Catalog of Basic Functions

P9-BJO-536

Linear Functions

$f(x) = mx + b$

Slope = m > 0

$f(x) = mx + b$

Slope = m < 0

Constant Function $f(x) = b$

Identity Function $f(x) = x$

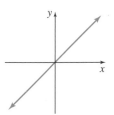

Square Function $f(x) = x^2$

Cube Function $f(x) = x^3$

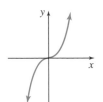

Square Root Function $f(x) = \sqrt{x}$

Greatest Integer Function $f(x) = [x]$

Absolute Value Function $f(x) = |x|$

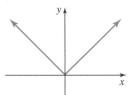

Power Functions

$f(x) = x^n$ (*n* even)

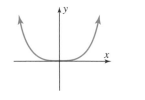

$f(x) = x^n$ (*n* odd)

Reciprocal Functions

$f(x) = \dfrac{1}{x}$

$f(x) = \dfrac{1}{x^2}$

Exponential Functions

$f(x) = b^x$ (**b > 1**)

$f(x) = b^x$ (**0 < b < 1**)

Logarithmic Functions

$f(x) = \log x$

$f(x) = \ln x$

Continues→

Catalog of Basic Functions   (*continued*)

Trigonometric Functions

$f(t) = \sin t$

$f(t) = \cos t$

$f(t) = \tan t$

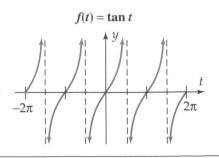

Rectangular and Parametric Equations for Conic Sections

Circles
Center (h, k), radius r

$$(x - h)^2 + (y - k)^2 = r^2$$

$\begin{aligned} x &= r \cos t + h \\ y &= r \sin t + k \end{aligned}$ $(0 \le t \le 2\pi)$

Ellipse
Center (h, k)

$$\frac{(x - h)^2}{a^2} + \frac{(y - k)^2}{b^2} = 1$$

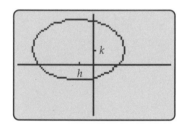

$\begin{aligned} x &= a \cos t + h \\ y &= b \sin t + k \end{aligned}$ $(0 \le t \le 2\pi)$

Parabola
Vertex (h, k)

$$(x - h)^2 = 4p(y - k)$$

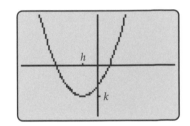

$\begin{aligned} x &= t \\ y &= \frac{(t - h)^2}{4p} + k \end{aligned}$ $(t \text{ any real})$

Parabola
Vertex (h, k)

$$(y - k)^2 = 4p(x - h)$$

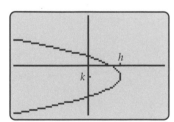

$\begin{aligned} x &= \frac{(t - k)^2}{4p} + h \\ y &= t \end{aligned}$ $(t \text{ any real})$

Hyperbola
Center (h, k)

$$\frac{(x - h)^2}{a^2} - \frac{(y - k)^2}{b^2} = 1$$

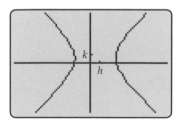

$\begin{aligned} x &= \frac{a}{\cos t} + h \\ y &= b \tan t + k \end{aligned}$ $(0 \le t \le 2\pi)$

Hyperbola
Center (h, k)

$$\frac{(y - k)^2}{a^2} - \frac{(x - h)^2}{b^2} = 1$$

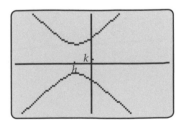

$\begin{aligned} x &= b \tan t + h \\ y &= \frac{a}{\cos t} + k \end{aligned}$ $(0 \le t \le 2\pi)$

www.brookscole.com

www.brookscole.com is the World Wide Web site for Brooks/Cole and is your direct source to dozens of online resources.

At *www.brookscole.com* you can find out about supplements, demonstration software, and student resources. You can also send email to many of our authors and preview new publications and exciting new technologies.

www.brookscole.com
Changing the way the world learns®

Contemporary Precalculus

A Graphing Approach

Fourth Edition

THOMAS W. HUNGERFORD
Cleveland State University

THOMSON
BROOKS/COLE

Australia • Canada • Mexico • Singapore • Spain
United Kingdom • United States

THOMSON

™

BROOKS/COLE

Mathematics Editor: *John-Paul Ramin*
Development Editor: *Leslie Lahr*
Assistant Editor: *Lisa Chow*
Project Manager, Editorial Production: *Janet Hill*
Technology Project Manager: *Star Mackenzie*
Marketing Manager: *Karin Sandberg*
Marketing Assistant: *Jennifer Gee*
Advertising Project Manager: *Bryan Vann*
Print/Media Buyer: *Kris Waller*
Production Service: *Hearthside Publishing Services/Laura Horowitz*

Text Designer: *Terri Wright*
Art Editor: *Hearthside Publishing Services*
Photo Researcher: *Terri Wright*
Copy Editor: *Barbara Willette*
Illustrator: *Hearthside Publishing Services*
Cover Designer: *Terri Wright*
Cover Image: *Jim Wehtje/Getty Images*
Cover Printer: *The Lehigh Press, Inc.*
Compositor: *Better Graphics, Inc.*
Printer: *R. R. Donnelley/Willard*

For more information about our products contact us at:
Thomson Learning Academic Resource Center
1-800-423-0563
For permission to use material from this text, contact us by:
Phone: 1-800-730-2214   **Fax:** 1-800-730-2215
Web: http://www.thomsonrights.com

Library of Congress Control Number: 2003103824

ISBN 0-534-40338-7

Instructor's Edition: ISBN 0-534-40339-5

Brooks/Cole–Thomson Learning
10 Davis Drive
Belmont, CA 94002
USA

Asia
Thomson Learning
5 Shenton Way #01-01
UIC Building
Singapore 068808

Australia/New Zealand
Thomson Learning
102 Dodds Street
Southbank, Victoria 3006
Australia

Canada
Nelson
1120 Birchmount Road
Toronto, Ontario M1K 5G4
Canada

Europe/Middle East/Africa
Thomson Learning
High Holborn House
50/51 Bedford Row
London WC1R 4LR
United Kingdom

Latin America
Thomson Learning
Seneca, 53
Colonia Polanco
11560 Mexico D.F.
Mexico

Spain/Portugal
Paraninfo
Calle/Magallanes, 25
28015 Madrid, Spain

PREFACE

This book provides the mathematical background needed in calculus for students who have had two or three years of high school mathematics. Mathematics is presented in an informal manner that stresses meaningful motivation, careful explanations, and numerous examples, with an ongoing focus on real-world problem solving. Technology is integrated into the text and students are expected to use it to participate actively in exploring various topics from algebraic, graphical, and numerical perspectives.

Changes in the Fourth Edition

The most noticeable differences from earlier editions are the following.

Variation   An optional Section on variation has been added to Chapter 1 for those instructors who want to review this topic.

Catalog of Functions   In response to user requests, the graphs of basic functions are now summarized in a catalog that appears in the front end-papers and at appropriate points in Chapters 3–6.

Trigonometry   As in previous editions, trigonometric functions of a real number are introduced first (Chapter 6) and triangle trigonometry is covered later (Chapter 8). In this edition, however, instructors who want to reverse this order may easily do so by using two alternate sections, as explained in To The Instructor on page x. In addition, Chapter 7 (Trigonometric Identities and Equations) has been reordered in response to user requests and placed before triangle trigonometry.

Conic Sections   The coverage of conic sections has been expanded and moved to the beginning of Chapter 10.

Limits   An optional chapter on limits appears at the end of the book.

Geometry of Triangles   The Geometry Appendix now contains full summary of the basic facts about congruent and similar triangles, including exercises.

A variety of other changes have been made, with many examples and exercises rewritten or replaced. A few sections have been extensively rewritten to improve coverage and clarity, including graphing with technology (Section 2.1), inverse functions (Section 3.7), and trigonometric equations (Section 7.5). Other improvements include the following.

Real-Data Applications   Significantly more exercises and examples are based on real data. All real-data examples and exercises from the previous edition have either been updated or replaced with more relevant ones.

Discovery Projects   Many of these have been replaced with more interesting topics. As before these investigative problems are suitable for small-group work.

Technology Tips   These have been updated, as necessary, with new ones added to clarify points that cause difficulties for students.

Artwork   In addition to an attractive multi-color format, the book now includes a number of photographs in the chapter openers and discovery projects.

Mathematical and Pedagogical Features

The mathematical approach remains the same: topics that are essential for success in calculus are thoroughly covered, including functional notation, graph reading, the natural exponential and logarithmic functions, average rates of change, trigonometric functions of a real variable, and limits.

All the helpful pedagogical features of earlier editions are retained here, such as:

Graphing Explorations, in which students discover and develop useful mathematics on their own;

Cautions that alert students to common misconceptions and mistakes;

Technology Tips that provide assistance in carrying out various procedures on specific calculators;

Calculator Investigations that encourage students to become familiar with the capabilities and limitations of graphing calculators;

Exercises that proceed from routine drill to those requiring some thought, including graph interpretation and applied problems, as well as *Thinkers* that challenge students to "think outside the box" (most of these are not difficult— just different);

Chapter Reviews that include a list of important concepts (referenced by section and page number), a summary of important facts and formulas, and a set of review questions;

Algebra Review Appendix for students who need a review of basic algebra;

Geometry Review Appendix that summarizes frequently used facts from plane geometry, with examples and exercises;

Program Appendix that provides a small number of programs that are useful for updating older calculator models or for more easily carrying out procedures discussed in the text.

Supplements for Instructors

Instructors who adopt this text may receive, free of charge, the following items.

INSTRUCTOR'S EDITION

This special version of the complete student text contains a Resource Integration Guide.   ISBN 0-534-40339-5

COMPLETE SOLUTIONS MANUAL

The *Complete Solutions Manual* provides worked out solutions to all of the problems in the text.   ISBN 0-534-40340-9

TEST BANK

The *Test Bank* includes 8 tests per chapter as well as 3 final exams. The tests are made up of a combination of multiple-choice, free-response, true/false, and fill-in-the-blank questions.   ISBN 0-534-40062-0

BCA INSTRUCTOR VERSION

With a balance of efficiency and high-performance, simplicity and versatility, *Brooks/Cole Assessment* gives you the power to transform the learning and teaching experience. *BCA Instructor Version* is made up of two components, *BCA Testing* and *BCA Tutorial*. *BCA Testing* is a revolutionary, internet-ready, text-specific testing suite that allows instructors to customize exams and track student progress in an accessible, browser-based format. BCA offers full algorithmic generation of problems and free response mathematics. *BCA Tutorial* is a text-specific, interactive tutorial software program, that is delivered via the web (at http://bca.brookscole.com) and is offered in both student and instructor versions. Like *BCA Testing*, it is browser-based, making it an intuitive mathematical guide even for students with little technological proficiency. So sophisticated, it's simple, *BCA Tutorial* allows students to work with real math notation in real time, providing instant analysis and feedback. The tracking program built into the instructor version of the software enables instructors to carefully monitor student progress. The complete integration of the testing, tutorial, and course management components simplifies your routine tasks. Results flow automatically to your gradebook and you can easily communicate to individuals, sections, or entire courses.   ISBN 0-534-40343-3

TEXT-SPECIFIC VIDEOTAPES

These text-specific videotape sets, available at no charge to qualified adopters of the text, feature 10- to 20-minute problem-solving lessons that cover each section of every chapter.   ISBN 0-534-40346-8

WEBTUTOR ON WEBCT

For online and distance learning needs, we offer *WebTutor on WebCT* with many of our texts! With *WebTutor's* text-specific, preformatted content and total flexibility, you can easily create and manage your own personal Web site! *WebTutor's* robust course management tool gives you the ability to provide virtual office hours, post syllabi, set up threaded discussions, track student progress with the quizzing material, and much more. For students, *WebTutor* offers real-time access to a full array of study tools, including chapter outlines, summaries, learning objectives, glossary, flashcards, practice quizzes, *InfoTrac*® *College Edition* exercises, and Web links.

Supplements for Students

STUDENT SOLUTIONS MANUAL

The *Student Solutions* manual provides worked out solutions to the odd-numbered problems in the text.   ISBN 0-534-40344-1

WEB SITE   http://mathematics.brookscole.com

When you adopt a Thomson-Brooks/Cole mathematics text, you and your students will have access to a variety of teaching and learning resources. This Web site features everything from book-specific resources to newsgroups. It's a great way to make teaching and learning an interactive and intriguing experience.

ON-LINE TUTORING WITH INFOTRAC

With Brooks/Cole's Assessment's On-Line Tutoring, your students have dynamic, flexible on-line tutorial resources at their fingertips. By entering a PIN code packaged with their textbook, students gain access to *BCA Tutorial,* a text-specific tutorial with step-by-step explanations, exercises and quizzes, and *vMentor,* live, one-on-one help from an experienced mathematics tutor. In addition to robust tutorial services, your students also receive anytime, anywhere access to *InfoTrac College Edition.* This on-line library offers the full text of articles from almost 4000 scholarly and popular publications, updated daily and going back as much as 22 years. Both adopters and their students receive unlimited access for 4 months.

BCA TUTORIAL STUDENT VERSION

This text-specific, interactive tutorial software is delivered via the Web (at http://bca.brookscole.com). It is browser-based, making it an intuitive mathematical guide, even for students with little technological proficiency. So sophisticated, it's simple. *BCA Tutorial* allows students to work with real math notation in real time, providing instant analysis and feedback. The tracking program built into the instructor version of the software enables instructors to carefully monitor student progress. *BCA Tutorial Student Version* is also available on CD-ROM for those students who want to use the tutorial locally on their computers and not via the Internet.
ISBN 0-534-40342-5

INTERACTIVE VIDEO SKILLBUILDER CD-ROM

The Interactive Video Skillbuilder CD-ROM contains more than 8 hours of video instruction. The problems worked during each video lesson are shown next to the viewing screen so that students can try working them before watching the solution. Examples and exercises in the text that are discussed in the lessons are marked with the icon 🎞 in the text. To help students evaluate their progress, each section contains a 10-question Web quiz (the results of which can be emailed to the instructor) and each chapter contains a chapter test, with the answer to each problem on each test.

The Skillbuilder CD-ROM also includes MathCue tutorial and testing software. MathCue is keyed to the text and includes these components:

- **MathCue Skillbuilder**   Presents problems to solve, evaluates answers, and tutors students by displaying complete solutions with step-by-step explanations.

- **MathCue Quiz**   Allows students to generate large numbers of quiz problems keyed to problem types from each section of the book.

- **MathCue Chapter Test**   Also provides large numbers of problems keyed to problem types from each chapter.

- **MathCue Solution Finder**   This unique tool allows students to enter their own basic problems and receive step-by-step help as if they were working with a tutor.

- Score reports for any MathCue session can be printed and handed in for credit or extra-credit.   ISBN 0-534-40347-6

Acknowledgments

I am particularly grateful to

Phil Embree, William Woods University

who checked all the examples and exercises for accuracy, proofread the book, and wrote several new Discovery Projects, as well as providing helpful suggestions on other aspects of the text. I also want to thank two Clevelend State students: Jonathan Landis, who prepared the answer section, and David Knapp, who assisted with manuscript preparation.

My sincere thanks go to the following reviewers, who provided many helpful suggestions that have improved this edition:

Kathleen A. Cantone, Onondaga Community College
Robert Fliess, West Liberty State College
Larry Howe, Rowan University
Conrad D. Krueger, San Antonio College
Anatoly S. Libgober, University of Illinois at Chicago
Trung G. Tran, Tacoma Community College
Cathleen M. Zucco-Teveloff, Trinity College

Thanks also go to reviewers from previous editions:
Deborah Adams, Jacksonville University
Kelly Bach, University of Kansas
David Blankenbaker,University of New Mexico
Bettyann Daley, University of Delaware
Margaret Donlan, University of Delaware
Patricia Dueck, Arizona State University
Betsy Farber, Buck County Community College
Alex Feldman, Boise State University
Betty Givan, Eastern Kentucky University
William Grimes, Central Missouri State University
Frances Gulick, University of Maryland
John Hamm, University of New Mexico
Lonnie Hass, North Dakota State University
Ann Lawrance, Wake Technical Community College
Charles Laws, Cleveland State Community College
Martha Lisle, Prince George's Community College
Matthew Liu, University of Wisconsin-Stevens Point
Sergey Lvin, University of Maine
George Matthews, Onandaga Community College
Nancy Matthews, University of Oklahoma
Ruth Meyering, Grand Valley State University

William Miller, Central Michigan University
Philip Montgomery, University of Kansas
Roger Nelsen, Lewis and Clark College
Jack Porter, University of Kansas
Robert Rogers, University of New Mexico
Barbara Sausen, Fresno City College
Hugo Sun, California State University at Fresno
Stuart Thomas, University of Oregon
Bettie Truitt, Black Hawk College
Jan Vandever, South Dakota State University
Judith Wolbert, Kettering University

I also want to thank the people who have prepared the various supplements that are available to instructors and students who use this book:

Fred Safier, City College of San Francisco; Complete Solutions Manual and Student Solution Manual
Margaret Donlan, University of Delaware; Test Bank.

It is a pleasure to acknowledge the invaluable assistance of the Brooks/Cole staff, including
John-Paul Ramin, Sponsoring Editor
Leslie Lahr, Development Editor
Karin Sandberg, Senior Marketing Manager
Janet Hill, Senior Project Manager, Editorial Production
Vernon Boes, Senior Art Director
Lisa Chow, Stacy Green, and Molly Nance, Assistant Editors.

Special thanks go to the outside production staff
Barbara Willette, Copyeditor
Terri Wright of Terri Wright Design

and especially to
Laura Horowitz of Hearthside Publishing Services who coordinated the production of the book.

The last word goes to the love of my life, my wife Mary Alice, who has provided understanding and support when it was most needed.

Thomas W. Hungerford

TO THE INSTRUCTOR

This book contains more than enough material for two semesters. By using the chart on the facing page (and the similar ones at the beginning of each chapter that show the interdependence of sections within the chapter) you can easily design a course to fit the needs of your students and the constraints of time. When planning your syllabus, two items are worth noting:

Special Topics Sections with this label are usually related to the immediately preceding section and are not prerequisites for other sections of the text. I am inclined to omit most of them (particularly in a short course), but I know that a number of people consider some of them to be essential material. So feel free to include as many or as few as you wish.

Trigonometry I personally think that a precalculus course should emphasize trigonometric functions of a real variable from the beginning, with triangle trigonometry treated later. If you agree, you should cover Chapters 6–8 in one of these orders: 6, 7, 8 or 6, 8, 7. A number of instructors, however, prefer to begin with triangle trigonometry. They can do so by covering Chapters 8, 6, and 7 in that order, using alternate sections 8.1 and 6.2 in place of the regular ones.

This text assumes the use of technology. You and your students will need either a graphing calculator or a computer with appropriate software. All discussions of calculators in the text apply (with obvious modifications) to computer software. You should be aware of the following facts.

Minimal Requirements Current calculator models that meet the minimal requirements include TI-82 and above, HP-38 and above, and Casio 9850 and above (including FX 2.0). A few of these may also require some programs from the Program Appendix to handle certain tasks

Technology Tips The Tips in the margin provide general information and advice, as well as listing the proper menus or keys needed to carry out procedures on specific calculators. Unless noted otherwise,

Tips for	also apply to
TI-83	TI-83 Plus, TI-82
TI-86	TI-85
TI-89	TI-92
Casio 9850	Casio 9970
HP-39	HP-38

Finally, you may want to explain to your students that the icon in the text indicates examples and exercises that are discussed in the Interactive Video Skillbuilder CD-ROM that accompanies the text.

INTERDEPENDENCE OF CHAPTERS

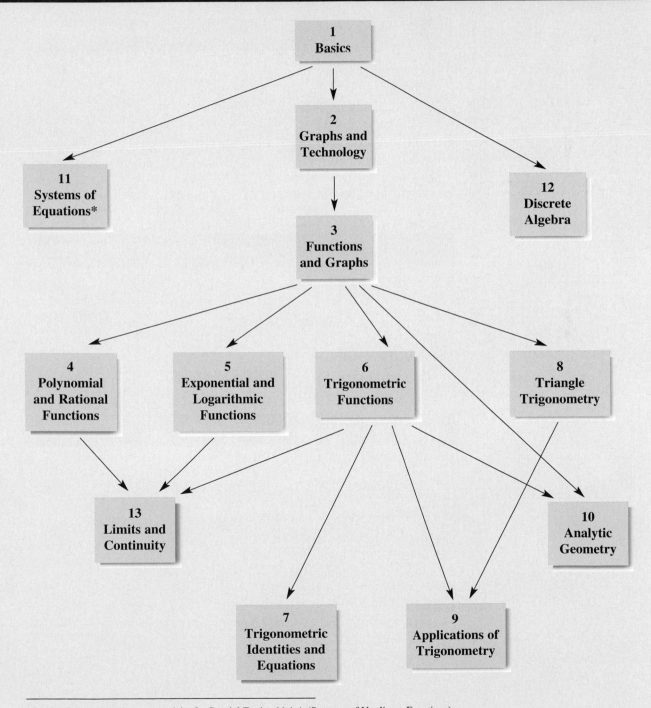

*Section 2.1 (graphs) is a prerequisite for Special Topics 11.1.A (Systems of Nonlinear Equations).

CONTENTS

1   *Basics*   *1*

1.1   The Real Number System   2
1.1.A   *SPECIAL TOPICS:* Decimal Representations of Real Numbers   16
1.2   Solving Equations Algebraically   18
1.2.A   *SPECIAL TOPICS:* Absolute Value Equations   28
1.2.B   *SPECIAL TOPICS:* Variation   30
1.3   The Coordinate Plane   36
1.4   Lines   48
     CHAPTER 1 REVIEW   63   •   DISCOVERY PROJECT 1   68

2   *Graphs and Technology*   *69*

2.1   Graphs   70
2.2   Solving Equations Graphically and Numerically   83
2.3   Applications of Equations   90
2.4   Optimization Applications   101
2.5   Linear Models   105
     CHAPTER 2 REVIEW   117   •   DISCOVERY PROJECT 2   121

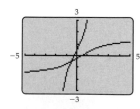

3   *Functions and Graphs*   *124*

3.1   Functions   125
3.2   Functional Notation   136
3.3   Graphs of Functions   146
3.3.A   *SPECIAL TOPICS:* Graph Reading   161
3.4   Graphs and Transformations   166
3.4.A   *SPECIAL TOPICS:* Symmetry   177
3.5   Operations on Functions   182
3.6   Rates of Change   192
3.7   Inverse Functions   203
     CHAPTER 3 REVIEW   213   DISCOVERY PROJECT 3   220

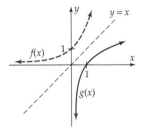

4 *Polynomial and Rational Functions* 224

4.1   Quadratic Functions   225
4.2   Polynomial Functions   235
4.2.A   *SPECIAL TOPICS:* Synthetic Division   243
4.3   Real Roots of Polynomials   247
4.4   Graphs of Polynomial Functions   254
4.4.A   *SPECIAL TOPICS:* Polynomial Models   267
4.5   Rational Functions   272
4.5.A   *SPECIAL TOPICS:* Other Rational Functions   288
4.6   Polynomial and Rational Inequalities   292
4.6.A   *SPECIAL TOPICS:* Absolute Value Inequalities   302
4.7   Complex Numbers   305
4.8   Theory of Equations   312
CHAPTER 4 REVIEW   318   •   DISCOVERY PROJECT 4   324

5 *Exponential and Logarithmic Functions* 325

5.1   Radicals and Rational Exponents   326
5.1.A   *SPECIAL TOPICS:* Radical Equations   334
5.2   Exponential Functions   341
5.2.A   *SPECIAL TOPICS:* Compound Interest and the Number *e*   354
5.3   Common and Natural Logarithmic Functions   360
5.4   Properties of Logarithms   369
5.4.A   *SPECIAL TOPICS:* Logarithmic Functions to Other Bases   375
5.5   Algebraic Solutions of Exponential and Logarithmic Equations   383
5.6   Exponential, Logarithmic, and Other Models   392
CHAPTER 5 REVIEW   403   •   DISCOVERY PROJECT 5   409

AM signal

6 *Trigonometric Functions* 411

6.1   Angles and Their Measurement   412
6.2   The Sine, Cosine, and Tangent Functions   423
6.2   *ALTERNATE:* The Sine, Cosine, and Tangent Functions   431
6.3   Algebra and Identities   437
6.4   Basic Graphs   445
6.5   Periodic Graphs and Simple Harmonic Motion   455
6.5.A   *SPECIAL TOPICS:* Other Trigonometric Graphs   468
6.6   Other Trigonometric Functions   474
CHAPTER 6 REVIEW   482   •   DISCOVERY PROJECT 6   487

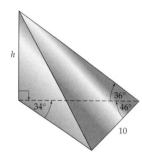

7 *Trigonometric Identities and Equations* 489

7.1   Basic Identities and Proofs   490
7.2   Addition and Subtraction Identities   498
7.2.A *SPECIAL TOPICS:* Lines and Angles   506
7.3   Other Identities   510
7.4   Inverse Trigonometric Functions   518
7.5   Trigonometric Equations   527
       CHAPTER 7 REVIEW   537   •   DISCOVERY PROJECT 7   540

8 *Triangle Trigonometry* 542

8.1   Trigonometric Functions of Angles   543
8.1   *ALTERNATE:* Trigonometric Functions of Angles   553
8.2   Applications of Right Triangle Trigonometry   558
8.3   The Law of Cosines   566
8.4   The Law of Sines   573
8.4.A *SPECIAL TOPICS:* The Area of a Triangle   582
       CHAPTER 8 REVIEW   586   •   DISCOVERY PROJECT 8   589

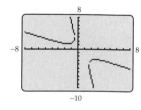

9 *Applications of Trigonometry* 591

9.1   The Complex Plane and Polar Form for Complex Numbers   592
9.2   DeMoivre's Theorem and the nth Roots of Complex Numbers   598
9.3   Vectors in the Plane   606
9.4   The Dot Product   621
       CHAPTER 9 REVIEW   631   •   DISCOVERY PROJECT 9   634

10 *Analytic Geometry* 636

10.1   Circles and Ellipses   638
10.2   Hyperbolas   652
10.3   Parabolas   663
10.4   Rotations and Second-Degree Equations   672
10.4.A *SPECIAL TOPICS:* Rotation of Axes   676
10.5   Plane Curves and Parametric Equations   681
10.6   Polar Coordinates   695
10.7   Polar Equations of Conics   704
        CHAPTER 10 REVIEW   715   •   DISCOVERY PROJECT 10   719

11 Systems of Equations 721

11.1 Systems of Linear Equations in Two Variables   722
11.1.A *SPECIAL TOPICS:* Systems of Nonlinear Equations   735
11.2 Large Systems of Linear Equations   742
11.3 Matrix Methods for Square Systems   758
CHAPTER 11 REVIEW   771 • **DISCOVERY PROJECT 11   774**

12 Discrete Algebra 776

12.1 Sequences and Sums   777
12.2 Arithmetic Sequences   788
12.3 Geometric Sequences   794
12.3.A *SPECIAL TOPICS:* Infinite Series   801
12.4 The Binomial Theorem   806
12.5 Mathematical Induction   813
CHAPTER 12 REVIEW   823 • **DISCOVERY PROJECT 12   826**

13 Limits and Continuity 828

13.1 Limits of Functions   829
13.2 Properties of Limits   840
13.2.A *SPECIAL TOPICS:* The Formal Definition of Limit   847
13.3 Continuity   854
13.4 Limits Involving Infinity   865
CHAPTER 13 REVIEW   876 • **DISCOVERY PROJECT 13   878**

APPENDIX 1: ALGEBRA REVIEW   882
APPENDIX 2: GEOMETRY REVIEW   902
APPENDIX 3: PROGRAMS   914
SELECTED ANSWERS   919
INDEX OF APPLICATIONS   996
INDEX   999

TO THE STUDENT

This text assumes the use of technology. You will need either a graphing calculator or a computer with appropriate software. All discussions of calculators in the text apply (with obvious modifications) to computer software. The following information should help you choose and effectively use a calculator.

Minimal Requirements Current calculator models that meet the minimal requirements include TI-82 and above, HP-38 and above, and Casio 9850 and above (including FX 2.0). A few of these may also require some programs from the Program Appendix to handle certain tasks.

Technology Tips The Tips in the margin provide general information and advice, as well as listing the proper menus or keys needed to carry out procedures on specific calculators. Unless noted otherwise,

Tips for	**also apply to**
TI-83	TI-83 Plus, TI-82
TI-86	TI-85
TI-89	TI-92
Casio 9850	Casio 9970
HP-39	HP-38

Calculator Investigations You may not be aware of the full capabilities of your calculator (or some of its limitations). The Calculator Investigations (which appear just before the exercise sets in some of the earlier sections of the book) will help you to become familiar with your calculator and to maximize the mathematical power it provides. Even if your instructor does not assign these investigations, you may want to look through them to be sure you are getting the most you can from your calculator.

Getting the Most Out of This Course

With all this talk about calculators, don't lose sight of this crucial fact:

Technology is only a *tool* for doing mathematics.

You can't build a house if you only use a hammer. A hammer is great for pounding nails, but useless for sawing boards. Similarly, a calculator is great for computations and graphing, but it is not the right tool for every mathematical task. To succeed in this course, you must develop and use your algebraic and geometric skills, your reasoning power and common sense, and you must be willing to work.

The key to success is to use all of the resources at your disposal: your instructor, your fellow students, your calculator (and its instruction manual), and this book and the Interactive Video Skillbuilder CD-ROM that accompanies it. Here are some tips for making the most of these resources.

Ask Questions   Remember the words of Hillel:

The bashful do not learn.

There is no such thing as a "dumb question" (assuming, of course, that you have attended class and read the text). Your instructor will welcome questions that arise from a serious effort on your part.

Read the Book   Not just the homework exercises, but the rest of the text as well. There is no way your instructor can possibly cover the essential topics, clarify ambiguities, explain the fine points, and answer all your questions during class time. You simply will not develop the level of understanding you need to succeed in this course and in calculus unless you read the text fully and carefully.

Be an Interactive Reader   You can't read a math book the way you read a novel or history book. You need pencil, paper, and your calculator at hand to work out the statements you don't understand and to make notes of things to ask your fellow students and/or your instructor.

Do the Graphing Explorations   When you come to a box labeled "Graphing Exploration," use your calculator as directed to complete the discussion. Typically, this will involve graphing one or more equations and answering some questions about the graphs. Doing these explorations as they arise will improve your understanding and clarify issues that might otherwise cause difficulties.

Do Your Homework   Remember that

Mathematics is not a spectator sport.

You can't expect to learn mathematics without doing mathematics, any more than you could learn to swim without getting wet. Like swimming or dancing or reading or any other skill, mathematics takes practice. Homework assignments are where you get the practice that is essential for passing this course and succeeding in calculus.

Basics

Stockphoto.com/Michael Howell

On a clear day, can you see forever?

If you are at the top of the Sears Tower in Chicago, how far can you see? In earlier centuries, the lookout on a sailing ship was posted atop the highest mast because he could see farther from there than from the deck. How much farther? These questions, and similar ones, can be answered (at least approximately) by using basic algebra and geometry. See Example 3 on page 8 and Excercise 72 on page 14.

1

Chapter Outline

Interdependence of Sections

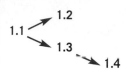

1.1 → 1.2
1.1 → 1.3
1.3 ⤵ 1.4

1.1	**The Real Number System**
1.1.A	*Special Topics:* Decimal Representation of Real Numbers
1.2	**Solving Equations Algebraically**
1.2.A	*Special Topics:* Absolute Value Equations
1.2.B	*Special Topics:* Variation
1.3	**The Coordinate Plane**
1.4	**Lines**

This chapter reviews the essential facts about real numbers, equations, the coordinate plane, and lines that are needed in this course and in calculus. The Algebra Review Appendix at the end of the book is a prerequisite for this material.

1.1 The Real Number System

You have been using **real numbers** most of your life. They include the **natural numbers** (or **positive integers**): 1, 2, 3, 4, . . . and the **integers:**

$$\ldots, -5, -4, -3, -2, -1, 0, 1, 2, 3, 4, 5, \ldots$$

A real number is said to be a **rational number** if it can be expressed as a fraction $\dfrac{r}{s}$, with r and s integers and $s \neq 0$; for instance,

$$\frac{1}{2}, \qquad -.983 = -\frac{983}{1000}, \qquad 47 = \frac{47}{1}, \qquad 8\tfrac{3}{5} = \frac{43}{5}.$$

Alternatively, rational numbers may be described as numbers that can be expressed as terminating decimals, such as $.25 = \dfrac{1}{4}$, or as nonterminating repeating decimals, in which a single digit or a block of digits repeats for ever, such as

$$\frac{5}{3} = 1.66666 \cdots \qquad \text{or} \qquad \frac{58}{333} = .174174174 \cdots .$$

A real number that cannot be expressed as a fraction with integer numerator and denominator is called an **irrational number.** Alternatively, an irrational number is one that can be expressed as a nonterminating, nonrepeating decimal (no block of digits repeats forever). For example, the number π, which is used to calculate the area of a circle, is irrational.[*] More information about decimal expansions of real numbers is given in Special Topics 1.1.A.

---

[*]This fact is difficult to prove. In the past you may have used $22/7$ as π; a calculator might display π as 3.141592654. However, these numbers are just *approximations* of π (close but not quite *equal* to π).

The real numbers are often represented geometrically as points on a **number line,** as in Figure 1–1. We shall assume that there is exactly one point on the line for every real number (and vice versa) and use phrases such as "the point 3.6" or "a number on the line." This mental identification of real numbers and points on the line is often extremely helpful.

Figure 1–1

:: SCIENTIFIC NOTATION

A positive real number is said to be in **scientific notation** when it is written in the form

$$a \times 10^n \qquad \text{where } 1 \le a < 10 \text{ and } n \text{ is an integer.}$$

EXAMPLE 1

Write each of these numbers in scientific notation:

(a) 356   (b) 1,564,000   (c) .072   (d) .00000087

SOLUTION   In each case, move the decimal point to the left or right to obtain a number between 1 and 10; then write the original number in scientific notation as follows.

(a) $356 = 3.56 \times 100 = 3.56 \times 10^2$ [Decimal point is moved 2 places to the *left,* and 10 is raised to the power of 2.]

(b) $1{,}564{,}000 = 1.564 \times 1{,}000{,}000 = 1.564 \times 10^6$ [Decimal point is moved 6 places to the *left,* and 10 is raised to the 6th power.]

(c) $.072 = 7.2 \times \dfrac{1}{100} = 7.2 \times 10^{-2}$ [Decimal point is moved 2 places to the *right,* and 10 is raised to -2.]*

(d) $.00000087 = 8.7 \times \dfrac{1}{10{,}000{,}000} = 8.7 \times 10^{-7}$ [Decimal point is moved 7 places to the *right,* and 10 is raised to the -7.]   ∎

Scientific notation is useful for computations with very large or very small numbers.

EXAMPLE 2   †

$$(.00000002)(4{,}300{,}000{,}000) = (2 \times 10^{-8})(4.3 \times 10^9)$$
$$= 2(4.3)10^{-8+9} = (8.6)10^1 = 86. \quad \blacksquare$$

---

*Negative exponents are explained in the first section of the Algebra Review Appendix.
†Examples and Exercises marked with this icon are discussed in the Interactive Video Skillbuilder CD-ROM that comes with this text.

When calculators display numbers in scientific notation, they omit the 10 and use an E to indicate the exponent. For instance,

$$7.235 \times 10^{-12} \quad \text{is displayed as} \quad 7.235 \text{ E} -12.$$

To enter a number in scientific notation, for example 5.6×10^{73}, type in 5.6, then press the EE key (labeled EEX or EXP or 10^x on some calculators) and type in 73. Calculators automatically switch to scientific notation whenever a number is too large or too small to be displayed in the standard way. If you try to enter a number with more digits than the calculator can handle, such as 45,000,000,333,222,111, a typical calculator will approximate it using scientific notation as 4.500000033 E 16, that is, as 45,000,000,330,000,000.

ORDER

The statement $c < d$, which is read **"c is less than d,"** and the statement $d > c$ (read **"d is greater than c"**) mean exactly the same thing:

c lies to the *left* of d on the number line.

For example, Figure 1–1 shows that $-5.78 < -2.2$ and $4 > \pi$.

The statement $c \le d$, which is read **"c is less than or equal to d,"** means

Either c is less than d or c is equal to d.

Only one part of an "either . . . or" statement needs to be true for the entire statement to be true. So the statement $5 \le 10$ is true because $5 < 10$, and the statement $5 \le 5$ is true because $5 = 5$. The statement $d \ge c$ (read **"d is greater than or equal to c"**) means exactly the same thing as $c \le d$.

The statement $b < c < d$ means

$b < c$ and simultaneously $c < d$.

For example, $3 < x < 7$ means that x is a number that is strictly between 3 and 7 on the number line (greater than 3 and less than 7). Similarly, $b \le c < d$ means

$b \le c$ and simultaneously $c < d$,

and so on.

Certain sets of numbers, defined in terms of the order relation, appear frequently enough to merit special notation. Let c and d be real numbers with $c < d$. Then

Interval Notation

$[c, d]$ denotes the set of all real numbers x such that $c \le x \le d$.

(c, d) denotes the set of all real numbers x such that $c < x < d$.

$[c, d)$ denotes the set of all real numbers x such that $c \le x < d$.

$(c, d]$ denotes the set of all real numbers x such that $c < x \le d$.

All four of these sets are called **intervals** from c to d. The numbers c and d are the **endpoints** of the interval. $[c, d]$ is called the **closed interval** from c to d (both endpoints included and *square* brackets), and (c, d) is called the **open inter-**

CAUTION

Your calculator may not always obey the laws of arithmetic when dealing with very large or very small numbers. For instance, we know that

$(1 + 10^{19}) - 10^{19}$

$= 1 + (10^{19} - 10^{19})$

$= 1 + 0$

$= 1,$

but a calculator may round off the number $1 + 10^{19}$ as 10^{19} and use this instead of the correct number

10,000,000,000,000,000,001.

So the calculator computes

$(1 + 10^{19}) - 10^{19}$

as $10^{19} - 10^{19} = 0$. (Try it!)

val from *c* to *d* (neither endpoint included and *round* brackets). Some examples are shown in Figure 1–2.[*]

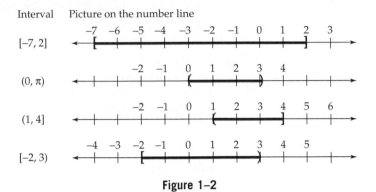

Figure 1–2

If *b* is a real number, then the half-line extending to the right or left of *b* is also called an **interval.** Depending on whether or not *b* is included, there are four possibilities.

Interval
Notation

> $[b, \infty)$ denotes the set of all real numbers *x* such that $x \geq b$.
>
> (b, ∞) denotes the set of all real numbers *x* such that $x > b$.
>
> $(-\infty, b]$ denotes the set of all real numbers *x* such that $x \leq b$.
>
> $(-\infty, b)$ denotes the set of all real numbers *x* such that $x < b$.

NOTE

The symbol ∞ is read "infinity," and we call the set $[b, \infty)$ "the interval from *b* to infinity." The symbol ∞ does *not* denote a real number; it is simply part of the notation used to label the first two sets of numbers defined in the previous box. Analogous remarks apply to the symbol $-\infty$, which is read "negative infinity."

Some examples are shown in Figure 1–3.

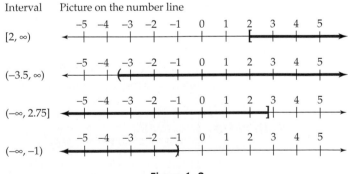

Figure 1–3

In a similar vein, $(-\infty, \infty)$ **denotes the set of all real numbers.**

---

[*]In Figures 1–2 and 1–3, a round bracket such as) or (indicates that the endpoint is *not* included, whereas a square bracket such as] or [indicates that the endpoint *is* included.

⬛⬛ ARITHMETIC

To avoid ambiguity when dealing with expressions such as $6 + 3 \times 5$, mathematicians have made the following agreement, which is also followed by your calculator.

Order of
Operations

> In an expression without parentheses, multiplication and division are performed first (from left to right). Addition and subtraction are performed last (from left to right).

In light of this convention, there is only one correct way to interpret $6 + 3 \times 5$:

$$6 + 3 \times 5 = 6 + 15 = 21. \qquad \text{[Multiplication first, addition last]}$$

On the other hand, if you want to "add $6 + 3$ and then multiply by 5," you must use parentheses:

$$(6 + 3) \cdot 5 = 9 \cdot 5 = 45.$$

This is an illustration of the first of two basic rules for dealing with parentheses.

Rules for
Parentheses

> 1. Do all computations inside the parentheses before doing any computations outside the parentheses.
>
> 2. When dealing with parentheses within parentheses, begin with the innermost pair and work outward.

For example,

$$8 + [11 - (6 \times 3)] = 8 + (11 - 18) = 8 + (-7) = 1.$$

Inside parentheses first

We assume that you are familiar with the basic properties of real number arithmetic, particularly the following fact.

Distributive
Law

> For all real numbers a, b, c,
>
> $$a(b + c) = ab + ac \qquad \text{and} \qquad (b + c)a = ba + ca.$$

The distributive law doesn't usually play a direct role in easy computations, such as $4(3 + 5)$. Most people don't say $4 \cdot 3 + 4 \cdot 5 = 12 + 20 = 32$. Instead, they mentally add the numbers in parentheses and say 4 times 8 is 32. But when symbols are involved, you can't do that, and the distributive law is essential. For example,

$$4(3 + x) = 4 \cdot 3 + 4x = 12 + 4x.$$

■■ NEGATIVE NUMBERS AND NEGATIVES OF NUMBERS

The **positive numbers** are those to the right of 0 on the number line, that is,

All numbers c with $c > 0$.

The **negative numbers** are those to the left of 0, that is,

All numbers c with $c < 0$.

The **nonnegative** numbers are the numbers c with $c \geq 0$.

The word "negative" has a second meaning in mathematics. The **negative *of a number*** c is the number $-c$. For example, the negative of 5 is -5, and the negative of -3 is $-(-3) = 3$. Thus the negative of a negative number is a positive number. Zero is its own negative, since $-0 = 0$. In summary,

Negatives

> The negative of the number c is $-c$.
>
> If c is a positive number, then $-c$ is a negative number.
>
> If c is a negative number, then $-c$ is a positive number.

■■ SQUARE ROOTS

The **square root** of a nonnegative real number d is defined as the nonnegative number whose square is d and is denoted $\sqrt{d}$. For instance,

$$\sqrt{25} = 5 \qquad \text{because} \qquad 5^2 = 25.$$

In the past you may have said that $\sqrt{25} = \pm 5$, since $(-5)^2$ is also 25. It is preferable, however, to have a single unambiguous meaning for the symbol $\sqrt{25}$. So in the real number system, the term "square root" and the radical symbol $\sqrt{\ }$ always denote a *nonnegative* number. To express -5 in terms of radicals, we write $-5 = -\sqrt{25}$.

Although $-\sqrt{25}$ is a real number, the expression $\sqrt{-25}$ is *not defined* in the real numbers because there is no real number whose square is -25. In fact, since the square of every real number is nonnegative,

No negative number has a square root in the real numbers.

Some square roots can be found (or verified) by hand, such as

$$\sqrt{225} = 15 \qquad \text{and} \qquad \sqrt{1.21} = 1.1.$$

Usually, however, a calculator is needed to obtain rational *approximations* of roots. For instance, we know that $\sqrt{87}$ is between 9 and 10 because $9^2 = 81$ and $10^2 = 100$. A calculator shows that $\sqrt{87} \approx 9.327379.^*$

You can easily convince yourself that $\sqrt{cd} = \sqrt{c}\sqrt{d}$ for any nonnegative real numbers c and d. Sums, however, are different.

$^*\approx$ means "approximately equal."

CAUTION

If c and d are positive real numbers, then

$$\sqrt{c+d} \neq \sqrt{c} + \sqrt{d}.$$

For example,

$$\sqrt{9+16} = \sqrt{25} = 5,$$

but

$$\sqrt{9} + \sqrt{16} = 3 + 4 = 7.$$

EXAMPLE 3

Suppose you are located h feet above the ground. Because of the curvature of the earth, the maximum distance you can see is approximately d miles, where

$$d = \sqrt{1.5h + (3.587 \times 10^{-8})h^2}.$$

How far can you see from the 500 foot-high Smith Tower in Seattle and from the 1454-foot-high Sears Tower in Chicago?

SOLUTION   For the Smith Tower, substitute 500 for h in the formula, and use your calculator:

$$d = \sqrt{1.5(500) + (3.587 \times 10^{-8})500^2} \approx 27.4 \text{ miles.}$$

For the Sears Tower, you can see almost 47 miles because

$$d = \sqrt{1.5(1454) + (3.587 \times 10^{-8})1454^2} \approx 46.7 \text{ miles.} \quad \blacksquare$$

ABSOLUTE VALUE

On an informal level, most students think of absolute value like this:

> The absolute value of a nonnegative number is the number itself.
>
> The absolute value of a negative number is found by "erasing the minus sign."

If $|c|$ denotes the absolute value of c, then, for example, $|5| = 5$ and $|-4| = 4$.

This informal approach is inadequate, however, for finding the absolute value of a number such as $\pi - 6$. It doesn't make sense to "erase the minus sign" here. So we must develop a more precise definition. The statement $|5| = 5$ suggests that the absolute value of a positive number ought to be the number itself. For negative numbers, such as -4, note that $|-4| = 4 = -(-4)$, that is, the absolute value of the negative number -4 is the *negative* of -4. These facts are the basis of the formal definition.

Absolute Value

> The **absolute value** of a real number c is denoted $|c|$ and is defined as follows.
>
> If $c \geq 0$, then $|c| = c$.
>
> If $c < 0$, then $|c| = -c$.

EXAMPLE 4

(a) $|3.5| = 3.5$ and $|-7/2| = -(-7/2) = 7/2$.

(b) To find $|\pi - 6|$, note that $\pi \approx 3.14$, so $\pi - 6 < 0$. Hence, $|\pi - 6|$ is defined to be the *negative* of $\pi - 6$, that is,

$$|\pi - 6| = -(\pi - 6) = -\pi + 6.$$

(c) $|5 - \sqrt{2}| = 5 - \sqrt{2}$ because $5 - \sqrt{2} \geq 0$. $\blacksquare$

Here are the important facts about absolute value:

Properties of Absolute Value

1. $|c| \geq 0$ and $|c| > 0$ when $c \neq 0$.
2. $|c| = |-c|$
3. $|cd| = |c| \cdot |d|$
4. $\left|\dfrac{c}{d}\right| = \dfrac{|c|}{|d|}$ $(d \neq 0)$

TECHNOLOGY TIP

To find $|9 - 3\pi|$ on a calculator, key in

$$ABS(9 - 3\pi).$$

The ABS key is located in this menu/submenu:

TI: MATH/NUM

Casio: OPTN/NUM

HP-39: Keyboard

EXAMPLE 5

Here are examples of some of the properties listed in the box.

2. $|3| = 3$ and $|-3| = 3$, so $|3| = |-3|$.
3. If $c = 6$ and $d = -2$, then

$$|cd| = |6(-2)| = |-12| = 12$$

and

$$|c| \cdot |d| = |6| \cdot |-2| = 6 \cdot 2 = 12,$$

so $|cd| = |c| \cdot |d|$.

4. If $c = -5$ and $d = 4$, then

$$\left|\frac{c}{d}\right| = \left|\frac{-5}{4}\right| = \left|-\frac{5}{4}\right| = \frac{5}{4} \quad \text{and} \quad \frac{|c|}{|d|} = \frac{|-5|}{|4|} = \frac{5}{4},$$

so $\left|\dfrac{c}{d}\right| = \dfrac{|c|}{|d|}$. ∎

When c is a positive number, then $\sqrt{c^2} = c$, but when c is negative, this is *false*. For example, if $c = -3$, then

$$\sqrt{c^2} = \sqrt{(-3)^2} = \sqrt{9} = 3 \quad (not \; -3),$$

so $\sqrt{c^2} \neq c$. In this case, however, $|c| = |-3| = 3$, so $\sqrt{c^2} = |c|$. The same thing is true for any negative number c. It is also true for positive numbers (since $|c| = c$ when c is positive). In other words,

Square Roots of Squares

For every real number c,

$$\sqrt{c^2} = |c|.$$

When dealing with long expressions inside absolute value bars, do the computations inside first, and then take the absolute value.

EXAMPLE 6

(a) $|5(2 - 4) + 7| = |5(-2) + 7| = |-10 + 7| = |-3| = 3$.

(b) $4 - |3 - 9| = 4 - |-6| = 4 - 6 = -2$. ∎

CAUTION

When c and d have opposite signs, $|c + d|$ is *not equal* to $|c| + |d|$. For example, when $c = -3$ and $d = 5$, then

$$|c + d| = |-3 + 5| = 2,$$

but

$$|c| + |d| = |-3| + |5| = 3 + 5 = 8.$$

The caution shows that $|c + d| < |c| + |d|$ when $c = -3$ and $d = 5$. In the general case, we have the following fact.

The Triangle Inequality

> For any real numbers c and d,
>
> $$|c + d| \le |c| + |d|.$$

■■ DISTANCE ON THE NUMBER LINE

Observe that the distance from -5 to 3 on the number line is 8 units:

Figure 1–4

This distance can be expressed in terms of absolute value by noting that

$$|(-5) - 3| = 8.$$

That is, the distance is the *absolute value of the difference* of the two numbers. Furthermore, the order in which you take the difference doesn't matter; $|3 - (-5)|$ is also 8. This reflects the geometric fact that the distance from -5 to 3 is the same as the distance from 3 to -5. The same thing is true in the general case.

Distance on the Number Line

> The distance between c and d on the number line is the number
>
> $$|c - d| = |d - c|.$$

EXAMPLE 7

The distance from 4.2 to 9 is $|4.2 - 9| = |-4.8| = 4.8$, and the distance from 6 to $\sqrt{2}$ is $|6 - \sqrt{2}|$. ■

When $d = 0$, the distance formula shows that $|c - 0| = |c|$. Hence,

Distance to Zero

$|c|$ is the distance between c and 0 on the number line.

Algebraic problems can sometimes be solved by translating them into equivalent geometric problems. The key is to interpret statements involving absolute value as statements about distance on the number line.

EXAMPLE 8

Solve the equation $|x + 5| = 3$ geometrically.

SOLUTION   We rewrite it as $|x - (-5)| = 3$. In this form it states that

The distance between x and −5 is 3 units.[*]

Figure 1–5 shows that -8 and -2 are the only two numbers whose distance to -5 is 3 units:

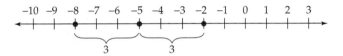

Figure 1–5

Thus $x = -8$ and $x = -2$ are the solutions of $|x + 5| = 3$.   ∎

EXAMPLE 9

The solutions of $|x - 1| \geq 2$ are all numbers x such that

The distance between x and 1 is greater than or equal to 2.

Figure 1–6 shows that the numbers 2 or more units away from 1 are the numbers x such that

$$x \leq -1 \qquad \text{or} \qquad x \geq 3.$$

So these numbers are the solutions of the inequality.   ∎

Figure 1–6

EXAMPLE 10

The solutions of $|x - 7| < 2.5$ are all numbers x such that

The distance between x and 7 is less than 2.5.

---

*It's necessary to rewrite the equation first because the distance formula involves the *difference* of two numbers, not their sum.

Figure 1–7 shows that the solutions of the inequality, that is, the numbers within 2.5 units of 7, are the numbers x such that $4.5 < x < 9.5$, that is, the interval $(4.5, 9.5)$. ■

Figure 1–7

CALCULATOR INVESTIGATIONS 1.1

1. EDIT AND REPLAY

Consider the equation $y = x^3 + 6x^2 - 5$.

(a) Find the value of y when $x = -7$ by keying in

$$(*) \qquad (-7)^3 + 6(-7)^2 - 5$$

and pressing ENTER.[†]

(b) To find the value of y when $x = 9$, without retyping the entire calculation, use the **replay** feature as follows:

Calculator	Procedure
TI	Press SECOND ENTER.
Casio	Press left or right arrow key *or* press AC and the up arrow key.
HP-39	Use up arrow key to shade the previous computation; then press COPY.

Your screen now returns to the previous calculation ($*$). Use the left/right arrow keys and the DEL(ete) key to move through the equation and replace -7 by 9. Press ENTER, and the result of the new computation is displayed.

(c) Use the replay feature again to find the value of y when $x = 108$. On some calculators, you might need to use the INS(ert) key when replacing 9 by 108 to avoid typing over some of the computation.

(d) Press the replay keys repeatedly [on Casio, press AC and then press the up arrow repeatedly]. The calculator will display all the preceding computations, in reverse order. Go back to the first one ($*$) and compute it again.

2. MATHEMATICAL OPERATIONS

(a) Key in each of the following, and explain why your answers are different. [See the Technology Tip on page 9.]

ABS(−) 9 + 2 ENTER   and   ABS ((−) 9 + 2) ENTER

(b) Find INT or FLOOR in the NUM or REAL submenu of the MATH or OPTN menu. Find out what this command does when you follow it by a number and ENTER. Try all kinds of numbers (positive, negative, decimals, fractions). Answers on Casio will sometimes differ from those on other calculators.

[†]Here and throughout the book, Casio users should read "EXE" in place of "ENTER."

3. SYMBOLIC CALCULATIONS

(a) To store the number 2 in memory A of a calculator, type

    2 STO ▶ A   ENTER   [TI-85/86] or

    2 STO ▶ ALPHA A ENTER   [TI-82/83/89, HP-39] or

    2 → ALPHA A EXE   [Casio].

If you now key in ALPHA A ENTER, what does the calculator display?

(b) In a similar fashion, store the number 5 in memory B and -10 in memory C. Then, using the ALPHA keys, display this expression on the screen: B + C/A. If you press ENTER, what happens? Explain what the calculator is doing.

(c) Experiment with other expressions, such as $B^2 - 4AC$.

4. INEQUALITIES

Find out what happens when you key in each of these statements and press ENTER:

$$8 < 9, \qquad 8 < 5, \qquad 9 > 2, \qquad 9 > 10.$$

[On TI and HP, the inequality symbols are located in the TEST(S) or INEQ (sub)menu; look on the keyboard or in the MATH menu. On Casio they are in the REL or LOGIC submenu of the PRGM menu, which is accessed from the home screen by pressing SHIFT VARS.]

✓ EXERCISES 1.1

1. Draw a number line and mark the location of each of these numbers: 0, -7, 8/3, 10, -1, -4.75, 1/2, -5, and 2.25.

2. Use your calculator to determine which of the following rational numbers is the best approximation of the irrational number π.

$$\frac{22}{7}, \qquad \frac{355}{113}, \qquad \frac{103{,}993}{33{,}102}, \qquad \frac{2{,}508{,}429{,}787}{798{,}458{,}000}.$$

If your calculator says that one of these numbers equals π, it's lying. All you can conclude is that the number agrees with π for as many decimal places as your calculator can handle (usually 12–14).

In Exercises 3–7, express the given numbers (based on 2002 estimates) in scientific notation.

3. Population of the world: 6,785,000,000

4. Population of the United States: 284,800,000

5. Mean distance from Earth to Pluto: 5,910,000,000,000 meters

6. Radius of a hydrogen atom: .00000000001 meter

7. Width of a DNA double helix: .000000002 meter

In Exercises 8–11, express the given number in normal decimal notation.

8. Speed of light in a vacuum: 2.9979×10^8 miles per second

9. Average distance from the earth to the sun: 1.50×10^{11} meters

10. Electron charge: 1.602×10^{-27} coulomb

11. Proton mass: 1.6726×10^{-19} kilogram

12. One light-year is the distance light travels in a 365-day year. The speed of light is about 186,282.4 miles per second.

    (a) How long is 1 light-year (in miles)? Express your answer in scientific notation.

    (b) Light from the North Star takes 680 years to reach the earth. How many miles is the North Star from the earth?

13. The gross federal debt was about 5626 billion dollars in 2000, when the U.S. population was approximately 281.3 million people.

    (a) Express the debt and the population in scientific notation.

    (b) At that time, what was each person's share of the federal debt?

In Exercises 14–22, express the given statement in symbols.

14. -4 is greater than -8.

15. -17 is less than 14.

16. π is less than 100.

17. x is nonnegative.

18. z is greater than or equal to -4.

19. t is positive.

20. d is not greater than 2.

21. c is at most 3.

22. z is at least -17.

In Exercises 23–27, fill the blank with $<$, $=$, or $>$ so that the resulting statement is true.

23. -6 ____ -2 **24.** 5 ____ -3

25. $3/4$ ____ $.75$ **26.** 3.1 ____ π

27. $1/3$ ____ $.33$

In Exercises 28–34, fill the blank so as to produce two equivalent statements. For example, the arithmetic statement "a is negative" is equivalent to the geometric statement "the point a lies to the left of the point 0."

Arithmetic Statement	Geometric Statement
28. $a \geq b$	_____
29. _____	a lies c units to the right of b
30. _____	a lies between b and c
31. $a - b > 0$	_____
32. a is positive	_____
33. _____	a lies to the left of b
34. $a + b < c$ $(b > 0)$	_____

In Exercises 35–40, draw a picture on the number line of the given interval.

35. $(0, 8]$ **36.** $(0, \infty)$ **37.** $[-2, 1]$

38. $(-1, 1)$ **39.** $(-\infty, 0]$ **40.** $[-2, 7)$

In Exercises 41–46, use interval notation to denote the set of all real numbers x that satisfy the given inequality.

41. $5 \leq x \leq 8$ **42.** $-2 \leq x \leq 7$

43. $-3 < x < 14$ **44.** $7 < x < 135$

45. $x \geq -8$ **46.** $x \geq 12$

In Exercises 47–56, simplify, and write the given number without using absolute values.

47. $|3 - 14|$ **48.** $|(-2)3|$ **49.** $3 - |2 - 5|$

50. $-2 - |-2|$ **51.** $|(-13)^2|$ **52.** $-|-5|^2$

53. $|\pi - \sqrt{2}|$ **54.** $|\sqrt{2} - 2|$ **55.** $|3 - \pi| + 3$

56. $|4 - \sqrt{2}| - 5$

In Exercises 57–62, fill the blank with $<$, $=$, or $>$ so that the resulting statement is true.

57. $|-2|$ ____ $|-5|$ **58.** 5 ____ $|-2|$

59. $|3|$ ____ $-|4|$ **60.** $|-3|$ ____ 0

61. -7 ____ $|-1|$ **62.** $-|-4|$ ____ 0

In Exercises 63–70, find the distance between the given numbers.

63. -3 and 4 **64.** 7 and 107

65. -7 and $15/2$ **66.** $-3/4$ and -10

67. π and 3 **68.** π and -3

69. $\sqrt{2}$ and $\sqrt{3}$ **70.** π and $\sqrt{2}$

71. Galileo discovered that the period of a pendulum depends only on the length of the pendulum and the acceleration of gravity. The period T of a pendulum (in seconds) is

$$T = 2\pi\sqrt{\frac{l}{g}},$$

where l is the length of the pendulum in feet and $g \approx 32.2$ ft/sec^2 is the acceleration due to gravity. Find the period of a pendulum whose length is 4 feet.

72. Suppose you are k miles (not feet) above the ground. The radius of the earth is approximately 3960 miles, and the point where your line of sight meets the earth is perpendicular to the radius of the earth at that point, as shown in the figure.

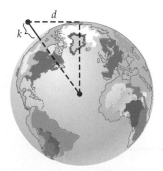

(a) Use the Pythagorean Theorem (see the Geometry Review Appendix) to show that

$$d = \sqrt{(3960 + k)^2 - 3960^2}.$$

(b) Show that the equation in part (a) simplifies to $d = \sqrt{7920k + k^2}$.

(c) If you are h feet above the ground, then you are $h/5280$ miles high (why?). Use this fact and the equation in part (b) to obtain the formula used in Example 3.

73. According to data from the Center for Science in the Public Interest, the healthy weight range for a person depends on the person's height. For example,

Height	Healthy Weight Range (lb)
5 ft 8 in.	143 ± 21
6 ft 0 in.	163 ± 26

Express each of these ranges as an absolute value inequality in which x is the weight of the person.

74. At Statewide Insurance, each department's expenses are reviewed monthly. A department can fail to pass the budget variance test in a category if either (i) the absolute value of the difference between actual expenses and the budget is more than $500 or (ii) the absolute value of the difference between the actual expenses and the budget is more than 5% of the budgeted amount. Which of the following items fail the budget variance test? Explain your answers.

Item	Budgeted Expense ($)	Actual Expense ($)
Wages	220,750	221,239
Overtime	10,500	11,018
Shipping and Postage	530	589

In Exercises 75–82, write the given expression without using absolute values.

75. $|t^2|$

76. $|-2 - y^2|$

77. $|b - 3|$ if $b \geq 3$

78. $|a - 5|$ if $a < 5$

79. $|c - d|$ if $c < d$

80. $|c - d|$ if $c \geq d$

81. $|u - v| - |v - u|$

82. $\dfrac{|u - v|}{|v - u|}$ if $u \neq v, u \neq 0, v \neq 0$

In Exercises 83 and 84, explain why the given statement is true for any numbers c and d. [Hint: Look at the properties of absolute value on page 9.]

83. $|(c - d)^2| = c^2 - 2cd + d^2$

84. $\sqrt{9c^2 - 18cd + 9d^2} = 3|c - d|$

In Exercises 85–90, express the given geometric statement about numbers on the number line algebraically, using absolute values.

85. The distance from x to 5 is less than 4.

86. x is more than 6 units from c.

87. x is at most 17 units from -4.

88. x is within 3 units of 7.

89. c is closer to 0 than b is.

90. x is closer to 1 than to 4.

In Exercises 91–94, translate the given algebraic statement into a geometric statement about numbers on the number line.

91. $|x - 3| < 2$

92. $|x - c| > 6$

93. $|x + 7| \leq 3$

94. $|u + v| \geq 2$

95. Match each of the following graphs with the appropriate absolute value equation or inequality.

(a) i. $|x - 17| = 7$

(b) ii. $|x - 17| = -7$

(c) iii. $|x - 17| \leq 7$

(d) iv. $|x - 17| \geq -7$

(e) v. $|x - 17| > 7$

96. Explain geometrically why this statement is always false:

$$|c - 1| < 2 \text{ and simultaneously } |c - 12| < 3.$$

In Exercises 97–108, use the geometric approach explained in the text to solve the given equation or inequality.

97. $|x| = 1$

98. $|x| = 3/2$

99. $|x - 2| = 1$

100. $|x + 3| = 2$

101. $|x + \pi| = 4$

102. $\left|x - \frac{3}{2}\right| = 5$

103. $|x| < 7$

104. $|x| \geq 5$

105. $|x - 5| < 2$

106. $|x - 6| > 2$

107. $|x + 2| \geq 3$

108. $|x + 4| \leq 2$

Thinkers

109. Explain why the statement $|a| + |b| + |c| > 0$ is algebraic shorthand for "at least one of the numbers a, b, c, is different from zero."

110. Find an algebraic shorthand version of the statement "none of the numbers a, b, c, is zero."

1.1.A   *SPECIAL TOPICS* Decimal Representation of Real Numbers

Every rational number can be expressed as a terminating or repeating decimal. For instance, $3/4 = 0.75$. To express $15/11$ as a decimal, divide the numerator by the denominator:

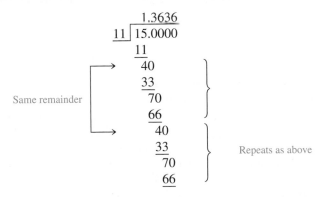

Since the remainder at the first step (namely, 4) occurs again at the third step, it is clear that the division process goes on forever with the two-digit block "36" repeating over and over in the quotient $15/11 = 1.3636363636 \cdots$.

The method used in the preceding example can be used to express any rational number as a decimal. During the division process, some remainder *necessarily repeats*. If the remainder at which this repetition starts is 0, the result is a repeating decimal ending in zeros—that is, a terminating decimal (for instance, $.75000 \cdots = .75$). If the remainder at which the repetition starts is nonzero, then the result is a nonterminating repeating decimal, as in the example above.

Conversely, there is a simple method for converting any repeating decimal into a rational number.

EXAMPLE 1

Write $d = .272727 \cdots$ as a rational number.

SOLUTION   Assuming that the usual rules of arithmetic hold, we see that

$$100d = 27.272727 \cdots \qquad \text{and} \qquad d = .272727 \cdots.$$

Now subtract d from $100d$:

$$
\begin{aligned}
100d &= 27.272727 \cdots \\
-d &= -.272727 \cdots \\
\hline
99d &= 27
\end{aligned}
$$

Dividing both sides of this last equation by 99 shows that $d = 27/99 = 3/11$. ∎

*The calculator denotes $12\frac{3}{4}$, for example, by 12⌐3⌐4 and denotes 15/29 by 15⌐29.

∷ IRRATIONAL NUMBERS

Many nonterminating decimals are *nonrepeating* (that is, no block of digits repeats forever), such as .202002000200002 · · · (where after each 2 there is one more zero than before). Although the proof is too long to give here, it is in fact true that every nonterminating and nonrepeating decimal represents an *irrational* real number. Conversely every irrational number can be expressed as a nonterminating and nonrepeating decimal (no proof to be given here).

A typical calculator can hold only the first 10–14 digits of a number in decimal form. Consequently, a calculator can contain the *exact* value only of those rational numbers whose decimal expansion terminates after 10–14 places. It must *approximate* all other real numbers.

Since every real number is either a rational number or an irrational one, the preceding discussion can be summarized as follows.

Decimal Representation

1. Every real number can be expressed as a decimal.

2. Every decimal represents a real number.

3. The terminating decimals and the nonterminating repeating decimals are the rational numbers.

4. The nonterminating, nonrepeating decimals are the irrational numbers.

CALCULATOR INVESTIGATIONS 1.1.A

1. FRAC KEY

If your calculator has a FRAC key or program key or program (see the Program Appendix), test its limitations by entering each of the following numbers and then pressing the FRAC key.

(a) .058823529411   (b) .0588235294117

(c) .058823529411724   (d) .0588235294117985

Which of your answers are correct? [*Hint:* Use Exercise 25 below to find the decimal expansion of 1/17.]

 EXERCISES 1.1.A

In Exercises 1–6, express the given rational number as a repeating decimal.

1. 7/9   **2.** 2/13   **3.** 23/14

4. 19/88   **5.** 1/19 (long)   **6.** 9/11

In Exercises 7–13, express the given repeating decimal as a fraction.

7. .373737 · · ·   **8.** .929292 · · ·

9. 76.63424242 · · · [*Hint:* Consider $10{,}000d - 100d$, where $d = 76.63424242 \cdots$.]

10. 13.513513 · · · [*Hint:* Consider $1000d - d$, where $d = 13.513513 \cdots$.]

11. .135135135 · · · [*Hint:* See Exercise 10.]

12. .33030303 · · ·   **13.** 52.31272727 · · ·

14. If two real numbers have the same decimal expansion through three decimal places, how far apart can they be on the number line?

In Exercises 15–22, state whether a calculator can express the given number exactly.

15. 2/3   **16.** 7/16   **17.** 1/64   **18.** 1/22

19. $3\pi/2$ **20.** $\pi - 3$ **21.** 1/.625   **22.** 1/.16

Thinkers

23. Use the methods in Exercises 7–13 to show that both .74999 · · · and .75000 · · · are decimal expansions of 3/4. [Every terminating decimal can also be expressed as a decimal ending in repeated 9's. It can be proved that these are the only real numbers with more than one decimal expansion.]

Finding remainders with a calculator

24. If you use long division to divide 369 by 7, you obtain:

$$
\begin{array}{r}
52 \quad \leftarrow \text{Quotient} \\
\text{Divisor} \rightarrow \quad 7\overline{)\,369} \quad \leftarrow \text{Dividend} \\
\underline{35} \\
19 \\
\underline{14} \\
5 \quad \leftarrow \text{Remainder}
\end{array}
$$

If you use a calculator to find 369 ÷ 7, the answer is displayed as 52.71428571. Observe that the integer part of this calculator answer, 52, is the quotient when you do the problem by long division. The usual "checking procedure" for long division shows that

$$7 \cdot 52 + 5 = 369 \quad \text{or, equivalently} \quad 369 - 7 \cdot 52 = 5.$$

Thus, the remainder is

$$\text{Dividend} - (\text{divisor})\left(\begin{array}{c}\text{integer part of} \\ \text{calculator answer}\end{array}\right).$$

Use this method to find the quotient and remainder in these problems:

(a) 5683 ÷ 9   (b) 1,000,000 ÷ 19
(c) 53,000,000 ÷ 37

In Exercises 25–30, find the decimal expansion of the given rational number. All these expansions are too long to fit in a calculator but can be readily found by using the hint in Exercise 25.

25. 1/17 [*Hint:* The first part of dividing 1 by 17 involves working this division problem: 1,000,000 ÷ 17. The method of Exercise 24 shows that the quotient is 58,823 and the remainder is 9. Thus the decimal expansion of 1/17 begins .058823, and the next block of digits in the expansion will be the quotient in the problem 9,000,000 ÷ 17. The remainder when 9,000,000 is divided by 17 is 13, so the next block of digits in the expansion of 1/17 is the quotient in the problem 13,000,000 ÷ 17. Continue in this way until the decimal expansion repeats.]

26. 3/19   **27.** 1/29   **28.** 3/43   **29.** 283/47

30. 768/59

31. (a) Show that there are at least as many irrational numbers (nonrepeating decimals) as there are terminating decimals. [*Hint:* With each terminating decimal associate a nonrepeating decimal.]

(b) Show that there are at least as many irrational numbers as there are repeating decimals. [*Hint:* With each repeating decimal, associate a nonrepeating decimal by inserting longer and longer strings of zeros: for instance, with .11111111 · · · associate the number .101001000100001 · · ·.]

1.2   Solving Equations Algebraically

This section deals with equations such as

$$3x - 6 = 7x + 4, \qquad x^2 - 5x + 6 = 0, \qquad 2x^4 - 13x^2 = 3.$$

A **solution** of an equation is a number that, when substituted for the variable *x,* produces a true statement.* For example, 5 is a solution of $3x + 2 = 17$ because $3 \cdot 5 + 2 = 17$ is a true statement. To **solve** an equation means to find all its solutions. Throughout this chapter, we shall deal only with **real solutions,** that is, solutions that are real numbers.

Two equations are said to be **equivalent** if they have the same solutions. For example, $3x + 2 = 17$ and $x - 2 = 3$ are equivalent because 5 is the only solution of each one.

---

*Any letter may be used for the variable.

Basic Principles for
Solving Equations

Performing any of the following operations on an equation produces an equivalent equation:

1. Add or subtract the same quantity from both sides of the equation.

2. Multiply or divide both sides of the equation by the same *nonzero* quantity.

The usual strategy in equation solving is to use these basic principles to transform a given equation into an equivalent one whose solutions are known.

A **first-degree,** or **linear, equation** is one that can be written in the form

$$ax + b = 0$$

for some constants *a, b,* with $a \neq 0$. Every first-degree equation has exactly one solution, which is easily found.

EXAMPLE 1

To solve $3x - 6 = 7x + 4$, we use the basic principles to transform this equation into an equivalent one whose solution is obvious:

$$3x - 6 = 7x + 4$$

Add 6 to both sides: $\qquad 3x = 7x + 10$

Subtract 7x from both sides: $\qquad -4x = 10$

Divide both sides by -4: $\qquad x = \dfrac{10}{-4} = -\dfrac{5}{2}.$

Since $-5/2$ is the only solution of this last equation, $-5/2$ is the only solution of the original equation, $3x - 6 = 7x + 4$. ■

EXAMPLE 2

The surface area *S* of the rectangular box in Figure 1–8 is given by

$$2lh + 2lw + 2wh = S.$$

Solve this equation for *h*.

SOLUTION   Treat *h* as the variable and all the other letters as constants. First, get all the terms involving the variable *h* on one side of the equation and everything else on the other side.

$$2lh + 2lw + 2wh = S$$

Sutract 2*lw* from both sides: $\qquad 2lh + 2wh = S - 2lw$

Factor out *h* on the left side: $\qquad (2l + 2w)h = S - 2lw$

Divide both sides by $(2l + 2w)$: $\qquad h = \dfrac{S - 2lw}{2l + 2w}.$ ■

Figure 1–8

⠿ QUADRATIC EQUATIONS

A **second-degree,** or **quadratic, equation** is an equation that can be written in the form

$$ax^2 + bx + c = 0$$

for some constants a, b, c, with $a \neq 0$. There are several techniques for solving such equations. We begin with the **factoring method,** which makes use of this property of the real numbers:

Zero Products

If a product of real numbers is zero, then at least one of the factors is zero; in other words,

If $cd = 0$, then $c = 0$ or $d = 0$ (or both).

EXAMPLE 3

To solve $3x^2 - x = 10$, we first rearrange the terms to make one side 0 and then factor:

Subtract 10 from each side: $3x^2 - x - 10 = 0$

Factor left side: $(3x + 5)(x - 2) = 0.$

If a product of real numbers is 0, then at least one of the factors must be 0. So this equation is equivalent to

$$3x + 5 = 0 \qquad \text{or} \qquad x - 2 = 0$$
$$3x = -5 \qquad\qquad x = 2$$
$$x = -5/3$$

CAUTION To guard against mistakes, check your solutions by substituting each one in the *original* equation to make sure it really *is* a solution.

Therefore the solutions are $-5/3$ and 2.   ∎

The solutions of $x^2 = 7$ are the numbers whose square is 7. Although 7 has just *one* square root, there are *two* numbers whose square is 7, namely, $\sqrt{7}$ and its negative, $-\sqrt{7}$. So the solutions of $x^2 = 7$ are $\sqrt{7}$ and $-\sqrt{7}$, or in abbreviated form, $\pm\sqrt{7}$. The same argument works for any positive real number d:

The solutions of $x^2 = d$ are $\sqrt{d}$ and $-\sqrt{d}$.

We now use a slight variation of this idea to develop a method for solving quadratic equations that don't readily factor. Consider, for example, the expression $x^2 + 6x$. If you add 9, the result is a perfect square:

$$x^2 + 6x + 9 = (x + 3)^2.$$

The process of adding a constant to produce a perfect square is called **completing the square.** Note that one-half the coefficient of x in $x^2 + 6x$ is $\dfrac{6}{2} = 3$. We added 9, which is 3^2, and the resulting perfect square was $(x + 3)^2$. The same idea works in the general case, as is proved in Exercise 78.

**Completing
the Square**

> To change $x^2 + bx$ into a perfect square, add $\left(\dfrac{b}{2}\right)^2$. The resulting polynomial factors as $\left(x + \dfrac{b}{2}\right)^2$.

The following example shows how completing the square can be used to solve quadratic equations.

EXAMPLE 4

To solve $x^2 + 6x + 1 = 0$, we first rewrite the equation as $x^2 + 6x = -1$. Next we complete the square on the left side by adding the square of half the coefficient of x, namely, $(6/2)^2 = 9$. To have an equivalent equation, we must add 9 to *both* sides:

$$x^2 + 6x + 9 = -1 + 9$$

Factor left side: $\qquad (x + 3)^2 = 8.$

Thus $x + 3$ is a number whose square is 8. The only numbers whose squares equal 8 are $\sqrt{8}$ and $-\sqrt{8}$. So we must have

$$x + 3 = \sqrt{8} \qquad \text{or} \qquad x + 3 = -\sqrt{8}$$
$$x = \sqrt{8} - 3 \qquad \text{or} \qquad x = -\sqrt{8} - 3.$$

Therefore the solutions of the original equation are $\sqrt{8} - 3$ and $-\sqrt{8} - 3$ or, in more compact notation, $\pm\sqrt{8} - 3$. ∎

CAUTION

Completing the square only works when the coefficient of x^2 is 1. In an equation such as

$$5x^2 - x + 2 = 0,$$

you must first divide every term on both sides by 5 and *then* complete the square.

We can use the completing-the-square method to solve *any* quadratic equation:*

$$ax^2 + bx + c = 0$$

Divide both sides by a: $\qquad x^2 + \dfrac{b}{a}x + \dfrac{c}{a} = 0$

Subtract $\dfrac{c}{a}$ from both sides: $\qquad x^2 + \dfrac{b}{a}x = -\dfrac{c}{a}$

Add $\left(\dfrac{b}{2a}\right)^2$ to both sides:[†] $\quad x^2 + \dfrac{b}{a}x + \left(\dfrac{b}{2a}\right)^2 = \left(\dfrac{b}{2a}\right)^2 - \dfrac{c}{a}$

Factor left side: $\qquad \left(x + \dfrac{b}{2a}\right)^2 = \left(\dfrac{b}{2a}\right)^2 - \dfrac{c}{a}$

Find common denominator for right side: $\quad \left(x + \dfrac{b}{2a}\right)^2 = \dfrac{b^2}{4a^2} - \dfrac{c}{a} = \dfrac{b^2 - 4ac}{4a^2}.$

---

*If you have trouble following any step here, do it for a numerical example, such as the case when $a = 3$, $b = 11$, $c = 5$.
[†]This is the square of half the coefficient of x.

Since the square of $x + \dfrac{b}{2a}$ equals $\dfrac{b^2 - 4ac}{4a^2}$, we must have

$$x + \frac{b}{2a} = \pm\sqrt{\frac{b^2 - 4ac}{4a^2}} = \pm\frac{\sqrt{b^2 - 4ac}}{2a}$$

Subtract $\dfrac{b}{2a}$ from both sides:

$$x = \frac{-b}{2a} \pm \frac{\sqrt{b^2 - 4ac}}{2a} = \frac{-b \pm \sqrt{b^2 - 4ac}}{2a}.$$

We have proved

The Quadratic Formula

> The solutions of the quadratic equation $ax^2 + bx + c = 0$ are
> $$x = \frac{-b \pm \sqrt{b^2 - 4ac}}{2a}.$$

You should memorize the quadratic formula.

EXAMPLE 5

Solve $x^2 + 3 = -8x$.

SOLUTION   Rewrite the equation as $x^2 + 8x + 3 = 0$, and apply the quadratic formula with $a = 1$, $b = 8$, and $c = 3$:

$$x = \frac{-b \pm \sqrt{b^2 - 4ac}}{2a} = \frac{-8 \pm \sqrt{8^2 - 4 \cdot 1 \cdot 3}}{2 \cdot 1}$$

$$= \frac{-8 \pm \sqrt{52}}{2} = \frac{-8 \pm \sqrt{4 \cdot 13}}{2}$$

$$= \frac{-8 \pm 2\sqrt{13}}{2} = -4 \pm \sqrt{13}.$$

The equation has two distinct real solutions, $-4 + \sqrt{13}$ and $-4 - \sqrt{13}$.   ∎

EXAMPLE 6

Solve $x^2 - 194x + 9409 = 0$.

SOLUTION   Use a calculator and the quadratic formula with $a = 1$, $b = -194$, and $c = 9409$:

$$x = \frac{-b \pm \sqrt{b^2 - 4ac}}{2a} = \frac{-(-194) \pm \sqrt{(-194)^2 - 4 \cdot 1 \cdot 9409}}{2 \cdot 1}$$

$$= \frac{194 \pm \sqrt{37636 - 37636}}{2} = \frac{194 \pm 0}{2} = 97.$$

Thus, 97 is the only solution of the equation.   ∎

EXAMPLE 7

Solve $2x^2 + x + 3 = 0$.

SOLUTION   Using the quadratic formula with $a = 2$, $b = 1$, and $c = 3$:

$$x = \frac{-b \pm \sqrt{b^2 - 4ac}}{2a} = \frac{-1 \pm \sqrt{1^2 - 4 \cdot 2 \cdot 3}}{2 \cdot 2} = \frac{-1 \pm \sqrt{1 - 24}}{4}$$

$$= \frac{-1 \pm \sqrt{-23}}{4}.$$

Since $\sqrt{-23}$ is not a real number, this equation has *no real solutions* (that is, no solutions in the real number system). ■

EXAMPLE 8

If an object is thrown upward, dropped, or thrown downward and travels in a straight line subject only to gravity (with wind resistance ignored), the height h of the object above the ground (in feet) after t seconds is given by

$$h = -16t^2 + v_0 t + h_0,$$

where h_0 is the height of the object when $t = 0$ and v_0 is the initial velocity at time $t = 0$. The value of v_0 is taken as positive if the object moves upward and negative if it moves downward. If a baseball is thrown down from the top of a 640-foot-high building with an initial velocity of 52 feet per second, how long does it take to reach the ground?

SOLUTION   In this case, v_0 is -52 and h_0 is 640, so that the height equation is

$$h = -16t^2 - 52t + 640.$$

The object is on the ground when $h = 0$, so we must solve the equation

$$0 = -16t^2 - 52t + 640.$$

Using the quadratic formula and a calculator, we see that

$$t = \frac{-(-52) \pm \sqrt{(-52)^2 - 4(-16)(640)}}{2(-16)} = \frac{52 \pm \sqrt{43{,}664}}{-32} = \begin{cases} -8.15 \\ \text{or} \\ 4.90. \end{cases}$$

Only the positive answer makes sense in this case. So it takes about 4.9 seconds for the baseball to reach the ground. ■

The expression $b^2 - 4ac$ in the quadratic formula is called the **discriminant.** As the last four examples illustrate, the discriminant determines the *number* of real solutions of the equation $ax^2 + bx + c = 0$.

Real Solutions of a
Quadratic Equation

Discriminant $b^2 - 4ac$	Number of Real Solutions of $ax^2 + bx + c = 0$	Example
> 0	Two distinct real solutions	$x^2 + 8x + 3 = 0$
$= 0$	One real solution	$x^2 - 194x + 9409 = 0$
< 0	No real solutions	$2x^2 + x + 3 = 0$

The quadratic formula and a calculator can be used to solve any quadratic equation with nonnegative discriminant. Experiment with your calculator to find the most efficient sequence of keystrokes for doing this.

EXAMPLE 9

Use a calculator to solve $3.2x^2 + 15.93x - 7.1 = 0$.

First compute $\sqrt{b^2 - 4ac} = \sqrt{15.93^2 - 4(3.2)(-7.1)}$ and store the result in memory D. Then the solutions of the equation are given by

$$x = \frac{-b \pm \sqrt{b^2 - 4ac}}{2a} = \frac{-15.93 \pm D}{2(3.2)}.$$

They are easily approximated on a calculator:

$$\frac{-15.93 + D}{2(3.2)} \approx .411658467 \qquad \text{and} \qquad \frac{-15.93 - D}{2(3.2)} \approx -5.389783347.$$

Remember that these answers are *approximations,* so they might not check exactly when substituted in the original equation. ∎

▓ HIGHER-DEGREE EQUATIONS

A **polynomial equation of degree n** is one that can be written in the form

$$a_n x^n + \cdots + a_3 x^3 + a_2 x^2 + a_1 x + a_0 = 0,$$

where n is a positive integer, each a_i is a constant, and $a_n \neq 0$. For instance,

$$4x^6 - 3x^5 + x^4 + 7x^3 - 8x^2 + 4x + 9 = 0$$

is a polynomial equation of degree 6. As a general rule, polynomial equations of degree 3 and above are best solved by the numerical or graphical methods presented in Section 2.2. However, some such equations can be solved algebraically by making a suitable substitution, as we now see.

TECHNOLOGY TIP

Most calculators have built-in polynomial equation solvers that will solve quadratic and other polynomial equations. See Calculator Investigation 1 at the end of this section. A quadratic formula program for other calculators is in the Program Appendix.

EXAMPLE 10

To solve $4x^4 - 13x^2 + 3 = 0$, substitute u for x^2 and solve the resulting quadratic equation:

$$4x^4 - 13x^2 + 3 = 0$$
$$4(x^2)^2 - 13x^2 + 3 = 0$$
$$4u^2 - 13u + 3 = 0$$
$$(u - 3)(4u - 1) = 0$$
$$u - 3 = 0 \quad \text{or} \quad 4u - 1 = 0$$
$$u = 3 \qquad\qquad 4u = 1$$
$$u = \frac{1}{4}.$$

Since $u = x^2$, we see that

$$x^2 = 3 \qquad \text{or} \qquad x^2 = \frac{1}{4}$$
$$x = \pm\sqrt{3} \qquad\qquad x = \pm\frac{1}{2}.$$

Hence, the original equation has four solutions: $-\sqrt{3}, \sqrt{3}, -1/2, 1/2$. ∎

EXAMPLE 11

To solve $x^4 - 4x^2 + 1 = 0$, let $u = x^2$:

$$x^4 - 4x^2 + 1 = 0$$
$$u^2 - 4u + 1 = 0.$$

The quadratic formula shows that

$$u = \frac{-(-4) \pm \sqrt{(-4)^2 - 4 \cdot 1 \cdot 1}}{2 \cdot 1} = \frac{4 \pm \sqrt{12}}{2}$$
$$= \frac{4 \pm \sqrt{4 \cdot 3}}{2} = \frac{4 \pm 2\sqrt{3}}{2} = 2 \pm \sqrt{3}.$$

Since $u = x^2$, we have the equivalent statements:

$$x^2 = 2 + \sqrt{3} \qquad \text{or} \qquad x^2 = 2 - \sqrt{3}$$
$$x = \pm\sqrt{2 + \sqrt{3}} \qquad\qquad x = \pm\sqrt{2 - \sqrt{3}}.$$

Therefore the original equation has four solutions. ∎

CALCULATOR

INVESTIGATIONS 1.2

1. POLYNOMIAL EQUATION SOLVERS

If you have one of the calculators mentioned below, use its polynomial equation solver to solve the equation $3.2x^2 + 15.93x - 7.1 = 0$. How do the answers you obtain compare with those in Example 9?

Solver Name and Location	Special Syntax (if needed)
TI-86 POLY on Keyboard	For "order," enter the degree of the polynomial.
TI-89 SOLVE in ALGEBRA menu	SOLVE $(3.2x^2 + 15.93x - 7.1 = 0, x)$
Casio EQUATION in main menu	
HP-39 POLYROOT in POLYNOM submenu of MATH menu	POLYROOT([3.2, 15.93, −7.1])

Note: TI-83 Plus and TI-89 users can download a polynomial solver similar to the one on TI-86.

Use the same technique to solve the following equations.

(a) $x^2 + 2x - 4 = 0$

(b) $x^2 + 5x + 2 = 0$

EXERCISES 1.2

In Exercises 1–6, solve the equation.

1. $3x + 2 = 26$

2. $\frac{y}{5} - 3 = 14$

3. $3x + 2 = 9x + 7$

4. $-7(t + 2) = 3(4t + 1)$

5. $\frac{3y}{4} - 6 = y + 2$

6. $2(1 + x) = 3x + 5$

In Exercises 7–12, solve the equation for the indicated variable.

7. $x = 3y - 5$ for y

8. $5x - 2y = 1$ for x

9. $A = \frac{h}{2}(b + c)$ for b

10. $V = \pi b^2 c$ for c

11. $V = \frac{\pi d^2 h}{4}$ for h

12. $\frac{1}{r} = \frac{1}{s} + \frac{1}{t}$ for r

In Exercises 13–22, solve the equation by factoring.

13. $x^2 - 8x + 15 = 0$

14. $x^2 + 5x + 6 = 0$

15. $x^2 - 5x = 14$

16. $x^2 + x = 20$

17. $2y^2 + 5y - 3 = 0$

18. $3t^2 - t - 2 = 0$

19. $4t^2 + 9t + 2 = 0$

20. $9t^2 + 2 = 11t$

21. $3u^2 + u = 4$

22. $5x^2 + 26x = -5$

In Exercises 23–26, solve the equation by completing the square.

23. $x^2 - 2x = 12$

24. $x^2 - 4x - 30 = 0$

25. $x^2 - x - 1 = 0$

26. $x^2 + 3x - 2 = 0$

In Exercises 27–32, find the number of real solutions of the equation by computing the discriminant.

27. $x^2 + 4x + 1 = 0$

28. $4x^2 - 4x - 3 = 0$

29. $9x^2 = 12x + 1$

30. $9t^2 + 15 = 30t$

31. $25t^2 + 49 = 70t$

32. $49t^2 + 5 = 42t$

In Exercises 33–42, use the quadratic formula to solve the equation.

33. $x^2 - 4x + 1 = 0$

34. $x^2 - 2x - 1 = 0$

35. $x^2 + 6x + 7 = 0$

36. $x^2 + 4x - 3 = 0$

37. $x^2 + 6 = 2x$

38. $x^2 + 11 = 6x$

39. $4x^2 - 4x = 7$

40. $4x^2 - 4x = 11$

41. $4x^2 - 8x + 1 = 0$

42. $2t^2 + 4t + 1 = 0$

In Exercises 43–52, solve the equation by any method.

43. $x^2 + 9x + 18 = 0$

44. $3t^2 - 11t - 20 = 0$

45. $4x(x + 1) = 1$

46. $25y^2 = 20y + 1$

47. $2x^2 = 7x + 15$

48. $2x^2 = 6x + 3$

49. $t^2 + 4t + 13 = 0$

50. $5x^2 + 2x = -2$

51. $\dfrac{7x^2}{3} = \dfrac{2x}{3} - 1$

52. $25x + \dfrac{4}{x} = 20$

In Exercises 53–56, use a calculator to find approximate solutions of the equation.

53. $4.42x^2 - 10.14x + 3.79 = 0$

54. $8.06x^2 + 25.8726x - 25.047256 = 0$

55. $3x^2 - 82.74x + 570.4923 = 0$

56. $7.63x^2 + 2.79x = 5.32$

In Exercises 57–64, find all real solutions of the equation exactly.

57. $y^4 - 7y^2 + 6 = 0$

58. $x^4 - 2x^2 + 1 = 0$

59. $x^4 - 2x^2 - 35 = 0$

60. $x^4 - 2x^2 - 24 = 0$

61. $2y^4 - 9y^2 + 4 = 0$

62. $6z^4 - 7z^2 + 2 = 0$

63. $10x^4 + 3x^2 = 1$

64. $6x^4 - 7x^2 = 3$

In Exercises 65–68, solve the equation and check your answers. [Hint: First, eliminate fractions by multiplying both sides by a common denominator.]

65. $1 - \dfrac{3}{x} = \dfrac{40}{x^2}$

66. $\dfrac{4x^2 + 5}{3x^2 + 5x - 2} = \dfrac{4}{3x - 1} - \dfrac{3}{x + 2}$

67. $\dfrac{2}{x^2} - \dfrac{5}{x} = 4$

68. $\dfrac{x}{x - 1} + \dfrac{x + 2}{x} = 3$

In a simple model of the economy (by John Maynard Keynes), equilibrium between national output and national expenditures is given by the equilibrium equation

$$Y = C + I + G + (X - M),$$

where Y is the national income, C is consumption (which depends on national income), I is the amount of investment, G is government spending, X is exports, and M is imports. In Exercises 69–70, solve the equilibrium equation for Y under the given conditions.

69. $C = 120 + .9Y$, $M = 20 + .2Y$, $I = 140$, $G = 150$, and $X = 60$

70. $C = 60 + .85Y$, $M = 35 + .2Y$, $I = 95$, $G = 145$, and $X = 50$

In Exercises 71–74, use the height equation in Example 8. Note that an object that is dropped (rather than thrown downward) has initial velocity $v_0 = 0$.

71. How long does it take a baseball to reach the ground if it is dropped from the top of a 640-foot-high building? Compare with Example 8.

72. You are standing on a cliff that is 200 feet high. How long will it take a rock to reach the ground if
   (a) you drop it?
   (b) you throw it downward at an initial velocity of 40 feet per second?
   (c) How far does the rock fall in 2 seconds if you throw it downward with an initial velocity of 40 feet per second?

73. A rocket is fired straight up from ground level with an initial velocity of 800 feet per second.
   (a) How long does it take the rocket to rise 3200 feet?
   (b) When will the rocket hit the ground?

74. A rocket loaded with fireworks is to be shot vertically upward from ground level with an initial velocity of 200 feet per second. When the rocket reaches a height of 400 feet on its upward trip the fireworks will be detonated. How many seconds after liftoff will this take place?

75. The atmospheric pressure a (in pounds per square foot) at height h thousand feet above sea level is approximately

$$a = .8315h^2 - 73.93h + 2116.1.$$

   (a) Find the atmospheric pressure at sea level and at the top of Mount Everest, the tallest mountain in the world (29,035 feet[*]). [Remember that h is measured in thousands.]
   (b) The atmospheric pressure at the top of Mount Rainier is 1223.43 pounds per square foot. How high is Mount Rainier?

76. Data from the U.S. Department of Health and Human Services indicates that the cumulative number N of reported cases of AIDS in the United States in year x can be approximated by the equation

$$N = 3362.1x^2 - 17,270.3x + 24,043,$$

where $x = 0$ corresponds to 1980. In what year did the total reach 550,000?

77. According to data from the U.S. Census Bureau, the population P of Cleveland, Ohio (in thousands) in year x can be approximated by $P = .08x^2 - 13.08x + 927$, where $x = 0$ corresponds to 1950. In what year in the past was the population about 804,200?

---

[*]Based on measurements in 1999 by climbers sponsored by the Boston Museum of Science and the National Geographic Society, using satellite-based technology.

78. (a) Let b be a real number. Multiply out the expression $\left(x + \dfrac{b}{2}\right)^2$.

(b) Explain why your computation in part (a) shows that this statement is true: If you add $\left(\dfrac{b}{2}\right)^2$ to the expression $x^2 + bx$, the resulting polynomial is a perfect square.

In Exercises 79–82, find a number k such that the given equation has exactly one real solution.

79. $x^2 + kx + 25 = 0$

80. $x^2 - kx + 49 = 0$

81. $kx^2 + 8x + 1 = 0$

82. $kx^2 + 24x + 16 = 0$

In Exercises 83–85, the discriminant of the equation $ax^2 + bx + c = 0$ (with a, b, c integers) is given. Use it to determine whether or not the solutions of the equation are rational numbers.

83. $b^2 - 4ac = 25$

84. $b^2 - 4ac = 0$

85. $b^2 - 4ac = 72$

86. Find the error in the following "proof" that $6 = 3$.

$$x = 3$$

Multiply both sides by x: $\quad x^2 = 3x$

Subtract 9 from both sides: $\quad x^2 - 9 = 3x - 9$

Factor each side: $\quad (x - 3)(x + 3) = 3(x - 3)$

Divide both sides by $x - 3$: $\quad x + 3 = 3$

Since $x = 3$: $\quad 3 + 3 = 3$

$$6 = 3$$

87. Find a number k such that 4 and 1 are the solutions of $x^2 - 5x + k = 0$.

88. Suppose a, b, c are fixed real numbers such that $b^2 - 4ac \geq 0$. Let r and s be the solutions of

$$ax^2 + bx + c = 0.$$

(a) Use the quadratic formula to show that $r + s = -b/a$ and $rs = c/a$.

(b) Use part (a) to verify that $ax^2 + bx + c = a(x - r)(x - s)$.

(c) Use part (b) to factor $x^2 - 2x - 1$ and $5x^2 + 8x + 2$.

1.2.A *SPECIAL TOPICS* Absolute Value Equations

If c is a real number, then by the definition of absolute value, $|c|$ is either c or $-c$ (whichever one is positive). This fact can be used to solve absolute value equations algebraically.

EXAMPLE 1

To solve $|3x - 4| = 8$, apply the fact stated above with $c = 3x - 4$. Then $|3x - 4|$ is either $3x - 4$ or $-(3x - 4)$, so

$$3x - 4 = 8 \qquad \text{or} \qquad -(3x - 4) = 8$$

$$3x = 12 \qquad\qquad\qquad -3x + 4 = 8$$

$$x = 4 \qquad\qquad\qquad\qquad -3x = 4$$

$$x = -4/3.$$

So there are two possible solutions of the original equation $|3x - 4| = 8$. You can readily verify that both 4 and $-4/3$ actually are solutions. ∎

EXAMPLE 2

Solve $|x + 4| = 5x - 2$.

SOLUTION   The left side of the equation is either $x + 4$ or $-(x + 4)$ (why?). Hence,

$$x + 4 = 5x - 2 \qquad \text{or} \qquad -(x + 4) = 5x - 2$$
$$-4x + 4 = -2 \qquad\qquad\qquad -x - 4 = 5x - 2$$
$$-4x = -6 \qquad\qquad\qquad\quad -6x = 2$$
$$x = \frac{-6}{-4} = \frac{3}{2} \qquad\qquad\qquad x = \frac{2}{-6} = -\frac{1}{3}.$$

We must check each of these possible solutions in the original equation,

$$|x + 4| = 5x - 2.$$

We see that $x = 3/2$ is a solution because

$$\left|\frac{3}{2} + 4\right| = \frac{11}{2} \qquad \text{and} \qquad 5\left(\frac{3}{2}\right) - 2 = \frac{11}{2}.$$

However, $x = -1/3$ is not a solution, since

$$\left|-\frac{1}{3} + 4\right| = \frac{11}{3} \qquad \text{but} \qquad 5\left(-\frac{1}{3}\right) - 2 = -\frac{11}{3}. \qquad \blacksquare$$

EXAMPLE 3

Solve the equation $|x^2 + 4x - 3| = 2$.

SOLUTION   The equation is equivalent to

$$x^2 + 4x - 3 = 2 \qquad \text{or} \qquad -(x^2 + 4x - 3) = 2$$
$$x^2 + 4x - 5 = 0 \qquad\qquad\qquad -x^2 - 4x + 3 = 2$$
$$-x^2 - 4x + 1 = 0$$
$$x^2 + 4x - 1 = 0.$$

The first of these equations can be solved by factoring and the second by the quadratic formula:

$$(x + 5)(x - 1) = 0 \qquad \text{or} \qquad x = \frac{-4 \pm \sqrt{4^2 - 4 \cdot 1 \cdot (-1)}}{2 \cdot 1}$$
$$x = -5 \quad \text{or} \quad x = 1 \qquad\qquad x = \frac{-4 \pm \sqrt{20}}{2} = \frac{-4 \pm 2\sqrt{5}}{2}$$
$$x = -2 + \sqrt{5} \quad \text{or} \quad x = -2 - \sqrt{5}.$$

Verify that all four of these numbers are solutions of the original equation. $\blacksquare$

✓ EXERCISES 1.2.A

In Exercises 1–12, find all real solutions of each equation.

1. $|2x + 3| = 9$

2. $|3x - 5| = 7$

3. $|6x - 9| = 0$

4. $|4x - 5| = -9$

5. $|2x + 3| = 4x - 1$

6. $|3x - 2| = 5x + 4$

7. $|x - 3| = x$

8. $|2x - 1| = 2x + 1$

9. $|x^2 + 4x - 1| = 4$

10. $|x^2 + 2x - 9| = 6$

11. $|x^2 - 5x + 1| = 3$

12. $|12x^2 + 5x - 7| = 4$

13. In statistical quality control, one needs to find the proportion of the product that is not acceptable. The upper and lower control limits are found by solving the following equation (in which $\bar{p}$ is the mean percent defective and n is the sample size) for CL.

$$|CL - \bar{p}| = 3\sqrt{\frac{\bar{p}(1 - \bar{p})}{n}}$$

Find the control limits when $\bar{p} = .02$ and $n = 200$.

1.2.B   *SPECIAL TOPICS* Variation

If a plane flies 400 mph for t hours, then it travels a distance of $400t$ miles. So the distance d is related to the time t by the equation $d = 400t$. We say that d *varies directly as t* and that the *constant of variation* is 400. Direct variation also occurs in other settings and is formally defined as follows.

Direct Variation

Suppose the quantities u and v are related by the equation

$$v = ku,$$

where k is a nonzero constant. We say that **v varies directly as u** or that **v is directly proportional to u.** In this case, k is called the **constant of variation** or the **constant of proportionality.**

EXAMPLE 1

When you swim underwater, the pressure p varies directly with the depth d at which you swim. At a depth of 20 feet, the pressure is 8.6 pounds per square inch. What is the pressure at 65 feet deep?

SOLUTION   The variation equation is $p = kd$ for some constant k. We first use the given information to find k. Since $p = 8.6$ when $d = 20$, we have

$$p = kd$$

$$8.6 = k(20)$$

$$k = \frac{8.6}{20} = .43.$$

Therefore, the variation equation is $p = .43d$. To find the pressure at 65 feet, substitute 65 for d in the equation:

$$p = .43d = .43(65) = 27.95 \text{ pounds per square inch.} \quad \blacksquare$$

The basic idea in direct variation is that the two quantities grow or shrink together. For instance, when $v = 3u$, then as u gets large, so does v. But two quantities can also be related in such a way that one grows as the other shrinks or vice versa. For example if $v = 5/u$, then when u is large (say, $u = 500$), v is small: $v = 5/500 = .01$. This is an example of *inverse variation*.

Inverse Variation

> Suppose the quantities u and v are related by the equation
>
> $$v = \frac{k}{u},$$
>
> where k is a nonzero constant. We say that v **varies inversely as u** or that v **is inversely proportional to u.** Once again, k is called the **constant of variation** or the **constant of proportionality.**

EXAMPLE 2

According to one of Parkinson's laws, the amount of time the Math Department spends discussing an item in its budget is inversely proportional to the cost of the item. If the department spent 40 minutes discussing the purchase of a copying machine for $450, how much time will it spend discussing a $100 appropriation for the annual picnic?

SOLUTION   Let t denote time (in minutes) and c the cost of an item. Then the variation equation is

$$t = \frac{k}{c}$$

for some constant k. We know that $t = 40$ when $c = 450$. Substituting these numbers in the equation, we see that

$$40 = \frac{k}{450}$$

$$k = 40 \cdot 450 = 18,000.$$

So the variation equation is

$$t = \frac{18,000}{c}.$$

To find the time spent discussing the picnic, let $c = 100$. Hence,

$$t = \frac{18,000}{100} = 180 \text{ minutes } (= 3 \text{ hours!}). \quad \blacksquare$$

The terminology of variation also applies to situations involving powers of variables, as illustrated in the table on the next page.

Example	Terminology	General Case*
$y = 7x^4$	y varies directly as the fourth power of x. y is directly proportional to the fourth power of x. (Constant of variation is 7.)	$v = ku^n$
$A = \pi r^2$	A varies directly as the square of r. A is directly proportional to the square for r. (Constant of variation is π).	
$y = \dfrac{10}{x^3}$	y varies inversely as the cube of x. y is inversely proportional to the cube of x. (Constant of variation is 10.)	$v = \dfrac{k}{u^n}$
$W = \dfrac{3.48 \times 10^9}{d^2}$	W varies inversely as the square of d. W is inversely proportional to the square of d. (Constant of variation is 3.48×10^9.)	

EXAMPLE 3

The density of light I falling on an object (measured in lumens per square foot) is inversely proportional to the square of the distance d from the light source. If an object 2 feet from a standard 100-watt light bulb receives 34.82 lumens per square foot, how much light falls on an object 10 feet from this bulb?

SOLUTION   The variation equation is

$$I = \frac{k}{d^2}$$

for some constant k. Since $I = 34.82$, when $d = 2$, we have

$$34.82 = \frac{k}{2^2}$$

$$k = 34.82(4) = 139.28.$$

Therefore, the variation equation is

$$I = \frac{139.28}{d^2}.$$

So the light density at 10 feet is

$$I = \frac{139.28}{10^2} = 1.3928 \text{ lumens per square foot.} \quad \blacksquare$$

In many situations, variation involves more than two variables. Typical terminology in such cases is illustrated in the following table.

---

*k is a nonzero constant; u and v are variables.

Example	Terminology
$I = 100rt$	I varies jointly as r and t. I is jointly proportional to r and t.
$H = .06sd^3$	H varies jointly as s and the cube of d. H is jointly proportional to s and the cube of d.
$P = \dfrac{3T}{V}$	P varies directly as T and inversely as V. P is directly proportional to T and inversely proportional to V.
$S = \dfrac{2.5wt^2}{\ell}$	S varies directly as w and the square of t and inversely as ℓ. S is directly proportional to w and the square of t and inversely proportional to ℓ.

EXAMPLE 4

The electrical resistance R of wire (of uniform material) varies directly as the length L and inversely as the square of the diameter d. A 2-meter-long piece of wire with a diameter of 4.4 millimeters has a resistance of 500 ohms. What diameter wire should be used if a 10-meter piece is to have a resistance of 1300 ohms?

SOLUTION   The variation equation is

$$R = \frac{kL}{d^2}$$

for some constant k. So we substitute the known information ($R = 500$ when $L = 2$ and $d = 4.4$) and solve for k:

$$R = \frac{kL}{d^2}$$

$$500 = \frac{k \cdot 2}{(4.4)^2}$$

$$k = \frac{500(4.4)^2}{2} = 4840.$$

Hence, the equation is

$$R = \frac{4840L}{d^2}.$$

We must find d when $L = 10$ and $R = 1300$:

$$1300 = \frac{4840 \cdot 10}{d^2}$$

$$1300d^2 = 48400$$

$$d^2 = \frac{48400}{1300} \approx 37.2308.$$

Since the diameter is positive, we must have $d \approx \sqrt{37.2308} \approx 6.1$ mm.   ■

✓ EXERCISES 1.2.B

In Exercises 1–6, express each geometric formula as a statement about variation by filling in the blanks in this sentence:

_____ varies _____ as _____; the constant of variation is _____.

1. Area of a circle of radius r: $A = \pi r^2$.

2. Volume of a sphere of radius r: $V = \dfrac{4}{3}\pi r^3$.

3. Area of a rectangle of length l and width w: $A = lw$.

4. Area of a triangle of base b and height h: $A = \dfrac{bh}{2}$.

5. Volume of a right circular cone of base radius r and height h: $V = \dfrac{\pi r^2 h}{3}$.

6. Volume of a triangular cylinder of base b, height h, and length l: $V = \dfrac{1}{2}bhl$.

In Exercises 7–12, find the variation equation. Use k for the constant of variation.

7. a varies inversely as b.

8. r is proportional to t.

9. z varies jointly as x, y, and w.

10. The weight w of an object varies inversely as the square of the distance d from the object to the center of the earth.

11. The distance d one can see to the horizon varies directly as the square root of the height h above sea level.

12. The pressure p exerted on the floor by a person's shoe heel is directly proportional to the weight w of the person and inversely proportional to the square of the width r of the heel.

In Exercises 13–22, express the given statement as an equation and find the constant of variation.

13. v varies directly as u; $v = 8$ when $u = 2$.

14. v is directly proportional to u; $v = .4$ when $u = .8$.

15. v varies inversely as u; $v = 8$ when $u = 2$.

16. v is inversely proportional to u; $v = .12$ when $u = .1$.

17. t varies jointly as r and s; $t = 24$ when $r = 2$ and $s = 3$.

18. B varies inversely as u and v; $B = 4$ when $u = 1$ and $v = 3$.

19. w varies jointly as x and y^2; $w = 96$ when $x = 3$ and $y = 4$.

20. p varies directly as the square of z and inversely as r; $p = 32/5$ when $z = 4$ and $r = 10$.

21. T varies jointly as p and the cube of v and inversely as the square of u; $T = 24$ when $p = 3$, $v = 2$, and $u = 4$.

22. D varies jointly as the square of r and the square of s and inversely as the cube of t; $D = 18$ when $r = 4$, $s = 3$, and $t = 2$.

In Exercises 23–30, use an appropriate variation equation to find the required quantity.

23. If r varies directly as t and $r = 6$ when $t = 3$, find r when $t = 2$.

24. If r is directly proportional to t and $r = 4$ when $t = 2$, find t when $r = 2$.

25. If b varies inversely as x and $b = 9$ when $x = 3$, find b when $x = 12$.

26. If b is inversely proportional to x and $b = 10$ when $x = 4$, find x when $b = 12$.

27. Suppose w is directly proportional to the sum of u and the square of v. If $w = 200$ when $u = 1$ and $v = 7$, then find u when $w = 300$ and $v = 5$.

28. Suppose z varies jointly as x and y. If $z = 30$ when $x = 5$ and $y = 2$, then find x when $z = 45$ and $y = 3$.

29. Suppose r varies inversely as s and t. If $r = 12$ when $s = 3$ and $t = 1$, then find r when $s = 6$ and $t = 2$.

30. Suppose u varies jointly as r and s and inversely as t. If $u = 1.5$ when $r = 2$, $s = 3$, and $t = 4$, then find r when $u = 27$, $s = 9$, and $t = 5$.

31. A resident of Michigan whose federal adjusted gross income (AGI) was $24,000 dollars paid state income tax of $1008 in 2001. Use the fact that state income tax varies directly with the AGI to determine the state tax paid by someone with an income of $39,000.

32. By experiment, you discover that the amount of water that comes from your garden hose varies directly with the water pressure. A pressure of 10 pounds per square inch is needed to produce a flow of 3 gallons per minute.
    (a) What pressure is needed to produce a flow of 4.2 gallons per minute?
    (b) If the pressure is 5 pounds per square inch, what is the flow rate?

33. According to Hooke's law, the distance d that a spring stretches when an object is attached to it is directly proportional to the weight of the object. Suppose that the string stretches 15.75 inches when a 7-pound weight is attached. What weight is required to stretch the spring 27 inches?

15.75 in.

7 lb

34. In a thunderstorm, there is a gap between the time you see the lightning and the time you hear the thunder. Your distance from the storm is directly proportional to the time interval between the lightning and the thunder. If you hear thunder from a storm that is 1.36 miles away 6.6 seconds after you see lightning, how far away is a storm for which the time gap between lightning and thunder is 12 seconds?

35. At a fixed temperature, the pressure of an enclosed gas is inversely proportional to its volume. The pressure is 50 kilograms per square centimeter when the volume is 200 cubic centimeters. If the gas is compressed to 125 cubic centimeters, what is the pressure?

36. The electrical resistance in a piece of wire of a given length and material varies inversely as the square of the diameter of the wire. If a wire of diameter .01 cm has a resistance of .4 ohm, what is the resistance of a wire of the same length and material but with diameter .025 cm?

37. The distance traveled by a falling object (subject only to gravity, with wind resistance ignored) is directly proportional to the square of the time it takes to fall that far. If an object falls 100 feet in 2.5 seconds, how far does it fall in 5 seconds?

38. The weight of an object varies inversely with the square of its distance from the center of the earth. Assume that the radius of the earth is 3960 miles. If an astronaut weighs 180 pounds on the surface of the earth, what does he weigh when traveling 300 miles above the surface of the earth?

39. The weight of a cylindrical can of glop varies jointly as the height and the square of the base radius. The weight is 250 ounces when the height is 20 inches and the base radius is 5 inches. What is the height when the weight is 960 ounces and the base radius is 8 inches?

40. The force of the wind blowing directly on a flat surface varies directly with the area of the surface and the square of the velocity of the wind. A 10 mile an hour (mph) wind blowing on a wooden gate that measures 3 by 6 feet exerts a force of 15 pounds. What is the force on a 1.5 by 2 foot traffic sign from a 50 mph wind?

41. The force needed to keep a car from skidding on a circular curve varies inversely as the radius of the curve and jointly as the weight of the car and the square of the speed. It takes 1500 kilograms of force to keep a 1000-kilogram car from skidding on a curve of radius 200 meters at a speed of 50 kilometers per hour. What force is needed to keep the same car from skidding on a curve of radius 320 meters at 100 kilometers per hour?

42. Boyle's law states that the volume of a given mass of gas varies directly as the temperature and inversely as the pressure. If a gas has a volume of 14 cubic inches when the temperature is 40° and the pressure is 280 pounds per square inch, what is the volume at a temperature of 50° and pressure of 175 pounds per square inch?

43. The maximum safe load that a rectangular beam can support varies directly with the width w and the square of the height h, and inversely with the length l of the beam. A 6-foot-long beam that is 2 inches high and 4 inches wide has a maximum safe load of 1000 pounds.
    (a) What is the maximum safe load for a 10-foot-long beam that is 4 inches high and 4 inches wide?
    (b) How long should a 4 inch by 4 inch beam be to safely support a 6000-pound load?

w

h

l

44. The period of a pendulum is the time it takes for the pendulum to make one complete swing (going both left and right) and return to its starting point. The period P varies directly with the *square root* of the length l of the pendulum.
    (a) Write the variation equation that describes this situation, using k for the constant of variation.
    (b) What happens to the period if the length of the pendulum is quadrupled?

1.3   The Coordinate Plane

Just as real numbers are identified with points on the number line, ordered *pairs* of real numbers can be identified with points in the plane. To do this, draw two number lines in the plane, one vertical and one horizontal, as in Figure 1–9. The horizontal line is usually called the **x-axis,** and the vertical line the **y-axis,** but other letters may be used if desired. The point where the axes intersect is the **origin.** The axes divide the plane into four regions, called **quadrants,** that are numbred as in Figure 1–9.

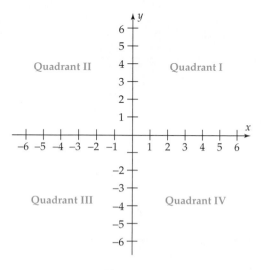

Figure 1–9

If *P* is a point in the plane, draw vertical and horizontal lines through *P* to the coordinate axes, as shown in Figure 1–10. These lines intersect the *x*-axis at some number *c* and the *y*-axis at *d*. We say that *P* has **coordinates** (c, d). The number *c* is the **x-coordinate** of *P,* and *d* is the **y-coordinate** of *P.* The plane is said to have a **rectangular** (or **Cartesian**) **coordinate system.**

Figure 1–10

You can think of the coordinates of a point as directions for locating it. For instance, to find $(4, -3)$, start at the origin and move 4 units to the right along the *x*-axis, then move 3 units downward, as shown in Figure 1–11, which also shows other points and their coordinates.

CAUTION

The coordinates of a point are an *ordered* pair. Figure 1–11 shows that the point *P* with coordinates $(-5, 2)$ is quite different from the point *Q* with coordinates $(2, -5)$. The same numbers (2 and -5) occur in both cases, but in *different order*.

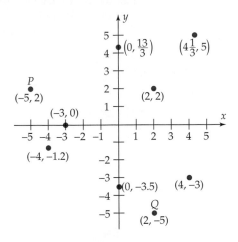

Figure 1–11

EXAMPLE 1

The following table, from the U.S. Department of Education shows the maximum Pell Grant for college students in selected years.

Year	1990	1992	1994	1996	1998	2000	2002
Amount	2300	2400	2300	2470	3000	3300	4000

One way to represent this data graphically is to represent each year's maximum by a point; for instance (1990, 2300) and (2002, 4000). Alternatively, to avoid using large numbers, we can let *x* be the number of years since 1990, so that $x = 0$ is 1990 and $x = 12$ is 2002. We can also list the dollar amounts in *hundreds,* so the points for 1990 and 2002 are (0, 23) and (12, 40). Plotting all the data in this way leads to the **scatter plot** in Figure 1–12. Connecting these data points with line segments produces the **line graph** in Figure 1–13. ■

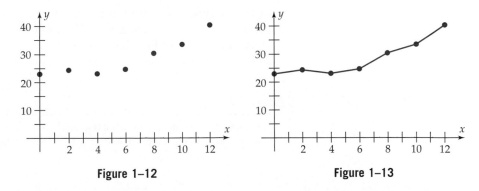

Figure 1–12   **Figure 1–13**

THE DISTANCE FORMULA

We shall often identify a point with its coordinates and refer, for example, to the point (2, 3). When dealing with several points simultaneously, it is customary to label the coordinates of the first point (x_1, y_1), the second point (x_2, y_2), the third point (x_3, y_3), and so on.* Once the plane is coordinatized, it's easy to compute the distance between any two points:

The Distance Formula

> The distance between points (x_1, y_1) and (x_2, y_2) is
> $$\sqrt{(x_1 - x_2)^2 + (y_1 - y_2)^2}.$$

Before proving the distance formula, we shall see how it is used.

EXAMPLE 2

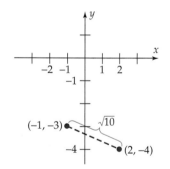

Figure 1–14

To find the distance between the points $(-1, -3)$ and $(2, -4)$ in Figure 1–14, substitute $(-1, -3)$ for (x_1, y_1) and $(2, -4)$ for (x_2, y_2) in the distance formula:

Distance formula:
$$\text{Distance} = \sqrt{(x_1 - x_2)^2 + (y_1 - y_2)^2}$$

Substitute:
$$= \sqrt{(-1 - 2)^2 + (-3 - (-4))^2}$$

Simplify:
$$= \sqrt{(-3)^2 + (-3 + 4)^2}$$
$$= \sqrt{9 + 1} = \sqrt{10}.$$

The order in which the points are used in the distance formula doesn't make a difference. If we substitute $(2, -4)$ for (x_1, y_1) and $(-1, -3)$ for (x_2, y_2), we get the same answer:

$$\sqrt{[2 - (-1)]^2 + [-4 - (-3)]^2} = \sqrt{3^2 + (-1)^2} = \sqrt{10}. \quad \blacksquare$$

EXAMPLE 3

In a Cubs game at Wrigley Field, a fielder catches the ball near the right-field corner and throws it to second base. The right-field corner is 353 feet from home plate along the right-field foul line. If the fielder is 5 feet from the outfield wall and 5 feet from the foul line, how far must he throw the ball?

---

*"x_1" is read "x-one" or "x-sub-one"; it is a *single symbol* denoting the first coordinate of the first point, just as c denotes the first coordinate of (c, d). Analogous remarks apply to y_1, x_2, and so on.

SOLUTION   Imagine that the playing field is placed on the coordinate plane, with home plate at the origin and the right-field foul line along the positive *x*-axis, as shown in Figure 1–15 (not to scale).

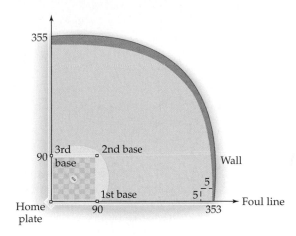

Figure 1–15

Since the four bases form a square whose sides measure 90 ft each, second base has coordinates (90, 90). The fielder is located 5 feet from the wall, so his *x*-coordinate is $353 - 5 = 348$. His *y*-coordinate is 5, since he is 5 feet from the foul line. Therefore the distance he throws is the distance from (348, 5) to (90, 90), which can be found as follows.

Distance formula: $\text{Distance} = \sqrt{(x_1 - x_2)^2 + (y_1 - y_2)^2}$

Substitute: $= \sqrt{(348 - 90)^2 + (5 - 90)^2}$

Simplify: $= \sqrt{258^2 + (-85)^2}$

$= \sqrt{73{,}789} \approx 271.6 \text{ feet.}$

Therefore, he must throw about 272 feet.   ■

EXAMPLE 4

To find the distance from (a, b) to $(2a, -b)$, where a and b are fixed real numbers, substitute a for x_1, b for y_1, $2a$ for x_2, and $-b$ for y_2 in the distance formula:

$$\sqrt{(x_1 - x_2)^2 + (y_1 - y_2)^2} = \sqrt{(a - 2a)^2 + (b - (-b))^2}$$
$$= \sqrt{(-a)^2 + (b + b)^2} = \sqrt{a^2 + (2b)^2}$$
$$= \sqrt{a^2 + 4b^2}. \quad ■$$

Proof of the Distance Formula   Figure 1–16 on the next page shows typical points *P* and *Q* in the plane. We must find length *d* of line segment *PQ*.

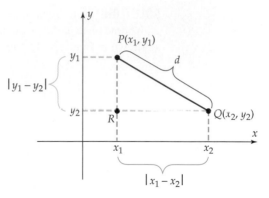

Figure 1–16

As shown in Figure 1–16, the length of RQ is the same as the distance from x_1 to x_2 on the x-axis (number line), namely, $|x_1 - x_2|$. Similarly, the length of PR is the same as the distance from y_1 to y_2 on the y-axis, namely, $|y_1 - y_2|$. According to the Pythagorean Theorem* the length d of PQ is given by

$$(\text{Length } PQ)^2 = (\text{length } RQ)^2 + (\text{length } PR)^2$$

$$d^2 = |x_1 - x_2|^2 + |y_1 - y_2|^2.$$

Since $|c|^2 = |c| \cdot |c| = |c^2| = c^2$ (because $c^2 \geq 0$), this equation becomes

$$d^2 = (x_1 - x_2)^2 + (y_1 - y_2)^2.$$

Since the length d is nonnegative, we must have

$$d = \sqrt{(x_1 - x_2)^2 + (y_1 - y_2)^2}. \qquad \blacksquare$$

The distance formula can be used to prove the following useful fact (see Exercise 86).

The Midpoint Formula

The midpoint of the line segment from (x_1, y_1) to (x_2, y_2) is

$$\left(\frac{x_1 + x_2}{2}, \frac{y_1 + y_2}{2} \right).$$

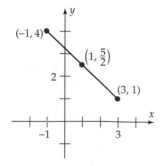

Figure 1–17

EXAMPLE 5

To find the midpoint of the segment joining $(-1, 4)$ and $(3, 1)$, use the formula in the box with $x_1 = -1$, $y_1 = 4$, $x_2 = 3$, and $y_2 = 1$. The midpoint is

$$\left(\frac{x_1 + x_2}{2}, \frac{y_1 + y_2}{2} \right) = \left(\frac{-1 + 3}{2}, \frac{4 + 1}{2} \right) = \left(1, \frac{5}{2} \right)$$

as shown in Figure 1–17. $\blacksquare$

---

*See the Geometry Review Appendix.

∷ GRAPHS

A **graph** is a set of points in the plane. Some graphs are based on data points, such as Figures 1–12 and 1–13. Other graphs arise from equations, as follows. A **solution** of an equation in variables x and y is a pair of numbers such that the substitution of the first number for x and the second for y produces a true statement. For instance, $(3, -2)$ is a solution of $5x + 7y = 1$ because

$$5 \cdot 3 + 7(-2) = 1,$$

and $(-2, 3)$ is *not* a solution because $5(-2) + 7 \cdot 3 \neq 1$. The **graph of an equation** in two variables is the set of points in the plane whose coordinates are solutions of the equation. Thus the graph is a *geometric picture of the solutions*.

EXAMPLE 6

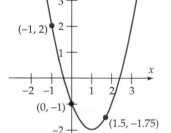

Figure 1–18

The graph of $y = x^2 - 2x - 1$ is shown in Figure 1–18. You can readily verify that each of the points whose coordinates are labeled is a solution of the equation. For instance, $(0, -1)$ is a solution because $-1 = 0^2 - 2(0) - 1$. ■

A graph may intersect the x- or y-axis at one or more points. The x-coordinate of a point where the graph intersects the x-axis is called an **x-intercept** of the graph. Similarly, the y-coordinate of a point where the graph intersects the y-axis is called a **y-intercept** of the graph.

EXAMPLE 7

Find the x- and y-intercepts of the graph of $y = x^2 - 2x - 1$ in Figure 1–18.

SOLUTION   The points where the graph intersects the x-axis have 0 as their y-coordinate (see Figure 1–18). We can find their x-coordinates by setting $y = 0$ and solving the resulting equation,

$$x^2 - 2x - 1 = 0.$$

By the quadratic formula

$$x = \frac{-(-2) \pm \sqrt{(-2)^2 - 4 \cdot 1 \cdot (-1)}}{2 \cdot 1} = \frac{2 \pm \sqrt{8}}{2} \approx \begin{cases} -.4142 \\ 2.4142. \end{cases}$$

So the x-intercepts are approximately $-.4142$ and 2.4142.

In this case, you can read the y-intercept from the graph; it is -1. Because points on the y-axis have 0 as their x-coordinate, the y-intercept can be found algebraically by setting $x = 0$ in the equation and solving for y. ■

The process in Example 7 can be summarized as follows.

x- and y-Intercepts

> To find the x-intercepts of the graph of an equation, set $y = 0$ and solve for x.
>
> To find the y-intercepts, set $x = 0$ and solve for y.

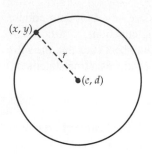

Figure 1–19

▚ CIRCLES

If (c, d) is a point in the plane and r a positive number, then the **circle with center (c, d) and radius r** consists of all points (x, y) that lie r units from (c, d), as shown in Figure 1–19. According to the distance formula, the statement that "the distance from (x, y) to (c, d) is r units" is equivalent to:

$$\sqrt{(x - c)^2 + (y - d)^2} = r$$

Squaring both sides shows that (x, y) satisfies this equation:

$$(x - c)^2 + (y - d)^2 = r^2$$

Reversing the procedure shows that any solution (x, y) of this equation is a point on the circle. Therefore

Circle
Equation

> The circle with center (c, d) and radius r is the graph of
> $$(x - c)^2 + (y - d)^2 = r^2.$$

We say that $(x - c)^2 + (y - d)^2 = r^2$ is the **equation of the circle** with center (c, d) and radius r.

EXAMPLE 8

(a) The graph of the equation $(x - 4)^2 + (y - 2)^2 = 9$ is of the form shown in the box, with $c = 4$, $d = 2$, and $r = 3$. So the graph is a circle with center $(4, 2)$ and radius 3, as shown in Figure 1–20.

(b) The circle with center $(-3, 2)$ and radius 2, shown in Figure 1–21, is the graph of the equation

$$(x - (-3))^2 + (y - 2)^2 = 2^2$$

or, equivalently

$$(x + 3)^2 + (y - 2)^2 = 4. \quad ■$$

Figure 1–20

Figure 1–21

If the center of a circle is at the origin (so that $(c, d) = (0, 0)$ in the preceding box), then its equation has a simple form.

Circle at the Origin

The circle with center $(0, 0)$ and radius r is the graph of

$$x^2 + y^2 = r^2.$$

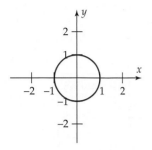

Figure 1–22

EXAMPLE 9

Letting $r = 1$ shows that the graph of $x^2 + y^2 = 1$ is the circle of radius 1 centered at the origin, as shown in Figure 1–22. This circle is called the **unit circle.** ∎

EXAMPLE 10

Find the equation of the circle with center $(3, -1)$ that passes through $(2, 4)$.

SOLUTION   We must first find the radius. Since $(2, 4)$ is on the circle, the radius is the distance from $(2, 4)$ to $(3, -1)$ as shown in Figure 1–23, namely,

$$\sqrt{(2 - 3)^2 + (4 - (-1))^2} = \sqrt{1 + 25} = \sqrt{26}.$$

The equation of the circle with center at $(3, -1)$ and radius $\sqrt{26}$ is

$$(x - 3)^2 + (y - (-1))^2 = (\sqrt{26})^2$$
$$(x - 3)^2 + (y + 1)^2 = 26$$
$$x^2 - 6x + 9 + y^2 + 2y + 1 = 26$$
$$x^2 + y^2 - 6x + 2y - 16 = 0. \quad ∎$$

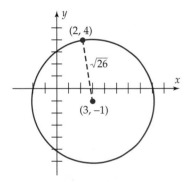

Figure 1–23

The equation of any circle can always be written in the form

$$x^2 + y^2 + Bx + Cy + D = 0$$

for some constants B, C, D, as in Example 10 (where $B = -6, C = 2, D = -16$). Conversely, the graph of such an equation can always be determined.

EXAMPLE 11

To find the graph of $3x^2 + 3y^2 - 12x - 30y + 45 = 0$, we divide both sides by 3 and rewrite the equation as

$$(x^2 - 4x) + (y^2 - 10y) = -15.$$

Next we complete the square in both expressions in parentheses (see page 21). To complete the square in $x^2 - 4x$, we add 4 (the square of half the coefficient of x), and to complete the square in $y^2 - 10y$, we add 25 (why?). To have an equivalent equation, we must add these numbers to *both* sides:

$$(x^2 - 4x + 4) + (y^2 - 10y + 25) = -15 + 4 + 25$$
$$(x - 2)^2 + (y - 5)^2 = 14$$

Since $14 = (\sqrt{14})^2$, this is the equation of the circle with center $(2, 5)$ and radius $\sqrt{14}$. ∎

EXERCISES 1.3

1. Find the coordinates of points A–I.

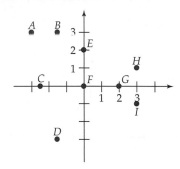

In Exercises 2–5, find the coordinates of the point P.

2. P lies 4 units to the left of the y-axis and 5 units below the x-axis.

3. P lies 3 units above the x-axis and on the same vertical line as $(-6, 7)$.

4. P lies 2 units below the x-axis, and its x-coordinate is three times its y-coordinate.

5. P lies 4 units to the right of the y-axis, and its y-coordinate is half its x-coordinate.

In Exercises 6–8, sketch a scatter plot and a line graph of the given data.

6. Tuition and fees at four-year public colleges in the fall of each year are shown in the table (*Source:* The College Board). Let $x = 0$ correspond to 1995.

Year	1995	1996	1997	1998	1999	2000
Tuition and Fees	2860	2966	3111	3247	3356	3510

7. The table shows projected sales of personal digital video recorders.[*] Let $x = 0$ correspond to 2000, and measure y in thousands.

Year	Number Sold
2000	257,000
2001	129,000
2002	143,000
2003	214,000
2004	315,000
2005	485,000

8. The maximum yearly contribution to an individual retirement account (IRA) is now $3000. It will change to $4000 in 2005 and change to $5000 in 2008. Assuming 3% inflation, however, the picture is somewhat different. The table shows the maximum IRA contribution in fixed 2003 dollars. Let $x = 0$ correspond to 2000.

Year	Maximum Contribution
2003	3000
2004	2910
2005	3764
2006	3651
2007	3541
2008	4294

9. (a) If the first coordinate of a point is greater than 3 and its second coordinate is negative, in what quadrant does it lie?
   (b) What is the answer in part (a) if the first coordinate is less than 3?

10. In what quadrant(s) does a point lie if the product of its coordinates is
   (a) positive?   (b) negative?

11. (a) Plot the points $(3, 2)$, $(4, -1)$, $(-2, 3)$, and $(-5, -4)$.
   (b) Change the sign of the y-coordinate in each of the points in part (a), and plot these new points.
   (c) Explain how the points (a, b) and $(a, -b)$ are related graphically. [*Hint:* What are their relative positions with respect to the x-axis?]

12. (a) Plot the points $(5, 3)$, $(4, -2)$, $(-1, 4)$, and $(-3, -5)$.
   (b) Change the sign of the x-coordinate in each of the points in part (a), and plot these new points.
   (c) Explain how the points (a, b) and $(-a, b)$ are related graphically. [*Hint:* What are their relative positions with respect to the y-axis?]

In Exercises 13–20, find the distance between the two points and the midpoint of the segment joining them.

13. $(-3, 5)$, $(2, -7)$ **14.** $(2, 4)$, $(1, 5)$

15. $(1, -5)$, $(2, -1)$ **16.** $(-2, 3)$, $(-3, 2)$

17. $(\sqrt{2}, 1)$, $(\sqrt{3}, 2)$ **18.** $(-1, \sqrt{5})$, $(\sqrt{2}, -\sqrt{3})$

19. (a, b), (b, a) **20.** (s, t), $(0, 0)$

---

*eBrain Market Research

21. Find the perimeter of the shaded area in the figure.

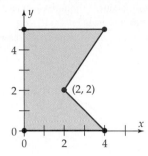

22. What is the perimeter of the triangle with vertices $(1, 1)$, $(5, 4)$, and $(-2, 5)$?

In Exercises 23–24, show that the three points are the ver-tices of a right triangle, and state the length of the hypotenuse. [You may assume that a triangle with sides of lengths a, b, c is a right triangle with hypotenuse c provided that $a^2 + b^2 = c^2$.]

23. $(0, 0), (1, 1), (2, -2)$

24. $(3, -2), (0, 4), (-2, 3)$

25. Sales of organically grown foods in the United States were $7.8 billion in 2000 and are projected to be $16 billion in 2004.[*]
  (a) Represent this data graphically by two points.
  (b) Find the midpoint of the line segment joining these two points.
  (c) How might this midpoint be intepreted? What assumptions, if any, are needed to make this interpretation?

26. Suppose a baseball playing field is placed on the coordinate plane, as in Example 3.
  (a) Find the coordinates of first and third base.
  (b) If the left fielder is at the point $(50, 325)$, how far is he from first base?
  (c) How far is the left fielder in part (b) from the right fielder, who is at the point $(280, 20)$?

27. A standard football field is 100 yards long and $53\frac{1}{3}$ yards wide. The quarterback, who is standing on the 10-yard line, 20 yards from the left sideline, throws the ball to a receiver who is on the 45-yard line, 5 yards from the right sideline, as shown in the figure.
  (a) How long was the pass? [*Hint*: Place the field in the first quadrant of the coordinate plane, with the left sideline on the *y*-axis and the goal line on the *x*-axis. What are the coordinates of the quarterback and the receiver?]

(b) A player is standing halfway between the quarterback and the receiver. What are his coordinates?

28. How far is the quarterback in Exercise 27 from a player who is on the 50-yard line, halfway between the sidelines?

In Exercises 29–34, determine whether the point is on the graph of the given equation.

29. $(1, -2); 3x - y - 5 = 0$

30. $(2, -1); x^2 + y^2 - 6x + 8y = -15$

31. $(6, 2); 3y + x = 12$

32. $(1, -2); 3x + y = 12$

33. $(3, 4); (x - 2)^2 + (y + 5)^2 = 4$

34. $(1, -1); \dfrac{x^2}{2} + \dfrac{y^2}{3} = 1$

In Exercises 35–40, find the x- and y-intercepts of the graph of the equation.

35. $x^2 - 6x + y + 5 = 0$

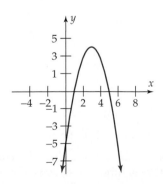

36. $x^2 - 2xy + 3y^2 = 1$

37. $(x - 2)^2 + y^2 = 9$

38. $(x + 1)^2 + (y - 2)^2 = 4$

39. $9x^2 + 24xy + 16y^2 + 90x - 128y = 0$

40. $2x^2 - 4xy + 2y^2 + 3x + 5y = 10$

In Exercises 41–44, find the equation of the circle with given center and radius r.

41. $(-3, 4)$; $r = 2$ **42.** $(-2, -1)$; $r = 3$

43. $(0, 0)$; $r = \sqrt{2}$ **44.** $(5, -2)$; $r = 1$

In Exercises 45–48, sketch the graph of the equation. Label the x- and y-intercepts.

45. $(x - 5)^2 + (y + 2)^2 = 5$

46. $(x + 6)^2 + y^2 = 4$

47. $(x + 1)^2 + (y - 3)^2 = 9$

48. $(x - 2)^2 + (y - 4)^2 = 1$

In Exercises 49–54, find the center and radius of the circle whose equation is given.

49. $x^2 + y^2 + 8x - 6y - 15 = 0$

50. $15x^2 + 15y^2 = 10$

51. $x^2 + y^2 + 6x - 4y - 15 = 0$

52. $x^2 + y^2 + 10x - 75 = 0$

53. $x^2 + y^2 + 25x + 10y = -12$

54. $3x^2 + 3y^2 + 12x + 12 = 18y$

55. The graph, which is based on data from the Actuarial Society of South Africa and assumes no changes in current behavior, shows the projected new cases of AIDS in South Africa (in millions) in coming years ($x = 0$ corresponds to 2000).

(a) Estimate the number of new cases in 2010.

(b) Estimate the year in which the largest number of new cases will occur. About how many new cases will there be in that year?

(c) In what years will the number of new cases be below 7,000,000?

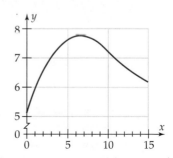

56. The graph shows the total number of alcohol-related car crashes in Ohio at a particular time of the day for the years 1991–2000.[*] Time is measured in hours after midnight. During what periods is the number of crashes

(a) below 5000?

(b) above 15,000?

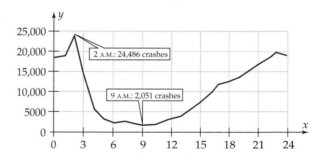

In Exercises 57–60, determine which of graphs A, B, C best describes the given situation.

57. You have a job that pays a fixed salary for the week. The graph shows your salary.

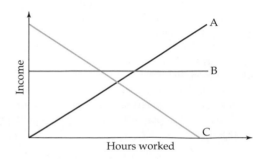

---

*The Cleveland *Plain Dealer*

58. You have a job that pays an hourly wage. The graph shows your salary.

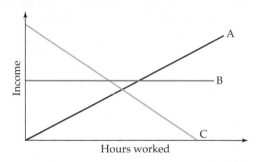

59. You inflate a birthday balloon. The graph shows the amount of air in the balloon.

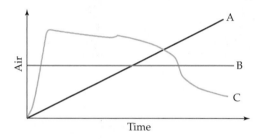

60. Alison's wading pool is filled with a hose by her big sister Emily, and Alison plays in the pool. When they are finished, Emily empties the pool. The graph shows the water level of the pool.

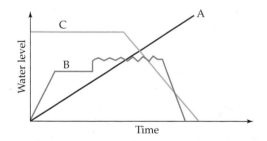

In Exercises 61–68, find the equation of the circle.

61. Center (2, 2); passes through the origin.

62. Center (−1, −3); passes through (−4, −2).

63. Center (1, 2); intersects *x*-axis at −1 and 3.

64. Center (3, 1); diameter 2.

65. Center (−5, 4); tangent (touching at one point) to the *x*-axis.

66. Center (2, −6); tangent to the *y*-axis.

67. Endpoints of diameter are (3, 3) and (1, −1).

68. Endpoints of diameter are (−3, 5) and (7, −5).

69. One diagonal of a square has endpoints (−3, 1) and (2, −4). Find the endpoints of the other diagonal.

70. Find the vertices of all possible squares with this property: Two of the vertices are (2, 1) and (2, 5). [*Hint:* There are three such squares.]

71. Do Exercise 70 with (*c, d*) and (*c, k*) in place of (2, 1) and (2, 5).

72. Find the three points that divide the line segment from (−4, 7) to (10, −9) into four parts of equal length.

73. Find all points *P* on the *x*-axis that are 5 units from (3, 4). [*Hint: P* must have coordinates (*x*, 0) for some *x* and the distance from *P* to (3, 4) is 5.]

74. Find all points on the *y*-axis that are 8 units from (−2, 4).

75. Find all points with first coordinate 3 that are 6 units from (−2, −5).

76. Find all points with second coordinate −1 that are 4 units from (2, 3).

77. Find a number *x* such that (0, 0), (3, 2), and (*x*, 0) are the vertices of an isosceles triangle, neither of whose two equal sides lie on the *x*-axis.

78. Do Exercise 77 if one of the two equal sides lies on the positive *x*-axis.

79. Show that the midpoint *M* of the hypotenuse of a right triangle is equidistant from the vertices of the triangle. [*Hint:* Place the triangle in the first quadrant of the plane, with right angle at the origin so that the situation looks like the figure.]

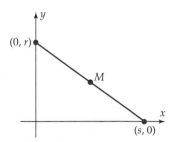

80. Show that the diagonals of a parallelogram bisect each other. [*Hint:* Place the parallelogram in the first quadrant with a vertex at the origin and one side along the *x*-axis so that the situation looks like the figure.]

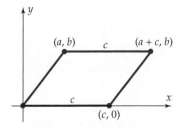

81. Show that the diagonals of a rectangle have the same length. [*Hint:* Place the rectangle in the first quadrant of the plane and label its vertices appropriately, as in Exercises 79–80.]

82. If the diagonals of a parallelogram have the same length, show that the parallelogram is actually a rectangle. [*Hint:* See Exercise 80.]

Thinkers

83. For each nonzero real number k, the graph of $(x - k)^2 + y^2 = k^2$ is a circle. Describe all possible such circles.

84. Suppose every point in the coordinate plane is moved 5 units straight up.
(a) To what point does each of these points go: $(0, -5)$, $(2, 2)$, $(5, 0)$, $(5, 5)$, $(4, 1)$?
(b) Which points go to each of the points in part (a)?
(c) To what point does (a, b) go?
(d) To what point does $(a, b - 5)$ go?
(e) What point goes to $(-4a, b)$?
(f) What points go to themselves?

85. Let (c, d) be any point in the plane with $c \neq 0$. Prove that (c, d) and $(-c, -d)$ lie on the same straight line through the origin, on opposite sides of the origin, the same distance from the origin. [*Hint:* Find the midpoint of the line segment joining (c, d) and $(-c, -d)$.]

86. *Proof of the Midpoint Formula* Let P and Q be the points (x_1, y_1) and (x_2, y_2), respectively, and let M be the point with coordinates

$$\left(\frac{x_1 + x_2}{2}, \frac{y_1 + y_2}{2}\right).$$

Use the distance formula to compute the following:
(a) The distance d from P to Q;
(b) The distance d_1 from M to P;
(c) The distance d_2 from M to Q.
(d) Verify that $d_1 = d_2$.
(e) Show that $d_1 + d_2 = d$. [*Hint:* Verify that $d_1 = \frac{1}{2}d$ and $d_2 = \frac{1}{2}d$.]
(f) Explain why parts (d) and (e) show that M is the midpoint of PQ.

1.4   Lines

When you move from a point P to a point Q on a line,[*] two numbers are involved, as illustrated in Figure 1–24:

(i) The vertical distance you move (the **change in y**);

(ii) The horizontal distance you move (the **change in x**).

Change in $x = 6$

$$\frac{\text{Change in } y}{\text{Change in } x} = \frac{4}{6} = \frac{2}{3}$$

(a)

Change in $x = 4$

$$\frac{\text{Change in } y}{\text{Change in } x} = \frac{4}{4} = 1$$

(b)

Change in $x = 1$

$$\frac{\text{Change in } y}{\text{Change in } x} = \frac{4}{1} = 4$$

(c)

Figure 1–24

---

[*]In this section, "line" means "straight line," and movement is from left to right.

Figure 1–25

The number $\dfrac{\text{change in } y}{\text{change in } x}$ measures the steepness of the line: the steeper the line, the larger the number. In Figure 1–24, the grid allowed us to measure the change in y and the change in x. When the coordinates of P and Q are given, then

The change in y is the difference of the y-coordinates of P and Q;

The change in x is the difference of the x-coordinates of P and Q;

as shown in Figure 1–25. Consequently, we have the following definition.

Slope of a Line

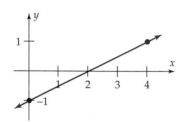

If (x_1, y_1) and (x_2, y_2) are points with $x_1 \neq x_2$, then the **slope** of the line through these points is the number

$$\frac{\text{change in } y}{\text{change in } x} = \frac{y_2 - y_1}{x_2 - x_1}.$$

EXAMPLE 1

Figure 1–26

To find the slope of the line through $(0, -1)$ and $(4, 1)$ (see Figure 1–26), we apply the formula in the previous box with $x_1 = 0$, $y_1 = -1$ and $x_2 = 4$, $y_2 = 1$:

$$\text{Slope} = \frac{y_2 - y_1}{x_2 - x_1} = \frac{1 - (-1)}{4 - 0} = \frac{2}{4} = \frac{1}{2}.$$

The order of the points makes no difference; if you use $(4, 1)$ for (x_1, y_1) and $(0, -1)$ for (x_2, y_2), you obtain the same number:

$$\text{Slope} = \frac{y_2 - y_1}{x_2 - x_1} = \frac{-1 - 1}{0 - 4} = \frac{-2}{-4} = \frac{1}{2}. \quad ■$$

CAUTION

When finding slopes, you must subtract the *y*-coordinates and *x*-coordinates in the same order. With the points (3, 4) and (1, 8), for instance, if you use $8 - 4$ in the numerator, you must use $1 - 3$ in the denominator (*not* $3 - 1$).

EXAMPLE 2

The lines shown in Figure 1–27 are determined by these points:

$$L_1\text{: } (-1, -1) \text{ and } (0, 2) \qquad L_2\text{: } (0, 2) \text{ and } (2, 4) \qquad L_3\text{: } (-6, 2) \text{ and } (3, 2)$$

$$L_4\text{: } (-3, 5) \text{ and } (3, -1) \qquad L_5\text{: } (1, 0) \text{ and } (2, -2).$$

Their slopes are as follows.

$$L_1\text{: } \frac{2 - (-1)}{0 - (-1)} = \frac{3}{1} = 3$$

$$L_2\text{: } \frac{4 - 2}{2 - 0} = \frac{2}{2} = 1$$

$$L_3\text{: } \frac{2 - 2}{3 - (-6)} = \frac{0}{9} = 0$$

$$L_4\text{: } \frac{-1 - 5}{3 - (-3)} = \frac{-6}{6} = -1$$

$$L_5\text{: } \frac{-2 - 0}{2 - 1} = \frac{-2}{1} = -2$$

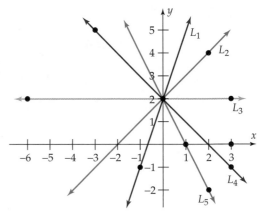

Figure 1–27   ∎

Example 2 illustrates how the slope measures the steepness of the line, as summarized below.

Properties of Slope

The slope of a nonvertical line is a number m that measures how steeply the line rises or falls.

If $m > 0$, the line rises from left to right; the larger m is, the more steeply the line rises.   [Lines L_1 and L_2 in Example 2]

If $m = 0$, the line is horizontal.   [Line L_3]

If $m < 0$, the line falls from left to right; the larger $|m|$ is, the more steeply the line falls.   [Lines L_4 and L_5]

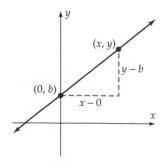

Figure 1–28

∷ SLOPE-INTERCEPT FORM

Let L be a nonvertical line with slope m and y-intercept b. Then $(0, b)$ is a point on L. Let (x, y) be any other point on L. Using the points $(0, b)$ and (x, y) to compute the slope of L (see Figure 1–28), we have

$$\text{Slope of } L = \frac{y - b}{x - 0}.$$

Since the slope of L is m, this equation becomes

$$m = \frac{y - b}{x}$$

$$mx = y - b$$

$$y = mx + b.$$

Thus the coordinates of any point on L satisfy the equation $y = mx + b$. So we have the following fact.

Slope-Intercept Form

> The line with slope m and y-intercept b is the graph of the equation
>
> $$y = mx + b.$$

EXAMPLE 3

Show that the graph of $2y - 5x = 2$ is a straight line. Find its slope, and graph the line.

SOLUTION   We begin by solving the equation for y:

$$2y = 5x + 2$$

$$y = 2.5x + 1.$$

The equation now has the form in the preceding box, with $m = 2.5$ and $b = 1$. Therefore, its graph is the line with slope 2.5 and y-intercept 1.

Since the y-intercept is 1, the point $(0, 1)$ is on the graph. To find another point on the line, choose a value for x, say, $x = 2$, and compute the corresponding value of y:

$$y = 2.5x + 1 = 2.5(2) + 1 = 6.$$

Hence, $(2, 6)$ is on the line. Plotting the line through $(0, 1)$ and $(2, 6)$ produces Figure 1–29. ∎

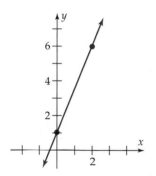

Figure 1–29

EXAMPLE 4

Describe and sketch the graph of the equation $y = 3$.

SOLUTION   We can write $y = 3$ as $y = 0x + 3$. So its graph is a line with slope 0, which means that the line is horizontal, and y-intercept 3, which means that the line crosses the y-axis at 3. This is sufficient information to obtain the graph in Figure 1–30. ∎

Example 4 is an illustration of this fact.

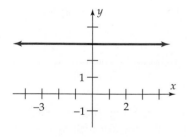

Figure 1–30

Horizontal Lines

> The horizontal line with y-intercept b is the graph of the equation
>
> $$y = b.$$

We have seen that the geometric interpretation of slope is that it measures the *steepness* or direction of the line. The next examples show that slope can also be interpreted as a *rate of change*.

EXAMPLE 5

According to the Kelley Blue Book, a Chevrolet Prizm that is now worth $13,500 will be worth $8500 in four years, provided that it is in good condition with average mileage. Assuming linear depreciation, what will the car be worth in two years? At what rate is the car depreciating?

SOLUTION   Linear depreciation means that the equation that gives the value y of the car in year x is linear. So the equation is of the form $y = mx + b$ for some constants m and b. Since the car is worth $13,500 now ($x = 0$), we have

$$y = mx + b$$
$$13,500 = m \cdot 0 + b$$
$$b = 13,500,$$

so that the equation is $y = mx + 13,500$. Since the car is worth $8,500 after four years (that is, $y = 8500$ when $x = 4$), we have

$$y = mx + 13,500$$
$$8500 = m \cdot 4 + 13,500$$
$$-5000 = 4m$$
$$m = -\frac{5000}{4} = -1250.$$

Consequently, the depreciation equation is $y = -1250x + 13,500$. The value of the car after two years ($x = 2$) is

$$y = -1250(2) + 13,500 = \$11,000.$$

The equation $y = -1250x + 13,500$ shows that the car depreciates $1250 each year (that is, each time x changes by 1). In other words, its value changes at the *rate* of -1250 per year. This rate is the *slope* of the graph of the depreciation equation.   ■

EXAMPLE 6

A factory that makes can openers has fixed costs (for building, fixtures, machinery, etc.) of $26,000. The variable cost (materials and labor) for making one can opener is $2.75.

(a) What is the total cost of making 1000 can openers? 20,000? 40,000?

(b) What is the average cost per can opener in each case?

SOLUTION

(a) Since each can opener costs $2.75, the variable costs for making x can openers is $2.75x$. The total cost y of making x can openers is

$$y = \text{variable costs} + \text{fixed costs} = 2.75x + 26{,}000.$$

The cost of making 1000 can openers is

$$y = 2.75x + 26{,}000 = 2.75(1000) + 26{,}000 = \$28{,}750.$$

Similarly, the cost of making 20,000 can openers is

$$y = 2.75(20{,}000) + 26{,}000 = \$81{,}000,$$

and the cost of 40,000 is

$$y = 2.75(40{,}000) + 26{,}000 = \$136{,}000.$$

The cost equation $y = 2.75x + 26{,}000$ shows that the cost of making x can openers increases at the rate of $2.75 per can opener. This rate of change is the slope of the graph of the cost equation.

(b) The average cost per can opener in each case is the total cost divided by the number of can openers. So the average cost per can opener is as follows.

For 1000: $28{,}750/1000 = \$28.75$ per can opener;

For 20,000: $81{,}000/20{,}000 = \$4.05$ per can opener;

For 40,000: $136{,}000/40{,}000 = \$3.40$ per can opener.   ■

Examples 5 and 6 illustrate this fact.

Linear Rate of Change

> The slope m of the line with equation
>
> $$y = mx + b$$
>
> is the rate of change of y with respect to x.

■■ POINT-SLOPE FORM

Suppose the line L passes through the point (x_1, y_1) and has slope m. Let (x, y) be any other point on L. Using the points (x_1, y_1) and (x, y) to compute the slope m of L (see Figure 1–31), we have

$$\frac{y - y_1}{x - x_1} = \text{slope of } L$$

$$\frac{y - y_1}{x - x_1} = m$$

$$y - y_1 = m(x - x_1).$$

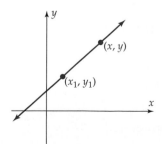

Figure 1–31

Thus the coordinates of every point on L satisfy the equation $y - y_1 = m(x - x_1)$, and we have this fact:

Point-Slope Form

> The line with slope m through the point (x_1, y_1) is the graph of the equation
>
> $$y - y_1 = m(x - x_1).$$

EXAMPLE 7

Figure 1–32

The tangent line to the graph of $y = 2x^2 - 8$ at the point $(1, -6)$ is shown in Figure 1–32. In calculus, it is shown that this line passes through $(1, -6)$ and has slope 4. Find the equation of the tangent line.

SOLUTION   Substitute 4 for m and $(1, -6)$ for (x_1, y_1) in the point-slope equation:

$$y - y_1 = m(x - x_1)$$

$$y - (-6) = 4(x - 1) \qquad \text{[point-slope form]}$$

$$y + 6 = 4x - 4$$

$$y = 4x - 10 \qquad \text{[slope-intercept form]} \qquad \blacksquare$$

EXAMPLE 8

The annual out-of-pocket spending (per person) on doctors and clinical services was approximately \$105 in 1997. According to projections from the Health Care Financing Committee, this cost is expected to rise linearly to \$196 in 2010, as indicated in Figure 1–33.

(a) Find an equation that gives the out-of-pocket cost y in year x.

(b) Use this equation to estimate the out-of-pocket costs in 2004 and 2006.

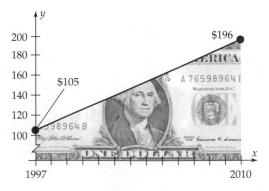

Figure 1–33

SOLUTION

(a) Let $x = 0$ correspond to 1997, so that $x = 13$ corresponds to 2010. Then the given information can be represented by the points (0, 105) and (13, 196). We must find the equation of the line through these points. Its slope is

$$\frac{196 - 105}{13 - 0} = \frac{91}{13} = 7.$$

Now we use the slope 7 and one of the points (0, 105) or (13, 196) to find the equation of the line. It doesn't matter which point, since both lead to the same equation.

$$y - y_1 = m(x - x_1) \qquad\qquad y - y_1 = m(x - x_1)$$
$$y - 105 = 7(x - 0) \qquad\qquad y - 196 = 7(x - 13)$$
$$y = 7x + 105 \qquad\qquad y - 196 = 7x - 91$$
$$y = 7x + 105.$$

(b) Since 2004 corresponds to $x = 7$, the projected out-of-pocket costs in 2004 are

$$y = 7x + 105 = 7 \cdot 7 + 105 = \$154.$$

The costs in 2006 ($x = 9$) are

$$y = 7x + 105 = 7 \cdot 9 + 105 = \$168. \qquad\blacksquare$$

▉ VERTICAL LINES AND GENERAL FORM

The preceding discussion does not apply to vertical lines, whose equations have a different form from those examined earlier.

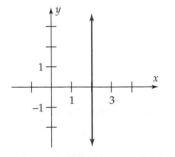

Figure 1–34

EXAMPLE 9

Every point on the vertical line in Figure 1–34 has first coordinate 2. Thus every point on the line satisfies $x + 0y = 2$. Thus the line is the graph of the equation $x = 2$. If you try to compute the slope of this line, say, using (2, 1) and (2, 4), you obtain $\dfrac{4 - 1}{2 - 2} = \dfrac{4}{0}$, which is not defined. $\qquad\blacksquare$

Example 9 illustrates these facts:

Vertical Lines

> The vertical line with *x*-intercept c is the graph of the equation
>
> $$x = c.$$
>
> The slope of this line is undefined.

By rearranging terms, if necessary, the equation of any line can be written in the **general form** $Ax + By = C$ for some constants A, B, and C. For instance,

This equation	can be written as
$y = 4x - 10$	$4x - 1y = 10$
$y - 196 = 7(x - 13)$	$-7x + 1y = 105$
$y = 3$	$0x + 1y = 3$
$x = 2$	$1x + 0y = 2$

In summary:

General Form

> Every line is the graph of an equation of the form
> $$Ax + By = C,$$
> where A and B are not both zero.

Figure 1–35

EXAMPLE 10

Graph $3x + 2y = 6$.

SOLUTION   From the preceding box, we know that the graph is a line. It is easily graphed by finding its intercepts. By setting $x = 0$ and solving for y, we see that the y-intercept is 3. Similarly, setting $y = 0$ and solving for x shows that the x-intercept is 2. This leads to the graph in Figure 1–35.   ■

▓ PARALLEL AND PERPENDICULAR LINES

The slope of a line measures how steeply it rises or falls. Since parallel lines rise or fall equally steeply, the following fact should be plausible (see Exercises 74–75 for a proof).

Parallel Lines

> Two nonvertical lines are parallel exactly when they have the same slope.

EXAMPLE 11

Find the equation of the line L through $(2, -1)$ that is parallel to the line M whose equation is $3x - 2y + 6 = 0$.

SOLUTION   First find the slope of M by rewriting its equation in slope-intercept form:

$$3x - 2y + 6 = 0$$
$$-2y = -3x - 6$$
$$y = \frac{3}{2}x + 3.$$

Therefore M has slope $3/2$. The parallel line L must have the same slope, $3/2$. Since $(2, -1)$ is on L, we can use the point-slope form to find its equation:

$$y - y_1 = m(x - x_1)$$

$$y - (-1) = \frac{3}{2}(x - 2) \qquad \text{[point-slope form]}$$

$$y + 1 = \frac{3}{2}x - 3$$

$$y = \frac{3}{2}x - 4 \qquad \text{[slope-intercept form]} \qquad \blacksquare$$

Two lines that meet in a right angle (90° angle) are said to be **perpendicular.** As you might suspect, there is a close relationship between the slopes of two perpendicular lines.

Perpendicular Lines

> Two nonvertical lines are perpendicular exactly when the product of their slopes is -1.

A proof of this fact is outlined in Exercise 76.

EXAMPLE 12

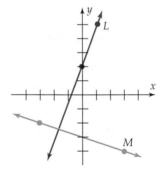

Figure 1–36

In Figure 1–36, the line L through $(0, 2)$ and $(1, 5)$ appears to be perpendicular to the line M through $(-3, -2)$ and $(3, -4)$. We can confirm this fact by computing the slopes of these lines:

$$\text{Slope } L = \frac{5 - 2}{1 - 0} = 3 \qquad \text{and} \qquad \text{slope } M = \frac{-4 - (-2)}{3 - (-3)} = \frac{-2}{6} = -\frac{1}{3}.$$

Since $3(-1/3) = -1$, the lines L and M are perpendicular. $\qquad \blacksquare$

EXAMPLE 13

Find the equation of the perpendicular bisector of the line segment with endpoints $(-5, -4)$ and $(7, 2)$.

SOLUTION The perpendicular bisector M goes through the midpoint of the line segment from $(-5, -4)$ and $(7, 2)$. The midpoint formula (page 40) shows that this midpoint is

$$\left(\frac{x_1 + x_2}{2}, \frac{y_1 + y_2}{2}\right) = \left(\frac{-5 + 7}{2}, \frac{-4 + 2}{2}\right) = (1, -1).$$

The line L through $(-5, -4)$ and $(7, 2)$ has slope

$$\frac{y_2 - y_1}{x_2 - x_1} = \frac{2 - (-4)}{7 - (-5)} = \frac{6}{12} = \frac{1}{2}.$$

Since M is perpendicular to L, we have (slope M)(slope L) = -1, so that

$$\text{Slope } M = \frac{-1}{\text{slope } L} = \frac{-1}{1/2} - -2.$$

Thus M is the line through $(1, -1)$ with slope -2, and its equation is

$$y - (-1) = -2(x - 1) \qquad \text{[point-slope form]}$$

$$y = -2x + 1. \qquad \text{[slope-intercept form]} \quad \blacksquare$$

✓ EXERCISES 1.4

1. For which of the line segments in the figure is the slope
   (a) largest?   (b) smallest?
   (c) largest in absolute value? (d) closest to zero?

2. The doorsill of a campus building is 5 feet above ground level. To allow wheelchair access, the steps in front of the door are to be replaced by a straight ramp with constant slope 1/12, as shown in the figure. How long must the ramp be? [The answer is *not* 60 feet.]

In Exercises 3–6, find the slope of the line through the given points.

3. $(1, 2)$; $(3, 7)$

4. $(-1, -2)$; $(2, -1)$

5. $(1/4, 0)$; $(3/4, 2)$

6. $(\sqrt{2}, -1)$; $(2, -9)$

In Exercises 7–10, find a number t such that the line passing through the two given points has slope -2.

7. $(0, t)$; $(9, 4)$

8. $(1, t)$; $(-3, 5)$

9. $(t + 1, 5)$; $(6, -3t + 7)$

10. (t, t); $(5, 9)$

11. Let L be a nonvertical straight line through the origin. L intersects the vertical line through $(1, 0)$ at a point P. Show that the second coordinate of P is the slope of L.

12. On one graph, sketch five line segments, not all meeting at a single point, whose slopes are five different positive numbers. Do this in such a way that the left-hand line has the largest slope, the second line from the left has the next largest slope, and so on.

In Exercises 13–16, find the slope and y-intercept of the line whose equation is given.

13. $2x - y + 5 = 0$ **14.** $3x + 4y = 7$

15. $3(x - 2) + y = 7 - 6(y + 4)$

16. $2(y - 3) + (x - 6) = 4(x + 1) - 2$

In Exercises 17–20, find the equation of the line with slope m that passes through the given point.

17. $m = 1$; $(3, 5)$ **18.** $m = 2$; $(-2, 1)$

19. $m = -1$; $(6, 2)$ **20.** $m = 0$; $(-4, -5)$

In Exercises 21–24, find the equation of the line through the given points.

21. $(0, -5)$ and $(-3, -2)$ **22.** $(4, 3)$ and $(2, -1)$

23. $(4/3, 2/3)$ and $(1/3, 3)$ **24.** $(6, 7)$ and $(6, 15)$

In Exercises 25–30, graph the equation. Label all intercepts.

25. $3x + 5y = 15$

26. $2x - 3y = 12$

27. $2y - x = 2$

28. $4x + 5y = -10$

29. $3x - 2y = 0$

30. $2x + 6y = 0$

In Exercises 31–34, determine whether the line through P and Q is parallel or perpendicular to the line through R and S or neither.

31. $P = (2, 5)$, $Q = (-1, -1)$ and $R = (4, 2)$, $S = (6, 1)$.

32. $P = (0, 3/2)$, $Q = (1, 1)$ and $R = (2, 7)$, $S = (3, 9)$.

33. $P = (-3, 1/3)$, $Q = (1, -1)$ and $R = (2, 0)$, $S = (4, -2/3)$.

34. $P = (3, 3)$, $Q = (-3, -1)$ and $R = (2, -2)$, $S = (4, -5)$.

In Exercises 35–37, determine whether the lines whose equations are given are parallel, perpendicular, or neither.

35. $2x + y - 2 = 0$ and $4x + 2y + 18 = 0$.

36. $3x + y - 3 = 0$ and $6x + 2y + 17 = 0$.

37. $y = 2x + 4$ and $.5x + y = -3$.

38. Use slopes to show that the points $(-4, 6)$, $(-1, 12)$, and $(-7, 0)$ all lie on the same straight line.

39. Use slopes to determine whether $(9, 6)$, $(-1, 2)$, and $(1, -3)$ are the vertices of a right triangle.

40. Use slopes to show that the points $(-5, -2)$ $(-3, 1)$, $(3, 0)$, and $(5, 3)$ are the vertices of a parallelogram.

In Exercises 41–48, find an equation for the line satisfying the given conditions.

41. Through $(-2, 1)$ with slope 3.

42. y-intercept -7 and slope 1.

43. Through $(2, 3)$ and parallel to $3x - 2y = 5$.

44. Through $(1, -2)$ and perpendicular to $y = 2x - 3$.

45. x-intercept 5 and y-intercept -5.

46. Through $(-5, 2)$ and parallel to the line through $(1, 2)$ and $(4, 3)$.

47. Through $(-1, 3)$ and perpendicular to the line through $(0, 1)$ and $(2, 3)$.

48. y-intercept 3 and perpendicular to $2x - y + 6 = 0$.

49. Find a real number k such that $(3, -2)$ is on the line $kx - 2y + 7 = 0$.

50. Find a real number k such that the line $3x - ky + 2 = 0$ has y-intercept -3.

If P is a point on a circle with center C, then the tangent line to the circle at P is the straight line through P that is perpendicular to the radius CP. In Exercises 51–54, find the equation of the tangent line to the circle at the given point.

51. $x^2 + y^2 = 25$ at $(3, 4)$ [*Hint:* Here C is $(0, 0)$ and P is $(3, 4)$; what is the slope of radius CP?]

52. $x^2 + y^2 = 169$ at $(-5, 12)$

53. $(x - 1)^2 + (y - 3)^2 = 5$ at $(2, 5)$

54. $x^2 + y^2 + 6x - 8y + 15 = 0$ at $(-2, 1)$

55. Let A, B, C, D be nonzero real numbers. Show that the lines $Ax + By + C = 0$ and $Ax + By + D = 0$ are parallel.

56. Let L be a line that is neither vertical nor horizontal and that does not pass through the origin. Show that L is the graph of $\dfrac{x}{a} + \dfrac{y}{b} = 1$, where a is the x-intercept and b is the y-intercept of L.

57. According to the March 31, 2001, issue of *The Economist*, worldwide automobile production was about 54 million vehicles in 2000 and is expected to be 63 million vehicles in 2008. This production growth is approximately linear.

(a) Let $x = 0$ correspond to 2000. Find a linear equation that gives the number y of vehicles (in millions) produced in year x.

(b) Estimate the number of vehicles produced in 2003.

(c) In what year will vehicle production first reach 60 million?

[*More exercises are on page 60.*]

58. Carbon dioxide (CO_2) concentration is measured regularly at the Mauna Loa observatory in Hawaii. The mean annual concentration in parts per million in various years is given in the table.[*]

Year	Concentration (ppm)
1970	325.7
1980	338.7
1990	354.2
2000	369.4

(a) Do the data points lie on a single line? How do you know?

(b) Write a linear equation that uses the data from 1970 and 2000 to model CO_2 concentration over time. What does this model say the concentration should have been in 1980? Does the model overestimate or underestimate the concentration?

(c) Write a linear equation that uses the data from 1980 and 2000 to model CO_2 concentration over time. What does this model say the concentration should have been in 1970? Does the model overestimate or underestimate the concentration?

(d) What do the two models say about the concentration in the year 2005? Which model do you think is the more accurate? Why?

59. The consumer price index (CPI), which measures the cost of a typical package of consumer goods, stood at 136.2 in 1991 and 177.1 in 2001.

(a) Find a linear equation that approximates the CPI y in year x, with $x = 0$ corresponding to 1991.

(b) Estimate the CPI in 2005.

(c) At what yearly rate is the CPI increasing?

60. At sea level, water boils at 212°F. At a height of 1100 feet, water boils at 210°F. The relationship between boiling point and height is linear.

(a) Find an equation that gives the boiling point y of water at a height of x feet.

Find the boiling point of water in each of the following cities (whose altitudes are given).

(b) Cincinnati, OH (550 feet)

(c) Springfield, MO (1300 feet)

(d) Billings, MT (3120 feet)

(e) Flagstaff, AZ (6900 feet)

61. According to the Center of Science in the Public Interest, the maximum healthy weight for a person who is 5 feet 5 inches tall is 150 pounds, and the maximum healthy weight for someone 6 feet 3 inches tall is 200 pounds. The relationship between weight and height here is linear.

(a) Find a linear equation that gives the maximum healthy weight y for a person whose height is x inches over 4 feet 10 inches. (Thus $x = 0$ corresponds to 4 feet 10 inches, $x = 2$ to 5 feet, etc.)

(b) What is the maximum healthy weight for a person whose height is 5 feet? 6 feet?

(c) How tall is a person who is at a maximum healthy weight of 220 pounds?

62. The age-adjusted death rate from heart disease was 307.2 per 100,000 people in 1950 and 126.6 per 100,000 in 1998.

(a) Assuming that the rate decreased linearly, find an equation that gives the number y of deaths per 100,000 in year x, with $x = 0$ corresponding to 1950.

(b) Use this equation to estimate the death rate in 1995 and 2005. [For comparison purposes, the actual rate was 138.3 in 1995.]

(c) Is this equation likely to remain accurate long into the future? [*Hint:* What does it give for a death rate in 2035?]

63. The number of unmarried couples living together in the United States has risen in an approximately linear fashion over the last quarter century. There were 1.1 million cohabiting adults in 1977 and 4.9 million in 1997.

(a) Assuming that this trend continues, use a linear equation to estimate the number of cohabiting adults in 2004.

(b) When will the number of cohabiting adults reach 7.2 million?

64. The Whismo Hat Company has fixed costs of $50,000 and variable costs of $8.50 per hat.

(a) Find an equation that gives the total cost y of producing x hats.

(b) What is the average cost per hat when 20,000 are made? 50,000? 100,000?

65. Suppose the cost of making x TV sets is given by $y = 145x + 120,000$.

(a) Write an equation that gives the average cost per set when x sets are made.

(b) How many sets should be made in order to have an average cost per set of $175?

---

[*]C. D. Keeling and T. P. Whorf, Scrips Institution of Oceanography.

66. The profit p (in thousands of dollars) on x thousand units of a specialty item is $p = .6x - 14.5$. The cost c of manufacturing x items is given by $c = .8x + 14.5$.
    (a) Find an equation that gives the revenue r from selling x items.
    (b) How many items must be sold for the company to break even (i.e., for revenue to equal cost)?

67. A publisher has fixed costs of $110,000 for a mathematics text. The variable costs are $50 per book. The book sells for $72. Find equations that give
    (a) The cost c of making x books
    (b) The revenue r from selling x books
    (c) The profit p from selling x books
    (d) What is the publisher's break-even point (see Exercise 66(b))?

68. The poverty level income for a family of four was $9287 in 1981. Because of inflation and other factors, the poverty level rose approximately linearly to $18,267 in 2001.[*]
    (a) At what rate is the poverty level increasing?
    (b) Estimate the poverty level in 1990 and 2005.

69. A 75-gallon water tank is being emptied. The graph shows the amount of water in the tank after x minutes.
    (a) At what rate is the tank emptying during the first 2 minutes? During the next 3 minutes? During the last minute?

    (b) Suppose the tank is emptied at a constant rate of 10 gallons per minute. Draw the graph that shows the amount of water after x minutes. What is the equation of the graph?

70. A Chrysler Concorde LXi that was worth $24,000 in 2002 will be worth $13,000 in 2006. Assuming linear depreciation, what is the rate of depreciation? What will the car be worth in 2009?

Use the graph and the following information for Exercises 71–73. Rocky is an "independent" ticket dealer who markets choice tickets for Los Angeles Lakers home games. (California currently has no laws against ticket scalping.) Each graph shows how many tickets will be demanded by buyers at a particular price. For instance, when the Lakers play the Chicago Bulls, the graph shows that at a price of $160, no tickets are demanded. As the price (y-coordinate) gets lower, the number of tickets demanded (x-coordinate) increases.

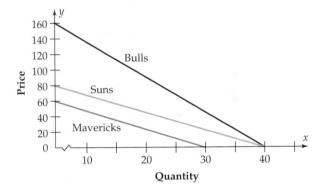

71. Write a linear equation that relates the quantity x of tickets demanded at price y when the Lakers play the
    (a) Dallas Mavericks
    (b) Phoenix Suns
    (c) Chicago Bulls
    [*Hint:* In each case, use the x- and y-intercepts to determine its slope.]

72. Use the equations from Exercise 71 to find the number of tickets Rocky would sell at a price of $40 for a game against the
    (a) Mavericks   (b) Bulls

73. Suppose Rocky has 20 tickets to sell. At what price could he sell them all when the Lakers play the
    (a) Mavericks   (b) Suns

---

[*]U.S. Census Bureau

Thinkers

74. Prove that nonvertical parallel lines L and M have the same slope, as follows. Suppose M lies above L, and choose two points (x_1, y_1) and (x_2, y_2) on L.

(a) Let P be the point on M with first coordinate x_1. Let b denote the vertical distance from P to (x_1, y_1). Show that the second coordinate of P is $y_1 + h$

(b) Let Q be the point on M with first coordinate x_2. Use the fact that L and M are parallel to show that the second coordinate of Q is $y_2 + b$.

(c) Compute the slope of L using (x_1, y_1) and (x_2, y_2). Compute the slope of M using the points P and Q. Verify that the two slopes are the same.

75. Show that two nonvertical lines with the same slope are parallel. [*Hint:* The equations of distinct lines with the same slope must be of the form $y = mx + b$ and $y = mx + c$ with $b \neq c$ (why?). If (x_1, y_1) were a point on both lines, its coordinates would satisfy both equations. Show that this leads to a contradiction, and conclude that the lines have no point in common.]

76. This exercise provides a proof of the statement about slopes of perpendicular lines in the box on page 57. First, assume that L and M are nonvertical perpendicular lines that both pass through the origin. L and M intersect the vertical line $x = 1$ at the points $(1, k)$ and $(1, m)$, respectively, as shown in the figure.

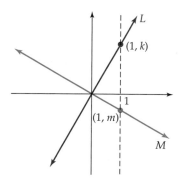

(a) Use $(0, 0)$ and $(1, k)$ to show that L has slope k. Use $(0, 0)$ and $(1, m)$ to show that M has slope m.

(b) Use the distance formula to compute the length of each side of the right triangle with vertices $(0, 0)$, $(1, k)$, and $(1, m)$.

(c) Use part (b) and the Pythagorean Theorem to find an equation involving k, m, and various constants. Show that this equation simplifies to $km = -1$. This proves half of the statement.

(d) To prove the other half, assume that $km = -1$, and show that L and M are perpendicular as follows. You may assume that a triangle whose sides a, b, c satisfy $a^2 + b^2 = c^2$ is a right triangle with hypotenuse c. Use this fact, and do the computation in part (b) in reverse (starting with $km = -1$) to show that the triangle with vertices $(0, 0)$, $(1, k)$, and $(1, m)$ is a right triangle, so that L and M are perpendicular.

(e) Finally, to prove the general case when L and M do not intersect at the origin, let L_1 be a line through the origin that is parallel to L, and let M_1 be a line through the origin that is parallel to M. Then L and L_1 have the same slope, and M and M_1 have the same slope (why?). Use this fact and parts (a)–(d) to prove that L is perpendicular to M exactly when $km = -1$.

77. Show that the diagonals of a square are perpendicular. [*Hint:* Place the square in the first quadrant of the plane, with one vertex at the origin and sides on the positive axes. Label the coordinates of the vertices appropriately.]

Chapter 1 Review

IMPORTANT CONCEPTS

Section 1.1

Real numbers, integers, rationals, irrationals 2
Number line 3
Scientific notation 3
Order ($<, \leq, >, \geq$) 4
Intervals, open intervals, closed intervals 4–5
Order of operations 6
Distributive law 6
Negatives 7
Square roots 7
Absolute value 8
Distance on the number line 10

Special Topics 1.1.A

Repeating and nonrepeating decimals 17

Section 1.2

Basic principles for solving equations 19

First-degree equations 19
Quadratic equations 20
Factoring 20
Completing the square 20–21
Quadratic formula 22
Discriminant 23–24
Higher-degree equations 24

Special Topics 1.2.A

Absolute value equations 28–29

Special Topics 1.2.B

Direct variation 30
Inverse variation 31
Constant of variation 30–31

Section 1.3

Coordinate plane, x-axis, y-axis, quadrants 36
Scatter plots and line graphs 37
Distance formula 38
Midpoint formula 40

Graph of an equation 41
x- and y-intercepts 41
Circle, center, radius 42
Equation of the circle 42
Unit circle 43

Section 1.4

Change in x, change in y 48
Slope 49
Properties of slope 50
Slope-intercept form of the equation of a line 50–51
Horizontal lines 51
Fixed and variable costs 52–53
Linear rate of change 53
Point-slope form of the equation of a line 53–54
Vertical lines 55
General form of the equation of a line 55–56
Slopes of parallel lines 56
Slopes of pependicular lines 57

IMPORTANT FACTS & FORMULAS

- $|c - d|$ = distance from c to d on the number line.

- *Quadratic Formula:* If $a \neq 0$, then the solutions of $ax^2 + bx + c = 0$ are
$$x = \frac{-b \pm \sqrt{b^2 - 4ac}}{2a}.$$

- If $a \neq 0$, then the number of real solutions of $ax^2 + bx + c = 0$ is 0, 1, or 2, depending on whether the discriminant $b^2 - 4ac$ is negative, zero, or positive.

- *Distance Formula:* The distance from (x_1, y_1) to (x_2, y_2) is
$$\sqrt{(x_1 - x_2)^2 + (y_1 - y_2)^2}.$$

- *Midpoint Formula:* The midpoint of the line segment from (x_1, y_1) to (x_2, y_2) is
$$\left(\frac{x_1 + x_2}{2}, \frac{y_1 + y_2}{2} \right).$$

- Equation of the circle with center (c, d) and radius r is
$$(x - c)^2 + (y - d)^2 = r^2.$$

■ The slope of the line through (x_1, y_1) and (x_2, y_2) (where $x_1 \neq x_2$) is

$$\frac{y_2 - y_1}{x_2 - x_1}.$$

■ The equation of the line with slope m and y-intercept b is

$$y = mx + b.$$

■ The equation of the line through (x_1, y_1) with slope m is

$$y - y_1 = m(x - x_1).$$

■ Two nonvertical lines are parallel exactly when they have the same slope.

■ Two nonvertical lines are perpendicular exactly when the product of their slopes is -1.

REVIEW QUESTIONS

1. Fill the blanks with one of the symbols $<$, $=$, or $>$ so that the resulting statement is true.
    (a) 142 ____ $|-51|$
    (b) $\sqrt{2}$ ____ $|-2|$
    (c) -1000 ____ $\dfrac{1}{10}$
    (d) $|-2|$ ____ $-|6|$
    (e) $|u - v|$ ____ $|v - u|$, where u and v are fixed real numbers.

2. List two real numbers that are *not* rational numbers.

3. Express in scientific notation:
    (a) $12{,}320{,}000{,}000{,}000{,}000$
    (b) $.0000000000789$

4. Express in decimal notation:
    (a) 4.78×10^8
    (b) 6.53×10^{-9}

5. Express in symbols:
    (a) y is negative, but greater than -10.
    (b) x is nonnegative and not greater than 10.

6. Express in symbols:
    (a) $c - 7$ is nonnegative.
    (b) $.6$ is greater than $|5x - 2|$.

7. Express in symbols:
    (a) x is less than 3 units from -7 on the number line.
    (b) y is farther from 0 than x is from 3 on the number line.

8. Simplify: $|b^2 - 2b + 1|$

9. Solve: $|x - 5| = 3$

10. Solve: $|x + 2| = 4$

11. Solve: $|x + 3| = \dfrac{5}{2}$

12. Solve: $|x - 5| \leq 2$

13. Solve: $|x + 2| \leq 2$

14. Solve: $|x - 1| > 4$

15. (a) $|\pi - 7| =$ ____   (b) $|\sqrt{23} - \sqrt{3}| =$ ____

16. If c and d are real numbers with $c \neq d$ what are the possible values of $\dfrac{c - d}{|c - d|}$?

17. Express in interval notation:
    (a) The set of all real numbers that are strictly greater than -8;
    (b) The set of all real numbers that are less than or equal to 5.

18. Express in interval notation:
    (a) The set of all real numbers that are strictly between -6 and 9;
    (b) The set of all real numbers that are greater than or equal to 5, but strictly less than 14.

19. Express $.282828 \cdots$ as a fraction.

20. Express $.362362362 \cdots$ as a fraction.

21. Solve for x:

$$2\left(\frac{x}{5} + 7\right) - 3x = \frac{x + 2}{5} - 4.$$

22. Solve for x in terms of y: $xy + 3 = x - 2y$

23. Solve for x: $3x^2 - 2x + 5 = 0$

24. Solve for y: $3y^2 - 2y = 5$

25. Solve for z: $5z^2 + 6z = 7$

26. The population P (in thousands) of St. Louis, Missouri, can be approximated by

$$P = .11x^2 - 15.95x + 864,$$

where $x = 0$ corresponds to 1950. When did the population first drop below 450,000?

REVIEW QUESTIONS

27. How many times larger is the viewing area on a 21-inch computer monitor than on a 14-inch monitor? Remember that the size of a monitor is its diagonal measurement, as shown in the figure, and assume that the height of the screen is three-fourths of its width.

21 in.

x

28. The median sales price S (in thousands of dollars) for a single-family home in the northeastern United States during the 1990s can be approximated by $S = .47x^2 + 1.52x + 158$, where $x = 0$ corresponds to 1990.[*] When did the median price reach $195,000?

29. Find the *number* of real solutions of the equation $20x^2 + 12 = 31x$.

30. For what value of k does the equation

$$kt^2 + 5t + 2 = 0$$

have exactly one real solution for t?

31. If x and y are directly proportional and $x = 12$ when $y = 36$, then what is y when $x = 2$?

32. Driving time varies inversely as speed. If it takes 3 hours to drive to the beach at an average speed of 48 mph, how long will it take to drive home at an average speed of 54 mph?

33. Find the constant of variation when T varies directly as the square of R and inversely as S if $T = .6$ when $R = 3$ and $S = 15$.

34. The statement "r varies directly as s and the square root of t and inversely as the cube of x" means that for some constant k,
    (a) $r = kstx^3$
    (b) $rx^3 = ks\sqrt{t}$
    (c) $r = \dfrac{kx^3\sqrt{t}}{s}$
    (d) $k = \dfrac{rs\sqrt{t}}{x^3}$
    (e) $kx^3s = k\sqrt{t}$

In Questions 35–38, find all real solutions of the equation. Do not approximate.

35. $x^4 - 11x^2 + 18 = 0$
36. $x^6 - 4x^3 + 4 = 0$
37. $|3x - 1| = 4$ 38. $|2x - 1| = x + 4$
39. Find the distance from $(1, -2)$ to $(4, 5)$.
40. Find the distance from $(3/2, 4)$ to $(3, 5/2)$.
41. Find the distance from (c, d) to $(c - d, c + d)$.
42. Find the midpoint of the line segment from $(-4, 7)$ to $(9, 5)$.
43. Find the midpoint of the line segment from (c, d) to $(2d - c, c + d)$.
44. Find the equation of the circle with center $(-3, 4)$ that passes through the origin.
45. (a) If $(1, 1)$ is on a circle with center $(2, -3)$, what is the radius of the circle?
    (b) Find the equation of the circle in part (a).
46. Sketch the graph of $3x^2 + 3y^2 = 12$.
47. Sketch the graph of $(x - 5)^2 + y^2 - 9 = 0$.
48. Find the center and radius of the circle whose equation is

$$x^2 + y^2 - 2x + 6y + 1 = 0.$$

49. Which of statements (a)–(d) are descriptions of the circle with center $(0, -2)$ and radius 5?
    (a) The set of all points (x, y) that satisfy $|x| + |y + 2| = 5$.
    (b) The set of all points whose distance from $(0, -2)$ is 5.
    (c) The set of all points (x, y) such that $x^2 + (y + 2)^2 = 5$.
    (d) The set of all points (x, y) such that $\sqrt{x^2 + (y + 2)^5} = 5$.

50. If the equation of a circle is $3x^2 + 3(y - 2)^2 = 12$, which of the following statements is true?
    (a) The circle has diameter 3.
    (b) The center of the circle is $(2, 0)$.
    (c) The point $(0, 0)$ is on the circle.
    (d) The circle has radius $\sqrt{12}$.
    (e) The point $(1, 1)$ is on the circle.

51. The graph of one of the equations below is *not* a circle. Which one?
    (a) $x^2 + (y + 5)^2 = \pi$
    (b) $7x^2 + 4y^2 - 14x + 3y^2 - 2 = 0$
    (c) $3x^2 + 6x + 3 = 3y^2 + 15$
    (d) $2(x - 1)^2 - 8 = -2(y + 3)^2$
    (e) $\dfrac{x^2}{4} + \dfrac{y^2}{4} = 1$

---

[*]Department of Housing and Urban Development

REVIEW QUESTIONS

52. The point $(7, -2)$ is on the circle whose center is on the midpoint of the segment joining $(3, 5)$ and $(-5, -1)$. Find the equation of this circle.

53. The table shows the number of Americans (in millions) who snowboarded each year from 1994 to 2000.[*]

Year	1994	1995	1996	1997	1998	1999	2000
Number	2	2.7	3.7	2.5	3.5	3.2	4.2

Sketch a scatter plot and a line graph for this data, letting $x = 0$ correspond to 1990.

54. The table shows the average speed (mph) of the winning car in the Indianapolis 500 race in selected years.[†]

Year	1990	1992	1994	1996	1998	2000	2002
Speed	186	134	161	148	145	168	166

Sketch a scatter plot and a line graph for this data, letting $x = 0$ correspond to 1990.

55. (a) What is the y-intercept of the graph of the line
$$y = x - \frac{x - 2}{5} + \frac{3}{5}?$$
(b) What is the slope of the line?

56. Find the equation of the line passing through $(1, 3)$ and $(2, 5)$.

57. Find the equation of the line passing through $(2, -1)$ with slope 3.

58. Find a point on the graph of $y = 3x$ whose distance to the origin is 2.

59. Find the equation of the line that crosses the y-axis at $y = 1$ and is perpendicular to the line $2y - x = 5$.

60. (a) Find the y-intercept of the line $2x + 3y - 4 = 0$.
(b) Find the equation of the line through $(1, 3)$ that has the same y-intercept as the line in part (a).

61. Find the equation of the line through $(-4, 5)$ that is parallel to the line through $(1, 3)$ and $(-4, 2)$.

62. Sketch the graph of the line $3x + y - 1 = 0$.

63. As a balloon is launched from the ground, the wind is blowing it due east. The conditions are such that the balloon is ascending along a straight line with slope $1/5$. After 1 hour the balloon is 5000 feet vertically above the ground. How far east has the balloon blown?

---

*National Sporting Goods Association, TransWorld Snowboarding Business.
†Indianapolis Motor Speedway Hall of Fame and Museum.

64. The point (u, v) lies on the line $y = 5x - 10$. What is the slope of the line passing through (u, v) and the point $(0, -10)$?

In Questions 65–71, determine whether the statement is true or false.

65. The graph of $x = 5y + 6$ has y-intercept 6.

66. The graph of $2y - 8 = 3x$ has y-intercept 4.

67. The lines $3x + 4y = 12$ and $4x + 3y = 12$ are perpendicular.

68. Slope is not defined for horizontal lines.

69. The line in the figure has positive slope.

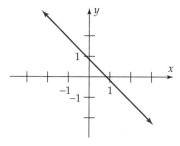

70. The line in the figure above does not pass through the third quadrant.

71. The y-intercept of the line in the figure above is negative.

72. Consider the *slopes* of the lines shown in the figure below. Which line has the slope with the largest *absolute value*?

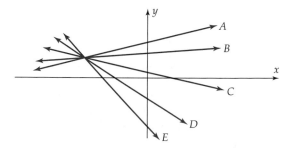

73. Which of the following lines rises most steeply from left to right?
(a) $y = -4x - 10$ (b) $y = 3x + 4$
(c) $20x + 2y - 20 = 0$ (d) $4x = y - 1$
(e) $4x = 1 - y$

74. Which of the following lines is *not* perpendicular to the line $y = x + 5$?
(a) $y = 4 - x$ (b) $y + x = -5$
(c) $4 - 2x - 2y = 0$ (d) $x = 1 - y$
(e) $y - x = \frac{1}{5}$

REVIEW QUESTIONS

75. Which of the following does *not* pass through the third quadrant?
(a) $y = x$ (b) $y = 4x - 7$
(c) $y = -2x - 5$ (d) $y = 4x + 7$
(e) $y = -2x + 5$

76. Let a, b be fixed real numbers. Where do the lines $x = a$ and $y = b$ intersect?
(a) Only at (b, a).   (b) Only at (a, b).
(c) These lines are parallel, so they don't intersect.
(d) If $a = b$, then these are the same line, so they have infinitely many points of intersection.
(e) Since these equations are not of the form $y = mx + b$, the graphs are not lines.

77. What is the y-intercept of the line $2x - 3y + 5 = 0$?

78. For what values of k will the graphs of $2y + x + 3 = 0$ and $3y + kx + 2 = 0$ be perpendicular lines?

79. Average life expectancy increased linearly from 70.8 years for a person born in 1970 to 76.8 years for a person born in 2000.
(a) Find a linear equation that gives the average life expectancy y of a person born in year x, with $x = 0$ corresponding to 1970.
(b) Use the equation in part (a) to estimate the average life expectancy of a person born in 1984.
(c) Assuming the equation remains valid, in what year will the average life expectancy be 80 years for people born in that year?

80. The population of San Diego, California, grew in an approximately linear fashion from 334,400 in 1950 to 1,223,400 in 2000.
(a) Find a linear equation that gives the population y of San Diego (in thousands) in year x, with $x = 0$ corresponding to 1950.
(b) Estimate the population of San Diego in 1995.
(c) Assuming the equation remains accurate, when will San Diego's population reach 1.5 million people?

In Exercises 81–84, match the given information with the graph, and determine the slope of the graph.

(a)

(b)

(c)

(d)

81. A salesman is paid $300 per week plus $75 for each unit sold.

82. A person is paying $25 per week to repay a $300 loan.

83. A gold coin that was purchased for $300 appreciates $20 per year.

84. A CD player that was purchased for $300 depreciates $80 per year.

DISCOVERY PROJECT 1　Modeling the Real World with Lines

Mathematical models are often used to represent the physical world. These models may take many forms, including equations, formulas, and graphs. When the model accurately reflects the real world, it may enable us to predict certain outcomes without actually using physical experimentation. Other models, such as the one presented here, result from different ways of measuring a particular quantity. They make it easy to compare the different measurement systems.

There are two common units that we (the average citizens of the world) use to measure temperature. One is the Fahrenheit scale, the usual one seen on U.S. television weather reports; the other is the Celsius, or centigrade, scale, the usual one seen on Canadian television. The two scales are linearly related and calibrated using the freezing and boiling points of water at sea level.

Temperature Scale	Fahrenheit Scale	Celsius Scale
Water Freezes	32°	0°
Water Boils	212°	100°

1. Write a formula that relates Fahrenheit temperature F to Celsius temperature C. Your answer should be in the form of $F = mC + b$.

2. Write a similar formula that relates Celsius to Fahrenheit.

3. If the temperature changes 1° Fahrenheit, how many degrees Celsius does the temperature change? What happens to the Fahrenheit temperature when the Celsius temperature changes 1°? What do the answers to these questions have to do with the models from questions 1 and 2?

4. Common wisdom (which is often neither) has it that the "normal" human body temperature is 98.6° Fahrenheit. What is this temperature in degrees Celsius?

5. Where do the two scales agree? That is, when is the temperature in degrees Fahrenheit equal to the temperature in degrees Celsius?

6. On the Kelvin scale, water freezes at 273°. One Kelvin is the same size as 1° Celsius. At what Kelvin temperature does water boil?

Graphs and Technology

AP/Wide World Photos

Are you on the Web?

The number of Internet users in the United States increased significantly from 1997 to 2002. Suppose this trend continues and you have data for the years 1997–2002. Then you can use technology to construct a linear model to predict the number of Internet users in future years. See Exercise 21 on page 115.

Chapter Outline

*Interdependence of
Sections*

2.1 ➤ 2.2 ➤ 2.3 ➤ 2.4
    ↘
      2.5

2.1   Graphs
2.2   Solving Equations Graphically and Numerically
2.3   Applications of Equations
2.4   Optimization Applications
2.5   Linear Models

Technology for graphing, solving equations, and building mathematical models is introduced in this chapter. It provides powerful tools for dealing with complicated mathematical situations and real-world problems that might otherwise be intractable. Keep in mind, however, that technology is not a substitute for mathematical knowledge or common sense. To use it effectively, you need a sound grounding in algebra and geometry.

2.1   Graphs

The traditional method of graphing an equation "by hand" is as follows: Construct a table of values with a reasonable number of entries, plot the corresponding points, and use whatever algebraic or other information is available to make an educated guess about the rest.

EXAMPLE 1

The graph of $y = x^2$ consists of all points (x, x^2), where x is a real number. You can easily construct a table of values and plot the corresponding points, as in Figure 2–1. These points suggest that the graph looks like the one in Figure 2–2, which is obtained by connecting the plotted points and extending the graph upward. ∎

x	$y = x^2$
-2.5	6.25
-2	4
-1.5	2.25
-1	1
$-.5$	.25
0	0
.5	.25
1	1
1.5	2.25
2	4
2.5	6.25

Figure 2–1

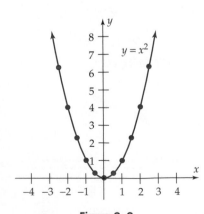

Figure 2–2

▖▖ GRAPHING WITH TECHNOLOGY

A graphing calculator or computer graphing program graphs in the same way you would graph by hand, but with much greater speed: It plots 95 or more points and simultaneously connects them with line segments. These graphs are generally quite accurate, but no technology is perfect. Algebra and geometry may be needed to interpret misleading (or incorrect) screen images. The next example illustrates the basics of graphing with technology.

EXAMPLE 2

Graph the equation

$$2x^3 - 8x - 2y + 4 = 0$$

(a) using a calculator;

(b) using a computer graphing program.

SOLUTION (a) For calculator graphing, proceed as follows:

Step 1 *Set the **viewing window**—*the portion of the coordinate plane that will appear on the screen. Since we don't know yet where the graph lies, we'll try the window with $-10 \le x \le 10$ and $-10 \le y \le 10$. We can change it later if necessary. Press the WINDOW (or RANGE or V-WINDOW or PLOT SET-UP) key,[*] and enter the appropriate numbers, as in Figure 2–3 (which shows a TI-83; other calculators are similar). Then the calculator will display the portion of the plane shown in Figure 2–4.

Figure 2–3

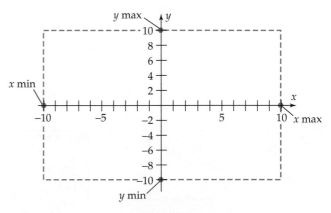

Figure 2–4

Setting Xscl $= 1$ in Figure 2–3 puts the tick marks 1 unit apart on the x-axis; similarly, Yscl $= 2$ puts the tick marks 2 units apart on the y-axis.[†]

---

[*]On TI-86, press GRAPH first, then WINDOW will appear as a menu choice.

[†]Xscl is labeled "X scale" on Casio and "Xtick" on HP-39. Some calculators do not have an Xres setting. On those that do, it should normally be set at 1 (or at "detail" on HP-39).

Step 2   *Enter the equation in the calculator.* To do this, you must first solve the equation for *y:*

$$2x^3 - 8x - 2y + 4 = 0$$

Rearrange terms: $-2y = -2x^3 + 8x - 4$

Divide by -2: $y = x^3 - 4x + 2.$

Now call up the **equation memory** by pressing Y= on TI (or SYMB on HP-39 or GRAPH (main menu) on Casio). Use the Technology Tip in the margin to enter the equation, as shown in Figure 2–5.

Step 3   *Graph the equation.* Press GRAPH (or PLOT or DRAW) to obtain Figure 2–6.

Figure 2–5

Figure 2–6

Because of the limited resolution of a calculator screen, the graph appears to consist of short adjacent line segments rather than a smooth unbroken curve.

Step 4   *If necessary, adjust the viewing window for a better view.* The graph in Figure 2–6 is scrunched up in the middle of the screen. So we change the viewing window (Figure 2–7) and press GRAPH again to obtain Figure 2–8.

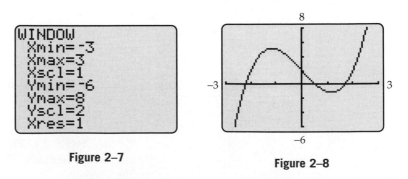

Figure 2–7

Figure 2–8

(b)   A typical computer graphing program requires the same information as does a calculator, namely, the equation to be graphed and the viewing window in which to graph it. The procedures for entering this information vary widely, so check your instruction manual or help key index. For example, the following command in Maple produces Figure 2–9, in which the range of y-values was chosen automatically by Maple.

$$\text{plot}(x\text{\textasciicircum}3 -4*x + 2, x = -3..3);$$

To duplicate Figure 2–8, we specify the range of y-values and the number of tick marks on each axis:

$$\text{plot}(x\text{\textasciicircum}3 -4*x + 2, x = -3..3, y = -6..8,$$
$$\text{xtickmarks} = 6, \text{ytickmarks} = 6);$$

The result is Figure 2–10.

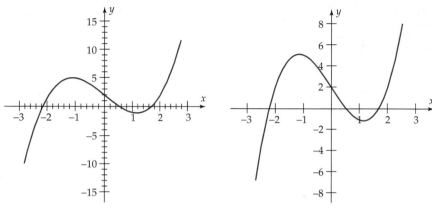

Figure 2–9   Figure 2–10

As is typical of computer-generated graphs, this one appears smooth and connected, as it should. ∎

> **NOTE** An equation stays in a calculator's equation memory until you delete it. When several equations are in the memory, you must turn "on" those you want graphed and turn "off" those you don't want graphed, as explained in the first Technology Tip in the margin.

TECHNOLOGY TIP

An equation is "on" if its equal sign is shaded or if there is a check mark next to it. Only equations that are "on" are graphed when you press GRAPH.

To turn an equation "on" or "off" on TI-83, move the cursor over the equal sign and press ENTER. On other calculators, move the cursor to the equation and press SELECT or CHECK.

▓ GRAPHING TOOLS: THE TRACE FEATURE

A calculator obtains a graph by plotting points and simultaneously connecting them. To see which points were actually plotted, press TRACE, and a flashing cursor appears on the graph. Use the left and right arrow keys to move the cursor along the graph. The coordinates of the point the cursor is on appear at the bottom of the screen. Figure 2–11 illustrates this for the graph of $y = x^3 - 4x + 2$.

TECHNOLOGY TIP

If TRACE is not on the keyboard, it will appear on screen after you press GRAPH (or PLOT or DRAW).

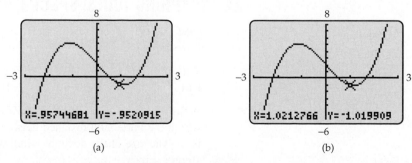

(a)   (b)

Figure 2–11

The trace cursor displays only the points that the calculator actually plotted. For instance, $(1, -1)$ is on the graph of $y = x^3 - 4x + 2$, as you can easily verify, but was not one of the points the calculator plotted to produce Figure 2–11. So the trace lands on two nearby points that *were* plotted, but skips $(1, -1)$.

■■ GRAPHING TOOLS: ZOOM IN/OUT

TECHNOLOGY TIP
If ZOOM is not on the keyboard, it will appear on screen after you press GRAPH (or PLOT or DRAW).
To set the zoom factors, look for FACT, ZFACT, or (SET) FACTORS in the ZOOM menu (or its MEMORY submenu).

The ZOOM-OUT and ZOOM-IN keys on the ZOOM menu make it easy to increase or decrease the size of the viewing window with a single keystroke. Figure 2–12 shows the graph of $y = x^3 - 2x^2 + 1$ in two windows. Window (b) was obtained from window (a) by zooming in at the origin by a factor of 4.

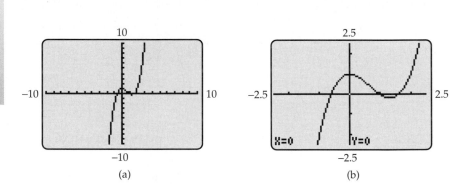

(a)   (b)

Figure 2–12

The calculator automatically changes the range of x- and y-values when zooming, but does not change the Xscl or Yscl settings. This may cause occasional viewing problems.

NOTE
The heading GRAPHING EXPLORATION indicates that you are to use your calculator or computer as directed to complete the discussion.

GRAPHING EXPLORATION

Graph $y = x^3 - 2x^2 + 1$ in the same window as Figure 2–12(a). Then zoom *out* from the origin by a factor of 4. Can you read the tick marks on the axes? Change the settings to Xscl $= 5$ and Yscl $= 5$, and regraph. Can you read them now?

■■ GRAPHING TOOLS: SPECIAL VIEWING WINDOWS

The ZOOM menu also has keys that enable you to obtain frequently used viewing windows in a single keystroke, including most or all of the following:

1. The **standard viewing window** (also labeled **Zstandard** or **Zstd** or **Default**) has $-10 \le x \le 10$ and $-10 \le y \le 10$.

2. A **decimal window** (also labeled **Zdecimal** or **Zdecm** and **Init**) is one in which the horizontal distance between two adjacent pixels on the screen is .1. The size of the decimal window depends on the width of your calculator screen.

EXPLORATION

Graph $y = x^4 - 2x^2 - 1$ in the standard window. Use the TRACE key, and watch the values of the x-coordinates. Now regraph by using DECIMAL in the ZOOM menu. Then use the TRACE key again. How do the x-coordinates change at each step? Finally, look in WINDOW to find the dimensions of your decimal window.

TECHNOLOGY TIP

For an approximately square window, the y-axis should be 2/3 as long as the x-axis on TI-83 and 3/5 as long on TI-86. It should be half as long on TI-89, Casio, and HP-39.

On calculators other than TI-86, the decimal window is a square window.

3. In a **square window,** a one-unit segment on the x-axis has the same length on the screen as a one-unit segment on the y-axis. Because calculator screens are wider than they are high, the y-axis in a square window must be shorter than the x-axis (see the Technology Tip in the margin). The next example illustrates the usefulness of square windows.

EXAMPLE 3

Graph the circle $x^2 + y^2 = 9$ on a calculator.

SOLUTION   First, we solve the equation for y:

$$y^2 = 9 - x^2$$
$$y = \sqrt{9 - x^2} \quad \text{or} \quad y = -\sqrt{9 - x^2}.$$

Graphing both of these equations on the same screen will produce the graph of the circle. In the standard viewing window (Figure 2–13), however, the graph doesn't look like a circle.

 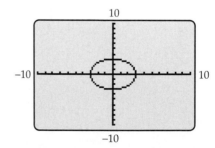

Figure 2–13

The reason is that this is *not* a square window (the 20 one-unit segments on the y-axis occupy less space than the 20 one-unit segments on the x-axis, which means that the ones on the x-axis are a tad longer than the ones on the y-axis). So we press SQUARE (or ZSQUARE or ZOOMSQR) on the ZOOM menu and obtain Figure 2–14 on the next page. The size of the x-axis has been changed to produce a square window in which the circle looks round, as it should.* ∎

---

*On some calculators, the SQUARE key changes the y-axis length instead of the x-axis. The result in either case is a square window.

Figure 2–14

Although any convenient viewing window is usually OK, you should use a square window when you want circles to look round and perpendicular lines to look perpendicular.

GRAPHING EXPLORATION

The lines $y = .5x$ and $y = -2x + 2$ are perpendicular (why?). Graph them in the standard viewing window. Do they look perpendicular? Now graph them in a square window. Do they look perpendicular?

The method used to graph the circle $x^2 + y^2 = 4$ in Example 3 can be used to graph any equation that can be solved for y.

EXAMPLE 4

To graph $12x^2 - 4y^2 + 16x + 12 = 0$, solve the equation for y:

$$4y^2 = 12x^2 + 16x + 12$$
$$y^2 = 3x^2 + 4x + 3$$
$$y = \pm\sqrt{3x^2 + 4x + 3}.$$

Every point on the graph of the equation is on the graph of either

$$y = \sqrt{3x^2 + 4x + 3} \qquad \text{or} \qquad y = -\sqrt{3x^2 + 4x + 3}.$$

GRAPHING EXPLORATION

Graph the previous two equations on the same screen. The result will be the graph of the original equation. ■

▪▪ GRAPHING TOOLS: THE MAXIMUM/MINIMUM FINDER

Many graphs have peaks and valleys (for instance, see Figure 2–12 on page 74 or Figure 2–15 on the next page). A calculator's maximum finder or minimum finder can locate these points with a high degree of accuracy, as illustrated in the next example.

EXAMPLE 5

The Cortopassi Computer Company can produce a maximum of 100,000 computers a year. Their annual profit is given by

$$y = -.003x^4 + .3x^3 - x^2 + 5x - 4000,$$

where y is the profit (in thousands of dollars) from selling x thousand computers. Use graphical methods to estimate how many computers should be sold to make the largest possible profit.

Figure 2–15

Figure 2–16

SOLUTION   We first choose a viewing window. The number x of computers is nonnegative, and no more than 100,000 can be produced, so that $0 \le x \le 100$ (because x is measured in thousands). The profit y may be positive or negative (the company could lose money). So we try a window with $-25,000 \le y \le 25,000$ and obtain Figure 2–15. For each point on the graph,

The x-coordinate is the number of thousands of computers produced;

The y-coordinate is the profit (in thousands) on that number of computers.

The largest possible profit occurs at the point with the largest y-coordinate, that is, the highest point in the window. The maximum finder on a TI-83 (see the Technology Tip in the margin) produced Figure 2–16. Since x and y are measured in thousands, we see that making about 72,789 computers results in a maximum profit of about $22,547,757. ■

TECHNOLOGY TIP

The graphical maximum finder is in the following menu/submenu:

TI-83: CALC

TI-86/89: GRAPH/MATH

Casio: GRAPH/G-SOLVE

HP-39: PLOT/FNC

It is labeled MAXIMUM, MAX, FMAX or EXTREMUM.

On some TI calculators, you must first select a left (or lower) bound, meaning an x-value to the left of the highest point, and a right (or upper) bound, meaning an x-value to its right, and make an initial guess. On other calculators, you may have to move the cursor near the point you are seeking.

GRAPHING EXPLORATION

Graph $y = .3x^3 + .8x^2 - 2x - 1$ in the window with $-5 \le x \le 5$ and $-5 \le y \le 5$. Use your maximum finder to approximate the coordinates of the highest point to the left of the y-axis. Then use your minimum finder (in the same menu) to approximate the coordinates of the *lowest* point to the right of the y-axis. How do these answers compare with the ones you get by using the trace feature?

▋▋ COMPLETE GRAPHS

A viewing window is said to display a **complete graph** if it shows all the important features of the graph (peaks, valleys, intercepts, etc.) and suggests the general shape of the portions of the graph that aren't in the window. Many different windows may show a complete graph. It's usually best to use a window that is small enough to show as much detail as possible.

In later chapters we shall develop algebraic facts that will enable us to know when certain graphs are complete. For the present, however, the best you may be able to do is try several different windows to see which, if any, appear to display a complete graph.

EXAMPLE 6

Sketch a complete graph of

$$y = .007x^5 - .2x^4 + 1.332x^3 - .004x^2 + 10.$$

SOLUTION   Four different viewing windows for this graph are shown in Figure 2–17. Graph (a) (the standard window) is certainly not complete, since it shows no points to the right of the y-axis. Graph (b) is not complete, since it indicates that parts of the graph lie outside the window.

(a)

(b)

(c)

(d)

Figure 2–17

Graph (d) tends to confirm what Graph (c) suggests: that the graph keeps climbing sharply forever as you move to the right and that it keeps falling sharply forever as you move to the left. Because of its large scale, however, graph (d) doesn't show the features of the graph near the origin. So we conclude that graph (c) is probably a complete graph, since it shows important features (twists and turns) near the origin, as well as suggesting the shape of the graph farther out. ■

EXAMPLE 7

If you graph $y = -2x^3 + 26x^2 + 18x + 50$ in the standard viewing window, you get a blank screen (try it!). In such cases, you can usually find at least one point on the graph by setting $x = 0$ and determining the corresponding value of y (the y-intercept of the graph). If $x = 0$ here, then $y = 50$, so the point $(0, 50)$ is on the graph. Consequently, the *y*-axis of our viewing window should extend well beyond 50. An alternative method of finding some points on the graph is in the second Technology Tip on page 78.

EXPLORATION

Find a complete graph of this equation. [*Hint:* The graph crosses the *x*-axis once and has one "peak" and one "valley."] ■

As a general rule, you should follow the directions in the following box when graphing equations.

Graphing Conventions

1. Unless directed otherwise, use technology for graphing.

2. Complete graphs are required unless a viewing window is specified or the context of a problem indicates that a partial graph is acceptable.

3. If the directions say "obtain the graph," "find the graph," or "graph the equation," you need not actually draw the graph on paper. For review purposes, however, it may be helpful to record the viewing window used.

4. The directions "sketch the graph" mean "draw the graph on paper, indicating the scale on each axis." This may involve simply copying the display on the screen, or it may require work if the display is misleading.

INVESTIGATIONS 2.1

1. TICK MARKS

(a) Set Xscl = 1 so that adjacent tick marks on the *x*-axis are one unit apart. Find the largest range of *x* values such that the tick marks on the *x*-axis are clearly distinguishable and appear to be equally spaced.

(b) Do part (a) with *y* in place of *x*.

2. VIEWING WINDOWS

Look in the ZOOM menu to find out how many built-in viewing windows your calculator has. Take a look at each one.

3. MAXIMUM/MINIMUM FINDERS

Use your minimum finder to approximate the *x*-coordinates of the lowest point on the graph of $y = x^3 - 2x + 5$ in the window with $0 \le x \le 5$ and $-3 \le y \le 8$. The correct answer is

$$x = \sqrt{\frac{2}{3}} \approx .816496580928.$$

How good is your approximation?

4. SQUARE WINDOWS

Find a square viewing window on your calculator that has $-10 \le x \le 10$.

5. DOT GRAPHING MODE

To see which points your calculator actually graphs (without any connecting line segments), change the graphing mode. Select DOT (or DRAW DOT or PLOT) in the appropriate menu/submenu:

TI-83: MODE   TI-89: Y= / STYLE

TI-86: GRAPH/FORMAT   Casio: SETUP/DRAWTYPE

On HP-39, uncheck "connect" on the second page of the PLOT SETUP menu.

Now graph $y = .5x^3 - 2x^2 + 1$ in the standard window. Try some other equations as well.

 EXERCISES 2.1

In Exercises 1–6, graph the equation by hand by plotting no more than six points and filling in the rest of the graph as best you can. Then use the calculator to graph the equation and compare the results.

1. $y = |x - 2|$

2. $y = \sqrt{x + 5}$

3. $y = x^2 - x$

4. $y = x^2 + x + 1$

5. $y = x^3 + 1$

6. $y = \dfrac{1}{x}$

In Exercises 7–12, find the graph of the equation in the standard window.

7. $3 + y = .5x$

8. $y - 2x = 4$

9. $y = x^2 - 5x + 2$

10. $y = .3x^2 + x - 4$

11. $y = .2x^3 + .1x^2 - 4x + 1$

12. $y = .2x^4 - .2x^3 - 2x^2 - 2x + 5$

13. (a) Graph

$$y = \frac{1}{x^2 + 1}$$

in the standard window.

(b) Does the graph appear to stop abruptly partway along the x-axis? Use the trace feature to explain why this happens. [*Hint:* In this viewing window, each pixel represents a rectangle that is approximately .32 unit high.]

(c) Find a viewing window with $-10 \le x \le 10$ that shows a complete graph that does not fade into the x-axis.

14. (a) Graph $y = x^3 - 2x^2 + x - 2$ in the standard window.

(b) Use the trace feature to show that the portion of the graph with $0 \le x \le 1.5$ is not actually horizontal. [*Hint:* All the points on a horizontal segment must have the same y-coordinate (why?).]

(c) Find a viewing window that clearly shows that the graph is not horizontal when $0 \le x \le 1.5$.

In Exercises 15–24, use the techniques of Examples 3 and 4 to graph the equation in a suitable square viewing window.

15. $x^2 + y^2 = 9$

16. $y^2 = x + 2$

17. $3x^2 + 2y^2 = 48$

18. $25(x - 5)^2 + 36(y + 4)^2 = 900$

19. $(x - 4)^2 + (y + 2)^2 = 25$

20. $9x^2 + 4y^2 = 36$

21. $4x^2 - 9y^2 = 36$

22. $9y^2 - x^2 = 9$

23. $9x^2 + 5y^2 = 45$

24. $x = y^2 - 2$

In Exercises 25–26, use zoom-in or a maximum/minimum finder to determine the highest and lowest point on the graph in the given window.

25. $y = .04x^4 + .01x^3 - 1.7x^2 + 2.5x + 3$ $(0 \le x \le 5$ and $-5 \le y \le 5)$

26. $y = .07x^5 - .3x^3 + 1.5x^2 - 2$ $(-3 \le x \le 2$ and $-6 \le y \le 6)$

27. The population y of New Orleans (in thousands) in year x of the twentieth century is approximated by

$$y = .000046685x^4 - .0108x^3 + .7194x^2$$
$$- 9.2426x + 305 \qquad (0 \le x \le 100),$$

where $x = 0$ corresponds to 1900. According to this model, in what year was the population largest?

28. The NASDAQ stock index between March 2000 and March 2002 is roughly approximated by

$$y = 11.19x^2 - 470x + 6358 \qquad (3 \le x \le 27),$$

where y is the index in month x, with $x = 3$ corresponding to March 2000. Determine the months during this period when the NASDAQ reached its high and its low. What were the approximate values of the NASDAQ at these times?

In Exercises 29–34, determine which of the following viewing windows gives the best view of the graph of the given equation.

(a) $-10 \le x \le 10; \quad -10 \le y \le 10$
(b) $-5 \le x \le 25; \quad 0 \le y \le 20$
(c) $-10 \le x \le 10; \quad -100 \le y \le 100$
(d) $-20 \le x \le 15; \quad -60 \le y \le 250$
(e) None of a, b, c, d gives a complete graph.

29. $y = 18x - 3x^2$ 30. $y = 4x^2 + 80x + 350$

31. $y = \frac{1}{3}x^3 - 25x + 100$ 32. $y = x^4 + x - 5$

33. $y = x^2 + 50x + 625$ 34. $y = .01(x - 15)^4$

In Exercises 35–40, use your algebraic knowledge to state whether or not the two equations have the same graph. Confirm your answer by graphing the equations in the standard window.

35. $y = |x + 3|$ and $y = |x| + 3$

36. $y = |x| - 4$ and $y = |x - 4|$

37. $y = \sqrt{x^2}$ and $y = |x|$

38. $y = \sqrt{x^2 + 6x + 9}$ and $y = |x + 3|$

39. $y = \sqrt{x^2 + 9}$ and $y = x + 3$

40. $y = \dfrac{1}{x^2 + 2}$ and $y = \dfrac{1}{x^2} + \dfrac{1}{2}$

41. (a) Confirm the accuracy of the factorization $x^2 - 5x + 6 = (x - 2)(x - 3)$ graphically. [*Hint:* Graph $y = x^2 - 5x + 6$ and $y = (x - 2)(x - 3)$ on the same screen. If the factorization is correct, the graphs will be identical (which means that you will see only a single graph on the screen).]
    (b) Show graphically that $(x + 5)^2 \ne x^2 + 5^2$. [*Hint:* Graph $y = (x + 5)^2$ and $y = x^2 + 5^2$ on the same screen. If the graphs are different, then the two expressions cannot be equal.]

True or False.   In Exercises 42–44, use the technique of Exercise 41 to determine graphically whether the given statement is possibly true or definitely false. (We say "possibly true" because two graphs that appear identical on a calculator screen may actually differ by small amounts or at places not shown in the window.)

42. $x^3 - 7x - 6 = (x + 1)(x + 2)(x - 3)$

43. $(1 - x)^6 = 1 - 6x + 15x^2 - 20x^3 + 15x^4 - 6x^5 + x^6$

44. $x^5 - 8x^4 + 16x^3 - 5x^2 + 4x - 20$
$$= (x - 2)^2(x - 5)(x^2 + x + 1)$$

45. A toy rocket is shot straight up from ground level and then falls back to earth; wind resistance is negligible. Use your calculator to determine which of the following equations has a graph whose portion above the x-axis provides the most plausible model of the path of the rocket.
    (a) $y = .1(x - 3)^3 - .1x^2 + 5$
    (b) $y = -x^4 + 16x^3 - 88x^2 + 192x$
    (c) $y = -16x^2 + 117x$
    (d) $y = .16x^2 - 3.2x + 16$
    (e) $y = -(.1x - 3)^6 + 600$

46. Monthly profits at DayGlo Tee Shirt Company appear to be given by the equation

$$y = -.00027(x - 15,000)^2 + 60,000,$$

where x is the number of shirts sold that month and y is the profit. DayGlo's maximum production capacity is 15,000 shirts per month.
    (a) If you plan to graph the profit equation, what range of x values should you use? [*Hint:* You can't make a negative number of shirts.]
    (b) The president of DayGlo wants to motivate the sales force (who are all in the profit-sharing plan), so he asks you to prepare a graph that shows DayGlo's profits increasing *dramatically* as sales increase. Using the profit equation and the x range from part (a), what viewing window would be suitable?
    (c) The City Council is talking about imposing more taxes. The president asks you to prepare a graph showing that DayGlo's profits are essentially flat. Using the profit equation and the x range from part (a), what viewing window would be suitable?

In each of the applied situations in Exercises 47–50, find an appropriate viewing window for the equation (that is, a window that includes all the points relevant to the problem but does not include large regions that are not relevant to the problem, and has easily readable tick marks on the axes). Explain why you chose this window. See the Hint in Exercise 46(a).

47. Beginning in 1905 the deer population in a region of Arizona rapidly increased because of a lack of natural predators. Eventually food resources were depleted to such a

degree that the deer population completely died out. In the equation $y = -.125x^5 + 3.125x^4 + 4000$, y is the number of deer in year x, where $x = 0$ corresponds to 1905.

48. A cardiac test measures the concentration y of a dye x seconds after a known amount is injected into a vein near the heart. In a normal heart

$$y = -.006x^4 + .14x^3 - .053x^2 + 179x.$$

49. The concentration of a certain medication in the bloodstream at time x hours is approximated by the equation

$$y = \frac{375x}{.1x^3 + 50},$$

where y is measured in milligrams per liter. After two days the medication has no effect.

50. A winery can produce x barrels of red wine and y barrels of white wine, where

$$y = \frac{200{,}000 - 50x}{2000 + x}.$$

In Exercises 51–53, use the viewing windows found in Exercises 47–49.

51. (a) Use the trace feature to estimate the year when the deer population in Exercise 47 first reached 40,000. In what later year was it also 40,000?
   (b) When was the deer population at its maximum? Approximately how many deer were there at that time?

52. (a) In the cardiac test of Exercise 48, use the trace feature to estimate the time at which all dye is gone from the body.
   (b) At what time during the test was the concentration of the dye the greatest?

53. (a) When was the concentration of the medication in Exercise 49 at its peak?
   (b) Use the trace feature to determine approximately when the concentration dropped below 10 milligrams per liter and stayed below that level.

54. Data from the EPA and the White House indicates that the balance in the Superfund Trust Fund (which is used to pay for cleaning up toxic waste sites) in year x can be roughly approximated by

$$y = -.0843x^2 + 1.081x + .3,$$

where $x = 0$ corresponds to 1990.
   (a) According to this model, in what year was the Superfund balance the largest? What was the balance then?
   (b) Estimate the year in which the balance in the Superfund reaches 0 (after which time cleanup costs will come from general tax revenues).

In Exercises 55–60, obtain a complete graph of the equation by trying various viewing windows. List a viewing window that produces this complete graph. (Many correct answers are possible; consider your answer to be correct if your window shows all the features in the window given in the answer section.)

55. $y = 7x^3 + 35x + 10$ **56.** $y = x^3 - 5x^2 + 5x - 6$

57. $y = \sqrt{x^2} - x$ **58.** $y = 1/x^2$

59. $y = -.1x^4 + x^3 + x^2 + x + 50$

60. $y = .002x^5 + .06x^4 - .001x^3 + .04x^2 - .2x + 15$

In Exercises 61–64, graph all four equations on the same screen, using a sufficiently large square viewing window, and answer this question: What is the geometric relationship of graphs (b), (c), and (d) to graph (a)?

61. (a) $y = x^2$ (b) $y = x^2 + 5$
    (c) $y = x^2 - 5$ (d) $y = x^2 - 2$

62. (a) $y = \sqrt{x}$ (b) $y = \sqrt{x - 3}$
    (c) $y = \sqrt{x + 3}$ (d) $y = \sqrt{x - 6}$

63. (a) $y = \sqrt{x}$ (b) $y = 2\sqrt{x}$
    (c) $y = 3\sqrt{x}$ (d) $y = \frac{1}{2}\sqrt{x}$

64. (a) $y = x^2$ (b) $y = -x^2$
    (c) $y = -\frac{1}{2}x^2$ (d) $y = -2x^2$

In Exercises 65–67, graph the two given equations and the equation $y = x$ on the same screen, using a sufficiently large square viewing window, and answer this question: What is the geometric relationship between graphs (a) and (b)?

65. (a) $y = x^3$ (b) $y = \sqrt[3]{x}$

66. (a) $y = \frac{1}{2}x^3 - 4$ (b) $y = \sqrt[3]{2x + 8}$

67. (a) $y = 5x - 15$ (b) $y = .2x + 3$

68. Put your calculator in *radian* mode, and use the viewing window given by $0 \le x \le 6.28$ and $-2 \le y \le 2$.[*]
   (a) Graph $y = \sin x$[*]
   (b) Graph $y = \sin (2x)$
   (c) Graph $y = \sin (3x)$
   (d) On the basis of parts (a)–(c), what do you think the graphs of $y = \sin (4x)$, $y = \sin (5x)$, $y = \sin (6x)$, and so on, will look like? Use the calculator to verify your answer.
   (e) On the basis of part (d), what do you think the graphs of $y = \sin (50x)$ and $y = \sin (100x)$ will look like? What does a calculator display instead? What might explain the graphs of the calculator?

---

[*]You don't need to know what radian mode is or what "sin" means. Just use the calculator key with this label.

2.2 Solving Equations Graphically and Numerically

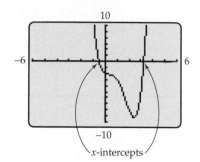

Figure 2–18

Algebraic techniques can be used to solve linear, quadratic, and some higher-degree equations, as we saw in Section 1.2. For many other equations, however, graphical and numerical approximation methods are the only practical alternatives. These methods are based on a connection between equations and graphs that we now examine.

Recall that the *x*-intercepts of the graph of

$$y = x^4 - 4x^3 + 3x^2 - x - 2$$

are the *x*-coordinates of the points where the graph intersects the *x*-axis (see Figure 2–18). Since points on the *x*-axis have 0 second coordinates, the *x*-intercepts are found algebraically by setting $y = 0$ and solving for *x*, that is, by solving the equation

$$x^4 - 4x^3 + 3x^2 - x - 2 = 0.$$

Graphical equation solving amounts to running this in the opposite direction: Solve this equation by finding the *x*-intercepts, as illustrated in the next example.

EXAMPLE 1

Solve $x^4 - 4x^3 + 3x^2 - x - 2 = 0$ graphically.

SOLUTION   As we have just seen, the solutions are the *x*-intercepts of the graph of

$$y = x^4 - 4x^3 + 3x^2 - x - 2.$$

These *x*-intercepts may be found by graphing the equation and using the graphical root finder, as in Figure 2–19. (See the Technology Tip in the margin.)

TIP

The graphical root finder is labeled ROOT or ZERO or X-INCPT and is in this menu/submenu:

TI-83: CALC

TI-86/89: GRAPH/MATH

Casio:DRAW/G-SOLV

HP-39: PLOT/FCN

On most TI calculators, you must select a left and a right bound (*x*-values to the left and right of the intercept) and make an initial guess.

Figure 2–19

Therefore, the approximate solutions of the original equation are $x \approx -.5242$ and $x \approx 3.2257$. ∎

The process in Example 1 works in the general case as well.

Solutions and
Intercepts

> The real solutions of a one-variable equation of the form
>
> $$\text{Expression in } x = 0$$
>
> are the x-intercepts of the graph of the two-variable equation
>
> $$y = \text{expression in } x.$$
>
> These x-intercepts (solutions) can be approximated by a graphical root finder.

EXAMPLE 2

Solve $x^3 - 4x^2 - 2x + 5 = 0$.

SOLUTION   We graph $y = x^3 - 4x^2 - 2x + 5$ (Figure 2–20) and see that it has three x-intercepts. So the equation has three real solutions. A root finder shows that one solutions is $x \approx -1.1926$ (Figure 2–21). ■

Figure 2–20

Figure 2–21

GRAPHING EXPLORATION

Use your graphical root finder to find the other two solutions (x-intercepts) of the equation in Example 2.

▪▪ THE INTERSECTION METHOD

The next examples illustrate an alternative graphical method, which is sometimes more convenient for solving equations.

EXAMPLE 3

Solve $|x^2 - 4x - 3| = x^3 + x - 6$.

SOLUTION   Let $y_1 = |x^2 - 4x - 3|$ and $y_2 = x^3 + x - 6$, and graph both equations on the same screen (Figure 2–22). Consider the point where the two graphs intersect. Since it is on the graph of y_1, its second coordinate is $|x^2 - 4x - 3|$, and since it is also on the graph of y_2, its second coordinate is $x^3 + x - 6$. So for this number x, we must have $|x^2 - 4x - 3| = x^3 + x - 6$. In other words, *the x-coordinate of the intersection point is the solution of the equation.*

Figure 2–22   Figure 2–23

TECHNOLOGY TIP

The graphical intersection finder is in the same menu/submenu as the root finder. It is labeled

INTERSECTION, INTERSECT, ISECT, ISCT, or INTSECT.

This coordinate can be approximated by using a graphical intersection finder (see the Technology Tip in the margin), as shown in Figure 2–23. Therefore, the solution of the original equation is $x \approx 2.207$. ∎

GRAPHING EXPLORATION

Show that the equation $x^2 - 2x - 6 = \sqrt{2x + 7}$ has two real solutions by graphing the left and right sides in the standard window and counting the number of intersection points. The positive solution is $x \approx 4.3094$. Find the negative solution.

EXAMPLE 4

According to data from the U.S. Census Bureau, the approximate population y (in millions) of Detroit is given by

$$y = (-6.985 \times 10^{-7})x^4 + (9.169 \times 10^{-5})x^3 - .00373x^2 + .0268x + 1.85,$$

where $x = 0$ corresponds to 1950. The approximate population of San Diego during the same period is given by

$$y = (-6.216 \times 10^{-6})x^3 + (4.569 \times 10^{-4})x^2 + .01035x + .346.$$

(a) Approximately when did the two cities have the same population?

(b) In what year was the population of Detroit about 1,500,000?

SOLUTION

(a) The populations of the two cities are equal when

$$(-6.985 \times 10^{-7})x^4 + (9.169 \times 10^{-5})x^3 - .00373x^2 + .0268x + 1.85$$
$$= (-6.216 \times 10^{-6})x^3 + (4.569 \times 10^{-4})x^2 + .01035x + .346.$$

We solve this equation by graphing the left side as y_1 and the right side as y_2 and finding the intersection point, as in Figure 2–24. Rounding the x-coordinate to the nearest year ($x \approx 38$), we see that the populations were equal around 1988.

Figure 2–24

(b) To find when the population of Detroit was 1,500,000 (that is, 1.5 million), we must solve the equation

$$(-6.985 \times 10^{-7})x^4 + (9.169 \times 10^{-5})x^3 - .00373x^2 + .0268x + 1.85 = 1.5.$$

Since the left side of this equation has already been graphed as y_1, we need only graph $y_3 = 1.5$ on the same screen and find the intersection point of y_1 and y_3, as in Figure 2–25. The solution $x \approx 20.45$ shows that Detroit had a population of 1.5 million in 1970. ■

Figure 2–25

▟▙ NUMERICAL METHODS

In addition to graphical tools for solving equations, calculators and computer algebra systems have **equation solvers** that can find or approximate the solutions of most equations. On most calculators, the solver finds one solution at a time. You must enter the equation and an initial guess and possibly an interval in which to search. A few calculators (such as TI-89) and most computer systems have **one-step equation solvers** that will find or approximate all the solutions in a single step. Check you calculator instruction manual or your computer's help menu for directions on using these solvers.

Several calculators and many computer algebra systems also have **polynomial solvers,** one-step solvers designed specifically for polynomial equations; you need only enter the degree and coefficients of the polynomial. A few of these (such as Casio 9850) are limited to equations of degree 2 and 3. Directions for using polynomial solvers on calculators are in Calculator Investigation 1 on page 26. Computer users should check the help menu for the proper syntax.

EXAMPLE 5

When asked to search the interval $-10 \leq x \leq 10$ with an initial guess of 3, the equation solver on a TI-83 found one solution of

$$4x^5 - 12x^3 + 8x - 1 = 0,$$

namely, $x \approx .128138691376$ (Figure 2–26). The POLY solver on a TI-86 produced that solution and four others (Figure 2–27).

TIP

The equation solver is labeled SOLVE or SOLVER. It is on the keyboard or in this menu:

TI-83: MATH

TI-89: ALGEBRA

HP-39: APLET

Casio: EQUA (Main Menu)

The syntax varies, so check your instruction manual.

Figure 2–26

Figure 2–27

On Maple, the command

$$\text{fsolve}(4*x^5 - 12*x^3 + 8*x - 1 = 0, x);$$

also produced all five solutions of the equation. ∎

⊞ STRATEGIES FOR SPECIAL CASES

Because of various technological shortcomings, some equations are easier to solve if an indirect approach is used.

EXAMPLE 6

Solve $\sqrt{x^4 + x^2 - 2x - 1} = 0$.

Figure 2–28

SOLUTION   The graph of $y = \sqrt{x^4 + x^2 - 2x - 1}$ in Figure 2–28 does not even appear to touch the x-axis (although it actually does touch at two places). Consequently, some solvers and graphical root finders will return an error message (check yours). Some computer solvers (such as "fsolve" on Maple) produce only one solution, unless told to search a specific interval. Even if your root finder or solver can handle this equation, it may fail in other similar situations. So the best approach is to use the following fact:

The only number whose square root is 0 is 0 itself.

Thus, $\sqrt{x^4 + x^2 - 2x - 1} = 0$ exactly when $x^4 + x^2 - 2x - 1 = 0$. So you need only solve the polynomial equation $x^4 + x^2 - 2x - 1 = 0$, which is easily done on any calculator or computer. In fact, if you have a polynomial or one-step solver, you can find all the solutions at once, which is faster than solving the original equation with a root finder or other solvers.

GRAPHING EXPLORATION

If you have a polynomial or one-step solver, use it to find all the real solutions of $x^4 + x^2 - 2x - 1 = 0$ at once. Otherwise, graph $y = x^4 + x^2 - 2x - 1$ and use the graphical root finder to obtain the solutions one at a time. Verify that the real solutions of this equation, and hence, of the original one, are $x \approx -.40469$ and $x \approx 1.18413$. ∎

The technique used in Example 6 is recommended for all equations of the form

$$\sqrt{\text{expression in } x} = 0:$$

Set the expression under the radical equal to 0 and solve.

EXAMPLE 7

Solve

$$\frac{2x^2 + x - 1}{9x^2 - 9x + 2} = 0.$$

SOLUTION   If you graph

$$y = \frac{2x^2 + x - 1}{9x^2 - 9x + 2},$$

you may get "garbage," as in Figure 2–29. You could experiment with other viewing windows, but it's easier to use this fact:

**A fraction is 0 exactly when its numerator is 0 and
its denominator is nonzero.**

Figure 2–29

To find where the numerator is 0, we need only solve $2x^2 + x - 1 = 0$. This is easily done algebraically.*

$$2x^2 + x - 1 = 0$$

$$(2x - 1)(x + 1) = 0$$

$$x = 1/2 \quad \text{or} \quad x = -1$$

You can readily verify that neither of these numbers make the denominator 0. Hence, the solutions of the original equation are $1/2$ and -1.   ■

The technique of Example 7 is recommended for equations of the form "fraction = 0": Set the numerator equal to 0, and solve. The solutions that do not make the denominator 0 are the solutions of the original equation. A number that makes *both* numerator and denominator 0 is *not* a solution because $0/0$ is not defined.

▪▪ CHOOSING A SOLUTION METHOD

We have seen that equations can be solved by algebraic, graphical, and numerical methods. Each method has both advantages and disadvantages, as summarized in the table on the next page.

The choice of solution method is up to you. In the rest of this book, we normally use algebraic means for solving linear and quadratic equations because this is often the fastest and most accurate method. Naturally, any such equation can also be solved graphically or numerically (and you may want to do that as a check against errors). Except in special cases, graphical and numerical methods will normally be used for more complicated equations.

---

*In other cases, you may need technology to solve the numerator equation.

Solution Method	Advantages	Possible Disadvantages
Algebraic	Produces exact solutions. Easiest method for most linear and quadratic equations.	May be difficult or impossible to use with complicated equations.
Graphical **Root Finder or** **Intersection Finder**	Works well for a large variety of equations. Gives visual picture of the location of the solutions.	Solutions may be approximations. Finding a useable viewing window may take a lot of time.
Numerical **Equation Solvers**		Solutions may be approximations.
Polynomial solver	Fast and easy.	Works only for polynomial equations.
One-step solver	Fast and easy.	May miss some solutions or be unable to solve certain equations.
Other solvers		May require a considerable amount of work to find the particular solution you want.

✓ **EXERCISES 2.2**

In Exercises 1–6, determine graphically the number of solutions of the equation, but don't solve the equation. You may need a viewing window other than the standard one to find all the x-intercepts.

1. $x^5 + 5 = 3x^4 + x$

2. $x^3 + 5 = 3x^2 + 24x$

3. $x^7 - 10x^5 + 15x + 10 = 0$

4. $x^5 + 36x + 25 = 13x^3$

5. $x^4 + 500x^2 - 8000x = 16x^3 - 32,000$

6. $6x^5 + 80x^3 + 45x^2 + 30 = 45x^4 + 86x$

In Exercises 7–20, use graphical approximation (a root finder or an intersection finder) to find a solution of the equation in the given interval.

7. $x^3 + 4x^2 + 10x + 15 = 0$; $(-3, -2)$

8. $x^3 + 9 = 3x^2 + 6x$; $(1, 2)$

9. $x^4 + x - 3 = 0$; $(0, \infty)$

10. $x^5 + 5 = 3x^4 + x$; $(-\infty, 1)$

11. $\sqrt{x^4 + x^3 - x - 3} = 0$; $(0, \infty)$

12. $\sqrt{8x^4 - 14x^3 - 9x^2 + 11x - 1} = 0$; $(-\infty, 0)$

13. $\sqrt{\dfrac{2}{5}x^5 + x^2 - 2x} = 0$; $(0, \infty)$

14. $\sqrt{x^4 + x^2 - 3x + 1} = 0$; $(0, 1)$

15. $x^2 = \sqrt{x + 5}$; $(-2, -1)$

16. $\sqrt{x^2 - 1} - \sqrt{x + 9} = 0$; $(3, 4)$

17. $\dfrac{2x^5 - 10x + 5}{x^3 + x^2 - 12x} = 0$; $(-\infty, 0)$

18. $\dfrac{3x^5 - 15x + 5}{x^7 - 8x^5 + 2x^2 - 5} = 0$; $(1, \infty)$

19. $\dfrac{x^3 - 4x + 1}{x^2 + x - 6} = 0$; $(1, \infty)$

20. $\dfrac{4}{x + 2} - \dfrac{3}{x + 1} = 0$; $(0, \infty)$ [*Hint:* Write the left side as a single fraction.]

In Exercises 21–34, use algebraic, graphical, or numerical methods to find all real solutions of the equation, approximating when necessary.

21. $2x^3 - 4x^2 + x - 3 = 0$

22. $6x^3 - 5x^2 + 3x - 2 = 0$

23. $x^5 - 6x + 6 = 0$ 24. $x^3 - 3x^2 + x - 1 = 0$

25. $10x^5 - 3x^2 + x - 6 = 0$ 26. $\dfrac{1}{4}x^4 - x - 4 = 0$

27. $2x - \dfrac{1}{2}x^2 - \dfrac{1}{12}x^4 = 0$

28. $\frac{1}{4}x^4 + \frac{1}{3}x^2 + 3x - 1 = 0$

29. $\frac{5x}{x^2 + 1} - 2x + 3 = 0$ **30.** $\frac{2x}{x + 5} = 1$

31. $|x^2 - 4| = 3x^2 - 2x + 1$ **32.** $|x^3 + 2| = 5 + x - x^2$

33. $\sqrt{x^2 + 3} = \sqrt{x - 2} + 5$

34. $\sqrt{x^3 + 2} = \sqrt{x + 5} + 4$

In Exercises 35–40, find an exact solution of the equation in the given interval. (For example, if the graphical approximation of a solution begins .3333, check to see whether 1/3 is the exact solution. Similarly, $\sqrt{2} \approx 1.414$; so if your approximation begins 1.414, check to see whether $\sqrt{2}$ is a solution.)

35. $3x^3 - 2x^2 + 3x - 2 = 0$; (0, 1)

36. $4x^3 - 3x^2 - 3x - 7 = 0$; (1, 2)

37. $12x^4 - x^3 - 12x^2 + 25x - 2 = 0$; (0, 1)

38. $8x^5 + 7x^4 - x^3 + 16x - 2 = 0$; (0, 1)

39. $4x^4 - 13x^2 + 3 = 0$; (1, 2)

40. $x^3 + x^2 - 2x - 2 = 0$; (1, 2)

Exercises 41–44 deal with exponential, logarithmic, and trigonometric equations, which will be dealt with in later chapters. If you are familiar with these concepts, solve each equation graphically or numerically.

41. $10^x - \frac{1}{4}x = 28$ **42.** $x + \sin\left(\frac{1}{2}x\right) = 4$

43. $\ln x - x^2 + 3 = 0$ **44.** $e^x - 6x = 5$

45. According to data from the U.S. Department of Education, the average cost y of tuition and fees at four-year public colleges and universities in year x is approximated by

$$y = .024x^4 - .87x^3 + 9.6x^2 + 97.2x + 2196,$$

where $x = 0$ corresponds to 1990. If this model continues to be accurate, in what year will tuition and fees reach $4000?

46. Use the information in Example 4 to determine the year in which the population of San Diego reached 1.1 million people.

47. According to data from the U.S. Department of Health and Human Services, the cumulative number y of AIDS cases (in thousands) as of year x is approximated by

$$y = .062x^4 - 1.54x^3 + 9.21x^2 + 57.54x + 199.36$$
$$(0 \le x \le 11),$$

where $x = 0$ corresponds to 1990.
(a) When did the cumulative number of cases reach 750,000?
(b) If this model remains accurate after 2001, in what year will the cumulative number of cases reach one million?

48. (a) How many real solutions does the equation

$$.2x^5 - 2x^3 + 1.8x + k = 0$$

have when $k = 0$?
(b) How many real solutions does it have when $k = 1$?
(c) Is there a value of k for which the equation has just one real solution?
(d) Is there a value of k for which the equation has no real solutions?

2.3   Applications of Equations

Actual problem situations are usually described verbally. To solve such problems, you must interpret this verbal information and express it as an equivalent mathematical problem. The following guidelines may be helpful.

Setting up Applied Problems

1. *Read* the problem carefully, and determine what is asked for.

2. *Label* the unknown quantities by letters (variables), and, if appropriate, draw a picture of the situation.

3. *Translate* the verbal statements in the problem and the relationships between the known and unknown quantities into mathematical language.

4. *Consolidate* the mathematical information into an equation in one variable that can be solved or an equation in two variables that can be graphed to determine at least one of the unknown quantities.

Here are some examples of how these guidelines are applied.

EXAMPLE 1

Set up the following problem: The average of two real numbers is 41.125, and their product is 1683. What are the numbers?

SOLUTION *Read:* We are asked for two numbers. *Label:* Call the numbers x and y. *Translate:*

English Language	Mathematical Language
Two numbers	x and y
Their average is 41.125.	$\dfrac{x + y}{2} = 41.125$
Their product is 1683.	$xy = 1683$

Consolidate: One technique to use when you have two unknowns is to express one in terms of the other and use this to obtain an equation in one variable. In this case, we can do that by solving the second equation for y:

$$xy = 1683$$

$$y = \frac{1683}{x}$$

and substituting the result in the first equation:

$$\frac{x + y}{2} = 41.125$$

$$\frac{x + \dfrac{1683}{x}}{2} = 41.125$$

Multiply both sides by 2: $\qquad x + \dfrac{1683}{x} = 82.25.$

The solution of this equation is one of the numbers, and $1683/x$ is the other. ■

EXAMPLE 2

Set up the following problem: A rectangle is twice as long as it is wide. If it has an area of 24.5 square inches, what are its dimensions?

SOLUTION *Read:* We are asked to find the length and width. *Label:* Let x denote the width and y the length, and draw a picture of the situation, as in Figure 2–30. *Translate:* Use the fact that the area of a rectangle is length $\times$ width.

Figure 2–30

English Language	Mathematical Language
The width and length of the rectangle	x and y
The length is twice the width.	$y = 2x$
The area is 24.5 square inches.	$xy = 24.5$

Consolidate: Substitute $y = 2x$ in the area equation:

$$xy = 24.5$$

$$x(2x) = 24.5.$$

So the equation to be solved is $2x^2 = 24.5$.   ■

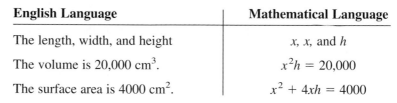

Set up this problem: A rectangular box with a square base and no top is to have a volume of 20,000 cubic centimeters. If the surface area of the box is 4000 square centimeters, what are its dimensions?

SOLUTION   *Read:* We must find the length, width, and height of the box. *Label:* Let x denote the length. Since the base is square, the length and width are the same. Let h denote the height, as in Figure 2–31. *Translate:* Recall that the volume of a box is given by the product length × width × height and that the surface area is the sum of the area of the base and the area of the four sides of the box. Then we have these translations:

English Language	Mathematical Language
The length, width, and height	x, x, and h
The volume is 20,000 cm³.	$x^2h = 20{,}000$
The surface area is 4000 cm².	$x^2 + 4xh = 4000$

Consolidate: We have two equations in two variables, so we solve the first equation for h

$$h = \frac{20{,}000}{x^2}$$

and substitute this result in the second equation:

$$x^2 + 4x\left(\frac{20{,}000}{x^2}\right) = 4000$$

$$x^2 + \frac{80{,}000}{x} = 4000.$$

The solution of this last equation will provide the solution of the problem.   ■

Once you are comfortable with the process for setting up problems, you can often do much of it mentally. In the rest of this section, the setup process will be simplified or shortened whenever doing so won't cause any confusion.

Figure 2–31

Setting up a problem is only half the job. You must then solve the equation you have obtained. Whenever possible, linear and quadratic equations should be solved algebraically, giving exact answers. Other equations may require graphical or numerical methods to find approximate solutions. Finally, you must check your answers in the original problem:

> **Interpret your answers in terms of the original problem.**
> **Do they make sense? Do they satisfy the required conditions?**

In particular, an equation may have several solutions, some of which might not make sense in the context of the problem. For instance, distance can't be negative, the number of people in a room cannot be a proper fraction, etc.

▪▪ APPLICATIONS

We begin with some problems involving interest. Recall that 8% means .08 and that "8% *of* 227" means ".08 *times* 227," that is, .08(227) = 18.16. The basic rule of annual simple interest is

$$\text{Interest} = \text{rate} \times \text{amount.}$$

EXAMPLE 4

A high-risk stock pays dividends at a rate of 12% per year, and a savings account pays 6% interest per year. How much of a $9000 investment should be put in the stock and how much in the savings account to obtain a return of 8% per year on the total investment?

SOLUTION   *Label:* Let x be the amount invested in stock. Then the rest of the $9000, namely, $(9000 - x)$ dollars, goes in the savings account. *Translate:* We want the total return on $9000 to be 8%, so we have

$$\left(\begin{array}{l}\text{Return on } x \text{ dollars} \\ \text{of stock at } 12\%\end{array}\right) + \left[\begin{array}{l}\text{Return on } (9000 - x) \\ \text{dollars of savings at } 6\%\end{array}\right] = 8\% \text{ of } \$9000$$

$$(12\% \text{ of } x \text{ dollars}) + [6\% \text{ of } (9000 - x) \text{ dollars}] = 8\% \text{ of } \$9000$$

$$.12x + .06(9000 - x) = .08(9000)$$
$$.12x + .06(9000) - .06x = .08(9000)$$
$$.12x + 540 - .06x = 720$$
$$.12x - .06x = 720 - 540$$
$$.06x = 180$$
$$x = \frac{180}{.06} = 3000.$$

Therefore, $3000 should be invested in stock and $(9000 - 3000) = 6000 in the savings account. If this is done, the total return will be 12% of $3000 ($360) plus 6% of $6000 ($360), a total of $720, which is precisely 8% of $9000. ∎

EXAMPLE 5

A car radiator contains 12 quarts of fluid, 20% of which is antifreeze. How much fluid should be drained and replaced with pure antifreeze in order that the resulting mixture be 50% antifreeze?

SOLUTION   Let x be the number of quarts of fluid to be replaced by pure antifreeze.* When x quarts are drained, there are $12 - x$ quarts of fluid left in the radiator, 20% of which is antifeeze. So we have

$$\left(\begin{matrix}\text{Amount of}\\\text{antifreeze in}\\\text{radiator after}\\\text{draining } x \text{ quarts}\\\text{of fluid}\end{matrix}\right) + \left(\begin{matrix}x \text{ quarts of}\\\text{antifreeze}\end{matrix}\right) = \left(\begin{matrix}\text{Amount of}\\\text{antifreeze in}\\\text{final mixture}\end{matrix}\right)$$

$$20\% \text{ of } (12 - x) + x = 50\% \text{ of } 12$$

$$.20(12 - x) + x = .50(12)$$
$$2.4 - .2x + x = 6$$
$$-.2x + x = 6 - 2.4$$
$$.8x = 3.6$$
$$x = \frac{3.6}{.8} = 4.5.$$

Therefore, 4.5 quarts should be drained and replaced with pure antifreeze. ∎

The basic formula for problems involving distance and a uniform rate of speed is

$$\textbf{Distance} = \textbf{rate} \times \textbf{time.}$$

For instance, if you drive at a rate of 55 mph for 2 hours, you travel a distance of $55 \cdot 2 = 110$ miles.

EXAMPLE 6

A pilot wants to make an 840-mile round trip from Cleveland to Peoria and back in 5 hours flying time. There will be a headwind of 30 mph going to Peoria, and

---

*Hereafter, we omit the headings *Label, Translate,* and so on.

it is estimated that there will be a 40-mph tailwind returning to Cleveland. At what constant air speed should the plane be flown?

SOLUTION   Let x be the engine speed of the plane. On the trip to Peoria, a distance of 420 miles, the actual speed will be $x - 30$ because of the headwind. Since Rate · time = distance, the time to Peoria will be

$$\frac{\text{Distance}}{\text{Rate}} = \frac{420}{x - 30}.$$

On the return trip, the actual speed will be $x + 40$ because of the tailwind and the time will be

$$\frac{\text{Distance}}{\text{Rate}} = \frac{420}{x + 40}.$$

Therefore,

$$5 = \left(\begin{array}{c}\text{Time from Cleveland}\\ \text{to Peoria}\end{array}\right) + \left(\begin{array}{c}\text{Time from Peoria}\\ \text{to Cleveland}\end{array}\right)$$

$$5 = \frac{420}{x - 30} + \frac{420}{x + 40}.$$

Multiplying both sides by the common denominator $(x - 30)(x + 40)$ and simplifying, we have

$$5(x - 30)(x + 40) = \frac{420}{x - 30} \cdot (x - 30)(x + 40) + \frac{420}{x + 40} \cdot (x - 30)(x + 40)$$

$$5(x - 30)(x + 40) = 420(x + 40) + 420(x - 30)$$

$$(x - 30)(x + 40) = 84(x + 40) + 84(x - 30)$$

$$x^2 + 10x - 1200 = 84x + 3360 + 84x - 2520$$

$$x^2 - 158x - 2040 = 0$$

$$(x - 170)(x + 12) = 0$$

$$x - 170 = 0 \quad \text{or} \quad x + 12 = 0$$

$$x = 170 \qquad\qquad x = -12.$$

Obviously, the negative doesn't apply. Since we multiplied both sides by a quantity involving the variable, we must check that 170 actually is a solution of the original equation. It is, so the plane should be flown at a speed of 170 mph. ■

The preceding examples used only algebraic models (equations in one variable). Sometimes a diagram (geometrical model) is helpful in visualizing the situation and setting up an appropriate equation.

EXAMPLE 7

A landscaper wants to put a cement walk of uniform width around a rectangular garden that measures 24 by 40 feet. She has enough cement to cover 660 square feet. How wide should the walk be to use all the cement?

Figure 2-32

SOLUTION   Let x denote the width of the walk (in feet), and draw a picture of the situation (Figure 2–32).

The length of the outer rectangle is $40 + 2x$ (the garden length plus walks on each end), and its width is $24 + 2x$.

$$\left(\begin{array}{c}\text{Area of outer}\\ \text{rectangle}\end{array}\right) - \left(\begin{array}{c}\text{Area of}\\ \text{garden}\end{array}\right) = \text{Area of walk}$$

$$\text{Length} \cdot \text{Width} - \text{Length} \cdot \text{Width} = 660$$

$$(40 + 2x)(24 + 2x) - 40 \cdot 24 = 660$$

$$960 + 128x + 4x^2 - 960 = 660$$

$$4x^2 + 128x - 660 = 0.$$

Dividing both sides by 4 and applying the quadratic formula yields

$$x^2 + 32x - 165 = 0$$

$$x = \frac{-32 \pm \sqrt{(32)^2 - 4 \cdot 1 \cdot (-165)}}{2.1}$$

$$x = \frac{-32 \pm \sqrt{1684}}{2} \approx \begin{cases} 4.5183 \\ \text{or} \\ -36.5183. \end{cases}$$

Only the positive solution makes sense in the context of this problem. The walk should be approximately 4.5 feet wide.   ■

EXAMPLE 8

A rectangular box with a square base and no top is to have a volume of 20,000 cubic centimeters. If the surface area of the box is 4000 square centimeters and the box is required to be higher than it is wide, what are its dimensions?

SOLUTION   This is essentially Example 3 with an extra condition (the box must be higher than it is wide). If the length, width, and height of the box are x, x, and h as shown in Figure 2–33, then as we saw in Example 3,

$$x^2 + \frac{80{,}000}{x} = 4000 \qquad \text{and} \qquad h = \frac{20{,}000}{x^2}.$$

Multiplying both sides of the first equation by x (which is nonzero because it is a length), we obtain

$$x^3 + 80{,}000 = 4000x$$

$$x^3 - 4000x + 80{,}000 = 0.$$

This equation can be solved in several ways.

Figure 2-33

Graphical: The graph of $y = x^3 - 4000x + 80{,}000$ in Figure 2–34 shows that the equation has three solutions. The negative solution does not apply here, nor does the one near 50 (because a value of x in that range makes the corresponding height $h = 20{,}000/x^2$ smaller than the width x, as indicated in Figure 2–35). The third solution, $x \approx 23.069$, is easily found (Figure 2–36).

Figure 2–34

Figure 2–35

Figure 2–36

Numerical: The polynomial solver on a TI-86 or the "fsolve" command on Maple produces the three solutions (Figure 2–37). As above, only one of them is applicable here, namely, $x \approx 23.069$.

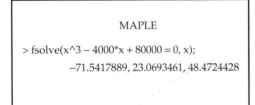

Figure 2–37

Therefore, the height is

$$h = \frac{20{,}000}{x^2} \approx \frac{20{,}000}{23.069^2} \approx 37.581.$$

So the approximate dimensions of the box are 23.069 by 23.069 by 37.581 cm.   ■

EXAMPLE 9

A box (with no top) of volume 1000 cubic inches is to be made from a 22 × 30 inch sheet of cardboard by cutting squares of equal size from each corner and folding up the flaps, as shown in Figure 2–38. If the box must be at least 4 inches high, what size square should be cut from each corner?

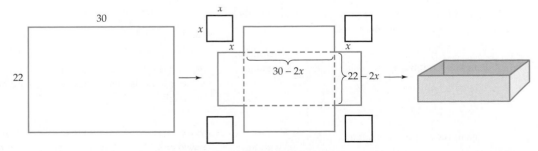

Figure 2–38

SOLUTION   Let x denote the length of the side of the square to be cut from each corner. The dashed rectangle in Figure 2–38 is the bottom of the box. Its length is $30 - 2x$ as shown in the figure. Similarly, the width of the box will be $22 - 2x$, and its height will be x inches. Therefore,

$$\text{Length} \times \text{Width} \times \text{Height} = \text{Volume of box}$$

$$(30 - 2x) \cdot (22 - 2x) \cdot x = 1000$$

$$(660 - 104x + 4x^2)x = 1000$$

$$4x^3 - 104x^2 + 660x - 1000 = 0.$$

Figure 2–39

Since the cardboard is 22 inches wide, x must be less than 11 (otherwise, you can't cut out two squares of length x). Since x is a length, it is positive. So we need only find solutions of the equation between 0 and 11. We graph

$$y = x^3 - 104x^2 + 660x - 1000$$

in a window with $0 \le x \le 11$ (Figure 2–39). A complete graph isn't needed here, only the x-intercepts (solutions). The one between 2 and 3 is not relevant here because x is the height of the box, which must be at least 4 inches.

GRAPHING EXPLORATION

Use a root finder or a polynomial solver on a calculator or computer to find the solution of the equation between 6 and 7. This is the side x of the square that should be cut from each corner. Round the value of x to two decimal places, and find the dimensions of the resulting box. ■

 EXERCISES 2.3

In Exercises 1–4, a problem situation is given.
(a) Decide what is being asked for, and label the unknown quantities.
(b) Translate the verbal statements in the problem and the relationships between the known and unknown quantities into mathematical language, using a table as in Examples 1–3. The table is provided in Exercises 1 and 2. You need not find an equation to be solved.

1. The sum of two numbers is 15, and the difference of their squares is 5. What are the numbers?

English Language	Mathematical Language
The two numbers	
Their sum is 15.	
The difference of their squares is 5.	

2. The sum of the squares of two consecutive integers is 4513. What are the integers?

English Language	Mathematical Language
The two integers	
The integers are consecutive.	
The sum of their squares is 4513.	

3. A rectangle has perimeter of 45 centimeters and an area of 112.5 square centimeters. What are its dimensions?

4. A triangle has area 96 square inches, and its height is two-thirds of its base. What are the base and height of the triangle?

In Exercises 5–8, set up the problem by labeling the unknowns, translating the given information into mathemati-

cal language, and finding an equation that will produce the solution to the problem. You need not solve this equation.

5. A worker gets an 8% pay raise and now makes $1600 per month. What was the worker's old salary?

6. A merchant has 5 pounds of mixed nuts that cost $30. He wants to add peanuts that cost $1.50 per pound and cashews that cost $4.50 per pound to obtain 50 pounds of a mixture that costs $2.90 per pound. How many pounds of peanuts are needed?

7. The diameter of a circle is 16 cm. By what amount must the radius be decreased to decrease the area by 48π square centimeters?

8. A corner lot has dimensions 25 by 40 yards. The city plans to take a strip of uniform width along the two sides bordering the streets to widen these roads. How wide should the strip be if the remainder of the lot is to have an area of 844 square yards?

In the remaining exercises, solve the applied problems.

9. You have already invested $550 in a stock with an annual return of 11%. How much of an additional $1100 should be invested at 12% and how much at 6% so that the total return on the entire $1650 is 9%?

10. If you borrow $500 from a credit union at 12% annual interest and $250 from a bank at 18% annual interest, what is the *effective annual interest rate* (that is, what single rate of interest on $750 would result in the same total amount of interest)?

11. A radiator contains 8 quarts of fluid, 40% of which is antifreeze. How much fluid should be drained and replaced with pure antifreeze so that the new mixture is 60% antifreeze?

12. A radiator contains 10 quarts of fluid, 30% of which is antifreeze. How much fluid should be drained and replaced with pure antifreeze so that the new mixture is 40% antifreeze?

13. An airplane flew with the wind for 2.5 hours and returned the same distance against the wind in 3.5 hours. If the cruising speed of the plane was a constant 360 mph in air, how fast was the wind blowing? [*Hint:* If the wind speed is r miles per hour, then the plane travels at $(360 + r)$ mph with the wind and at $(360 - r)$ mph against the wind.]

14. A train leaves New York for Boston, 200 miles away, at 3:00 P.M. and averages 75 mph. Another train leaves Boston for New York on an adjacent set of tracks at 5:00 P.M. and averages 45 mph. At what time will the trains meet?

15. The average of two real numbers is 41.125, and their product is 1683. What are the numbers? [*Hint:* See Example 1.]

16. A rectangle is twice as long as it is wide. If it has an area of 24.5 inches, what are its dimensions? [*Hint:* See Example 2.]

17. Two cars leave a gas station at the same time, one traveling north and the other south. The northbound car travels at 50 mph. After 3 hours, the cars are 345 miles apart. How fast is the southbound car traveling?

18. A student leaves the university at noon, bicycling south at a constant rate. At 12:30 P.M., a second student leaves the same point and heads west, bicycling 7 mph faster than the first student. At 2:00 P.M., they are 30 miles apart. How fast is each one going?

19. A 13-foot-long ladder leans on a wall, as shown in the figure. The bottom of the ladder is 5 feet from the wall. If the bottom is pulled out 3 feet farther from the wall, how far does the top of the ladder move down the wall? [*Hint:* Draw pictures of the right triangle formed by the ladder, the ground, and the wall before and after the ladder is moved. In each case, use the Pythagorean Theorem to find the distance from the top of the ladder to the ground.]

5 ft

20. A 15-foot-long pole leans against a wall. The bottom is 9 feet from the wall. How much farther should the bottom be pulled away from the wall so that the top moves the same amount down the wall?

21. A concrete walk of uniform width is to be built around a circular pool, as shown in the figure. The radius of the pool is 12 meters, and enough concrete is available to cover 52π square meters. If all the concrete is to be used, how wide should the walk be?

12

22. In Example 6 of Section 1.4, how many can openers must be produced to have an average cost per can opener of $3?

23. A rock is dropped into a well, and 3 seconds later the sound of its splash is heard. How deep is the well? Assume that sound travels at 1100 feet per second and that an object falls a distance of $16t^2$ feet in t seconds. [*Hint:* Let t be the time it takes for the rock to fall to the water. Then it takes $3 - t$ seconds for the sound of its splash to rise to the top (why?). Both the rock and the sound travel the same distance.]

Rock falls ———→   ←——— Sound rises

24. A group of homeowners are to share equally in the $210 cost of repairing a bus-stop shelter near their homes. At the last moment, two members of the group decide not to participate, and this raises the share of each remaining person by $28. How many people were in the group at the beginning?

25. Red Riding Hood drives the 432 miles to Grandmother's house in 1 hour less than it takes the Wolf to drive the same route. Her average speed is 6 mph faster than the Wolf's average speed. How fast does each drive?

26. To get to work, Sam jogs 3 kilometers to the train and then rides the remaining 5 kilometers. If the train goes 40 kilometers per hour faster than Sam's constant rate of jogging and the entire trip takes 30 minutes, how fast does Sam jog?

27. The dimensions of a rectangular box are consecutive integers. If the box has volume of 13,800 cubic centimeters, what are its dimensions?

28. Find a real number that exceeds its cube by 2.

29. The surface area S of the right circular cone in the figure below is given by $S = \pi r \sqrt{r^2 + h^2}$. What radius should be used to produce a cone of height 5 inches and surface area 100 square inches?

30. The surface area of the right square pyramid in the figure below is given by $S = b \sqrt{b^2 + 4h^2}$. If the pyramid has height 10 feet and surface area 100 square feet, what is the length of a side b of its base?

31. Suppose that the open-top box being made from a sheet of cardboard in Example 9 is required to have at least one of its dimensions *greater* than 18 inches. What size square should be cut from each corner?

32. A homemade loaf of bread turns out to be a perfect cube. Five slices of bread, each .6 inch thick, are cut from one end of the loaf. The remainder of the loaf now has a volume of 235 cubic inches. What were the dimensions of the original loaf?

33. A rectangular bin with an open top and volume of 38.72 cubic feet is to be built. The length of its base must be twice the width, and the bin must be at least 3 feet high. Material for the base of the bin costs $12 per square foot, and material for the sides costs $8 per square foot. If it costs $538.56 to build the bin, what are its dimensions?

Thinker

34. One corner of an 8.5 × 11 inch piece of paper is folded over to the opposite side, as shown in the figure. The area of the darkly shaded triangle at the lower left is 6 square inches and we want to find the length x.
  (a) Take a piece of paper this size and experiment. Approximately, what is the largest value x could have (and still have the paper look like the figure)? With this value of x, what is the approximate area of the triangle? Try some other possibilities.
  (b) Now find an exact answer by constructing and solving a suitable equation. Explain why one of the solutions to the equation is not an answer to this problem.

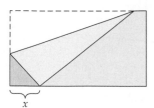

2.4 Optimization Applications*

Many real-life situations require you to find the largest or smallest quantity that satisfies certain conditions. For instance, automotive engineers want to design engines with maximum fuel efficiency. Similarly, a cereal manufacturer who needs a box of volume 300 cubic inches might want to know the dimensions of the box that requires the least amount of cardboard (and hence is cheapest). The exact solutions of such minimum/maximum problems require calculus. However, graphing technology can provide very accurate approximate solutions, as we shall see. Before reading the following examples, you should review the guidelines for setting up applied problems (page 90).

EXAMPLE 1

Find two negative numbers whose product is 50 and whose sum is as large as possible.

SOLUTION   Let x and z be the two negative numbers, and let y be their sum. Then $xz = 50$ and $y = x + z$. Solving $xz = 50$ for z, we have $z = 50/x$ so that

$$y = x + z = x + 50/x.$$

We must find the value of x that makes y as large as possible. Since x must be negative, we graph $y = x + 50/x$ in a window with $-20 \leq x \leq 0$ (Figure 2–40).

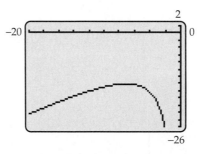

Figure 2–40

Each point (x, y) on this graph represents one of the possibilities:

   x represents one of the two negative numbers ($50/x$ is the other);

   y is the sum of the two negative numbers.

Since we want y to be as large as possible, we must find the point on the graph with largest y-coordinate, that is, the highest point on this part of the graph.

---

*This section may be omitted or postponed until Chapter 4 (Polynomial and Rational Functions). It will be used only in occasional examples and in clearly identifiable exercises.

GRAPHING EXPLORATION

Use a maximum finder to approximate the highest point. Its first coordinate x is one of the two numbers, and $50/x$ is the other. ■

EXAMPLE 2

A box with no top is to be made from a 22×30 inch sheet of cardboard by cutting squares of equal size from each corner and bending up the flaps, as shown in Figure 2–41. To the nearest hundredth of an inch, what size square should be cut from each corner in order to obtain a box with the largest possible volume, and what is the volume of this box?

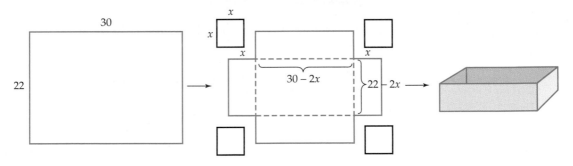

Figure 2–41

SOLUTION   Let x denote the length of the side of the square to be cut from each corner. Then,

$$\text{Volume of box} = \text{Length} \times \text{Width} \times \text{Height}$$

$$= (30 - 2x) \cdot (22 - 2x) \cdot x$$

$$= 4x^3 - 104x^2 + 660x.$$

Thus, the equation $y = 4x^3 - 104x^2 + 660x$ gives the volume y of the box that results from cutting a square of side x from each corner. Since the shortest side of the cardboard is 22 inches, the length x of the side of the cut-out square must be less than 11 (why?).

Each point on the graph of $y = 4x^3 - 104x^2 + 660x$ $(0 < x < 11)$ in Figure 2–42 represents one of the possibilities:

The x-coordinate is the size of the square to be cut from each corner;

The y-coordinate is the volume of the resulting box.

The box with the largest volume corresponds to the point with the largest y-coordinate, that is, the highest point in the viewing window. A maximum finder (Figure 2–43) shows that this point is approximately $(4.182, 1233.809)$. Therefore, a square measuring approximately 4.18×4.18 inches should be cut from each corner, producing a box of volume approximately 1233.81 cubic inches. ■

Figure 2–42

Figure 2–43

Figure 2–44

EXAMPLE 3

A cylindrical can of volume 58 cubic inches (approximately 1 quart) is to be designed. For convenient handling, it must be at least 1 inch high and 2 inches in diameter. What dimensions will use the least amount of material?

SOLUTION   We can construct a can by rolling a rectangular sheet of metal into a tube and then attaching the top and bottom, as shown in Figure 2–44. The surface area of the can (which determines the amount of material) is

$$\underset{\text{sheet}}{\text{Area of rectangular}} + \underset{\text{top}}{\text{Area of}} + \underset{\text{bottom}}{\text{Area of}}$$

$$ch + \pi r^2 + \pi r^2 = ch + 2\pi r^2.$$

When the sheet is rolled into a tube, the width c of the sheet is the circumference of the end of the can, so that $c = 2\pi r$, and hence,

$$\text{Surface area} = ch + 2\pi r^2 = 2\pi rh + 2\pi r^2.$$

The volume of a cylinder of radius r and height h is $\pi r^2 h$. Since the can is to have volume 58 cubic inches, we have

$$\pi r^2 h = 58, \qquad \text{or equivalently,} \qquad h = \frac{58}{\pi r^2}.$$

Therefore,

$$\text{Surface area} = 2\pi rh + 2\pi r^2 = 2\pi r\left(\frac{58}{\pi r^2}\right) + 2\pi r^2 = \frac{116}{r} + 2\pi r^2.$$

Note that r must be greater than 1 (since the diameter $2r$ must be at least 2). Furthermore, r cannot be more than 5 (if $r > 5$ and $h \geq 1$, then the volume $\pi r^2 h$ would be at least $\pi \cdot 25 \cdot 1$, which is greater than 58).

The situation can be represented by the graph of the equation

$$y = 116/x + 2\pi x^2,$$

as in Figure 2–45. The x-coordinate of each point represents a possible radius, and the y-coordinate represents the surface area of the corresponding can. We must find the point with the smallest y-coordinate, that is, the lowest point on the graph. A graphical minimum finder (Figure 2–45) shows that the coordinates of this point are approximately (2.09773, 82.946845). If the radius is 2.09773, then the height is $58/(\pi 2.09773^2) \approx 4.1955$. As a practical matter, it would probably be best to round to one decimal place and construct a can of radius 2.1 and height 4.2 inches.   ■

200

```
Minimum
X=2.09773        Y=82.946845
```
1   5

−50

Figure 2–45

EXERCISES 2.4

In Exercises 1–6, find the coordinates of the highest or lowest point on the part of the graph of the equation in the given viewing window. Only the range of x-coordinates for the window are given; you must choose an appropriate range of y-coordinates.

1. $y = 2x^3 - 3x^2 - 12x + 1$;   highest point when $-3 \leq x \leq 3$

2. $y = 2x^6 + 3x^5 + 3x^3 - 2x^2$;   lowest point when $-3 \leq x \leq 3$

3. $y = \dfrac{4}{x^2} - \dfrac{7}{x} + 1$;   lowest point when $-10 \leq x \leq 10$

4. $y = \dfrac{1}{x^2 + 2x + 2}$;   highest point when $-5 \leq x \leq 5$

5. $y = \dfrac{x^2(x + 1)^3}{(x - 2)^2(x - 4)^2}$;   highest point when $-1 \leq x \leq 0$

[*Hint:* Think small.]

6. $y = \dfrac{x^2(x + 1)^3}{(x - 2)(x - 4)^2}$;   lowest point when $-10 \leq x \leq -1$

7. Find the highest point on the part of the graph of $y = x^3 - 3x + 2$ that is shown in the given window. The answers are not all the same.
   (a) $-2 \leq x \leq 0$ (b) $-2 \leq x \leq 2$
   (c) $-2 \leq x \leq 3$

8. Find the lowest point on the part of the graph of $y = x^3 - 3x + 2$ that is shown in the given window.
   (a) $0 \leq x \leq 2$ (b) $-2 \leq x \leq 2$
   (c) $-3 \leq x \leq 2$

9. Find the dimensions of the rectangular box with a square base and no top that has volume 20,000 cubic centimeters and the smallest possible surface area. [*Hint:* See Example 3 in Section 2.3.]

10. An open-top box with a square base is to be constructed from 120 square centimeters of material. What dimensions will produce a box
    (a) of volume 100 cubic centimeters?
    (b) with largest possible volume?

11. A 20-inch square piece of metal is to be used to make an open-top box by cutting equal-sized squares from each corner and folding up the sides (as in Example 2). The length, width, and height of the box are each to be less than 14 inches. What size squares should be cut out to produce a box with
    (a) volume 550 cubic inches?
    (b) largest possible volume?

12. A cylindrical waste container with no top, a diameter of at least 2 feet, and a volume of 25 cubic feet is to be constructed. What should its radius be if
    (a) 65 square feet of material are to be used to construct it?
    (b) the smallest possible amount of material is to be used to construct it? In this case, how much material is needed?

13. If $c(x)$ is the cost of producing x units, then $c(x)/x$ is the *average cost* per unit.[*] Suppose the cost of producing x units is given by $c(x) = .13x^3 - 70x^2 + 10{,}000x$ and that no more than 300 units can be produced per week.
    (a) If the average cost is $1100 per unit, how many units are being produced?
    (b) What production level should be used in order to minimize the average cost per unit? What is the minimum average cost?

14. A manufacturer's revenue (in cents) from selling x items per week is given by $200x - .02x^2$. It costs $60x + 30{,}000$ cents to make x items.
    (a) Approximately how many items should be made each week to make a profit of $1100? (Don't confuse cents and dollars.)
    (b) How many items should be made each week to have the largest possible profit? What is that profit?

15. (a) A company makes novelty bookmarks that sell for $142 per hundred. The cost (in dollars) of making x hundred bookmarks is $x^3 - 8x^2 + 20x + 40$. Because of other projects, a maximum of 600 bookmarks per day can be manufactured. Assuming that the company can sell all the bookmarks it makes, how many should it make each day to maximize profits?
    (b) Owing to a change in other orders, as many as 1600 bookmarks can now be manufactured each day. How many should be made to maximize profits?

16. If the cost of material to make the can in Example 3 is 5 cents per square inch for the top and bottom and 3 cents per square inch for the sides, what dimensions should be used to minimize the cost of making the can? [The answer is not the same as in Example 3.]

17. A certain type of fencing comes in rigid 10-foot-long segments. Four uncut segments are used to fence in a garden on the side of a building, as shown in the figure. What value of x will result in a garden of the largest possible area, and what is that area?

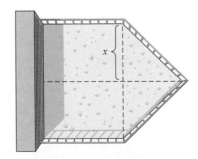

18. A rectangle is to be inscribed in a semicircle of radius 2, as shown in the figure. What is the largest possible area of such a rectangle? [*Hint:* The width of the rectangle is the second coordinate of the point P (why?), and P is on the top half of the circle $x^2 + y^2 = 4$.]

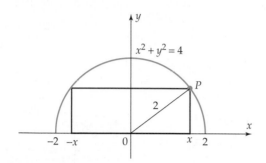

19. Find the point on the graph of $y = 5 - x^2$ that is closest to the point $(0, 1)$ and has positive coordinates. [*Hint:* The distance from the point (x, y) on the graph to $(0, 1)$ is $\sqrt{(x - 0)^2 + (y - 1)^2}$; express y in terms of x.]

20. A manufacturer's cost (in thousands of dollars) of producing x thousand units is $x^3 - 6x^2 + 15x$ dollars, and the revenue (in thousands) from x thousand units is $9x$ dollars. What production level(s) will result in the largest possible profit?

21. A hardware store sells ladders throughout the year. It costs $20 every time an order for ladders is placed and $10 to store a ladder until it is sold. When ladders are ordered x times per year, then an average of $300/x$ ladders are in storage at any given time. How often should the company order ladders each year to minimize its total ordering and storage costs? [*Be careful:* The answer must be an integer.]

22. A mathematics book has 36 square inches of print per page. Each page has a left side margin of 1.5 inches and top, bottom, and right side margins of .5 inch. If a page cannot be wider than 7.5 inches, what should its length and width be to use the least amount of paper?

2.5 Linear Models*

People working in business, medicine, agriculture, and other fields often need to make judgments based on past data. For instance, a stock analyst might use the past profits of a company to estimate next year's profits, or a doctor might use data on previous patients to determine the ideal dosage of a drug for a new patient. In such situations, the available data can sometimes be used to construct a **mathematical model,** such as an equation or graph, that approximates the likely outcome in cases that are not included in the data. In this section, we consider applications in which the data can be modeled by a linear equation.

EXAMPLE 1

The profits of the General Electric Company (in billions of dollars) over a five-year period are shown in the table.[†]

Year	1997	1998	1999	2000	2001
Profit	8	9	11	13	14

---

*This section is optional. It will be used only in clearly identifiable exercises or (sub)sections of the text that can be omitted by those not interested.

[†]GE Annual Report 2001, March 2002. The profits (net earnings) are rounded to the nearest billion.

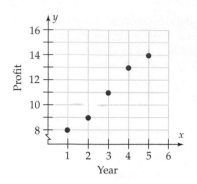

Figure 2–46

Let $x = 1$ correspond to 1997, $x = 2$ to 1998, and so on. Then the data points given by the table are (1, 8), (2, 9), (3, 11), (4, 13), and (5, 14). They are plotted in Figure 2–46, which shows that they lie roughly in a straight line. One way to construct a linear model is to use the line determined by two of the data points. Here are two such models:

Points: (1, 8) and (5, 14)

Slope of line: $\dfrac{14 - 8}{5 - 1} = \dfrac{6}{4} = 1.5$

Equation:
$$y - 8 = 1.5(x - 1)$$
$$y = 1.5x - 1.5 + 8$$
$$y = 1.5x = 6.5.$$

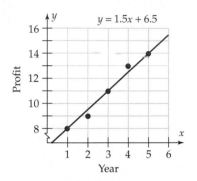

Figure 2–47

Points: (2, 9) and (4, 13)

Slope of line: $\dfrac{13 - 9}{4 - 2} = \dfrac{4}{2} = 2$

Equation:
$$y - 9 = 2(x - 2)$$
$$y = 2x - 4 + 9$$
$$y = 2x + 5.$$

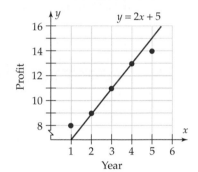

Figure 2–48

Choosing two data points is not the only way to construct a linear model. Figure 2–49 below shows another model for the data, $y = 1.6x + 6.2$, that was obtained by a process that will be described later. Visual inspection of the graphs suggests that the first and third models (Figures 2–47 and 2–49) fit the data better than the second model (Figure 2–48). The first model goes through three of the data points but misses the other two by a noticeable amount. The third model is very close to *all* the data points, although it passes through only one of them.

Figure 2–49

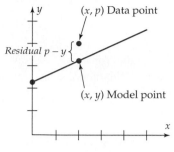

Figure 2–50

To determine the best model, we compute for each one the difference between the actual profit p and the amount y given by the model. If the data point is (x, p) and the corresponding point on the line is (x, y), then the difference $p - y$ measures the error in the model for that particular value of x. The number $p - y$ is called a **residual.** As shown in Figure 2–50 above, the residual $p - y$ is the vertical distance from the data point to the line (positive when the data point is above the line, negative when it is below the line, and 0 when it is on the line). The following tables compute the residuals for the three models, as well as some additional information that is discussed below.

$y = 1.5x + 6.5$			
Data Point (x, p)	Model Point (x, y)	Residual $p - y$	Squared Residual $(p - y)^2$
(1, 8)	(1, 8)	0	0
(2, 9)	(2, 9.5)	−.5	.25
(3, 11)	(3, 11)	0	0
(4, 13)	(4, 12.5)	.5	.25
(5, 14)	(5, 14)	0	0
		Sum: 0	*Sum:* .5

$y = 2x + 5$			
Data Point (x, p)	Model Point (x, y)	Residual $p - y$	Squared Residual $(p - y)^2$
(1, 8)	(1, 7)	1	1
(2, 9)	(2, 9)	0	0
(3, 11)	(3, 11)	0	0
(4, 13)	(4, 13)	0	0
(5, 14)	(5, 15)	−1	1
		Sum: 0	*Sum:* 2

$y = 1.6x + 6.2$			
Data Point (x, p)	Model Point (x, y)	Residual $p - y$	Squared Residual $(p - y)^2$
(1, 8)	(1, 7.8)	.2	.04
(2, 9)	(2, 9.4)	−.4	.16
(3, 11)	(3, 11)	0	0
(4, 13)	(4, 12.6)	.4	.16
(5, 14)	(5, 14.2)	−.2	.04
		Sum: 0	*Sum:* .4

The tables show that the sum of the residuals in each case is 0. This indicates that positive and negative errors cancel each other out (suggesting that the models are reasonable), but it doesn't help us to decide whether one model is better than another.

So we try another accuracy measure: Compute the sum of the *squares* of the residuals, as shown in the last columns of the preceding tables. Now there is no canceling, and each model has a different sum. Using this number as a measure of accuracy has the effect of emphasizing large errors (those with absolute value greater than 1) because the square is greater than the residual and minimizing small errors (those with absolute value less than 1) because the square is less than the residual. So a smaller sum means that the line has less overall error and fits the data better. By this measure, the best of the three models is the third one, $y = 1.6x + 6.2$, because the sum of the squares of its residuals is the smallest. ■

The following fact, which requires multivariate calculus for its proof, shows that there is always a best possible model for linear data.

Linear Regression Theorem

For any set of data points, there is one and only one line for which the sum of the squares of the residuals is as small as possible. This line is called the **least squares regression line.**

The computational process for finding the least squares regression line is called **linear regression.** The linear regression formulas are quite complicated and can be tiresome to use with a large data set. Fortunately, however, linear regression routines are built into most calculators and are also available in spreadsheet and other computer programs. A calculator's linear regression feature was used to obtain the third model in Example 1. According to the Linear Regression Theorem, it is the best possible linear model for that data.

EXAMPLE 2

The following table from the U.S. Census Bureau shows the poverty level for a family of four in selected years (families whose income is below this level are considered to be in poverty).

Figure 2–51

Year	1994	1996	1997	1998	1999	2000	2001
Income	$15,141	16,036	16,400	16,660	17,029	17,603	18,267

(a) Use linear regression to find an equation that models this data.

(b) Use the equation to estimate the poverty level in 1995 and 2005.

SOLUTION

(a) Let $x = 0$ correspond to 1990, and write the income in thousands (for instance, 15.141 in place of 15,141). Then the data points are (4, 15.141), (6, 16.036), . . . , (11, 18.267). We display the calculator's statistics editor and enter the data points: x-coordinates in the first list and y-coordinates (in the same order) in the second list, as in Figure 2–51.* Using the statistical plotting feature to plot the data points, we obtain Figure 2–52, which shows that the data is approximately linear.

Figure 2–52

*Consult the Technology Tip at the end of this section for details on how to carry out this and subsequent steps.

Now we use the linear regression feature to obtain Figure 2–53, which shows that the equation of the least squares regression line is

$$y = .4226885246x + 13.41259016.$$

When you do this on most calculators, you can simultaneously store the regression equation in the equation memory so that it can be graphed along with the data points, as in Figure 2–54.

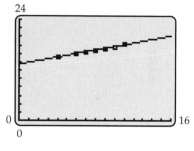

Figure 2–53   Figure 2–54

Figure 2–54 suggests that the regression line provides a good model of the data. This is also indicated in Figure 2–53 by the number r, as is explained after the example.

(b) The year 1995 corresponds to $x = 5$. If $x = 5$, then

$$y = .4226885246(5) + 13.41259016 \approx 15.526.$$

So the regression model estimates the poverty level at $15,526 in 1995. The actual level in that year was $15,569, a difference of only $43. Similarly, the model predicts the poverty level in 2005 ($x = 15$) to be about $19,753 because

$$y = .4226885246(15) + 13.41259016 \approx 19.753.$$

Because 2005 lies outside of the range of the data points, this figure might not prove to be as accurate as the estimate for 1995, which lies within the data range. ∎

TECHNOLOGY TIP

The correlation coefficient r is labeled "corr" on TI-86/89, and also on HP-39 (where it is in the NUM/STATS menu).

If r fails to appear on TI-83, go to CATALOG, select "Diagnostics On," and press ENTER twice. Then re-run the regression calculation.

Figure 2–53 contains a number r (and its square) in addition to the coefficients of the regression line. The number r, which is called the **correlation coefficient,** is a statistical measure of how well the regression line fits the data. It is always between -1 and 1. The closer the absolute value of r is to 1, the better the fit. For instance, the regression line in Example 2 is an excellent fit because $r \approx .99$, as shown in Figure 2–53. When $|r| = 1$, the fit is perfect: All the data points are on the regression line. Conversely, a regression coefficient near 0 indicates a poor fit.

EXAMPLE 3

The table shows the number of accidental deaths per 100,000 population in the United States in selected years.

Year	Death Rate	Year	Death Rate	Year	Death Rate
1910	84.4	1960	52.1	1995	35.5
1920	71.2	1970	56.2	1998	34.1
1930	80.5	1980	46.5	1999	35.5
1940	73.4	1990	36.9	2000	33.9
1950	60.3				

Find a linear model for this data, and graph it. If the model fits the data well, use it to estimate the accidental death rate in 2004.

SOLUTION   We let $x = 0$ correspond to 1900 and enter the data points in the statistical editor of a calculator. Then we proceed as above to find the least squares regression line (Figure 2–55), which is graphed in Figure 2–56.

Figure 2–55

Figure 2–56

The correlation coefficient $r \approx -.97$ indicates that the model fits the data well, and the graph confirms this. To estimate the rate in 2004, substitute $x = 104$ in the regression equation:

$$y = -.5661507143(104) + 90.55376165 \approx 31.7.$$

In 2004, the accidental death rate should be about 31.7 per 100,000 population.

The correlation coefficient always has the same sign as the slope of the least squares regression line. So when r is negative, as in Example 3, the regression line moves downward from left to right (Figure 2–56). In other words, as x increases, y decreases. In such cases, we say that the data has a **negative correlation.** When r is positive, as in Example 2, the regression line slopes upward from left to right (Figure 2–54), and we say that the data has a **positive correlation:** As x increases, y also increases. When r is close to 0 (regardless of sign), we say that there is **no correlation.**

EXAMPLE 4

The number of unemployed people in the labor force (in millions) for 1988–2002 is as follows.*

Year	Unemployed	Year	Unemployed	Year	Unemployed
1988	6.701	1993	8.940	1998	6.210
1989	6.528	1994	7.996	1999	5.880
1990	7.047	1995	7.404	2000	5.655
1991	8.628	1996	7.236	2001	6.742
1992	9.613	1997	6.739	2002	7.991

Is a linear equation a good model for this data?

SOLUTION   After entering the data as two lists in the statistics editor (with $x = 0$ corresponding to 1980), you can test it graphically or analytically.

> *Graphical:* Plotting the data points (Figure 2–57) shows that they do not form a linear pattern (unemployment tends to rise and fall).

> *Analytical:* Linear regression (Figure 2–58) produces an equation whose correlation coefficient is $r \approx -.32$, a number rather close to 0. This indicates that there is no correlation, that is, the regression line is a very poor fit for the data.

Therefore, a linear equation is not a good model for this data.   ■

Figure 2–57

Figure 2–58

GRAPHING EXPLORATION

Enter the data from Example 4 in the statistics editor of your calculator. Graph the data points. Graph the least squares regression line on the same screen to see how poorly it fits the data.

The following Technology Tip may be helpful for learning how to use the regression feature of your calculator.

---

*U.S. Bureau of Labor Statistics; 2002 data is for the first quarter.

TIP

Most calculators allow you to store three or more statistics graphs (identified by number); the following directions assume that the first one is used.

To call up the statistics editor, use these commands:

TI-83: STAT   EDIT   (lists are L_1, L_2, . . .);

TI-85/86: STAT   EDIT   (built-in lists are x-stat, y-stat; it is usually better to create your own lists; we use L1 and L2 as list names here);*

TI-89: APPS   DATA/MATRIX EDITOR   NEW; then choose DATA as the TYPE, enter a VARIABLE name (we use L here), and key in ENTER (lists are C_1, C_2, . . .);

Casio 9850: STAT   (lists are List 1, List 2, . . .);

HP-39: APLET   STATISTICS   (lists are C_1, C_2, . . .).

To graph the data points that have been entered as lists (x-coordinates in the first list, corresponding y-coordinates in the second), choose an appropriate viewing window on TI calculators. On TI-85, use STAT   DRAW   SCAT   to plot the points determined by the lists currently chosen in the STAT   EDIT screen. On other calculators, use these commands to enter the setup screen:

TI-83: STAT-PLOT (on keyboard); choose PLOT 1;

TI-86: STAT; choose PLOT and PLOT 1;

TI-89: from the Data Editor, choose PLOT-SETUP (F_2) and DEFINE (F_1);

Casio: STAT   GRPH   SET;

HP-39: APLET   STATISTICS   PLOT-SETUP   and   SYMB.

Key in the appropriate information [on/off (TI-83/86), graph number (Casio and HP), lists to be used, graph type (scatter plot), the mark to be used for data points, and, on HP-39, the viewing window]. Then press GRAPH (on TI) or GPH 1 (on Casio) or PLOT (on HP-39).

To produce the least squares regression line, store its equation as y_1 in the equation memory, and graph it, use these commands:

TI-83: STAT   CALC   LinReg ($ax + b$) L_1, L_2, Y_1;

TI-85: STAT   CALC   L1, L2   LINR; then press STREG and enter y_1 as the name;

TI-86: STAT   CALC   Lin R   L1, L2, y_1;

TI-89: From the Data Editor, choose CALC (F_5); enter LIN REG as CALCULATION TYPE, C_1 as x, C_2 as y; and choose $y_1(x)$ in STORE REGEQ;

Then press GRAPH or DRAW to obtain the graph of the equation and the data points (assuming that PLOT 1 has not been turned off). On HP-39, after plotting the data points, use MENU FIT to graph the regression line and SYMB to see its equation. On Casio 9850, after plotting the data points, press X for the equation of the regression line and DRAW for its graph.

---

*When statistical computations are run on TI-85/86, the lists used are automatically copied into the x-stat and y-stat lists, replacing whatever was there before. So use the x-stat and y-stat lists only if you don't want to save them. See your instruction manual to find out how to create new lists in the statistics editor.

EXERCISES 2.5

In Exercises 1–4, two linear models are given for the data. For each model,
(a) Find the residuals and their sum;
(b) Find the sum of the squares of the residuals;
(c) Determine which model is the better fit.

1. The weekly amount spent on advertising and the weekly sales revenue of a small store over a five-week period are shown in the table. Two models are $y = x$ and $y = .5x + 1.5$.

Advertising Expenditures x (in hundreds of dollars)	1	2	3	4	5
Sales Revenue y (in thousands of dollars)	2	2	3	3	5

2. The table gives the consumer price index (CPI) in April of selected years.[*] Two models are $y = 4x + 131$ and $y = 4.05x + 131.2$, where $x = 0$ corresponds to 1990.

Year	1994	1998	2000	2002
CPI	147.4	162.5	171.3	179.8

3. Advertising expenditures in the United States (in billions of dollars) in selected years are shown in the table.[†] Two models are $y = 11x + 121$ and $y = 11.5x + 129$, where $x = 0$ corresponds to 1990.

Year	1990	1995	1999	2000
Amount	129	161	215	244

4. Revenues for U.S. public elementary and secondary schools (in billions of dollars) in the fall of selected years are shown in the table.[‡] Two models are

$$y = 17.3x + 205.8 \quad \text{and} \quad y = 17.45x + 203.6,$$

where $x = 0$ corresponds to 1990.

Year	1994	1996	1998	2000
Revenues	275	306	343	379

In Exercises 5–8, determine whether the given scatter plot of the data indicates that there is a positive correlation, a negative correlation, or very little correlation.

5.

6.

7.

8.

In Exercises 9–14, construct a scatter diagram for the data and answer these questions: (a) Does the data appear to be linear? (b) If so, is there a positive or negative correlation?

9. Banking fees for various services have become a significant part of a typical bank's income. The table shows fee income as a percent of net operating revenue for U.S. banks in the past two decades.[§] Let $x = 0$ correspond to 1980.

Year	1980	1985	1990	1995	2000
Fees as a Percent of Income	20	25	32	35	43

[*]U.S. Bureau of Labor Statistics.
[†]*Advertising Age.*
[‡]*Statistical Abstract of the United States: 2001.*

[§]U.S. Federal Reserve.

10. The U.S. gross domestic product (GDP) is the total value of all goods and services produced in the United States. The table shows the GDP in billions of 1996 dollars.[*] Let $x = 0$ correspond to 1990.

Year	1990	1992	1994	1996	1998	2000
GDP	6707.9	6880.0	7347.7	7813.2	8495.7	9318.5

11. In mid-2002, Grange Life Insurance advertised the following monthly premium rates for a $50,000 term policy for a female nonsmoker. Let x represent age and y the monthly premium.

Age	30	35	40	45	50
Premium	6.48	6.52	6.78	7.88	9.93

Age	55	60	65	70
Premium	12.69	16.71	23.67	36.79

12. The vapor pressure y of water depends on the temperature x, as given in the table.

Temperature (°C)	Pressure (mm Hg)
0	4.6
10	9.2
20	17.5
30	31.8
40	55.3
50	92.5
60	149.4
70	233.7
80	355.1
90	525.8
100	760

13. The table shows the U.S. Bureau of Census population data for St. Louis, Missouri. Let $x = 0$ correspond to 1950.

Year	Population
1950	856,796
1970	622,236
1980	452,801
1990	396,685
2000	348,189

14. The table shows the U.S. disposable income (personal income less personal taxes) in billions of dollars.[†] Let $x = 0$ correspond to 1990.

Year	Disposal Personal Income
1990	4166.8
1992	4613.7
1994	5018.9
1996	5534.7
1998	6320.0
1999	6618.0
2000	7031.0

In Exercises 15–24, use linear regression to find the requested linear model.

15. Expenditures for research and development at Intel Corporation (in millions of dollars) are shown in the table.[‡]

Year	1993	1995	1997	1999	2000	2001
Expenditures	970	1296	2347	3111	3897	3796

(a) Make a scatter plot of the data, with $x = 0$ corresponding to 1990.
(b) Find a linear model for the data.
(c) Use the model to estimate the expenditures in 1994 and 1998. [For comparison, the actual expenditures were $1111 million and $2509 million, respectively.]
(d) Assuming that the model remains accurate after 2001, estimate the expenditures in 2004.

16. The table on the next page shows the median time (in months after the application has been made) for the U.S. Food and Drug Administration to approve a new drug.[§]

---

[*]U.S. Bureau of Economic Analysis.

[†]Bureau of Economics Analysis U.S. Department of Commerce.
[‡]Intel Corporation, 2001 Annual Report.
[§]U.S. Food and Drug Administration.

Year	Median Time to Approval
1990	24.3
1991	22.1
1992	22.6
1993	23.0
1994	17.5
1995	15.9
1996	14.3
1997	13.4
1998	12.0
1999	11.6
2000	15.6

(a) Make a scatter plot of the data, with $x = 0$ corresponding to 1990.

(b) Find a linear model for the data.

(c) What are the limitations of this model? [*Hint:* What does it say about approval time in 2009?]

17. National expenditures on health care (in billions of dollars) in various years are shown in the table.[*]

Year	1990	1995	1997	1999
Amount	699.4	993.7	1092.4	1228.5

Year	2000	2001	2002	2003
Amount	1316.2	1403.6	1495.5	1590.4

(a) Find a linear model for this data, with $x = 0$ corresponding to 1990.

(b) Interpret the meaning of the slope and the y-intercept.

(c) If the model remains accurate in the future, what will health care expenditures be in 2007?

18. Enrollment in public colleges (in thousands) in selected years is shown in the table.[†]

Year	1980	1985	1990	1995	2000	2003
Enrollment	9457	9479	10,845	11,092	11,750	11,894

(a) Find a linear model for this data, with $x = 0$ corresponding to 1980.

(b) If the model remains accurate in the future, estimate public college enrollment in 2006 and 2009.

(c) According to this model, when will public college enrollment reach 13 million?

In Exercises 19 and 20, use the following table, which gives the median weekly earnings of full-time workers 25 years and older by the amount of education.[‡]

Year	No High School Diploma	High School Graduate	Some College	College Graduate
1996	$317	$443	$518	$758
1997	$321	$461	$535	$779
1998	$337	$479	$558	$821
1999	$346	$490	$580	$860
2000	$360	$506	$598	$896
2001	$378	$520	$621	$924

19. (a) Find linear models for the median weekly earnings of full-time workers 25 years and older who did not graduate from high school and for those who graduated from high school, but did not attend college. Let $x = 0$ correspond to 1990.

(b) Estimate the median weekly income for each group in part (a) in 2006.

20. (a) Find linear models for the median weekly earnings of full-time workers 25 years and older who attended college, but did not graduate, and those who graduated from college. Let $x = 0$ correspond to 1990.

(b) Use the models in Exercises 19(a) and 20(a) to determine the approximate yearly rate at which the median weekly earnings of each of the four groups in the table are increasing. What does this say about the value of education?

21. The table shows the number of Internet users in the United States (in millions) in recent years.[§]

Year	1997	1998	1999	2000	2001	2002
Users	60	83	104	122	138	153

(a) Find a linear model for this data (with $x = 0$ corresponding to 1997), and use it to predict the number of Internet users in 2004.

(b) For how long do you think this model might remain accurate? [*Note:* The U.S. population is approximately 283 million in 2003 and is expected to increase by about 2.5 million per year during the next decade.]

---

[*]Data and projections from the U.S. Health Care Financing Administration.

[†]Data and projections from the U.S. National Center for Education Statistics.

[‡]U.S. Bureau of Labor Statistics.

[§]U.S. Department of Commerce; Jupiter Media Metrix.

22. The projected number of new cases of Alzheimer's disease (in thousands) in the United States in selected years is shown in the table.[*]

Year	2000	2010	2020	2030	2040	2050
New Cases	400	467	489	600	800	956

(a) Find a linear model for this data, with $x = 0$ corresponding to 2000, and use it to estimate the number of new cases in the years 2005, 2015, and 2025.

(b) Use the model to predict the years in which the number of new cases will be 750,000 and 1,000,000.

23. The table shows the maximum Pell Grant for college students in recent years.[†]

Year	1995	1996	1997	1998	1999
Maximum Grant	$2340	$2470	$2700	$3000	$3125

Year	2000	2001	2002	2003
Maximum Grant	$3300	$3750	$4000	$4000

(a) Find a linear model for this data, with $x = 0$ corresponding to 1990.

(b) According to this model, when will the maximum Pell Grant reach $5000?

24. During the current decade, the amount of money each American spends annually on prescription drugs (in addition to the amounts paid by insurance) is projected to increase, as shown in the table.[‡]

Year	2000	2002	2004	2006	2008	2010
Amount	$143	$175	$207	$231	$287	$351

(a) Find a linear model for this data, with $x = 0$ corresponding to 2000.

(b) Use the model to estimate the amount each person will spend on prescription drugs in 2005, 2007, and 2011.

(c) According to this model, in what year will each American spend $400 on prescription drugs?

---

[*]L. Hebert, L. Beckett, P. Scherr, D. Evans, "Annual Incidence of Alzheimer Disease in the United States Projected to the Years 2000 Through 2050," in *Alzheimer Disease and Associated Disorders*, Vol. 15, No. 4, 2001.

[†]U.S. Education Department.

[‡]Centers for Medicare and Medicaid Services.

Chapter 2 Review

IMPORTANT CONCEPTS

Section 2.1

Viewing window   71
Equation memory   72
Trace   73
Zoom   74
Standard viewing window   74
Decimal window   74
Square window   75
Maximum/minimum finder   76
Complete graph   77

Solving equations graphically   83
Graphical root finder   83
Graphical intersection finder   85
Equation solvers   86
Solution methods   89

Section 2.3

Guidelines for setting up applied
   problems   90
Solving applied problems with
   equations   93–98

Section 2.2

x-Intercepts and equation solutions
   83–84

Section 2.4

Solving optimization problems
   101–103

Section 2.5

Mathematical model   105
Residual   106
Least squares regression line   107
Linear regression   108
Correlation coefficient   109
Positive and negative correlation
   110

REVIEW QUESTIONS

In Questions 1–6,
(a) *Determine which of the viewing windows a–e shows a complete graph of the equation.*
(b) *For each viewing window that does not show a complete graph, explain why.*
(c) *Find a viewing window that gives a "better" complete graph than windows a–e (meaning that the window is small enough to show as much detail as possible, yet large enough to show a complete graph).*

(a) Standard viewing window
(b) $-10 \le x \le 10, -200 \le y \le 200$
(c) $-20 \le x \le 20, -500 \le y \le 500$
(d) $-50 \le x \le 50, -50 \le y \le 50$
(e) $-1000 \le x \le 1000, -1000 \le y \le 1000$

1. $y = .2x^3 - .8x^2 - 2.2x + 6$

2. $y = x^3 - 11x^2 - 25x + 275$

3. $y = x^4 - 7x^3 - 48x^2 + 180x + 200$

4. $y = x^3 - 6x^2 - 4x + 24$

5. $y = .03x^5 - 3x^3 + 69.12x$

6. $y = .00000002x^6 - .0000014x^5 - .00017x^4 + .0107x^3 + .2568x^2 - 12.096x$

In Questions 7–10, sketch a complete graph of the equation, and give reasons why it is complete.

7. $y = x^2 - 10$

8. $y = x^3 + x + 4$

9. $y = \sqrt{x - 5}$

10. $y = x^4 + x^2 - 6$

In Questions 11–14, sketch a complete graph of the equation.

11. $y = x^2 - 13x + 43$

12. $y = |x|$

13. $y = |x + 5|$

14. $y = 1/x$

In Questions 15–22, solve the equation graphically. You need only find solutions in the given interval.

15. $x^3 + 2x^2 = 11x + 6; \quad [0, \infty)$

16. $x^3 + 2x^2 = 11x + 6; \quad (-\infty, 0)$

17. $x^4 + x^3 - 10x^2 = 8x + 16; \quad [0, \infty)$

18. $2x^4 + x^3 - 2x^2 + 6x + 2 = 0; \quad (-\infty, -1)$

19. $\dfrac{x^3 + 2x^2 - 3x + 4}{x^2 + 2x - 15} = 0; \quad (-10, \infty)$

20. $\dfrac{3x^4 + x^3 - 6x^2 - 2x}{x^5 + x^3 + 2} = 0; \quad [0, \infty)$

21. $\sqrt{x^3 + 2x^2 - 3x - 5} = 0; \quad [0, \infty)$

REVIEW **QUESTIONS**

22. $\sqrt{1 + 2x - 3x^2 + 4x^3 - x^4} = 0$; $(-5, 5)$

23. A jeweler wants to make a 1-ounce ring consisting of gold and silver, using \$200 worth of metal. If gold costs \$600 per ounce and silver costs \$50 per ounce, how much of each metal should she use?

24. A calculator is on sale for 15% less than the list price. The sale price, plus a 5% shipping charge, totals \$210. What is the list price?

25. Karen can do a job in 5 hours, and Claire can do the same job in 4 hours. How long will it take them to do the job together?

26. A car leaves the city traveling at 54 mph. One-half hour later, a second car leaves from the same place and travels at 63 mph along the same road. How long will it take for the second car to catch up with the first?

27. A 12-foot-long rectangular board is cut in two pieces so that one piece is four times as long as the other. How long is the bigger piece?

28. George owns 200 shares of stock, 40% of which are in the computer industry. How many more shares must he buy to have 50% of his total shares in computers?

29. A square region is changed into a rectangular one by making it 2 feet wider and twice as long. If the area of the rectangular region is three times larger than the area of the original square region, what was the length of a side of the square before it was changed?

30. The radius of a circle is 10 inches. By how many inches should the radius be increased so that the area increases by 5π square inches?

31. The cost of manufacturing x caseloads of ballpoint pens is

$$\frac{600x^2 + 600x}{x^2 + 1}$$

dollars. How many caseloads should be manufactured to have an *average cost* of \$25? [Average cost was defined in Exercise 13 of Section 2.4.]

32. An open-top box with a rectangular base is to be constructed. The box is to be at least 2 inches wide and twice as long as it is wide and is to have a volume of 150 cubic inches. What should the dimensions of the box be if the surface area is to be

    (a) 90 square inches?   (b) as small as possible?

33. A farmer has 120 yards of fencing and wants to construct a rectangular pen, divided in two parts by an interior fence, as shown in the figure. What should the dimensions of the pen be to enclose the maximum possible area?

34. The top and bottom margins of a rectangular poster are each 5 inches, and each side margin is 3 inches. The printed material on the poster occupies an area of 400 square inches. Find the dimensions that will use the least possible amount of posterboard.

35. A rectangle has one side on the x-axis, and its other two corners sit on the graph of $y = 9 - x^2$, as shown in the figure. What value of x gives a rectangle of maximum area?

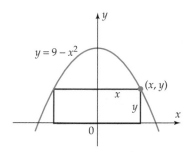

36. The window in the figure has a rectangular bottom, with a semicircle of radius r lying on top of it, and a perimeter of 40 feet. In order that the window have the maximum possible area, what should r and h be?

37. The table on the next page shows monthly premiums (as of 2002) for a \$100,000 term life insurance policy from Grange Life Insurance for a male smoker.

Age	Premium
25	$20.30
30	$20.39
35	$20.39
40	$22.05
45	$28.61
50	$41.65
55	$60.90
60	$85.58
65	$132.91

(a) Make a scatter plot of the data, using x for age and y for premiums.

(b) Does the data appear to be linear?

38. For which of the following scatter plots would a linear model be reasonable? Which sets of data show positive correlation, and which show negative correlation?

(a)

(b)

(c)

(d)

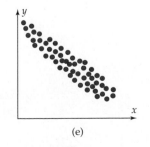

(e)

In Exercises 39–42, use linear regression to find the requested linear model.

39. The table shows the average hourly earnings of production workers in manufacturing.[*]

Year	1991	1993	1995	1997	1999	2001
Hourly Earnings	$11.18	$11.74	$12.37	$13.17	$13.90	$14.84

(a) Find a linear model for the data, with $x = 0$ corresponding to 1990.

(b) Use the model to estimate the average hourly earnings in 1993 and 2000. How do the estimates of the model compare with the actual figures of $11.74 and $14.38, respectively.

(c) Estimate the average hourly earnings in 2004.

40. The amount spent on marketing prescription drugs (in billions of dollars) in recent years is shown in the table.[†]

Year	1996	1997	1998	1999	2000	2001
Amount	9.3	11	12.5	13.8	15.8	19.0

(a) Find a linear model for the data, with $x = 0$ corresponding to 1996.

(b) Estimate the amount spent on marketing in 2002.

41. The table shows the total amount of charitable giving (in billions of dollars) in the United States in recent years.[‡]

Year	Charitable Giving
1992	110.4
1993	116.5
1994	119.2
1995	124.0
1996	138.6
1997	153.8
1998	172.1
1999	190.8
2000	203.5

---

[*]U.S. Bureau of Labor Statistics.
[†]*IMS Health.*
[‡]*Statistical Abstract of the United States: 2001.*

(a) Find a linear model for the data, with $x = 0$ corresponding to 1990.
(b) Estimate the amount of charitable giving in 2002 and 2005.
(c) According to the model, in what year will charitable giving reach $220 billion?

42. The table shows, for selected states, the percent of high school students in the class of 2001 who took the SAT and the average SAT math score.*

(a) Make a scatter plot of average SAT math score y and percent x of students who took the SAT.
(b) Find a linear model for the data.
(c) What is the slope of your linear model? What does this mean in the context of the problem?
(d) Here are the data on four additional states. How well does the model match the actual figures for these states?

State	Students Who Took SAT (%)	Average Math Score
Connecticut	82	510
Delaware	67	499
Georgia	63	489
Idaho	17	542
Indiana	60	501
Iowa	5	603
Montana	23	539
Nevada	33	515
New Jersey	81	513
New Mexico	13	542
North Dakota	4	599
Ohio	26	539
Pennsylvania	71	499
South Carolina	57	488
Washington	53	527

State	Students Who Took SAT (%)	Average Math Score
Alaska	51	510
Arizona	34	525
Hawaii	52	515
Oklahoma	8	561

*The College Board.

DISCOVERY PROJECT 2   Supply and Demand

Claus Meyer/Black Star Publishing/PictureQuest

Economists study the forces at play as buyers and sellers interact in what we call the **market** for a product. Because their money is limited, consumers tend to buy less of a particular product if the price is higher and more if the price is lower. This is the essence of the **demand curve,** which is just a depiction of the relationship between the price of an item and the maximum quantity that people will buy at that price. The demand curve represents the aggregate demand of all consumers in a market.

Manufacturers, however, have other considerations. Production costs and limited resources (materials, labor, technology, and capital, for instance) factor into the decision to produce at different levels. Ultimately, as the ones supplying (selling) the product, they want to sell more of them if the price is high and less of them if the price is low. This is the essence of the **supply curve,** a depiction of the relationship between the price of an item and the number of those items the manufacturers are willing to sell at that price. Like the demand curve, the supply curve represents the aggregate of all suppliers of a given product.

Here is a simple example of these ideas. The table below shows the quantity demanded for a hypothetical product at several different prices. You can easily verify that the equation $p = 5.85 - .005q$ describes this relationship.

Price p (in dollars)	Quantity q (in thousands)
3.25	520
3.00	570
2.75	620
2.50	670
2.25	720

Suppose the supply curve for this product is given by the equation $p = .01q - 3.75$ and that the producer would like to set the price of the product at \$2.80. To find the supply and demand at this price we must solve the appropriate equation for q when $p = 2.80$.[*]

Supply	**Demand**
$.01q - 3.75 = p$	$5.85 - .005q = p$
$.01q - 3.75 = 2.80$	$5.85 - .005q = 2.80$
$.01q = 6.55$	$-.005q = -3.05$
$q = \dfrac{6.55}{.01} = 655$	$q = \dfrac{-3.05}{-.005} = 610.$

---

[*]We use algebra to solve the equation, but you could also use graphical means or an equation solver. The technological methods are often the best choice for nonlinear supply and demand equations.

These results pose a problem: At a price of $2.80 the producer would be willing to sell 655,000 items, but the demand curve says that only 610,000 items would be demanded by consumers at this price. In technical terms, there is a **surplus** because supply exceeds demand.

You can readily verify that the situation is reversed when the price is $2.40. Solving the supply and demand equations when $p = 2.40$ shows that consumer demand is 690,000 items, but producers are only willing to sell 615,000. In this case, there is a **shortage** because demand exceeds supply.

When there is a surplus, sellers are inclined to lower the price because storing unsold items (or having to discard perishable ones) is expensive. When there is a shortage, consumers are willing to pay more in order to get a product they want. These tendencies push the market toward the **equilibrium point,** at which buyers and sellers agree on both price and quantity. This occurs when

$$\text{quantity demanded} = \text{quantity supplied.}$$

The price p at this point is called the **equilibrium price,** the price at which supply equals demand.

The equilibrium point can be found graphically by finding the intersection point of the supply curve and the demand curve. Because economists normally put price (p) on the vertical axis and quantity (q) on the horizontal axis, we follow the same convention here (using X in place of q and Y in place of p when graphing on a calculator). Figure 1 shows that the equilibrium point for our example is (640, 2.65), which means that the equilibrium price is $2.65 and that 640,000 items will be demanded and supplied at that price.

Supply: $Y = .01X - 3.75$

Demand: $Y = -.005X + 5.85$

Intersection
X=640   Y=2.65

Figure 1

Supply and demand curves provide quantities that correspond to a limited collection of feasible prices at a given point in time. So the corresponding equations are valid only in this range and may not be valid outside of it. In our example, the equations are valid when $520 \leq q \leq 720$, which determines the viewing window in Figure 1.

When these ideas are used to make business decisions, the first step is to determine the supply and demand equations. This is often done by using statistical data from a trial market, as in the following exercises.

1. An economist obtained the following data about the demand for oranges.

Price per pound (dollars)	0.93	0.90	0.84	0.70	0.65	0.63	0.60
Quantity demanded (in 100,000 tons)	10	10.6	11.5	13.8	14.7	15.1	15.5

Find a linear demand equation of the form $p = mq + b$ either by using linear regression *or* by using the first and last data points [(10, .93) and (15.5, .60)]. If you use regression, round the coefficients to four decimal places.

2. Use the model you found in Exercise 1 to find the quantity demanded at a price of 75¢ per pound.

3. The economist in Exercise 1 also obtained this data about the supply of oranges.

Price per pound (dollars)	0.20	0.36	0.45	0.55	0.65	0.75	1.00
Quantity demanded (in 100,000 tons)	10	11.4	12.4	13.5	14.6	15.5	18

Find a linear supply equation of the form $p = mq + b$ either by using linear regression *or* by using the first and last data points, as in Exercise 1.

4. Use the model you found in Exercise 3 to find the price at which producers are willing to supply 1,300,000 tons of oranges.

5. Find the equilibrium point for the situation in Exercises 1 and 3. What is the equilibrium price? How many oranges will be supplied/demanded at this price?

6. Here are the equations of the curves for the sale of apples:

$$p = .06q - .2 \quad \text{and} \quad p = -.04q + .7,$$

where p is in dollars and q is in 100,000 tons.

(a) Which curve is supply and which one is demand? How can you tell?

(b) Find the equilibrium point. What is the equilibrium price? How many apples will be supplied/demanded at this price?

7. Graph $p = .28$ together with the demand and supply curves of Exercise 6. Is this price above or below the equilibrium price? Would this price lead to a surplus, a shortage, or neither? Explain.

Functions and Graphs

Spencer Grant/PhotoEdit

Looking for a house?

If you buy a house, you'll probably need a mortgage. If you can get a low interest rate, your monthly payments will be lower (or, alternatively, you can afford a more expensive house). The timing of your purchase can make a difference because mortgage interest rates constantly fluctuate. In mathematical terms, rates are a function of time. The graph of this function provides a picture of how interest rates change. See Exercise 47 on page 159.

Chapter Outline

Interdependence of Sections

3.1 → 3.2

3.3 → 3.4
3.3 → 3.6
3.2 → 3.7
3.5

3.1   **Functions**
3.2   **Functional Notation**
3.3   **Graphs of Functions**
3.3.A   *Special Topics:* Graph Reading
3.4   **Graphs and Transformations**
3.4.A   *Special Topics:* Symmetry
3.5   **Operations on Functions**
3.6   **Rates of Change**
3.7   **Inverse Functions**

The concept of a function and functional notation are central to modern mathematics and its applications. In this chapter, you will be introduced to functions and operations on functions, learn how to use functional notation, and develop skill in constructing and interpreting graphs of functions.

3.1   Functions

To understand the origin of the concept of function, it may help to consider some real-life situations in which one numerical quantity depends on, corresponds to, or determines another.

EXAMPLE 1

The amount of state income tax Louisiana residents pay depends on their income. The way that the income determines the tax is given by the following tax law.*

Income		Tax
At Least	**But Less Than**	
0	$4,500	0
$4,500	$10,000	2% of amount over $4,500
$10,000	$50,000	$110 + 4% of amount over $10,000
$50,000		$1710 + 6% of amount over $50,000

■

---

*2002 rates for a single person with one exemption and no deductions; actual tax amount may vary slightly from this formula when tax tables are used.

EXAMPLE 2

The graph in Figure 3–1 shows the temperatures in Cleveland, Ohio, on April 11, 2001, as recorded by the U.S. Weather Bureau at Hopkins Airport. The graph indicates the temperature that corresponds to each given time.   ■

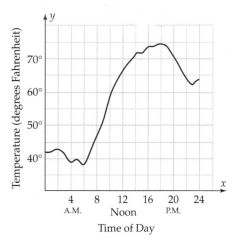

Figure 3–1

EXAMPLE 3

Suppose a rock is dropped straight down from a high place. Physics tells us that the distance traveled by the rock in t seconds is $16t^2$ feet. So the distance depends on the time.   ■

These examples share several common features. Each involves two sets of numbers, which we can think of as inputs and outputs. In each case, there is a rule by which each input determines an output, as summarized here.

	Set of Inputs	Set of Outputs	Rule
Example 1	All incomes	All tax amounts	The tax law
Example 2	Hours since midnight	Temperatures during the day	Time/temperature graph
Example 3	Seconds elapsed after dropping the rock	Distance rock travels	Distance $= 16t^2$

Each of these examples may be mentally represented by an idealized calculator that has a single operation key: A number is entered [*input*], the rule key is pushed [*rule*], and an answer is displayed [*output*]. The formal definition of function incorporates these common features (input/rule/output), with a slight change in terminology.

Functions

> A **function** consists of
>
> A set of inputs (called the **domain**);
>
> A **rule** by which each input determines one and only one output;
>
> A set of outputs (called the **range**).

The phrase "one and only one" in the definition of the rule of a function may need some clarification. In Example 2, for each time of day (input), there is one and only one temperature (output). But it is quite possible to have the same temperature (output) at different times (inputs). In general,

For each input, the rule of a function determines exactly one output. But different inputs may produce the same output.

Although real-world situations, such as Examples 1–3, are the motivation for functions, much of the emphasis in mathematics courses is on the functions themselves, independent of possible interpretations in specific situations, as illustrated in the following examples.

EXAMPLE 4

Could either of the following be the table of values of a function?

(a)

Input	−4	−2	0	2	4
Output	21	7	1	3	7

(b)

Input	3	2	1	3	5
Output	4	0	2	6	9

SOLUTION In table (a), two different inputs (−2 and 4) produce the same output, but that's OK because each input produces exactly one output. So the table (a) could be the table of values of a function. In table (b), however, the input 3 produces two different outputs (4 and 6), so this table cannot possibly represent a function. ■

EXAMPLE 5

For each real number *s* that is not an integer, let [*s*] denote the *integer* that is closest to *s* on the *left* side of *s* on the number line; if *s* is itself an integer, we define [*s*] = *s*. Some examples are shown in Figure 3-2 on the next page.

$$[-4.7] = -5, \qquad [-3] = -3, \qquad [-1.5] = -2,$$

$$[0] = 0, \qquad \left[\frac{5}{3}\right] = 1, \quad [\pi] = 3.$$

Figure 3–2

The **greatest integer function** is the function whose domain is the set of all real numbers, whose range is the set of integers, and whose rule is

> For each input (real number) x, the output is the integer $[x]$.   ■

TECHNOLOGY TIP

The greatest integer function is denoted INT or FLOOR on TI and HP-39, and INTG on Casio. It is in this menu/submenu:

TI: MATH/NUM

HP-39: MATH/REAL

Casio: OPTN/NUM

∷ FUNCTIONS DEFINED BY EQUATIONS AND GRAPHS

Equations in two variables are *not* the same things as functions. However, many equations can be used to define functions.

EXAMPLE 6

The equation $4x - 2y^3 + 5 = 0$ can be solved uniquely for y:

$$2y^3 = 4x + 5$$

$$y^3 = 2x + \frac{5}{2}$$

$$y = \sqrt[3]{2x + \frac{5}{2}}.$$

If a number is substituted for x in this equation, then exactly one value of y is produced. So we can define a function whose domain is the set of all real numbers and whose rule is

> The input x produces the output $\sqrt[3]{2x + 5/2}$.

In this situation, we say that the equation defines y **as a function of** x.

The original equation can also be solved for x:

$$4x = 2y^3 - 5$$

$$x = \frac{2y^3 - 5}{4}.$$

Now if a number is substituted for y, exactly one value of x is produced. So we can think of y as the input and the corresponding x as the output and say that the equation defines x **as a function of** y.   ■

EXAMPLE 7

If you solve the equation

$$y^2 - x + 1 = 0$$

for y, you obtain

$$y^2 = x - 1$$
$$y = \pm\sqrt{x - 1}.$$

This equation does *not* define y as a function of x because, for example, the input $x = 5$ produces two outputs: $y = \pm 2$. ■

EXAMPLE 8

A group of students drives from Cleveland to Seattle, a distance of 2350 miles, at an average speed of 52 mph.

(a) Express their distance from Cleveland as a function of time.

(b) Express their distance from Seattle as a function of time.

SOLUTION

(a) Let t denote the time traveled in hours after leaving Cleveland, and let D be the distance from Cleveland at time t. Then the equation that expresses D as a function of t is

$$D = \text{Distance traveled in } t \text{ hours at 52 mph} = 52t.$$

(b) At time t, the car has traveled $52t$ miles of the 2350-mile journey, so the distance K remaining to Seattle is given by $K = 2350 - 52t$. This equation expresses K as a function of t. ■

Graphing calculators are designed to deal with equations that define y as a function of x. Calculators can evaluate such functions (that is, produce the outputs from various inputs). One method is illustrated in Calculator Investigation 1 at the end of this section, and another in the next example.

EXAMPLE 9

The equation $y = x^3 - 2x + 3$ defines y as a function of x. Use the table feature of a calculator[*] to find the outputs for each of the following inputs:

(a) $-3, -2, -1, 0, 1, 2, 3, 4, 5$ (b) $-5, -11, 8, 7.2, -.44$

---

[*]TI-85 does not have a built-in table feature, but a program to proivde one is in the Program Appendix.

TECHNOLOGY

TIP

To find the table setup screen, look for TBLSET (or RANG or NUM SETUP) on the keyboard or in the TABLE menu.

The increment is called ∆TBL on TI (and PITCH or NUM-STEP on others).

The table type is called INDPNT on TI, and NUMTYPE on HP-39.

SOLUTION

(a) To use the table feature, we first enter $y = x^3 - 2x + 3$ in the equation memory, say, as y_1. Then we call up the setup screen (see the Technology Tip in the margin and Figure 3–3) and enter the *starting number* (-3), the *increment* (the amount the input changes for each subsequent entry, which is 1 here), and the *table type* (AUTO, which means the calculator will compute all the outputs at once).* Then press TABLE to obtain the table in Figure 3–4. To find values that don't appear on the screen in Figure 3–4, use the up and down arrow keys to scroll through the table.

(b) With an apparently random list of inputs, as here, we change the table type to ASK (or USER or BUILD YOUR OWN).† Then key in each value of x, and hit ENTER. This produces the table one line at a time, as in Figure 3–5. ∎

Figure 3–3

Figure 3–4

Figure 3–5

EXPLORATION

Construct a table of values for the function in Example 9 that shows the outputs for these inputs: 2, 2.4, 2.8, 3.2, 3.6, and 4. What is the increment here?

EXAMPLE 10

The revenues y of MTV in year x can be approximated by the equation

$$y = .257x^3 + 11.6x^2 + 4.5x + 177.5 \qquad (0 \le x \le 14),$$

where $x = 0$ corresponds to 1987 and y is in millions of dollars.‡ So the revenue y is a function of the year x. In what year did revenues first exceed one billion dollars?

Figure 3–6

SOLUTION   Since y is in millions, one billion dollars corresponds to $y = 1000$. Make a table of values for the function (Figure 3–6). It shows that the revenue was approximately $865,550,000 in 1994 ($x = 7$) and $1,087,500,000 in 1995 ($x = 8$). ∎

---

*There is no table type selection on Casio, but you must enter a maximum value for x.
†Casio users should follow the directions in Calculator Investigation 2 at the end of this section. This type of table is not available on TI-85, but there is an effective alternative (see Calculator Investigation 3).
‡Based on data from MTV Networks.

EXAMPLE 11

The graph in Figure 3–7 defines a function whose rule is as follows:

**For input *x*, the output is the unique
number *y* such that (*x*, *y*) is on the graph.**

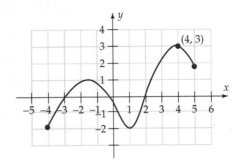

Figure 3–7

Input 4, for example, produces output 3 because (4, 3) is on the graph. Similarly, (−3, 0) is on the graph, which means that input −3 produces output 0. Since the first coordinates of all points on the graph (the inputs) lie between −4 and 5, the domain of this function is the interval [−4, 5]. The range is the interval [−2, 3] because all the second coordinates of points on the graph (the outputs) lie between −2 and 3. ∎

CALCULATOR INVESTIGATIONS 3.1

1. FUNCTION EVALUATION

The equation $y = x^3 - 4x + 1$ defines *y* as a function of *x*. On calculators other than HP-39, the function may be evaluated as follows.

(a) Enter this equation as y_1 in the equation memory, and return to the home screen. To evaluate *y* at *x* = 5, store 5 in memory X*, where X* denotes the "variable" key that is used to enter the equation y_1 in the function memory. Then choose y_1, which is found in the following menu/submenu:

> TI-83: VARS/Y-VARS
>
> Casio: VARS/GRPH [choose Y and type in 1]
>
> TI-86/89: Type in y_1 on the keyboard.

Now press ENTER and the calculator displays the value of *y* at *x* = 5.

(b) Evaluate *y* at the following values of *x*: 3.5, 2, −1, −4.7.

(c) Enter the equation $y = x^2 + x - 3$ as y_2 in the equation memory.

Evaluate this function at *x* = 7 and *x* = −5.5.

2. USER-CONTROLLED TABLES ON CASIO

To construct the table in Example 9(b), proceed as follows.

(a) Enter the equation $y = x^3 - 2x + 3$ in the equation memory. This can be done by selecting either TABLE or GRAPH in the MAIN menu.

(b) Return to the MAIN menu and select LIST. Enter the numbers at which you want to evaluate this function [in this case, $-5, -11, 8, 7.2, -.44$] as List 1.

(c) Return to the MAIN menu and select TABLE. Then press SET-UP [that is, 2nd MENU] and select LIST as the Variable; on the LIST menu, choose List 1. Press EXIT and then press TABL to produce the table.

(d) Use the up/down arrow key to scroll through the table. If you change an entry in the X column, the corresponding y_1 value will automatically change. Find the value of y_1 at $x = 11$ and $x = -6$.

3. FUNCTION EVALUATION ON TI-85/86

The equation $y = x^2 - 3.7x + 4.5$ defines y as a function of x.

(a) Enter this equation as y_1 in the equation memory, and return to the home screen by pressing 2nd QUIT. To determine the value of y when $x = 29$, key in EVAL   29   ENTER (you will find EVAL in the MISC submenu of the MATH menu). Use this method to find the value of y when $x = .45$ and 611.

(b) Enter two more equations in the equation memory: $y_2 = x^3 - 6$ and $y_3 = 5x + 2$. Now when you key in EVAL   29   ENTER, the value of all three functions at $x = 29$ will be displayed as a list, which you can scroll through by using the arrow keys.

(c) In the equation memory, turn off y_2, and return to the home screen. If you use EVAL now, what happens?

4. CUSTOM MENU ON TI-85/86

If you press CUSTOM, the calculator displays a menu that may be blank. You may fill in any entries you want, so that frequently used operations (such as EVAL and FRAC) may be accessed quickly from the CUSTOM menu, rather than searching for them in varius menus and submenus. Check your instruction manual for details.

 EXERCISES 3.1

In Exercises 1–4, determine whether or not the given table could possibly be a table of values of a function. Give reasons for your answer.

1.

Input	-2	0	3	1	-5
Output	2	3	-2.5	2	14

2.

Input	-5	3	0	-3	5
Output	7	3	0	5	-3

3.

Input	-5	1	3	-5	7
Output	0	2	4	6	8

4.

Input	1	-1	2	-2	3
Output	1	-2	± 5	-6	8

Exercises 5–10 deal with the greatest integer function of Example 4, which is given by the equation $y = [x]$. Compute the following values of the function:

5. $[6.75]$ **6.** $[.75]$

7. $[-4/3]$ **8.** $[5/3]$

9. $[-16.0001]$

10. Does the equation $y = [x]$ define x as a function of y? Give reasons for your answer.

In Exercises 11–18, determine whether the equation defines y as a function of x or defines x as a function of y.

11. $y = 3x^2 - 12$

12. $y = 2x^4 + 3x^2 - 2$

13. $y^2 = 4x + 1$

14. $5x - 4y^4 + 64 = 0$

15. $3x + 2y = 12$

16. $y - 4x^3 - 14 = 0$

17. $x^2 + y^2 = 9$

18. $y^2 - 3x^4 + 8 = 0$

In Exercises 19–22, each equation defines y as a function of x. Create a table that shows the values of the function for the given values of x.

19. $y = x^2 + x - 4$; $x = -2, -1.5, -1, \ldots, 3, 3.5, 4$.

20. $y = x^3 - x^2 + 4x + 1$; $x = 3, 3.1, 3.2, \ldots, 3.9, 4$

21. $y = \sqrt{4 - x^2}$; $x = -2, -1.2, -.04, .04, 1.2, 2$

22. $y = |x^2 - 5|$; $x = -8, -6, \ldots, 8, 10, 12$

Exercises 23–26 refer to the tax law in Example 1.

23. Find the output (tax amount) that is produced by each of the following inputs (incomes):

$500	$1509	$3754
$6783	$12,500	$55,342

24. Find four different numbers in the domain of this function that produce the same output (number in the range).

25. Explain why your answer in Exercise 24 does *not* contradict the definition of a function (in the box on page 127).

26. Is it possible to do Exercise 24 if all four numbers in the domain are required to be greater than 4500? Why or why not?

27. The amount of postage required to mail a first-class letter is determined by its weight. In this situation, is weight a function of postage? Or vice versa? Or both?

28. Could the following statement ever be the rule of a function?

> For input *x*, the output is the
> number whose square is *x*.

Why or why not? If there is a function with this rule, what is its domain and range?

29. (a) Use the following chart to make two tables of values (one for an average man and one for an average woman) in which the inputs are the number of drinks per hour and the outputs are the corresponding blood alcohol contents.*

Blood alcohol content

A look at the number of drinks consumed and blood alcohol content in one hour under optimum conditions:

(b) Does each of these tables define a function? If so, what are the domain and range of each function? [Remember that you can have part of a drink.]

30. The prime rate is the rate that large banks charge their best corporate customers for loans. The graph shows how the prime rate charged by a particular bank has varied in recent years.[†] Answer the following questions by reading the graph as best you can.

(a) What was prime rate in January 1997? In January 2002? In mid-2000?

(b) In what time periods was the prime rate 6%?

(c) On the basis of the data provided by this graph, can the prime rate be considered a function of time? Can time be considered a function of the prime rate?

31. Find an equation that expresses the area *A* of a circle as a function of its
   (a) radius *r*   (b) diameter *d*

32. Find an equation that expresses the area of a square as a function of its
   (a) side *x*   (b) diagonal *d*

33. A box with a square base of side *x* is four times higher than it is wide. Express the volume *V* of the box as a function of *x*.

34. The surface area of a cylindrical can of radius *r* and height *h* is $2\pi r^2 + 2\pi rh$. If the can is twice as high as the diameter of its top, express its surface area *S* as a function of *r*.

---

*National Highway Traffic Safety Administration. Art by AP/Amy Kranz.

[†]Federal Reserve Board.

35. Suppose you drop a rock from the top of a 400-foot-high building. Express the distance D from the rock to the ground as a function of time t. What is the range of this function? [*Hint:* See Example 3.]

36. A bicycle factory has weekly fixed costs of $26,000. In addition, the material and labor costs for each bicycle are $125. Express the total weekly cost C as a function of the number x of bicycles that are made.

37. The table below shows the percentage of single-parent families in various years.*

Year	Percent
1960	12.8
1970	13.2
1980	17.5
1990	20.8
2000	23.2

(a) The equation

$$y = (2.0284 \times 10^{-7})x^4 - (2.0237 \times 10^{-5})x^3$$
$$+ (6.584 \times 10^{-4})x^2 - .00434x + .1282,$$

in which $x = 0$ corresponds to 1960, defines y as a function of x. Make a table of values that includes the x-values corresponding to the years in the Census Bureau table.

(b) How do the values of y in your table compare with the percentages in the Census Bureau table? Does this equation seem to provide a reasonable model of the Census Bureau data?

(c) Use the equation to estimate the percentage of single-parent families in 1995 and 2005.

(d) Assuming that this model remains reasonably accurate, in what year will 30% of families be single-parent families?

38. The table shows the amount spent on student scholarships (in millions of dollars) by Oberlin College in recent years. [1995 indicates the school year 1995–96, and so on.]

Year	Scholarships
1995	19.8
1996	22.0
1997	25.7
1998	27.5
1999	28.7
2000	31.1
2001	34.3

(a) Use linear regression to find an equation that expresses the amount of scholarships y as a function of the year x, with $x = 0$ corresponding to 1995.

(b) Assuming that the function in part (a) remains accurate, estimate the amount spent on scholarships in 2004.

Use the following figures for Exercises 39–44. Each of the graphs in the figure defines a function, as in Example 11.

(a)

(b)

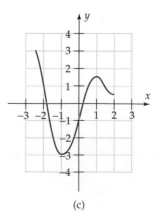

(c)

---

*U.S. Census Bureau.

39. State the domain and range of the function defined by graph (a).

40. State the output (number in the range) that the function of Exercise 39 produces from the following inputs (numbers in the domain): $-2, -1, 0, 1$.

41. State the domain and range of the function defined by graph (b).

42. State the output (number in the range) that the function of Exercise 41 produces from the following inputs (numbers in the domain): $-2, 0, 1, 2.5, -1.5$.

43. State the domain and range of the function defined by graph (c).

44. State the output (number in the range) that the function of Exercise 43 produces from the following inputs (numbers in the domain): $-2, -1, 0, 1/2, 1$.

45. Explain why none of the graphs in the figure below defines a function according to the procedure in Example 11. What goes wrong?

Thinkers

46. Consider the function whose rule uses a calculator as follows: "Press COS, and then press LN; then enter a number in the domain, and press ENTER."[*] Experiment with this function, then answer the following questions. You may not be able to prove your answers—just make the best estimate you can based on the evidence from your experiments.

    (a) What is the largest set of real numbers that could be used for the domain of this function? [If applying the rule to a number produces an error message or a complex number, that number cannot be in the domain.]

    (b) Using the domain in part (a), what is the range of this function?

47. Do Exercise 46 for the function whose rule is "Press 10^x, and then press TAN; then enter a number in the domain, and press ENTER."

48. The *integer part* function has the set of all real numbers (written as decimals) as its domain. The rule is "For each input number, the output is the part of the number to the left of the decimal point." For instance, the input 37.986 produces the output 37, and the input -1.5 produces the output -1. On most calculators, the integer part function is denoted "iPart." On calculators that use "Intg" or "Floor" for the greatest integer function, the integer part function is denoted by "INT."

    (a) For each nonnegative real number input, explain why both the integer part function and the greatest integer function [Example 5] produce the same output.

    (b) For which negative numbers do the two functions produce the same output?

    (c) For which negative numbers do the two functions produce different outputs?

(a)   (b)

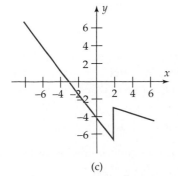

(c)

---

[*]You don't need to know what these keys mean to do this exercise.

3.2   Functional Notation

Functional notation is a convenient shorthand language that facilitates the analysis of mathematical problems involving functions. It arises from real-life situations, such as the following.

EXAMPLE 1

In Section 3.1, we saw that the 2001 Louisiana state income tax rates (for a single person with one exemption and no deductions) were as follows.

Income		Tax
At Least	But Less Than	
0	$4,500	0
$4,500	$10,000	2% of amount over $4,500
$10,000	$50,000	$110 + 4% of amount over $10,000
$50,000		$1710 + 6% of amount over $50,000

Let I denote income, and write $T(I)$ (read "T of I") to denote the amount of tax on income I. In this shorthand language, $T(7500)$ denotes "the tax on an income of 7500." The sentence "The tax on an income of $7500 is $60" is abbreviated as $T(7500) = 60$. Similarly, $T(25,000) = 710$ says that the tax on an income of $25,000 is 710. There is nothing that forces us to use the letters T and I here:

Any choice of letters will do, provided that we make clear what is meant by these letters. ■

EXAMPLE 2

Recall that a falling rock travels $16t^2$ feet after t seconds. Let $d(t)$ stand for the phrase "the distance the rock has traveled after t seconds." Then the sentence "The distance the rock has traveled after t seconds is $16t^2$ feet" can be abbreviated as $d(t) = 16t^2$. For instance,

$$d(1) = 16 \cdot 1^2 = 16$$

means "the distance the rock has traveled after 1 second is 16 feet," and

$$d(4) = 16 \cdot 4^2 = 256$$

means "the distance the rock has traveled after 4 seconds is 256 feet." ■

Functional notation is easily adapted to mathematical settings, in which the particulars of time, distance, etc., are not mentioned. Suppose a function is given. Denote the function by f and let x denote a number in the domain. Then

$f(x)$ **denotes the output produced by input** x.

CAUTION

The parentheses in $d(t)$ do *not* denote multiplication as in the algebraic equation $3(a + b) = 3a + 3b$. The entire symbol $d(t)$ is part of a *shorthand language.* In particular

$d(1 + 4)$ is *not* equal to $d(1) + d(4)$.

We saw above that $d(1) = 16$ and $d(4) = 256$, so $d(1) + d(4) = 16 + 256 = 272$. But $d(1 + 4)$ is "the distance traveled after $1 + 4$ seconds," that is, the distance after 5 seconds, namely, $16 \cdot 5^2 = 400$. In general,

Functional notation is a convenient shorthand for phrases and sentences in the English language. It is *not* the same as ordinary algebraic notation.

For example, $f(6)$ is the output produced by the input 6. The sentence

"*y* is the output produced by input *x* according to the rule of the function *f*"

is abbreviated

$$y = f(x),$$

which is read "*y* equals *f* of *x*." The output $f(x)$ is sometimes called the **value** of the function *f* at *x*.

In actual practice, functions are seldom presented in the style of domain, rule, range, as they have been here. Usually, you will be given a phrase such as "the function $f(x) = \sqrt{x^2 + 1}$." This should be understood as a set of directions:

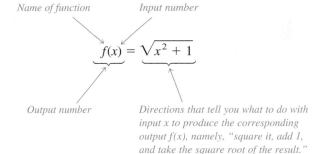

Name of function   *Input number*

$$f(x) = \sqrt{x^2 + 1}$$

Output number   *Directions that tell you what to do with input x to produce the corresponding output f(x), namely, "square it, add 1, and take the square root of the result."*

For example, to find $f(3)$, the output of the function *f* for input 3, simply replace *x* by 3 in the formula:

$$f(x) = \sqrt{x^2 + 1}$$
$$f(3) = \sqrt{3^2 + 1} = \sqrt{10}.$$

Similarly, replacing *x* by -5 and 0 shows that

$$f(-5) = \sqrt{(-5)^2 + 1} = \sqrt{26} \qquad \text{and} \qquad f(0) = \sqrt{0^2 + 1} = 1.$$

EXAMPLE 3

The expression

$$h(x) = \frac{x^2 + 5}{x - 1}$$

defines the function *h* whose rule is

For input *x*, the output is the number $\dfrac{x^2 + 5}{x - 1}$.

Find each of the following:

$$h(\sqrt{3}), \qquad h(-2), \qquad h(-a), \qquad h(r^2 + 3), \qquad h(\sqrt{c + 2}).$$

SOLUTION   To find $h(\sqrt{3})$ and $h(-2)$, replace *x* by $\sqrt{3}$ and -2, respectively, in the rule of *h*:

$$h(\sqrt{3}) = \frac{(\sqrt{3})^2 + 5}{\sqrt{3} - 1} = \frac{8}{\sqrt{3} - 1} \qquad \text{and} \qquad h(-2) = \frac{(-2)^2 + 5}{-2 - 1} = -3.$$

The value of the function h at any quantity, such as $-a$, $r^2 + 3$, etc., can be found by using the same procedure: *Replace x in the formula for $h(x)$ by that quantity:*

$$h(-a) = \frac{(-a)^2 + 5}{-a - 1} = \frac{a^2 + 5}{-a - 1}$$

$$h(r^2 + 3) = \frac{(r^2 + 3)^2 + 5}{(r^2 + 3) - 1} = \frac{r^4 + 6r^2 + 9 + 5}{r^2 + 2} = \frac{r^4 + 6r^2 + 14}{r^2 + 2}$$

$$h(\sqrt{c + 2}) = \frac{(\sqrt{c + 2})^2 + 5}{\sqrt{c + 2} - 1} = \frac{c + 2 + 5}{\sqrt{c + 2} - 1} = \frac{c + 7}{\sqrt{c + 2} - 1}. \quad \blacksquare$$

When functional notation is used in expressions such as $f(-x)$ or $f(x + h)$, the same basic rule applies: Replace x in the formula by the *entire* expression in parentheses.

TECHNOLOGY TIP

One way to evaluate a function $f(x)$ is to enter its rule as an equation $y = f(x)$ in the equation memory and use TABLE or (on TI-85/86) EVAL; see Example 9 or Calculator Investigation 3 in Section 3.1.

EXAMPLE 4

If $f(x) = x^2 + x - 2$, then

$$f(-x) = (-x)^2 + (-x) - 2 = x^2 - x - 2.$$

Note that in this case, $f(-x)$ is *not* the same as $-f(x)$, because $-f(x)$ is the negative of the number $f(x)$, that is,

$$-f(x) = -(x^2 + x - 2) = -x^2 - x + 2. \quad \blacksquare$$

EXAMPLE 5

If $f(x) = x^2 - x + 2$ and $h \neq 0$, find

(a) $f(x + h)$ (b) $f(x + h) - f(x)$ (c) $\dfrac{f(x + h) - f(x)}{h}$.

SOLUTION

(a) Replace x by $x + h$ in the rule of the function:

$$f(x) = x^2 - x + 2$$
$$f(x + h) = (x + h)^2 - (x + h) + 2 = x^2 + 2xh + h^2 - x - h + 2.$$

(b) By part (a),

$$f(x + h) - f(x) = [(x + h)^2 - (x + h) + 2] - [x^2 - x + 2]$$
$$= [x^2 + 2xh + h^2 - x - h + 2] - [x^2 - x + 2]$$
$$= x^2 + 2xh + h^2 - x - h + 2 - x^2 + x - 2$$
$$= 2xh + h^2 - h.$$

(c) By part (b), we have

$$\frac{f(x + h) - f(x)}{h} = \frac{2xh + h^2 - h}{h} = \frac{h(2x + h - 1)}{h} = 2x + h - 1. \quad \blacksquare$$

If f is a function, then the quantity $\dfrac{f(x + h) - f(x)}{h}$, as in Example 5($c$), is called the **difference quotient** of f. Difference quotients, whose significance is explained in Section 3.6, play an important role in calculus.

EXAMPLE 6

Compute and simplify the difference quotient for the function $f(x) = x^3 + 2x$.

SOLUTION

$$\frac{f(x + h) - f(x)}{h} = \frac{\overbrace{[(x + h)^3 + 2(x + h)]}^{f(x+h)} - \overbrace{[x^3 + 2x]}^{f(x)}}{h}$$

$$= \frac{[x^3 + 3x^2h + 3xh^2 + h^3 + 2x + 2h] - [x^3 + 2x]}{h}$$

$$= \frac{x^3 + 3x^2h + 3xh^2 + h^3 + 2x + 2h - x^3 - 2x}{h}$$

$$= \frac{3x^2h + 3xh^2 + h^3 + 2h}{h}$$

$$= \frac{h(3x^2 + 3xh + h^2 + 2)}{h} = 3x^2 + 3xh + h^2 + 2. \quad \blacksquare$$

As the preceding examples illustrate, functional notation is a specialized shorthand language. Treating it as ordinary algebraic notation may lead to mistakes.

EXAMPLE 7

⊘ CAUTION

Common Mistakes with Functional Notation

Each of the following statements may be FALSE:

1. $f(a + b) = f(a) + f(b)$
2. $f(a - b) = f(a) - f(b)$
3. $f(ab) = f(a)f(b)$
4. $f(ab) = af(b)$
5. $f(ab) = f(a)b$

Here are examples of three of the errors listed in the Caution box.

1. If $f(x) = x^2$, then
$$f(3 + 2) = f(5) = 5^2 = 25.$$
But
$$f(3) + f(2) = 3^2 + 2^2 = 9 + 4 = 13.$$
So $f(3 + 2) \neq f(3) + f(2)$.

3. If $f(x) = x + 7$, then
$$f(3 \cdot 4) = f(12) = 12 + 7 = 19.$$
But
$$f(3)f(4) = (3 + 7)(4 + 7) = 10 \cdot 11 = 110.$$
So $f(3 \cdot 4) \neq f(3)f(4)$.

5. If $f(x) = x^2 + 1$, then

$$f(2 \cdot 3) = (2 \cdot 3)^2 + 1 = 36 + 1 = 37.$$

But

$$f(2) \cdot 3 = (2^2 + 1)3 = 5 \cdot 3 = 15.$$

So $f(2 \cdot 3) \neq f(2) \cdot 3.$ ■

⠃⠃ DOMAINS

When the rule of a function is given by a formula, as in Examples 3–7, its domain (set of inputs) is determined by the following convention.

Domain Convention

> Unless specific information to the contrary is given, the domain of a function f includes every real number (input) for which the rule of the function produces a real number as output.

Thus, the domain of a polynomial function such as $f(x) = x^3 - 4x + 1$ is the set of all real numbers, since $f(x)$ is defined for every value of x. In cases in which applying the rule of a function leads to division by zero or to the square root of a negative number, however, the domain may not consist of all real numbers.

EXAMPLE 8

Find the domain of the function given by

(a) $k(x) = \dfrac{x^2 - 6x}{x - 1}$ (b) $f(u) = \sqrt{u + 2}$

SOLUTION

(a) When $x = 1$, the denominator of $\dfrac{x^2 - 6x}{x - 1}$ is 0, and the fraction is not defined. When $x \neq 1$, however, the denominator is nonzero and the fraction *is* defined. Therefore, the domain of the function k consists of all real numbers *except* 1.

(b) Since negative numbers do not have real square roots, $\sqrt{u + 2}$ is a real number only when $u + 2 \geq 0$, that is, when $u \geq -2$. Therefore, the domain of f consists of all real numbers greater than or equal to -2, that is, the interval $[-2, \infty)$.   ■

EXAMPLE 9

A **piecewise-defined** function is one whose rule includes several formulas, such as

$$f(x) = \begin{cases} 2x + 3 & \text{if } x < 4 \\ x^2 - 1 & \text{if } 4 \leq x \leq 10. \end{cases}$$

Find each of the following.

(a) $f(-5)$ (b) $f(8)$ (c) $f(k)$

(d) The domain of f.

SOLUTION

(a) Since $-5 < 4$, the first part of the rule applies:

$$f(-5) = 2(-5) + 3 = -7.$$

(b) Since 8 is between 4 and 10, the second part of the rule applies:

$$f(8) = 8^2 - 1 = 63.$$

(c) We cannot find $f(k)$ unless we know whether $k < 4$ or $4 \le k \le 10$.

(d) The rule of f gives no directions when $x > 10$, so the domain of f consists of all real numbers x with $x \le 10$, that is, $(-\infty, 10]$. ■

EXAMPLE 10

Use Example 1 to write the rule of the piecewise-defined function T that gives the Louisiana state income tax $T(x)$ on an income of x dollars.

SOLUTION   By translating the information in the table in Example 1 into functional notation, we obtain

$$T(x) = \begin{cases} 0 & \text{if } 0 \le x < 4500 \\ .02(x - 4500) & \text{if } 4500 \le x < 10{,}000 \\ 110 + .04(x - 10{,}000) & \text{if } 10{,}000 \le x < 50{,}000 \\ 1710 + .06(x - 50{,}000) & \text{if } x \ge 50{,}000. \end{cases}$$ ■

■■ APPLICATIONS

The domain convention does not always apply when dealing with applications. Consider, for example, the distance function for falling objects, $d(t) = 16t^2$ (see Example 2). Since t represents time, only nonnegative values of t make sense here, even though the rule of the function is defined for all values of t. Analogous comments apply to other applications.

> **A real-life situation may lead to a function whose domain does not include all the numbers for which the rule of the function is defined.**

EXAMPLE 11

A glassware factory has fixed expenses (mortgage, taxes, machinery, etc.) of $12,000 per week. It costs 80 cents to make one cup (labor, materials, shipping). A cup sells for $1.95. At most 18,000 cups can be manufactured and sold each week.

(a) Express the weekly revenue as a function of the number x of cups made.

(b) Express the weekly costs as a function of x.

(c) Find the domain and the rule of the weekly profit function.

SOLUTION

(a) If $R(x)$ is the weekly revenue from selling x cups, then

$$R(x) = (\text{price per cup}) \times (\text{number sold})$$

$$R(x) = 1.95x.$$

(b) If $C(x)$ is the weekly cost of manufacturing x cups, then

$$C(x) = (\text{cost per cup}) \times (\text{number sold}) + (\text{fixed expenses})$$

$$C(x) = .80x + 12,000.$$

(c) If $P(x)$ is the weekly profit from selling x cups, then

$$P(x) = \text{Revenue} - \text{Cost}$$

$$P(x) = R(x) - C(x)$$

$$P(x) = 1.95x - (.80x + 12,000) = 1.95x - .80x - 12,000$$

$$P(x) = 1.15x - 12,000.$$

Although this rule is defined for all real numbers x, the domain of the function P consists of the possible number of cups that can be made each week. Since you can make only whole cups and the maximum production is 18,000, the domain of P consists of all integers from 0 to 18,000.   ■

EXAMPLE 12

Let P be the profit function in Example 11.

(a) What is the profit from selling 5000 cups? From 14,000 cups?

(b) What is the break-even point?

SOLUTION

(a) We evaluate the function $P(x) = 1.15x - 12,000$ at the required values of x:

$$P(5000) = 1.15(5000) - 12,000 = -\$6250$$

$$P(14,000) = 1.15(14,000) - 12,000 = \$4100.$$

Thus, sales of 5000 cups produce a loss of $6250, while sales of 14,000 produce a profit of $4100.

(b) The break-even point occurs when revenue equals costs (that is, when profit is 0). So we set $P(x) = 0$ and solve for x:

$$1.15x - 12,000 = 0$$

$$1.15x = 12,000$$

$$x = \frac{12,000}{1.15} \approx 10,434.78.$$

Thus, the break-even point occurs between 10,434 and 10,435 cups. There is a slight loss from selling 10,434 cups and a slight profit from selling 10,435.   ■

EXERCISES 3.2

In Exercises 1 and 2, find the indicated values of the function **by hand** *and by using the table feature of a calculator (or the EVAL key on TI-85/86). If your answers do not agree with each other or with those at the back of the book, you are either making algebraic mistakes or incorrectly entering the function in the equation memory.*

1. $f(x) = \dfrac{x - 3}{x^2 + 4}$

   (a) $f(-1)$ (b) $f(0)$ (c) $f(1)$ (d) $f(2)$ (e) $f(3)$

2. $g(x) = \sqrt{x + 4} - 2$

   (a) $g(-2)$ (b) $g(0)$ (c) $g(4)$ (d) $g(5)$ (e) $g(12)$

Exercises 3–22 refer to these three functions:

$$f(x) = \sqrt{x + 3} - x + 1$$

$$g(t) = t^2 - 1$$

$$h(x) = x^2 + \frac{1}{x} + 2.$$

In each case, find the indicated value of the function.

3. $f(0)$ **4.** $f(1)$

5. $f(\sqrt{2})$ **6.** $f(\sqrt{2} - 1)$

7. $f(-2)$ **8.** $f(-3/2)$

9. $h(3)$ **10.** $h(-4)$

11. $h(3/2)$ **12.** $h(\pi + 1)$

13. $h(a + k)$ **14.** $h(-x)$

15. $h(2 - x)$ **16.** $h(x - 3)$

17. $g(3)$ **18.** $g(-2)$

19. $g(s + 1)$ **20.** $g(1 - r)$

21. $g(-t)$ **22.** $g(t + h)$

In Exercises 23–26, compute:

(a) $f(r)$ (b) $f(r) - f(x)$ (c) $\dfrac{f(r) - f(x)}{r - x}$

In part (c), assume that $r \neq x$ and simplify your answer.
Example: If $f(x) = x^2$, then

$$\frac{f(r) - f(x)}{r - x} = \frac{r^2 - x^2}{r - x} = \frac{(r + x)(r - x)}{r - x} = r + x.$$

23. $f(x) = 3x + 7$ **24.** $f(x) = x^3$

25. $f(x) = x - x^2$ **26.** $f(x) = x^2 + 1$

In Exercises 27–36, assume $h \neq 0$. Compute and simplify the difference quotient

$$\frac{f(x + h) - f(x)}{h}$$

27. $f(x) = x + 1$ **28.** $f(x) = -10x$

29. $f(x) = 3x + 7$ **30.** $f(x) = x^2$

31. $f(x) = x - x^2$ **32.** $f(x) = x^3$

33. $f(x) = \sqrt{x}$ **34.** $f(x) = 1/x$

35. $f(x) = x^2 + 3$ **36.** $f(x) = x^2 + 2x + 1$

37. In each part, compute $f(a)$, $f(b)$, and $f(a + b)$, and determine whether the statement "$f(a + b) = f(a) + f(b)$" is true or false for the given function.
   (a) $f(x) = x^2$ (b) $f(x) = 3x$

38. In each part, compute $g(a)$, $g(b)$, and $g(ab)$, and determine whether the satement "$g(ab) = g(a) \cdot g(b)$" is true or false for the given function.
   (a) $g(x) = x^3$ (b) $g(x) = 5x$

39. If $f(x) = x^3 + cx^2 + 4x - 1$ for some constant c and $f(1) = 2$, find c. [*Hint:* Use the rule of f to compute $f(1)$.]

40. If $f(x) = \dfrac{dx - 5}{x - 3}$ and $f(4) = 3$, find d.

41. The rule of the function f is given by the graph, as in Example 11 of Section 3.1. Find
   (a) The domain of f
   (b) The range of f
   (c) $f(-3)$
   (d) $f(-1)$
   (e) $f(1)$
   (f) $f(2)$

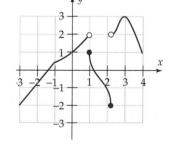

42. The rule of the function g is given by the graph, as in Example 11 of Section 3.1. Find
   (a) The domain of g
   (b) The range of g
   (c) $g(-3)$
   (d) $g(-1)$
   (e) $g(1)$
   (f) $g(4)$

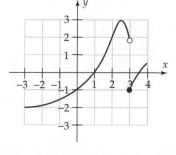

43. If $f(x) = \begin{cases} x^2 + 2x & \text{if } x < 2 \\ 3x - 5 & \text{if } 2 \le x \le 20 \end{cases}$ find

(a) The domain of f

(b) $f(-3)$ (c) $f(-1)$ (d) $f(2)$ (e) $f(7/3)$

44. If $g(x) = \begin{cases} 2x - 3 & \text{if } x < -1 \\ |x| - 5 & \text{if } -1 \le x \le 2 \\ x^2 & \text{if } x > 2 \end{cases}$ find

(a) The domain of g

(b) $g(-2.5)$ (c) $g(-1)$ (d) $g(2)$ (e) $g(4)$

In Exercises 45–58, determine the domain of the function according to the usual convention.

45. $f(x) = x^2$

46. $g(x) = \dfrac{1}{x^2} + 2$

47. $h(t) = |t| - 1$

48. $k(u) = \sqrt{u}$

49. $k(x) = |x| + \sqrt{x - 1}$

50. $h(x) = \sqrt{(x + 1)^2}$

51. $g(u) = \dfrac{|u|}{u}$

52. $h(x) = \dfrac{\sqrt{x - 1}}{x^2 - 1}$

53. $g(y) = [-y]$

54. $f(t) = \sqrt{-t}$

55. $g(u) = \dfrac{u^2 + 1}{u^2 - u - 6}$

56. $f(t) = \sqrt{4 - t^2}$

57. $f(x) = -\sqrt{9 - (x - 9)^2}$ **58.** $f(x) = \sqrt{-x} + \dfrac{2}{x + 1}$

59. Give an example of two different functions f and g that have all of the following properties:

$$f(-1) = 1 = g(-1) \quad \text{and} \quad f(0) = 0 = g(0)$$

$$\text{and} \quad f(1) = 1 = g(1).$$

60. Give an example of a function g with the property that $g(x) = g(-x)$ for every real number x.

In Exercises 61–64, find the values of x for which f(x) = g(x).

61. $f(x) = 2x^2 + 4x - 3; \quad g(x) = x^2 + 12x + 7$

62. $f(x) = 2x^2 + 12x - 14; \quad g(x) = 7x - 2$

63. $f(x) = 3x^2 - x + 5; \quad g(x) = x^2 - 2x + 26$

64. $f(x) = 2x^2 - x + 1; \quad g(x) = x^2 - 4x + 4$

In Exercises 65–67, the rule of a function f is given. Write an algebraic formula for f(x).

65. Double the input, subtract 5, and take the square root of the result.

66. Square the input, multiply by 3, and subtract the result from 8.

67. Cube the input, add 6, and divide the result by 5.

68. A potato chip factory has a daily overhead from salaries and building costs of $1800. The cost of ingredients and packaging to produce a pound of potato chips is 50¢. A pound of potato chips sells for $1.20. Show that the factory's daily profit is a function of the number of pounds of potato chips sold, and find the rule of this function. (Assume that the factory sells all the potato chips it produces each day.)

69. Jack and Jill are salespersons in the suit department of a clothing store. Jack is paid $200 per week plus $5 for each suit he sells, whereas Jill is paid $10 for every suit she sells.

(a) Let $f(x)$ denote Jack's weekly income, and let $g(x)$ denote Jill's weekly income from selling x suits. Find the rules of the functions f and g.

(b) Use algebra or a table to find $f(20)$ and $g(20)$, $f(35)$ and $g(35)$, $f(50)$ and $g(50)$.

(c) If Jack sells 50 suits a week, how many must Jill sell to have the same income as Jack?

70. A person who needs crutches can determine the correct length as follows: a 50-inch-tall person needs a 38-inch-long crutch. For each additional inch in the person's height, add .72 inch to the crutch length.

(a) If a person is y inches taller than 50 inches, write an expression for the proper crutch length.

(b) Write the rule of a function f such that $f(x)$ is the proper crutch length (in inches) for a person who is x inches tall. [*Hint:* Replace y in your answer to part (a) with an expression in x. How are x and y related?]

71. The table shows the 2002 federal income tax rates for a single person.

Taxable Income Over	But Not Over	Tax
0	$6,000	10% of income
$6,000	$27,950	$600 + 15% of amount over $6,000
$27,950	$67,700	$3892.50 + 27% of amount over $27,950
$67,700	$141,250	$14,625 + 30% of amount over $67,700
$141,250	$307,050	$36,690 + 35% of amount over $141,250
$307,050		$94,720 + 38.6% of amount over $307,050

(a) Write the rule of a piecewise-defined function T such that $T(x)$ is the tax due on a taxable income of x dollars.

(b) Find $T(24,000)$, $T(35,000)$, and $T(100,000)$.

72. Suppose a car travels at a constant rate of 55 mph for 2 hours and travels at 45 mph thereafter. Show that distance traveled is a function of time, and find the rule of the function.

73. A man walks for 45 minutes at a rate of 3 mph, then jogs for 75 minutes at a rate of 5 mph, then sits and rests for 30 minutes, and finally walks for $1\frac{1}{2}$ hours. Find the rule of the function that expresses his distance traveled as a function of time. [*Caution:* Don't mix up the units of time; use either minutes or hours, not both.]

74. Suppose that the width and height of the box in the figure are equal and that the sum of the length and the girth is 108 (the maximum size allowed by the post office).

(a) Express the length y as a function of the width x. [*Hint:* Use the girth.]
(b) Express the volume V of the box as a function of the width x. [*Hint:* Find a formula for the volume and use part (a).]

75. A rectangular region of 6000 square feet is to be fenced in on three sides with fencing costing $3.75 per foot and on the fourth side with fencing costing $2.00 per foot. Express the cost of the fence as a function of the length x of the fourth side.

76. A box with a square base measuring $t \times t$ ft is to be made of three kinds of wood. The cost of the wood for the base is 85¢ per square foot; the wood for the sides costs 50¢ per square foot, and the wood for the top $1.15 per square foot. The volume of the box is to be 10 cubic feet. Express the total cost of the box as a function of the length t.

77. Average tuition and fees in private four-year colleges in recent years were as follows.[*]

Year	Tuition and Fees	Year	Tuition and Fees
1995	$12,432	1998	$14,709
1996	$12,823	1999	$15,380
1997	$13,664	2000	$16,332

(a) Use linear regression to find the rule of a function f that gives the approximate average tuition in year x, where $x = 0$ corresponds to 1990.
(b) Find $f(6)$, $f(8)$, and $f(10)$. How do they compare with the actual figures?
(c) Use f to estimate tuition in 2003.

78. The number of U.S. commercial radio stations whose primary format is top 40 hits has been increasing in recent years, as shown in the table.[†]

Year	Number of Stations
1995	318
1996	333
1997	358
1998	379
1999	401
2001	468

(a) Use linear regression to find the rule of a function g that gives the number of top-40 stations in year x, where $x = 0$ corresponds to 1990.
(b) Find $g(7)$ and $g(11)$. How do they compare with the actual figures?
(c) Data for the year 2000 is missing. Estimate the number of stations in 2000.
(d) Assuming that this function remains accurate, estimate the number of stations in 2005.

---

[*]The College Board.
[†]World Almanac and Book of Facts 2002.

3.3   Graphs of Functions

The graph of a function f is the graph of the equation $y = f(x)$. Hence

The graph of the function f consists of the points $(x, f(x))$ for every number x in the domain of f.

When the rule of a function is given by an algebraic formula, the graph is easily obtained with technology. However, machine-generated graphs can sometimes be incomplete or misleading. So the emphasis here is on using your algebraic knowledge *before* reaching for a calculator. Doing so will often tell you that a calculator is inappropriate or help you to interpret screen images when a calculator is used.

Some functions appear so frequently that you should memorize the shapes of their graphs, which are easily obtained by hand-graphing or by using technology. Regardless of how you first find these graphs, you should be able to reproduce them without looking them up or resorting to technology. These basic graphs are summarized in the **catalog of functions** at the end of this section. A title on an example (such as "linear function" in Example 1) indicates a function that is in the catalog.

EXAMPLE 1

Linear Functions   The graph of a function of the form

$$f(x) = mx + b \qquad \text{(with m and b constants)}$$

is the graph of the equation $y = mx + b$. As we saw in Section 1.4, the graph is a straight line with slope m and y-intercept b that can easily be obtained by hand. Some typical linear functions are graphed in Figure 3–8 (several of them have special names). On a calculator, these graphs will usually look a bit bumpy and jagged.   ■

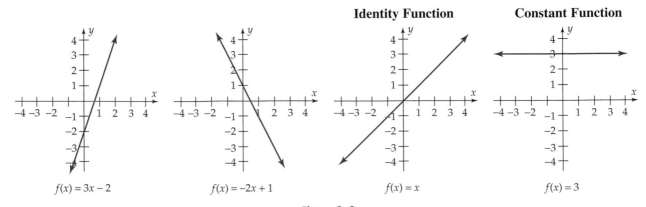

$f(x) = 3x - 2$ $f(x) = -2x + 1$ Identity Function $f(x) = x$ Constant Function $f(x) = 3$

Figure 3–8

EXAMPLE 2

Square and Cube Functions   Figure 3–9 shows the graphs of $f(x) = x^2$ and $g(x) = x^3$. They can be obtained by plotting points (as was done for f in Example 1 of Section 2.1) or by using technology.   ■

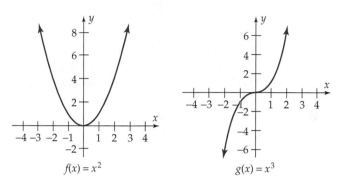

$f(x) = x^2$ $g(x) = x^3$

Figure 3–9

Figure 3–10

EXAMPLE 3

Square Root Function   The graph of $f(x) = \sqrt{x}$ in Figure 3–10 is easily found. Why does it lie entirely in the first quadrant?   ■

▦ STEP FUNCTIONS

The greatest integer function $f(x) = [x]$ was introduced in Example 5 of Section 3.1. It can easily be graphed by hand, as in the next example.

EXAMPLE 4

Greatest Integer Function   The function $f(x) = [x]$ was introduced in Example 5 of Section 3.1. It can easily be graphed by hand by considering the values of the function between each two consecutive integers. For instance,

Figure 3–11

x	$-2 \le x < -1$	$-1 \le x < 0$	$0 \le x < 1$	$1 \le x < 2$	$2 \le x < 3$
$[x]$	-2	-1	0	1	2

Thus, between $x = -2$ and $x = -1$, the value of $f(x) = [x]$ is always -2, so the graph there is a horizontal line segment, all of whose points have second coordinate -2. The rest of the graph is obtained similarly (Figure 3–11). An open circle in Figure 3–11 indicates that the endpoint of the segment is *not* on the graph, whereas a closed circle indicates that the endpoint is on the graph.   ■

A function whose graph consists of horizontal line segments, such as Figure 3–11, is called a **step function.** Graphing step functions with reasonable accuracy on a calculator requires some care. Even then, some features of the graph might not be shown.

EXPLORATION

Graph the greatest integer function $f(x) = [x]$ on your calculator (see the Technology Tip on page 128). Does your graph look like Figure 3–11, or does it include vertical segments? Now change the graphing mode of your calculator to "dot" rather than "connected" (see the Technology Tip in the margin), and graph again. How does this graph compare with Figure 3–11? Can you tell from the graph which endpoints are included and which are excluded?

TECHNOLOGY TIP

Directions for switching to dot graphing mode are in Calculator Investigation 5 at the end of Section 2.1.

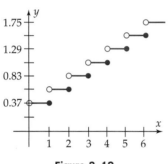

Figure 3–12

EXAMPLE 5

At this writing, first-class postage rates are 37¢ for the first ounce and 23¢ for each additional ounce or fraction thereof. Verify that the postage $P(x)$ for a letter weighing x ounces is given by $P(x) = .37 - .23[1 - x]$. For instance, the postage for a 2.5 ounce letter is

$$P(2.5) = .37 - .23[1 - 2.5] = .37 - .23(-2) = .83.$$

Although the rule of P makes sense for all real numbers, the domain of the function consists of positive numbers (why?). The graph of P is in Figure 3–12. ∎

PIECEWISE-DEFINED FUNCTIONS

Piecewise-defined functions were introduced in Example 9 of Section 3.2. Graphing them correctly requires some care.

EXAMPLE 6

The graph of the piecewise-defined function

$$f(x) = \begin{cases} x^2 & \text{if } x \le 1 \\ x + 2 & \text{if } 1 < x \le 4 \end{cases}$$

is made up of *parts* of two graphs, corresponding to the different parts of the rule of the function:

$x \le 1$	For these values of x, the graph of f coincides with the graph of $y = x^2$, which was sketched in Figure 3–9.
$1 < x \le 4$	For these values of x, the graph of f coincides with the graph of $y = x + 2$, which is a straight line.

Therefore, we must graph

$$y = x^2 \quad \text{when } x \le 1 \qquad \text{and} \qquad y = x + 2 \quad \text{when } 1 < x \le 4.$$

Combining these partial graphs produces the graph of f in Figure 3–13. ■

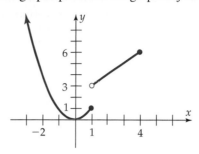

Figure 3–13

Piecewise-defined functions can be graphed on a calculator, provided that you use the correct syntax. Once again, however, the screen does not show which endpoints are included or excluded from the graph.

TECHNOLOGY TIP

Inequality symbols can be found in the following menu/submenu:

TI-83/86: TEST

TI-89: MATH/TEST

HP-39: MATH/TESTS

GRAPHING EXPLORATION

Use the Technology Tip in the margin and the directions here to graph the function f of Example 6 on a calculator. On TI-83/86 and HP-39, graph these two equations on the same screen:

$$y_1 = \frac{x^2}{(x \le 1)}$$

$$y_2 = \frac{(x + 2)}{(x > 1)(x \le 4)}.$$

On TI-89, graph the following equations on the same screen (the symbol | is on the keyboard; "and" is in the TESTS submenu of the MATH menu):

$$y_1 = x^2 | x \le 1$$

$$y_2 = x + 2 | x > 1 \quad \text{and} \quad x \le 4.$$

To graph f on Casio, with the viewing window of Figure 3–13, graph these equations on the same screen (including commas and square brackets):

$$y_1 = x^2, [-6, 1]$$

$$y_2 = x + 2, [1, 4].$$

How does your graph compare with Figure 3–13?

EXAMPLE 7

Absolute Value Function   The absolute value function $f(x) = |x|$ is also a piecewise-defined function, since by definition,

$$|x| = \begin{cases} x & \text{if } x \ge 0 \\ -x & \text{if } x < 0. \end{cases}$$

Its graph can be obtained by drawing the part of the line $y = x$ to the right of the origin and the part of the line $y = -x$ to the left of the origin (Figure 3–14) or by graphing $y = $ ABS x on a calculator (Figure 3–15).   ■

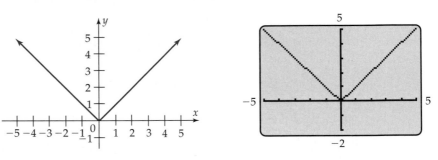

Figure 3–14   **Figure 3–15**

▋▋ LOCAL MAXIMA AND MINIMA

The graph of a function may include some peaks and valleys (Figure 3–16). A peak is not necessarily the highest point on the graph, but it is the highest point in its neighborhood. Similarly, a valley is the lowest point in the neighborhood but not necessarily the lowest point on the graph.

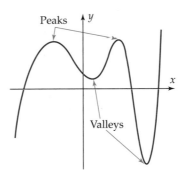

Figure 3–16

More formally, we say that a function f has a **local maximum** at $x = c$ if the graph of f has a peak at the point $(c, f(c))$. This means that all nearby points $(x, f(x))$ have smaller y-coordinates, that is,

$$f(x) \leq f(c) \quad \text{for all } x \text{ near } c.$$

Similarly, a function has a **local minimum** at $x = d$ provided that

$$f(x) \geq f(d) \quad \text{for all } x \text{ near } d.$$

In other words, the graph of f has a valley at $(d, f(d))$ because all nearby points $(x, f(x))$ have larger y-coordinates.

Calculus is usually needed to find the exact location of local maxima and minima (the plural forms of maximum and minimum). However, they can be accurately approximated by the maximum finder or minimum finder of a calculator.

EXAMPLE 8

The graph of $f(x) = x^3 - 1.8x^2 + x + 1$ in Figure 3–17 does not appear to have any local maxima or minima. However, if you use the trace feature to move along the flat segment to the right of the *y*-axis, you find that the *y*-coordinates increase, then decrease, then increase (try it!). To see what's really going on, we change viewing windows (Figure 3–18) and see that the function actually has a local maximum and a local minimum (Figure 3–19). The calculator's minimum finder shows that the local minimum occurs when $x \approx .7633$. ∎

Figure 3–17

Figure 3–18

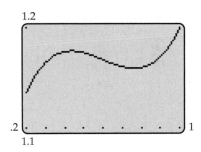

Figure 3–19

GRAPHING EXPLORATION

Graph the function in Example 8 in the viewing window of Figure 3–18. Use the maximum finder to approximate the location of the local maximum.

EXAMPLE 9

U.S. military troop levels (in millions) during 1980–2003 can be approximated by the function

$$g(x) = (2.063 \times 10^{-5})x^4 - (5.9 \times 10^{-4})x^3$$
$$- (4.88 \times 10^{-4})x^2 + .047x + 2.052,$$

where $x = 0$ corresponds to 1980. During this period, when were troop levels at their highest?

Figure 3–20

SOLUTION We graph the function and use the maximum finder to determine that the local maximum occurs when $x \approx 5.640$ and $g(5.640) \approx 2.217$, as shown in Figure 3–20. Therefore, troop levels reached their highest level (about 2,217,000) in late 1985. ∎

▪▪ INCREASING AND DECREASING FUNCTIONS

A function is said to be **increasing on an interval** if its graph always rises as you move from left to right over the interval. It is **decreasing on an interval** if its graph always falls as you move from left to right over the interval. A function is said to be **constant on an interval** if its graph is horizontal over the interval.

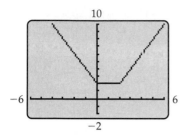

Figure 3–21

EXAMPLE 10

Figure 3–21 suggests that $f(x) = |x| + |x - 2|$ is decreasing on the interval $(-\infty, 0)$, increasing on $(2, \infty)$, and constant on $[0, 2]$. You can confirm that the function is actually constant between 0 and 2 by using the trace feature to move along the graph there (the y coordinates remain the same, as they should on a horizontal segment). For an algebraic proof that f is constant on $[0, 2]$, see Exercise 20. ∎

 CAUTION

A horizontal segment on a calculator graph does not always mean that the function is constant there. There may be **hidden behavior**, as was the case in Example 8. When in doubt, use the trace feature to see if the y-coordinates remain constant as you move along the "horizontal" segment, or change the viewing window.

EXAMPLE 11

Figure 3–22

On what (approximate) intervals is the function $g(x) = .5x^3 - 3x$ increasing or decreasing?

SOLUTION The (complete) graph of g in Figure 3–22 shows that g has a local maximum at P and a local minimum at Q. The maximum and minimum finders show that the approximate coordinates of P and Q are

$$P = (-1.4142, 2.8284) \qquad \text{and} \qquad Q = (1.4142, -2.8284).$$

Therefore, f is increasing on $(-\infty, -1.4142)$ and $(1.4142, \infty)$. It is decreasing on $(-1.4142, 1.4142)$. ∎

▪▪ THE VERTICAL LINE TEST

The following fact, which distinguishes graphs of functions from other graphs, can also be used to interpret some calculator-generated graphs.

Vertical Line Test

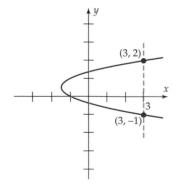

Figure 3–23

The graph of a function $y = f(x)$ has this property:

No vertical line intersects the graph more than once.

Conversely, any graph with this property is the graph of a function.

To see why this is true, consider Figure 3–23, in which the graph intersects the vertical line at two points. If this were the graph of a function f, then we would have $f(3) = 2$ [because $(3, 2)$ is on the graph] *and* $f(3) = -1$ [because $(3, -1)$ is on the graph]. This means that the input 3 produces two different outputs, which is impossible for a function. Therefore, Figure 3–23 is not the graph of a function. A similar argument works in the general case.

Care must be used when applying the Vertical Line Test to a calculator graph, as the following Exploration illustrates.

GRAPHING EXPLORATION

Graph $g(x) = x^{15} + 2$ in the standard window. Does the graph appear to be vertical near $x = 1$? Now graph g in the window with $.7 \leq x \leq 1.3$ and $-2 \leq y \leq 50$. Does the graph of g pass the Vertical Line Test near $x = 1$? Find a viewing window that shows that the function g also passes the Vertical Line Test near $x = -1$.

▪▪ PARAMETRIC GRAPHING*

Most of the functions graphed above can be described by equations in which y is a function of x and, hence, are easily graphed on a calculator. When you must graph an equation that expresses x as a function of y, a different calculator graphing technique is needed.

EXAMPLE 12

Graph $x = y^3 - 3y^2 - 4y + 7$.

SOLUTION   Let t be any real number. If $y = t$, then

$$x = y^3 - 3y^2 - 4y + 7 = t^3 - 3t^2 - 4t + 7.$$

Thus, the graph consists of all points (x, y) such that

$$x = t^3 - 3t^2 - 4t + 7 \qquad \text{and} \qquad y = t \quad (t \text{ any real number}).$$

When written in this form, the equation can be graphed as follows.

Change the graphing mode to **parametric mode** (see the Technology Tip in the margin), and enter the equations

$$x_{1t} = t^3 - 3t^2 - 4t + 7$$

$$y_{1t} = t$$

in the equation memory.† Next set the viewing window so that

$$-6 \leq t \leq 6, \qquad -10 \leq x \leq 10, \qquad -6 \leq y \leq 6,$$

TECHNOLOGY TIP

To change to parametric graphing mode, select PAR(AMETRIC) in the following menu/submenu:

TI: MODE

Casio: GRAPH/TYPE

HP-39: APLET

---

*This material will be used only in Sections 3.7 and 6.2, Chapter 10, and occasional exercises. It may be omitted or postponed until then, if desired.

†On some calculators x_{1t} is denoted x_{t1} or $x_1(T)$, and similarly for y_{1t}.

as partially shown in Figure 3–24 (scroll down to see the rest). We use the same range for t and y here because $y = t$. Note that we must also set "t-step" (or "t pitch"), which determines how much t changes each time a point is plotted. A t-step between .05 and .15 usually produces a relatively smooth graph in a reasonable amount of time. Finally, pressing GRAPH (or PLOT or DRAW) produces the graph in Figure 3–25.   ∎

Figure 3–24   **Figure 3–25**

TECHNOLOGY TIP

If you have trouble finding appropriate ranges for t, x, and y, it might help to use the TABLE feature to display a table of t-x-y values produced by the parametric equations.

As is illustrated in Example 12, the underlying idea of **parametric graphing** is to express both x and y as functions of a third variable t. The equations that define x and y are called **parametric equations,** and the variable t is called the **parameter.** Example 12 illustrates just one of the many applications of parametric graphing. It can also be used to graph curves that are not graphs of a single equation in x and y.

EXAMPLE 13

Graph the curve given by

$$x = t^2 - t - 1 \quad \text{and} \quad y = t^3 - 4t - 6 \quad (-2 \le t \le 3).$$

SOLUTION   Using the standard viewing window, we obtain the graph in Figure 3–26. Note that the graph crosses over itself at one point and that it does not extend forever to the left and right but has endpoints.

GRAPHING EXPLORATION

Graph these same parametric equations, but set the range of t values so that $-4 \le t \le 4$. What happens to the graph? Now change the range of t values so that $-10 \le t \le 10$. Find a viewing window large enough to show the entire graph, including endpoints.   ∎

Figure 3–26

Any function of the form $y = f(x)$ can be expressed in terms of parametric equations and graphed that way. For instance, to graph $f(x) = x^2 + 1$, let $x = t$ and $y = f(x) = t^2 + 1$. Parametric graphing will be used hereafter whenever it is convenient and will be studied more thoroughly in Section 10.5.

▦ CATALOG OF BASIC FUNCTIONS

As was noted at the beginning of this section, there are a number of functions whose graphs you should know by heart. The ones in this section are listed below; others will be added as we go along. The entire catalog appears on the inside front cover of this book.

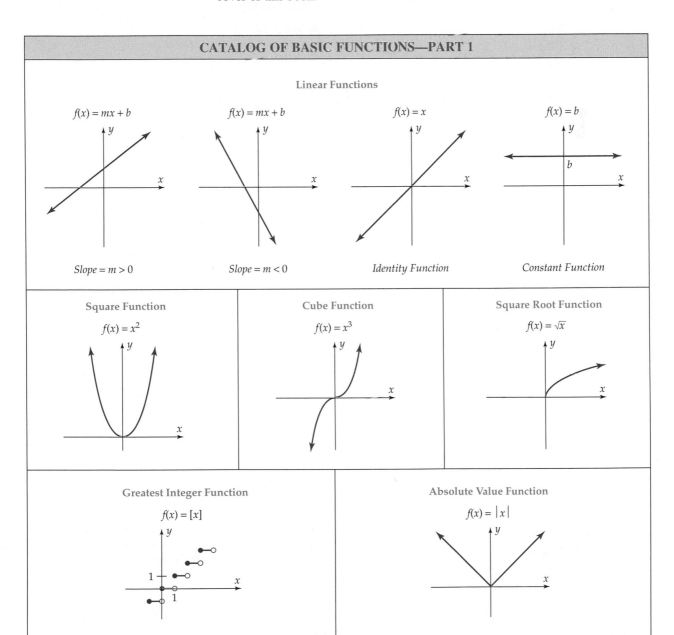

CATALOG OF BASIC FUNCTIONS—PART 1

Linear Functions

$f(x) = mx + b$

$f(x) = mx + b$

$f(x) = x$

$f(x) = b$

Slope = m > 0

Slope = m < 0

Identity Function

Constant Function

Square Function

$f(x) = x^2$

Cube Function

$f(x) = x^3$

Square Root Function

$f(x) = \sqrt{x}$

Greatest Integer Function

$f(x) = [x]$

Absolute Value Function

$f(x) = |x|$

✓ EXERCISES 3.3

In Exercises 1–4, state whether or not the graph is the graph of a function. If it is, find f(3).

1.

2.

3.

4.

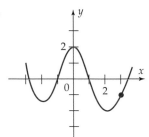

In Exercises 5–12, sketch the graph of the function, being sure to indicate which endpoints are included and which ones are excluded.

5. $f(x) = 2[x]$

6. $f(x) = -[x]$

7. $g(x) = [-x]$ [This is *not* the same function as in Exercisee 6.]

8. $h(x) = [x] + [-x]$

9. $f(x) = \begin{cases} x^2 & \text{if } x \geq -1 \\ 2x + 3 & \text{if } x < -1 \end{cases}$

10. $g(x) = \begin{cases} |x| & \text{if } x < 1 \\ -3x + 4 & \text{if } x \geq 1 \end{cases}$

11. $k(u) = \begin{cases} -2u - 2 & \text{if } u < -3 \\ u - [u] & \text{if } -3 \leq u \leq 1 \\ 2u^2 & \text{if } u > 1 \end{cases}$

12. $f(x) = \begin{cases} x^2 & \text{if } x < -2 \\ x & \text{if } -2 \leq x < 4 \\ \sqrt{x} & \text{if } x \geq 4 \end{cases}$

13. At this writing, first-class postage rates are 37¢ for the first ounce or fraction thereof plus 23¢ for each additional ounce or fraction thereof. Assume that each first-class letter carries one 37¢ stamp and as many 23¢ stamps as are necessary. Then the *number* of stamps required for a first-class letter is a function of the weight of the letter in ounces. Call this function the *postage stamp function.*
(a) Describe the rule of the postage stamp function algebraically.
(b) Sketch the graph of the postage stamp function.
(c) Sketch the graph of the function whose rule is

$$f(x) = p(x) - [x],$$

where *p* is the postage stamp function.

14. A common mistake is to graph the function *f* in Example 6 by graphing both $y = x^2$ and $y = x + 2$ on the same screen (with no restrictions on *x*). Explain why this graph could not possibly be the graph of a function.

In Exercises 15–19, (a) Use the fact that the absolute value function is piecewise-defined (see Example 7) to write the rule of the given function as a piecewise-defined function whose rule does not include any absolute value bars. (b) Graph the function.

15. $f(x) = |x| + 2$

16. $g(x) = |x| - 4$

17. $h(x) = |x|/2 - 2$

18. $g(x) = |x + 3|$

19. $f(x) = |x - 5|$

20. Show that the function $f(x) = |x| + |x - 2|$ is constant on the interval [0, 2]. [*Hint:* Use the definition of absolute value (see Example 7) to compute $f(x)$ when $0 \leq x \leq 2$.]

In Exercises 21–26, find the approximate location of all local maxima and minima of the function.

21. $f(x) = x^3 - x$

22. $g(t) = -\sqrt{16 - t^2}$

23. $h(x) = \dfrac{x}{x^2 + 1}$

24. $k(x) = x^3 - 3x + 1$

25. $f(x) = x^3 - 1.8x^2 + x + 2$

26. $g(x) = 2x^3 + x^2 + 1$

In Exercises 27–28, find the approximate intervals on which the function whose graph is shown is increasing and those on which it is decreasing.

27.

28.

In Exercises 29–34, find the approximate intervals on which the function is increasing, those on which it is decreasing, and those on which it is constant.

29. $f(x) = |x - 1| - |x + 1|$

30. $g(x) = |x - 1| + |x + 2|$

31. $f(x) = -x^3 - 8x^2 + 8x + 5$

32. $f(x) = x^4 - .7x^3 - .6x^2 + 1$

33. $g(x) = .2x^4 - x^3 + x^2 - 2$

34. $g(x) = x^4 + x^3 - 4x^2 + x - 1$

35. Find the dimensions of the rectangle with perimeter 100 inches and largest possible area, as follows.

   (a) Use the figure to write an equation in x and z that expresses the fact that the perimeter of the rectangle is 100.

   (b) The area A of the rectangle is given by $A = xz$ (why?). Write an equation that expresses A as a function of x. [*Hint:* Solve the equation in part (a) for z, and substitute the result in the area equation.]

   (c) Graph the function in part (b), and find the value of x that produces the largest possible value of A. What is z in this case?

36. Find the dimensions of the rectangle with area 240 square inches and smallest possible perimeter, as follows.

   (a) Using the figure for Exercise 35, write an equation for the perimeter P of the rectangle in terms of x and z.

   (b) Write an equation in x and z that expresses the fact that the area of the rectangle is 240.

   (c) Write an equation that expresses P as a function of x. [*Hint:* Solve the equation in part (b) for z, and substitute the result in the equation of part (a).]

   (d) Graph the function in part (c), and find the value of x that produces the smallest possible value of P. What is z in this case?

37. Find the dimensions of a box with a square base that has a volume of 867 cubic inches and the smallest possible surface area, as follows.

   (a) Write an equation for the surface area S of the box in terms of x and h. [Be sure to include all four sides, the top, and the bottom of the box.]

   (b) Write an equation in x and h that expresses the fact that the volume of the box is 867.

   (c) Write an equation that expresses S as a function of x. [*Hint:* Solve the equation in part (b) for h, and substitute the result in the equation of part (a).]

   (d) Graph the function in part (c), and find the value of x that produces the smallest possible value of S. What is h in this case?

38. Find the radius *r* and height *h* of a cylindrical can with a surface area of 60 square inches and the largest possible volume, as follows.

(a) Write an equation for the volume *V* of the can in terms of *r* and *h*.

(b) Write an equation in *r* and *h* that expresses the fact that the surface area of the can is 60. [*Hint:* Think of cutting the top and bottom off the can; then cut the side of the can lengthwise and roll it out flat; it's now a rectangle. The surface area is the area of the top and bottom plus the area of this rectangle. The length of the rectangle is the same as the circumference of the original can (why?).]

(c) Write an equation that expresses *V* as a function of *r*. [*Hint:* Solve the equation in part (b) for *h*, and substitute the result in the equation of part (a).]

(d) Graph the function in part (c), and find the value of *r* that produces the largest possible value of *V*. What is *h* in this case?

39. Match each of the following functions with the graph that best fits the situation.

(a) The phases of the moon as a function of time;

(b) The demand for a product as a function of its price;

(c) The height of a ball thrown from the top of a building as a function of time;

(d) The distance a woman runs at constant speed as a function of time;

(e) The temperature of an oven turned on and set to 350° as a function of time.

(i)

(ii)

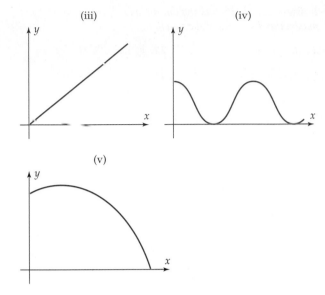

(iii)

(iv)

(v)

In Exercises 40–42, sketch a plausible graph of the given function. Label the axes and specify a reasonable domain and range.

40. The distance from the top of your head to the ground as you jump on a trampoline is a function of time.

41. The amount you spend on gas each week is a function of the number of gallons you put in your car.

42. The temperature of an oven that is turned on, set to 350°, and 45 minutes later turned off is a function of time.

43. A bacteria population in a laboratory culture contains about three million bacteria at 8 A.M. The culture grows very rapidly until noon, when a bactericide is introduced and the bacteria population plunges. By 4 P.M., the bacteria have adapted to the bactericide, and the culture slowly increases in population until 10 P.M., when the culture is accidentally destroyed by the cleanup crew. Let *g(t)* denote the bacteria population at time *t*, and draw a plausible graph of the function *g*. [Many correct answers are possible.]

44. A plane flies from Austin, Texas, to Cleveland, Ohio, a distance of 1200 miles. Let *f* be the function whose rule is *f(t)* = distance (in miles) from Austin at time *t* hours. Draw a plausible graph of *f* under the given circumstances. [There are many possible correct answers for each part.]

(a) The flight is nonstop and takes less than 4 hours.

(b) Bad weather forces the plane to land in Dallas (about 200 miles from Austin), remain overnight (for 8 hours), and continue the next day.

(c) The flight is nonstop, but owing to heavy traffic, the plane must fly in a holding pattern over Cincinnati (about 200 miles from Cleveland) for an hour before going on to Cleveland.

In Exercises 45–46, the graph of a function f is shown. Find and label the given points on the graph.

45. (a) $(k, f(k))$

(b) $(-k, f(-k))$

(c) $(k, -f(k))$

46. (a) $(k, f(k))$

(b) $(k, .5f(k))$

(c) $(.5k, f(.5k))$

(d) $(2k, f(2k))$

47. The graph of the function f, whose rule is $f(x)$ = average interest rate on a 30-year fixed-rate mortgage in year x, is shown in the figure.* Use it to answer these questions (reasonable approximations are OK).

(a) $f(1977) = ?$ (b) $f(1982) = ?$ (c) $f(2000) = ?$

(d) In what year between 1990 and 1996 were rates the lowest? The highest?

(e) During what three-year period were rates changing the fastest? How do you determine this from the graph?

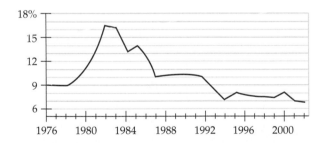

48. The figure shows the graph of the function g whose rule is $g(x)$ = the Pentagon's budget in year x (in billions of 2001 dollars, adjusted for inflation), with $x = 0$ corresponding to 1975.† Use it to answer the following questions.

(a) Over what intervals was this function approximately constant?

(b) Over what intervals was this function decreasing?

(c) When did the Pentagon budget reach a local maximum?

49. Draw the graph of a function f that satisfies the following four conditions:

(i) domain $f = [-2, 4]$

(ii) range $f = [-5, 6]$

(iii) $f(-1) = f(3)$

(iv) $f\left(\dfrac{1}{2}\right) = 0$

50. Sketch the graph of a function f that satisfies these five conditions:

(i) $f(-1) = 2$

(ii) $f(x) \geq 2$ when x is in the interval $\left(-1, \frac{1}{2}\right)$

(iii) $f(x)$ starts decreasing when $x = 1$

(iv) $f(3) = 3 = f(0)$

(v) $f(x)$ starts increasing when $x = 5$

[*Note:* The function whose graph you sketch need not be given by an algebraic formula.]

In Exercises 51–56, use parametric graphing. Find a viewing window that shows a complete graph of the equation.

51. $x = y^3 + 5y^2 - 4y - 5$

52. $\sqrt[3]{y^2 - y - 1} - x + 2 = 0$

53. $xy^2 + xy + x = y^3 - 2y^2 + 4$

[*Hint:* First solve for x.]

54. $2y = xy^2 + 180x$ **55.** $x - \sqrt{y} + y^2 + 8 = 0$

56. $y^2 - x - \sqrt{y + 5} + 4 = 0$

In Exercises 57–62, find a viewing window that shows a complete graph of the curve determined by the parametric equations.

57. $x = 3t^2 - 5$ and $y = t^2$ $(-4 \leq t \leq 4)$

---

*Federal Home Mortgage Corporation.

†Department of Defense; figures for 2002–2005 are estimates.

58. The Zorro curve: $x = .1t^3 - .2t^2 - 2t + 4$ and $y = 1 - t$ $(-5 \le t \le 6)$

59. $x = t^2 - 3t + 2$ and $y = 8 - t^3$ $(-4 \le t \le 4)$

60. $x = t^2 - 6t$ and $y = \sqrt{t + 7}$ $(-5 \le t \le 9)$

61. $x - 1 - t^2$ and $y = t^3 - t - 1$ $(-4 \le t \le 4)$

62. $x = t^2 - t - 1$ and $y = 1 - t - t^7$

63. Wireless telephone services are growing rapidly. The table shows the industry's revenue (in billions of dollars) over a five-year period.[*]

Year	Revenue
1998	40.265
1999	46.751
2000	57.552
2001	74.978
2002	87.967

(a) Make a scatter plot of the data, with $x = 0$ corresponding to 1998.

(b) Use linear regression to find a function that models this data. Assume that the model remains accurate.

(c) What is the revenue in 2004?

(d) When will revenue reach $130 billion?

64. The percentage of adults in the United States who smoke has been decreasing, as shown in the table.[†]

Year	Percent Who Smoke
1965	42.5
1974	37.0
1980	33.3
1987	29.2
1994	25.4
2000	22.9

(a) Make a scatter plot of the data, with $x = 0$ corresponding to 1965.

(b) Use linear regression to find a function that models this data.

(c) Use the model to estimate the percentage of smokers in 1991 and 2003. [For comparison purposes, the actual figure for 1991 is 25.8%]

(d) If this model remains accurate, when will less than 20% of adults smoke?

(e) According to this model, will smoking even disappear entirely? If so, when?

Thinkers

65. Graph the curve given by

$$x = (t^2 - 1)(t^2 - 4)(t + 5) + t + 3$$
$$y = (t^2 - 1)(t^2 - 4)(t^3 + 4) + t - 1$$
$$(-2.5 \le t \le 2.5)$$

How many times does this curve cross itself?

66. Use parametric equations to describe a curve that crosses itself more times than the curve in Exercise 65. [Many correct answers are possible.]

In Exercises 67 and 68, sketch the graph of the equation.

67. $|x| + |y| = 1$ **68.** $|y| = x^2$

---

[*]*New York Times 2002 Almanac.*
[†]Centers for Disease Control and Prevention.

3.3.A *SPECIAL TOPICS* Graph Reading

Until now, we have concentrated on translating statements into functional notation and functional notation into graphs. It is just as important, however, to be
able to translate graphical information into equivalent statements in English or
functional notation.

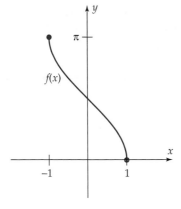

Figure 3–27

EXAMPLE 1

The entire graph of a function f is shown in Figure 3–27. Find the domain and
range of f.

SOLUTION   The graph of f consists of all points of the form $(x, f(x))$. Thus, the
first coordinates of points on the graph are the inputs (numbers in the domain
of f) and the second coordinates are the outputs (the numbers in the range of f).
Figure 3–27 shows that the first coordinates of points on the graph all satisfy
$-1 \le x \le 1$, so these numbers are the domain of f. Similarly, the range of f consists of all numbers y such that $0 \le y \le \pi$, because these are the second coordinates of points on the graph. ∎

EXAMPLE 2

The consumer confidence level reflects people's feelings about their employment
opportunities and income prospects. Let $C(t)$ be the consumer confidence level at
time t (with $t = 0$ corresponding to 1970) and consider the graph of the function
C in Figure 3–28. *

Figure 3–28

(a) How did the consumer confidence level vary in the 1980s?

(b) What was the lowest level of consumer confidence during the 1990s?

(c) During what time periods was the confidence level above 110?

SOLUTION

(a) The 1980s correspond to the interval $10 \le t \le 20$, so we consider the part of
the graph that lies between the vertical lines $t = 10$ and $t = 20$. The second

---

*The consumer confidence level is scaled to be 100 in 1985.

coordinates of these points range from approximately 60 to 117. So the consumer confidence level varied from a low of 60 to a high of 117 during the 1980s.

(b) The 1990s correspond to the interval $20 \leq t \leq 30$. Figure 3–28 shows that the graph has local minimums at $t = 22$ and $t = 29$. The lowest of these three points is the one at $t = 22$. Hence, the lowest level of consumer confidence in the 1990s occurred at the beginning of 1992.

(c) We must find the values of t for which the graph lies above the horizontal line through 110. Figure 3–28 shows that this occurs approximately when $17.5 \leq t \leq 19.5$ and when $26.5 \leq t \leq 31.5$. Thus, the confidence level was above 110 from the middle of 1987 to the middle of 1989 and from the middle of 1996 to the fall of 2001. ■

EXAMPLE 3

Use the graphs of the functions g and h in Figure 3–29 to find:

(a) All numbers x such that $h(x) < 0$;

(b) All numbers x with $-3 \leq x \leq 3$ such that $g(x) = 2$;

(c) The largest interval over which g is increasing and h is deceasing and $h(x) \geq g(x)$ for every x in the interval.

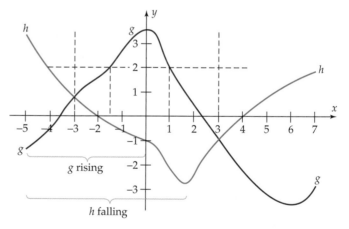

Figure 3–29

SOLUTION

(a) The graph of h consists of all points $(x, h(x))$. The numbers such that $h(x) < 0$ correspond to the points on the graph with negative y-coordinates, that is, points that lie below the x-axis. The graph of h shows that these points are the ones with $-2 < x < 4$.

(b) The graph of g consists of the points $(x, g(x))$. The points on the graph that lie *between* the vertical lines through -3 and 3 and *on* the horizontal line through 2 have x-coordinates satisfying $-3 \leq x \leq 3$ and y-coordinate $g(x) = 2$. Figure 3–29 shows that the only such points are $(-1.5, 2)$ and $(1, 2)$. So the answer is $x = -1.5$ and $x = 1$.

(c) Figure 3–29 shows that the only place where the graph of h is falling *and* the graph of g is rising is where $-5 \leq x \leq 0$. Now $h(x) \geq g(x)$ when the point $(x, h(x))$ lies above the point $(x, g(x))$. This occurs when $-5 \leq x \leq -3$. ■

EXERCISES 3.3.A

In Exercise 1–4, use your calculator to determine the domain and range of the function by reading its graph as best you can (the trace feature may be helpful). Careful approximate answers are acceptable.

1. $f(x) = 3x - 2$

2. $g(x) = x^2 - 4$

3. $h(x) = \sqrt{x^2 - 4}$

4. $k(x) = \sqrt{x^2 + 4}$

Exercises 5–16 deal with the function g whose entire graph is shown in the figure.

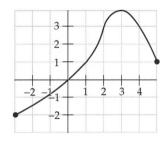

5. What is the domain of g?

6. What is the range of g?

7. If $t = 1.5$, then $g(2t) = ?$

8. If $t = 1.5$, then $2g(t) = ?$

9. If $y = 2$, then $g(y + 1.5) = ?$

10. If $y = 2$, then $g(y) + g(1.5) = ?$

11. If $y = 2$, then $g(y) + 1.5 = ?$

12. For what values of x is $g(x) < 0$?

13. If $v = 1.5$, then $g(3v - 1.5) = ?$

14. If $s = 2$, then $g(-s) = ?$

15. For what values of z is $g(z) = 1$?

16. For what values of z is $g(z) = -1$?

Exercises 17–24 deal with the function f whose entire graph is shown in the figure.

17. What is the domain of f?

18. What is the range of f?

19. Find all numbers x such that $f(x) = 2$.

20. Find all numbers x such that $f(x) > 2$.

21. Find at least three numbers x such that $f(x) = f(-x)$.

22. Find all numbers x such that $f(x) = f(7)$.

23. Find a number x such that $f(x + 1) = 0$.

24. Find two numbers x such that $f(x - 2) = 4$.

25. Let $F(x) =$ the U.S. federal debt in year x, and let $p(x) =$ the federal debt as a percent of the gross domestic product in year x. The graphs of these functions appear below.[*] Explain why the graph of F is increasing after 1996, while the graph of p is decreasing during that period.

Gross Federal Debt

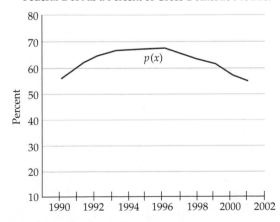

Federal Debt as a Percent of Gross Domestic Product

---

*Graphs prepared by U.S. Census Bureau, based on data from the U.S. Office of Management and Budget.

26. The annual percentage changes in various consumer price indexes (CPIs) are shown in the figure.[*] Use it to answer the following questions. In each case, explain how you got your answer from the graph.

(a) Did the CPI for medical care increase or decrease from 1990 to 1996?

(b) During what time intervals was the CPI for fuel oil increasing?

(c) If the CPI for fuel oil stood at 91 at the beginning of 1999, approximately what was it at the beginning of 2000?

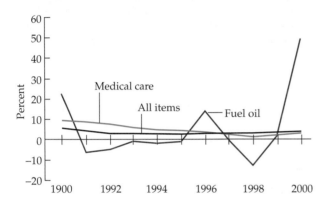

For Exercises 27–30, use the Cleveland temperature graph from Example 2 of Section 3.1, which is reproduced below. Let $T(x)$ denote the temperature at time x hours after midnight.

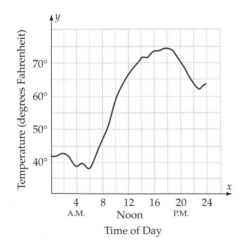

Time of Day

27. Find $T(4)$ and $T(7 + 11)$.

28. Is $T(8)$ larger than, equal to, or less than $T(16)$?

29. Find an eight-hour period in which $T(x) > 65°$ for all x in this period. [There are many possible correct answers.]

30. Determine whether the following statements are true or false.

(a) $T(4 \cdot 3) = T(4) \cdot T(3)$

(b) $T(4 \cdot 3) = 4 \cdot T(3)$

(c) $T(4 + 14) = T(4) + T(14)$

Exercises 31–38 deal with the two functions f and g whose entire graphs are shown in the figure.

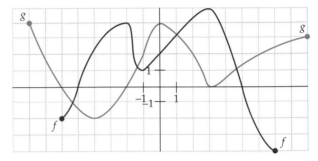

31. What is the domain of f? The domain of g?

32. What is the range of f? The range of g?

33. Find all numbers x in the interval $[-3, 1]$ such that
$$f(x) = 2.$$

34. Find all numbers x in the interval $[-3, 3]$ such that
$$g(x) \geq 2.$$

35. Find the number x for which $f(x) - g(x)$ is largest?

36. For how many values of x is it true that $f(x) = g(x)$?

37. Find all intervals over which both functions are defined, f is decreasing, and g is increasing.

38. Find all intervals over which g is decreasing.

---

[*]Graph prepared by U.S. Census Bureau, based on data from the Bureau of Labor Statistics.

Exercises 39–42 deal with this situation: The owners of the Melville & Pluth Hammer Factory have determined that both their weekly manufacturing expenses and their weekly sales income are functions of the number of hammers manufactured each week. The figure shows the graphs of these two functions.

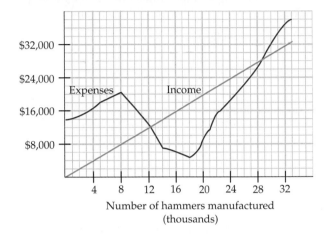

Number of hammers manufactured
(thousands)

39. Use careful measurement on the graph and the fact that profit = income − expenses to determine the weekly profit if 5000 hammers are manufactured.

40. Do the same if 10,000, 14,000, 18,000, or 22,000 hammers are manufactured.

41. What is the smallest number of hammers that can be manufactured each week without losing money?

42. What is the largest number of hammers that can be manufactured without losing money?

In Exercises 43–46, express the length h as a function of x.

43.

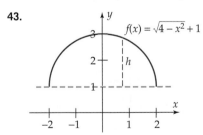

$f(x) = \sqrt{4 - x^2} + 1$

44.

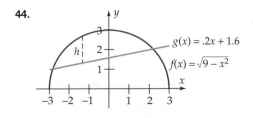

$g(x) = .2x + 1.6$

$f(x) = \sqrt{9 - x^2}$

45.

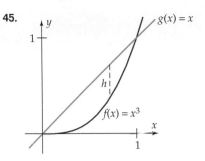

$g(x) = x$

$f(x) = x^3$

46.

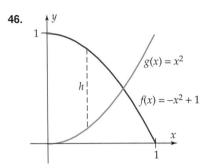

$g(x) = x^2$

$f(x) = -x^2 + 1$

Thinkers

47. A jogger begins her daily run from her home. The graph shows her distance from home at time t minutes. The graph shows, for example, that she ran at a slow but steady pace for 10 minutes, then increased her pace for 5 minutes, all the time moving farther from home. Describe the rest of her run.

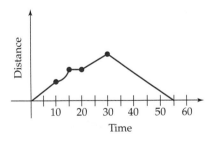

48. The graph shows the speed (in mph) at which a driver is going at time t minutes. Describe his journey.

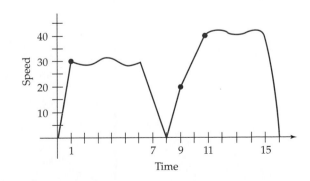

3.4   Graphs and Transformations

In this section, we shall see that when the rule of a function is algebraically changed in certain ways, so as to produce a new function, then the graph of the new function can be obtained from the graph of the original function by a simple geometric transformation. The same format will be used for each topic:

First   You will be asked to assemble some evidence by doing a graphing exploration.

Next   General conclusions deduced from the evidence will be summarized in the boxes.

Last   There may be a discussion of how these conclusions can be proved in particular cases and possibly some additional examples.

▒ VERTICAL SHIFTS

GRAPHING EXPLORATION

Using the standard viewing window, graph these three functions on the same screen:

$$f(x) = x^2 \qquad g(x) = x^2 + 5 \qquad h(x) = x^2 - 7,$$

and answer these questions:

Do the graphs of g and h look very similar to the graph of f in *shape*?

How do their vertical positions differ?

Where would you predict that the graph of $k(x) = x^2 - 9$ is located relative to the graph $f(x) = x^2$, and what is its shape?

Confirm your prediction by graphing k on the same screen as f, g, and h.

The results of this Exploration should make the following statements plausible.

Vertical Shifts

Let f be a function and c a positive constant.

The graph of $g(x) = f(x) + c$ is the graph of f shifted c units upward.

The graph of $h(x) = f(x) - c$ is the graph of f shifted c units downward.

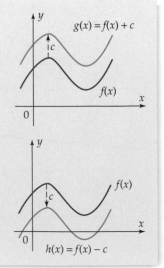

To see why these statements are true, suppose $f(x) = x^2$, $c = 5$, and $g(x) = x^2 + 5$. For any given value of x, consider the points

$P = (x, x^2)$ on the graph of f and $Q = (x, x^2 + 5)$ on the graph of g.

The x-coordinataes show that P and Q lie in the same vertical line. The y-coordinates show that Q lies 5 units directly above P. Thus, the graph of $g(x) = f(x) + 5$ is just the graph of f shifted 5 units upward.

EXAMPLE 1

A calculator was used to obtain a complete graph of $f(x) = .04x^3 - x - 3$ in Figure 3–30. The graph of

$$h(x) = f(x) - 4 = (.04x^3 - x - 3) - 4 = .04x^3 - x - 7$$

is the graph of f shifted 4 units downward, as shown in Figure 3–31.

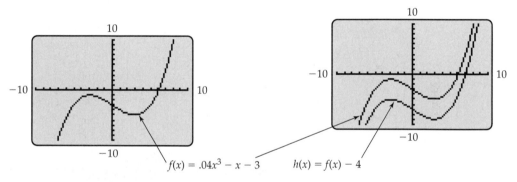

Figure 3–30   **Figure 3–31**

Although it may appear that the graph of h is closer to the graph of f at the outer edges of Figure 3–31 than in the center, this is an optical illusion. The *vertical* distance between the graphs is always 4 units.

EXPLORATION

Use the trace feature of your calculator as follows to confirm that the vertical distance is always 4:[*]

Move the cursor to any point on the graph of f, and note its coordinates.

Use the down arrow to drop the cursor to the graph of h, and note the coordinates of the cursor in its new position.

The x-coordinates will be the same in both cases, and the new y-coordinate will be 4 less than the original y-coordinate. ∎

---

*The trace cursor can be moved vertically from graph to graph by using the up and down arrows.

◼◼ HORIZONTAL SHIFTS

GRAPHING EXPLORATION

Using the standard viewing window, graph these three functions on the same screen:

$$f(x) = 2x^3 \qquad g(x) = 2(x + 6)^3 \qquad h(x) = 2(x - 8)^3,$$

and answer these questions:

Do the graphs of g and h look very similar to the graph of f in *shape*?

How do their horizontal positions differ?

Where would you predict that the graph of $k(x) = 2(x + 2)^3$ is located relative to the graph of $f(x) = 2x^3$, and what is its shape?

Confirm your prediction by graphing k on the same screen as f, g, and h.

The results of this Exploration should make the following statements, plausible.

Horizontal Shifts

Let f be a function and c a positive constant.

The graph of $g(x) = f(x + c)$ is the graph of f shifted horizontally c units to the left.

The graph of $h(x) = f(x - c)$ is the graph of f shifted horizontally c units to the right.

To see why the first statement is true, suppose $c = 4$ and $g(x) = f(x + 4)$. Then the value of g at x is the same as the value of f at $x + 4$, which is 4 units to the right of x on the horizontal axis. So the graph of f is the graph of g shifted 4 units to the *right,* which means that the graph of g is the graph of f shifted 4 units to the *left.* An analogous argument works for the second statement in the box.

EXAMPLE 2

In some cases, shifting the graph of a function f horizontally may produce a graph that overlaps the graph of f. For instance, a complete graph of $f(x) = x^2 - 7$ is shown in red in Figure 3–32. The graph of

$$g(x) = f(x + 5) = (x + 5)^2 - 7 = x^2 + 10x + 25 - 7 = x^2 + 10x + 18$$

is the graph of f shifted 5 units to the left, and the graph of

$$h(x) = f(x - 4) = (x - 4)^2 - 7 = x^2 - 8x + 16 - 7 = x^2 - 8x + 9$$

is the graph of f shifted 4 units to the right, as shown in Figure 3–32. ∎

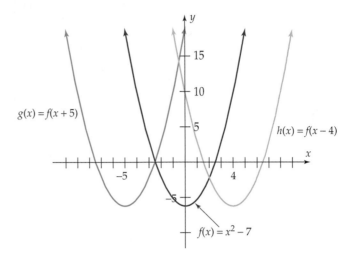

Figure 3–32

■■ EXPANSIONS AND CONTRACTIONS

GRAPHING EXPLORATION

In the viewing window with $-5 \le x \le 5$ and $-15 \le y \le 15$, graph these functions on the same screen:

$$f(x) = x^2 - 4 \qquad g(x) = 3f(x) = 3(x^2 - 4).$$

The table of values in Figure 3–33 shows that the y-coordinates on the graph of $Y_2 = g(x)$ are always 3 times the y-coordinates of the corresponding points on the graph of $Y_1 = f(x)$. To translate this into visual terms, imagine that the graph of f is nailed to the x-axis at its intercepts (± 2). The graph of g is then obtained by

X	Y1	Y2
-3	5	15
-2	0	0
-1	-3	-9
0	-4	-12
1	-3	-9
2	0	0
3	5	15

X=3

Figure 3–33

"stretching" the graph of *f* away from the *x*-axis (with the nails holding the *x*-intercepts in place) by a factor of 3, as shown in Figure 3–34.

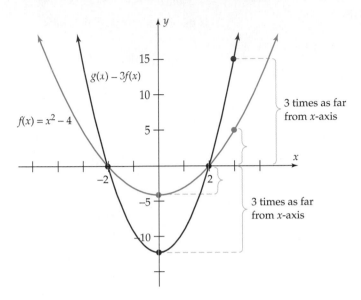

Figure 3–34

GRAPHING EXPLORATION

In the viewing window with $-5 \leq x \leq 5$ and $-5 \leq y \leq 10$, graph these functions on the same screen:

$$f(x) = x^2 - 4 \qquad h(x) = \frac{1}{4}(x^2 - 4).$$

Your screen should suggest that the graph of *h* is the graph of *f* "shrunk" vertically toward the *x*-axis by a factor of 1/4.

Analogous facts are true in the general case.

Expansions and Contractions

If $c > 1$, then the graph of $g(x) = cf(x)$ is the graph of *f* stretched vertically away from the *x*-axis by a factor of *c*.

If $0 < c < 1$, then the graph of $h(x) = cf(x)$ is the graph of *f* shrunk vertically toward the *x*-axis by a factor of *c*.

▚ REFLECTIONS

GRAPHING **EXPLORATION**

In the standard viewing window, graph these functions on the same screen.

$$f(x) = .04x^3 - x \qquad g(x) = -f(x) = -(.04x^3 - x).$$

By moving your trace cursor from graph to graph, verify that for every point on the graph of f, there is a point on the graph of g with the same first coordinate that is on the opposite side of the x-axis, the same distance from the x-axis.

This Exploration shows that the graph of g is the mirror image (reflection) of the graph of f, with the x-axis being the (two-way) mirror. The same thing is true in the general case.

Reflections

Let f be a function. The graph of $g(x) = -f(x)$ is the graph of f reflected in the x-axis.

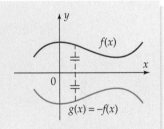

EXAMPLE 3

If $f(x) = x^2 - 3$, then the graph of

$$g(x) = -f(x) = -(x^2 - 3)$$

is the reflection of the graph of f in the x-axis, as shown in Figure 3–35. ■

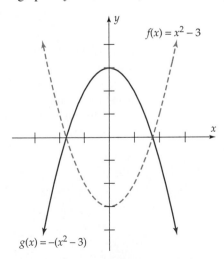

Figure 3–35

GRAPHING EXPLORATION

In the standard viewing window, graph these functions on the same screen:

$$f(x) = \sqrt{5x + 10} \qquad \text{and} \qquad h(x) = f(-x) = \sqrt{5(-x) + 10}.$$

Reflect carefully: How are the two graphs related to the y-axis? Now graph these two functions on the same screen:

$$f(x) = x^2 + 3x - 3$$

$$h(x) = f(-x) = (-x)^2 + 3(-x) - 3 = x^2 - 3x - 3.$$

Are the graphs of f and h related in the same way as the first pair?

This Exploration shows that the graph of h in each case is the mirror image (reflection) of the graph of f, with the y-axis as the mirror. The same thing is true in the general case.

Reflections

Let f be a function. The graph of $h(x) = f(-x)$ is the graph of f reflected in the y-axis.

To see why this is true, let a be any number. Then

$(a, f(a))$ is on the graph of f and $(-a, h(-a))$ is on the graph of h.

However, $h(-a) = f(-(-a)) = f(a)$, so the two points are

$(a, f(a))$ on the graph of f and $(-a, f(a))$ on the graph of h.

These points lie on opposite sides of the y-axis, the same distance from the axis, because their first coordinates are negatives of each other. The points are on the same horizontal line because their second coordinates are the same. Thus, every point on the graph of f has a mirror-image point on the graph of h, the y-axis being the mirror.

Other algebraic operations and their graphical effects are considered in Exercises 47–58.

⠃⠃ COMBINING TRANSFORMATIONS

The transformations described above may be used in sequence to analyze the graphs of functions whose rules are algebraically complicated.

EXAMPLE 4

To understand the graph of $g(x) = 2(x - 3)^2 - 1$, note that the rule of g may be obtained from the rule of $f(x) = x^2$ in three steps:

$$f(x) = x^2 \xrightarrow{\text{Step 1}} (x - 3)^2 \xrightarrow{\text{Step 2}} 2(x - 3)^2 \xrightarrow{\text{Step 3}} 2(x - 3)^2 - 1 = g(x).$$

Step 1 shifts the graph of f horizontally 3 units to the right; step 2 stretches the resulting graph away from the x-axis by a factor of 2; step 3 shifts this graph 1 unit downward, thus producing the graph of g in Figure 3–36.   ■

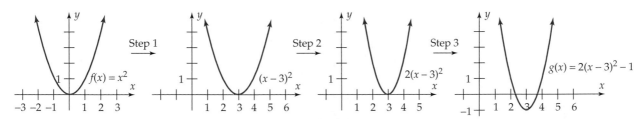

Figure 3–36

INVESTIGATIONS 3.4

Investigations 1 and 2 show that, because of its small screen size, a calculator might not always display clearly what it should.

1. Graph $f(x) = x^2 - x - 6$ in the standard viewing window. Describe verbally what the graph of $h(x) = f(x - 1000)$ should look like (see the box on page 168). Now find an appropriate viewing window, and graph h. Can you find a viewing window that clearly displays both the graph of f and the graph of h?

2. Graph $f(x) = x^2 - x - 6$ in the standard viewing window. Describe verbally what the graph of $g(x) = 1000f(x)$ should look like (see the box on page 170). Now find an appropriate viewing window, and graph g. Can you find a viewing window that clearly displays both the graph of f and the graph of g?

EXERCISES 3.4

In Exercises 1–8, use the catalog of functions at the end of Section 3.3 and information from this section to match each function with its graph, which is one of those shown here.

A.

B.

C.

D.

E.

F.

G.

H.

I.

J.

K.

L.
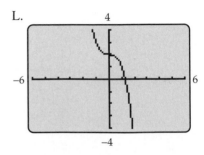

1. $f(x) = x^2 + 2$

2. $f(x) = \sqrt{x - 2}$

3. $g(x) = (x - 2)^3$

4. $g(x) = (x - 2)^2$

5. $f(x) = -\sqrt{x}$

6. $f(x) = (x + 1)^2 - 1$

7. $g(x) = -x^2 + 2$

8. $g(x) = -x^3 + 2$

In Exercises 9–12, use the catalog of functions and information from this section (but not a calculator) to sketch the graph of the function.

9. $f(x) = |x - 2|$

10. $(g)|x| = |x| - 2$

11. $g(x) = -|x|$

12. $f(x) = |x + 2| - 2$

In Exercises 13–16, find a single viewing window that shows complete graphs of the functions f, g, and h.

13. $f(x) = .25x^3 - 9x + 5$; $g(x) = f(x) + 15$;
 $h(x) = f(x) - 20$

14. $f(x) = \sqrt{x^2 - 9} - 5$; $g(x) = 3f(x)$;
 $h(x) = .5f(x)$

15. $f(x) = |x^2 - 5|$; $g(x) = f(x + 8)$;
 $h(x) = f(x - 6)$

16. $f(x) = .125x^3 - .25x^2 - 1.5x + 5$;
 $g(x) = f(x) - 5$; $h(x) = 5 - f(x)$

In Exercises 17 and 18, find complete graphs of the functions f and g in the same viewing window.

17. $f(x) = \dfrac{4 - 5x^2}{x^2 + 1}$; $g(x) = -f(x)$

18. $f(x) = x^4 - 4x^3 + 2x^2 + 3$; $g(x) = f(-x)$

In Exercises 19–24, describe a sequence of transformations that will transform the graph of the function f into the graph of the function g.

19. $f(x) = x^2 + x$; $g(x) = (x - 3)^2 + (x - 3) + 2$

20. $f(x) = x^2 + 5$; $g(x) = (x + 2)^2 + 10$

21. $f(x) = \sqrt{x^3 + 5}$; $g(x) = -\dfrac{1}{2}\sqrt{x^3 + 5} - 6$

22. $f(x) = \sqrt{x^4 + x^2 + 1}$; $g(x) = 10 - \sqrt{4x^4 + 4x^2 + 4}$

23. $f(x) = \dfrac{3x}{x^2 + 10}$; $g(x) = \dfrac{-6x + 12}{(x - 2)^2 + 10}$

24. $f(x) = \dfrac{1}{2x^2 + 2}$; $g(x) = \dfrac{-1}{(x - 1)^2 + 1}$

In Exercises 25–28, write the rule of a function g whose graph can be obtained from the graph of the function f by performing the transformations in the order given.

25. $f(x) = x^2 + 2$; shift the graph horizontally 5 units to the left and then vertically upward 4 units.

26. $f(x) = x^2 - x + 1$; reflect the graph in the *x*-axis, then shift it vertically upward 3 units.

27. $f(x) = \sqrt{x}$; shift the graph horizontally 6 units to the right, stretch it away from the *x*-axis by a factor of 2, and shift it vertically downward 3 units.

28. $f(x) = \sqrt{-x}$; shift the graph horizontally 3 units to the left, then reflect it in the *x*-axis, and shrink it toward the *x*-axis by a factor of $1/2$.

29. Let $f(x) = x^2 + 3x$, and let $g(x) = f(x) + 2$.
 (a) Write the rule of $g(x)$.
 (b) Find the difference quotients of $f(x)$ and $g(x)$. How are they related?

30. Let $f(x) = x^2 + 5$, and let $g(x) = f(x - 1)$.
 (a) Write the rule of $g(x)$ and simplify.
 (b) Find the difference quotients of $f(x)$ and $g(x)$.
 (c) Let $d(x)$ denote the difference quotient of $f(x)$. Show that the difference quotient of $g(x)$ is $d(x - 1)$.

In Exercises 31–34, use the graph of the function f in the figure to sketch the graph of the function g.

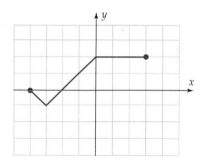

31. $g(x) = f(x) + 3$

32. $g(x) = f(x) - 1$

33. $g(x) = 3f(x)$

34. $g(x) = .25f(x)$

In Exercises 35–38, use the graph of the function f in the figure to sketch the graph of the function h.

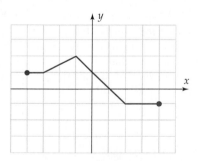

35. $h(x) = -f(x)$

36. $h(x) = -4f(x)$

37. $h(x) = f(-x)$

38. $h(x) = f(-x) + 2$

In Exercises 39–44, use the graph of the function f in the figure to sketch the graph of the function g.

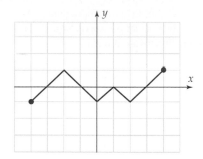

39. $g(x) = f(x + 3)$

40. $g(x) = f(x - 2)$

41. $g(x) = f(x - 2) + 3$

42. $g(x) = f(x + 1) - 3$

43. $g(x) = 2 - f(x)$

44. $g(x) = f(-x) + 2$

45. Graph $f(x) = -|x - 3| - |x - 17| + 20$ in the window with $0 \le x \le 20$ and $-2 \le y \le 12$. Think of the x-axis as a table and the graph as a side view of a fast-food carton placed upside down on the table (the flat part of the graph is the bottom of the carton). Find the rule of a function g whose graph (in this viewing window) looks like another fast-food carton, which has been placed right side up on top of the first one.

46. A factory has a linear cost function $c(x) = ax + b$, where b represents fixed costs and a represents the variable costs (labor and materials) of making one item, both in thousands of dollars.
   (a) If property taxes (part of the fixed costs) are increased by $28,000 per year, what effect does this have on the graph of the cost function?
   (b) If variable costs increase by 12 cents per item, what effect does this have on the graph of the cost function?

In Exercises 47–49, assume $f(x) = (.2x)^6 - 4$. Use the standard viewing window to graph the functions f and g on the same screen.

47. $g(x) = f(2x)$ **48.** $g(x) = f(3x)$ **49.** $g(x) = f(4x)$

50. On the basis of the results of Exercises 47–49, describe the transformation that transforms the graph of a function $f(x)$ into the graph of the function $f(cx)$, where c is a constant with $c > 1$. [*Hint:* How are the two graphs related to the y-axis? Stretch your mind.]

In Exercises 51–53, assume $f(x) = x^2 - 3$. Use the standard viewing window to graph the functions f and g on the same screen.

51. $g(x) = f\left(\frac{1}{2}x\right)$ **52.** $g(x) = f\left(\frac{1}{3}x\right)$

53. $g(x) = f\left(\frac{1}{4}x\right)$

54. On the basis of the results of Exercises 51–53, describe the transformation that transforms the graph of a function $f(x)$ into the graph of the function $f(cx)$, where c is a constant with $0 < c < 1$. [*Hint:* How are the two graphs related to the y-axis?]

In Exercises 55–57, use the standard viewing window to graph the function f and the function $g(x) = f(|x|)$ on the same screen.

55. $f(x) = x - 4$ **56.** $f(x) = x^3 - 3$

57. $f(x) = .5(x - 4)^2 - 9$

58. On the basis of the results of Exercises 55–57, describe the relationship between the graph of a function $f(x)$ and the graph of the function $f(|x|)$.

In Exercises 59–61, use the standard viewing window to graph the function f and the function $g(x) = |f(x)|$ on the same screen. Exercise 62 may be helpful for interpreting the results.

59. $f(x) = .5x^2 - 5$ **60.** $f(x) = x^3 - 4x^2 + x + 3$

61. $f(x) = x + 3$

62. (a) Let f be a function, and let g be the function defined by $g(x) = |f(x)|$. Use the definition of absolute value (page 8) to explain why the following statement is true:

$$g(x) = \begin{cases} f(x) & \text{if } f(x) \ge 0 \\ -f(x) & \text{if } f(x) < 0 \end{cases}$$

   (b) Use part (a) and your knowledge of transformations to explain why the graph of g consists of those parts of the graph of f that lie above the x-axis together with the reflection in the x-axis of those parts of the graph of f that lie below the x-axis.

3.4.A *SPECIAL TOPICS* Symmetry

A graph is **symmetric with respect to the y-axis** if the part of the graph on the right side of the y-axis is the mirror image of the part on the left side of the y-axis (with the y-axis being the mirror), as shown in Figure 3–37.

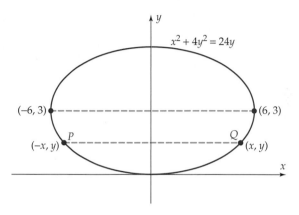

Figure 3–37

Each point P on the left side of the graph has a mirror image point Q on the right side of the graph, as indicated by the dashed lines. Note that:

Their second coordinates are the same (P and Q are on the same side of the x-axis and the same distance from it);

Their first coordinates are negatives of each other (P and Q lie on opposite sides of the y-axis and the same distance from it).

Thus, a graph is symmetric with respect to the y-axis provided that

Whenever (x, y) is on the graph, then $(-x, y)$ is also on it.

In algebraic terms, this means that replacing x by $-x$ in the equation leads to the same number y. In other words, replacing x by $-x$ produces an equivalent equation.

EXAMPLE 1

Replacing x by $-x$ in the equation $y = x^4 - 5x^2 + 3$ produces

$$(-x)^4 - 5(-x)^2 + 3,$$

which is the same equation because $(-x)^2 = x^2$ and $(-x)^4 = x^4$. Therefore, the graph is symmetric with respect to the y-axis.

GRAPHING EXPLORATION

Confirm this fact by graphing the equation. ■

▐▐ x-AXIS SYMMETRY

A graph is **symmetric with respect to the x-axis** if the part of the graph above the x-axis is the mirror image of the part below the x-axis (the x-axis being the mirror), as shown in Figure 3–38.

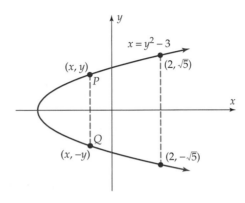

Figure 3–38

Using Figure 3–38 and argument analogous to the one preceding Example 1, we see that a graph is symmetric with respect to the x-axis provided that

Whenever (x, y) is on the graph, then $(x, -y)$ is also on it.

In algebraic terms, this means that replacing y by $-y$ in the equation leads to the same number x. In other words, replacing y by $-y$ produces an equivalent equation.

EXAMPLE 2

Replacing y by $-y$ in the equation $y^2 = 4x - 12$ produces $(-y)^2 = 4x - 12$, which is the same equation, so the graph is symmetric with respect to the x-axis.

GRAPHING EXPLORATION

Confirm this fact by graphing the equation. To do this, note that every point on the graph of $y^2 = 4x - 12$ is also on the graph of either $y = \sqrt{4x - 12}$ or $y = -\sqrt{4x - 12}$. Each of these latter equations defines a function; graph them both on the same screen. ■

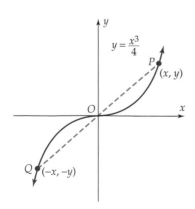

Figure 3–39

▐▐ ORIGIN SYMMETRY

A graph is **symmetric with respect to the origin** if a straight line through the origin and any point P on the graph also intersects the graph at a point Q such that the origin is the midpoint of segment PQ, as shown in Figure 3–39.

Using Figure 3–39, we can also describe symmetry with respect to the origin in terms of coordinates and equations (as proved in Exercise 34):

Whenever (x, y) is on the graph, then $(-x, -y)$ is also on it.

In algebraic terms, this means that replacing x by $-x$ and y by $-y$ in the equation produces an equivalent equation.

EXAMPLE 3

Replacing x by $-x$ and y by $-y$ in the equation $y = \dfrac{x^3}{10} - x$ yields

$$-y = \frac{(-x)^3}{10} - (-x), \qquad \text{that is,} \qquad -y = \frac{-x^3}{10} + x.$$

This equation is equivalent to

$$y = \frac{x^3}{10} - x,$$

since it can be obtained from it by multiplying by -1. Therefore, the graph of

$$y = \frac{x^3}{10} - x$$

is symmetric with respect to the origin.

GRAPHING EXPLORATION

Confirm this fact by graphing the equation.   ■

Here is a summary of the various tests for symmetry:

Symmetry Tests

Symmetry with Respect to	Coordinate Test for Symmetry	Algebraic Test for Symmetry
y-axis	(x, y) on graph implies $(-x, y)$ on graph.	Replacing x by $-x$ produces an equivalent equation.
x-axis	(x, y) on graph implies $(x, -y)$ on graph.	Replacing y by $-y$ produces an equivalent equation.
origin	(x, y) on graph implies $(-x, -y)$ on graph.	Replacing x by $-x$ and y by $-y$ produces an equivalent equation.

▟ EVEN AND ODD FUNCTIONS

For *functions,* the algebraic description of symmetry takes a different form. A function f whose graph is symmetric with respect to the y-axis is called an **even function.** To say that the graph of $y = f(x)$ is symmetric with respect to the y-axis means that replacing x by $-x$ produces the same y value. In other words, the function takes the same value at both x and $-x$. Therefore,

Even
Functions

> A function f is even provided that
>
> $$f(x) = f(-x) \text{ for every number } x \text{ in the domain of } f.$$
>
> The graph of an even function is symmetric with respect to the y-axis.

For example, $f(x) = x^4 + x^2$ is even because

$$f(-x) = (-x)^4 + (-x)^2 = x^4 + x^2 = f(x).$$

Thus, the graph of f is symmetric with respect to the y-axis, as you can easily verify with your calculator (do it!).

Except for zero functions ($f(x) = 0$ for every x in the domain), *the graph of a function is never symmetric with respect to the x-axis.* The reason is the vertical line test: The graph of a function never contains two points with the same first coordinate. If both $(5, 3)$ and $(5, -3)$, for instance, were on the graph, this would say that $f(5) = 3$ and $f(5) = -3$, which is impossible when f is a function.

A function whose graph is symmetric with respect to the origin is called an **odd function.** If both (x, y) and $(-x, -y)$ are on the graph of such a function f, then we must have both

$$y = f(x) \qquad \text{and} \qquad -y = f(-x),$$

so $f(-x) = -y = -f(x)$. Therefore,

Odd
Functions

> A function f is odd provided that
>
> $$f(-x) = -f(x) \text{ for every number } x \text{ in the domain of } f.$$
>
> The graph of an odd function is symmetric with respect to the origin.

For example, $f(x) = x^3$ is an odd function because

$$f(-x) = (-x)^3 = -x^3 = -f(x).$$

Hence, the graph of f is symmetric with respect to the origin (verify this with your calculator).

✓ EXERCISES 3.4.A

In Exercises 1–4, find the graph of the equation. If the graph is symmetric with respect to the x-axis, the y-axis, or the origin, say so.

1. $y = x^2 + 2$

2. $x = (y - 3)^2$

3. $y = x^3 + 2$

4. $y = (x + 2)^3$

In Exercises 5–14, determine whether the given function is even, odd, or neither.

5. $f(x) = 4x$

6. $k(t) = -5t$

7. $f(x) = x^2 - |x|$

8. $h(u) = |3u|$

9. $k(t) = t^4 - 6t^2 + 5$

10. $f(x) = x(x^4 - x^2) + 4$

11. $f(t) = \sqrt{t^2 - 5}$ **12.** $h(x) = \sqrt{7 - 2x^2}$

13. $f(x) = \dfrac{x^2 + 2}{x - 7}$ **14.** $g(x) = \dfrac{x^2 + 1}{x^2 - 1}$

In Exercises 15–18, determine algebraically whether or not the graph of the given equation is symmetric with respect to the x-axis.

15. $x^2 - 6x + y^2 + 8 = 0$ **16.** $x^2 + 8x + y^2 = -15$

17. $x^2 - 2x + y^2 + 2y = 2$ **18.** $x^2 - x + y^2 - y = 0$

In Exercises 19–24, determine whether the given graph is symmetric with respect to the y-axis, the x-axis, or the origin.

19.

20.

21.

22.

23.

24.

In Exercises 25–28, complete the graph of the given function, assuming that it satisfies the given symmetry condition.

25. Even

26. Even

27. Odd

28. Odd

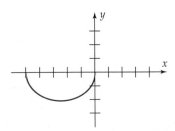

29. (a) Draw some coordinate axes, and plot the points $(0, 1)$, $(1, -3)$, $(-5, 2)$, $(-3, 5)$, $(2, 3)$, and $(4, 1)$.

  (b) Suppose the points in part (a) lie on the graph of an *even* function f. Plot the points $(0, f(0))$, $(-1, f(-1))$, $(5, f(5))$, $(3, f(3))$, $(-2, f(-2))$, and $(-4, f(-4))$.

30. Draw the graph of an *even* function that includes the points $(0, -3)$, $(-3, 0)$, $(2, 0)$, $(1, -4)$, $(2.5, -1)$, $(-4, 3)$, and $(-5, 3)$.*

31. (a) Plot the points $(0, 0)$; $(2, 3)$; $(3, 4)$; $(5, 0)$; $(7, -3)$; $(-1, -1)$; $(-4, -1)$; $(-6, 1)$.
   (b) Suppose the points in part (a) lie on the graph of an *odd* function f. Plot the points $(-2, f(-2))$; $(-3, f(-3))$; $(-5, f(-5))$; $(-7, f(-7))$; $(1, f(1))$; $(4, f(4))$; $(6, f(6))$.
   (c) Draw the graph of an odd function f that includes all the points plotted in parts (a) and (b).*

32. Draw the graph of an odd function that includes the points $(-3, 5)$, $(-1, 1)$, $(2, -6)$, $(4, -9)$, and $(5, -5)$.*

33. Show that any graph that has two of the three types of symmetry (x-axis, y-axis, origin) necessarily has the third type also.

34. Use the midpoint formula to show that $(0, 0)$ is the midpoint of the segment joining (x, y) and $(-x, -y)$. Conclude that the coordinate test for symmetry with respect to the origin (page 179) is correct.

---

*There are many correct answers.

3.5 | Operations on Functions

We now examine ways in which two or more given functions can be used to create new functions. If f and g are functions, then their **sum** is the function h defined by the rule

$$h(x) = f(x) + g(x).$$

For example, if $f(x) = 3x^2 + x$ and $g(x) = 4x - 2$, then

$$h(x) = f(x) + g(x) = (3x^2 + x) + (4x - 2) = 3x^2 + 5x - 2.$$

Instead of using a different letter h for the sum function, we shall usually denote it by $f + g$. Thus, the sum $f + g$ is defined by the rule

$$(f + g)(x) = f(x) + g(x).$$

This rule is *not* just a formal manipulation of symbols. If x is a number, then so are $f(x)$ and $g(x)$. The plus sign in $f(x) + g(x)$ is addition of *numbers,* and the result is a number. But the plus sign in $f + g$ is addition of *functions,* and the result is a new function.

The **difference** $f - g$ is the function defined by the rule

$$(f - g)(x) = f(x) - g(x).$$

The domain of the sum and difference functions is the set of all real numbers that are in both the domain of f and the domain of g.

EXAMPLE 1

If $f(x) = \sqrt{9 - x^2}$ and $g(x) = \sqrt{x - 2}$, find the rules of the functions $f + g$ and $f - g$ and their domains.

SOLUTION   We have

$$(f + g)(x) = f(x) + g(x) = \sqrt{9 - x^2} + \sqrt{x - 2};$$
$$(f - g)(x) = f(x) - g(x) = \sqrt{9 - x^2} - \sqrt{x - 2}.$$

The domain of f consists of all x such that $9 - x^2 \geq 0$ (so that the square root will be defined), that is, all x with $-3 \leq x \leq 3$. Similarly, the domain of g consists of all x such that $x \geq 2$. The domain of $f + g$ and $f - g$ consists of all real numbers in both the domain of f and the domain of g, namely, all x such that $2 \leq x \leq 3$. ∎

The **product** and **quotient** of functions f and g are the functions defined by the rules

$$(fg)(x) = f(x)g(x) \qquad \text{and} \qquad \left(\frac{f}{g}\right)(x) = \frac{f(x)}{g(x)}.$$

The domain of fg consists of all real numbers in both the domain of f and the domain of g. The domain of f/g consists of all real numbers x in both the domain of f and the domain of g such that $g(x) \neq 0$.

EXAMPLE 2

If $f(x) = \sqrt{3x}$ and $g(x) = x^2 - 1$, find the rules of the functions fg and f/g and their domains.

SOLUTION The rules are

$$(fg)(x) = f(x)g(x) = \sqrt{3x}\,(x^2 - 1) = \sqrt{3x}\,x^2 - \sqrt{3x};$$

$$\left(\frac{f}{g}\right)(x) = \frac{f(x)}{g(x)} = \frac{\sqrt{3x}}{x^2 - 1}.$$

The domain of fg consists of all numbers x in both the domain of f (all nonnegative real numbers) and the domain of g (all real numbers), that is, all $x \geq 0$. The domain of f/g consists of all these x for which $g(x) \neq 0$, that is, all nonnegative real numbers *except* $x = 1$. ∎

If c is a real number and f is a function, then the product of f and the constant function $g(x) = c$ is usually denoted cf. For example, if the function

$$f(x) = x^3 - x + 2,$$

and $c = 5$, then $5f$ is the function given by

$$(5f)(x) = 5 \cdot f(x) = 5(x^3 - x + 2) = 5x^3 - 5x + 10.$$

■■ COMPOSITION OF FUNCTIONS

Another way of combining functions is illustrated by the function $h(x) = \sqrt{x^3}$. To compute $h(4)$, for example, you first find $4^3 = 64$ and then take the square root $\sqrt{64} = 8$. So the rule of h may be rephrased as follows:

First apply the function $f(x) = x^3$,

Then apply the function $g(t) = \sqrt{t}$ to the result.

The same idea can be expressed in functional notation like this:

$$x \xrightarrow{\textit{first apply } f} f(x) \xrightarrow{\textit{then apply } g \textit{ to the result}} g(f(x)).$$
$$x \qquad\qquad x^3 \qquad\qquad\qquad\qquad \sqrt{x^3}$$
$$\underset{\textit{apply } h}{\underline{\hspace{8cm}}\uparrow}$$

So the rule of h may be written as $h(x) = g(f(x))$, where $f(x) = x^3$ and $g(t) = \sqrt{t}$. We can think of h as being made up of two simpler functions f and g, or we can think of f and g being "composed" to create the function h. Both viewpoints are useful.

EXAMPLE 3

Suppose $f(x) = 4x^2 + 1$ and $g(t) = \dfrac{1}{t + 2}$. Define a new function h whose rule is "first apply f; then apply g to the result." In functional notation,

$$x \xrightarrow{\textit{first apply } f} f(x) \xrightarrow{\textit{then apply } g \textit{ to the result}} g(f(x)).$$

So the rule of the function h is $h(x) = g(f(x))$. Evaluating $g(f(x))$ means that whenever t appears in the formula for $g(t)$, we must replace it by $f(x) = 4x^2 + 1$:

$$h(x) = g(f(x)) = \frac{1}{f(x) + 2} = \frac{1}{(4x^2 + 1) + 2} = \frac{1}{4x^2 + 3}. \qquad \blacksquare$$

The function h in Example 3 is an illustration of the following definition.

Composite Functions

Let f and g be functions. The **composite function** of f and g defined as follows.

For input x, the output is $g(f(x))$.

This composite function is denoted $g \circ f$.

The symbol "$g \circ f$" is read "g circle f" or "f followed by g." (Note the order carefully; the functions are applied *right* to *left*.) So the rule of the composite function is

$$(g \circ f)(x) = g(f(x)).$$

EXAMPLE 4

If $f(x) = 2x + 5$ and $g(t) = 3t^2 + 2t + 4$, then find

$$(f \circ g)(2), \qquad (g \circ f)(-1), \qquad (g \circ f)(5).$$

SOLUTION

$$(f \circ g)(2) = f(g(2)) = f(3 \cdot 2^2 + 2 \cdot 2 + 4) = f(20) = 2 \cdot 20 + 5 = 45.$$

Similarly,

$$(g \circ f)(-1) = g(f(-1)) = g(2(-1) + 5) = g(3) = 3 \cdot 3^2 + 2 \cdot 3 + 4 = 37.$$

The value of a composite function can also be computed like this:

$$(g \circ f)(5) = g(f(5)) = 3(f(5)^2) + 2(f(5)) + 4 = 3(15^2) + 2(15) + 4 = 709. \quad \blacksquare$$

The domain of $g \circ f$ is determined by this convention.

Domain of
$g \circ f$

> The domain of the composite function $g \circ f$ is the set of all real numbers x such that x is in the domain of f and $f(x)$ is in the domain of g.

EXAMPLE 5

Find the rule and domain of $g \circ f$, when $f(x) = \sqrt{x}$ and $g(t) = t^2 - 5$.

SOLUTION

$$(g \circ f)(x) = g(f(x)) = (f(x))^2 - 5 = (\sqrt{x})^2 - 5 = x - 5.$$

Although $x - 5$ is defined for every real number x, the domain of $g \circ f$ is *not* the set of all real numbers. The domain of g is the set of all real numbers, but the function $f(x) = \sqrt{x}$ is defined only when $x \geq 0$. So the domain of $g \circ f$ is the set of nonnegative real numbers, that is, the interval $[0, \infty)$. $\blacksquare$

TECHNOLOGY TIP

Evaluating composite functions is easy on calculators other than TI-85 and most Casio calculators. If the functions are entered in the equation memory as $y_1 = g(x)$ and $y_2 = h(x)$ (with f in place of y on HP-39), then keying in $y_2(y_1(5))$ ENTER produces the number $h(g(5))$.

On TI-85 and most Casio calculators, this syntax does *not* produce $h(g(5))$; it produces

$$h(x) \cdot g(x) \cdot 5$$

for whatever number is stored in the x-memory.

EXAMPLE 6

Write the function $h(x) = \sqrt{3x^2 + 1}$ in two different ways as the composite of two functions.

SOLUTION   Let $f(x) = 3x^2 + 1$ and $g(x) = \sqrt{x}.$* Then

$$(g \circ f)(x) = g(f(x)) = g(3x^2 + 1) = \sqrt{3x^2 + 1} = h(x).$$

Similarly, h is also the composite $j \circ k$, where $j(x) = \sqrt{x + 1}$ and $k(x) = 3x^2$:

$$(j \circ k)(x) = j(k(x)) = j(3x^2) = \sqrt{3x^2 + 1} = h(x). \quad \blacksquare$$

EXAMPLE 7

If $k(x) = (x^2 - 2x + \sqrt{x})^3$, then k is $g \circ f$, where $f(x) = x^2 - 2x + \sqrt{x}$ and $g(x) = x^3$ because

$$(g \circ f)(x) = g(f(x)) = g(x^2 - 2x + \sqrt{x}) = (x^2 - 2x + \sqrt{x})^3 = k(x). \quad \blacksquare$$

---

*Now that you have the idea of composite functions, we'll use the same letter for the variable in both functions.

By using the operations above, a complicated function may be considered as being built up from simple parts.

EXAMPLE 8

The function

$$f(x) = \sqrt{\frac{3x^2 - 4x + 5}{x^3 + 1}}$$

may be considered as the composite $f = g \circ h$, where

$$h(x) = \frac{3x^2 - 4x + 5}{x^3 + 1} \quad \text{and} \quad g(x) = \sqrt{x},$$

since

$$(g \circ h)(x) = g(h(x)) = g\left(\frac{3x^2 - 4x + 5}{x^3 + 1}\right) = \sqrt{\frac{3x^2 - 4x + 5}{x^3 + 1}} = f(x).$$

The function

$$h(x) = \frac{3x^2 - 4x + 5}{x^3 + 1}$$

is the quotient $\dfrac{p}{q}$, where

$$p(x) = 3x^2 - 4x + 5 \quad \text{and} \quad q(x) = x^3 + 1.$$

The function $p(x) = 3x^2 - 4x + 5$ may be written $p = k - s + r$, where

$$k(x) = 3x^2, \quad s(x) = 4x, \quad r(x) = 5.$$

The function k, in turn, can be considered as the product $3I^2$, where I is the *identity function* [whose rule is $I(x) = x$]:

$$(3I^2)(x) = 3(I^2(x)) = 3(I(x)I(x)) = 3 \cdot x \cdot x = 3x^2 = k(x).$$

Similarly, $s(x) = (4I)(x) = 4I(x) = 4x$. The function $q(x) = x^3 + 1$ may be "decomposed" in the same way.

Thus, the complicated function f is just the result of performing suitable operations on the identity function I and various constant functions.   ■

As you may have noticed, there are two possible ways to form a composite function from two given functions. If f and g are functions, we can consider either

$$(g \circ f)(x) = g(f(x)), \qquad \text{[the composite of } f \text{ and } g\text{]}$$

$$(f \circ g)(x) = f(g(x)). \qquad \text{[the composite of } g \text{ and } f\text{]}$$

The *order is important,* as we shall now see:

$g \circ f$ and $f \circ g$ usually are *not* the same function.

EXAMPLE 9

If $f(x) = x^2$ and $g(x) = x + 3$, then

$$(g \circ f)(x) = g(f(x)) = g(x^2) = x^2 + 3$$

but

$$(f \circ g)(x) = f(g(x)) = f(x + 3) = (x + 3)^2 = x^2 + 6x + 9.$$

Obviously, $g \circ f \neq f \circ g$, since, for example, they have different values at $x = 0$. ∎

CAUTION

Don't confuse the product function fg with the composite function $f \circ g$ (g followed by f). For instance, if $f(x) = 2x^2$ and $g(x) = x - 3$, then the product fg is given by

$$(fg)(x) = f(x)g(x) = 2x^2(x - 3) = 2x^3 - 6x^2.$$

It is *not* the same as the composite $f \circ g$ because

$$(f \circ g)(x) = f(g(x)) = f(x - 3) = 2(x - 3)^2 = 2x^2 - 12x + 18.$$

■ APPLICATIONS

Composition of functions arises in applications involving several functional relationships simultaneously. In such cases, one quantity may have to be expressed as a function of another.

EXAMPLE 10

A circular puddle of liquid is evaporating and slowly shrinking in size. After t minutes, the radius r of the puddle measures $\dfrac{18}{2t + 3}$ inches; in other words, the radius is a function of time. The area A of the puddle is given by $A = \pi r^2$, that is, area is a function of the radius r. We can express the area as a function of time by substituting $r = \dfrac{18}{2t + 3}$ in the area equation:

$$A = \pi r^2 = \pi\left(\frac{18}{2t + 3}\right)^2.$$

This amounts to forming the composite function $f \circ g$, where $f(r) = \pi r^2$ and $g(t) = \dfrac{18}{2t + 3}$:

$$(f \circ g)(t) = f(g(t)) = f\left(\frac{18}{2t + 3}\right) = \pi\left(\frac{18}{2t + 3}\right)^2.$$

When area is expressed as a function of time, it is easy to compute the area of the puddle at any time. For instance, after 12 minutes, the area of the puddle is

$$A = \pi\left(\frac{18}{2t + 3}\right)^2 = \pi\left(\frac{18}{2 \cdot 12 + 3}\right)^2 = \frac{4\pi}{9} \approx 1.396 \text{ square inches.} \quad ∎$$

EXAMPLE 11

At noon, a car leaves Podunk on a straight road, heading south at 45 mph, and a plane 3 miles above the ground passes over Podunk heading east at 350 mph.

(a) Express the distance r traveled by the car and the distance s traveled by the plane as functions of time.

(b) Express the distance d between the plane and the car in terms of r and s.

(c) Express d as a function of time.

(d) How far apart were the plane and the car at 1:30 P.M.?

SOLUTION

(a) Traveling at 45 mph for t hours, the car will go a distance of $45t$ miles. Hence, the equation $r = 45t$ expresses the distance r as a function of the time t. Similarly, the equation $s = 350t$ expresses the distance s as a function of the time t.

(b) To express the distance d as a function of r and s, consider Figure 3–40.

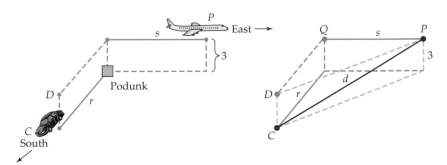

Figure 3–40

Right triangle PQD and the Pythagorean Theorem show that $(PD)^2 = r^2 + s^2$; hence, $PD = \sqrt{r^2 + s^2}$. Applying the Pythagorean Theorem to right triangle PDC, we have

$$d^2 = 3^2 + (PD)^2$$
$$d^2 = 3^2 + (\sqrt{r^2 + s^2})^2$$
$$d^2 = 9 + r^2 + s^2$$
$$d = \sqrt{9 + r^2 + s^2}.$$

(c) The preceding equation expresses d in terms of r and s. By substituting $r = 45t$ and $s = 350t$ in this equation, we can express d as a function of the time t:

$$d = \sqrt{9 + r^2 + s^2}$$
$$d = \sqrt{9 + (45t)^2 + (350t)^2}$$
$$d = \sqrt{9 + 2025t^2 + 122{,}500t^2} = \sqrt{9 + 124{,}525t^2}.$$

(d) At 1:30 P.M., we have $t = 1.5$ (since noon is $t = 0$). At this time,

$$d = \sqrt{9 + 124{,}525t^2} = \sqrt{9 + 124{,}525(1.5)^2} = \sqrt{280{,}190.25}$$
$$\approx 529.33 \text{ miles.} \quad \blacksquare$$

EXERCISES 3.5

In Exercises 1–4, find $(f + g)(x)$, $(f - g)(x)$, and $(g - f)(x)$.

1. $f(x) = -3x + 2$, $g(x) = x^3$

2. $f(x) = x^2 + 2$, $g(x) = -4x + 7$

3. $f(x) = 1/x$, $g(x) = x^2 + 2x - 5$

4. $f(x) = \sqrt{x}$, $g(x) = x^2 + 1$

In Exercises 5–8, find $(fg)(x)$, $(f/g)(x)$, and $(g/f)(x)$.

5. $f(x) = -3x + 2$, $g(x) = x^3$

6. $f(x) = 4x^2 + x^4$, $g(x) = \sqrt{x^2 + 4}$

7. $f(x) = x^2 - 3$, $g(x) = \sqrt{x - 3}$

8. $f(x) = \sqrt{x^2 - 1}$, $g(x) = \sqrt{x - 1}$

In Exercises 9–12, find the domains of fg and f/g.

9. $f(x) = x^2 + 1$, $g(x) = 1/x$

10. $f(x) = x + 2$. $g(x) = \dfrac{1}{x + 2}$

11. $f(x) = \sqrt{4 - x^2}$, $g(x) = \sqrt{3x + 4}$

12. $f(x) = 3x^2 + x^4 + 2$, $g(x) = 4x - 3$

In Exercises 13–16, find the indicated values, where

$$g(t) = t^2 - t \text{ and } f(x) = 1 + x.$$

13. $g(f(0))$

14. $(f \circ g)(3)$

15. $g(f(2) + 3)$

16. $f(2g(1))$

In Exercises 17–20, find $(g \circ f)(3)$, $(f \circ g)(1)$, and $(f \circ f)(0)$.

17. $f(x) = 3x - 2$, $g(x) = x^2$

18. $f(x) = |x + 2|$, $g(x) = -x^2$

19. $f(x) = x$, $g(x) = -3$

20. $f(x) = x^2 - 1$, $g(x) = \sqrt{x}$

In Exercises 21–24, find the rule of the function $f \circ g$, the domain of $f \circ g$, the rule of $g \circ f$, and the domain of $g \circ f$.

21. $f(x) = x^2$, $g(x) = x + 3$

22. $f(x) = -3x + 2$, $g(x) = x^3$

23. $f(x) = 1/x$, $g(x) = \sqrt{x}$

24. $f(x) = \dfrac{1}{2x + 1}$, $g(x) = x^2 - 1$

In Exercises 25–28, find the rules of the functions ff and $f \circ f$.

25. $f(x) = x^3$

26. $f(x) = (x - 1)^2$

27. $f(x) = 1/x$

28. $f(x) = \dfrac{1}{x - 1}$

In Exercises 29–32, verify that $(f \circ g)(x) = x$ and $(g \circ f)(x) = x$ for every x.

29. $f(x) = 9x + 2$, $g(x) = \dfrac{x - 2}{9}$

30. $f(x) = \sqrt[3]{x - 1}$, $g(x) = x^3 + 1$

31. $f(x) = \sqrt[3]{x} + 2$, $g(x) = (x - 2)^3$

32. $f(x) = 2x^3 - 5$, $g(x) = \sqrt[3]{\dfrac{x + 5}{2}}$

Exercises 33 and 34 refer to the function f whose graph is shown in the figure.

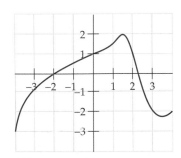

33. Let g be the composite function $f \circ f$, that is,

$$g(x) = (f \circ f)(x) = f(f(x)).$$

Use the graph of f to fill in the following table (approximate where necessary).

x	$f(x)$	$g(x) = f(f(x))$
-4		
-3		
-2	0	1
-1		
0		
1		
2		
3		
4		

34. Use the information obtained in Exercise 33 to sketch the graph of the function g.

In Exercises 35–38, fill the blanks in the given table. In each case the values of the functions f and g are given by these tables:

x	$f(x)$
1	3
2	5
3	1
4	2
5	3

t	$g(t)$
1	5
2	4
3	4
4	3
5	2

35.

x	$(g \circ f)(x)$
1	4
2	
3	5
4	
5	

36.

t	$(f \circ g)(t)$
1	
2	2
3	
4	
5	

37.

x	$(f \circ f)(x)$
1	
2	
3	3
4	
5	

38.

t	$(g \circ g)(t)$
1	
2	
3	
4	4
5	

In Exercises 39–44, write the given function as the composite of two functions, neither of which is the identity function, as in Examples 6 and 7. (There may be more than one way to do this.)

39. $f(x) = \sqrt[3]{x^2 + 2}$

40. $g(x) = \sqrt{x + 3} - \sqrt[3]{x + 3}$

41. $h(x) = (7x^3 - 10x + 17)^7$

42. $k(x) = \sqrt[3]{(7x - 3)^2}$

43. $f(x) = \dfrac{1}{3x^2 + 5x - 7}$

44. $g(t) = \dfrac{3}{\sqrt{t - 3}} + 7$

45. If $f(x) = x + 1$ and $g(t) = t^2$, then

$$(g \circ f)(x) = g(f(x)) = g(x + 1) = (x + 1)^2$$
$$= x^2 + 2x + 1$$

Find two other functions $h(x)$ and $k(t)$ such that $(k \circ h)(x) = x^2 + 2x + 1$.

46. If f is any function and I is the identity function, what are $f \circ I$ and $I \circ f$?

In Exercises 47–50, determine whether the functions $f \circ g$ and $g \circ f$ are defined. If a composite is defined, find its domain.

47. $f(x) = x^3$, $g(x) = \sqrt{x}$

48. $f(x) = x^2 + 1$, $g(x) = \sqrt{x}$

49. $f(x) = \sqrt{x + 10}$, $g(x) = 5x$

50. $f(x) = -x^2$, $g(x) = \sqrt{x}$

51. (a) If $f(x) = 2x^3 + 5x - 1$, find $f(x^2)$.
 (b) If $f(x) = 2x^3 + 5x - 1$, find $(f(x))^2$.
 (c) Are the answers in parts (a) and (b) the same? What can you conclude about $f(x^2)$ and $(f(x))^2$?

52. Give an example of a function f such that

$$f\left(\frac{1}{x}\right) \neq \frac{1}{f(x)}.$$

In Exercises 53 and 54, graph both $f \circ g$ and $g \circ f$ on the same screen. Use the graphs to determine whether $f \circ g$ is the same function as $g \circ f$.

53. $f(x) = x^5 - x^3 - x$; $g(x) = x - 2$

54. $f(x) = x^3 + x$; $g(x) = \sqrt[3]{x - 1}$

In Exercises 55–58, find $g \circ f$, and find the difference quotient of the function $g \circ f$.

55. $f(x) = x + 3$; $g(x) = x^2 + 1$

56. $f(x) = 2x$; $g(x) = x^2 + x$

57. $f(x) = x + 1$; $g(x) = 1/x$

58. $f(x) = x^3$; $g(x) = x + 2$

59. (a) What is the area of the puddle in Example 10 after one day? After a week? After a month?
 (b) Does the puddle ever totally evaporate? Is this realistic? Under what circumstances might this area function be an accurate model of reality?

60. In a laboratory culture, the number $N(d)$ of bacteria (in thousands) at temperature d degrees Celsius is given by the function

$$N(d) = \frac{-90}{d + 1} + 20 \quad (4 \leq d \leq 32).$$

The temperature $D(t)$ at time t hours is given by the function $D(t) = 2t + 4$ $(0 \leq t \leq 14)$.
 (a) What does the composite function $N \circ D$ represent?
 (b) How many bacteria are in the culture after 4 hours? After 10 hours?

61. A certain fungus grows in a circular shape. Its diameter after t weeks is $6 - \dfrac{50}{t^2 + 10}$ inches.
   (a) Express the area covered by the fungus as a function of time.
   (b) What is the area covered by the fungus when $t = 0$? What area does it cover at the end of 8 weeks?
   (c) When is its area 25 square inches?

62. Tom left point P at 6 A.M. walking south at 4 mph. Anne left point P at 8 A.M. walking west at 3.2 mph.
   (a) Express the distance between Tom and Anne as a function of the time t elapsed since 6 A.M.
   (b) How far apart are Tom and Anne at noon?
   (c) At what time are they 35 miles apart?

63. As a weather balloon is inflated, its radius increases at the rate of 4 centimeters per second. Express the volume of the balloon as a function of time and determine the volume of the balloon after 4 seconds. [*Hint:* The volume of a sphere of radius r is $4\pi r^3/3$.]

64. Express the surface area of the weather balloon in Exercise 63 as a function of time. [*Hint:* The surface area of a sphere of radius r is $4\pi r^2$.]

65. Charlie, who is 6 feet tall, walks away from a streetlight that is 15 feet high at a rate of 5 feet per second, as shown in the figure. Express the length s of Charlie's shadow as a function of time. [*Hint:* First use similar triangles to express s as a function of the distance d from the streetlight to Charlie.]

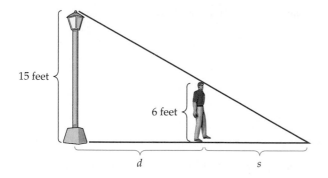

15 feet   6 feet d s

66. A water-filled balloon is dropped from a window 120 feet above the ground. Its height above the ground after t seconds is $120 - 16t^2$ feet. Laura is standing on the ground 40 feet from the point where the balloon will hit the ground, as shown in the figure.
   (a) Express the distance d between Laura and the balloon as a function of time.
   (b) When is the balloon exactly 90 feet from Laura?

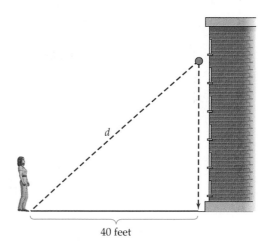

d

40 feet

Thinker

67. Find a function f (other than the identity function) such that $(f \circ f \circ f)(x) = x$ for every x in the domain of f. [Several correct answers are possible.]

3.6 | Rates of Change

Rates of change play a central role in the analysis of many real-world situations. To understand the basic ideas involved in rates of change, we take another look at the falling rock from Sections 3.1 and 3.2. We saw that when the rock is dropped from a high place, then the distance it travels (ignoring wind resistance) is given by the function

$$d(t) = 16t^2$$

with distance $d(t)$ measured in feet and time t in seconds. The following table shows the distance the rock has fallen at various times:

Time t	0	1	2	3	3.5	4	4.5	5
Distance $d(t)$	0	16	64	144	196	256	324	400

To find the distance the rock falls from time $t = 1$ to $t = 3$, we note that at the end of three seconds, the rock has fallen $d(3) = 144$ feet, whereas it had fallen only $d(1) = 16$ feet at the end of one second. So during this time interval, the rock traveled

$$d(3) - d(1) = 144 - 16 = 128 \text{ feet.}$$

The distance traveled by the rock during other time intervals can be found similarly:

Time Interval	Distance Traveled
$t = 1$ to $t = 4$	$d(4) - d(1) = 256 - 16 = 240$
$t = 2$ to $t = 3.5$	$d(3.5) - d(2) = 196 - 64 = 132$
$t = 2$ to $t = 4.5$	$d(4.5) - d(2) = 324 - 64 = 260$

The same procedure works in general:

The distance traveled from time $t = a$ to time $t = b$ is $d(b) - d(a)$ feet.

In the preceding chart, the length of each time interval can be computed by taking the difference between the two times. For example, from $t = 1$ to $t = 4$ is a time interval of length $4 - 1 = 3$ seconds. Similarly, the interval from $t = 2$ to $t = 3.5$ is of length $3.5 - 2 = 1.5$ seconds, and in general,

The time interval from $t = a$ to $t = b$ is an interval of $b - a$ seconds.

Since Distance = Average speed × Time,

$$\text{Average speed} = \frac{\text{Distance traveled}}{\text{Time interval}}.$$

Hence, the average speed over the time interval from $t = a$ to $t = b$ is

$$\text{Average speed} = \frac{\text{Distance traveled}}{\text{Time interval}} = \frac{d(b) - d(a)}{b - a}.$$

For example, to find the average speed from $t = 1$ to $t = 4$, apply the preceding formula with $a = 1$ and $b = 4$:

$$\text{Average speed} = \frac{d(4) - d(1)}{4 - 1} = \frac{256 - 16}{4 - 1} = \frac{240}{3} = 80 \text{ ft per second.}$$

Similarly, the average speed from $t = 2$ to $t = 4.5$ is

$$\frac{d(4.5) - d(2)}{4.5 - 2} = \frac{324 - 64}{4.5 - 2} = \frac{260}{2.5} = 104 \text{ ft per second.}$$

The units in which average speed is measured here (feet per second) indicate the number of units of distance traveled during each unit of time, that is, the *rate of change* of distance (feet) with respect to time (seconds). The preceding discussion can be summarized by saying that the average speed (rate of change of distance with respect to time) as time changes from $t = a$ to $t = b$ is given by

$$\text{Average speed} = \text{Average rate of change}$$

$$= \frac{\text{Change in distance}}{\text{Change in time}} = \frac{d(b) - d(a)}{b - a}.$$

Although speed is the most familiar example, rates of change play a role in many other situations as well, as illustrated in Examples 1–3 below. Consequently, we define the average rate of change of any function as follows.

Average Rate of Change

> Let f be a function. The **average rate of change of $f(x)$** with respect to x as x changes from a to b is the number
>
> $$\frac{\text{Change in } f(x)}{\text{Change in } x} = \frac{f(b) - f(a)}{b - a}.$$

EXAMPLE 1

A large heavy-duty balloon is being filled with water. Its approximate volume (in gallons) is given by

$$V(x) = \frac{x^3}{55}$$

where x is the radius of the balloon (in inches). Find the average rate of change of the volume of the balloon as the radius increases from 5 to 10 inches.

SOLUTION

$$\frac{\text{Change in volume}}{\text{Change in radius}} = \frac{V(10) - V(5)}{10 - 5} \approx \frac{18.18 - 2.27}{10 - 5} = \frac{15.91}{5}$$

$$= 3.182 \text{ gallons per inch.} \quad \blacksquare$$

EXAMPLE 2

According to U.S. Census Bureau data, the population of Cleveland, Ohio, can be approximated by

$$f(x) = (4.063 \times 10^{-5})x^4 - .00587x^3 - .0274x^2 + 21.8x + 377$$
$$(0 \le x \le 100),$$

where $x = 0$ corresponds to 1900 and $f(x)$ is in thousands. Find the average rate of change of the population during the following time periods.

(a) 1900–1950   (b) 1960–2000

SOLUTION

(a) Now 1900 corresponds to $x = 0$ and 1950 to $x = 50$, so

$$\text{Average rate of change} = \frac{f(50) - f(0)}{50 - 0} \approx \frac{918.69 - 377}{50} \approx 10.83.$$

Since $f(x)$ is measured in thousands, the population was increasing at an average rate of approximately 10,830 people per year during this period.

(b) During this period ($x = 60$ to $x = 100$), the average rate of change was

$$\frac{f(100) - f(60)}{100 - 60} \approx \frac{476 - 845}{40} \approx -9.23.$$

The rate is negative, which means that the population was *decreasing* at an average rate of approximately 9,230 people per year during this period. ∎

EXAMPLE 3

Figure 3–41 is the graph of the temperature function f during a particular day; $f(x)$ is the temperature at x hours after midnight. What is the average rate of change of the temperature (a) from 4 A.M. to noon? (b) from 3 P.M. to 8 P.M.?

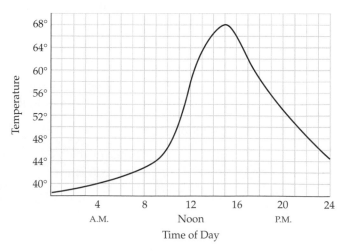

Figure 3–41

SOLUTION

(a) The graph shows that the temperature at 4 A.M. is $f(4) = 40°$ and the temperature at noon is $f(12) = 58°$. The average rate of change of temperature is

$$\frac{\text{Change in temperature}}{\text{Change in time}} = \frac{f(12) - f(4)}{12 - 4} = \frac{58 - 40}{12 - 4} = \frac{18}{8}$$

$$= 2.25° \text{ per hour.}$$

The rate of change is positive because the temperature is increasing at an average rate of 2.25° per hour.

(b) Now 3 P.M. corresponds to $x = 15$ and 8 P.M. to $x = 20$. The graph shows that $f(15) = 68°$ and $f(20) = 53°$. Hence, the average rate of change of temperature is

$$\frac{\text{Change in temperature}}{\text{Change in time}} = \frac{f(20) - f(15)}{20 - 15} = \frac{53 - 68}{20 - 15} = \frac{-15}{5}$$

$$= -3° \text{ per hour.}$$

The rate of change is negative because the temperature is decreasing at an average rate of 3° per hour.   ■

⁙ GEOMETRIC INTERPRETATION OF AVERAGE RATE OF CHANGE

If P and Q are points on the graph of a function f, then the straight line determined by P and Q is called a **secant line.** Figure 3–42 shows the secant line joining the points (4, 40) and (12, 58) on the graph of the temperature function f of Example 3.

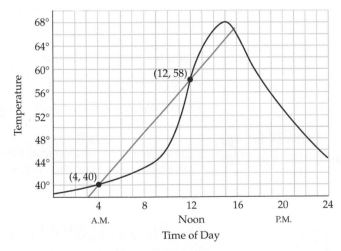

Figure 3–42

Using the points (4, 40) and (12, 58), we see that the slope of this secant line is

$$\frac{58 - 40}{12 - 4} = \frac{18}{8} = 2.25.$$

To say that (4, 40) and (12, 58) are on the graph of f means that $f(4) = 40$ and $f(12) = 58$. Thus,

$$\text{Slope of secant line} = 2.25 = \frac{58 - 40}{12 - 4} = \frac{f(12) - f(4)}{12 - 4}$$

$$= \text{Average rate of change as } x \text{ goes from 4 to 12.}$$

The same thing happens in the general case.

Secant Lines and Average Rates of Change

If f is a function, then the average rate of change of $f(x)$ with respect to x as x changes from $x = a$ to $x = b$ is the slope of the secant line joining the points $(a, f(a))$ and $(b, f(b))$ on the graph of f.

EXAMPLE 4

Current data and projections from the Office of Management and Budget show that between 1990 and 2004, the national debt (the amount the U.S. government owes) can be approximated by

$$f(x) = 199x + 3760 \quad (0 \le x \le 14),$$

where $x = 0$ corresponds to 1990 and $f(x)$ is in billions of dollars. At what average rate is the natioal debt changing from year to year during this period?

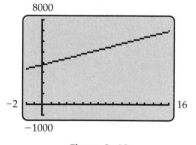

Figure 3–43

SOLUTION   The graph of $f(x) = 199x + 3760$ is a straight line (Figure 3–43). The secant line joining any two points on the graph is just *the graph itself,* that is, the line $y = 199x + 3760$. As we saw in Section 1.4, the slope of this line is 199. Therefore, the average rate of change of the national debt function between any two values of x is 199. In other words, at any time during this period, the national debt is increasing at a constant rate of \$199 billion per year. ∎

The argument used in Example 4 works for any function whose graph is a straight line and leads to this conclusion.

Rates of Change of Linear Functions

The average rate of change of a linear function $f(x) = mx + b$. as x changes from one value to another, is the slope m of the line.

⠶ THE DIFFERENCE QUOTIENT

Average rates of change are often computed for very small intervals. For instance, we might compute the rate from 4 to 4.01 or from 4 to 4.001. Since $4.01 = 4 + .01$ and $4.001 = 4 + .001$, we are doing essentially the same thing in both cases: computing the rate of change over the interval from 4 to $4 + h$ for some small nonzero quantity h. Furthermore, it's often possible to use a single calculation to determine the average rate for all possible values of h.

EXAMPLE 5

Consider the falling rock with which this section began. The distance the rock has traveled at time t is given by $d(t) = 16t^2$, and its average speed (rate of change) from $t = 4$ to $t = 4 + h$ is

$$\text{Average speed} = \frac{d(4 + h) - d(4)}{(4 + h) - 4} = \frac{16(4 + h)^2 - 16 \cdot 4^2}{h}$$

$$= \frac{16(16 + 8h + h^2) - 256}{h} = \frac{256 + 128h + 16h^2 - 256}{h}$$

$$= \frac{128h + 16h^2}{h} = \frac{h(128 + 16h)}{h} = 128 + 16h.$$

Thus, we can quickly compute the average speed over the interval from 4 to $4 + h$ seconds for any value of h by using the formula

$$\text{Average speed} = 128 + 16h.$$

For example, the average speed from 4 seconds to 4.001 seconds (here $h = .001$) is

$$128 + 16h = 128 + 16(.001) = 128 + .016 = 128.016 \text{ feet per second.} \quad ■$$

Similar calculations can be done with any number in place of 4. In each such case, we are dealing with an interval from x to $x + h$ for some number x. As in Example 5, a single computation can often be used for all possible x and h.

EXAMPLE 6

The average speed of the falling rock of Example 4 from time x to time $x + h$ is:[*]

$$\text{Average speed} = \frac{d(x + h) - d(x)}{(x + h) - x} = \frac{16(x + h)^2 - 16x^2}{h}$$

$$= \frac{16(x^2 + 2xh + h^2) - 16x^2}{h} = \frac{16x^2 + 32xh + 16h^2 - 16x^2}{h}$$

$$= \frac{32xh + 16h^2}{h} = \frac{h(32x + 16h)}{h} = 32x + 16h.$$

When $x = 4$, then this result states that the average speed from 4 to $4 + h$ is $32(4) + 16h = 128 + 16h$, which is exactly what we found in Example 4. To find the average speed from 3 to 3.1 seconds, apply the formula

$$\text{Average speed} = 32x + 16h$$

with $x = 3$ and $h = .1$:

$$\text{Average speed} = 32 \cdot 3 + 16(.1) = 96 + 1.6 = 97.6 \text{ feet per second.} \quad ■$$

---

[*]Note that this calculation is the same as in Example 5 except that 4 has been replaced by x.

More generally, we can compute the average rate of change of any function f over the interval from x to $x + h$ just as we did in Example 6: Apply the definition of average rate of change in the box on page 193 with x in place of a and $x + h$ in place of b:

$$\text{Average rate of change} = \frac{f(b) - f(a)}{b - a} = \frac{f(x + h) - f(x)}{(x + h) - x}$$

$$= \frac{f(x + h) - f(x)}{h}.$$

This last quantity is just the difference quotient of f (see page 139). Therefore,

Difference Quotients and Rates of Change

> If f is a function, then the average rate of change of f over the interval from x to $x + h$ is given by the difference quotient
>
> $$\frac{f(x + h) - f(x)}{h}.$$

EXAMPLE 7

Find the difference quotient of $V(x) = x^3/55$, and use it to find the average rate of change of V as x changes from 8 to 8.01.

SOLUTION   Use the definition of the difference quotient and algebra:

$$\frac{V(x + h) - V(x)}{h} = \frac{\overbrace{\frac{(x + h)^3}{55}}^{V(x+h)} - \overbrace{\frac{x^3}{55}}^{V(x)}}{h} = \frac{\frac{1}{55}[(x + h)^3 - x^3]}{h}$$

$$= \frac{1}{55} \cdot \frac{(x + h)^3 - x^3}{h} = \frac{1}{55} \cdot \frac{x^3 + 3x^2h + 3xh^2 + h^3 - x^3}{h}$$

$$= \frac{1}{55} \cdot \frac{3x^2h + 3xh^2 + h^3}{h} = \frac{1}{55} \cdot \frac{h(3x^2 + 3xh + h^2)}{h}$$

$$= \frac{3x^2 + 3xh + h^2}{55}.$$

When x changes from 8 to $8.01 = 8 + .01$, we have $x = 8$ and $h = .01$. So the average rate of change is

$$\frac{3x^2 + 3xh + h^2}{55} = \frac{3 \cdot 8^2 + 3 \cdot 8(.01) + (.01)^2}{55} \approx 3.495. \quad\blacksquare$$

▪▪ INSTANTANEOUS RATE OF CHANGE

Rates of change are a major theme in calculus—not just the average rate of change discussed above, but also the *instantaneous rate of change* of a function (that is, its rate of change at a particular instant). Even without calculus, however, we can obtain quite accurate approximations of instantaneous rates of change by using average rates appropriately.

EXAMPLE 8

A rock is dropped from a high place. What is its speed exactly 3 seconds after it is dropped?

SOLUTION   The distance the rock has fallen at time t is given by the function $d(t) = 16t^2$. The exact speed at $t = 3$ can be approximated by finding the average speed over very small time intervals, say, 3 to 3.01 or even shorter intervals. Over a very short time span, such as a hundredth of a second, the rock cannot change speed very much, so these average speeds should be a reasonable approximation of its speed at the instant $t = 3$. Example 5 shows that the average speed is given by the difference quotient $32x + 16h$. When $x = 3$, the difference quotient is $32 \cdot 3 + 16h = 96 + 16h$, and we have the following:

Change in Time 3 to $3 + h$	h	Average Speed [Difference Quotient at $x = 3$] $96 + 16h$
3 to 3.1	.1	$96 + 16(.1) = 97.6$ ft per second
3 to 3.01	.01	$96 + 16(.01) = 96.16$ ft per second
3 to 3.005	.005	$96 + 16(.005) = 96.08$ ft per second
3 to 3.00001	.00001	$96 + 16(.00001) = 96.00016$ ft per second

The table suggests that the exact speed of the rock at the instant $t = 3$ seconds is very close to 96 feet per second.   ∎

EXAMPLE 9

A balloon is being filled with water in such a way that when its radius is x inches, then its volume is $V(x) = x^3/55$ gallons. In Example 1, we saw that the average rate of change of the volume as the radius increases from 5 inches to 10 inches is 3.182 gallons per inch. What is the rate of change at the instant when the radius is 7 inches?

SOLUTION   The average rate of change when the radius goes from x to $x + h$ inches is given by the difference quotient of $V(x)$, which was found in Example 7:

$$\frac{V(x + h) - V(x)}{h} = \frac{3x^2 + 3xh + h^2}{55}.$$

Therefore, when $x = 7$, the difference quotient is

$$\frac{3 \cdot 7^2 + 3 \cdot 7 \cdot h + h^2}{55} = \frac{147 + 21h + h^2}{55},$$

and we have these average rates of change over small intervals near 7:

Change in Radius 7 to 7 + h	h	Average Rate of Change of Volume [Difference Quotient at $x = 7$] $\dfrac{147 + 21h + h^2}{55}$
7 to 7.01	.01	2.6765 gallons per inch
7 to 7.001	.001	2.6731 gallons per inch
7 to 7.0001	.0001	2.6728 gallons per inch
7 to 7.00001	.00001	2.6727 gallons per inch

The chart suggests that at the instant the radius is 7 inches, the volume is changing at a rate of approximately 2.673 gallons per inch. ■

  EXERCISES 3.6

1. A car moves along a straight test track. The distance traveled by the car at various times is shown in this table:

Time (seconds)	0	5	10	15	20	25	30
Distance (feet)	0	20	140	400	680	1400	1800

Find the average speed of the car over the interval from
(a) 0 to 10 seconds   (b) 10 to 20 seconds
(c) 20 to 30 seconds   (d) 15 to 30 seconds

2. Find the average rate of change of the volume of the balloon in Example 1 as the radius increases from
(a) 2 to 5 inches   (b) 4 to 8 inches

3. Use the function in Example 2 to find the average rate of change of Cleveland's population from
(a) 1920 to 1960   (b) 1990 to 2000.

4. The graph shows the total amount spent on advertising (in millions of dollars) in the United States, as estimated by a leading advertising publication.[*] Find the average rate of change in advertising expenditures over the following time periods:
(a) 1950–1970   (b) 1970–1980
(c) 1980–2000   (d) 1950–2000

(e) During which of these periods were expenditures increasing at the fastest rate?

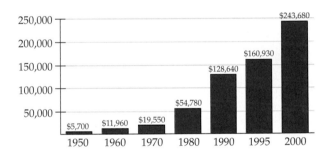

5. The table below shows the total projected elementary and secondary school enrollment (in thousands) for selected years.[†]

Find the average rate of change of enrollment from
(a) 1980 to 1985   (b) 1985 to 1995
(c) 1995 to 2005   (d) 2005 to 2010
(e) During which of these periods was enrollment increasing at the fastest rate? At the slowest rate?
(f) During which of these periods was enrollment decreasing at the fastest rate? at the slowest rate?

Year	1980	1985	1990	1995	2000	2005	2010
Enrollment	46,208	44,979	46,448	50,503	52,989	53,465	53,017

*Advertising Age.

†U.S. National Center for Education Statistics.

6. The graph shows the minimum wage (in constant 2000 dollars), with $x = 0$ corresponding to 1950.[*] Find the average rate of change in the minimum wage from
(a) 1950 to 1976   (b) 1976 to 1995
(c) 1995 to 2000   (d) 1976 to 2000

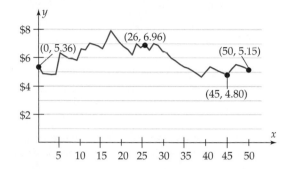

7. The table shows the total number of shares traded (in millions) on the New York Stock Exchange in selected years.[†]

Year	1991	1993	1995	1997	1999	2001
Volume	45.3	66.9	87.2	133.3	203.9	307.5

Find the average rate of change in share volume from
(a) 1991 to 1995   (b) 1995 to 1999
(c) 1999 to 2001   (d) 1991 to 2001
(e) During which of these periods did share volume increase at the fastest rate?

8. The graph in the figure shows the annual sales of floral pattern ties (in thousands of ties) made by Neckwear, Inc., over a 48-month period. Sales are very low when the ties are first introduced, increase significantly, hold steady for a while, and then drop off as the ties go out of fashion. Find the average rate of change of sales (in ties per month) over the interval:
(a) 0 to 12   (b) 8 to 24   (c) 12 to 24
(d) 20 to 28   (e) 28 to 36   (f) 32 to 44
(g) 36 to 40   (h) 40 to 48

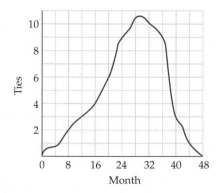

---

[*]U.S. Employment Standards Administration.
[†]*NYSE Factbook.*

9. The XYZ Company has found that its sales are related to the amount of advertising it does in trade magazines. The graph in the figure shows the sales (in thousands of dollars) as a function of the amount of advertising (in number of magazine ad pages). Find the average rate of change of sales when the number of ad pages increases from
(a) 10 to 20   (b) 20 to 60
(c) 60 to 100   (d) 0 to 100
(e) Is it worthwhile to buy more than 70 pages of ads, if the cost of a one-page ad is $2000? If the cost is $5000? If the cost is $8000?

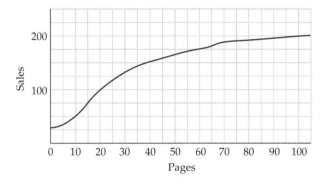

10. When blood flows through an artery (which can be thought of as a cylindrical tube) its velocity is greatest at the center of the artery. Because of friction along the walls of the tube, the blood's velocity decreases as the distance r from the center of the artery increases, finally becoming 0 at the wall of the artery. The velocity (in centimeters per second) is given by the function $v = 18{,}500(.000065 - r^2)$, where r is measured in centimeters. Find the average rate of change of the velocity as the distance from the center changes from
(a) $r = .001$ to $r = .002$ (b) $r = .002$ to $r = .003$
(c) $r = 0$ to $r = .005$

In Exercises 11–16, find the average rate of change of the function f over the given interval.

11. $f(x) = 2 - x^2$ from $x = 0$ to $x = 2$

12. $f(x) = .25x^4 - x^2 - 2x + 4$ from $x = -1$ to $x = 4$

13. $f(x) = x^3 - 3x^2 - 2x + 6$ from $x = -1$ to $x = 3$

14. $f(x) = -\sqrt{x^4 - x^3 + 2x^2 - x + 4}$ from $x = 0$ to $x = 3$

15. $f(x) = \sqrt{x^3 + 2x^2 - 6x + 5}$ from $x = 1$ to $x = 2$

16. $f(x) = \dfrac{x^2 - 3}{2x - 4}$ from $x = 3$ to $x = 6$

In Exercises 17–24, compute the difference quotient of the function.

17. $f(x) = x + 5$ **18.** $f(x) = 7x + 2$

19. $f(x) = x^2 + 3$ **20.** $f(x) = x^2 + 3x - 1$

21. $f(t) = 160,000 - 8000t + t^2$

22. $V(x) = x^3$

23. $A(r) = \pi r^2$ **24.** $V(p) = 5/p$

25. Water is draining from a large tank. After t minutes, there are $160,000 - 8000t + t^2$ gallons of water in the tank.
   (a) Use the results of Exercise 21 to find the average rate at which the water runs out in the interval from 10 to 10.1 minutes.
   (b) Do the same for the interval from 10 to 10.01 minutes.
   (c) Estimate the rate at which the water runs out after exactly 10 minutes.

26. Use the results of Exercise 22 to find the average rate of change of the volume of a cube whose side has length x as x changes from
   (a) 4 to 4.1   (b) 4 to 4.01   (c) 4 to 4.001
   (d) Estimate the rate of change of the volume at the instant when $x = 4$.

27. Use the results of Exercise 23 to find the average rate of change of the area of a circle of radius r as r changes from
   (a) 3 to 3.5   (b) 3 to 3.2   (c) 3 to 3.1
   (d) Estimate the rate of change at the instant when $r = 3$.
   (e) How is your answer in part (d) related to the circumference of a circle of radius 3?

28. Under certain conditions, the volume V of a quantity of air is related to the pressure p (which is measured in kilopascals) by the equation $V = 5/p$. Use the results of Exercise 24 to estimate the rate at which the volume is changing at the instant when the pressure is 50 kilopascals.

29. Two cars race on a straight track, beginning from a dead stop. The distance (in feet) each car has covered at each time during the first 16 seconds is shown in the figure.

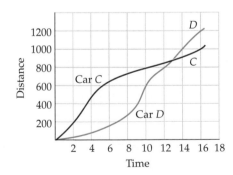

   (a) What is the average speed of each car during this 16-second interval?
   (b) Find an interval beginning at $t = 4$ during which the average speed of car D was approximately the same as the average speed of car C from $t = 2$ to $t = 10$.
   (c) Use secant lines and slopes to justify the statement "car D traveled at a higher average speed than car C from $t = 4$ to $t = 10$."

30. The figure shows the profits earned by a certain company during the last quarters of three consecutive years.

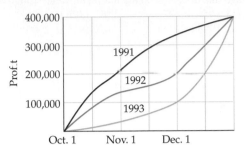

   (a) Explain why the average rate of change of profits from October 1 to December 31 was the same in all three years.
   (b) During what month in what year was the average rate of change of profits the greatest?

31. The graph in the figure shows the chipmunk population in a certain wilderness area. The population increases as the chipmunks reproduce but then decreases sharply as predators move into the area.

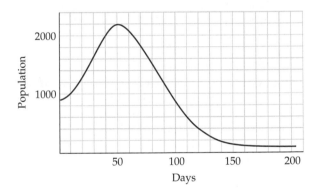

   (a) During what approximate time period (beginning on day 0) is the average growth rate of the chipmunk population positive?
   (b) During what approximate time period, beginning on day 0, is the average growth rate of the chipmunk population 0?
   (c) What is the average growth rate of the chipmunk population from day 50 to day 100? What does this number mean?
   (d) What is the average growth rate from day 45 to day 50? From day 50 to day 55? What is the approximate average growth rate from day 49 to day 51?

32. Lucy has a viral flu. How bad she feels depends primarily on how fast her temperature is rising at that time. The figure shows her temperature during the first day of the flu.

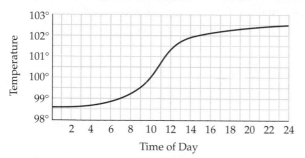

(a) At what average rate does her temperature rise during the entire day?

(b) During what two-hour period during the day does she feel worst?

(c) Find two time intervals, one in the morning and one in the afternoon, during which she feels about the same (that is, during which her temperature is rising at the same average rate).

33. The table shows the average weekly earnings (including overtime) of production workers and nonsupervisory employees in industry (excluding agriculture) in selected years.*

Year	1980	1985	1990	1995	2000
Weekly Earnings	$191	$257	$319	$369	$454

(a) Use linear regression to find a function that models this data, with $x = 0$ corresponding to 1980.

(b) According to your function, what is the average rate of change in earnings over any time period between 1980 and 2000?

(c) Use the data in the table to find the average rate of change in earnings from 1980 to 1990 and from 1990 to 2000. How do these rates compare with the ones given by the model?

(d) If the model remains accurate, when will average weekly earnings reach $550?

34. The estimated number of 15- to 24-year-old people worldwide (in millions) who are living with HIV/AIDS in selected years is given in the table.[†]

Year	2001	2003	2005	2007	2009
15- to 24-year-olds with HIV/AIDS	12	14.5	17	19	20.5

(a) Use linear regression to find a function that models this data, with $x = 0$ corresponding to 2000.

(b) According to your function, what is the average rate of change in this HIV/AIDS population over any time period between 2001 and 2009?

(c) Use the data in the table to find the average rate of change in this HIV/AIDS population from 2001 to 2009. How does this rate compare with the one given by the model?

(d) If the model remains accurate, when will the number of people in this age group with HIV/AIDS reach 25 million?

---

*U.S. Bureau of Labor Statistics.

[†]Kaiser Family Foundation, UNICEF, U.S. Census Bureau.

3.7   Inverse Functions*

Consider the functions f and h given by these tables:

f-input	−2	−1	0	1	2
f-output	−3	−2	1	4	5

h-input	1	2	3	4	5
h-output	−1	3	0	3	2

With the function h, two different inputs (2 and 4) produce the same output 3. With the function f, however, different inputs always produce different outputs. Functions with this property have a special name. A function f is said to be **one-to-one** if distinct inputs always produce distinct outputs, that is,

$$\text{if } a \neq b, \text{ then } f(a) \neq f(b)$$

---

*This section is used only in Section 5.3, Special Topics 5.3.A, and Section 7.4. It may be postponed until then.

In graphical terms, this means that two points on the graph, $(a, f(a))$ and $(b, f(b))$, that have different x-coordinates $[a \neq b]$ must also have different y-coordinates $[[f(a) \neq f(b)]]$. Consequently, these points cannot lie on the same horizontal line because all points on a horizontal line have the same y-coordinate. Therefore, we have this geometric test to determine whether a function is one-to-one.

The Horizontal Line Test

If a function f is one-to-one, then it has this property:

No horizontal line intersects the graph of f more than once.

Conversely, if the graph of a function has this property, then the function is one-to-one.

EXAMPLE 1

Which of the following functions are one-to-one?

(a) $f(x) = 7x^5 + 3x^4 - 2x^3 + 2x + 1$

(b) $g(x) = x^3 - 3x - 1$

(c) $h(x) = 1 - .2x^3$

SOLUTION   Complete graphs of each function are shown in Figure 3–44.

(a)

(b)

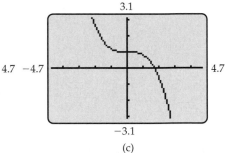
(c)

Figure 3–44

(a) The graph of f in Figure 3–44(a) passes the horizontal line test, since no horizontal line intersects the graph more than once. Hence, f is one-to-one.

(b) The graph of g in Figure 3–44(b) obviously fails the horizontal line test because many horizontal lines (including the x-axis) intersect the graph more than once. Therefore, g is not one-to-one.

(c) The graph of h in Figure 3–44(c) appears to contain a horizontal line segment. So h appears to fail the Horizontal Line Test because the horizontal line through $(0, 1)$ seems to intersect the graph infinitely many times. But appearances are deceiving.

TECHNOLOGY TIP

Although a horizontal segment may appear on a calculator screen when the graph is actually rising or falling, there is another possibility. The graph may have a tiny wiggle (less than the height of a pixel) and thus fail the horizontal line test:

You can usually detect such a wiggle by zooming in to magnify that portion of the graph or by using the trace feature to see whether the *y*-coordinates increase and then decrease (or vice versa) along the "horizontal" segment.

GRAPHING EXPLORATION

Graph $h(x) = 1 - .2x^3$ and use the trace feature to move from left to right along the "horizontal" segment. Do the *y*-coordinates stay the same, or do they decrease?

The Exploration shows that the graph is actually falling from left to right, so that each horizontal line intersects it only once. (It appears to have a horizontal segment because the amount the graph falls there is less than the height of a pixel on the screen.) Therefore, *h* is a one-to-one function. ■

The function *f* in Example 1 is an **increasing function** (its graph is always rising from left to right), and the function *h* is a **decreasing function** (its graph is always falling from left to right). Every increasing or decreasing function is necessarily one-to-one because its graph can never touch the same horizontal line twice (it would have to change from rising to falling, or vice versa, to do so).

▦ INVERSE FUNCTIONS

We begin with a simple example that illustrates the basic idea of an inverse function. Consider the one-to-one function *f* introduced at the beginning of this section.

f-input	−2	−1	0	1	2
f-output	−3	−2	1	4	5

Now define a new function *g* by the following table (which simply *reverses* the rows in the *f* table).

g-input	−3	−2	1	4	5
g-output	−2	−1	0	1	2

Note that the inputs of *f* are the outputs of *g* and the outputs of *f* are the inputs of *g*. In other words,

Domain of *f* = Range of *g*   and   Range of *f* = Domain of *g*.

The rule of *g* reverses the action of *f* by taking each output of *f* back to the input it came from. For instance,

$$g(4) = 1 \qquad \text{and} \qquad f(1) = 4$$
$$g(-3) = -2 \qquad \text{and} \qquad f(-2) = -3$$

and in general,

$$g(y) = x \qquad \text{exactly when} \qquad f(x) = y.$$

We say that *g* is the *inverse function* of *f*.

The preceding construction works for any one-to-one function f. Each output of f comes from exactly one input (because different inputs produce different outputs). Consequently, we can define a new function g that reverses the action of f by sending each output back to the unique input it came from. For instance, if $f(7) = 11$, then $g(11) = 7$. Thus, the outputs of f become the inputs of g, and we have this definition.

Inverse Functions

Let f be a one-to-one function. Then the **inverse function** of f is the function g whose rule is

$$g(y) = x \qquad \text{exactly when} \qquad f(x) = y.$$

The domain of g is the range of f and the range of g is the domain of f.

EXAMPLE 2

The graph of $f(x) = 3x - 2$ is a straight line that certainly passes the Horizontal Line Test, so f is one-to-one and has an inverse function g. From the definition of g we know that

$$g(y) = x \qquad \text{exactly when} \qquad f(x) = y$$

that is,

$$g(y) = x \qquad \text{exactly when} \qquad 3x - 2 = y.$$

To find the rule of g, we need only solve this last equation for x:

$$3x - 2 = y$$

Add 2 to both sides: $\qquad 3x = y + 2$

Divide both sides by 3: $\qquad x = \dfrac{y + 2}{3}$

Since $g(y) = x$, we see that the rule of g is $g(y) = \dfrac{y + 2}{3}$. ∎

Recall that the letter used for the variable of a function doesn't matter. For instance, $h(x) = x^2$ and $h(t) = t^2$ and $h(u) = u^2$ all describe the same function, whose rule is "square the input." When dealing with inverse functions, it is customary to use the same variable for both f and its inverse g. Consequently, the inverse function in Example 2 would normally be written as

$$g(x) = \frac{x + 2}{3}.$$

We can summarize this procedure as follows.

Finding Inverse Functions Algebraically

To find the inverse function of a one-to-one function f:

1. Solve the equation $f(x) = y$ for x.

2. The solution is an expression in y, which is the rule of the inverse function g, that is, $x = g(y)$.

3. Rewrite the rule of $x = g(y)$ by interchanging x and y.

EXAMPLE 3

Use your calculator to verify that the function $f(x) = x^3 + 5$ passes the Horizontal Line Test and hence is one-to-one. Its inverse can be found by solving for x in the equation $x^3 + 5 = y$:

Subtract 5 from both sides: $\qquad x^3 = y - 5$

Take cube roots on both sides: $\qquad x = \sqrt[3]{y - 5}.$

Therefore, $g(y) = \sqrt[3]{y - 5}$ is the inverse function of f. Interchanging x and y, we write this rule as $g(x) = \sqrt[3]{x - 5}.$ ∎

EXAMPLE 4

The function $f(x) = \sqrt{x - 3}$ is one-to-one, as you can verify with your calculator. To find its inverse, we solve the equation:

$$y = \sqrt{x - 3}$$

Square both sides: $\qquad y^2 = x - 3$

Add 3 to both sides: $\qquad x = y^2 + 3.$

Although this last equation is defined for all real numbers y, the original equation $y = \sqrt{x - 3}$ has $y \geq 0$ (since square roots are nonnegative). In other words, the range of the function f (the possible values of y) consists of all nonnegative real numbers. Consequently, the domain of the inverse function g is the set of all nonnegative real numbers, and its rule is

$$g(y) = y^2 + 3 \quad (y \geq 0).$$

Once again, it's customary to use the same variable to describe both f and its inverse function, so we write the rule of g as $g(x) = x^2 + 3 \ (x \geq 0).$ ∎

⊞ THE ROUND-TRIP PROPERTIES

The inverse function g of a function f was designed to send each output of f back to the input it came from. Consequently, if you first apply f and then apply g to the result, you obtain the number you started with, as illustrated in the next example.

EXAMPLE 5

As we saw in Example 2, the inverse function of $f(x) = 3x - 2$ is

$$g(x) = \frac{x + 2}{3}.$$

If we start with a number c and apply f, we obtain $f(c) = 3c - 2$. If we now apply g to this result, we obtain

$$g(f(c)) = g(3c - 2) = \frac{(3c - 2) + 2}{3} = c.$$

So we are back where we started. Similarly, if we first apply g and then apply f to a number, we end up where we started:

$$f(g(c)) = f\left(\frac{c+2}{3}\right) = 3\left(\frac{c+2}{3}\right) - 2 = c.$$

The function $f(x) = x^3 + 5$ of Example 3 and its inverse function

$$g(x) = \sqrt[3]{x - 5}$$

also have these "round-trip" properties. If you apply one function and then the other, you wind up at the number you started with:

$$g(f(x)) = g(x^3 + 5) = \sqrt[3]{(x^3 + 5) - 5} = \sqrt[3]{x^3} = x$$

and

$$f(g(x)) = f(\sqrt[3]{x - 5}) = (\sqrt[3]{x - 5})^3 + 5 = (x - 5) + 5 = x. \quad\blacksquare$$

Not only do a function and its inverse have the round-trip properties illustrated in Example 5, but somewhat more is true.

Round-Trip Theorem

A one-to-one function f and its inverse function g have these properties:

$$g(f(x)) = x \quad \text{for every } x \text{ in the domain of } f;$$

$$f(g(x)) = x \quad \text{for every } x \text{ in the domain of } g.$$

Conversely, if f and g are functions having these properties, then f is one-to-one and its inverse is g.

Proof By the definition of inverse function,

$$g(d) = c \quad \text{exactly when} \quad f(c) = d.$$

Consequently, for any c in the domain of f.

$$g(f(c)) = g(d) \quad (\text{because } f(c) = d)$$

$$= c \quad (\text{because } g(d) = c).$$

A similar argument shows that $f(g(d)) = d$ for any d in the domain of g. The last statement in the Theorem is proved in Exercise 53. $\blacksquare$

EXAMPLE 6

If

$$f(x) = \frac{5}{2x - 4} \quad \text{and} \quad g(x) = \frac{4x + 5}{2x},$$

then for every x in the domain of f (that is, all $x \neq 2$),

$$g(f(x)) = g\!\left(\frac{5}{2x - 4}\right) = \frac{4\!\left(\dfrac{5}{2x - 4}\right) + 5}{2\!\left(\dfrac{5}{2x - 4}\right)} = \frac{\dfrac{20 + 5(2x - 4)}{2x - 4}}{\dfrac{10}{2x - 4}}$$

$$= \frac{20 + 5(2x - 4)}{10} = \frac{20 + 10x - 20}{10} = \frac{10x}{10} = x,$$

and for every x in the domain of g (all $x \neq 0$),

$$f(g(x)) = f\!\left(\frac{4x + 5}{2x}\right) = \frac{5}{2\!\left(\dfrac{4x + 5}{2x}\right) - 4} = \frac{5}{\dfrac{4x + 5}{x} - 4}$$

$$= \frac{5}{\dfrac{4x + 5 - 4x}{x}} = \frac{5}{\dfrac{5}{x}} = x.$$

By the Round-Trip Theorem, f is a one-to-one function with inverse g. ∎

⠿ GRAPHS OF INVERSE FUNCTIONS

Finding the rule of the inverse function g of a one-to-one function f by solving the equation $y = f(x)$ for x, as in the preceding examples, is not always possible (some equations are hard to solve). But even if you don't know the rule of g, you can always find its graph, as shown below.

Suppose f is a one-to-one function and g is its inverse function. Then by the definition of inverse function:

$$f(a) = b \qquad \text{exactly when} \qquad g(b) = a.$$

But $f(a) = b$ means that (a, b) is on the graph of f and $g(b) = a$ means that (b, a) is on the graph of g. Therefore,

Inverse Function Graphs

> If f is a one-to-one function and g is its inverse function, then
>
> (a, b) is on the graph of f exactly when (b, a) is on the graph of g.

Therefore, the graph of the inverse function g can be obtained by reversing the coordinates of each point on the graph of f. There are two practical ways of doing this, each of which is illustrated below.

EXAMPLE 7

Verify that $f(x) = .7x^5 + .3x^4 - .2x^3 + 2x + .5$ has an inverse function g. Use parametric graphing to graph both f and g.

SOLUTION   First, we graph f in parametric mode by letting

$$x = t \qquad \text{and} \qquad y = f(t) = .7t^5 + .3t^4 - .2t^3 + 2t + .5.$$

The complete graph in Figure 3–45 shows that f is one-to-one (why?) and hence has an inverse function g. According to the preceding box, the graph of g can be obtained by taking each point on the graph of f and reversing its coordinates. Thus, g can be graphed parametrically by letting

$$x = f(t) = .7t^5 + .3t^4 - .2t^3 + 2t + .5 \qquad \text{and} \qquad y = t.$$

Figure 3–46 shows the graphs of g and f on the same screen. ∎

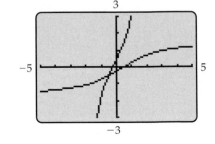

Figure 3–45   **Figure 3–46**

The second method of graphing inverse functions by reversing coordinates depends on this geometric fact, which is proved in Exercise 51:

> The line $y = x$ is the perpendicular bisector of the line segment from (a, b) to (b, a),

as shown in Figure 3–47 when $a = 7$, $b = 2$. Thus (a, b) and (b, a) lie on opposite sides of $y = x$, the same distance from it: They are mirror images of each other, with the line $y = x$ being the mirror.* Consequently, the graph of the inverse function g is the mirror image of the graph of f. In formal terms,

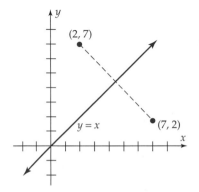

Figure 3–47

Inverse Function Graphs

> If g is the inverse function of f, then the graph of g is the reflection of the graph of f in the line $y = x$.

GRAPHING EXPLORATION

Illustrate this fact by graphing the line $y = x$, the function

$$f(x) = x^3 + 5$$

of Example 3, and its inverse $g(x) = \sqrt[3]{x - 5}$ on the same screen (use a square viewing window so that the mirror effect won't be distorted).

NOTE: In many texts, the inverse function of a function f is denoted f^{-1}. In this notation, for instance, the inverse of the function $f(x) = x^3 + 5$ in Example 3 would be written as $f^{-1}(x) = \sqrt[3]{x - 5}$. Similarly, the reversal properties of inverse functions become

$$f^{-1}(f(x)) = x \text{ for every } x \text{ in the domain of } f; \text{ and}$$
$$f(f^{-1}(x)) = x \text{ for every } x \text{ in the domain of } f^{-1}.$$

In this context, f^{-1} does *not* mean $1/f$ (see Exercise 47).

---

*In technical terms, (a, b) and (b, a) are **symmetric with respect to the line** $y = x$.

EXERCISES 3.7

In Exercises 1–8, use a calculator and the Horizontal Line Test to determine whether or not the function f is one-to-one.

1. $f(x) = x^4 - 4x^2 + 3$

2. $f(x) = x^4 - 4x + 3$

3. $f(x) = x^3 + x - 5$

4. $f(x) = \begin{cases} x - 3 & \text{for } x \le 3 \\ 2x - 6 & \text{for } x > 3 \end{cases}$

5. $f(x) = x^5 + 2x^4 - x^2 + 4x - 5$

6. $f(x) = x^3 - 4x^2 + x - 10$

7. $f(x) = .1x^3 - .1x^2 - .005x + 1$

8. $f(x) = .1x^3 + .005x + 1$

In Exercises 9–22, use algebra to find the inverse of the given one-to-one function.

9. $f(x) = -x$ **10.** $f(x) = -x + 1$

11. $f(x) = 5x - 4$ **12.** $f(x) = -3x + 5$

13. $f(x) = 5 - 2x^3$ **14.** $f(x) = (x^5 + 1)^3$

15. $f(x) = \sqrt{4x - 7}$ **16.** $f(x) = 5 + \sqrt{3x - 2}$

17. $f(x) = 1/x$ **18.** $f(x) = 1/\sqrt{x}$

19. $f(x) = \dfrac{1}{2x + 1}$ **20.** $f(x) = \dfrac{x}{x + 1}$

21. $f(x) = \dfrac{x^3 - 1}{x^3 + 5}$ **22.** $f(x) = \sqrt[5]{\dfrac{3x - 1}{x - 2}}$

In Exercises 23–28, use the Round-Trip Theorem on page 208 to show that g is the inverse of f.

23. $f(x) = x + 1,$ $g(x) = x - 1$

24. $f(x) = 2x - 6,$ $g(x) = \dfrac{x}{2} + 3$

25. $f(x) = \dfrac{1}{x + 1},$ $g(x) = \dfrac{1 - x}{x}$

26. $f(x) = \dfrac{-3}{2x + 5},$ $g(x) = \dfrac{-3 - 5x}{2x}$

27. $f(x) = x^5,$ $g(x) = \sqrt[5]{x}$

28. $f(x) = x^3 - 1,$ $g(x) = \sqrt[3]{x + 1}$

29. Show that the inverse function of the function f whose rule is $f(x) = \dfrac{2x + 1}{3x - 2}$ is f itself.

30. List three different functions (other than the one in Exercise 29), each of which is its own inverse. [Many correct answers are possible.]

In Exercises 31 and 32, the graph of a function f is given. Sketch the graph of the inverse function of f. [Reflect carefully.]

31.

32.

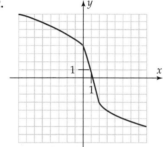

In Exercises 33–38, each given function has an inverse function. Sketch the graph of the inverse function.

33. $f(x) = \sqrt{x + 3}$ **34.** $f(x) = \sqrt{3x - 2}$

35. $f(x) = .3x^5 + 2$ **36.** $f(x) = \sqrt[3]{x + 3}$

37. $f(x) = \sqrt[5]{x^3 + x - 2}$

38. $f(x) = \begin{cases} x^2 - 1 & \text{for } x \le 0 \\ -.5x - 1 & \text{for } x > 0 \end{cases}$

In Exercises 39–46, none of the functions has an inverse. State at least one way of restricting the domain of the function (that is, find a function with the same rule and a smaller domain) so that the restricted function has an inverse. Then find the rule of the inverse function.

Example: $f(x) = x^2$ has no inverse. But the function h with domain all $x \ge 0$ and rule $h(x) = x^2$ is increasing (its graph is the right half of the graph of f—see Figure 2–2 on page 70)—and therefore has an inverse.

39. $f(x) = |x|$ **40.** $f(x) = |x - 3|$

41. $f(x) = -x^2$ **42.** $f(x) = x^2 + 4$

43. $f(x) = \dfrac{x^2 + 6}{2}$ **44.** $f(x) = \sqrt{4 - x^2}$

45. $f(x) = \dfrac{1}{x^2 + 1}$ **46.** $f(x) = 3(x + 5)^2 + 2$

47. (a) Using the f^{-1} notation for inverse functions, find $f^{-1}(x)$ when $f(x) = 3x + 2$.
(b) Find $f^{-1}(1)$ and $1/f(1)$. Conclude that f^{-1} is not the same function as $1/f$.

48. Let C be the temperature in degrees Celsius. Then the temperature in degrees Fahrenheit is given by $f(C) = \frac{9}{5}C + 32$. Let g be the function that converts degrees Fahrenheit to degrees Celsius. Show that g is the inverse function of f and find the rule of g.

Thinkers

49. Let m and b be constants with $m \neq 0$. Show that the function $f(x) = mx + b$ has an inverse function g and find the rule of g.

50. Prove that the function $h(x) = 1 - .2x^3$ of Example 1(c) is one-to-one by showing that it satisfies the definition:

$$\text{If } a \neq b, \text{ then } h(a) \neq h(b).$$

[*Hint:* Use the rule of h to show that when $h(a) = h(b)$, then $a = b$. If this is the case, then it is impossible to have $h(a) = h(b)$ when $a \neq b$.]

51. Show that the points $P = (a, b)$ and $Q = (b, a)$ are symmetric with respect to the line $y = x$ as follows.
(a) Find the slope of the line through P and Q.
(b) Use slopes to show that the line through P and Q is perpendicular to $y = x$.
(c) Let R be the point where the line $y = x$ intersects line segment PQ. Since R is on $y = x$, it has coordinates (c, c) for some number c, as shown in the figure. Use the distance formula to show that segment PR has the same length as segment RQ. Conclude that the line $y = x$ is the perpendicular bisector of segment PQ. Therefore, P and Q are symmetric with respect to the line $y = x$.

52. (a) Experiment with your calculator or use some of the preceding exercises to find four different increasing functions. For each function, sketch the graph of the function and the graphs of its inverse on the same set of axes.
(b) On the basis of the evidence in part (a), do you think the following statement true or false: The inverse function of every increasing function is also an increasing function.
(c) Do parts (a) and (b) with "increasing" replaced by "decreasing."

53. Suppose that functions f and g have these round-trip properties:
(1) $g(f(x)) = x$ for every x in the domain of f.
(2) $f(g(y)) = y$ for every y in the domain of g.
To complete the proof of the Round-Trip Theorem, we must show that g is the inverse function of f. Do this as follows.
(a) Prove that f is one-to-one by showing that

$$\text{if } \quad a \neq b, \quad \text{then} \quad f(a) \neq f(b).$$

[*Hint:* If $f(a) = f(b)$, apply g to both sides and use (1) to show that $a = b$. Consequently, if $a \neq b$, it is impossible to have $f(a) = f(b)$.]
(b) If $g(y) = x$, show that $f(x) = y$. [*Hint:* Use (2).]
(c) If $f(x) = y$, show that $g(y) = x$. [*Hint:* Use (1).]
Parts (b) and (c) prove that

$$g(y) = x \quad \text{exactly when} \quad f(x) = y.$$

Hence, g is the inverse function of f (see page 206).

54. Prove that every function f that has an inverse function g is one-to-one. [*Hint:* The proof of the Round-Trip Theorem on page 208 shows that f and g have the round-trip properties; use Exercise 53(a).]

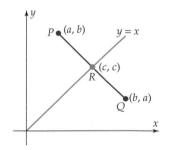

Chapter 3 Review

IMPORTANT CONCEPTS

Section 3.1

Function   127
Domain   127
Range   127
Rule   127
Greatest integer function   127–128
Functions defined by equations and graphs   128

Section 3.2

Functional notation   137
Difference quotient   139
Common mistakes with functional notation   139
Domain convention   140
Piecewise-defined functions   140–141

Section 3.3

Graphs of functions   146
Step functions   147
Graphs of piecewise-defined functions   148
Local maximum and local minimum   150

Increasing and decreasing functions   151
Vertical Line Test   152
Parametric graphing   153

Special Topics 3.3.A

Graph reading   161

Section 3.4

Vertical shifts   166
Horizontal shifts   168
Expansions and contractions   170
Reflections in the x- and y-axis   171–172

Special Topics 3.4.A

Symmetry with respect to the y-axis   177
Symmetry with respect to the x-axis   178
Symmetry with respect to the origin   178
Coordinate and algebraic tests for symmetry   179

Even and odd functions   179–180

Section 3.5

Sums, differences, products, and quotients of functions   182–183
Composition of functions   184–185

Section 3.6

Average rate of change   193
Secant lines and average rate of change   196
Difference quotient   196
Estimating instantaneous rates of change   199

Section 3.7

One-to-one functions   203
Horizontal Line Test   204
Inverse functions   206
Round-Trip Theorem   208
Graphs of inverse functions   209–210

IMPORTANT FACTS & FORMULAS

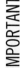

- The average rate of change of a function f as x changes from a to b is the number

$$\frac{f(b) - f(a)}{b - a}.$$

- The average rate of change of a function f as x changes from a to b is the slope of the secant line joining the points $(a, f(a))$ and $(b, f(b))$.

- The difference quotient of the function f is the quantity

$$\frac{f(x + h) - f(x)}{h}.$$

IMPORTANT **FACTS & FORMULAS**

CATALOG OF BASIC FUNCTIONS—PART 1

1. Let $[x]$ denote the greatest integer function and evaluate
   (a) $[-5/2] = $ ____.
   (b) $[1755] = $ ____.
   (c) $[18.7] + [-15.7] = $ ____.
   (d) $[-7] - [7] = $ ____.

2. If $f(x) = x + |x| + [x]$, then find $f(0), f(-1), f(1/2)$, and $f(-3/2)$.

3. Let f be the function given by the rule $f(x) = 7 - 2x$. Complete this table:

x	0	1	2	-4	t	k	$b - 1$	$1 - b$	$6 - 2u$
$f(x)$	7								

4. What is the domain of the function g given by
$$g(t) = \frac{\sqrt{t - 2}}{t - 3} ?$$

5. In each case, give a *specific* example of a function and numbers a, b to show that the given statement may be *false*.
   (a) $f(a + b) = f(a) + f(b)$
   (b) $f(ab) = f(a)f(b)$

6. If $f(x) = |3 - x| \sqrt{x - 3} + 7$, then
$$f(7) - f(4) = \underline{\qquad}.$$

7. What is the domain of the function given by
$$g(r) = \sqrt{r - 4} + \sqrt{r - 2}?$$

8. What is the domain of the function $f(x) = \sqrt{-x + 2}$?

9. If $h(x) = x^2 - 3x$, then $h(t + 2) = $ ____.

10. Which of the following statements about the greatest integer function $f(x) = [x]$ is true for *every* real number x?
    (a) $x - [x] = 0$
    (b) $x - [x] \le 0$
    (c) $[x] + [-x] \le 0$
    (d) $[-x] \ge [x]$
    (e) $3[x] = [3x]$

11. If $f(x) = 2x^3 + x + 1$, then $f(x/2) = $ ____.

12. If $g(x) = x^2 - 1$, then $g(x - 1) - g(x + 1) = $ ____.

13. The radius of an oil spill (in meters) is 50 times the square root of the time t (in hours).
    (a) Write the rule of a function f that gives the radius of the spill at time t.
    (b) Write the rule of a function g that gives the area of the spill at time t.
    (c) What are the radius and area of the spill after 9 hours?
    (d) When will the spill have an area of 100,000 square meters?

14. The cost of renting a limousine for 24 hours is given by
$$C(x) = \begin{cases} 150 & \text{if } 0 < x \le 25 \\ 1.75x + 150 & \text{if } x > 25, \end{cases}$$
where x is the number of miles driven.
    (a) What is the cost if the limo is driven 20 miles? 30 miles?
    (b) If the cost is $218.25, how many miles were driven?

15. Sketch the graph of the function f given by
$$f(x) = \begin{cases} x^2 & \text{if } x \le 0 \\ x + 1 & \text{if } 0 < x < 4 \\ \sqrt{x} & \text{if } x \ge 4. \end{cases}$$

16. U.S. Express Mail rates in 2003 are shown in the following table. Sketch the graph of the function e, whose rule is $e(x) = $ cost of sending a package weighing x pounds by Express Mail.

EXPRESS MAIL Letter Rate—Post Office to Addressee Service	
Up to 8 ounces	$13.65
Over 8 ounces to 2 pounds	$17.85
Up to 3 pounds	$21.05
Up to 4 pounds	$24.20
Up to 5 pounds	$27.30
Up to 6 pounds	$30.40
Up to 7 pounds	$33.45

17. Which of the following are graphs of functions of x?
    (a)

    (b)

REVIEW QUESTIONS

18. The function whose graph is shown gives the number of Americans who snowboard (in millions) in year x where $x = 0$ corresponds to 1988.*

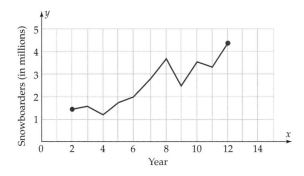

(a) What is the domain of the function?

(b) Approximately, what is the range of the function?

(c) Find the average rate of change of the function from $x = 6$ to $x = 8$ and from $x = 9$ to $x = 10$.

In Questions 19–22, determine the local maxima and minima of the function, the intervals on which the function is increasing, and the intervals on which it is decreasing.

19. $g(x) = \sqrt{x^2 + x + 1}$

20. $f(x) = 2x^3 - 5x^2 + 4x - 3$

21. $g(x) = x^3 + 8x^2 + 4x - 3$

22. $f(x) = .5x^4 + 2x^3 - 6x^2 - 16x + 2$

In Questions 23 and 24, sketch the graph of the curve given by the parametric equations.

23. $x = t^2 - 4$ and $y = 2t + 1$ $(-3 \leq t \leq 3)$

24. $x = t^3 + 3t^2 - 1$ and $y = t^2 + 1$ $(-3 \leq t \leq 2)$

25. Sketch a graph that is symmetric with respect to both the x-axis and the y-axis. (There are many correct answers and your graph need not be the graph of an equation.)

26. Sketch the graph of a function that is symmetric with respect to the origin. (There are many correct answers and you don't have to state the rule of your function.)

In Questions 27 and 28, determine algebraically whether the graph of the given equation is symmetric with respect to the x-axis, the y-axis, or the origin.

27. $x^2 = y^2 + 2$

28. $5y = 7x^2 - 2x$

In Questions 29–31, determine whether the given function is even, odd, or neither.

29. $g(x) = 9 - x^2$ **30.** $f(x) = |x|x + 1$

31. $h(x) = 3x^5 - x(x^4 - x^2)$

32. (a) Draw some coordinate axes and plot the points $(-2, 1), (-1, 3), (0, 1), (3, 2), (4, 1)$.

(b) Suppose the points plotted in part (a) lie on the graph of an *even* function f. Plot these points: $(2, f(2)), (1, f(1)), (0, f(0)), (-3, f(-3)), (-4, f(-4))$.

33. Determine whether the circle with equation

$$x^2 + y^2 + 6y = -5$$

is symmetric with respect to the x-axis, the y-axis, or the origin.

34. Sketch the graph of a function f that satisfies all of these conditions:

(i) domain of $f = [-3, 4]$

(ii) range of $f = [-2, 5]$

(iii) $f(-2) = 0$

(iv) $f(1) > 2$

[*Note:* There are many possible correct answers and the function whose graph you sketch need *not* have an algebraic rule.]

35. Sketch the graph of $g(x) = 5 + \dfrac{4}{x - 5}$.

Use the graph of the function f in the figure to answer Questions 36–39.

36. What is the domain of f?

37. What is the range of f?

38. Find all numbers x such that $f(x) = 1$.

39. Find a number x such that $f(x + 1) < f(x)$. (Many correct answers are possible.)

---

*National Sporting Goods Association, TransWorld Snowboarding Business.

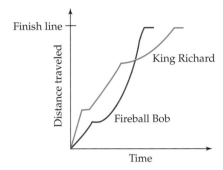

Use the graph of the function f in the figure to answer Questions 40–46.

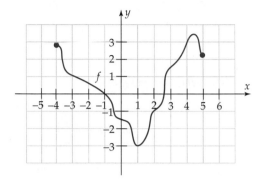

40. What is the domain of f?   **41.** $f(-3) =$ _____.

42. $f(2 + 2) =$ _____.

43. $f(-1) + f(1) =$ _____.

44. True or false: $2f(2) = f(4)$.

45. True or false: $3f(2) = -f(4)$.

46. True or false: $f(x) = 3$ for exactly one number x.

Use the graphs of the functions f and g in the figure to answer Questions 47–52.

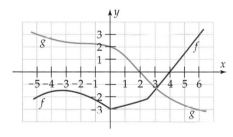

47. For which values of x is $f(x) = 0$?

48. True or false: If a and b are numbers such that $-5 \leq a < b \leq 6$, then $g(a) < g(b)$.

49. For which values of x is $g(x) \geq f(x)$?

50. Find $f(0) - g(0)$.

51. For which values of x is $f(x + 1) < 0$?

52. What is the distance from the point $(-5, g(-5))$ to the point $(6, g(6))$?

53. Fireball Bob and King Richard are two NASCAR racers. The graph shows their distance traveled in a recent race as a function of time.
  (a) Which car made the most pit stops?
  (b) Which car started out the fastest?
  (c) Which car won the race?

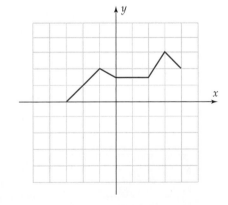

In Questions 54–57, list the transformations, in the order in which they should be performed on the graph of $g(x) = x^2$, so as to produce a complete graph of the function f.

54. $f(x) = (x - 2)^2$ **55.** $f(x) = .25x^2 + 2$

56. $f(x) = -(x + 4)^2 - 5$

57. $f(x) = -3(x - 7)^2 + 2$

58. The graph of a function f is shown in the figure. On the same coordinate plane, carefully draw the graphs of the functions g and h whose rules are:

$$g(x) = -f(x) \qquad \text{and} \qquad h(x) = 1 - f(x)$$

REVIEW QUESTIONS

59. The figure shows the graph of a function f. If g is the function given by $g(x) = f(x + 2)$, then which of these statements about the graph of g is true?
(a) It does not cross the x-axis.
(b) It does not cross the y-axis.
(c) It crosses the y-axis at $y = 4$.
(d) It crosses the y-axis at the origin.
(e) It crosses the x-axis at $x = -3$.

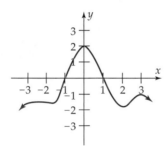

60. If $f(x) = 3x + 2$ and $g(x) = x^3 + 1$, find:
(a) $(f + g)(-1)$ (b) $(f - g)(2)$ (c) $(fg)(0)$

61. If $f(x) = \dfrac{1}{x - 1}$ and $g(x) = \sqrt{x^2 + 5}$, find:
(a) $(f/g)(2)$ (b) $(g/f)(x)$
(c) $(fg)(c + 1)$ $(c \neq 1)$

62. Find two functions f and g such that neither is the identity function, and
$$(f \circ g)(x) = (2x + 1)^2.$$

63. Use the graph of the function g in the figure to fill in the following table, in which h is the composite function $g \circ g$.

x	-4	-3	-2	-1	0	1	2	3	4
$g(x)$					-1				
$h(x) = g(g(x))$									

Questions 64–69 refer to the functions $f(x) = \dfrac{1}{x + 1}$ *and* $g(t) = t^3 + 3$.

64. $(f \circ g)(1) = $ _____.

65. $(g \circ f)(2) = $ _____.

66. $g(f(-2)) = $ _____.

67. $(g \circ f)(x - 1) = $ _____.

68. $g(2 + f(0)) = $ _____.

69. $f(g(1) - 1) = $ _____.

70. Let f and g be the functions given by
$$f(x) = 4x + x^4 \quad \text{and} \quad g(x) = \sqrt{x^2 + 1}$$
(a) $(f \circ g)(x) = $ _____.   (b) $(g - f)(x) = $ _____.

71. If $f(x) = \dfrac{1}{x}$ and $g(x) = x^2 - 1$, then
$$(f \circ g)(x) = \text{\_\_\_\_} \quad \text{and} \quad (g \circ f)(x) = \text{\_\_\_\_}.$$

72. Let $f(x) = x^2$. Give an example of a function g with domain all real numbers such that $g \circ f \neq f \circ g$.

73. If $f(x) = \dfrac{1}{1 - x}$ and $g(x) = \sqrt{x}$, then find the domain of the composite function $f \circ g$.

74. These tables show the values of the functions f and g at certain numbers:

x	-1	0	1	2	3
$f(x)$	1	0	1	3	5

and

t	0	1	2	3	4
$g(t)$	-1	0	1	2	5

Which of the following statements are *true*?
(a) $(g - f)(1) = 1$ (b) $(f \circ g)(2) = (f - g)(0)$
(c) $f(1) + f(2) = f(3)$ (d) $(g \circ f)(2) = 1$
(e) None of the above is true.

75. Find the average rate of change of the function
$$g(x) = \frac{x^3 - x + 1}{x + 2}$$
as x changes from
(a) -1 to 1   (b) 0 to 2

76. Find the average rate of change of the function $f(x) = \sqrt{x^2 - x + 1}$ as x changes from
(a) -3 to 0   (b) -3 to 3.5   (c) -3 to 5

77. If $f(x) = 2x + 1$ and $g(x) = 3x - 2$, find the average rate of change of the composite function $f \circ g$ as x changes from 3 to 5.

78. If $f(x) = x^2 + 1$ and $g(x) = x - 2$, find the average rate of change of the composite function $f \circ g$ as x changes from -1 to 1.

In Questions 79–82, find the difference quotient of the function.

79. $f(x) = 3x + 4$

80. $g(x) = \sqrt{x}$

REVIEW QUESTIONS

81. $g(x) = x^2 - 1$

82. $f(x) = x^2 + x$

83. The profit (in hundreds of dollars) from selling x tons of Wonderchem is given by

$$P(x) = .2x^2 + .5x - 1.$$

What is the average rate of change of profit when the number of tons of Wonderchem sold increases from (a) 4 to 8 tons? (b) 4 to 5 tons? (c) 4 to 4.1 tons?

84. On the planet Mars, the distance traveled by a falling rock (ignoring atmospheric resistance) in t seconds is $6.1t^2$ feet. How far must a rock fall in order to have an average speed of 25 feet per second over that time interval?

85. The graph in the figure shows the population of fruit flies during a 50-day experiment in a controlled atmosphere.

(a) During what 5-day period is the average rate of population growth the slowest?

(b) During what 10-day period is the average rate of population growth the fastest?

(c) Find an interval beginning at the 30th day during which the average rate of population growth is the same as the average rate from day 10 to day 20.

86. The graph of the function g in the figure consists of straight line segments. Find an interval over which the average rate of change of g is

(a) 0   (b) -3 (c) .5

(d) Explain why the average rate of change of g is the same from -3 to -1 as it is from -2.5 to 0.

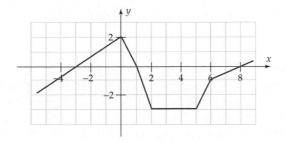

87. The table shows the median sale price for single-family homes in the western United States in selected years.

Year	1985	1990	1995	1998	2000	2001
Median Price	$95,400	$129,600	$141,000	$164,800	$183,000	$194,500

(a) Find the average rate of change of the median price from 1985 to 1998, from 1998 to 2001, and from 1985 to 2001.

(b) If the average rate of change from 1998 to 2001 continues in years after 2001, what is the median price in 2005?

88. Find the inverse of the function $f(x) = 2x + 1$.

89. Find the inverse of the function $f(x) = \sqrt{5 - x} + 7$.

90. Find the inverse of the function $f(x) = \sqrt[5]{x^3 + 1}$.

91. The graph of a function f is shown in the figure. Sketch the graph of the inverse function of f.

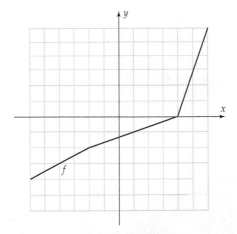

92. Which of the following functions have inverse functions (give reasons for your answers):

(a) $f(x) = x^3$ (b) $f(x) = 1 - x^2$, $x \le 0$

(c) $f(x) = |x|$

In Exercises 93–95, determine whether or not the given function has an inverse function (give reasons for your answer). If it does, find the graph of the inverse function.

93. $f(x) = 1/x$

94. $f(x) = .02x^3 - .04x^2 + .6x - 4$

95. $f(x) = .2x^3 - 4x^2 + 6x - 15$

DISCOVERY PROJECT 3   Feedback—Good and Bad

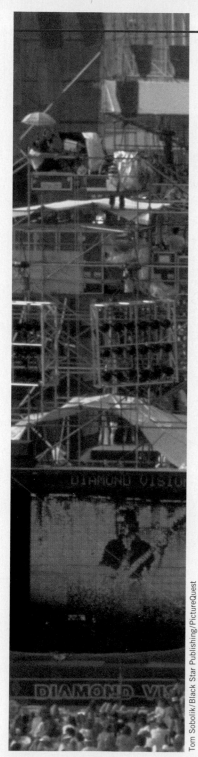

Whether it's for a concert or a worship service, a sporting event or a graduation ceremony, a shopping mall or a lecture hall, audio engineers are concerned to avoid audio feedback. If the system is not set up correctly, then sound from the speakers reaches the microphone with enough clarity that it is fed back into the audio system and amplified. This feedback cycle repeats again and again. Each time the amplified sound follows very quickly behind the previous one so that the audience hears an unpleasant hum that rapidly becomes a loud screech. This is feedback we'd prefer to avoid.

Other types of feedback are considerably more pleasing. Consider, for example, an investment of x dollars that earns 4% interest compounded annually. Then the amount in the account (principal plus interest) after one year is $1.04x$. Assume the initial deposit is left in the account without adding deposits or withdrawing funds. If $1000 is invested, then the balance (rounded to the nearest penny) is

Beginning	1000
End of year 1	$1.04 \cdot 1000 = 1081.60$
End of year 2	$1.04 \cdot 1081.60 = 1124.86$
End of year 3	$1.04 \cdot 1124.86 = 1169.86$

and so on. You can easily carry out this process on a calculator as follows: key in 1000 and press ENTER; then key in "$\times 1.04$" and press ENTER repeatedly. Each time you press ENTER, the calculator computes

$$(\text{Answer from preceding calculation}) \times 1.04$$

as shown in Figure 1. The result is an ever-increasing bank balance.

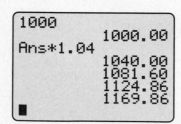

Figure 1

In each of the preceding examples, the output of the process was fed back as input and the process repeated to generate the result—unpleasant in the first case and quite pleasant in the second. Now we strip away the specifics and look at the process in more general terms. Suppose we have a function f and x is in its domain. We find $f(x)$ and feed that value back to the function to find

$f(f(x))$. Then feed that value back to find $f(f(f(x)))$. The sequence of values obtained by continuing this process

$$f(x), f(f(x)), f(f(f(x))), f(f(f(f(x)))), \ldots$$

is called the **orbit** of x and the feedback process is called **iteration.** The functions themselves $(f, f^2 = f \circ f, f^3 = f \circ f \circ f, \ldots)$ are called **iterates** of f.

Technology makes it easy to compute orbits. For example, let $f(x) = \sqrt{x}$ and enter f as Y_1 in your calculator. Key in 8 and press ENTER. Then key in "STO➡ X : Y$_1$" and press ENTER repeatedly.* Each time you press ENTER, the calculator stores the answer from the preceding calculation as X and evaluates Y_1 at X, as shown in Figure 2.

 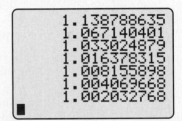

Figure 2

Thus, the (approximate) orbit of 8 is

$$\{8, 2.828, 1.682, 1.297, 1.139, 1.067, 1.033, \ldots\}.$$

Note that the terms of the orbit are getting closer and closer to 1.

1. Find the orbit of .2 and two other numbers of your choice (except 0 and 1). Do the terms of these orbits get closer and closer to 1?

For the function $f(x) = \sqrt{x}$ we can see that $f(1) = 1$, so that iteration produces the orbit $\{1, 1, 1, \ldots\}$; we say that 1 is a **fixed point** for this function. Another fixed point is zero because $f(0) = 0$. A fixed point a is called an **attracting fixed point** if nearby numbers have orbits that approach a or a **repelling fixed point** if nearby numbers have orbits that move farther away from a. The following exercises illustrate these ideas for the function $f(x) = \sqrt{x}$.

2. Confirm numerically that 1 is an attracting fixed point by computing the orbits of .5, 2, and 11.†

---

*The colon (:) is on the TI and HP-39 keyboards and in the Casio PRGM menu (press SHIFT VARS).

†Your calculator may tell you that the terms of the orbits eventually are equal to 1, but this is due to round-off error.

3. Explain mathematically why 1 is an attracting fixed point. [*Hint:* You may assume that if $u < v$, then $\sqrt{u} < \sqrt{v}$. Is $\sqrt{x}$ larger or smaller than x when $x < 1$? When $x > 1$?]

4. Determine whether zero is an attracting fixed point or a repelling fixed point.

5. Do you think this function has any other fixed points? Why or why not?

Generalizations are interesting in their own right but you may be wondering whether these ideas have any relevance to the real world. So let's look at another application. Ecologists want to model the growth of the population of some animal in an ecosystem. To simplify notation, let's let the variable x represent a fraction of the theoretical maximum population, so that $0 \le x \le 1$. [If the population were the maximum the ecosystem could sustain, then $x = 1$ and if there are none of these animals in the ecosystem, then $x = 0$. We're interested in the usual situation where x is somewhere in between.] With a growth rate r, the function $f(x) = rx$ seems to fill the bill.[*] But logistics, the limitation of resources (food, nesting locations, etc.), demands an adjustment. The **logistic function** $f(x) = rx(1 - x)$ gives a more realistic model of the growth, which is nearly exponential when x is small (and $1 - x$ is nearly 0). It also reflects the influence of logistics in that $(1 - x)$ approaches 0 as x approaches 1.

Do these exercises for the logistic function $f(x) = rx(1 - x)$ and the given value of r.

6. Use $r = 2$ and different initial values (such as .25, .8, and .1) to confirm that $f(x) = 2x(1 - x)$ has fixed point .5.

7. Find the fixed point (to the nearest hundredth) when $r =$
   (a) 2.5 (b) 2.8 (c) 3 (d) 3.2

8. Verify that for $r = 3.3$, the orbit of any point x (with $0 < x < 1$) seems to (eventually) jump back and forth between *two* values. Somewhere between 3.2 and 3.3 a **bifurcation** has occurred. Instead of a single fixed point we now have a periodic oscillation, with **period 2.** Find the two values.

9. Verify that periodic oscillation also occurs when $r = 3.4$. Is the period still 2?

10. Verify that $r = 3.5$ eventually oscillates among *four* values (that is, the period is $4 = 2^2$). Another bifurcation has occurred. Find the four values.

---

[*]Notice the similarity to the 4% interest example, where the "balance function" was $f(x) = 1.04x$. Notice also, that if r is less than 1, the population declines to extinction.

11. Verify that for $r = 3.55$ the period is still four but at $r = 3.56$ there are *eight* values (period is $8 = 2^3$). Find them.

12. Is it possible to determine exactly where bifurcations occur? When r is very large, do you think that the system breaks down to "chaos" or that there are values of r that have period 2^n for large values of n?

13. If you are able to consult a naturalist, determine whether some animal populations follow yearly patterns consistent with the logistic model.

Polynomial and Rational Functions

Rob Crandall/Stock, Boston Inc./PictureQuest

Can you afford to go to college?

As you (and your parents) know, the cost of a four-year college education (tuition, fees, room, board, books) has steadily increased. It is expected to continue to do so in the foreseeable future, according to projections by a large insurance company. This growth can be modeled by a fourth-degree polynomial function. See Exercise 46 on page 322.

Chapter Outline

Interdependence of Sections

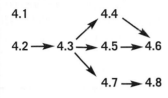

4.1   Quadratic Functions

4.2   Polynomial Functions

4.2.A   *Special Topics:* Synthetic Division

4.3   Real Roots of Polynomials

4.4   Graphs of Polynomial Functions

4.4.A   *Special Topics:* Polynomial Models

4.5   Rational Functions

4.5.A   *Special Topics:* Other Rational Functions

4.6   Polynomial and Rational Inequalities

4.6.A   *Special Topics:* Absolute Value Inequalities

4.7   Complex Numbers

4.8   Theory of Equations

Polynomial functions arise naturally in many applications. Many complicated functions in applied mathematics can be approximated by polynomial functions or their quotients (rational functions).

4.1   Quadratic Functions*

A **quadratic function** is a function whose rule can be written in the form

$$f(x) = ax^2 + bx + c$$

for some constants a, b, c, with $a \neq 0$. The graph of a quadratic function is called a **parabola.**

GRAPHING EXPLORATION

Using the standard viewing window, graph the following quadratic functions on the same screen:

$$f(x) = x^2, \qquad f(x) = 3x^2 + 30x + 77, \qquad f(x) = -x^2 + 4x,$$

$$f(x) = -.2x^2 + 1.5x - 5$$

As the Exploration illustrates, all parabolas have the same basic "cup" shape, though the cup may be broad or narrow. The parabola opens upward when the coefficient of x^2 is positive and downward when this coefficient is negative.

---

*This section may be omitted or postponed if desired. Section 3.4 (Graphs and Transformations) is a prerequisite.

225

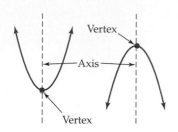

Figure 4–1

If a parabola opens upward, its **vertex** is the lowest point on the graph and if a parabola opens downward, its **vertex** is the highest point on the graph, as shown in Figure 4–1. Every parabola is symmetric with respect to the vertical line through its vertex; this line is called the **axis** of the parabola.

Parabolas are easily graphed on a calculator. The vertex can always be approximated by using the trace feature or a maximum/minimum finder. However, algebraic techniques can be used to find the vertex precisely and to graph the parabola.

EXAMPLE 1

Show that the function $g(x) = 2(x - 3)^2 + 1$ is quadratic, graph the function, and find its vertex.

SOLUTION   The function g is quadratic because its rule can be written in the form $g(x) = ax^2 + bx + c$:

$$g(x) = 2(x - 3)^2 + 1 = 2(x^2 - 6x + 9) + 1 = 2x^2 - 12x + 19.$$

Figure 4–2

Graphing the function in the standard viewing window (Figure 4–2) and using the minimum finder, we see that the vertex is approximately $(2.999, 1)$. To find the vertex exactly, we use the techniques of Section 3.4. The graph of

$$g(x) = 2(x - 3)^2 + 1$$

can be obtained from the graph of $y = x^2$ as follows:

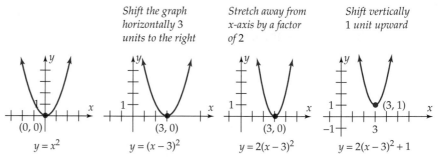

Figure 4–3

Figure 4–3 shows that when the vertex $(0, 0)$ of $y = x^2$ is shifted 3 units to the right and 1 unit upward, it moves to $(3, 1)$. Therefore, $(3, 1)$ is the vertex of $g(x) = 2(x - 3)^2 + 1$. Note how the coordinates of the vertex are related to the rule of the function g:

The vertex of the function g in Example 1 was easily determined because the rule of g had a special algebraic form. The vertex of the graph of any quadratic function can be determined in a similar fashion by first rewriting its rule.

EXAMPLE 2

Find the vertex of the graph of $g(x) = 3x^2 + 30x + 77$ algebraically.

SOLUTION   We rewrite the rule of g as follows, being careful at each step not to change its value:

$$g(x) = 3x^2 + 30x + 77$$

Factor out 3: $$= 3(x^2 + 10x) + 77$$

Next, we want to complete the square in the expression in parentheses by adding 25 (the square of half the coefficient of x).[*] In order not to change the rule of the function, however, we must also *subtract* 25:

Add $25 - 25$ inside parentheses: $$= 3(x^2 + 10x + 25 - 25) + 77$$

Use the distributive law: $$= 3(x^2 + 10x + 25) - 3 \cdot 25 + 77$$

Simplify: $$= 3(x^2 + 10x + 25) + 2$$

Factor expression in parentheses: $$= 3(x + 5)^2 + 2.$$

As we saw in Section 3.4, the graph of $g(x)$ is the graph of $f(x) = x^2$ shifted horizontally 5 units to the left, stretched by a factor of 3, and shifted 2 units upward, as shown in Figure 4–4. In this process, the vertex $(0, 0)$ of f moves to $(-5, 2)$. Therefore, $(-5, 2)$ is the vertex of $g(x) = 3(x + 5)^2 + 2$. Once again, note how the coordinates of the vertex are related to the rule of the function.

Negatives

$$g(x) = 3(x + 5)^2 + 2 \qquad\qquad \text{vertex } (-5, 2)$$

Same

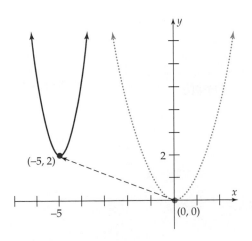

Figure 4–4

---

[*]Completing the square was discussed on page 21.

The technique used to find the vertex in Example 2 works for any quadratic function $f(x) = ax^2 + bx + c$.* First, rewrite the rule of f as follows.

$$f(x) = ax^2 + bx + c$$

Factor out a:
$$= a\left(x^2 + \frac{b}{a}x\right) + c$$

Next, complete the square in the expression in parentheses, as in Example 2. Note that half the coefficient of x is $\dfrac{b}{2a}$ and the square of this number is $\dfrac{b^2}{4a^2}$.

Add $\dfrac{b^2}{4a^2} - \dfrac{b^2}{4a^2}$ inside parentheses.
$$= a\left(x^2 + \frac{b}{a}x + \frac{b^2}{4a^2} - \frac{b^2}{4a^2}\right) + c$$

Use the distributive law:
$$= a\left(x^2 + \frac{b}{a}x + \frac{b^2}{4a^2}\right) - a \cdot \frac{b^2}{4a^2} + c$$

Simplify and rearrange:
$$= a\left(x^2 + \frac{b}{a}x + \frac{b^2}{4a^2}\right) + \left(c - \frac{b^2}{4a}\right)$$

Factor first expression in parentheses:†
$$= a\left(x + \frac{b}{2a}\right)^2 + \left(c - \frac{b^2}{4a}\right).$$

As in the preceding examples, the graph of f is just the graph of x^2 shifted horizontally, stretched by a factor of a, and shifted vertically. As above, the vertex of this parabola can be read from the rule of the function:

$$\underset{\text{Same}}{\underbrace{f(x = a\left(x + \overset{\text{Negatives}}{\frac{b}{2a}}\right)^2 + \left(c - \frac{b^2}{4a}\right)}} \qquad \text{vertex} \left(\frac{-b}{2a}, c - \frac{b^2}{4a}\right)$$

If we let $h = -\dfrac{b}{2a}$ and $k = c - \dfrac{b^2}{4a}$, then we have these useful facts.

Quadratic Functions

> The rule of the quadratic function $f(x) = ax^2 + bx + c$ can be rewritten in the form
>
> $$f(x) = a(x - h)^2 + k,$$
>
> where $h = -b/2a$. The graph of f is a parabola with vertex (h, k). It opens upward if $a > 0$ and downward if $a < 0$.

---

*The following argument is exactly the one used in Example 2, with a in place of 3, b in place of 30, c in place of 77, and $b^2/4a^2$ in place of 25.

†Verify that this factorization is correct by multiplying out $\left(x + \dfrac{b}{2a}\right)^2$.

EXAMPLE 3

Describe the graph of

$$f(x) = -x^2 + 5x + 1.$$

SOLUTION The graph is a downward-opening parabola (because f is a quadratic function and the coefficient of x^2 is negative). According to the preceding box, the x-coordinate of its vertex is

$$-\frac{b}{2a} = -\frac{5}{2(-1)} = \frac{-5}{-2} = \frac{5}{2} = 2.5.$$

To find the y-coordinate of the vertex, we need only evaluate f at this number:

$$f(2.5) = -(2.5^2) + 5(2.5) + 1 = 7.25.$$

Therefore, the vertex is (2.5, 7.25). ■

CALCULATOR EXPLORATION

Graphically confirm the results of Example 3 by graphing $y = -x^2 + 5x + 1$ and using your maximum finder to approximate the vertex.

EXAMPLE 4

Find the rule of the quadratic function whose graph is a parabola with vertex (3, 4) that passes through the point $(-1, 36)$.

SOLUTION The rule of a quadratic function can be written in the form $f(x) = a(x - h)^2 + k$, and its graph is a parabola with vertex (h, k). In this case, the vertex is (3, 4), so we have $h = 3$ and $k = 4$. Hence, the rule of f is $f(x) = a(x - 3)^2 + 4$. Since $(-1, 36)$ is on the graph, we have

$$f(-1) = 36$$

Substitute -1 for x in the rule of f: $\quad a(-1 - 3)^2 + 4 = 36$

Simplify: $\quad 16a + 4 = 36$

Subtract 4 from both sides. $\quad 16a = 32$

Divide both sides by 16: $\quad a = 2.$

Therefore, the rule of the function is $f(x) = 2(x - 3)^2 + 4.$ ■

APPLICATIONS

The solution of many applied problems depends on finding the vertex of a parabola.

EXAMPLE 5

Find the area and dimensions of the largest rectangular field that can be enclosed with 3000 feet of fence.

SOLUTION   Let x denote the length and y the width of the field, as shown in Figure 4–5.

$$\text{Perimeter} = x + y + x + y$$
$$= 2x + 2y$$
$$\text{Area} = xy$$

Figure 4–5

Since the perimeter is the length of the fence, $2x + 2y = 3000$. Hence,

$$2y = 3000 - 2x$$

and $y = 1500 - x$. Consequently, the area is

$$A = xy = x(1500 - x) = 1500x - x^2 = -x^2 + 1500x.$$

The largest possible area is just the maximum value of the quadratic function $A(x) = -x^2 + 1500x$. This maximum occurs at the vertex of the graph of $A(x)$ (which is a downward-opening parabola because the coefficient of x^2 is negative). The vertex may be found by using the fact in the box on page 228 (with $a = -1$ and $b = 1500$):

$$\text{The } x\text{-coordinate of the vertex is } -\frac{1500}{2(-1)} = 750 \text{ feet.}$$

Hence, the y-coordinate of the vertex, the maximum value of $A(x)$, is

$$A(750) = -750^2 + 1500 \cdot 750 = 562{,}500 \text{ square feet.}$$

It occurs when the length is $x = 750$. In this case the width is

$$y = 1500 - x = 1500 - 750 = 750. \quad\blacksquare$$

EXAMPLE 6

The owner of a 20-unit apartment complex has found that each $50 increase in monthly rent results in a vacant apartment. All units are now rented at $400 per month. How many $50 increases in rent will produce the largest possible income for the owner?

SOLUTION   Let *x* represent the number of $50 increases. Then the monthly rent will be $400 + 50x$ dollars. Since one apartment goes vacant for each increase, the number of occupied apartments will be $20 - x$. Then the owner's monthly income $R(x)$ is given by

$$R(x) = \text{(number of apartments rented)} \times \text{(rent per apartment)}$$

$$R(x) = (20 - x)(400 + 50x)$$

$$R(x) = -50x^2 + 600x + 8000$$

There are three ways to find the maximum possible income.

Algebraic Method.   The graph of $R(x) = -50x^2 + 600x + 8000$ is a downward-opening parabola (why?). Maximum income occurs at the vertex of this parabola, that is, when

$$x = \frac{-b}{2a} = \frac{-600}{2(-50)} = \frac{-600}{-100} = 6.$$

Therefore, six increases of $50 will produce maximum income.

Figure 4–6

Graphical Method.   Graphing $R(x) = -50x^2 + 600x + 8000$ and using a maximum finder to determine the coordinates of the vertex, as in Figure 4–6, shows that maximum income of $9800 occurs when there are six rent increases.

Table Method.   Make a table of values of $R(x)$ for $0 \leq x \leq 20$, as in Figure 4–7.* The table shows that the maximum income of $9800 occurs when $x = 6$. In this case, there will be $20 - 6 = 14$ apartments rented at a monthly rent of $400 + 6(50) = \$700$. ∎

Figure 4–7

---

*This method is feasible here because there are only 20 apartments, but it cannot be used when the number of possibilities is very large or infinite.

EXERCISES 4.1

In Exercises 1–8, use the catalog of functions at the end of Section 3.3 and the information in this section to match each function with its graph, which is one of those shown here.

A.

B.

C.

D.

E.

F.

G.

H.

I.

J.

K.

L.

1. $f(x) = x^2 + 2$

2. $g(x) = x^2 - 2$

3. $g(x) = (x - 2)^2$

4. $f(x) = -(x + 2)^2$

5. $f(x) = 2(x - 2)^2 + 2$

6. $g(x) = -2(x - 2)^2 - 2$

7. $g(x) = -2(x + 2)^2 + 2$

8. $f(x) = 2(x + 2)^2 - 2$

In Exercises 9–20, without graphing, *determine the vertex of the given parabola and state whether it opens upward or downward.*

9. $f(x) = 3(x - 5)^2 + 2$

10. $g(x) = -6(x - 2)^2 - 5$

11. $y = -(x - 1)^2 + 2$

12. $h(x) = -x^2 + 1$

13. $f(x) = x^2 - 6x + 3$

14. $g(x) = x^2 + 8x - 1$

15. $h(x) = x^2 + 3x + 6$

16. $f(x) = x^2 - 5x - 7$

17. $y = 2x^2 + 12x - 3$

18. $y = 3x^2 + 6x + 1$

19. $f(x) = -x^2 + 8x - 2$

20. $g(x) = -x^2 - 6x + 4$

In Exercises 21–24, find

(a) The difference quotient of the functions;

(b) The vertex of the function's graph;

(c) The value of the difference quotient at the x-coordinate of the vertex.

21. $f(x) = -x^2 + x$

22. $g(x) = 2x^2 - x - 1$

23. $f(x) = -2x^2 + 2x - 1$

24. $g(x) = -3x^2 + 4x + 5$

25. The graph of the quadratic function g is obtained from the graph of $f(x) = x^2$ by vertically stretching it by a factor of 2 and then shifting vertically 5 units downward. What is the rule of the function g? What is the vertex of its graph?

26. The graph of the quadratic function g is obtained from the graph of $f(x) = x^2$ by shifting it horizontally 4 units to the left, then vertically stretching it by a factor of 3, and then shifting vertically 2 units upward. What is the rule of the function g? What is the vertex of its graph?

27. If the graph of the quadratic function h is shifted vertically 4 units downward, then shrunk by a factor of $1/2$, and then shifted horizontally 3 units to the left, the resulting graph is the parabola $f(x) = x^2$. What is the rule of the function h? What is the vertex of its graph?

28. If the graph of the quadratic function h is shifted vertically 3 units upward, then reflected in the x-axis, and then shifted horizontally 5 units to the right, the resulting graph is the parabola $f(x) = x^2$. What is the rule of the function h? What is the vertex of its graph?

In Exercises 29–32, find the rule of the quadratic function whose graph satisfies the given conditions.

29. Vertex at $(0, 0)$; passes through $(2, 12)$

30. Vertex at $(0, 1)$; passes through $(2, -7)$

31. Vertex at $(2, 5)$; passes through $(-3, 80)$

32. Vertex at $(4, 1)$; passes through $(2, -11)$

33. Find the number b such that the vertex of the parabola $y = x^2 + bx + c$ lies on the y-axis.

34. Find the number c such that the vertex of the parabola $y = x^2 + 8x + c$ lies on the x-axis.

35. If the vertex of the parabola $f(x) = x^2 + bx + c$ is at $(2, 4)$, find b and c.

36. If the vertex of the parabola $f(x) = -x^2 + bx + 8$ has second coordinate 17 and is in the second quadrant, find b.

37. The Leslie Lahr Luggage Company has determined that its profit on its Luxury ensemble is given by

$$p(x) = 1600x - 4x^2 - 50{,}000,$$

where x is the number of units sold.

(a) What is the profit on 50 units? On 250 units?

(b) How many units should be sold to maximize profit? In that case, what will be the profit on each unit?

38. On the basis of data from past years, a consultant informs Bob's Bicycles that their profit from selling x bicycles is given by the function

$$p(x) = 250x - x^2/4 - 15{,}000.$$

(a) How much profit do they make by selling 100 bicycles? By selling 400 bicycles?

(b) How many bicycles should be sold to maximize profit? In that case, what will be the profit per bicycle?

39. During the Civil War, the standard heavy gun for coastal artillery was the 15-inch Rodman cannon, which fired a 330-pound shell. If one of these guns is fired from the top of a 50-foot-high shoreline embankment, then the height of the shell above the water (in feet) can be approximated by the function

$$p(x) = -.0000167x^2 + .23x + 50,$$

where x is the horizontal distance (in feet) from the foot of the embankment to a point directly under the shell. How high does the shell go, and how far away does it hit the water?

40. The Golden Gate Bridge is supported by two huge cables strung between the towers at each end of the bridge. The function

$$f(x) = .0001193x^2 - .50106x + 526.113$$

gives the approximate height of the cables above the roadway at a point on the road x feet from one of the towers. The cables touch the road halfway between the two towers. How far apart are the towers?

41. The braking distance (in meters) for a car with excellent brakes on a good road with an alert driver can be modeled by the quadratic function $B(s) = .01s^2 + .7s$, where s is the car's speed in kilometers per hour.

(a) What is the braking distance for a car traveling 30 kilometers per hour? For one traveling 100 kilometers per hour?

(b) If the car takes 60 meters to come to a complete stop, what was its speed?

42. Jack throws a baseball. Its height above the gound (in feet) is given by

$$h(x) = -.0013x^2 + .26x + 5,$$

where x is the distance (in feet) from Jack to a point on the ground directly below the ball.
(a) How far from Jack is the ball when it reaches the highest point on its flight? How high is the ball at that point?
(b) How far from Jack does the ball hit the ground?

In Exercises 43–46, use the formula for the height h of an object that is traveling vertically (subject only to gravity) at time t:

$$h = -16t^2 + v_0 t + h_0,$$

where h_0 is the initial height and v_0 is the initial velocity; t is measured in seconds and h in feet.

43. A ball is thrown upward from the top of a 96-foot-high tower with an initial velocity of 80 feet per second. When does the ball reach its maximum height and how high is it at that time?

44. A rocket is fired upward from ground level with an initial velocity of 1600 feet per second. When does it attain its maximum height and what is that height?

45. A ball is thrown upward from a height of 6 feet with an initial velocity of 32 feet per second. Find its maximum height.

46. A bullet is fired upward from ground level with an initial velocity of 1500 feet per second. How high does it go?

47. The sum of the height h and the base b of a triangle is 30. What height and base will produce a triangle of maximum area?

48. A gutter is to be made by bending up the edge of a 15-inch-wide piece of aluminum. What depth should the gutter be to have the maximum possible cross-sectional area?

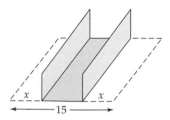

49. A field bounded on one side by a river is to be fenced on three sides so as to form a rectangular enclosure. If 200 feet of fencing is to be used, what dimensions will yield an enclosure of the largest possible area?

50. A rectangular box (with top) has a square base. The sum of the lengths of its 12 edges is 8 feet. What dimensions should the box have so that its surface area is as large as possible?

51. A gardener wants to use 130 feet of fencing to enclose a rectangular garden and divide it into two plots, as shown in the figure. What is the largest possible area for such a garden?

52. A rectangular garden next to a building is to be fenced on three sides. Fencing for the side parallel to the building costs $80 per foot, and material for the other two sides costs $20 per foot. If $1800 is to be spent on fencing, what are the dimensions of the garden with the largest possible area?

53. At Middleton Place, a plantation near Charleston, South Carolina, there is a "joggling board" that was once used for courting. A young girl would sit at one end, her suitor at the other end, and her mother in the center. The mother would bounce on the board, thus causing the girl and her suitor to move closer together. A joggling board is 8 feet long and an average mother sitting at its center causes the board to deflect 2 inches, as shown in the figure. The shape of the deflected board is parabolic.

2 inches

8 feet

(a) Find the equation of the parabola, assuming that the joggling board lies on the *x*-axis with its center at the origin.
(b) How far from the center of the board is the deflection 1 inch?

54. A salesperson finds that her sales average 40 cases per store when she visits 20 stores a week. Each time she visits an additional store per week, the average sales per store decreases by 1 case. How many stores should she visit each week if she wants to maximize her sales?

55. A potter can sell 120 bowls per week at $4 per bowl. For each 50¢ decrease in price, 20 more bowls are sold. What price should be charged to maximize sales income?

56. A vendor can sell 200 souvenirs per day at a price of $2 each. Each 10¢ price increase decreases the number of sales by 25 per day. Souvenirs cost the vendor $1.50 each. What price should be charged to maximize the profit?

57. When a basketball team charges $4 per ticket, average attendance is 500 people. For each 20¢ decrease in ticket price, average attendance increases by 30 people. What should the ticket price be to ensure maximum income?

Thinker

58. The *discriminant* of a quadratic function

$$f(x) = ax^2 + bx + c$$

is the number $b^2 - 4ac$. For each of the discriminants listed, state which graphs could possibly be the graph of f.
(a) $b^2 - 4ac = 25$
(b) $b^2 - 4ac = 0$
(c) $b^2 - 4ac = -49$
(d) $b^2 - 4ac = 72$

(i)   (ii)

(iii)   (iv)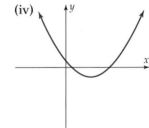

4.2 Polynomial Functions

Informally, a **polynomial** is an algebraic expression such as

$$x^3 - 6x^2 + \tfrac{1}{2} \quad \text{or} \quad x^{15} + x^{10} + 7 \quad \text{or} \quad x - 6.7 \quad \text{or} \quad 12.$$

Formally, a **polynomial in *x*** is an algebraic expression that can be written in the form

$$a_nx^n + a_{n-1}x^{n-1} + \cdots + a_3x^3 + a_2x^2 + a_1x + a_0,$$

where *n* is a nonnegative integer, *x* is a variable, and each of $a_0, a_1, \ldots, a_n$ is a constant, called a **coefficient.** The coefficient a_0 is called the **constant term.** A polynomial that consists only of a constant term, such as 12, is called a **constant polynomial.** The **zero polynomial** is the constant polynomial 0.

---

*Any letter may be used as the variable in a polynomial.

The *exponent* of the highest power of x that appears with *nonzero* coefficient is the **degree** of the polynomial, and the nonzero coefficient of this highest power of x is the **leading coefficient.** For example,

Polynomial	Degree	Leading Coefficient	Constant Term
$6x^7 + 4x^3 + 5x^2 - 7x + 10$	7	6	10
x^3	3	1	0
12 (think of this as $12x^0$)	0	12	12
$0x^9 + 2x^6 + 3x^7 + x^8 - 2x - 4$	8	1	-4

The degree of the zero polynomial is *not defined* since no exponent of x occurs with nonzero coefficient.

A **polynomial function** is a function whose rule is given by a polynomial, such as $f(x) = x^5 + 3x^2 - 2$. First-degree polynomial functions, such as $g(x) = 3x - 4$, are called **linear functions,** and, as we saw in Section 4.1, second-degree polynomial functions are called **quadratic functions.**

▗▖ POLYNOMIAL ARITHMETIC

You should be familiar with addition, subtraction, and multiplication of polynomials, which are presented in the Algebra Review Appendix. Long division of polynomials is quite similar to long division of numbers, as we now see.

EXAMPLE 1

Divide $8x^3 + 2x^2 + 1$ by $2x^2 - x$.

SOLUTION   We set up the division in the same way that is used for numbers.

$$\text{Divisor} \rightarrow \quad 2x^2 - x \,\overline{\smash{\big)}\, 8x^3 + 2x^2 + 1} \quad \leftarrow \text{Dividend}$$

Begin by dividing the first term of the divisor ($2x^2$) into the first term of the dividend ($8x^3$) and putting the result $\left(\text{namely, } \dfrac{8x^3}{2x^2} = 4x\right)$ on the top line, as shown below. Then multiply $4x$ times the entire divisor, put the result on the third line, and subtract

$$
\begin{array}{r}
4x \qquad\qquad\quad \leftarrow \text{Partial Quotient} \\
2x^2 - x \,\overline{\smash{\big)}\, 8x^3 + 2x^2 + 1} \\
\underline{8x^3 - 4x^2} \qquad \leftarrow 4x(2x^2 - x) \\
6x^2 + 1 \quad \leftarrow \text{Subtraction}^*
\end{array}
$$

Now divide the first term of the divisor ($2x^2$) into $6x^2$ and put the result $\left(\dfrac{6x^2}{2x^2} = 3\right)$ on the top line, as shown on the next page.

---

*If this subtraction is confusing, write it out horizontally and watch the signs:

$$(8x^3 + 2x^2 + 1) - (8x^3 - 4x^2) = 8x^3 + 2x^2 + 1 - 8x^3 + 4x^2 = 6x^2 + 1.$$

Then multiply 3 times the entire divisor, put the result on the fifth line, and subtract

$$
\begin{array}{r}
4x + 3 \qquad\qquad\quad \leftarrow \text{Quotient} \\
2x^2 - x\,\overline{\big)\,8x^3 + 2x^2 \qquad\quad + 1} \\
\underline{8x^3 - 4x^2 \qquad\qquad}\quad \leftarrow 4x(2x^2 - x) \\
6x^2 \qquad + 1 \quad \leftarrow \text{Subtraction} \\
\underline{6x^2 - 3x \qquad}\quad \leftarrow 3(2x^2 - x) \\
\text{Remainder} \rightarrow\; 3x + 1 \quad \leftarrow \text{Subtraction}
\end{array}
$$

The division process stops when the remainder is 0 or has smaller degree than the divisor, which is the case here. ■

Recall how you check a long division problem with numbers.

$$
\begin{array}{r}
145 \\
31\,\overline{\big)\,4509} \qquad\qquad \text{Check:} \\
\underline{31} \\
140 \\
\underline{124} \\
169 \\
\underline{155} \\
14
\end{array}
\qquad
\begin{array}{r}
145 \quad \leftarrow \text{Quotient} \\
\underline{\times\ 31} \quad \leftarrow \text{Divisor} \\
145 \\
\underline{435} \\
4495 \\
\underline{+\ 14} \quad \leftarrow \text{Remainder} \\
4509 \quad \leftarrow \text{Dividend}
\end{array}
$$

We can summarize this process in one line:

$$\text{Divisor} \cdot \text{Quotient} + \text{Remainder} = \text{Dividend}.$$

The same thing works for division of polynomials, as you can see by examining the division problem from Example 1.

$$
\overset{\text{Divisor} \cdot\ \text{Quotient}\ +\ \text{Remainder}}{(2x^2 - x) \cdot (4x + 3) + (3x + 1)} = (8x^3 + 2x^2 - 3x) + (3x + 1)
$$

$$
= \underset{\text{Dividend}}{8x^3 + 2x^2 + 1}
$$

This fact is so important that it is given a special name and a formal statement.

The Division Algorithm

If a polynomial $f(x)$ is divided by a nonzero polynomial $h(x)$, then there is a quotient polynomial $q(x)$ and a remainder polynomial $r(x)$ such that

$$\text{Dividend} = \text{Divisor} \cdot \text{Quotient} + \text{Remainder}$$

$$f(x) = h(x)\, q(x) + r(x),$$

where either $r(x) = 0$ or $r(x)$ has degree less than the degree of the divisor $h(x)$.

EXAMPLE 2

Show that $2x^2 + 1$ is a factor of $6x^3 - 4x^2 + 3x - 2$.

SOLUTION   We divide $6x^3 - 4x^2 + 3x - 2$ by $2x^2 + 1$ and find that the remainder is 0.

$$
\begin{array}{r}
3x - 2 \\
2x^2 + 1 \overline{\smash{\big)}\, 6x^3 - 4x^2 + 3x - 2} \\
\underline{6x^3 + 3x } \\
-4x^2 - 2 \\
\underline{-4x^2 - 2} \\
0.
\end{array}
$$

Since the remainder is 0, the Division Algorithm tells us that

$$\text{Dividend} = \text{Divisor} \cdot \text{Quotient} + \text{Remainder}$$
$$6x^3 - 4x^2 + 3x - 2 = (2x^2 + 1)(3x - 2) + 0$$
$$= (2x^2 + 1)(3x - 2).$$

Therefore, $2x^2 + 1$ is a factor of $6x^3 - 4x^2 + 3x - 2$, and the other factor is the quotient $3x - 2$.   ■

Example 2 illustrates this fact.

Remainders and Factors

The remainder in polynomial division is 0 exactly when the divisor is a factor of the dividend. In this case, the quotient is the other factor.

■■ REMAINDERS AND ROOTS

When a polynomial $f(x)$ is divided by a first-degree polynomial, such as $x - 3$ or $x + 5$, the remainder is a constant (because constants are the only polynomials of degree less than 1). This remainder has an interesting connection with the values of the polynomial function $f(x)$.

EXAMPLE 3

Let $f(x) = x^3 - 2x^2 - 4x + 5$.

(a) Find the quotient and remainder when $f(x)$ is divided by $x - 3$.

(b) Find $f(3)$.

NOTE

When the divisor is a first-degree polynomial such as $x - 3$, there is a convenient shorthand method of division, called **synthetic division.** See Special Topics 4.2.A for details.

SOLUTION

(a) Using long division, we have

$$
\begin{array}{r}
x^2 + x - 1 \\
x - 3 \overline{\smash{\big)}\, x^3 - 2x^2 - 4x + 5} \\
\underline{x^3 - 3x^2} \\
x^2 - 4x + 5 \\
\underline{x^2 - 3x} \\
-x + 5 \\
\underline{-x + 3} \\
2.
\end{array}
$$

Therefore, the quotient is $x^2 + x - 1$, and the remainder is 2.

(b) Using the Division Algorithm, we can write the dividend $f(x) = x^3 - 2x^2 - 4x + 5$ as

Dividend = Divisor · Quotient + Remainder

$$f(x) = (x - 3)(x^2 + x - 1) + 2.$$

Hence,

$$f(3) = (3 - 3)(3^2 + 3 - 1) + 2 = 0 + 2 = 2.$$

Note that the number $f(3)$ is the same as the remainder when $f(x)$ is divided by $x - 3$. ■

The argument used in Example 3 to show that $f(3)$ is the remainder when $f(x)$ is divided by $x - 3$ also works in the general case and proves this fact.

Remainder Theorem

If a polynomial $f(x)$ is divided by $x - c$, then the remainder is the number $f(c)$.

EXAMPLE 4

To find the remainder when $f(x) = x^{79} + 3x^{24} + 5$ is divided by $x - 1$, we apply the Remainder Theorem with $c = 1$. The remainder is

$$f(1) = 1^{79} + 3 \cdot 1^{24} + 5 = 1 + 3 + 5 = 9. \quad ■$$

EXAMPLE 5

To find the remainder when $f(x) = 3x^4 - 8x^2 + 11x + 1$ is divided by $x + 2$, we must apply the Remainder Theorem *carefully*. The divisor in the theorem is $x - c$, not $x + c$. So we rewrite $x + 2$ as $x - (-2)$ and apply the theorem with $c = -2$. The remainder is

$$f(-2) = 3(-2)^4 - 8(-2)^2 + 11(-2) + 1 = 48 - 32 - 22 + 1 = -5. \quad ■$$

If $f(x)$ is a polynomial, then a solution of the equation $f(x) = 0$ is called a **root** or **zero** of $f(x)$. Thus, a number c is a root of $f(x)$ if $f(c) = 0$. A root that is a real number is called a **real root**. For example, 4 is a real root of the polynomial $f(x) = 3x - 12$ because $f(4) = 3 \cdot 4 - 12 = 0$. There is an interesting connection between the roots of a polynomial and its factors.

EXAMPLE 6

Let $f(x) = x^3 - 4x^2 + 2x + 3$.

(a) Show that 3 is a root of $f(x)$.

(b) Show that $x - 3$ is a factor of $f(x)$.

SOLUTION

(a) Evaluating $f(x)$ at 3 shows that

$$f(3) = 3^3 - 4(3^2) + 2(3) + 3 = 0.$$

Therefore, 3 is a root of $f(x)$.

(b) If $f(x)$ is divided by $x - 3$, then by the Division Algorithm, there is a quotient polynomial $q(x)$ such that

$$f(x) = (x - 3)q(x) + \text{remainder}.$$

The remainder Theorem shows that the remainder when $f(x)$ is divided by $x - 3$ is the number $f(3)$, which is 0, as we saw in part (a). Therefore,

$$f(x) = (x - 3)q(x) + 0 = (x - 3)q(x).$$

Thus, $x - 3$ is a factor of $f(x)$. [To determine the other factor, the quotient $q(x)$, you have to perform the division.] ∎

Example 6 illustrates this fact, which can be proved by the same argument used in the example.

Factor Theorem

The number c is a root of the polynomial $f(x)$ exactly when $x - c$ is a factor of $f(x)$.

EXAMPLE 7

The graph of $f(x) = 15x^3 - x^2 - 114x + 72$ in the standard viewing window (Figure 4–8) is obviously not complete but suggests that -3 is an x-intercept, and hence a root of $f(x)$. It is easy to verify that this is indeed the case.

$$f(-3) = 15(-3)^3 - (-3)^2 - 114(-3) + 72 = -405 - 9 + 342 + 72 = 0.$$

Since -3 is a root, $x - (-3) = x + 3$ is a factor of $f(x)$. Use synthetic or long division to verify that the other factor is $15x^2 - 46x + 24$. By factoring this quadratic, we obtain a complete factorization of $f(x)$.

$$f(x) = (x + 3)(15x^2 - 46x + 24) = (x + 3)(3x - 2)(5x - 12). \quad ∎$$

Figure 4–8

> ### EXAMPLE 8
>
> Find three polynomials of different degrees that have 1, 2, 3, and -5 as roots.
>
> **SOLUTION**   A polynomial that has 1, 2, 3, and -5 as roots must have $x - 1$, $x - 2$, $x - 3$, and $x - (-5) = x + 5$ as factors. Many polynomials satisfy these conditions, such as
>
> $$g(x) = (x - 1)(x - 2)(x - 3)(x + 5) = x^4 - x^3 - 19x^2 + 49x - 30$$
>
> $$h(x) = 8(x - 1)(x - 2)(x - 3)^2(x + 5)$$
>
> $$k(x) = 2(x + 4)^2(x - 1)(x - 2)(x - 3)(x + 5)(x^2 + x + 1).$$
>
> Note that g has degree 4. When h is multiplied out, its leading term is $8x^5$, so h has degree 5. Similarly, k has degree 8 since its leading term is $2x^8$.   ∎

If a polynomial $f(x)$ has four roots, say a, b, c, d, then by the same argument used in Example 8, it must have

$$(x - a)(x - b)(x - c)(x - d)$$

as a factor. Since $(x - a)(x - b)(x - c)(x - d)$ has degree 4 (multiply it out—its leading term is x^4), $f(x)$ must have degree at least 4. In particular, this means that no polynomial of degree 3 can have four or more roots. A similar argument works in the general case.

Number	
of Roots	A polynomial of degree n has at most n distinct roots.

EXERCISES 4.2

In Exercises 1–8, determine whether the given algebraic expression is a polynomial. If it is, list its leading coefficient, constant term, and degree.

1. $1 + x^3$

2. -7

3. $(x - 1)(x^2 + 1)$

4. $7^x + 2x + 1$

5. $(x + \sqrt{3})(x - \sqrt{3})$

6. $4x^2 + 3\sqrt{x} + 5$

7. $\dfrac{7}{x^2} + \dfrac{5}{x} - 15$

8. $(x - 1)^k$ (where k is a fixed positive integer)

In Exercises 9–14, state the quotient and remainder when the first polynomial is divided by the second. Check your division by calculating (Divisor)(Quotient) + Remainder.

9. $3x^4 + 2x^2 - 6x + 1$; $x + 1$

10. $x^5 - x^3 + x - 5$; $x - 2$

11. $x^5 + 2x^4 - 6x^3 + x^2 - 5x + 1$; $x^3 + 1$

12. $3x^4 - 3x^3 - 11x^2 + 6x - 1$; $x^3 + x^2 - 2$

13. $5x^4 + 5x^2 + 5$; $x^2 - x + 1$

14. $x^5 - 1$; $x - 1$

In Exercises 15–18, determine whether the first polynomial is a factor of the second.

15. $x^2 + 3x - 1$; $x^3 + 2x^2 - 5x - 6$

16. $x^2 + 9$; $x^5 + x^4 - 81x - 81$

17. $x^2 + 3x - 1$; $x^4 + 3x^3 - 2x^2 - 3x + 1$

18. $x^2 - 5x + 7$; $x^3 - 3x^2 - 3x + 9$

In Exercises 19–22, determine which of the given numbers are roots of the given polynomial.

19. 2, 3, 0, −1; $g(x) = x^4 + 6x^3 - x^2 - 30x$

20. 1, 1/2, 2, −1/2, 3; $f(x) = 6x^2 + x - 1$

21. 2√2, √2, −√2, 1, −1; $h(x) - x^3 + x^2 \quad 8x \quad 8$

22. √3, −√3, 1, −1; $k(x) = 8x^3 - 12x^2 - 6x + 9$

In Exercises 23–32, find the remainder when f(x) is divided by g(x), without using division.

23. $f(x) = x^{10} + x^8$; $g(x) = x - 1$

24. $f(x) = x^6 - 10$; $g(x) = x - 2$

25. $f(x) = 3x^4 - 6x^3 + 2x - 1$; $g(x) = x + 1$

26. $f(x) = x^5 - 3x^2 + 2x - 1$; $g(x) = x - 2$

27. $f(x) = x^3 - 2x^2 + 5x - 4$; $g(x) = x + 2$

28. $f(x) = 10x^{75} - 8x^{65} + 6x^{45} + 4x^{32} - 2x^{15} + 5$;
    $g(x) = x - 1$

29. $f(x) = 2x^5 - 3x^4 + x^3 - 2x^2 + x - 8$;
    $g(x) = x - 10$

30. $f(x) = x^3 + 8x^2 - 29x + 44$; $g(x) = x + 11$

31. $f(x) = 2x^5 - 3x^4 + 2x^3 - 8x - 8$; $g(x) = x - 20$

32. $f(x) = x^5 - 10x^4 + 20x^3 - 5x - 95$; $g(x) = x + 10$

In Exercises 33–38, use the Factor Theorem to determine whether or not h(x) is a factor of f(x).

33. $h(x) = x - 1$; $f(x) = x^5 + 1$

34. $h(x) = x - 1/2$; $f(x) = 2x^4 + x^3 + x - 3/4$

35. $h(x) = x + 2$; $f(x) = x^3 - 3x^2 - 4x - 12$

36. $h(x) = x + 1$; $f(x) = x^3 - 4x^2 + 3x + 8$

37. $h(x) = x - 1$; $f(x) = 14x^{99} - 65x^{56} + 51$

38. $h(x) = x - 2$; $f(x) = x^3 + x^2 - 4x + 4$

In Exercises 39–42, use the Factor Theorem and a calculator to factor the polynomial, as in Example 7.

39. $f(x) = 6x^3 - 7x^2 - 89x + 140$

40. $g(x) = x^3 - 5x^2 - 5x - 6$

41. $h(x) = 4x^4 + 4x^3 - 35x^2 - 36x - 9$

42. $f(x) = x^5 - 5x^4 - 5x^3 + 25x^2 + 6x - 30$

In Exercises 43–46, each graph is of a polynomial function f(x) of degree 5 whose leading coefficient is 1. The graph is not drawn to scale. Use the Factor Theorem to find the poly-

nomial. [Hint: *What are the roots of f(x)? What does the Factor Theorem tell you?*]

43.

44.

45.

46.

In Exercises 47–50, find a polynomial with the given degree n, the given roots, and no other roots.

47. $n = 3$ roots 1, 7, −4

48. $n = 3$;   roots 1, −1

49. $n = 6$;   roots 1, 2, π

50. $n = 5$;   root 2

51. Find a polynomial function f of degree 3 such that

$$f(10) = 17$$

and the roots of $f(x)$ are 0, 5, and 8.

52. Find a polynomial function g of degree 4 such that the roots of g are 0, -1, 2, -3, and $g(3) = 288$.

In Exercises 53–56, find a number k satisfying the given condition.

53. $x + 2$ is a factor of $x^3 + 3x^2 + kx - 2$.

54. $x - 3$ is a factor of $x^4 - 5x^3 - kx^2 + 18x + 18$.

55. $x - 1$ is a factor of $k^2x^4 - 2kx^2 + 1$.

56. $x + 2$ is a factor of $x^3 - kx^2 + 3x + 7k$.

57. Use the Factor Theorem to show that for every real number c, $x - c$ is *not* a factor of $x^4 + x^2 + 1$.

58. Let c be a real number and n a positive integer.
(a) Show that $x - c$ is a factor of $x^n - c^n$.
(b) If n is even, show that $x + c$ is a factor of $x^n - c^n$.
[*Remember:* $x + c = x - (-c)$.]

59. (a) If c is a real number and n an odd positive integer, give an example to show that $x + c$ may not be a factor of $x^n - c^n$.
(b) If c and n are as in part (a), show that $x + c$ is a factor of $x^n + c^n$.

Thinkers

60. For what value of k is the difference quotient of
$$g(x) = kx^2 + 2x + 1$$
equal to $7x + 2 + 3.5h$?

61. For what value of k is the difference quotient of
$$f(x) = x^2 + kx$$
equal to $2x + 5 + h$?

4.2.A   *SPECIAL TOPICS* Synthetic Division

Synthetic division is a fast method of doing polynomial division when the divisor is a first-degree polynomial of the form $x - c$ for some real number c. To see how it works, we first consider an example of ordinary long division.

$$
\begin{array}{r}
3x^3 + 6x^2 + 4x - 3 \quad \leftarrow \text{Quotient} \\
\text{Divisor} \rightarrow x - 2 \overline{\smash{\big)}\ 3x^4 \quad\quad - 8x^2 - 11x + 1} \quad \leftarrow \text{Dividend} \\
\underline{3x^4 - 6x^3} \quad\quad\quad\quad\quad\quad\quad \\
6x^3 - 8x^2 \quad\quad\quad\quad\quad \\
\underline{6x^3 - 12x^2} \quad\quad\quad\quad \\
4x^2 - 11x \quad\quad \\
\underline{4x^2 - 8x} \quad\quad \\
- 3x + 1 \\
\underline{- 3x + 6} \\
- 5 \quad \leftarrow \text{Remainder}
\end{array}
$$

This calculation obviously involves a lot of repetitions. If we insert 0 coefficients for terms that don't appear above and keep the various coefficients in the proper columns, we can eliminate the repetition and all the x's.

$$
\begin{array}{r}
3 \quad 6 \quad 4 \quad -3 \quad \leftarrow \text{Quotient} \\
\text{Divisor} \rightarrow 1 - 2 \overline{\smash{\big)}\ 3 \quad 0 \quad -8 \quad -11 \quad 1} \quad \leftarrow \text{Dividend} \\
\underline{-6} \quad\quad\quad\quad\quad\quad \\
6 \quad\quad\quad\quad\quad \\
\underline{-12} \quad\quad\quad \\
4 \quad\quad\quad \\
\underline{-8} \quad\quad \\
-3 \\
\underline{+6} \\
-5 \quad \leftarrow \text{Remainder}
\end{array}
$$

We can save space by moving the lower lines upward and writing 2 in the divisor position (since that's enough to remind us that the divisor is $x - 2$).

$$
\begin{array}{r}
3 \quad\ \ 6 \quad\ \ 4 \quad -3 \qquad\qquad \leftarrow \text{Quotient} \\
\text{Divisor} \rightarrow 2\,|\,3 \quad\ \ 0 \quad -8 \quad -11 \quad\ \ 1 \leftarrow \text{Dividend} \\
\underline{-6 \quad -12 \quad -8 \quad\ \ 6} \\
6 \quad\ \ 4 \quad -3 \ \boxed{-5} \leftarrow \text{Remainder}
\end{array}
$$

Since the last line contains most of the quotient line, we can save more space and still preserve the essential information by inserting a 3 in the last line and omitting the top line.

$$
\begin{array}{r}
\text{Divisor} \rightarrow 2\,|\,3 \quad\ \ 0 \quad -8 \quad -11 \quad\ \ 1 \leftarrow \text{Dividend} \\
\underline{-6 \quad -12 \quad -8 \quad\ \ 6} \\
3 \quad\ \ 6 \quad\ \ 4 \quad -3 \ \boxed{-5} \leftarrow \text{Remainder}
\end{array}
$$

$$\underbrace{}_{\text{Quotient}}$$

Synthetic division is a quick method for obtaining the last row of this array. Here is a step-by-step explanation of the division of $3x^4 - 8x^2 - 11x + 1$ by $x - 2$.

Step 1 In the first row, list the 2 from the divisor and the coefficients of the dividend in order of decreasing powers of x (insert 0 coefficients for missing powers of x).

$$2\,|\ \ 3 \quad\ \ 0 \quad -8 \quad -11 \quad\ \ 1$$

Step 2 Bring down the first dividend coefficient (namely, 3) to the third row

$$2\,|\ \ 3 \quad\ \ 0 \quad -8 \quad -11 \quad\ \ 1$$
$$\qquad 3$$

Step 3 Multiply $2 \cdot 3$ and insert the answer 6 in the second row, in the position shown here.

$$2\,|\ \ 3 \quad\ \ 0 \quad -8 \quad -11 \quad\ \ 1$$
$$\qquad\quad 6$$
$$\qquad 3$$

Step 4 Add $0 + 6$ and write the answer 6 in the third row.

$$2\,|\ \ 3 \quad\ \ 0 \quad -8 \quad -11 \quad\ \ 1$$
$$\qquad\quad 6$$
$$\qquad 3 \quad\ \ 6$$

Step 5 Multiply $2 \cdot 6$ and insert the answer 12 in the second row.

$$2\,|\ \ 3 \quad\ \ 0 \quad -8 \quad -11 \quad\ \ 1$$
$$\qquad\quad 6 \quad 12$$
$$\qquad 3 \quad\ \ 6$$

Step 6 Add $-8 + 12$ and write the answer 4 in the third row.

$$2\,|\ \ 3 \quad\ \ 0 \quad -8 \quad -11 \quad\ \ 1$$
$$\qquad\quad 6 \quad 12$$
$$\qquad 3 \quad\ \ 6 \quad\ \ 4$$

Step 7 Multiply $2 \cdot 4$ and insert the answer 8 in the second row.

$$
\begin{array}{r|rrrrr}
2 & 3 & 0 & -8 & -11 & 1 \\
  &   & 6 & 12 & 8 & \\
\hline
  & 3 & 6 & 4 & & \\
\end{array}
$$

Step 8 Add $-11 + 8$ and write the answer -3 in the third row.

$$
\begin{array}{r|rrrrr}
2 & 3 & 0 & -8 & -11 & 1 \\
  &   & 6 & 12 & 8 & \\
\hline
  & 3 & 6 & 4 & -3 & \\
\end{array}
$$

Step 9 Multiply $2 \cdot (-3)$ and insert the answer -6 in the second row.

$$
\begin{array}{r|rrrrr}
2 & 3 & 0 & -8 & -11 & 1 \\
  &   & 6 & 12 & 8 & -6 \\
\hline
  & 3 & 6 & 4 & -3 & \\
\end{array}
$$

Step 10 Add $1 + (-6)$ and write the answer -5 in the third row.

$$
\begin{array}{r|rrrrr}
2 & 3 & 0 & -8 & -11 & 1 \\
  &   & 6 & 12 & 8 & -6 \\
\hline
  & 3 & 6 & 4 & -3 & \underline{-5} \\
\end{array}
$$

Except for the signs in the second row, this last array is the same as the array obtained from the long division process, and we can read off the quotient and remainder:

The last number in the third row is the remainder.

The other numbers in the third row are the coefficients of the quotient (arranged in order of decreasing powers of x).

Since we are dividing the *fourth*-degree polynomial $3x^4 - 8x^2 - 11x + 1$ by the *first*-degree polynomial $x - 2$, the quotient must be a polynomial of degree *three* with coefficients 3, 6, 4, -3, namely, $3x^3 + 6x^2 + 4x - 3$. The remainder is -5.

CAUTION

Synthetic division can be used *only* when the divisor is a first-degree polynomial of the form $x - c$. In the example above, $c = 2$. If you want to use synthetic, division with a divisor such as $x + 3$, you must write it as $x - (-3)$, which is of the form $x - c$ with $c = -3$.

EXAMPLE 1

TECHNOLOGY TIP
Synthetic division programs are in the Program Appendix.

To divide $x^5 + 5x^4 + 6x^3 - x^2 + 4x + 29$ by $x + 3$, we write the divisor as $x - (-3)$ and proceed as above.

$$
\begin{array}{r|rrrrrr}
-3 & 1 & 5 & 6 & -1 & 4 & 29 \\
   &   & -3 & -6 & 0 & 3 & -21 \\
\hline
   & 1 & 2 & 0 & -1 & 7 & \underline{8} \\
\end{array}
$$

The last row shows that the quotient is $x^4 + 2x^3 - x + 7$ and the remainder is 8. ■

EXAMPLE 2

Show that $x - 7$ is a factor of $8x^5 - 52x^4 + 2x^3 - 198x^2 - 86x + 14$ and find the other factor.

SOLUTION $x - 7$ is a factor exactly when division by $x - 7$ leaves remainder 0, in which case the quotient is the other factor. Using synthetic division, we have

$$
\begin{array}{r|rrrrrr}
7 & 8 & -52 & 2 & -198 & -86 & 14 \\
  &   & 56 & 28 & 210 & 84 & -14 \\
\hline
  & 8 & 4 & 30 & 12 & -2 & 0
\end{array}
$$

Since the remainder is 0, the divisor $x - 7$ and the quotient

$$8x^4 + 4x^3 + 30x^2 + 12x - 2$$

are factors.

$$8x^5 - 52x^4 + 2x^3 - 198x^2 - 86x + 14$$
$$= (x - 7)(8x^4 + 4x^3 + 30x^2 + 12x - 2). \quad \blacksquare$$

✓ EXERCISES 4.2.A

In Exercises 1–8, use synthetic division to find the quotient and remainder.

1. $(3x^4 - 8x^3 + 9x + 5) \div (x - 2)$

2. $(4x^3 - 3x^2 + x + 7) \div (x - 2)$

3. $(2x^4 + 5x^3 - 2x - 8) \div (x + 3)$

4. $(3x^3 - 2x^2 - 8) \div (x + 5)$

5. $(5x^4 - 3x^2 - 4x + 6) \div (x - 7)$

6. $(3x^4 - 2x^3 + 7x - 4) \div (x - 3)$

7. $(x^4 - 6x^3 + 4x^2 + 2x - 7) \div (x - 2)$

8. $(x^6 - x^5 + x^4 - x^3 + x^2 - x + 1) \div (x + 3)$

In Exercises 9–12, use synthetic division to find the quotient and the remainder. In each divisor $x - c$, the number c is not an integer, but the same technique will work.

9. $(3x^4 - 2x^2 + 2) \div \left(x - \dfrac{1}{4}\right)$

10. $(2x^4 - 3x^2 + 1) \div \left(x - \dfrac{1}{2}\right)$

11. $(2x^4 - 5x^3 - x^2 + 3x + 2) \div \left(x + \dfrac{1}{2}\right)$

12. $\left(10x^5 - 3x^4 + 14x^3 + 13x^2 - \dfrac{4}{3}x + \dfrac{7}{3}\right) \div \left(x + \dfrac{1}{5}\right)$

In Exercises 13–16, use synthetic division to show that the first polynomial is a factor of the second and find the other factor.

13. $x + 4;$ $3x^3 + 9x^2 - 11x + 4$

14. $x - 5;$ $x^5 - 8x^4 + 17x^2 + 293x - 15$

15. $x - 1/2;$ $2x^5 - 7x^4 + 15x^3 - 6x^2 - 10x + 5$

16. $x + 1/3;$ $3x^6 + x^5 - 6x^4 + 7x^3 + 3x^2 - 15x - 5$

In Exercises 17 and 18, use a calculator and synthetic division to find the quotient and remainder.

17. $(x^3 - 5.27x^2 + 10.708x - 10.23) \div (x - 3.12)$

18. $(2.79x^4 + 4.8325x^3 - 6.73865x^2 + .9255x - 8.125)$
$\div (x - 1.35)$

Thinkers

19. When $x^3 + cx + 4$ is divided by $x + 2$, the remainder is 4. Find c.

20. If $x - d$ is a factor of $2x^3 - dx^2 + (1 - d^2)x + 5$, what is d?

4.3   Real Roots of Polynomials

Finding the real roots of polynomials is the same as solving polynomial equations. The root of a first-degree polynomial, such as $5x - 3$, can be found by solving the equation $5x - 3 = 0$. Similarly, the roots of any second-degree polynomial can be found by using the quadratic formula (Section 1.2). Although the roots of higher-degree polynomials can always be approximated graphically as in Section 2.2, it is better to find exact solutions, if possible.

▪▪ RATIONAL ROOTS

When a polynomial has integer coefficients, all of its **rational roots** (roots that are rational numbers) can be found exactly by using the following result.

The Rational Root Test

If a rational number r/s (in lowest terms) is a root of the polynomial

$$a_n x^n + \cdots + a_1 x + a_0,$$

where the coefficients $a_n, \ldots, a_1, a_0$ are integers with $a_n \neq 0$, $a_0 \neq 0$, then

r is a factor of the constant term a_0 and

s is a factor of the leading coefficient a_n.

The test states that every rational root must satisfy certain conditions.[*] By finding all the numbers that satisfy these conditions, we produce a list of *possible* rational roots. Then we must evaluate the polynomial at each number on the list to see whether the number actually is a root. This testing process can be considerably shortened by using a calculator, as in the next example.

EXAMPLE 1

Find the rational roots of

$$f(x) = 2x^4 + x^3 - 17x^2 - 4x + 6.$$

SOLUTION   If $f(x)$ has a rational root r/s, then by the Rational Root Test r must be a factor of the constant term 6. Therefore, r must be one of ± 1, ± 2, ± 3, or ± 6 (the only factors of 6). Similarly, s must be a factor of the leading coefficient 2, so s must be one of ± 1 or ± 2 (the only factors of 2). Consequently, the only *possibilities* for r/s are

$$\frac{\pm 1}{\pm 1}, \frac{\pm 2}{\pm 1}, \frac{\pm 3}{\pm 1}, \frac{\pm 6}{\pm 1}, \frac{\pm 1}{\pm 2}, \frac{\pm 2}{\pm 2}, \frac{\pm 3}{\pm 2}, \frac{\pm 6}{\pm 2}.$$

---

[*]Since the proof of the Rational Root Test sheds no light on how the test is actually used to solve equations, it will be omitted.

Eliminating duplications from this list, we see that the only *possible* rational roots are

$$1, -1, 2, -2, 3, -3, 6, -6, \frac{1}{2}, -\frac{1}{2}, \frac{3}{2}, -\frac{3}{2}.$$

Now graph $f(x)$ in a viewing window that includes all of these numbers on the x-axis, say $-7 \le x \le 7$ and $-5 \le y \le 5$ (Figure 4–9). A complete graph isn't necessary, since we are interested only in the x-intercepts.

| Figure 4–9 | Figure 4–10 |

Figure 4–9 shows that the only numbers on our list that could possibly be roots (x-intercepts) are -3, $-1/2$, and $1/2$, so these are the only ones that need be tested. We use the table feature to evaluate $f(x)$ at these three numbers (Figure 4–10). The table shows that -3 and $1/2$ are the only rational roots of $f(x)$. Its other roots (x-intercepts) in Figure 4–9 must be irrational numbers.   ■

ROOTS AND THE FACTOR THEOREM

Once some roots of a polynomial have been found, the Factor Theorem can be used to factor the polynomial, which may lead to additional roots.

EXAMPLE 2

Find all the roots of $f(x) = 2x^4 + x^3 - 17x^2 - 4x + 6$.

SOLUTION   In Example 1, we saw that -3 and $1/2$ are the rational roots of $f(x)$. By the Factor Theorem, $x - (-3) = x + 3$ and $x - 1/2$ are factors of $f(x)$. Using synthetic or long division twice, we have

$$2x^4 + x^3 - 17x^2 - 4x + 6 = (x + 3)(2x^3 - 5x^2 - 2x + 2)$$

$$= (x + 3)(x - .5)(2x^2 - 4x - 4).$$

The remaining roots of $f(x)$ are the roots of $2x^2 - 4x - 4$, that is, the solutions of

$$2x^2 - 4x - 4 = 0$$

$$x^2 - 2x - 2 = 0.$$

They are easily found by the quadratic formula.

$$x = \frac{-(-2) \pm \sqrt{(-2)^2 - 4 \cdot 1 \cdot (-2)}}{2 \cdot 1}$$

$$= \frac{2 \pm \sqrt{12}}{2} = \frac{2 \pm 2\sqrt{3}}{2} = 1 \pm \sqrt{3}.$$

Therefore, $f(x)$ has rational roots -3 and $1/2$, and irrational roots $1 + \sqrt{3}$ and $1 - \sqrt{3}$. ■

EXAMPLE 3

Factor $f(x) = 2x^5 - 10x^4 + 7x^3 + 13x^2 + 3x + 9$ completely.

SOLUTION   We begin by finding as many roots of $f(x)$ as we can. By the Rational Root Test, every rational root is of the form r/s where $r = \pm 1, \pm 3,$ or ± 9 and $s = \pm 1$ or ± 2. Thus, the possible rational roots are

$$\pm 1, \quad \pm 3, \quad \pm 9, \quad \pm\frac{1}{2}, \quad \pm\frac{3}{2}, \quad \pm\frac{9}{2}.$$

Figure 4–11

The partial graph of $f(x)$ in Figure 4–11 shows that the only possible roots (x-intercepts) are -1 and 3. You can easily verify that both -1 and 3 are roots of $f(x)$.

Since -1 and 3 are roots, $x - (-1) = x + 1$ and $x - 3$ are factors of $f(x)$ by the Factor Theorem. Division shows that

$$f(x) = 2x^5 - 10x^4 + 7x^3 + 13x^2 - 6x + 9$$

$$= (x + 1)(2x^4 - 12x^3 + 19x^2 - 6x + 9)$$

$$= (x + 1)(x - 3)(2x^3 - 6x^2 + x - 3).$$

The other roots of $f(x)$ are the roots of $g(x) = 2x^3 - 6x^2 + x - 3$. We first check for rational roots of $g(x)$. Since every root of $g(x)$ is also a root of $f(x)$ (why?), the only possible rational roots of $g(x)$ are -1 and 3 [the rational roots of $f(x)$]. We have

$$g(-1) = 2(-1)^3 - 6(-1)^2 + (-1) - 3 = -12;$$

$$g(3) = 2(3^3) - 6(3^2) + 3 - 3 = 0.$$

So -1 is not a root, but 3 is a root of $g(x)$. By the Factor Theorem, $x - 3$ is a factor of $g(x)$. Division shows that

$$f(x) = (x + 1)(x - 3)(2x^3 - 6x^2 + x - 3)$$

$$= (x + 1)(x - 3)(x - 3)(2x^2 + 1).$$

Since $2x^2 + 1$ has no real roots, it cannot be factored. So the factorization of $f(x)$ is complete. ■

▪▪ BOUNDS

The polynomial $f(x)$ in Example 2 had degree 4 and had four real roots. Since a polynomial of degree n has at most n roots, we know that we found all the roots of $f(x)$. In other cases, however, special techniques may be needed to guarantee that we have found all the roots.

EXAMPLE 4

Prove that all the real roots of

$$g(x) = x^5 - 2x^4 - x^3 + 3x + 1$$

lie between -1 and 3. Then find all the real roots of $g(x)$.

SOLUTION   We first prove that $g(x)$ has no root larger than 3, as follows. Use synthetic division to divide $g(x)$ by $x - 3$.*

$$
\begin{array}{r|rrrrrr}
3 & 1 & -2 & -1 & 0 & 3 & 1 \\
  &   & 3 & 3 & 6 & 18 & 63 \\
\hline
  & 1 & 1 & 2 & 6 & 21 & \underline{64}
\end{array}
$$

Thus, the quotient is $x^4 + x^3 + 2x^2 + 6x + 21$, and the remainder is 64. Applying the Division Algorithm, we have

$$f(x) = (x - 3)(x^4 + x^3 + 2x^2 + 6x + 21) + 64.$$

When $x > 3$, then the factor $x - 3$ is positive, and the quotient

$$x^4 + x^3 + 2x^2 + 6x + 21$$

is also positive (because all its coefficients are). The remainder 64 is also positive. Therefore, $f(x)$ is positive whenever $x > 3$. In particular, $f(x)$ is never zero when $x > 3$, and so there are no roots of $f(x)$ greater than 3.

Now we show that $g(x)$ has no root less than -1. Divide $g(x)$ by

$$x - (-1) = x + 1:$$

$$
\begin{array}{r|rrrrrr}
-1 & 1 & -2 & -1 & 0 & 3 & 1 \\
   &   & -1 & 3 & -2 & 2 & -5 \\
\hline
   & 1 & -3 & 2 & -2 & 5 & \underline{-4}
\end{array}
$$

Read off the quotient and remainder and apply the Division Algorithm:

$$f(x) = (x + 1)(x^4 - 3x^3 + 2x^2 - 2x + 5) - 4.$$

When $x < -1$, then the factor $x + 1$ is negative. When x is negative, its odd powers are negative and its even powers are positive. Consequently, the quotient $x^4 - 3x^3 + 2x^2 - 2x + 5$ is positive (because the odd powers of x are multiplied by negative coefficients). The product of the positive quotient with the negative factor $x + 1$ is negative. The remainder -4 is also negative. Hence, $f(x)$ is negative whenever $x < -1$. So there are no real roots less than -1. Therefore, all the real roots of $g(x)$ lie between -1 and 3.

Finally, we find the roots of $g(x) = x^5 - 2x^4 - x^3 + 3x + 1$. The only possible rational roots are ± 1 (why?) and it is easy to verify that neither is actually a root. The graph of $g(x)$ in Figure 4–12 shows that there are exactly three real roots (x-intercepts) between -1 and 3. Since all the real roots of $g(x)$ lie between -1 and 3, $g(x)$ has only these three real roots. They are readily approximated by a root finder:

$$x \approx -.3361, \qquad x \approx 1.4268, \qquad \text{and} \qquad x \approx 2.2012. \qquad \blacksquare$$

Figure 4–12

---

*If you haven't read Special Topics 4.2.A, use long division to find the quotient and remainder.

Suppose $f(x)$ is a polynomial and r and s are real numbers with $r < s$. If all the real roots of $f(x)$ are between r and s, we say that r is a **lower bound** and s is an **upper bound** for the real roots of $f(x)$.* Example 4 shows that -1 is a lower bound and 3 is an upper bound for the real roots of $g(x) = x^5 - 2x^4 - x^3 + 3x + 1$.

If you know lower and upper bounds for the real roots of a polynomial, you can usually determine the number of real roots the polynomial has, as we did in Example 4. The technique used in Example 4 to test possible lower and upper bounds works in the general case:

Bounds Test

Let $f(x)$ be a polynomial with positive leading coefficient.

If $d > 0$ and every number in the last row in the synthetic division of $f(x)$ by $x - d$ is nonnegative,† then d is an upper bound for the real roots of $f(x)$.

If $c < 0$ and the numbers in the last row of the synthetic division of $f(x)$ by $x - c$ are alternately positive and negative [with 0 considered as either]‡ then c is a lower bound for the real roots of $f(x)$.

NOTE

When the last row of the synthetic division by $x - d$ is nonnegative, then d is definitely an upper bound. But if the last row contains some negative entries, no conclusion can be drawn from the Bounds Test [d might or might not be an upper bound]. In such a case, try a larger number in place of d.

Analogous remarks apply to lower bounds.

EXAMPLE 5

Find all real roots of

$$f(x) = x^7 - 6x^6 + 9x^5 + 7x^4 - 28x^3 + 33x^2 - 36x + 20.$$

SOLUTION By the Rational Root Test, the only possible roots are

$$\pm 1, \quad \pm 2, \quad \pm 4, \quad \pm 5, \quad \pm 10, \quad \text{and} \quad \pm 20.$$

The graph of $f(x)$ in Figure 4–13 is hard to read but shows that the possible roots are quite close to the origin. Changing the window (Figure 4–14), we see that the only numbers on the list that could possibly be roots (x-intercepts) are 1 and 2. You can easily verify that both 1 and 2 are roots of $f(x)$.

Figure 4–13

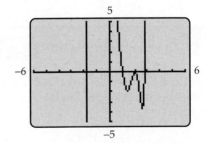

Figure 4–14

---

*The bounds are not unique. Any number smaller than r is also a lower bound, and any number larger than s is also an upper bound.
†Equivalently, all the coefficients of the quotient and the remainder are nonnegative.
‡Equivalently, the coefficients of the quotient are alternately positive and negative, with the last one and the remainder having opposite signs.

In Figures 4–13 and 4–14, all the real roots of $f(x)$ lie between -2 and 6, which suggests that these numbers might be lower and upper bounds for the real roots of $f(x)$.* The Bounds Test shows that this is indeed the case:

-2	1	-6	9	7	-28	33	-36	20
		2	16	-50	86	-116	172	-272
	1	-8	25	-43	58	-86	136	$\boxed{-252}$

Alternating Signs
-2 is a lower bound

6	1	-6	9	7	-28	33	-36	20
		6	0	54	366	2028	12,366	73,980
	1	0	9	61	338	2061	12,330	$\boxed{74,000}$

All Nonnegative
6 is an upper bound

Therefore, the four x-intercepts in Figure 4–14 are the only real roots of $f(x)$. We have seen that two of these are the rational roots, 1 and 2. A root finder shows that the other roots are $x \approx -1.7913$ and $x \approx 2.7913$. ∎

⊞ SUMMARY

The examples above illustrate the following guidelines for finding all the real roots of a polynomial $f(x)$.

1. Use the Rational Root Test to find all the rational roots of $f(x)$. [*Examples 1, 3, 5*]

2. Write $f(x)$ as the product of linear factors (one for each rational root) and another factor $g(x)$. [*Examples 2, 3*]

3. If $g(x)$ has degree 2, find its roots by factoring or the quadratic formula. [*Example 2*]

4. If $g(x)$ has degree 3 or more, use the Bounds Test, if possible, to find lower and upper bounds for the roots of $g(x)$ and approximate the remaining roots graphically. [*Examples 4, 5*]

Shortcuts and variations are always possible. For instance, if the graph of a cubic shows three x-intercepts, then it has three real roots (the maximum possible) and there is no point in finding bounds on the roots. In order to find as many roots as possible exactly in guideline 4, check to see if the rational roots of $f(x)$ are also roots of $g(x)$ and factor $g(x)$ accordingly, as in Example 3.

TECHNOLOGY TIP

The polynomial solvers on TI-86/89, Casio FX 2.0, and HP-39 can find or approximate all the roots of a polynomial simultaneously. The solver on Casio 9850 is limited to polynomials of degree 2 or 3. See the Calculator Investigation on page 26 for details.

✓ **EXERCISES 4.3**

Directions: When asked to find the roots of a polynomial, find exact roots whenever possible and approximate the other roots.

In Exercises 1–12, find all the rational roots of the polynomial.

1. $x^3 + 3x^2 - x - 3$

2. $x^3 - x^2 - 3x + 3$

3. $x^3 + 5x^2 - x - 5$

4. $3x^3 + 8x^2 - x - 20$

5. $f(x) = 2x^5 + 5x^4 - 11x^3 + 4x^2$ [*Hint:* The Rational Root Test can only be used on polynomials with nonzero con-

stant terms. Factor $f(x)$ as a product of a power of x and a polynomial $g(x)$ with nonzero constant term. Then use the Rational Root Test on $g(x)$.]

6. $2x^6 - 3x^5 - 7x^4 - 6x^3$

7. $f(x) = \frac{1}{12}x^3 - \frac{1}{12}x^2 - \frac{2}{3}x + 1$ [*Hint:* The Rational Root Test can only be used on polynomials with integer coefficients. Note that $f(x)$ and $12f(x)$ have the same roots (why?).]

---

*If you are wondering why we don't test 3, 4, or 5 as upper bounds, we did—but the Bounds Test is inconclusive for these numbers, as you can easily verify. See the Note on the preceding page.

8. $\frac{2}{3}x^4 + \frac{1}{2}x^3 - \frac{5}{4}x^2 - x - \frac{1}{6}$

9. $\frac{1}{3}x^4 - x^3 - x^2 + \frac{13}{3}x - 2$

10. $\frac{1}{3}x^7 - \frac{1}{2}x^6 - \frac{1}{6}x^5 + \frac{1}{6}x^4$

11. $.1x^3 - 1.9x + 3$ **12.** $.05x^3 + .45x^2 - .4x + 1$

In Exercises 13–18, factor the polynomial as a product of linear factors and a factor $g(x)$ such that $g(x)$ is either a constant or a polynomial that has no rational roots.

13. $2x^3 - 4x^2 + x - 2$ **14.** $6x^3 - 5x^2 + 3x - 1$

15. $x^6 + 2x^5 + 3x^4 + 6x^3$

16. $x^5 - 2x^4 + 2x^3 - 3x + 2$

17. $x^5 - 4x^4 + 8x^3 - 14x^2 + 15x - 6$

18. $x^5 + 4x^3 + x^2 + 6x$

In Exercises 19–22, use the Bounds Test to find lower and upper bounds for the real roots of the polynomial.

19. $x^3 + 2x^2 - 7x + 20$ **20.** $x^3 - 15x^2 - 16x + 12$

21. $-x^5 - 5x^4 + 9x^3 + 18x^2 - 68x + 176$ [*Hint:* The Bounds Test applies only to polynomials with positive leading coefficient. The polynomial $f(x)$ has the same roots as $-f(x)$ (why?).]

22. $-.002x^3 - 5x^2 + 8x - 3$

In Exercises 23–36, find all real roots of the polynomial.

23. $2x^3 - 5x^2 + x + 2$ **24.** $t^4 - t^3 + 2t^2 - 4t - 8$

25. $6x^3 - 11x^2 + 6x - 1$ **26.** $z^3 + z^2 + 2z + 2$

27. $x^4 + x^3 - 19x^2 + 32x - 12$

28. $3x^5 + 2x^4 - 7x^3 + 2x^2$

29. $2x^5 - x^4 - 10x^3 + 5x^2 + 12x - 6$

30. $x^5 - x^3 + x$

31. $x^6 - 4x^5 - 5x^4 - 9x^2 + 36x + 45$

32. $x^5 + 3x^4 - 4x^3 - 11x^2 - 3x + 2$

33. $3x^4 + 2x^3 - 4x^2 + 4x - 1$

34. $x^5 + 8x^4 + 20x^3 + 9x^2 - 27x - 27$

35. $x^4 - 48x^3 - 101x^2 + 49x + 50$

36. $3x^7 + 8x^6 - 13x^5 - 36x^4 - 10x^3 + 21x^2 + 41x + 10$

37. (a) Show that $\sqrt{2}$ is an irrational number. [*Hint:* $\sqrt{2}$ is a root of $x^2 - 2$. Does this polynomial have any rational roots?]
    (b) Show that $\sqrt{3}$ is irrational.

38. Graph $f(x) = .001x^3 - .199x^2 - .23x + 6$ in the standard viewing window.
    (a) How many roots does $f(x)$ appear to have? Without changing the viewing window, explain why $f(x)$ must have an additional root. [*Hint:* Each root corresponds to a factor of $f(x)$. What does the rest of the factorization consist of?]
    (b) Find all the roots of $f(x)$.

39. According to data from the FBI, the number of people murdered each year per 100,000 population can be approximated by the polynomial function

$$f(x) = .00038x^4 + .0037x^3 - .1528x^2 + .3879x + 9.42$$
$$(0 \le x \le 11),$$

where $x = 0$ corresponds to 1990.
    (a) What was the murder rate in 1995 and in 2000?
    (b) In what year was the murder rate 7 people per 100,000?
    (c) In what year between 1990 and 2001 was the murder rate the highest?

40. During the first 150 hours of an experiment, the growth rate of a bacteria population at time t hours is

$$g(t) = -.0003t^3 + .04t^2 + .3t + .2 \text{ bacteria per hour.}$$

    (a) What is the growth rate at 50 hours? At 100 hours?
    (b) What is the growth rate at 145 hours? What does this mean?
    (c) At what time is the growth rate 0?
    (d) At what time is the growth rate -50 bacteria per hour?
    (e) At what time does the highest growth rate occur?

41. An open-top reinforced box is to be made from a 12- by 36-inch piece of cardboard by cutting along the marked lines, discarding the shaded pieces, and folding as shown in the figure. If the box must be less than 2.5 inches high, what size squares should be cut from the corners in order for the box to have a volume of 448 cubic inches?

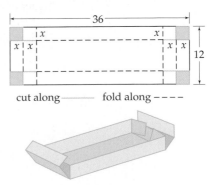

cut along ——— fold along - - - -

42. A box with a lid is to be made from a 48- by 24-inch piece of cardboard by cutting and folding, as shown in the figure. If the box must be at least 6 inches high, what size squares should be cut from the two corners in order for the box to have a volume of 1000 cubic inches?

43. In a sealed chamber where the temperature varies, the instantaneous rate of change of temperature with respect to time over an 11-day period is given by

$$F(t) = .0035t^4 - .4t^2 - .2t + 6,$$

where time is measured in days and temperature in degrees Fahrenheit (so that rate of change is in degrees per day).

(a) At what rate is the temperature changing at the beginning of the period ($t = 0$)? At the end of the period ($t - 11$)?

(b) When is the temperature increasing at a rate of 4°F per day?

(c) When is the temperature decreasing at a rate of 3°F per day?

(d) When is the temperature decreasing at the fastest rate?

44. (a) If c is a root of

$$f(x) = 5x^4 - 4x^3 + 3x^2 - 4x + 5,$$

show that $1/c$ is also a root.

(b) Do part (a) with $f(x)$ replaced by

$$g(x) = 2x^6 + 3x^5 + 4x^4 - 5x^3 + 4x^2 + 3x + 2.$$

(c) Let $f(x) = a_{12}x^{12} + a_{11}x^{11} + \cdots + a_2x^2 + a_1x + a_0$. What conditions must the coefficients a_i satisfy in order that this statement be true: If c is a root of $f(x)$, then $1/c$ is also a root?

4.4   Graphs of Polynomial Functions

The graphs of first- and second-degree polynomial functions are straight lines and parabolas respectively (Sections 1.4 and 4.1). The emphasis here will be on higher degree polynomial functions.

The simplest ones are those of the form $f(x) = ax^n$ (where a is a constant). Their graphs are of four types, as shown in the following chart.

GRAPH OF $f(x) = ax^n$

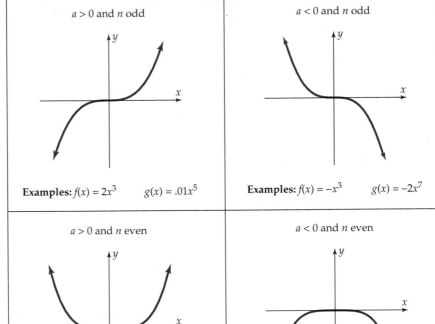

$a > 0$ and n odd

Examples: $f(x) = 2x^3$ $g(x) = .01x^5$

$a < 0$ and n odd

Examples: $f(x) = -x^3$ $g(x) = -2x^7$

$a > 0$ and n even

Examples: $f(x) = 2x^4$ $g(x) = 2x^6$

$a < 0$ and n even

Examples: $f(x) = -2x^2$ $g(x) = -3x^4$

GRAPHING EXPLORATION

Verify the accuracy of the preceding summary by graphing each of the examples in the window with $-5 \le x \le 5$ and $-30 \le y \le 30$.

The graphs of more complicated polynomial functions can vary considerably in shape. Understanding the properties discussed below should assist you to interpret screen images correctly and to determine when a polynomial graph is complete.

▉▉ CONTINUITY

Every polynomial graph is **continuous,** meaning that it is an unbroken curve, with no jumps, gaps, or holes. Furthermore, polynomial graphs have no sharp corners. Thus, neither of the graphs in Figure 4–15 is the graph of a polynomial function. On a calculator screen, however, a polynomial graph may look like a series of juxtaposed line segments, rather than a smooth, continuous curve.

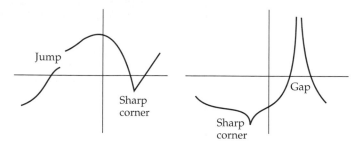

Figure 4–15

▉▉ SHAPE OF THE GRAPH WHEN $|x|$ IS LARGE

The shape of a polynomial graph at the far left and far right of the coordinate plane is easily determined by using our knowledge of graphs of functions of the form $f(x) = ax^n$.

EXAMPLE 1

Consider the function $f(x) = 2x^3 + x^2 - 6x$ and the function determined by its leading term $g(x) = 2x^3$.

EXPLORATION

Using the standard viewing window, graph f and g on the same screen.

Do the graphs look different? Now graph f and g in the viewing window with $-20 \le x \le 20$ and $-10{,}000 \le y \le 10{,}000$. Do the graphs look almost the same?

Finally, graph f and g in the viewing window with

$$-100 \le x \le 100 \text{ and } -1{,}000{,}000 \le y \le 1{,}000{,}000.$$

Do the graphs look virtually identical?

The reason the answer to the last question is "yes" can be understood from this table.

x	-100	-50	70	100
$-6x$	600	300	-420	-600
x^2	10,000	2,500	4,900	10,000
$g(x) = 2x^3$	$-2{,}000{,}000$	$-250{,}000$	686,000	2,000,000
$f(x) = 2x^3 + x^2 - 6x$	$-1{,}989{,}400$	$-247{,}200$	690,480	2,009,400

It shows that when $|x|$ is large, the terms x^2 and $-6x$ are insignificant compared with $2x^3$ and play a very minor role in determining the value of $f(x)$. Hence, the values of $f(x)$ and $g(x)$ are relatively close. ■

Example 1 is typical of what happens in every case: When $|x|$ is very large, the highest power of x totally overwhelms all lower powers and plays the greatest role in determining the value of the function.

Behavior When $|x|$ Is Large

When $|x|$ is very large, the graph of a polynomial function closely resembles the graph of its highest degree term.

In particular, when the polynomial function has odd degree, one end of its graph shoots upward and the other end downward.

When the polynomial function has even degree, both ends of its graph shoot upward or both ends shoot downward.

⊞ *x*-INTERCEPTS

As we saw in Section 4.3, the *x*-intercepts of the graph of a polynomial function are the real roots of the polynomial. Since a polynomial of degree n has at most n distinct roots (page 241), we have the following fact.

x-Intercepts

The graph of a polynomial function of degree n meets the *x*-axis at most n times.

There is another connection between roots and graphs. For example, it is easy to see that the roots of

$$f(x) = (x + 3)^2(x + 1)(x - 1)^3$$

are -3, -1, and 1. We say that

$$-3 \text{ is a root of multiplicity 2;}$$

$$-1 \text{ is a root of multiplicity 1;}$$

$$1 \text{ is a root of multiplicity 3.}$$

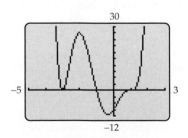

Figure 4–16

Observe that the graph of $f(x)$ in Figure 4–16 does not cross the *x*-axis at -3 (a root whose multiplicity is an *even* number) but does cross the *x*-axis at -1 and 1 (roots of *odd* multiplicity). More generally, a number c is a **root of multiplicity** k of a polynomial $f(x)$ if $(x - c)^k$ is a factor of $f(x)$ and no higher power of $(x - c)$ is a factor, and we have this fact.

Multiplicity and Graphs

Let c be a root of multiplicity k of a polynomial function f.

If k is odd, the graph of f crosses the *x*-axis at c.

If k is even, the graph of f touches, but does not cross, the *x*-axis at c.

▪▪ LOCAL EXTREMA

The term **local extremum** (plural, extrema) refers to either a local maximum or a local minimum, that is, a point where the graph is a peak or a valley.

GRAPHING EXPLORATION

Graph $f(x) = x^3 + 2x^2 - 4x - 3$ in the standard viewing window. What is the total number of peaks and valleys on the graph? What is the degree of $f(x)$?

Now graph $g(x) = x^4 - 3x^3 - 2x^2 + 4x + 5$ in the standard viewing window. What is the total number of peaks and valleys on the graph? What is the degree of $g(x)$? ■

The two polynomials you have just graphed are illustrations of the following fact, which is proved in calculus.

Local Extrema

> A polynomial function of degree n has at most $n - 1$ local extrema. In other words, the total number of peaks and valleys on the graph is at most $n - 1$.

▪▪ BENDING

A polynomial graph may bend upward or downward as indicated here by the vertical arrows (Figure 4–17):

Figure 4–17

TECHNOLOGY TIP

Points of inflection may be found by using INFLC in the TI-86/89 GRAPH MATH menu.

A point at which the graph changes from bending downward to bending upward (or vice versa) is called a **point of inflection.** The direction in which a graph bends may not always be clear on a calculator screen, and calculus is usually required to determine the exact location of points of inflection. The number of inflection points and hence the amount of bending in the graph are governed by these facts, which are proved in calculus.

Points of Inflection

> The graph of a polynomial function of degree n (with $n \geq 2$) has at most $n - 2$ points of inflection.
>
> The graph of a polynomial function of odd degree has at least one point of inflection.

Thus, the graph of a quadratic function (degree 2) has no points of inflection ($n - 2 = 2 - 2 = 0$), and the graph of a cubic has exactly one (since it has at least one and at most $3 - 2 = 1$).

∷ COMPLETE GRAPHS OF POLYNOMIAL FUNCTIONS

By using the facts discussed earlier, you can often determine whether or not the graph of a polynomial function is complete (that is, shows all the important features).

EXAMPLE 2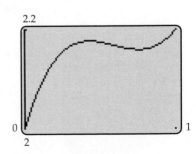

Find a complete graph of

$$f(x) = x^4 + 10x^3 + 21x^2 - 40x - 80.$$

SOLUTION   Since $f(0) = -80$, the standard viewing window probably won't show a complete graph, so we try the window with

$$-10 \le x \le 10 \qquad \text{and} \qquad -100 \le y \le 100$$

Figure 4–18

and obtain Figure 4–18. The three peaks and valleys shown here are the only ones because a fourth-degree polynomial graph has at most three local extrema. There cannot be more x-intercepts than the two shown here because if the graph turned toward the x-axis farther out, there would be an additional peak, which is impossible. Finally, the outer ends of the graph resemble the graph of x^4, the highest-degree term (see the chart on page 255). Hence, Figure 4–18 includes all the important features of the graph and is therefore complete.   ∎

EXAMPLE 3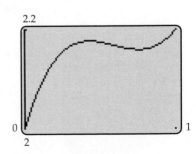

Find a complete graph of $f(x) = x^3 - 1.8x^2 + x + 2$.

SOLUTION   We first try the standard window (Figure 4–19). The graph is similar to the graph of the leading term $y = x^3$ but does not appear to have any local extrema. However, if you use the trace feature on the flat portion of the graph to the right of the x-axis, you see that the y-coordinates increase, then decrease, then increase (try it!). Zooming in on the portion of the graph between 0 and 1 (Figure 4–20), we see that the graph actually has a tiny peak and valley (the maximum possible number of local extrema for a cubic). So Figures 4–19 and 4–20 together provide a complete graph of f.   ∎

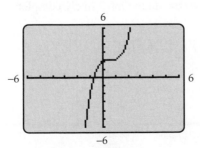

Figure 4–19   **Figure 4–20**

The following fact is proved in Exercise 63.

No polynomial graph contains any horizontal line segments.

However, a calculator may erroneously show horizontal segments, as in Figure 4–19. So always investigate such segments, by using trace or zoom-in, to determine any hidden behavior, such as that in Example 3.

EXAMPLE 4

The graph of $f(x) = .01x^5 + x^4 - x^3 - 6x^2 + 5x + 4$ in the standard window is shown in Figure 4–21. Explain why this is *not* a complete graph and find a complete graph of f.

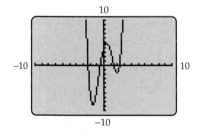

Figure 4–21

SOLUTION   When $|x|$ is large, the graph of f must resemble the graph of $y = .01x^5$, whose left end goes downward (see the chart on page 255). Since Figure 4–21 does not show this, it is not a complete graph. To have the same shape as the graph of $y = .01x^5$, the graph of f must turn downward and cross the x-axis somewhere to the left of the origin. Figure 4–21 shows three local extrema. Even without graphing, we can see that there must be one more peak (where the graph turns downward on the left), making a total of four local extrema (the most a fifth-degree polynomial can have), and another x-intercept, for a total of five. When these additional features are shown, we will have a complete graph.

GRAPHING EXPLORATION

Find a viewing window that includes the local maximum and x-intercept not shown in Figure 4–21. When you do, the scale will be such that the local extrema and x-intercepts shown in Figure 4–21 will no longer be visible.

Consequently, a complete graph of $f(x)$ requires several viewing windows in order to see all the important features. ∎

The graphs obtained in Examples 2–4 were known to be complete because in each case, they included the maximum possible number of local extrema. In many cases, however, a graph may not have the largest possible number of peaks and valleys. In such cases, use any available information and try several viewing windows to obtain the most likely complete graph.

■■ APPLICATIONS

The solution of many applied problems reduces to finding a local extremum of a polynomial function.

EXAMPLE 5

A rectangular box with a square base (Figure 4–22) is to be mailed. The sum of the height of the box and the perimeter of the base is to be 84 inches, the maxi-

Figure 4–22

mum allowable under postal regulations. What are the dimensions of the box with largest possible volume that meets these conditions?

SOLUTION   If the length of one side of the base is x, then the perimeter of the base (the sum of the length of its four sides) is $4x$. If the height of the box is d, then $4x + d = 84$, so $d = 84 - 4x$, and hence, the volume is

$$V = x \cdot x \cdot d = x \cdot x \cdot (84 - 4x) = 84x^2 - 4x^3.$$

The graph of the polynomial function $V(x) = 84x^2 - 4x^3$ in Figure 4–23 is complete (why?). However, the only relevant part of the graph in this situation is the portion with x and $V(x)$ positive (because x is a length and $V(x)$ is a volume). The graph of $V(x)$ has a local maximum between 10 and 20, and this local maximum value is the largest possible volume for the box.

5500

−10   24

−1000

Figure 4–23

 EXPLORATION

GRAPHING
Use a maximum finder to find the x-value at which the local maximum occurs. State the dimensions of the box in this case.   ■

✓ **EXERCISES 4.4**

In Exercises 1–6, decide whether the given graph could possibly be the graph of a polynomial function.

1.

2.

3.

4.

5.

6.

In Exercises 7–12, determine whether the given graph could possibly be the graph of a polynomial function of degree 3, of degree 4, or of degree 5.

7.

8.

9.

10.

11.

12.

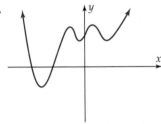

In Exercises 13 and 14, find a viewing window in which the graph of the given polynomial function f appears to have the same general shape as the graph of its leading term.

13. $f(x) = x^4 - 6x^3 + 9x^2 - 3$

14. $f(x) = x^3 - 5x^2 + 4x - 2$

In Exercises 15–18, the graph of a polynomial function is shown. List each root of the polynomial and state whether its multiplicity is even or odd.

15.

16.

17.

18.

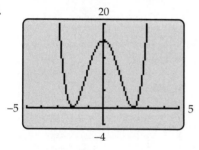

In Exercises 19–24, use your knowledge of polynomial graphs, not a calculator, to match the given function with its graph, which is one of those shown here.

(a)

(b)

(c)

(d)

(e)

(f)

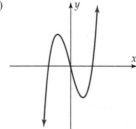

19. $f(x) = 2x - 3$ **20.** $g(x) = x^2 - 4x + 7$

21. $g(x) = x^3 - 4x$ **22.** $f(x) = x^4 - 5x^2 + 4$

23. $f(x) = -x^4 + 6x^3 - 9x^2 + 2$

24. $g(x) = -2x^2 + 3x + 1$

In Exercises 25–28, graph the function in the standard viewing window and explain why that graph cannot possibly be complete.

25. $f(x) = .01x^3 - .2x^2 - .4x + 7$

26. $g(x) = .01x^4 + .1x^3 - .8x^2 - .7x + 9$

27. $h(x) = .005x^4 - x^2 + 5$

28. $f(x) = .001x^5 - .01x^4 - .2x^3 + x^2 + x - 5$

In Exercises 29–34, find a single viewing window that shows a complete graph of the function.

29. $f(x) = x^3 + 8x^2 + 5x - 14$

30. $g(x) = x^3 - 3x^2 - 4x - 5$

31. $g(x) = -x^4 - 3x^3 + 24x^2 + 80x + 15$

32. $f(x) = x^4 - 10x^3 + 35x^2 - 50x + 24$

33. $f(x) = 2x^5 - 3.5x^4 - 10x^3 + 5x^2 + 12x + 6$

34. $g(x) = x^5 + 8x^4 + 20x^3 + 9x^2 - 27x - 7$

In Exercises 35–40, find a complete graph of the function and list the viewing window(s) that show this graph.

35. $f(x) = .1x^5 + 3x^4 - 4x^3 - 11x^2 + 3x + 2$

36. $g(x) = x^4 - 48x^3 - 101x^2 + 49x + 50$

37. $g(x) = .03x^3 - 1.5x^2 - 200x + 5$

38. $f(x) = .25x^6 + .25x^5 - 35x^4 - 7x^3 + 823x^2 + 25x - 2750$

39. $g(x) = 2x^3 - .33x^2 - .006x + 5$

40. $f(x) = .3x^5 + 2x^4 - 7x^3 + 2x^2$

41. (a) Explain why the graph of a cubic polynomial function has either two local extrema or none at all. [*Hint:* If it had only one, what would the graph look like when $|x|$ is very large?]

  (b) Explain why the general shape of the graph of a cubic polynomial function must be one of the following:

(a)   (b)   (c)   (d)

42. The figure shows an incomplete graph of a fourth-degree even polynomial function f. (Even functions were defined in Special Topics 3.4.A.)

  (a) Find the roots of f.

  (b) Explain why

$$f(x) = k(x - a)(x - b)(x - c)(x - d),$$

  where a, b, c, d are the roots of f.

  (c) Experiment with your calculator to find the value of k that produces the graph in the figure.

  (d) Find all local extrema of f.

  (e) List the approximate intervals on which f is increasing and those on which it is decreasing.

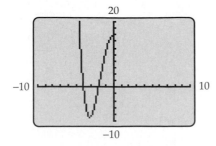

43. A complete graph of a polynomial function g is shown below.

  (a) Is the degree of $g(x)$ even or odd?

  (b) Is the leading coefficient of $g(x)$ positive or negative?

  (c) What are the real roots of $g(x)$?

  (d) What is the smallest possible degree of $g(x)$?

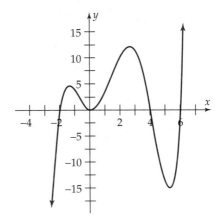

44. Do Exercise 43 for the polynomial function g whose complete graph is shown here.

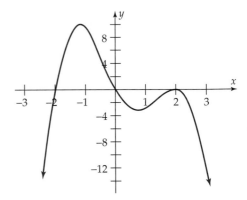

45. The figure is a partial view of the graph of a cubic polynomial whose leading coefficient is negative. Which of the patterns shown in Exercise 41 does this graph have?

46. The figure is a partial view of the graph of a fourth-degree polynomial. Sketch the general shape of the graph and state whether the leading coefficient is positive or negative.

In Exercises 47–54, sketch a complete graph of the function. Label each x-intercept and the coordinates of each local extremum; find intercepts and coordinates exactly when possible and otherwise approximate them.

47. $f(x) = x^3 - 3x^2 + 4$ **48.** $g(x) = 4x - 4x^3/3$

49. $h(x) = .25x^4 - 2x^3 + 4x^2$

50. $f(x) = .25x^4 - 2x^3/3$

51. $g(x) = 3x^3 - 18.5x^2 - 4.5x - 45$

52. $h(x) = 2x^3 + x^2 - 4x - 2$

53. $f(x) = x^5 - 3x^3 + x + 1$

54. $g(x) = .25x^4 - x^2 + .5$

55. The sales $f(x)$ of a certain product (in dollars) are related to the amount x (in thousands of dollars) spent on advertising by

$$f(x) = -3x^3 + 135x^2 + 3600x + 12,000$$
$$(0 \le x \le 40).$$

(a) Graph f in the window with $0 \le x \le 40$ and $0 \le y \le 180,000$ and verify that f is concave upward near the origin and concave downward near $x = 40$.

(b) Compute the average rate of change of $f(x)$ from $x = 0$ to $x = 15$ and from $x = 15$ to $x = 40$. What do these numbers tell you about the rate at which sales are increasing in each interval?

(c) This function has an inflection point at $x = 15$ (a fact you might want to verify if your calculator can find inflection points). Use the results of part (b) to explain why the inflection point is sometimes called the **point of diminishing returns.**

56. The profits (in thousands of dollars) from producing x hundred thousand items are given by

$$g(x) = -x^3 + 27x^2 + 20x - 60 \qquad (0 \le x \le 20).$$

(a) Graph g in a window with $0 \le x \le 20$. If you have an appropriate calculator, verify that there is a point of inflection when $x = 9$.

(b) Verify that the point of inflection is the point of diminishing returns (see Exercise 55) by computing the average rate of change of profit from $x = 0$ to $x = 9$ and from $x = 9$ to $x = 20$.

57. An auto parts manufacturer makes radiators that sell for $347 each. The cost (in thousands of dollars) to produce x thousand radiators is given by

$$C(x) = .0001x^5 - .29x^3 + 8x^2 + 1600$$

(a) What is the revenue function in this situation? What is the profit function?

(b) How many radiators must be sold to make any profit?

(c) What is the maximum possible number of radiators that can be sold without losing money?

(d) What number of radiators will produce the largest possible profit?

58. When there are 22 apple trees per acre, the average yield has been found to be 500 apples per tree. For each additional tree planted per acre, the yield per tree decreases by 15 apples per tree. How many additional trees per acre should be planted to maximize the yield?

59. Name tags can be sold for $29 per thousand. The cost of manufacturing x thousand tags is $.001x^3 + .06x^2 - 1.5x$ dollars. Assuming that all tags manufactured are sold,

(a) What number of tags should be made to guarantee a maximum profit? What will that profit be?

(b) What is the largest number of tags that can be made without losing money?

60. The top of a 12-ounce can of soda pop is three times thicker than the sides and bottom (so that the flip-top opener will work properly), and the can has a volume of 355 cubic centimeters. What should the radius and height of the can be in order to use the least possible amount of metal? [Assume that the entire can is made from a single sheet of metal, with three layers being used for the top. Example 3 in Section 2.4 may be helpful.]

61. An open-top reinforced box is to be made from a 12-by-36-inch piece of cardboard as in Exercise 41 of Section 4.3. What size squares should be cut from the corners in order to have a box with maximum volume?

62. A box with a lid is to be made from a 48-by-24-inch piece of cardboard by cutting and folding, as in Exercise 42 of Section 4.3. What size squares should be cut from the two corners in order to have a box of maximum volume?

Thinkers

63. (a) Graph $g(x) = .01x^3 - .06x^2 + .12x + 3.92$ in the viewing window with $-3 \leq x \leq 3$ and $0 \leq y \leq 6$ and verify that the graph appears to coincide with the horizontal line $y = 4$ between $x = 1$ and $x = 3$. In other words, it appears that every x with $1 \leq x \leq 3$ is a solution of the equation

$$.01x^3 - .06x^2 + .12x + 3.92 = 4.$$

Explain why this is impossible. Conclude that the actual graph is not horizontal between $x = 1$ and $x = 3$.

(b) Use the trace feature to verify that the graph is actually rising from left to right between $x = 1$ and $x = 3$. Find a viewing window that shows this.

(c) Show that it is not possible for the graph of a polynomial $f(x)$ to contain a horizontal segment. [*Hint:* A horizontal line segment is part of the horizontal line $y = k$ for some constant k. Adapt the argument in part (a), which is the case $k = 4$.]

64. (a) Let $f(x)$ be a polynomial of odd degree. Explain why $f(x)$ must have at least one real root. [*Hint:* Why must the graph of f cross the x-axis, and what does this mean?]

(b) Let $g(x)$ be a polynomial of even degree, with a negative leading coefficient and a positive constant term. Explain why $g(x)$ must have at least one positive and at least one negative root.

65. For x-values in a particular interval, a nonpolynomial function $g(x)$ can often be approximated by a polynomial function, meaning that there is some polynomial $f(x)$ such that $g(x) \approx f(x)$ for every x in the interval.

(a) In the standard viewing window, graph both

$$g(x) = \sqrt{x}$$

(which is not a polynomial function) and the polynomial function

$$f(x) = .26705x^3 - .78875x^2 + 1.3021x + .22033.$$

Are the graphs similar?

(b) Graph $f(x)$ and $g(x)$ in the viewing window with $0.26 \leq x \leq 1$ and $0 \leq y \leq 1$. Does it now appear that $f(x)$ is a good approximation of $g(x)$ over the interval $[.26, 1]$?

(c) For any particular value of x, the error in this approximation is the difference between $f(x)$ and $g(x)$. In other words, $h(x) = f(x) - g(x)$ measures the error in the approximation. Graph the function $h(x)$ in the viewing window with

$$.26 \leq x \leq 1 \quad \text{and} \quad -.001 \leq y \leq .001$$

and use the trace feature to determine the maximum error in the approximation.

66. (a) Graph $f(x) = x^3 - 4x$ in the viewing window with $-3 \leq x \leq 3$ and $-5 \leq y \leq 5$.

(b) Graph the difference quotient of $f(x)$ (with $h = .01$) on the same screen.

(c) Find the x-coordinates of the relative extrema of $f(x)$. How do these numbers compare with the x-intercepts of the difference quotient?

67. The graph of

$$f(x) = (x + 18)(x^2 - 20)(x - 2)^2(x - 10)$$

has x-intercepts at each of its roots, that is, at $x = -18$, $\pm\sqrt{20} \approx \pm 4.472$, 2, and 10. It is also true that $f(x)$ has a relative minimum at $x = 2$.

(a) Draw the x-axis and mark the roots of $f(x)$. Then use the fact that $f(x)$ has degree 6 (why?) to sketch the general shape of the graph (as was done for cubics in Exercise 41).

(b) Now graph $f(x)$ in the standard viewing window. Does the graph resemble your sketch? Does it even show all the x-intercepts between -10 and 10?

(c) Graph $f(x)$ in the viewing window with $-19 \leq x \leq 11$ and $-10 \leq y \leq 10$. Does this window include all the x-intercepts as it should?

(d) List viewing windows that give a complete graph of $f(x)$.

4.4.A   *SPECIAL TOPICS* **Polynomial Models***

Linear regression was used in Section 2.5 to construct a linear function that modeled a set of data points. When the scatter plot of the data points looks more like a higher-degree polynomial graph than a straight line, similar least squares regression procedures are available on most calculators for constructing quadratic, cubic, and quartic (fourth-degree) polynomial functions to model the data.

TECHNOLOGY TIP

The quadratic, cubic, and quartic regression commands are in the same menu as the linear regression command and are labeled as follows:

TI-83/89: QuadReg, CubicReg, QuartReg

TI-86: P2Reg, P3Reg, P4Reg.

Casio 9850: $x \wedge 2$, $x \wedge 3$, $x \wedge 4$,

Casio FX2.0: Quad, Cubic, Quart

HP-39: quadratic, cubic

Quartic regression is not available on HP-39.

EXAMPLE 1

The table shows the population of San Francisco in selected years.[†]

Year	1950	1960	1970	1980	1990	2000
Population	775,357	740,316	715,674	678,974	723,959	776,733

(a) Find a polynomial model for this data.

(b) Use the model to estimate the population of San Francisco in 1995 and 2004.

SOLUTION

(a) Let $x = 0$ correspond to 1950 and plot the data points, as in Figure 4–24. We use quartic regression to obtain the following function:

$$f(x) = -.1072x^4 + 13.2x^3 - 387.24x^2 - 168.65x + 774,231.\text{[‡]}$$

(The procedure is the same as for linear regression; see the Technology Tip on page 112 and the one in the margin here. The graph of f in Figure 4–25 appears to fit the data well.

(b) To estimate the population in 1995 and 2004, we evaluate f at $x = 45$ and $x = 54$, as shown in Figure 4–26. According to this model, the 1995 population was about 745,844 and the 2004 population is about 803,155. ■

Figure 4–24

Figure 4–25

Figure 4–26

---

*Section 2.5 is a prerequisite for this optional section. This material will be used in the optional Section 5.5 and in clearly identified exercises but not elsewhere.

[†]U.S. Census Bureau.

[‡]Here and later, coefficients are rounded for convenient reading, but the full coefficients are used to produce the graphs and estimates.

Use your minimum finder and the population function in Example 1 to estimate the year since 1950 when the population of San Francisco was smallest.

EXAMPLE 2

The table below, which is based on statistics from the Department of Health and Human Services, gives the cumulative number of reported cases of AIDS in the United States from 1982 through 2000. It shows, for example, that 41,662 cases were reported from 1982 through 1986.

Year	Cases	Year	Cases	Year	Cases
1982	1,563	1991	232,383	1996	595,559
1984	10,845	1992	278,189	1997	652,439
1986	41,662	1993	380,601	1998	698,527
1988	105,489	1994	457,789	1999	743,418
1990	188,872	1995	528,421	2000	784,518

Find a suitable polynomial model for this data.

SOLUTION   Letting $x = 0$ correspond to 1980 and plotting the data points (2, 1563), (4, 10845), etc., we obtain the scatter plot in Figure 4–27. The points are not in a straight line but could be part of a polynomial graph of higher degree.

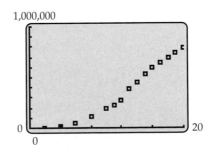

Figure 4–27

Since the data points suggest a curve that is convex upward on the left and convex downward on the right, a third- or fourth-degree polynomial might provide a reasonable model. We use the regression feature of a calculator to find these models.

Cubic: $f(x) = -218.6x^3 + 9094.1x^2 - 59,841.6x + 102,952.9$

Quartic: $g(x) = -20.2x^4 + 678.3x^3 - 4273.6x^2 + 15,349.3x - 17,615.7$

The graphs of both f and g are shown in Figure 4–28; as you can see, they are virtually identical in this viewing window. If you change the viewing window, however, you see that g is the better model in the 1980s (Figure 4–29), although there is little difference after 1990 (Figure 4–30).

Figure 4–28   Figure 4–29   Figure 4–30

Although both functions provide reasonable models after 1990, our knowledge of polynomial graphs suggests that they may not be accurate in the future. As x gets larger, the graphs of f and g will resemble the graphs of $y = -218.6x^3$ and $y = -20.2x^4$, respectively (why?), both of which turn downward to the right (see the chart on page 255). However, the cumulative number of cases can't decrease. (Even when there are no new cases, the cumulative total stays the same.)

GRAPHING EXPLORATION

Graph f and g in a window with $10 \leq x \leq 28$. Which graph provides a plausible model for any years after 2000? How long will this model remain reasonable? ∎

NOTE: You must have at least three data points for quadratic regression, at least four for cubic regression, and at least five for quartic regression. If you have exactly the required minimum number data points, no two of them can have the same first coordinate. In this case, the polynomial regression function will pass through all of the data points (an exact fit). When you have more than the minimum number of data points required, the fit will generally be approximate rather than exact.

EXERCISES 4.4.A

In Exercises 1–4, a scatter plot of data is shown. State the type of polynomial model that seems most appropriate for the data (linear, quadratic, cubic, or quartic). If none of them is likely to provide a reasonable model, say so.

1.

2.

3.

4.

5. The table, which is based on the FBI Uniform Crime Reports, shows the rate of property crime per 100,000 population.

Year	Crimes	Year	Crimes
1984	4492.1	1996	4450.1
1986	4862.6	1997	4318.7
1988	5027.1	1998	4051.8
1990	5088.5	1999	3743.6
1992	4902.7	2000	3617.9
1994	4660.0	2001	3697.5

(a) Use cubic regression to find a polynomial function that models this data, with $x = 0$ corresponding to 1980.

(b) According to this model, what was the property crime rate in 1987 and 1993?

(c) The actual crime rate was about 4591 in 1995. What does the model predict?

(d) For how many years in the future is this model likely to be a reasonable one?

6. The table shows actual and projected enrollment (in millions) in public high schools in selected years.[*]

Year	Enrollment	Year	Enrollment
1975	14.3	1995	12.5
1980	13.2	2000	13.5
1985	12.4	2005	14.4
1990	11.3	2010	14.1

(a) Use quartic regression to find a polynomial function that models this data, with $x = 0$ corresponding to 1975.

(b) According to the model, what was enrollment in 1998 and in 1999?

(c) Estimate the year between 1975 and 2000 in which enrollment was the lowest. Does this estimate appear to be accurate?

7. The table shows the air temperature at various times during a spring day in Gainesville, Florida.

Time	Temp (F°)	Time	Temp (F°)
6 A.M.	52	1 P.M.	82
7 A.M.	56	2 P.M.	86
8 A.M.	61	3 P.M.	85
9 A.M.	67	4 P.M.	83
10 A.M.	72	5 P.M.	78
11 A.M.	77	6 P.M.	72
noon	80		

(a) Sketch a scatter plot of the data, with $x = 0$ corresponding to midnight.

(b) Find a quadratic polynomial model for the data.

(c) What is the predicted temperature for noon? For 9 A.M.? For 2 P.M.?

_____

*U.S. National Center for Educational Statistics.

8. The fall enrollment in college level Spanish classes (in thousands) is shown in the table.[*]

Year	Enrollment	Year	Enrollment
1970	389.2	1986	411.3
1974	362.2	1990	533.9
1977	376.7	1995	606.3
1980	379.4	1998	656.6
1983	386.2		

(a) Sketch a scatter plot of the data, with $x = 0$ corresponding to 1970.

(b) Find a cubic polynomial model for this data.

(c) Use your model to estimate enrollment in college level Spanish classes in 2003.

(d) Estimate the year between 1970 and 2000 in which enrollment was at its lowest level.

Use the following table in Exercises 9 and 10. It shows the median income of U.S. households in constant 1999 dollars.[†]

Year	Median Income	Year	Median Income
1985	$36,568	1993	$36,019
1987	$38,220	1995	$37,251
1989	$38,836	1997	$38,411
1991	$36,850	1999	$40,816

9. (a) Sketch a scatter plot of the data from 1985 to 1999, with $x = 0$ corresponding to 1985 and y measured in thousands.

(b) Decide whether a quadratic or a quartic model seems more appropriate and state its rule.

(c) Use the model to predict the median income in 2002.

(d) Does this model seem reasonable after 2002?

10. (a) Sketch a scatter plot of the data from 1989 to 1999, with $x = 0$ corresponding to 1989 and y measured in thousands.

(b) Find both cubic and quartic models for this data.

(c) Is there any significant difference between the models from 1989 to 1999? What about from 1999 to 2005?

(d) According to these models, when will the median income reach $45,000?

11. The table shows the percentage of male high school students who currently smoke cigarettes in various years.[‡]

Year	Percent	Year	Percent
1991	27.6	1997	37.7
1993	29.8	1999	34.7
1995	35.4	2001	29.2

(a) Sketch a scatter plot of the data, with $x = 0$ corresponding to 1990.

(b) Find quadratic, cubic, and quartic polynomial models for the data.

(c) Which model seems to fit this data best? Which one seems most reasonable for future years.

12. The table shows the US. public debt per person (in dollars) in selected years.[§]

Year	Debt	Year	Debt
1981	$4,338	1993	$17,105
1983	$5,870	1995	$18,930
1985	$7,598	1997	$20,026
1987	$9,615	1999	$20,746
1989	$11,545	2001	$20,353
1991	$14,436		

(a) Sketch a scatter plot of the data.

(b) Find a quartic polynomial model for the data.

(c) Use the model to estimate the public debt per person in 1996. How does your estimate compare with the actual figure of $19,805?

(d) Is this model likely to be accurate after 2001? Why?

13. (a) Find both a cubic and a quartic model for the data on the number of unemployed people in the labor force in Example 4 of Section 2.5.

(b) Does either model seem likely to be accurate in the future?

14. The table shows the total advertising expenditures (in billions of dollars) in selected years.[¶]

Year	1990	1992	1994	1996	1998	2000
Expenditures	129.59	132.65	151.68	175.23	201.59	236.33

(a) Sketch a scatter plot of the data, with $x = 0$ corresponding to 1990.

(b) Find a quadratic model for the data.

(c) Use the model to estimate expenditures in 1995 and 2002.

(d) If this model remains accurate, when will expenditures reach $350 billion?

---

[*]Association of Departments of Foreign Languages.
[†]U.S. Census Bureau.
[‡]Youth at Risk Behavior Survey.

[§]U.S. Bureau of Public Debt.
[¶]*Statistical Abstract of the United States: 2001.*

4.5 Rational Functions

A **rational function** is a function whose rule is the quotient of two polynomials, such as

$$f(x) = \frac{1}{x}, \qquad t(x) = \frac{4x - 3}{2x + 1}, \qquad k(x) = \frac{2x^3 + 5x + 2}{x^2 - 7x + 6}.$$

A polynomial function is defined for every real number, but the rational function $f(x) = g(x)/h(x)$ is defined only when its denominator is nonzero. Hence,

Domain

> The domain of the rational function $f(x) = \dfrac{g(x)}{h(x)}$ is the set of all real numbers that are *not* roots of the denominator $h(x)$.

For instance, the domain of

$$f(x) = \frac{x^2 + 3x + 1}{x^2 - x - 6}$$

is the set of all real numbers except -2 and 3 [the roots of $x^2 - x - 6 = (x + 2)(x - 3)$]. The key to understanding the behavior of rational functions is this fact from arithmetic.

The Big-Little Principle

> If c is a number far from 0, then $1/c$ is a number close to 0. Conversely, if c is close to 0, then $1/c$ is far from 0. In less precise but more suggestive terms:
>
> $$\frac{1}{\text{big}} = \text{little} \qquad \text{and} \qquad \frac{1}{\text{little}} = \text{big.}$$

For example, 5000 is big (far from 0), and $1/5000$ is little (close to 0). Similarly, $-1/1000$ is very close to 0, but

$$\frac{1}{-1/1000} = -1000$$

is far from 0. To see the role played by the Big-Little Principle, we examine two rational functions that are part of the catalog of basic functions.

EXAMPLE 1

Reciprocal Functions   Graph

$$f(x) = \frac{1}{x} \quad \text{and} \quad g(x) = \frac{1}{x^2}.$$

SOLUTION   Note that f and g are not defined when $x = 0$ (root of the denominator). The graphs are easily obtained, either by hand or by calculator (see Figures 4–31 and 4–32), and you should know their shapes by heart. ■

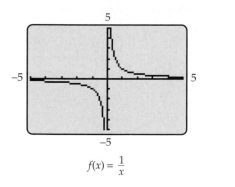

$$f(x) = \frac{1}{x}$$

Figure 4–31

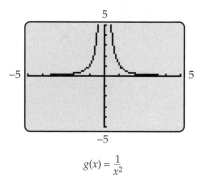

$$g(x) = \frac{1}{x^2}$$

Figure 4–32

The Big-Little Principle explains why these graphs have the shapes they do. When x is very close to 0, both $1/x$ and $1/x^2$ are very far from 0. That's why the graphs "explode" near the undefined place at $x = 0$, shooting sharply upward or downward on either side of the y-axis. We say that the y-axis is the **vertical asymptote** of the graph: The graph gets closer and closer to this line, but never touches it. A vertical asymptote can occur only where the function is not defined.

Similarly, when x is very far from 0, then $1/x$ and $1/x^2$ are very small numbers. So the farther you go from the origin, the closer the graphs get to the line $y = 0$ (the x-axis), which is called the **horizontal asymptote** of the graph.

EXAMPLE 2

Without using technology, describe the graphs of

(a) $h(x) = \dfrac{1}{x + 3}$ and   (b) $k(x) = \dfrac{1}{x^2 - 4x + 4}.$

SOLUTION

(a) Recall the graph of $f(x) = \dfrac{1}{x}$ in Figure 4–31, and note that

$$f(x + 3) = \frac{1}{x + 3} = h(x)$$

From Section 3.4, we know that the graph of h is the graph of f shifted horizontally 3 units to the left, as shown in Figure 4–33(a). The vertical asymptote of this graph is at $x = -3$ [the root of the denominator of $h(x)$].

(b) To understand the graph of k, recall the graph of $g(x) = \dfrac{1}{x^2}$ in Figure 4–32 and factor the denominator of $k(x)$:

$$k(x) = \frac{1}{x^2 - 4x + 4} = \frac{1}{(x-2)^2} = g(x-2).$$

Thus, the graph of k is the graph of g shifted horizontally 2 units to the right, as shown in Figure 4–33(b). The vertical asymptote is at $x = 2$ [the root of the denominator of $k(x)$]. ■

> **NOTE**
>
> The asymptotes are shown as dashed lines in Figure 4–33 and in the figures below. The asymptotes are included for easier visualization, but they are *not* part of the graph.

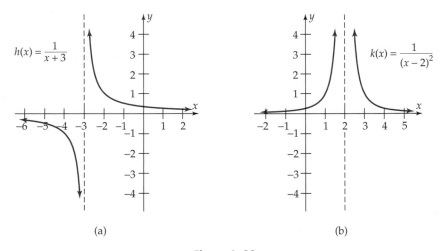

(a)   (b)

Figure 4–33

▚ LINEAR RATIONAL FUNCTIONS

Next we consider the graphs of rational functions, where both the numerator and denominator are either constants or first-degree polynomials.

EXAMPLE 3

Find the asymptotes of the graph of

$$f(x) = \frac{x + 1}{2x - 4}$$

and graph the function.

SOLUTION   The function is not defined when $x = 2$ because the denominator is 0 there. When x is a number very close to 2, then

The numerator $x + 1$ is very close to 3;

The denominator $2x - 4$ is very close to 0.

Therefore,

$$f(x) = \frac{x + 1}{2x - 4} \approx \frac{3}{\text{little}} = \text{BIG (far from 0)}.$$

Consequently, the graph explodes near $x = 2$, as shown in the table and partial graph in Figure 4–34. The vertical line $x = 2$ is the vertical asymptote of the graph.

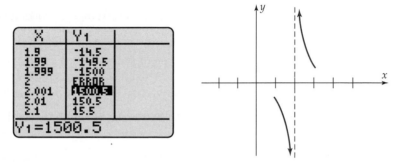

Figure 4–34

To determine the horizontal asymptote of the graph, we rewrite the rule of f like this:

$$f(x) = \frac{x + 1}{2x - 4} = \frac{\dfrac{x + 1}{x}}{\dfrac{2x - 4}{x}} = \frac{1 + \dfrac{1}{x}}{2 - \dfrac{4}{x}} \cdot$$

As x gets larger in absolute value (far from 0), both $1/x$ and $4/x$ get very close to 0 by the Big-Little Principle. Consequently,

$$f(x) = \frac{1 + (1/x)}{2 - (4/x)}$$

gets very close to

$$\frac{1 + 0}{2 - 0} = \frac{1}{2}.$$

So when $|x|$ is large, the graph gets closer and closer to the horizontal line $y = 1/2$, but never touches it, as shown in Figure 4–35 on the next page. Thus, the line $y = 1/2$ is the horizontal asymptote of the graph.

The preceding information, together with a few hand-plotted points, produces the graph in Figure 4–35.   ■

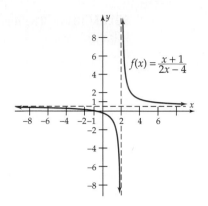

Figure 4–35

The analysis in Example 3 works for all linear rational functions.

**Linear Rational
Functions**

The graph of $f(x) = \dfrac{ax + b}{cx + d}$ (with $c \neq 0$ and $ad \neq bc$) has two asymptotes:

The vertical asymptote occurs at the root of the denominator.

The horizontal asymptote is the line $y = a/c$.

Example 3 is the case where $a = 1$ and $c = 2$. Similarly, in Example 1, we saw that the horizontal asymptote of

$$f(x) = \frac{1}{x} = \frac{0x + 1}{1x + 0}$$

was the line

$$y = \frac{0}{1} = 0.$$

Figure 4–36 shows some additional examples.

$$f(x) = \frac{-5x + 12}{2x - 4} \qquad\qquad k(x) = \frac{3x + 6}{x} = \frac{3x + 6}{1x + 0}$$

Vertical asymptote $x = 2$ Vertical asymptote $x = 0$

Horizontal asymptote $y = -\frac{5}{2}$ Horizontal asymptote $y = \frac{3}{1} = 3$

(a)

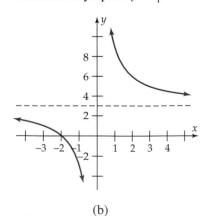

(b)

Figure 4–36

▐▐ RATIONAL FUNCTIONS AND TECHNOLOGY

Getting an accurate graph of a rational function on a calculator often depends on choosing an appropriate viewing window. For example, a TI-83 produced the following graphs of

$$f(x) = \frac{x + 1}{2x - 4}$$

in Figure 4–37, two of which do not look like Figure 4–35 as they should.

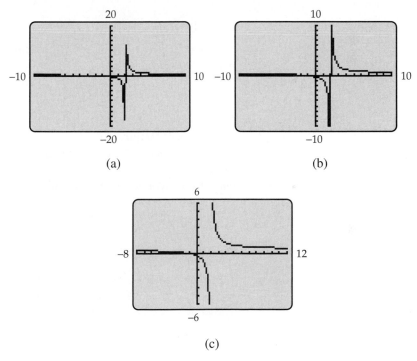

(a)   (b)

(c)

Figure 4–37

The vertical segments in graphs (a) and (b) are *not* representations of the vertical asymptote. They are a result of the calculator evaluating $f(x)$ just to the left of $x = 2$ and just to the right of $x = 2$ but not at $x = 2$ and then erroneously connecting these points with a near vertical segment that looks like an asymptote.

In the accurate graph (c), the calculator attempted to plot a point with $x = 2$ and, when it found that $f(2)$ was not defined, skipped a pixel and did not join the points on either side of the skipped one.

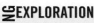

TECHNOLOGY TIP

To avoid erroneous vertical lines, use a window with the vertical asymptote in the center, as in Figure 4–37(c) (where the asymptote at $x = 2$ is half-way between −8 and 12). Also see the Technology Tip on page 282.

GRAPHING EXPLORATION

Find a viewing window that displays the graph of $f(x)$ in Figure 4–36(a), without any erroneous vertical line segments being shown. The Technology Tip in the margin may be helpful.

▪▪ PROPERTIES OF RATIONAL GRAPHS

Here is a summary of the important characteristics of graphs of more complicated rational functions.

CONTINUITY

There will be breaks in the graph of a rational function wherever the function is not defined. Except for breaks at these undefined points, the graph is a continuous unbroken curve. In addition, the graph has no sharp corners.

LOCAL MAXIMA AND MINIMA

The graph may have some local extrema (peaks and valleys), and calculus is needed to determine their exact location. There are no simple rules for the possible number of peaks and valleys as there were with polynomial functions.

INTERCEPTS

As with any function, the y-intercept of the graph of a rational function f occurs at $f(0)$, provided that f is defined at $x = 0$. The x-intercepts of the graph of any function f occur at each number c for which $f(c) = 0$. Now a fraction is 0 only when its numerator is 0 and its denominator is nonzero (since division by 0 is not defined). Thus,

Intercepts

> The x-intercepts of the graph of the rational function $f(x) = \dfrac{g(x)}{h(x)}$ occur at the numbers that are roots of the numerator $g(x)$ but *not* of the denominator $h(x)$.
>
> If f has a y-intercept, it occurs at $f(0)$.

For example, the graph of

$$f(x) = \frac{x^2 - x - 2}{x - 5}$$

has x-intercepts at $x = -1$ and $x = 2$ [which are the roots of $x^2 - x - 2 = (x + 1)(x - 2)$, but not of $x - 5$] and y-intercept at $y = 2/5$ (the value of f at $x = 0$).

VERTICAL ASYMPTOTES

In Example 3, we saw that the graph of

$$f(x) = \frac{x + 1}{2x - 4}$$

had a vertical asymptote at $x = 2$. Note that $x = 2$ is a root of the denominator $2x - 4$, but not of the numerator $x + 1$. The same thing occurs in the general case.

Vertical Asymptotes

> The function $f(x) = \dfrac{g(x)}{h(x)}$ has a vertical asymptote at every number that is a root of the denominator $h(x)$, but *not* of the numerator $g(x)$.

Figure 4–38

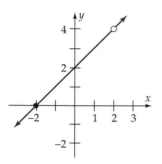

Figure 4–39

HOLES

You have often cancelled factors, as here:

$$\frac{x^2 - 4}{x - 2} = \frac{(x + 2)(x - 2)}{x - 2} = x + 2.$$

But the functions given by

$$p(x) = \frac{x^2 - 4}{x - 2} \qquad \text{and} \qquad q(x) = x + 2$$

are *not* the same, because when $x = 2$,

$$q(2) = 2 + 2 = 4, \qquad \text{but} \qquad p(2) = \frac{2^2 - 2}{2 - 2} = \frac{0}{0},$$

which is not defined. For any number other than 2, the two functions do have the same value, and hence, the same graph. The graph of $q(x) = x + 2$ is a straight line that includes the point $(2, 4)$, as shown in Figure 4–38. The graph of $p(x)$ is the same straight line, but with the point $(2, 4)$ omitted, so that the graph of p has a **hole** at $x = 2$ (indicated by an open circle in Figure 4–39). Note that the hole occurs at $x = 2$, which is a root of multiplicity 1 in both the numerator and the denominator of

$$p(x) = \frac{x^2 - 4}{x - 2} = \frac{(x + 2)(x - 2)}{x - 2}.*$$

Similarly, the graph of $g(x) = \frac{x^2}{x^3}$ is the same as the graph of $f(x) = \frac{1}{x}$, except at $x = 0$, where neither function is defined. In this case, however, there is a vertical asymptote rather than a hole at $x = 0$ (see Figure 4–31). Note that the vertical asymptote occurs at $x = 0$, which is a root of multiplicity 2 in the numerator, but of larger multiplicity 3 in the denominator. In general,

Holes

> Let $f(x) = \frac{g(x)}{h(x)}$ be a rational function and d a root of both $g(x)$ and $h(x)$. If the multiplicity of d as a root of $g(x)$ is greater than or equal to its multiplicity as a root of $h(x)$, then the graph of f has a hole at $x = d$. Otherwise, the graph has a vertical asymptote at $x = d$.

A calculator-drawn graph may not show holes where it should. If the calculator actually attempts to compute an undefined quantity, it indicates a hole by skipping a pixel; otherwise, it may erroneously show a continuous graph with no hole.

BEHAVIOR WHEN $|x|$ IS LARGE

The shape of a rational graph at the far left and far right (that is, when $|x|$ is large) can usually be found by algebraic analysis, as in the next example.

---

*Multiplicity of roots was discussed on page 257.

EXAMPLE 4

Determine the shape of the graph when $|x|$ is large for the following functions.

(a) $f(x) = \dfrac{7x^4 - 6x^3 + 4}{2x^4 + x^2}$

(b) $g(x) = \dfrac{x^2 - 2}{x^3 - 3x^2 + x - 3}$

SOLUTION

(a) When $|x|$ is very large, a polynomial function behaves in essentially the same way as its highest degree term, as we saw on page 257. Consequently, we have this approximation:

$$f(x) = \frac{7x^4 - 6x^3 + 4}{2x^4 + x^2} \approx \frac{7x^4}{2x^4} = \frac{7}{2} = 3.5.$$

Thus, when $|x|$ is large, the graph of $f(x)$ is very close to the horizontal line $y = 3.5$, which is a horizontal asymptote of the graph.

EXPLORATION

Confirm the last statement by graphing $f(x)$ and $y = 3.5$ in the window with $-10 \le x \le 10$ and $-2 \le y \le 12$.

(b) When $|x|$ is large, the graph of g closely resembles the graph of

$$y = \frac{x^2}{x^3} = \frac{1}{x}.$$

By the Big-Little Principle, $1/x$ is very close to 0 when $|x|$ is large. So the line $y = 0$ (that is, the x-axis) is the horizontal asymptote. ■

Arguments similar to those in the preceding example, using the highest-degree terms in the numerator and denominator, carry over to the general case and lead to this conclusion.

Horizontal Asymptotes

> Let $f(x) = \dfrac{ax^n + \cdots}{cx^k + \cdots}$ be a rational function whose numerator has degree n and whose denominator has degree k.
>
> If $n = k$, then the line $y = a/c$ is a horizontal asymptote.
>
> If $n < k$, then the x-axis (the line $y = 0$) is a horizontal asymptote.

The asymptotes of rational functions in which the denominator has smaller degree than the numerator are discussed in Special Topics 4.5.A.

▪▪ GRAPHS OF RATIONAL FUNCTIONS

The procedure for finding accurate graphs of rational functions is summarized here.

Graphing

$$f(x) = \frac{g(x)}{h(x)} \text{ When}$$

Degree g(x) ≤
Degree h(x)

1. Analyze the function algebraically to determine its vertical asymptotes, holes, and intercepts.

2. Determine the horizontal asymptote of the graph when $|x|$ is large by using the facts in the box on the opposite page.

3. Use the preceding information to select an appropriate viewing window (or windows), to interpret the calculator's version of the graph (if necessary), and to sketch an accurate graph.

Figure 4–40

EXAMPLE 5

If you ignore the preceding advice and simply graph $f(x) = \dfrac{x - 1}{x^2 - x - 6}$ in the standard viewing window, you get garbage (Figure 4–40). So let's try analyzing the function. We begin by factoring.

$$f(x) = \frac{x - 1}{x^2 - x - 6} = \frac{x - 1}{(x + 2)(x - 3)}.$$

The factored form allows us to read off the necessary information:

Vertical Asymptotes: $x = -2$ and $x = 3$ (roots of the denominator but not of the numerator).

Horizontal Asymptote: *x*-axis (because denominator has larger degree than the numerator).

Intercepts: *y*-intercept at $f(0) = \dfrac{0 - 1}{0^2 - 0 - 6} = \dfrac{1}{6}$; *x*-intercept at $x = 1$ (root of the numerator but not of the denominator).

Interpreting Figure 4–40 in light of this information suggests that a complete graph of *f* looks something like Figure 4–41.

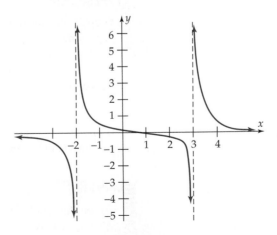

Figure 4–41

TECHNOLOGY **TIP**

A decimal window (with the y-range suitably adjusted) will usually produce an accurate graph of a rational function whose vertical asymptotes occur at numbers such as $x = -2.0$ or 3.7 or 4.1 that are within the x-range.

GRAPHING EXPLORATION

Find a viewing window in which the graph of f looks similar to Figure 4–41. The Tip Technology in the margin may be helpful.   ■

NOTE: The graph of a rational function never touches a horizontal asymptote when x is large in absolute value. For smaller values of x, however, the graph may cross the asymptote, as in Example 5.

EXAMPLE 6

To graph

$$f(x) = \frac{2x^2}{x^2 + x - 2},$$

we factor and then read off the necessary information:

$$f(x) = \frac{2x^2}{x^2 + x - 2} = \frac{2x^2}{(x + 2)(x - 1)}.$$

Vertical Asymptotes: $x = -2$ and $x = 1$ (roots of denominator).

Horizontal Asymptote: $y = 2/1 = 2$ (because numerator and denominator have the same degree; see the box on page 280).

Intercepts: x-intercept at $x = 0$ (root of numerator); y-intercept at $f(0) = 0$.

Using this information and selecting a decimal viewing window that will accurately portray the graph near the vertical asymptotes, we obtain what seems to be a reasonably complete graph in Figure 4–42. The graph appears to be falling to the right of $x = 1$, but this is deceptive.

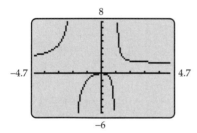

Figure 4–42

GRAPHING EXPLORATION

Graph f in this same viewing window and use the trace feature, beginning at approximately $x = 1.1$ and moving to the right. For what values of x is the graph above the horizontal asymptote $y = 2$? For what values of x is the graph below the horizontal asymptote?

This use of the trace feature indicates that there is some *hidden behavior* of the graph that is not visible in Figure 4–42.

GRAPHING EXPLORATION

To see this hidden behavior, graph both f and the line $y = 2$ in the viewing window with $1 \le x \le 50$ and $1.7 \le y \le 2.1$.

This Exploration shows that the graph has a local minimum near $x = 4$ and then stays below the asymptote, moving closer and closer to it as x takes larger values. ∎

APPLICATIONS

Several applications of rational functions were considered in Section 2.4. Here is another one.

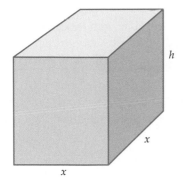

Figure 4–43

EXAMPLE 7

A cardboard box with a square base and a volume of 1000 cubic inches is to be constructed (Figure 4–43). The box must be at least 2 inches in height.

(a) What are the possible lengths for a side of the base if no more than 1100 square inches of cardboard can be used to construct the box?

(b) What is the least possible amount of cardboard that can be used?

(c) What are the dimensions of the box that uses the least possible amount of cardboard?

SOLUTION   The amount of cardboard needed is given by the surface area S of the box. From Figure 4–43, we have

$$S = \begin{array}{c} area \ of \\ top \end{array} + \begin{array}{c} area \ of \\ bottom \end{array} + \begin{array}{c} area \ of \ each \\ side \end{array}$$

$$S = x^2 + x^2 + xh + xh + xh + xh = 2x^2 + 4xh.$$

Since the volume of the box is given by

$$\text{Length} \times \text{Width} \times \text{Height} = \text{Volume},$$

we have

$$x \cdot x \cdot h = 1000 \qquad \text{or, equivalently,} \qquad h = \frac{1000}{x^2}.$$

Substituting this into the surface area formula allows us to express the surface area as a function of x:

$$S(x) = 2x^2 + 4xh = 2x^2 + 4x\left(\frac{1000}{x^2}\right) = 2x^2 + \frac{4000}{x} = \frac{2x^3 + 4000}{x}.$$

Although the rational function $S(x)$ is defined for all nonzero real numbers, x is a length here and must be positive. Furthermore, $x^2 \leq 500$ because if $x^2 > 500$, then $h = \dfrac{1000}{x^2}$ would be less than 2, contrary to specifications. Hence, the only values of x that make sense in this context are those with $0 < x \leq \sqrt{500}$. Since $\sqrt{500} \approx 22.4$, we choose the viewing window in Figure 4–44. For each point (x, y) on the graph, x is a possible side length for the base of the box, and y is the corresponding surface area.

Figure 4–44

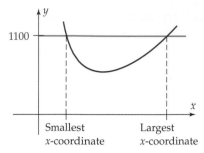

Figure 4–45

(a) The points on the graph corresponding to the requirement that no more than 1100 square inches of cardboard be used are those whose y-coordinates are less than or equal to 1100. The x-coordinates of these points are the possible side lengths. The x-coordinates of the points where the graph of S meets the horizontal line $y = 1100$ are the smallest and largest possible values for x, as indicated schematically in Figure 4–45.

EXPLORATION

Graph $S(x)$ and $y = 1100$ on the same screen. Use an intersection finder to show that the possible side lengths that use no more than 1100 square inches of cardboard are those with $3.73 \leq x \leq 21.36$.

(b) The least possible amount of cardboard corresponds to the point on the graph of $S(x)$ with the smallest y-coordinate.

EXPLORATION

Show that the graph of S has a local minimum at the point (10.00, 600.00). Consequently, the least possible amount of cardboard is 600 square inches and this occurs when $x = 10$.

(c) When $x = 10$, $h = 1000/10^2 = 10$. So the dimensions of the box using the least amount of cardboard are $10 \times 10 \times 10$. ∎

EXERCISES 4.5

In Exercises 1–6, find the domain of the function. You may need to use some of the techniques of Section 4.3.

1. $f(x) = \dfrac{-3x}{2x + 5}$

2. $g(x) = \dfrac{x^3 + x + 1}{2x^2 - 5x - 3}$

3. $h(x) = \dfrac{6x - 5}{x^2 - 6x + 4}$

4. $g(x) = \dfrac{x^3 - x^2 - x - 1}{x^5 - 36x}$

5. $f(x) = \dfrac{x^5 - 2x^3 + 7}{x^3 - x^2 - 2x + 2}$

6. $h(x) = \dfrac{x^5 - 5}{x^4 + 12x^3 + 60x^2 + 50x - 125}$

In Exercises 7–10, use the graphs in Example 1 and the information in Section 3.4 to match the function with its graph, which is one of those shown here.

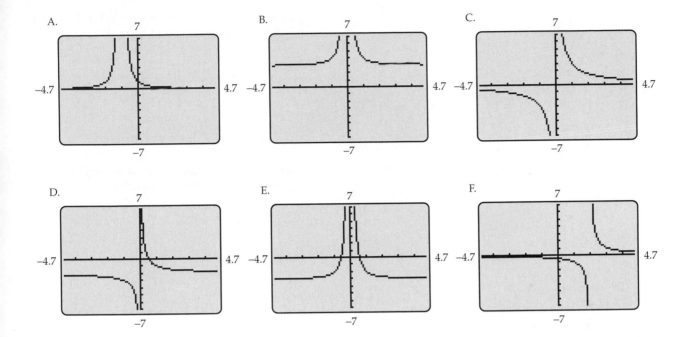

A.   B.   C.
D.   E.   F.

7. $f(x) = \dfrac{1}{x-2}$

8. $g(x) = \dfrac{3}{x}$

9. $h(x) = \dfrac{1}{(x+1)^2}$

10. $f(x) = \dfrac{1}{x^2} - 3$

21. $g(x) = \dfrac{5x^3 - 8x^2 + 4}{2x^3 + 2x}$

22. $h(x) = \dfrac{8x^5 - 6x^3 + 2x - 1}{.5x^5 + x^4 + 3x^2 + x}$

In Exercises 11–16, use algebra to determine the location of the vertical asymptotes and holes in the graph of the function.

11. $f(x) = \dfrac{x^2 + 4}{x^2 - 5x - 6}$

12. $g(x) = \dfrac{x - 5}{x^3 + 7x^2 + 2x}$

13. $f(x) = \dfrac{x}{x^3 + 2x^2 + x}$

14. $g(x) = \dfrac{x}{x^3 + 5x}$

15. $f(x) = \dfrac{x^2 - 4x + 4}{(x + 2)(x - 2)^3}$

16. $h(x) = \dfrac{x - 3}{x^2 - x - 6}$

In Exercises 23–40, analyze the function algebraically: List its vertical asymptotes, holes, and horizontal asymptote. Then sketch a complete graph of the function.

23. $f(x) = \dfrac{1}{x + 5}$

24. $q(x) = \dfrac{-7}{x - 6}$

25. $f(x) = \dfrac{3x}{x - 1}$

26. $p(x) = \dfrac{x - 2}{x}$

27. $f(x) = \dfrac{2 - x}{x - 3}$

28. $g(x) = \dfrac{3x - 2}{x + 3}$

29. $f(x) = \dfrac{1}{x(x + 1)^2}$

30. $g(x) = \dfrac{x}{2x^2 - 5x - 3}$

31. $f(x) = \dfrac{x - 3}{x^2 + x - 2}$

32. $g(x) = \dfrac{x + 2}{x^2 - 1}$

33. $h(x) = \dfrac{(x^2 + 6x + 5)(x + 5)}{(x + 5)^3(x - 1)}$

34. $f(x) = \dfrac{x^2 - 1}{x^3 - 2x^2 + x}$

35. $f(x) = \dfrac{-4x^2 + 1}{x^2}$

In Exercises 17–22, find the horizontal asymptote of the graph of the function when $|x|$ is large and find a viewing window in which the ends of the graph are within .1 of this asymptote.

17. $f(x) = \dfrac{3x - 2}{x + 3}$

18. $g(x) = \dfrac{3x^2 + x}{2x^2 - 2x + 4}$

19. $h(x) = \dfrac{5 - x}{x - 2}$

20. $f(x) = \dfrac{4x^2 - 5}{2x^3 - 3x^2 + x}$

36. $k(x) = \dfrac{x^2 + 1}{x^2 - 1}$

37. $q(x) = \dfrac{x^2 + 2x}{x^2 - 4x - 5}$

38. $F(x) = \dfrac{x^2 + x}{x^2 - 2x + 4}$

39. $p(x) = \dfrac{(x + 3)(x - 3)}{(x - 5)(x + 4)(x + 3)}$

40. $p(x) = \dfrac{x^3 + 3x^2}{x^4 - 4x^2}$

In Exercises 41–48, find a viewing window, or windows, that shows a complete graph of the function (if possible, with no erroneous vertical line segments). Be alert for hidden behavior, such as that in Example 6.

41. $f(x) = \dfrac{x^3 + 4x^2 - 5x}{(x^2 - 4)(x^2 - 9)}$

42. $g(x) = \dfrac{x^2 + x - 6}{x^3 - 19x + 30}$

43. $h(x) = \dfrac{2x^2 - x - 6}{x^3 + x^2 - 6x}$

44. $f(x) = \dfrac{x^3 - x + 1}{x^4 - 2x^3 - 2x^2 + x - 1}$

45. $h(x) = \dfrac{3x^2 + x - 4}{2x^2 - 5x}$

46. $f(x) = \dfrac{2x^2 - 1}{3x^3 + 2x + 1}$

47. $g(x) = \dfrac{x - 4}{2x^3 - 5x^2 - 4x + 12}$

48. $h(x) = \dfrac{x^2 - 9}{x^3 + 2x^2 - 23x - 60}$

In Exercises 49–53, find the difference quotient of the function. [See Section 3.6.]

49. $f(x) = \dfrac{1}{x}$

50. $h(x) = \dfrac{1}{x + 5}$

51. $g(x) = \dfrac{1}{x^2}$

52. $g(x) = \dfrac{3}{2x^2}$

53. (a) Use the difference quotient in Exercise 49 to determine the average rate of change of $f(x) = 1/x$ as x changes from 2 to 2.1, from 2 to 2.01, and from 2 to 2.001. Estimate the instantaneous rate of change of f at $x = 2$.

   (b) Determine the average rate of change of $f(x) = 1/x$ as x changes from 3 to 3.1, from 3 to 3.01, and from 3 to 3.001. Estimate the instantaneous rate of change of f at $x = 3$.

   (c) How are the instantaneous rates of change of f at $x = 2$ and $x = 3$ related to the values of the function $g(x) = -1/x^2$ at $x = 2$ and $x = 3$?

54. The graph of

$$f(x) = \dfrac{2x^3 - 2x^2 - x + 1}{3x^3 - 3x^2 + 2x - 1}$$

has a vertical asymptote. Find a viewing window that demonstrates this fact.

55. (a) When $x \geq 0$, what rational function has the same graph as

$$f(x) = \dfrac{x - 1}{|x| - 2}?$$

   [*Hint:* Use the definition of absolute value on page 8.]

   (b) When $x < 0$, what rational function has the same graph as

$$f(x) = \dfrac{x - 1}{|x| - 2}?$$

   [See the hint for part (a).]

   (c) Use parts (a) and (b) to explain why the graph of

$$f(x) = \dfrac{x - 1}{|x| - 2}$$

   has two vertical asymptotes. What are they? Confirm your answer by graphing the function.

56. The percentage c of a drug in a person's bloodstream t hours after its injection is approximated by

$$c(t) = \dfrac{5t}{4t^2 + 5}.$$

   (a) Approximately what percentage of the drug is in the person's bloodstream after four and a half hours?

   (b) Graph the function c in an appropriate window for this situation.

   (c) What is the horizontal asymptote of the graph? What does it tell you about the amount of the drug in the bloodstream?

   (d) At what time is the percentage the highest? What is the percentage at that time?

57. It costs 2.5 cents per square inch to make the top and bottom of the box in Example 7. The sides cost 1.5 cents per square inch. What are the dimensions of the cheapest possible box?

58. A box with a square base and a volume of 1000 cubic inches is to be constructed. The material for the top and bottom of the box costs $3 per 100 square inches, and the material for the sides costs $1.25 per 100 square inches.

   (a) If x is the length of a side of the base, express the cost of constructing the box as a function of x.

   (b) If the side of the base must be at least 6 inches long, for what value of x will the cost of the box be $7.50?

59. A truck traveling at a constant speed on a reasonably straight, level road burns fuel at the rate of $g(x)$ gallons per mile, where x is the speed of the truck (in miles per hour) and $g(x)$ is given by

$$g(x) = \dfrac{800 + x^2}{200x}.$$

   (a) If fuel costs $1.40 per gallon, find the rule of the cost function $c(x)$ that expresses the cost of fuel for a 500-mile trip as a function of the speed. [*Hint:* $500g(x)$ gallons of fuel are needed to go 500 miles. (Why?)]

(b) What driving speed will make the cost of fuel for the trip $250?

(c) What driving speed will minimize the cost of fuel for the trip?

60. Pure alcohol is being added to 50 gallons of a coolant mixture that is 40% alcohol.

(a) Find the rule of the concentration function $c(x)$ that expresses the percentage of alcohol in the resulting mixture as a function of the number x of gallons of pure alcohol that are added. [*Hint:* The final mixture contains $50 + x$ gallons (why?). So $c(x)$ is the amount of alcohol in the final mixture divided by the total amount $50 + x$. How much alcohol is in the original 50-gallon mixture? How much is in the final mixture?]

(b) How many gallons of pure alcohol should be added to produce a mixture that is at least 60% alcohol and no more than 80% alcohol?

(c) Determine algebraically the exact amount of pure alcohol that must be added to produce a mixture that is 70% alcohol.

61. A rectangular garden with an area of 250 square meters is to be located next to a building and fenced on three sides, with the building acting as a fence on the fourth side.

(a) If the side of the garden parallel to the building has length x meters, express the amount of fencing needed as a function of x.

(b) For what values of x will less than 60 meters of fencing be needed?

(c) What value of x will result in the least possible amount of fencing being used? What are the dimensions of the garden in this case?

62. A certain company has fixed costs of $40,000 and variable costs of $2.60 per unit.

(a) Let x be the number of units produced. Find the rule of the average cost function. [The average cost is the cost of the units divided by the number of units.]

(b) Graph the average cost function in a window with $0 \le x \le 100{,}000$ and $0 \le y \le 20$.

(c) Find the horizontal asymptote of the average cost function. Explain what the asymptote means in this situation. [How low can the average cost possibly be?]

63. Radioactive waste is stored in a cylindrical tank, whose exterior has radius r and height h as shown in the figure. The sides, top, and bottom of the tank are 1 foot thick, and the tank has a volume of 150 cubic feet (including top, bottom, and walls).

(a) Express the interior height h_1 (that is, the height of the storage area) as a function of h.

(b) Express the interior height as a function of r.

(c) Express the volume of the interior as a function of r.

(d) Explain why r must be greater than 1.

(e) What should the dimensions of the tank be in order for it to hold as much as possible?

64. The relationship between the fixed focal length F of a camera, the distance u from the object being photographed to the lens, and the distance v from the lens to the film is given by

$$\frac{1}{F} = \frac{1}{u} + \frac{1}{v}.$$

(a) If the focal length is 50 millimeters, express v as a function of u.

(b) What is the horizontal asymptote of the graph of the function in part (a)?

(c) Graph the function in part (a) when 50 millimeters $< u <$ 35,000 millimeters.

(d) When you focus the camera on an object, the distance between the lens and the film is changed. If the distance from the lens to the film changes by less than .1 millimeter, the object will remain in focus. Explain why you have more latitude in focusing on distant objects than on very close ones.

65. The formula for the gravitational acceleration (in units of meters per second squared) of an object relative to the earth is

$$g(r) = \frac{3.987 \times 10^{14}}{(6.378 \times 10^6 + r)^2},$$

where r is the distance in meters above the earth's surface.

(a) What is the gravitational acceleration at the earth's surface?

(b) Graph the function $g(r)$ for $r \ge 0$.

(c) Can you ever escape the pull of gravity? [Does the graph have any r-intercepts?]

4.5.A   *SPECIAL TOPICS* Other Rational Functions

We now examine the graphs of rational functions in which the degree of the denominator is smaller than the degree of the numerator. Such a graph has no horizontal asymptote. However, it does have some polynomial curve as an asymptote, which means that the graph will get very close to this curve when $|x|$ is very large.

EXAMPLE 1

To graph

$$f(x) = \frac{x^3 + 3x^2 + x + 1}{x^2 + 2x - 1},$$

we begin by finding the vertical asymptotes and the x- and y-intercepts. The quadratic formula can be used to find the roots of the denominator.

$$x = \frac{-2 \pm \sqrt{2^2 - 4 \cdot 1(-1)}}{2 \cdot 1} = \frac{-2 \pm \sqrt{8}}{2} = \frac{-2 \pm 2\sqrt{2}}{2} = -1 \pm \sqrt{2}.$$

It is easy to verify that neither of these numbers is a root of the numerator, so the graph has vertical asymptotes at $x = -1 - \sqrt{2}$ and $x = -1 + \sqrt{2}$. The y-intercept is $f(0) = -1$. The x-intercepts are the roots of the numerator.

GRAPHING EXPLORATION

Use a calculator to verify that $x^3 + 3x^2 + x + 1$ has exactly one real root, located between -3 and -2.

Therefore, the graph of $f(x)$ has one x-intercept. Using this information and the calculator graph in Figure 4–46 (which erroneously shows some vertical segments), we conclude that the graph looks approximately like Figure 4–47.

Figure 4–46

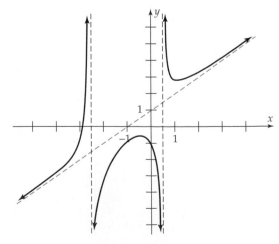

Figure 4–47

At the left and right ends, the graph moves away from the x-axis. To understand the behavior of the graph when $|x|$ is large, divide the numerator $f(x)$ by its denominator.

$$
\begin{array}{r}
x + 1 \\
x^2 + 2x - 1 \overline{\smash{\big)}\, x^3 + 3x^2 + x + 1} \\
\underline{x^3 + 2x^2 - x } \\
x^2 + 2x + 1 \\
\underline{x^2 + 2x - 1} \\
2.
\end{array}
$$

By the Division Algorithm,

$$x^3 + 3x^2 + x + 1 = (x^2 + 2x - 1)(x + 1) + 2.$$

Dividing both sides by $x^2 + 2x - 1$, we have

$$\frac{x^3 + 3x^2 + x + 1}{x^2 + 2x - 1} = \frac{(x^2 + 2x - 1)(x + 1) + 2}{x^2 + 2x - 1}$$

$$f(x) = (x + 1) + \frac{2}{x^2 + 2x - 1}.$$

Now when x is very large in absolute value, so is $x^2 + 2x - 1$. Hence, $2/(x^2 + 2x - 1)$ is very close to 0 by the Big-Little Principle, and $f(x)$ is very close to $(x + 1) + 0$. Therefore, as x gets larger in absolute value, the graph of $f(x)$ gets closer and closer to the line $y = x + 1$ (the dashed blue line in Figure 4–49) and this line is an asymptote of the graph.[*] Note that $x + 1$ is just the quotient obtained in the long division above. ∎

It is instructive to examine the graph in Example 1 further to see that the asymptote accurately indicates the behavior of the function when $|x|$ is large.

GRAPHING EXPLORATION

Using the viewing window with $-20 \le x \le 20$ and $-20 \le y \le 20$, graph both $f(x)$ and $y = x + 1$ on the same screen.

Except near the vertical asymptotes of $f(x)$, the two graphs are virtually identical.

EXAMPLE 2

To graph

$$g(x) = \frac{x^3 + 2x^2 - 7x + 5}{x - 1},$$

we first note that there is a vertical asymptote at $x = 1$ (root of the denominator, but not the numerator). The y-intercept is at $g(0) = -5$. By carefully choosing a viewing window that accurately portrays the behavior of $g(x)$ near its vertical asymptote, we obtain Figure 4–48.

Figure 4–48

(graph with window labels: 11, −4.7, 4.7, −11)

---

[*]An asymptote that is a nonvertical and nonhorizontal straight line is called an **oblique asymptote**.

GRAPHING**EXPLORATION**

Verify that the x-intercept near $x = -4$ is the only one by showing graphically that the numerator of $g(x)$ has exactly one real root.

To confirm that Figure 4–48 is a complete graph, we find its asymptote when $|x|$ is large. Divide the numerator by the denominator.

$$
\begin{array}{r}
x^2 + 3x - 4 \\
x - 1 \overline{\smash{\big)}\, x^3 + 2x^2 - 7x + 5} \\
\underline{x^3 - x^2} \\
3x^2 - 7x + 5 \\
\underline{3x^2 - 3x} \\
-4x + 5 \\
\underline{-4x + 4} \\
1
\end{array}
$$

Hence, by the Division Algorithm,

$$x^3 + 2x^2 - 7x + 5 = (x - 1)(x^2 + 3x - 4) + 1$$

$$\frac{x^3 + 2x^2 - 7x + 5}{x - 1} = \frac{(x - 1)(x^2 + 3x - 4) + 1}{x - 1}$$

$$g(x) = (x^2 + 3x - 4) + \frac{1}{x - 1}.$$

When $|x|$ is large, $1/(x - 1)$ is very close to 0 (why?), so that $y = x^2 + 3x - 4$ is the asymptote. Once again, the asymptote is given by the quotient of the division.

EXPLORATION

Graph $g(x)$ and $y = x^2 + 3x - 4$ on the same screen to show that the graph of $g(x)$ does get very close to the asymptote when $|x|$ is large. Then find a large enough viewing window that the two graphs appear to be identical. ■

The procedures used in the preceding examples may be summarized as follows.

Graphing
$$f(x) = \frac{g(x)}{h(x)} \text{ When}$$
Degree $g(x)$ >
Degree $h(x)$

1. Analyze the function algebraically to determine its vertical asymptotes, holes, and intercepts.

2. Divide the numerator $g(x)$ by the denominator $h(x)$. The quotient $q(x)$ is the nonvertical asymptote of the graph, which describes the behavior of the graph when $|x|$ is large.

3. Use the preceding information to select an appropriate viewing window (or windows), to interpret the calculator's version of the graph (if necessary), and to sketch an accurate graph.

✔ **EXERCISES 4.5.A**

In Exercises 1–4, find the nonvertical asymptote of the graph of the function when $|x|$ is large and find a viewing window in which the ends of the graph are within .1 of this asymptote.

1. $f(x) = \dfrac{x^3 - 1}{x^2 - 4}$

2. $g(x) = \dfrac{x^3 - 4x^2 + 6x + 5}{x - 2}$

3. $h(x) = \dfrac{x^3 + 3x^2 - 4x + 1}{x + 4}$

4. $f(x) = \dfrac{x^3 + 3x^2 - 4x + 1}{x^2 - x}$

In Exercises 5–12, analyze the function algebraically. List its vertical asymptotes and holes, and determine its nonvertical asymptote. Then sketch a complete graph of the function.

5. $f(x) = \dfrac{x^2 - x - 6}{x - 2}$

6. $k(x) = \dfrac{x^2 + x - 2}{x}$

7. $Q(x) = \dfrac{4x^2 + 4x - 3}{2x - 5}$

8. $K(x) = \dfrac{3x^2 - 12x + 15}{3x + 6}$

9. $f(x) = \dfrac{x^3 - 2}{x - 1}$

10. $p(x) = \dfrac{x^3 + 8}{x + 1}$

11. $q(x) = \dfrac{x^3 - 1}{x - 2}$

12. $f(x) = \dfrac{x^4 - 1}{x^2}$

In Exercises 13–18, find a viewing window (or windows) that shows a complete graph of the function (if possible, with no erroneous vertical line segments). Be alert for hidden behavior.

13. $f(x) = \dfrac{2x^2 + 5x + 2}{2x + 7}$

14. $g(x) = \dfrac{2x^3 + 1}{x^2 - 1}$

15. $h(x) = \dfrac{x^3 - 2x^2 + x - 2}{x^2 - 1}$

16. $f(x) = \dfrac{3x^3 - 11x - 1}{x^2 - 4}$

17. $g(x) = \dfrac{2x^4 + 7x^3 + 7x^2 + 2x}{x^3 - x + 50}$

18. $h(x) = \dfrac{2x^3 + 7x^2 - 4}{x^2 + 2x - 3}$

19. (a) Show that when $0 < x < 4$, the rational function

$$r(x) = \frac{4096x^3 + 34{,}560x^2 + 19{,}440x + 729}{18{,}432x^2 + 34{,}560x + 5832}$$

is a good approximation of the function $s(x) = \sqrt{x}$ by graphing both functions in the viewing window with $0 \le x \le 4$ and $0 \le y \le 2$.

(b) For what values of x is $r(x)$ within .01 of $s(x)$?

20. Find a rational function f that has these properties:
   (i) The curve $y = x^3 - 8$ is an asymptote of the graph of f.
   (ii) $f(2) = 1$.
   (iii) The line $x = 1$ is a vertical asymptote of the graph of f.

4.6   Polynomial and Rational Inequalities

Inequalities may be solved by using algebraic or geometric methods, both of which are discussed here. Whenever possible, we shall use algebra to obtain exact solutions. When algebraic methods are too difficult, approximate graphical solutions will be found. The basic tools for working with inequalities are the following principles.

Basic Principles for Solving Inequalities

Performing any of the following operations on an inequality produces an equivalent inequality:*

1. Add or subtract the same quantity on both sides of the inequality.

2. Multiply or divide both sides of the inequality by the same *positive* quantity.

3. Multiply or divide both sides of the inequality by the same *negative* quantity and *reverse the direction of the inequality.*

Note principle 3 carefully. It says, for example, that if you multiply both sides of $-3 < 5$ by -2, the equivalent inequality is $6 > -10$ (direction of inequality is reversed).

▘▘ LINEAR INEQUALITIES

EXAMPLE 1

Solve $5x + 3 \leq 6 + 7x$.

SOLUTION   We use the basic principles to transform the inequality into one whose solutions are obvious.

Subtract $7x$ from both sides: $\qquad -2x + 3 \leq 6$

Subtract 3 from both sides: $\qquad -2x \leq 3$

Divide both sides by -2 and reverse the direction of the inequality: $\qquad x \geq -3/2$

Therefore, the solutions are all real numbers greater than or equal to $-3/2$, that is, the interval $[-3/2, \infty)$. ■

---

*Two inequalities are **equivalent** if they have the same solutions.

EXAMPLE 2

A solution of the inequality $2 \leq 3x + 5 < 2x + 11$ is any number that is a solution of *both* of these inequalities:

$$2 \leq 3x + 5 \qquad \text{and} \qquad 3x + 5 < 2x + 11.$$

Each of these inequalities can be solved by the methods used earlier. For the first one, we have

$$2 \leq 3x + 5$$

Subtract 5 from both sides: $\qquad -3 \leq 3x$

Divide both sides by 3: $\qquad -1 \leq x.$

The second inequality is solved similarly:

$$3x + 5 < 2x + 11$$

Subtract 5 from both sides. $\qquad 3x < 2x + 6$

Subtract $2x$ from both sides: $\qquad x < 6.$

The solutions of the original inequality are the numbers x that satisfy *both* $-1 \leq x$ *and* $x < 6$, that is, all x with $-1 \leq x < 6$. Thus, the solutions are the numbers in the interval $[-1, 6)$. ∎

EXAMPLE 3

When solving the inequality $4 < 3 - 5x < 18$, in which the variable appears only in the middle part, you can proceed as follows.

$$4 < 3 - 5x < 18$$

Subtract 3 from each part: $\qquad 1 < -5x < 15$

Divide each part by -5 and reverse the directions of the inequalities: $\qquad -\dfrac{1}{5} > x > -3.$

CAUTION
All inequality signs in an inequality should point in the same direction. *Don't* write things like $4 < x > 2$ or $-3 \geq x < 5$.

Reading this last inequality from right to left we see that

$$-3 < x < -1/5,$$

so the solutions are the numbers in the interval $(-3, -1/5)$. ∎

▪▪ POLYNOMIAL INEQUALITIES

Although the basic principles play a role in the solution of nonlinear inequalities, the key to solving such inequalities is this geometric fact.

The graph of $y = f(x)$ lies above the x-axis exactly when $f(x) > 0$ and below the x-axis exactly when $f(x) < 0$.

Consequently, the solutions of $f(x) > 0$ are the numbers x for which the graph of f lies above the x-axis and the solutions $f(x) < 0$ are the numbers x for which the graph of f lies below the x-axis.

EXAMPLE 4

Solve $2x^3 - 15x < x^2$.

SOLUTION   We replace the inequality by an equivalent one,

$$2x^3 - x^2 - 15x < 0,$$

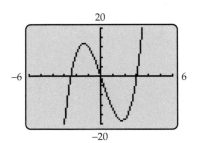

20

−6   6

−20

Figure 4–49

and consider the graph of the function $f(x) = 2x^3 - x^2 - 15x$ (Figure 4–49). Since $f(x)$ factors as

$$f(x) = 2x^3 - x^2 - 15x = x(2x^2 - x - 15) = x(2x + 5)(x - 3),$$

its roots (the x-intercepts of its graph) are $x = 0$, $x = -5/2$, and $x = 3$. The graph of $f(x) = 2x^3 - x^2 - 15x$ in Figure 4–49 is complete (why?) and lies below the x-axis when $x < -5/2$ or $0 < x < 3$. Therefore, the solutions of

$$2x^3 - x^2 - 15x < 0,$$

and hence of the original inequality, are all numbers x such that $x < -5/2$ or $0 < x < 3$.   ∎

EXAMPLE 5

Solve $2x^3 - x^2 - 15x \geq 0$.

SOLUTION   Figure 4–49 shows that the solutions of $2x^3 - x^2 - 15x > 0$ (that is, the numbers x for which the graph of $f(x) = 2x^3 - x^2 - 15x$ lies above the x-axis) are all x such that $-5/2 < x < 0$ or $x > 3$. The solutions of the equation $2x^3 - x^2 - 15x = 0$ are the roots of $f(x) = 2x^3 - x^2 - 15x$, namely, 0, $-5/2$, and 3 as we saw in Example 4. Therefore, the solutions of the given inequality are all numbers x such that $-5/2 \leq x \leq 0$ or $x \geq 3$.   ∎

When the roots of a polynomial $f(x)$ cannot be determined exactly, a root finder can be used to approximate them and to find approximate solutions of the inequalities $f(x) > 0$ and $f(x) < 0$.

EXAMPLE 6

Solve $x^4 + 10x^3 + 21x^2 + 8 > 40x + 88$.

SOLUTION   This inequality is equivalent to

$$x^4 + 10x^3 + 21x^2 - 40x - 80 > 0.$$

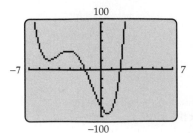

100

−7   7

−100

Figure 4–50

The graph $f(x) = x^4 + 10x^3 + 21x^2 - 40x - 80$ in Figure 4–50 is complete (why?) and shows that $f(x)$ has two roots, one between -2 and -1 and the other near 2.

GRAPHING EXPLORATION

Use a root finder to show that the approximate roots of $f(x)$ are -1.53 and 1.89.

Therefore, the approximate solutions of the inequality (the numbers x for which the graph is above the x-axis) are all numbers x such that $x < -1.53$ or $x > 1.89$. ∎

CAUTION

Do not attempt to write the solution in Example 6, namely, "$x < -1.53$ or $x > 1.89$" as a single inequality. If you do, the result will be a *nonsense statement* such as $-1.53 > x > 1.89$ (which says, among other things, that $-1.53 > 1.89$).

▪▪ QUADRATIC AND FACTORABLE INEQUALITIES

The preceding examples show that solving a polynomial inequality depends only on knowing the roots of a polynomial and the places where its graph is above or below the x-axis. In the case of quadratic inequalities or completely factored polynomial inequalities, a calculator is not needed to determine this information.

EXAMPLE 7

The solutions of $2x^2 + 3x - 4 \leq 0$ are the numbers x at which the graph of $f(x) = 2x^2 + 3x - 4$ lies on or below the x-axis. The points where the graph meets the x-axis are the roots of $f(x) = 2x^2 + 3x - 4$, which can be found by means of the quadratic formula:

$$x = \frac{-3 \pm \sqrt{3^2 - 4 \cdot 2(-4)}}{2 \cdot 2} = \frac{-3 \pm \sqrt{41}}{4}.$$

From Section 4.1, we know that the graph of $f(x)$ is an upward-opening parabola, so the graph must have the general shape shown in Figure 4–51.

Figure 4–51

The graph lies below the x-axis between the two roots. Therefore, the solutions of the original inequality are all numbers x such that

$$\frac{-3 - \sqrt{41}}{4} \leq x \leq \frac{-3 + \sqrt{41}}{4}. \qquad ∎$$

EXAMPLE 8

Solve $(x + 15)(x - 2)^6(x - 10) \leq 0$.

SOLUTION   The roots of $f(x) = (x + 15)(x - 2)^6(x - 10)$ are easily read from the factored form: -15, 2, and 10. So we need only determine where the graph of $f(x)$ is on or below the x-axis. To do this without a calculator, note that the three roots of $f(x)$ divide the x-axis into four intervals:

$$x < -15, \qquad -15 < x < 2, \qquad 2 < x < 10, \qquad x > 10.$$

For each of these intervals, we shall determine whether the graph is above or below the x-axis.

Consider, for example, the interval between the roots 2 and 10. The graph of $f(x)$ touches the x-axis at $x = 2$ and $x = 10$ but does not touch the axis at any point in between, since the only other root (x-intercept) is -15. Furthermore, the graph of $f(x)$ is continuous, which means that between the points $(2, 0)$ and $(10, 0)$ it can be drawn without lifting the pencil from the paper. If you experiment with Figure 4–52, you will find that the only way to draw a graph from $(2, 0)$ to $(10, 0)$ without lifting your pencil and without touching the x-axis is to stay entirely on one side of the x-axis (try it!). In other words, between the two adjacent roots 2 and 10, the graph of $f(x)$ is either entirely above the x-axis or entirely below it.

To determine which is the case, choose any number between 2 and 10, say, $x = 4$, and evaluate $f(4)$.

$$f(4) = (4 + 15)(4 - 2)^6(4 - 10) = 19(2^6)(-6).$$

You don't even have to finish the computation to see that $f(4)$ is a negative number. Therefore, the point $(4, f(4))$ on the graph of $f(x)$ lies below the x-axis. Since one point of the graph between 2 and 10 lies below the x-axis, the entire graph must be below the x-axis between 2 and 10.

The location of the graph on the other intervals can be determined similarly, by choosing a test number in each interval, as summarized in this chart.

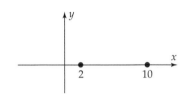

Figure 4–52

Interval	$x < -15$	$-15 < x < 2$	$2 < x < 10$	$x > 10$
Test number in this interval	-20	0	4	11
Value of $f(x)$ at test number	$(-5)(-22)^6(-30)$	$15(-2)^6(-10)$	$19(2^6)(-6)$	$26(9^6)(1)$
Sign of $f(x)$ at test number	$+$	$-$	$-$	$+$
Graph	Above x-axis	Below x-axis	Below x-axis	Above x-axis

The last line of the chart shows that the intervals where the graph is below the x-axis are $-15 < x < 2$ and $2 < x < 10$. Since the graph touches the x-axis at the roots -15, 2, and 10, the solutions of the original inequality (the numbers x for which the graph is on or below the x-axis) are all numbers x such that $-15 \leq x \leq 10$. ∎

The procedures used in Examples 4–8 may be summarized as follows.

Solving Polynomial Inequalities

1. Write the inequality in one of these forms:
$$f(x) > 0, \qquad f(x) \geq 0, \qquad f(x) < 0. \qquad f(x) \leq 0.$$

2. Determine the roots of $f(x)$, exactly if possible, approximately otherwise.

3. Use a calculator (as in Examples 4–6), your knowledge of quadratic functions (as in Example 7), or a sign chart (as in Example 8) to determine whether the graph of $f(x)$ is above or below the x-axis on each of the intervals determined by the roots.

4. Use the information in step 3 to find the solutions of the inequality.

▚ RATIONAL INEQUALITIES

Rational inequalities are solved in essentially the same way that polynomial inequalities are solved, with one difference. The graph of a rational function may cross the x-axis at an x-intercept, but there is another possibility: The graph may be above the x-axis on one side of a vertical asymptote and below it on the other side (see, for instance, Examples 5–6 in Section 4.5). Since the x-intercepts of the graph of the rational function $g(x)/h(x)$ are determined by the roots of its numerator $g(x)$ and the vertical asymptotes by the roots of its denominator $h(x)$, all of these roots must be considered in determining the solution of an inequality involving $g(x)/h(x)$.

EXAMPLE 9

Solve $\dfrac{x}{x-1} > -6$.

SOLUTION   There are three ways to solve this inequality.

Geometric: The fastest way to get an approximate solution is to replace the given inequality by an equivalent one,

$$\frac{x}{x-1} + 6 > 0.$$

and graph the function $f(x) = \dfrac{x}{x-1} + 6$ as in Figure 4–53.

The graph is above the x-axis everywhere except between the x-intercept and the vertical asymptote $x = 1$. Using a root finder, we see that the x-intercept is approximately .857. Therefore, the approximate solutions of the original inequality are all numbers x such that $x < .857$ or $x > 1$.

Figure 4–53

Algebraic/Geometric: Proceed as above, but rewrite the rule of the function f as a single rational expression before graphing.

$$f(x) = \frac{x}{x-1} + 6 = \frac{x}{x-1} + \frac{6(x-1)}{x-1} = \frac{x + 6x - 6}{x-1} = \frac{7x-6}{x-1}.$$

When the rule of f is written in this form, it is easy to see that the x-intercept of the graph (the root of the numerator) is $x = 6/7$ (whose decimal approximation begins .857). Therefore, the exact solutions of the original inequality (the numbers x for which the graph in Figure 4–53 is above the x-axis) are all numbers x such that $x < 6/7$ or $x > 1$.

Algebraic: Write the rule of the function f as a single rational expression $f(x) = \frac{7x-6}{x-1}$. The roots of the numerator and denominator (6/7 and 1) divide the x-axis into three intervals. Use test numbers and a sign chart instead of graphing to determine the location of the graph on each interval:[*]

Interval	$x < 6/7$	$6/7 < x < 1$	$x > 1$
Test number in this interval	0	.9	2
Value of $f(x)$ at test number	$\dfrac{7 \cdot 0 - 6}{}$	$\dfrac{7(.9) - 6}{.9 - 1}$	$\dfrac{7 \cdot 2 - 6}{}$
Sign of $f(x)$ at test number	+	−	+
Graph	Above x-axis	Below x-axis	Above x-axis

The last line of the chart shows that the solutions of the original inequality (the numbers x for which the graph is above the x-axis) are all such that $x < 6/7$ or $x > 1$. ∎

   The algebraic technique of writing the left side of the inequality as a single rational expression is useful whenever the resulting numerator has low degree (so that its roots can be found exactly), but can usually be omitted when the roots of the numerator must be approximated.

---

[*]The justification for this approach is essentially the same as that in Example 8: Because f is continuous everywhere that it is defined, the graph can change from one side of the x-axis to the other only at x-intercepts or vertical asymptotes, so testing one number in each interval is sufficient to determine the side on which the graph lies.

CAUTION

Don't treat rational inequalities as if they are equations, as in this *incorrect* "solution" of the preceding example:

$$\frac{x}{x-1} > -6$$

$$x > -6(x-1) \qquad \text{[Both sides multiplied by } x-1]$$

$$x > -6x + 6$$

$$7x > 6$$

$$x > \frac{6}{7}$$

According to this, the inequality has no negative solution and $x = 1$ is a solution, but as we saw in Example 9, *every* negative number is a solution and $x = 1$ is not.*

:: APPLICATIONS

EXAMPLE 10

A computer store has determined that the cost C of ordering and storing x laser printers is given by

$$C = 2x + \frac{300{,}000}{x}.$$

If the delivery truck can bring at most 450 printers per order, how many printers should be ordered at a time to keep the cost below $1600?

SOLUTION   To find the values of x that make C less than 1600, we must solve the inequality

$$2x + \frac{300{,}000}{x} < 1600 \qquad \text{or, equivalently,} \qquad 2x + \frac{300{,}000}{x} - 1600 < 0.$$

We shall solve this inequality graphically, although it can also be solved algebraically. In this context, the only solutions that make sense are those between 0 and 450. So we choose the viewing window in Figure 4–54 and graph

$$f(x) = 2x + \frac{300{,}000}{x} - 1600.$$

Figure 4–54 is consistent with the fact that $f(x)$ has a vertical asymptote at $x = 0$ and shows that the desired solutions (numbers where the graph is below the x-axis) are all numbers x between the root and 450. A root finder shows that the root is $x \approx 300$. In fact, this is the exact root, since a simple computation shows that $f(300) = 0$. (Do it!) Therefore, to keep costs under $1600, x printers should be ordered each time, with $300 < x \le 450$.   ■

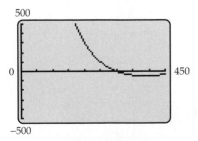

Figure 4–54

---

*The source of the error is multiplying by $x - 1$. This quantity is negative for some values of x and positive for others. To do this calculation correctly, you must consider two separate cases and reverse the direction of the inequality when $x - 1$ is negative.

✓ **EXERCISES 4.6**

In Exercises 1–20, solve the inequality and express your answer in interval notation.

1. $2x + 4 \leq 7$ **2.** $3x - 5 > -6$

3. $3 - 5x < 13$ **4.** $2 - 3x < 11$

5. $6x + 3 \leq x - 5$ **6.** $5x + 3 \leq 2x + 7$

7. $5 - 7x < 2x - 4$ **8.** $5 - 3x > 7x - 3$

9. $2 < 3x - 4 < 8$ **10.** $1 < 5x + 6 < 9$

11. $0 < 5 - 2x \leq 11$ **12.** $-4 \leq 7 - 3x < 0$

13. $2x + 7(3x - 2) < 2(x - 1)$

14. $x + 3(x - 5) \geq 3x + 2(x + 1)$

15. $\dfrac{x + 1}{2} - 3x \leq \dfrac{x + 5}{3}$

16. $\dfrac{x - 1}{4} + 2x \geq \dfrac{2x - 1}{3} + 2$

17. $2x + 3 \leq 5x + 6 < -3x + 7$

18. $4x - 2 < x + 8 < 9x + 1$

19. $3 - x < 2x + 1 \leq 3x - 4$

20. $2x + 5 \leq 4 - 3x < 1 - 4x$

In Exercises 21–24, a, b, c, and d are positive constants. Solve the inequality for x.

21. $ax - b < c$ **22.** $d - cx > a$

23. $0 < x - c < a$ **24.** $-d < x - c < d$

In Exercises 25–54, solve the inequality. Find exact solutions when possible and approximate ones otherwise.

25. $x^2 - 4x + 3 \leq 0$ **26.** $x^2 - 7x + 10 \leq 0$

27. $x^2 + 9x + 15 \geq 0$ **28.** $x^2 + 8x + 20 \geq 0$

29. $8 + x - x^2 \leq 0$ **30.** $4 - 3x - x^2 \geq 0$

31. $x^3 - x \geq 0$ **32.** $x^3 + 2x^2 + x > 0$

33. $x^3 - 2x^2 - 3x < 0$ **34.** $x^4 - 14x^3 + 48x^2 \geq 0$

35. $x^4 - 5x^2 + 4 < 0$ **36.** $x^4 - 10x^2 + 9 \leq 0$

37. $x^3 - 2x^2 - 5x + 7 \geq 2x + 1$

38. $x^4 - 6x^3 + 2x^2 < 5x - 2$

39. $2x^4 + 3x^3 < 2x^2 + 4x - 2$

40. $x^5 + 5x^4 > 4x^3 - 3x^2 + 2$

41. $\dfrac{3x + 1}{2x - 4} > 0$ **42.** $\dfrac{2x - 1}{5x + 3} \geq 0$

43. $\dfrac{x^2 + x - 2}{x^2 - 2x - 3} < 0$ **44.** $\dfrac{2x^2 + x - 1}{x^2 - 4x + 4} \geq 0$

45. $\dfrac{x - 2}{x - 1} < 1$ **46.** $\dfrac{-x + 5}{2x + 3} \geq 2$

47. $\dfrac{x - 3}{x + 3} \leq 5$ **48.** $\dfrac{2x + 1}{x - 4} > 3$

49. $\dfrac{2}{r + 3} \geq \dfrac{1}{x - 1}$ **50.** $\dfrac{1}{x - 1} < \dfrac{-1}{x + 2}$

51. $\dfrac{x^3 - 3x^2 + 5x - 29}{x^2 - 7} > 3$

52. $\dfrac{x^4 - 3x^3 + 2x^2 + 2}{x - 2} > 15$

53. $\dfrac{2x^2 + 6x - 8}{2x^2 + 5x - 3} < 1$ [Be alert for hidden behavior.]

54. $\dfrac{1}{x^2 + x - 6} + \dfrac{x - 2}{x + 3} > \dfrac{x + 3}{x - 2}$

In Exercises 55–57, read the solution of the inequality from the given graph.

55. $3 - 2x < .8x + 7$

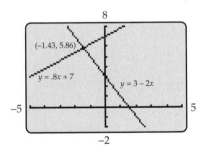

56. $8 - |7 - 5x| > 3$

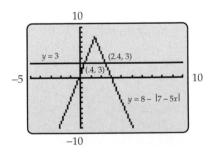

57. $x^2 + 3x + 1 \geq 4$

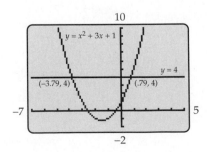

58. The graphs of the revenue and cost functions for a manufacturing firm are shown in the figure.
   (a) What is the break-even point?
   (b) Shade in the region representing profit.
   (c) What does the *y*-intercept of the cost graph represent? Why is the *y*-intercept of the revenue graph 0?

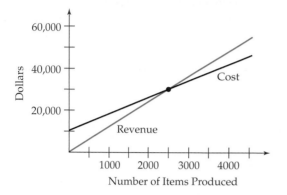

59. One freezer costs $623.95 and uses 90 kilowatt hours (kwh) of electricity each month. A second freezer costs $500 and uses 100 kwh of electricity each month. The expected life of each freezer is 12 years. What is the minimum electric rate (in *cents* per kwh) for which the 12-year total cost (purchase price + electricity costs) will be less for the first freezer?

60. A business executive leases a car for $300 per month. She decides to lease another brand for $250 per month but has to pay a penalty of $1000 for breaking the first lease. How long must she keep the second car to come out ahead?

61. One salesperson is paid a salary of $1000 per month plus a commission of 2% of her total sales. A second salesperson receives no salary but is paid a commission of 10% of her total sales. What dollar amount of sales must the second salesperson have to earn more per month than the first?

62. A developer subdivided 60 acres of a 100-acre tract, leaving 20% of the 60 acres as a park. Zoning laws require that at least 25% of the total tract be set aside for parks. For financial reasons, the developer wants to have no more than 30% of the tract as parks. How many one-quarter-acre lots can the developer sell in the remaining 40 acres and still meet the requirements for the whole tract?

63. If $5000 is invested at 8%, how much more should be invested at 10% to guarantee a total annual interest income between $800 and $940?

64. How many gallons of a 12% salt solution should be added to 10 gallons of an 18% salt solution to produce a solution whose salt content is between 14% and 16%?

65. Find all pairs of numbers that satisfy these two conditions: Their sum is 20, and the sum of their squares is less than 362.

66. The length of a rectangle is 6 inches longer than its width. What are the possible widths if the area of the rectangle is at least 667 square inches?

67. It costs a craftsman $5 in materials to make a medallion. He has found that if he sells the medallions for $50 - x$ dollars each, where x is the number of medallions produced each week, then he can sell all that he makes. His fixed costs are $350 per week. If he wants to sell all he makes and show a profit each week, what are the possible numbers of medallions he should make?

68. A retailer sells file cabinets for $80 - x$ dollars each, where x is the number of cabinets she receives from the supplier each week. She pays $10 for each file cabinet and has fixed costs of $600 per week. How many file cabinets should she order from the supplier each week to guarantee that she makes a profit?

In Exercises 69–72, you will need the formula for the height h of an object above the ground at time t seconds:

$$h = -16t^2 + v_0t + h_0;$$

this formula was explained on page 234.

69. A toy rocket is fired straight up from ground level with an initial velocity of 80 feet per second. During what time interval will it be at least 64 feet above the ground?

70. A projectile is fired straight up from ground level with an initial velocity of 72 feet per second. During what time interval is it at least 37 feet above the ground?

71. A ball is dropped from the roof of a 120-foot-high building. During what time period will it be strictly between 56 feet and 39 feet above the ground?

72. A ball is thrown straight up from a 40-foot-high tower with an initial velocity of 56 feet per second.
   (a) During what time interval is the ball at least 8 feet above the ground?
   (b) During what time interval is the ball between 53 feet and 80 feet above the ground?

73. (a) Solve the inequalities $x^2 < x$ and $x^2 > x$.
   (b) Use the results of part (a) to show that for any nonzero real number c with $|c| < 1$, it is always true that $c^2 < |c|$.
   (c) Use the results of part (a) to show that for any nonzero real number c with $|c| > 1$, it is always true that $c^2 > c$.

74. (a) If $0 < a \le b$, prove that $1/a \ge 1/b$.
   (b) If $a \le b < 0$, prove that $1/a \ge 1/b$.
   (c) If $a < 0 < b$, how are $1/a$ and $1/b$ related?

4.6.A *SPECIAL TOPICS* Absolute Value Inequalities

Polynomial and rational inequalities involving absolute value can be solved graphically, just as was done earlier: Rewrite the inequality in an equivalent form that has 0 on the right side of the inequality sign; then graph the function whose rule is given by the left side and determine where the graph is above or below the *x*-axis.

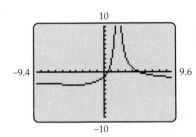

Figure 4–55

EXAMPLE 1

To solve $\left|\dfrac{x+4}{x-2}\right| > 3$, use the equivalent inequality

$$\left|\frac{x+4}{x-2}\right| - 3 > 0$$

and graph the function $f(x) = \left|\dfrac{x+4}{x-2}\right| - 3$ (Figure 4–55). The graph is above the *x*-axis between the two *x*-intercepts, which can be found algebraically or graphically.

GRAPHING EXPLORATION

Verify that the *x*-intercepts are $x = 1/2$ and $x = 5$.

Since $f(x)$ is not defined at $x = 2$ (where the graph has a vertical asymptote), the solutions of the original inequality are all *x* such that $1/2 < x < 2$ or $2 < x < 5$. ∎

EXAMPLE 2

The solutions of $\left|x^4 + 2x^2 - x + 2\right| < 11x$ can be found by determining the numbers for which the graph of

$$f(x) = \left|x^4 + 2x^2 - x + 2\right| - 11x$$

lies below the *x*-axis. (Why?) Convince yourself that the graph of $f(x)$ in Figure 4–56 is complete.

A root finder shows that the approximate *x*-intercepts are $x = .17$ and $x = 1.92$. Therefore, the approximate solutions of the original inequality (the numbers where the graph is below the *x*-axis) are all *x* such that $.17 < x < 1.92$. ∎

Figure 4–56

ALGEBRAIC METHODS

Most linear and quadratic inequalities involving absolute values can be solved exactly by algebraic means. In fact, this is often the easiest way to solve such inequalities. The key to the algebraic method is the fact that the absolute value of a number can be interpreted as distance on the number line. For example, the inequality $|r| \leq 5$ states that the distance from r to 0 (namely, $|r|$) is 5 units or less. A glance at the number line in Figure 4–57 shows that these are the numbers r with $-5 \leq r \leq 5$.

Figure 4–57

Similarly, the numbers r such that $|r| \geq 5$ are those whose distance to 0 is 5 or more units, that is, the numbers r with $r \leq -5$ or $r \geq 5$. This argument works with any positive number k in place of 5 and proves the following facts (which are also true with $<$ and $>$ in place of $\leq$ and $\geq$).

Absolute Value Inequalities

> Let k be a positive number and r any real number.
>
> $$|r| \leq k \qquad \text{is equivalent to} \qquad -k \leq r \leq k.$$
> $$|r| \geq k \qquad \text{is equivalent to} \qquad r \leq -k \quad \text{or} \quad r \geq k.$$

EXAMPLE 3

To solve $|3x - 7| \leq 11$, apply the first fact in the box, with $3x - 7$ in place of r and 11 in place of k, and obtain this equivalent inequality: $-11 \leq 3x - 7 \leq 11$. Then

Add 7 to each part: $\qquad -4 \leq 3x \leq 18$

Divide each part by 3: $\qquad -4/3 \leq x \leq 6$,

Therefore, the solutions of the original inequality are all numbers in the interval $[-4/3, 6]$. ■

EXAMPLE 4

To solve $|5x + 2| > 3$, apply the second fact in the box, with $5x + 2$ in place of r, and 3 in place of k, and $>$ in place of $\geq$. This produces the equivalent statement:

$$5x + 2 < -3 \qquad \text{or} \qquad 5x + 2 > 3$$
$$5x < -5 \qquad\qquad\qquad 5x > 1$$
$$x < -1 \qquad \text{or} \qquad x > 1/5.$$

Therefore, the solutions of the original inequality are the numbers in *either* of the intervals $(-\infty, -1)$ or $(1/5, \infty)$. ■

EXAMPLE 5

If a and δ are real numbers with δ positive, then the inequality $|x - a| < \delta$ is equivalent to $-\delta < x - a < \delta$. Adding a to each part shows that

$$a - \delta < x < a + \delta. \quad \blacksquare$$

EXAMPLE 6

To solve $|x^2 - x - 4| \geq 2$, we use the fact in the box on the preceding page to replace it by an equivalent inequality:

$$x^2 - x - 4 \leq -2 \qquad \text{or} \qquad x^2 - x - 4 \geq 2,$$

which is the same as

$$x^2 - x - 2 \leq 0 \qquad \text{or} \qquad x^2 - x - 6 \geq 0.$$

The solutions are all numbers that are solutions of *either one* of the two inequalities.

To solve the first of these inequalities, note that the graph of

$$f(x) = x^2 - x - 2 = (x + 1)(x - 2)$$

is an upward-opening parabola that crosses the x-axis at -1 and 2 (the roots of $f(x)$). Therefore, the solutions of

$$x^2 - x - 2 \leq 0$$

(the numbers for which the graph of $f(x)$ is on or below the x-axis) are all x with $-1 \leq x \leq 2$. The second inequality above, $x^2 - x - 6 \geq 0$, is solved similarly.

GRAPHING EXPLORATION

What is the shape of the graph of $g(x) = x^2 - x - 6$ and what are its x-intercepts?

This Graphing Exploration shows that the solutions of the second inequality (the numbers for which the graph of $g(x)$ is on or above the x-axis) are all x with $x \leq -2$ or $x \geq 3$.

Consequently, the solutions of the original inequality are all numbers x such that $x \leq -2$ or $-1 \leq x \leq 2$ or $x \geq 3$, as shown in Figure 4–58. $\blacksquare$

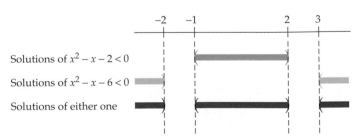

Figure 4–58

EXERCISES 4.6.A

In Exercises 1–32, solve the inequality. Find exact solutions when possible and approximate ones otherwise.

1. $|3x + 2| \le 2$

2. $|5x - 1| < 3$

3. $|3 - 2x| < 2/3$

4. $|4 - 5x| \le 4$

5. $|2x + 3| > 1$

6. $|3x - 1| \ge 2$

7. $|5x + 2| \ge \dfrac{3}{4}$

8. $|2 - 3x| > 4$

9. $\left|\dfrac{12}{5} + 2x\right| > \dfrac{1}{4}$

10. $\left|\dfrac{5}{6} + 3x\right| < \dfrac{7}{6}$

11. $\left|\dfrac{x - 1}{x + 2}\right| \le 3$

12. $\left|\dfrac{x + 1}{3x + 5}\right| < 2$

13. $\left|\dfrac{2x - 1}{x + 5}\right| > 1$

14. $\left|\dfrac{x + 1}{x + 2}\right| \ge 2$

15. $\left|\dfrac{1 - 4x}{2 + 3x}\right| < 1$

16. $\left|\dfrac{3x + 1}{1 - 2x}\right| \ge 2$

17. $|x^2 - 2| < 1$

18. $|x^2 - 4| \le 3$

19. $|x^2 - 2| > 4$

20. $\left|\dfrac{1}{x^2 - 1}\right| \le 2$

21. $|x^2 + x - 1| \ge 1$

22. $|x^2 + x - 4| \le 2$

23. $|3x^2 - 8x + 2| < 2$

24. $|x^2 + 3x - 4| < 6$

25. $|x^5 - x^3 + 1| < 2$

26. $|4x - x^3| > 1$

27. $|x^4 - x^3 + x^2 - x + 1| > 4$

28. $|x^3 - 6x^2 + 4x - 5| < 3$

29. $\dfrac{x + 2}{|x - 3|} \le 4$

30. $\dfrac{x^2 - 9}{|x^2 - 4|} < -2$

31. $\left|\dfrac{2x^2 + 2x - 12}{x^3 - x^2 + x - 2}\right| > 2$

32. $\left|\dfrac{x^2 - x - 2}{x^2 + x - 2}\right| > 3$

Thinkers

33. Let E be a fixed real number. Show that every solution of $|x - 3| < E/5$ is also a solution of $|(5x - 4) - 11| < E$.

34. Let a and b be fixed real numbers with $a < b$. Show that the solutions of

$$\left|x - \frac{a + b}{2}\right| < \frac{b - a}{2}$$

are all x with $a < x < b$.

4.7 Complex Numbers

The equation $x^2 = 2$ has no solutions in the rational number system but has $\sqrt{2}$ and $-\sqrt{2}$ as solutions in the real number system. So the idea of enlarging a number system to solve an equation that can't be solved in the present system is a natural one.

Equations such as $x^2 = -1$ and $x^2 = -4$ have no solutions in the real number system because $\sqrt{-1}$ and $\sqrt{-4}$ are not real numbers. To solve such equations (or, equivalently, to find square roots of negative numbers), we must enlarge the number system again. We claim that there is a number system, called the **complex number system,** with the following properties.

Properties of the Complex Number System

1. The complex number system contains all real numbers.

2. Addition, subtraction, multiplication, and division of complex numbers obey the same rules of arithmetic that hold in the real number system, with one exception: The exponent laws hold for *integer* exponents, but not necessarily for fractional ones.

3. The complex number system contains a number (usually denoted by i) such that $i^2 = -1$.

4. Every complex number can be written in the **standard form** $a + bi$, where a and b are real numbers.[*]

5. Two complex numbers $a + bi$ and $c + di$ are equal exactly when $a = c$ and $b = d$.

In view of our past experience with enlarging the number system, this claim *ought* to appear plausible. But the mathematicians who invented the complex numbers in the seventeenth century were very uneasy about a number i such that $i^2 = -1$ (that is, $i = \sqrt{-1}$). Consequently, they called numbers of the form bi (b any real number), such as $5i$ and $-\frac{1}{4}i$, **imaginary numbers.** The old familiar numbers (integers, rationals, irrationals) were called **real numbers.** Sums of real and imaginary numbers, numbers of the form $a + bi$, such as

$$5 + 2i, \qquad 7 - 4i, \qquad 18 + \frac{3}{2}i, \qquad \sqrt{3} - 12i$$

were called **complex numbers.**[†]

Every real number is a complex number; for instance, $7 = 7 + 0i$. Similarly, every imaginary number bi is a complex number, since $bi = 0 + bi$. Since the usual laws of arithmetic still hold, it's easy to add, subtract, and multiply complex numbers. As the following examples demonstrate, *all symbols can be treated as if they were real numbers, provided that i^2 is replaced by* -1. Unless directed otherwise, express your answers in the standard form $a + bi$.

EXAMPLE 1

(a) $(1 + i) + (3 - 7i) = 1 + i + 3 - 7i$
$$= (1 + 3) + (i - 7i) = 4 - 6i.$$

(b) $(4 + 3i) - (8 - 6i) = 4 + 3i - 8 - (-6i)$
$$= (4 - 8) + (3i + 6i) = -4 + 9i.$$

---

[*]Hereafter, whenever we write $a + bi$ or $c + di$, it is assumed that a, b, c, d are real numbers and $i^2 = -1$.

[†]This terminology is still used, even though there is nothing complicated, unreal, or imaginary about complex numbers—they are just as valid mathematically as are real numbers. See Exercise 80 for a formal construction of the complex numbers and proofs of the claims made here.

(c) $4i\left(2 + \dfrac{1}{2}i\right) = 4i \cdot 2 + 4i\left(\dfrac{1}{2}i\right) = 8i + 4 \cdot \dfrac{1}{2} \cdot i^2$

$\qquad\qquad = 8i + 2i^2 = 8i + 2(-1) = -2 + 8i.$

(d) $(2 + i)(3 - 4i) = 2 \cdot 3 + 2(-4i) + i \cdot 3 + i(-4i)$

$\qquad\qquad\qquad = 6 - 8i + 3i - 4i^2 = 6 - 8i + 3i - 4(-1)$

$\qquad\qquad\qquad = (6 + 4) + (-8i + 3i) = 10 - 5i.$ ■

The familiar multiplication patterns and exponent laws for integer exponents hold in the complex number system.

EXAMPLE 2

(a) $(3 + 2i)(3 - 2i) = 3^2 - (2i)^2$

$\qquad\qquad\qquad\quad = 9 - 4i^2 = 9 - 4(-1) = 9 + 4 = 13.$

(b) $(4 + i)^2 = 4^2 + 2 \cdot 4 \cdot i + i^2 = 16 + 8i + (-1) = 15 + 8i.$

(c) To find i^{54}, we first note that $i^4 = i^2 i^2 = (-1)(-1) = 1$ and that

$$54 = 52 + 2 = 4 \cdot 13 + 2.$$

Consequently,

$$i^{54} = i^{52+2} = i^{52} i^2 = i^{4 \cdot 13} i^2 = (i^4)^{13} i^2 = 1^{13}(-1) = -1. \quad ■$$

TECHNOLOGY TIP

To do complex arithmetic on TI-86 and HP-39, enter $a + bi$ as (a, b). On other calculators, use the special *i* key whose location is

TI-83/89: keyboard

Casio FX 2.0: keyboard

Casio: 9850: OPTN/CPLX

The **conjugate** of the complex number $a + bi$ is the number $a - bi$, and the conjugate of $a - bi$ is $a + bi$. For example, the conjugate of $3 + 4i$ is $3 - 4i$ and the conjugate of $-3i = 0 - 3i$ is $0 + 3i = 3i$. *Every real number is its own conjugate;* for instance, the conjugate of $17 = 17 + 0i$ is $17 - 0i = 17$.

For any complex number $a + bi$, we have

$$(a + bi)(a - bi) = a^2 - (bi)^2 = a^2 - b^2 i^2 = a^2 - b^2(-1) = a^2 + b^2.$$

Since a^2 and b^2 are nonnegative real numbers, so is $a^2 + b^2$. Therefore, *the product of a complex number and its conjugate is a nonnegative real number.* This fact enables us to express quotients of complex numbers in standard form.

EXAMPLE 3

To express $\dfrac{3 + 4i}{1 + 2i}$ in the form $a + bi$, *multiply both numerator and denominator by the conjugate of the denominator,* namely, $1 - 2i$:

$$\frac{3 + 4i}{1 + 2i} = \frac{3 + 4i}{1 + 2i} \cdot \frac{1 - 2i}{1 - 2i} = \frac{(3 + 4i)(1 - 2i)}{(1 + 2i)(1 - 2i)}$$

$$= \frac{3 + 4i - 6i - 8i^2}{1^2 - (2i)^2} = \frac{3 + 4i - 6i - 8(-1)}{1 - 4i^2} = \frac{11 - 2i}{1 - 4(-1)}$$

$$= \frac{11 - 2i}{5} = \frac{11}{5} - \frac{2}{5}i.$$

This is the form $a + bi$ with $a = 11/5$ and $b = -2/5$. ■

EXAMPLE 4

To express $\dfrac{1}{1-i}$ in standard form, note that the conjugate of the denominator is $1 + i$, and therefore

$$\frac{1}{1-i} = \frac{1 \cdot (1+i)}{(1-i)(1+i)} = \frac{1+i}{1^2 - i^2} = \frac{1+i}{1-(-1)} = \frac{1+i}{2} = \frac{1}{2} + \frac{1}{2}i.$$

We can check this result by multiplying $\dfrac{1}{2} + \dfrac{1}{2}i$ by $1 - i$ to see whether the product is 1 $\left(\text{which it should be if } \dfrac{1}{2} + \dfrac{1}{2}i = \dfrac{1}{1-i}\right)$:

$$\left(\frac{1}{2} + \frac{1}{2}i\right)(1-i) = \frac{1}{2} \cdot 1 - \frac{1}{2}i + \frac{1}{2}i \cdot 1 - \frac{1}{2}i^2 = \frac{1}{2} - \frac{1}{2}(-1) = 1. \quad \blacksquare$$

Since $i^2 = -1$, we define $\sqrt{-1}$ to be the complex number i. Similarly, since $(5i)^2 = 5^2 i^2 = 25(-1) = -25$, we define $\sqrt{-25}$ to be $5i$. In general,

Square Roots of Negative Numbers

> Let b be a positive real number.
>
> $$\sqrt{-b} \text{ is defined to be } \sqrt{b}\,i$$
>
> because $(\sqrt{b}\,i)^2 = (\sqrt{b})^2 i^2 = b(-1) = -b.$

CAUTION

$\sqrt{b}\,i$ is *not* the same as $\sqrt{bi}$. To avoid confusion it may help to write $\sqrt{b}\,i$ as $i\sqrt{b}$.

EXAMPLE 5

(a) $\sqrt{-3} = \sqrt{3}\,i = i\sqrt{3}.$

(b) $\dfrac{1 - \sqrt{-7}}{3} = \dfrac{1 - \sqrt{7}\,i}{3} = \dfrac{1}{3} - \dfrac{\sqrt{7}}{3}i. \quad \blacksquare$

CAUTION

The property $\sqrt{cd} = \sqrt{c}\,\sqrt{d}$ (or equivalently in exponential notation, $(cd)^{1/2} = c^{1/2}d^{1/2}$), which is valid for positive real numbers, does *not hold* when both c and d are negative.

$$\sqrt{-20}\,\sqrt{-5} = \sqrt{20}i \cdot \sqrt{5}i = \sqrt{20}\,\sqrt{5} \cdot i^2 = \sqrt{20 \cdot 5}(-1)$$

$$= \sqrt{100}(-1) = -10.$$

But $\sqrt{(-20)(-5)} = \sqrt{100} = 10$, so that

$$\sqrt{(-20)(-5)} \neq \sqrt{-20}\,\sqrt{-5}.$$

To avoid difficulty, *always write square roots of negative numbers in terms of i before doing any simplification.*

EXAMPLE 6

$$(7 - \sqrt{-4})(5 + \sqrt{-9}) = (7 - \sqrt{4}i)(5 + \sqrt{9}i)$$
$$= (7 - 2i)(5 + 3i)$$
$$= 35 + 21i - 10i - 6i^2$$
$$= 35 + 11i - 6(-1) = 41 + 11i. \quad \blacksquare$$

Since every negative real number has a square root in the complex number system, we can now find complex solutions for equations that have no real solutions. For example, the solutions of $x^2 = -25$ are $x = \pm\sqrt{-25} = \pm 5i$. In fact,

Every quadratic equation with real coefficients has solutions in the complex number system.

EXAMPLE 7

To solve the equation $2x^2 + x + 3 = 0$, we apply the quadratic formula.

$$x = \frac{-1 \pm \sqrt{1^2 - 4 \cdot 2 \cdot 3}}{2 \cdot 2} = \frac{-1 \pm \sqrt{-23}}{4}.$$

Since $\sqrt{-23}$ is not a real number, this equation has no real number solutions. But $\sqrt{-23}$ *is* a complex number, namely, $\sqrt{-23} = \sqrt{23}i$. Thus, the equation does have solutions in the complex number system.

$$x = \frac{-1 \pm \sqrt{-23}}{4} = \frac{-1 \pm \sqrt{23}i}{4} = -\frac{1}{4} \pm \frac{\sqrt{23}}{4}i.$$

Note that the two solutions, $-\dfrac{1}{4} + \dfrac{\sqrt{23}}{4}i$ and $-\dfrac{1}{4} - \dfrac{\sqrt{23}}{4}i$, are conjugates of each other. $\quad \blacksquare$

EXAMPLE 8

To find *all* solutions of $x^3 = 1$, we rewrite the equation and use the Difference of Cubes pattern (see the Algebra Review Appendix) to factor:

$$x^3 = 1$$
$$x^3 - 1 = 0$$
$$(x - 1)(x^2 + x + 1) = 0$$
$$x - 1 = 0 \quad \text{or} \quad x^2 + x + 1 = 0.$$

The solution of the first equation is $x = 1$. The solutions of the second can be obtained from the quadratic formula.

$$x = \frac{-1 \pm \sqrt{1^2 - 4 \cdot 1 \cdot 1}}{2 \cdot 1} = \frac{-1 \pm \sqrt{-3}}{2} = \frac{-1 \pm \sqrt{3}i}{2} = -\frac{1}{2} \pm \frac{\sqrt{3}}{2}i.$$

Therefore, the equation $x^3 = 1$ has one real solution ($x = 1$) and two nonreal complex solutions [$x = -1/2 + (\sqrt{3}/2)i$ and $x = -1/2 - (\sqrt{3}/2)i$]. Each of these solutions is said to be a **cube root of 1** or a **cube root of unity.** Observe that the two nonreal complex cube roots of unity are conjugates of each other. ∎

The preceding examples illustrate this useful fact (whose proof is discussed in Section 4.8).

Conjugate
Solutions

If $a + bi$ is a solution of a polynomial equation with *real* coefficients, then its conjugate $a - bi$ is also a solution of this equation.

✓ EXERCISES 4.7

In Exercises 1–54, perform the indicated operation and write the result in the form $a + bi$.

1. $(2 + 3i) + (6 - i)$ **2.** $(-5 + 7i) + (14 + 3i)$

3. $(2 - 8i) - (4 + 2i)$ **4.** $(3 + 5i) - (3 - 7i)$

5. $\dfrac{5}{4} - \left(\dfrac{7}{4} + 2i\right)$ **6.** $(\sqrt{3} + i) + (\sqrt{5} - 2i)$

7. $\left(\dfrac{\sqrt{2}}{2} + i\right) - \left(\dfrac{\sqrt{3}}{2} - i\right)$

8. $\left(\dfrac{1}{2} + \dfrac{\sqrt{3}i}{2}\right) + \left(\dfrac{3}{4} - \dfrac{5\sqrt{3}i}{2}\right)$

9. $(2 + i)(3 + 5i)$ **10.** $(2 - i)(5 + 2i)$

11. $(-3 + 2i)(4 - i)$ **12.** $(4 + 3i)(4 - 3i)$

13. $(2 - 5i)^2$ **14.** $(1 + i)(2 - i)i$

15. $(\sqrt{3} + i)(\sqrt{3} - i)$ **16.** $\left(\dfrac{1}{2} - i\right)\left(\dfrac{1}{4} + 2i\right)$

17. i^{15} **18.** i^{26} **19.** i^{33} **20.** $(-i)^{53}$

21. $(-i)^{107}$ **22.** $(-i)^{213}$ **23.** $\dfrac{1}{5 - 2i}$ **24.** $\dfrac{1}{i}$

25. $\dfrac{1}{3i}$ **26.** $\dfrac{i}{2 + i}$ **27.** $\dfrac{3}{4 + 5i}$ **28.** $\dfrac{2 + 3i}{i}$

29. $\dfrac{1}{i(4 + 5i)}$ **30.** $\dfrac{1}{(2 - i)(2 + i)}$ **31.** $\dfrac{2 + 3i}{i(4 + i)}$

32. $\dfrac{2}{(2 + 3i)(4 + i)}$ **33.** $\dfrac{2 + i}{1 - i} + \dfrac{1}{1 + 2i}$

34. $\dfrac{1}{2 - i} + \dfrac{3 + i}{2 + 3i}$ **35.** $\dfrac{i}{3 + i} - \dfrac{3 + i}{4 + i}$

36. $6 + \dfrac{2i}{3 + i}$ **37.** $\sqrt{-36}$ **38.** $\sqrt{-81}$

39. $\sqrt{-14}$ **40.** $\sqrt{-50}$ **41.** $-\sqrt{-16}$

42. $-\sqrt{-12}$ **43.** $\sqrt{-16} + \sqrt{-49}$

44. $\sqrt{-25} - \sqrt{-9}$ **45.** $\sqrt{-15} - \sqrt{-18}$

46. $\sqrt{-12} \sqrt{-3}$ **47.** $\sqrt{-16}/\sqrt{-36}$

48. $-\sqrt{-64}/\sqrt{-4}$ **49.** $(\sqrt{-25} + 2)(\sqrt{-49} - 3)$

50. $(5 - \sqrt{-3})(-1 + \sqrt{-9})$

51. $(2 + \sqrt{-5})(1 - \sqrt{-10})$

52. $\sqrt{-3}(3 - \sqrt{-27})$ **53.** $1/(1 + \sqrt{-2})$

54. $(1 + \sqrt{-4})(3 - \sqrt{-9})$

In Exercises 55–58, find x and y. Remember that

$$a + bi = c + di$$

exactly when $a = c$ and $b = d$.

55. $3x - 4i = 6 + 2yi$ **56.** $8 - 2yi = 4x + 12i$

57. $3 + 4xi = 2y - 3i$ **58.** $8 - xi = \dfrac{1}{2}y + 2i$

In Exercises 59–70, solve the equation and express each solution in the form $a + bi$.

59. $3x^2 - 2x + 5 = 0$ **60.** $5x^2 + 2x + 1 = 0$

61. $x^2 + x + 2 = 0$ **62.** $5x^2 - 6x + 2 = 0$

63. $2x^2 - x = -4$ **64.** $x^2 + 1 = 4x$

65. $2x^2 + 3 = 6x$ **66.** $3x^2 + 4 = -5x$

67. $x^3 - 8 = 0$ **68.** $x^3 + 125 = 0$

69. $x^4 - 1 = 0$ **70.** $x^4 - 81 = 0$

71. Simplify: $i + i^2 + i^3 + \cdots + i^{15}$

72. Simplify: $i - i^2 + i^3 - i^4 + i^5 - \cdots + i^{15}$

Thinkers

If $z = a + bi$ is a complex number, then its conjugate is usually denoted $\bar{z}$, that is, $\bar{z} = a - bi$. In Exercises 73–77, prove that for any complex numbers $z = a + bi$ and $w = c + di$:

73. $\overline{z + w} = \bar{z} + \bar{w}$

74. $\overline{zw} = \bar{z} \cdot \bar{w}$

75. $\overline{\left(\dfrac{z}{w}\right)} = \dfrac{\bar{z}}{\bar{w}}$

76. $\bar{\bar{z}} = z$

77. z is a real number exactly when $\bar{z} = z$.

78. The **real part** of the complex number $a + bi$ is defined to be the real number a. The **imaginary part** of $a + bi$ is defined to the real number b (*not bi*).

  (a) Show that the real part of $z = a + bi$ is $\dfrac{z + \bar{z}}{2}$.

  (b) Show that the imaginary part of $z = a + bi$ is $\dfrac{z - \bar{z}}{2i}$.

79. If $z = a + bi$ (with a, b real numbers, not both 0), express $1/z$ in standard form.

80. Construction of the Complex Numbers. We assume that the real number system is known. To construct a new number system with the desired properties, we must do the following:

  (i) Define a set C (whose elements will be called complex numbers).

  (ii) Ensure that the set C contains the real numbers or at least a copy of them.

  (iii) Define addition and multiplication in the set C in such a way that the usual laws of arithmetic are valid.

  (iv) Show that C has the other properties listed in the box on page 306.

We begin by defining C to be the set of all ordered pairs of real numbers. Thus, $(1, 5)$, $(-6, 0)$, $(4/3, -17)$, and $(\sqrt{2}, 12/5)$ are some of the elements of the set C. More generally, a complex number (= element of C) is any pair (a, b), where a and b are real numbers. By definition, two complex numbers are *equal* exactly when they have the same first and the same second coordinate.

(a) *Addition in C is defined by this rule:*

$$(a, b) + (c, d) = (a + c, b + d)$$

For example,

$$(3, 2) + (5, 4) = (3 + 5, 2 + 4) = (8, 6).$$

Verify that this addition has the following properties. For any complex numbers (a, b), (c, d), (e, f) in C:

  (i) $(a, b) + (c, d) = (c, d) + (a, b)$

  (ii) $[(a, b) + (c, d)] + (e, f) =$
  $$(a, b) + [(c, d) + (e, f)]$$

  (iii) $(a, b) + (0, 0) = (a, b)$

  (iv) $(a, b) + (-a, -b) = (0, 0)$

(b) *Multiplication in C is defined by this rule:*

$$(a, b)(c, d) = (ac - bd, bc + ad)$$

For example,

$$(3, 2)(4, 5) = (3 \cdot 4 - 2 \cdot 5, 2 \cdot 4 + 3 \cdot 5)$$
$$= (12 - 10, 8 + 15) = (2, 23).$$

Verify that this multiplication has the following properties. For any complex numbers (a, b), (c, d), (e, f) in C:

  (i) $(a, b)(c, d) = (c, d)(a, b)$

  (ii) $[(a, b)(c, d)](e, f) = (a, b)[(c, d)(e, f)]$

  (iii) $(a, b)(1, 0) = (a, b)$

  (iv) $(a, b)(0, 0) = (0, 0)$

(c) Verify that for any two elements of C with second coordinate zero:

  (i) $(a, 0) + (c, 0) = (a + c, 0)$

  (ii) $(a, 0)(c, 0) = (ac, 0)$

Identify $(t, 0)$ with the real number t. Statements (i) and (ii) show that when addition or multiplication in C is performed on two real numbers (that is, elements of C with second coordinate 0), the result is the usual sum or product of real numbers. Thus, C contains (a copy of) the real number system.

(d) *New Notation.* Since we are identifying the complex number $(a, 0)$ with the real number a, we shall hereafter denote $(a, 0)$ simply by the symbol a. Also, let i denote the complex number $(0, 1)$.

  (i) Show that $i^2 = -1$, that is,

$$(0, 1)(0, 1) = (-1, 0).$$

  (ii) Show that for any complex number $(0, b)$, $(0, b) = bi$ [that is, $(0, b) = (b, 0)(0, 1)$].

  (iii) Show that any complex number (a, b) can be written: $(a, b) = a + bi$, that is,

$$(a, b) = (a, 0) + (b, 0)(0, 1).$$

In this new notation, every complex number is of the form $a + bi$ with a, b real and $i^2 = -1$, and our construction is finished.

4.8   Theory of Equations*

The complex numbers were constructed to obtain a solution for the equation $x^2 = -1$, that is, a root of the polynomial $x^2 + 1$. In Section 4.7, we saw that *every* quadratic polynomial with real coefficients has roots in the complex number system. A natural question now arises:

> Do we have to enlarge the complex number system (perhaps many times) to find roots for higher degree polynomials?

In this section, we shall see that the somewhat surprising answer is no.

To give the full answer, we shall consider not just polynomials with real coefficients, but also those with complex coefficients, such as

$$x^3 - ix^2 + (4 - 3i)x + 1 \quad \text{or} \quad (-3 + 2i)x^6 - 3x + (5 - 4i).$$

The discussion of polynomial division in Section 4.2 can easily be extended to include polynomials with complex coefficients. In fact, *all of the results in Section 4.2 are valid for polynomials with complex coefficients.* For example, you can check that i is a root of $f(x) = x^2 + (i - 1)x + (2 + i)$ and that $x - i$ is a factor of $f(x)$:

$$f(x) = x^2 + (i - 1)x + (2 + i) = (x - i)[x - (1 - 2i)].$$

Since every real number is also a complex number, polynomials with real coefficients are just special cases of polynomials with complex coefficients. So in the rest of this section, "polynomial" means "polynomial with complex (possibly real) coefficients" unless specified otherwise. We can now answer the question posed in the first paragraph.

Fundamental Theorem of Algebra

> Every nonconstant polynomial has a root in the complex number system.

Although this is obviously a powerful result, neither the Fundamental Theorem nor its proof provides a practical method for *finding* a root of a given polynomial.[†] The proof of the Fundamental Theorem is beyond the scope of this book, but we shall explore some of the useful implications of the theorem, such as this one.

Factorization over the Complex Numbers

> Let $f(x)$ be a polynomial of degree $n > 0$ with leading coefficient d. Then there are (not necessarily distinct) complex numbers $c_1, c_2, \ldots, c_n$ such that
> $$f(x) = d(x - c_1)(x - c_2)(x - c_3) \cdots (x - c_n).$$
> Furthermore, $c_1, c_2, \ldots, c_n$ are the only roots of $f(x)$.

---

*Section 4.7 is a prerequisite for this section.
[†]It might seem strange that you can prove that a root exists without actually exhibiting one. But such "existence theorems" are quite common. A rough analogy is the situation that occurs when someone is killed by a sniper's bullet. The police know that there is a killer, but *finding* the killer may be impossible.

Proof By the Fundamental Theorem, $f(x)$ has a complex root c_1. The Factor Theorem shows that $x - c_1$ must be a factor of $f(x)$, say,

$$f(x) = (x - c_1)g(x),$$

where $g(x)$ has degree $n - 1$.[†] If $g(x)$ is nonconstant, then it has a complex root c_2 by the Fundamental Theorem. Hence, $x - c_2$ is a factor of $g(x)$, so

$$f(x) = (x - c_1)(x - c_2)h(x)$$

for some $h(x)$ of degree $n - 2$ [1 less than the degree of $g(x)$]. If $h(x)$ is nonconstant, then it has a complex root c_3, and the argument can be repeated. Continuing in this way, with the degree of the last factor going down by 1 at each step, we reach a factorization in which the last factor is a constant (degree 0 polynomial):

(∗) $$f(x) = (x - c_1)(x - c_2)(x - c_3) \cdots (x - c_n)d.$$

If the right side were multiplied out, it would look like

$$dx^n + \text{lower degree terms.}$$

So the constant factor d is the leading coefficient of $f(x)$.

It is easy to see from the factored form (∗) that the numbers $c_1, c_2, \ldots, c_n$ are roots of $f(x)$. If k is *any* root of $f(x)$, then

$$0 = f(k) = d(k - c_1)(k - c_2)(k - c_3) \cdots (k - c_n).$$

The product on the right is 0 only when one of the factors is 0. Since the leading coefficient d is nonzero, we must have

$$k - c_1 = 0 \quad \text{or} \quad k - c_2 = 0 \quad \text{or} \quad \cdots \quad k - c_n = 0$$

$$k = c_1 \quad \text{or} \quad k = c_2 \quad \text{or} \quad \cdots \quad k = c_n.$$

Therefore, k is one of the c's, and $c_1, \ldots, c_n$ are the only roots of $f(x)$. This completes the proof. ■

Since the n roots $c_1, \ldots, c_n$ of $f(x)$ might not all be distinct, we see that

TECHNOLOGY TIP

To find all roots of a polynomial, see the second Tip on page 309.

To factor a polynomial as a product of linear factors on TI-89, try

ALGEBRA/COMPLEX/cFACTOR.

In the Casio FX 2.0 main menu, try

CAS/TRANS/rFACTOR.

The Answer Type should be set to "complex" in the Casio SET-UP menu.

Number of Roots

Every polynomial of degree $n > 0$ has at most n different roots in the complex number system.

Suppose $f(x)$ has repeated roots, meaning that some of the $c_1, c_2, \ldots, c_n$ are the same in factorization (∗). Recall that a root c is said to have multiplicity k if $(x - c)^k$ is a factor of $f(x)$ but no higher power of $(x - c)$ is a factor. Consequently, if every root is counted as many times as its multiplicity, then the statement in the preceding box implies that

A polynomial of degree n has exactly n roots.

[†]The degree of $g(x)$ is 1 less than the degree n of $f(x)$ because $f(x)$ is the product of $g(x)$ and $x - c_1$ (which has degree 1).

EXAMPLE 1

Find a polynomial $f(x)$ of degree 5 such that 1, -2, and 5 are roots, 1 is a root of multiplicity 3 and $f(2) = -24$.

SOLUTION   Since 1 is a root of multiplicity 3, $(x - 1)^3$ must be a factor of $f(x)$. There are at least two other factors corresponding to the roots -2 and 5: $x - (-2) = x + 2$ and $x - 5$. The product of these factors $(x - 1)^3(x + 2)(x - 5)$ has degree 5, as does $f(x)$, so $f(x)$ must look like this:

$$f(x) = d(x - 1)^3(x + 2)(x - 5),$$

where d is the leading coefficient. Since $f(2) = -24$, we have

$$d(2 - 1)^3(2 + 2)(2 - 5) = f(2) = -24,$$

which reduces to $-12d = -24$. Therefore, $d = (-24)/(-12) = 2$, and

$$f(x) = 2(x - 1)^3(x + 2)(x - 5)$$
$$= 2x^5 - 12x^4 + 4x^3 + 40x^2 - 54x + 20. \quad \blacksquare$$

▪▪ POLYNOMIALS WITH REAL COEFFICIENTS

Recall that the **conjugate** of the complex number $a + bi$ is the number $a - bi$ (see page 307). We usually write a complex number as a single letter, say z, and indicate its conjugate by $\bar{z}$ (sometimes read "z bar"). For instance, if $z = 3 + 7i$, then $\bar{z} = 3 - 7i$. Conjugates play a role whenever a quadratic polynomial with real coefficients has complex roots.

EXAMPLE 2

The quadratic formula shows that $x^2 - 6x + 13$ has two complex roots.

$$\frac{-(-6) \pm \sqrt{(-6)^2 - 4 \cdot 1 \cdot 13}}{2 \cdot 1} = \frac{6 \pm \sqrt{-16}}{2} = \frac{6 \pm 4i}{2} = 3 \pm 2i.$$

The complex roots are $z = 3 + 2i$ and its conjugate $\bar{z} = 3 - 2i$. $\blacksquare$

The preceding example is a special case of a more general theorem, whose proof is outlined in Exercises 59 and 60.

Conjugate Roots
Theorem

Let $f(x)$ be a polynomial with *real* coefficients. If the complex number z is a root of $f(x)$, then its conjugate $\bar{z}$ is also a root of $f(x)$.

EXAMPLE 3

Find a polynomial with real coefficients whose roots include the numbers 2 and $3 + i$.

SOLUTION   Since $3 + i$ is a root, its conjugate $3 - i$ must also be a root. Consider the polynomial

$$f(x) = (x - 2)[(x - (3 + i)][x - (3 - i)].$$

Obviously, $2, 3 + i$. and $3 - i$ are roots of $f(x)$. Multiplying out this factored form shows that $f(x)$ *does* have real coefficients.

$$
\begin{aligned}
f(x) &= (x - 2)[x^2 - (3 - i)x - (3 + i)x + (3 + i)(3 - i)] \\
&= (x - 2)(x^2 - 3x + ix - 3x - ix + 9 - i^2) \\
&= (x - 2)(x^2 - 6x + 10) \\
&= x^3 - 8x^2 + 22x - 20.
\end{aligned}
$$

The next-to-last line of this calculation also shows that $f(x)$ can be factored as a product of a linear and a quadratic polynomial, each with *real* coefficients.   ■

The technique in Example 3 works because the polynomial

$$[x - (3 + i)][x - (3 - i)]$$

turns out to have real coefficients. The proof of the following result shows why this must always be the case.

Factorization over the Real Numbers

> Every nonconstant polynomial with real coefficients can be factored as a product of linear and quadratic polynomials with real coefficients in such a way that the quadratic factors, if any, have no real roots.

Proof   The second box on page 312 shows that

$$f(x) = d(x - c_1)(x - c_2) \cdots (x - c_n),$$

where $c_1. \ldots, c_n$ are the roots of $f(x)$. If some c_i is a real number, then the factor $x - c_i$ is a linear polynomial with real coefficients.[*] If some c_j is a nonreal complex root, then its conjugate must also be a root. Thus some c_k is the conjugate of c_j, say, $c_j = a + bi$ (with a, b real) and $c_k = a - bi$.[†] In this case,

$$
\begin{aligned}
(x - c_j)(x - c_k) &= [x - (a + bi)][x - (a - bi)] \\
&= x^2 - (a - bi)x - (a + bi)x + (a + bi)(a - bi) \\
&= x^2 - ax + bix - ax - bix + a^2 - (bi)^2 \\
&= x^2 - 2ax + (a^2 + b^2).
\end{aligned}
$$

---

[*]In Example 3, for instance, 2 is a real root and $x - 2$ a linear factor.
[†]In Example 3, for instance, $c_j = 3 + i$ and $c_k = 3 - i$ are conjugate roots.

Therefore, the factor $(x - c_j)(x - c_k)$ of $f(x)$ is a quadratic with real coefficients (because a and b are real numbers). Its roots (c_j and c_k) are nonreal. By taking the real roots of $f(x)$ one at a time and the nonreal ones in conjugate pairs in this fashion, we obtain the desired factorization of $f(x)$.   ■

EXAMPLE 4

Given that $1 + i$ is a root of $f(x) = x^4 - 2x^3 - x^2 + 6x - 6$, factor $f(x)$ completely over the real numbers.

SOLUTION   Since $1 + i$ is a root of $f(x)$, so is its conjugate $1 - i$, and hence $f(x)$ has this quadratic factor:

$$[x - (1 + i)][x - (1 - i)] = x^2 - 2x + 2.$$

Dividing $f(x)$ by $x^2 - 2x + 2$ shows that the other factor is $x^2 - 3$, which factors as $(x + \sqrt{3})(x - \sqrt{3})$. Therefore,

$$f(x) = (x + \sqrt{3})(x - \sqrt{3})(x^2 - 2x + 2). \quad ■$$

✔ EXERCISES 4.8

In Exercises 1–6, find the remainder when $f(x)$ is divided by $g(x)$ without using synthetic or long division.

1. $f(x) = x^{10} + x^8$; $g(x) = x - 1$

2. $f(x) = x^6 - 10$; $g(x) = x - 2$

3. $f(x) = 3x^4 - 6x^3 + 2x - 1$; $g(x) = x + 1$

4. $f(x) = x^5 - 3x^2 + 2x - 1$; $g(x) = x - 2$

5. $f(x) = x^3 - 2x^2 + 5x - 4$; $g(x) = x + 2$

6. $f(x) = 10x^{75} - 8x^{65} + 6x^{45} + 4x^{32} - 2x^{15} + 5$;
   $g(x) = x - 1$

In Exercises 7–10, list the roots of the polynomial and state the multiplicity of each root.

7. $f(x) = x^{54}\left(x + \dfrac{4}{5}\right)$

8. $g(x) = 3\left(x + \dfrac{1}{6}\right)\left(x - \dfrac{1}{5}\right)\left(x + \dfrac{1}{4}\right)$

9. $h(x) = 2x^{15}(x - \pi)^{14}[x - (\pi + 1)]^{13}$

10. $k(x) = (x - \sqrt{7})^7(x - \sqrt{5})^5(2x - 1)$

In Exercises 11–22, find all the roots of $f(x)$ in the complex number system; then write $f(x)$ as a product of linear factors.

11. $f(x) = x^2 - 2x + 5$ **12.** $f(x) = x^2 - 4x + 13$

13. $f(x) = 3x^2 + 2x + 7$ **14.** $f(x) = 3x^2 - 5x + 2$

15. $f(x) = x^3 - 27$ [*Hint:* Factor first.]

16. $f(x) = x^3 + 125$ **17.** $f(x) = x^3 + 8$

18. $f(x) = x^6 - 64$ [*Hint:* Let $u = x^3$ and factor $u^2 - 64$ first.]

19. $f(x) = x^4 - 1$ **20.** $f(x) = x^4 - x^2 - 6$

21. $f(x) = x^4 - 3x^2 - 10$ **22.** $f(x) = 2x^4 - 7x^2 - 4$

In Exercises 23–44, find a polynomial $f(x)$ with real coefficients that satisfies the given conditions. Some of these problems have many correct answers.

23. Degree 3; only roots are 1, 7, −4.

24. Degree 3; only roots are 1 and −1.

25. Degree 6; only roots are 1, 2, π.

26. Degree 5; only root is 2.

27. Degree 3; roots $-3, 0, 4$; $f(5) = 80$.

28. Degree 3; roots $-1, 1/2, 2$; $f(0) = 2$.

29. Roots include $2 + i$ and $2 - i$.

30. Roots include $1 + 3i$ and $1 - 3i$.

31. Roots include 2 and $2 + i$.

32. Roots include 3 and $4i - 1$.

33. Roots include $-3, 1 - i, 1 + 2i$.

34. Roots include $1, 2 + i, 3i, - 1$.

35. Degree 2; roots $1 + 2i$ and $1 - 2i$.

36. Degree 4; roots $3i$ and $-3i$, each of multiplicity 2.

37. Degree 4; only roots are $4, 3 + i$, and $3 - i$.

38. Degree 5; roots 2 (of multiplicity 3), i, and $-i$.

39. Degree 6; roots 0 (of multiplicity 3) and $3, 1 + i, 1 - i$, each of multiplicity 1.

40. Degree 6; roots include i (of multiplicity 2) and 3.

41. Degree 2; roots include $1 + i$; $f(0) = 6$.

42. Degree 2; roots include $3 + i$; $f(2) = 3$.

43. Degree 3; roots include i and 1; $f(-1) = 8$.

44. Degree 3; roots include $2 + 3i$ and -2; $f(2) = -3$.

In Exercises 45–48, find a polynomial with complex coefficients that satisfies the given conditions.

45. Degree 2; roots i and $1 - 2i$.

46. Degree 2; roots $2i$ and $1 + i$.

47. Degree 3; roots $3, i$, and $2 - i$.

48. Degree 4; roots $\sqrt{2}, -\sqrt{2}, 1 + i$, and $1 - i$.

In Exercises 49–56, one root of the polynomial is given; find all the roots.

49. $x^3 - 2x^2 - 2x - 3$; root 3.

50. $x^3 + x^2 + x + 1$; root i.

51. $x^4 + 3x^3 + 3x^2 + 3x + 2$; root i.

52. $x^4 - x^3 - 5x^2 - x - 6$; root i.

53. $x^4 - 2x^3 + 5x^2 - 8x + 4$; root 1 of multiplicity 2.

54. $x^4 - 6x^3 + 29x^2 - 76x + 68$; root 2 of multiplicity 2.

55. $x^4 - 4x^3 + 6x^2 - 4x + 5$; root $2 - i$.

56. $x^4 - 5x^3 + 10x^2 - 20x + 24$; root $2i$.

57. Let $z = a + bi$ and $w = c + di$ be complex numbers (a, b, c, d are real numbers). Prove the given equality by computing each side and comparing the results.

    (a) $\overline{z + w} = \bar{z} + \bar{w}$ (The left side says: First find $z + w$ and then take the conjugate. The right side says: First take the conjugates of z and w and then add.)

    (b) $\overline{z \cdot w} = \bar{z} \cdot \bar{w}$

58. Let $g(x)$ and $h(x)$ be polynomials of degree n and assume that there are $n + 1$ numbers $c_1, c_2, \ldots, c_n, c_{n+1}$ such that

    $$g(c_i) = h(c_i) \quad \text{for every } i.$$

    Prove that $g(x) = h(x)$. [*Hint:* Show that each c_i is a root of $f(x) = g(x) - h(x)$. If $f(x)$ is nonzero, what is its largest possible degree? To avoid a contradiction, conclude that $f(x) = 0$.]

59. Suppose $f(x) = ax^3 + bx^2 + cx + d$ has real coefficients and z is a complex root of $f(x)$.

    (a) Use Exercise 57 and the fact that $\bar{r} = r$, when r is a real number, to show that

    $$\overline{f(x)} = \overline{az^3 + bz^2 + cz + d}$$
    $$= a\bar{z}^3 + b\bar{z}^2 + c\bar{z} + d = f(\bar{z}).$$

    (b) Conclude that $\bar{z}$ is also a root of $f(x)$. [*Note:* $f(\bar{z}) = \overline{f(z)} = \bar{0} = 0$.]

60. Let $f(x)$ be a polynomial with real coefficients and z a complex root of $f(x)$. Prove that the conjugate $\bar{z}$ is also a root of $f(x)$. [*Hint:* Exercise 59 is the case when $f(x)$ has degree 3; the proof in the general case is similar.]

61. Use the statement in the box on page 315 to show that every polynomial with real coefficients and *odd* degree must have at least one real root.

62. Give an example of a polynomial $f(x)$ with complex, non-real coefficients and a complex number z such that z is a root of $f(x)$, but its conjugate is not. Hence, the conclusion of the Conjugate Roots Theorem (page 314) may be *false* if $f(x)$ doesn't have real coefficients.

Chapter 4 Review

IMPORTANT CONCEPTS

Section 4.1

Quadratic function   225
Parabola   225
Vertex   226
Applications of quadratic
   functions   230

Section 4.2

Polynomial   235
Coefficient   235
Constant term   235
Zero polynomial   235
Degree of a polynomial   236
Leading coefficient   236
Polynomial function   236
Division of polynomials   236
Division Algorithm   237
Remainder Theorem   239
Root or zero of a
   polynomial   240
Real root   240
Factor Theorem   240
Number of roots of a
   polynomial   241

Special Topics 4.2.A

Synthetic division   243

Section 4.3

The Rational Root Test   247

Factoring polynomials   249
Bounds Test   251

Section 4.4

Graph of $y = ax^n$ 255
Continuity   255
Behavior when $|x|$ is large   256
x-Intercepts   257
Multiplicity   257
Local extrema   258
Points of inflection   258

Special Topics 4.4.A

Quadratic, cubic, and quartic
   regression   267

Section 4.5

Rational function   272
Domain   272
The Big-Little Principle   272
Linear rational functions   276
Intercepts   278
Vertical asymptotes   278
Holes   279
Horizontal asymptotes   280

Special Topics 4.5.A

Nonvertical asymptotes   288–290

Section 4.6

Basic principles for solving
   inequalities   292
Linear inequalities   292
Polynomial inequalities   293
Rational inequalities   297

Special Topics 4.6.A

Absolute value inequalities   303

Section 4.7

Complex number   306
Imaginary number   306
Real numbers   306
Conjugate   307
Square roots of negative
   numbers   308

Section 4.8

The Fundamental Theorem of
   Algebra   312
Factorization over the complex
   numbers   312
Multiplicity and number of
   roots   313
Conjugate Roots Theorem   314
Factorization over the real
   numbers   315

IMPORTANT FACTS & FORMULAS

■ The graph of $f(x) = ax^2 + bx + c$ is a parabola whose vertex has x-coordinate $-b/2a$.

CATALOG OF BASIC FUNCTIONS—PART 2

REVIEW QUESTIONS

In Questions 1–5, find the vertex of the graph of the quadratic function.

1. $f(x) = (x - 2)^2 + 3$

2. $f(x) = 2(x + 1)^2 - 1$

3. $f(x) = x^2 - 8x + 12$

4. $f(x) = x^2 - 7x + 6$

5. $f(x) = 3x^2 - 9x + 1$

6. Which of the following statements about the functions

$$f(x) = 3x^2 + 2 \quad \text{and} \quad g(x) = -3x^2 + 2$$

is *false*?

(a) The graphs of f and g are parabolas.

(b) The graphs of f and g have the same vertex.

(c) The graphs of f and g open in opposite directions.

(d) The graph of f is the graph of $y = 3x^2$ shifted 2 units to the right.

7. A preschool wants to construct a fenced playground. The fence will be attached to the building at two corners, as shown in the figure. There is 400 feet of fencing available, all of it to be used.
   (a) Write an equation in x and y that gives the amount of fencing to be used. Solve the equation for y.
   (b) Write the area of the playground as a function of x. [Part (a) may be helpful.]
   (c) What are the dimensions of the playground with the largest possible area?

8. A model rocket is launched straight up from a platform at time $t = 0$ (where t is time measured in seconds). The altitude $h(t)$ of the rocket above the ground at given time (t) is given by

$$h(t) = 10 + 112t - 16t^2$$

   (where $h(t)$ is measured in feet).
   (a) What is the altitude of the rocket the instant it is launched?
   (b) What is the altitude of the rocket 2 seconds after launching?
   (c) What is the maximum altitude attained by the rocket?
   (d) At what time does the rocket return to the altitude at which it was launched?

9. A rectangular garden next to a building is to be fenced with 120 feet of fencing. The side against the building will not be fenced. What should the lengths of the other three sides be to ensure the largest possible area?

10. A factory offers 100 calculators to a retailer at a price of $20 each. The price per calculator on the entire order will be reduced 5¢ for each additional calculator over 100. What number of calculators will produce the largest possible sales revenue for the factory?

11. Which of the following are polynomials?
   (a) $2^3 + x^2$
   (b) $x + \dfrac{1}{x}$

   (c) $x^3 - \dfrac{1}{\sqrt{2}}$
   (d) $\sqrt[3]{x^4}$
   (e) $\pi^3 - x$
   (f) $\sqrt{2} + 2x^2$
   (g) $\sqrt{x} + 2x^2$
   (h) $|x|$

12. What is the remainder when $x^4 + 3x^3 + 1$ is divided by $x^2 + 1$?

13. What is the remainder when

$$x^{112} - 2x^8 + 9x^5 - 4x^4 + x - 5$$

   is divided by $x - 1$?

14. Is $x - 1$ a factor of $f(x) = 14x^{87} - 65x^{56} + 51$? Justify your answer.

15. Use synthetic division to show that $x - 2$ is a factor of $x^6 - 5x^5 + 8x^4 + x^3 - 17x^2 + 16x - 4$ and find the other factor.

16. List the roots of this polynomial and the multiplicity of each root:

$$f(x) = 5(x - 4)^3(x - 2)(x + 17)^3(x^2 - 4).$$

17. Find a polynomial f of degree 3 such that $f(-1) = 0$, $f(1) = 0$, and $f(0) = 5$.

18. Find the root(s) of $2\left(\dfrac{x}{5} + 7\right) - 3x - \dfrac{x + 2}{5} + 4$.

19. Find the roots of $3x^2 - 2x - 5$.

20. Factor the polynomial $x^3 - 8x^2 + 9x + 6$. [*Hint:* 2 is a root.]

21. Find all real roots of $x^6 - 4x^3 + 4$.

22. Find all real roots of $9x^3 - 6x^2 - 35x + 26$. [*Hint:* Try $x = -2$].

23. Find all real roots of $3y^3(y^4 - y^2 - 5)$.

24. Find the rational roots of $x^4 - 2x^3 - 4x^2 + 1$.

25. Consider the polynomial $2x^3 - 8x^2 + 5x + 3$.
   (a) List the only *possible* rational roots.
   (b) Find one rational root.
   (c) Find all the roots of the polynomial.

26. (a) Find all rational roots of $x^3 + 2x^2 - 2x - 2$.
   (b) Find two consecutive integers such that an irrational root of $x^3 + 2x^2 - 2x - 2$ lies between them.

27. How many distinct real roots does $x^3 + 4x$ have?

28. How many distinct real roots does $x^3 - 6x^2 + 11x - 6$ have?

29. Find the roots of $x^4 - 11x^2 + 18$.

30. The polynomial $x^3 - 2x + 1$ has
   (a) no real roots.
   (b) only one real root.
   (c) three rational roots.
   (d) only one rational root.
   (e) none of the above.

31. Show that 5 is an upper bound for the real roots of $x^4 - 4x^3 + 16x - 16$.

32. Show that -1 is a lower bound for the real roots of $x^4 - 4x^3 + 15$.

In Questions 33 and 34, find the real roots of the polynomial.

33. $x^6 - 2x^5 - x^4 + 3x^3 - x^2 - x + 1$

34. $x^5 - 3x^4 - 2x^3 - x^2 - 23x - 20$

In Questions 35 and 36, compute and simplify the difference quotient of the function.

35. $f(x) = x^2 + x$ **36.** $g(x) = x^3 - x + 1$

37. Draw the graph of a function that could not possibly be the graph of a polynomial function and explain why.

38. Draw a graph that could be the graph of a polynomial function of degree 5. You need not list a specific polynomial, nor do any computation.

39. Which of the statements below is *not* true about the polynomial function f whose graph is shown in the figure?

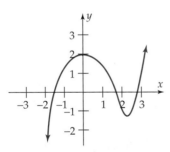

(a) f has three roots between -2 and 3.
(b) $f(x)$ could possibly be a fifth-degree polynomial.
(c) $(f \circ f)(0) > 0$.
(d) $f(2) - f(-1) < 3$.
(e) $f(x)$ is positive for all x in the interval $[-1, 0]$.

40. Which of the statements (i)–(v) about the polynomial function f whose graph is shown in the figure are *false?*

(i) f has 2 roots in the interval $[-6, -3)$.
(ii) $f(-3) - f(-6) < 0$.
(iii) $f(0) < f(1)$.
(iv) $f(2) - 2 = 0$.
(v) f has degree ≤ 4.

In Questions 41–44, find a viewing window (or windows) that shows a complete graph of the function. Be alert for hidden behavior.

41. $f(x) = .5x^3 - 4x^2 + x + 1$

42. $g(x) = .3x^5 - 4x^4 + x^3 - 4x^2 + 5x + 1$

43. $h(x) = 4x^3 - 100x^2 + 600x$

44. $f(x) = 32x^3 - 99x^2 + 100x + 2$

45. HomeArt makes plastic replicas of famous statues. Their total cost to produce copies of a particular statue are shown in the table below.

(a) Sketch a scatter plot of the data.
(b) Use cubic regression to find a function $C(x)$ that models the data [that is, $C(x)$ is the cost of making x statues]. Assume that C is reasonably accurate when $x \leq 100$.
(c) Use C to estimate the cost of making the seventy-first statue.
(d) Use C to approximate the average cost per statue when 35 are made and when 75 are made. [Recall that the average cost of x statues is $C(x)/x$.]

Number of Statues	Total Cost
0	$2,000
10	2,519
20	2,745
30	2,938
40	3,021
50	3,117
60	3,269
70	3,425

46. The table gives the estimated cost of a college education at a public institution. Costs include tuition, fees, books, and room and board for four years. (*Source:* Teachers Insurance and Annuity Association College Retirement Equities Fund)

Enrollment Year	Costs	Enrollment Year	Costs
1998	$46,691	2008	$ 83,616
2000	52,462	2010	93,951
2002	58,946	2012	105,564
2004	66,232	2014	118,611
2006	74,418		

(a) Sketch a scatter plot of the data (with $x = 0$ corresponding to 1990).
(b) Use quartic regression to find a function C that models the data.
(c) Estimate the cost of a college education in 2007 and in 2015.

In Questions 47–54, sketch a complete graph of the function. Label the x-intercepts, all local extrema, holes, and asymptotes.

47. $f(x) = x^3 - 9x$

48. $g(x) = x^3 - 2x^2 + 3$

49. $h(x) = x^4 - x^3 - 4x^2 + 4x + 2$

50. $f(x) = x^4 - 3x - 2$

51. $g(x) = \dfrac{-2}{x + 4}$

52. $h(x) = \dfrac{3 - x}{x - 2}$

53. $k(x) = \dfrac{4x + 10}{3x - 9}$

54. $f(x) = \dfrac{x + 1}{x^2 - 1}$

In Questions 55 and 56, list all asymptotes of the graph of the function.

55. $f(x) = \dfrac{x^2 - 1}{x^3 - 2x^2 - 5x + 6}$

56. $g(x) = \dfrac{x^4 - 6x^3 + 2x^2 - 6x + 2}{x^2 - 3}$

In Questions 57–60, find a viewing window (or windows) that shows a complete graph of the function. Be alert for hidden behavior.

57. $f(x) = \dfrac{x - 3}{x^2 + x - 2}$

58. $g(x) = \dfrac{x^2 - x - 6}{x^3 - 3x^2 + 3x - 1}$

59. $h(x) = \dfrac{x^4 + 4}{x^4 - 99x^2 - 100}$

60. $k(x) = \dfrac{x^3 - 2x^2 - 4x + 8}{x - 10}$

61. It costs the Junkfood Company 50¢ to produce a bag of Munchies. There are fixed costs of $500 per day for building, equipment, etc. The company has found that if the price of a bag of Munchies is set at $1.95 - \dfrac{x}{2000}$ dollars, where x is the number of bags produced per day, then all the bags that are produced will be sold. What number of bags can be produced each day if all are to be sold and the company is to make a profit? What are the possible prices?

62. Highway engineers have found that a good model of the relationship between the density of automobile traffic and the speed at which it moves along a particular section of highway is given by the function

$$s = \frac{100}{1 + d^2} \qquad (0 \le d \le 3)$$

where the density of traffic d is measured in hundreds of cars per mile and the speed s at which traffic moves is measured in miles per hour. Then the traffic flow q is given by the product of the density and the speed, that is, $q = ds$, with q being measured in hundreds of cars per hour.
(a) Express traffic flow as a function of traffic density.
(b) For what densities will traffic flow be at least 3000 cars per hour?
(c) What traffic density will maximize traffic flow? What is the maximum flow?

63. Charlie lives 150 miles from the city. He drives 40 miles to the station (obeying the 55 mph speed limit) and catches a train to the city. The average speed of the train is 25 mph faster than the average speed of the car.
(a) Express the total time for the journey as a function of the speed of the car. What speeds make sense in this context?
(b) How fast should Charlie drive if the entire journey is to take no more than 2.5 hours?

64. The survival rate s of seedlings in the vicinity of a parent tree is given by

$$s = \frac{.5x}{1 + .4x^2},$$

where x is the distance from the seedling to the tree (in meters) and $0 < x \le 10$.
(a) For what distances is the survival rate at least .21?
(b) What distance produces the maximum survival rate?

REVIEW QUESTIONS

In Questions 65 and 66, find the average rate of change of the function between x and x + h.

65. $f(x) = \dfrac{x}{x + 1}$ **66.** $g(x) = \dfrac{1}{x^2 + 1}$

67. Which of these statements about the graph of

$$f(x) = \frac{(x - 1)(x + 3)}{(x^2 + 1)(x^2 - 1)}$$

is *true*?

(a) The graph has two vertical asymptotes.
(b) The graph touches the *x*-axis at $x = 3$.
(c) The graph lies above the *x*-axis when $x < -1$.
(d) The graph has a hole at $x = 1$.
(e) The graph has no horizontal asymptotes.

68. Solve for *x:* $-3(x - 4) \leq 5 + x$.

69. Solve for *y:* $\left| \dfrac{y + 2}{3} \right| \geq 5$.

70. Solve for *x:* $-4 < 2x + 5 < 9$.

71. On which intervals is $\dfrac{2x - 1}{3x + 1} < 1$?

72. On which intervals is $\dfrac{2}{x + 1} < x$?

73. Solve for *x:* $\left| \dfrac{1}{1 - x^2} \right| \geq \dfrac{1}{2}$.

74. Solve for *x:* $x^2 + x > 12$.

75. Solve for *x:* $(x - 1)^2(x^2 - 1)x \leq 0$.

76. If $0 < r \leq s - t$, then which of these statements is *false*?

(a) $s \geq r + t$ (b) $t - s \leq -r$
(c) $-r \geq s - t$ (d) $\dfrac{s - t}{r} > 0$
(e) $s - r \geq t$

77. If $\dfrac{x + 3}{2x + 3} > 1$, then which of these statements is *true*?

(a) $\dfrac{x - 3}{2x - 3} > 1$ (b) $\dfrac{2x + 3}{x + 3} < -1$
(c) $\dfrac{3 - 2x}{x + 3} < -1$ (d) $2x + 3 < x - 3$
(e) None of these

78. Solve and express your answer in interval notation:

$$2x - 3 \leq 5x + 9 < -3x + 4.$$

In Questions 79–86, solve the inequality.

79. $|3x + 2| \geq 2$ **80.** $x^2 + x - 20 > 0$

81. $\dfrac{x - 2}{x + 4} \leq 3$

82. $(x + 1)^2(x - 3)^4(x + 2)^3(x - 7)^5 > 0$

83. $\dfrac{x^2 + x - 9}{x + 3} < 1$ **84.** $\dfrac{x^2 - x - 6}{x - 3} > 1$

85. $\dfrac{x^2 - x - 5}{x^2 + 2} > -2$

86. $\dfrac{x^4 - 3x^2 + 2x - 3}{x^2 - 4} < -1$

In Questions 87–94, solve the equation in the complex number system.

87. $x^2 + 3x + 10 = 0$ **88.** $x^2 + 2x + 5 = 0$

89. $5x^2 + 2 = 3x$

90. $-3x^2 + 4x - 5 = 0$

91. $3x^4 + x^2 - 2 = 0$

92. $8x^4 + 10x^2 + 3 = 0$

93. $x^3 + 8 = 0$ **94.** $x^3 - 27 = 0$

95. One root of $x^4 - x^3 - x^2 - x - 2$ is *i*. Find all the roots.

96. One root of $x^4 + x^3 - 5x^2 + x - 6$ is *i*. Find all the roots.

97. Give an example of a fourth-degree polynomial with real coefficients whose roots include 0 and $1 + i$.

98. Find a fourth-degree polynomial *f* whose only roots are $2 + i$ and $2 - i$, such that $f(-1) = 50$.

DISCOVERY PROJECT 4   Architectural Arches

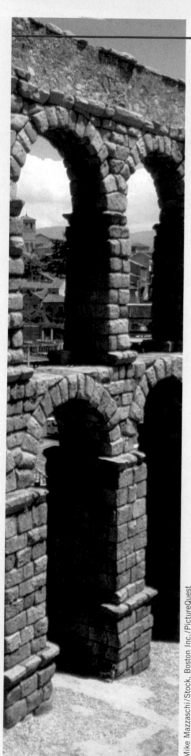

You can see arches almost everywhere you look—in windows, entryways, tunnels, and bridges. Common arch shapes are semicircles, semiellipses, and parabolas. When constructed on a level base, arches are symmetric from left to right. This means that a mathematical function that describes an arch must be an even function.

A *semicircular arch* always has the property that its base is twice as wide as its height. This ratio can be modified by placing a rectangular area under the semicircle, giving a shape known as a *Norman arch.* This approach gives a tunnel or room a vaulted ceiling. *Parabolic arches* can also be created to give a more vaulted appearance.

Semicircular   Norman   Parabolic

For the following exercises in arch modeling, you should always set the origin of your coordinate system to be the center of the base of the arch.

1. Show that the function that models a semicircular arch of radius r is
   $$h(x) = \sqrt{r^2 - x^2}.$$

2. Write a function $h(x)$ that models a semicircular arch that is 15 feet tall. How wide is the arch?

3. Write a function $n(x)$ that models a Norman arch that is 15 feet tall and 16 feet wide at the base.

4. Parabolic arches are typically modeled by using the function
   $$p(x) = H - ax^2,$$
   where H is the height of the arch. Write a function $p(x)$ for an arch that is 15 feet tall and 16 feet wide at the base.

5. Would a truck that is 12 feet tall and 9 feet wide fit through all three arches? How could you fix any of the arches that is too small so that the truck would fit through?

C H A P T E R 5

Exponential and Logarithmic Functions

Walter Stuart/Index Stock Imagery/PictureQuest

How old is that dinosaur?

Population growth (of humans, fish, bacteria, etc.), compound interest, radioactive decay, and a host of other phenomena can be mathematically described by exponential functions (see Exercises 44, 57 and 59 on pages 351 and 353). Archeologists sometimes use carbon-14 dating to determine the approximate age of an artifact (such as a dinosaur skeleton, a mummy, or a wooden statue). This involves using logarithms to solve an appropriate exponential equation. See Exercise 61 on page 391.

325

Chapter Outline

Interdependence of Sections

5.4.A can be covered before 5.3 if desired.

5.1	**Radicals and Rational Exponents**
5.1.A	*Special Topics:* Radical Equations
5.2	**Exponential Functions**
5.2.A	*Special Topics:* Compound Interest and the Number e
5.3	**Common and Natural Logarithmic Functions**
5.4	**Properties of Logarithms**
5.4.A	*Special Topics:* Logarithmic Functions to Other Bases
5.5	**Algebraic Solutions of Exponential and Logarithmic Equations**
5.6	**Exponential, Logarithmic, and Other Models**

Exponential and logarithmic functions are essential for the mathematical description of a variety of phenomena in the physical sciences, engineering, and economics. Although a calculator is necessary to evaluate these functions at most numbers, you won't be able to use your calculator efficiently or interpret its answers unless you understand the properties of these functions. When calculations can readily be done by hand, you will be expected to do them without a calculator.

5.1 Radicals and Rational Exponents

When $c \geq 0$, the *square root* of c is defined to be the nonnegative solution of the equation $x^2 = c$. For $n > 2$, *nth roots* are defined as the solutions of the equation $x^n = c$. The graphs on the next page show that this equation may have two, one, or no solutions, depending on whether n is even or odd and whether c is positive or negative.[*]

---

[*]The solutions of $x^n = c$ are the x-coordinates of the intersection points of the graph of $y = x^n$ and the horizontal line $y = c$. The graph of $y = x^n$ was discussed on page 255.

$x^n = c$ n odd	$x^n = c$ n even		
Exactly one solution for any c	$c > 0$ One positive and one negative solution	$c = 0$ One solution $x = 0$	$c < 0$ No solution

Consequently, we have the following definition.

nth Roots

Let c be a real number and n a positive integer. The ***n*th root of c** is denoted by either of the symbols

$$\sqrt[n]{c} \qquad \text{or} \qquad c^{1/n}$$

and is defined to be

The solution of $x^n = c$, when n is odd;

The nonnegative solution of $x^n = c$, when n is even and $c \geq 0$.

As before, square roots are denoted $\sqrt{}$ rather than $\sqrt[2]{}$.

EXAMPLE 1

Find the following roots.

(a) $\sqrt[3]{-8}$ (b) $81^{1/4}$ (c) $\sqrt[5]{32}$

SOLUTION

(a) $\sqrt[3]{-8} = (-8)^{1/3} = -2$ because -2 is the solution of $x^3 = -8$.

(b) $81^{1/4} = \sqrt[4]{81} = 3$ because 3 is the positive solution of $x^4 = 81$.

(c) $\sqrt[5]{32} = 32^{1/5} = 2$ because 2 is the solution of $x^5 = 32$. ∎

EXAMPLE 2

Use a calculator to approximate

(a) $40^{1/5}$ (b) $225^{1/11}$

Figure 5–1

SOLUTION

(a) Since $1/5 = .2$, you can compute $40^{.2}$, as in Figure 5–1.

(b) $1/11$ is the infinite decimal $.090909 \ldots$. In such cases, it is best to leave the exponent in fractional form and use parentheses, as shown in Figure 5–1. If you round off the decimal equivalent of $1/11$, say as $.0909$, you will not get the same answer, as Figure 5–1 shows. ■

▪▪ RATIONAL EXPONENTS

The next step is to give a meaning to fractional exponents for any fraction, not just those of the form $1/n$. If possible, they should be defined in such a way that the various exponent rules continue to hold. Consider, for example, how $4^{3/2}$ might be defined. The exponent $3/2$ can be written as either

$$3 \cdot \left(\frac{1}{2}\right) \qquad \text{or} \qquad \left(\frac{1}{2}\right) \cdot 3.$$

If the power of a power property $(c^m)^n = c^{mn}$ is to hold, we might define $4^{3/2}$ as either $(4^3)^{1/2}$ or $(4^{1/2})^3$. The result is the same in both cases:

$$(4^3)^{1/2} = 64^{1/2} = \sqrt{64} = 8 \qquad \text{and} \qquad (4^{1/2})^3 = (\sqrt{4})^3 = 2^3 = 8.$$

It can be proved that the same thing is true in the general case, which leads to this definition.

Rational Exponents

> Let c be a positive real number and let t/k be a rational number with positive denominator. Then,
>
> $$c^{t/k} \text{ is defined to be the number } (c^t)^{1/k} = (c^{1/k})^t.$$

Since every terminating decimal is a rational number, expressions such as $13^{3.78}$ now have a meaning, namely, $13^{378/100}$. (Actually, we used this fact earlier when we computed $225^{.0909}$ in Figure 5–1.)

TECHNOLOGY TIP

TI-89 and Casio FX 2.0 will produce real number answers to computations like $(-8)^{2/3}$ if you select "Real" in the appropriate menu/submenu:

TI-89: MODE/COMPLEX FORMAT

Casio FX 2.0: SET-UP/ COMPLEX MODE

NOTE: The preceding definition is also valid when c is negative, *provided that* the exponent is in lowest terms, with an odd denominator, such as $(-8)^{2/3}$ (see Exercise 61). Nevertheless, if you try to compute $(-8)^{2/3}$ on your calculator, you may get an error message or a complex number (indicated by an ordered pair or an expression involving i) instead of the correct answer,

$$(-8)^{2/3} = [(-8)^2]^{1/3} = \sqrt[3]{(-8)^2} = \sqrt[3]{64} = 4.$$

If this happens, you can probably get the correct answer by keying in either $[(-8)^2]^{1/3}$ or $[(-8)^{1/3}]^2$, each of which is equal to $(-8)^{2/3}$.

Rational exponents were defined in a way that guaranteed that one of the familiar exponent laws would remain valid. In fact, all exponent laws developed for integer exponents are valid for rational exponents, as summarized here and illustrated in Examples 3 through 5.

Exponent Laws

Let c and d be nonnegative real numbers, and let r and s be any rational numbers. Then

1. $c^r c^s = c^{r+s}$

2. $\dfrac{c^r}{c^s} = c^{r-s}$ $(c \neq 0)$

3. $(c^r)^s = c^{rs}$

4. $(cd)^r = c^r d^r$

5. $\left(\dfrac{c}{d}\right)^r = \dfrac{c^r}{d^r}$ $(d \neq 0)$

6. $c^{-r} = \dfrac{1}{c^r}$ $(c \neq 0)$

CAUTION

The exponent laws deal only with products and quotients. There are no analogous properties for sums. In particular, if both c and d are nonzero, then

$(c + d)^r$ is **not** equal to $c^r + d^r$.

EXAMPLE 3

Compute the product $x^{1/2}(x^{3/4} - x^{3/2})$.

SOLUTION

Distributive law:

$$x^{1/2}(x^{3/4} - x^{3/2}) = x^{1/2}x^{3/4} - x^{1/2}x^{3/2}$$

Multiplication with exponents (law 1):

$$= x^{1/2+3/4} - x^{1/2+3/2}$$

$\dfrac{1}{2} + \dfrac{3}{4} = \dfrac{5}{4}$ and $\dfrac{1}{2} + \dfrac{3}{2} = 2$:

$$= x^{5/4} - x^2. \quad\blacksquare$$

EXAMPLE 4

Simplify $(8r^{3/4}s^{-3})^{2/3}$, and express it without negative exponents.

SOLUTION

Product to a power (law 4):

$$(8r^{3/4}s^{-3})^{2/3} = 8^{2/3}(r^{3/4})^{2/3}(s^{-3})^{2/3}$$

Power of a power (law 3):

$$= 8^{2/3}r^{(3/4)(2/3)}s^{(-3)(2/3)}$$

$\dfrac{3}{4} \cdot \dfrac{2}{3} = \dfrac{1}{2}$ and $(-3)\dfrac{2}{3} = -2$:

$$= 8^{2/3}r^{1/2}s^{-2}$$

Negative exponents (law 6):

$$= \frac{8^{2/3}r^{1/2}}{s^2}$$

$8^{2/3} = \sqrt[3]{8^2} = \sqrt[3]{64} = 4$:

$$= \frac{4r^{1/2}}{s^2} \quad \text{or} \quad \frac{4\sqrt{r}}{s^2}. \blacksquare$$

EXAMPLE 5

Simplify $\dfrac{x^{7/2}y^3}{(xy^{7/4})^2}$ and express it without negative exponents.

SOLUTION

Product to a power (law 4):
$$\frac{x^{7/2}y^3}{(xy^{7/4})^2} = \frac{x^{7/2}y^3}{x^2(y^{7/4})^2}$$

$\dfrac{7}{4} \cdot 2 = \dfrac{7}{2}$:
$$= \frac{x^{7/2}y^3}{x^2 y^{7/2}}$$

Division with exponents (law 2):
$$= x^{7/2-2}y^{3-7/2}$$

$\dfrac{7}{2} - 2 = \dfrac{3}{2}$ and $3 - \dfrac{7}{2} = -\dfrac{1}{2}$:
$$= x^{3/2}y^{-1/2}$$

Negative exponents (law 6):
$$= \frac{x^{3/2}}{y^{1/2}} \quad \blacksquare$$

▪▪ RADICAL NOTATION

As we saw above, there are two notations for nth roots: $\sqrt[n]{c}$ and $c^{1/n}$. Similarly, $c^{t/k}$ can be expressed in terms of radicals.

$$c^{t/k} = (c^t)^{1/k} = \sqrt[k]{c^t} \quad \text{and} \quad c^{t/k} = (c^{1/k})^t = (\sqrt[k]{c})^t.$$

When radical notation is used, the exponent laws take on a different form. For instance,

$$(cd)^{1/n} = c^{1/n}d^{1/n} \quad \text{means the same thing as} \quad \sqrt[n]{cd} = \sqrt[n]{c}\,\sqrt[n]{d};$$

$$\left(\frac{c}{d}\right)^{1/n} = \frac{c^{1/n}}{d^{1/n}} \quad \text{means the same thing as} \quad \sqrt[n]{\frac{c}{d}} = \frac{\sqrt[n]{c}}{\sqrt[n]{d}}.$$

When simplifying expression involving radicals, either you can change to rational exponents and proceed as in Examples 3–5, or you can use the exponent laws in their radical notation, as in the next example.

EXAMPLE 6

Simplify each of the following.

(a) $\sqrt{63}$ (b) $\sqrt{12} - \sqrt{75}$ (c) $\sqrt[3]{40k^4}$

SOLUTION

(a) Factor 63 as the product of 9 (a perfect square) and 7 and then use the product of roots property.

$$\sqrt{63} = \sqrt{9 \cdot 7} = \sqrt{9}\,\sqrt{7} = 3\sqrt{7}$$

(b) Look for perfect squares.

Factor perfect squares out of 12 and 75:
$$\sqrt{12} - \sqrt{75} = \sqrt{4 \cdot 3} - \sqrt{25 \cdot 3}$$

Product of roots:
$$= \sqrt{4}\,\sqrt{3} - \sqrt{25}\,\sqrt{3}$$
$$= 2\sqrt{3} - 5\sqrt{3} = -3\sqrt{3}.$$

(c) Note that $8k^3$ is a factor of $40k^4$ and that $8k^3 = (2k)^3$ is a perfect cube.

Factor out $8k^3$: $\qquad \sqrt[3]{40k^4} = \sqrt[3]{8k^3 \cdot 5k}$

Product of roots: $\qquad\qquad = \sqrt[3]{8k^3}\,\sqrt[3]{5k}$

$$= \sqrt[3]{(2k)^3}\,\sqrt[3]{5k} = 2k\,\sqrt[3]{5k}. \qquad\blacksquare$$

▘▘ RATIONALIZING DENOMINATORS AND NUMERATORS

When dealing with fractions in the days before calculators, it was customary to *rationalize the denominators,* that is, write equivalent fractions with no radicals in the denominator, because this made many computations easier. With calculators, of course, there is no computational advantage to rationalizing denominators. Nevertheless, rationalizing denominators or numerators is sometimes needed to simplify expressions and to derive useful formulas.

EXAMPLE 7

(a) To rationalize the denominator of $\dfrac{7}{\sqrt{5}}$, multiply it by 1, with 1 written as a suitable radical fraction.

$$\frac{7}{\sqrt{5}} = \frac{7}{\sqrt{5}} \cdot 1 = \frac{7}{\sqrt{5}} \cdot \frac{\sqrt{5}}{\sqrt{5}} = \frac{7\sqrt{5}}{5}.$$

(b) To rationalize the denominator of $\dfrac{2}{3 + \sqrt{6}}$, multiply by 1 written as a radical fraction and use the multiplication pattern $(a + b)(a - b) = a^2 - b^2$.

$$\frac{2}{3 + \sqrt{6}} = \frac{2}{3 + \sqrt{6}} \cdot 1$$

$$= \frac{2}{3 + \sqrt{6}} \cdot \frac{3 - \sqrt{6}}{3 - \sqrt{6}} = \frac{2(3 - \sqrt{6})}{(3 + \sqrt{6})(3 - \sqrt{6})}$$

$$= \frac{6 - 2\sqrt{6}}{3^2 - (\sqrt{6})^2} = \frac{6 - 2\sqrt{6}}{9 - 6} = \frac{6 - 2\sqrt{6}}{3}. \qquad\blacksquare$$

EXAMPLE 8

Assume that $h \neq 0$, and rationalize the numerator of $\dfrac{\sqrt{x + h} - \sqrt{x}}{h}$; that is, write an equivalent fraction with no radicals in the numerator.

SOLUTION   Again, the technique is to multiply the fraction by 1, with 1 written as a suitable radical fraction:

$$\frac{\sqrt{x+h}-\sqrt{x}}{h} = \frac{\sqrt{x+h}-\sqrt{x}}{h} \cdot 1$$

$$= \frac{\sqrt{x+h}-\sqrt{x}}{h} \cdot \frac{\sqrt{x+h}+\sqrt{x}}{\sqrt{x+h}+\sqrt{x}} = \frac{(\sqrt{x+h})^2-(\sqrt{x})^2}{h(\sqrt{x+h}+\sqrt{x})}$$

$$= \frac{x+h-x}{h(\sqrt{x+h}+\sqrt{x})} = \frac{h}{h(\sqrt{x+h}+\sqrt{x})} = \frac{1}{\sqrt{x+h}+\sqrt{x}}. \qquad \blacksquare$$

■■ IRRATIONAL EXPONENTS

An example will illustrate how a^t is defined when t is an irrational number.[*] To compute $10^{\sqrt{2}}$ we use the infinite decimal expansion $\sqrt{2} \approx 1.414213562 \ldots$ (see Special Topics 1.1.A). Each of

$$1.4, \; 1.41, \; 1.414, \; 1.4142, \; 1.41421, \ldots$$

is a rational number approximation of $\sqrt{2}$, and each is a more accurate approximation than the preceding one. We know how to raise 10 to each of these rational numbers:

$$10^{1.4} \approx 25.1189 \qquad 10^{1.4142} \approx 25.9537$$

$$10^{1.41} \approx 25.7040 \qquad 10^{1.41421} \approx 25.9543$$

$$10^{1.414} \approx 25.9418 \qquad 10^{1.414213} \approx 25.9545$$

It appears that as the exponent r gets closer and closer to $\sqrt{2}$, 10^r gets closer and closer to a real number whose decimal expansion begins 25.954 . . . . We define $10^{\sqrt{2}}$ to be this number.

Similarly, for any $a > 0$,

a^t is a well-defined *positive* number for each real exponent t.

We shall also assume this fact.

The exponent laws (page 329) are valid for *all* real exponents.

✓ EXERCISES 5.1

Note: *Unless directed otherwise, assume that all letters represent positive real numbers.*

In Exercises 1–10, simplify the expression.

1. $(25k^2)^{3/2}(16k^{1/3})^{3/4}$

2. $(3x^{2/3})(4y^{3/4})(2x^{-1/3})(3y^{13/4})$

3. $(c^{2/5}d^{-2/3})(c^6d^3)^{4/3}$ 4. $(x^{1/2}y^3)(x^0y^7)^{-2}$

5. $\dfrac{(x^2)^{1/3}(y^2)^{2/3}}{3x^{2/3}y^2}$

6. $\dfrac{(c^{1/2})^3(d^3)^{1/2}}{(c^3)^{1/4}(d^{1/4})^3}$

7. $\dfrac{(7a)^2(5b)^{3/2}}{(5a)^{3/2}(7b)^4}$

8. $\dfrac{(6a)^{1/2}\sqrt{ab}}{a^2b^{3/2}}$

9. $(a^{x^2})^{1/x}$

10. $\dfrac{(b^x)^{x-1}}{b^{-x}}$

___

[*]This example is not a proof but should make the idea plausible. Calculus is required for a rigorous proof.

In Exercises 11–16, compute and simplify.

11. $x^{1/2}(x^{2/3} - x^{4/3})$ **12.** $x^{1/2}(3x^{3/2} + 2x^{-1/2})$

13. $(x^{1/2} + y^{1/2})(x^{1/2} - y^{1/2})$

14. $(x^{1/3} + y^{1/2})(2x^{1/3} - y^{3/2})$

15. $(x + y)^{1/2}[(x + y)^{1/2} - (x + y)]$

16. $(x^{1/3} + y^{1/3})(x^{2/3} - x^{1/3}y^{1/3} + y^{2/3})$

In Exercises 17–22, factor the given expression. For example,

$$x - x^{1/2} - 2 = (x^{1/2} - 2)(x^{1/2} + 1).$$

17. $x^{2/3} + x^{1/3} - 6$ **18.** $x^{2/5} + 11x^{1/5} + 30$

19. $x + 4x^{1/2} + 3$ **20.** $x^{1/3} + 7x^{1/6} + 10$

21. $x^{4/5} - 81$ **22.** $x^{2/3} - 6x^{1/3} + 9$

In Exercises 23–28, write the given expression without using radicals.

23. $\sqrt[3]{a^2 + b^2}$ **24.** $\sqrt[4]{a^{-3} - b^3}$ **25.** $\sqrt[4]{\sqrt[4]{a^3}}$

26. $\sqrt{\sqrt[3]{a^3b^4}}$ **27.** $\sqrt[5]{t}\sqrt{16t^5}$ **28.** $\sqrt{x}(\sqrt[3]{x^2})(\sqrt[4]{x^3})$

In Exercises 29–42, simplify the expression without using a calculator.

29. $\sqrt{80}$ **30.** $\sqrt{96}$

31. $\sqrt{6}\sqrt{12}$ **32.** $\sqrt{8}\sqrt{96}$

33. $\dfrac{-6 + \sqrt{99}}{15}$ **34.** $\dfrac{5 - \sqrt{175}}{10}$

35. $\sqrt{50} - \sqrt{72}$ **36.** $\sqrt{75} + \sqrt{192}$

37. $5\sqrt{20} - \sqrt{45} + 2\sqrt{80}$

38. $\sqrt[3]{40} + 2 \cdot \sqrt[3]{135} - 5 \cdot \sqrt[3]{320}$

39. $\sqrt{16a^8b^{-2}}$ **40.** $\sqrt{24x^6y^{-4}}$

41. $\dfrac{\sqrt{c^2d^6}}{\sqrt{4c^3d^{-4}}}$ **42.** $\dfrac{\sqrt{a^{-10}b^{-12}}}{\sqrt{a^{14}d^{-4}}}$

In Exercises 43–48, rationalize the denominator and simplify your answer.

43. $\dfrac{3}{\sqrt{8}}$ **44.** $\dfrac{2}{\sqrt{6}}$

45. $\dfrac{3}{2 + \sqrt{12}}$ **46.** $\dfrac{1 + \sqrt{3}}{5 + \sqrt{10}}$

47. $\dfrac{2}{\sqrt{x} + 2}$ **48.** $\dfrac{\sqrt{x}}{\sqrt{x} - \sqrt{c}}$

In Exercises 49–52, find the difference quotient of the given function. Then rationalize its numerator and simplify.

49. $f(x) = \sqrt{x + 1}$

50. $g(x) = 2\sqrt{x + 3}$

51. $f(x) = \sqrt{x^2 + 1}$

52. $g(x) = \sqrt{x^2 - x}$

In Exercises 53–56, use the equation $y = 92.8935 \cdot x^{.6669}$ which gives the approximate distance y (in millions of miles) from the sun to a planet that takes x earth years to complete one orbit of the sun. Find the distance from the sun to the planet whose orbit time is given.

53. Mercury (.24 years)   **54.** Mars (1.88 years)

55. Saturn (29.46 years)   **56.** Pluto (247.69 years)

Between 1790 and 1860, the population y of the United States (in millions) in year x was given by $y = 3.9572(1.0299^x)$, where $x = 0$ corresponds to 1790. In Exercises 57–60, find the U.S. population in the given year.

57. 1800   **58.** 1817

59. 1845   **60.** 1859

61. Here are some of the reasons why restrictions are necessary when defining fractional powers of a negative number.
(a) Explain why the equations $x^2 = -4$, $x^4 = -4$, $x^6 = -4$, etc., have no real solutions. Hence, we cannot define $c^{1/2}$, $c^{1/4}$, $c^{1/6}$ when $c = -4$.
(b) Since $1/3$ is the same as $2/6$, it should be true that $c^{1/3} = c^{2/6}$, that is, that $\sqrt[3]{c} = \sqrt[6]{c^2}$. Show that this is false when $c = -8$.

62. Use a calculator to find a *six*-place decimal approximation of $(311)^{-4.2}$. Explain why your answer cannot possibly be the number $(311)^{-4.2}$.

63. (a) Graph $f(x) = x^5$ and explain why this function has an inverse function.
(b) Show algebraically that the inverse function is $g(x) = x^{1/5}$.

64. If n is an odd positive integer, show that $f(x) = x^n$ has an inverse function and find the rule of the inverse function. [*Hint:* Exercise 63 is the case when $n = 5$.]

In Exercises 65–67, use the catalog of basic functions (page 214) and Section 3.4 to describe the graph of the given function.

65. $g(x) = \sqrt{x + 3}$

66. $h(x) = \sqrt{x} - 2$

67. $k(x) = \sqrt{x + 4} - 4$

68. (a) Suppose r is a solution of the equation $x^n = c$ and s is a solution of $x^n = d$. Verify that rs is a solution of $x^n = cd$.
(b) Explain why part (a) shows that $\sqrt[n]{cd} = \sqrt[n]{c}\sqrt[n]{d}$.

69. The output Q of an industry depends on labor L and capital C according to the equation

$$Q = L^{1/4}C^{3/4}.$$

(a) Use a calculator to determine the output for the following resource combinations.

L	C	$Q = L^{1/4}C^{3/4}$
10	7	
20	14	
30	21	
40	28	
60	42	

(b) When you double both labor and capital, what happens to the output? When you triple both labor and capital, what happens to the output?

70. Do Exercise 69 when the equation relating output to resources is $Q = L^{1/4}C^{1/2}$.

71. Do Exercise 69 when the equation relating output to resources is $Q = L^{1/2}C^{3/4}$.

72. In Exercises 69–71, how does the sum of the exponents on L and C affect the increase in output?

5.1.A   *SPECIAL TOPICS* Radical Equations

The algebraic solution of equations involving radicals depends on this fact: If two quantities are equal, say,

$$x - 2 = 3,$$

then their squares are also equal:

$$(x - 2)^2 = 9.$$

Thus,

every solution of $x - 2 = 3$ is also a solution of $(x - 2)^2 = 9$.

But *be careful:* This works only in one direction. For instance, -1 is a solution of $(x - 2)^2 = 9$, but not of $x - 2 = 3$. This is an example of the Power Principle.

Power Principle

> If both sides of an equation are raised to the same positive integer power, then every solution of the original equation is also a solution of the new equation. But the new equation may have solutions that are *not* solutions of the original one.

Consequently, if you raise both sides of an equation to a power, you must *check your solutions* in the *original* equation. Graphing provides a quick way to eliminate most extraneous solutions. But only an algebraic computation can confirm an exact solution.

EXAMPLE 1

Solve $5 + \sqrt{3x - 11} = x$.

SOLUTION We first rearrange terms to get the radical expression alone on one side.

$$\sqrt{3x - 11} = x - 5.$$

Then we square both sides and solve the resulting equation.

$$(\sqrt{3x - 11})^2 = (x - 5)^2$$
$$3x - 11 = x^2 - 10x + 25$$
$$0 = x^2 - 13x + 36$$
$$0 = (x - 4)(x - 9)$$
$$x - 4 = 0 \quad \text{or} \quad x - 9 = 0$$
$$x = 4 \qquad\qquad x = 9.$$

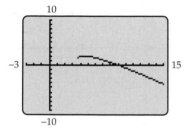

10

−3   15

−10

Figure 5–2

If these numbers are solutions of the original equation, they should be x-intercepts of the graph of $y = \sqrt{3x - 11} - x + 5$ (why?). But the graph of f in Figure 5–2 doesn't have an x-intercept at $x = 4$, so 4 is not a solution of the original equation. The graph appears to show that $x = 9$ is a solution of the original equation, and direct calculation confirms this.

Left side: $\quad 5 + \sqrt{3x - 11}$ Right side: $\quad x$

$$5 + \sqrt{3 \cdot 9 - 11} \qquad\qquad 9$$
$$5 + \sqrt{16}$$
$$9$$

Hence, 9 is the only solution of the original equation. ∎

EXAMPLE 2

Solve $\sqrt{2x - 3} - \sqrt{x + 7} = 2$.

SOLUTION We first rearrange terms so that one side contains only a single radical term.

$$\sqrt{2x - 3} = \sqrt{x + 7} + 2.$$

Then we square both sides and simplify.

$$(\sqrt{2x - 3})^2 = (\sqrt{x + 7} + 2)^2$$
$$2x - 3 = (\sqrt{x + 7})^2 + 2 \cdot 2 \cdot \sqrt{x + 7} + 2^2$$
$$2x - 3 = x + 7 + 4\sqrt{x + 7} + 4$$
$$x - 14 = 4\sqrt{x + 7}.$$

Now we square both sides and solve the resulting equation.

$$(x - 14)^2 = (4\sqrt{x + 7})^2$$
$$x^2 - 28x + 196 = 4^2 \cdot (\sqrt{x + 7})^2$$
$$x^2 - 28x + 196 = 16(x + 7)$$
$$x^2 - 28x + 196 = 16x + 112$$
$$x^2 - 44x + 84 = 0$$
$$(x - 2)(x - 42) = 0$$
$$x - 2 = 0 \quad \text{or} \quad x - 42 = 0$$
$$x = 2 \qquad\qquad x = 42.$$

Substituting 2 and 42 in the left side of the original equation shows that

$$\sqrt{2 \cdot 2 - 3} - \sqrt{2 + 7} = \sqrt{1} - \sqrt{9} = 1 - 3 = -2;$$
$$\sqrt{2 \cdot 42 - 3} - \sqrt{42 + 7} = \sqrt{81} - \sqrt{49} = 9 - 7 = 2.$$

Therefore, 42 is the only solution of the equation.   ■

Many radical equations are not amenable to algebraic techniques. In such cases, graphical or numerical means must be used to approximate the solutions.

EXAMPLE 3

Solve $\sqrt[5]{x^2 - 6x + 2} = x - 4$.

SOLUTION   Raising both sides of the equation to the fifth power would eliminate the radical, but produce a fifth-degree polynomial equation that is not readily solvable. So we use graphical methods to approximate the solutions. If you rewrite the equation as

$$\sqrt[5]{x^2 - 6x + 2} - x + 4 = 0,$$

then the solutions are the x-intercepts of the graph of

$$h(x) = \sqrt[5]{x^2 - 6x + 2} - x + 4$$

(see Figure 5–3). Alternatively, you can graph $f(x) = \sqrt[5]{x^2 - 6x + 2}$ and

$$g(x) = x - 4$$

on the same screen and find the x-coordinate of their intersection (Figure 5–4).

Figure 5–3

Figure 5–4

GRAPHING EXPLORATION

Use a root finder in Figure 5–3 or an intersection finder in Figure 5–4 to verify that $x \approx 2.534$ is the solution of the equation. ■

Equations involving rational exponents can be solved graphically, provided that the function to be graphed is entered properly. Many such equations can also be solved algebraically by making an appropriate substitution.

EXAMPLE 4

Solve $x^{2/3} - 2x^{1/3} - 15 = 0$ algebraically and graphically.

SOLUTION

Algebraic: Let $u = x^{1/3}$, rewrite the equation, and solve:

$$x^{2/3} - 2x^{1/3} - 15 = 0$$
$$(x^{1/3})^2 - 2x^{1/3} - 15 = 0$$
$$u^2 - 2u - 15 = 0$$
$$(u + 3)(u - 5) = 0$$

$$u + 3 = 0 \qquad \text{or} \qquad u - 5 = 0$$
$$u = -3 \qquad\qquad\qquad u = 5$$
$$x^{1/3} = -3 \qquad\qquad\quad x^{1/3} = 5.$$

Cubing both sides of these last equations shows that

$$(x^{1/3})^3 = (-3)^3 \qquad \text{or} \qquad (x^{1/3})^3 = 5^3$$
$$x = -27 \qquad\qquad\qquad x = 125.$$

Since we cubed both sides, we must check these numbers in the original equation. Verify that both *are* solutions.

Graphical: Graph the function $f(x) = x^{2/3} - 2x^{1/3} - 15$ and find the *x*-intercepts, namely, $x = -27$ and $x = 125$. The only difficulty is the one mentioned in the Note on page 328. Depending on your calculator, you might have to enter the function f in one of these forms:

$$f(x) = (x^2)^{1/3} - 2x^{1/3} - 15 \qquad \text{or} \qquad f(x) = (x^{1/3})^2 - 2x^{1/3} - 15.$$

Otherwise, the calculator might not produce a graph when x is negative. ■

EXAMPLE 5

Leslie, who is standing at point A on the bank of a 2.5-kilometer-wide river wants to reach point B, 15 kilometers downstream on the opposite bank. She plans to row to a point C on the opposite shore and then run to B, as shown in Figure 5–5. She can row at a rate of 4 kilometers per hour and can run at 8 kilometers per hour.

(a) If her trip is to take 3 hours, how far from B should she land?

(b) How far from B should she land to make the time for the trip as short as possible?

Figure 5–5

SOLUTION   Let x be the distance that Leslie must run from C to B. Using the basic formula for distance, we have

$$\text{Rate} \times \text{Time} = \text{Distance}$$

$$\text{Time} = \frac{\text{Distance}}{\text{Rate}} = \frac{x}{8}.$$

Similarly, the time required to row distance d is

$$\text{Time} = \frac{\text{Distance}}{\text{Rate}} = \frac{d}{4}.$$

Since $15 - x$ is the distance from D to C, the Pythagorean Theorem applied to right triangle ADC shows that

$$d^2 = (15 - x)^2 + 2.5^2 \qquad \text{or, equivalently,} \qquad d = \sqrt{(15 - x)^2 + 6.25}.$$

Therefore, the total time for the trip is given by

$$T(x) = \text{Rowing time} + \text{Running time} = \frac{d}{4} + \frac{x}{8} = \frac{\sqrt{(x - 15)^2 + 6.25}}{4} + \frac{x}{8}.$$

(a) If the trip is to take 3 hours, then $T(x) = 3$, and we must solve the equation

$$\frac{\sqrt{(x - 15)^2 + 6.25}}{4} + \frac{x}{8} = 3.$$

GRAPHING EXPLORATION

Using the viewing window with $0 \le x \le 15$ and $-2 \le y \le 2$, graph the function

$$f(x) = \frac{\sqrt{(x - 15)^2 + 6.25}}{4} + \frac{x}{8} - 3$$

and use a root finder to find its x-intercept (the solution of the equation).

This Graphing Exploration shows that Leslie should land approximately 6.74 kilometers from B to make the trip in 3 hours.

(b) To find the shortest possible time, we must find the value of x that makes

$$T(x) = \frac{\sqrt{(x - 15)^2 + 6.25}}{4} + \frac{x}{8}$$

as small as possible.

GRAPHING EXPLORATION

Using the viewing window with $0 \le x \le 15$ and $0 \le y \le 4$, graph $T(x)$ and use a minimum finder to verify that the lowest point on the graph (that is, the point with the y-coordinate $T(x)$ as small as possible) is approximately $(13.56, 2.42)$.

Therefore, the shortest time for the trip will be 2.42 hours and will occur if Leslie rows to a point 13.56 kilometers from B. ∎

✓ EXERCISES 5.1.A

In Exercises 1–26, find all real solutions of each equation. Find exact solutions when possible and approximate ones otherwise.

1. $\sqrt{x + 2} = 3$
2. $\sqrt{x - 7} = 4$
3. $\sqrt{4x + 9} = 5$
4. $\sqrt{3x - 2} = 7$
5. $\sqrt[3]{5 - 11x} = 3$
6. $\sqrt[3]{6x - 10} = 2$
7. $\sqrt[3]{x^2 - 1} = 2$
8. $(x + 1)^{2/3} = 4$
9. $\sqrt{x^2 - x - 1} = 1$
10. $\sqrt{x^2 - 5x + 4} = 2$
11. $\sqrt{x + 7} = x - 5$
12. $\sqrt{x + 5} = x - 1$
13. $\sqrt{3x^2 + 7x - 2} = x + 1$
14. $\sqrt{4x^2 - 10x + 5} = x - 3$
15. $\sqrt[3]{x^3 + x^2 - 4x + 5} = x + 1$
16. $\sqrt[3]{x^3 - 6x^2 + 2x + 3} = x - 1$
17. $\sqrt[5]{9 - x^2} = x^2 + 1$
18. $\sqrt[4]{x^3 - x + 1} = x^2 - 1$
19. $\sqrt[3]{x^5 - x^3 - x} = x + 2$
20. $\sqrt{x^3 + 2x^2 - 1} = x^3 + 2x - 1$
21. $\sqrt{x^2 + 3x - 6} = x^4 - 3x^2 + 2$
22. $\sqrt[3]{x^4 + x^2 + 1} = x^2 - x - 5$

23. $\sqrt{5x + 6} = 3 + \sqrt{x + 3}$
24. $\sqrt{3y + 1} - 1 = \sqrt{y + 4}$
25. $\sqrt{2x - 5} = 1 + \sqrt{x - 3}$
26. $\sqrt{x - 3} + \sqrt{x + 5} = 4$

27. The surface area S of the right circular cone in the figure is given by $S = \pi r \sqrt{r^2 + h^2}$. What radius should be used to produce a cone of height 5 inches and surface area 100 square inches?

28. What is the radius of the base of a cone whose surface area is 18π square centimeters and whose height is 4 cm?

29. Find the radius of the base of a conical container whose height is 1/3 of the radius and whose volume is 180 cubic inches. [*Note:* The volume of a cone of radius r and height h is $\pi r^2 h / 3$.]

30. The surface area of the right square pyramid in the figure is given by $S = b\sqrt{b^2 + 4h^2}$. If the pyramid has height 10 feet and surface area 100 square feet, what is the length of a side b of its base?

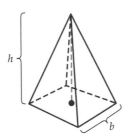

In Exercises 31–34, assume that all letters represent positive numbers and solve each equation for the required letter.

31. $A = \sqrt{1 + \dfrac{a^2}{b^2}}$ for b

32. $T = 2\pi\sqrt{\dfrac{m}{g}}$ for g

33. $K = \sqrt{1 - \dfrac{x^2}{u^2}}$ for u

34. $R = \sqrt{d^2 + k^2}$ for d

In Exercises 35–42, solve each equation algebraically.

35. $x - 4x^{1/2} + 4 = 0$ [*Hint:* Let $u = x^{1/2}$.]

36. $x - x^{1/2} - 12 = 0$

37. $2x - \sqrt{x} - 6 = 0$

38. $3x - 11\sqrt{x} - 4 = 0$

39. $x^{2/3} + 3x^{1/3} + 2 = 0$ [*Hint:* Let $u = x^{1/3}$.]

40. $x^{2/3} - 4x^{1/3} + 3 = 0$

41. $x^{1/2} - x^{1/4} - 2 = 0$ [*Hint:* Let $u = x^{1/4}$.]

42. $x^{1/3} + x^{1/6} - 2 = 0$

In Exercises 43–46, solve each equation graphically.

43. $x^{3/5} - 2x^{2/5} + x^{1/5} - 6 = 0$

44. $x^{5/3} + x^{4/3} - 3x^{2/3} + x = 5$

45. $x^{-3} + 2x^{-2} - 4x^{-1} + 5 = 0$

46. $x^{-2/3} - 3x^{-1/3} = 4$

47. A rope is to be stretched at uniform height from a tree to a 35-foot-long fence, which is 20 feet from the tree, and then to the side of a building at a point 30 feet from the fence, as shown in the figure.
   (a) If 63 feet of rope is to be used, how far from the building wall should the rope meet the fence?
   (b) How far from the building wall should the rope meet the fence if as little rope as possible is to be used?

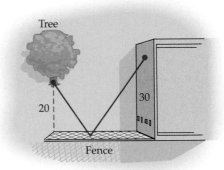

48. Anne is standing on a straight road and wants to reach her helicopter, which is located 2 miles down the road from her, a mile from the road in a field (see the figure). She can run 5 miles per hour on the road and 3 miles per hour in the field. She plans to run down the road, then cut diagonally across the field to reach the helicopter.
   (a) Where should she leave the road to reach the helicopter in exactly 42 minutes (.7 hour)?
   (b) Where should she leave the road to reach the helicopter as soon as possible?

49. A power plant is located on the bank of a river that is $\frac{1}{2}$ mile wide. Wiring is to be laid across the river and then along the shore to a substation 8 miles downstream, as shown in the figure. It costs $12,000 per mile for underwater wiring and $8000 per mile for wiring on land. If $72,000 is to be spent on the project, how far from the substation should the wiring come to shore?

50. A spotlight is to be placed on a building wall to illuminate a bench that is 32 feet from the base of the wall. The intensity I of the light at the bench is known to be x/d^3, where x is the height of the spotlight above the ground and d is the distance from the bench to the spotlight.

(a) Express I as a function of x. [It may help to draw a picture.]

(b) How high should the spotlight be in order to provide maximum illumination at the bench?

5.2 Exponential Functions

For each positive real number a there is a function (called the **exponential function with base a**) whose domain is all real numbers and whose rule is $f(x) = a^x$. For example,

$$f(x) = 10^x, \qquad g(x) = 2^x, \qquad h(x) = \left(\frac{1}{2}\right)^x, \qquad k(x) = \left(\frac{3}{2}\right)^x.$$

The graph of $f(x) = a^x$ is the next entry in the catalog of basic functions. To see how its graph depends on the size of the base a, do the following Graphing Exploration.

GRAPHING EXPLORATION

Using viewing window with $-3 \le x \le 7$ and $-2 \le y \le 18$, graph

$$f(x) = 1.3^x, \qquad g(x) = 2^x, \qquad \text{and} \qquad h(x) = 10^x$$

on the same screen. How is the steepness of the graph of $f(x) = a^x$ related to the size of a?

The Graphing Exploration illustrates these facts.

The Exponential Function
$$f(x) = a^x\,(a > 1)$$

When $a > 1$, the graph of $f(x) = a^x$ has the shape shown here and the properties listed below.

The graph is above the x-axis.

The y-intercept is 1.

$f(x)$ is an increasing function.

The negative x-axis is a horizontal asymptote.

The larger the base a, the more steeply the graph rises to the right.

The figure above doesn't really show how explosively an exponential graph rises.

EXAMPLE 1

Graph $f(x) = 2^x$ and estimate the height of the graph when $x = 50$.

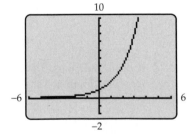

Figure 5–6

SOLUTION   A small portion of the graph is shown in Figure 5–6. If the x-axis were to be extended with the same scale, $x = 50$ would be at the right edge of the page. At that point, the height of the graph is $f(50) = 2^{50}$. Now the y-axis scale in Figure 5–6 is approximately 12 units to the inch, which is equivalent to 760,320 units per mile, as you can readily verify. Therefore, the height of the graph at $x = 50$ is

$$\frac{2^{50}}{760,320} = 1{,}480{,}823{,}741 \text{ MILES},$$

which would put that part of the graph well beyond the planet Saturn! ■

When the base a is between 0 and 1, then the graph of $f(x) = a^x$ has a different shape.

GRAPHING EXPLORATION

Using viewing window with $-4 \le x \le 4$ and $-1 \le y \le 4$, graph

$$f(x) = .2^x, \qquad g(x) = .4^x, \qquad h(x) = .6^x, \qquad \text{and} \qquad k(x) = .8^x$$

on the same screen. How is the steepness of the graph of $f(x) = a^x$ related to the size of a?

The exploration supports this conclusion.

The Exponential Function
$$f(x) = a^x \, (0 < a < 1)$$

When $0 < a < 1$, the graph of $f(x) = a^x$ has the shape shown here and the properties listed below.

The graph is above the x-axis.

The y-intercept is 1.

$f(x)$ is a decreasing function.

The positive x-axis is a horizontal asymptote.

The closer the base a is to 0, the more steeply the graph falls to the right.

EXAMPLE 2

Without graphing, describe the graph of $g(x) = 3^{-x}$.

SOLUTION   Note that

$$g(x) = 3^{-x} = (3^{-1})^x = \left(\frac{1}{3}\right)^x.$$

So $g(x)$ is an exponential function with a positive base less than 1. Its graph has the shape shown in the preceding box: It falls quickly to the right and rises very steeply to the left of the y-axis. ■

GRAPHING EXPLORATION

Verify the analysis in Example 2 by graphing $g(x) = 3^{-x}$ in the viewing window with $-4 \le x \le 4$ and $-2 \le y \le 18$.

Since most quantities that grow exponentially don't increase or decrease as dramatically as the graphs of $f(x) = a^x$, exponential functions that model real-life situations generally have the form $f(x) = Pa^{kx}$, such as

$$f(x) = 5 \cdot 2^{.45x}, \qquad g(x) = 3.5(10^{-.03x}), \qquad h(x) = (-6)(1.076^{2x}).$$

Their graphs have the same basic shape as the graph of $f(x) = a^x$, but rise or fall at different rates, depending on the constants P, a, and k.

EXAMPLE 3

Figure 5–7 shows the graphs of

$$f(x) = 3^x, \qquad g(x) = 3^{.15x}, \qquad h(x) = 3^{.35x}, \qquad k(x) = 3^{-x}, \qquad p(x) = 3^{-.4x}.$$

Note how the coefficient of x determines the steepness of the graph. When this coefficient is positive, the graph rises, and when it is negative, the graph falls from left to right.

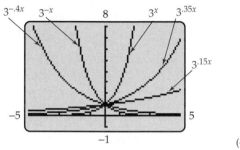

Figure 5–7   **Figure 5–8**

Figure 5–8 shows the graphs of

$$p(x) = 3^{-.4x}, \qquad q(x) = 4 \cdot 3^{-.4x}, \qquad r(x) = (-2)3^{-.4x}.$$

As we saw in Section 3.4, the graph of $q(x) = 4 \cdot 3^{-.4x}$ is the graph of $p(x) = 3^{-.4x}$ stretched away from the x-axis by a factor of 4. The graph of $r(x) = (-2)3^{-.4x}$ is the graph of $p(x) = 3^{-.4x}$ stretched away from the x-axis by a factor of 2 *and* reflected in the x-axis. ∎

⬛ EXPONENTIAL GROWTH

Exponential functions are useful for modeling situations in which a quantity increases by a fixed factor.

EXAMPLE 4

If you deposit $5000 in a savings account that pays 3% interest, compounded annually, how much money is in the account after nine years?

SOLUTION   After one year, the account balance is

$$5000 + 3\% \text{ of } 5000 = 5000 + (.03)5000 = 5000(1 + .03)$$

$$= 5000(1.03) = \$5150.$$

The initial balance has grown by a factor of 1.03. If you leave the $5150 in the account, then at the end of the second year, the balance is

$$5150 + 3\% \text{ of } 5150 = 5150 + (.03)5150 = 5150(1 + .03) = 5150(1.03).$$

Once again, the amount at the beginning of the year has grown by a factor of 1.03. The same thing happens every year. A balance of P dollars at the beginning of the year grows to $P(1.03)$. So the balance grows like this:

$$5000 \rightarrow \underset{\text{Year 1}}{5000(1.03)} \rightarrow \underbrace{\underset{\text{Year 2}}{[5000(1.03)](1.03)}}_{5000(1.03)^2} \rightarrow \underbrace{\underset{\text{Year 3}}{[5000(1.03)(1.03)](1.03)}}_{5000(1.03)^3} \rightarrow \cdots$$

Consequently, the balance at the end of year x is given by

$$f(x) = 5000 \cdot 1.03^x.$$

The balance at the end of nine years is $f(9) = 5000(1.03^9) = \$6523.87$ (rounded to the nearest penny). ■

EXAMPLE 5

The world population in 1980 was about 4.5 billion people and has been increasing at approximately 1.5% per year.

(a) Estimate the world population in 2005.

(b) In what year will the population be double what it is in 2005?

SOLUTION

(a) The world population in 1981 was

$$4.5 + 1.5\% \text{ of } 4.5 = 4.5 + .015(4.5) = 4.5(1 + .015) = 4.5(1.015).$$

Similarly, in each successive year, the population increased by a factor of 1.015, so the population (in billions) in year x is given by

$$g(x) = 4.5(1.015^x),$$

where $x = 0$ corresponds to 1980. The year 2005 corresponds to $x = 25$, so the population then is $g(25) = 4.5(1.015^{25}) \approx 6.53$ billion people.

(b) Twice the population in 2005 is $2(6.53) = 13.06$ billion. We must find the number x such that $g(x) = 13.06$; that is, we must solve the equation

$$4.5(1.015^x) = 13.06.$$

This can be done with an equation solver or by graphical means, as in Figure 5–9, which shows the intersection point of $y = 4.5(1.015^x)$ and $y = 13.06$. The solution is $x \approx 71.6$, which corresponds to the year 2051. Thus, the world population will double in your lifetime. This is what is meant by the population explosion. ■

Figure 5–9

Examples 4 and 5 illustrate **exponential growth.** The functions developed there, $f(x) = 5000(1.03^x)$ and $g(x) = 4.5(1.015^x)$, are typical of the general case.

Exponential
Growth

Exponential growth can be described by a function of the form

$$f(x) = Pa^x,$$

where $f(x)$ is the quantity at time x, P is the initial quantity (when $x = 0$) and $a > 1$ is the factor by which the quantity changes when x increases by 1.

If the quantity is growing at the rate r per time period, then $a = 1 + r$, and

$$f(x) = Pa^x = P(1 + r)^x.$$

EXAMPLE 6

At the beginning of an experiment, a culture contains 1000 bacteria. Five hours later, there are 7600 bacteria. Assuming that the bacteria grow exponentially, how many will there be after 24 hours?

SOLUTION   The bacterial population is given by $f(x) = Pa^x$, where P is the initial population, a is the change factor, and x is the time in hours. We are given that $P = 1000$, so $f(x) = 1000a^x$. The next step is to determine a. Since there are 7600 bacteria when $x = 5$, we have

$$7600 = f(5) = 1000a^5,$$

so

$$1000\, a^5 = 7600$$

$$a^5 = 7.6$$

$$a = \sqrt[5]{7.6} = (7.6)^{.2}.$$

Therefore, the population function is $f(x) = 1000(7.6^{.2})^x = 1000 \cdot (7.6)^{.2x}$. After 24 hours, the bacteria population will be

$$f(24) = 1000(7.6)^{.2(24)} \approx 16,900,721. \qquad \blacksquare$$

██ EXPONENTIAL DECAY

In some situations, a quantity *decreases* by a fixed factor as time goes on.

EXAMPLE 7

When tap water is filtered through a layer of charcoal and other purifying agents, 30% of the chemical impurities in the water are removed, and 70% remain. If the water is filtered through a second purifying layer, then the amount of impurities remaining is 70% of 70%, that is, $(.7)(.7) = .7^2 = .49$ or 49%. A third layer results in $.7^3$ of the impurities remaining. Thus, the function

$$f(x) = .7^x$$

Figure 5–10

gives the percentage of impurities remaining in the water after it passes through x layers of purifying material. How many layers are needed to ensure that 95% of the impurities are removed from the water?

SOLUTION   If 95% of the impurities are removed, then 5% will remain. Hence, we must find x such that $f(x) = .05$, that is, we must solve the equation $.7^x = .05$. This can be done numerically or graphically. Figure 5–10 shows that the solution is $x \approx 8.4$, so 8.4 layers of material are needed.

GRAPHING EXPLORATION

How many layers are needed to ensure that 99% of the impurities are removed? ■

Example 7 illustrates **exponential decay.** Note that the impurities were removed at a rate of 30% = .3 and that the amount of impurities remaining in the water was changing by a factor of $1 - .30 = .7$. The same thing is true in the general case.

Exponential Decay

Exponential decay can be described by a function of the form

$$f(x) = Pa^x,$$

where $f(x)$ is the quantity at time x, P is the initial quantity (when $x = 0$) and $0 < a < 1$. Here, a is the factor by which the quantity changes when x increases by 1.

If the quantity is decaying at the rate r per time period, then $a = 1 - r$, and

$$f(x) = Pa^x = P(1 - r)^x.$$

One of the important uses of exponential functions is to describe radioactive decay. The **half-life** of a radioactive element is the time it takes a given quantity to decay to one-half of its original mass. The half-life depends only on the substance and not on the size of the sample. Exercise 74 proves the following result.

Radioactive Decay

The mass $M(x)$ of a radioactive element at time x is given by

$$M(x) = c(.5^{x/h}),$$

where c is the original mass and h is the half-life of the element.

EXAMPLE 8

Plutonium (^{239}Pu) has a half-life of 24,360 years. So the amount remaining from 1 kilogram after x years is given by

$$M(x) = 1(.5^{x/24360}) = (.5^{1/24360})^x \approx .99997^x.$$

Thus, M is an exponential function whose base is very close to 1. Its graph falls *very slowly* from left to right, as you can easily verify by graphing M in a window with $0 \leq x \leq 2000$. This means that even after an extremely long time, a substantial amount of plutonium will remain. In fact, most of the original kilogram is still there after *ten thousand years* because $M(10{,}000) \approx .75$ kg. This is the reason that nuclear waste disposal is such a serious problem. ■

▪▪ THE NUMBER e AND THE NATURAL EXPONENTIAL FUNCTION

There is an irrational number, denoted e, that arises naturally in a variety of phenomena and plays a central role in the mathematical description of the physical universe. Its decimal expansion begins

$$e = 2.718281828459045 \ldots .$$

Your calculator has an e^x key that can be used to evaluate the **natural exponential function** $f(x) = e^x$. If you key in e^1, the calculator will display the first part of the decimal expansion of e.

The graph of $f(x) = e^x$ has the same shape as the graph of $g(x) = 2^x$ in Figure 5–6 but climbs more steeply.

TECHNOLOGY TIP

On most calculators, you use the e^x key, but not the x^y or $\wedge$ keys to enter the function $f(x) = e^x$.

GRAPHING EXPLORATION

Graph $f(x) = e^x$, $g(x) = 2^x$, and $h(x) = 3^x$ on the same screen in a window with $-5 \leq x \leq 5$. The Technology Tip in the margin may be helpful.

EXAMPLE 9

Population Growth   If the population of the United States continues to grow as it has recently, then the approximate population of the United States (in millions) in year t will be given by the function

$$P(t) = 227e^{.0093t},$$

where 1980 corresponds to $t = 0$.

(a) Estimate the population in 2015.

(b) When will the population reach half a billion?

SOLUTION

(a) The population in 2015 (that is, $t = 35$) will be approximately

$$P(35) = 227e^{.0093(35)} \approx 314.3 \text{ million people.}$$

(b) Half a billion is 500 million people. So we must find the value of t for which $P(t) = 500$, that is, we must solve the equation

$$227e^{.0093t} = 500.$$

Figure 5–11

This can be done graphically by finding the intersection of the graph of $P(t)$ and the horizontal line $y = 500$, which occurs when $t \approx 84.9$ (Figure 5–11). Therefore, the population will reach half a billion late in the year 2064. ■

▌▌ OTHER EXPONENTIAL FUNCTIONS

The population growth models in earlier examples do not take into account factors that may limit population growth in the future (wars, new diseases, etc.). Example 10 illustrates a function, called a **logistic model,** that is designed to model such situations more accurately.

EXAMPLE 10

Inhibited Population Growth   There is an upper limit on the fish population in a certain lake due to the oxygen supply, available food, etc. The population of fish in this lake at time t months is given by the function

$$p(t) = \frac{20{,}000}{1 + 24e^{-t/4}} \qquad (t \geq 0).$$

What is the upper limit on the fish population?

SOLUTION   The graph of $p(t)$ in Figure 5–12 suggests that the horizontal line $y = 20{,}000$ is a horizontal asymptote of the graph.

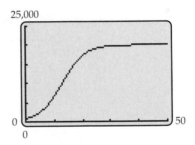

Figure 5–12

In other words, the fish population never goes above 20,000. You can confirm this algebraically by rewriting the rule of p in this form:

$$p(t) = \frac{20{,}000}{1 + 24e^{-t/4}} = \frac{20{,}000}{1 + \dfrac{24}{e^{t/4}}}.$$

When t is very large, so is $t/4$, which means that $e^{t/4}$ is huge. Hence, by the Big-Little Principle (page 272), $\dfrac{24}{e^{t/4}}$ is very close to 0, and $p(t)$ is very close to $\dfrac{20{,}000}{1 + 0} = 20{,}000$. Since $e^{t/4}$ is positive, the denominator of $p(t)$ is slightly bigger than 1, so $p(t)$ is always less than 20,000.   ■

When a cable, such as a power line, is suspended between towers of equal height, it forms a curve called a **catenary,** which is the graph of a function of the form

$$f(x) = A(e^{kx} + e^{-kx})$$

for suitable constants A and k. The Gateway Arch in St. Louis (Figure 5–13) has the shape of an inverted catenary, which was chosen because it evenly distributes the internal structural forces.

Figure 5–13

GRAPHING EXPLORATION

Graph each of the following functions in the window with $-5 \le x \le 5$ and $-10 \le y \le 80$.

$$y_1 = 10(e^{.4x} + e^{-.4x}), \qquad y_2 = 10(e^{2x} + e^{-2x}),$$
$$y_3 = 10(e^{3x} + e^{-3x}).$$

How does the coefficient of x affect the shape of the graph?

Predict the shape of the graph of $y = -y_1 + 80$. Confirm your answer by graphing.

 EXERCISES 5.2

In Exercises 1–6, sketch a complete graph of the function.

1. $f(x) = 4^{-x}$

2. $f(x) = (5/2)^{-x}$

3. $f(x) = 2^{3x}$

4. $g(x) = 3^{x/2}$

5. $f(x) = 2^{x^2}$

6. $g(x) = 2^{-x^2}$

In Exercises 7–12, list the transformations needed to transform the graph of $h(x) = 2^x$ into the graph of the given function. [Section 3.4 may be helpful.]

7. $f(x) = 2^x - 5$

8. $g(x) = -(2^x)$

9. $k(x) = 3(2^x)$

10. $g(x) = 2^{x-1}$

11. $f(x) = 2^{x+2} - 5$

12. $g(x) = -5(2^{x-1}) + 7$

In Exercises 13 and 14, match the functions to the graphs. Assume $a > 1$ and $c > 1$.

13. $f(x) = a^x$
  $g(x) = a^x + 3$
  $h(x) = a^{x+5}$

14. $f(x) = c^x$
$g(x) = -3c^x$
$h(x) = c^{x+5}$
$k(x) = -3c^x - 2$

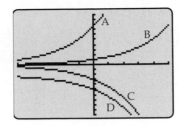

In Exercises 15–19, determine whether the function is even, odd, or neither (see Special Topics 3.4.A).

15. $f(x) = 10^x$

16. $g(x) = 2^x - x$

17. $f(x) = \dfrac{e^x + e^{-x}}{2}$

18. $f(x) = \dfrac{e^x - e^{-x}}{2}$

19. $f(x) = e^{-x^2}$

20. Use the Big-Little Principle to explain why $e^x + e^{-x}$ is approximately equal to e^x when x is large.

In Exercises 21–24, find the average rate of change of the function.

21. $f(x) = x2^x$ as x goes from 1 to 3

22. $g(x) = 3^{x^2 - x}$ as x goes from -1 to 1

23. $h(x) = 5^{-x^2}$ as x goes from -1 to 0

24. $f(x) = e^x - e^{-x}$ as x goes from -3 to -1

In Exercises 25–28, find the difference quotient of the function.

25. $f(x) = 10^x$

26. $g(x) = 5^{x^2}$

27. $f(x) = 2^x + 2^{-x}$

28. $f(x) = e^x - e^{-x}$

In Exercises 29–36, find a viewing window (or windows) that shows a complete graph of the function.

29. $k(x) = e^{-x}$

30. $f(x) = e^{-x^2}$

31. $f(x) = \dfrac{e^x + e^{-x}}{2}$

32. $h(x) = \dfrac{e^x - e^{-x}}{2}$

33. $g(x) = 2^x - x$

34. $k(x) = \dfrac{2}{e^x + e^{-x}}$

35. $f(x) = \dfrac{5}{1 + e^{-x}}$

36. $g(x) = \dfrac{10}{1 + 9e^{-x/2}}$

In Exercises 37–42, list all asymptotes of the graph of the function and the approximate coordinates of each local extremum.

37. $f(x) = x2^x$

38. $g(x) = x2^{-x}$

39. $h(x) = e^{x^2/2}$

40. $k(x) = 2^{x^2 - 6x + 2}$

41. $f(x) = e^{-x^2}$

42. $g(x) = -xe^{x^2/20}$

43. The population of a colony of fruit flies t days from now is given by the function $p(t) = 100 \cdot 3^{t/10}$.
  (a) What will the population be in 15 days? In 25 days?
  (b) When will the population reach 2500?

44. If current rates of deforestation and fossil fuel consumption continue, then the amount of atmospheric carbon dioxide in parts per million (ppm) will be given by $f(x) = 375e^{.00609x}$, where $x = 0$ corresponds to 2000.[*]
  (a) What is the amount of carbon dioxide in 2003? In 2022?
  (b) In what year will the amount of carbon dioxide reach 500 ppm?

45. The pressure of the atmosphere $p(x)$ (in pounds per square inch) is given by

$$p(x) = ke^{-.0000425x},$$

where x is the height above sea level (in feet) and k is a constant.
  (a) Use the fact that the pressure at sea level is 15 pounds per square inch to find k.
  (b) What is the pressure at 5000 feet?
  (c) If you were in a spaceship at an altitude of 160,000 feet, what would the pressure be?

46. (a) The function $g(t) = 1 - e^{-.0479t}$ gives the percentage of the population (expressed as a decimal) that has seen a new TV show t weeks after it goes on the air. What percentage of people have seen the show after 24 weeks?
  (b) Approximately when will 90% of the people have seen it?

47. According to data from the National Center for Health Statistics, the life expectancy at birth for a person born in a year x is approximated by the function

$$D(x) = \frac{79.257}{1 + 9.7135 \times 10^{24} \cdot e^{-.0304x}}$$

$$(1900 \le x \le 2050).$$

  (a) What is the life expectancy of someone born in 1980? in 2000?
  (b) In what year was life expectancy at birth 60 years?

*Based on projections from the International Panel on Climate Change.

48. The number of subscribers to basic cable TV (in millions) can be approximated by

$$g(x) = \frac{76.7}{1 + 16(.8444^x)},$$

where $x = 0$ corresponds to 1970.[*]
(a) Estimate the number of subscribers in 1995 and 2005.
(b) When does the number of subscribers reach 70 million?
(c) According to this model, will the number of subscribers ever each 90 million?

49. (a) The beaver population near a certain lake in year t is approximately

$$p(t) = \frac{2000}{1 + 199e^{-.5544t}}.$$

What is the population now ($t = 0$) and what will it be in 5 years?
(b) Approximately when will there be 1000 beavers?

50. The Gateway Arch (Figure 5–13) is 630 feet high and 630 feet wide at ground level. Suppose it were placed on a coordinate plane with the x-axis at ground level and the y-axis going through the center of the arch. Find a catenary function $g(x) = A(e^{kx} + e^{-kx})$ and a constant C such that the graph of the function $f(x) = g(x) + C$ provides a model of the arch. [*Hint:* Experiment with various values of A, k, C as in the Graphing Exploration on page 350. Many correct answers are possible.]

51. (a) A genetic engineer is growing cells in a fermenter. The cells multiply by splitting in half every 15 minutes. The new cells have the same DNA as the original ones. Complete the following table.

Time (hours)	Number of Cells
0	1
.25	2
.5	4
.75	
1	

(b) Write the rule of the function that gives the number of C cells at time t hours.

52. Do Exercise 51, using the following table.

Time (hours)	Number of Cells
0	300
.25	600
.5	1200
.75	
1	

53. A weekly census of the tree-frog population in Frog Hollow State Park produces the following results.

Week	1	2	3	4	5	6
Population	18	54	162	486	1458	4374

(a) Find a function of the form $f(x) = Pa^x$ that describes the frog population at time x weeks.
(b) What is the growth factor in this situation (that is, by what number must this week's population be multiplied to obtain next week's population)?
(c) Each tree frog requires 10 square feet of space and the park has an area of 6.2 square miles. Will the space required by the frog population exceed the size of the park in 12 weeks? In 14 weeks? [Remember: 1 square mile = 5280^2 square feet.]

54. An eccentric billionaire offers you a job for the month of September. She says that she will pay you 2¢ on the first day, 4¢ on the second day, 8¢ on the third day, and so on, doubling your pay on each successive day.
(a) Let $P(x)$ denote your salary in *dollars* on day x. Find the rule of the function P.
(b) Would you be better off financially if instead you were paid $10,000 per day? [*Hint:* Consider $P(30)$.]

55. Take an ordinary piece of typing paper and fold it in half; then the folded sheet is twice as thick as the single sheet was. Fold it in half again so that it is twice as thick as before. Keep folding it in half as long as you can. Soon the folded paper will be so thick and small that you will be unable to continue, but suppose you could keep folding the paper as many times as you wanted. Assume that the paper is .002 inches thick.
(a) Make a table showing the thickness of the folded paper for the first four folds (with fold 0 being the thickness of the original unfolded paper).
(b) Find a function of the form $f(x) = Pa^x$ that describes the thickness of the folded paper after x folds.
(c) How thick would the paper be after 20 folds?
(d) How many folds would it take to reach the moon (which is 243,000 miles from the earth)? [*Hint:* One mile is 5280 feet.]

[*]Based on data from *The Cable TV Financial Datebook* and *The Pay TV Newsletter*.

56. The figure is the graph of an exponential growth function $f(x) = Pa^x$.

(a) In this case, what is P? [*Hint:* What is $f(0)$?]

(b) Find the rule of the function f by finding a. [*Hint:* What is $f(2)$?]

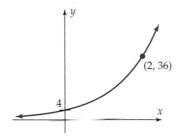

57. Suppose you invest $1200 in an account that pays 4% interest, compounded annually and paid from date of deposit to date of withdrawal.

(a) Find the rule of the function f that gives the amount you would receive if you closed the account after x years.

(b) How much would you receive after 3 years? After 5 years and 9 months?

(c) When should you close the account to receive $1850?

58. Anne now has a balance of $800 on her credit card, on which 1.5% interest per month is charged. Assume that she makes no further purchases or payments (and that the credit card company doesn't turn her account over to a bill collector).

(a) Find the rule of the function g that gives Anne's total credit card debt after x months.

(b) How much will Anne owe after one year? After two years?

(c) When will she owe twice the amount she owes now?

59. The population of Mexico was 100.4 million in 2000 and is expected to grow at the rate of 1.4% per year.

(a) Find the rule of the function f that gives Mexico's population (in millions) in year x, with $x = 0$ corresponding to 2000.

(b) Estimate Mexico's population in 2010.

(c) When will the population reach 125 million people?

60. The number of digital devices (such as MP3 players, handheld computers, cell phones, and PCs) in the world was approximately .94 billion in 1999 and is growing at a rate of 28.3% a year.*

(a) Find the rule of a function that gives the number of digital devices (in billions) in year x, with $x = 0$ corresponding to 1999.

(b) Approximately how many digital devices will be in use in 2004?

(c) If this model remains accurate, when will the number of digital devices reach 6 billion?

61. The U.S. Census Bureau estimates that the Hispanic population in the United States will increase from 32.44 million in 2000 to 98.23 million in 2050.[†]

(a) Find an exponential function that gives the Hispanic population in year x, with $x = 0$ corresponding to 2000.

(b) What is the projected Hispanic population in 2010 and 2025?

(c) In what year will the Hispanic population reach 55 million?

62. The U.S. Department of Commerce estimated that there were 54 million Internet users in the United States in 1999 and 85 million in 2002.

(a) Find an exponential function that models the number of Internet users in year x, with $x = 0$ corresponding to 1999.

(b) For how long is this model likely to remain accurate? [*Hint:* The current U.S. population is about 230 million.]

63. At the beginning of an experiment, a culture contains 200 *H. pylori* bacteria. An hour later there are 205 bacteria. Assuming that the *H. pylori* bacteria grow exponentially, how many will there be after 10 hours? After 2 days?

64. The population of India was approximately 1030 million in 2001 and was 865 million a decade earlier. If the population continues to grow exponentially at the same rate, what will it be in 2006?

65. Kerosene is passed through a pipe filled with clay to remove various pollutants. Each foot of pipe removes 25% of the pollutants.

(a) Write the rule of a function that gives the percentage of pollutants remaining in the kerosene after it has passed through x feet of pipe. [See Example 7.]

(b) How many feet of pipe are needed to ensure that 90% of the pollutants have been removed from the kerosene?

66. If inflation runs at a steady 3% per year, then the amount a dollar is worth decreases by 3% each year.

(a) Write the rule of a function that gives the value of a dollar in year x.

(b) How much will the dollar be worth in 5 years? In 10 years?

(c) How many years will it take before today's dollar is worth only a dime?

67. You have 5 grams of carbon-14, whose half-life is 5730 years.

(a) Write the rule of the function that gives the amount of carbon-14 remaining after x years. [See the box preceding Example 8.]

(b) How much carbon-14 will be left after 4000 years? After 8000 years?

(c) When will there be just 1 gram left?

---

*Based on data and projections from IDC.

[†]*Statistical Abstract of the United States: 2001.*

68. (a) The half-life of radium is 1620 years. If you start with 100 milligrams of radium, what is the rule of the function that gives the amount remaining after t years?

   (b) How much radium is left after 800 years? After 1600 years? After 3200 years?

Thinkers

69. Find a function $f(x)$ with the property $f(r + s) = f(r)f(s)$ for all real numbers r and s. [*Hint:* Think exponential.]

70. Find a function $g(x)$ with the property $g(2x) = (g(x))^2$ for every real number x.

71. (a) Using the viewing window with $-4 \le x \le 4$ and $-1 \le y \le 8$, graph $f(x) = \left(\frac{1}{2}\right)^x$ and $g(x) = 2^x$ on the same screen. If you think of the y-axis as a mirror, how would you describe the relationship between the two graphs?

   (b) Without graphing, explain how the graphs of $g(x) = 2^x$ and $k(x) = 2^{-x}$ are related.

72. Look back at Section 4.4, where the basic properties of graphs of polynomial functions were discussed. Then review the basic properties of the graph of $f(x) = a^x$ discussed in this section. Using these various properties, give an argument to show that for any fixed positive number $a(\ne 1)$, it is *not* possible to find a polynomial function $g(x) = c_n x^n + \cdots + c_1 x + c_0$ such that $a^x = g(x)$ for *all* numbers x. In other words, *no exponential function is a polynomial function*. However, see Exercise 73.

73. **Approximating exponential functions by polynomials.** For each positive integer n, let f_n be the polynomial function whose rule is

$$f_n(x) = 1 + x + \frac{x^2}{2!} + \frac{x^3}{3!} + \frac{x^4}{4!} + \cdots + \frac{x^n}{n!},$$

where $k!$ is the product $1 \cdot 2 \cdot 3 \cdots k$.

   (a) Using the viewing window with $-4 \le x \le 4$ and $-5 \le y \le 55$, graph $g(x) = e^x$ and $f_4(x)$ on the same screen. Do the graphs appear to coincide?

   (b) Replace the graph of $f_4(x)$ by that of $f_5(x)$, then by $f_6(x), f_7(x)$, and so on until you find a polynomial $f_n(x)$ whose graph appears to coincide with the graph of $g(x) = e^x$ In this viewing window. Use the trace feature to move from graph to graph at the same value of x to see how accurate this approximation is.

   (c) Change the viewing window so that $-6 \le x \le 6$ and $-10 \le y \le 400$. Is the polynomial you found in part (b) a good approximation for $g(x)$ in this viewing window? What polynomial is?

74. This exercise provides a justification for the claim that the function $M(x) = c(.5)^{x/h}$ gives the mass after x years of a radioactive element with half-life h years. Suppose we have c grams of an element that has a half-life of 50 years. Then after 50 years, we would have $c\left(\frac{1}{2}\right)$ grams. After another 50 years, we would have half of that, namely, $c\left(\frac{1}{2}\right)\left(\frac{1}{2}\right) = c\left(\frac{1}{2}\right)^2$.

   (a) How much remains after a third 50-year period? After a fourth 50-year period?

   (b) How much remains after t 50-year periods?

   (c) If x is the number of years, then $x/50$ is the number of 50-year periods. By replacing the number of periods t in part (b) by $x/50$, you obtain the amount remaining after x years. This gives the function $M(x)$ when $h = 50$. The same argument works in the general case (just replace 50 by h).

5.2.A   *SPECIAL TOPICS* Compound Interest and the Number *e*

When money earns compound interest, as in Example 4 on page 344, the exponential growth function can be described as follows.

Compound Interest Formula

If P dollars is invested at interest rate r per time period (expressed as a decimal), then the amount A after t periods is

$$A = P(1 + r)^t$$

EXAMPLE 1

Suppose you borrow $50 from your friendly neighborhood loan shark, who charges 18% interest per week. How much do you owe after one year (assuming that he lets you wait that long to pay)?

SOLUTION   You use the compound interest formula with $P = 50$, $r = .18$, and $t = 52$ (because interest is compounded weekly and there are 52 weeks in a year). So you figure that you owe

$$A = P(1 + r)^t = 50(1 + .18)^{52} = 50 \cdot 1.18^{52} = \$273,422.58.^*$$

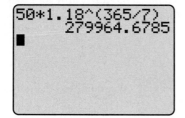

Figure 5–14

When you try to pay the loan shark this amount, however, he points out that a 365-day year has more than 52 weeks, namely, $\dfrac{365}{7} = 52\dfrac{1}{7}$ weeks. So you recalculate with $t = 365/7$ (and careful use of parentheses, as shown in Figure 5–14) and find that you actually owe

$$A = P(1 + r)^t = 50(1 + .18)^{365/7} = 50 \cdot 1.18^{365/7} = \$279,964.68.$$

Ouch!   ■

As Example 1 illustrates, the compound interest formula can be used even when the number of periods t is not an integer. You must also learn how to read "financial language" to apply the formula correctly, as shown in the following example.

EXAMPLE 2

Determine the amount a $3500 investment is worth after three and a half years at the following interest rates:

(a) 6.4% compounded annually;

(b) 6.4% compounded quarterly;

(c) 6.4% compounded monthly.

SOLUTION

(a) Using the compound interest formula with $P = 3500$, $r = .064$, and $t = 3.5$, we have

$$A = 3500(1 + .064)^{3.5} = \$4348.74.$$

(b) "6.4% interest, compounded quarterly" means that the interest period is one-fourth of a year and the interest rate per period is $.064/4 = .016$. Since there are four interest periods per year, the number of periods in 3.5 years is $4(3.5) = 14$, so

$$A = 3500\left(1 + \frac{.064}{4}\right)^{14} = 3500(1 + .016)^{14} = \$4370.99.$$

---

*Here and below, all financial answers are rounded to the nearest penny.

(c) Similarly, "6.4% compounded monthly" means that the interest period is one month (1/12 of a year) and the interest rate per period is .064/12. The number of periods (months) in 3.5 years is 42, so

$$A = 3500\left(1 + \frac{.064}{12}\right)^{42} = \$4376.14.$$

Note that the more often interest is compounded, the larger the final amount. ■

EXAMPLE 3

If \$5000 is invested at 6.5% annual interest, compounded monthly, how long will it take for the investment to double?

SOLUTION
The compound interest formula (with $P = 5000$ and $r = .065/12$) shows that the amount in the account after t months is $5000\left(1 + \frac{.065}{12}\right)^t$. We must find the value of t such that

$$5000\left(1 + \frac{.065}{12}\right)^t = 10{,}000.$$

Algebraic methods for solving this equation will be considered in Section 5.5. For now, we use technology.

GRAPHING EXPLORATION

Solve the equation, either by using an equation solver, or by graphical means as follows. Graph

$$y = 5000\left(1 + \frac{.065}{12}\right)^t - 10{,}000$$

in a viewing window with $0 \le t \le 240$ (that's 20 years) and find the t-intercept.

The exploration shows that it will take 128.3 months (approximately 10.7 years) for the investment to double.   ■

EXAMPLE 4

What interest rate, compounded annually, is needed for a \$16,000 investment to grow to \$50,000 in 18 years?

SOLUTION   In the compound interest formula, we have $A = 50{,}000$, $P = 16{,}000$ and $t = 18$. We must find r in the equation

$$16{,}000(1 + r)^{18} = 50{,}000.$$

The equation can be solved numerically with an equation solver or by one of the following methods.

Figure 5–15

Graphical: Rewrite the equation as $16,000(1 + r)^{18} - 50,000 = 0$. Then the solution is the r-intercept of the graph of $y = 16,000(1 + r)^{18} - 50,000$, as shown in Figure 5–15.

Algebraic:

$$16,000(1 + r)^{18} = 50,000$$

Divide both sides by 16,000: $$(1 + r)^{18} = \frac{50,000}{16,000} = 3.125$$

Take 18th roots on both sides: $$\sqrt[18]{(1 + r)^{18}} = \sqrt[18]{3.125}$$

$$1 + r = \sqrt[18]{3.125}$$

$$r - \sqrt[18]{3.125} - 1 \approx .06535.$$

So the necessary interest rate is about 6.535%. ■

⚏ CONTINUOUS COMPOUNDING AND THE NUMBER *e*

As a general rule, the more often interest is compounded, the better off you are, as we saw in Example 2. But there is, alas, a limit.

EXAMPLE 5

The Number *e* You have \$1 to invest for 1 year. The Exponential Bank offers to pay 100% annual interest, compounded *n* times per year and rounded to the nearest penny. You may pick any value you want for *n*. Can you choose *n* so large that your \$1 will grow to some huge amount?

SOLUTION Since interest rate is compounded *n* times per year and the annual rate is 100% (= 1.00), the interest rate per period is $r = 1/n$ and the number of periods in 1 year is *n*. According to the formula, the amount at the end of the year will be $A = \left(1 + \dfrac{1}{n}\right)^n$. Here's what happens for various values of *n*:

Interest Is Compounded	$n =$	$\left(1 + \dfrac{1}{n}\right)^n =$
Annually	1	$\left(1 + \frac{1}{1}\right)^1 = 2$
Semiannually	2	$\left(1 + \frac{1}{2}\right)^2 = 2.25$
Quarterly	4	$\left(1 + \frac{1}{4}\right)^4 \approx 2.4414$
Monthly	12	$\left(1 + \frac{1}{12}\right)^{12} \approx 2.6130$
Daily	365	$\left(1 + \frac{1}{365}\right)^{365} \approx 2.71457$
Hourly	8760	$\left(1 + \frac{1}{8760}\right)^{8760} \approx 2.718127$
Every minute	525,600	$\left(1 + \frac{1}{525,600}\right)^{525,600} \approx 2.7182792$
Every second	31,536,000	$\left(1 + \frac{1}{31,536,000}\right)^{31,536,000} \approx 2.7182818$

Since interest is rounded to the nearest penny, your dollar will grow no larger than \$2.72, no matter how big *n* is. ■

The last entry in the preceding table, 2.7182818, is the number e to seven decimal places. This is just one example of how e arises naturally in real-world situations. In calculus, it is provided that e is the *limit* of $\left(1 + \dfrac{1}{n}\right)^n$, meaning that as n gets larger and larger, $\left(1 + \dfrac{1}{n}\right)^n$ gets closer and closer to e.

GRAPHING EXPLORATION

Confirm this fact graphically by graphing the function

$$f(x) = \left(1 + \frac{1}{x}\right)^x$$

and the horizontal line $y = e$ in the viewing window with $0 \le x \le 5000$ and $-1 \le y \le 4$ and noting that the two graphs appear to be identical.

When interest is compounded n times per year for larger and larger values of n, as in Example 5, we say that the interest is **continuously compounded.** In this terminology, Example 5 says that \$1 will grow to \$2.72 in 1 year at an interest rate of 100% compounded continuously. A similar argument with more realistic interest rates (see Exercise 30) produces the following result (Example 5 is the case when $P = 1$, $r = 1$, and $t = 1$).

Continuous Compounding

If P dollars is invested at interest rate r, compounded continuously, then the amount A after t years is

$$A = Pe^{rt}.$$

EXAMPLE 6

If \$3800 is invested at 9.2% interest, compounded continuously, find:

(a) The amount in the account after seven and a half years.

(b) The number of years for the account balance to reach \$5000.

SOLUTION

(a) Apply the continuous compounding formula with $P = 3800$, $r = .092$, and $t = 7.5$.

$$A = 3800e^{(.092)7.5} = 3800e^{.69} \approx \$7576.12.$$

(b) We must solve the equation

$$3800e^{.092t} = 5000, \qquad \text{or, equivalently,} \qquad 3800e^{.092t} - 5000 = 0.$$

GRAPHING EXPLORATION

Solve the equation graphically and verify that it will take just a few days less than three years for the investment to be worth \$5000.

∎

EXERCISES 5.2.A

1. If $1,000 is invested at 8%, find the value of the investment after 5 years if interest is compounded
   (a) annually.   (b) quarterly.   (c) monthly.
   (d) weekly.

2. If $2500 is invested at 11.5%, what is the value of the investment after 10 years if interest is compounded
   (a) annually?   (b) monthly?   (c) daily?

In Exercises 3–12, determine how much money will be in a savings account if the initial deposit was $500 and the interest rate is:

3. 2% compounded annually for 8 years.

4. 2% compounded annually for 10 years.

5. 2% compounded quarterly for 10 years.

6. 2.3% compounded monthly for 9 years.

7. 2.9% compounded daily for 8.5 years.

8. 3.5% compounded weekly for 7 years and 7 months.

9. 3% compounded continuously for 4 years.

10. 3.5% compounded continuously for 10 years.

11. 2.45% compounded continuously for 6.2 years.

12. 3.25% compounded continuously for 11.6 years.

*A sum of money P that can be deposited today to yield some larger amount A in the future is called the **present value** of A. In Exercises 13–16, find the present value of the given amount A. [Hint: Substitute A, the interest rate per period r, and the number t of periods in the compound interest formula and solve for P.]*

13. $5000 at 6% compounded annually for 7 years.

14. $3500 at 5.5% compounded annually for 4 years.

15. $4800 at 7.2% compounded quarterly for 5 years.

16. $7400 at 5.9% compounded quarterly for 8 years.

17. You have $10,000 to invest for two years. Fund *A* pays 13.2% interest, compounded annually. Fund *B* pays 12.7% interest, compounded quarterly. Fund *C* pays 12.6% interest, compounded monthly. Which fund will return the most money?

18. If you invest $7400 for five years, are you better off with an interest rate of 5% compounded quarterly or 4.8% compounded continuously?

19. If you borrow $1200 at 14% interest, compounded monthly, and pay off the loan (principal and interest) at the end of two years, how much interest will you have paid?

20. A developer borrows $150,000 at 6.5% interest, compounded quarterly, and agrees to pay off the loan in four years. How much interest will she owe?

21. A manufacturer has settled a lawsuit out of court by agreeing to pay $1.5 million four years from now. At this time, how much should the company put in an account paying 6.4% annual interest, compounded monthly, to have $1.5 million in four years? [*Hint:* Exercises 13–16.]

22. Lisa Chow wants to have $30,000 available in five years for a down payment on a house. She has inherited $25,000. How much of the inheritance should be invested at 5.7% annual interest, compounded quarterly, to accumulate the $30,000?

23. If an investment of $1000 grows to $1407.10 in seven years with interest compounded annually, what is the interest rate?

24. If an investment of $2000 grows to $2700 in three and a half years, with an annual interest rate that is compounded quarterly, what is the annual interest rate?

25. If you put $3000 in a savings account today, what interest rate (compounded annually) must you receive in order to have $4000 after five years?

26. If interest is compounded continuously, what annual rate must you receive if your investment of $1500 is to grow to $2100 in six years?

27. At an interest rate of 8% compounded annually, how long will it take to double an investment of
   (a) $100   (b) $500   (c) $1200?
   (d) What conclusion about doubling time do parts (a)–(c) suggest?

28. At an interest rate of 6% compounded annually, how long will it take to double an investment of *P* dollars?

29. How long will it take to double an investment of $500 at 7% annual interest, compounded continuously?

30. This exercise provides an illustration of why the continuous compounding formula (page 358) is valid, using a realistic interest rate. We shall determine the value of $4000 deposited for three years at 5% interest compounded *n* times per year for larger and larger values of *n*. In this case, the interest rate per period is $.05/n$ and the number of periods in three years is $3n$. So the amount in the account at the end of three years is:

$$A = 4000\left(1 + \frac{.05}{n}\right)^{3n} = 4000\left[\left(1 + \frac{.05}{n}\right)^{n}\right]^{3}.$$

continues

(a) Fill in the missing entries in the following table.

n	$\left(1 + \dfrac{.05}{n}\right)^n$
1,000	
10,000	
500,000	
1,000,000	
5,000,000	
10,000,000	

(b) Compare the entries in the second volume of the table with the number $e^{.05}$ and fill the blank in the following sentence:

As n gets larger and larger, the value of $\left(1 + \dfrac{.05}{n}\right)^n$ gets closer and closer to the number _____.

(c) Use you answer to part (b) to fill the blank in the following sentence:

As n gets larger and larger, the value of

$$A = 4000\left[\left(1 + \frac{.05}{n}\right)^n\right]^3$$

gets closer and closer to _____.

(d) Compare your answer in part (c) to the value of the investment given by the continuous compounding formula.

5.3   Common and Natural Logarithmic Functions*

Roadmap

We begin with the only logarithms that are in widespread use, common and natural logarithms. Natural logarithms are emphasized because of their central role in calculus. Those who prefer to begin with logarithms to an arbitrary base b should cover Special Topics 5.4.A before reading this section.

From their invention in the seventeenth century until the development of computers and calculators, logarithms were the only effective tool for numerical computation in astronomy, chemistry, physics, and engineering. Although they are no longer needed for computation, logarithmic functions still play an important role in the sciences and engineering.

▪▪ COMMON LOGARITHMS

The exponential function $f(x) = 10^x$, whose graph is shown in Figure 5–16, is an increasing function and hence is one-to-one (as explained on page 205). Therefore, f has an inverse function g whose graph is the reflection of the graph of f in the line $y = x$ (see page 210), as shown in Figure 5–17.[†]

Figure 5–16   Figure 5–17

*Section 3.7 (Inverse Functions) is a prerequisite for this section.
[†]The graph of $f(x) = 10^x$ can be obtained in parametric mode by letting

$$x = t \quad \text{and} \quad y = 10^t \quad (t \text{ any real number}).$$

As explained on pages 209–210, the graph of the inverse function g can then be obtained by letting

$$x = 10^t \quad \text{and} \quad y = t \quad (t \text{ any real number}).$$

This inverse function *g* is called the **common logarithmic function.** The value of this function at the number *x* is denoted **log *x*** and called the **common logarithm** of the number *x*. Every calculator has a LOG key for evaluating the function $g(x) = \log x$. For instance,

$$\log .6 = -.2218 \qquad \text{and} \qquad \log 327 = 2.5145.^*$$

As we saw in Section 3.7, the relationship between a function *f* and its inverse function *g* is given by

$$g(v) = u \qquad \text{exactly when} \qquad f(u) = v.$$

When $f(x) = 10^x$ and $g(x) = \log x$, this statement takes the following form.

Definition of Common Logarithms

> Let *u* and *v* be real numbers, with $v > 0$. Then
>
> $$\log v = u \qquad \text{exactly when} \qquad 10^u = v.$$
>
> In other words,
>
> $\qquad$ log *v* is the exponent to which 10 must be raised to produce *v*.

EXAMPLE 1

Without using a calculator, find

(a) log 1000 $\qquad\qquad$ (b) log 1 $\qquad\qquad$ (c) $\log \sqrt{10}$

SOLUTION

(a) To find log 1000, ask yourself, "What power of 10 equals 1000?" The answer is 3 because $10^3 = 1000$. Therefore, log 1000 = 3.

(b) What power must 10 be raised to in order to produce 1? Since $10^0 = 1$, we conclude that log 1 = 0.

(c) Log $\sqrt{10} = 1/2$ because 1/2 is the exponent to which 10 must be raised to produce $\sqrt{10}$, that is $10^{1/2} = \sqrt{10}$. ∎

EXAMPLE 2

Translate each of the following logarithmic statements into an equivalent exponential statement.

$$\log 29 = 1.4624 \qquad \log .47 = -.3279 \qquad \log (k + t) = d.$$

---

*Here and below, all logarithms are rounded to four decimal places, and an equal sign is used rather than the more correct "approximately equal." The word "common" will be omitted except when it is necessary to distinguish these logarithms from other types that are introduced below.

SOLUTION   Using the preceding box, we have these translations.

Logarithmic Statement	**Equivalent Exponential Statement**
$\log 29 = 1.4624$	$10^{1.4624} = 29$
$\log .47 = -.3279$	$10^{-.3279} = .47$
$\log (k + t) = d$	$10^d = k + t$ ∎

EXAMPLE 3

Translate each of the following exponential statements into an equivalent logarithmic statement.

$$10^{5.5} = 316{,}227.766 \qquad 10^{.66} = 4.5708819 \qquad 10^{rs} = t$$

SOLUTION   Translate as follows.

Exponential Statement	**Equivalent Logarithmic Statement**
$10^{5.5} = 316{,}277.766$	$\log 316{,}277.766 = 5.5$
$10^{.66} = 4.5708819$	$\log 4.5708819 = .66$
$10^{rs} = t$	$\log t = rs$ ∎

EXAMPLE 4

Solve the equation $\log x = 4$.

SOLUTION $\log x = 4$ is equivalent to $10^4 = x$. So the solution is $x = 10{,}000$.   ∎

▟ NATURAL LOGARITHMS

Common logarithms are closely related to the exponential function $f(x) = 10^x$. With the advent of calculus, however, it became clear that the most useful exponential function in science and engineering is $g(x) = e^x$. Consequently, a new type of logarithm, based on the number e instead of 10, was developed. This development is essentially a carbon copy of what was done above, with some minor changes in notation.

The exponential function $f(x) = e^x$ whose graph is shown in Figure 5–18 is increasing and hence one-to-one, so f has an inverse function g whose graph is the reflection of the graph of f in the line $y = x$, as shown in Figure 5–19.

Figure 5–18   Figure 5–19

This inverse function g is called the **natural logarithmic function.** The value $g(x)$ of this function at a number x is denoted **ln x** and called the **natural logarithm** of the number x. Every calculator has an LN key for evaluating natural logarithms. For instance,

$$\ln .15 = -1.8971 \qquad \text{and} \qquad \ln 186 = 5.2257.$$

When the relationship of inverse functions (Section 3.7)

$$g(v) = u \qquad \text{exactly when} \qquad f(u) = v$$

is applied to the function $f(x) = e^x$ and its inverse $g(x) = \ln x$, it says the following.

Definition of Natural Logarithms

Let u and v be real numbers, with $v > 0$. Then

$$\ln v = u \qquad \text{exactly when} \qquad e^u = v.$$

In other words,

$\ln v$ is the exponent to which e must be raised to produce v.

EXAMPLE 5

Translate:

(a) $\ln 14 = 2.6391$ into an equivalent exponential statement.

(b) $e^{5.0626} = 158$ into an equivalent logarithmic statement.

SOLUTION

(a) Using the preceding definition, we see that $\ln 14 = 2.6391$ is equivalent to $e^{2.6391} = 14$.

(b) Similarly, $e^{5.0626} = 158$ is equivalent to $\ln 158 = 5.0626$. ∎

⠿ PROPERTIES OF LOGARITHMS

Since common and natural logarithms have almost identical definitions (just replace 10 by e), it is not surprising that they share the same essential properties. You don't need a calculator to understand these properties. You need only use the definition of logarithms or translate logarithmic statements into equivalent exponential ones (or vice versa).

EXAMPLE 6

What is $\ln(-10)$?

Translation: To what power must e be raised to produce -10?

Answer: The graph of $f(x) = e^x$ in Figure 5–18 shows that every power of e is *positive*. So e^x can *never* be -10 or any negative number or zero, and hence, ln (-10) is not defined. Similarly, log(-10) is not defined because every power of 10 is positive. Therefore,

ln v and log v are defined only when $v > 0$. ■

EXAMPLE 7

What is ln 1?

Translation: To what power must e be raised to produce 1?

Answer: We know that $e^0 = 1$, which means that ln 1 = 0. Combining this fact with Example 1(b), we have

ln 1 = 0   and   log 1 = 0. ■

EXAMPLE 8

What is ln e^9?

Translation: To what power must e be raised to produce e^9?

Answer: Obviously, the answer is 9. So ln $e^9 = 9$ and in general

ln $e^k = k$ for every real number k.

Similarly,

log $10^k = k$ for every real number k

because k is the exponent to which 10 must be raised to produce 10^k. In particular, when $k = 1$, we have

ln $e = 1$ and   log 10 = 1. ■

EXAMPLE 9

Find $10^{\log 678}$ and $e^{\ln 678}$.

SOLUTION   By definition, log 678 is the exponent to which 10 must be raised to produce 678. So if you raise 10 to this exponent, the answer will be 678, that is, $10^{\log 678} = 678$. Similarly, ln 678 is the exponent to which e must be raised to produce 678, so that $e^{\ln 678} = 678$. The same argument works with any positive number v in place of 678:

$e^{\ln v} = v$ and $10^{\log v} = v$ for every $v > 0$. ■

The facts presented in the preceding examples may be summarized as follows.

Properties of Logarithms

Natural Logarithms	Common Logarithms
1. $\ln v$ is defined only when $v > 0$;	$\log v$ is defined only when $v > 0$.
2. $\ln 1 = 0$ and $\ln e = 1$;	$\log 1 = 0$ and $\log 10 = 1$.
3. $\ln e^k = k$ for every real number k;	$\log 10^k = k$ for every real number k.
4. $e^{\ln v} = v$ for every $v > 0$;	$10^{\log v} = v$ for every $v > 0$.

EXAMPLE 10

Applying Property 3 with $k = 2x^2 + 7x + 9$ shows that

$$\ln e^{2x^2+7x+9} = 2x^2 + 7x + 9. \quad \blacksquare$$

EXAMPLE 11

Solve the equation $\ln(x + 1) = 2$.

SOLUTION   Since $\ln(x + 1) = 2$, we have

$$e^{\ln(x+1)} = e^2.$$

Applying Property 4 with $v = x + 1$ shows that

$$x + 1 = e^{\ln(x+1)} = e^2$$

$$x = e^2 - 1 \approx 6.3891. \quad \blacksquare$$

Property 4 has another interesting consequence. If a is any positive number, then $e^{\ln a} = a$. Hence, the rule of the exponential function $f(x) = a^x$ can be written as

$$f(x) = a^x = (e^{\ln a})^x = e^{(\ln a)x}.$$

For example, $f(x) = 2^x = e^{(\ln 2)x} \approx e^{.6931x}$. Thus, we have this useful result.

Exponential Functions

Every exponential growth or decay function can be written in the form

$$f(x) = Pe^{kx},$$

where $f(x)$ is the amount at time x, P is the initial quantity, and k is positive for growth and negative for decay.

▚ GRAPHS OF LOGARITHMIC FUNCTIONS

Figure 5–20 shows the graphs of two more entries in the catalog of basic functions, $f(x) = \log x$ and $g(x) = \ln x$. Both are increasing functions with these four properties:

Domain: all positive real numbers   ***x*-intercept:** 1
Range: all real numbers   **Vertical Asymptote:** *y*-axis

 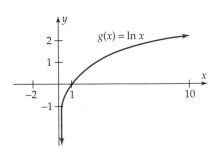

Figure 5–20

Calculators and computers do not accurately show that the *y*-axis is a vertical asymptote of these graphs. By evaluating the functions at very small numbers (such as $x = 1/10^{500}$), you can see that the graphs go lower and lower as x gets closer to 0. On a calculator, however, the graph will appear to end abruptly near the *y*-axis (try it!).

EXAMPLE 12

Sketch the graph of $f(x) = \ln (x - 2)$.

SOLUTION   Using a calculator to graph $f(x) = \ln (x - 2)$, we obtain Figure 5–21, in which the graph appears to end abruptly near $x = 2$. Fortunately, however, we have read Section 3.4, so we know that this is *not* how the graph looks. From Section 3.4, we know that the graph of $f(x) = \ln (x - 2)$ is the graph of $g(x) = \ln x$ shifted horizontally 2 units to the right, as shown in Figure 5–22. In particular, the graph of f has a vertical asymptote at $x = 2$ and drops sharply downward there.   ∎

Figure 5–21   **Figure 5–22**

EXPLORATION

Graph $y_1 = \ln (x - 2)$ and $y_2 = -5$ in the same viewing window and verify that the graphs do not appear to intersect, as they should. Nevertheless, try to solve the equation $\ln (x - 2) = -5$ by finding the intersection point of y_1 and y_2. Some calculators will find the intersection point even through it does not show on the screen. Others produce an error message, in which case the SOLVER feature should be used instead of a graphical solution.

Although logarithms are only defined for positive numbers, many logarithmic functions include negative numbers in their domains.

EXAMPLE 13

Find the domain of each of the following functions.

(a) $f(x) = \ln (x + 4)$ (b) $g(x) = \log x^2$

SOLUTION

(a) $f(x) = \ln (x + 4)$ is defined only when $x + 4 > 0$, that is, when $x > -4$. So the domain of f consists of all real numbers greater than -4.

(b) Since $x^2 > 0$ for all nonzero x, the domain of $g(x) = \log x^2$ consists of all real numbers except 0. ■

EXPLORATION

Verify the conclusions of Example 13 by graphing each of the functions. What is the vertical asymptote of each graph?

EXERCISES 5.3

Unless stated otherwise, all letters represent positive numbers.

In Exercises 1–4, find the logarithm, without using a calculator.

1. $\log 10{,}000$ **2.** $\log .001$

3. $\log \dfrac{\sqrt{10}}{1000}$ **4.** $\log \sqrt[3]{.01}$

In Exercises 5–14, translate the given logarithmic statement into an equivalent exponential statement.

5. $\log 1000 = 3$ **6.** $\log .001 = -3$

7. $\log 750 = 2.88$ **8.** $\log (.8) = -.097$

9. $\ln 3 = 1.0986$ **10.** $\ln 10 = 2.3026$

11. $\ln .01 = -4.6052$ **12.** $\ln s = r$

13. $\ln (x^2 + 2y) = z + w$ **14.** $\log (a + c) = d$

In Exercises 15–24, translate the given exponential statement into an equivalent logarithmic one.

15. $10^{-2} = .01$ **16.** $10^3 = 1000$

17. $10^{.4771} = 3$ **18.** $10^{7k} = r$

19. $e^{3.25} = 25.79$ **20.** $e^{-4} = .0183$

21. $e^{12/7} = 5.5527$ **22.** $e^k = t$

23. $e^{2/r} = w$ **24.** $e^{4uv} = m$

In Exercises 25–36, evaluate the given expression without using a calculator.

25. $\log 10^{\sqrt{43}}$

26. $\log 10^{\sqrt{x^2+y^2}}$

27. $\ln e^{15}$

28. $\ln e^{3.78}$

29. $\ln \sqrt{e}$

30. $\ln \sqrt[5]{e}$

31. $e^{\ln 931}$

32. $e^{\ln 34.17}$

33. $\ln e^{x+y}$

34. $\ln e^{x^2+2y}$

35. $e^{\ln x^2}$

36. $e^{\ln \sqrt{x+3}}$

In Exercises 37–40, write the rule of the function in the form $f(x) = Pe^{kx}$. (See the discussion and box after Example 11.)

37. $f(x) = 4(25^x)$

38. $g(x) = 3.9(1.03^x)$

39. $g(x) = -16(30.5^x)$

40. $f(x) = -2.2(.75^x)$

In Exercises 41–44, find the domain of the given function (that is, the largest set of real numbers for which the rule produces well-defined real numbers).

41. $f(x) = \ln (x + 1)$

42. $g(x) = \ln (x + 2)$

43. $h(x) = \log (-x)$

44. $k(x) = \log (2 - x)$

45. (a) Graph $y = x$ and $y = e^{\ln x}$ in separate viewing windows [or use a split-screen if your calculator has that feature]. For what values of x are the graphs identical?
  (b) Use the properties of logarithms to explain your answer in part (a).

46. (a) Graph $y = x$ and $y = \ln (e^x)$ in separate viewing windows [or a split-screen if your calculator has that feature]. For what values of x are the graphs identical?
  (b) Use the properties of logarithms to explain your answer in part (a).

47. Do the graphs of $f(x) = \log x^2$ and $g(x) = 2 \log x$ appear to be the same? How do they differ?

48. Do the graphs of $h(x) = \log x^3$ and $k(x) = 3 \log x$ appear to be the same?

In Exercises 49–54, list the transformations that will change the graph of $g(x) = \ln x$ into the graph of the given function. [Section 3.4 may be helpful.]

49. $f(x) = 2 \cdot \ln x$

50. $f(x) = \ln x - 7$

51. $h(x) = \ln (x - 4)$

52. $k(x) = \ln (x + 2)$

53. $h(x) = \ln (x + 3) - 4$

54. $k(x) = \ln (x - 2) + 2$

In Exercises 55–58, sketch the graph of the function.

55. $f(x) = \log (x - 3)$

56. $g(x) = 2 \ln x + 3$

57. $h(x) = -2 \log x$

58. $f(x) = \ln (-x) - 3$

In Exercises 59–64, find a viewing window (or windows) that shows a complete graph of the function.

59. $f(x) = \dfrac{x}{\ln x}$

60. $g(x) = \dfrac{\ln x}{x}$

61. $h(x) = \dfrac{\ln x^2}{x}$

62. $k(x) = e^{2/\ln x}$

63. $f(x) = 10 \log x - x$

64. $f(x) = \dfrac{\log x}{x}$

In Exercises 65–68, find the average rate of change of the function.

65. $f(x) = \ln (x - 2)$, as x goes from 3 to 5.

66. $g(x) = x - \ln x$, as x goes from .5 to 1.

67. $g(x) = \log (x^2 + x + 1)$, as x goes from -5 to -3.

68. $f(x) = x \log |x|$, as x goes from 1 to 4.

69. (a) What is the average rate of change of $f(x) = \ln x$, as x goes from 3 to $3 + h$?
  (b) What is the value of h when the average rate of $f(x) = \ln x$, as x goes from 3 to $3 + h$, is .25?

70. (a) Find the average rate of change of $f(x) = \ln x^2$, as x goes from .5 to 2.
  (b) Find the average rate of change of $g(x) = \ln (x - 3)^2$, as x goes from 3.5 to 5.
  (c) What is the relationship between your answers in parts (a) and (b) and why is this so?

71. Show that $g(x) = \ln \left(\dfrac{x}{1 - x} \right)$ is the inverse function of $f(x) = \dfrac{1}{1 + e^{-x}}$. (See Section 3.7.)

72. The doubling function $D(x) = \dfrac{\ln 2}{\ln (1 + x)}$ gives the years required to double your money when it is invested at interest rate x (expressed as a decimal), compounded annually.
  (a) Find the time it takes to double your money at each of these interest rates: 4%, 6%, 8%, 12%, 18%, 24%, 36%.
  (b) Round the answers in part (a) to the nearest year and compare them with these numbers: $72/4, 72/6, 72/8, 72/12, 72/18, 72/24, 72/36$. Use this evidence to state a rule of thumb for determining approximate doubling time, without using the function D. This rule of thumb, which has long been used by bankers, is called the **rule of 72.**

73. Suppose $f(x) = A \ln x + B$, where A and B are constants. If $f(1) = 10$ and $f(e) = 1$, what are A and B?

74. If $f(x) = A \ln x + B$ and $f(e) = 5$ and $f(e^2) = 8$, what are A and B?

75. The height h above sea level (in meters) is related to air temperature t (in degrees Celsius), the atmospheric pressure p (in centimeters of mercury at height h), and the atmospheric pressure c at sea level by

$$h = (30t + 8000) \ln (c/p).$$

If the pressure at the top of Mount Rainier is 44 centimeters on a day when sea level pressure is 75.126 centimeters and the temperature is 7°C, what is the height of Mount Rainier?

76. Mount Everest is 8850 meters high. What is the atmospheric pressure at the top of the mountain on a day when the temperature is −25°C and the atmospheric pressure at sea level is 75 centimeters? [See Exercise 75.]

77. Beef consumption in the United States (in billions of pounds) in year x can be approximated by the function

$$f(x) = -154.41 + 39.38 \ln x \qquad (x \ge 90).$$

where $x = 90$ corresponds to 1990.*
(a) How much beef was consumed in 1999 and in 2002?
(b) When will beef consumption reach 30 billion pounds per year?

78. Students in a precalculus class were given a final exam. Each month thereafter, they took an equivalent exam. The class average on the exam taken after t months is given by

$$F(t) = 82 - 8 \cdot \ln (t + 1).$$

(a) What was the class average after six months?
(b) After a year?
(c) When did the class average drop below 55?

79. One person with a flu virus visited the campus. The number T of days it took for the virus to infect x people was given by:

$$T = -.93 \ln \left[\frac{7000 - x}{6999x} \right].$$

(a) How many days did it take for 6000 people to become infected?
(b) After two weeks, how many people were infected?

80. The population of St. Petersburg, Florida (in thousands) can be approximated by the function

$$g(x) = -127.9 + 81.91 \ln x \qquad (x \ge 70),$$

where $x = 70$ corresponds to 1970.
(a) Estimate the population in 1995 and 2003.
(b) If this model remains accurate, when will the population be 260,000?

81. A bicycle store finds that the number N of bikes sold is related to the number d of dollars spent on advertising by $N = 51 + 100 \cdot \ln (d/100 + 2)$.
(a) How many bikes will be sold if nothing is spent on advertising? If $1000 is spent? If $10,000 is spent?
(b) If the average profit is $25 per bike, is it worthwhile to spend $1000 on advertising? What about $10,000?
(c) What are the answers in part (b) if the average profit per bike is $35?

82. Approximating Logarithmic Functions by Polynomials. For each positive integer n, let f_n be the polynomial function whose rule is

$$f_n(x) = x - \frac{x^2}{2} + \frac{x^3}{3} - \frac{x^4}{4} + \frac{x^5}{5} - \cdots \pm \frac{x^n}{n}$$

where the sign of the last term is $+$ if n is odd and $-$ if n is even. In the viewing window with $-1 \le x \le 1$ and $-4 \le y \le 1$, graph $g(x) = \ln (1 + x)$ and $f_4(x)$ on the same screen. For what values of x does f_4 appear to be a good approximation of g?

83. Using the viewing window in Exercise 82, find a value of n for which the graph of the function f_n (as defined in Exercise 82) appears to coincide with the graph of $g(x) = \ln (1 + x)$. Use the trace feature to move from graph to graph to see how good this approximation actually is.

---

*Based on data from the U.S. Department of Agriculture.

5.4   Properties of Logarithms

Logarithms have several important properties in addition to those presented in Section 5.3. These properties, which we shall call *logarithm laws,* arise from the fact that logarithms are exponents. Essentially, they are properties of exponents translated into logarithmic language.

The first law of exponents says that $b^m b^n = b^{m+n}$, or in words,

The exponent of a product is the sum of the exponents of the factors.

Since logarithms are just particular kinds of exponents, this statement translates as follows.

The logarithm of a product is the sum of the logarithms of the factors.

Here is the same statement in symbolic language.

Product Law
for Logarithms

For all $v, w > 0$,

$$\ln(vw) = \ln v + \ln w$$

and

$$\log(vw) = \log v + \log w.$$

Proof   According to Property 4 of logarithms (in the box on page 365),

$$e^{\ln v} = v \qquad \text{and} \qquad e^{\ln w} = w.$$

Therefore, by the first law of exponents (with $m = \ln v$ and $n = \ln w$),

$$vw = e^{\ln v}e^{\ln w} = e^{\ln v + \ln w}.$$

So raising e to the exponent $(\ln v + \ln w)$ produces vw. But the definition of logarithm says that $\ln vw$ is the exponent to which e must be raised to produce vw. Therefore, we must have $\ln vw = \ln v + \ln w$. A similar argument works for common logarithms (just replace e by 10 and "ln" by "log"). ■

EXAMPLE 1

A calculator shows that $\ln 7 = 1.9459$ and $\ln 9 = 2.1972$. Therefore,

$$\ln 63 = \ln (7 \cdot 9) = \ln 7 + \ln 9 = 1.9459 + 2.1972 = 4.1341. \qquad ■$$

CALCULATOR EXPLORATION

We know that $5 \cdot 7 = 35$. Key in LOG(35) ENTER. Then key in LOG(5) + LOG(7) ENTER. The answers are the same by the Product Law. Do you get the same answer if you key in LOG(5) $\times$ LOG(7) ENTER?

EXAMPLE 2

Use the Product Law to write

(a) $\log (7xy)$ as a sum of three logarithms.

(b) $\log x^2 + \log y + 1$ as a single logarithm.

SOLUTION

(a) $\log (7xy) = \log 7x + \log y = \log 7 + \log x + \log y$

(b) Note that $\log 10 = 1$ (why?). Hence,

$$\log x^2 + \log y + 1 = \log x^2 + \log y + \log 10$$

$$= \log (x^2y) + \log 10$$

$$= \log (10x^2y). \qquad ■$$

CAUTION

A common error in applying the Product Law for Logarithms is to write the *false* statement

$$\ln 7 + \ln 9 = \ln (7 + 9)$$

$$= \ln 16$$

instead of the correct statement

$$\ln 7 + \ln 9 = \ln (7 \cdot 9)$$

$$= \ln 63.$$

GRAPHING EXPLORATION

Illustrate the Caution in the margin graphically by graphing both

$$f(x) = \ln x + \ln 9 \qquad \text{and} \qquad g(x) = \ln (x + 9)$$

in the standard viewing window and verifying that the graphs are not the same. In particular, the functions have different values at $x = 7$.

The second law of exponents, namely, $b^m / b^n = b^{m-n}$, may be roughly stated in words as follows.

> The exponent of the quotient is the difference of exponents.

When the exponents are logarithms, this says

The logarithm of a quotient is the difference of the logarithms.

In other words,

Quotient Law for Logarithms

For all $v, w > 0$,

$$\ln \left(\frac{v}{w} \right) = \ln v - \ln w$$

and

$$\log \left(\frac{v}{w} \right) = \log v - \log w.$$

The proof of the Quotient Law is very similar to the proof of the Product Law (see Exercise 26).

```
log(297/39)
         .8816918423
log(297)-log(39)
         .8816918423
■
```

Figure 5–23

EXAMPLE 3

Figure 5–23 illustrates the Quotient Law by showing that

$$\log \left(\frac{297}{39} \right) = \log 297 - \log 39. \qquad ■$$

EXAMPLE 4

For any $w > 0$,

$$\ln \left(\frac{1}{w} \right) = \ln 1 - \ln w = 0 - \ln w = -\ln w$$

and

$$\log \left(\frac{1}{w} \right) = \log 1 - \log w = 0 - \log w = -\log w. \qquad ■$$

CAUTION

Do not confuse $\ln\left(\dfrac{v}{w}\right)$ with the quotient $\dfrac{\ln v}{\ln w}$. They are *different* numbers. For example,

$$\ln\left(\frac{36}{3}\right) = \ln(12) = 2.4849, \qquad \text{but} \qquad \frac{\ln 36}{\ln 3} = \frac{3.5835}{1.0986} = 3.2619.$$

EXPLORATION

Illustrate the preceding Caution graphically by graphing both $f(x) = \ln(x/3)$ and $g(x) = (\ln x)/(\ln 3)$ and verifying that the graphs are not the same at $x = 36$ (or anywhere else, for that matter).

The third law of exponents, namely, $(b^m)^k = b^{mk}$, can also be translated into logarithmic language.

Power Law for Logarithms

For all k and all $v > 0$,

$$\ln(v^k) = k(\ln v)$$

and

$$\log(v^k) = k(\log v).$$

Proof Since $v = 10^{\log v}$ (why?), the third law of exponents (with $b = 10$ and $m = \log v$) shows that

$$v^k = (10^{\log v})^k = 10^{(\log v)k} = 10^{k(\log v)}.$$

So raising 10 to the exponent $k(\log v)$ produces v^k. But the exponent to which 10 must be raised to produce v^k is, by definition, $\log(v^k)$. Therefore, $\log(v^k) = k(\log v)$, and the proof is complete. A similar argument with e in place of 10 and "ln" in place of "log" works for natural logarithms. ■

EXAMPLE 5

Express $\ln\sqrt{19}$ without radicals or exponents.

SOLUTION First write $\sqrt{19}$ in exponent notation, then use the Power Law:

$$\ln\sqrt{19} = \ln 19^{1/2}$$

$$= \frac{1}{2}\ln 19 \qquad \text{or} \qquad \frac{\ln 19}{2}. \quad ■$$

EXAMPLE 6

Express as a single logarithm: $\dfrac{\log(x^2 + 1)}{3} - \log x.$

SOLUTION

$$\frac{\log(x^2 + 1)}{3} - \log x = \frac{1}{3}\log(x^2 + 1) - \log x$$

$$= \log(x^2 + 1)^{1/3} - \log x \qquad \text{[Power Law]}$$

$$= \log \sqrt[3]{x^2 + 1} - \log x$$

$$= \log\left(\frac{\sqrt[3]{x^2 + 1}}{x}\right) \qquad \text{[Quotient Law]} \qquad \blacksquare$$

EXAMPLE 7

Express as a single logarithm: $\ln 3x + 4\ln x - \ln 3xy$.

SOLUTION

$$\ln 3x + 4 \cdot \ln x - \ln 3xy = \ln 3x + \ln x^4 - \ln 3xy \qquad \text{[Power Law]}$$

$$= \ln(3x \cdot x^4) - \ln 3xy \qquad \text{[Product Law]}$$

$$= \ln \frac{3x^5}{3xy} \qquad \text{[Quotient Law]}$$

$$= \ln \frac{x^4}{y} \qquad \text{[Cancel } 3x] \qquad \blacksquare$$

EXAMPLE 8

Simplify: $\ln\left(\frac{\sqrt{x}}{x}\right) + \ln \sqrt[4]{ex^2}$.

SOLUTION   Begin by changing to exponential notation.

$$\ln\left(\frac{x^{1/2}}{x}\right) + \ln(ex^2)^{1/4} = \ln(x^{-1/2}) + \ln(ex^2)^{1/4}$$

$$= -\frac{1}{2} \cdot \ln x + \frac{1}{4} \cdot \ln ex^2 \qquad \text{[Power Law]}$$

$$= -\frac{1}{2} \cdot \ln x + \frac{1}{4}(\ln e + \ln x^2) \qquad \text{[Product Law]}$$

$$= -\frac{1}{2} \cdot \ln x + \frac{1}{4}(\ln e + 2 \cdot \ln x) \qquad \text{[Power Law]}$$

$$= -\frac{1}{2} \cdot \ln x + \frac{1}{4} \cdot \ln e + \frac{1}{2} \cdot \ln x$$

$$= \frac{1}{4} \cdot \ln e = \frac{1}{4} \qquad [\ln e = 1] \qquad \blacksquare$$

▪▪ APPLICATIONS

Because logarithmic growth is slow, measurements on a logarithmic scale (that is, on a scale determined by a logarithmic function) can sometimes be deceptive.

EXAMPLE 9

Earthquakes   The magnitude $R(i)$ of an earthquake on the Richter scale is given by $R(i) = \log(i/i_0)$, where i is the amplitude of the ground motion of the earthquake and i_0 is the amplitude of the ground motion of the so-called zero earthquake.[*] A moderate earthquake might have 1000 times the ground motion of the zero earthquake (that is, $i = 1000i_0$). So its magnitude would be

$$\log(1000i_0/i_0) = \log 1000 = \log 10^3 = 3.$$

An earthquake with 10 times this ground motion (that is, $i = 10 \cdot 1000i_0 = 10{,}000i_0$) would have a magnitude of

$$\log(10{,}000i_0/i_0) = \log 10{,}000 = \log 10^4 = 4.$$

So a *tenfold* increase in ground motion produces only a one-point change on the Richter scale. In general,

Increasing the ground motion by a factor of 10^k increases the Richter magnitude by k units.[†]

For instance, the 1989 World Series earthquake in San Francisco measured 7.0 on the Richter scale, and the great earthquake of 1906 measured 8.3. The difference of 1.3 points means that the 1906 quake was $10^{1.3} \approx 20$ times more intense than the 1989 one in terms of ground motion.   ■

---

[*]The zero earthquake has ground motion amplitude of less than 1 micron on a standard seismograph 100 kilometers from the epicenter.
[†]*Proof:* If one quake has ground motion amplitude i and the other $10^k i$, then

$$R(10^k i) = \log(10^k i/i_0) = \log 10^k + \log(i/i_0)$$
$$= k + \log(i/i_0) = k + R(i).$$

✓ EXERCISES 5.4

In Exercises 1–10, write the given expression as a single logarithm.

1. $\ln x^2 + 3 \ln y$ **2.** $\ln 2x + 2(\ln x) - \ln 3y$

3. $\log(x^2 - 9) - \log(x + 3)$

4. $\log 3x - 2[\log x - \log(2 + y)]$

5. $2(\ln x) - 3(\ln x^2 + \ln x)$

6. $\ln(e/\sqrt{x}) - \ln\sqrt{ex}$

7. $3 \ln(e^2 - e) - 3$ **8.** $2 - 2 \log(20)$

9. $\log(10x) + \log(20y) - 1$

10. $\ln(e^2 x) + \ln(ey) - 3$

In Exercises 11–16, let $u = \ln x$ and $v = \ln y$. Write the given expression in terms of u and v. For example,

$$\ln x^3 y = \ln x^3 + \ln y = 3 \ln x + \ln y = 3u + v.$$

11. $\ln(x^2 y^5)$ **12.** $\ln(x^3 y^2)$ **13.** $\ln(\sqrt{x} \cdot y^2)$

14. $\ln\left(\dfrac{\sqrt{x}}{y}\right)$ **15.** $\ln\left(\sqrt[3]{x^2}\sqrt{y}\right)$ **16.** $\ln\left(\dfrac{\sqrt{x^2 y}}{\sqrt[3]{y}}\right)$

In Exercises 17–22, use graphical or algebraic means to determine whether the statement is true or false.

17. $\ln |x| = |\ln x|$?

18. $\ln\left(\dfrac{1}{x}\right) = \dfrac{1}{\ln x}$?

19. $\log x^5 = 5(\log x)$?

20. $e^{x \ln x} = x^x$ $(x > 0)$?

21. $\ln x^3 = (\ln x)^3$?

22. $\log \sqrt{x} = \sqrt{\log x}$?

In Exercises 23–24, find values of a and b for which the statement is false.

23. $\dfrac{\log a}{\log b} = \log\left(\dfrac{a}{b}\right)$

24. $\log (a + b) = \log a + \log b$

25. If $\ln b^7 = 7$, what is b?

26. Prove the Quotient Law for Logarithms: For $v, w > 0$, $\ln\left(\dfrac{v}{w}\right) = \ln v - \ln w$. (Use properties of exponents and the fact that $v = e^{\ln v}$ and $w = e^{\ln w}$.)

In Exercises 27–30, state the magnitude on the Richter scale of an earthquake that satisfies the given condition.

27. 100 times stronger than the zero quake.

28. $10^{4.7}$ times stronger than the zero quake.

29. 350 times stronger than the zero quake.

30. 2500 times stronger than the zero quake.

Exercises 31–34 deal with the energy intensity i of a sound, which is related to the loudness of the sound by the function $L(i) = 10 \cdot \log (i/i_0)$, where i_0 is the minimum intensity detectable by the human ear and $L(i)$ is measured in decibels. Find the decibel measure of the sound.

31. Ticking watch (intensity is 100 times i_0).

32. Soft music (intensity is 10,000 times i_0).

33. Loud conversation (intensity is 4 million times i_0).

34. Victoria Falls in Africa (intensity is 10 billion times i_0).

35. How much louder is the sound in Exercise 32 than the sound in Exercise 31?

36. The perceived loudness L of a sound of intensity I is given by $L = k \cdot \ln I$, where k is a certain constant. By how much must the intensity be increased to double the loudness? (That is, what must be done to I to produce $2L$?)

Thinkers

37. Compute each of the following pairs of numbers.

(a) $\log 18$ and $\dfrac{\ln 18}{\ln 10}$

(b) $\log 456$ and $\dfrac{\ln 456}{\ln 10}$

(c) $\log 8950$ and $\dfrac{\ln 8950}{\ln 10}$

(d) What do these results suggest?

38. Prove that for any positive number c, $\log c = \dfrac{\ln c}{\ln 10}$. [*Hint:* We know that $10^{\log c} = c$ (why?). Take natural logarithms on both sides and use a logarithm law to simplify and solve for $\log c$.]

39. Find each of the following logarithms.

(a) $\log 8.753$ (b) $\log 87.53$ (c) $\log 875.3$

(d) $\log 8753$ (e) $\log 87,530$

(f) How are the numbers 8.753, 87.53, . . . , 87,530 related to one another? How are their logarithms related? State a general conclusion that this evidence suggests.

40. Prove that for every positive number c, $\log c$ can be written in the form $k + \log b$, where k is an integer and $1 < b < 10$. [*Hint:* Write c in scientific notation and use logarithm laws to express $\log c$ in the required form.]

5.4.A SPECIAL TOPICS Logarithmic Functions to Other Bases*

The same procedure used in Sections 5.3–5.4 can be carried out with any positive number b in place of 10 and e.

> **Throughout this section, b is a fixed positive number with $b > 1$.†**

---

*This material is not needed in the sequel and may be read before Section 5.3 if desired. Section 3.7 (Inverse Functions) is a prerequisite for this section, which replicates the discussion of Sections 5.3 and 5.4 in a more general context.

†The discussion is also valid when $0 < b < 1$, but in that case, the graphs have a different shape.

The exponential function $f(x) = b^x$, whose graph is shown in Figure 5–24, is an increasing function and hence is one-to-one (as explained on page 205). Therefore, f has an inverse function g whose graph is the reflection of the graph of f in the line $y = x$ (see page 210), as shown in Figure 5–25.

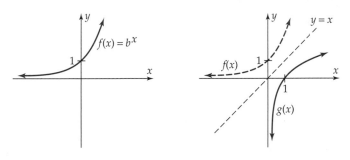

Figure 5–24   **Figure 5–25**

This inverse function g is called the **logarithmic function to the base b.** The value $g(x)$ of this function at a number x is denoted **$\log_b x$** and called the **logarithm to the base b** of the number x.

In Section 3.7, we saw that the relationship between a function f and its inverse function g is given by

$$g(v) = u \qquad \text{exactly when} \qquad f(u) = v.$$

When $f(x) = b^x$ and $g(x) = \log_b x$, this statement takes the following form.

Definition of Logarithms to Base b

> Let u and v be real numbers, with $v > 0$. Then
> $$\log_b v = u \qquad \text{exactly when} \qquad b^u = v.$$
> In other words,
> $$\log_b v \text{ is the exponent to which } b \text{ must be raised to produce } v.$$

EXAMPLE 1

To find $\log_2 16$, ask yourself, "What power of 2 equals 16?" Since $2^4 = 16$, we see that $\log_2 16 = 4$. Similarly, $\log_2 (1/8) = -3$ because $2^{-3} = 1/8$.   ■

EXAMPLE 2

Since logarithms are just exponents, every logarithmic statement can be translated into exponential language.

Logarithmic Statement	Equivalent Exponential Statement
$\log_3 81 = 4$	$3^4 = 81$
$\log_4 64 = 3$	$4^3 = 64$
$\log_{125} 5 = \dfrac{1}{3}$	$125^{1/3} = 5^*$
$\log_8 \left(\dfrac{1}{4}\right) = -\dfrac{2}{3}$	$8^{-2/3} = \dfrac{1}{4}$ (verify!) ∎

EXAMPLE 3

Solve $\log_5 x = 3$.

SOLUTION The equation $\log_5 x = 3$ is equivalent to the exponential statement $5^3 = x$, so the solution is $x = 125$. ∎

EXAMPLE 4

Logarithms to the base 10 are called **common logarithms.** Is it customary to write $\log v$ instead of $\log_{10} v$. Then

$$\log 100 = 2 \quad \text{because} \quad 10^2 = 100;$$

$$\log .001 = -3 \quad \text{because} \quad 10^{-3} = \frac{1}{10^3} = \frac{1}{1000} = .001.$$

Calculators have a LOG key for evaluating common logarithms. For instance,[†]

$$\log .4 = -0.3979, \quad \log 45.3 = 1.6561, \quad \log 685 = 2.8357. \quad ∎$$

EXAMPLE 5

The most frequently used base for logarithms in modern applications is the number e ($\approx 2.71828 \ldots$). Logarithms to the base e are called **natural logarithms** and use a different notation: We write $\ln v$ instead of $\log_e v$. Calculators also have an LN key for evaluating natural logarithms. For example,

$$\ln .5 = -0.6931, \quad \ln 65 = 4.1744, \quad \ln 158 = 5.0626. \quad ∎$$

You *don't* need a calculator to understand the essential properties of logarithms. You need only translate logarithmic statements into exponential ones (or vice versa).

---

*Because $125^{1/3} = \sqrt[3]{125} = 5$.

[†]Here and below, all logarithms are rounded to four decimal places. So strictly speaking, the equal sign should be replaced by an "approximately equal" sign ($\approx$).

EXAMPLE 6

What is $\log(-25)$?

Translation: To what power must 10 be raised to produce -25? The graph of $f(x) = 10^x$ lies entirely above the x-axis (use your calculator), which means that *every* power of 10 is *positive*. So 10^x can *never* be -25, or any negative number, or zero. The same argument works for any base b.

$$\log_b v \text{ is defined only when } v > 0. \quad \blacksquare$$

EXAMPLE 7

What is $\log_5 1$?

Translation: To what power must 5 be raised to produce 1? The answer, of course, is $5^0 = 1$. So $\log_5 1 = 0$. Similarly, $\log_5 5 = 1$ because 1 is the answer to "what power of 5 equals 5?" In general,

$$\log_b 1 = 0 \quad \text{and} \quad \log_b b = 1. \quad \blacksquare$$

EXAMPLE 8

What is $\log_2 2^9$?

Translation: To what power must 2 be raised to produce 2^9? Obviously, the answer is 9. So $\log_2 2^9 = 9$ and, in general,

$$\log_b b^k = k \quad \text{for every real number } k.$$

This property holds even when k is a complicated expression. For instance, if x and y are positive, then

$$\log_6 6^{\sqrt{3x+y}} = \sqrt{3x + y} \quad \text{(here } k = \sqrt{3x + y}\text{).} \quad \blacksquare$$

EXAMPLE 9

What is $10^{\log 439}$? Well, $\log 439$ is the power to which 10 must be raised to produce 439, that is $10^{\log 439} = 439$. Similarly,

$$b^{\log_b v} = v \quad \text{for every } v > 0. \quad \blacksquare$$

Here is a summary of the facts illustrated in the preceding examples.

Properties of Logarithms

1. $\log_b v$ is defined only when $v > 0$.

2. $\log_b 1 = 0$ and $\log_b b = 1$.

3. $\log_b (b^k) = k$ for every real number k.

4. $b^{\log_b v} = v$ for every $v > 0$.

▪▪ LOGARITHM LAWS

The first law of exponents states that $b^m b^n = b^{m+n}$, or in words,

> The exponent of a product is the sum of the exponents of the factors.

Since logarithms are just particular kinds of exponents, this statement translates as follows.

> **The logarithm of a product is the sum of the logarithms of the factors.**

The second and third laws of exponents, namely, $b^m/b^n = b^{m-n}$ and $(b^m)^k = b^{mk}$, can also be translated into logarithmic language.

Logarithm Laws

> Let b, v, w, k be real numbers, with b, v, w positive and $b \neq 1$.
>
> *Product Law:* $\log_b (vw) = \log_b v + \log_b w$.
>
> *Quotient Law:* $\log_b\left(\dfrac{v}{w}\right) = \log_b v - \log_b w$.
>
> *Power Law:* $\log_b (v^k) = k(\log_b v)$.

Proof of the Quotient Law   According to Property 4 in the box on page 378,

$$b^{\log_b v} = v \quad \text{and} \quad b^{\log_b w} = w.$$

Therefore, by the second law of exponents (with $m = \log_b v$ and $n = \log_b w$), we have

$$\frac{v}{w} = \frac{b^{\log_b v}}{b^{\log_b w}} = b^{\log_b v - \log_b w}.$$

Since $\log_b (v/w)$ is the exponent to which b must be raised to produce v/w, we must have $\log_b (v/w) = \log_b v - \log_b w$. This proves the Quotient Law. The Product and Power Laws are proved in a similar fashion. ▪

EXAMPLE 10

Simplify and write as a single logarithm.

(a) $\log_3 (x + 2) + \log_3 y - \log_3 (x^2 - 4)$

(b) $3 - \log_5 (125x)$

SOLUTION

(a) $\log_3 (x + 2) + \log_3 y - \log_3 (x^2 - 4)$

$$= \log_3[(x + 2)y] - \log_3 (x^2 - 4) \qquad \text{[Product Law]}$$

$$= \log_3\left(\frac{(x + 2)y}{x^2 - 4}\right) \qquad \text{[Quotient Law]}$$

$$= \log_3\left(\frac{(x + 2)y}{(x + 2)(x - 2)}\right) \qquad \text{[Factor denominator]}$$

$$= \log_3\left(\frac{y}{x - 2}\right) \qquad \text{[Cancel common factor]}$$

(b) $3 - \log_5 (125x)$

$$= 3 - (\log_5 125 + \log_5 x) \qquad \text{[Product Law]}$$
$$= 3 - \log_5 125 - \log_5 x$$
$$= 3 - 3 - \log_5 x \qquad \text{[}\log_5 125 = 3 \text{ because } 5^3 = 125\text{]}$$
$$= -\log_5 x$$
$$= \log_5 x^{-1} = \log_5 \left(\frac{1}{x}\right) \qquad \text{[Power Law]} \qquad \blacksquare$$

 CAUTION

1. A common error in using the Product Law is to write something like
   $\log 6 + \log 7 = \log (6 + 7) = \log 13$ instead of the correct statement
   $\log 6 + \log 7 = \log (6 \cdot 7) = \log 42$.

2. Do not confuse $\log_b \left(\dfrac{v}{w}\right)$ with the quotient $\dfrac{\log_b v}{\log_b w}$. They are *different*

   numbers. For example, when $b = 10$

   $$\log\left(\frac{48}{4}\right) = \log 12 = 1.0792 \qquad \text{but} \qquad \frac{\log 48}{\log 4} = \frac{1.6812}{0.6021} = 2.7922.$$

For graphic illustrations of the errors mentioned in the Caution, see Exercises 82 and 83.

EXAMPLE 11

Given that

$$\log_7 2 = .3562, \qquad \log_7 3 = .5646, \qquad \text{and} \qquad \log_7 5 = .8271,$$

find:

(a) $\log_7 10$;   (b) $\log_7 2.5$;   (c) $\log_7 48$.

SOLUTION

(a) By the Product Law,

$$\log_7 10 = \log_7 (2 \cdot 5) = \log_7 2 + \log_7 5 = .3562 + .8271 = 1.1833.$$

(b) By the Quotient Law,

$$\log_7 2.5 = \log_7 \left(\frac{5}{2}\right) = \log_7 5 - \log_7 2 = .8271 - .3562 = .4709.$$

(c) By the Product and Power Laws,

$$\log_7 48 = \log_7 (3 \cdot 16) = \log_7 3 + \log_7 16 = \log_7 3 + \log_7 2^4$$
$$= \log_7 3 + 4 \cdot \log_7 2 = .5646 + 4(.3562)$$
$$= 1.9894. \qquad \blacksquare$$

Example 11 worked because we were *given* several logarithms to base 7. But there's no $\log_7$ key on the calculator, so how do you find logarithms to base 7 or to any base other than e or 10? *Answer:* Use the LN key on the calculator and the following formula.

Change of Base Formula

For any positive numbers b and v,

$$\log_b v = \frac{\ln v}{\ln b}.$$

Proof   By Property 4 in the box on page 378, $b^{\log_b v} = v$. Take the natural logarithm of each side of this equation:

$$\ln (b^{\log_b v}) = \ln v.$$

Apply the Power Law for natural logarithms on the left side.

$$(\log_b v)(\ln b) = \ln v.$$

Dividing both sides by $\ln b$ finishes the proof.

$$\log_b v = \frac{\ln v}{\ln b}. \quad \blacksquare$$

EXAMPLE 12

To find $\log_7 3$, apply the change of base formula with $b = 7$.

$$\log_7 3 = \frac{\ln 3}{\ln 7} = \frac{1.0986}{1.9459} = .5646. \quad \blacksquare$$

EXERCISES 5.4.A

Note: *Unless stated otherwise, all letters represent positive numbers and $b \neq 1$.*

In Exercises 1–8, fill in the missing entries in each table.

1.

x	0	1	2	4
$f(x) = \log_4 x$				

2.

x	1/25	5	25	$\sqrt{5}$
$g(x) = \log_5 x$				

3.

x		1/6	1	216
$h(x) = \log_6 x$	-2			

4.

x	10/3	4	6	12
$k(x) = \log_3 (x - 3)$				

5.

x	0	1/7	$\sqrt{7}$	49
$f(x) = 2 \log_7 x$				

6.

x			100	1000
$g(x) = 3 \log x$	6	3		

7.

x	-2.75	-1	1	29
$h(x) = 3 \log_2 (x + 3)$				

8.

x	1/e	1	e	e^2
$k(x) = 2 \ln x$				

In Exercises 9–18, translate the given exponential statement into an equivalent logarithmic one.

9. $10^{-2} = .01$

10. $10^3 = 1000$

11. $\sqrt[3]{10} = 10^{1/3}$

12. $10^{.4771} \approx 3$

13. $10^{7k} = r$

14. $10^{(a+b)} = c$

15. $7^8 = 5{,}764{,}801$

16. $2^{-3} = 1/8$

17. $3^{-2} = 1/9$

18. $b^{14} = 3379$

In Exercises 19–28, translate the given logarithmic statement into an equivalent exponential one.

19. $\log 10{,}000 = 4$

20. $\log .001 = -3$

21. $\log 750 \approx 2.88$

22. $\log (.8) = -.097$

23. $\log_5 125 = 3$

24. $\log_8 (1/4) = -2/3$

25. $\log_2 (1/4) = -2$

26. $\log_2 \sqrt{2} = 1/2$

27. $\log (x^2 + 2y) = z + w$

28. $\log (a + c) = d$

In Exercises 29–36, evaluate the given expression without using a calculator.

29. $\log 10^{\sqrt{97}}$

30. $\log_{17} (17^{17})$

31. $\log 10^{x^2 + y^2}$

32. $\log_{3.5} [3.5^{(x^2-1)}]$

33. $\log_{16} 4$

34. $\log_2 64$

35. $\log_{\sqrt{3}} (27)$

36. $\log_{\sqrt{3}} (1/9)$

In Exercises 37–40, a graph or a table of values for the function $f(x) = \log_b x$ is given. Find b.

37.

38.

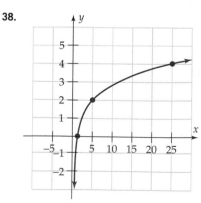

39.

x	.05	1	400	$2\sqrt{5}$
$f(x)$	-1	0	2	$1/2$

40.

x	$1/25$	1	5	125
$f(x)$	-4	0	2	6

In Exercises 41–46, find x.

41. $\log_3 243 = x$

42. $\log_{81} 27 = x$

43. $\log_{27} x = 1/3$

44. $\log_5 x = -4$

45. $\log_x 64 = 3$

46. $\log_x (1/9) = -2/3$

In Exercises 47–60, write the given expression as the logarithm of a single quantity, as in Example 10.

47. $2 \log x + 3 \log y - 6 \log z$

48. $5 \log_8 x - 3 \log_8 y + 2 \log_8 z$

49. $\log x - \log (x + 3) + \log (x^2 - 9)$

50. $\log_3 (y + 2) + \log_3 (y - 3) - \log_3 y$

51. $\frac{1}{2} \log_2 (25c^2)$

52. $\frac{1}{3} \log_2 (27b^6)$

53. $-2 \log_4 (7c)$

54. $\frac{1}{3} \log_5 (x + 1)$

55. $2 \ln (x + 1) - \ln (x + 2)$

56. $\ln (z - 3) + 2 \ln (z + 3)$

57. $\log_2 (2x) - 1$

58. $2 - \log_5 (25z)$

59. $2 \ln (e^2 - e) - 2$

60. $4 - 4 \log_5 (20)$

In Exercises 61–68, use a calculator and the change of base formula to find the logarithm.

61. $\log_2 10$

62. $\log_2 22$

63. $\log_7 5$

64. $\log_5 7$

65. $\log_{500} 1000$

66. $\log_{500} 250$

67. $\log_{12} 56$

68. $\log_{12} 725$

In Exercises 69–74, answer true or false and give reasons for your answer.

69. $\log_b (r/5) = \log_b r - \log_b 5$

70. $\dfrac{\log_b a}{\log_b c} = \log_b \left(\dfrac{a}{c}\right)$

71. $(\log_b r)/t = \log_b (r^{1/t})$

72. $\log_b (cd) = \log_b c + \log_b d$

73. $\log_5 (5x) = 5(\log_5 x)$

74. $\log_b (ab)^t = t(\log_b a) + t$

Thinkers

75. Which is larger: 397^{398} or 398^{397}? [*Hint:* $\log 397 \approx 2.5988$ and $\log 398 \approx 2.5999$ and $f(x) = 10^x$ is an increasing function.]

76. If $\log_b 9.21 = 7.4$ and $\log_b 359.62 = 19.61$, then what is $\log_b 359.62 / \log_b 9.21$?

In Exercises 77–80, assume that a and b are positive, with $a \neq 1$ and $b \neq 1$.

77. Express $\log_b u$ in terms of logarithms to the base a.

78. Show that $\log_b a = 1/\log_a b$.

79. How are $\log_{10} u$ and $\log_{100} u$ related?

80. Show that $a^{\log b} = b^{\log a}$.

81. If $\log_b x = \frac{1}{2}\log_b v + 3$, show that $x = (b^3)\sqrt{v}$.

82. Graph the functions $f(x) = \log x + \log 7$ and $g(x) = \log(x + 7)$ on the same screen. For what values of x is it true that $f(x) = g(x)$? What do you conclude about the statement

$$\log 6 + \log 7 = \log(6 + 7)?$$

83. Graph the functions $f(x) = \log(x/4)$ and $g(x) = (\log x)/(\log 4)$. Are they the same? What does this say about a statement such as

$$\log\left(\frac{48}{4}\right) = \frac{\log 48}{\log 4}?$$

In Exercises 84–86, sketch a complete graph of the function, labeling any holes, asymptotes, or local extrema.

84. $f(x) = \log_5 x + 2$ **85.** $h(x) = x \log x^2$

86. $g(x) = \log_{20} x^2$

5.5 Algebraic Solutions of Exponential and Logarithmic Equations

Most of the exponential and logarithmic equations solved by graphical means earlier in this chapter could also have been solved algebraically. The algebraic techniques for solving such equations are based on the properties of logarithms.

EXPONENTIAL EQUATIONS

The easiest exponential equations to solve are those in which both sides are powers of the same base.

EXAMPLE 1

Solve $8^x = 2^{x+1}$.

SOLUTION   Using the fact that $8 = 2^3$, we rewrite the equation as follows.

$$8^x = 2^{x+1}$$
$$(2^3)^x = 2^{x+1}$$
$$2^{3x} = 2^{x+1}$$

Since the powers of 2 are equal, the exponents must be the same, that is,

$$3x = x + 1$$
$$2x = 1$$
$$x = \frac{1}{2}. \quad \blacksquare$$

When different bases are involved in an exponential equation, a different solution technique is needed.

EXAMPLE 2

To solve $5^x = 2$,

Take logarithms on each side:[*]	$\ln 5^x = \ln 2$
Use the Power Law:	$x(\ln 5) = \ln 2$
Divide both sides by ln 5:	$x = \dfrac{\ln 2}{\ln 5} \approx \dfrac{.6931}{1.6094} \approx .4307.$

Remember: $\dfrac{\ln 2}{\ln 5}$ is *not* $\ln \dfrac{2}{5}$ or $\ln 2 - \ln 5.$ ■

EXAMPLE 3

To solve $2^{4x-1} = 3^{1-x}$,

Take logarithms of each side:	$\ln 2^{4x-1} = \ln 3^{1-x}$
Use the Power Law:	$(4x - 1)(\ln 2) = (1 - x)(\ln 3)$
Multiply out both sides:	$4x(\ln 2) - \ln 2 = \ln 3 - x(\ln 3)$
Rearrange terms:	$4x(\ln 2) + x(\ln 3) = \ln 2 + \ln 3$
Factor left side:	$(4 \cdot \ln 2 + \ln 3)x = \ln 2 + \ln 3$
Divide both sides by $(4 \cdot \ln 2 + \ln 3)$:	$x = \dfrac{\ln 2 + \ln 3}{4 \cdot \ln 2 + \ln 3} \approx .4628.$ ■

■■ APPLICATIONS OF EXPONENTIAL EQUATIONS

As we saw in Section 5.2, the mass of a radioactive element at time x is given by

$$M(x) = c(.5^{x/h}),$$

where c is the initial mass and h is the half-life of the element.

EXAMPLE 4

After 43 years, a 20-milligram sample of strontium-90 (^{90}Sr) decays to 6.071 mg. What is the half-life of strontium-90?

SOLUTION   The mass of the sample at time x is given by

$$f(x) = 20(.5^{x/h}),$$

---

[*]We shall use natural logarithms, but the same techniques are valid for logarithms to other bases (Exercise 34).

where h is the half-life of strontium-90. We know that $f(x) = 6.071$ when $x = 43$, that is, $6.071 = 20(.5^{43/h})$. We must solve this equation for h.

Divide both sides by 20:
$$\frac{6.071}{20} = .5^{43/h}$$

Take logarithms on both sides:
$$\ln \frac{6.071}{20} = \ln .5^{43/h}$$

Use the Power Law:
$$\ln \frac{6.071}{20} = \frac{43}{h} \ln .5$$

Multiply both sides by h:
$$h \ln \frac{6.071}{20} = 43 \ln .5$$

Divide both sides by $\ln \frac{6.071}{20}$:
$$h = \frac{43 \ln .5}{\ln(6.071/20)} \approx 25.$$

Therefore, strontium-90 has a half-life of 25 years. ∎

EXAMPLE 5

When a living organism dies, its carbon-14 decays. The half-life of carbon-14 is 5730 years. If the skeleton of a mastodon has lost 58% of its original carbon-14, when did the mastodon die?[*]

SOLUTION Time is measured from the death of the mastodon. The amount of carbon-14 left in the skeleton at time x is given by

$$M(x) = c(.5^{x/5730}),$$

where c is the original mass of carbon-14. The skeleton has lost 58% of c, that is, $.58c$. So the present value of $M(x)$ is $c - .58c = .42c$, and we have

$$M(x) = c(.5^{x/5730})$$
$$.42c = c(.5^{x/5730})$$
$$.42 = .5^{x/5730}.$$

The solution of this equation is the time elapsed from the mastodon's death to the present. It can be solved as above.

$$\ln .42 = \ln (.5)^{x/5730}$$

$$\ln .42 = \frac{x}{5730} (\ln .5)$$

$$5730(\ln .42) = x(\ln .5)$$

$$x = \frac{5730(\ln .42)}{\ln .5} \approx 7171.32.$$

Therefore, the mastodon died approximately 7200 years ago. ∎

---

[*]Archeologists can determine how much carbon-14 has been lost by a technique that involves measuring the ratio of carbon-14 to carbon-12 in the skeleton.

EXAMPLE 6

A certain bacteria is known to grow exponentially, with the population at time t given by a function of the form $g(t) = Pe^{kt}$, where P is the original population and k is the continuous growth rate. A culture shows 1000 bacteria present. Seven hours later, there are 5000.

(a) Find the continuous growth rate k.

(b) Determine when the population will reach one billion.

SOLUTION

(a) The original population is $P = 1000$, so the growth function is $g(t) = 1000e^{kt}$. We know that $g(7) = 5000$, that is,

$$1000e^{k \cdot 7} = 5000.$$

To determine the growth rate, we solve this equation for k.

Divide both sides by 1000: $e^{7k} = 5$

Take logarithms of both sides: $\ln e^{7k} = \ln 5$

Use the Power Law: $7k \ln e = \ln 5.$

Since $\ln e = 1$ (why?), this equation becomes

$$7k = \ln 5$$

Divide both sides by 7: $k = \dfrac{\ln 5}{7} \approx .22992.$

Therefore, the growth function is $g(t) \approx 1000e^{.22992t}$.

(b) The population will reach one billion when $g(t) = 1,000,000,000$, that is, when

$$1000e^{.22992t} = 1,000,000,000.$$

So we solve this equation for t:

Divide both sides by 1000: $e^{.22992t} = 1,000,000$

Take logarithms on both sides: $\ln e^{.22992t} = \ln 1,000,000$

Use the Power Law: $.22992t \ln e = \ln 1,000,000$

Remember $\ln e = 1$: $.22992t = \ln 1,000,000$

Divide both sides by .22992: $t = \dfrac{\ln 1,000,000}{.22992} \approx 60.09.$

Therefore, it will take a bit more than 60 hours for the culture to grow to one billion.   ■

EXAMPLE 7

Inhibited Population Growth The population of fish in a lake at time t months is given by the function

$$p(t) = \frac{20{,}000}{1 + 24e^{-t/4}}.$$

How long will it take for the population to reach 15,000?

SOLUTION We must solve this equation for t.

$$15{,}000 = \frac{20{,}000}{1 + 24e^{-t/4}}$$

$$15{,}000(1 + 24e^{-t/4}) = 20{,}000$$

$$1 + 24e^{-t/4} = \frac{20{,}000}{15{,}000} = \frac{4}{3}$$

$$24e^{-t/4} = \frac{1}{3}$$

$$e^{-t/4} = \frac{1}{3} \cdot \frac{1}{24} = \frac{1}{72}$$

$$\ln e^{-t/4} = \ln\left(\frac{1}{72}\right)$$

$$\left(-\frac{t}{4}\right)(\ln e) = \ln 1 - \ln 72$$

$$-\frac{t}{4} = -\ln 72 \qquad\qquad [\ln e = 1 \text{ and } \ln 1 = 0]$$

$$t = 4(\ln 72) \approx 17.1067.$$

So the population reaches 15,000 in a little over 17 months. ■

⠿ LOGARITHMIC EQUATIONS

Equations that involve only logarithmic terms may be solved by using this fact, which is proved in Exercise 33 (and is valid with log replaced by ln).

If $\log u = \log v$, then $u = v$.

EXAMPLE 8

Solve $\log (3x + 2) + \log (x + 2) = \log (7x + 6)$.

SOLUTION First we write the left side as a single logarithm.

$$\log (3x + 2) + \log (x + 2) = \log (7x + 6)$$

Use the Product Law: $\qquad \log[(3x + 2)(x + 2)] = \log (7x + 6)$

Multiply out left side: $\qquad \log (3x^2 + 8x + 4) = \log (7x + 6).$

Since the logarithms are equal, we must have

$$3x^2 + 8x + 4 = 7x + 6$$

Subtract $7x + 6$ from both sides: $3x^2 + x - 2 = 0$

Factor: $(3x - 2)(x + 1) = 0$

$$3x - 2 = 0 \qquad \text{or} \qquad x + 1 = 0$$

$$3x = 2 \qquad\qquad\qquad x = -1$$

$$x = \frac{2}{3}$$

Figure 5–26

Thus, $x = 2/3$ and $x = -1$ are the *possible* solutions and must be checked in the *original* equation. When $x = 2/3$, both sides of the original equation have the same value, as shown in Figure 5–26. So $2/3$ is a solution. When $x = -1$, however, the right side of the equation is

$$\log (7x + 6) = \log [7(-1) + 6] = \log (-1),$$

which is not defined. So -1 is not a solution.   ∎

Equations that involve both logarithmic and constant terms may be solved by using the basic property of logarithms (see page 365).

$(*)$ $\qquad\qquad 10^{\log v} = v \qquad \text{and} \qquad e^{\ln v} = v.$

EXAMPLE 9

Solve $7 + 2 \log 5x = 11$.

SOLUTION   We start by getting all the logarithmic terms on one side and the constant on the other.

Subtract 7 from both sides: $2 \log 5x = 4$

Divide both sides by 2: $\log 5x = 2.$

We know that if two quantities are equal, say $a = b$, then $10^a = 10^b$. We use this fact here, with the two sides of the preceding equation as a and b.

Exponentiate both sides: $10^{\log 5x} = 10^2$

Use the basic logarithm property $(*)$: $5x = 100$

Divide both sides by 5: $x = 20.$

Verify that 20 is actually a solution of the original equation.   ∎

EXAMPLE 10

Solve $\ln (x - 3) = 5 - \ln (x - 3)$.

SOLUTION   We proceed as in Example 9, but since the base for these logarithms is e, we use e rather than 10 when we exponentiate.

$$\ln (x - 3) = 5 - \ln (x - 3)$$

Add $\ln (x - 3)$ to both sides: $$2 \ln (x - 3) = 5$$

Divide both sides by 2: $$\ln (x - 3) = \frac{5}{2}$$

Exponentiate both sides: $$e^{\ln(x-3)} = e^{5/2}$$

Use the basic property of logarithms (∗): $$x - 3 = e^{5/2}$$

Add 3 to both sides: $$x = e^{5/2} + 3 \approx 15.1825.$$

This is the only possibility for a solution. A calculator shows that it actually is a solution of the original equation.   ■

EXAMPLE 11

Solve $\log (x - 16) = 2 - \log (x - 1)$.

SOLUTION

$$\log (x - 16) = 2 - \log (x - 1)$$

Add $\log (x - 1)$ to both sides: $$\log (x - 16) + \log (x - 1) = 2$$

Use the Product Law: $$\log [(x - 16)(x - 1)] = 2$$

Multiply out left side: $$\log (x^2 - 17x + 16) = 2$$

Exponentiate both sides: $$10^{\log (x^2-17x+16)} = 10^2$$

Use the basic logarithm property (∗): $$x^2 - 17x + 16 = 100$$

Subtract 100 from both sides: $$x^2 - 17x - 84 = 0$$

Factor: $$(x + 4)(x - 21) = 0$$

$$x + 4 = 0 \quad \text{or} \quad x - 21 = 0$$

$$x = -4 \quad \text{or} \quad x = 21.$$

You can easily verify that 21 is a solution of the original equation, but -4 is not [when $x = -4$, then $\log (x - 16) = \log (-20)$, which is not defined].   ■

EXAMPLE 12

To solve $\log (x + 5) = 1 - \log (x - 2)$,

Rearrange terms: $$\log (x + 5) + \log (x - 2) = 1$$

Use the Product Law: $$\log [(x + 5)(x - 2)] = 1$$

$$\log (x^2 + 3x - 10) = 1$$

Exponentiate both sides: $$10^{\log (x^2+3x-10)} = 10^1$$

Use the basic logarithm property (∗): $$x^2 + 3x - 10 = 10$$

$$x^2 + 3x - 20 = 0.$$

This equation can be solved with the quadratic formula.

$$x = \frac{-3 \pm \sqrt{3^2 - 4 \cdot 1 \cdot (-20)}}{2 \cdot 1} = \frac{-3 \pm \sqrt{89}}{2}.$$

An easy way to verify that

$$x = \frac{-3 + \sqrt{89}}{2}$$

is a solution is to store this number in your calculator as A and then evaluate both sides of the original equation at $x = A$, as shown in Figure 5–27. The other possibility, however, is not a solution because

$$x = \frac{-3 - \sqrt{89}}{2}$$

Figure 5–27

is negative, so log $(x - 2)$ is not defined. ■

 EXERCISES 5.5

In Exercises 1–8, solve the equation without using logarithms.

1. $3^x = 81$ **2.** $3^x + 3 = 30$ **3.** $3^{x+1} = 9^{5x}$

4. $4^{5x} = 16^{2x-1}$ **5.** $3^{5x}9^{x^2} = 27$

6. $2^{x^2+5x} = 1/16$ **7.** $9^{x^2} = 3^{-5x-2}$

8. $4^{x^2-1} = 8^x$

In Exercises 9–22, solve the equation. First express your answer in terms of natural logarithms (for instance, $x = (2 + \ln 5)/(\ln 3)$). Then use a calculator to find an approximation for the answer.

9. $3^x = 5$ **10.** $5^x = 4$ **11.** $2^x = 3^{x-1}$

12. $4^{x+2} = 2^{x-1}$ **13.** $3^{1-2x} = 5^{x+5}$

14. $4^{3x-1} = 3^{x-2}$ **15.** $2^{1-3x} = 3^{x+1}$

16. $3^{z+3} = 2^z$ **17.** $e^{2x} = 5$

18. $e^{-3x} = 2$ **19.** $6e^{-1.4x} = 21$

20. $3.4e^{-x/3} = 5.6$ **21.** $2.1e^{(x/2)\ln 3} = 5$

22. $7.8e^{(x/3)\ln 5} = 14$

In Exercises 23–29, solve the equation for x by first making an appropriate substitution, as in the Hint for Exercise 23.

23. $9^x - 4 \cdot 3^x + 3 = 0$ *[Hint: Let $u = 3^x$ and note that $9^x = (3^2)^x = 3^{2x} = (3^x)^2$. Hence, the equation becomes $u^2 - 4u + 3 = 0$. Solve this equation for u. In each solution, replace u by 3^x and solve for x.]*

24. $4^x - 6 \cdot 2^x = -8$

25. $e^{2x} - 5e^x + 6 = 0$ *[Hint: Let $u = e^x$.]*

26. $2e^{2x} - 9e^x + 4 = 0$ **27.** $6e^{2x} - 16e^x = 6$

28. $8e^{2x} + 8e^x = 6$ **29.** $4^x + 6 \cdot 4^{-x} = 5$

In Exercises 30–32, solve the equation for x.

30. $\dfrac{e^x + e^{-x}}{e^x - e^{-x}} = t$ **31.** $\dfrac{e^x - e^{-x}}{2} = t$

32. $\dfrac{e^x - e^{-x}}{e^x + e^{-x}} = t$

33. Prove that if $\ln u = \ln v$, then $u = v$. *[Hint: Property (∗) on page 388.]*

34. (a) Solve $7^x = 3$, using natural logarithms. Leave your answer in logarithmic form; don't approximate with a calculator.

(b) Solve $7^x = 3$, using common (base 10) logarithms. Leave your answer in logarithmic form.

(c) Use the change of base formula in Special Topics 5.4.A to show that your answers in parts (a) and (b) are the same.

In Exercises 35–44, solve the equation as in Example 8.

35. $\ln (3x - 5) = \ln 11 + \ln 2$

36. $\log (4x - 1) = \log (x + 1) + \log 2$

37. $\log (3x - 1) + \log 2 = \log 4 + \log (x + 2)$

38. $\ln (x + 6) - \ln 10 = \ln (x - 1) - \ln 2$

39. $2 \ln x = \ln 36$

40. $2 \log x = 3 \log 4$

41. $\ln x + \ln (x + 1) = \ln 3 + \ln 4$

42. $\ln (6x - 1) + \ln x = \frac{1}{2}\ln 4$

43. $\ln x = \ln 3 - \ln (x + 5)$

44. $\ln (2x + 3) + \ln x = \ln e$

In Exercises 45–52, solve the equation.

45. $\ln (x + 9) - \ln x = 1$

46. $\ln (2x + 1) - 1 = \ln (x - 2)$

47. $\log x + \log (x - 3) = 1$

48. $\log (x - 1) + \log (x + 2) = 1$

49. $\log \sqrt{x^2 - 1} = 2$

50. $\log \sqrt[3]{x^2 + 21x} = 2/3$

51. $\ln (x^2 + 1) - \ln (x - 1) = 1 + \ln (x + 1)$

52. $\dfrac{\ln (x + 1)}{\ln (x - 1)} = 2$

Exercises 53–62 deal with radioactive decay and the function $M(x) = c(.5^{x/h})$; see Examples 4 and 5.

53. A sample of 300 grams of uranium decays to 200 grams in .26 billion years. Find the half-life of uranium.

54. It takes 1000 years for a sample of 100 mg of radium-226 to decay to 65 mg. Find the half-life of radium-226.

55. A 3-gram sample of an isotope of sodium decays to 1 gram in 23.7 days. Find the half-life of the isotope of sodium.

56. The half-life of cobalt-60 is 4.945 years. How long will it take for 25 grams to decay to 15 grams?

57. After six days a sample of radon-222 decayed to 33.6% of its original mass. Find the half-life of radon-222. [*Hint:* When $x = 6$, then $M(x) = .336P$.]

58. Krypton-85 loses 6.44% of its mass each year. What is its half-life?

59. How old is a piece of ivory that has lost 36% of its carbon-14?

60. How old is a mummy that has lost 49% of its carbon-14?

61. A Native American mummy was found recently. If it has lost 26.4% of its carbon-14, approximately how long ago did the Native American die?

62. How old is a wooden statue that has only one-third of its original carbon-14?

Exercises 63–68 deal with the compound interest formula $A = P(1 + r)^t$, which was discussed in Special Topics 5.2.A.

63. At what annual rate of interest should $1000 be invested so that it will double in 10 years if interest is compounded quarterly?

64. How long does it take $500 to triple if it is invested at 6% compounded: (a) annually, (b) quarterly, (c) daily?

65. (a) How long will it take to triple your money if you invest $500 at a rate of 5% per year compounded annually?
(b) How long will it take at 5% compounded quarterly?

66. At what rate of interest (compounded annually) should you invest $500 if you want to have $1500 in 12 years?

67. How much money should be invested at 5% interest, compounded quarterly, so that 9 years later the investment will be worth $5000? This amount is called the **present value** of $5000 at 5% interest.

68. Find a formula that gives the time needed for an investment of P dollars to double, if the interest rate is r% compounded annually. [*Hint:* Solve the compound interest formula for t, when $A = 2P$.]

Exercises 69–76 deal with functions of the form $f(x) = Pe^{kx}$, where k is the continuous exponential growth rate (see Example 6).

69. The present concentration of carbon dioxide in the atmosphere is 364 parts per million (ppm) and is increasing exponentially at a continuous yearly rate of .4% (that is, $k = .004$). How many years will it take for the concentration to reach 500 ppm?

70. The amount P of ozone in the atmosphere is currently decaying exponentially each year at a continuous rate of $\frac{1}{4}$% (that is, $k = -.0025$). How long will it take for half the ozone to disappear (that is, when will the amount be $P/2$)? [Your answer is the half-life of ozone.]

71. The population of Brazil increased from 151 million in 1990 to 173 million in 2000.[*]
(a) At what continuous rate was the population growing during this period?
(b) Assuming that Brazil's population continues to increase at this rate, when will it reach 250 million?

72. Outstanding consumer debt increased exponentially from $781.5 billion in 1990 to $1765.5 billion in 2002.[†]
(a) At what continuous growth rate is consumer debt growing?
(b) Assuming that this rate continues, when will consumer debt reach $2500 billion?

73. The probability P percent of having an accident while driving a car is related to the alcohol level of the driver's blood by the formula $P = e^{kt}$, where k is a constant. Accident statistics show that the probability of an accident is 25% when the blood alcohol level is $t = .15$.
(a) Find k. [Use $P = 25$, not .25.]
(b) At what blood alcohol level is the probability of having an accident 50%?

---

[*]U.S. Census Bureau, International Data Base.
[†]*Federal Reserve Bulletin.*

74. Under normal conditions, the atmospheric pressure (in millibars) at height h feet above sea level is given by $P(h) = 1015e^{-kh}$, where k is a positive constant.
    (a) If the pressure at 18,000 feet is half the pressure at sea level, find k.
    (b) Using the information from part (a), find the atmospheric pressure at 1000 feet, 5000 feet, and 15,000 feet.

75. One hour after an experiment begins, the number of bacteria in a culture is 100. An hour later, there are 500.
    (a) Find the number of bacteria at the beginning of the experiment and the number three hours later.
    (b) How long does it take the number of bacteria at any given time to double?

76. If the population at time t is given by $S(t) = ce^{kt}$, find a formula that gives the time it takes for the population to double.

77. The spread of a flu virus in a community of 45,000 people is given by the function
$$f(t) = \frac{45,000}{1 + 224e^{-.899t}},$$
where $f(t)$ is the number of people infected in week t.
    (a) How many people had the flu at the outbreak of the epidemic? After three weeks?
    (b) When will half the town be infected?

78. The beaver population near a certain lake in year t is approximately
$$p(t) = \frac{2000}{1 + 199e^{-.5544t}}.$$
    (a) When will the beaver population reach 1000?
    (b) Will the population ever reach 2000? Why?

Thinkers

79. According to one theory of learning, the number of words per minute N that a person can type after t weeks of practice is given by $N = c(1 - e^{-kt})$, where c is an upper limit that N cannot exceed and k is a constant that must be determined experimentally for each person.
    (a) If a person can type 50 wpm (words per minute) after four weeks of practice and 70 wpm after eight weeks, find the values of k and c for this person. According to the theory, this person will never type faster than c wpm.
    (b) Another person can type 50 wpm after four weeks of practice and 90 wpm after eight weeks. How many weeks must this person practice to be able to type 125 wpm?

80. Kate has been offered two jobs, each with the same starting salary of $32,000 and identical benefits. Assuming satisfactory performance, she will receive a $1600 raise each year at the Great Gizmo Company, whereas the Wonder Widget Company will give her a 4% raise each year.
    (a) In what year (after the first year) would her salary be the same at either company? Until then, which company pays better? After that, which company pays better?
    (b) Answer the questions in part (a) assuming that the annual raise at Great Gizmo is $2000.

5.6   Exponential, Logarithmic, and Other Models*

Many data sets can be modeled by suitable exponential, logarithmic, and related functions. Most calculators have regression procedures for constructing the following models.

Model	Equation	Examples	
Power	$y = ax^r$	$y = 5x^{2.7}$	$y = 3.5x^{-.045}$
Exponential	$y = ab^x$ or $y = ae^{kx}$	$y = 2(1.64)^x$	$y = 2 \cdot e^{.4947x}$
Logistic	$y = \dfrac{a}{1 + be^{-kx}}$	$y = \dfrac{20,000}{1 + 24e^{-.25x}}$	$y = \dfrac{650}{1 + 6e^{.3x}}$
Logarithmic	$y = a + b \ln x$	$y = 5 + 4.2 \ln x$	$y = 2 - 3 \ln x$

*This section is optional; its prerequisites are Section 2.5 and Special Topics 4.4.A. It will be used in clearly identifiable exercises but not elsewhere in the text.

We begin by examining exponential models, such as $y = 3 \cdot 2^x$. A table of values for this model is shown below. Look carefully at the ratio of successive entries (that is, each entry divided by its predecessor).

x	$y = 3 \cdot 2^x$
0	3
4	48
8	768
12	12,288
16	196,608

$$\frac{48}{3} = 16$$
$$\frac{768}{48} = 16$$
$$\frac{12,288}{768} = 16$$
$$\frac{196,608}{12,228} = 16$$

It should not be a surprise that the ratio of successive entries is constant. For at each step, x changes from x to $x + 4$ (from 0 to 4, from 4 to 8, and so on) and y changes from $3 \cdot 2^x$ to $3 \cdot 2^{x+4}$. Hence, the ratio of successive terms is always

$$\frac{3 \cdot 2^{x+4}}{3 \cdot 2^x} = \frac{3 \cdot 2^x \cdot 2^4}{3 \cdot 2^x} = 2^4 = 16.$$

A similar argument applies to any exponential model $y = ab^x$ and shows that if x changes by a fixed amount k, then the ratio of the corresponding y values is the constant b^k (in our example b was 2 and k was 4). This suggests that

when the ratio of successive entries in a table of data is approximately constant, an exponential model is appropriate.

EXAMPLE 1

In the years before the Civil War, the population of the United States grew rapidly, as shown in the following table from the U.S. Bureau of the Census. Find a model for this growth.

Year	Population in Millions	Year	Population in Millions
1790	3.93	1830	12.86
1800	5.31	1840	17.07
1810	7.24	1850	23.19
1820	9.64	1860	31.44

Figure 5–28

SOLUTION   The data points (with $x = 0$ corresponding to 1790) are shown in Figure 5–28. Their shape suggests either a polynomial graph of even degree or an exponential graph. Since populations generally grow exponentially, an exponential model is likely to be a good choice. We can confirm this by looking at the ratios of successive entries in the table.

Year	Population		Year	Population	
1790	3.93		1830	12.86	
		$\dfrac{5.31}{3.93} \approx 1.351$			$\dfrac{17.07}{12.86} \approx 1.327$
1800	5.31		1840	17.07	
		$\dfrac{7.24}{5.31} \approx 1.363$			$\dfrac{23.19}{17.07} \approx 1.359$
1810	7.24		1850	23.19	
		$\dfrac{9.64}{7.24} \approx 1.331$			$\dfrac{31.44}{23.19} \approx 1.356$
1820	9.64		1860	31.44	
		$\dfrac{12.86}{9.64} \approx 1.334$			
1830	12.86				

The ratios are almost constant, as they would be in an exponential model. So we use regression to find such an exponential model. The procedure is the same as for linear and polynomial regression (see the Tips on pages 112 and 267). It produces this model:[*]

$$y = 3.9572\,(1.0299^x).$$

Figure 5–29

The graph in Figure 5–29 appears to fit the data quite well. In fact, you can readily verify that the model has an error of less than 1% for each of the data points. Furthermore, as discussed before the example, when x changes by 10, the value of y changes by approximately $1.0299^{10} \approx 1.343$, which is very close to the successive ratios of the data that were computed above.   ∎

EXAMPLE 2

After the Civil War, the U.S. population continued to increase, as shown below.

Year	Population in Millions	Year	Population in Millions	Year	Population in Millions
1870	38.56	1920	106.02	1960	179.32
1880	50.19	1930	123.20	1970	202.30
1890	62.98	1940	132.16	1980	226.54
1900	76.21	1950	151.33	1990	248.72
1910	92.23			2000	281.42

---

[*]Throughout this section, coefficients are rounded for convenient reading, but the full expansion is used for calculations and graphs.

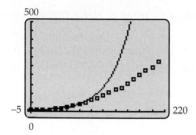

Figure 5–30

However, the model from Example 1 does not remain valid, as can be seen in Figure 5–30, which shows its graph together with all the data points from 1790 through 2000 ($x = 0$ corresponds to 1790).

The problem is that the *rate* of growth has steadily decreased since the Civil War. For instance, the ratio of the first two entries in the preceding table is

$$\frac{50.19}{38.56} \approx 1.302,$$

and the ratio of the last two is

$$\frac{281.42}{248.72} \approx 1.13.$$

So an exponential model may not be the best choice now. Other possibilities are polynomial models (which grow at a slower rate) or logistic models (in which the growth rate decreases with time). Figure 5–31 shows three possible models, each obtained by using the appropriate regression program on a calculator, with all the data points from 1790 through 2000.

Exponential Model

$$y = 6.06616 \cdot 1.02039^x$$

Polynominal Model

$$y = (7.94 \times 10^{-8})x^4 - (2.76 \times 10^{-5})x^3$$
$$+ .0093x^2 - .1621x + 5.462$$

Logistic Model*

$$y = \frac{442.1}{1 + 56.33e^{-.0216x}}$$

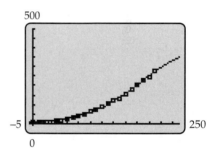

Figure 5–31

As expected, the polynomial and logistic models fit the data better than does the exponential model. The main difference between them is that the polynomial model indicates unlimited future growth, whereas the logistic model has the population growing more slowly in the future (and eventually leveling off—see Exercise 13). ∎

In Example 1, we used the ratios of successive entries of the data table to determine that an exponential model was appropriate. Here is another way to make that determination. Consider the exponential function $y = ab^x$. Taking natural logarithms of both sides and using the logarithm laws on the right side shows that

$$\ln y = \ln (ab^x) = \ln a + \ln b^x = \ln a + x \ln b.$$

Now $\ln a$ and $\ln b$ are constants, say, $k = \ln a$ and $m = \ln b$, so

$$\ln y = mx + k.$$

---

*This model was obtained on a TI-83. Other calculators may produce a slightly different model or an error message.

Thus, the points $(x, \ln y)$ lie on the straight line with slope m and y-intercept k. Consequently, we have this guideline.

> **If (x, y) are data points and if the points $(x, \ln y)$ are approximately linear, then an exponential model may be appropriate for the data.**

Similarly, if $y = ax^r$ is a power function, then

$$\ln y = \ln (ax^r) = \ln a + r \ln x.$$

Since $\ln a$ is a constant, say $k = \ln a$, we have

$$\ln y = r \ln x + k,$$

which means that the points $(\ln x, \ln y)$ lie on a straight line with slope r and y-intercept k. Consequently, we have this guideline.

> **If (x, y) are data points and if the points $(\ln x, \ln y)$ are approximately linear, then a power model may be appropriate for the data.**

EXAMPLE 3

The length of time that a planet takes to make one complete rotation around the sun is its year. The table shows the length (in earth years) of each planet's year and the distance of that planet from the sun (in millions of miles).[*] Find a model for this data in which x is the length of the year and y the distance from the sun.

Planet	Year	Distance	Planet	Year	Distance
Mercury	.24	36.0	Saturn	29.46	886.7
Venus	.62	67.2	Uranus	84.01	1783.0
Earth	1	92.9	Neptune	164.79	2794.0
Mars	1.88	141.6	Pluto	247.69	3674.5
Jupiter	11.86	483.6			

SOLUTION   Figure 5–32 shows the data points for the five planets with the shortest years. Figure 5–33 shows all the data points, but on this scale, the first four points look like a single large one near the origin.

Figure 5–32

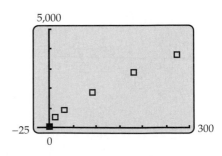

Figure 5–33

---

*Since the orbit of a planet around the sun is not circular, its distance from the sun varies through the year. Each given number is the average of its maximum and minimum distances from the sun.

Plotting the point $(x, \ln y)$ for each data point (x, y) (see the Technology Tip in the margin) produces Figure 5–34. Its points do not form a linear pattern (four of them are almost vertical near the *y*-axis and the other five almost horizontal), so an exponential function is not an appropriate model. On the other hand, the points $(\ln x, \ln y)$ in Figure 5–35 do form a linear pattern, which suggests that a power model will work.

Figure 5–34

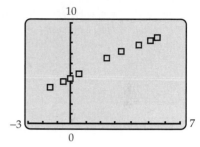

Figure 5–35

A calculator's power regression feature produces this model.

$$y = 92.8935 \, x^{.6669}.$$

Its graph in Figure 5–36 shows that it fits the original data points quite well. ■

Figure 5–36

If $y = a + b \ln x$ is a logarithmic model, then the points $(\ln x, y)$ lie on the straight line with slope *b* and *y*-intercept *a* (why?). Thus, we have this guideline.

If (x, y) are data points and if the points $(\ln x, y)$ are approximately linear, then a logarithmic model may be appropriate for the data.

EXAMPLE 4

Find a model for population growth in El Paso, Texas, given the following data.[*]

Year	1950	1970	1980	1990	2000
Population	130,485	322,261	425,259	515,342	563,662

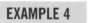

*U.S. Bureau of the Census.

SOLUTION   The scatter plot of the data points (with $x = 50$ corresponding to 1950) in Figure 5–37 bends slightly, suggesting a logarithmic curve. So we plot the points $(\ln x, y)$, that is,

$$(\ln 50, 130485), (\ln 70, 322261), \ldots, (\ln 100, 563662),$$

in Figure 5–38. Since these points lie approximately on a straight line, a logarithmic model is appropriate. Using logarithmic regression on a calculator, we obtain this model.

$$-2{,}382{,}368.345 + 640{,}666.815 \ln x.$$

Its graph in Figure 5–39 is a good fit for the data.   ■

CAUTION
When using logarithmic models, you must have data points with positive first coordinates (since logarithms of negative numbers and 0 are not defined).

Figure 5–37

Figure 5—38

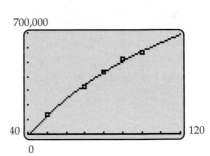

Figure 5–39

EXERCISES 5.6

In Exercises 1–10, state which of the following models might be appropriate for the given scatter plot of data (more than one model may be appropriate).

Model	Corresponding Function
A. Linear	$y = ax + b$
B. Quadratic	$y = ax^2 + bx + c$
C. Power	$y = ax^r$
D. Cubic	$y = ax^3 + bx^2 + cx + d$
E. Exponential	$y = ab^x$
F. Logarithmic	$y = a + b \ln x$
G. Logistic	$y = \dfrac{a}{1 + be^{-kx}}$

1.

2.

3.

4.

5.

6.

7.

8.

9.

10.

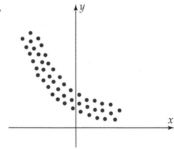

In Exercises 11 and 12, compute the ratios of successive entries in the table to determine whether or not an exponential model is appropriate for the data.

11.

x	0	2	4	6	8	10
y	3	15.2	76.9	389.2	1975.5	9975.8

12.

x	1	3	5	7	9	11
y	3	21	55	105	171	253

13. (a) Show algebraically that in the logistic model for the U.S. population in Example 2, the population can never exceed 442.1 million people.

(b) Confirm your answer in part (a) by graphing the logistic model in a window that includes the next three centuries.

14. According to estimates by the U.S. Bureau of the Census, the U.S. population was 287.7 million in 2002. On the basis of this information, which of the models in Example 2 appears to be the most accurate predictor?

15. Graph each of the following power functions in a window with $0 \le x \le 20$.
(a) $f(x) = x^{-1.5}$ (b) $g(x) = x^{.75}$ (c) $h(x) = x^{2.4}$

16. On the basis of your graphs in Exercise 15, describe the general shape of the graph of $y = ax^r$ when $a > 0$ and
(a) $r < 0$ (b) $0 < r < 1$ (c) $r > 1$

In Exercises 17–20, determine whether an exponential, power, or logarithmic model (or none or several of these) is appropriate for the data by determining which (if any) of the following sets of points are approximately linear:

$$\{(x, \ln y)\}, \qquad \{(\ln x, \ln y)\}, \qquad \{(\ln x, y)\}.$$

where the given data set consists of the points $\{(x, y)\}$.

17.

x	1	3	5	7	9	11
y	2	25	81	175	310	497

18.

x	3	6	9	12	15	18
y	385	74	14	2.75	.5	.1

19.

x	5	10	15	20	25	30
y	17	27	35	40	43	48

20.

x	5	10	15	20	25	30
y	2	110	460	1200	2500	4525

21. The table shows the number of babies born as twins, triplets, quadruplets, etc., in recent years.

Year	Multiple Births
1989	92,916
1990	96,893
1991	98,125
1992	99,255
1993	100,613
1994	101,658
1995	101,709

(a) Sketch a scatter plot of the data, with $x = 1$ corresponding to 1989.

(b) Plot each of the following models on the same screen as the scatter plot.

$$f(x) = 93{,}201.973 + 4{,}545.977 \ln x;$$
$$g(x) = \frac{102{,}519.98}{1 + .1536e^{-.4263x}}.$$

(c) Use the table feature to estimate the number of multiple births in 2000 and 2005.

(d) Over the long run, which model do you think is the better predictor?

22. The graph shows the U.S. Census Bureau estimates of future U.S. population.

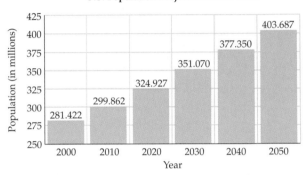

U.S. Population Projections: 2000–2050

(a) How well do the projections in the graph compare with those given by the logistic model in Example 2?

(b) Find a logistic model of the U.S. population, using the data given in Example 2 for the years from 1900 to 2000.

(c) How well do the projections in the graph compare with those given by the model in part (b)?

23. Infant mortality rates in the United States are shown in the table.

Year	Infant Mortality Rate*	Year	Infant Mortality Rate*
1920	76.7	1980	12.6
1930	60.4	1985	10.6
1940	47.0	1990	9.2
1950	29.2	1995	7.6
1960	26.0	2000	6.9
1970	20.0		

(a) Sketch a scatter plot of the data, with $x = 0$ corresponding to 1900.

(b) Verify that the set of points $(x, \ln y)$, where (x, y) are the original data points, is approximately linear.

(c) On the basis of part (b), what type of model would be appropriate for this data? Find such a model.

---

*Rates are infant (under 1 year) deaths per 1000 live births.

24. The number of children who were home schooled in the United States in selected years is shown in the table. *
  (a) Sketch a scatter plot of the data, with $x = 0$ corresponding to 1980.
  (b) Find a quadratic model for the data.
  (c) Find a logistic model for the data.
  (d) What is the number of home-schooled children predicted by each model for the year 2003?
  (e) What are the limitations of each model?

Fall of School Year	Number of Children (in thousands)
1985	183
1988	225
1990	301
1992	470
1993	588
1994	735
1995	800
1996	920
1997	1100
1999	1400
2000	1700

25. The average number of students per computer in the U.S. public schools (elementary through high school) is shown in the table at the top of the next column.
  (a) Sketch a scatter plot of the data, with $x = 7$ corresponding to 1987.
  (b) Find an exponential model for the data.
  (c) Use the model to estimate the number of students per computer in 2003.
  (d) In what year, according to this model, will each student have his or her own computer in school?
  (e) What are the limitations of this model?

---

*National Home Education Research Institute

Fall of School Year	Students/Computer
1987	32
1988	25
1989	22
1990	20
1991	18
1992	16
1993	14
1994	10.5
1995	10
1996	7.8
1997	6.1
1998	5.7
1999	5.4

26. (a) Find an exponential model for the federal debt, based on the data in the table. Let $x = 0$ correspond to 1960.

ACCUMULATED GROSS FEDERAL DEBT	
Year	Amount (in billions of dollars)
1960	284.1
1965	313.8
1970	370.1
1975	533.2
1980	907.7
1985	1823.1
1990	3233.3
1995	4974.0
2000	5674.2

  (b) Use the model to estimate the federal debt in 2003.

27. The number of U.S. households (in millions) from 1995 to 2000 are shown in the first two columns of the following table.[*]

Year	Households	Predicted Number of Households	Ratio
1995	99.0		
1996	99.6		
1997	101.0		
1998	102.5		
1999	103.6		
2000	104.8		

(a) Sketch a scatter plot of the data with $x = 0$ corresponding to 1995.
(b) Fill in column 4 by dividing each entry in column 2 by the preceding one; round your answers to two decimal places. In view of column 4, what type of model is appropriate here?
(c) Find an appropriate model for the data.
(d) Use the model to complete column 3 in the table. How well does the model fit the data?
(e) If the model remains accurate, how many U.S. households are there in 2003?

28. In the past two decades, more women than men have been entering college. The table shows the percentage of male first-year college students in selected years.[†]

Year	1985	1990	1995	1997	1998	1999	2000
Percent	48.9	46.9	45.6	45.5	45.5	45.3	45.2

(a) Find three models for this data: exponential, logarithmic, and power, with $x = 5$ corresponding to 1985.
(b) For the years 1985–2000, is there any significant difference among the models?
(c) Assume that the models remain accurate. What year does each predict as the first year in which fewer than 43% of first-year college students will be male?

29. The table gives the life expectancy (at birth) of a woman born in the given year.[‡]

Year	Life Expectancy (in years)
1910	51.8
1920	54.6
1930	61.6
1940	65.2
1950	71.1
1960	73.1
1970	74.7
1980	77.5
1990	78.8
2000	79.4

(a) Find a logarithmic model for the data, with $x = 10$ corresponding to 1910.
(b) Use the model to find the life expectancy of a woman born in 1986. [For comparison, the actual expectancy is 78.3 years.]
(c) Assume that the model remains accurate. In what year will the life expectancy of a woman born in that year be at least 81 years?

30. The table gives the death rate in motor vehicle accidents (per 100,000 population) in selected years.[‡]

Year	1970	1980	1985	1990	1995	2000
Death Rate	26.8	23.4	19.3	18.8	16.5	15.6

(a) Find an exponential model for the data, with $x = 0$ corresponding to 1970.
(b) What was the death rate in 1998 and in 2002?
(c) Assume that the model remains accurate, when will the death rate drop to 13 per 100,000?

---

[*]U.S. Department of Commerce.
[†]Higher Education Research Institute at UCLA.
[‡]National Center for Health Statistics.

Chapter 5 Review

IMPORTANT CONCEPTS

Section 5.1

nth root   327
Rational exponents   328
Exponent Laws   329
Radical notation   330
Rationalizing numerators and
   denominators   331
Irrational exponents   332

Special Topics 5.1.A

Power Principle   334

Section 5.2

Exponential functions   342–343
Exponential growth and decay
   346–347
The number e 348

Special Topics 5.2.A

Compound interest formula   354
Continuous compounding   358

Section 5.3

Common logarithmic functions
   361
Natural logarithmic functions   363
Properties of logarithms   365

Section 5.4

Product Law for Logarithms   370
Quotient Law for Logarithms   371
Power Law for Logarithms   372
Richter scale   374

Special Topics 5.4.A

Logarithmic functions to base b
   376
Properties of logarithms   378
Logarithm Laws   379
Change of base formula   381

Section 5.5

Exponential equations   383
Logarithmic equations   387

Section 5.6

Exponential, logarithmic, power,
   and logistic models   392

IMPORTANT FACTS & FORMULAS

■ *Laws of Exponents:*

$$c^r c^s = c^{r+s} \qquad (cd)^r = c^r d^r$$

$$\frac{c^r}{c^s} = c^{r-s} \qquad \left(\frac{c}{d}\right)^r = \frac{c^r}{d^r}$$

$$(c^r)^s = c^{rs} \qquad c^{-r} = \frac{1}{c^r}$$

■ $g(x) = \log x$ is the inverse function of $f(x) = 10^x$:

$$10^{\log v} = v \text{ for all } v > 0 \qquad \text{and} \qquad \log 10^u = u \text{ for all } u.$$

■ $g(x) = \ln x$ is the inverse function of $f(x) = e^x$:

$$e^{\ln v} = v \text{ for all } v > 0 \qquad \text{and} \qquad \ln(e^u) = u \text{ for all } u.$$

■ $h(x) = \log_b x$ is the inverse function of $k(x) = b^x$:

$$b^{\log_b v} = v \text{ for all } v > 0 \qquad \text{and} \qquad \log_b(b^u) = u \text{ for all } u.$$

■ *Logarithm Laws:* For all v, $w > 0$ and any k:

$$\ln (vw) = \ln v + \ln w \qquad \log_b (vw) = \log_b v + \log_b w$$

$$\ln \left(\frac{v}{w} \right) = \ln v - \ln w \qquad \log_b \left(\frac{v}{w} \right) = \log_b v - \log_b w$$

$$\ln (v^k) = k(\ln v) \qquad \log_b (v^k) = k(\log_b v)$$

■ *Exponential Growth Functions:*

$$f(x) = P(1 + r)^x \quad (0 < r < 1),$$

$$f(x) = Pa^x \quad (a > 1),$$

$$f(x) = Pe^{kx} \quad (k > 0)$$

■ *Exponential Decay Functions:*

$$f(x) = P(1 - r)^x \quad (0 < r < 1),$$

$$f(x) = Pa^x \quad (0 < a < 1),$$

$$f(x) = Pe^{kx} \quad (k < 0)$$

■ *Compound Interest Formula:* $A = P(1 + r)^t$

■ *Change of Base Formula:* $\log_b v = \dfrac{\ln v}{\ln b}$

CATALOG OF BASIC FUNCTIONS—PART 3

Exponential Functions

$f(x) = b^x \quad (b > 1)$

$f(x) = b^x \quad (0 < b < 1)$

Logarithmic Functions

$f(x) = \log x$

$f(x) = \ln x$

REVIEW QUESTIONS

In Questions 1–6, simplify the expression.

1. $\sqrt[3]{\sqrt{c^{12}}}$

2. $(\sqrt[3]{4}c^3d^2)^3(c\sqrt{d})^2$

3. $(a^{-2/3}b^{2/5})(a^3b^6)^{4/3}$

4. $\dfrac{(3c)^{3/5}(2d)^{-2}(4c)^{1/2}}{(4c)^{1/5}(2d)^4(2c)^{-3/2}}$

5. $(u^{1/4} - v^{1/4})(u^{1/4} + v^{1/4})$

6. $c^{3/2}(2c^{1/2} + 3c^{-3/2})$

In Exercises 7 and 8, simplify and write the expression without radicals or negative exponents:

7. $\dfrac{\sqrt[3]{6c^4d^{14}}}{\sqrt[3]{48c^{-2}d^2}}$

8. $\dfrac{(8u^5)^{1/4}2^{-1}u^{-3}}{2u^8}$

9. Rationalize the numerator and simplify:

$$\frac{\sqrt{2x + 2h + 1} - \sqrt{2x + 1}}{h}.$$

10. Rationalize the denominator: $\dfrac{5}{\sqrt{x} - 3}$.

In Questions 11–16, find all real solutions of the equation.

11. $\sqrt{x - 1} = 2 - x$

12. $\sqrt[3]{1 - t^2} = -2$

13. $\sqrt{x + 1} + \sqrt{x - 1} = 1$

14. $x^{1/4} - 5x^{1/2} + 6 = 0$

15. $\sqrt[3]{x^4 - 2x^3 + 6x - 7} = x + 3$

16. $x^{4/3} + x - 2x^{2/3} + x^{1/3} = 4$

In Questions 17 and 18, find a viewing window (or windows) that shows a complete graph of the function.

17. $f(x) = 2^{x^3 - x - 2}$

18. $g(x) = \ln\left(\dfrac{x}{x - 2}\right)$

In Questions 19–22, sketch a complete graph of the function. Indicate all asymptotes clearly.

19. $g(x) = 2^x - 1$

20. $f(x) = 2^{x-1}$

21. $h(x) = \ln(x + 4) - 2$

22. $k(x) = \ln\left(\dfrac{x + 4}{x}\right)$

23. A computer software company claims the following function models the "learning curve" for their mathematical software.

$$P(t) = \frac{100}{1 + 48.2e^{-.52t}},$$

where t is measured in months and $P(t)$ is the average percent of the software program's capabilities mastered after t months.

(a) Initially, what percent of the program is mastered?

(b) After six months, what percent of the program is mastered?

(c) Roughly, when can a person expect to "learn the most in the least amount of time"?

(d) If the company's claim is true, how many months will it take to have completely mastered the program?

24. Compunote has offered you a starting salary of $60,000 with $1000 yearly raises. Calcuplay offers you an initial salary of $30,000 and a guaranteed 6% raise each year.

(a) Complete the following table for each company.

Year	Compunote		Year	Calcuplay
1	$60,000		1	$30,000
2	$61,000		2	$31,800
3			3	
4			4	
5			5	

(b) For each company, write a function that gives your salary in terms of years employed.

(c) If you plan on staying with the company for only five years, which job should you take to earn the most money?

(d) If you plan on staying with the company for 20 years, which is your better choice?

(e) In what year does the salary at Calcuplay exceed the salary at Compunote?

In Questions 25–30, translate the given exponential statement into an equivalent logarithmic one.

25. $e^{6.628} = 756$

26. $e^{5.8972} = 364$

27. $e^{r^2 - 1} = u + v$

28. $e^{a-b} = c$

29. $10^{2.8785} = 756$

30. $10^{c+d} = t$

In Questions 31–36, translate the given logarithmic statement into an equivalent exponential one.

31. $\ln 1234 = 7.118$

32. $\ln(ax + b) = y$

33. $\ln(rs) = t$

34. $\log 1234 = 3.0913$

35. $\log_5(cd - k) = u$

36. $\log_d(uv) = w$

In Questions 37–40, evaluate the given expression without using a calculator.

37. $\ln e^3$

38. $\ln \sqrt[3]{e}$

39. $e^{\ln 3/4}$

40. $e^{\ln (x+2y)}$

41. Simplify: $3 \ln \sqrt{x} + (1/2)\ln x$

42. Simplify: $\ln (e^{4e})^{-1} + 4e$

In Questions 43–45, write the given expression as a single logarithm.

43. $\ln 3x - 3 \ln x + \ln 3y$

44. $\log_7 7x + \log_7 y - 1$

45. $4 \ln x - 2(\ln x^3 + 4 \ln x)$

46. $\log (-.01) = ?$

47. $\log_{20} 400 = ?$

48. You are conducting an experiment about memory. The people who participate agree to take a test at the end of your course and every month thereafter for a period of two years. The average score for the group is given by the model

$$M(t) = 91 - 14 \ln (t + 1) \qquad (0 \le t \le 24)$$

where t is time in months after the first test.
(a) What is the average score on the initial exam?
(b) What is the average score after three months?
(c) When will the average drop below 50%?
(d) Is the magnitude of the rate of memory loss greater in the first month after the course (from $t = 0$ to $t = 1$) or after the first year (from $t = 12$ to $t = 13$)?
(e) Hypothetically, if the model could be extended past $t = 24$ months, would it be possible for the average score to be 0%?

49. Which of the following statements are *true*?
(a) $\ln 10 = (\ln 2)(\ln 5)$ (b) $\ln (e/6) = \ln e + \ln 6$
(c) $\ln (1/7) + \ln 7 = 0$ (d) $\ln (-e) = -1$
(e) None of the above is true.

50. Which of the following statements are *false*?
(a) $10 (\log 5) = \log 50$ (b) $\log 100 + 3 = \log 10^5$
(c) $\log 1 = \ln 1$ (d) $\log 6/\log 3 = \log 2$
(e) All of the above are false.

Use the following six graphs for Questions 51 and 52.

REVIEW QUESTIONS

51. If $b > 1$, then the graph of $f(x) = -\log_b x$ could possibly be:
(a) I
(b) IV
(c) V
(d) VI
(e) none of these

52. If $0 < b < 1$, then the graph of $g(x) = b^x + 1$ could possibly be:
(a) II
(b) III
(c) IV
(d) VI
(e) none of these

53. If $\log_3 9^{x^2} = 4$, what is x?

54. What is the domain of the function $f(x) = \ln\left(\dfrac{x}{x-1}\right)$?

In Questions 55–63, solve the equation for x.

55. $8^x = 4^{x^2-3}$

56. $e^{3x} = 4$

57. $2 \cdot 4^x - 5 = -4$

58. $725e^{-4x} = 1500$

59. $u = c + d \ln x$

60. $2^x = 3^{x+3}$

61. $\ln x + \ln(3x - 5) = \ln 2$

62. $\ln(x + 8) - \ln x = 1$

63. $\log(x^2 - 1) = 2 + \log(x + 1)$

64. At a small community college the spread of a rumor through the population of 500 faculty and students can be modeled by.

$$\ln(n) - \ln(1000 - 2n) = .65t - \ln 998,$$

where n is the number of people who have heard the rumor after t days.
(a) How many people know the rumor initially? (at $t = 0$)
(b) How many people have heard the rumor after four days?
(c) Roughly, in how many weeks will the entire population have heard the rumor?
(d) Use the properties of logarithms to write n as a function of t; in other words solve the model above for n in terms of t.
(e) Enter the function you found in part (d) into your calculator and use the table feature to check your answers to parts (a), (b), and (c). Do they agree?
(f) Now graph the function. Roughly over what time interval does the rumor seem to "spread" the fastest?

65. The half-life of polonium (^{210}Po) is 140 days. If you start with 10 milligrams, how much will be left at the end of a year?

66. An insect colony grows exponentially from 200 to 2000 in three months' time. How long will it take for the insect population to reach 50,000?

67. Hydrogen-3 decays at a rate of 5.59% per year. Find its half-life.

68. The half-life of radium-88 is 1590 years. How long will it take for 10 grams to decay to 1 gram?

69. How much money should be invested at 8% per year, compounded quarterly, in order to have $1000 in 10 years?

70. At what annual interest rate should you invest your money if you want to double it in six years?

71. One earthquake measures 4.6 on the Richter scale. A second earthquake is 1000 times more intense than the first. What does it measure on the Richter scale?

72. The table gives the population of Austin, Texas.[*]

Year	Population
1950	132,459
1970	253,539
1980	345,890
1990	465,622
2000	656,562

(a) Sketch a scatter plot of the data, with $x = 0$ corresponding to 1950.
(b) Find an exponential model for the data.
(c) Use the model to estimate the population of Austin in 1960 and 2005.

---

*U.S. Census Bureau.

73. The wind-chill factor is the temperature that would produce the same cooling effect on a person's skin if there were no wind. The table shows the wind-chill factors for various wind speeds when the temperature is 25°F.[*]

Wind Speed (mph)	Wind Chill Temperature (in °F)
0	25
5	19
10	15
15	13
20	11
25	9
30	8
35	7
40	6
45	5

(a) What does a 20-mph wind make 25°F feel like?
(b) Sketch a scatter plot of the data.
(c) Explain why an exponential model would be appropriate.
(d) Find an exponential model for the data.
(e) According to the model, what is the wind-chill factor for a 23-mph wind?

74. Cigarette consumption in the United States has been decreasing for some time, as shown in the table (in which the number of cigarettes each year is in billions).[†]

Year	Cigarettes
1980	631.5
1990	525
1997	480
1998	470
1999	435
2000	430

(a) Let $x = 10$ correspond to 1980 and find exponential, logarithmic, and power models for the data.
(b) Which of the three models do you think is most appropriate? Justify your answer
(c) If you were a doctor, which model would you prefer for the future? Which one would you prefer if you manufactured cigarettes?

----

*National Weather Service.

[†]Centers for Disease Control and Prevention; USDA Economic Research Services.

Exponential and Logistic Modeling of Diseases

Diseases that are contagious and are transmitted homogeneously through a population often appear to be spreading exponentially. That is, the rate of spread is proportional to the number of people in the population who are already infected. This is a reasonable model as long as the number of infected people is relatively small in comparison with the number of people in the population who can be infected. The standard exponential model looks like this:

$$f(t) = Y_0 e^{rt}.$$

Y_0 is the initial number of infected people (the number on the arbitrarily decided day 0), and r is the rate by which the disease spreads through the population. If the time t is measured in days, then r is the ratio of new infections to current infections each day.

Suppose that in Big City, population 3855, there is an outbreak of dingbat disease. On the first Monday after the outbreak was discovered (day 0), 72 people have dingbat disease. On the following Monday (day 7), 193 people have dingbat disease.

1. Using the exponential model, Y_0 is clearly 72. Calculate the value of r.
2. Using your values of Y_0 and r, predict the number of cases of dingbat disease that will be reported on day 14.

It turns out that eventually, the spread of disease must slow as the number of infected people approaches the number of susceptible people. What happens is that some of the people to whom the disease would spread are already infected. As time goes on, the spread of the disease becomes proportional to the number of susceptible and uninfected people. The disease then follows the logistic model:

$$g(t) = \frac{rY_0}{aY_0 + (r - aY_0)e^{-rt}}.$$

Y_0 is still the initial value, and r serves the same function as before, at least at the initial time. The extra parameter a is not so obvious, but it is inversely related to the number of people susceptible to the disease. Unfortunately, the algebra to solve for a is quite complicated. It is much easier to approximate a using the same r from the exponential model.

3. On day 14 in Big City, 481 people have dingbat disease. Using the values of Y_0 and r from Exercise 1 in the rule of the function g, determine the value of a. [Hint: $g(14) = 481$.] Does g overestimate or underestimate the number of people with dingbat disease on day 7?
4. Use the function g from Exercise 3 to approximate the number of people in Big City who are susceptible to the disease. Does this model make sense? [Remember, as time goes on, the number of people infected approaches the number of people susceptible.]

Greg Pease/Getty Images

5. In the logistic model, the rate at which the disease spreads tends to fall over time. This means that the value of r you calculated in Exercise 1 is a little low. Raise the value of r and find the new value of a as in Exercise 3. Experiment until you find a value of r for which $g(7) = 193$ (meaning that the model g matches the data on day 7).

6. Using the function g from Exercise 5, repeat Exercise 4.

CHAPTER 6

Trigonometric Functions

Don Smetzer/Getty Images

Don't touch that dial!

Radio stations transmit by sending out a signal in the form of an electromagnetic wave that can be described by a trigonometric function. The shape of this signal is modified by the sounds being transmitted. AM radio signals are modified by varying the "height," or amplitude, of the waves, whereas FM signals are modified by varying the frequency of the waves. See Exercise 43 on page 465.

AM signal

FM signal

Chapter Outline

Interdependence of Sections

6.5

6.1 → 6.2 → 6.3 → 6.4

6.6*

6.1   **Angles and Their Measurement**
6.2   **The Sine, Cosine, and Tangent Functions**
6.2   **ALTERNATE   The Sine, Cosine, and Tangent Functions**
6.3   **Algebra and Identities**
6.4   **Basic Graphs**
6.5   **Periodic Graphs and Simple Harmonic Motion**
6.5.A   *Special Topics:* **Other Trigonometric Graphs**
6.6   **Other Trigonometric Functions**

Roadmap

Instructors who want to introduce triangle trigonometry before doing trigonometric functions of a real number should cover Chapter 8 before Chapter 6.

In this chapter, those who have read Chapter 8 should replace Section 6.2 by Alternate Section 6.2.

The ancient Greeks developed trigonometry for measuring angles and sides of triangles to solve problems in astronomy, navigation, and surveying.[†] But with the invention of calculus in the seventeenth century and the subsequent explosion of knowledge in the physical sciences, a different viewpoint toward trigonometry arose.

Whereas the ancients dealt only with *angles*, the classical trigonometric concepts of sine and cosine are now considered as *functions* with domain the set of all *real numbers*. The advantage of this switch in viewpoint is that almost any phenomenon involving rotation or vibration can be described in terms of trigonometric functions, including light rays, sound waves, electron orbitals, planetary orbits, radio transmission, vibrating strings, pendulums, and many more.

The presentation of trigonometry here reflects this modern viewpoint. Nevertheless, angles still play an important role in defining the trigonometric functions, so the chapter begins with them.

6.1   Angles and Their Measurement

In trigonometry an **angle** is formed by rotating a half-line around its endpoint (the **vertex**), as shown in Figure 6–1, where the arrow indicates the direction of rotation. The position of the half-line at the beginning is the **initial side,** and its final position is the **terminal side** of the angle.

---

*Parts of Section 6.6 may be covered much earlier; see the Roadmap at the beginning of Section 6.2.
[†]In fact, "trigonometry" means "triangle measurement."

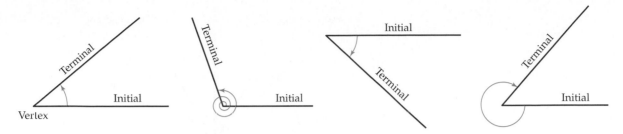

Figure 6–1

Figure 6–2 shows that different angles (that is, angles obtained by different rotations) may have the same initial and terminal side.* Such angles are said to be **coterminal.**

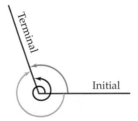

Figure 6–2

An angle in the coordinate plane is said to be in **standard position** if its vertex is at the origin and its initial side on the positive x-axis, as in Figure 6–3. When measuring angles in standard position, we use positive numbers for angles obtained by counterclockwise rotation (**positive angles**) and negative numbers for ones obtained by clockwise rotation (**negative angles**).

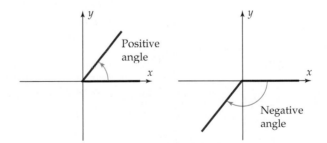

Figure 6–3

The classical unit for angle measurement is the **degree** (in symbols, °), as explained in the Geometry Review Appendix. You should be familiar with the positive angles in standard position shown in Figure 6–4 on the next page. Note that a 360° angle corresponds to one full revolution and thus is coterminal with an angle of 0°.

---

*They are *not* the same angle, however. For instance, both $\frac{1}{2}$ turn and $1\frac{1}{2}$ turns put a circular faucet handle in the same position, but the water flow is quite different.

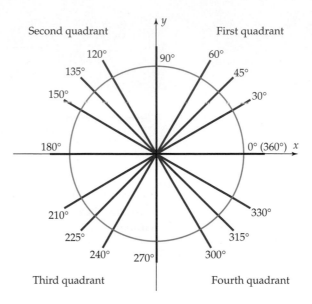

Figure 6–4

:: RADIAN MEASURE

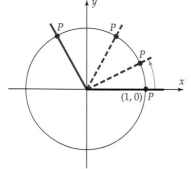

Figure 6–5

Because it simplifies many formulas in calculus and physics, a different unit of angle measurement is used in mathematical and scientific applications. Recall that the unit circle is the circle of radius 1 with center at the origin, whose equation is $x^2 + y^2 = 1$. When a positive angle in standard position is formed by rotating the initial side (the positive x-axis) counterclockwise, then the point $P = (1, 0)$ moves along the unit circle, as in Figure 6–5. The **radian measure** of the angle is defined to be

> **the distance traveled along the unit circle by the point P as it moves from its starting position on the initial side to its final position on the terminal side of the angle.**

The radian measure of a negative angle in standard position is found in the same way, except that you move clockwise along the unit circle. Figure 6–6 shows angles of 3.75, 7, and −2 radians, respectively.

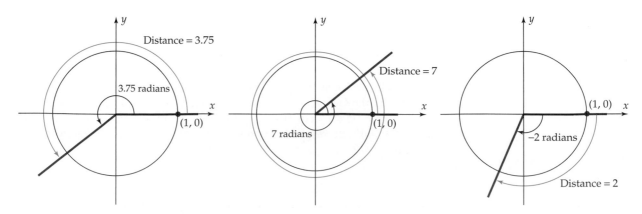

Figure 6–6

To become comfortable with radian measure, think of the terminal side of the angle revolving around the origin: When it makes one full revolution, it produces an angle of 2π radians (because the circumference of the unit circle is 2π). When it makes half a revolution, it forms an angle whose radian measure is $1/2$ of 2π, that is, π radians, and so on, as illustrated in Figure 6–7 and the table below.

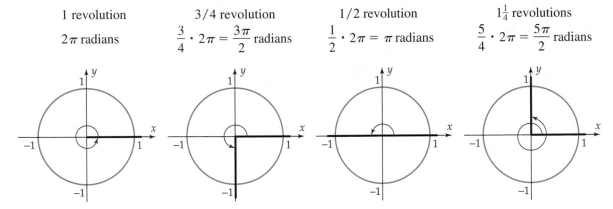

1 revolution

2π radians

3/4 revolution

$\dfrac{3}{4} \cdot 2\pi = \dfrac{3\pi}{2}$ radians

1/2 revolution

$\dfrac{1}{2} \cdot 2\pi = \pi$ radians

$1\frac{1}{4}$ revolutions

$\dfrac{5}{4} \cdot 2\pi = \dfrac{5\pi}{2}$ radians

Figure 6–7

Terminal Side	Radian Measure of Angle	Equivalent Degree Measure
1 revolution	2π	$360°$
$\dfrac{7}{8}$ revolution	$\dfrac{7}{8} \cdot 2\pi = \dfrac{7\pi}{4}$	$\dfrac{7}{8} \cdot 360 = 315°$
$\dfrac{3}{4}$ revolution	$\dfrac{3}{4} \cdot 2\pi = \dfrac{3\pi}{2}$	$\dfrac{3}{4} \cdot 360 = 270°$
$\dfrac{2}{3}$ revolution	$\dfrac{2}{3} \cdot 2\pi = \dfrac{4\pi}{3}$	$\dfrac{2}{3} \cdot 360 = 240°$
$\dfrac{1}{2}$ revolution	$\dfrac{1}{2} \cdot 2\pi = \pi$	$\dfrac{1}{2} \cdot 360 = 180°$
$\dfrac{1}{3}$ revolution	$\dfrac{1}{3} \cdot 2\pi = \dfrac{2\pi}{3}$	$\dfrac{1}{3} \cdot 360 = 120°$
$\dfrac{1}{4}$ revolution	$\dfrac{1}{4} \cdot 2\pi = \dfrac{\pi}{2}$	$\dfrac{1}{4} \cdot 360 = 90°$
$\dfrac{1}{6}$ revolution	$\dfrac{1}{6} \cdot 2\pi = \dfrac{\pi}{3}$	$\dfrac{1}{6} \cdot 360 = 60°$
$\dfrac{1}{8}$ revolution	$\dfrac{1}{8} \cdot 2\pi = \dfrac{\pi}{4}$	$\dfrac{1}{8} \cdot 360 = 45°$
$\dfrac{1}{12}$ revolution	$\dfrac{1}{12} \cdot 2\pi = \dfrac{\pi}{6}$	$\dfrac{1}{12} \cdot 360 = 30°$

Although equivalent degree measures are given in the table, you should learn to "think radian" as much as possible rather than mentally translating from radians to degrees.

Figure 6–8

$$\frac{16\pi}{3} = 2\pi + 2\pi + \frac{4\pi}{3}$$

EXAMPLE 1

To construct an angle of $16\pi/3$ radians in standard position, note that

$$\frac{16\pi}{3} = \frac{12\pi}{3} + \frac{4\pi}{3} = 4\pi + \frac{4\pi}{3} = 2\pi + 2\pi + \frac{4\pi}{3}.$$

So the terminal side must be rotated counterclockwise through two complete revolutions (each full-circle revolution is 2π radians) and then rotated an additional $2/3$ of a revolution (since $4\pi/3$ is $2/3$ of a complete revolution of 2π radians), as shown in Figure 6–8. ∎

EXAMPLE 2

Since $-5\pi/4 = -\pi - \pi/4$, an angle of $-5\pi/4$ radians in standard position is obtained by rotating the terminal side *clockwise* for half a revolution (π radians) plus an additional $1/8$ of a revolution (since $\pi/4$ is $1/8$ of a full-circle revolution of 2π radians), as shown in Figure 6–9. ∎

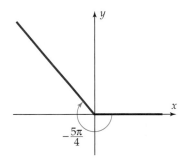

Figure 6–9

Consider an angle of t radians in standard position (Figure 6–10). Since 2π radians corresponds to a full revolution of the terminal side, this angle has the same terminal side as an angle of $t + 2\pi$ radians or $t - 2\pi$ radians or $t + 4\pi$ radians.

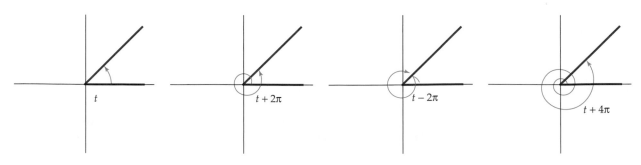

Figure 6–10

The same thing is true in general.

Coterminal Angles

> Increasing or decreasing the radian measure of an angle by an integer multiple of 2π results in a coterminal angle.

EXAMPLE 3

Find angles in standard position that are coterminal with an angle of

(a) $23\pi/5$ radians   (b) $-\pi/12$ radians.

SOLUTION

(a) We can subtract 2π to obtain a coterminal angle whose measure is

$$\frac{23\pi}{5} - 2\pi = \frac{23\pi}{5} - \frac{10\pi}{5} = \frac{13\pi}{5} \text{ radians,}$$

or we can subtract 4π to obtain a coterminal angle of measure

$$\frac{23\pi}{5} - 4\pi = \frac{3\pi}{5} \text{ radians.}$$

Subtracting 6π produces a coterminal angle of

$$\frac{23\pi}{5} - 6\pi = -\frac{7\pi}{5} \text{ radians.}$$

(b) An angle of $\dfrac{-\pi}{12}$ radians is coterminal with an angle of

$$-\frac{\pi}{12} + 2\pi = \frac{23\pi}{12} \text{ radians}$$

and with an angle of

$$-\frac{\pi}{12} - 2\pi = -\frac{25\pi}{12} \text{ radians.} \quad \blacksquare$$

⊞ RADIAN/DEGREE CONVERSION

Although we shall generally work with radians, it may occasionally be necessary to convert from radian to degree measure or vice versa. The key to doing this is the fact that

(∗) π *radians* $= 180°.$

Dividing both sides of (∗) by π shows that

$$1 \text{ radian} = \frac{180}{\pi} \text{ degrees} \approx 57.3°,$$

and dividing both sides of (∗) by 180 shows that

$$1° = \frac{\pi}{180} \text{ radians} \approx .0175 \text{ radians.}$$

Consequently, we have these rules.

Radian/Degree Conversion

> To convert radians to degrees, multiply by $\dfrac{180}{\pi}$.
>
> To convert degrees to radians, multiply by $\dfrac{\pi}{180}$.

EXAMPLE 4

(a) To find the degree measure of an angle of 2.4 radians, multiply by $\dfrac{180}{\pi}$.

$$(2.4)\left(\frac{180}{\pi}\right) = \frac{432}{\pi} \approx 137.51°.$$

(b) An angle of $-.3$ radians has degree measure

$$(-.3)\left(\frac{180}{\pi}\right) = \frac{-54}{\pi} \approx -17.19°. \quad ■$$

Angle conversions from radian to degree measure or vice versa can be done directly on a calculator. Since procedures vary from one model to another, check your instruction manual.

▪▪ ARC LENGTH*

Consider a circle of radius r, with center at the origin, and an angle of θ radians in standard position, as shown in Figure 6–11. As you can see, the sides of the angle of θ radians determine an arc length s along the circle. We say that the **central angle** of θ radians **subtends an arc** of length s on the circle. It can be shown that the ratio of the arc length s to the circumference of the entire circle (namely, $2\pi r$) is the same as the ratio of the angle of θ radians to the full-circle angle of 2π radians; that is,

$$\frac{s}{2\pi r} = \frac{\theta}{2\pi}.$$

Solving this equation for s, we obtain the following fact.

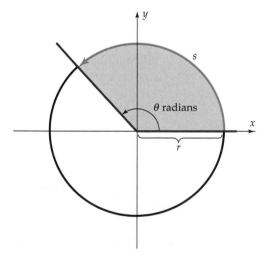

Figure 6–11

---

*The remainder of this section is not needed in the sequel and may be omitted if desired.

Arc Length

> A central angle of θ radians in a circle of radius r subtends an arc of length
>
> $$s = \theta r,$$
>
> that is, the length s of the arc is the product of the radian measure of the angle and the radius of the circle.

EXAMPLE 5

The second hand on a large clock is 6 inches long. How far does the tip of the second hand move in 15 seconds?

SOLUTION   The second hand makes a full revolution every 60 seconds, that is, it moves through an angle of 2π radians. During a 15-second interval it will make $\frac{15}{60} = \frac{1}{4}$ of a revolution, moving through an angle of $\pi/2$ radians (Figure 6–12). If we think of the second hand as the radius of a circle, then during a 15-second interval, its tip travels along the arc subtended by an angle of $\pi/2$ radians. Therefore, the distance (arc length) traveled by the tip of the second hand is

$$s = \theta r = \frac{\pi}{2} \cdot 6 = 3\pi \approx 9.425 \text{ inches.} \qquad \blacksquare$$

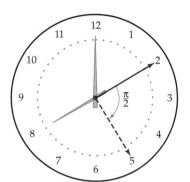

Figure 6–12

EXAMPLE 6

Suppose the circle in Figure 6–11 has radius 6 and the arc length s is 14.64. We can find the measure of the central angle that subtends the arc s by solving $s = \theta r$ for θ.

$$\theta = \frac{s}{r} = \frac{14.64}{6} = 2.44 \text{ radians.} \qquad \blacksquare$$

■■ ANGULAR SPEED

Suppose a point P moves along a circle of radius r at a constant rate, tracing out an arc of length s, as in Figure 6–13. As P moves, the radius OP traces out an angle of θ radians. The *linear speed* of the point P is given by

$$\textbf{Linear speed} = \frac{\text{Distance traveled by } P}{\text{Times elapsed}} = \frac{s}{t} = \frac{\theta r}{t}.$$

Similarly, the *angular speed* of P is defined by

$$\textbf{Angular speed} = \frac{\text{Angle traced out by } OP}{\text{Time elapsed}} = \frac{\theta}{t}.$$

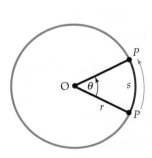

Figure 6–13

By rewriting the linear speed equation, we obtain the relationship between linear and angular speed.

$$\frac{\theta r}{t} = r \cdot \frac{\theta}{t}$$

Linear speed = r (Angular speed).

420 CHAPTER 6 Trigonometric Functions

EXAMPLE 7

A merry-go-round makes eight revolutions per minute.

(a) What is the angular speed of the merry-go-round in radians per minute?

(b) How fast is a horse 12 feet from the center traveling?

(c) How fast is a horse 4 feet from the center traveling?

SOLUTION

(a) Each revolution of the merry-go-round corresponds to a central angle of 2π radians, so it travels through an angle of $8 \cdot 2\pi = 16\pi$ radians in one minute. Therefore, its

$$\text{Angular speed} = \frac{\theta}{t} = \frac{16\pi}{1} = 16\pi \text{ radians per minute.}$$

(b) The horse that is 12 feet from the center travels along a circle of radius 12 and travels through an angle of 16π radians in 1 minute. Therefore, its

$$\text{Linear speed} = \frac{\theta r}{t} = \frac{16\pi \cdot 12}{1} = 192\pi \text{ feet per minute,}$$

which is approximataely 6.9 miles per hour.

(c) For the horse that is 4 feet from the center, $r = 4$, $\theta = 16\pi$, and $t = 1$. Hence,

$$\text{Linear speed} = \frac{\theta r}{t} = \frac{16\pi \cdot 4}{1} = 64\pi \text{ feet per minute (about 2.28 mph).} \blacksquare$$

EXAMPLE 8

A bicycle, each of whose wheels is 26 inches in diameter, travels at a constant 14 mph.

(a) What is the angular speed of one of its wheels (in radians per hour)?

(b) How many revolutions per minute does each wheel make?

SOLUTION   Within each computation, we must be careful to use comparable units.

(a) Since the wheel size is given in inches, we also express the speed in terms of inches. Recall that there are 5280 feet in a mile, which means that

$$1 \text{ mile} = 5280 \cdot 12 = 63,360 \text{ inches.}$$

Thus, the bike's linear speed is

$$14 \text{ miles per hour} = 14(63,360) \text{ inches per hour.}$$

As we saw above, linear speed = r (angular speed), so

$$\text{Angular speed} = \frac{\text{Linear speed}}{r} = \frac{14(63,360)}{13} \approx 68,233.85 \text{ radians per hour.}$$

(b) Each revolution corresponds to an angle of 2π radians, so the number of revolutions per hour is $\dfrac{68{,}233.85}{2\pi}$. Dividing this number by 60 gives the revolutions per minute.

$$\frac{1}{60}\left(\frac{68{,}233.85}{2\pi}\right) \approx 181 \text{ revolutions per minute.} \quad \blacksquare$$

✓ EXERCISES 6.1

In Exercises 1–5, find the radian measure of the angle in standard position formed by rotating the terminal side the given amount.

1. 1/9 of a circle   **2.** 1/24 of a circle

3. 1/18 of a circle   **4.** 1/72 of a circle

5. 1/36 of a circle

6. State the radian measure of *every* standard position angle in the figure.*

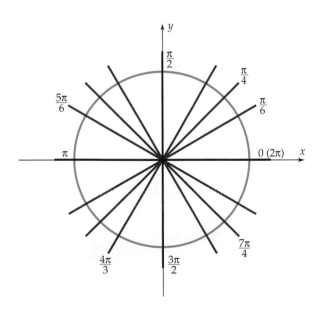

In Exercises 7–10, find the radian measure of four angles in standard position that are coterminal with the angle in standard position whose measure is given.

7. $\pi/4$ **8.** $7\pi/5$

9. $-\pi/6$ **10.** $-9\pi/7$

This is the same diagram that appears in Figure 6–4 on page 414, showing positive angles in standard position.

In Exercises 11–18, find the radian measure of an angle in standard position that has measure between 0 and 2π and is coterminal with the angle in standard position whose measure is given.

11. $-\pi/3$ **12.** $-3\pi/4$ **13.** $19\pi/4$

14. $16\pi/3$ **15.** $-7\pi/5$ **16.** $45\pi/8$

17. 7   **18.** 18.5

In Exercises 19–30, convert the given degree measure to radians.

19. $6°$ **20.** $-10°$ **21.** $-12°$ **22.** $36°$

23. $75°$ **24.** $-105°$ **25.** $135°$ **26.** $-165°$

27. $-225°$ **28.** $252°$ **29.** $930°$ **30.** $-585°$

In Exercises 31–42, convert the given radian measure to degrees.

31. $\pi/5$ **32.** $-\pi/6$ **33.** $-\pi/10$

34. $2\pi/5$ **35.** $3\pi/4$ **36.** $-5\pi/3$

37. $\pi/45$ **38.** $-\pi/60$ **39.** $-5\pi/12$

40. $7\pi/15$ **41.** $27\pi/5$ **42.** $-41\pi/6$

In Exercises 43–48, determine the positive radian measure of the angle that the second hand of a clock traces out in the given time.

43. 40 seconds   **44.** 50 seconds

45. 35 seconds

46. 2 minutes and 15 seconds.

47. 3 minutes and 25 seconds

48. 1 minute and 55 seconds

49. The second hand on a clock is 6 centimeters long. How far does its tip travel in 40 seconds? (See Exercise 43.)

50. The second hand on a clock is 5 centimeters long. How far does its tip travel in 2 minutes and 15 seconds? (See Exercise 46.)

51. If the radius of the circle in the figure is 20 centimeters and the length of arc *s* is 85 centimeters, what is the radian measure of the angle θ?

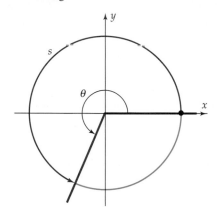

52. Find the radian measure of the angle θ in the preceding figure if the *diameter* of the circle is 150 centimeters and *s* has length 360 centimeters.

In Exercises 53–56, assume that a wheel on a car has radius 36 centimeters. Find the angle (in radians) that the wheel turns while the car travels the given distance.

53. 2 meters (= 200 centimeters)   54. 5 meters

55. 720 meters   56. 1 kilometer (= 1000 meters)

In Exercises 57–60, find the length of the circular arc subtended by the central angle whose radian measure is given. Assume the circle has diameter 10.

57. 1 radian   58. 2 radians   59. 1.75 radians

60. 2.2 radians

*In Exercises 61–64, the latitudes of a pair of cities are given. Assume that one city is directly south of the other and that the earth is a perfect sphere of radius 4000 miles. Find the distance between the two cities. [The **latitude** of a point P on the earth is the degree measure of the angle θ between the point and the plane of the equator (with the center of the earth being the vertex), as shown in the figure. Remember that angles are measured in radians in the arc length formula.]*

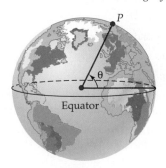

61. The North Pole (latitude 90° north) and Springfield, Illinois (latitude 40° north).

62. San Antonio, Texas (latitude 29.5° north), and Mexico City, Mexico (latitude 20° north).

63. Cleveland, Ohio (latitude 41.5° north), and Tampa, Florida (latitude 28° north).

64. Copenhagen, Denmark (latitude 54.3° north), and Rome, Italy (latitude 42° north).

In Exercises 65–72, a wheel is rotating around its axle. Find the angle (in radians) through which the wheel turns in the given time when it rotates at the given number of revolutions per minute (rpm). Assume that t > 0 and k > 0.

65. 3.5 minutes, 1 rpm   66. *t* minutes, 1 rpm

67. 1 minute, 2 rpm   68. 3.5 minutes, 2 rpm

69. 4.25 minutes, 5 rpm   70. *t* minutes, 5 rpm

71. 1 minute, *k* rpm   72. *t* minutes, *k* rpm

73. One end of a rope is attached to a winch (circular drum) of radius 2 feet and the other to a steel beam on the ground. When the winch is rotated, the rope wraps around the drum and pulls the object upward (see the figure). Through what angle must the winch be rotated to raise the beam 6 feet above the ground?

74. A circular saw blade has an angular speed of 15,000 radians per minute.
    (a) How many revolutions per minute does the saw make?
    (b) How long will it take the saw to make 6000 revolutions?

75. A circular gear rotates at the rate of 200 revolutions per minute (rpm).
    (a) What is the angular speed of the gear in radians per minute?
    (b) What is the linear speed of a point on the gear 2 inches from the center in inches per minute and in feet per minute?

76. A wheel in a large machine is 2.8 feet in diameter and rotates at 1200 rpm.
  (a) What is the angular speed of the wheel?
  (b) How fast is a point on the circumference of the wheel traveling in feet per minute? In miles per hour?

77. A riding lawn mower has wheels that are 15 inches in diameter. If the wheels are making 2.5 revolutions per second
  (a) What is the angular speed of a wheel?
  (b) How fast is the lawn mower traveling in miles per hour?

78. A weight is attached to the end of a 4-foot-long leather cord. A boy swings the cord in a circular motion over his head. If the weight makes 12 revolutions every 10 seconds, what is its linear speed?

79. A merry-go-round horse is traveling at 10 feet per second, and the merry-go-round is making 6 revolutions per minute. How far is the horse from the center of the merry-go-round?

80. The pedal sprocket of a bicycle has radius 4.5 inches, and the rear wheel sprocket has radius 1.5 inches (see figure). If the rear wheel has a radius of 13.5 inches and the cyclist is pedaling at the rate of 80 rpm, how fast is the bicycle traveling in feet per minute? In miles per hour?

6.2   The Sine, Cosine, and Tangent Functions

NOTE

If you have read Chapter 8, use Alternate Section 6.2 on page 431 in place of this section.

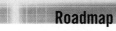
Roadmap

Instructors who wish to cover all six trigonometric functions simultaneously should incorporate Section 6.6 into Sections 6.2–6.4, as follows.

Subsection of Section 6.6	Cover at the end of
Definitions; Alternate Descriptions	Section 6.2
Algebra and Identities	Section 6.3
Graphs	Section 6.4

Instructors who prefer to introduce triangle trigonometry early should cover Section 8.1 between Sections 6.2 and 6.3.

Unlike most of the functions seen thus far, the definitions of the sine and cosine functions do not involve algebraic formulas. Instead, these functions are defined geometrically, using the unit circle.* Recall that the unit circle is the circle of radius 1 with center at the origin, whose equation is $x^2 + y^2 = 1$.

Both the sine and cosine functions have the set of all real numbers as domain. Their rules are given by the following three-step geometric process:

1. Given a real number t, construct an angle of t radians in standard position.

2. Find the coordinates of the point P where the terminal side of this angle meets the unit circle $x^2 + y^2 = 1$, say $P = (a, b)$.

3. The value of the **cosine function** at t (denoted $\cos t$) is the x-coordinate of P:

$$\cos t = a.$$

The value of the **sine function** at t (denoted $\sin t$) is the y-coordinate of P:

$$\sin t = b.$$

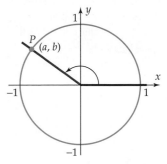

---

*If you have previously studied the trigonometry of triangles, the definition given here may not look familiar. If this is the case, just concentrate on this definition and don't worry about relating it to any definition you remember from the past. The connection between this definition and the trigonometry of triangles will be explained in Chapter 8.

In other words,

Sine and Cosine

> If P is the point where the terminal side of an angle of t radians in standard position meets the unit circle, then
>
> $$P \text{ has coordinates } (\cos t, \sin t).$$

GRAPHING EXPLORATION

With your calculator in radian mode and parametric graphing mode, set the range values as follows:

$$0 \le t \le 2\pi, \qquad -1.8 \le x \le 1.8, \qquad -1.2 \le y \le 1.2^*$$

Then graph the curve given by the parametric equations

$$x = \cos t \qquad \text{and} \qquad y = \sin t.$$

The graph is the unit circle. Use the trace to move around the circle. At each point, the screen will display three numbers: the values of t, x, and y. For each t, the cursor is on the point where the terminal side of an angle of t radians meets the unit circle, so the corresponding x is the number $\cos t$ and the corresponding y is the number $\sin t$.

The **tangent function** is defined as the quotient of the sine and cosine functions. Its value at the number t, denoted $\tan t$, is given by

$$\tan t = \frac{\sin t}{\cos t}$$

Every calculator has SIN, COS, and TAN keys for evaluating the sine, cosine, and tangent functions. For instance, a calculator (in radian mode) gives these approximations (Figure 6–14).

```
sin(2.5)
       .5984721441
cos(-6)
       .9601702867
tan(15)
      -.8559934009
```

Figure 6–14

Nevertheless, you should know how to use the definition to find exact values of these functions when possible, as illustrated in Examples 1–4.

---

*These settings give a square viewing window on calculators with a screen measuring approximately 95 by 63 pixels, and hence the unit circle will look like a circle. For wider screens, adjust the x range settings to obtain a square window.

EXAMPLE 1

Evaluate the three trigonometric functions at

(a) $t = \pi$ 　　　　　　　　　　(b) $t = \pi/2$.

SOLUTION

(a) Construct an angle of π radians, as in Figure 6–15. Its terminal side lies on the negative x-axis and intersects the unit circle at $P = (-1, 0)$. Hence,

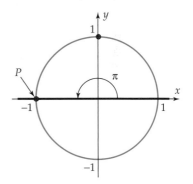

$\sin \pi = y$-coordinate of $P = 0$

$\cos \pi = x$-coordinate of $P = -1$

$\tan \pi = \dfrac{\sin \pi}{\cos \pi} = \dfrac{0}{-1} = 0$

Figure 6–15

(b) An angle of $\pi/2$ radians (Figure 6–16) has its terminal side on the positive y-axis and intersects the unit circle at $P = (0, 1)$.

$\cos \dfrac{\pi}{2} = x$-coordinate of $P = 0$

$\sin \dfrac{\pi}{2} = y$-coordinate of $P = 1$

$\tan \dfrac{\pi}{2} = \dfrac{\sin (\pi/2)}{\cos (\pi/2)} = \dfrac{1}{0}$ *undefined*　■

Figure 6–16

Example 1(b) shows that the domain of the tangent function must exclude numbers for which $\cos t = 0$. These occur when the point P has coordinates $(0, 1)$ or $(0, -1)$. Therefore,

the domain of the tangent function consists of all real numbers except $t = \pm\pi/2, \pm 3\pi/2, \pm 5\pi/2, \ldots$

▗▖ SPECIAL VALUES

The trigonometric functions can be evaluated exactly at $t = \pi/3$, $t = \pi/4$, $t = \pi/6$, and any integer multiples of these numbers by using the following facts (which are explained in the Geometry Review Appendix):

A right triangle with hypotenuse 1 and angles of $\pi/6$ and $\pi/3$ radians has sides of lengths $1/2$ (opposite the angle of $\pi/6$) and $\sqrt{3}/2$ (opposite the angle of $\pi/3$).

A right triangle with hypotenuse 1 and two angles of $\pi/4$ radians has two sides of length $\sqrt{2}/2$.

Figure 6–17

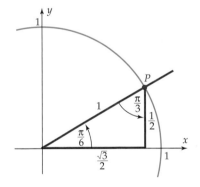

Figure 6–18

EXAMPLE 2

Evaluate the three trigonometric functions at $t = \pi/6$.

SOLUTION Construct an angle of $\pi/6$ radians in standard position and let P be the point where its terminal side intersects the unit circle (Figure 6–18). Draw a vertical line from P to the x-axis, forming a right triangle with hypotenuse 1, angles of $\pi/6$ and $\pi/3$ radians, and sides of lengths $1/2$ and $\sqrt{3}/2$, as described above. Then P has coordinates $(\sqrt{3}/2, 1/2)$, and by definition,

$$\sin \frac{\pi}{6} = y\text{-coordinate of } P = \frac{1}{2}$$

$$\cos \frac{\pi}{6} = x\text{-coordinate of } P = \frac{\sqrt{3}}{2}$$

$$\tan \frac{\pi}{6} = \frac{\sin(\pi/6)}{\cos(\pi/6)} = \frac{1/2}{\sqrt{3}/2} = \frac{1}{\sqrt{3}} = \frac{\sqrt{3}}{3}. \quad ■$$

EXAMPLE 3

Evaluate the trigonometric functions at $-5\pi/4$.

SOLUTION Construct an angle of $-5\pi/4$ radians in standard position and let P be the point where the terminal side intersects the unit circle. Draw a vertical line from P to the x-axis, forming a right triangle with hypotenuse 1 and two angles of $\pi/4$ radians, as shown in Figure 6–19.

Figure 6–19

Each side of this triangle has length $\sqrt{2}/2$, as explained above. Therefore, the coordinates of P are $(-\sqrt{2}/2,\ \sqrt{2}/2)$, so

$$\sin\frac{-5\pi}{4} = y\text{-coordinate of } P = \frac{\sqrt{2}}{2}$$

$$\cos\frac{-5\pi}{4} = x\text{-coordinate of } P = -\frac{\sqrt{2}}{2}$$

$$\tan\frac{-5\pi}{4} = \frac{\sin t}{\cos t} = \frac{\sqrt{2}/2}{-\sqrt{2}/2} = -1. \quad\blacksquare$$

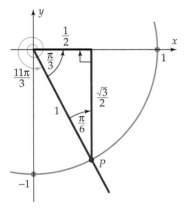

Figure 6–20

EXAMPLE 4

Evaluate the trigonometric functions at $11\pi/3$.

SOLUTION   Construct an angle of $11\pi/3$ radians in standard position and draw a vertical line from the x-axis to the point P where the terminal side of the angle meets the unit circle, as shown in Figure 6–20. The right triangle formed in this way has hypotenuse 1 and angles of $\pi/6$ and $\pi/3$ radians. The sides of the triangle must have lengths $1/2$ and $\sqrt{3}/2$, so the coordinates of P are $(1/2,\ -\sqrt{3}/2)$, and

$$\sin t = y\text{-coordinate of } P = -\frac{\sqrt{3}}{2}$$

$$\cos t = x\text{-coordinate of } P = \frac{1}{2}$$

$$\tan t = \frac{\sin t}{\cos t} = \frac{-\sqrt{3}/2}{1/2} = -\sqrt{3}. \quad\blacksquare$$

■■ POINT-IN-THE-PLANE DESCRIPTION OF TRIGONOMETRIC FUNCTIONS

In evaluating $\sin t$, $\cos t$ and $\tan t$, from the definition, we use the point where the unit circle intersects the terminal side of an angle of t radians in standard position. Here is an alternative method of evaluating the trigonometric functions that uses *any* point on the terminal side of the angle (except the origin).

Point-in-the-Plane Description

Let t be a real number. Let (x, y) be any point (except the origin) on the terminal side of an angle of t radians in standard position. Then,

$$\sin t = \frac{y}{r} \qquad \cos t = \frac{x}{r} \qquad \tan t = \frac{y}{x}$$

where $r = \sqrt{x^2 + y^2}$ is the distance from (x, y) to the origin.

Proof   Let Q be the point on the terminal side of the standard position angle of t radians and let P be the point where the terminal side meets the unit circle, as in Figure 6–21. The definition of sine and cosine shows that P has coordinates $(\cos t, \sin t)$. The distance formula shows that the segment OQ has length $\sqrt{x^2 + y^2}$, which we denote by r.

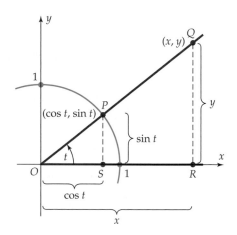

Figure 6–21

Both triangles QOR and POS are right triangles containing an angle of t radians. Therefore, these triangles are *similar.** Consequently,

$$\frac{\text{length } OP}{\text{length } OQ} = \frac{\text{length } PS}{\text{length } QR} \quad \text{and} \quad \frac{\text{length } OP}{\text{length } OQ} = \frac{\text{length } OS}{\text{length } OR}.$$

Figure 6–21 shows what each of these lengths is. Hence,

$$\frac{1}{r} = \frac{\sin t}{y} \quad \text{and} \quad \frac{1}{r} = \frac{\cos t}{x}$$

$$r \sin t = y \qquad\qquad r \cos t = x$$

$$\sin t = \frac{y}{r} \qquad\qquad \cos t = \frac{x}{r}$$

Similar arguments work when the terminal side is not in the first quadrant. In every case, $\tan t = \dfrac{\sin t}{\cos t} = \dfrac{y/r}{x/r} = \dfrac{y}{x}$. This completes the proof of the statements in the box.   ■

EXAMPLE 5

Figure 6–22 shows an angle of t radians in standard position. Evaluate the three trigonometric functions at t.

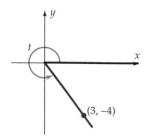

Figure 6–22

---

*See the Geometry Review Appendix for the basic facts about similar triangles.

SOLUTION   Apply the facts in the box above with $(x, y) = (3, -4)$ and

$$r = \sqrt{x^2 + y^2} = \sqrt{3^2 + (-4)^2} = \sqrt{25} = 5.$$

Then we have

$$\sin t = \frac{y}{r} = \frac{-4}{5} = -\frac{4}{5}, \qquad \cos t = \frac{x}{r} = \frac{3}{5}, \qquad \tan t = \frac{y}{x} = \frac{-4}{3} = -\frac{4}{3}. \quad \blacksquare$$

Figure 6–23

EXAMPLE 6

The terminal side of a first-quadrant angle of t radians in standard position lies on the line with equation $2x - 3y = 0$. Evaluate the three trigonometric functions at t.

SOLUTION   Verify that the point $(3, 2)$ satisfies the equation and hence lies on the terminal side of the angle (Figure 6–23). Now we have

$$(x, y) = (3, 2) \qquad \text{and} \qquad r = \sqrt{x^2 + y^2} = \sqrt{3^2 + 2^2} = \sqrt{13}.$$

Therefore,

$$\sin t = \frac{y}{r} = \frac{2}{\sqrt{13}}, \qquad \cos t = \frac{x}{r} = \frac{3}{\sqrt{13}}, \qquad \tan t = \frac{y}{x} = \frac{2/\sqrt{13}}{3/\sqrt{13}} = \frac{2}{3}. \quad \blacksquare$$

✓ EXERCISES 6.2

Note: Unless stated otherwise, all angles are in standard position.

In Exercises 1–10, use the definition (not a calculator) to find the function value.

1. $\sin 3\pi/2$

2. $\sin(-\pi)$

3. $\cos 3\pi/2$

4. $\cos(-\pi/2)$

5. $\tan 4\pi$

6. $\tan(-\pi)$

7. $\cos(-3\pi/2)$

8. $\sin 9\pi/2$

9. $\cos(-11\pi/2)$

10. $\tan(-13\pi)$

In Exercises 11–14, assume that the terminal side of an angle of t radians passes through the given point on the unit circle. Find $\sin t$, $\cos t$, $\tan t$.

11. $(-2/\sqrt{5}, 1/\sqrt{5})$

12. $(1/\sqrt{10}, -3/\sqrt{10})$

13. $(-3/5, -4/5)$

14. $(.6, -.8)$

In Exercises 15–29, find the exact value of the sine, cosine, and tangent of the number, without using a calculator.

15. $\pi/3$

16. $2\pi/3$

17. $\pi/4$

18. $5\pi/4$

19. $3\pi/4$

20. $-7\pi/3$

21. $5\pi/6$

22. 3π

23. $-23\pi/6$

24. $11\pi/6$

25. $-19\pi/3$

26. $-10\pi/3$

27. $-15\pi/4$

28. $-25\pi/4$

29. $-17\pi/2$

30. Fill the blanks in the following table. Write each entry as a fraction with denominator 2 and with a radical in the numerator. For example,

$$\sin \frac{\pi}{2} = 1 = \frac{\sqrt{4}}{2}.$$

Some students find the resulting pattern an easy way to remember these functional values.

t	0	$\pi/6$	$\pi/4$	$\pi/3$	$\pi/2$
$\sin t$					
$\cos t$					

In Exercises 31–36, write the expression as a single real number. Do not use decimal approximations. [Hint: Exercises 15–21 may be helpful.]

31. $\sin(\pi/3) \cos(\pi + \sin \pi \cos(\pi/3)$

32. $\sin(\pi/6) \cos(\pi/2) - \cos(\pi/6) \sin(\pi/2)$

33. $\cos(\pi/2)\cos(\pi/4) - \sin(\pi/2)\sin(\pi/4)$

34. $\cos(2\pi/3)\cos\pi + \sin(2\pi/3)\sin\pi$

35. $\sin(3\pi/4)\cos(5\pi/6) - \cos(3\pi/4)\sin(5\pi/6)$

36. $\sin(-7\pi/3)\cos(5\pi/4) + \text{co}\,(-7\pi/3)\sin(5\pi/4)$

In Exercises 37–42, find sin t, cos t, tan t when the terminal side of an angle of t radians in standard position passes through the given point.

37. $(2, 7)$ **38.** $(-3, 2)$ **39.** $(-5, -6)$

40. $(4, -3)$ **41.** $(\sqrt{3}, -10)$ **42.** $(-\pi, 2)$

In Exercises 43–48, assume that the terminal side of an angle of t radians in standard position lies in the given quadrant on the given straight line. Find sin t, cos t, tan t. [Hint: Find a point on the terminal side of the angle.]

43. Quadrant IV; line with equation $y = -3x$.

44. Quadrant III; line with equation $2y - 4x = 0$.

45. Quadrant IV; line through $(-3, 5)$ and $(-9, 15)$.

46. Quadrant III; line through the origin parallel to

$$7x - 2y = -6.$$

47. Quadrant II; line through the origin parallel to

$$2y + x = 6.$$

48. Quadrant I; line through the origin perpendicular to

$$3y + x = 6.$$

49. The values of $\sin t$, $\cos t$, and $\tan t$ are determined by the point (x, y) where the terminal side of an angle of t radians in standard position intersects the unit circle. The coordinates x and y are positive or negative, depending on what quadrant (x, y) lies in. For instance, in the second quadrant x is negative and y is positive, so that $\cos t$ (which is x by definition) is negative. Fill the blanks in this chart with the appropriate sign ($+$ or $-$).

Quadrant II $\pi/2 < t < \pi$		Quadrant I $0 < t < \pi/2$	
$\sin t$		$\sin t$	$+$
$\cos t$	$-$	$\cos t$	
$\tan t$		$\tan t$	

Quadrant III $\pi < t < 3\pi/2$		Quadrant IV $3\pi/2 < t < 2\pi$	
$\sin t$		$\sin t$	
$\cos t$		$\cos t$	
$\tan t$		$\tan t$	

50. (a) Find two numbers c and d such that

$$\sin(c + d) \neq \sin c + \sin d.$$

(b) Find two numbers c and d such that

$$\cos(c + d) \neq \cos c + \cos d.$$

In Exercises 51–56, draw a rough sketch to determine if the given number is positive.

51. sin 1 [*Hint:* The terminal side of an angle of 1 radian lies in the first quadrant (why?), so any point on it will have a positive y-coordinate.]

52. cos 2   **53.** tan 3   **54.** (cos 2)(sin 2)

55. tan 1.5   **56.** cos 3 + sin 3

In Exercises 57–62, find all the solutions of the equation.

57. $\sin t = 1$ **58.** $\cos t = -1$ **59.** $\tan t = 0$

60. $\sin t = -1$ **61.** $|\sin t| = 1$ **62.** $|\cos t| = 1$

Thinkers

63. Using only the definition and no calculator, determine which number is larger: $\sin(\cos 0)$ or $\cos(\sin 0)$.

64. With your calculator in radian mode and function graphing mode, graph the following functions on the same screen, using the viewing window with $0 \leq x \leq 2\pi$ and $-3 \leq y \leq 3$: $f(x) = \cos x^3$ and $g(x) = (\cos x)^3$. Are the graphs the same? What do you conclude about the statement $\cos x^3 = (\cos x)^3$?

65. Figure R is a diagram of a merry-go-round that includes horses A through F. The distance from the center P to A is 1 unit and the distance from P to D is 5 units. Define six functions as follows:

$A(t) = $ vertical distance from horse A to the x-axis at time t minutes;

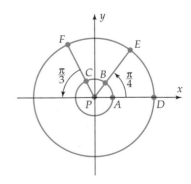

Figure R

and similarly for $B(t)$, $C(t)$, $D(t)$, $E(t)$, $F(t)$. The merry-go-round rotates counterclockwise at a rate of 1 revolution per minute, and the horses are in the positions shown in Figure R at the starting time $t = 0$. As the merry-go-round rotates, the horses move around the circles shown in Figure R.

(a) Show that $B(t) = A(t + 1/8)$ for every t.
(b) In a similar manner, express $C(t)$ in terms of the function $A(t)$.
(c) Express $E(t)$ and $F(t)$ in terms of the function $D(t)$.
(d) Explain why Figure S is valid and use it and similar triangles to express $D(t)$ in terms of $A(t)$.
(e) In a similar manner, express $E(t)$ and $F(t)$ in terms of $A(t)$.
(f) Show that $A(t) = \sin(2\pi t)$ for every t. [*Hint:* Exercises 65–72 in Section 6.1 may be helpful.]

(g) Use parts (a), (b), and (f) to express $B(t)$ and $C(t)$ in terms of the sine function.
(h) Use parts (d), (e), and (f) to express $D(t)$, $E(t)$, and $F(t)$ in terms of the sine function.

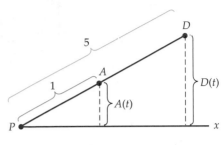

Figure S

6.2   *ALTERNATE*   The Sine, Cosine, and Tangent Functions

> **NOTE**
>
> If you have not read Chapter 8, omit this section. If you have read Chapter 8, use this section in place of Section 6.2.

Trigonometric functions of any angle are defined in Alternate Section 8.1, using a point on the terminal side of the angle. According to that definition, the domains of the sine, cosine, and tangent functions consist of *angles*. We now define trigonometric functions whose domains consist of *real numbers*. The basic idea is quite simple: If t is a real number, then

sin t is defined to be the sine of an angle of t radians.

cos t is defined to be the cosine of an angle of t radians.

tan t is defined to be the tangent of an angle of t radians.

Roadmap

Instructors who wish to cover all six trigonometric functions simultaneously should incorporate Section 6.6 into Sections 6.2–6.4, as follows.

Subsection of Section 6.6	Cover at the end of
Definitions; Alternate Descriptions	Section 6.2
Algebra and Identities	Section 6.3
Graphs	Section 6.4

In other words, instead of starting with an angle as in Chapter 8, we now start with a number t and then move to an angle of t radians, as summarized here:

Adapting the definition of Alternate Section 8.1 to this new viewpoint, we have the following.

Point-in-the-Plane Description

Let t be a real number. Let (x, y) be any point (except the origin) on the terminal side of an angle of t radians in standard position. Then

$$\sin t = \frac{y}{r}, \qquad \cos t = \frac{x}{r}, \qquad \tan t = \frac{y}{x},$$

where $r = \sqrt{x^2 + y^2}$ is the distance from (x, y) to the origin.

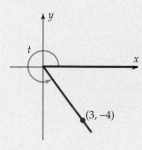

Figure 6–24

EXAMPLE 1

Figure 6–24 shows an angle of t radians in standard position. Find $\sin t$, $\cos t$, and $\tan t$.

SOLUTION   Apply the facts in the box above with $(x, y) = (3, -4)$ and
$$r = \sqrt{x^2 + y^2} = \sqrt{3^2 + (-4)^2} = \sqrt{25} = 5.$$

Then we have

$$\sin t = \frac{y}{r} = \frac{-4}{5} = -\frac{4}{5}. \qquad \cos t = \frac{x}{r} = \frac{3}{5}, \qquad \tan t = \frac{y}{x} = \frac{-4}{3} = -\frac{4}{3}. \quad\blacksquare$$

EXAMPLE 2

The terminal side of a first-quadrant angle of t radians in standard position lies on the line with equation $2x - 3y = 0$. Evaluate the three trigonometric functions at t.

SOLUTION   Verify that the point $(3, 2)$ satisfies the equation and hence lies on the terminal side of the angle (Figure 6–25). Now we have $(x, y) = (3, 2)$ and $r = \sqrt{x^2 + y^2} = \sqrt{3^2 + 2^2} = \sqrt{13}$. Therefore,

$$\sin t = \frac{y}{r} = \frac{2}{\sqrt{13}}, \qquad \cos t = \frac{x}{r} = \frac{3}{\sqrt{13}}, \qquad \tan t = \frac{y}{x} = \frac{2/\sqrt{13}}{3/\sqrt{13}} = \frac{2}{3}. \quad\blacksquare$$

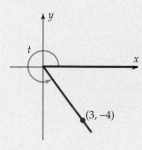

Figure 6–25

In most cases, evaluating trigonometric functions is not as simple as it was in the preceding examples. Usually, you must use a calculator (in radian mode) to approximate the values of these functions, as illustrated in Figure 6–26.

TECHNOLOGY TIP

Throughout this chapter, be sure your calculator is set for RADIAN mode. Use the MODE(S) menu on TI and HP and the SETUP menu Casio.

```
sin(2.5)
          .5984721441
cos(-6)
          .9601702867
tan(15)
         -.8559934009
■
```

Figure 6–26

However, for a few numbers, we can use our knowledge of special angles to evaluate the trigonometric functions exactly.

EXAMPLE 3

Find the exact value of

(a) $\sin \pi/6$ (b) $\cos \pi/6$ (c) $\tan 3\pi/4$

SOLUTION

(a) An angle of $\pi/6$ radians is the same as an angle of $30°$ and from Example 3(a) of Section 8.1, we know that $\sin 30° = 1/2$. Therefore,

$$\sin \pi/6 = \sin 30° = \frac{1}{2}.$$

(b) In Example 3(a) of Section 8.1, we also saw that $\cos 30° = \sqrt{3}/2$. Hence,

$$\cos \pi/6 = \cos 30° = \frac{\sqrt{3}}{2}.$$

(c) An angle of $3\pi/4$ is the same as an angle of $135°$, so from Example 11 of Alternate Section 8.1, we now have

$$\tan 3\pi/4 = \tan 135° = -1. \quad ■$$

▪▪ THE UNIT CIRCLE DESCRIPTION

We now develop a description of sine, cosine, and tangent that is based on the **unit circle,** which is the circle of radius 1 with center at the origin, whose equation is $x^2 + y^2 = 1$. Let t be any real number and construct an angle of t radians in standard position. Let $P = (x, y)$ be the point where the terminal side of this angle intersects the unit circle, as shown in Figure 6–27.

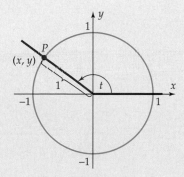

Figure 6–27

The distance from (x, y) to the origin is 1 unit because the radius of the unit circle is 1. Using the point (x, y) and $r = 1$, we see that

$$\cos t = \frac{x}{r} = \frac{x}{1} = x \qquad \text{and} \qquad \sin t = \frac{y}{r} = \frac{y}{1} = y.$$

In other words,

Sine and Cosine

> If P is the point where the terminal side of an angle of t radians in standard position meets the unit circle, then
>
> $$P \text{ has coordinates } (\cos t, \sin t).$$

This description is often used as a definition of sine and cosine. To get a better feel for it, do the following Graphing Exploration.

GRAPHING EXPLORATION

With your calculator in radian mode and parametric graphing mode, set the range values as follows:

$$0 \le t \le 2\pi, \qquad -1.8 \le x \le 1.8, \qquad -1.2 \le y \le 1.2^*$$

Then graph the curve given by the parametric equations.

$$x = \cos t \qquad \text{and} \qquad y = \sin t.$$

The graph is the unit circle. Use the trace to move around the circle. At each point, the screen will display three numbers: the values of t, x, and y. For each t, the cursor is on the point where the terminal side of an angle of t radians meets the unit circle, so the corresponding x is the number $\cos t$ and the corresponding y is the number $\sin t$.

Suppose an angle of t radians in standard position intersects the unit circle at the point (x, y), as in Figure 6–27. From our earlier definition, $\tan t = y/x$ and by the unit circle description, we have $x = \cos t$ and $y = \sin t$. Therefore, we have another description of the tangent function:

$$\tan t = \frac{\sin t}{\cos t}.$$

EXAMPLE 4

Use the unit circle description and the preceding equation to find the exact values of $\sin t$, $\cos t$, and $\tan t$ when

(a) $t = \pi$ (b) $t = \pi/2$.

---

*These settings give a square viewing window on calculators with a screen measuring approximately 95 by 63 pixels, and hence the unit circle will look like a circle. For wider screens, adjust the x range settings to obtain a square window.

SOLUTION

(a) Construct an angle of π radians, as in Figure 6–28. Its terminal side lies on the negative x-axis and intersects the unit circle at $P = (-1, 0)$. Hence,

$\sin \pi = y$-coordinate of $P = 0$

$\cos \pi = x$-coordinate of $P = -1$

$\tan \pi = \dfrac{\sin \pi}{\cos \pi} = \dfrac{0}{-1} = 0.$

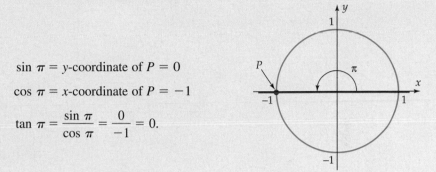

Figure 6–28

(b) An angle of $\pi/2$ radians (Figure 6–29) has its terminal side on the positive y-axis and intersects the unit circle at $P = (0, 1)$.

$\cos \dfrac{\pi}{2} = x$-coordinate of $P = 0$

$\sin \dfrac{\pi}{2} = y$-coordinate of $P = 1$

$\tan \dfrac{\pi}{2} = \dfrac{\sin(\pi/2)}{\cos(\pi/2)} = \dfrac{1}{0} \; undefined.$ ∎

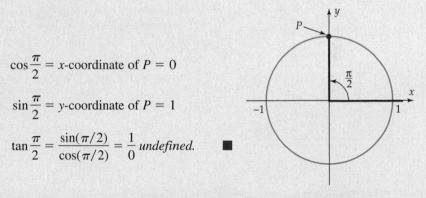

Figure 6–29

Example 4(b) shows that the domain of the tangent function must exclude numbers for which $\cos t = 0$. These occur when the point P has coordinates $(0, 1)$ or $(0, -1)$. Therefore,

The domain of the tangent function consists of all real numbers except $t = \pm\pi/2, \pm3\pi/2, \pm5\pi/2, \ldots.$

⊞ SPECIAL VALUES

The trigonometric functions can be evaluated exactly at $t = \pi/3$, $t = \pi/4$, $t = \pi/6$, and any integer multiples of these numbers by using the following facts (which are explained in the Geometry Review Appendix):

A right triangle with hypotenuse 1 and angles of $\pi/6$ and $\pi/3$ radians has sides of lengths $1/2$ (opposite the angle of $\pi/6$) and $\sqrt{3}/2$ (opposite the angle of $\pi/3$).

A right triangle with hypotenuse 1 and two angles of $\pi/4$ radians has two sides of length $\sqrt{2}/2$.

Figure 6–30

EXAMPLE 5

Find the exact values of the three trigonometric functions at $t = -5\pi/4$.

SOLUTION   Construct an angle of $-5\pi/4$ radians in standard position and let P be the point where the terminal side intersects the unit circle (Figure 6–31). Draw a vertical line from P to the x-axis, forming a right triangle with hypotenuse 1 and two angles of $\pi/4$ radians (45°), as shown in Figure 6–31.

Each side of this triangle has length $\sqrt{2}/2$, as explained above. Therefore, the coordinates of P are $(-\sqrt{2}/2, \sqrt{2}/2)$, so

Figure 6–31

$$\sin \frac{-5\pi}{4} = y\text{-coordinate of } P = \frac{\sqrt{2}}{2}$$

$$\cos \frac{-5\pi}{4} = x\text{-coordinate of } P = -\frac{\sqrt{2}}{2}$$

$$\tan \frac{-5\pi}{4} = \frac{\sin t}{\cos t} = \frac{\sqrt{2}/2}{-\sqrt{2}/2} = -1. \quad \blacksquare$$

EXAMPLE 6

Evaluate the trigonometric functions exactly at $t = 11\pi/3$.

SOLUTION   Construct an angle of $11\pi/3$ radians in standard position and draw a vertical line from the x-axis to the point P where the terminal side of the angle meets the unit circle, as shown in Figure 6–32. The right triangle formed in this way has hypotenuse 1 and angles of $\pi/6$ and $\pi/3$ radians. The sides of the triangle must have lengths $1/2$ and $\sqrt{3}/2$, as explained above. So the coordinates of P are $(1/2, -\sqrt{3}/2)$, and

Figure 6–32

$$\sin t = y\text{-coordinate of } P = -\frac{\sqrt{3}}{2}$$

$$\cos t = x\text{-coordinate of } P = \frac{1}{2}$$

$$\tan t = \frac{\sin t}{\cos t} = \frac{-\sqrt{3}/2}{1/2} = -\sqrt{3}. \quad \blacksquare$$

EXERCISES ALTERNATE 6.2

Use the exercises for Section 6.2 on page 429.

6.3   Algebra and Identities

Roadmap

Section 8.1 may be covered at this point by instructors who prefer to introduce right triangle trigonometry early.

In the previous section, we concentrated on evaluating the trigonometric functions. In this section, the emphasis is on the algebra of such functions. When dealing with trigonometric functions, two notational conventions are normally observed:

1. **Parentheses are omitted whenever no confusion can result.**

   For example,

$\sin(t)$	is written	$\sin t$
$-(\cos(5t))$	is written	$-\cos 5t$
$4(\tan t)$	is written	$4 \tan t$.

   On the other hand, parentheses *are* needed to distinguish

   $$\cos(t + 3) \quad \text{from} \quad \cos t + 3,$$

   because the first one says, "Add 3 to t and take the cosine of the result," but the second one says, "Take the cosine of t and add 3 to the result." When $t = 5$, for example, these are different numbers, as shown in Figure 6–33.

TECHNOLOGY TIP

TI-83 and HP-39 automatically insert an opening parenthesis when the COS key is pushed. The display COS(5 + 3 is interpreted as COS(5 + 3). If you want cos 5 + 3, you must insert a parenthesis after the 5: COS(5) + 3.

```
cos(5+3)
         -.14500033809
cos 5+3
          3.2836621854G
■
```

Figure 6–33

2. **When dealing with powers of trigonometric functions, exponents (other than −1) are written between the function symbol and the variable.**

For example,

$$(\cos t)^3 \qquad \text{is written} \qquad \cos^3 t$$

$$(\sin t)^4(\tan 7t)^2 \qquad \text{is written} \qquad \sin^4 t \tan^2 7t.$$

Furthermore,

$$\sin t^3 \qquad \text{means} \qquad \sin(t^3) \qquad [not\ (\sin t)^3 \qquad \text{or} \qquad \sin^3 t]$$

For instance, when $t = 4$, we have Figure 6–34.

```
sin 4^3
        .920026038197
sin(4^3)
        .920026038197
(sin 4)^3
       -.433458641981
■
```

Figure 6–34

TECHNOLOGY TIP

Calculators do not use convention 2. To obtain $\sin^3 4$, you must enter $(\sin 4)^3$.

CAUTION

Convention 2 is not used when the exponent is −1. For example, $\sin^{-1} t$ does *not* mean $(\sin t)^{-1}$ or $1/(\sin t)$; it has an entirely different meaning that will be discussed in Section 7.4. Similar remarks apply to $\cos^{-1} t$ and $\tan^{-1} t$.

Except for these two conventions and the Caution in the margin, the algebra of trigonometric functions is just like the algebra of other functions. They may be added, subtracted, multiplied, composed, etc.

EXAMPLE 1

If $f(t) = \sin^2 t + \tan t$ and $g(t) = \tan^3 t + 5$, then the product function fg is given by the rule

$$(fg)(t) = f(t)g(t) = (\sin^2 t + \tan t)(\tan^3 t + 5)$$

$$= \sin^2 t \tan^3 t + 5 \sin^2 t + \tan^4 t + 5 \tan t. \qquad ■$$

EXAMPLE 2

Factor $2 \cos^2 t - 5 \cos t - 3$.

SOLUTION   You can do this directly, but it may be easier to understand if you make a substitution. Let $u = \cos t$, then

$$2 \cos^2 t - 5 \cos t - 3 = 2(\cos t)^2 - 5 \cos t - 3$$

$$= 2u^2 - 5u - 3$$

$$= (2u + 1)(u - 3)$$

$$= (2 \cos t + 1)(\cos t - 3). \qquad ■$$

EXAMPLE 3

If $f(t) = \cos^2 t - 9$ and $g(t) = \cos t + 3$, then the quotient function f/g is given by the rule

$$\left(\frac{f}{g}\right)(t) = \frac{f(t)}{g(t)} = \frac{\cos^2 t - 9}{\cos t + 3} = \frac{(\cos t + 3)(\cos t - 3)}{\cos t + 3} = \cos t - 3. \quad \blacksquare$$

CAUTION

You are dealing with *functional notation* here, so the symbol sin *t* is a *single entity*, as are cos *t* and tan *t*. Don't try some nonsensical "canceling" operation, such as

$$\frac{\sin t}{\cos t} = \frac{\sin}{\cos} \qquad \text{or} \qquad \frac{\cos t^2}{\cos t} = \frac{\cos t}{\cos} = t.$$

EXAMPLE 4

If $f(t) = \sin t$ and $g(t) = t^2 + 3$, then the composite function $g \circ f$ is given by the rule

$$(g \circ f)(t) = g(f(t)) = g(\sin t) = \sin^2 t + 3.$$

The composite function $f \circ g$ is given by the rule

$$(f \circ g)(t) = f(g(t)) = f(t^2 + 3) = \sin(t^2 + 3).$$

The parentheses are absolutely necessary here because $\sin(t^2 + 3)$ is *not* the same function as $\sin t^2 + 3$. For instance, a calculator in radian mode shows that for $t = 5$,

$$\sin(5^2 + 3) = \sin(25 + 3) = \sin 28 \approx .2709,$$

whereas

$$\sin 5^2 + 3 = \sin 25 + 3 \approx (-.1324) + 3 = 2.8676. \quad \blacksquare$$

▌▌ THE PYTHAGOREAN IDENTITY

Trigonometric functions have numerous interrelationships that are usually expressed as *identities*. An **identity** is an equation that is true for all values of the variable for which every term of the equation is defined. Here is one of the most important trigonometric identities.

Pythagorean Identity

For every real number t,

$$\sin^2 t + \cos^2 t = 1.$$

Proof For each real number t, the point P, where the terminal side of an angle of t radians intersects the unit circle, has coordinates $(\cos t, \sin t)$, as in

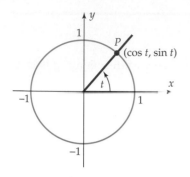

Figure 6–35

Figure 6–35. Since P lies on the unit circle, its coordinates must satisfy the equation of the unit circle: $x^2 + y^2 = 1$, that is, $\cos^2 t + \sin^2 t = 1$. ■

GRAPHING EXPLORATION

Recall that the graph of $y = 1$ is a horizontal line through $(0, 1)$. Verify the Pythagorean identity by graphing the equation

$$y = (\sin x)^2 + (\cos x)^2$$

in the window with $-10 \le x \le 10$ and $-3 \le y \le 3$ and using the trace feature.

EXAMPLE 5

If $\pi/2 < t < \pi$ and $\sin t = 2/3$, find $\cos t$ and $\tan t$.

SOLUTION By the Pythagorean identity,

$$\cos^2 t = 1 - \sin^2 t = 1 - \left(\frac{2}{3}\right)^2 = 1 - \frac{4}{9} = \frac{5}{9}.$$

So there are two possibilities:

$$\cos t = \sqrt{5/9} = \sqrt{5}/3 \qquad \text{or} \qquad \cos t = -\sqrt{5/9} = -\sqrt{5}/3.$$

Since $\pi/2 < t < \pi$, $\cos t$ is negative (see Exercise 49 on page 430). Therefore, $\cos t = -\sqrt{5}/3$, and

$$\tan t = \frac{\sin t}{\cos t} = \frac{2/3}{-\sqrt{5}/3} = \frac{-2}{\sqrt{5}} = \frac{-2\sqrt{5}}{5}.$$ ■

EXAMPLE 6

The Pythagorean identity is valid for *any* number t. For instance, if $t = 3k + 7$, then $\sin^2(3k + 7) + \cos^2(3k + 7) = 1$. ■

EXAMPLE 7

To simplify the expression $\tan^2 t \cos^2 t + \cos^2 t$, we use the definition of tangent and the Pythagorean identity:

$$\tan^2 t \cos^2 t + \cos^2 t = \frac{\sin^2 t}{\cos^2 t}\cos^2 t + \cos^2 t$$

$$= \sin^2 t + \cos^2 t = 1.$$ ■

⠿ PERIODICITY IDENTITIES

Let t be any real number and construct two angles in standard position of measure t and $t + 2\pi$ radians, respectively, as shown in Figure 6–36. As we saw in Section 6.1, both of these angles have the same terminal side. Therefore, the point P where the terminal side meets the unit circle is the *same* in both cases.

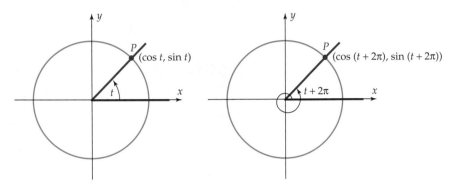

Figure 6–36

Therefore, the coordinates of P are the same, that is,

$$\sin t = \sin(t + 2\pi) \qquad \text{and} \qquad \cos t = \cos(t + 2\pi).$$

Furthermore, since an angle of t radians has the same terminal side as angles of radian measure $t \pm 2\pi, t \pm 4\pi, t \pm 6\pi$, and so forth, the same argument shows that

$$\sin t = \sin(t \pm 2\pi) = \sin(t \pm 4\pi) = \sin(t \pm 6\pi) = \cdots$$

$$\cos t = \cos(t \pm 2\pi) = \cos(t \pm 4\pi) = \cos(t \pm 6\pi) = \cdots$$

Figure 6–37

as illustrated (for $t = 5$) in Figure 6–37.

There is a special name for functions that repeat their values at regular intervals. A function f is said to be **periodic** if there is a positive constant k such that $f(t) = f(t + k)$ for every number t in the domain of f. There will be more than one constant k with this property; the smallest one is called the **period** of the function f. We have just seen that sine and cosine are periodic with $k = 2\pi$. Exercises 65 and 66 show that 2π is the smallest such positive constant k. Therefore, we have the following.

Period of Sine and Cosine

> The sine and cosine functions are periodic with period 2π: For every real number t,
>
> $$\sin t = \sin(t \pm 2\pi) \qquad \text{and} \qquad \cos t = \cos(t \pm 2\pi)$$

EXAMPLE 8

Find $\sin \dfrac{13\pi}{6}$.

SOLUTION   The periodicity identity shows that

$$\sin \frac{13\pi}{6} = \sin\left(\frac{\pi}{6} + \frac{12\pi}{6}\right) = \sin\left(\frac{\pi}{6} + 2\pi\right) = \sin \frac{\pi}{6} = \frac{1}{2}. \qquad ■$$

The tangent function is also periodic (see Exercise 36), but its period is π rather than 2π, that is,

$$\tan(t + \pi) = \tan t \quad \text{for every real number } t,$$

as we shall see in Section 6.4.

⊞ NEGATIVE ANGLE IDENTITIES

GRAPHING EXPLORATION

(a) In a viewing window with $-2\pi \le x \le 2\pi$, graph $y_1 = \sin x$ and $y_2 = \sin(-x)$ on the same screen. Use trace to move along $y_1 = \sin x$. Stop at a point and note its y-coordinate. Use the up or down arrow to move vertically to the graph of $y_2 = \sin(-x)$. The x-coordinate remains the same, but the y-coordinate is different. How are the two y-coordinates related? Is one the negative of the other? Repeat the procedure for other points. Are the results the same?

(b) Now graph $y_1 = \cos x$ and $y_2 = \cos(-x)$ on the same screen. How do the graphs compare?

(c) Repeat part (a) for $y_1 = \tan x$ and $y_2 = \tan(-x)$. Are the results similar to those for sine?

The preceding Graphing Exploration suggests the truth of the following statement.

Negative Angle Identities

For every real number t,

$$\sin(-t) = -\sin t$$

$$\cos(-t) = \cos t$$

$$\tan(-t) = -\tan t.$$

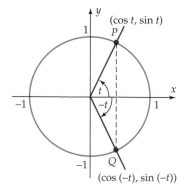

Figure 6–38

Proof   Consider angles of t radians and $-t$ radians in standard position, as in Figure 6–38. By the definition of sine and cosine, P has coordinates $(\cos t, \sin t)$, and Q has coordinates $(\cos(-t), \sin(-t))$. As Figure 6–38 suggests, P and Q lie on the same vertical line. Therefore, they have the same first coordinate, that is, $\cos(-t) = \cos t$. As the figure also suggests, P and Q lie at equal distances from the x-axis.* So the y-coordinate of Q must be the negative of the y-coordinate of P, that is, $\sin(-t) = -\sin t$. Finally, by the definition of the tangent function and the two identities just proved, we have

$$\tan(-t) = \frac{\sin(-t)}{\cos(-t)} = \frac{-\sin t}{\cos t} = -\frac{\sin t}{\cos t} = -\tan t. \quad \blacksquare$$

---

*These facts can be proved by using congruent triangles. (See Exercise 13 in the Geometry Review Appendix.)

EXAMPLE 9

In Example 2 of Section 6.2, we showed that

$$\sin\frac{\pi}{6} = \frac{1}{2}, \qquad \cos\frac{\pi}{6} = \frac{\sqrt{3}}{2}, \qquad \tan\frac{\pi}{6} = \frac{\sqrt{3}}{3}.$$

Using the negative angle identities, we have

$$\sin\left(-\frac{\pi}{6}\right) = -\sin\frac{\pi}{6} = -\frac{1}{2}, \qquad \cos\left(-\frac{\pi}{6}\right) = \cos\frac{\pi}{6} = \frac{\sqrt{3}}{2},$$

$$\tan\left(-\frac{\pi}{6}\right) = -\tan\frac{\pi}{6} = -\frac{\sqrt{3}}{3}. \qquad ■$$

EXAMPLE 10

To simplify $(1 + \sin t)(1 + \sin(-t))$, we use the negative angle identity and the Pythagorean identity.

$$(1 + \sin t)(1 + \sin(-t)) = (1 + \sin t)(1 - \sin t)$$

$$= 1 - \sin^2 t$$

$$= \cos^2 t. \qquad ■$$

✓ EXERCISES 6.3

In Exercises 1–4, find the rule of the product function fg.

1. $f(t) = 3 \sin t; \qquad g(t) = \sin t + 2 \cos t$

2. $f(t) = 5 \tan t; \qquad g(t) = \tan^3 t - 1$

3. $f(t) = 3 \sin^2 t; \qquad g(t) = \sin t + \tan t$

4. $f(t) = \sin 2t + \cos^4 t; \qquad g(t) = \cos 2t + \cos^2 t$

In Exercises 5–14, factor the given expression.

5. $\cos^2 t - 4$

6. $25 - \tan^2 t$

7. $\sin^2 t - \cos^2 t$

8. $\sin^3 t - \sin t$

9. $\tan^2 t + 6 \tan t + 9$

10. $\cos^2 t - \cos t - 2$

11. $6 \sin^2 t - \sin t - 1$

12. $\tan t \cos t + \cos^2 t$

13. $\cos^4 t + 4 \cos^2 t - 5$

14. $3 \tan^2 t + 5 \tan t - 2$

In Exercises 15–18, find the rules of the composite functions f ∘ g and g ∘ f.

15. $f(t) = \cos t; \qquad g(t) = 2t + 4$

16. $f(t) = \sin t + 2; \qquad g(t) = t^2$

17. $f(t) = \tan(t + 3); \qquad g(t) = t^2 - 1$

18. $f(t) = \cos^2(t - 2); \qquad g(t) = 5t + 2$

In Exercises 19–24, determine if it is possible for a number t to satisfy the given conditions. [Hint: Think Pythagorean.]

19. $\sin t = 5/13$ and $\cos t = 12/13$

20. $\sin t = -2$ and $\cos t = 1$

21. $\sin t = -1$ and $\cos t = 1$

22. $\sin t = 1/\sqrt{2}$ and $\cos t = -1/\sqrt{2}$

23. $\sin t = 1$ and $\tan t = 1$

24. $\cos t = 8/17$ and $\tan t = 15/8$

In Exercises 25–28, use the Pythagorean identity to find sin t.

25. $\cos t = -.5 \qquad$ and $\qquad \pi < t < 3\pi/2$

26. $\cos t = -3/\sqrt{10} \qquad$ and $\qquad \pi/2 < t < \pi$

27. $\cos t = 1/2 \qquad$ and $\qquad 0 < t < \pi/2$

28. $\cos t = 2/\sqrt{5} \qquad$ and $\qquad 3\pi/2 < t < 2\pi$

In Exercises 29–35, assume that sin t = 3/5 and 0 < t < π/2. Use identities in the text to find the number.

29. $\sin(-t)$ **30.** $\sin(t + 10\pi)$ **31.** $\sin(2\pi - t)$

32. $\cos t$ **33.** $\tan t$ **34.** $\cos(-t)$

35. $\tan(2\pi - t)$

36. (a) Show that $\tan(t + 2\pi) = \tan t$ for every t in the domain of tan t. [*Hint:* Use the definition of tangent and some identities proved in the text.]

   (b) Verify that it appears true that $\tan(x + \pi) = \tan x$ for every t in the domain by using your calculator's table feature to make a table of values for $y_1 = \tan(x + \pi)$ and $y_2 = \tan x$.

In Exercises 37–42, assume that

$$\cos t = -2/5 \quad and \quad \pi < t < 3\pi/2.$$

Use identities to find the number.

37. $\sin t$ **38.** $\tan t$ **39.** $\cos(2\pi - t)$

40. $\cos(-t)$ **41.** $\sin(4\pi + t)$ **42.** $\tan(4\pi - t)$

In Exercises 43–46, assume that

$$\sin(\pi/8) = \frac{\sqrt{2 - \sqrt{2}}}{2}$$

and use identities to find the exact functional value.

43. $\cos(\pi/8)$ **44.** $\tan(\pi/8)$

45. $\sin(17\pi/8)$ **46.** $\tan(-15\pi/8)$

In Exercises 47–58, use algebra and identities in the text to simplify the expression. Assume all denominators are nonzero.

47. $(\sin t + \cos t)(\sin t - \cos t)$

48. $(\sin t - \cos t)^2$ **49.** $\tan t \cos t$

50. $(\sin t)/(\tan t)$ **51.** $\sqrt{\sin^3 t \cos t}\sqrt{\cos t}$

52. $(\tan t + 2)(\tan t - 3) - (6 - \tan t) + 2\tan t$

53. $\left(\dfrac{4\cos^2 t}{\sin^2 t}\right)\left(\dfrac{\sin t}{4\cos t}\right)^2$

54. $\dfrac{5\cos t}{\sin^2 t} \cdot \dfrac{\sin^2 t - \sin t \cos t}{\sin^2 t - \cos^2 t}$

55. $\dfrac{\cos^2 t + 4\cos t + 4}{\cos t + 2}$

56. $\dfrac{\sin^2 t - 2\sin t + 1}{\sin t - 1}$

57. $\dfrac{1}{\cos t} - \sin t \tan t$

58. $\dfrac{1 - \tan^2 t}{1 + \tan^2 t} + 2\sin^2 t$

In Exercises 59–63, show that the given function is periodic with period less than 2π. [Hint: Find a positive number k with $k < 2\pi$ such that $f(t + k) = f(t)$ for every t in the domain of f.]

59. $f(t) = \sin 2t$

60. $f(t) = \cos 3t$

61. $f(t) = \sin 4t$

62. $f(t) = \cos(3\pi t/2)$

63. $f(t) = \tan 2t$

64. Fill the blanks with "even" or "odd" so that the resulting statement is true. Then prove the statement by using an appropriate identity. [*Hint: Special Topics* 3.4.A may be helpful.]

   (a) $f(t) = \sin t$ is an _____ function.
   (b) $g(t) = \cos t$ is an _____ function.
   (c) $h(t) = \tan t$ is an _____ function.
   (d) $f(t) = t \sin t$ is an _____ function.
   (e) $g(t) = t + \tan t$ is an _____ function.

65. Here is a proof that the cosine function has period 2π. We saw in the text that $\cos(t + 2\pi) = \cos t$ for every t. We must show that there is no positive number smaller than 2π with this property. Do this as follows:

   (a) Find all numbers k such that $0 < k < 2\pi$ and $\cos k = 1$. [*Hint:* Draw a picture and use the definition of the cosine function.]

   (b) Suppose k is a number such that $\cos(t + k) = \cos t$ for every number t. Show that $\cos k = 1$. [*Hint:* Consider $t = 0$.]

   (c) Use parts (a) and (b) to show that there is no positive number k less than 2π with the property that $\cos(t + k) = \cos t$ for *every* number t. Therefore, $k = 2\pi$ is the smallest such number, and the cosine function has period 2π.

66. Here is proof that the sine function has period 2π. We saw in the text that $\sin(t + 2\pi) = \sin t$ for every t. We must show that there is no positive number smaller than 2π with this property. Do this as follows:

   (a) Find a number t such that $\sin(t + \pi) \neq \sin t$.

   (b) Find all numbers k such that $0 < k < 2\pi$ and $\sin k = 0$. [*Hint:* Draw a picture and use the definition of the sine function.]

   (c) Suppose k is a number such that $\sin(t + k) = \sin t$ for every number t. Show that $\sin k = 0$. [*Hint:* Consider $t = 0$.]

   (d) Use parts (a)–(c) to show that there is no positive number k less than 2π with the property that $\sin(t + k) = \sin t$ for *every* number t. Therefore, $k = 2\pi$ is the smallest such number, and the sine function has period 2π.

6.4 Basic Graphs

Although a graphing calculator will quickly sketch the graphs of the sine, cosine, and tangent functions, it will not give you much insight into why these graphs have the shapes they do and why these shapes are important. So the emphasis here is on the connection between the definition of these functions and their graphs.

If P is the point where the unit circle meets the terminal side of an angle of t radians, then the y-coordinate of P is the number $\sin t$. We can get a rough sketch of the graph of $f(t) = \sin t$ by watching the y-coordinate of P.

As t Increases	The Point P Moves	The y-coordinate of $P\,(=\sin t)$	Rough Sketch of the Graph
from 0 to $\dfrac{\pi}{2}$	from (1, 0) to (0, 1)	increases from 0 to 1	
from $\dfrac{\pi}{2}$ to π	from (0, 1) to (−1, 0)	decreases from 1 to 0	
from π to $\dfrac{3\pi}{2}$	from (−1, 0) to (0, −1)	decreases from 0 to −1	
from $\dfrac{3\pi}{2}$ to 2π	from (0, −1) to (1, 0)	increases from −1 to 0	

GRAPHING **EXPLORATION**

Your calculator can provide a dynamic simulation of this process. Put it in parametric graphing mode and set the range values as follows:

$$0 \le t \le 6.28 \qquad -1 \le x \le 6.28 \qquad -2.5 \le y \le 2.5.$$

On the same screen, graph the two functions given by

$$x_1 = \cos t, \quad y_1 = \sin t \qquad \text{and} \qquad x_2 = t, \quad y_2 = \sin t.$$

Using the trace feature, move the cursor along the first graph (the unit circle). Stop at a point on the circle, and note the value of t and the y-coordinate of the point. Then switch the trace to the second graph (the sine function) by using the up or down cursor arrows. The value of t remains the same. What are the x- and y-coordinates of the new point? How does the y-coordinate of the new point compare with the y-coordinate of the original point on the unit circle?

TECHNOLOGY TIP

Calculators have built-in windows for trigonometric functions, in which the x-axis tick marks are at intervals of $\pi/2$.
Choose TRIG or ZTRIG in this menu:

TI: ZOOM

HP-39: VIEWS

Casio: V-WINDOW

To complete the graph of the sine function, note that as t goes from 2π to 4π, the point P on the unit circle *retraces* the path it took from 0 to 2π, so *the same wave shape will repeat* on the graph. The same thing happens when t goes from 4π to 6π, or from -2π to 0, and so on. This repetition of the same pattern is simply the graphical expression of the fact that the sine function has period 2π: For any number t, the points

$$(t, \sin t) \qquad \text{and} \qquad (t + 2\pi, \sin(t + 2\pi))$$

on the graph have the same second coordinate.

A graphing calculator or some point plotting with an ordinary calculator now produces the graph of $f(t) = \sin t$ (Figure 6–39).

Figure 6–39

NOTE: Throughout this chapter, we use t as the variable for trigonometric functions to avoid any confusion with the x's and y's that are part of the definition of these functions. For calculator graphing in "function mode," however, you must use x as the variable: $f(x) = \sin x$, $g(x) = \cos x$, etc.

The graph of the sine function and the techniques of Section 3.4 can be used to graph other trigonometric functions.

EXAMPLE 1

The graph of $h(t) = 3 \sin t$ is the graph of $f(t) = \sin t$ stretched away from the horizontal axis by a factor of 3, as shown in Figure 6–40. ∎

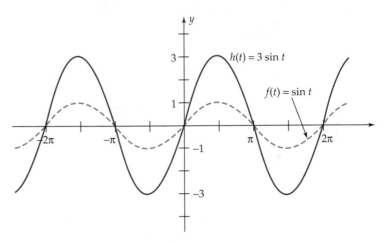

Figure 6–40

EXAMPLE 2

The graph of $k(t) = -\frac{1}{2} \sin t$ is the graph of $f(t) = \sin t$ shrunk by a factor of $1/2$ toward the horizontal axis and then reflected in the horizontal axis, as shown in Figure 6–41. ∎

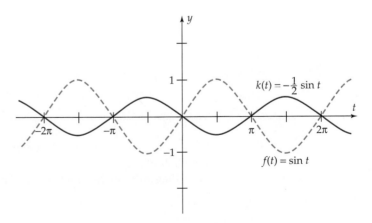

Figure 6–41

▦ GRAPH OF THE COSINE FUNCTION

To obtain the graph of $g(t) = \cos t$, we follow the same procedure as with sine, except that we now watch the x-coordinate of P (which is $\cos t$).

As t Increases	The Point P Moves	The x-coordinate of P $(= \cos t)$	Rough Sketch of the Graph
from 0 to $\frac{\pi}{2}$	from $(1, 0)$ to $(0, 1)$	decreases from 1 to 0	
from $\frac{\pi}{2}$ to π	from $(0, 1)$ to $(-1, 0)$	decreases from 0 to -1	
from π to $\frac{3\pi}{2}$	from $(-1, 0)$ to $(0, -1)$	increases from -1 to 0	
from $\frac{3\pi}{2}$ to 2π	from $(0, -1)$ to $(1, 0)$	increases from 0 to 1	

As t takes larger values, P begins to retrace its path around the unit circle, so the graph of $g(t) = \cos t$ repeats the same wave pattern, and similarly for negative values of t. So the graph looks like Figure 6–42.

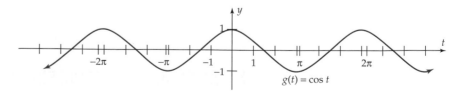

$g(t) = \cos t$

Figure 6–42

For a dynamic simulation of the cosine graphing process described above, see Exercise 51.

The graphs of the cosine function in Figure 6–42 and the sine function in Figure 6–39 lie between -1 and 1 on the y-axis. This is a graphical illustration of the following fact.

Range of Sine and Cosine

For every real number t

$$-1 \le \sin t \le 1$$

and

$$-1 \le \cos t \le 1.$$

The techniques of Section 3.4 can be used to graph variations of the cosine function.

EXAMPLE 3

The graph of $h(t) = 4 \cos t$ is the graph of $g(t) = \cos t$ stretched away from the horizontal axis by a factor of 4, as shown in Figure 6–43. ■

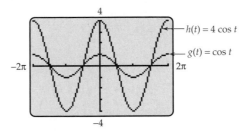

Figure 6–43

EXAMPLE 4

The graph of $k(t) = -2 \cos t + 3$ is the graph of $g(t) = \cos t$ stretched away from the horizontal axis by a factor of 2, reflected in the horizontal axis, and shifted vertically 3 units upward as shown in Figure 6–44. ■

Figure 6–44

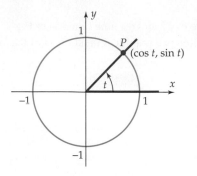

Figure 6–45

▪▪ GRAPH OF THE TANGENT FUNCTION

To determine the shape of the graph of $h(t) = \tan t$, we use an interesting connection between the tangent function and straight lines. As shown in Figure 6–45, the point P where the terminal side of an angle of t radians in standard position meets the unit circle has coordinates $(\cos t, \sin t)$. We can use this point and the point $(0, 0)$ to compute the *slope* of the terminal side.

$$\text{slope} = \frac{\sin t - 0}{\cos t - 0} = \frac{\sin t}{\cos t} = \tan t$$

Therefore, we have the following.

Slope and Tangent

> The slope of the terminal side of an angle of t radians in standard position is the number $\tan t$.

The graph of $h(x) = \tan t$ can now be sketched by watching the slope of the terminal side of an angle of t radians, as t takes different values. Recall that the more steeply a line rises from left to right, the larger its slope. Similarly, lines that fall from left to right have negative slopes that increase in absolute value as the line falls more steeply.

As *t* Changes	The Terminal Side of the Angle Moves	Its Slope (tan *t*)	Rough Sketch of the Graph
from 0 to $\frac{\pi}{2}$	from horizontal upward toward vertical	increases from 0 in the positive direction and keeps getting larger	
from 0 to $-\frac{\pi}{2}$	from horizontal downward toward vertical	decreases from 0 in the negative direction and keeps getting larger in absolute value	

When $t = \pm \pi/2$, the terminal side of the angle is vertical, and hence its slope is not defined. This corresponds to the fact that the tangent function is not defined when $t = \pm \pi/2$. The vertical lines through $\pm \pi/2$ are vertical asymptotes of the graph: It gets closer and closer to these lines but never touches them.

As t goes from $\pi/2$ to $3\pi/2$, the terminal side goes from almost vertical with negative slope to horizontal to almost vertical with positive slope (draw a picture), exactly as it does between $-\pi/2$ and $\pi/2$. So the graph repeats the same pattern. The same thing happens between $3\pi/2$ and $5\pi/2$, between $-3\pi/2$ and $-\pi/2$, etc. Therefore, the entire graph looks like Figure 6–46.

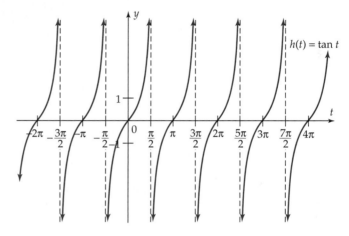

Figure 6–46

Because calculators sometimes do not graph accurately across vertical asymptotes, the graph may look slightly different on a calculator screen (with vertical line segments where the asymptotes should be).

The graph of the tangent function repeats the same pattern at intervals of length π. This means that the tangent function repeats its values at intervals of π.

Period of Tangent

> The tangent function is periodic with period π: For every real number t in its domain,
>
> $$\tan(t \pm \pi) = \tan t.$$

EXAMPLE 5

As we saw in Section 3.4, the graph of

$$k(t) = \tan\left(t - \frac{\pi}{2}\right)$$

is the graph of $h(t) = \tan t$ shifted horizontally $\pi/2$ units to the right (Figure 6–47).   ∎

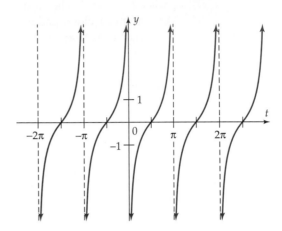

Figure 6–47

▪▪ GRAPHS AND IDENTITIES

Graphing calculators can be used to identify equations that could possibly be identities. A calculator cannot *prove* that such an equation is an identity; but it can provide evidence that it *might* be one. On the other hand, a calculator *can* prove that a particular equation is not an identity.

___

EXAMPLE 6

Which of the following equations could possibly be an identity?

(a) $\cos\left(\dfrac{\pi}{2} + t\right) = \sin t$ \qquad\qquad (b) $\cos\left(\dfrac{\pi}{2} - t\right) = \sin t$

SOLUTION

(a) Consider the functions $f(t) = \cos\left(\dfrac{\pi}{2} + t\right)$ and $g(t) = \sin t$, whose rules are given by the two sides of the equation

$$\cos\left(\frac{\pi}{2} + t\right) = \sin t.$$

If this equation is an identity, then $f(t) = g(t)$ for every real number t, and hence, f and g have the same graph. But the graphs of f and g on the interval $[-2\pi, 2\pi]$ (Figure 6–48) are obviously different. Therefore, this equation is *not* an identity.

(b) We can test this equation in the same manner. The graph of the left side, that is, the graph of

$$h(t) = \cos\left(\frac{\pi}{2} - t\right),$$

Figure 6–48

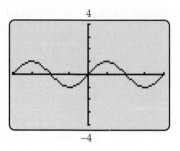

Figure 6–49

in Figure 6–49 appears to be the same as the graph of $g(t) = \sin t$ on the interval $[-2\pi, 2\pi]$ (Figure 6–48). So let's explore further.

GRAPHING EXPLORATION

Graph $h(t) = \cos\left(\dfrac{\pi}{2} - t\right)$ and $g(t) = \sin t$ on the same screen and use the trace feature to confirm that the graphs appear to be identical.

The fact that the graphs appear to be identical means that the two functions have the same value at every number t that the calculator computed in making the graphs (at least 95 numbers). This evidence strongly suggests that the equation $\cos\left(\dfrac{\pi}{2} - t\right) = \sin t$ is an identity, but does not prove it. All we can say at this point is that the equation could possibly be an identity. ■

CAUTION

Do not assume that two graphs that look the same on a calculator screen actually are the same. Depending on the viewing window, two graphs that are actually quite different may appear to be identical. See Exercises 43–50 for some examples.

EXERCISES 6.4

In Exercises 1–6, use the graphs of the sine and cosine functions to find all the solutions of the equation.

1. $\sin t = 0$ **2.** $\cos t = 0$ **3.** $\sin t = 1$

4. $\sin t = -1$ **5.** $\cos t = -1$ **6.** $\cos t = 1$

In Exercises 7–10, find tan t, where the terminal side of an angle of t radians lies on the given line.

7. $y = 11x$ **8.** $y = 1.5x$ **9.** $y = 1.4x$

10. $y = .32x$

In Exercises 11–22, list the transformations needed to change the graph of f(t) into the graph of g(t). [See Section 3.4.]

11. $f(t) = \sin t$; $g(t) = \sin t + 3$

12. $f(t) = \cos t$; $g(t) = \cos t - 2$

13. $f(t) = \cos t$; $g(t) = -\cos t$

14. $f(x) = \sin t$; $g(t) = -3 \sin t$

15. $f(t) = \tan t$; $g(t) = \tan t + 5$

16. $f(t) = \tan t$; $g(t) = -\tan t$

17. $f(t) = \cos t$; $g(t) = 3 \cos t$

18. $f(t) = \sin t$; $g(t) = -2 \sin t$

19. $f(t) = \sin t$; $g(t) = 3 \sin t + 2$

20. $f(t) = \cos t$; $g(t) = 5 \cos t + 3$

21. $f(t) = \sin t$; $g(t) = \sin(t - 2)$

22. $f(t) = \cos t$; $g(t) = 3 \cos(t + 2) - 3$

In Exercises 23–30, use the graphs of the trigonometric functions to determine the number *of solutions of the equation between* 0 *and* 2π.

23. $\sin t = 3/5$ [*Hint:* How many points on the graph of $f(t) = \sin t$ between $t = 0$ and $t = 2\pi$ have second coordinate $3/5$?]

24. $\cos t = -1/4$ **25.** $\tan t = 4$ **26.** $\cos t = 2/3$

27. $\sin t = -1/2$

28. $\sin t = k$, where k is a nonzero constant such that $-1 < k < 1$.

29. $\cos t = k$, where k is a constant such that $-1 < k < 1$.

30. $\tan t = k$, where k is any constant.

In Exercises 31–42, use graphs to determine whether the equation could possibly be an identity or definitely is not an identity.

31. $\sin(-t) = -\sin t$ **32.** $\cos(-t) = \cos t$

33. $\sin^2 t + \cos^2 t - 1$ **34.** $\sin(t + u) - = -\sin t$

35. $\sin t = \cos(t - \pi/2)$

36. $\sin^2 t - \tan^2 t = -(\sin^2 t)(\tan^2 t)$

37. $\dfrac{\sin t}{1 + \cos t} = \tan t$

38. $\dfrac{\cos t}{1 - \sin t} = \dfrac{1}{\cos t} + \tan t$

39. $\cos\left(\dfrac{\pi}{2} + t\right) = -\sin t$

40. $\sin\left(\dfrac{\pi}{2} + t\right) = -\cos t$

41. $(1 + \tan t)^2 = \dfrac{1}{\cos t}$

42. $(\cos^2 t - 1)(\tan^2 t + 1) = -\tan^2 t$

Thinkers

Exercises 43–46 explore various ways in which a calculator can produce inaccurate graphs of trigonometric functions. These exercises also provide examples of two functions, with different graphs, whose graphs appear identical in certain viewing windows.

43. Choose a viewing window with $-3 \le y \le 3$ and $0 \le x \le k$, where k is chosen as follows.

Width of Screen	k
95 pixels (TI-83)	188π
127 pixels (TI-86, Casio)	252π
131 pixels (HP-39)	260π
159 pixels (TI-89)	316π

(a) Graph $y = \cos x$ and the constant function $y = 1$ on the same screen. Do the graphs look identical? Are the functions the same?

(b) Use the trace feature to move the cursor along the graph of $y = \cos x$, starting at $x = 0$. For what values of x did the calculator plot points? [*Hint:* $2\pi \approx 6.28$.] Use this information to explain why the two graphs look identical.

44. Using the viewing window in Exercise 43, graph $y = \tan x + 2$ and $y = 2$ on the same screen. Explain why the graphs look identical even though the functions are not the same.

45. The graph of $g(x) = \cos x$ is a series of repeated waves (see Figure 6–42). A full wave (from the peak, down to the trough, and up to the peak again) starts at $x = 0$ and finishes at $x = 2\pi$.

(a) How many full waves will the graph make between $x = 0$ and $x = 502.65 \ (\approx 80 \cdot 2\pi)$?

(b) Graph $g(t) = \cos t$ in a viewing window with $0 \le t \le 502.65$. How many full waves are shown on the graph? Is your answer the same as in part (a)? What's going on?

46. Find a viewing window in which the graphs of $y = \cos x$ and $y = .54$ appear identical. [*Hint:* See the chart in Exercise 43 and note that $\cos 1 \approx .54$.]

Exercises 47–50 provide further examples of functions with different graphs, whose graphs appear identical in certain viewing windows.

47. *Approximating trigonometric functions by polynomials.* For each odd positive integer n, let f_n be the function whose rule is

$$f_n(t) = t - \frac{t^3}{3!} + \frac{t^5}{5!} - \frac{t^7}{7!} + \cdots - \frac{t^n}{n!}.$$

Since the signs alternate, the sign of the last term might be $+$ instead of $-$, depending on what n is. Recall that $n!$ is the product of all integers from 1 to n; for instance, $5! = 1 \cdot 2 \cdot 3 \cdot 4 \cdot 5 = 120$.

(a) Graph $f_7(t)$ and $g(t) = \sin t$ on the same screen in a viewing window with $-2\pi \le t \le 2\pi$. For what values of t does f_7 appear to be a good approximation of g?

(b) What is the smallest value of n for which the graphs of f_n and g appear to coincide in this window? In this case, determine how accurate the approximation is by finding $f_n(2)$ and $g(2)$.

48. For each even positive integer n, let f_n be the function whose rule is

$$f_n(t) = 1 - \frac{t^2}{2!} + \frac{t^4}{4!} - \frac{t^6}{6!} + \frac{t^8}{8!} - \cdots + \frac{t^n}{n!}.$$

(The sign of the last term may be $-$ instead of $+$, depending on what n is.)

(a) In a viewing window with $-2\pi \le t \le 2\pi$, graph f_6, f_{10}, and f_{12}.

(b) Find a value of n for which the graph of f_n appears to coincide (in this window) with the graph of a well-known trigonometric function. What is the function?

49. Find a rational function whose graph appears to coincide with the graph of $h(t) = \tan t$ when

$$-2\pi \le t \le 2\pi.$$

[*Hint:* Exercises 47 and 48.]

50. Find a periodic function whose graph consists of "square waves." [*Hint:* Consider the sum

$$\sin \pi t + \frac{1}{3}\sin 3\pi t + \frac{1}{5}\sin 5\pi t + \frac{1}{7}\sin 7\pi t + \cdots.]$$

51. With your calculator in parametric graphing mode and the range values

$$0 \le t \le 6.28 \qquad -1 \le x \le 6.28 \qquad -2.5 \le y \le 2.5,$$

graph the following two functions on the same screen:

$$x_1 = \cos t, y_1 = \sin t \qquad \text{and} \qquad x_2 = t, y_2 = \cos t.$$

Using the trace feature, move the cursor along the first graph (the unit circle). Stop at a point on the circle, note

the value of t and the x-coordinate of the point. Then switch the trace to the second graph (the cosine function) by using the up or down cursor arrows. The value of t remains the same. How does the y-coordinate of the new point compare with the x-coordinate of the original point on the unit circle? Explain what's going on.

52. (a) Judging from their graphs, which of the functions $f(t) = \sin t$, $g(t) = \cos t$, and $h(t) = \tan t$ appear to be even functions? Which appear to be odd functions?
(b) Confirm your answers in part (a) algebraically by using appropriate identities from Section 6.3.

6.5   Periodic Graphs and Simple Harmonic Motion

We now analyze functions whose rule is of the form

$$f(t) = A\,\sin(bt + c) \qquad \text{or} \qquad g(t) = A\,\cos(bt + c),$$

where A, b, and c are constants. Many periodic phenomena can be modeled by such functions, as we shall see below.

▪▪ PERIOD

The functions $f(t) = \sin t$ and $g(t) = \cos t$ have period 2π, so each of their graphs makes one full wave between 0 and 2π. The sine wave begins on the horizontal axis, rises to height 1, falls to -1, and returns to the axis (Figure 6–50). The cosine wave between 0 and 2π begins at height 1, falls to -1, and rises to height 1 again (Figure 6–51).

Figure 6–50

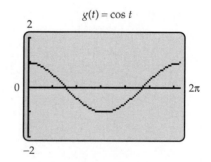

Figure 6–51

GRAPHING EXPLORATION

Graph the following functions, one at a time, in a viewing window with $0 \le t \le 2\pi$. Determine the number of complete waves in each graph and the period of the function (the length of one wave).

$$f(t) = \sin 2t, \qquad g(t) = \cos 3t, \qquad h(x) = \sin 4t, \qquad k(x) = \cos 5t.$$

This exploration suggests the following.

Period

> If $b > 0$, then the graph of either
>
> $$f(t) = \sin bt \qquad \text{or} \qquad g(t) = \cos bt$$
>
> makes b complete waves between 0 and 2π. Hence, each function has period $2\pi/b$.

Although we arrived at this statement by generalizing from several graphs, it can also be explained algebraically.

EXAMPLE 1

The graph of $g(t) = \cos t$ makes one complete wave as t takes values from 0 to 2π. Similarly, the graph of $k(t) = \cos 3t$ will complete one wave as the quantity $3t$ takes values from 0 to 2π. However,

$$3t = 0 \text{ when } t = 0 \qquad \text{and} \qquad 3t = 2\pi \text{ when } t = 2\pi/3.$$

So the graph of $k(t) = \cos 3t$ makes one complete wave between $t = 0$ and $t = 2\pi/3$, as shown in Figure 6–52, and hence k has period $2\pi/3$. Similarly, the graph makes a complete wave from $t = 2\pi/3$ to $t = 4\pi/3$ and another one from $t = 4\pi/3$ to $t = 2\pi$, as shown in Figure 6–52. ■

Figure 6–52

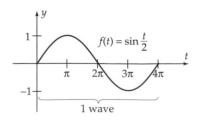

Figure 6–53

EXAMPLE 2

According to the box above, the function $f(t) = \sin \frac{1}{2}t$ has period $\dfrac{2\pi}{1/2} = 4\pi$. Its graph makes *half* a wave from $t = 0$ to $t = 2\pi$ (just as $\sin t$ does from $t = 0$ to $t = \pi$) and the other half of the wave from $t = 2\pi$ to $t = 4\pi$, as shown in Figure 6–53. ■

EXAMPLE 3

Your calculator is incapable of accurately graphing $f(t) = \sin bt$ or $g(t) = \cos bt$ when b is large. For instance, we know that the graph of

$$f(t) = \sin 500t$$

should show 500 complete waves between 0 and 2π. Depending on the model, however, your calculator will produce either garbage (Figure 6–54) or a graph with far fewer than 500 waves (Figure 6–55). For the reason why, see Exercises 55 and 56. ■

Figure 6–54

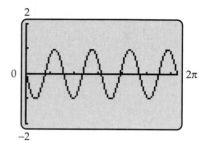

Figure 6–55

▪▪ AMPLITUDE

As we saw in Section 3.4, multiplying the rule of a function by a positive constant has the effect of stretching its graph away from or shrinking it toward the horizontal axis.

EXAMPLE 4

The function $g(t) = 7 \cos 3t$ is just the function $k(t) = \cos 3t$ multiplied by 7. Consequently, the graph of g is just the graph of k (which was obtained in Example 1) stretched away from the horizontal axis by a factor of 7, as shown in Figure 6–56.

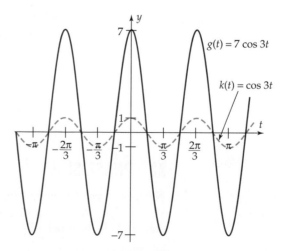

Figure 6–56

Stretching the graph affects only the height of the waves, not the period of the function: Both graphs have period $2\pi/3$, and each full wave has length $2\pi/3$. ∎

The waves of the graph of $g(t) = 7 \cos 3t$ in Figure 6–56 rise 7 units above the t-axis and drop 7 units below the axis. More generally, the waves of the graph of $f(t) = A \sin bt$ or $g(t) = A \cos bt$ move a distance of $|A|$ units above and below the t-axis, and we say that these functions have **amplitude** $|A|$. In summary, we have the following.

Amplitude and Period

If $A \neq 0$ and $b > 0$, then each of the functions

$$f(t) = A \sin bt \qquad \text{or} \qquad g(t) = A \cos bt$$

has amplitude $|A|$ and period $2\pi/b$.

EXAMPLE 5

The function $f(t) = -2 \sin 4t$ has amplitude $|-2| = 2$ and period $2\pi/4 = \pi/2$. So the graph consists of waves of length $\pi/2$ that rise and fall between -2 and 2. But be careful: The waves in the graph of $2 \sin 4t$ (like the waves of $\sin t$) begin at height 0, rise, and then fall. But the graph of $f(t) = -2 \sin 4t$ is the graph of $2 \sin 4t$ reflected in the horizontal axis (see page 171). So its waves start at height 0, move *downward,* and then rise, as shown in Figure 6–57. ∎

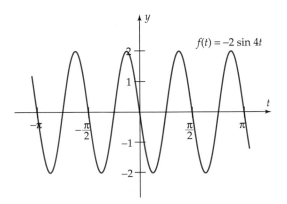

Figure 6–57

⠿ PHASE SHIFT

Next, we consider horizontal shifts. As we saw in Section 3.4, the graph of $\sin(t - 3)$ is the graph of $\sin t$ shifted 3 units to the right, and the graph of $\sin(t + 3)$ is the graph of $\sin t$ shifted 3 units to the left.

EXAMPLE 6

(a) Find a sine function whose graph looks like Figure 6–58.

(b) Find a cosine function whose graph looks like Figure 6–58.

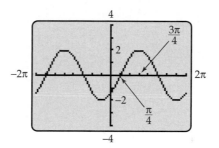

Figure 6–58

SOLUTION

(a) Since each wave has height 2, Figure 6–58 looks like the graph of $2 \sin t$ shifted $\pi/4$ units to the right (so that a sine wave starts at $t = \pi/4$). Since the graph of $2 \sin(t - \pi/4)$ is the graph of $2 \sin t$ shifted $\pi/4$ units to the right (see page 168), we conclude that Figure 6–58 closely resembles the graph of $f(t) = 2 \sin(t - \pi/4)$.

(b) Figure 6–58 also looks like the graph of $2 \cos t$ shifted $3\pi/4$ units to the right (so that a cosine wave starts at $t = 3\pi/4$). Hence, Figure 6–58 could also be the graph of $g(t) = 2 \cos(t - 3\pi/4)$. ■

EXAMPLE 7

(a) Find the amplitude and the period of

$$f(t) = 3 \sin(2t + 5).$$

(b) Do the same for the function $f(t) = A \sin(bt + c)$, where A, b, c are constants.

SOLUTION   The analysis of $f(t) = 3 \sin(2t + 5)$ is in the left-hand column below, and the analysis of the general case $f(t) = A \sin(bt + c)$ is in the right-hand column. Observe that exactly the same procedure is used in both cases: Just change 3 to A, 2 to b, and 5 to c.

(a) Rewrite the rule of $f(t) = 3 \sin(2t + 5)$ as

$$f(t) = 3 \sin(2t + 5) = 3 \sin\left(2\left(t + \frac{5}{2}\right)\right).$$

Thus, the rule of f can be obtained from the rule of the function $k(t) = 3 \sin 2t$ by replacing t with $t + \frac{5}{2}$. Therefore, the graph of f is just the graph of k shifted horizontally $5/2$ units to the left, as shown in Figure 6–59 on the next page.

(b) Rewrite the rule of $f(t) = A \sin(bt + c)$ as

$$f(t) = A \sin(bt + c) = A \sin\left(b\left(t + \frac{c}{b}\right)\right).$$

Thus, the rule of f can be obtained from the rule of the function $k(t) = A \sin bt$ by replacing t with $t + \frac{c}{b}$. Therefore, the graph of f is just the graph of k shifted horizontally by c/b units.

Hence, $f(t) = 3 \sin(2t + 5)$ has the same amplitude as $k(t) = 3 \sin 2t$, namely, 3, and the same period, namely, $2\pi/2 = \pi$.

On the graph of $k(t) = 3 \sin 2t$, a wave begins when $t = 0$. On the graph of

$$f(t) = 3 \sin 2\left(t + \frac{5}{2}\right),$$

the shifted wave begins when $t + 5/2 = 0$, that is, when $t = -5/2$.

Hence, $f(t) = A \sin(bt + c)$ has the same amplitude as $k(t) = A \sin bt$, namely, $|A|$, and the same period, namely, $2\pi/b$.

On the graph of $k(t) = A \sin bt$, a wave begins when $t = 0$. On the graph of

$$f(t) = A \sin b\left(t + \frac{c}{b}\right),$$

the shifted wave begins when $t + c/b = 0$, that is, when $t = -c/b$. ∎

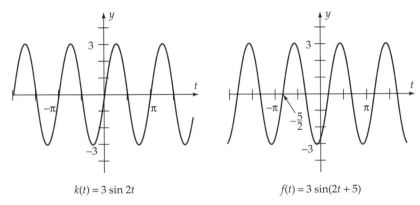

$$k(t) = 3 \sin 2t \qquad\qquad f(t) = 3 \sin(2t + 5)$$

Figure 6–59

We say that the function $f(t) = A \sin(bt + c)$ has **phase shift** $-c/b$. A similar analysis applies to the function $g(t) = \cos(bt + c)$ and leads to this conclusion.

Amplitude, Period, and Phase Shift

If $A \neq 0$ and $b > 0$, then each of the functions

$$f(t) = A \sin(bt + c) \qquad \text{and} \qquad g(t) = A \cos(bt + c)$$

has

amplitude $|A|$,   period $2\pi/b$,   phase shift $-c/b$.

A wave of the graph begins at $t = -c/b$.

EXAMPLE 8

Describe the graph of $g(t) = 2 \cos(3t - 4)$.

SOLUTION   The rule of g can be rewritten as

$$g(t) = 2 \cos(3t + (-4)).$$

This is the case described in the preceding box with $A = 2$, $b = 3$, and $c = -4$. Therefore, the function g has

$$\text{amplitude } |A| = |2| = 2, \qquad \text{period } \frac{2\pi}{b} = \frac{2\pi}{3},$$

$$\text{phase shift } -\frac{c}{b} = -\frac{-4}{3} = \frac{4}{3}.$$

Hence, the graph of g consists of waves of length of $2\pi/3$ that run vertically between 2 and -2. A wave begins at $t = 4/3$.

EXPLORATION

Verify the accuracy of this analysis by graphing $y = 2 \cos(3t - 4)$ in the viewing window with $-2\pi \le t \le 2\pi$ and $-3 \le y \le 3$. ■

Many other types of trigonometric graphs, including those consisting of waves of varying height and length, are considered in Special Topics 6.5.A.

■■ APPLICATIONS

The sine and cosine functions, or variations of them, can be used to describe many different phenomena.

EXAMPLE 9

A wheel of radius 2 centimeters is rotating counterclockwise at 3 radians per second. A free-hanging rod 10 centimeters long is connected to the edge of the wheel at point P and remains vertical as the wheel rotates (Figure 6–60). Assuming that the center of the wheel is at the origin and that P is at $(2, 0)$ at time $t = 0$, find a function that describes the y-coordinate of the tip E of the rod at time t.

SOLUTION   The wheel is rotating at 3 radians per second, so after t seconds, the point P has moved through an angle of $3t$ radians and is 2 units from the origin, as shown in Figure 6–61. By the point-in-the-plane description, the coordinates (x, y) of P satisfy

$$\frac{x}{2} = \cos 3t \qquad \frac{y}{2} = \sin 3t$$

$$x = 2 \cos 3t \qquad y = 2 \sin 3t.$$

Figure 6–61

Figure 6–60

Since E lies 10 centimeters directly below P, its y-coordinate is 10 less than the y-coordinate of P. Hence, the function giving the y-coordinate of E at time t is

$$f(t) = y - 10 = 2 \sin 3t - 10. \qquad ■$$

EXAMPLE 10

Suppose that a weight hanging from a spring is set in motion by an upward push (Figure 6–62) and that it takes 5 seconds for it to move from its equilibrium position to 8 centimeters above, then drop to 8 centimeters below, and finally return to its equilibrium position. [We consider an idealized situation in which the spring has perfect elasticity and friction, air resistance, etc., are negligible.]

Equilibrium position

Figure 6–62

Let $h(t)$ denote the distance of the weight above $(+)$ or below $(-)$ its equilibrium position at time t. Then $h(t)$ is 0 when $t = 0$. As t runs from 0 to 5, $h(t)$ increases from 0 to 8, decreases to -8, and increases again to 0. In the next 5 seconds, it repeats the same pattern, and so on. Thus, the graph of h has some kind of wave shape. Two possibilities are shown in Figure 6–63.

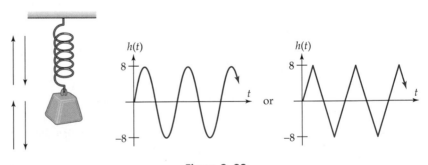

Figure 6–63

Careful physical experiment suggests that the left-hand curve in Figure 6–63, which resembles the sine graphs studied above, is a reasonably accurate model of this process. Facts from physics, calculus, and differential equations show that the rule of the function h is the form $h(t) = A \sin(bt + c)$ for some constants A, b, c. Since the amplitude of h is 8, its period is 5, and its phase shift is 0, the constants A, b, and c must satisfy

$$A = 8, \qquad \frac{2\pi}{b} = 5, \qquad -\frac{c}{b} = 0$$

or, equivalently,

$$A = 8 \qquad b = \frac{2\pi}{5}, \qquad c = 0.$$

Therefore, the motion of the moving spring can be described by the function

$$h(t) = A \sin(bt + c) = 8 \sin\left(\frac{2\pi}{5}t + 0\right) = 8 \sin\frac{2\pi t}{5}. \qquad \blacksquare$$

Motion that can be described by a function of the form $f(t) = A \sin(bt + c)$ or $f(t) = A \cos(bt + c)$ is called **simple harmonic motion.** Many kinds of physical motion are simple harmonic motions. Other periodic phenomena, such as sound waves, are more complicated to describe. Their graphs consist of waves of varying amplitude. Such graphs are discussed in Special Topics 6.5.A.

EXAMPLE 11[*]

The table shows the average monthly temperature in Cleveland, OH, based on data from 1971 to 2000.[†] Since average temperatures are not likely to vary much from year to year, the data essentially repeats the same pattern in subsequent years. So a periodic model is appropriate.

Month	Temperature (°F)	Month	Temperature (°F)
Jan	25.7	Jul	71.9
Feb	28.4	Aug	70.2
Mar	37.5	Sep	63.3
Apr	47.6	Oct	52.2
May	58.5	Nov	41.8
Jun	67.5	Dec	31.1

Figure 6–64

The data for a two-year period is plotted in Figure 6–64 (with $x = 1$ corresponding to January, $x = 2$ to February, and so on). The sine regression feature on a calculator produces this model from the 24 data points:

$$y = 22.7 \sin(.5219x - 2.1842) + 49.5731.$$

The period of this function is $2\pi/.5219 \approx 12.04$ slightly off from the 12-month period we would expect. However, its graph in Figure 6–65 appears to fit the data well. $\blacksquare$

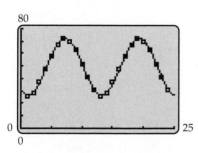

Figure 6–65

---

*Skip this example if you haven't read Sections 2.5 and 5.5 on regression.
†National Climatic Data Center.

✓ EXERCISES 6.5

In Exercises 1–7, state the amplitude, period, and phase shift of the function.

1. $g(t) = 3 \sin(2t - \pi)$ **2.** $h(t) = -4 \cos(3t - \pi/6)$

3. $q(t) = -7 \sin(7t + 1/7)$

4. $g(t) = 97 \cos(14t + 5)$

5. $f(t) = \cos 2\pi t$ **6.** $k(t) = \cos(2\pi t/3)$

7. $p(t) = 6 \cos(3\pi t + 1)$

8. (a) What is the period of $f(t) = \sin 2\pi t$?
  (b) For what values of t (with $0 \le t \le 2\pi$) is $f(t) = 0$?
  (c) For what values of t (with $0 \le t \le 2\pi$) is $f(t) = 1$? or $f(t) = -1$?

In Exercises 9–14, give the rule of a periodic function with the given numbers as amplitude, period, and phase shift (in this order).

9. $3, \pi/4, \pi/5$ **10.** $2, 3, 0$ **11.** $2/3, 1, 0$

12. $4/5, 2, 3$ **13.** $7, 5/3, -\pi/2$ **14.** $19, 4, -5$

In Exercises 15–18, state the rule of a function of the form $f(t) = A \sin bt$ or $g(t) = A \cos bt$ whose graph appears to be identical to the given graph.

15.

16.

17.

18.

In Exercises 19–22,
(a) *State the period of the function.*
(b) *Describe the graph of the function between 0 and 2π.*
(c) *Find a viewing window that accurately shows exactly four complete waves of the graph.*

19. $f(t) = \sin 300t$ **20.** $f(t) = \sin 1200t$

21. $g(t) = \cos 900t$ **22.** $g(t) = \cos 575t$

In Exercises 23–26,
(a) *State the rule of a function of the form*

$$f(t) = A \sin(bt + c)$$

whose graph appears to be identical with the given graph.
(b) *State the rule of a function of the form*

$$g(t) = A \cos(bt + c)$$

whose graph appears to be identical with the given graph.

23.

24.

25.

26.

In Exercises 27–32, sketch a complete graph of the function.

27. $k(t) = -3 \sin t$ **28.** $y(t) = -2 \cos 3t$

29. $p(t) = -\dfrac{1}{2}\sin 2t$ **30.** $q(t) = \dfrac{2}{3} \cos \dfrac{3}{2}t$

31. $h(t) = 3 \sin(2t + \pi/2)$ **32.** $p(t) = 3 \cos(3t - \pi)$

In Exercises 33–36, graph the function over the interval $[0, 2\pi)$ and determine the location of all local maxima and minima. [This can be done either graphically or algebraically.]

33. $f(t) = \dfrac{1}{2} \sin\left(t - \dfrac{\pi}{3}\right)$

34. $g(t) = 2 \sin(2t/3 - \pi/9)$

35. $f(t) = -2 \sin(3t - \pi)$

36. $h(t) = \dfrac{1}{2} \cos\left(\dfrac{\pi}{2}t - \dfrac{\pi}{8}\right) + 1$

In Exercises 37–40, graph $f(t)$ in a viewing window with $-2\pi \le t \le 2\pi$. Use the trace feature to determine constants A, b, c such that the graph of $f(t)$ appears to coincide with the graph of $g(t) = A \sin(bt + c)$.

37. $f(t) = 2 \sin t + 5 \cos t$

38. $f(t) = -3 \sin t + 2 \cos t$

39. $f(t) = 3 \sin(4t + 2) + 2 \cos(4t - 1)$

40. $f(t) = 2 \sin(3t - 5) - 3 \cos(3t + 2)$

In Exercises 41 and 42, explain why there could not possibly be constants $A, b,$ and c such that the graph of $g(t) = A \sin(bt + c)$ coincides with the graph of $f(t)$.

41. $f(t) = \sin 2t + \cos 3t$

42. $f(t) = 2 \sin(3t - 1) + 3 \cos(4t + 1)$

43. The current generated by an AM radio transmitter is given by a function of the form $f(t) = A \sin 2000\pi mt$, where $550 \le m \le 1600$ is the location on the broadcast dial and t is measured in seconds. For example, a station at 980 on the AM dial has a function of the form

$$f(t) = A \sin 2000\pi(980)t = A \sin 1,960,000\pi t.$$

Sound information is added to this signal by varying (modulating) A, that is, by changing the amplitude of the waves being transmitted. (*AM* means "amplitude modulation.") For a station at 980 on the dial, what is the period of function f? What is the frequency (number of complete waves per second)?

44. Find the function f (as in Exercise 43), its period, and its frequency for a station at 1440 on the dial.

45. The original Ferris wheel, built by George Ferris for the Columbian Exposition of 1893, was much larger and slower than its modern counterparts: It had a diameter of 250 feet and contained 36 cars, each of which held 60 people; it made one revolution every 10 minutes. Imagine that the Ferris wheel revolves counterclockwise in the x-y plane with its center at the origin. A car had coordinates $(125, 0)$ at time $t = 0$. Find the rule of a function that gives the y-coordinate of the car at time t.

46. Do Exercise 45 if the wheel turns at 2 radians per minute and the car is at $(0, -125)$ at time $t = 0$.

47. A circular wheel of radius 1 foot rotates counterclockwise. A 4-foot-long rod has one end attached to the edge of this wheel and the other end to the base of a piston (see the figure). It transfers the rotary motion of the wheel into a back-and-forth linear motion of the piston. If the wheel is rotating at 10 revolutions per second, point W is at $(1, 0)$ at time $t = 0$, and point P is always on the x-axis, find the rule of a function that gives the x-coordinate of P at time t.

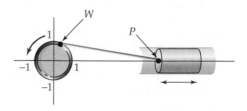

48. Do Exercise 47 if the wheel has a radius of 2 feet, rotates at 50 revolutions per second, and is at $(2, 0)$ when $t = 0$.

In Exercises 49–52, suppose there is a weight hanging from a spring (under the same idealized conditions as described in Example 10). The weight is given a push to start it moving. At any time t, let h(t) be the height (or depth) of the weight above (or below) its equilibrium point. Assume that the maximum distance the weight moves in either direction from the equilibrium point is 6 centimeters and that it moves through a complete cycle every 4 seconds. Express h(t) in terms of the sine or cosine function under the stated conditions.

49. Initial push is *upward* from the equilibrium point.

50. Initial push is *downward* from the equilibrium point. [*Hint:* What does the graph of $A \sin bt$ look like when $A < 0$?]

51. Weight is pulled 6 centimeters above equilibrium, and the initial movement (at $t = 0$) is downward. [*Hint:* Think cosine.]

52. Weight is pulled 6 centimeters below equilibrium, and the initial movement is upward.

53. A pendulum swings uniformly back and forth, taking 2 seconds to move from the position directly above point A to the position directly above point B.

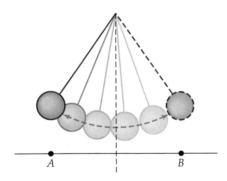

The distance from A to B is 20 centimeters. Let $d(t)$ be the horizontal distance from the pendulum to the (dashed) center line at time t seconds (with distances to the right of the line measured by positive numbers and distances to the left by negative ones). Assume that the pendulum is on the center line at time $t = 0$ and moving to the right. Assume that the motion of the pendulum is simple harmonic motion. Find the rule of the function $d(t)$.

54. The diagram shows a merry-go-round that is turning counterclockwise at a constant rate, making 2 revolutions in 1 minute. On the merry-go-round are horses A, B, C, and D at 4 meters from the center and horses E, F, and G at 8 meters from the center. There is a function $a(t)$ that gives the distance the horse A is from the y-axis (this is the x-coordinate of the position A is in) as a function of time t

(measured in minutes). Similarly, $b(t)$ gives the x-coordinate for B as a function of time, and so on. Assume that the diagram shows the situation at time $t = 0$.

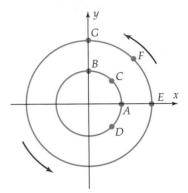

(a) Which of the following functions does $a(t)$ equal?

$$4 \cos t, \quad 4 \cos \pi t, \quad 4 \cos 2t, \quad 4 \cos 2\pi t,$$
$$4 \cos \left(\tfrac{1}{2}t\right), \quad 4 \cos ((\pi/2)t), \quad 4 \cos 4\pi t$$

Explain.

(b) Describe the functions $b(t)$, $c(t)$, $d(t)$, and so on using the cosine function:

$$b(t) = \underline{\quad}, \ c(t) = \underline{\quad}, \ d(t) = \underline{\quad}.$$
$$e(t) = \underline{\quad}, f(t) = \underline{\quad}, \ g(t) = \underline{\quad}.$$

(c) Suppose the x-coordinate of a horse S is given by the function $4 \cos(4\pi t - (5\pi/6))$ and the x-coordinate of another horse T is given by $8 \cos(4\pi t - (\pi/3))$. Where are these horses located in relation to the rest of the horses? Mark the positions of T and S at $t = 0$ into the figure.

Exercises 55–56 explore various ways in which a calculator can produce inaccurate or misleading graphs of trigonometric functions.

55. (a) If you were going to draw a rough picture of a full wave of the sine function by plotting some points and connecting them with straight-line segments, approximately how many points would you have to plot?

(b) If you were drawing a rough sketch of the graph of $f(t) = \sin 100t$ when $0 \le t \le 2\pi$, according to the method in part (a), approximately how many points would have to be plotted?

(c) How wide (in pixels) is your calculator screen? Your answer to this question is the maximum number of points that your calculator plots when graphing any function.

(d) Use parts (a)–(c) to explain why your calculator cannot possibly produce an accurate graph of $f(t) = \sin 100t$ in any viewing window with $0 \le t \le 2\pi$.

56. (a) Using a viewing window with $0 \leq t \leq 2\pi$, use the trace feature to move the cursor along the horizontal axis. [On some calculators, it may be necessary to graph $y = 0$ to do this.] What is the distance between one pixel and the next (to the nearest hundredth)?

(b) What is the period of $f(t) = \sin 300t$? Since the period is the length of one full wave of the graph, approximately how many waves should there be between two adjacent pixels? What does this say about the possibility of your calculator's producing an accurate graph of this function between 0 and 2π?

57. The table below shows the number of unemployed people in the labor force (in millions) for 1984–2002.[*]

(a) Sketch a scatter plot of the data, with $x = 0$ corresponding to 1980.

(b) Does the data appear to be periodic? If so, find an appropriate model.

(c) Do you think this model is likely to be accurate much beyond 2002? Why?

Year	Unemployed	Year	Unemployed
1984	8.539	1994	7.996
1985	8.312	1995	7.404
1986	8.237	1996	7.236
1987	7.425	1997	6.739
1988	6.701	1998	6.210
1989	6.528	1999	5.880
1990	7.047	2000	5.655
1991	8.628	2001	6.742
1992	9.613	2002	8.234
1993	8.940		

In Exercises 58 and 59, do the following.

(a) Use 12 data points (with $x = 1$ corresponding to January) to find a periodic model of the data.

(b) What is the period of the function found in part (a)? Is this reasonable?

(c) Plot 24 data points (two years) and graph the function from part (a) on the same screen. Is the function a good model in the second year?

(d) Use the 24 data points in part (c) to find another periodic model for the data.

(e) What is the period of the function in part (d)? Does its graph fit the data well?

58. The table shows the average monthly temperature in Chicago, IL, based on data from 1971 to 2000.[†]

Month	Temperature (°F)
Jan	22.0
Feb	27.0
Mar	37.3
Apr	47.8
May	58.7
Jun	68.2
Jul	73.3
Aug	71.7
Sep	63.8
Oct	52.1
Nov	39.3
Dec	27.4

59. The table shows the average monthly precipitation (in inches) in San Francisco, CA, based on data from 1971 to 2000.[‡]

Month	Precipitation
Jan	4.45
Feb	4.01
Mar	3.26
Apr	1.17
May	.38
Jun	.11
Jul	.03
Aug	.07
Sep	.2
Oct	1.04
Nov	2.49
Dec	2.89

[†]National Climatic Data Center.
[‡]National Climatic Data Center.

---

[*]U.S. Bureau of Labor Statistics.

Thinkers

60. On the basis of the results of Exercises 37–40, under what conditions on the constants a, k, h, d, r, s does it appear that the graph of

$$f(t) = a \sin(kt + h) + d \cos(rt + s)$$

coincides with the graph of the function

$$g(t) = A \sin(bt + c)?$$

61. A grandfather clock has a pendulum length of k meters and its swing is given (as in Exercise 53) by the function $f(t) = .25 \sin(\omega t)$, where

$$\omega = \sqrt{\frac{9.8}{k}}.$$

(a) Find k such that the period of the pendulum is 2 seconds.

(b) The temperature in the summer months causes the pendulum to increase its length by .01%. How much time will the clock lose in June, July, and August? [*Hint:* These three months have a total of 92 days (7,948,800 seconds). If k is increased by .01%, what is $f(2)$?]

6.5.A *SPECIAL TOPICS* Other Trigonometric Graphs

A graphing calculator or computer enables you to explore with ease a wide variety of trigonometric functions.

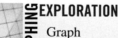**GRAPHING EXPLORATION**

Graph

$$g(t) = \cos t \quad \text{and} \quad f(t) = \sin(t + \pi/2)$$

on the same screen. Is there any apparent difference between the two graphs?

This exploration suggests that the equation $\cos t = \sin(t + \pi/2)$ is an identity and hence that the graph of the cosine function can be obtained by horizontally shifting the graph of the sine function. This is indeed the case, as will be proved in Section 7.2. Consequently, every graph in Section 6.5 is actually the graph of a function of the form $f(t) = A \sin(bt + c)$. In fact, considerably more is true.

Sinusoidal Graphs

If b, D, E, r, s are constants, then the graph of the function

$$g(t) = D \sin(bt + r) + E \cos(bt + s)$$

is a sine curve: There exist constants A and c such that

$$D \sin(bt + r) + E \cos(bt + s) = A \sin(bt + c).$$

EXAMPLE 1

Estimate the constants A, b, c such that

$$A \sin(bt + c) = 4 \sin(3t + 2) + 2 \cos(3t - 4).$$

SOLUTION   The function $g(t) = 4 \sin(3t + 2) + 2 \cos(3t - 4)$ has period $2\pi/3$ because this is the period of both $\sin(3t + 2)$ and $\cos(3t - 4)$. The function $f(t) = A \sin(bt + c)$ has period $2\pi/b$. So we must have

$$\frac{2\pi}{b} = \frac{2\pi}{3}, \quad \text{or equivalently,} \quad b = 3.$$

Using a maximum finder and a root finder on the graph of

$$g(t) = 4 \sin(3t + 2) + 2 \cos(3t - 4)$$

in Figure 6–66, we see that the maximum height (amplitude) of a wave is approximately 3.94 and that a sine wave begins at approximately $t = -.84$. Therefore, the graph has (approximate) amplitude 3.94 and phase shift $-.84$. Since $b = 3$ and $f(t) = A \sin(bt + c)$ has amplitude $|A|$ and phase shift $-c/b = -c/3$, we have $A \approx 3.94$ and

$$-\frac{c}{3} \approx -.84 \quad \text{or, equivalently,} \quad c \approx 3(.84) = 2.52.$$

Therefore,

$$3.94 \sin(3t + 2.52) \approx 4 \sin(3t + 2) + 2 \cos(3t - 4).$$

Figure 6–66

EXPLORATION

Graphically confirm this fact by graphing

$$f(t) = 3.94 \sin(3t + 2.52)$$

and

$$g(t) = 4 \sin(3t + 2) + 2 \cos(3t - 4)$$

on the same screen. Do the graphs appear identical?   ∎

In the preceding examples, the variable t had the same coefficient in both the sine and cosine term of the function's rule. When this is not the case, the graph will consist of waves of varying size and shape, as you can readily illustrate.

EXPLORATION

Graph each of the following functions separately in the viewing window with $-2\pi \le t \le 2\pi$ and $-6 \le y \le 6$.

$$f(t) = \sin 3t + \cos 2t, \quad g(t) = -2 \sin(3t + 5) + 4 \cos(t + 2),$$

$$h(t) = 2 \sin 2t - 3 \cos 3t.$$

EXAMPLE 2

Find a complete graph of

$$f(t) = 4 \sin 100\pi t + 2 \cos 40\pi t.$$

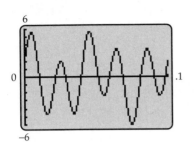

Figure 6–67

SOLUTION   If you graph f in a window with $-2\pi \leq t \leq 2\pi$, you will get "garbage" on the screen (try it!). Trial and error might lead to a viewing window that shows a readable graph, but the graph might not be accurate. A better procedure is to note that this is a periodic function. Hence, we need only graph it over one period to have a complete graph. The period of $4 \sin 100\pi t$ is

$$\frac{2\pi}{100\pi} = \frac{1}{50} = .02,$$

and the period of $2 \cos 40\pi t$ is $2\pi/40\pi = 1/20 = .05$. So the period of their sum will be the least common multiple of .02 and .05, which is .1.[*] By graphing f in the viewing window with $0 \leq t \leq .1$ and $-6 \leq y \leq 6$, we obtain the complete graph in Figure 6–67.   ■

■■   DAMPED AND COMPRESSED TRIGONOMETRIC GRAPHS

Many physical situations can be described by functions whose graphs consist of waves of different heights. Other situations (for instance, sound waves in FM radio transmission) are modeled by functions whose graphs consist of waves of uniform height and varying frequency. Here are some examples of such functions.

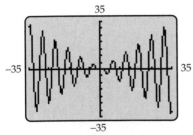

Figure 6–68

EXAMPLE 3

Explain why the graph of $f(t) = t \cos t$ in Figure 6–68 has the shape it does.

SOLUTION   The graph appears to consist of waves that get larger and larger as you move away from the origin. To explain this situation, we analyze the situation algebraically. We know that

$$-1 \leq \cos t \leq 1 \quad \text{for every } t.$$

If we multiply each term of this inequality by t and remember the rules for changing the direction of inequalities when multiplying by negatives, we see that

$$-t \leq t \cos t \leq t \quad \text{when } t \geq 0$$

and

$$-t \geq t \cos t \geq t \quad \text{when } t < 0.$$

In graphical terms, this means that the graph of $f(x) = t \cos t$ lies between the straight lines $y = t$ and $y = -t$, with the waves growing larger or smaller to fill this space. The graph touches the lines $y = \pm t$ exactly when $t \cos t = \pm t$, that is, when $\cos t = \pm 1$. This occurs when $t = 0, \pm\pi, \pm2\pi, \pm3\pi, \ldots$.

---

[*]The multiples of .02 are .02, .04, .06, .08, .10, . . . , and the multiples of .05 are .05, .10, . . .
Hence, the smallest common multiple is .10.

GRAPHING EXPLORATION

Illustrate this analysis by graphing $f(t) = t \cos t$, $y = t$, and $y = -t$ on the same screen. ■

EXAMPLE 4

No single viewing window gives a completely readable graph of $g(t) = .5^t \sin t$ (try some). To the left of the *y*-axis, the graph gets quite large, but to the right, it almost coincides with the horizontal axis. To get a better mental picture, note that $.5^t > 0$ for every *t*. Multiplying each term of the known inequality $-1 \le \sin t \le 1$ by $.5^t$, we see that

$$-.5^t \le .5^t \sin t \le .5^t \quad \text{for every } t.$$

Hence, the graph of *g* lies between the graphs of the exponential functions $y = -.5^t$ and $y = .5^t$, which are shown in Figure 6–69. The graph of *g* will consist of sine waves rising and falling between those exponential graphs, as indicated in the sketch in Figure 6–70 (which is not to scale).

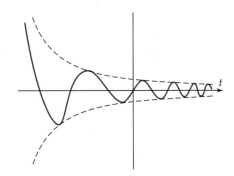

Figure 6–69   Figure 6–70

The best you can do with a calculator is to look at various viewing windows in which a portion of the graph is readable.

GRAPHING EXPLORATION

Find viewing windows that clearly show the graph of $g(t) = .5^t \sin t$ in each of these ranges.

$$-2\pi \le t \le 0, \qquad 0 \le t \le 2\pi, \qquad 2\pi \le t \le 4\pi. \quad ■$$

EXAMPLE 5

If you graph $f(t) = \sin(\pi/t)$ in a wide viewing window such as Figure 6–71, it is clear that the horizontal axis is an asymptote of the graph.[*] Near the origin, however, the graph is not very readable, even in a very narrow viewing window like Figure 6–72.

Figure 6–71   Figure 6–72

To understand the behavior of f near the origin, consider what happens as you move left from $t = 1/2$ to $t = 0$:

$$\text{As } t \text{ goes from } \frac{1}{2} \text{ to } \frac{1}{4}, \text{ then } \frac{\pi}{t} \text{ goes from } \frac{\pi}{1/2} = 2\pi \text{ to } \frac{\pi}{1/4} = 4\pi.$$

As π/t takes all values from 2π to 4π, the graph of $f(t) = \sin(\pi/t)$ makes one complete sine wave. Similarly,

$$\text{As } t \text{ goes from } \frac{1}{4} \text{ to } \frac{1}{6}, \text{ then } \frac{\pi}{t} \text{ goes from } \frac{\pi}{1/4} = 4\pi \text{ to } \frac{\pi}{1/6} = 6\pi.$$

As π/t takes all values from 4π to 6π, the graph of $f(t) = \sin(\pi/t)$ makes another complete sine wave. The same pattern continues, so the graph of f makes a complete wave from $t = 1/2$ to $t = 1/4$, another from $t = 1/4$ to $t = 1/6$, another from $t = 1/6$ to $t = 1/8$, and so on. A similar phenomenon occurs as t takes values between $-1/2$ and 0. Consequently, the graph of f near 0 oscillates infinitely often between -1 and 1, with the waves becoming more and more compressed as t gets closer to 0, as indicated in Figure 6–73. Since the function is not defined at $t = 0$, the left and right halves of the graph are not connected. ∎

Graph oscillates
infinitely often here

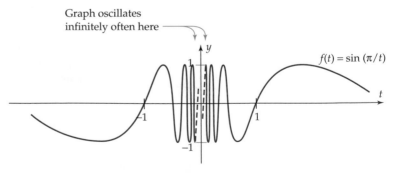

Figure 6–73

---

[*]This can also be demonstrated algebraically: When t is very large in absolute value, then π/t is very close to 0 by the Big-Little Principle, and hence, $\sin(\pi/t)$ is very close to 0 as well.

EXAMPLE 6

Describe the graph of $g(t) = \cos e^t$.

SOLUTION When t is negative, then e^t is very close to 0 (why?), and hence, $\cos e^t$ is very close to 1. Therefore, the horizontal line $y = 1$ is an asymptote of the half of the graph to the left of the origin. As t takes increasing positive values, the corresponding values of e^t increase at a much faster rate (remember exponential growth). For instance, as t goes from 0 to 2π,

$$e^t \text{ goes from } e^0 = 1 \text{ to } e^{2\pi} \approx 535.5 \approx 170\pi = 85(2\pi).$$

Consequently, $\cos e^t$ runs through 85 periods, that is, the graph of g makes 85 full waves between 0 and 2π. As t gets larger, the graph of g makes waves at a faster and faster rate.

GRAPHING EXPLORATION

To see how compressed the waves become, graph $g(t)$ in three viewing windows, with

$$0 \le t \le 3.5, \qquad 4.5 \le t \le 5, \qquad 6 \le t \le 6.2,$$

and note how the number of waves increases in each succeeding window, even though the widths of the windows are getting smaller. ∎

EXERCISES 6.5.A

In Exercises 1–6, estimate constants A, b, c such that

$$f(t) = A \sin(bt + c).$$

1. $f(t) = \sin t + 2 \cos t$ **2.** $f(t) = 3 \sin t + 2 \cos t$

3. $f(t) = 2 \sin 4t - 5 \cos 4t$

4. $f(t) = 3 \sin(2t - 1) + 4 \cos(2t + 3)$

5. $f(t) = -5 \sin(3t + 2) + 2 \cos(3t - 1)$

6. $f(t) = .3 \sin(2t + 4) - .4 \cos(2t - 3)$

In Exercises 7–16, find a viewing window that shows a complete graph of the function.

7. $g(t) = (5 \sin 2t)(\cos 5t)$ **8.** $h(t) = e^{\sin t}$

9. $f(t) = t/2 + \cos 2t$

10. $g(t) = \sin\left(\dfrac{t}{3} - 2\right) + 2 \cos\left(\dfrac{t}{4} - 2\right)$

11. $h(t) = \sin 300t + \cos 500t$

12. $f(t) = 3 \sin(200t + 1) - 2 \cos(300t + 2)$

13. $g(t) = -5 \sin(250\pi t + 5) + 2 \cos(400\pi t - 7)$

14. $h(t) = 4 \sin(600\pi t + 3) - 6 \cos(500\pi t - 3)$

15. $f(t) = 4 \sin .2\pi t - 5 \cos .4\pi t$

16. $g(t) = 6 \sin .05\pi t + 2 \cos .04\pi t$

In Exercises 17–24, describe the graph of the function verbally (including such features as asymptotes, undefined points, amplitude and number of waves between 0 and 2π, etc.) as in Examples 4–6. Find viewing windows that illustrate the main features of the graph.

17. $g(t) = \sin e^t$ **18.** $h(t) = \dfrac{\cos 2t}{1 + t^2}$

19. $f(t) = \sqrt{|t|} \cos t$ **20.** $g(t) = e^{-t^2/8} \sin 2\pi t$

21. $h(t) = \dfrac{1}{t} \sin t$ **22.** $f(t) = t \sin \dfrac{1}{t}$

23. $g(t) = \ln |\cos t|$ **24.** $h(t) = \ln |\sin t + 1|$

6.6   Other Trigonometric Functions

This section is divided into three parts, each of which may be covered earlier, as shown in the table.

Subsection of Section 6.6	May be covered at the end of
Definitions, Alternate Descriptions	Section 6.2
Algebra and Identities	Section 6.3
Graphs	Section 6.4

■■ DEFINITIONS

The three remaining trigonometric functions are defined in terms of sine and cosine, as follows.

Definition of Cotangent, Secant, and Cosecant Functions

Name of Function	Value of Function at t Is Denoted	Rule of Function
cotangent	$\cot t$	$\cot t = \dfrac{\cos t}{\sin t}$
secant	$\sec t$	$\sec t = \dfrac{1}{\cos t}$
cosecant	$\csc t$	$\csc t = \dfrac{1}{\sin t}$

The domain of each function consists of all real numbers for which the denominator is not 0. The graphs of the sine and cosine function in Section 6.4 show that $\sin t = 0$ only when t is an integer multiple of π and that $\cos t = 0$ only when t is an odd integer multiple of $\pi/2$. Therefore,

The domain of the cotangent and cosecant functions consists of all real numbers *except* $0, \pm\pi, \pm2\pi, \pm3\pi$, and so on.

The domain of the secant function consists of all real numbers *except* $\pm\pi/2, \pm3\pi/2, \pm5\pi/2$, and so on.

The values of the secant and cosecant functions may be approximated on a calculator by using the SIN and COS keys. For instance,

$$\sec 7 = \frac{1}{\cos 7} \approx 1.3264 \qquad \text{and} \qquad \csc 18.5 = \frac{1}{\sin 18.5} \approx -2.9199.$$

To evaluate the cotangent function, either use the SIN or COS keys or use the TAN key and this identity (which is proved below): $\cot t = 1/\tan t$. For example,

$$\cot(-5) = \frac{\cos(-5)}{\sin(-5)} \approx .2958 \qquad \text{or} \qquad \cot(-5) = \frac{1}{\tan(-5)} \approx .2958.$$

CAUTION

The calculator keys labeled SIN^{-1}, COS^{-1}, and TAN^{-1} do *not* denote the functions $1/\sin t$, $1/\cos t$, and $1/\tan t$. For instance, if you key in

  COS^{-1} 7 ENTER

you will get an error message, not the number sec 7, and if you key in

  TAN^{-1} −5 ENTER

you will obtain −1.3734, which is *not* cot(−5).

These new trigonometric functions may be evaluated exactly at any integer multiple of $\pi/3$, $\pi/4$, or $\pi/6$.

EXAMPLE 1

Evaluate the cotangent, secant, and cosecant functions at $t = \pi/3$.

SOLUTION   Let P be the point where the terminal side of an angle of $\pi/3$ radians in standard position meets the unit circle (Figure 6–74). Draw the vertical line from P to the x-axis, forming a right triangle with hypotenuse 1, angles of $\pi/3$ and $\pi/6$ radians, and sides of lengths of $1/2$ and $\sqrt{3}/2$ as explained on page 426. Then P has coordinates $(1/2, \sqrt{3}/2)$, and by definition,

$$\sin \frac{\pi}{3} = y\text{-coordinate of } P = \sqrt{3}/2,$$

$$\cos \frac{\pi}{3} = x\text{-coordinate of } P = 1/2.$$

Therefore,

$$\csc \frac{\pi}{3} = \frac{1}{\sin(\pi/3)} = \frac{1}{\sqrt{3}/2} = \frac{2}{\sqrt{3}} = \frac{2\sqrt{3}}{3},$$

$$\sec \frac{\pi}{3} = \frac{1}{\cos(\pi/3)} = \frac{1}{1/2} = 2,$$

$$\cot \frac{\pi}{3} = \frac{\cos(\pi/3)}{\sin(\pi/3)} = \frac{1/2}{\sqrt{3}/2} = \frac{1}{\sqrt{3}} = \frac{\sqrt{3}}{3}. \quad \blacksquare$$

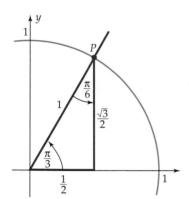

Figure 6–74

▐▐ ALTERNATE DESCRIPTIONS

The point-in-the-plane description of sine, cosine, and tangent readily extends to these new functions.

Point-in-the-Plane Description

Let t be a real number and (x, y) any point (except the origin) on the terminal side of an angle of t radians in standard position. Let

$$r = \sqrt{x^2 + y^2}.$$

Then,

$$\cot t = \frac{x}{y}, \qquad \sec t = \frac{r}{x}, \qquad \csc t = \frac{r}{y}$$

for each number t in the domain of the given function.

These statements are proved by using the similar descriptions of sine and cosine. For instance,

$$\cot t = \frac{\cos t}{\sin t} = \frac{x/r}{y/r} = \frac{x}{y}.$$

The proofs of the other statements are similar.

EXAMPLE 2

Evaluate all six trigonometric functions at $t = 3\pi/4$.

SOLUTION   The terminal side of an angle of $3\pi/4$ radians in standard position lies on the line $y = -x$, as shown in Figure 6–75. We shall use the point $(-1, 1)$ on this line to compute the function values. In this case,

$$r = \sqrt{x^2 + y^2} = \sqrt{(-1)^2 + 1^2} = \sqrt{2}.$$

Therefore,

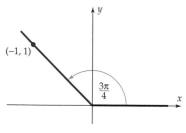

(-1, 1)

$\dfrac{3\pi}{4}$

Figure 6–75

$$\sin \frac{3\pi}{4} = \frac{y}{r} = \frac{1}{\sqrt{2}} = \frac{\sqrt{2}}{2} \qquad \cos \frac{3\pi}{4} = \frac{x}{r} = \frac{-1}{\sqrt{2}} = \frac{-\sqrt{2}}{2}$$

$$\tan \frac{3\pi}{4} = \frac{y}{x} = \frac{1}{-1} = -1 \qquad \csc \frac{3\pi}{4} = \frac{r}{y} = \frac{\sqrt{2}}{1} = \sqrt{2}$$

$$\sec \frac{3\pi}{4} = \frac{r}{x} = \frac{\sqrt{2}}{-1} = -\sqrt{2} \qquad \cot \frac{3\pi}{4} = \frac{x}{y} = \frac{-1}{1} = -1. \qquad ∎$$

■■ ALGEBRA AND IDENTITIES

We begin by noting the relationship between the cotangent and tangent functions.

***Reciprocal
Identities***

The cotangent and tangent functions are reciprocals; that is,

$$\cot t = \frac{1}{\tan t} \qquad \text{and} \qquad \tan t = \frac{1}{\cot t}$$

for every number t in the domain of both functions.

The proof of these facts comes directly from the definitions; for instance,

$$\cot t = \frac{\cos t}{\sin t} = \frac{1}{\dfrac{\sin t}{\cos t}} = \frac{1}{\tan t}.$$

***Period of Secant,
Cosecant, Cotangent***

The secant and cosecant functions are periodic with period 2π and the cotangent function is periodic with period π. In symbols,

$$\sec(t + 2\pi) = \sec t, \qquad \csc(t + 2\pi) = \csc t,$$

$$\cot(t + \pi) = \cot t$$

for every number t in the domain of the given function.

The proof of these statements uses the fact that each of these functions is the reciprocal of a function whose period is known. For instance,

$$\csc(t + 2\pi) = \frac{1}{\sin(t + 2\pi)} = \frac{1}{\sin t} = \csc t,$$

$$\cot(t + \pi) = \frac{1}{\tan(t + \pi)} = \frac{1}{\tan t} = \cot t.$$

The other details are left as an exercise.

Pythagorean
Identities

For every number t in the domain of both functions,

$$1 + \tan^2 t = \sec^2 t$$

and

$$1 + \cot^2 t = \csc^2 t.$$

Proof   By the definitions of the functions and the Pythagorean identity, $\sin^2 t + \cos^2 t = 1$.

$$1 + \tan^2 t = 1 + \frac{\sin^2 t}{\cos^2 t} = \frac{\cos^2 t + \sin^2 t}{\cos^2 t} = \frac{1}{\cos^2 t} = \left(\frac{1}{\cos t}\right)^2 = \sec^2 t.$$

The second identity is proved similarly.   ■

EXAMPLE 3

Simplify the expression $\dfrac{30 \cos^3 t \sin t}{6 \sin^2 t \cos t}$, assuming that $\sin t \neq 0$, $\cos t \neq 0$.

SOLUTION

$$\frac{30 \cos^3 t \sin t}{6 \sin^2 t \cos t} = \frac{5 \cos^3 t \sin t}{\cos t \sin^2 t} = \frac{5 \cos^2 t}{\sin t} = 5\frac{\cos t}{\sin t}\cos t = 5 \cot t \cos t.$$ ■

EXAMPLE 4

Assume that $\cos t \neq 0$ and simplify $\cos^2 t + \cos^2 t \tan^2 t$.

SOLUTION

$$\cos^2 t + \cos^2 t \tan^2 t = \cos^2 t(1 + \tan^2 t) = \cos^2 t \sec^2 t = \cos^2 t \cdot \frac{1}{\cos^2 t} = 1.$$ ■

EXAMPLE 5

If $\tan t = 3/4$ and $\sin t < 0$, find $\cot t$, $\cos t$, $\sin t$, $\sec t$, and $\csc t$.

SOLUTION   First we have $\cot t = 1/\tan t = 1/(3/4) = 4/3$. Next we use the Pythagorean identity to obtain

$$\sec^2 t = 1 + \tan^2 t = 1 + \left(\frac{3}{4}\right)^2 = 1 + \frac{9}{16} = \frac{25}{16}$$

$$\sec t = \pm\sqrt{\frac{25}{16}} = \pm\frac{5}{4}$$

$$\frac{1}{\cos t} = \pm\frac{5}{4} \quad \text{or, equivalently,} \quad \cos t = \pm\frac{4}{5}.$$

Since $\sin t$ is given as negative and $\tan t = \sin t/\cos t$ is positive, $\cos t$ must be negative. Hence, $\cos t = -4/5$. Consequently,

$$\frac{3}{4} = \tan t = \frac{\sin t}{\cos t} = \frac{\sin t}{(-4/5)}$$

so

$$\sin t = \left(-\frac{4}{5}\right)\left(\frac{3}{4}\right) = -\frac{3}{5}.$$

Therefore,

$$\sec t = \frac{1}{\cos t} = \frac{1}{(-4/5)} = -\frac{5}{4} \quad \text{and} \quad \csc t = \frac{1}{\sin t} = \frac{1}{(-3/5)} = -\frac{5}{3}. \quad \blacksquare$$

⊞ GRAPHS

The general shape of the graph of $g(t) = \sec t$ can be determined by using the graph of the cosine function and the fact that $\sec t = 1/\cos t$. First of all, $\sec t$ is not defined whenever $\cos t = 0$, that is, when $t = \pm\pi/2, \pm3\pi/2, \pm5\pi/2$, etc. When $\cos t$ is a number near 1 or -1, then so is $\sec t = 1/\cos t$ and their graphs are close together. When $\cos t$ is near 0 (so its graph is close to the x-axis), then $\sec t = 1/\cos t$ is very large in absolute value[*] (so its graph is far from the x-axis), as shown in Figure 6–76. The graph of $g(t) = \sec t$ has vertical asymptotes at those points where the function is not defined, that is $x = \pi/2$, $x = -\pi/2$, $x = 3\pi/2$, $x = -3\pi/2$, etc.

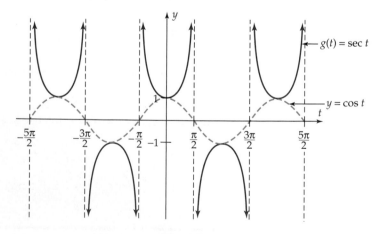

Figure 6–76

---

*See the Big-Little Principle on page 272.

The graphs of $h(t) = \csc t = 1/\sin t$ and $f(t) = \cot t = 1/\tan t$ can be obtained in a similar fashion (Figure 6–77).

Figure 6–77

 EXERCISES 6.6

Note: *The arrangement of the exercises corresponds to the subsections of this section.*

Definitions and Alternate Descriptions

In Exercises 1–6, determine the quadrant containing the terminal side of an angle of t radians in standard position under the given conditions.

1. $\cos t > 0$ and $\sin t < 0$

2. $\sin t < 0$ and $\tan t > 0$

3. $\sec t < 0$ and $\cot t < 0$

4. $\csc t < 0$ and $\sec t > 0$

5. $\sec t > 0$ and $\cot t < 0$

6. $\sin t > 0$ and $\sec t < 0$

In Exercises 7–16, evaluate all six trigonometric functions at t, where the given point lies on the terminal side of an angle of t radians in standard position.

7. $(3, 4)$

8. $(0, 6)$

9. $(-5, 12)$

10. $(-2, -3)$

11. $(-1/5, 1)$

12. $(4/5, -3/5)$

13. $(\sqrt{2}, \sqrt{3})$

14. $(-2\sqrt{3}, \sqrt{3})$

15. $(1 + \sqrt{2}, 3)$

16. $(1 + \sqrt{3}, 1 - \sqrt{3})$

In Exercises 17–19, evaluate all six trigonometric functions at the given number without using a calculator.

17. $\dfrac{4\pi}{3}$ **18.** $-\dfrac{7\pi}{6}$ **19.** $\dfrac{7\pi}{4}$

20. Fill in the missing entries in the following table. Give exact answers, not decimal approximations.

t	0	$\dfrac{\pi}{6}$	$\dfrac{\pi}{4}$	$\dfrac{\pi}{3}$	$\dfrac{\pi}{2}$	$\dfrac{2\pi}{3}$	$\dfrac{3\pi}{4}$	$\dfrac{5\pi}{6}$	π	$\dfrac{3\pi}{2}$
$\sin t$										
$\cos t$										
$\tan t$				—						—
$\cot t$	—							—		
$\sec t$					—					—
$\csc t$	—								—	

Algebra and Identities

21. Find the average rate of change of $f(t) = \cot t$ from $t = 1$ to $t = 3$.

22. Find the average rate of change of $g(t) = \csc t$ from $t = 2$ to $t = 3$.

23. (a) Find the average rate of change of $f(t) = \tan t$ from $t = 2$ to $t = 2 + h$, for each of these values of h: .01, .001, .0001, and .00001.
 (b) Compare your answers in part (a) with the number $(\sec 2)^2$. What would you guess that the instantaneous rate of change of $f(t) = \tan t$ is at $t = 2$?

In Exercises 24–30, perform the indicated operations, then simplify your answers by using appropriate definitions and identities.

24. $\tan t(\cos t - \csc t)$

25. $\cos t \sin t(\csc t + \sec t)$

26. $(1 + \cot t)^2$

27. $(1 - \sec t)^2$

28. $(\sin t - \csc t)^2$

29. $(\cot t - \tan t)(\cot^2 t + 1 + \tan^2 t)$

30. $(\sin t + \csc t)(\sin^2 t + \csc^2 t - 1)$

In Exercises 31–36, factor and simplify the given expression.

31. $\sec t \csc t - \csc^2 t$

32. $\tan^2 t - \cot^2 t$

33. $\tan^4 t - \sec^4 t$

34. $4\sec^2 t + 8\sec t + 4$

35. $\cos^3 t - \sec^3 t$

36. $\csc^4 t + 4\csc^2 t - 5$

In Exercises 37–42, simplify the given expression. Assume that all denominators are nonzero and all quantities under radicals are nonnegative.

37. $\dfrac{\cos^2 t \sin t}{\sin^2 t \cos t}$

38. $\dfrac{\sec^2 t + 2\sec t + 1}{\sec t}$

39. $\dfrac{4\tan t \sec t + 2\sec t}{6\sin t \sec t + 2\sec t}$

40. $\dfrac{\sec^2 t \csc t}{\csc^2 t \sec t}$

41. $(2 + \sqrt{\tan t})(2 - \sqrt{\tan t})$

42. $\dfrac{6\tan t \sin t - 3\sin t}{9\sin^2 t + 3\sin t}$

In Exercises 43–46, prove the given identity.

43. $1 + \cot^2 t = \csc^2 t$ [*Hint:* Look at the proof of the similar identity on page 477]

44. $\cot(-t) = -\cot t$ [*Hint:* Express the left side in terms of sine and cosine; then use the negative angle identities and express the result in terms of cotangent.]

45. $\sec(-t) = \sec t$ [Adapt the hint for Exercise 44.]

46. $\csc(-t) = -\csc t$

In Exercises 47–52, find the values of all six trigonometric functions at t if the given conditions are true.

47. $\cos t = -1/2$ and $\sin t > 0$
 [*Hint:* $\sin^2 t + \cos^2 t = 1$.]

48. $\cos t = \dfrac{1}{2}$ and $\sin t < 0$

49. $\cos t = 0$ and $\sin t = 1$

50. $\sin t = -2/3$ and $\sec t > 0$

51. $\sec t = -13/5$ and $\tan t < 0$

52. $\csc t = 8$ and $\cos t < 0$

Graphs

In Exercises 53–56, use graphs to determine whether the equation could possibly be an identity or is definitely not an identity.

53. $\tan t = \cot\left(\dfrac{\pi}{2} - t\right)$

54. $\dfrac{\cos t}{\cos(t - \pi/2)} = \cot t$

55. $\dfrac{\sin t}{1 - \cos t} = \cot t$

56. $\dfrac{\sec t + \csc t}{1 + \tan t} = \csc t$

57. Show graphically that the equation sec $t = t$ has infinitely many solutions, but none between $-\pi/2$ and $\pi/2$.

Thinkers

58. In the diagram of the unit circle in the figure, find six line segments whose respective lengths are sin t, cos t, tan t, cot t, sec t, csc t. [*Hint:* sin t = length CA. Why? Note that OC has length 1 and various right triangles in the figure are similar.]

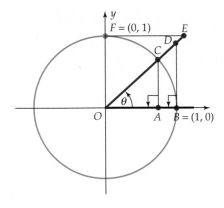

59. In the figure for Exercise 58, find the following areas in terms of θ.
(a) triangle OCA
(b) triangle ODB
(c) circular segment OCB

Chapter 6 Review

IMPORTANT CONCEPTS

Section 6.1

Angle 412
Vertex 412
Initial side 412
Terminal side 412
Coterminal angles 413
Standard position 413
Degree measure 413
Radian measure 414
Arc length 419
Angular speed 419

Section 6.2
[Alternate Section 6.2]

Sine function 423 [431, 434]
Cosine function 423 [431, 434]
Tangent function 424 [431, 434]
Special values 426 [436]

Point-in-the-plane description 427
[432]

Section 6.3

Algebra with trigonometric
  functions 437
Pythagorean identity 439
Periodicity 441
Negative angle identities 442

Section 6.4

Graphs of sine, cosine, and tangent
  functions 445–451
Graphs and identities 452

Section 6.5

Period 456

Amplitude 458
Phase shift 460
Simple harmonic motion 463

Special Topics 6.5.A

Sinusoidal graphs 468
Damped and compressed
  graphs 470

Section 6.6

Cotangent function 474
Secant function 474
Cosecant function 474
Alternative descriptions 475
Identities 476–477
Graphs of cotangent, secant, and
  cosecant functions 478–479

IMPORTANT FACTS & FORMULAS

- *Conversion Rules:* To convert radians to degrees, multiply by $180/\pi$. To convert degrees to radians, multiply by $\pi/180$.

- *Definition of Trigonometric Functions:* If P is the point where the terminal side of an angle of t radians in standard position meets the unit circle, then

$$\sin t = y\text{-coordinate of } P,$$

$$\cos t = x\text{-coordinate of } P,$$

$$\tan t = \frac{\sin t}{\cos t}, \quad \cot t = \frac{\cos t}{\sin t}, \quad \sec t = \frac{1}{\cos t}, \quad \csc t = \frac{1}{\sin t}.$$

- *Point-in-the-Plane Description:* If (x, y) is any point other than the origin on the terminal side of an angle of t radians in standard position and $r = \sqrt{x^2 + y^2}$, then

$$\sin t = \frac{y}{r}, \quad \cos t = \frac{x}{r},$$

$$\tan t = \frac{y}{x}, \quad \cot t = \frac{x}{y},$$

$$\sec t = \frac{r}{x}, \quad \csc t = \frac{r}{y}.$$

■ *Basic Identities:*

$$\sin^2 t + \cos^2 t = 1 \qquad\qquad \sin(-t) = -\sin t$$

$$1 + \tan^2 t = \sec^2 t \qquad\qquad \cos(-t) = \cos t$$

$$1 + \cot^2 t = \csc^2 t \qquad\qquad \tan(-t) = -\tan t$$

$$\tan t = \frac{1}{\cot t} \qquad\qquad \cot t = \frac{1}{\tan t}$$

$$\sin(t \pm 2\pi) = \sin t \qquad \csc(t \pm 2\pi) = \csc t$$

$$\cos(t \pm 2\pi) = \cos t \qquad \sec(t \pm 2\pi) = \sec t$$

$$\tan(t \pm \pi) = \tan t \qquad \cot(t \pm \pi) = \cot t$$

■ If $A \neq 0$ and $b > 0$, then each of the $f(t) = A \sin(bt + c)$ and $g(t) = A \cos(bt + c)$ has

amplitude $|A|$,   period $2\pi/b$,   phase shift $-c/b$.

CATALOG OF BASIC FUNCTIONS—PART 4

REVIEW QUESTIONS

1. Find a number t between 0 and 2π such that an angle of t radians in standard position is coterminal with an angle of $-23\pi/3$ radians in standard position.

2. Through how many radians does the second hand of a clock move in 2 minutes and 40 seconds?

3. $\dfrac{9\pi}{5}$ radians = _____ degrees.

4. 36 degrees = _____ radians.

5. $220°$ = _____ radians.

6. $\dfrac{17\pi}{12}$ radians = _____ degrees.

7. $-\dfrac{11\pi}{4}$ radians = _____ degrees.

8. $-135°$ = _____ radians.

9. If an angle of v radians has its terminal side in the second quadrant and $\sin v = \sqrt{8/9}$, then find $\cos v$.

10. $\cos \dfrac{47\pi}{2} = ?$

11. $\sin(-13\pi) = ?$

12. Simplify: $\dfrac{\tan(t + \pi)}{\sin(t + 2\pi)}$

REVIEW

Use the figure below in Questions 13–17.

13. $\cos\dfrac{\pi}{5} = ?$

14. $\sin\left(\dfrac{7\pi}{6}\right) = ?$

15. $\cos\left(\dfrac{-5\pi}{6}\right) = ?$

16. $\sin\left(\dfrac{16\pi}{6}\right) = ?$

17. $\sin\left(\dfrac{-4\pi}{3}\right) = ?$

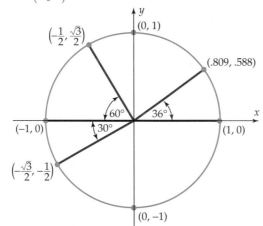

18. $\left[3\sin\left(\dfrac{\pi}{5^{500}}\right)\right]^2 + \left[3\cos\left(\dfrac{\pi}{5^{500}}\right)\right]^2 = ?$

19. Fill in the blanks (approximations not allowed):

t	0	$\dfrac{\pi}{6}$	$\dfrac{\pi}{4}$	$\dfrac{\pi}{3}$	$\dfrac{\pi}{2}$
$\sin t$					
$\cos t$					

20. Express as a single real number:

$$\cos\dfrac{3\pi}{4}\sin\dfrac{5\pi}{6} - \sin\dfrac{3\pi}{4}\cos\dfrac{5\pi}{6}$$

21. $\left(\sin\dfrac{\pi}{6} + 1\right)^2 = ?$

22. $\sin(\pi/2) + \sin 0 + \cos 0 = ?$

23. If $f(x) = \log_{10} x$ and $g(t) = -\cos t$, then $(f \circ g)(\pi) = ?$

24. Cos t is negative when the terminal side of an angle of t radians in standard position lies in which quadrants?

25. If $\sin t = 1/\sqrt{3}$ and the terminal side of an angle of t radians in standard position lies in the second quadrant, then $\cos t = ?$

26. Which of the following could possibly be a true statement about a real number t?
   (a) $\sin t = -2$ and $\cos t = 1$
   (b) $\sin t = 1/2$ and $\cos t = \sqrt{2}/2$
   (c) $\sin t = -1$ and $\cos t = 1$
   (d) $\sin t = \pi/2$ and $\cos t = 1 - (\pi/2)$
   (e) $\sin t = 3/5$ and $\cos t = 4/5$

27. If $\sin t = -4/5$ and the terminal side of an angle of t radians in standard position lies in the third quadrant, then $\cos t = $ _____.

28. If $\sin(-101\pi/2) = -1$, then $\sin(-105\pi/2) = ?$

29. If $\pi/2 < t < \pi$ and $\sin t = 5/13$, then $\cos t = ?$

30. $\cos\left(-\dfrac{\pi}{6}\right) = ?$ **31.** $\cos\left(\dfrac{2\pi}{3}\right) = ?$

32. $\sin\left(-\dfrac{11\pi}{6}\right) = ?$ **33.** $\sin\left(\dfrac{\pi}{3}\right) = ?$

34. $\tan(5\pi/3) = ?$

35. Which of the following is *not* true about the graph of $f(t) = \sin t$?
   (a) It has no sharp corners.
   (b) It crosses the horizontal axis more than once.
   (c) It rises higher and higher as t gets larger.
   (d) It is periodic.
   (e) It has no vertical asymptotes.

36. Which of the following functions has the graph in the figure between $-\pi$ and π?
   (a) $f(x) = \begin{cases} \sin x, & \text{if } x \geq 0 \\ \cos x, & \text{if } x < 0 \end{cases}$
   (b) $g(x) = \cos x - 1$
   (c) $h(x) = \begin{cases} \sin x, & \text{if } x \geq 0 \\ \sin(-x), & \text{if } x < 0 \end{cases}$
   (d) $k(x) = |\cos x|$
   (e) $p(x) = \sqrt{1 - \sin^2 x}$

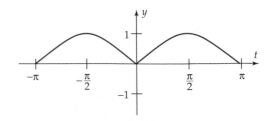

The point $(-3/\sqrt{50},\ 7/\sqrt{50})$ lies on the terminal side of an angle of t radians (in standard position). Find:

37. $\sin t$ **38.** $\cos t$

39. tan *t*

40. sec *t*

41. Find the equation of the straight line containing the terminal side of an angle of $5\pi/3$ radians (in standard position).

Suppose that an angle of w radians has its terminal side in the fourth quadrant and $\cos w = 2/\sqrt{13}$. *Find:*

42. sin *w*

43. tan *w*

44. csc *w*

45. $\cos(-w)$

46. Fill in the blanks (approximations not allowed):

t	sin *t*	tan *t*	sec *t*
$\pi/4$			
$2\pi/3$			
$5\pi/6$			

47. Sketch the graphs of $f(t) = \sin t$ and $h(t) = \csc t$ on the same set of coordinate axes $(-2\pi \le t \le 2\pi)$.

48. *Let* θ *be the angle shown in the figure. Which of the following statements is true?*

(a) $\sin \theta = \dfrac{\sqrt{2}}{2}$

(b) $\cos \theta = \dfrac{\sqrt{2}}{2}$

(c) $\tan \theta = 1$

(d) $\cos \theta = \sqrt{2}$

(e) $\tan \theta = -1$

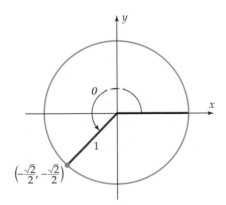

$\left(-\dfrac{\sqrt{2}}{2}, -\dfrac{\sqrt{2}}{2}\right)$

49. Let θ be as indicated in the figure in the next column. Which of the statements **(i)–(iii)** are *true?*

(i) $\cos \theta = -\dfrac{1}{3}$

(ii) $\tan \theta = \dfrac{2\sqrt{2}}{9}$

(iii) $\sin \theta = -\dfrac{2\sqrt{2}}{3}$

(a) only ii

(b) only ii and iii

(c) all of them

(d) only i and iii

(e) none of them

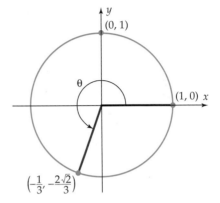

$\left(-\dfrac{1}{3}, -\dfrac{2\sqrt{2}}{3}\right)$

50. Between (and including) 0 and 2π, the function $h(t) = \tan t$ has

(a) 3 roots and is undefined at 2 places.

(b) 2 roots and is undefined at 3 places.

(c) 2 roots and is undefined at 2 places.

(d) 3 roots and is defined everywhere.

(e) no roots and is undefined at 3 places.

51. If the terminal side of an angle of θ radians in standard position passes through the point $(-2, 3)$, then $\tan \theta =$ _____.

52. Which of the statements **(i)–(iii)** are true?

(i) $\sin(-x) = -\sin x$

(ii) $\cos(-x) = -\cos x$

(iii) $\tan(-x) = -\tan x$

(a) (i) and (ii) only

(b) (ii) only

(c) (i) and (iii) only

(d) all of them

(e) none of them

53. If $\sec x = 1$ and $-\pi/2 < x < \pi/2$, then $x = ?$

54. If $\tan t = 4/3$ and $0 < t < \pi$, what is $\cos t$?

55. Which of the following is true about sec *t*?

(a) $\sec(0) = 0$

(b) $\sec t = 1/\sin t$

(c) Its graph has no asymptotes.

(d) It is a periodic function.

(e) It is never negative.

56. If $\cot t = 0$ and $0 < t \le \pi$, then $t =$ _____.

57. What is $\cot\left(\dfrac{2\pi}{3}\right)$?

58. Which of the following functions has the graph in the figure?
(a) $f(t) = \tan t$
(b) $g(t) = \tan\left(t + \dfrac{\pi}{2}\right)$
(c) $h(t) = 1 + \tan t$
(d) $k(t) = 3 \tan t$
(e) $p(t) = -\tan t$

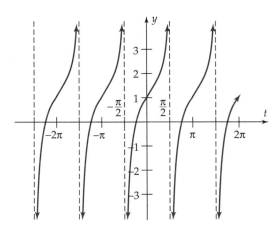

59. Let $f(t) = \frac{3}{2} \sin 5t$.
(a) What is the largest possible value of $f(x)$?
(b) Find the smallest positive number t such that $f(t) = 0$.

60. Sketch the graph of $g(t) = -2 \cos t$.

61. Sketch the graph of $f(t) = -\frac{1}{2} \sin 2t \ (-2\pi \le t \le 2\pi)$.

62. Sketch the graph of $f(t) = \sin 4t \ (0 \le t \le 2\pi)$.

In Questions 63–66, determine graphically whether the given equation could possibly be an identity.

63. $\cos t = \sin\left(t - \dfrac{\pi}{2}\right)$

64. $\tan \dfrac{t}{2} = \dfrac{\sin t}{1 + \cos t}$

65. $\dfrac{\sin t - \sin 3t}{\cos t + \cos 3t} = -\tan t$

66. $\cos 2t = \dfrac{1}{1 - 2\sin^2 t}$

67. What is the period of the function $g(t) = \sin 4\pi t$?

68. What are the amplitude, period, and phase shift of the function
$$h(t) = 13 \cos(14t + 15)?$$

69. State the rule of a periodic function whose graph from $t = 0$ to $t = 2\pi$ closely resembles the one in the figure.

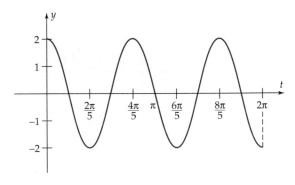

70. State the rule of a periodic function with amplitude 3, period π, and phase shift $\pi/3$.

71. State the rule of a periodic function with amplitude 8, period 5, and phase shift 14.

72. If $g(t) = 20 \sin(200t)$, for how many values of t, with $0 \le t \le 2\pi$, is it true that $g(t) = 1$?

In Exercises 73 and 74, estimate constants A, b, c such that $f(t) = A \sin(bt + c)$.

73. $f(t) = 6 \sin(4t + 7) - 5 \cos(4t + 8)$

74. $f(t) = -5 \sin(5t - 3) + 2 \cos(5t + 2)$

In Exercises 75 and 76, find a viewing window that shows a complete graph of the function.

75. $f(t) = 3 \sin(300t + 5) - 2 \cos(500t + 8)$

76. $g(t) = -5 \sin(400\pi t + 1) + 2 \cos(150\pi t - 6)$

DISCOVERY PROJECT 6   Pistons and Flywheels

A common and well-proven piece of technology is the piston and flywheel combination. It is clearly visible in photographs of steam locomotives from the middle of the nineteenth century. The structure involves a wheel (or crankshaft) connected to a sliding plug in a cylinder by a rigid arm. The axis of rotation of the wheel is perpendicular to the central axis of the cylinder. The sliding plug, the piston, moves in one dimension, in and out of the cylinder. The motion of the piston is periodic, like the basic trigonometric functions, but doesn't have the same elegant symmetry.

It is quite easy to superimpose a coordinate plane on the flywheel-piston system so that the center of the flywheel is the origin, the flywheel rotates counterclockwise, and the piston moves along the x-axis. As you can see in the diagram, the flywheel typically has a larger radius than the radial distance from the center to the point where the arm attaches. In this particular figure, the radius of the flywheel is 50 centimeters, and the radial distance to the attachment point Q is 45 centimeters. The arm is 150 centimeters long measured from the base of the piston to the attachment point.

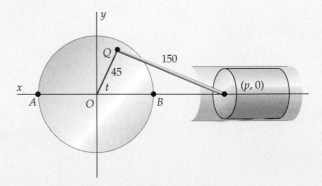

1. What are the coordinates of A and B in the figure? How close does the base of the piston come to the flywheel? What is the length of the piston stroke?

2. Show that Q has coordinates $(45 \cos t, 45 \sin t)$, where t is the radian measure of angle BOQ. [*Hint:* If Q has coordinates (x, y), use the point-in-the-plane description to compute $\cos t$ and solve for x; find y similarly.]

3. Let p be the x-coordinate of the center point of the base of the piston. Express p as a function of t. [*Hint:* Use the distance formula to express the distance from Q to $(p, 0)$ in terms of p and t. Set this expression equal to 150 (why?) and solve for p.]

4. Let $p(t)$ be the function found in Question 3 (that is, $p(t)$ = the x-coordinate of the base of the piston when angle BOQ measures t radians). What is the range of this function? How does this relate to Question 1?

5. Approximate the values of $p(0)$, $p(\pi/2)$, and $p(\pi)$. Note that $p(\pi/2)$ is not halfway between $p(0)$ and $p(\pi)$. Which side of the halfway point is it on? Does this mean that the piston moves faster on the average when it is to the right or left of the halfway point? Find the value of t that places the base of the piston at the halfway point of its motion.

6. If the flywheel spins at a constant speed, does the piston move back and forth at a constant speed? How do you know?

Trigonometric Identities and Equations

Photodisc Collection/Getty Images

Beam me up, Scotty!

When a light beam passes from one medium to another (for instance, from air to water), it changes both its speed and its direction. If you know what some of these numbers are (say, the speed of light in air or the angle at which a light beam hits the water), then you can determine the unknown ones by solving a trigonometric equation. See Exercises 37–40 on page 535.

Chapter Outline

Interdependence of Sections

Chapters 7–9 are independent of each other and may be read in any order. The interdependence of the sections of this chapter is shown below. Sections 7.1, 7.4, and 7.5 are independent of one another and may be read in any order.

7.1 ⟶ 7.2 ⟶ 7.3

7.4

7.5

7.1   **Basic Identities and Proofs**

7.2   **Addition and Subtraction Identities**

7.2.A   *Special Topics:* **Lines and Angles**

7.3   **Other Identities**

7.4   **Inverse Trigonometric Functions**

7.5   **Trigonometric Equations**

Until now, the variable t has been used for trigonometric functions, to avoid confusion with the x's and y's that appear in their definitions. Now that you are comfortable with these functions, we shall usually use the letter x (or occasionally y) for the variable. Unless stated, all trigonometric functions in this chapter are considered to be functions of real numbers, rather than functions of angles in degree measure.

Two kinds of trigonometric equations are considered here. *Identities* (Sections 7.1–7.3) are equations that are valid for all values of the variable for which the equation is defined, such as

$$\sin^2 x + \cos^2 x = 1 \qquad \text{and} \qquad \cot x = \frac{1}{\tan x}.$$

Identities can be used for simplifying expressions, rewriting the rule of a trigonometric function, performing numerical computations, and in other ways. *Conditional equations* (Section 7.5) are valid only for certain values of the variable, such as

$$\sin x = 0 \qquad \text{(true only when } x \text{ is an integer multiple of } \pi\text{)}.$$

Inverse trigonometric functions, which have a number of uses, are considered in Section 7.4.

7.1   Basic Identities and Proofs

When you suspect that an equation might be an identity, it's a good idea to see whether there is any graphical evidence to support this conclusion.

EXAMPLE 1

Is either of the following equations an identity?

(a) $2\sin^2 x - \cos x = 2\cos^2 x + \sin x$

(b) $\dfrac{1 + \sin x - \sin^2 x}{\cos x} = \cos x + \tan x$

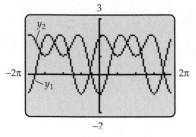

Figure 7–1

SOLUTION

(a) Test the equation by graphing these two equations on the same screen:

$$y_1 = 2 \sin^2 x - \cos x \qquad [\textit{left side of equation being tested}]$$

$$y_2 = 2 \cos^2 x + \sin x \qquad [\textit{right side of equation being tested}]$$

If the given equation is an identity (meaning that $y_1 = y_2$), then the two graphs will be identical. As Figure 7–1 shows, however, the graphs are quite different. Hence, this equation is not an identity.

(b) Test the second equation graphically.

GRAPHING EXPLORATION

Graph the functions

$$f(x) = \frac{1 + \sin x - \sin^2 x}{\cos x} \qquad \text{and} \qquad g(x) = \cos x + \tan x$$

on the same screen, using a viewing window with $-2\pi \le x \le 2\pi$. Do the graphs appear to be identical?

The exploration suggests that the equation *may* be an identity because the two graphs *appear* to be identical. However, this graphical evidence is *not* a proof—that must be done algebraically. ■

Any equation can be graphically tested, as in Example 1, to see whether it might be an identity. If the left-side and right-side graphs are different, then it definitely is *not* an identity. If the graphs appear to be the same, then it is *possible,* but not certain, that the equation is an identity.

> **The fact that two graphs appear identical on a calculator screen does *not* prove that they are actually the same, as the following Graphing Exploration shows.**

GRAPHING EXPLORATION

In the viewing window with $-\pi \le x \le \pi$ and $-2 \le y \le 2$, graph both sides of the equation

$$\cos x = 1 - \frac{x^2}{2} + \frac{x^4}{24} - \frac{x^6}{720} + \frac{x^8}{40,320}.$$

Do the graphs appear to be identical? Now change the viewing window so that $-2\pi \le x \le 2\pi$. Is the equation an identity?

▪▪ PROVING IDENTITIES

The phrases "prove the identity" and "verify the identity" mean "prove that the given equation is an identity." In the proofs here, we shall assume the elementary identities that were proved in Chapter 6 and are summarized here.

Basic Trigonometric
Identities

Reciprocal Identities

$$\sec x = \frac{1}{\cos x} \qquad \csc x = \frac{1}{\sin x}$$

$$\tan x = \frac{\sin x}{\cos x} \qquad \cot x = \frac{\cos x}{\sin x}$$

$$\tan x = \frac{1}{\cot x} \qquad \cot x = \frac{1}{\tan x}$$

Periodicity Identities

$$\sin(x + 2\pi) = \sin x \qquad \cos(x + 2\pi) = \cos x$$

$$\sec(x + 2\pi) = \sec x \qquad \csc(x + 2\pi) = \csc x$$

$$\tan(x + \pi) = \tan x \qquad \cot(x + \pi) = \cot x$$

Pythagorean Identities

$$\sin^2 x + \cos^2 x = 1 \qquad 1 + \tan^2 x = \sec^2 x \qquad 1 + \cot^2 x = \csc^2 x$$

Negative Angle Identities

$$\sin(-x) = -\sin x \qquad \cos(-x) = \cos x \qquad \tan(-x) = -\tan x$$

There are no cut-and-dried rules for simplifying trigonometric expressions or proving identities, but there are some common strategies that are often helpful. Four of these strategies are illustrated in the following examples. There are often a variety of ways to proceed, and it will take some practice before you can easily decide which strategies are likely to be the most efficient in a particular case.

Strategy 1

Express everything in terms of sine and cosine.

EXAMPLE 2

Simplify $(\csc x + \cot x)(1 - \cos x)$.

SOLUTION Using Strategy 1, we have

$$(\csc x + \cot x)(1 - \cos x) = \left(\frac{1}{\sin x} + \frac{\cos x}{\sin x}\right)(1 - \cos x) \qquad \text{[Reciprocal identities]}$$

$$= \frac{(1 + \cos x)}{\sin x}(1 - \cos x)$$

$$= \frac{(1 + \cos x)(1 - \cos x)}{\sin x}$$

$$= \frac{1 - \cos^2 x}{\sin x} = \frac{\sin^2 x}{\sin x} \qquad \text{[Pythagorean identity]}$$

$$= \sin x. \qquad \blacksquare$$

Strategy 2

Use algebra and identities to transform the expression on one side of the equal sign into the expression on the other side.[*]

EXAMPLE 3

In Example 1, we verified graphically that the equation

$$\frac{1 + \sin x - \sin^2 x}{\cos x} = \cos x + \tan x$$

might be an identity. Prove that it is.

SOLUTION We use Strategy 2, beginning with the left side of the equation:

$$\frac{1 + \sin x - \sin^2 x}{\cos x} = \frac{(1 - \sin^2 x) + \sin x}{\cos x}$$

$$= \frac{\cos^2 x + \sin x}{\cos x} \qquad \text{[Pythagorean identity]}$$

$$= \frac{\cos^2 x}{\cos x} + \frac{\sin x}{\cos x}$$

$$= \cos x + \frac{\sin x}{\cos x}$$

$$= \cos x + \tan x. \qquad \blacksquare$$

The strategies presented above and those to be considered below are "plans of attack." By themselves they are not much help unless you also have some techniques for carrying out these plans. In the previous examples, we used the techniques of basic algebra and the use of known identities to change trigonometric expressions to equivalent ones. Here is another technique that is often useful when dealing with fractions:

Rewrite a fraction in equivalent form by multiplying its numerator and denominator by the same quantity.

EXAMPLE 4

Prove that

$$\frac{\sin x}{1 + \cos x} = \frac{1 - \cos x}{\sin x}.$$

---

[*]That is, start with expression A on one side and use identities and algebra to produce a sequence of equalities $A = B$, $B = C$, $C = D$, $D = E$, where E is the other side of the identity to be proved; conclude that $A = E$.

SOLUTION We shall use Strategy 2, beginning with the left side, whose denominator is $1 + \cos x$. Multiply its numerator and denominator by $1 - \cos x$:†

$$\frac{\sin x}{1 + \cos x} = \frac{\sin x}{1 + \cos x} \cdot \frac{1 - \cos x}{1 - \cos x} = \frac{\sin x(1 - \cos x)}{(1 + \cos x)(1 - \cos x)}$$

$$= \frac{\sin x(1 - \cos x)}{1 - \cos^2 x}$$

$$= \frac{\sin x(1 - \cos x)}{\sin^2 x} \qquad \text{[Pythagorean identity]}$$

$$= \frac{1 - \cos x}{\sin x}.$$

ALTERNATE SOLUTION The numerators of the given equation look similar to the Pythagorean identity—with the squares missing. So we begin with the left side and introduce some squares by multiplying it by $\frac{\sin x}{\sin x} = 1$.

$$\frac{\sin x}{1 + \cos x} = 1 \cdot \frac{\sin x}{1 + \cos x} = \frac{\sin x}{\sin x} \cdot \frac{\sin x}{1 + \cos x} = \frac{\sin^2 x}{\sin x(1 + \cos x)}$$

$$= \frac{1 - \cos^2 x}{\sin x(1 + \cos x)} \qquad \text{[Pythagorean identity]}$$

$$= \frac{(1 - \cos x)(1 + \cos x)}{\sin x(1 + \cos x)} \qquad \text{[Factor numerator]}$$

$$= \frac{1 - \cos x}{\sin x}. \qquad ■$$

Strategy 3

Deal separately with each side of the equation $A = B$. First use identities and algebra to transform A into some expression C (so that $A = C$). Then use (possibly different) identities and algebra to transform B into the *same* expression C (so that $B = C$). Conclude that $A = B$.

EXAMPLE 5

Prove that $\csc x - \cot x = \dfrac{\sin x}{1 + \cos x}$.

SOLUTION We use Strategy 3, together with Strategy 1, beginning with the left side.

$$(*) \qquad \csc x - \cot x = \frac{1}{\sin x} - \frac{\cos x}{\sin x} = \frac{1 - \cos x}{\sin x}.$$

---

†This is analogous to the process used to rationalize the denominator of a fraction by multiplying its numerator and denominator by the conjugate of the denominator, as in this example:
$$\frac{1}{3 + \sqrt{2}} = \frac{1}{3 + \sqrt{2}} \cdot \frac{3 - \sqrt{2}}{3 - \sqrt{2}} = \frac{3 - \sqrt{2}}{3^2 - (\sqrt{2})^2} = \frac{3 - \sqrt{2}}{7}.$$

Example 4 shows that the right side of the identity to be proved can also be transformed into this same expression.

$$(**) \qquad \frac{\sin x}{1 + \cos x} = \frac{1 - \cos x}{\sin x}.$$

Combining the equalities $(*)$ and $(**)$ proves the identity.

$$\csc x - \cot x = \frac{1 - \cos x}{\sin x} = \frac{\sin x}{1 + \cos x}. \qquad \blacksquare$$

Proving identities involving fractions can sometimes be quite complicated. It often helps to approach a fractional identity indirectly, as in the following example.

EXAMPLE 6

Prove these identities.

(a) $\sec x(\sec x - \cos x) = \tan^2 x$

(b) $\dfrac{\sec x}{\tan x} = \dfrac{\tan x}{\sec x - \cos x}.$

SOLUTION

(a) Beginning with the left side and using Strategy 1, we have

$$\sec x(\sec x - \cos x) = \sec^2 x - \sec x \cos x$$

$$= \sec^2 x - \frac{1}{\cos x} \cos x$$

$$= \sec^2 x - 1$$

$$= \tan^2 x. \qquad \text{[Pythagorean identity]}$$

Therefore, $\sec x(\sec x - \cos x) = \tan^2 x$.

(b) By part (a), we know that

$$\sec x(\sec x - \cos x) = \tan^2 x$$

Dividing both sides of this equation by $\tan x(\sec x - \cos x)$ shows that

$$\frac{\sec x(\sec x - \cos x)}{\tan x(\sec x - \cos x)} = \frac{\tan^2 x}{\tan x(\sec x - \cos x)}$$

$$\frac{\sec x(\sec x - \cos x)}{\tan x(\sec x - \cos x)} = \frac{\tan x \tan x}{\tan x(\sec x - \cos x)}$$

$$\frac{\sec x}{\tan x} = \frac{\tan x}{\sec x - \cos x}. \qquad \blacksquare$$

Look carefully at how identity (b) was proved in Example 6. We first proved identity (a), which is of the form $AD = BC$ (with $A = \sec x$, $B = \tan x$, $C = \tan x$, and $D = \sec x - \cos x$). Then we divided both sides by BD (that is, by $\tan x(\sec x - \cos x)$) to conclude that $A/B = C/D$. The same argument works in general and provides the following useful strategy for dealing with identities involving fractions.

Strategy 4

> If you can prove that $AD = BC$, with $B \neq 0$ and $D \neq 0$, then you can conclude that
>
> $$\frac{A}{B} = \frac{C}{D}.$$

Many students misunderstand Strategy 4: It does *not* say that you begin with a fractional equation $A/B = C/D$ and cross-multiply to eliminate the fractions. If you did that, you would be assuming what has to be proved. What the strategy says is that to prove an identity involving fractions, you need only prove a different identity that does not involve fractions. In other words, if you prove that $AD = BD$ whenever $B \neq 0$ and $D \neq 0$, then you can conclude that $A/B = C/D$. Note that you do not *assume* that $AD = BC$; you use Strategy 2 or 3 or some other means to *prove* this statement.

EXAMPLE 7

Prove that $\dfrac{\cot x - 1}{\cot x + 1} = \dfrac{1 - \tan x}{1 + \tan x}$.

SOLUTION   We use Strategy 4, with $A = \cot x - 1$, $B = \cot x + 1$, $C = 1 - \tan x$, and $D = 1 + \tan x$. We must prove that this equation is an identity:

$$AD = BC$$

$(\ast\ast\ast)$ $$(\cot x - 1)(1 + \tan x) = (\cot x + 1)(1 - \tan x).$$

Strategy 3 will be used. Multiplying out the left side shows that

$$(\cot x - 1)(1 + \tan x) = \cot x - 1 + \cot x \tan x - \tan x$$

$$= \cot x - 1 + \frac{1}{\tan x}\tan x - \tan x$$

$$= \cot x - 1 + 1 - \tan x$$

$$= \cot x - \tan x.$$

Similarly, on the right side of $(\ast\ast\ast)$,

$$(\cot x + 1)(1 - \tan x) = \cot x + 1 - \cot x \tan x - \tan x$$

$$= \cot x + 1 - 1 - \tan x$$

$$= \cot x - \tan x.$$

Since the left and right sides are equal to the same expression, we have proved that $(\ast\ast\ast)$ is an identity. Therefore, by Strategy 4, we conclude that

$$\frac{\cot x - 1}{\cot x + 1} = \frac{1 - \tan x}{1 + \tan x}$$

is also an identity.   ■

TECHNOLOGY TIP

Using SOLVE in the TI-89 ALGEBRA menu to solve an equation that might be an identity produces one of three responses. "True" means the equation *probably* is an identity [algebraic proof is required for certainty]. "False" means the equation is not an identity. A numerical answer is inconclusive [the equation may or may not be an identity].

It takes a good deal of practice, as well as *much* trial and error, to become proficient in proving identities. The more practice you have, the easier it will get. Since there are many correct methods, your proofs may be quite different from those of your instructor or the text answers.

If you don't see what to do immediately, try something and see where it leads: Multiply out or factor or multiply numerator and denominator by the same nonzero quantity. Even if this doesn't lead anywhere, it might give you some ideas on other things to try. When you do obtain a proof, check to see whether it can be done more efficiently. In your final proof, don't include the side trips that may have given you some ideas but aren't themselves part of the proof.

EXERCISES 7.1

In Exercises 1–4, test the equation graphically to determine whether it might be an identity. You need not prove those equations that seem to be identities.

1. $\dfrac{\sec x - \cos x}{\sec x} = \sin^2 x$

2. $\tan x + \cot x = (\sin x)(\cos x)$

3. $\dfrac{1 - \cos(2x)}{2} = \sin^2 x$

4. $\dfrac{\tan x + \cot x}{\csc x} = \sec x$

In Exercises 5–8, insert one of A–F on the right of the equal sign so that the resulting equation appears to be an identity when you test it graphically. You need not prove the identity.

A. $\cos x$ **B.** $\sec x$ **C.** $\sin^2 x$

D. $\sec^2 x$ **E.** $\sin x - \cos x$ **F.** $\dfrac{1}{\sin x \cos x}$

5. $\csc x \tan x = \underline{\hspace{1cm}}$

6. $\dfrac{\sin x}{\tan x} = \underline{\hspace{1cm}}$

7. $\dfrac{\sin^4 x - \cos^4 x}{\sin x + \cos x} = \underline{\hspace{1cm}}$

8. $\tan^2(-x) - \dfrac{\sin(-x)}{\sin x} = \underline{\hspace{1cm}}$

In Exercises 9–18, prove the identity.

9. $\tan x \cos x = \sin x$

10. $\cot x \sin x = \cos x$

11. $\cos x \sec x = 1$

12. $\sin x \csc x = 1$

13. $\tan x \csc x = \sec x$

14. $\sec x \cot x = \csc x$

15. $\dfrac{\tan x}{\sec x} = \sin x$

16. $\dfrac{\cot x}{\csc x} = \cos x$

17. $(1 + \cos x)(1 - \cos x) = \sin^2 x$

18. $(\csc x - 1)(\csc x + 1) = \cot^2 x$

In Exercises 19–48, state whether or not the equation is an identity. If it is an identity, prove it.

19. $\sin x = \sqrt{1 - \cos^2 x}$

20. $\cot x = \dfrac{\csc x}{\sec x}$

21. $\dfrac{\sin(-x)}{\cos(-x)} = -\tan x$

22. $\tan x = \sqrt{\sec^2 x - 1}$

23. $\cot(-x) = -\cot x$

24. $\sec(-x) = \sec x$

25. $1 + \sec^2 x = \tan^2 x$

26. $\sec^4 x - \tan^4 x = 1 + 2\tan^2 x$

27. $\sec^2 x - \csc^2 x = \tan^2 x - \cot^2 x$

28. $\sec^2 x + \csc^2 x = \sec^2 x \csc^2 x$

29. $\sin^2 x(\cot x + 1)^2 = \cos^2 x(\tan x + 1)^2$

30. $\cos^2 x(\sec x + 1)^2 = (1 + \cos x)^2$

31. $\sin^2 x - \tan^2 x = -\sin^2 x \tan^2 x$

32. $\cot^2 x - 1 = \csc^2 x$

33. $(\cos^2 x - 1)(\tan^2 x + 1) = -\tan^2 x$

34. $(1 - \cos^2 x)\csc x = \sin x$

35. $\tan x = \dfrac{\sec x}{\csc x}$

36. $\dfrac{\cos(-x)}{\sin(-x)} = -\cot x$

37. $\cos^4 x - \sin^4 x = \cos^2 x - \sin^2 x$

38. $\cot^2 x - \cos^2 x = \cos^2 x \cot^2 x$

39. $(\sin x + \cos x)^2 = \sin^2 x + \cos^2 x$

40. $(1 + \tan x)^2 = \sec^2 x$

41. $\dfrac{\sec x}{\csc x} + \dfrac{\sin x}{\cos x} = 2 \tan x$

42. $\dfrac{1 + \cos x}{\sin x} + \dfrac{\sin x}{1 + \cos x} = 2 \csc x$

43. $\dfrac{\sec x + \csc x}{1 + \tan x} = \csc x$ **44.** $\dfrac{\cot x - 1}{1 - \tan x} = \dfrac{\csc x}{\sec x}$

45. $\dfrac{1}{\csc x - \sin x} = \sec x \tan x$

46. $\dfrac{1 + \csc x}{\csc x} = \dfrac{\cos^2 x}{1 - \sin x}$

47. $\dfrac{\sin x - \cos x}{\tan x} = \dfrac{\tan x}{\sin x + \cos x}$

48. $\dfrac{\cot x}{\csc x - 1} = \dfrac{\csc x + 1}{\cot x}$

In Exercises 49–52, half of an identity is given. Graph this half in a viewing window with $-2\pi \le x \le 2\pi$ and make a conjecture as to what the right side of the identity is. Then prove your conjecture.

49. $1 - \dfrac{\sin^2 x}{1 + \cos x} = \,?$ [*Hint:* What familiar function has a graph that looks like this?]

50. $\dfrac{1 + \cos x - \cos^2 x}{\sin x} - \cot x = \,?$

51. $(\sin x + \cos x)(\sec x + \csc x) - \cot x - 2 = \,?$

52. $\cos^3 x(1 - \tan^4 x + \sec^4 x) = \,?$

In Exercises 53–66, prove the identity.

53. $\dfrac{1 - \sin x}{\sec x} = \dfrac{\cos^3 x}{1 + \sin x}$

54. $\dfrac{\sin x}{1 - \cot x} + \dfrac{\cos x}{1 - \tan x} = \cos x + \sin x$

55. $\dfrac{\cos x}{1 - \sin x} = \sec x + \tan x$

56. $\dfrac{1 + \sec x}{\tan x + \sin x} = \csc x$

57. $\dfrac{\cos x \cot x}{\cot x - \cos x} = \dfrac{\cot x + \cos x}{\cos x \cot x}$

58. $\dfrac{\cos^3 x - \sin^3 x}{\cos x - \sin x} = 1 + \sin x \cos x$

59. $\log_{10}(\cot x) = -\log_{10}(\tan x)$

60. $\log_{10}(\sec x) = -\log_{10}(\cos x)$

61. $\log_{10}(\csc x + \cot x) = -\log_{10}(\csc x - \cot x)$

62. $\log_{10}(\sec x + \tan x) = -\log_{10}(\sec x - \tan x)$

63. $\tan x - \tan y = -\tan x \tan y(\cot x - \cot y)$

64. $\dfrac{\tan x - \tan y}{\cot x - \cot y} = -\tan x \tan y$

65. $\dfrac{\cos x - \sin y}{\cos y - \sin x} = \dfrac{\cos y + \sin x}{\cos x + \sin y}$

66. $\dfrac{\tan x + \tan y}{\cot x + \cot y} = \dfrac{\tan x \tan y - 1}{1 - \cot x \cot y}$

7.2   Addition and Subtraction Identities

A common student ERROR is to write

$$\sin\left(x + \frac{\pi}{6}\right) = \sin x + \sin\frac{\pi}{6} = \sin x + \frac{1}{2}.$$

GRAPHING EXPLORATION

Verify graphically that the equation above is NOT an identity by graphing $y = \sin(x + \pi/6)$ and $y = \sin x + 1/2$ on the same screen.

The exploration shows that "$\sin(x + y) = \sin x + \sin y$" is NOT an identity (because it's false when $y = \pi/6$). There is an identity that relates these quantities, but it is a bit more complicated, as we now see.

**Addition and
Subtraction Identities**

$$\sin(x + y) = \sin x \cos y + \cos x \sin y$$

$$\sin(x - y) = \sin x \cos y - \cos x \sin y$$

$$\cos(x + y) = \cos x \cos y - \sin x \sin y$$

$$\cos(x - y) = \cos x \cos y + \sin x \sin y$$

The addition and subtraction identities are probably the most important of all the trigonometric identities. Before reading their proofs at the end of this section, you should become familiar with the examples and special cases below.

EXAMPLE 1

Use the addition identities to find the *exact* values of $\sin(5\pi/12)$ and $\cos(5\pi/12)$.

SOLUTION   Since

$$\frac{5\pi}{12} = \frac{2\pi}{12} + \frac{3\pi}{12} = \frac{\pi}{6} + \frac{\pi}{4},$$

we apply the addition identities with $x = \pi/6$ and $y = \pi/4$.

$$\sin\frac{5\pi}{12} = \sin\left(\frac{\pi}{6} + \frac{\pi}{4}\right) = \sin\frac{\pi}{6}\cos\frac{\pi}{4} + \cos\frac{\pi}{6}\sin\frac{\pi}{4}$$

$$= \frac{1}{2} \cdot \frac{\sqrt{2}}{2} + \frac{\sqrt{3}}{2} \cdot \frac{\sqrt{2}}{2} = \frac{\sqrt{2}(\sqrt{3} + 1)}{4},$$

$$\cos\frac{5\pi}{12} = \cos\left(\frac{\pi}{6} + \frac{\pi}{4}\right) = \cos\frac{\pi}{6}\cos\frac{\pi}{4} - \sin\frac{\pi}{6}\sin\frac{\pi}{4}$$

$$= \frac{\sqrt{3}}{2} \cdot \frac{\sqrt{2}}{2} - \frac{1}{2} \cdot \frac{\sqrt{2}}{2} = \frac{\sqrt{2}(\sqrt{3} - 1)}{4}. \qquad \blacksquare$$

EXAMPLE 2

Find $\sin(\pi - y)$.

SOLUTION   Apply the subtraction identity with $x = \pi$.

$$\sin(\pi - y) = \sin \pi \cos y - \cos \pi \sin y$$

$$= (0)(\cos y) - (-1)(\sin y)$$

$$= \sin y. \qquad \blacksquare$$

EXAMPLE 3

Show that for the function $f(x) = \sin x$ and any number $h \neq 0$,

$$\frac{f(x + h) - f(x)}{h} = \sin x\left(\frac{\cos h - 1}{h}\right) + \cos x\left(\frac{\sin h}{h}\right).$$

[This fact is needed in calculus.]

SOLUTION   Use the addition identity for $\sin(x + y)$ with $y = h$.

$$\frac{f(x + h) - f(x)}{h} = \frac{\sin(x + h) - \sin x}{h}$$

$$= \frac{\sin x \cos h + \cos x \sin h - \sin x}{h}$$

$$= \frac{\sin x(\cos h - 1) + \cos x \sin h}{h}$$

$$= \sin x\left(\frac{\cos h - 1}{h}\right) + \cos x\left(\frac{\sin h}{h}\right). \quad \blacksquare$$

EXAMPLE 4

Prove that

$$\cos x \cos y = \frac{1}{2}[\cos(x + y) + \cos(x - y)].$$

SOLUTION   We begin with the more complicated right side and use the addition and subtraction identities for cosine to transform it into the left side.

$$\frac{1}{2}[\cos(x + y) + \cos(x - y)] = \frac{1}{2}[(\cos x \cos y - \sin x \sin y)$$
$$+ (\cos x \cos y + \sin x \sin y)]$$

$$= \frac{1}{2}(\cos x \cos y + \cos x \cos y)$$

$$= \frac{1}{2}(2 \cos x \cos y) = \cos x \cos y. \quad \blacksquare$$

The addition and subtraction identities for sine and cosine can be used to obtain the following identities, as outlined in Exercise 36.

Addition and Subtraction Identities for Tangent

$$\tan(x + y) = \frac{\tan x + \tan y}{1 - \tan x \tan y}$$

$$\tan(x - y) = \frac{\tan x - \tan y}{1 + \tan x \tan y}$$

It is sometimes convenient to say that x is a number *in the first quadrant* if $0 < x < \pi/2$, that x is a number *in the second quadrant* if $\pi/2 < x < \pi$, and so on.

EXAMPLE 5

Suppose x is a number in the first quadrant and y is a number in the third quadrant. If $\sin x = 3/4$ and $\cos y = -1/3$, find the exact values of $\sin(x + y)$ and $\tan(x + y)$ and determine in which quadrant $x + y$ lies.

SOLUTION   Using the Pythagorean identity and the fact that $\cos x$ and $\tan x$ are positive when $0 < x < \pi/2$, we have

$$\cos x = \sqrt{1 - \sin^2 x} = \sqrt{1 - \left(\frac{3}{4}\right)^2} = \sqrt{1 - \frac{9}{16}} = \sqrt{\frac{7}{16}} = \frac{\sqrt{7}}{4},$$

$$\tan x = \frac{\sin x}{\cos x} = \frac{3/4}{\sqrt{7}/4} = \frac{3}{4} \cdot \frac{4}{\sqrt{7}} = \frac{3}{\sqrt{7}} = \frac{3\sqrt{7}}{7}.$$

Since y lies between π and $3\pi/2$, its sine is negative; hence,

$$\sin y = -\sqrt{1 - \cos^2 y} = \sqrt{1 - \left(-\frac{1}{3}\right)^2} = -\sqrt{\frac{8}{9}} = -\frac{\sqrt{8}}{3} = -\frac{2\sqrt{2}}{3},$$

$$\tan y = \frac{\sin y}{\cos y} = \frac{-2\sqrt{2}/3}{-1/3} = \frac{-2\sqrt{2}}{3} \cdot \frac{3}{-1} = 2\sqrt{2}.$$

The addition identities for sine and tangent now show that

$$\sin(x + y) = \sin x \cos y + \cos x \sin y$$

$$= \frac{3}{4} \cdot \frac{-1}{3} + \frac{\sqrt{7}}{4} \cdot \frac{-2\sqrt{2}}{3} = \frac{-3}{12} - \frac{2\sqrt{14}}{12} = \frac{-3 - 2\sqrt{14}}{12},$$

$$\tan(x + y) = \frac{\tan x + \tan y}{1 - \tan x \tan y}$$

$$= \frac{\dfrac{3\sqrt{7}}{7} + 2\sqrt{2}}{1 - \left(\dfrac{3\sqrt{7}}{7}\right)(2\sqrt{2})} = \frac{\dfrac{3\sqrt{7} + 14\sqrt{2}}{7}}{\dfrac{7 - 6\sqrt{14}}{7}} = \frac{3\sqrt{7} + 14\sqrt{2}}{7 - 6\sqrt{14}}.$$

So both the sine and tangent of $x + y$ are negative numbers. Therefore, $x + y$ must be in the interval $(3\pi/2, 2\pi)$ since the sign chart in Exercise 49 on page 430 shows that this is the only one of the four quadrants in which both sine and tangent are negative. ∎

■■ COFUNCTION IDENTITIES

Other special cases of the addition and subtraction identities are the cofunction identities:

Cofunction Identities

$$\sin x = \cos\left(\frac{\pi}{2} - x\right) \qquad \cos x = \sin\left(\frac{\pi}{2} - x\right)$$

$$\tan x = \cot\left(\frac{\pi}{2} - x\right) \qquad \cot x = \tan\left(\frac{\pi}{2} - x\right)$$

$$\sec x = \csc\left(\frac{\pi}{2} - x\right) \qquad \csc x = \sec\left(\frac{\pi}{2} - x\right)$$

The first confunction identity is proved by using the identity for $\cos(x - y)$ with $\pi/2$ in place of x and x in place of y.

$$\cos\left(\frac{\pi}{2} - x\right) = \cos\frac{\pi}{2}\cos x + \sin\frac{\pi}{2}\sin x = (0)(\cos x) + (1)(\sin x) = \sin x.$$

Since the first cofunction identity is valid for *every* number x, it is also valid with the number $\pi/2 - x$ in place of x.

$$\sin\left(\frac{\pi}{2} - x\right) = \cos\left[\frac{\pi}{2} - \left(\frac{\pi}{2} - x\right)\right] = \cos x.$$

Thus, we have proved the second cofunction identity. The others now follow from these two. For instance,

$$\tan\left(\frac{\pi}{2} - x\right) = \frac{\sin[(\pi/2) - x]}{\cos[(\pi/2) - x]} = \frac{\cos x}{\sin x} = \cot x.$$

EXAMPLE 6

Verify that $\dfrac{\cos(x - \pi/2)}{\cos x} = \tan x$.

SOLUTION   Beginning on the left side, we see that the term $\cos(x - \pi/2)$ looks almost, but not quite, like the term $\cos(\pi/2 - x)$ in the cofunction identity. But note that $-(x - \pi/2) = \pi/2 - x$. Therefore,

$$\frac{\cos\left(x - \dfrac{\pi}{2}\right)}{\cos x} = \frac{\cos\left[-\left(x - \dfrac{\pi}{2}\right)\right]}{\cos x}$$

$$= \frac{\cos\left(\dfrac{\pi}{2} - x\right)}{\cos x} \qquad \text{[Negative angle identity with } x - \frac{\pi}{2} \text{ in place of } x\text{]}$$

$$= \frac{\sin x}{\cos x} \qquad \text{[Cofunction identity]}$$

$$= \tan x. \qquad \text{[Reciprocal identity]}$$

▪▪ PROOF OF THE ADDITION AND SUBTRACTION IDENTITIES

We first prove that

$$\cos(x - y) = \cos x \cos y + \sin x \sin y.$$

If $x = y$, then this is true by the Pythagorean identity:

$$\cos(x - x) = \cos 0 = 1 = \cos^2 x + \sin^2 x = \cos x \cos x + \sin x \sin x.$$

Next we prove the identity in the case when $x > y$. Let P be the point where the terminal side of an angle of x radians in standard position meets the unit circle and let Q be the point where the terminal side of an angle of y radians in standard position meets the circle, as shown in Figure 7–2. According to the definitions of sine and cosine, P has coordinates $(\cos x, \sin x)$ and Q has coordinates $(\cos y, \sin y)$.

The angle QOP formed by the two terminal sides has radian measure $x - y$. Rotate this angle clockwise until side OQ lies on the horizontal axis, as shown in Figure 7–3. Angle QOP is now in standard position, and its terminal side meets the unit circle at P. Since angle QOP has radian measure $x - y$, the definitions of sine and cosine show that the point P, in this new location, has coordinates $(\cos(x - y), \sin(x - y))$. Q now has coordinates $(1, 0)$.

Using the coordinates of P and Q *before* the angle was rotated and the distance formula, we have

Distance from P to Q

$$= \sqrt{(\cos x - \cos y)^2 + (\sin x - \sin y)^2}$$
$$= \sqrt{\cos^2 x - 2\cos x \cos y + \cos^2 y + \sin^2 x - 2\sin x \sin y + \sin^2 y}$$
$$= \sqrt{(\cos^2 x + \sin^2 x) + (\cos^2 y + \sin^2 y) - 2\cos x \cos y - 2\sin x \sin y}$$
$$= \sqrt{1 + 1 - 2\cos x \cos y - 2\sin x \sin y}$$
$$= \sqrt{2 - 2\cos x \cos y - 2\sin x \sin y}.$$

Figure 7–2

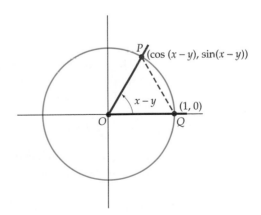

Figure 7–3

But using the coordinates of P and Q *after* the angle is rotated shows that

$$\text{Distance from } P \text{ to } Q = \sqrt{[\cos(x-y) - 1]^2 + [\sin(x-y) - 0]^2}$$
$$= \sqrt{\cos^2(x-y) - 2\cos(x-y) + 1 + \sin^2(x-y)}$$
$$= \sqrt{\cos^2(x-y) + \sin^2(x-y) - 2\cos(x-y) + 1}$$
$$= \sqrt{1 - 2\cos(x-y) + 1} \qquad \text{[Pythagorean identity]}$$
$$= \sqrt{2 - 2\cos(x-y)}.$$

The two expressions for the distance from P to Q must be equal. Hence,

$$\sqrt{2 - 2\cos(x-y)} = \sqrt{2 - 2\cos x \cos y - 2\sin x \sin y}.$$

Squaring both sides of this equation and simplifying the result yields

$$2 - 2\cos(x-y) = 2 - 2\cos x \cos y - 2\sin x \sin y$$
$$-2\cos(x-y) = -2(\cos x \cos y + \sin x \sin y)$$
$$\cos(x-y) = \cos x \cos y + \sin x \sin y.$$

This completes the proof of the last addition identity when $x > y$. If $y > x$, then the proof just given is valid with the roles of x and y interchanged; it shows that

$$\cos(y-x) = \cos y \cos x + \sin y \sin x$$
$$= \cos x \cos y + \sin x \sin y.$$

The negative angle identity with $x - y$ in place of x shows that

$$\cos(x-y) = \cos[-(x-y)] = \cos(y-x).$$

Combining this fact with the previous one shows that

$$\cos(x-y) = \cos x \cos y + \sin x \sin y$$

in this case also. Therefore, the last addition identity is proved.

The identity for $\cos(x+y)$ now follows from one for $\cos(x-y)$ by using the negative angle identities for sine and cosine:

$$\cos(x+y) = \cos[x - (-y)] = \cos x \cos(-y) + \sin x \sin(-y)$$
$$= \cos x \cos y + (\sin x)(-\sin y) = \cos x \cos y - \sin x \sin y.$$

The proof of the first two cofunction identities on page 502 depended only on the addition identity for $\cos(x-y)$. Since that has been proved, we can validly use the first two cofunction identities in the remainder of the proof. In particular,

$$\sin(x-y) = \cos\left[\frac{\pi}{2} - (x-y)\right] = \cos\left[\left(\frac{\pi}{2} - x\right) + y\right].$$

Applying the proven identity for $\cos(x+y)$ with $(\pi/2) - x$ in place of x and the two cofunction identities now yields

$$\sin(x-y) = \cos\left[\left(\frac{\pi}{2} - x\right) + y\right]$$
$$= \cos\left(\frac{\pi}{2} - x\right)\cos y - \sin\left(\frac{\pi}{2} - x\right)\sin y$$
$$= \sin x \cos y - \cos x \sin y.$$

This proves the second of the addition and subtraction identities. The first is obtained from the second in the same way the third was obtained from the last.

EXERCISES 7.2

In Exercises 1–12, find the exact value.

1. $\sin \dfrac{\pi}{12}$

2. $\cos \dfrac{\pi}{12}$

3. $\tan \dfrac{\pi}{12}$

4. $\sin \dfrac{5\pi}{12}$

5. $\cot \dfrac{5\pi}{12}$

6. $\cos \dfrac{7\pi}{12}$

7. $\tan \dfrac{7\pi}{12}$

8. $\cos \dfrac{11\pi}{12}$

9. $\cot \dfrac{11\pi}{12}$

10. $\sin 75°$ [*Hint:* $75° = 45° + 30°$.]*

11. $\sin 105°$*

12. $\cos 165°$*

In Exercises 13–18, rewrite the given expression in terms of $\sin x$ *and* $\cos x$.

13. $\sin\left(\dfrac{\pi}{2} + x\right)$

14. $\cos\left(x + \dfrac{\pi}{2}\right)$

15. $\cos\left(x - \dfrac{3\pi}{2}\right)$

16. $\csc\left(x + \dfrac{\pi}{2}\right)$

17. $\sec(x - \pi)$

18. $\cot(x + \pi)$

In Exercises 19–24, simplify the given expression.

19. $\sin 3 \cos 5 - \cos 3 \sin 5$

20. $\sin 37° \sin 53° - \cos 37° \cos 53°$

21. $\cos(x + y) \cos y + \sin(x + y) \sin y$

22. $\sin(x - y) \cos y + \cos(x - y) \sin y$

23. $\cos(x + y) - \cos(x - y)$

24. $\sin(x + y) - \sin(x - y)$

25. If $\sin = \dfrac{1}{3}$ and $0 < x < \dfrac{\pi}{2}$, then $\sin\left(\dfrac{\pi}{4} + x\right) = ?$

26. If $\cos x = -\dfrac{1}{4}$ and $\dfrac{\pi}{2} < x < \pi$, then $\cos\left(\dfrac{\pi}{6} - x\right) = ?$

27. If $\cos x = -\dfrac{1}{5}$ and $\pi < x < \dfrac{3\pi}{2}$, then $\sin\left(\dfrac{\pi}{3} - x\right) = ?$

28. If $\sin x = -\dfrac{3}{4}$ and $\dfrac{3\pi}{2} < x < 2\pi$, then $\cos\left(\dfrac{\pi}{4} + x\right) = ?$

In Exercises 29–32, assume that $\sin x = .8$ *and* $\sin y = \sqrt{.75}$ *and that* x, y *lie between* 0 *and* $\pi/2$. *Evaluate the given expressions.*

29. $\sin(x + y)$

30. $\cos(x - y)$

31. $\sin(x - y)$

32. $\tan(x + y)$

33. The figure shows an angle of t radians. Prove that for any number x,
$$5 \sin(x + t) = 3 \sin x + 4 \cos x.$$

34. The figure shows an angle of t radians. Prove that for any number y,
$$13 \cos(t - y) = 12 \cos y + 5 \sin y.$$

35. If $f(x) = \cos x$ and h is a fixed nonzero number, prove that:
$$\dfrac{f(x + h) - f(x)}{h} = \cos x\left(\dfrac{\cos h - 1}{h}\right) - \sin x\left(\dfrac{\sin h}{h}\right).$$

36. Prove the addition and subtraction identities for the tangent function (page 500). [*Hint:*
$$\tan(x + y) = \dfrac{\sin(x + y)}{\cos(x + y)}.$$
Use the addition identities on the numerator and denominator; then divide both numerator and denominator by $\cos x \cos y$ and simplify.]

37. If x is in the first and y is in the second quadrant, $\sin x = 24/25$, and $\sin y = 4/5$, find the exact value of $\sin(x + y)$ and $\tan(x + y)$ and the quadrant in which $x + y$ lies.

38. If x and y are in the second quadrant, $\sin x = 1/3$, and $\cos y = -3/4$, find the exact value of $\sin(x + y)$, $\cos(x + y)$, $\tan(x + y)$, and find the quadrant in which $x + y$ lies.

39. If x is in the first and y is in the second quadrant, $\sin x = 4/5$, and $\cos y = -12/13$, find the exact value of $\cos(x + y)$ and $\tan(x + y)$ and the quadrant in which $x + y$ lies.

40. If x is in the fourth and y is in the first quadrant, $\cos x = 1/3$, and $\cos y = 2/3$, find the exact value of $\sin(x - y)$ and $\tan(x - y)$ and the quadrant in which $x - y$ lies.

*Skip Exercises 10–12 if you haven't read Section 8.1.

41. Express $\sin(u + v + w)$ in terms of sines and cosines of u, v, and w. [*Hint:* First apply the addition identity with $x = u + v$ and $y = w$.]

42. Express $\cos(x + y + z)$ in terms of sines and cosines of x, y, and z.

43. If $x + y = \pi/2$, show that $\sin^2 x + \sin^2 y = 1$.

44. Prove that $\cot(x + y) = \dfrac{\cot x \cot y - 1}{\cot x + \cot y}$.

In Exercises 45–56, prove the identity.

45. $\sin(x - \pi) = -\sin x$ 46. $\cos(x - \pi) = -\cos x$

47. $\cos(\pi - x) = -\cos x$ 48. $\tan(\pi - x) = -\tan x$

49. $\sin(x + \pi) = -\sin x$ 50. $\cos(x + \pi) = -\cos x$

51. $\tan(x + \pi) = \tan x$

52. $\sin x \cos y = \frac{1}{2}[\sin(x + y) + \sin(x - y)]$

53. $\sin x \sin y = \frac{1}{2}[\cos(x - y) - \cos(x + y)]$

54. $\cos x \sin y = \frac{1}{2}[\sin(x + y) - \sin(x - y)]$

55. $\cos(x + y)\cos(x - y) = \cos^2 x \cos^2 y - \sin^2 x \sin^2 y$

56. $\sin(x + y)\sin(x - y) = \sin^2 x \cos^2 y - \cos^2 x \sin^2 y$

In Exercises 57–66, determine graphically whether the equation could possibly be an identity (by choosing a numerical value for y and graphing both sides). If it could, prove that it is.

57. $\dfrac{\cos(x - y)}{\sin x \cos y} = \cot x + \tan y$

58. $\dfrac{\cos(x + y)}{\sin x \cos y} = \cot x - \tan y$

59. $\sin(x - y) = \sin x - \sin y$

60. $\cos(x + y) = \cos x + \cos y$

61. $\dfrac{\sin(x + y)}{\sin(x - y)} = \dfrac{\tan x + \tan y}{\tan x - \tan y}$

62. $\dfrac{\sin(x + y)}{\sin(x - y)} = \dfrac{\cot y + \cot x}{\cot y - \cot x}$

63. $\dfrac{\cos(x + y)}{\cos(x - y)} = \dfrac{\cot x + \tan y}{\cot x - \tan y}$

64. $\dfrac{\cos(x - y)}{\cos(x + y)} = \dfrac{\cot y + \tan x}{\cot y - \tan x}$

65. $\tan(x + y) = \tan x + \tan y$

66. $\cot(x - y) = \cot x - \cot y$

7.2.A   *SPECIAL TOPICS*   Lines and Angles

If L is a nonhorizontal straight line, the **angle of inclination** of L is the angle θ formed by the part of L above the x-axis and the x-axis in the positive direction, as shown in Figure 7–4.

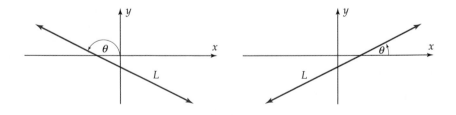

Figure 7–4

The angle of inclination of a horizontal line is defined to be $\theta = 0$. Thus, the radian measure of the angle of inclination of any line satisfies $0 \leq \theta < \pi$. Furthermore,

Angle of Inclination

If L is a nonvertical line with an angle of inclination of θ radians, then

$$\tan \theta = \text{slope of } L.$$

Proof   First, suppose L is horizontal. Then L has slope 0 and angle of inclination $\theta = 0$. Hence,

$$\tan \theta = \tan 0 = 0 = \text{slope } L.$$

Next, suppose L is not horizontal. L is parallel to a line M through the origin, as shown in Figure 7–5.[*]

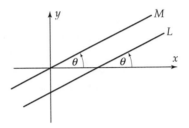

Figure 7–5

Basic facts about parallel lines show that M has the same angle of inclination θ as L. Furthermore, M lies on the terminal side of an angle of θ radians in standard position. Therefore, as we proved in Section 6.4,

$$\text{slope } M = \tan \theta.$$

Since parallel lines have the same slope, we have

$$\text{slope } L = \text{slope } M = \tan \theta. \qquad \blacksquare$$

▉▉ ANGLES BETWEEN TWO LINES

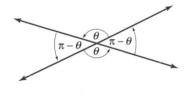

Figure 7–6

If two lines intersect, then they determine four angles with vertex at the point of intersection, as shown in Figure 7–6. If one of these angles measures θ radians, then each of the two angles adjacent to it measures $\pi - \theta$ radians. (Why?) The fourth angle also measures θ radians by the vertical angle theorem from plane geometry.

The angles between intersecting lines can be determined from the angles of inclination of the lines. Suppose L and M have angles of inclination α and β, respectively, such that $\beta \geq \alpha$. Basic facts about parallel lines, as illustrated in Figure 7–7, show that $\beta - \alpha$ is one angle between L and M and $\pi - (\beta - \alpha)$ is the other one.

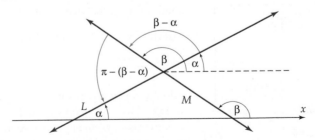

Figure 7–7

---

[*]Figure 7–5 illustrates the case when θ is acute and L lies to the right of M. The pictures are different in the other possible cases, but the argument is the same.

The angle between two lines can also be found from their slopes by using this fact.

Angle Between Two Lines

> If two nonvertical, nonperpendicular lines have slopes m and k, then one angle θ between them satisfies
>
> $$\tan \theta = \left| \frac{m - k}{1 + mk} \right|.$$

Proof   Suppose the line with slope k has angle of inclination α and the line with slope m has angle of inclination β. Then

$$\tan \alpha = k \qquad \text{and} \qquad \tan \beta = m.$$

If $\beta \geq \alpha$, then $\theta = \beta - \alpha$ is one angle between the lines. By the subtraction identity for tangent,

$$\tan \theta = \tan(\beta - \alpha) = \frac{\tan \beta - \tan \alpha}{1 + \tan \beta \tan \alpha} = \frac{m - k}{1 + mk}.$$

The other angle between the lines is $\pi - \theta$, and

$$\tan(\pi - \theta) = \tan(-\theta) \qquad \text{[tangent has period } \pi\text{]}$$

$$= -\tan \theta \qquad \text{[negative angle identity]}$$

$$= -\frac{m - k}{1 + mk}.$$

By the definition of absolute value,

$$\left| \frac{m - k}{1 + mk} \right| = \pm \frac{m - k}{1 + mk},$$

whichever is positive. Thus, the tangent of one of the angles θ or $\pi - \theta$ is

$$\left| \frac{m - k}{1 + mk} \right|.$$

This completes the proof when $\beta \geq \alpha$. The proof when $\alpha \geq \beta$ is similar.   ■

EXAMPLE 1

Find the angle between a line L with slope 8 and a line M of slope -3.

SOLUTION   One angle θ between the lines satisfies

$$\tan \theta = \left| \frac{8 - (-3)}{1 + 8(-3)} \right| = \left| \frac{11}{-23} \right| = \frac{11}{23}.$$

Although you could solve the equation $\tan \theta = 11/23$ graphically, it is easier to use the TAN^{-1} key on your calculator. When you key in TAN$^{-1}(11/23)$, the calculator displays an angle between 0 and $\pi/2$ radians such that $\tan \theta = 11/23$, as in Figure 7–8.* Therefore, $\theta \approx .4461$ radians.   ■

```
tan⁻¹(11/23)
           .4461055489
■
```

Figure 7–8

---

*Tan$^{-1}(x)$ denotes the inverse tangent function, which is explained in Section 7.4. In this context, using the TAN^{-1} key is the electronic equivalent of searching through a table of tangent values until you find a number whose tangent is $11/23$.

We can now prove the following fact, which was first presented in Section 1.4.

Slope Theorem for Perpendicular Lines

Let L be a line with slope k and M a line with slope m. Then

$$L \text{ and } M \text{ are perpendicular exactly when } km = -1.$$

Proof First, suppose L and M are perpendicular. We must show that $km = -1$. If α and β (with $\beta \geq \alpha$) are the angles of inclination of L and M, then $\beta - \alpha$ is the angle between L and M, so $\beta - \alpha = \pi/2$, or, equivalently, $\beta = \alpha + \pi/2$. Therefore, by the addition identities for sine and cosine,

$$m = \tan \beta = \tan\left(\alpha + \frac{\pi}{2}\right) = \frac{\sin[\alpha + (\pi/2)]}{\cos[\alpha + (\pi/2)]}$$

$$= \frac{\sin \alpha \cos(\pi/2) + \cos \alpha \sin(\pi/2)}{\cos \alpha \cos(\pi/2) - \sin \alpha \sin(\pi/2)}$$

$$= \frac{\sin \alpha(0) + \cos \alpha(1)}{\cos \alpha(0) - \sin \alpha(1)}$$

$$= -\frac{\cos \alpha}{\sin \alpha} = -\cot \alpha = \frac{-1}{\tan \alpha} = \frac{-1}{k}.$$

Thus, $m = -1/k$, and hence, $mk = -1$.

Now suppose that $mk = -1$. We must show that L and M are perpendicular. If L and M are *not* perpendicular, then neither of the angles between them is $\pi/2$. In this case, if θ is either of the angles between L and M, then $\tan \theta$ is a well-defined real number. But we know that one of these angles must satisfy

$$\tan \theta = \left| \frac{m - k}{1 + mk} \right| = \left| \frac{m - k}{1 + (-1)} \right| = \frac{|m - k|}{0},$$

which is *not* defined. This contradiction shows that L and M must be perpendicular. ∎

EXERCISES 7.2.A

In Exercises 1–6, find $\tan \theta$*, where* θ *is the angle of inclination of the line through the given points.*

1. $(-1, 2)$, $(3, 5)$

2. $(0, 4)$, $(5, -1)$

3. $(1, 4)$, $(6, 0)$

4. $(4, 2)$, $(-3, -2)$

5. $(3, -7)$, $(3, 5)$

6. $(0, 0)$, $(-4, -5)$

In Exercises 7–13, find one of the angles between the straight lines L and M.

7. L has slope $3/2$ and M has slope -1.

8. L has slope 1 and M has slope 3.

9. L has slope -1 and M has slope 0.

10. L has slope -2 and M has slope -3.

11. $(3, 2)$ and $(5, 6)$ are on L; $(0, 3)$ and $(4, 0)$ are on M.

12. $(-1, 2)$ and $(3, -3)$ are on L; $(3, -3)$ and $(6, 1)$ are on M.

13. L is parallel to the line with equation $y = 3x - 2$ and M is perpendicular to the line with equation $y = -.5x + 1$.

14. If θ is an angle between two nonperpendicular lines with slopes m and k, respectively, and $\tan \theta = \left| \dfrac{m - k}{1 + mk} \right|$, explain why θ is an acute angle. [*Hint*: For what values of θ is $\tan \theta$ positive?]

7.3   Other Identities

We now present a variety of identities that are special cases of the addition and subtraction identities of Section 7.2, beginning with

Double-Angle Identities

$$\sin 2x = 2 \sin x \cos x$$

$$\cos 2x = \cos^2 x - \sin^2 x$$

$$\tan 2x = \frac{2 \tan x}{1 - \tan^2 x}$$

Proof   Let $x = y$ in the addition identities:

$$\sin 2x = \sin(x + x) = \sin x \cos x + \cos x \sin x = 2 \sin x \cos x,$$

$$\cos 2x = \cos(x + x) = \cos x \cos x - \sin x \sin x = \cos^2 x - \sin^2 x,$$

$$\tan 2x = \tan(x + x) = \frac{\tan x + \tan x}{1 - \tan x \tan x} = \frac{2 \tan x}{1 - \tan^2 x}. \qquad \blacksquare$$

EXAMPLE 1

If $\pi < x < 3\pi/2$ and $\cos x = -8/17$, find $\sin 2x$ and $\cos 2x$, and show that $5\pi/2 < 2x < 3\pi$.

SOLUTION   To use the double-angle identities, we first must determine $\sin x$. It can be found by using the Pythagorean identities.

$$\sin^2 x = 1 - \cos^2 x = 1 - \left(-\frac{8}{17}\right)^2 = 1 - \frac{64}{289} = \frac{225}{289}.$$

Since $\pi < x < 3\pi/2$, we know $\sin x$ is negative. Therefore,

$$\sin x = -\sqrt{\frac{225}{289}} = -\frac{15}{17}.$$

We now substitute these values in the double-angle identities.

$$\sin 2x = 2 \sin x \cos x = 2\left(-\frac{15}{17}\right)\left(-\frac{8}{17}\right) = \frac{240}{289} \approx .83$$

$$\cos 2x = \cos^2 x - \sin^2 x = \left(-\frac{8}{17}\right)^2 - \left(-\frac{15}{17}\right)^2$$

$$= \frac{64}{289} - \frac{225}{289} = -\frac{161}{289} \approx -.56.$$

Since $\pi < x < 3\pi/2$, we know that $2\pi < 2x < 3\pi$. The calculations above show that at $2x$, sine is positive and cosine is negative. This can occur only if $2x$ lies between $5\pi/2$ and 3π. $\blacksquare$

EXAMPLE 2

Express the rule of the function $f(x) = \sin 3x$ in terms of $\sin x$ and constants.

SOLUTION   We first use the addition identity for $\sin(x + y)$ with $y = 2x$.

$$f(x) = \sin 3x = \sin(x + 2x) = \sin x \cos 2x + \cos x \sin 2x.$$

Next apply the double-angle identities for $\cos 2x$ and $\sin 2x$.

$$\begin{aligned}
f(x) = \sin 3x &= \sin x \cos 2x + \cos x \sin 2x \\
&= \sin x(\cos^2 x - \sin^2 x) + \cos x(2 \sin x \cos x) \\
&= \sin x \cos^2 x - \sin^3 x + 2 \sin x \cos^2 x \\
&= 3 \sin x \cos^2 x - \sin^3 x.
\end{aligned}$$

Finally, use the Pythagorean identity.

$$\begin{aligned}
f(x) = \sin 3x = 3 \sin x \cos^2 x - \sin^3 x &= 3 \sin x(1 - \sin^2 x) - \sin^3 x \\
&= 3 \sin x - 3 \sin^3 x - \sin^3 x = 3 \sin x - 4 \sin^3 x. \quad \blacksquare
\end{aligned}$$

The double-angle identity for $\cos 2x$ can be rewritten in several useful ways. For instance, we can use the Pythagorean identity in the form of $\cos^2 x = 1 - \sin^2 x$ to obtain:

$$\cos 2x = \cos^2 x - \sin^2 x = (1 - \sin^2 x) - \sin^2 x = 1 - 2 \sin^2 x.$$

Similarly, using the Pythagorean identity in the form $\sin^2 x = 1 - \cos^2 x$, we have:

$$\cos 2x = \cos^2 x - \sin^2 x = \cos^2 x - (1 - \cos^2 x) = 2 \cos^2 x - 1.$$

In summary:

More Double-Angle
Identities

$$\cos 2x = 1 - 2 \sin^2 x$$
$$\cos 2x = 2 \cos^2 x - 1$$

EXAMPLE 3

Prove that

$$\frac{1 - \cos 2x}{\sin 2x} = \tan x.$$

SOLUTION   The first identity in the preceding box and the double-angle identity for sine show that

$$\frac{1 - \cos 2x}{\sin 2x} = \frac{1 - (1 - 2 \sin^2 x)}{2 \sin x \cos x} = \frac{2 \sin^2 x}{2 \sin x \cos x} = \frac{\sin x}{\cos x} = \tan x. \quad \blacksquare$$

If we solve the first equation in the preceding box for $\sin^2 x$ and the second one for $\cos^2 x$, we obtain a useful alternate form for these identities.

Power-Reducing
Identities

$$\sin^2 x = \frac{1 - \cos 2x}{2}$$

$$\cos^2 x = \frac{1 + \cos 2x}{2}$$

EXAMPLE 4

Express the rule of the function $f(x) = \sin^4 x$ in terms of constants and first powers of the cosine function.

SOLUTION   We begin by applying the power-reducing identity.

$$f(x) = \sin^4 x = \sin^2 x \sin^2 x = \frac{1 - \cos 2x}{2} \cdot \frac{1 - \cos 2x}{2}$$

$$= \frac{1 - 2\cos 2x + \cos^2 2x}{4}.$$

Next we apply the power-reducing identity for cosine to $\cos^2 2x$. Note that this means using $2x$ in place of x in the identity.

$$\cos^2 2x = \frac{1 + \cos 2(2x)}{2} = \frac{1 + \cos 4x}{2}.$$

Finally, we substitute this last result in the expression for $\sin^4 x$ above.

$$f(x) = \sin^4 x = \frac{1 - 2\cos 2x + \cos^2 2x}{4} = \frac{1 - 2\cos 2x + \dfrac{1 + \cos 4x}{2}}{4}$$

$$= \frac{1}{4} - \frac{1}{2}\cos 2x + \frac{1}{8}(1 + \cos 4x)$$

$$= \frac{3}{8} - \frac{1}{2}\cos 2x + \frac{1}{8}\cos 4x. \quad \blacksquare$$

▪▪ HALF-ANGLE IDENTITIES

If we use the power-reducing identity with $x/2$ in place of x, we obtain

$$\sin^2\left(\frac{x}{2}\right) = \frac{1 - \cos 2\left(\dfrac{x}{2}\right)}{2} = \frac{1 - \cos x}{2}.$$

Consequently, we must have

$$\sin\left(\frac{x}{2}\right) = \pm\sqrt{\frac{1 - \cos x}{2}}.$$

This proves the first of the half-angle identities.

Half-Angle Identities

$$\sin\frac{x}{2} = \pm\sqrt{\frac{1-\cos x}{2}} \qquad \cos\frac{x}{2} = \pm\sqrt{\frac{1+\cos x}{2}}$$

$$\tan\frac{x}{2} = \pm\sqrt{\frac{1-\cos x}{1+\cos x}}$$

The half-angle identity for cosine is derived from a power-reducing identity, as was the half-angle identity for sine. The half-angle identity for tangent then follows immediately since $\tan(x/2) = \sin(x/2)/\cos(x/2)$. In all cases, *the sign in front of the radical depends on the quadrant in which $x/2$ lies.*

EXAMPLE 5

Find the exact value of

(a) $\cos 5\pi/8$ (b) $\sin \pi/12$.

SOLUTION

(a) Since $5\pi/8 = \frac{1}{2}(5\pi/4)$, we use the half-angle identity with $x = 5\pi/4$ and the fact that $\cos(5\pi/4) = -\sqrt{2}/2$. The sign chart in Exercise 49 on page 430 shows that $\cos(5\pi/8)$ is negative because $5\pi/8$ is in the second quadrant. So we use the negative sign in front of the radical.

$$\cos\frac{5\pi}{8} = \cos\frac{5\pi/4}{2} = -\sqrt{\frac{1+\cos(5\pi/4)}{2}}$$

$$= -\sqrt{\frac{1+(-\sqrt{2}/2)}{2}} = -\sqrt{\frac{(2-\sqrt{2})/2}{2}}$$

$$= -\sqrt{\frac{2-\sqrt{2}}{4}}$$

$$= \frac{-\sqrt{2-\sqrt{2}}}{2}.$$

(b) Since $\pi/12 = \frac{1}{2}(\pi/6)$ and $\pi/12$ is in the first quadrant, where sine is positive, we have

$$\sin\frac{\pi}{12} = \sin\frac{\pi/6}{2} = \sqrt{\frac{1-\cos(\pi/6)}{2}}$$

$$= \sqrt{\frac{1-\sqrt{3}/2}{2}} = \sqrt{\frac{(2-\sqrt{3})/2}{2}} = \sqrt{\frac{2-\sqrt{3}}{4}}$$

$$= \frac{\sqrt{2-\sqrt{3}}}{2} \qquad \blacksquare$$

The problem of determining signs in the half-angle formulas can be eliminated with tangent by using these identities.

Half-Angle Identities for Tangent

$$\tan\frac{x}{2} = \frac{1 - \cos x}{\sin x}$$

$$\tan\frac{x}{2} = \frac{\sin x}{1 + \cos x}$$

Proof   In the identity

$$\tan x = \frac{1 - \cos 2x}{\sin 2x},$$

which was proved in Example 3, replace x by $x/2$.

$$\tan\left(\frac{x}{2}\right) = \frac{1 - \cos 2(x/2)}{\sin 2(x/2)} = \frac{1 - \cos x}{\sin x}.$$

The second identity in the box is proved in Exercise 71.   ■

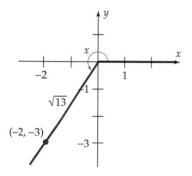

Figure 7–9

EXAMPLE 6

If $\tan x = \dfrac{3}{2}$ and $\pi < x < \dfrac{3\pi}{2}$, find $\tan\dfrac{x}{2}$.

SOLUTION   The terminal side of an angle of x radians in standard position lies
in the third quadrant, as shown in Figure 7–9. The tangent of the angle in stan-
dard position whose terminal side passes through the point $(-2, -3)$ is $\dfrac{-3}{-2} = \dfrac{3}{2}$.
Since there is only one angle in the third quadrant with tangent $3/2$, the point
$(-2, -3)$ must lie on the terminal side of the angle of x radians.
    Since the distance from $(-2, -3)$ to the origin is

$$\sqrt{(-2 - 0)^2 + (-3 - 0)^2} = \sqrt{13},$$

we have

$$\sin x = \frac{-3}{\sqrt{13}} \qquad \text{and} \qquad \cos x = \frac{-2}{\sqrt{13}}.$$

Therefore, by the first of the half-angle identities for tangent

$$\tan\frac{x}{2} = \frac{1 - \cos x}{\sin x} = \frac{1 - \left(\dfrac{-2}{\sqrt{13}}\right)}{\dfrac{-3}{\sqrt{13}}} = \frac{\dfrac{\sqrt{13} + 2}{\sqrt{13}}}{\dfrac{-3}{\sqrt{13}}} = -\frac{\sqrt{13} + 2}{3}. \qquad ■$$

▪▪ PRODUCT AND FACTORING IDENTITIES

Using the addition and subtraction identities to compute

$$\sin(x + y) + \sin(x - y),$$

we see that

$$\sin(x + y) + \sin(x - y) = \sin x \cos y + \cos x \sin y + \sin x \cos y - \cos x \sin y$$

$$= 2 \sin x \cos y.$$

Dividing both sides of this last equation by 2 produces the first of the following identities.

Product Identities

$$\sin x \cos y = \frac{1}{2}[\sin(x + y) + \sin(x - y)]$$

$$\sin x \sin y = \frac{1}{2}[\cos(x - y) - \cos(x + y)]$$

$$\cos x \cos y = \frac{1}{2}[\cos(x + y) + \cos(x - y)]$$

$$\cos x \sin y = \frac{1}{2}[\sin(x + y) - \sin(x - y)]$$

The proofs of the second and fourth product identities are similar to the proof of the first. The third product identity was proved in Example 4 of Section 7.2.

If we use the first product identity with $\frac{1}{2}(x + y)$ in place of x and $\frac{1}{2}(x - y)$ in place of y, we obtain

$$\sin\left[\frac{1}{2}(x + y)\right] \cos\left[\frac{1}{2}(x - y)\right] = \frac{1}{2}\left[\sin\left(\frac{1}{2}(x + y) + \frac{1}{2}(x - y)\right)\right.$$
$$\left. + \sin\left(\frac{1}{2}(x + y) - \frac{1}{2}(x - y)\right)\right]$$

$$= \frac{1}{2}(\sin x + \sin y).$$

Multiplying both sides of the last equation by 2 produces the first of the following identities.

Factoring Identities

$$\sin x + \sin y = 2 \sin\left(\frac{x + y}{2}\right)\cos\left(\frac{x - y}{2}\right)$$

$$\sin x - \sin y = 2 \cos\left(\frac{x + y}{2}\right)\sin\left(\frac{x - y}{2}\right)$$

$$\cos x + \cos y = 2 \cos\left(\frac{x + y}{2}\right)\cos\left(\frac{x - y}{2}\right)$$

$$\cos x - \cos y = -2 \sin\left(\frac{x + y}{2}\right)\sin\left(\frac{x - y}{2}\right)$$

The last three factoring identities are proved in the same way as the first.

EXAMPLE 7

Prove the identity

$$\frac{\sin t + \sin 3t}{\cos t + \cos 3t} = \tan 2t.$$

SOLUTION   Using the first factoring identity with $x = t$ and $y = 3t$ yields

$$\sin t + \sin 3t = 2 \sin\left(\frac{t + 3t}{2}\right)\cos\left(\frac{t - 3t}{2}\right) = 2 \sin 2t \cos(-t).$$

Similarly,

$$\cos t + \cos 3t = 2 \cos\left(\frac{t + 3t}{2}\right)\cos\left(\frac{t - 3t}{2}\right) = 2 \cos 2t \cos(-t),$$

so

$$\frac{\sin t + \sin 3t}{\cos t + \cos 3t} = \frac{2 \sin 2t \cos(-t)}{2 \cos 2t \cos(-t)} = \frac{\sin 2t}{\cos 2t} = \tan 2t. \quad \blacksquare$$

EXERCISES 7.3

In Exercises 1–12, use the half-angle identities to evaluate the given expression exactly.

1. $\cos\dfrac{\pi}{8}$ **2.** $\tan\dfrac{\pi}{8}$ **3.** $\sin\dfrac{3\pi}{8}$ **4.** $\cos\dfrac{3\pi}{8}$

5. $\tan\dfrac{\pi}{12}$ **6.** $\sin\dfrac{5\pi}{8}$ **7.** $\cos\dfrac{\pi}{12}$ **8.** $\tan\dfrac{5\pi}{8}$

9. $\sin\dfrac{7\pi}{8}$ **10.** $\cos\dfrac{7\pi}{8}$ **11.** $\tan\dfrac{7\pi}{8}$ **12.** $\cot\dfrac{\pi}{8}$

In Exercises 13–18, write each expression as a sum or difference.

13. $\sin 4x \cos 6x$ **14.** $\sin 5x \sin 7x$

15. $\cos 2x \cos 4x$ **16.** $\sin 3x \cos 5x$

17. $\sin 17x \sin(-3x)$ **18.** $\cos 13x \cos(-5x)$

In Exercises 19–22, write each expression as a product.

19. $\sin 3x + \sin 5x$ **20.** $\cos 2x + \cos 6x$

21. $\sin 9x - \sin 5x$ **22.** $\cos 5x - \cos 7x$

In Exercises 23–29, find $\sin 2x$, $\cos 2x$, and $\tan 2x$ under the given conditions.

23. $\sin x = \dfrac{5}{13} \quad \left(0 < x < \dfrac{\pi}{2}\right)$

24. $\sin x = -\dfrac{4}{5} \quad \left(\pi < x < \dfrac{3\pi}{2}\right)$

25. $\cos x = -\dfrac{3}{5} \quad \left(\pi < x < \dfrac{3\pi}{2}\right)$

26. $\cos x = -\dfrac{1}{3} \quad \left(\dfrac{\pi}{2} < x < \pi\right)$

27. $\tan x = \dfrac{3}{4} \quad \left(\pi < x < \dfrac{3\pi}{2}\right)$

28. $\tan x = -\dfrac{3}{2} \quad \left(\dfrac{\pi}{2} < x < \pi\right)$

29. $\csc x = 4 \quad \left(0 < x < \dfrac{\pi}{2}\right)$

30. A batter hits a baseball that is caught by a fielder. If the ball leaves the bat at an angle of θ radians to the horizontal, with an initial velocity of v feet per second, then the approximate horizontal distance d traveled by the ball is given by

$$d = \frac{v^2 \sin \theta \cos \theta}{16}.$$

(a) If the initial velocity is 90 feet per second, find the horizontal distances traveled by the ball when $\theta = .5$ radian, $\theta = .75$ radian, and $\theta = 1$ radian.

(b) Do part (a) for an initial velocity of 105 feet per second.

(c) Use an identity to show that

$$d = \frac{v^2 \sin 2\theta}{32}.$$

(d) If the initial velocity is 115 ft/second, what angle θ will produce the maximum distance? [*Hint:* Use part (c). For what value of θ is $\sin 2\theta$ as large as possible?]

In Exercises 31–36, find $\sin \frac{x}{2}$, $\cos \frac{x}{2}$, *and* $\tan \frac{x}{2}$ *under the given conditions.*

31. $\cos x = .4 \quad \left(0 < x < \frac{\pi}{2}\right)$

32. $\sin x = .6 \quad \left(\frac{\pi}{2} < x < \pi\right)$

33. $\sin x = -\frac{3}{5} \quad \left(\frac{3\pi}{2} < x < 2\pi\right)$

34. $\cos x = .8 \quad \left(\frac{3\pi}{2} < x < 2\pi\right)$

35. $\tan x = \frac{1}{2} \quad \left(\pi < x < \frac{3\pi}{2}\right)$

36. $\cot x = 1 \quad \left(-\pi < x < -\frac{\pi}{2}\right)$

In Exercises 37–42, assume $\sin = .6$ *and* $0 < x < \pi/2$ *and evaluate the given expression.*

37. $\sin 2x$ **38.** $\cos 4x$ **39.** $\cos 2x$ **40.** $\sin 4x$

41. $\sin\dfrac{x}{2}$ **42.** $\cos\dfrac{x}{2}$

43. Express $\cos 3x$ in terms of $\cos x$.

44. (a) Express the rule of the function $f(x) = \cos^3 x$ in terms of constants and first powers of the cosine function as in Example 4.

(b) Do the same for $f(x) = \cos^4 x$.

In Exercises 45–50, simplify the given expression.

45. $\dfrac{\sin 2x}{2 \sin x}$ **46.** $1 - 2 \sin^2\left(\dfrac{x}{2}\right)$

47. $2 \cos 2y \sin 2y$ (Think!)

48. $\cos^2\left(\dfrac{x}{2}\right) - \sin^2\left(\dfrac{x}{2}\right)$ **49.** $(\sin x + \cos x)^2 - \sin 2x$

50. $2 \sin x \cos^3 x - 2 \sin^3 x \cos x$

In Exercises 51–64, determine graphically whether the equation could possibly be an identity. If it could, prove that it is.

51. $\sin 16x = 2 \sin 8x \cos 8x$

52. $\cos 8x = \cos^2 4x - \sin^2 4x$

53. $\cos^4 x - \sin^4 x = \cos 2x$

54. $\sec 2x = \dfrac{1}{1 - 2 \sin^2 x}$

55. $\cos 4x = 2 \cos 2x - 1$

56. $\sin^2 x = \cos^2 x - 2 \sin x$

57. $\dfrac{1 + \cos 2x}{\sin 2x} = \cot x$ **58.** $\sin 2x = \dfrac{2 \cot x}{\csc^2 x}$

59. $\sin 3x = (\sin x)(3 - 4 \sin^2 x)$

60. $\sin 4x = (4 \cos x \sin x)(1 - 2 \sin^2 x)$

61. $\cos 2x = \dfrac{2 \tan x}{\sec^2 x}$

62. $\cos 3x = (\cos x)(3 - 4 \cos^2 x)$

63. $\csc^2\left(\dfrac{x}{2}\right) = \dfrac{2}{1 - \cos x}$ **64.** $\sec^2\left(\dfrac{x}{2}\right) = \dfrac{2}{1 + \cos x}$

In Exercises 65–70, prove the identity.

65. $\dfrac{\sin x - \sin 3x}{\cos x + \cos 3x} = -\tan x$

66. $\dfrac{\sin x - \sin 3x}{\cos x - \cos 3x} = -\cot 2x$

67. $\dfrac{\sin 4x + \sin 6x}{\cos 4x - \cos 6x} = \cot x$

68. $\dfrac{\cos 8x + \cos 4x}{\cos 8x - \cos 4x} = -\cot 6x \cot 2x$

69. $\dfrac{\sin x + \sin y}{\cos x - \cos y} = -\cot\left(\dfrac{x - y}{2}\right)$

70. $\dfrac{\sin x - \sin y}{\cos x + \cos y} = \tan\left(\dfrac{x - y}{2}\right)$

71. (a) Prove that $\dfrac{1 - \cos x}{\sin x} = \dfrac{\sin x}{1 + \cos x}$.

(b) Use part (a) and the half-angle identity proved in the text to prove that

$$\tan \frac{x}{2} = \frac{\sin x}{1 + \cos x}.$$

7.4   Inverse Trigonometric Functions

Before reading this section, you should review the concept of an inverse function (Section 3.7). As explained there, a function f has an inverse function only when its graph passes the Horizontal Line Test:

No horizontal line intersects the graph of f more than once.

The graphs of the sine, cosine, and tangent functions certainly do not have this property. However, functions that are closely related to them (same rules but smaller domains) *do* have inverse functions.

The **restricted sine function** is defined as follows:

$$\text{Domain: } [-\pi/2, \pi/2] \qquad \text{Rule: } f(x) = \sin x.$$

Its graph in Figure 7–10 shows that for each number v between -1 and 1, there is exactly one number u between $-\pi/2$ and $\pi/2$ such that $\sin u = v$.

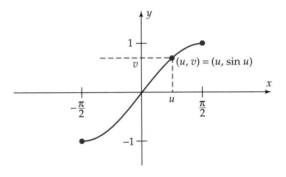

Figure 7–10

Since the graph of the restricted sine function passes the horizontal line test, we know that it has an inverse function. This inverse function is called the **inverse sine** (or **arcsine**) **function** and is denoted by $g(x) = \sin^{-1}x$ **or** $g(x) = $ **arcsin x.** The domain of the inverse sine function is the interval $[-1, 1]$, and its rule is as follows.

Inverse Sine Function

For each v with $-1 \le v \le 1$,

$\sin^{-1}v = $ the unique number u between $-\pi/2$ and $\pi/2$ whose sine is v;

that is,

$$\sin^{-1}v = u \qquad \text{exactly when} \qquad \sin u = v.$$

EXAMPLE 1

Find

(a) $\sin^{-1}(1/2)$

(b) $\sin^{-1}(-\sqrt{2}/2)$.

SOLUTION

(a) $\text{Sin}^{-1}(1/2)$ is the one number between $-\pi/2$ and $\pi/2$ whose sine is $1/2$. From our study of special values, we know that $\sin \pi/6 = 1/2$, *and* $\pi/6$ is between $-\pi/2$ and $\pi/2$. Hence, $\sin^{-1}(1/2) = \pi/6$.

(b) $\text{Sin}^{-1}(-\sqrt{2}/2) = -\pi/4$ because $\sin(-\pi/4) = -\sqrt{2}/2$ and $-\pi/4$ is between $-\pi/2$ and $\pi/2$. ■

EXAMPLE 2

Except for special values, you should use the SIN^{-1} key (labeled ASIN on some calculators) in *radian mode* to evaluate the inverse sine function. For instance,

$$\sin^{-1}(-.67) = -.7342 \qquad \text{and} \qquad \sin^{-1}(.42) = .4334. \quad ■$$

EXAMPLE 3

If you key in SIN^{-1} 2 ENTER, you will get an error message, because 2 is not in the domain of the inverse sine function.* ■

 CAUTION

The notation $\sin^{-1}x$ is *not* exponential notation. It does *not* mean either $(\sin x)^{-1}$ or $\dfrac{1}{\sin x}$. For instance, Example 1 shows that

$$\sin^{-1}(1/2) = \pi/6 \approx .5236,$$

but

$$\left(\sin \frac{1}{2}\right)^{-1} = \frac{1}{\sin \dfrac{1}{2}} \approx \frac{1}{.4794} \approx 2.0858.$$

Suppose $-1 \le v \le 1$ and $\sin^{-1}v = u$. Then by the definition of the inverse sine function, we know that $-\pi/2 \le u \le \pi/2$ and $\sin u = v$. Therefore,

$$\sin^{-1}(\sin u) = \sin^{-1}(v) = u \qquad \text{and} \qquad \sin(\sin^{-1}v) = \sin u = v.$$

This shows that the restricted sine function and the inverse sine function have the usual "round-trip properties" of inverse functions. In summary,

Properties of Inverse Sine

$$\sin^{-1}(\sin u) = u \qquad \text{if} \qquad -\frac{\pi}{2} \le u \le \frac{\pi}{2}$$

$$\sin(\sin^{-1}v) = v \qquad \text{if} \qquad -1 \le v \le 1$$

---

*TI-85/86 and HP-39 display the complex number $(1.5707\cdots, -1.3169\cdots)$ for $\sin^{-1}(2)$. Casio FX 2.0 does the same when it is in complex mode. For our purposes, this is equivalent to an error message, since we deal only with functions whose values are real numbers.

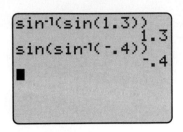

Figure 7–11

A calculator can illustrate the identities in the preceding box, as shown in Figure 7–11. Nevertheless, when special values are involved, you should be able to deal with them by hand.

EXAMPLE 4

Find   (a) $\sin^{-1}(\sin \pi/6)$ (b) $\sin^{-1}(\sin 5\pi/6)$.

SOLUTION

(a) We know that $\sin \pi/6 = 1/2$. Hence,

$$\sin^{-1}\left(\sin \frac{\pi}{6}\right) = \sin^{-1}\left(\frac{1}{2}\right) = \frac{\pi}{6}$$

because $\pi/6$ is the number between $-\pi/2$ and $\pi/2$ whose sine is $1/2$.

(b) We also have $\sin 5\pi/6 = 1/2$, so the expression $\sin^{-1}(\sin 5\pi/6)$ is defined. However,

$$\sin^{-1}\left(\sin \frac{5\pi}{6}\right) \qquad \text{is NOT equal to} \qquad \frac{5\pi}{6}$$

because the identity in the box on page 519 is valid only when u is between $-\pi/2$ and $\pi/2$. Using the result of part (a), we see that

$$\sin^{-1}\left(\sin \frac{5\pi}{6}\right) = \sin^{-1}\left(\frac{1}{2}\right) = \frac{\pi}{6}. \qquad \blacksquare$$

Recall that the graph of the inverse function of f can be obtained in two ways: Reverse the coordinates of each point on the graph of f or, equivalently, reflect the graph of f in the line $y = x$, as explained on pages 209–210. In summary,

Restricted Sine Function	Inverse Sine Function
$f(x) = \sin x$	$g(x) = \sin^{-1}x$
Domain: $[-\pi/2, \pi/2]$	Domain: $[-1, 1]$
Range: $[-1, 1]$	Range: $[-\pi/2, \pi/2]$

▪▪ THE INVERSE COSINE FUNCTION

The **restricted cosine function** is defined as follows:

$$\text{Domain: } [0, \pi] \qquad \text{Rule: } f(x) = \cos x.$$

Its graph in Figure 7–12 shows that for each number v between -1 and 1, there is exactly one number u between 0 and π such that $\cos u = v$.

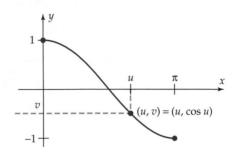

Figure 7–12

Since the graph of the restricted cosine function passes the Horizontal Line Test, we know that it has an inverse function. This inverse function is called the **inverse cosine** (or **arccosine**) **function** and is denoted by $h(x) = \cos^{-1}x$ or $h(x) = \arccos x$. The domain of the inverse cosine function is the interval $[-1, 1]$, and its rule is as follows.

Inverse Cosine Function

For each v with $-1 \le v \le 1$,

$$\cos^{-1}v = \text{the unique number } u \text{ between 0 and } \pi \text{ whose cosine is } v;$$

that is,

$$\cos^{-1}v = u \qquad \text{exactly when} \qquad \cos u = v.$$

The inverse cosine function has these properties:

$$\cos^{-1}(\cos u) = u \qquad \text{if} \qquad 0 \le u \le \pi;$$
$$\cos(\cos^{-1}v) = v \qquad \text{if} \qquad -1 \le v \le 1.$$

EXAMPLE 5

Find   (a) $\cos^{-1}(1/2)$ (b) $\cos^{-1}(0)$ (c) $\cos^{-1}(-.63)$.

SOLUTION

(a) $\text{Cos}^{-1}(1/2) = \pi/3$ since $\pi/3$ is the unique number between 0 and π whose cosine is $1/2$.

(b) $\text{Cos}^{-1}(0) = \pi/2$ because $\cos \pi/2 = 0$ and $0 \le \pi/2 \le \pi$.

(c) The COS^{-1} key on a calculator in *radian mode* shows that
$\cos^{-1}(-.63) = 2.2523.$ ■

CAUTION

$\text{Cos}^{-1}x$ does *not* mean $(\cos x)^{-1}$, or $1/\cos x$.

EXAMPLE 6

Write $\sin(\cos^{-1} v)$ as an algebraic expression in v.

SOLUTION $\text{Cos}^{-1} v = u$, where $\cos u = v$ and $0 \le u \le \pi$. Hence, $\sin u$ is non-negative, and by the Pythagorean identity, $\sin u = \sqrt{\sin^2 u} = \sqrt{1 - \cos^2 u}$. Also, $\cos^2 u = v^2$. Therefore,

$$\sin(\cos^{-1} v) = \sin u = \sqrt{1 - \cos^2 u} = \sqrt{1 - v^2}. \quad \blacksquare$$

EXAMPLE 7

Prove the identity $\sin^{-1} x + \cos^{-1} x = \pi/2$.

SOLUTION   Suppose $\sin^{-1} x = u$, with $-\pi/2 \le u \le \pi/2$. Verify that $0 \le \pi/2 - u \le \pi$ (Exercise 24). Then we have

$$\sin u = x \qquad \text{[Definition of inverse sine]}$$

$$\cos\left(\frac{\pi}{2} - u\right) = x \qquad \text{[Cofunction identity]}$$

$$\cos^{-1} x = \frac{\pi}{2} - u. \qquad \text{[Definition of inverse cosine]}$$

Therefore,

$$\sin^{-1} x + \cos^{-1} x = u + \left(\frac{\pi}{2} - u\right) = \frac{\pi}{2}. \quad \blacksquare$$

The graph of the inverse cosine function is the reflection of the graph of the restricted cosine function in the line $y = x$, as shown below.

Restricted Cosine Function	Inverse Cosine Function
$f(x) = \cos x$	$g(x) = \cos^{-1} x$
Domain: $[0, \pi]$	Domain: $[-1, 1]$
Range: $[-1, 1]$	Range: $[0, \pi]$

▪▪ THE INVERSE TANGENT FUNCTION

The **restricted tangent function** is defined as follows:

$$\text{Domain: } (-\pi/2,\ \pi/2) \qquad \text{Rule: } f(x) = \tan x.$$

Its graph in Figure 7–13 shows that for every real number v, there is exactly one number u between $-\pi/2$ and $\pi/2$ such that $\tan u = v$.

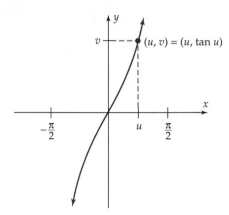

Figure 7–13

Since the graph of the restricted tangent function passes the horizontal line test, we know that it has an inverse function. This inverse function is called the **inverse tangent** (or **arctangent**) **function** and is denoted by $g(x) = \tan^{-1}x$ or $g(x) =$ **arctan x.** The domain of the inverse tangent function is the set of all real numbers, and its rule is as follows.

Inverse Tangent Function

> For each real number v,
>
> $\tan^{-1}v =$ the unique number u between $-\pi/2$ and $\pi/2$ whose tangent is v;
>
> that is,
>
> $$\tan^{-1}v = u \qquad \text{exactly when} \qquad \tan u = v.$$
>
> The inverse tangent function has these properties:
>
> $$\tan^{-1}(\tan u) = u \qquad \text{if} \qquad -\frac{\pi}{2} < u < \frac{\pi}{2};$$
>
> $$\tan(\tan^{-1}v) = v \qquad \text{for every number } v.$$

CAUTION

$\text{Tan}^{-1}x$ does *not* mean $(\tan x)^{-1}$, or $1/\tan x$.

EXAMPLE 8

$\text{Tan}^{-1}1 = \pi/4$ because $\pi/4$ is the unique number between $-\pi/2$ and $\pi/2$ such that $\tan \pi/4 = 1$. A calculator in *radian mode* shows that $\tan^{-1}(136) = 1.5634$. ∎

EXAMPLE 9

Find the exact value of $\cos[\tan^{-1}(\sqrt{5}/2)]$.

SOLUTION   Consider an angle of u radians in standard position whose terminal side passes through $(2, \sqrt{5})$, as in Figure 7–14. By the point-in-the-plane description,

$$\tan u = \sqrt{5}/2.$$

Since u is between $-\pi/2$ and $\pi/2$ and $\tan u = \sqrt{5}/2$, we must have

$$u = \tan^{-1}(\sqrt{5}/2).$$

Furthermore, the distance from $(2, \sqrt{5})$ to the origin is

$$\sqrt{(2 - 0)^2 + (\sqrt{5} - 0)^2} = \sqrt{4 + 5} = 3,$$

so

$$\cos u = 2/3.$$

Therefore,

$$\cos[\tan^{-1}(\sqrt{5}/2)] = \cos u = 2/3. \quad ∎$$

The graph of the inverse tangent function is the reflection of the graph of the restricted tangent function in the line $y = x$, as shown below.

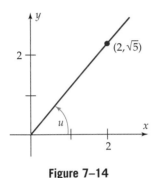

Figure 7–14

Restricted Tangent Function	Inverse Tangent Function
$f(x) = \tan x$	$g(x) = \tan^{-1}x$
Domain: $(-\pi/2, \pi/2)$	Domain: All real numbers
Range: All real numbers	Range: $(-\pi/2, \pi/2)$

EXERCISES 7.4

In Exercises 1–14, find the exact functional value without using a calculator.

1. $\sin^{-1} 1$ **2.** $\cos^{-1} 0$ **3.** $\tan^{-1}(-1)$

4. $\sin^{-1}(-1)$ **5.** $\cos^{-1} 1$ **6.** $\tan^{-1} 1$

7. $\tan^{-1}(\sqrt{3}/3)$ **8.** $\cos^{-1}(\sqrt{3}/2)$

9. $\sin^{-1}(-\sqrt{2}/2)$ **10.** $\sin^{-1}(\sqrt{3}/2)$

11. $\tan^{-1}(-\sqrt{3})$ **12.** $\cos^{-1}(-\sqrt{2}/2)$

13. $\cos^{-1}\left(-\dfrac{1}{2}\right)$ **14.** $\sin^{-1}\left(-\dfrac{1}{2}\right)$

In Exercises 15–23, use a calculator in radian mode to approximate the functional value.

15. $\sin^{-1} .35$ **16.** $\cos^{-1} .76$

17. $\tan^{-1}(-3.256)$ **18.** $\sin^{-1}(-.795)$

19. $\sin^{-1}(\sin 7)$ [The answer is *not* 7.]

20. $\cos^{-1}(\cos 3.5)$ **21.** $\tan^{-1}[\tan(-4)]$

22. $\sin^{-1}[\sin(-2)]$ **23.** $\cos^{-1}[\cos(-8.5)]$

24. Let u be a number such that

$$-\frac{\pi}{2} \le u \le \frac{\pi}{2}.$$

Prove that

$$0 \le \frac{\pi}{2} - u \le \pi.$$

25. Given that $u = \sin^{-1}(-\sqrt{3}/2)$, find the exact value of $\cos u$ and $\tan u$.

26. Given that $u = \tan^{-1}(4/3)$, find the exact value of $\sin u$ and $\sec u$.

In Exercises 27–42, find the exact functional vlaue without using a calculator.

27. $\sin^{-1}(\cos 0)$ **28.** $\cos^{-1}(\sin \pi/6)$

29. $\cos^{-1}(\sin 4\pi/3)$ **30.** $\tan^{-1}(\cos \pi)$

31. $\sin^{-1}(\cos 7\pi/6)$ **32.** $\cos^{-1}(\tan 7\pi/4)$

33. $\sin^{-1}(\sin 2\pi/3)$ (See Exercise 19.)

34. $\cos^{-1}(\cos 5\pi/4)$ **35.** $\cos^{-1}[\cos(-\pi/6)]$

36. $\tan^{-1}[\tan(-4\pi/3)]$

37. $\sin[\cos^{-1}(3/5)]$ (See Example 9.)

38. $\tan[\sin^{-1}(3/5)]$ **39.** $\cos[\tan^{-1}(-3/4)]$

40. $\cos[\sin^{-1}(\sqrt{3}/5)]$ **41.** $\tan[\sin^{-1}(5/13)]$

42. $\sin[\cos^{-1}(3/\sqrt{13})]$

In Exercises 43–46, write the expression as an algebraic expression in v, as in Example 6.

43. $\cos(\sin^{-1} v)$ **44.** $\cot(\cos^{-1} v)$

45. $\tan(\sin^{-1} v)$ **46.** $\sin(2 \sin^{-1} v)$

In Exercises 47–50, graph the function.

47. $f(x) = \cos^{-1}(x + 1)$ **48.** $g(x) = \tan^{-1} x + \pi$

49. $h(x) = \sin^{-1}(\sin x)$ **50.** $k(x) = \sin(\sin^{-1} x)$

51. In an alternating current circuit, the voltage is given by the formula

$$V = V_{max} \cdot \sin(2\pi ft + \phi),$$

where V_{max} is the maximum voltage, f is the frequency (in cycles per second), t is the time in seconds, and ϕ is the phase angle.

(a) If the phase angle is 0, solve the voltage equation for t.

(b) If $\phi = 0$, $V_{max} = 20$, $V = 8.5$, and $f = 120$, find the smallest positive value of t.

52. A model plane 40 feet above the ground is flying away from an observer.

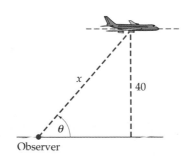

(a) Express the angle of elevation θ of the plane as a function of the distance x from the observer to the plane.

(b) What is θ when the plane is 250 feet away from the observer?

53. A 15-foot-wide highway sign is placed 10 feet from a road, perpendicular to the road (see figure). A spotlight at the edge of the road is aimed at the sign.

(a) Express θ as a function of the distance x from point A to the spotlight.

(b) How far from point A should the spotlight be placed so that the angle θ is as large as possible?

54. A camera on a 5-foot-high tripod is placed in front of a 6-foot-high picture that is mounted 3 feet above the floor.

(a) Express angle θ as a function of the distance x from the camera to the wall.

(b) The photographer wants to use a particular lens, for which $\theta = 36°$ ($\pi/5$ radians). How far should she place the camera from the wall to be sure that the entire picture will show in the photograph?

55. Show that the restricted secant function, whose domain consists of all numbers x such that $0 \le x \le \pi$ and $x \ne \pi/2$, has an inverse function. Sketch its graph.

56. Show that the restricted cosecant function, whose domain consists of all numbers x such that $-\pi/2 \le x \le \pi/2$ and $x \ne 0$, has an inverse function. Sketch its graph.

57. Show that the restricted cotangent function, whose domain is the interval $(0, \pi)$, has an inverse function. Sketch its graph.

58. (a) Show that the inverse cosine function actually has the two properties listed in the box on page 521.

(b) Show that the inverse tangent function actually has the two properties listed in the box on page 523.

In Exercises 59–66, prove the identity.

59. $\sin^{-1}(-x) = -\sin^{-1}x$ [*Hint:* Let $u = \sin^{-1}(-x)$ and show that $\sin^{-1}x = -u$.]

60. $\tan^{-1}(-x) = -\tan^{-1}x$

61. $\cos^{-1}(-x) = \pi - \cos^{-1}x$ [*Hint:* Let $u = \cos^{-1}(-x)$ and show that $0 \le \pi - u \le \pi$; use the identity
$$\cos(\pi - u) = -\cos u.]$$

62. $\sin^{-1}(\cos x) = \pi/2 - x \quad (0 \le x \le \pi)$

63. $\tan^{-1}(\cot x) = \pi/2 - x \quad (0 < x < \pi)$

64. $\tan^{-1}x + \tan^{-1}\left(\dfrac{1}{x}\right) = \dfrac{\pi}{2}$

65. $\sin^{-1}x = \tan^{-1}\left(\dfrac{x}{\sqrt{1 - x^2}}\right) \quad (-1 < x < 1)$

[*Hint:* Let $u = \sin^{-1}x$ and show that $\tan u = x/\sqrt{1 - x^2}$. Since $\sin u = x$, $\cos u = \pm\sqrt{1 - x^2}$. Show that in this case, $\cos u = \sqrt{1 - x^2}$.]

66. $\cos^{-1}x = \dfrac{\pi}{2} - \tan^{-1}\left(\dfrac{x}{\sqrt{1 - x^2}}\right) \quad (-1 < x < 1)$

[*Hint:* See Example 7 and Exercise 65.]

67. Is it true that $\tan^{-1}x = \dfrac{\sin^{-1}x}{\cos^{-1}x}$? Justify your answer.

68. Using the viewing window with $-2\pi \le x \le 2\pi$ and $-4 \le y \le 4$ graph the functions $f(x) = \cos(\cos^{-1}x)$ and $g(x) = \cos^{-1}(\cos x)$. How do you explain the shapes of the two graphs?

7.5 Trigonometric Equations

Any equation that involves trigonometric functions can be solved graphically, and many can be solved algebraically. Unlike the equations solved previously, trigonometric equations typically have an infinite number of solutions. In most cases, these solutions can be systematically determined by using periodicity, as we now see.

■■ BASIC EQUATIONS

We begin with **basic equations,** such as

$$\sin x = .39, \qquad \cos x = .2, \qquad \tan x = -3.$$

Basic equations can be solved by the methods illustrated in Examples 1–3.

EXAMPLE 1

Solve $\tan x = 2$.

SOLUTION   The equation can be solved graphically by graphing $y = \tan x$ and $y = 2$ on the same coordinate axes and finding the intersection points. The x-coordinate of every such point is a number whose tangent is 2, that is, a solution of the equation.

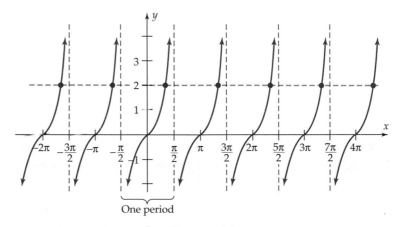

Figure 7–15

Figure 7–15 shows that there is exactly one solution in each period of $\tan x$ (for instance, between $-\pi/2$ and $\pi/2$). This solution could be found graphically, but it's faster to key in

$$\text{TAN}^{-1}2 \quad \text{ENTER}^*$$

---

*Unless stated otherwise, radian mode is used throughout this section.

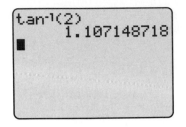

Figure 7–16

The calculator then displays the number between $-\pi/2$ and $\pi/2$ whose tangent is 2, namely, $x = 1.1071$, as shown in Figure 7–16.* Since the tangent graph repeats its pattern with period π, all the other solutions differ from this one by a multiple of π. Thus, all the solutions are

$$1.1071, \quad 1.1071 \pm \pi, \quad 1.1071 \pm 2\pi, \quad 1.1071 \pm 3\pi, \quad \text{etc.}$$

These solutions are customarily written like this:

$$x = 1.1071 + k\pi \quad (k = 0, \pm 1, \pm 2, \pm 3, \ldots). \quad \blacksquare$$

EXAMPLE 2

Solve $\sin x = -.75$

SOLUTION   The solutions are the x-coordinates of the points where the graphs of $y = \sin x$ and $y = -.75$ intersect (why?). Note that there are exactly two solutions in every period of $\sin x$ (for instance, between $-\pi/2$ and $3\pi/2$).

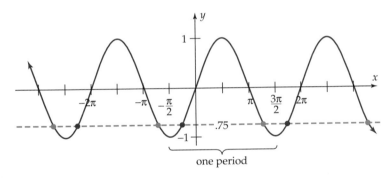

Figure 7–17

Figure 7–17 shows that there is one solution between $-\pi/2$ and $\pi/2$. By keying in

$$\text{SIN}^{-1} - .75 \quad \text{ENTER},$$

we find that this solution is $x = -.8481$, as shown in Figure 7–18. Since the sine graph repeats its pattern with period 2π, all of the following numbers are also solutions:

$$-.8481, \quad -.8481 \pm 2\pi, \quad -.8481 \pm 4\pi, \quad -.8481 \pm 6\pi, \quad \text{etc.}$$

These solutions correspond to the red intersection points in Figure 7–17, each of which is 2π units from the next red point. As you can see, there are still more solutions (corresponding to the blue points). One of them can be found by using the identity that was proved in Example 2 of Section 7.2:

$$\sin(\pi - x) = \sin x.$$

Figure 7–18

---

*For convenient reading, all solutions in the text are rounded to four or fewer decimal places, but the full decimal expansion given by the calculator is used in all computations. We write $=$ rather than $\approx$ even though these calculator solutions are approximations of the actual solutions.

Applying this identity with the solution $x = -.8481$ shows that

$$\sin[\pi - (-.8481] = \sin(-.8481) = -.75.$$

In other words, $\pi - (-.8481) = 3.9897$ is also a solution of $\sin x = -.75$. The other solutions of the equation are

$$3.9897, \qquad 3.9897 \pm 2\pi, \qquad 3.9897 \pm 4\pi, \qquad 3.9897 \pm 6\pi, \qquad \text{etc.,}$$

corresponding to the blue intersection points in Figure 7–17, each of which is 2π units from the next blue point. Therefore, all the solutions of $\sin x = -0.75$ are

$$x = -.8481 + 2k\pi \qquad \text{and} \qquad x = 3.9897 + 2k\pi$$

$$(k = 0, \pm 1, \pm 2, \pm 3, \ldots). \qquad \blacksquare$$

EXAMPLE 3

Solve $\cos x = \sqrt{3}/2$.

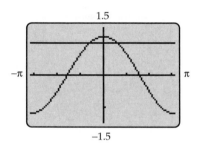

1.5

$-\pi$ π

-1.5

Figure 7–19

SOLUTION   By graphing $y = \cos x$ and $y = \sqrt{3}/2$ on the same screen (Figure 7–19), we see that there are two solutions of the equation between $-\pi$ and π (one full period of cosine). The positive solution could be approximated by the COS^{-1} key on a calculator. In this case, however, our knowledge of special values provides an exact solution. As we saw in Section 6.2, $\cos \pi/6 = \sqrt{3}/2$. Therefore, one solution of the equation is $x = \pi/6$. The negative angle identity $\cos(-x) = \cos x$ shows that the second solution is $x = -\pi/6$, because

$$\cos\left(-\frac{\pi}{6}\right) = \cos\left(\frac{\pi}{6}\right) = \frac{\sqrt{3}}{2}.$$

Since the interval $[-\pi, \pi]$ is one full period of cosine, all the solutions of the equation are

$$x = \frac{\pi}{6} + 2k\pi \qquad \text{and} \qquad x = -\frac{\pi}{6} + 2k\pi \qquad (k = 0, \pm 1, \pm 2, \pm 3, \ldots). \qquad \blacksquare$$

The procedures in Examples 1–3 work in the general case and lead to the following conclusion.

Solution Algorithm for Basic Trigonometric Equations

If an equation of the form

$$\sin x = c, \qquad \cos x = c, \qquad \tan x = c$$

has solutions, then one solution may be found by using the appropriate inverse function $[\sin^{-1}(x), \cos^{-1}(x), \text{ or } \tan^{-1}(x)]$ or by using your knowledge of special values.

When one solution u has been found, then the remaining solutions are as follows.

Equation	All Solutions
$\sin x = c$	$u + 2k\pi$ and $(\pi - u) + 2k\pi$ $(k = 0, \pm 1, \pm 2, \pm 3, \ldots)$
$\cos x = c$	$u + 2k\pi$ and $-u + 2k\pi$ $(k = 0, \pm 1, \pm 2, \pm 3, \ldots)$
$\tan x = c$	$u + k\pi$ $(k = 0, \pm 1, \pm 2, \pm 3, \ldots)$

EXAMPLE 4

Find all solutions of $\sec x = 8$ in the interval $[0, 2\pi)$.

SOLUTION Note that $\sec x = 8$ exactly when

$$\frac{1}{\cos x} = 8 \quad \text{or, equivalently,} \quad \cos x = \frac{1}{8} = .125.$$

Since $\cos^{-1}(.125) = 1.4455$, the solutions of $\cos x = .125$, and hence of $\sec x = 8$, are

$$x = 1.4455 + 2k\pi \quad \text{and} \quad x = -1.4455 + 2k\pi$$

$$(k = 0, \pm 1, \pm 2, \pm 3, \ldots).$$

Of these solutions, the two between 0 and 2π are

$$x = 1.4455 \quad \text{and} \quad x = -1.4455 + 2\pi = 4.8377. \quad \blacksquare$$

EXAMPLE 5

Solve $\sin u = \sqrt{2}/2$ without using a calculator.

SOLUTION Our knowledge of special values shows that $u = \pi/4$ is one solution. Hence, a second solution is

$$u = \pi - \frac{\pi}{4} = \frac{3\pi}{4},$$

and all solutions are

$$u = \frac{\pi}{4} + 2k\pi \quad \text{and} \quad u = \frac{3\pi}{4} + 2k\pi \quad (k = 0, \pm 1, \pm 2, \pm 3, \ldots). \quad \blacksquare$$

EXAMPLE 6

Solve exactly: $\sin 2x = \sqrt{2}/2$.

SOLUTION First, let $u = 2x$ and solve the basic equation $\sin u = \sqrt{2}/2$. As we saw in Example 5, the solutions are

$$u = \frac{\pi}{4} + 2k\pi \quad \text{and} \quad u = \frac{3\pi}{4} + 2k\pi \quad (k = 0, \pm 1, \pm 2, \pm 3, \ldots).$$

Since $u = 2x$, each of these solutions leads to a solution of the original equation.

$$2x = u = \frac{\pi}{4} + 2k\pi \quad \text{or, equivalently,} \quad x = \frac{1}{2}\left(\frac{\pi}{4} + 2k\pi\right) = \frac{\pi}{8} + k\pi.$$

Similarly,

$$2x = u = \frac{3\pi}{4} + 2k\pi \quad \text{or, equivalently,} \quad x = \frac{1}{2}\left(\frac{3\pi}{4} + 2k\pi\right) = \frac{3\pi}{8} + k\pi.$$

Therefore, all solutions of $\sin 2x = \sqrt{2}/2$ are given by

$$x = \frac{\pi}{8} + k\pi \quad \text{and} \quad x = \frac{3\pi}{8} + k\pi \quad (k = 0, \pm 1, \pm 2, \pm 3, \ldots).$$

The fact that the solutions are obtained by adding multiples of π rather than 2π is a reflection of the fact that the period of $\sin 2x$ is π. ■

ALGEBRAIC SOLUTION OF OTHER TRIGONOMETRIC EQUATIONS

Many trigonometric equations can be solved algebraically by using factoring, the quadratic formula, and identities to reduce the problem to an equivalent one that involves only basic equations.

EXAMPLE 7

Solve $-10 \cos^2 x - 3 \sin x + 9 = 0$.

SOLUTION   We first use the Pythagorean identity to rewrite the equation in terms of the sine function.

$$-10 \cos^2 x - 3 \sin x + 9 = 0$$
$$-10(1 - \sin^2 x) - 3 \sin x + 9 = 0$$
$$-10 + 10 \sin^2 x - 3 \sin x + 9 = 0$$
$$10 \sin^2 x - 3 \sin x - 1 = 0.$$

Now factor the left side:[*]

$$(2 \sin x - 1)(5 \sin x + 1) = 0$$
$$2 \sin x - 1 = 0 \quad \text{or} \quad 5 \sin x + 1 = 0$$
$$2 \sin x = 1 \qquad\qquad 5 \sin x = -1$$
$$\sin x = 1/2 \qquad\qquad \sin x = -1/5 = -.2.$$

Each of these basic equations is readily solved. We note that $\sin(\pi/6) = 1/2$, so $x = \pi/6$ and $x = \pi - \pi/6 = 5\pi/6$ are solutions of the first one. Since $\sin^{-1}(-.2) = -.2014$, both $x = -.2014$ and $x = \pi - (-.2014) = 3.3430$ are solutions of the second equation. Therefore, all solutions of the original equation are given by

$$x = \frac{\pi}{6} + 2k\pi, \quad x = \frac{5\pi}{6} + 2k\pi, \quad x = -.2014 + 2k\pi,$$

$$x = 3.3430 + 2k\pi,$$

where $k = 0, \pm 1, \pm 2, \pm 3, \ldots$. ■

---

[*]The factorization may be easier to see if you first substitute v for $\sin x$, so that $10 \sin^2 x - 3 \sin x - 1$ becomes $10v^2 - 3v - 1 = (2v - 1)(5v + 1)$.

EXAMPLE 8

Solve $\sec^2 x + 5 \tan x = -2$.

SOLUTION   We use the Pythagorean identity $\sec^2 x = 1 + \tan^2 x$ to obtain an equivalent equation.

$$\sec^2 x + 5 \tan x = -2$$

$$\sec^2 x + 5 \tan x + 2 = 0$$

$$(1 + \tan^2 x) + 5 \tan x + 2 = 0$$

$$\tan^2 x + 5 \tan x + 3 = 0.$$

If we let $u = \tan x$, this last equation becomes $u^2 + 5u + 3 = 0$. Since the left side does not readily factor, we use the quadratic formula to solve the equation.

$$u = \frac{-5 \pm \sqrt{5^2 - 4 \cdot 1 \cdot 3}}{2} = \frac{-5 \pm \sqrt{13}}{2}.$$

Since $u = \tan x$, the original equation is equivalent to

$$\tan x = \frac{-5 + \sqrt{13}}{2} \approx -.6972 \quad \text{or} \quad \tan x = \frac{-5 - \sqrt{13}}{2} \approx -4.3028.$$

Solving these basic equations as above, we find that $x = -.6089$ is a solution of the first and $x = -1.3424$ is a solution of the second. Hence, the solutions of the original equation are

$$x = -.6089 + k\pi \quad \text{and} \quad x = -1.3424 + k\pi$$

$$(k = 0, \pm 1, \pm 2, \pm 3, \ldots). \quad \blacksquare$$

EXAMPLE 9

Solve $5 \cos x + 3 \cos 2x = 3$.

SOLUTION   We use the double-angle identity: $\cos 2x = 2 \cos^2 x - 1$ as follows.

$$5 \cos x + 3 \cos 2x = 3$$

Use double-angle identity: $\qquad 5 \cos x + 3(2 \cos^2 x - 1) = 3$

Multiply out left side: $\qquad 5 \cos x + 6 \cos^2 x - 3 = 3$

Rearrange terms: $\qquad 6 \cos^2 x + 5 \cos x - 6 = 0$

Factor left side: $\qquad (2 \cos x + 3)(3 \cos x - 2) = 0$

$$2 \cos x + 3 = 0 \quad \text{or} \quad 3 \cos x - 2 = 0$$

$$2 \cos x = -3 \qquad\qquad 3 \cos x = 2$$

$$\cos x = -\frac{3}{2} \qquad\qquad \cos x = \frac{2}{3}.$$

The equation $\cos x = -3/2$ has no solutions because $\cos x$ always lies between -1 and 1. A calculator shows that the solutions of $\cos x = 2/3$ are

$$x = .8411 + 2k\pi \qquad \text{and} \qquad x = -.8411 + 2k\pi$$

$$(k = 0, \pm 1, \pm 2, \pm 3, \ldots). \qquad \blacksquare$$

▟▙ GRAPHICAL SOLUTION METHOD

When the techniques of the preceding examples are inadequate, trigonometric equations may be solved by the following graphical procedure.

Graphical Method for Solving Trigonometric Equations

1. Write the equation in the form $f(x) = 0$.

2. Determine the period of p of $f(x)$.

3. Graph $f(x)$ over an interval of length p.

4. Use a graphical root finder to determine the x-intercepts of the graph in this interval.

5. For each x-intercept u, all of the numbers
$$u + kp \qquad (k = 0, \pm 1, \pm 2, \pm 3, \ldots)$$
are solutions of the equation.

EXAMPLE 10

Solve $3 \sin^2 x - \cos x - 2 = 0$,

SOLUTION   Both sine and cosine have period 2π, so $f(x) = 3 \sin^2 x - \cos x - 2$ also has period 2π. Figure 7–20 shows one full period of the graph of f. A graphical root finder shows that the four x-intercepts (solutions of the equation) in this window are

$$x = 1.1216, \qquad x = 2.4459, \qquad x = 3.8373, \qquad x = 5.1616.$$

Since the graph repeats its pattern to the left and right, the other x-intercepts (solutions) will differ from these four by multiples of 2π. For instance, in addition to the solution $x = 1.1216$, each of the following is a solution.

$$x = 1.1216 \pm 2\pi, \qquad x = 1.1216 \pm 4\pi, \qquad x = 1.1216 \pm 6\pi, \text{ etc.}$$

A similar analysis applies to the other solutions between 0 and 2π. Hence, all solutions of the equation are given by

$$x = 1.1216 + 2k\pi, \qquad x = 2.4459 + 2k\pi, \qquad x = 3.8373 + 2k\pi,$$

$$x = 5.1616 + 2k\pi, \qquad \text{where } k = 0, \pm 1, \pm 2, \pm 3, \ldots. \qquad \blacksquare$$

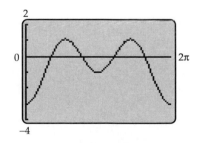

Figure 7–20

EXAMPLE 11

Solve $\tan x = 3 \sin 2x$.

SOLUTION   We first rewrite the equation as

$$\tan x - 3 \sin 2x = 0.$$

Both $\tan x$ and $\sin 2x$ have period π (see pages 451 and 456). Hence, the function given by the left side of the equation, $f(x) = \tan x - 3 \sin 2x$, also has period π. The graph of f on the interval $[0, \pi)$ (Figure 7–21) shows an erroneous vertical line segment at $x = \pi/2$, where tangent is not defined, as well as x-intercepts at the endpoints of the interval. Consequently, we use the more easily read graph f in Figure 7–22, which uses the interval $(-\pi/2, \pi/2)$.

Figure 7–21

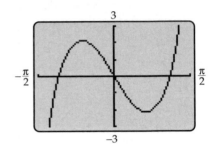

Figure 7–22

Even without the graph, we can verify that there is an x-intercept at the origin because

$$f(0) = \tan 0 - 3 \sin(2 \cdot 0) = 0.$$

A root finder shows that the other two x-intercepts in Figure 7–22 are

$$x = -1.1503 \qquad \text{and} \qquad x = 1.1503.$$

Since $f(x)$ has period π, all solutions of the equation are given by

$$x = -1.1503 + k\pi, \qquad x = 0 + k\pi, \qquad x = 1.1503 + k\pi$$

$$(k = 0, \pm 1, \pm 2, \pm 3, \ldots). \qquad \blacksquare$$

✓ **EXERCISES 7.5**

In all exercises, find exact solutions if possible (as in Examples 3, 5, and 6) and approximate ones otherwise. When a calculator is used, round your answers (but not any intermediate results) to four decimal places.

In Exercises 1–10, find all the solutions of the equation.

1. $\sin x = -.465$

2. $\sin x = -.682$

3. $\cos x = -.564$

4. $\cos x = -.371$

5. $\tan x = -.237$

6. $\tan x = -12.45$

7. $\cot x = 2.3$ [*Remember:* $\cot x = 1/\tan x.$]

8. $\cot x = -3.5$ 9. $\sec x = -2.65$

10. $\csc x = 5.27$

In Exercises 11–14, approximate all solutions in $[0, 2\pi)$ of the given equation.

11. $\sin x = .119$ **12.** $\cos x = .958$

13. $\tan x = 5$ **14.** $\tan x = 17.65$

In Exercises 15–22, use your knowledge of special values to find the exact solutions of the equation.

15. $\sin x = \sqrt{3}/2$ **16.** $2 \cos x = \sqrt{2}$

17. $\tan x = -\sqrt{3}$ **18.** $\tan x = 1$

19. $2 \cos x = -\sqrt{3}$ **20.** $\sin x = 0$

21. $2 \sin x + 1 = 0$ **22.** $\csc x = \sqrt{2}$

In Exercises 23–32, find all the solutions of the equation.

23. $\sin 2x = -\sqrt{3}/2$ **24.** $\cos 2x = \sqrt{2}/2$

25. $2 \cos \dfrac{x}{2} = \sqrt{2}$ **26.** $2 \sin \dfrac{x}{3} = 1$

27. $\tan 3x = -\sqrt{3}$ **28.** $5 \sin 2x = 2$

29. $5 \cos 3x = -3$ **30.** $2 \tan 4x = 16$

31. $4 \tan \dfrac{x}{2} = 8$ **32.** $5 \sin \dfrac{x}{4} = 4$

At the instant you hear a sonic boom from an airplane overhead, your angle of elevation α to the plane is given by the equation $\sin \alpha = 1/m$, where m is the Mach number for the speed of the plane (Mach 1 is the speed of sound, Mach 2.5 is 2.5 times the speed of sound, etc.). In Exercises 33–36, find the angle of elevation (in degrees) for the given Mach number. Remember that an angle of elevation must be between $0°$ and $90°$.

33. $m = 1.1$ **34.** $m = 1.6$

35. $m = 2$ **36.** $m = 2.4$

When a light beam passes from one medium to another (for instance, from air to water), it changes both its speed and direction. According to Snell's Law of Refraction,

$$\frac{\sin \theta_1}{\sin \theta_2} = \frac{v_1}{v_2},$$

where v_1 is the speed of light in the first medium, v_2 its speed in the second medium, θ_1 the angle of incidence, and θ_2 the angle of refraction, as shown in the figure. The number v_1/v_2 is called the index of refraction. *Use this information to do Exercises 37–40.*

Angle of incidence

Incident ray, speed v_1

θ_1

Refracted ray, speed v_2

θ_2

Angle of refraction

37. The index of refraction of light passing from air to water is 1.33. If the angle of incidence is $38°$, find the angle of refraction.

38. The index of refraction of light passing from air to ordinary glass is 1.52. If the angle of incidence is $17°$, find the angle of refraction.

39. The index of refraction of light passing from air to dense glass is 1.66. If the angle of incidence is $24°$, find the angle of refraction.

40. The index of refraction of light passing from air to quartz is 1.46. If the angle of incidence is $50°$, find the angle of refraction.

In Exercises 41–66, use factoring, the quadratic formula, or identities to solve the equation. Find all solutions in the interval $[0, 2\pi)$.

41. $3 \sin^2 x - 8 \sin x - 3 = 0$

42. $5 \cos^2 x + 6 \cos x = 8$

43. $2 \tan^2 x + 5 \tan x + 3 = 0$

44. $3 \sin^2 x + 2 \sin x = 5$

45. $\cot x \cos x = \cos x$ [Be careful; see Exercise 97.]

46. $\tan x \cos x = \cos x$ **47.** $\cos x \csc x = 2 \cos x$

48. $\tan x \sec x + 3 \tan x = 0$

49. $4 \sin x \tan x - 3 \tan x + 20 \sin x - 15 = 0$
[*Hint:* One factor is $\tan x + 5$.]

50. $25 \sin x \cos x - 5 \sin x + 20 \cos x = 4$

51. $\sin^2 x + 2 \sin x - 2 = 0$ **52.** $\cos^2 x + 5 \cos x = 1$

53. $\tan^2 x + 1 = 3 \tan x$ **54.** $4 \cos^2 x - 2 \cos x = 1$

55. $2 \tan^2 x - 1 = 3 \tan x$ **56.** $6 \sin^2 x + 4 \sin x = 1$

57. $\sin^2 x + 3 \cos^2 x = 0$ **58.** $\sec^2 x - 2 \tan^2 x = 0$

59. $\sin 2x + \cos x = 0$ **60.** $\cos 2x - \sin x = 1$

61. $9 - 12 \sin x = 4 \cos^2 x$ **62.** $\sec^2 x + \tan x = 3$

63. $\cos^2 x - \sin^2 x + \sin x = 0$

64. $2 \tan^2 x + \tan x = 5 - \sec^2 x$

65. $\sin \dfrac{x}{2} = 1 - \cos x$ **66.** $4 \sin^2\left(\dfrac{x}{2}\right) + \cos^2 x = 2$

In Exercises 67–76, find all angles θ with $0° \le \theta < 360°$ that are solutions of the given equation. [Hint: Put your calculator in degree mode and replace π by $180°$ in the solution algorithms for basic equations.]

67. $\tan \theta = 7.95$ **68.** $\tan \theta = 69.4$

69. $\cos \theta = -.42$ **70.** $\cot \theta = -2.4$

71. $2 \sin^2\theta + 3 \sin \theta + 1 = 0$

72. $4 \cos^2\theta + 4 \cos \theta - 3 = 0$

73. $\tan^2\theta - 3 = 0$ **74.** $2 \sin^2\theta = 1$

75. $4 \cos^2\theta + 4 \cos \theta + 1 = 0$

76. $\sin^2\theta - 3 \sin \theta = 10$

In Exercises 77–88, solve the equation graphically.

77. $4 \sin 2x - 3 \cos 2x = 2$

78. $5 \sin 3x + 6 \cos 3x = 1$

79. $3 \sin^3 2x = 2 \cos x$

80. $\sin^2 2x - 3 \cos 2x + 2 = 0$

81. $\tan x + 5 \sin x = 1$

82. $2 \cos^2 x + \sin x + 1 = 0$

83. $\cos^3 x - 3 \cos x + 1 = 0$

84. $\tan x = 3 \cos x$

85. $\cos^4 x - 3 \cos^3 x + \cos x = 1$

86. $\sec x + \tan x = 3$

87. $\sin^3 x + 2 \sin^2 x - 3 \cos x + 2 = 0$

88. $\csc^2 x + \sec x = 1$

89. The number of hours of daylight in Detroit on day t of a non–leap year (with $t = 0$ being January 1) is given by the function

$$d(t) = 3 \sin\left[\dfrac{2\pi}{365}(t - 80)\right] + 12.$$

(a) On what days of the year are there exactly 11 hours of daylight?
(b) What day has the maximum amount of daylight?

90. A weight hanging from a spring is set into motion (see Figure 6–62 on page 462), moving up and down. Its distance (in centimeters) above or below the equilibrium point at time t seconds is given by

$$d = 5(\sin 6t - 4 \cos 6t).$$

At what times during the first 2 seconds is the weight at the equilibrium position ($d = 0$)?

In Exercises 91–94, use the following fact: When a projectile (such as a ball or a bullet) leaves its starting point at angle of elevation θ with velocity v, the horizontal distance d it travels is given by the equation

$$d = \dfrac{v^2}{32} \sin 2\theta,$$

where d is measured in feet and v in feet per second. Note that the horizontal distance traveled may be the same for two different angles of elevation, so some of these exercises may have more than one correct answer.

(a)   (b)

91. If muzzle velocity of a rifle is 300 feet per second, at what angle of elevation (in radians) should it be aimed for the bullet to hit a target 2500 feet away?

92. Is it possible for the rifle in Exercise 91 to hit a target that is 3000 feet away? [At what angle of elevation would it have to be aimed?]

93. A fly ball leaves the bat at a velocity of 98 mph and is caught by an outfielder 288 feet away. At what angle of elevation (in degrees) did the ball leave that bat?

94. An outfielder throws the ball at a speed of 75 mph to the catcher who is 200 feet away. At what angle of elevation was the ball thrown?

Thinkers

95. Under what conditions (on the constant) does a basic equation involving the sine and cosine function have *no* solutions?

96. Do Exercise 95 for the secant and cosecant functions.

97. What is wrong with this so-called solution?

$$\sin x \tan x = \sin x$$

$$\tan x = 1$$

$$x = \dfrac{\pi}{4} \quad \text{or} \quad \dfrac{5\pi}{4}.$$

[*Hint:* Solve the original equation by moving all terms to one side and factoring. Compare your answers with the ones above.]

98. Let n be a fixed positive integer. Describe *all* solutions of the equation $\sin nx = 1/2$. [*Hint:* See Exercises 23–32.]

Chapter 7 Review

IMPORTANT CONCEPTS

Section 7.1

Reciprocal identities   492
Periodicity identities   492
Pythagorean identities   492
Negative angle identities   492
Strategies and proof techniques
   for identities   492–496

Section 7.2

Addition and subtraction identities
   499–500
Cofunction identities   502

Special Topics 7.2.A

Angle of inclination   506
Angle between two lines   508
Slope theorem for perpendicular
   lines   509

Section 7.3

Double-angle identities   510–511
Power-reducing identities   512
Half-angle identities   513–514
Product identities   515
Factoring identities   515

Section 7.4

Inverse sine function   518
Inverse cosine function   521
Inverse tangent function   523

Section 7.5

Solution algorithms for basic
   trigonometric equations   529
Algebraic solution of trigonometric
   equations   531–532
Graphical solution of trigonometric
   equations   533

IMPORTANT FACTS & FORMULAS

- All identities in the Chapter 6 Review
- *Addition and Subtraction Identities:*

$$\sin(x + y) = \sin x \cos y + \cos x \sin y$$

$$\sin(x - y) = \sin x \cos y - \cos x \sin y$$

$$\cos(x + y) = \cos x \cos y - \sin x \sin y$$

$$\cos(x - y) = \cos x \cos y + \sin x \sin y$$

$$\tan(x + y) = \frac{\tan x + \tan y}{1 - \tan x \tan y}$$

$$\tan(x - y) = \frac{\tan x - \tan y}{1 + \tan x \tan y}$$

- *Cofunction Identities:*

$$\sin x = \cos\left(\frac{\pi}{2} - x\right) \qquad \cos x = \sin\left(\frac{\pi}{2} - x\right)$$

$$\tan x = \cot\left(\frac{\pi}{2} - x\right) \qquad \cot x = \tan\left(\frac{\pi}{2} - x\right)$$

$$\sec x = \csc\left(\frac{\pi}{2} - x\right) \qquad \csc x = \sec\left(\frac{\pi}{2} - x\right)$$

■ *Double-Angle Identities:*

$$\sin 2x = 2 \sin x \cos x$$

$$\cos 2x = \cos^2 x - \sin^2 x$$

$$\cos 2x = 2 \cos^2 x - 1 \qquad \cos 2x = 1 - 2 \sin^2 x$$

$$\tan 2x = \frac{2 \tan x}{1 - \tan^2 x}$$

■ *Half-Angle Identities:*

$$\sin \frac{x}{2} = \pm \sqrt{\frac{1 - \cos x}{2}}$$

$$\cos \frac{x}{2} = \pm \sqrt{\frac{1 + \cos x}{2}}$$

$$\tan \frac{x}{2} = \frac{1 - \cos x}{\sin x} \qquad \tan \frac{x}{2} = \frac{\sin x}{1 + \cos x} \qquad \tan \frac{x}{2} = \pm \sqrt{\frac{1 - \cos x}{1 + \cos x}}$$

■ $\sin^{-1} v = u$ exactly when $\sin u = v \quad \left(-\dfrac{\pi}{2} \le u \le \dfrac{\pi}{2}, -1 \le v \le 1\right)$

■ $\cos^{-1} v = u$ exactly when $\cos u = v \quad (0 \le u \le \pi, -1 \le v \le 1)$

■ $\tan^{-1} v = u$ exactly when $\tan u = v \quad \left(-\dfrac{\pi}{2} < u < \dfrac{\pi}{2}, \text{any } v\right)$

REVIEW QUESTIONS

In Questions 1–4, simplify the given expression.

1. $\dfrac{\sin^2 t + (\tan^2 t + 2 \tan t - 4) + \cos^2 t}{3 \tan^2 t - 3 \tan t}$

2. $\dfrac{\sec^2 t \csc t}{\csc^2 t \sec t}$

3. $\dfrac{\tan^2 x - \sin^2 x}{\sec^2 x}$

4. $\dfrac{(\sin x + \cos x)(\sin x - \cos x) + 1}{\sin^2 x}$

In Questions 5–12, determine graphically whether the equation could possibly be an identity. If it could, prove that it is.

5. $\sin^4 t - \cos^4 t = 2 \sin^2 t - 1$

6. $1 + 2 \cos^2 t + \cos^4 t = \sin^4 t$

7. $\dfrac{\sin t}{1 - \cos t} = \dfrac{1 + \cos t}{\sin t}$

8. $\dfrac{\sin^2 t}{\cos^2 t} + 1 = \dfrac{1}{\cos^2 t}$

9. $\dfrac{\cos^2(\pi + t)}{\sin^2(\pi + t)} - 1 = \dfrac{1}{\sin^2 t}$

10. $\tan x + \cot x = \sec x \csc x$

11. $(\sin x + \cos x)^2 - \sin 2x = 1$

12. $\dfrac{1 - \cos 2x}{\tan x} = \sin 2x$

In Questions 13–22, prove the given identity.

13. $\dfrac{\tan x - \sin x}{2 \tan x} = \sin^2\left(\dfrac{x}{2}\right)$

14. $2 \cos x - 2 \cos^3 x = \sin x \sin 2x$

15. $\cos(x + y)\cos(x - y) = \cos^2 x - \sin^2 y$

16. $\dfrac{\cos(x - y)}{\cos x \cos y} = 1 + \tan x \tan y$

17. $\dfrac{\sec x + 1}{\tan x} = \dfrac{\tan x}{\sec x - 1}$

18. $\dfrac{\cos^4 x - \sin^4 x}{1 - \tan^4 x} = \cos^4 x$

REVIEW QUESTIONS 💿

19. $\dfrac{1 + \tan^2 x}{\tan^2 x} = \csc^2 x$

20. $\sec x - \cos x = \sin x \tan x$

21. $\tan^2 x - \sec^2 x = \cot^2 x - \csc^2 x$

22. $\sin 2x = \dfrac{1}{\tan x + \cot 2x}$

23. If $\tan x = 5/12$ and $\sin x > 0$, find $\sin 2x$.

24. If $\cos x = 15/17$ and $0 < x < \pi/2$, find $\sin(x/2)$.

25. If $\tan x = 4/3$ with $\pi < x < 3\pi/2$, and $\cot y = -5/12$ with $3\pi/2 < y < 2\pi$, find $\sin(x - y)$.

26. If $\sin x = -12/13$ with $\pi < x < 3\pi/2$, and $\sec y = 13/12$ with $3\pi/2 < y < 2\pi$, find $\cos(x + y)$.

27. If $\sin x = 1/4$ and $0 < x < \pi/2$, then $\sin(\pi/3 + x) = ?$

28. If $\sin x = -2/5$ and $3\pi/2 < x < 2\pi$, then $\cos(\pi/4 + x) = ?$

29. If $\sin x = 0$, is it true that $\sin 2x = 0$? Justify your answer.

30. If $\cos x = 0$, is it true that $\cos 2x = 0$? Justify your answer.

31. Show that
$$\sqrt{2 + \sqrt{3}} = \dfrac{\sqrt{2} + \sqrt{6}}{2}$$
by computing $\cos(\pi/12)$ in two ways, using the half-angle identity and the subtraction identity for cosine.

32. True or false: $2 \sin x = \sin 2x$. Justify your answer.

33. $\sin(5\pi/12) = ?$

34. Express $\sec(x - \pi)$ in terms of $\sin x$ and $\cos x$.

35. $\sqrt{\dfrac{1 - \cos^2 x}{1 - \sin^2 x}} = \underline{\quad}$.

    (a) $|\tan x|$ (b) $|\cot x|$

    (c) $\sqrt{\dfrac{1 - \sin^2 x}{1 - \cos^2 x}}$ (d) $\sec x$

    (e) undefined

36. $\dfrac{1}{(\csc x)(\sec^2 x)} = \underline{\quad}$.

    (a) $\dfrac{1}{(\sin x)(\cos^2 x)}$ (b) $\sin x - \sin^3 x$

    (c) $\dfrac{1}{(\sin x)(1 + \tan^2 x)}$ (d) $\sin x - \dfrac{1}{1 + \tan^2 x}$

    (e) $1 + \tan^3 x$

37. If $\sin x = .6$ and $0 < x < \pi/2$, find $\sin 2x$.

38. If $\sin x = .6$ and $0 < x < \pi/2$, find $\sin(x/2)$.

39. Find the angle of inclination of the straight line through the points $(2, 6)$ and $(-2, 2)$.

40. Find one of the angles between the line L through the points $(-3, 2)$ and $(5, 1)$ and the line M, which has slope 2.

41. $\cos^{-1}(\sqrt{2}/2) = ?$

42. $\sin^{-1}(\sqrt{3}/2) = ?$

43. $\tan^{-1}\sqrt{3} = ?$

44. $\sin^{-1}(\cos 11\pi/6) = ?$

45. $\cos^{-1}(\sin 5\pi/3) = ?$

46. $\tan^{-1}(\cos 7\pi/2) = ?$

47. $\sin^{-1}(\sin .75) = ?$

48. $\cos^{-1}(\cos 2) = ?$

49. $\sin^{-1}(\sin 8\pi/3) = ?$

50. $\cos^{-1}(\cos 13\pi/4) = ?$

51. Sketch the graph of $f(x) = \tan^{-1} x - \pi$.

52. Sketch the graph of $g(x) = \sin^{-1}(x - 2)$.

53. Find the exact value of $\sin[\cos^{-1}(1/4)]$.

54. Find the exact value of $\sin[\tan^{-1}(1/2) - \cos^{-1}(4/5)]$.

In Questions 55–70, solve the equation by any means. Find exact solutions when possible and approximate ones otherwise.

55. $2 \sin x = 1$

56. $\cos x = \sqrt{3}/2$

57. $\tan x = -1$

58. $\sin 3x = -\sqrt{3}/2$

59. $\sin x = .7$

60. $\cos x = -.8$

61. $\tan x = 13$

62. $\cot x = .4$

63. $2 \sin^2 x + 5 \sin x = 3$

64. $4 \cos^2 x - 2 = 0$

65. $2 \sin^2 x - 3 \sin x = 2$

66. $\cos 2x = \cos x$ [*Hint:* First use an identity.]

67. $\sec^2 x + 3 \tan^2 x = 13$

68. $\sec^2 x = 4 \tan x - 2$

69. $2 \sin^2 x + \sin x - 2 = 0$

70. $\cos^2 x - 3 \cos x - 2 = 0$

In Questions 71–74, solve the equation graphically.

71. $5 \tan x = 2 \sin 2x$

72. $\sin^3 x + \cos^2 x - \tan x = 2$

73. $\sin x + \sec^2 x = 3$

74. $\cos^2 x - \csc^2 x + \tan(x - \pi/2) + 5 = 0$

75. Find all angles θ with $0° \le \theta \le 360°$ such that $\sin \theta = -.7133$.

76. Find all angles θ with $0° \le \theta \le 360°$ such that $\tan \theta = 3.7321$.

77. A cannon has a muzzle velocity of 600 feet per second. At what angle of elevation should it be fired in order to hit a target 3500 feet away? [*Hint:* Use the projectile equation preceding Exercise 91 of Section 7.5.]

78. A weight hanging from a spring is set into motion (see Figure 6–62 on page 462), moving up and down. Its distance (in centimeters) above or below the equilibrium point at time t seconds is given by $d = 5 \sin 3t - 3 \cos 3t$. At what times during the first 2 seconds is the weight at the equilibrium position ($d = 0$)?

DISCOVERY PROJECT 7 The Sun and the Moon

It has long been known that the cycles of the sun and the moon are periodic; that is, the moon is full at regular intervals and new at regular intervals. The same is true of the sun; the interval between the summer and winter solstices is also regular. It is therefore quite natural to use the sun and the moon to keep track of time, and to study the interaction between the two to predict events such as full moons, new moons, solstices, equinoxes, and eclipses. Indeed, these solar and lunar events have consequences on the earth, including the succession of the seasons and the severity of tides.

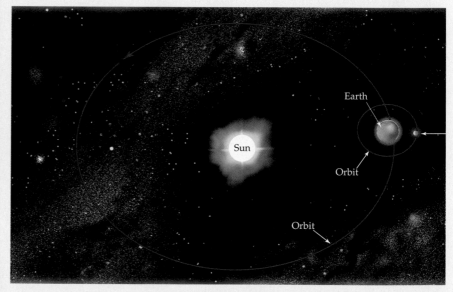

1. The following is a list of the days in 1999 when the moon was full. The length of the lunar month is the length of time between full moons. Use the data to approximate the length of the lunar month.

January 2	May 30	October 24
January 31	June 28	November 23
March 2	July 28	December 22
March 31	August 26	
April 30	September 25	

2. Write a function that has value 1 when the moon is full and 0 when the moon is new. Measure the independent variable in days with January 2, 1999, set as time 0. Use a function of the form

$$m(x) = \frac{\cos kt + 1}{2}$$

with period equal to the length of the lunar month.

3. Use your function to predict the date of the first full moon of the 21st century (in January of the year 2001). Does your function agree with the actual date of January 9? If not, what could have caused the discrepancy?

4. The solar year is approximately 365.24 days long. Write a function $s(x)$ with period equal to the length of the solar year so that on the date of the summer solstice, $s(x) = 1$ and on the day of the winter solstice, $s(x) = 0$. The summer solstice in 1999 was June 21 and the winter solstice falls midway between summer solstices.

5. If your functions $s(x)$ and $m(x)$ were accurate, when would you expect to see the next full moon on the summer solstice?

6. How would you go about making your models $s(x)$ and $m(x)$ more accurate?

CHAPTER 8

Triangle Trigonometry

Chris Alan Wilton/Getty Images

Where are we?

Navigators at sea must determine their location. Surveyors need to determine the height of a mountain or the width of a canyon when direct measurement isn't feasible. A fighter plane's computer must set the course of a missile so that it will hit a moving target. These and many similar problems can be solved by using triangle trigonometry. See Exercise 33 on page 572 and Exercise 33 on page 580.

Chapter Outline

Interdependence of Sections

8.1 ⟶ 8.2 ⟶ 8.3
           ⟶ 8.4

Roadmap

This Chapter may be read before Chapter 6. To do so, replace Section 8.1 by Alternate Section 8.1

For those who have read Chapter 6, Chapters 8 and 9 are independent of each other and may be read in either order.

8.1   Trigonometric Functions of Angles

8.1   ALTERNATE   Trigonometric Functions of Angles

8.2   Applications of Right Triangle Trigonometry

8.3   The Law of Cosines

8.4   The Law of Sines

8.4.A   *Special Topics:* The Area of a Triangle

Trigonometry was first used by the ancients to solve practical problems in astronomy, navigation, and surveying that involved triangles. Trigonometric functions, as presented in Chapter 6, came much later. The early mathematicians took a somewhat different viewpoint than we have used up to now, but their approach is often the best way to deal with problems involving triangles.

8.1   Trigonometric Functions of Angles

Trigonometric functions were defined in Chapter 6 as functions whose domains consist of real numbers. In the classical approach, however, the domains of the trigonometric functions consist of *angles.* In other words, instead of starting with a number t and then moving to an angle of t radians, we start directly with the angle, as summarized here.

In this chapter (and hereafter, whenever convenient), we shall take this classical approach and begin with angles. From there on, everything is essentially the same. The *values* of the trigonometric functions are still *numbers* and are obtained as before. For example, the point-in-the-plane description now reads as follows.

Point-in-the-Plane
Description

Let θ be an angle in standard position and let (x, y) be any point (except the origin) on the terminal side of θ. Let $r = \sqrt{x^2 + y^2}$. Then, the values of the six trigonometric functions of the angle θ are given by

$$\sin \theta = \frac{y}{r} \qquad \cos \theta = \frac{x}{r} \qquad \tan \theta = \frac{y}{x}$$

$$\csc \theta = \frac{r}{y} \qquad \sec \theta = \frac{r}{x} \qquad \cot \theta = \frac{x}{y}$$

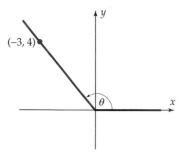

Figure 8–1

EXAMPLE 1

Evaluate the six trigonometric functions at the angle θ shown in Figure 8–1.

SOLUTION   We use $(-3, 4)$ as the point (x, y), so
$$r = \sqrt{x^2 + y^2} = \sqrt{9 + 16} = \sqrt{25} = 5.$$

Thus,

$$\sin \theta = \frac{y}{r} = \frac{4}{5} \qquad \cos \theta = \frac{x}{r} = \frac{-3}{5} \qquad \tan \theta = \frac{y}{x} = \frac{4}{-3}$$

$$\csc \theta = \frac{r}{y} = \frac{5}{4} \qquad \sec \theta = \frac{r}{x} = \frac{5}{-3} \qquad \cot \theta = \frac{x}{y} = \frac{-3}{4}. \qquad \blacksquare$$

▮▮ DEGREES AND RADIANS

Angles can be measured in either degrees or radians. If radian measure is used (as was the case in Chapter 6), then everything is the same as before. For example, sin 30 denotes the sine of an angle of 30 radians.

But when angles are measured in degrees (as will be done in the rest of this chapter), new notation is needed. To denote the value of the sine function at an angle of 30 *degrees,* we write

$$\sin 30° \qquad \text{[note the degree symbol]}$$

The degree symbol here is essential for avoiding error. For example, an angle of 30 degrees is the same as an angle of $\pi/6$ radians. Therefore,

$$\sin 30° = \sin \pi/6 = 1/2$$

This is *not* the same as sin 30 (the sine of an angle of 30 *radians*); a calculator in radian mode shows that sin 30 $\approx -.988$.

The various identities proved in earlier sections are valid for angles measured in degrees, provided that π radians is replaced by 180°. For any angle θ measured in degrees for which the functions are defined, the following identities hold.

*Identities for
Angles Measured
in Degrees*

Periodicity Identities

$$\sin(\theta + 360°) = \sin\theta \qquad \csc(\theta + 360°) = \csc\theta$$
$$\cos(\theta + 360°) = \cos\theta \qquad \sec(\theta + 360°) = \sec\theta$$
$$\tan(\theta + 180°) = \tan\theta \qquad \cot(\theta + 180°) = \cot\theta$$

Pythagorean Identities

$$\sin^2\theta + \cos^2\theta = 1 \quad 1 + \tan^2\theta = \sec^2\theta \quad 1 + \cot^2\theta = \csc^2\theta$$

Negative Angle Identities

$$\sin(-\theta) = -\sin\theta \qquad \cos(-\theta) = \cos\theta \qquad \tan(-\theta) = -\tan\theta$$

▉▉ RIGHT TRIANGLE DESCRIPTION OF TRIGONOMETRIC FUNCTIONS

For angles between $0°$ and $90°$, the trigonometric functions may be evaluated by using right triangles as follows. Suppose θ is an angle in a right triangle. Place the triangle so that angle θ is in standard position, with the hypotenuse as its terminal side, as shown in Figure 8–2.

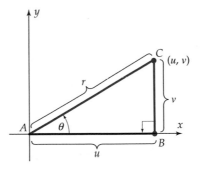

Figure 8–2

Denote the length of the side AB (the one *adjacent* to angle θ) by u and the length of side BC (the one *opposite* angle θ) by v. Then the coordinates of C are (u, v). Let r be the length of the *hypotenuse* AC (the distance from (u, v) to the origin). Then the point-in-the-plane description shows that

$$\sin\theta = \frac{v}{r} = \frac{\text{length of opposite side}}{\text{length of hypotenuse}}, \qquad \cos\theta = \frac{u}{r} = \frac{\text{length of adjacent side}}{\text{length of hypotenuse}},$$

$$\tan\theta = \frac{v}{u} = \frac{\text{length of opposite side}}{\text{length of adjacent side}},$$

and similarly for the other trigonometric functions. These facts can be summarized as follows.

Right Triangle Description

Consider a right triangle containing an angle θ.

The values of the six trigonometric functions of the angle θ are given by

$$\sin \theta = \frac{\text{opposite}}{\text{hypotenuse}} \qquad \cos \theta = \frac{\text{adjacent}}{\text{hypotenuse}} \qquad \tan \theta = \frac{\text{opposite}}{\text{adjacent}}$$

$$\csc \theta = \frac{\text{hypotenuse}}{\text{opposite}} \qquad \sec \theta = \frac{\text{hypotenuse}}{\text{adjacent}} \qquad \cot \theta = \frac{\text{adjacent}}{\text{opposite}}$$

This is the description of the trigonometric functions that is usually presented first when trigonometry is studied from the viewpoint of angles and triangles, rather than from the viewpoint of functions. It has the advantage of being independent of both the unit circle and the coordinate system in the plane.

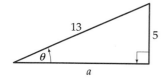

Figure 8–3

EXAMPLE 2

Evaluate $\sin \theta$, $\cos \theta$, and $\tan \theta$ when θ is the angle shown in Figure 8–3.

SOLUTION First, we find the length of the third side a by using the Pythagorean Theorem.

$$a^2 + 5^2 = 13^2$$

Multiply out terms: $\qquad a^2 + 25 = 169$

Subtract 25 from both sides: $\qquad a^2 = 144$

Take square roots on both sides.* $\qquad a = \sqrt{144} = 12.$

Now we can calculate the values of the trigonometric functions.

$$\sin \theta = \frac{\text{opposite}}{\text{hypotenuse}} = \frac{5}{13}, \qquad \cos \theta = \frac{\text{adjacent}}{\text{hypotenuse}} = \frac{12}{13},$$

$$\tan \theta = \frac{\text{opposite}}{\text{adjacent}} = \frac{5}{12}. \qquad \blacksquare$$

---

*The equation $a^2 = 144$ has two solutions, 12 and -12, but only the positive one applies here since a is the side of a triangle.

Unless you are given an appropriate triangle, whose sides are known (or can be computed) as in Example 2, it may be difficult to find the exact values of the trigonometric functions at an angle θ. Fortunately, however, your calculator can provide good approximations, as illustrated in Figure 8–4.

```
sin(52)
          .7880107536
cos(27)
          .8910065242
tan(75)
          3.732050808
■
```

Figure 8–4

In a few cases, however, we can find the exact values of sine, cosine, and tangent.

EXAMPLE 3

Evaluate sin θ, cos θ, tan θ when (a) $\theta = 30°$ (b) $\theta = 60°$.

SOLUTION

(a) Consider a 30°-60°-90° triangle whose hypotenuse has length 2. As explained in Example 3 of the Geometry Review Appendix, the side opposite the 30° angle must have length 1 (half the hypotenuse) and the side adjacent to this angle must have length $\sqrt{3}$, as shown in Figure 8–5. According to the right triangle description,

$$\sin 30° = \frac{\text{opposite}}{\text{hypotenuse}} = \frac{1}{2}, \qquad \cos 30° = \frac{\text{adjacent}}{\text{hypotenuse}} = \frac{\sqrt{3}}{2},$$

$$\tan 30° = \frac{\text{opposite}}{\text{adjacent}} = \frac{1}{\sqrt{3}} = \frac{\sqrt{3}}{3}.$$

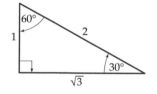

Figure 8–5

(b) The same triangle can be used to evaluate the trigonometric functions at 60°. In this case, the opposite side has length $\sqrt{3}$ and the adjacent side has length 1. Therefore,

$$\sin 60° = \frac{\text{opposite}}{\text{hypotenuse}} = \frac{\sqrt{3}}{2}, \qquad \cos 60° = \frac{\text{adjacent}}{\text{hypotenuse}} = \frac{1}{2},$$

$$\tan 60° = \frac{\text{opposite}}{\text{adjacent}} = \frac{\sqrt{3}}{1} = \sqrt{3}. \qquad ■$$

EXAMPLE 4

Evaluate sin 45°, cos 45°, tan 45°.

SOLUTION   Consider a 45°-45°-90° triangle whose sides each have length 3 (Figure 8–6). According to the Pythagorean Theorem, the hypotenuse d satisfies

$$d^2 = 3^2 + 3^2 = 18,$$

Figure 8–6

so
$$d = \sqrt{18} = \sqrt{9 \cdot 2} = \sqrt{9}\,\sqrt{2} = 3\,\sqrt{2}.$$

Therefore,

$$\sin 45° = \frac{\text{opposite}}{\text{hypotenuse}} = \frac{3}{3\,\sqrt{2}} = \frac{1}{\sqrt{2}} = \frac{\sqrt{2}}{2}$$

$$\cos 45° = \frac{\text{adjacent}}{\text{hypotenuse}} = \frac{3}{3\,\sqrt{2}} = \frac{1}{\sqrt{2}} = \frac{\sqrt{2}}{2}$$

$$\tan 45° = \frac{\text{opposite}}{\text{adjacent}} = \frac{3}{3} = 1. \quad\blacksquare$$

▪▪ SOLVING RIGHT TRIANGLES

Many applications of trigonometry involve **"solving a triangle."** This means finding the lengths of all three sides and the measures of all three angles when only some of these quantities are given. Solving the right triangles depends on this fact.

> **The right angle description of a trigonometric function (such as sin θ = opposite/hypotenuse) relates three quantities: the angle θ and two sides of the right triangle.**

When two of these three quantities are known, then the third can always be found.

Figure 8–7

EXAMPLE 5

Find the lengths of sides b and c in the right triangle shown in Figure 8–7.

SOLUTION   Since the side c is opposite the 75° angle and the hypotenuse is 17, we have

$$\sin 75° = \frac{\text{opposite}}{\text{hypotenuse}} = \frac{c}{17}.$$

We can solve this equation for c.

$$\frac{c}{17} = \sin 75°$$

Multiply both sides by 17: $c = 17 \sin 75°$

Use a calculator to evaluate sin 75°: $c \approx 17(.9659) \approx 16.42.$

Side b can now be found by the Pythagorean Theorem or by using the fact that

$$\cos 75° = \frac{\text{adjacent}}{\text{hypotenuse}} = \frac{b}{17}.$$

Solving this equation and using a calculator shows that

$$\frac{b}{17} = \cos 75°$$

$$b = 17 \cos 75° \approx 17(.2588) \approx 4.40. \quad\blacksquare$$

Figure 8–8

EXAMPLE 6

Find the degree measure of the angle θ in Figure 8–8.

SOLUTION   We first note that

$$\cos \theta = \frac{\text{adjacent}}{\text{hypotenuse}} = \frac{4}{5} = .8.$$

Before calculators were available, θ was found by using a table of cosine values, as follows: Look through the column of cosine values for the closest one to .8, then look in the first column for the corresponding value of θ. You can do the same thing by having your calculator generate a table for $y_1 = \cos x$, as in Figure 8–9. The closest entry to .8 in the cosine (y_1) column is .79968, which corresponds to an angle of 36.9°. Hence, $\theta \approx 36.9°$.

Figure 8–9   **Figure 8–10**

A faster and more accurate method of finding θ is to use the COS^{-1} key on your calculator (labeled ACOS on some models). When you key in COS^{-1} .8, as in Figure 8–10, the calculator produces an acute angle whose cosine is .8, namely, $\theta \approx 36.8699°$. Thus, the COS^{-1} key provides the electronic equivalent of searching the cosine table, without actually having to construct the table.   ∎

NOTE: In this chapter, we shall use the COS^{-1} key, and the analogous keys SIN^{-1} and TAN^{-1}, as they were used in the preceding example: as a way to find an angle θ in a triangle, when $\sin \theta$ or $\cos \theta$ or $\tan \theta$ is known. The other uses of these keys are discussed in Section 7.4, which deals with the inverse functions of sine, cosine, and tangent.

Figure 8–11

EXAMPLE 7

Without using the Pythagorean Theorem, find angles α and β and side c of the triangle in Figure 8–11.

SOLUTION   We first use the fact that

$$\tan \alpha = \frac{\text{opposite}}{\text{adjacent}} = \frac{10}{7}.$$

The TAN^{-1} key on a calculator (Figure 8–12) shows that $\alpha \approx 55.01°$. Since the sum of the angles of a triangle is 180°, we have

$$\alpha + \beta + 90° = 180°$$

$$55.01 + \beta + 90° \approx 180°$$

$$\beta \approx 180° - 90° - 55.01° = 34.99°.$$

Next, we note that

$$\sin \alpha = \frac{\text{opposite}}{\text{hypotenuse}} = \frac{10}{c}$$

$$\sin 55.01° = \frac{10}{c}.$$

Multiplying both sides of this equation by c shows that

$$c \sin 55.01° = 10$$

$$c = \frac{10}{\sin 55.01°}.$$

A calculator shows that $c \approx 12.21$ (Figure 8–13).   ■

Figure 8–12

Figure 8–13

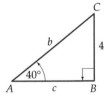

Figure 8–14

EXAMPLE 8

Solve the right triangle in Figure 8–14.

SOLUTION   We must find the measure of $\angle C$ and the lengths of sides b and c. Since the sum of the angles of a triangle is 180°, we have

$$40° + 90° + \angle C = 180°$$

$$\angle C = 180° - 40° - 90° = 50°.$$

Furthermore, Figure 8–14 shows that

$$\sin A = \frac{4}{b}$$

Since $A = 40°$: $$\sin 40° = \frac{4}{b}$$

Multiply both sides by b: $$b \sin 40° = 4$$

Divide both sides by $\sin 40°$: $$b = \frac{4}{\sin 40°} \approx 6.22.$$

Figure 8–15

Now side c can be found by using the Pythagorean Theorem.

$$4^2 + c^2 = b^2$$

$$c^2 = b^2 - 16$$

$$c = \sqrt{b^2 - 16} = \sqrt{\left(\frac{4}{\sin 40°}\right)^2 - 16}.$$

A calculator shows that $c \approx 4.77$ (Figure 8–15). ∎

CALCULATOR EXPLORATION

In Example 8, use the approximation $b \approx 6.22$ and the Pythagorean Theorem to find c. Is your answer the same as the length of c found in Figure 8–15? Why not? The moral here is: Don't use approximations in intermediate steps if you can avoid it. However, rounding your final answer is usually appropriate.

EXERCISES 8.1

Directions: *When solving triangles here, all decimal approximations should be rounded off to one decimal place at the end of the computation.*

In Exercises 1–6, evaluate the trigonometric functions at the angle (in standard position) whose terminal side contains the given point.

1. $(2, 3)$ **2.** $(4, -2)$ **3.** $(-5, 6)$

4. $(\sqrt{2}, \sqrt{3})$ **5.** $(-3, -\sqrt{2})$ **6.** $(-5, 3)$

In Exercises 7–12, find sin θ, cos θ, tan θ.

7.

8.

9.

10.

11.

12.

In Exercises 13–18, find side c of the right triangle in the figure under the given conditions.

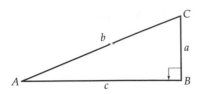

13. $\cos A = 12/13$ and $b = 39$

14. $\sin C = 3/4$ and $b = 12$

15. $\tan A = 5/12$ and $a = 15$

16. $\sec A = 2$ and $b = 8$

17. $\cot A = 6$ and $a = 1.4$

18. $\csc C = 1.5$ and $b = 4.5$

In Exercises 19–24, find the length h of the side of the right triangle, without using a calculator.

19.

20.

21.

22.

23.

24.

In Exercises 25–28, find the required side without using a calculator.

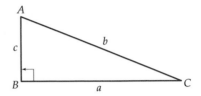

25. $a = 4$ and angle A measures $60°$; find c.

26. $c = 5$ and angle A measures $60°$; find a.

27. $c = 10$ and angle A measures $30°$; find a.

28. $a = 12$ and angle A measures $30°$; find c.

In Exercises 29–36, use the figure for Exercises 25–28. Solve the right triangle under the given conditions.

29. $b = 10$ and $\angle C = 50°$

30. $c = 12$ and $\angle C = 37°$

31. $a = 6$ and $\angle A = 14°$

32. $a = 8$ and $\angle A = 40°$

33. $c = 5$ and $\angle A = 65°$

34. $c = 4$ and $\angle C = 28°$

35. $b = 3.5$ and $\angle A = 72°$

36. $a = 4.2$ and $\angle C = 33°$

In Exercises 37–40, find angle θ.

37.

38.

39.

40.

In Exercises 41–48, use the figure for Exercises 25–28 to find angles A and C under the given conditions.

41. $a = 4$ and $c = 6$

42. $b = 14$ and $c = 5$

43. $a = 7$ and $b = 10$

44. $a = 5$ and $c = 3$

45. $b = 18$ and $c = 12$

46. $a = 4$ and $b = 9$

47. $a = 2.5$ and $c = 1.4$

48. $b = 3.7$ and $c = 2.2$

49. Let θ and α be acute angles of a right triangle, as shown in the figure.
  (a) Find $\sin \theta$ and $\cos \alpha$.
  (b) Explain why $\theta + \alpha = 90°$.
  (c) Use parts (a) and (b) to conclude that for any acute angle θ,

$$\cos(90° - \theta) = \sin \theta.$$

50. (a) Using the figure for Exercise 49, find $\cos \theta$ and $\sin \alpha$.
  (b) Use part (a) and part (b) of Exercise 49 to show that for any acute angle θ,

$$\sin(90° - \theta) = \cos \theta.$$

This equation and the one in Exercise 49(c) are called *cofunction identities.*

8.1   *ALTERNATE*   Trigonometric Functions of Angles

NOTE

If you have read Chapter 6, omit this section. If you have not read Chapter 6, use this section in place of Section 8.1.

Before reading this section, it might be a good idea to read the Geometry Review Appendix, which presents the basic facts about angles and triangles that frequently are used here. In particular, recall that a **right triangle** is one that contains a **right angle,** that is, an angle of 90°. An **acute angle** is an angle whose measure is less than 90°. Consider the right triangles in Figure 8–1, each of which has an acute angle of θ degrees.

Figure 8–16

Since the sum of the angles of any triangle is 180°, we see that the third angle in each of these triangles has the same measure, namely, $180° - 90° - \theta$. Thus, all three triangles have equal corresponding angles and, therefore, are similar.

Consequently, by the Ratios Theorem of the Geometry Review Appendix, we know that the ratio of corresponding sides is the same, that is

$$\frac{a}{c} = \frac{a'}{c'} = \frac{a''}{c''}.$$

Each of these fractions is the ratio

$$\frac{\text{length of the side opposite angle } \theta}{\text{length of the hypotenuse}},$$

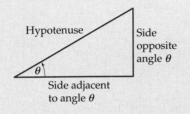

Figure 8–17

as indicated in Figure 8–17. Consequently, *this ratio depends only on the angle θ* and *not* on the size of the triangle. Similar remarks apply to the ratios of other sides of the triangle in Figure 8–17 and make it possible to define three new functions. For each function, the input is an acute angle θ, and the corresponding output is a ratio of sides in any right triangle containing angle θ, as summarized here.

Trigonometric Functions of Acute Angles

Name of Function	Abbreviation	Rule of Function
sine	sin	$\sin \theta = \dfrac{\text{length of side opposite angle } \theta}{\text{length of hypotenuse}}$
cosine	cos	$\cos \theta = \dfrac{\text{length of side adjacent to angle } \theta}{\text{length of hypotenuse}}$
tangent	tan	$\tan \theta = \dfrac{\text{length of side opposite angle } \theta}{\text{length of side adjacent to angle } \theta}$

EXAMPLE 9

Evaluate sin θ, cos θ, and tan θ for the angle θ shown in Figure 8–18.

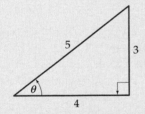

Figure 8–18

SOLUTION   The side opposite angle θ has length 3, and the side adjacent to angle θ has length 4. Hence,

$$\sin \theta = \frac{\text{length of side opposite angle } \theta}{\text{length of hypotenuse}} = \frac{3}{5},$$

$$\cos \theta = \frac{\text{length of side adjacent to angle } \theta}{\text{length of hypotenuse}} = \frac{4}{5},$$

$$\tan \theta = \frac{\text{length of side opposite angle } \theta}{\text{length of side adjacent to angle } \theta} = \frac{3}{4}. \quad \blacksquare$$

Now that you have the idea, we shall abbreviate the ratio

$$\frac{\text{length of the side adjacent to angle } \theta}{\text{length of the hypotenuse}} \quad \text{as} \quad \frac{\text{adjacent}}{\text{hypotenuse}}$$

and similarly for the others.

NOTE: Now turn to page 546 and begin reading at Example 2. When you have finished Example 8 on page 550, return here and continue reading below. [If you prefer, you can delay reading the material below until Section 8.3, where it will first be used.]

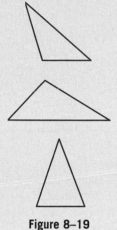

Figure 8–19

The preceding discussion applies only to right triangles. The next step is to learn how to solve other triangles, such as those in Figure 8–19. To do this, we must find a description of the trigonometric functions that is not limited to acute angles of a right triangle. First, we introduce some terminology.

An angle in the coordinate plane is said to be in **standard position** if its vertex is at the origin and one of its sides (which we call the **initial side**) is on the positive *x*-axis, as shown in Figure 8–20. The other side of the angle will be called its **terminal side.**

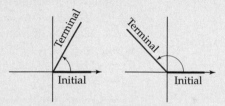

Figure 8–20

Let θ be an acute angle in standard position. Choose any point (x, y) on the terminal side of θ (except the origin) and consider the right triangle with vertices $(0, 0)$, $(x, 0)$, and (x, y) shown in Figure 8–21.

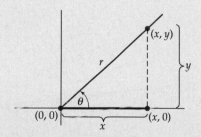

Figure 8–21

The legs of this triangle have lengths *x* and *y*, respectively. The distance formula shows that the hypotenuse has length

$$\sqrt{(x - 0)^2 + (y - 0)^2} = \sqrt{x^2 + y^2},$$

which we denote by *r*. The triangle shows that

$$\sin \theta = \frac{y}{r} \qquad \cos \theta = \frac{x}{r} \qquad \tan \theta = \frac{y}{x}.$$

We now have a description of the trigonometric functions in terms of the coordinate plane rather than right triangles. Furthermore, this description makes sense for *any* angle. Consequently, we make the following definition.

Point-in-the-Plane Description

Let θ be an angle in standard position and (x, y) any point (except the origin) on its terminal side. Then the trigonometric functions are defined by these rules:

$$\sin \theta = \frac{y}{r}, \qquad \cos \theta = \frac{x}{r}, \qquad \tan \theta = \frac{y}{x} \quad (x \neq 0),$$

where $r = \sqrt{x^2 + y^2}$ is the distance from (x, y) to the origin.

The discussion preceding the box shows that when θ is an acute angle, these definitions produce the same numbers for $\sin \theta$, $\cos \theta$, and $\tan \theta$ as does the right triangle definition on page 554. It can be shown that the values of the trigonometric functions of θ are independent of the point that is chosen on the terminal side (just as the definition for acute angles is independent of the size of the right triangle).

EXAMPLE 10

Find $\sin \theta$, $\cos \theta$, and $\tan \theta$ for the angle θ shown in Figure 8–22.

SOLUTION   Since $(-5, 7)$ is on the terminal side of θ, we apply the definitions in the preceding box with $(x, y) = (-5, 7)$ and

$$r = \sqrt{x^2 + y^2} = \sqrt{(-5)^2 + 7^2} = \sqrt{74}:$$

$$\sin \theta = \frac{y}{r} = \frac{7}{\sqrt{74}}$$

$$\cos \theta = \frac{x}{r} = \frac{-5}{\sqrt{74}}$$

$$\tan \theta = \frac{y}{x} = \frac{7}{-5} = -\frac{7}{5}.$$

Figure 8–22

For most angles, we use a calculator to approximate the values of the trigonometric functions, as in Figure 8–23. But there are a few angles for which we can compute the exact values of the trigonometric functions.

```
sin(125)
        .8191520443
cos(178)
        -.999390827
tan(98)
        -7.115369722
■
```

Figure 8–23

EXAMPLE 11

Find the exact values of the trigonometric functions at $\theta = 90°$.

SOLUTION   We use the point $(0, 1)$ on the terminal side of an angle of $90°$ in standard position (Figure 8–24). In this case, $r = \sqrt{x^2 + y^2} = \sqrt{0^2 + 1^2} = 1$, so

$$\sin 90° = \frac{y}{r} = \frac{1}{1} = 1$$

$$\cos 90° = \frac{x}{r} = \frac{0}{1} = 0$$

$$\tan 90° = \frac{y}{x} = \frac{1}{0} \quad (undefined). \quad \blacksquare$$

Figure 8–24

The last part of Example 11 shows that the domain of the tangent function excludes $90°$, whereas the domains of sine and cosine include all angles.

___

EXAMPLE 12

Find the exact values of $\sin 135°$, $\cos 135°$, and $\tan 135°$.

SOLUTION   Construct an angle of $135°$ in standard position and let P be the point on the terminal side that is 1 unit from the origin (Figure 8–25). Draw a vertical line from P to the x-axis, forming a right triangle with hypotenuse 1 and two angles of $45°$. Each side of this triangle has length $\sqrt{2}/2$, as explained in Example 2 of the Geometry Review Appendix. Therefore, the coordinates of P are $(-\sqrt{2}/2, \sqrt{2}/2)$, and we have

$$\sin 135° = \frac{y}{r} = \frac{\sqrt{2}/2}{1} = \frac{\sqrt{2}}{2}$$

$$\cos 135° = \frac{x}{r} = \frac{-\sqrt{2}/2}{1} = -\frac{\sqrt{2}}{2}$$

$$\tan 135° = \frac{y}{x} = \frac{\sqrt{2}/2}{-\sqrt{2}/2} = -1. \quad \blacksquare$$

Figure 8–25

 EXERCISES ALTERNATE 8.1

Use the exercises for Section 8.1 on page 551.

8.2   Applications of Right Triangle Trigonometry

The following examples illustrate a variety of practical applications of triangle trigonometry.

EXAMPLE 1

A hiker sees a tree on the river's edge directly opposite her. She walks along the riverbank for 120 feet and notes that the angle formed by her path and a line to the tree measures 70°, as indicated in Figure 8–26. How wide is the river?

Figure 8–26

SOLUTION   Using the right triangle whose one leg is the width w of the river and whose other leg is the 120-foot path on this side of the river, we see that

$$\tan 70° = \frac{\text{opposite}}{\text{adjacent}} = \frac{w}{120}.$$

Solving this equation for w, we have

$$w = 120 \tan 70° \approx 329.7.$$

So the river is about 330 feet wide.   ∎

EXAMPLE 2

A straight road leads from an ocean beach into the nearby hills. The road has a constant upward grade of 3°. After taking this road from the beach for 1 mile, how high above sea level are you?

SOLUTION   Figure 8–27 shows the situation [*Remember:* 1 mile = 5280 feet]:

Figure 8–27

Figure 8–27 shows that

$$\sin 3° = \frac{\text{opposite}}{\text{hypotenuse}} = \frac{h}{5280}$$

$$h = 5280 \sin 3° \approx 276.33 \text{ feet.} \quad \blacksquare$$

EXAMPLE 3

According to the safety sticker on an extension ladder, the distance from the foot of the ladder to the base of the wall on which it leans should be one-fourth of the length of the ladder. If the ladder is in this position, what angle does it make with the ground?

SOLUTION Let c be the distance from the foot of the ladder to the base of the wall. Then the ladder's length is $4c$, as shown in Figure 8–28. If θ is the angle the ladder makes with the ground, then

$$\cos \theta = \frac{\text{adjacent}}{\text{hypotenuse}} = \frac{c}{4c} = \frac{1}{4}.$$

Using the $\cos^{-1}$ key, we find that $\theta \approx 75.52°$, as shown in Figure 8–29. $\quad \blacksquare$

Figure 8–28

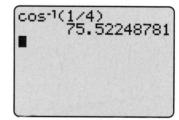

Figure 8–29

In many practical applications, one uses the angle between the horizontal and some other line (for instance, the line of sight from an observer to a distant object). This angle is called the **angle of elevation** or the **angle of depression,** depending on whether the line is above or below the horizontal, as shown in Figure 8–30.

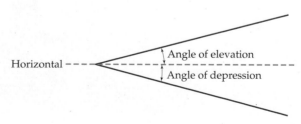

Figure 8–30

EXAMPLE 4

A wire is to be stretched from the top of a 10-meter-high building to a point on the ground. From the top of the building, the angle of depression to the ground point is 22°. How long must the wire be?

Figure 8–31

SOLUTION   Figure 8–31 shows that the sum of the angle of depression and the angle α is 90°. Hence, α measures $90° - 22° = 68°$. We know the length of the side of the triangle adjacent to the angle α and must find the hypotenuse w (the length of the wire). Using the cosine function, we see that

$$\cos 68° = \frac{\text{adjacent}}{\text{hypotenuse}} = \frac{10}{w}$$

$$w = \frac{10}{\cos 68°} \approx 26.7 \text{ meters.} \quad \blacksquare$$

EXAMPLE 5

A large American flag flies from a pole on top of the Terminal Tower in Cleveland (Figure 8–32). At a point 203 feet from the base of the tower, the angle of elevation to the bottom of the flag pole is 74°, and the angle of elevation to the top of the pole is 75.285°. To the nearest foot, how long is the flagpole?

Figure 8–32   **Figure 8–33**

SOLUTION   By abstracting the given information, we see that there are two right triangles, as shown in Figure 8–33 (which is not to scale). The length of the flagpole is $c - d$. We can use the two triangles to find c and d.

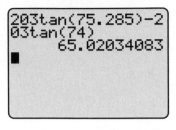

Figure 8–34

Larger Triangle	**Smaller Triangle**
$\dfrac{c}{203} = \dfrac{\text{opposite}}{\text{adjacent}} = \tan 75.285°$	$\dfrac{d}{203} = \dfrac{\text{opposite}}{\text{adjacent}} = \tan 74°$
$c = 203 \tan 75.285° \approx 773$	$d = 203 \tan 74° \approx 708$

As shown in Figure 8–34, the length of the flagpole is

$$c - d \approx 773 - 708 = 65 \text{ feet.} \qquad \blacksquare$$

 EXAMPLE 6

A person standing on the edge of one bank of a canal observes a lamp post on the edge of the other bank of the canal. The person's eye level is 152 centimeters above the ground (approximately 5 feet). The angle of elevation from eye level to the top of the lamp post is 12°, and the angle of depression from eye level to the bottom of the lamp post is 7°, as shown in Figure 8–35. How wide is the canal? How high is the lamp post?

Figure 8–35

SOLUTION   Abstracting the essential information, we obtain the diagram in Figure 8–36.

Figure 8–36

We must find the height of the lamp post BD and the width of the canal AC (or ED). The eye level height AE of the observer is 152 centimeters. Since AC and ED are parallel, CD also has length 152 centimeters. In right triangle ACD, we know the angle of 7° and the side *CD* opposite it. We must find the adjacent side *AC*. The tangent function is needed.

$$\tan 7° = \frac{\text{opposite}}{\text{adjacent}} = \frac{152}{AC} \quad \text{or, equivalently,} \quad AC = \frac{152}{\tan 7°}$$

$$AC = \frac{152}{\tan 7°} \approx 1237.94 \text{ centimeters.}$$

So the canal is approximately 12.3794 meters[*] wide (about 40.6 feet). Now using right triangle *ACB*, we see that

$$\tan 12° = \frac{\text{opposite}}{\text{adjacent}} = \frac{BC}{AC} \approx \frac{BC}{1237.94}$$

or, equivalently,

$$BC \approx 1237.94(\tan 12°) \approx 263.13 \text{ centimeters.}$$

Therefore, the height of the lamp post *BD* is $BC + CD \approx 263.13 + 152 = 415.13$ centimeters or, equivalently, 4.1513 meters. ∎

---

*Remember, 100 centimeters = 1 meter.

  EXERCISES 8.2

1. What is the width of the river in Example 1 if the angle to the tree is 40° (and all other information is the same)?

2. If you travel the road in Example 2 for a mile and a half, how high above sea level are you?

3. Suppose you have a 24-foot-long ladder and you ignore the safety advice in Example 3 by placing the foot of the ladder 9 feet from the base of the wall. What angle does the ladder make with the ground?

4. How long a wire is needed in Example 4 if the angle of depression is 25.8°?

5. Suppose that the flagpole on the Terminal Tower (Example 5) has been replaced. Now from a point 240 feet from the base of the tower, the angle of elevation to the bottom of the flagpole is 71.3°, and the angle of elevation to the top of the pole is 72.9°. To the nearest foot, how long is the new flagpole?

6. A 150-foot-long ramp connects a ground-level parking lot with the entrance of a building. If the entrance is 8 feet above the ground, what angle does the ramp make with the ground?

7. A 20-foot-long ladder leans on a wall of a building. The foot of the ladder makes an angle of 50° with the ground. How far above the ground does the top of the ladder touch the wall?

8. A guy wire stretches from the top of an antenna tower to a point on level ground 18 feet from the base of the tower. The angle between the wire and the ground is 63°. How high is the tower?

9. Consider a 16-foot-long drawbridge on a medieval castle, as shown in the figure. The royal army is engaged in ignominious retreat. The king would like to raise the end of the drawbridge 8 feet off the ground so that Sir Rodney can jump onto the drawbridge and scramble into the castle while the enemy's cavalry are held at bay. Through how much of an angle must the drawbridge be raised for the end of it to be 8 feet off the ground?

16 ft

10. Through what angle must the drawbridge in Exercise 9 be raised in order that its end be directly above the center of the moat?

11. A plane takes off at an angle of 5°. After traveling 1 mile along this flight path, how high (in feet) is the plane above the ground?

12. A plane takes off at an angle of 6° traveling at the rate of 200 feet/second. If it continues on this flight path at the same speed, how many minutes will it take to reach an altitude of 8000 feet?

13. It is claimed that the Ohio Turnpike never has an uphill grade of more than 3°. How long must a straight uphill segment of the road be to allow a vertical rise of 450 feet?

14. Ruth is flying a kite. Her hand is 3 feet above ground level and is holding the end of a 300-foot-long kite string, which makes an angle of 57° with the horizontal. How high is the kite above the ground?

15. If you stand upright on a mountainside that makes a 62° angle with the horizontal and stretch your arm straight out at shoulder height, you may be able to touch the mountain (as shown in the figure). Can a person with an arm reach of 27 inches, whose shoulder is 5 feet above the ground, touch the mountain?

16. A swimming pool is 3 feet deep in the shallow end. The bottom of the pool has a steady downward drop of 12°. If the pool is 50 feet long, how deep is it at the deep end?

17. Batman is on the edge of a 200-foot-deep chasm and wants to jump to the other side. A tree on the edge of the chasm is directly across from him. He walks 20 feet to his right and notes that the angle to the tree is 54°. His jet belt enables him to jump a maximum of 24 feet. How wide is the chasm and is it safe for Batman to jump?

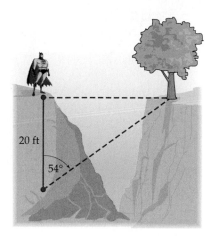

18. A wire from the top of a TV tower to the ground makes an angle of 49.5° with the ground and touches ground 225 feet from the base of the tower. How high is the tower?

19. A woman 5.5 feet tall stands 10 feet from a streetlight and casts a 4-foot-long shadow. How tall is the streetlight? What is angle θ?

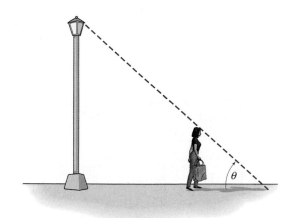

20. A plane flies a straight course. On the ground directly below the flight path, observers 2 miles apart spot the plane at the same time. The plane's angle of elevation is 46° from one observation point and 71° from the other. How high is the plane?

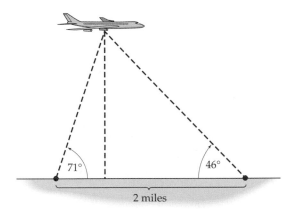

21. A buoy in the ocean is observed from the top of a 40-meter-high radar tower on shore. The angle of depression from the top of the tower to the base of the buoy is 6.5°. How far is the buoy from the base of the radar tower?

22. A plane passes directly over your head at an altitude of 500 feet. Two seconds later, you observe that its angle of elevation is 42°. How far did the plane travel during those 2 seconds?

23. A man stands 12 feet from a statue. The angle of elevation from eye level to the top of the statue is 30°, and the angle of depression to the base of the statue is 15°. How tall is the statue?

24. Two boats lie on a straight line with the base of a lighthouse. From the top of the lighthouse (21 meters above water level), it is observed that the angle of depression of the nearest boat is 53° and the angle of depression of the farthest boat is 27°. How far apart are the boats?

25. A rocket shoots straight up from the launchpad. Five seconds after liftoff, an observer 2 miles away notes that the rocket's angle of elevation is 3.5°. Four seconds later, the angle of elevation is 41°. How far did the rocket rise during those 4 seconds?

26. From a 35-meter-high window, the angle of depression to the top of a nearby streetlight is 55°. The angle of depression to the base of the streetlight is 57.8°. How high is the streetlight?

27. A closed 60-foot-long drawbridge is 24 feet above water level. When open, the bridge makes an angle of 33° with the horizontal.

(a) How high is the tip *P* of the open bridge above the water?

(b) When the bridge is open, what is the distance from *P* to *Q*?

28. One plane flies straight east at an altitude of 31,000 feet. A second plane is flying west at an altitude of 14,000 feet on a course that lies directly below that of the first plane and directly above the straight road from Thomasville to Johnsburg. As the first plane passes over Thomasville, the second is passing over Johnsburg. At that instant both planes spot a beacon next to the road between Thomasville to Johnsburg. The angle of depression from the first plane to the beacon is 61°, and the angle of depression from the second plane to the beacon is 34°. How far is Thomasville from Johnsburg?

29. A schematic diagram of a pedestrian overpass is shown in the figure. If you walk on the overpass from one end to the other, how far have you walked?

30. A 5-inch-high plastic beverage glass has a 2.5-inch-diameter base. Its sides slope outward at a 4° angle as shown. What is the diameter of the top of the glass?

31. In aerial navigation, directions are given in degrees clockwise from north. Thus, east is 90°, south is 180°, and so on, as shown in the figure. A plane travels from an airport for 200 miles in the direction 300°. How far west of the airport is the plane then?

32. A plane travels at a constant 300 mph in the direction 65° (see Exercise 31).
(a) How far east of its starting point is the plane after half an hour?
(b) How far north of its starting point is the plane after 2 hours and 24 minutes?

33. A car on a straight road passes under a bridge. Two seconds later an observer on the bridge, 20 feet above the road, notes that the angle of depression to the car is 7.4°. How fast (in miles per hour) is the car traveling? [*Note:* 60 mph is equivalent to 88 feet/second.]

Thinkers

34. A gutter is to be made from a strip of metal 24 inches wide by bending up the sides to form a trapezoid.

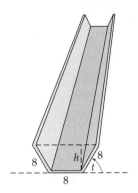

(a) Express the area of the cross section of the gutter as a function of the angle *t*. [*Hint:* The area of a trapezoid with bases *b* and *b′* and height *h* is $h(b + b')/2$.]
(b) For what value of *t* will this area be as large as possible?

35. The cross section of a tunnel is a semicircle with radius 10 meters. The interior walls of the tunnel form a rectangle.

(a) Express the area of the rectangular cross section of the tunnel opening as a function of angle *t*.
(b) For what value of *t* is the cross-sectional area of the tunnel opening as large as possible? What are the dimensions of the tunnel opening in this case?

36. A spy plane on a practice run over the Midwest takes a picture that shows Cleveland, Ohio, on the eastern horizon and St. Louis, Missouri, 520 miles away, on the western horizon (the figure is not to scale). Assuming that the radius of the earth is 3950 miles, how high was the plane when the picture was taken? [*Hint:* The sight lines from the plane to the horizons are tangent to the earth, and a tangent line to a circle is perpendicular to the radius at that point. The arc of the earth between St. Louis and Cleveland is 520 miles long. Use this fact and the arc length formula to find angle θ (your answers will be in radians). Note that $\alpha = \theta/2$ (why?).]

St. Louis θ α Cleveland

37. A 50-foot-high flagpole stands on top of a building. From a point on the ground, the angle of elevation of the top of the pole is 43°, and the angle of elevation of the bottom of the pole is 40°. How high is the building?

38. Two points on level ground are 500 meters apart. The angles of elevation from these points to the top of a nearby hill are 52° and 67°, respectively. The two points and the ground level point directly below the top of the hill lie on a straight line. How high is the hill?

8.3   The Law of Cosines

Figure 8–37

We now consider the solution of *oblique* triangles (ones that don't contain a right angle). We shall use **standard notation** for triangles: Each vertex is labeled with a capital letter, and the length of the side opposite that vertex is denoted by the same letter in lower case, as shown in Figure 8–37. The letter *A* will also be used to label the *angle* at vertex *A* and similarly for *B* and *C*. So we shall make statements such as $A = 37°$ or $\cos B = .326$.

The first fact needed to solve oblique triangles is the Law of Cosines, whose proof is given at the end of this section.

Law of Cosines

In any triangle ABC, with sides of lengths a, b, c, as in Figure 8–37,

$$a^2 = b^2 + c^2 - 2bc \cos A$$

$$b^2 = a^2 + c^2 - 2ac \cos B$$

$$c^2 = a^2 + b^2 - 2ab \cos C$$

You need only memorize one of these equations since each of them provides essentially the same information: a description of one side of a triangle in terms of the angle opposite it and the other two sides.

> NOTE: When *C* is a right angle, then *c* is the hypotenuse and
>
> $$\cos C = \cos 90° = 0,$$
>
> so the third equation in the Law of Cosines becomes the Pythagorean Theorem:
>
> $$c^2 = a^2 + b^2$$

Solving the first equation in the Law of Cosines for cos *A*, we obtain the following.

Law of Cosines:
Alternate Form

In any triangle *ABC*, with sides of lengths *a, b, c*, as in Figure 8–37,

$$\cos A = \frac{b^2 + c^2 - a^2}{2bc}.$$

The other two equations can be similarly rewritten. In this form, the Law of Cosines provides a description of each angle of a triangle in terms of the three sides. Consequently, the Law of Cosines can be used to solve triangles in these cases:

1. Two sides and the angle between them are known (SAS).

2. Three sides are known (SSS).

EXAMPLE 1

SAS   Solve triangle *ABC* in Figure 8–38.

SOLUTION   We have *a* = 16, *b* = 10, and *C* = 110°. The right side of the third equation in the Law of Cosines involves only these known quantities. Hence,

$$c^2 = a^2 + b^2 - 2ab \cos C$$

$$= 16^2 + 10^2 - 2 \cdot 16 \cdot 10 \cos 110°$$

$$\approx 256 + 100 - 320(-.342) \approx 465.4.^*$$

Therefore, $c \approx \sqrt{465.4} \approx 21.6$. Now use the alternate form of the Law of Cosines.

$$\cos A = \frac{b^2 + c^2 - a^2}{2bc} \approx \frac{10^2 + (21.6)^2 - 16^2}{2 \cdot 10 \cdot 21.6} \approx .7172.$$

The COS^{-1} key (in degree mode) produces an angle with cosine .7172: *A* ≈ 44.2°. Hence, *B* ≈ 180° − (44.2° + 110°) = 25.8°.   ■

Figure 8–38

---

*Throughout this chapter, all decimals are printed in rounded-off form for reading convenience, but no rounding is done in the actual computation until the final answer is obtained.

Figure 8–39

EXAMPLE 2

SSS   Find the angles of triangle ABC in Figure 8–39.

SOLUTION   In this case, $a = 20$, $b = 15$, and $c = 8.3$. By the alternate form of the Law of Cosines,

$$\cos A = \frac{b^2 + c^2 - a^2}{2bc} = \frac{15^2 + 8.3^2 - 20^2}{2 \cdot 15 \cdot 8.3} = \frac{-106.11}{249} \approx -.4261.$$

The $\cos^{-1}$ key shows that $A \approx 115.2°$. Similarly, the alternate form of the Law of Cosines yields

$$\cos B \approx \frac{a^2 + c^2 - b^2}{2ac} = \frac{20^2 + 8.3^2 - 15^2}{2 \cdot 20 \cdot 8.3} = \frac{243.89}{332} \approx .7346$$

$$B \approx 42.7°.$$

Therefore, $C \approx 180° - (115.2° + 42.7°) = 180° - 157.9° = 22.1°.$ ■

EXAMPLE 3

Two trains leave a station on different tracks. The tracks make an angle of 125° with the station as vertex. The first train travels at an average speed of 100 kilometers per hour, and the second travels at an average speed of 65 kilometers per hour. How far apart are the trains after 2 hours?

SOLUTION   The first train A traveling at 100 kilometers per hour for 2 hours goes a distance of $100 \times 2 = 200$ kilometers. The second train B travels a distance of $65 \times 2 = 130$ kilometers. So we have the situation shown in Figure 8–40.

Figure 8–40

By the Law of Cosines,

$$c^2 = a^2 + b^2 - 2ab \cos C$$

$$= 130^2 + 200^2 - 2 \cdot 130 \cdot 200 \cos 125°$$

$$= 56,900 - 52,000 \cos 125° \approx 86,725.97$$

$$c \approx \sqrt{86,725.97} = 294.5 \text{ kilometers}.$$

The trains are 294.5 kilometers apart after 2 hours.   ■

EXAMPLE 4

A 100-foot-tall antenna tower is to be placed on a hillside that makes an angle of 12° with the horizontal. It is to be anchored by two cables from the top of the tower to points 85 feet uphill and 95 feet downhill from the base. How much cable is needed?

SOLUTION   The situation is shown in Figure 8–41.

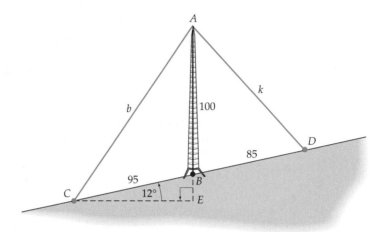

Figure 8–41

In triangle *BEC*, angle *E* is a right angle and by hypothesis, angle *C* measures 12°. Since the sum of the angles of a triangle is 180°, we must have

$$\angle CBE = 180° - (90° + 12°) = 78°.$$

As shown in the figure, the sum of angles *CBE* and *CBA* is a straight angle (180°). Hence,

$$\angle CBA = 180° - 78° = 102°.$$

Apply the Law of Cosines to triangle *ABC*.

$$b^2 = 95^2 + 100^2 - 2 \cdot 95 \cdot 100 \cos 102°$$

$$= 9025 + 10,000 - 19,000 \cos 102°$$

$$\approx 22,975.32.$$

Therefore, the length of the downhill cable is $b \approx \sqrt{22,975.32} \approx 151.58$ feet.

    To find the length of the uphill cable, note that the sum of angles *CBA* and *DBA* is a straight angle, so

$$\angle DBA = 180° - \angle CBA = 180° - 102° = 78°.$$

Applying the Law of Cosines to triangle *DBA*, we have

$$k^2 = 85^2 + 100^2 - 2 \cdot 85 \cdot 100 \cos 78°$$

$$= 7225 + 10,000 - 17,000 \cos 78° \approx 13,690.50.$$

Hence, the length of the uphill cable is $k = \sqrt{13,690.50} \approx 117.01$ feet. ■

▜▜ PROOF OF THE LAW OF COSINES

Given triangle ABC, position it on a coordinate plane so that angle A is in standard position with initial side c and terminal side b. Depending on the size of angle A, there are two possibilities, as shown in Figure 8–42.

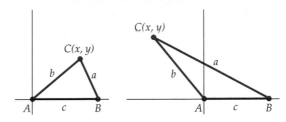

Figure 8–42

The coordinates of B are $(c, 0)$. Let (x, y) be the coordinates of C. Now C is a point on the terminal side of angle A, and the distance from C to the origin A is obviously b. Therefore, according to the point-in-the-plane description of sine and cosine, we have

$$\frac{x}{b} = \cos A \qquad \text{or, equivalently,} \qquad x = b \cos A,$$

$$\frac{y}{b} = \sin A \qquad \text{or, equivalently,} \qquad y = b \sin A.$$

Using the distance formula on the coordinates of B and C, we have

$$a = \text{distance from } C \text{ to } B$$
$$= \sqrt{(x - c)^2 + (y - 0)^2} = \sqrt{(b \cos A - c)^2 + (b \sin A - 0)^2}.$$

Squaring both sides of this last equation and simplifying, using the Pythagorean identity, yields

$$a^2 = (b \cos A - c)^2 + (b \sin A)^2$$
$$a^2 = b^2 \cos^2 A - 2bc \cos A + c^2 + b^2 \sin^2 A$$
$$a^2 = b^2(\sin^2 A + \cos^2 A) + c^2 - 2bc \cos A$$
$$a^2 = b^2 + c^2 - 2bc \cos A.$$

This proves the first equation in the Law of Cosines. Similar arguments beginning with angle B or C in standard position prove the other two equations.

✓ EXERCISES 8.3

Directions: *Standard notation for triangle ABC is used throughout. Use a calculator and round off your answers to one decimal place at the end of the computation.*

In Exercises 1–16, solve the triangle ABC under the given conditions.

1. $A = 20°$, $b = 10$, $c = 7$

2. $B = 40°$, $a = 12$, $c = 20$

3. $C = 118°$, $a = 6$, $b = 10$

4. $C = 52.5°$, $a = 6.5$, $b = 9$

5. $A = 140°$, $b = 12$, $c = 14$

6. $B = 25.4°$, $a = 6.8$, $c = 10.5$

7. $C = 78.6°, a = 12.1, b = 20.3$

8. $A = 118.2°, b = 16.5, c = 10.7$

9. $a = 7, b = 3, c = 5$

10. $a = 8, b = 5, c = 10$

11. $a = 16, b = 20, c = 32$

12. $a = 5.3, b = 7.2, c = 10$

13. $a = 7.2, b = 6.5, c = 11$

14. $a = 6.8, b = 12.4, c = 15.1$

15. $a = 12, b = 16.5, c = 21.3$

16. $a = 5.7, b = 20.4, c = 16.8$

17. Find the angles of the triangle whose vertices are $(0, 0)$, $(5, -2)$, $(1, -4)$.

18. Find the angles of the triangle whose vertices are $(-3, 4)$, $(6, 1)$, $(2, -1)$.

19. Two ships leave port, one traveling in a straight course at 22 mph and the other traveling a straight course at 31 mph. Their courses diverge by 38°. How far apart are they after 3 hours?

20. One plane flies west from Cleveland at 350 mph. A second plane leaves Cleveland at the same time and flies southeast at 200 mph. How far apart are the planes after 1 hour and 36 minutes?

21. A 400-foot-high tower stands on level ground, anchored by two cables on the west side. The end of the cable closest to the tower makes an angle of 70° with the horizontal. The two cable ends are 100 feet apart, as shown in the figure. How long are the cables?

22. The side of a hill makes an angle of 12° with the horizontal. A wire is to be run from the top of a 175-foot tower on the top of the hill to a stake located 120 feet down the hillside from the base of the tower. How long a wire is needed?

23. The pitcher's mound on a standard baseball diamond (which is actually a square) is 60.5 feet from home plate (see the figure). How far is the pitcher's mound from first base?

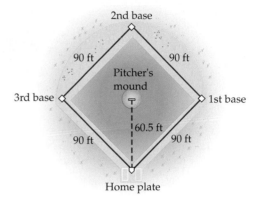

24. At Wrigley Field in Chicago, the straight-line distance from home plate over second base to the center field wall is 400 feet. How far is it from first base to the same point at the center field wall? [*Hint:* Adapt and extend the figure from Exercise 23.]

25. A stake is located 10.8 feet from the end of a closed gate that is 8 feet long. The gate swings open, and its end hits the stake. Through what angle did the gate swing?

26. The distance from Chicago to St. Louis is 440 kilometers, that from St. Louis to Atlanta 795 kilometers, and that from Atlanta to Chicago 950 kilometers. What are the angles in the triangle with these three cities as vertices?

27. A boat runs in a straight line for 3 kilometers, then makes a 45° turn and goes for another 6 kilometers (see the figure). How far is the boat from its starting point?

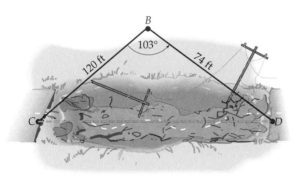

28. A plane flies in a straight line at 400 mph for 1 hour and 12 minutes. It makes a 15° turn and flies at 375 mph for 2 hours and 27 minutes. How far is it from its starting point?

29. An engineer wants to measure the width *CD* of a sinkhole. So he places a stake *B* and determines the measurements shown in the figure. How wide is the sinkhole?

30. A straight tunnel is to be dug through a hill. Two people stand on opposite sides of the hill where the tunnel entrances are to be located. Both can see a stake located 530 meters from the first person and 755 meters from the second. The angle determined by the two people and the stake (vertex) is 77°. How long must the tunnel be?

31. One diagonal of a parallelogram is 6 centimeters long, and the other is 13 centimeters long. They form an angle of 42° with each other. How long are the sides of the parallelogram? [*Hint:* The diagonals of a parallelogram bisect each other.]

32. A parallelogram has diagonals of lengths 12 and 15 inches that intersect at an angle of 63.7°. How long are the sides of the parallelogram?

33. A ship is traveling at 18 mph from Corsica to Barcelona, a distance of 350 miles. To avoid bad weather, the ship leaves Corsica on a route 22° south of the direct route (see the figure). After 7 hours, the bad weather has been bypassed. Through what angle should the ship now turn to head directly to Barcelona?

34. A plane leaves South Bend for Buffalo, 400 miles away, intending to fly a straight course in the direction 70° (aerial navigation is explained in Exercise 31 of Section 8.2). After flying 180 miles, the pilot realizes that an error has been made and that he has actually been flying in the direction 55°.
(a) At that time, how far is the plane from Buffalo?
(b) In what direction should the plane now go to reach Buffalo?

35. Assume that the earth is a sphere of radius 3980 miles. A satellite travels in a circular orbit around the earth, 900 miles above the equator, making one full orbit every 6 hours. If it passes directly over a tracking station at 2 P.M., what is the distance from the satellite to the tracking station at 2:05 P.M.?

36. Two planes at the same altitude approach an airport. One plane is 16 miles from the control tower and the other is 22 miles from the tower. The angle determined by the planes and the tower, with the tower as vertex, is 11°. How far apart are the planes?

37. Assuming that the circles in the figure are mutually tangent, find the lengths of the sides and the measures of the angles in triangle *ABC*.

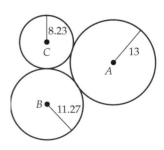

38. Assuming that the circles in the figure are mutually tangent, find the lengths of the sides and the measures of the angles in triangle *ABC*.

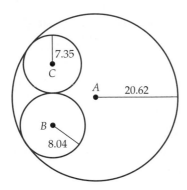

Thinkers

39. A rope is attached at points *A* and *B* and taut around a pulley whose center is at *C*, as shown in the figure (in which *AC* has length 8 and *BC* length 7). The rope lies on the pulley from *D* to *E* and the radius of the pulley is 1 meter. How long is the rope?

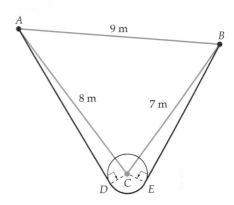

40. Use the Law of Cosines to prove that the sum of the squares of the lengths of the two diagonals of a parallelogram equals the sum of the squares of the lengths of the four sides.

8.4   The Law of Sines

To solve oblique triangles in cases in which the Law of Cosines cannot be used, we need this fact.

Law of Sines

In any triangle *ABC* (in standard notation),

$$\frac{a}{\sin A} = \frac{b}{\sin B} = \frac{c}{\sin C}.^*$$

Proof   Position triangle *ABC* on a coordinate plane so that angle *C* is in standard position, with initial side *b* and terminal side *a*, as shown in Figure 8–43.

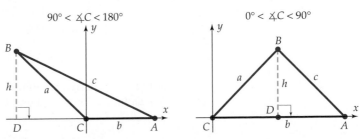

Figure 8–43

---

*An equality of the form $u = v = w$ is shorthand for the statement $u = v$ and $v = w$ and $w = u$.

In each case, we can compute sin C by using the point B on the terminal side of angle C. The second coordinate of B is h, and the distance from B to the origin is a. Therefore, by the point-in-the-plane description of sine,

$$\sin C = \frac{h}{a} \qquad \text{or, equivalently,} \qquad h = a \sin C.$$

In each case, right triangle ADB shows that

$$\sin A = \frac{\text{opposite}}{\text{hypotenuse}} = \frac{h}{c} \qquad \text{or, equivalently,} \qquad h = c \sin A.$$

Combining this with the fact that $h = a \sin C$, we have

$$a \sin C = c \sin A.$$

Since angles in a triangle are nonzero, $\sin A \neq 0$ and $\sin C \neq 0$. Dividing both sides of the last equation by $\sin A \sin C$ yields

$$\frac{a}{\sin A} = \frac{c}{\sin C}.$$

This proves one equation in the Law of Sines. Similar arguments beginning with angles A or B in standard position prove the other equations.   ■

The Law of Sines can be used to solve triangles in these cases:

1. Two angles and one side are known (AAS).

2. Two sides and the angle opposite one of them are known (SSA).

EXAMPLE 1

AAS   If $B = 20°$, $C = 31°$, and $b = 210$ in Figure 8–44, find the other angles and sides.

Figure 8–44

SOLUTION   Since the sum of the angles of a triangle is $180°$,

$$A = 180° - (20° + 31°) = 180° - 51° = 129°.$$

To find side a, we observe that we know three of the four quantities in one of the equations given by the Law of Sines.

$$\frac{a}{\sin A} = \frac{b}{\sin B}$$

$$\frac{a}{\sin 129°} = \frac{210}{\sin 20°}.$$

Multiplying both sides by sin $129°$, we obtain

$$a = \frac{210(\sin 129°)}{\sin 20°} \approx 477.2.$$

Side c is found similarly. Beginning with an equation from the Law of Sines involving c and three known quantities, we have

$$\frac{c}{\sin C} = \frac{b}{\sin B}$$

$$\frac{c}{\sin 31°} = \frac{210}{\sin 20°}$$

$$c = \frac{210 \sin 31°}{\sin 20°} \approx 316.2 \qquad ■$$

In the AAS case, there is exactly one triangle that satisfies the given data.[*] But when two sides of a triangle and the angle opposite one of them are known (SSA), there may be one, two, or no triangles that satisfy the given data (the **ambiguous case**). To see why this can happen, suppose sides a and b and angle A are given. Place angle A in standard position with terminal side b. If angle A is less than 90°, then there are four possibilities for side a:

(i) Side a is too short to reach the third side: *no solution.*

(ii) Side a just reaches the third side and is perpendicular to it: *one solution.*

(iii) An arc of radius a meets the third side at 2 points to the right of A: *two solutions.*

(iv) $a \geq b$, so that an arc of radius a meets the third side at just one point to the right of A: *one solution.*

When angle A is more than 90°, then there are only two possibilities:

(i) $a \leq b$, so that side a is too short to reach the third side: *no solution.*

(ii) $a > b$, so that an arc of radius a meets the third side at just one point to the right of A: *one solution.*

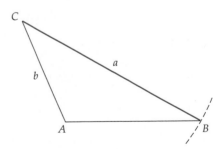

---

[*]Once you know two angles, you know all three (their sum must be 180°). Hence, you know two angles and the included side. Any two triangles satisfying these conditions will be congruent by the ASA Theorem of plane geometry.

To deal with the case of two solutions, we need this identity.

Supplementary
Angle Identity

If $0° \leq \theta \leq 90°$, then

$$\sin \theta = \sin(180° - \theta).$$

Proof*   Place the angle $180° - \theta$ in standard position and choose a point D on its terminal side. Let r be the distance from D to the origin. The situation looks like Figure 8–45.

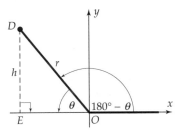

Figure 8–45

Since h is the second coordinate of D, we have $\sin(180° - \theta) = h/r$. Right triangle OED shows that

$$\sin \theta = \frac{\text{opposite}}{\text{hypotenuse}} = \frac{h}{r} = \sin(180° - \theta). \qquad \blacksquare$$

EXAMPLE 2

SSA   Given triangle ABC with $a = 6$, $b = 7$, and $A = 65°$, find angle B.

SOLUTION   We use an equation from the Law of Sines involving the known quantities.

$$\frac{b}{\sin B} = \frac{a}{\sin A}$$

$$\frac{7}{\sin B} = \frac{6}{\sin 65°}$$

$$\sin B = \frac{7 \sin 65°}{6} \approx 1.06.$$

There is no angle B whose sine is greater than 1. Therefore, there is no triangle satisfying the given data. $\blacksquare$

---

*You may skip this proof if you have read Example 2 in Section 7.2, where the identity [in the form $\sin y = \sin(\pi - y)$] is proved for every angle.

EXAMPLE 3

An airplane A takes off from carrier B and flies in a straight line for 12 kilometers. At that instant, an observer on destroyer C, located 5 kilometers from the carrier, notes that the angle determined by the carrier, the destroyer (vertex), and the plane is 37°. How far is the plane from the destroyer?

SOLUTION   The given data provide Figure 8–46.

Figure 8–46

We must find side b. To do this, we first use the Law of Sines to find angle A.

$$\frac{a}{\sin A} = \frac{c}{\sin C}$$

$$\frac{5}{\sin A} = \frac{12}{\sin 37°}$$

$$\sin A = \frac{5 \sin 37°}{12} \approx .2508.$$

The SIN^{-1} key on a calculator in degree mode shows that 14.5° is an angle whose sine is .2508. The supplementary angle identity shows that $180° - 14.5° = 165.5°$ is also an angle with sine .2508. But if $A = 165.5°$ and $C = 37°$, the sum of angles A, B, C would be greater than 180°. Since this is impossible, $A = 14.5°$ is the only solution here. Therefore,

$$B = 180° - (37° + 14.5°) = 180° - 51.5° = 128.5°.$$

Using the Law of Sines again, we have

$$\frac{b}{\sin B} = \frac{c}{\sin C}$$

$$\frac{b}{\sin 128.5°} = \frac{12}{\sin 37°}$$

$$b = \frac{12 \sin 128.5°}{\sin 37°} \approx 15.6.$$

Thus, the plane is approximately 15.6 kilometers from the destroyer.   ■

EXAMPLE 4

SSA Solve triangle ABC when $a = 7.5$, $b = 12$, and $A = 35°$.

SOLUTION The Law of Sines shows that

$$\frac{b}{\sin B} = \frac{a}{\sin A}$$

$$\frac{12}{\sin B} = \frac{7.5}{\sin 35°}$$

$$\sin B = \frac{12 \sin 35°}{7.5} \approx .9177.$$

The SIN^{-1} key shows that $66.6°$ is a solution of $\sin B = .9177$. Therefore, $180° - 66.6° = 113.4°$ is also a solution of $\sin B = .9177$ by the supplementary angle identity. In each case the sum of angles A and B is less than $180°$, so there are two triangles ABC satisfying the given data, as shown in Figure 8–47.

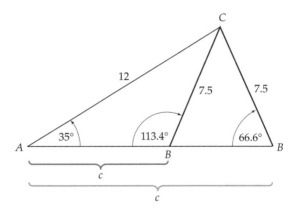

Figure 8–47

Case 1. If $B = 66.6°$, then $C = 180° - (35° + 66.6°) = 78.4°$. By the Law of Sines,

$$\frac{c}{\sin C} = \frac{a}{\sin A}$$

$$\frac{c}{\sin 78.4°} = \frac{7.5}{\sin 35°}$$

$$c = \frac{7.5 \sin 78.4°}{\sin 35°} \approx \frac{7.5(.9796)}{.5736} \approx 12.8.$$

Case 2. If $B = 113.4°$, then $C = 180° - (35° + 113.4°) = 31.6°$. Consequently,

$$\frac{c}{\sin C} = \frac{a}{\sin A}$$

$$\frac{c}{\sin 31.6°} = \frac{7.5}{\sin 35°}$$

$$c = \frac{7.5 \sin 31.6°}{\sin 35°} \approx \frac{7.5(.5240)}{.5736} \approx 6.9. \quad ■$$

EXAMPLE 5

A plane flying in a straight line passes directly over point A on the ground and later directly over point B, which is 3 miles from A. A few minutes after the plane passes over B, the angle of elevation from A to the plane is 43° and the angle of elevation from B to the plane is 67°. How high is the plane at that moment?

SOLUTION   If C represents the plane, then the situation is represented in Figure 8–48. We must find the length of h.

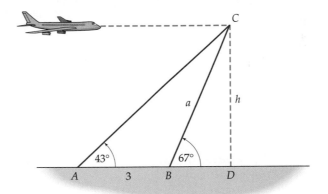

Figure 8–48

Note that angle ABC measures $180° - 67° = 113°$, and hence,

$$\angle BCA = 180° - (43° + 113°) = 24°.$$

Use the Law of Sines to find side a of triangle ABC.

$$\frac{a}{\sin 43°} = \frac{3}{\sin 24°}$$

$$a = \frac{3 \sin 43°}{\sin 24°} \approx 5.03.$$

Now in the right triangle CBD, we have

$$\sin 67° = \frac{\text{opposite}}{\text{hypotenuse}} = \frac{h}{a} \approx \frac{h}{5.03}.$$

Therefore, $h \approx 5.03 \sin 67° \approx 4.63$ miles.   ■

✓ EXERCISES 8.4

Directions: *Standard notation for triangle ABC is used throughout. Use a calculator and round off your answers to one decimal place at the end of the computation.*

In Exercises 1–8, solve triangle ABC under the given conditions.

1. $A = 48°, B = 22°, a = 5$

2. $B = 33°, C = 46°, b = 4$

3. $A = 116°, C = 50°, a = 8$

4. $A = 105°, B = 27°, b = 10$

5. $B = 44°, C = 48°, b = 12$

6. $A = 67°, C = 28°, a = 9$

7. $A = 102.3°, B = 36.2°, a = 16$

8. $B = 97.5°, C = 42.5°, b = 7$

In Exercises 9–32, solve the triangle. The Law of Cosines may be needed in Exercises 19–32.

9. $b = 15, c = 25, B = 47°$

10. $b = 30, c = 50, C = 60°$

11. $a = 12, b = 5, B = 20°$

12. $b = 12.5, c = 20.1, B = 37.3°$

13. $a = 5, c = 12, A = 102°$

14. $a = 9, b = 14, B = 95°$

15. $b = 11, c = 10, C = 56°$

16. $a = 12.4, c = 6.2, A = 72°$

17. $A = 41°, B = 67°, a = 10.5$

18. $a = 30, b = 40, A = 30°$

19. $b = 4, c = 10, A = 75°$

20. $a = 50, c = 80, C = 45°$

21. $a = 6, b = 12, c = 16$

22. $B = 20.67°, C = 34°, b = 185$

23. $a = 16.5, b = 18.2, C = 47°$

24. $a = 21, c = 15.8, B = 71°$

25. $b = 17.2, c = 12.4, B = 62.5°$

26. $b = 24.1, c = 10.5, C = 26.3°$

27. $a = 10.1, b = 18.2, A = 50.7°$

28. $b = 14.6, c = 7.8, B = 40.4°$

29. $b = 12.2, c = 20, A = 65°$

30. $a = 44, c = 84, C = 42.2°$

31. $A = 19°, B = 35°, a = 110$

32. $b = 15.4, c = 19.3, A = 42°$

33. A surveyor marks points A and B 200 meters apart on one bank of a river. She sights a point C on the opposite bank and determines the angles shown in the figure. What is the distance from A to C?

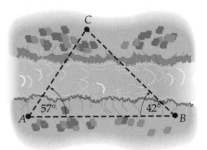

34. A forest fire is spotted from two fire towers. The triangle determined by the two towers and the fire has angles of 28° and 37° at the tower vertices. If the towers are 3000 meters apart, which one is closer to the fire?

35. A visitor to the Leaning Tower of Pisa observed that the tower's shadow was 40 meters long and that the angle of elevation from the tip of the shadow to the top of the tower was 57°. The tower is now 54 meters tall (measured from the ground to the top along the center line of the tower). Approximate the angle α that the center line of the tower makes with the vertical.

36. A pole tilts at an angle 9° from the vertical, away from the sun, and casts a shadow 24 feet long. The angle of elevation from the end of the pole's shadow to the top of the pole is 53°. How long is the pole?

37. A side view of a bus shelter is shown in the figure. The brace *d* makes an angle of 37.25° with the back and an angle of 34.85° with the top of the shelter. How long is this brace?

8 ft

5 ft

d

38. A straight path makes an angle of 6° with the horizontal. A statue at the higher end of the path casts a 6.5-meter-long shadow straight down the path. The angle of elevation from the end of the shadow to the top of the statue is 32°. How tall is the statue?

39. A vertical statue 6.3 meters high stands on top of a hill. At a point on the side of the hill 35 meters from the statue's base, the angle between the hillside and a line from the top of the statue is 10°. What angle does the side of the hill make with the horizontal?

40. A fence post is located 50 feet from one corner of a building and 40 feet from the adjacent corner. Fences are put up between the post and the building corners to form a triangular garden area. The 40-foot fence makes a 58° angle with the building. How long is the building wall?

41. Two straight roads meet at an angle of 40° in Harville, one leading to Eastview and the other to Wellston. Eastview is 18 kilometers from Harville and 20 kilometers from Wellston. What is the distance from Harville to Wellston?

Harville 40° Eastview

Wellston

42. Each of two observers 400 feet apart measures the angle of elevation to the top of a tree that sits on the straight line between them. These angles are 51° and 65°, respectively. How tall is the tree? How far is the base of its trunk from each observer?

43. From the top of the 800-foot-tall Cartalk Tower, Tom sees a plane; the angle of elevation is 67°. At the same instant, Ray, who is on the ground, 1 mile from the building, notes that his angle of elevation to the plane is 81° and that his angle of elevation to the top of Cartalk Tower is 8.6°. Assuming that Tom and Ray and the airplane are in a plane perpendicular to the ground, how high is the airplane?

800 ft   67°   81°   8.6°

1 mile

44. A plane flies in a direction of 105° from airport *A*. After a time, it turns and proceeds in a direction of 267°. Finally, it lands at airport *B*, 120 miles directly south of airport *A*. How far has the plane traveled? [*Note:* Aerial navigation directions are explained in Exercise 31 of Section 8.2.]

45. Charlie is afraid of water; he can't swim and refuses to get in a boat. However, he must measure the width of a river for his geography class. He has a long tape measure but no way to measure angles. While pondering what to do, he paces along the side of the river using the five paths joining points A, B, C, and D. If he can't determine the width of the river, he will flunk the course.

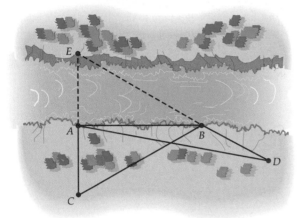

(a) Save Charlie from disaster by explaining how he can determine the width AE simply by measuring the lengths AB, AC, AD, BC, and BD and using trigonometry.

(b) Charlie determines that $AB = 75$ feet, $AC = 25$ feet, $AD = 90$ feet, $BC = 80$ feet, and $BD = 22$ feet. How wide is the river between A and E?

46. A plane flies in a direction of 85° from Chicago. It then turns and flies in the direction of 200° for 150 miles. It is then 195 miles from its starting point. How far did the plane fly in the direction of 85°? (See the note in Exercise 44.)

47. A hinged crane makes an angle of 50° with the ground. A malfunction causes the lock on the hinge to fail and the top part of the crane swings down. How far from the base of the crane does the top hit the ground?

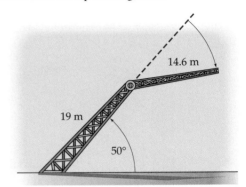

8.4.A SPECIAL TOPICS The Area of a Triangle

The proof of the Law of Sines enables us to prove a useful fact.

Area of a Triangle

The area of a triangle containing an angle C with sides of lengths a and b is

$$\frac{1}{2}ab \sin C.$$

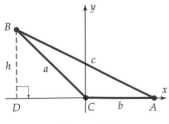

Figure 8–49

Proof Place the vertex of angle C at the origin, with side b on the positive x-axis (Figure 8–49).* Then b is the base and h is the altitude of the triangle so that

$$\text{area of triangle } ABC = \frac{1}{2} \times \text{base} \times \text{altitude} = \frac{1}{2} \cdot b \cdot h.$$

*Figure 8–49 is the case when C is larger than 90°; the argument when C is less than 90° is similar.

The proof of the Law of Sines on page 573 shows that $h = a \sin C$. Therefore,

$$\text{area of triangle } ABC = \frac{1}{2} \cdot b \cdot h = \frac{1}{2} \cdot b \cdot a \sin C = \frac{1}{2}ab \sin C. \quad \blacksquare$$

EXAMPLE 1

Find the area of the triangle shown in Figure 8–50.

SOLUTION

$$\frac{1}{2} \cdot 8 \cdot 13 \sin 130° \approx 39.83 \text{ square centimeters.} \quad \blacksquare$$

8 cm   130°   13 cm

Figure 8–50

Here is a useful formula for the area of a triangle in terms of its sides.

Heron's Formula

> The area of a triangle with sides a, b, c is
> $$\sqrt{s(s - a)(s - b)(s - c)},$$
> where $s = \frac{1}{2}(a + b + c)$.

Proof   The preceding area formula and the Pythagorean identity,

$$\sin^2 C = 1 - \cos^2 C = (1 + \cos C)(1 - \cos C),$$

show that the area of triangle ABC (standard notation) is

$$\frac{1}{2}ab \sin C = \sqrt{\left(\frac{1}{2}ab \sin C\right)^2} = \sqrt{\frac{1}{4}a^2b^2 \sin^2 C}$$

$$= \sqrt{\frac{1}{4}a^2b^2(1 - \cos^2 C)}$$

$$= \sqrt{\frac{1}{2}ab(1 + \cos C)\frac{1}{2}ab(1 - \cos C)}.$$

Exercise 20 uses the Law of Cosines to show that

$$\frac{1}{2}ab(1 + \cos C) = \frac{(a + b)^2 - c^2}{4} = \frac{(a + b) + c}{2} \cdot \frac{(a + b) - c}{2}$$

$$= s(s - c)$$

and

$$\frac{1}{2}\, ab(1 - \cos C) = \frac{c^2 - (a - b)^2}{4} = \frac{c - (a - b)}{2} \cdot \frac{c + (a - b)}{2}$$

$$= (s - a)(s - b).$$

Combining these facts completes the proof.

$$\text{Area} = \frac{1}{2}\, ab \sin C = \sqrt{\frac{1}{2}\, ab(1 + \cos C)\frac{1}{2}\, ab(1 - \cos C)}$$

$$= \sqrt{s(s - a)(s - b)(s - c)}. \quad \blacksquare$$

EXAMPLE 2

Find the area of the triangle whose sides have lengths 7, 9, and 12.

SOLUTION Apply Heron's Formula with $a = 7$, $b = 9$, $c = 12$, and

$$s = \frac{1}{2}(a + b + c) = \frac{1}{2}(7 + 9 + 12) = 14.$$

The area is

$$\sqrt{s(s - a)(s - b)(s - c)} = \sqrt{14(14 - 7)(14 - 9)(14 - 12)}$$

$$= \sqrt{980} \approx 31.3 \text{ square units.} \quad \blacksquare$$

EXERCISES 8.4.A

In Exercises 1–8, find the area of triangle ABC (standard notation) under the given conditions.

1. $a = 4$, $b = 8$, $C = 27°$

2. $b = 10$, $c = 14$, $A = 36°$

3. $c = 7$, $a = 10$, $B = 68°$

4. $a = 9$, $b = 13$, $C = 75°$

5. $a = 11$, $b = 15$, $c = 18$

6. $a = 4$, $b = 12$, $c = 14$

7. $a = 7$, $b = 9$, $c = 11$

8. $a = 17$, $b = 27$, $c = 40$

In Exercises 9–11, find the area of the triangle with the given vertices.

9. $(0, 0)$, $(2, -5)$, $(-3, 1)$ **10.** $(-4, 2)$, $(5, 7)$, $(3, 0)$

In Exercises 11 and 12, find the area of the polygonal region. [Hint: Divide the region into triangles.]

11.

12.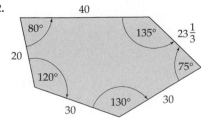

13. A triangular banner is hung from a window along the side of a building. The edges that touch the window are 20 and 24 feet long respectively. The third side is parallel to the ground. The angle between the 20-foot side and the third side is 44°. What is the area of the banner?

14. A triangular lot has sides of length 120 feet and 160 feet. The angle between these sides is 42°. Adjacent to this lot is a rectangular lot whose longest side has length 200 feet and whose shortest side is the same length as the shortest side of the triangular lot. What is the total area of both lots?

15. If a gallon of paint covers 400 square feet, how many gallons are needed to paint a triangular deck with sides of lengths 65 feet, 72 feet, and 88 feet?

16. Find the volume of the prism in the figure. The volume is given by the formula $V = \dfrac{1}{3} Bh$, where B is the area of the base and h is the height.

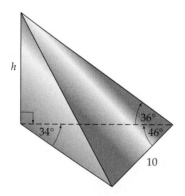

17. A rigid plastic triangle *ABC* rests on three vertical rods, as shown in the figure. What is its area?

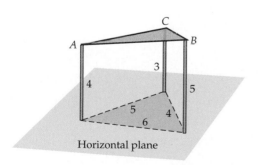

18. Prove that the area of triangle *ABC* (standard notation) is given by $\dfrac{a^2 \sin B \sin C}{2 \sin A}$.

19. What is the area of a triangle whose sides have lengths 12, 20, and 36? (If your answer turns out strangely, try drawing a picture.)

20. Complete the proof of Heron's Formula as follows. Let $s = \dfrac{1}{2}(a + b + c)$.

(a) Show that

$$\frac{1}{2}ab(1 + \cos C) = \frac{(a + b)^2 - c^2}{4}$$

$$= \frac{(a + b) + c}{2} \cdot \frac{(a + b) - c}{2}$$

$$= s(s - c).$$

[*Hint:* Use the Law of Cosines to express cos *C* in terms of *a, b, c*; then simplify.]

(b) Show that

$$\frac{1}{2}ab(1 - \cos C) = \frac{c^2 - (a - b)^2}{4}$$

$$= \frac{c - (a - b)}{2} \cdot \frac{c + (a - b)}{2}$$

$$= (s - a)(s - b).$$

Chapter 8 Review

IMPORTANT CONCEPTS

Section 8.1
[Alternate Section 8.1]

Trigonometric functions of
  angles   543 [554]
Point-in-the-plane description   554
  [556]
Right triangle description   546
  [554]
Solving right triangles   548

Section 8.2

Applications of right triangle
  trigonometry   558
Angles of elevation and
  depression   559

Section 8.3

Standard notation for
  triangles   566
Law of Cosines   566–567

Solution of oblique triangles
  567–569

Section 8.4

Law of Sines   573
Ambiguous case   575

Special Topics 8.4.A

Area formulas for triangles
  582–583

IMPORTANT FACTS & FORMULAS

■ *Right Triangle Description:* In a right triangle containing an angle θ,

$$\sin \theta = \frac{\text{opposite}}{\text{hypotenuse}} \qquad \cos \theta = \frac{\text{adjacent}}{\text{hypotenuse}}$$

$$\tan \theta = \frac{\text{opposite}}{\text{adjacent}} \qquad \cot \theta = \frac{\text{adjacent}}{\text{opposite}}$$

$$\sec \theta = \frac{\text{hypotenuse}}{\text{adjacent}} \qquad \csc \theta = \frac{\text{hypotenuse}}{\text{opposite}}$$

■ *Law of Cosines:* $a^2 = b^2 + c^2 - 2bc \cos A$

■ *Law of Cosines—Alternate Form:* $\cos A = \dfrac{b^2 + c^2 - a^2}{2bc}$

■ *Law of Sines:* $\dfrac{a}{\sin A} = \dfrac{b}{\sin B} = \dfrac{c}{\sin C}$

■ $\sin D = \sin(180° - D)$

■ Area of triangle $ABC = \dfrac{ab \sin C}{2}$

■ *Heron's Formula:*

$$\text{Area of triangle } ABC = \sqrt{s(s - a)(s - b)(s - c)}$$

where $s = \dfrac{1}{2}(a + b + c)$.

REVIEW QUESTIONS

Note: *Standard notation is used for triangles.*

1. Which of the following statements about the angle θ is true?
   (a) $\sin \theta = 3/4$
   (b) $\cos \theta = 5/4$
   (c) $\tan \theta = 3/5$
   (d) $\sin \theta = 4/5$
   (e) $\sin \theta = 4/3$

2. Suppose θ is a real number. Consider the right triangle with sides as shown in the figure. Then
   (a) $x = 1$
   (b) $x = 2$
   (c) $x = 4$
   (d) $x = 2 (\cos \theta + \sin \theta)$
   (e) none of the above

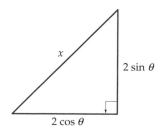

3. Use the right triangle in the figure to find $\sec \theta$.
   (a) $\sec \theta = 7/4$
   (b) $\sec \theta = 4/\sqrt{65}$
   (c) $\sec \theta = 7/\sqrt{65}$
   (d) $\sec \theta = \sqrt{65}/7$
   (e) $\sec \theta = \sqrt{65}/4$

4. Find the length of side h in the triangle, where angle A measures $40°$ and the distance from C to A is 25.

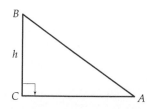

In Questions 5–8, angle B is a right angle. Solve triangle ABC.

5. $a = 12$, $c = 13$

6. $A = 40°$, $b = 10$

7. $C = 35°$, $a = 12$

8. $A = 56°$, $a = 11$

9. From a point on level ground 145 feet from the base of a tower, the angle of elevation to the top of the tower is $57.3.°$. How high is the tower?

10. A pilot in a plane at an altitude of 22,000 feet observes that the angle of depression to a nearby airport is $26°$. How many *miles* is the airport from a point on the ground directly below the plane?

11. A road rises 140 feet per horizontal mile. What angle does the road make with the horizontal?

12. A lighthouse keeper 100 feet above the water sees a boat sailing in a straight line directly toward her. As she watches, the angle of depression to the boat changes from $25°$ to $40°$. How far has the boat traveled during this time?

In Questions 13–16, use the Law of Cosines to solve triangle ABC.

13. $a = 12$, $b = 10$, $c = 15$

14. $a = 7.5$, $b = 3.2$, $c = 6.4$

15. $a = 10$, $c = 14$, $B = 130°$

16. $a = 7$, $b = 8.6$, $C = 72.4°$

17. Two trains depart simultaneously from the same station. The angle between the two tracks on which they leave is $120°$. One train travels at an average speed of 45 mph and the other at 70 mph. How far apart are the trains after 3 hours?

18. A 40-foot-high flagpole sits on the side of a hill. The hillside makes a $17°$ angle with the horizontal. How long is a wire that runs from the top of the pole to a point 72 feet downhill from the base of the pole?

In Questions 19–24, use the Law of Sines to solve triangle ABC.

19. $B = 124°$, $C = 31°$, $c = 3.5$

20. $A = 96°$, $B = 44°$, $b = 12$

21. $a = 75$, $c = 84$, $C = 62°$

22. $a = 5$, $c = 2.5$, $C = 30°$

23. $a = 3.5$, $b = 4$, $A = 60°$

24. $a = 3.8$, $c = 2.8$, $C = 41°$

25. Find the area of triangle ABC if $b = 24$, $c = 15$, and $A = 55°$.

26. Find the area of triangle ABC if $a = 10$, $c = 14$, and $B = 75°$.

27. A boat travels for 8 kilometers in a straight line from the dock. It is then sighted from a lighthouse that is 6.5 kilometers from the dock. The angle determined by the dock, the lighthouse (vertex), and the boat is 25°. How far is the boat from the lighthouse?

28. A pole tilts 12° from the vertical, away from the sun, and casts a 34-foot-long shadow on level ground. The angle of elevation from the end of the shadow to the top of the pole is 64°. How long is the pole?

In Questions 29–32, solve triangle ABC.

29. $A = 48°$, $B = 57°$, $b = 47$

30. $A = 67°$, $c = 125$, $a = 100$

31. $a = 5$, $c = 8$, $B = 76°$

32. $a = 90$, $b = 70$, $c = 40$

33. Two surveyors, Joe and Alice, are 240 meters apart on a riverbank. Each sights a flagpole on the opposite bank. The angle from the pole to Joe (vertex) to Alice is 63°. The angle from the pole to Alice (vertex) to Joe is 54°. How far are Joe and Alice from the pole?

34. A surveyor stakes out points A and B on opposite sides of a building. Point C on the side of the building is 300 feet from A and 440 feet from B. Angle ACB measures 38°. What is the distance from A to B?

35. A woman on the top of a 448-foot-high building spots a small plane. As she views the plane, its angle of elevation is 62°. At the same instant, a man at the ground-level entrance to the building sees the plane and notes that its angle of elevation is 65°.
    (a) How far is the woman from the plane?
    (b) How far is the man from the plane?
    (c) How high is the plane?

36. A straight road slopes at an angle of 10° with the horizontal. When the angle of elevation of the sun (from horizontal) is 62.5°, a telephone pole at the side of the road casts a 15-foot shadow downhill, parallel to the road. How high is the telephone pole?

37. Find angle ABC.

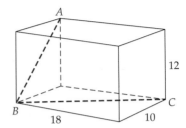

38. Use the Law of Sines to prove Engelsohn's equations: For any triangle ABC (standard notation),

$$\frac{a + b}{c} = \frac{\sin A + \sin B}{\sin C}$$

and

$$\frac{a - b}{c} = \frac{\sin A - \sin B}{\sin C}.$$

In Questions 39–42, find the area of triangle ABC under the given conditions.

39. There is an angle of 30°, the sides of which have lengths 5 and 8.

40. There is an angle of 40°, the sides of which have lengths 3 and 12.

41. The sides have lengths 7, 11, and 14.

42. The sides have lengths 4, 8, and 10.

DISCOVERY PROJECT 8   Life on a Sphere

Russell Illig/Getty Images

Although we experience the surface of the earth as a level surface, it is not. The earth, of course, is a sphere whose radius is about 6370 kilometers. When you stand on a sphere, strange things happen, especially when the surface is particularly smooth.

Granted, the horizon is often interfered with by hills, forests, buildings, and such. However, if you are out in an area of flat land like that found in eastern Kansas or northern Ontario, the hills are very slight, and the ground is very much like a sphere. The same is true of large lakes and oceans.

Generally speaking, you are not as interested in seeing the horizon as you are in objects that the horizon might hide. On the ocean, for instance, you might want to see another ship. On land, you might be looking for a particular building or vehicle.

1. Suppose that your average human is standing on a highway and a 3.5-meter-high truck passes by. Assume also that your average human has eyes that are 1.55 meters off the ground. If the road is flat and straight, how far away is the truck when it disappears below the horizon?

1.55 m   6370 km   6370 km   3.5 m   6370 km

2. Engineers building large bridges must also account for the curvature of the earth. In the figure, a bridge with towers 200 meters tall and 900 meters apart (straight-line distance) at the base is constructed. How much farther apart are the tops of the towers than the bases?

200 m   200 m

900 m

6370 km

6370 km

3. How much difference does it make if the distance between the towers is measured along the curve of the earth?

C H A P T E R 9

Applications of Trigonometry

Tomas Muscionico/Contact Press Images/PictureQuest

Is this bridge safe?

When planning a bridge or a building, architects and engineers must determine the stress on cables and other parts of the structure to be sure that all parts are adequately supported. Problems like these can be modeled and solved by using vectors. See Exercise 73 on page 620.

Chapter Outline

Interdependence of Sections

The chapter is divided into two independent parts, each consisting of two sections. The parts may be read in either order:

9.1 ———————▶ 9.2

9.3 ———————▶ 9.4

9.1   The Complex Plane and Polar Form for Complex Numbers

9.2   DeMoivre's Theorem and nth Roots of Complex Numbers

9.3   Vectors in the Plane

9.4   The Dot Product

Trigonometry has a variety of useful applications in geometry, algebra, and the physical sciences, several of which are discussed in this chapter.

9.1   The Complex Plane and Polar Form for Complex Numbers*

The complex number system can be represented geometrically by the coordinate plane:

The complex number $a + bi$ corresponds to the point (a, b) in the plane.

For example, the point $(2, 3)$ in Figure 9–1 is labeled by $2 + 3i$, and similarly for the other points shown:

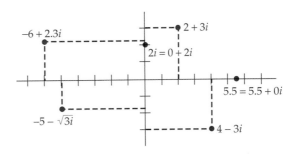

Figure 9–1

> ### TECHNOLOGY TIP
>
> To do complex arithmetic on TI-86 and HP-39, enter $a + bi$ as (a, b). On other calculators, use the special i key whose location is:
>
> TI-83/89: keyboard
>
> Casio FX 2.0: keyboard
>
> Casio 9850: OPTN/CPLX

When the coordinate plane is labeled by complex numbers in this way, it is called the **complex plane.** Each real number $a = a + 0i$ corresponds to the point $(a, 0)$ on the horizontal axis; so this axis is called the **real axis.** The vertical axis is called the **imaginary axis** because every imaginary number $bi = 0 + bi$ corresponds to the point $(0, b)$ on the vertical axis.

---

*Section 4.7 is a prerequisite for this section.

The absolute value of a real number c is the distance from c to 0 on the number line (see page 11). So we define the **absolute value** (or **modulus**) of the complex number $a + bi$ to be the distance from $a + bi$ to the origin in the complex plane:

$$|a + bi| = \text{distance from } (a, b) \text{ to } (0, 0) = \sqrt{(a - 0)^2 + (b - 0)^2}.$$

Therefore, we have the following.

Absolute Value

> The **absolute value** (or **modulus**) of the complex number $a + bi$ is
> $$|a + bi| = \sqrt{a^2 + b^2}.$$

TECHNOLOGY TIP

To find the absolute value of a complex number, use the ABS key, which is in this menu/submenu:

TI-83: MATH/CPX

TI-86: CPLX

TI-89: MATH/COMPLEX

Casio: OPTN/CPLX

HP-39: Keyboard

EXAMPLE 1

(a) $|3 + 2i| = \sqrt{3^2 + 2^2} = \sqrt{13}$.

(b) $|4 - 5i| = \sqrt{4^2 + (-5)^2} = \sqrt{41}$. ∎

Let $a + bi$ be a nonzero complex number, and denote $|a + bi|$ by r. Then r is the length of the line segment joining (a, b) and $(0, 0)$ in the plane. Let θ be the angle in standard position with this line segment as its terminal side (Figure 9–2).

Figure 9–2

According to the point-in-the-plane description of sine and cosine,

$$\cos \theta = \frac{a}{r} \qquad \text{and} \qquad \sin \theta = \frac{b}{r},$$

so

$$a = r \cos \theta \qquad \text{and} \qquad b = r \sin \theta.$$

Consequently,

$$a + bi = r \cos \theta + (r \sin \theta)i = r(\cos \theta + i \sin \theta).^*$$

When a complex number $a + bi$ is written in this way, it is said to be in **polar form** or **trigonometric form.** The angle θ is called the **argument** and is usually expressed in radian measure. The number $r = |a + bi|$ is sometimes called the

TECHNOLOGY TIP

To find the argument θ, use the ANGLE or ARG key in the same menu/submenu as the ABS key [except on HP-39, where it is in MATH/COMPLEX].

---

*It is customary to place i in front of $\sin \theta$ rather than after it. Some books abbreviate $r(\cos \theta + i \sin \theta)$ as r cis θ.

modulus (plural, moduli). The number 0 can also be written in polar notation by letting $r = 0$ and θ be any angle. Thus, we have the following.

Polar Form

> Every complex number $a + bi$ can be written in polar form:
> $$r(\cos \theta + i \sin \theta),$$
> where $r = |a + bi| = \sqrt{a^2 + b^2}$, $a = r \cos \theta$, and $b = r \sin \theta$.

When a complex number is written in polar form, the argument θ is not uniquely determined, since θ, $\theta \pm 2\pi$, $\theta \pm 4\pi$, etc., all satisfy the conditions in the box.

EXAMPLE 2

Express $-\sqrt{3} + i$ in polar form.

SOLUTION   Here $a = -\sqrt{3}$ and $b = 1$, so
$$r = \sqrt{a^2 + b^2} = \sqrt{(-\sqrt{3})^2 + 1^2} = \sqrt{3 + 1} = 2.$$

The angle θ must satisfy

$$\cos \theta = \frac{a}{r} = \frac{-\sqrt{3}}{2} \qquad \text{and} \qquad \sin \theta = \frac{b}{r} = \frac{1}{2}.$$

Since $-\sqrt{3} + i$ lies in the second quadrant (Figure 9–3), θ must be a second-quadrant angle. Our knowledge of special angles and Figure 9–3 show that $\theta = 5\pi/6$ satisfies these conditions. Hence,

$$-\sqrt{3} + i = 2\left(\cos \frac{5\pi}{6} + i \sin \frac{5\pi}{6}\right). \qquad \blacksquare$$

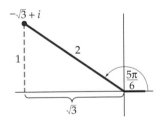

Figure 9–3

EXAMPLE 3*

Express $-2 + 5i$ in polar form.

---

*Omit this example if you haven't read Section 7.5.

SOLUTION   Since $a = -2$ and $b = 5$, $r = \sqrt{(-2)^2 + 5^2} = \sqrt{29}$. The angle θ must satisfy

$$\cos \theta = \frac{a}{r} = \frac{-2}{\sqrt{29}} \qquad \text{and} \qquad \sin \theta = \frac{b}{r} = \frac{5}{\sqrt{29}},$$

so

$$\tan \theta = \frac{\sin \theta}{\cos \theta} = \frac{5/\sqrt{29}}{-2/\sqrt{29}} = -\frac{5}{2} = -2.5.$$

Figure 9–4

Since $-2 + 5i$ lies in the second quadrant (Figure 9–4), θ lies between $\pi/2$ and π. As we saw in Section 7.5, the only solution of the equation $\tan \theta = -2.5$ that lies between $\pi/2$ and π is $\theta \approx -1.1903 + \pi = 1.9513$. Therefore,

$$-2 + 5i \approx \sqrt{29}(\cos 1.9513 + i \sin 1.9513). \qquad \blacksquare$$

Multiplication and division of complex numbers in polar form are done by the following rules, which are proved at the end of the section.

Polar Multiplication and Division Rules

> If $z_1 = r_1(\cos \theta_1 + i \sin \theta_1)$ and $z_2 = r_2(\cos \theta_2 + i \sin \theta_2)$ are any two complex numbers, then
>
> $$z_1 z_2 = r_1 r_2[\cos(\theta_1 + \theta_2) + i \sin(\theta_1 + \theta_2)]$$
>
> and
>
> $$\frac{z_1}{z_2} = \frac{r_1}{r_2}[\cos(\theta_1 - \theta_2) + i \sin(\theta_1 - \theta_2)] \quad (z_2 \neq 0).$$

In other words, to multiply two numbers in polar form, just *multiply the moduli and add the arguments.* To divide, just *divide the moduli and subtract the arguments.* Before proving the statements in the box, we will illustrate them with some examples.

EXAMPLE 4

Find $z_1 z_2$, when

$$z_1 = 2[\cos(5\pi/6) + i \sin(5\pi/6)] \quad \text{and} \quad z_2 = 3[\cos(7\pi/4) + i \sin(7\pi/4)].$$

SOLUTION   Here r_1 is the number 2, and $\theta_1 = 5\pi/6$; similarly, $r_2 = 3$, and $\theta_2 = 7\pi/4$, and we have

$$z_1 z_2 = r_1 r_2[\cos(\theta_1 + \theta_2) + i \sin(\theta_1 + \theta_2)]$$

$$= 2 \cdot 3\left[\cos\left(\frac{5\pi}{6} + \frac{7\pi}{4}\right) + i \sin\left(\frac{5\pi}{6} + \frac{7\pi}{4}\right)\right]$$

$$= 6\left[\cos\left(\frac{10\pi}{12} + \frac{21\pi}{12}\right) + i \sin\left(\frac{10\pi}{12} + \frac{21\pi}{12}\right)\right]$$

$$= 6\left(\cos\frac{31\pi}{12} + i \sin\frac{31\pi}{12}\right). \qquad \blacksquare$$

TECHNOLOGY TIP
Complex arithmetic can be done with numbers in polar form on TI calculators and Casio FX 2.0. Some answers may be expressed in rectangular form.

EXAMPLE 5

Find z_1/z_2, where

$$z_1 = 10[\cos(\pi/3) + i\sin(\pi/3)] \qquad \text{and} \qquad z_2 = 2[\cos(\pi/4) + i\sin(\pi/4)].$$

SOLUTION

$$\frac{z_1}{z_2} = \frac{10\left(\cos\dfrac{\pi}{3} + i\sin\dfrac{\pi}{3}\right)}{2\left(\cos\dfrac{\pi}{4} + i\sin\dfrac{\pi}{4}\right)} = \frac{10}{2}\left[\cos\left(\dfrac{\pi}{3} - \dfrac{\pi}{4}\right) + i\sin\left(\dfrac{\pi}{3} - \dfrac{\pi}{4}\right)\right]$$

$$= 5\left(\cos\dfrac{\pi}{12} + i\sin\dfrac{\pi}{12}\right). \qquad \blacksquare$$

▪▪ PROOF OF THE POLAR MULTIPLICATION RULE

If $z_1 = r_1(\cos\theta_1 + i\sin\theta_1)$ and $z_2 = r_2(\cos\theta_2 + i\sin\theta_2)$, then

$$z_1z_2 = r_1(\cos\theta_1 + i\sin\theta_1)r_2(\cos\theta_2 + i\sin\theta_2)$$

$$= r_1r_2(\cos\theta_1 + i\sin\theta_1)(\cos\theta_2 + i\sin\theta_2)$$

$$= r_1r_2(\cos\theta_1\cos\theta_2 + i\sin\theta_1\cos\theta_2 + i\cos\theta_1\sin\theta_2 + i^2\sin\theta_1\sin\theta_2)$$

$$= r_1r_2[(\cos\theta_1\cos\theta_2 - \sin\theta_1\sin\theta_2) + i(\sin\theta_1\cos\theta_2 + \cos\theta_1\sin\theta_2)].$$

But the addition identities for sine and cosine (page 499) show that

$$\cos\theta_1\cos\theta_2 - \sin\theta_1\sin\theta_2 = \cos(\theta_1 + \theta_2)$$

$$\sin\theta_1\cos\theta_2 + \cos\theta_1\sin\theta_2 = \sin(\theta_1 + \theta_2).$$

Therefore,

$$z_1z_2 = r_1r_2[(\cos\theta_1\cos\theta_2 - \sin\theta_1\sin\theta_2) + i(\sin\theta_1\cos\theta_2 + \cos\theta_1\sin\theta_2)]$$

$$= r_1r_2[\cos(\theta_1 + \theta_2) + i\sin(\theta_1 + \theta_2)].$$

This completes the proof of the multiplication rule. The division rule is proved similarly (Exercise 51).

✓ EXERCISES 9.1

In Exercises 1–8, plot the point in the complex plane corresponding to the number.

1. $3 + 2i$

2. $-7 + 6i$

3. $-\dfrac{8}{3} - \dfrac{5}{3}i$

4. $\sqrt{2} - 7i$

5. $(1 + i)(1 - i)$

6. $(2 + i)(1 - 2i)$

7. $2i\left(3 - \dfrac{5}{2}i\right)$

8. $\dfrac{4i}{3}(-6 - 3i)$

In Exercises 9–14, find the absolute value.

9. $|5 - 12i|$

10. $|2i|$

11. $|1 + \sqrt{2}i|$

12. $|2 - 3i|$

13. $|-12i|$

14. $|i^7|$

15. Give an example of complex numbers z and w such that $|z + w| \neq |z| + |w|$.

16. If $z = 3 - 4i$, find $|z|^2$ and $z\bar{z}$, where $\bar{z}$ is the conjugate of z (see page 307).

In Exercises 17–24, sketch the graph of the equation in the complex plane (z denotes a complex number of the form $a + bi$).

17. $|z| = 4$ [*Hint:* The graph consists of all points that lie 4 units from the origin.]

18. $|z| = 1$

19. $|z - 1| = 10$ [*Hint:* 1 corresponds to $(1, 0)$ in the complex plane. What does the equation say about the distance from z to 1?]

20. $|z + 3| = 1$ **21.** $|z - 2i| = 4$

22. $|z - 3i + 2| = 9$
[*Hint:* Rewrite it as $|z - (-2 + 3i)| = 9$.]

23. $\text{Re}(z) = 2$ [The **real part** of the complex number $z = a + bi$ is defined to be the number a and is denoted $\text{Re}(z)$.]

24. $\text{Im}(z) = -5/2$ [The **imaginary part** of $z = a + bi$ is defined to be the number b *(not bi)* and is denoted $\text{Im}(z)$.]

In Exercises 25–32, express the number in polar form.

25. $3 + 4i$ **26.** $-4 + 3i$ **27.** $5 - 12i$

28. $-\sqrt{7} - 3i$ **29.** $1 + 2i$ **30.** $3 - 5i$

31. $-\dfrac{5}{2} + \dfrac{7}{2}i$ **32.** $\sqrt{5} + \sqrt{11}i$

In Exercises 33–38, perform the indicated multiplication or division; express your answer in both rectangular form $a + bi$ and polar form $r(\cos\theta + i\sin\theta)$.

33. $\left(\cos\dfrac{\pi}{12} + i\sin\dfrac{\pi}{12}\right) \cdot 2\left(\cos\dfrac{7\pi}{12} + i\sin\dfrac{7\pi}{12}\right)$

34. $3\left(\cos\dfrac{\pi}{8} + i\sin\dfrac{\pi}{8}\right) \cdot 12\left(\cos\dfrac{3\pi}{8} + i\sin\dfrac{3\pi}{8}\right)$

35. $12\left(\cos\dfrac{11\pi}{12} + i\sin\dfrac{11\pi}{12}\right) \cdot \dfrac{7}{2}\left(\cos\dfrac{\pi}{4} + i\sin\dfrac{\pi}{4}\right)$

36. $\dfrac{8\left(\cos\dfrac{5\pi}{18} + i\sin\dfrac{5\pi}{18}\right)}{4\left(\cos\dfrac{\pi}{9} + i\sin\dfrac{\pi}{9}\right)}$

37. $\dfrac{6\left(\cos\dfrac{7\pi}{20} + i\sin\dfrac{7\pi}{20}\right)}{4\left(\cos\dfrac{\pi}{10} + i\sin\dfrac{\pi}{10}\right)}$

38. $\dfrac{\sqrt{54}\left(\cos\dfrac{9\pi}{4} + i\sin\dfrac{9\pi}{4}\right)}{\sqrt{6}\left(\cos\dfrac{7\pi}{12} + i\sin\dfrac{7\pi}{12}\right)}$

In Exercises 39–46, convert to polar form and then multiply or divide. Express your answer in polar form.

39. $(1 + i)(1 + \sqrt{3}i)$ **40.** $(1 - i)(3 - 3i)$

41. $\dfrac{1 + i}{1 - i}$ **42.** $\dfrac{2 - 2i}{-1 - i}$

43. $3i(2\sqrt{3} + 2i)$ **44.** $\dfrac{-4i}{\sqrt{3} + i}$

45. $i(i + 1)(-\sqrt{3} + i)$

46. $(1 - i)(2\sqrt{3} - 2i)(-4 - 4\sqrt{3}i)$

47. Explain what is meant by saying that multiplying a complex number $z = r(\cos\theta + i\sin\theta)$ by i amounts to rotating z 90° counterclockwise around the origin. [*Hint:* Express i and iz in polar form. What are their relative positions in the complex plane?]

48. Describe what happens geometrically when you multiply a complex number by 2.

Thinkers

49. The sum of two distinct complex numbers, $a + bi$ and $c + di$, can be found geometrically by means of the so-called **parallelogram rule:** Plot the points $a + bi$ and $c + di$ in the complex plane, and form the parallelogram, three of whose vertices are 0, $a + bi$, and $c + di$, as in the figure. Then the fourth vertex of the parallelogram is the point whose coordinate is the sum

$$(a + bi) + (c + di) = (a + c) + (b + d)i.$$

Complete the following *proof* of the parallelogram rule when $a \neq 0$ and $c \neq 0$.
(a) Find the *slope* of the line K from 0 to $a + bi$. [*Hint:* K contains the points $(0, 0)$ and (a, b).]
(b) Find the *slope* of the line N from 0 to $c + di$.
(c) Find the *equation* of the line L through $a + bi$ and parallel to line N of part (b). [*Hint:* The point (a, b) is on L; find the slope of L by using part (b) and facts about the slope of parallel lines.]
(d) Find the *equation* of the line M through $c + di$ and parallel to line K of part (a).
(e) Label the lines K, L, M, and N in the figure.
(f) Show by using substitution that the point $(a + c, b + d)$ satisfies both the equation of line L and the equation of line M. Therefore, $(a + c, b + d)$ lies on both L and M. Since the only point on both L and M is the fourth vertex of the parallelogram (see the figure), this vertex must be $(a + c, b + d)$. Hence, this vertex has coordinate

$$(a + c) + (b + d)i = (a + bi) + (c + di).$$

50. Let $z = a + bi$ be a complex number and denote its conjugate $a - bi$ by $\bar{z}$. Prove that $|z|^2 = z\bar{z}$.

51. *Proof of the polar division rule.* Let
$z_1 = r_1(\cos\theta_1 + i\sin\theta_1)$ and $z_2 = r_2(\cos\theta_2 + i\sin\theta_2)$.
Then

$$\frac{z_1}{z_2} = \frac{r_1(\cos\theta_1 + i\sin\theta_1)}{r_2(\cos\theta_2 + i\sin\theta_2)}$$

$$= \frac{r_1(\cos\theta_1 + i\sin\theta_1)}{r_2(\cos\theta_2 + i\sin\theta_2)} \cdot \frac{\cos\theta_2 - i\sin\theta_2}{\cos\theta_2 - i\sin\theta_2}.$$

(a) Multiply out the denominator on the right side and use the Pythagorean identity to show that it is just the number r_2.

(b) Multiply out the numerator on the right side; use the subtraction identities for sine and cosine (page 499) to show that it is

$$r_1[\cos(\theta_1 - \theta_2) + i\sin(\theta_1 - \theta_2)].$$

Therefore,

$$\frac{z_1}{z_2} = \left(\frac{r_1}{r_2}\right)[\cos(\theta_1 - \theta_2) + i\sin(\theta_1 - \theta_2)].$$

52. (a) If $s(\cos\beta + i\sin\beta) = r(\cos\theta + i\sin\theta)$, explain why we must have $s = r$. [*Hint:* Think distance.]

(b) If $r(\cos\beta + i\sin\beta) = r(\cos\theta + i\sin\theta)$, explain why $\cos\beta = \cos\theta$ and $\sin\beta = \sin\theta$. [*Hint:* See property 5 of the complex numbers on page 306.]

(c) If $\cos\beta = \cos\theta$ and $\sin\beta = \sin\theta$, show that angles β and θ in standard position have the same terminal side. [*Hint:* $(\cos\beta, \sin\beta)$ and $(\cos\theta, \sin\theta)$ are points on the unit circle.]

(d) Use parts (a)–(c) to prove this **equality rule for polar form:**

$$s(\cos\beta + i\sin\beta) = r(\cos\theta + i\sin\theta)$$

exactly when $s = r$ and $\beta = \theta + 2k\pi$ for some integer k. [*Hint:* Angles with the same terminal side must differ by an integer multiple of 2π.]

9.2 DeMoivre's Theorem and *n*th Roots of Complex Numbers

Polar form provides a convenient way to calculate both powers and roots of complex numbers. If $z = r(\cos\theta + i\sin\theta)$, then the multiplication formula on page 595 shows that

$$z^2 = z \cdot z = r \cdot r[\cos(\theta + \theta) + i\sin(\theta + \theta)]$$
$$= r^2(\cos 2\theta + i\sin 2\theta).$$

Similarly,

$$z^3 = z^2 \cdot z = r^2 \cdot r[\cos(2\theta + \theta) + i\sin(2\theta + \theta)]$$
$$= r^3(\cos 3\theta + i\sin 3\theta).$$

Repeated application of the multiplication formula proves the following theorem.

DeMoivre's Theorem

For any complex number $z = r(\cos\theta + i\sin\theta)$ and any positive integer n,

$$z^n = r^n(\cos n\theta + i\sin n\theta).$$

EXAMPLE 1

Compute $(-\sqrt{3} + i)^5$.

SOLUTION   We first express $-\sqrt{3} + i$ in polar form (as in Example 2 on page 594):

$$-\sqrt{3} + i = 2\left(\cos\frac{5\pi}{6} + i\sin\frac{5\pi}{6}\right).$$

By DeMoivre's Theorem,

$$(-\sqrt{3} + i)^5 = 2^5\left[\cos\left(5\cdot\frac{5\pi}{6}\right) + i\sin\left(5\cdot\frac{5\pi}{6}\right)\right] = 32\left(\cos\frac{25\pi}{6} + i\sin\frac{25\pi}{6}\right).$$

Since $25\pi/6 = (\pi/6) + (24\pi/6) = (\pi/6) + 4\pi$, we have

$$(-\sqrt{3} + i)^5 = 32\left(\cos\frac{25\pi}{6} + i\sin\frac{25\pi}{6}\right) = 32\left(\cos\frac{\pi}{6} + i\sin\frac{\pi}{6}\right)$$
$$= 32\left(\frac{\sqrt{3}}{2} + \frac{1}{2}i\right) = 16\sqrt{3} + 16i. \quad\blacksquare$$

EXAMPLE 2

Find $(1 + i)^{10}$.

SOLUTION   First verify that the polar form of $1 + i$ is

$$1 + i = \sqrt{2}\left(\cos\frac{\pi}{4} + i\sin\frac{\pi}{4}\right).$$

Therefore, by DeMoivre's Theorem,

$$(1 + i)^{10} = (\sqrt{2})^{10}\left(\cos\frac{10\pi}{4} + i\sin\frac{10\pi}{4}\right)$$
$$= (2^{1/2})^{10}\left(\cos\frac{5\pi}{2} + i\sin\frac{5\pi}{2}\right) = 2^5(0 + i\cdot 1) = 32i.$$

Figure 9–5

A calculator that can do complex arithmetic confirms this result (Figure 9–5). $\blacksquare$

⬛ *n*TH ROOTS

Recall that for a positive real number c, we called the positive solution of the equation $x^n = c$ the nth root of c. In the complex number system, however, "positive" and "negative" are not meaningful terms (for instance, should $3 - 2i$ be called positive or negative?). Consequently, if $a + bi$ is a complex number, then *any* solution of the equation

$$z^n = a + bi$$

is called an **nth root** of $a + bi$.

Every real number is, of course, a complex number. When the definition of an *n*th root of a complex number is applied to a real number, we must change our previous terminology. For instance, 16 now has *four* fourth roots, since each of 2, −2, 2*i*, and −2*i* is a solution of $z^4 = 16$, whereas we previously said that 2 was *the* fourth root of 16. In the context of complex numbers, this change will not cause any confusion.

Although *n*th roots are no longer unique, the radical symbol will be used only for nonnegative real numbers and will have the same meaning as before: If *r* is a nonnegative real number, then $\sqrt[n]{r}$ denotes the unique nonnegative real number whose *n*th power is *r*.

All *n*th roots of a complex number $a + bi$ can easily be found if $a + bi$ is written in polar form, as illustrated in the next example.

EXAMPLE 3

Find the fourth roots of $-8 + 8\sqrt{3}i$.

SOLUTION   To solve $z^4 = -8 + 8\sqrt{3}i$, first verify that the polar form of $-8 + 8\sqrt{3}i$ is $16\left(\cos\dfrac{2\pi}{3} + i\sin\dfrac{2\pi}{3}\right)$. We must find numbers *s* and β such that

$$[s(\cos\beta + i\sin\beta)]^4 = 16\left(\cos\frac{2\pi}{3} + i\sin\frac{2\pi}{3}\right).$$

By DeMoivre's Theorem, we must have

$$s^4(\cos 4\beta + i\sin 4\beta) = 16\left(\cos\frac{2\pi}{3} + i\sin\frac{2\pi}{3}\right).$$

The equality rules for complex numbers (Exercise 52 in Section 9.1) show that this can happen only when

$$s^4 = 16 \qquad\text{and}\qquad 4\beta = \frac{2\pi}{3} + 2k\pi \quad (k\text{ an integer})$$

$$s = \sqrt[4]{16} = 2 \qquad\qquad \beta = \frac{2\pi/3 + 2k\pi}{4}.$$

Substituting these values in $s(\cos\beta + i\sin\beta)$ shows that the solutions of $z^4 = 16\left(\cos\dfrac{2\pi}{3} + i\sin\dfrac{2\pi}{3}\right)$ are

$$z = 2\left(\cos\frac{2\pi/3 + 2k\pi}{4} + i\sin\frac{2\pi/3 + 2k\pi}{4}\right) \quad (k = 0, \pm 1, \pm 2, \pm 3, \ldots).$$

which can be simplified as

$$z = 2\left[\cos\left(\frac{\pi}{6} + \frac{k\pi}{2}\right) + i\sin\left(\frac{\pi}{6} + \frac{k\pi}{2}\right)\right] \quad (k = 0, \pm 1, \pm 2, \pm 3, \ldots).$$

Letting $k = 0, 1, 2, 3$, produces four distinct solutions:

$k = 0$: $z = 2\left(\cos\dfrac{\pi}{6} + i\sin\dfrac{\pi}{6}\right) = \sqrt{3} + i.$

$k = 1$: $z = 2\left[\cos\left(\dfrac{\pi}{6} + \dfrac{\pi}{2}\right) + i\sin\left(\dfrac{\pi}{6} + \dfrac{\pi}{2}\right)\right] = 2\left(\cos\dfrac{2\pi}{3} + i\sin\dfrac{2\pi}{3}\right)$
$$= -1 + \sqrt{3}i.$$

$k = 2$: $z = 2\left[\cos\left(\dfrac{\pi}{6} + \pi\right) + i\sin\left(\dfrac{\pi}{6} + \pi\right)\right] = 2\left(\cos\dfrac{7\pi}{6} + i\sin\dfrac{7\pi}{6}\right)$
$$= -\sqrt{3} - i.$$

$k = 3$: $z = 2\left[\cos\left(\dfrac{\pi}{6} + \dfrac{3\pi}{2}\right) + i\sin\left(\dfrac{\pi}{6} + \dfrac{3\pi}{2}\right)\right] = 2\left(\cos\dfrac{5\pi}{3} + i\sin\dfrac{5\pi}{3}\right).$
$$= 1 - \sqrt{3}i.$$

Any other value of k produces an angle β with the same terminal side as one of the four angles used above, and hence leads to the same solution. For instance, when $k = 4$, then $\beta = \dfrac{\pi}{6} + \dfrac{4\pi}{2} = \dfrac{\pi}{6} + 2\pi$ and β has the same terminal side as $\pi/6$. Therefore, we have found *all* the solutions—the four fourth roots of $-8 + 8\sqrt{3}i.$* ■

The general equation $z^n = r(\cos\theta + i\sin\theta)$ can be solved by exactly the same method used in the preceding example—just substitute n for 4, r for 16, and θ for $2\pi/3$, as follows. A solution is a number $s(\cos\beta + i\sin\beta)$ such that

$$[s(\cos\beta + i\sin\beta)]^n = r(\cos\theta + i\sin\theta)$$

$$s^n(\cos n\beta + i\sin n\beta) = r(\cos\theta + i\sin\theta).$$

Therefore,

$$s^n = r \qquad \text{and} \qquad n\beta = \theta + 2k\pi \quad (k \text{ any integer})$$

$$s = \sqrt[n]{r} \qquad\qquad \beta = \frac{\theta + 2k\pi}{n}$$

Taking $k = 0, 1, 2, \ldots, n - 1$ produces n distinct angles β. Any other value of k leads to an angle β with the same terminal side as one of these. Hence,

Formula for nth Roots

> For each positive integer n, nonzero complex number
>
> $$r(\cos\theta + i\sin\theta)$$
>
> has exactly n distinct nth roots. They are given by
>
> $$\sqrt[n]{r}\left[\cos\left(\frac{\theta + 2k\pi}{n}\right) + i\sin\left(\frac{\theta + 2k\pi}{n}\right)\right],$$
>
> where $k = 0, 1, 2, 3, \ldots n - 1.$

*Alternatively, page 313 shows that a fourth-degree equation, such as $z^4 = -8 + 8\sqrt{3}i$, has at most four distinct solutions.

EXAMPLE 4

Find the fifth roots of $4 + 4i$.

SOLUTION First write it in polar form as $4\sqrt{2}\left(\cos\dfrac{\pi}{4} + i\sin\dfrac{\pi}{4}\right)$. Now apply the root formula with $n = 5$, $r = 4\sqrt{2}$, $\theta = \pi/4$, and $k = 0, 1, 2, 3, 4$. Note that

$$\sqrt[5]{r} = \sqrt[5]{4\sqrt{2}} = (4\sqrt{2})^{1/5} = (2^2 2^{1/2})^{1/5} = (2^{5/2})^{1/5} = 2^{5/10} = 2^{1/2} = \sqrt{2}.$$

Therefore, the fifth roots are

$$\sqrt{2}\left[\cos\left(\frac{\pi/4 + 2k\pi}{5}\right) + i\sin\left(\frac{\pi/4 + 2k\pi}{5}\right)\right] \quad k = 0, 1, 2, 3, 4,$$

that is,

$$k = 0: \sqrt{2}\left[\cos\left(\frac{\pi/4 + 0}{5}\right) + i\sin\left(\frac{\pi/4 + 0}{5}\right)\right] = \sqrt{2}\left(\cos\frac{\pi}{20} + i\sin\frac{\pi}{20}\right),$$

$$k = 1: \sqrt{2}\left[\cos\left(\frac{\pi/4 + 2\pi}{5}\right) + i\sin\left(\frac{\pi/4 + 2\pi}{5}\right)\right] = \sqrt{2}\left(\cos\frac{9\pi}{20} + i\sin\frac{9\pi}{20}\right),$$

$$k = 2: \sqrt{2}\left[\cos\left(\frac{\pi/4 + 4\pi}{5}\right) + i\sin\left(\frac{\pi/4 + 4\pi}{5}\right)\right] = \sqrt{2}\left(\cos\frac{17\pi}{20} + i\sin\frac{17\pi}{20}\right),$$

$$k = 3: \sqrt{2}\left[\cos\left(\frac{\pi/4 + 6\pi}{5}\right) + i\sin\left(\frac{\pi/4 + 6\pi}{5}\right)\right] = \sqrt{2}\left(\cos\frac{25\pi}{20} + i\sin\frac{25\pi}{20}\right),$$

$$k = 4: \sqrt{2}\left[\cos\left(\frac{\pi/4 + 8\pi}{5}\right) + i\sin\left(\frac{\pi/4 + 8\pi}{5}\right)\right] = \sqrt{2}\left(\cos\frac{33\pi}{20} + i\sin\frac{33\pi}{20}\right).$$ ■

■■ ROOTS OF UNITY

The n distinct nth roots of 1 (the solutions of $z^n = 1$) are called the **nth roots of unity.** Since $\cos 0 = 1$ and $\sin 0 = 0$, the polar form of the number 1 is $\cos 0 + i\sin 0$. Applying the root formula with $r = 1$ and $\theta = 0$ shows that

Roots of Unity

> For each positive integer n, there are n distinct nth roots of unity:
>
> $$\cos\frac{2k\pi}{n} + i\sin\frac{2k\pi}{n} \quad (k = 0, 1, 2, \ldots, n - 1).$$

EXAMPLE 5

Find the cube roots of unity.

SOLUTION   Apply the formula with $n = 3$ and $k = 0, 1, 2$:

$$k = 0: \qquad \cos 0 + i \sin 0 = 1,$$

$$k = 1: \qquad \cos \frac{2\pi}{3} + i \sin \frac{2\pi}{3} = -\frac{1}{2} + \frac{\sqrt{3}}{2} i,$$

$$k = 2: \qquad \cos \frac{4\pi}{3} + i \sin \frac{4\pi}{3} = -\frac{1}{2} - \frac{\sqrt{3}}{2} i. \qquad \blacksquare$$

Denote by ω the first complex cube root of unity obtained in Example 5:

$$\omega = \cos \frac{2\pi}{3} + i \sin \frac{2\pi}{3}.$$

If we use DeMoivre's Theorem to find ω^2 and ω^3, we see that these numbers are the other two cube roots of unity found in Example 5:

$$\omega^2 = \left(\cos \frac{2\pi}{3} + i \sin \frac{2\pi}{3} \right)^2 = \cos \frac{4\pi}{3} + i \sin \frac{4\pi}{3},$$

$$\omega^3 = \left(\cos \frac{2\pi}{3} + i \sin \frac{2\pi}{3} \right)^3 = \cos \frac{6\pi}{3} + i \sin \frac{6\pi}{3} = \cos 2\pi + i \sin 2\pi$$

$$= 1 + 0 \cdot i = 1.$$

In other words, all the cube roots of unity are powers of ω. The same thing is true in the general case.

Roots of Unity

Let n be a positive integer with $n > 1$. Then the number

$$z = \cos \frac{2\pi}{n} + i \sin \frac{2\pi}{n}$$

is an nth root of unity and all the nth roots of unity are

$$z, z^2, z^3, z^4, \ldots z^{n-1}, z^n = 1.$$

The nth roots of unity have an interesting geometric interpretation. Every nth root of unity has absolute value 1 by the Pythagorean identity.

$$\left| \cos \frac{2k\pi}{n} + i \sin \frac{2k\pi}{n} \right| = \left(\cos \frac{2k\pi}{n} \right)^2 + \left(\sin \frac{2k\pi}{n} \right)^2$$

$$= \cos^2 \left(\frac{2k\pi}{n} \right) + \sin^2 \left(\frac{2k\pi}{n} \right) = 1.$$

Therefore, in the complex plane, every nth root of unity is exactly 1 unit from the origin. In other words, the nth roots of unity all lie on the unit circle.

EXAMPLE 6

Find the fifth roots of unity.

SOLUTION   They are

$$\cos \frac{2k\pi}{5} + i \sin \frac{2k\pi}{5} \quad (k = 0, 1, 2, 3, 4),$$

that is,

$$\cos 0 + i \sin 0 = 1, \qquad \cos \frac{2\pi}{5} + i \sin \frac{2\pi}{5}, \qquad \cos \frac{4\pi}{5} + i \sin \frac{4\pi}{5},$$

$$\cos \frac{6\pi}{5} + i \sin \frac{6\pi}{5}, \qquad \cos \frac{8\pi}{5} + i \sin \frac{8\pi}{5}.$$

These five roots can be plotted in the complex plane by starting at $1 = 1 + 0i$ and moving counterclockwise around the unit circle, moving through an angle of $2\pi/5$ at each step, as shown in Figure 9–6. If you connect these five roots, they form the vertices of a regular pentagon (Figure 9–7). ■

Figure 9–6

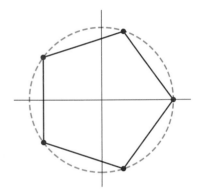

Figure 9–7

⊞ EXPLORATION

GRAPHING

With your calculator in parametric graphing mode, set these range values:

$$0 \le t \le 2\pi, \qquad t\text{-step} \approx .067,$$

$$-1.5 \le x \le 1.5, \qquad -1 \le y \le 1$$

and graph the unit circle, whose parametric equations are

$$x = \cos t \qquad \text{and} \qquad y = \sin t.^*$$

Reset the t-step to be $2\pi/5$ and graph again. Your screen now looks exactly like the red lines in Figure 9–7 because the calculator plotted only the five points corresponding to $t = 0, 2\pi/5, 4\pi/5, 6\pi/5, 8\pi/5$[†] and connected them with the shortest possible segments. Use the trace feature to move along the graph. The cursor will jump from vertex to vertex, that is, it will move from one fifth root of unity to the next.

---

*On wide-screen calculators, use $-2 \le x \le 2$ or $-1.7 \le x \le 1.7$ so that the unit circle looks like a circle.

[†]The point corresponding to $t = 10\pi/5 = 2\pi$ is the same as the one corresponding to $t = 0$.

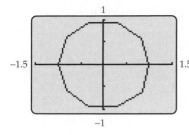

Figure 9–8

EXAMPLE 7

Find the tenth roots of unity graphically.

SOLUTION Graph the unit circle as in the preceding exploration, but use $2\pi/10$ as the *t*-step. The result (Figure 9–8) is a regular decagon whose vertices are the tenth roots of unity. By using the trace feature, you can approximate each of them.

GRAPHING EXPLORATION

Verify that the two tenth roots of unity in the first quadrant are (approximately) $.8090 + .5878i$ and $.3090 + .9511i$. ∎

EXERCISES 9.2

In Exercises 1–10, calculate the given product and express your answer in the form a + bi.

1. $\left(\cos\dfrac{\pi}{12} + i\sin\dfrac{\pi}{12}\right)^6$ **2.** $\left(\cos\dfrac{\pi}{5} + i\sin\dfrac{\pi}{5}\right)^{20}$

3. $\left[3\left(\cos\dfrac{7\pi}{30} + i\sin\dfrac{7\pi}{30}\right)\right]^5$ **4.** $\left[\sqrt[3]{4}\left(\cos\dfrac{7\pi}{36} + i\sin\dfrac{7\pi}{36}\right)\right]^{12}$

5. $(1 - i)^{12}$ [*Hint:* Use polar form and DeMoivre's Theorem.]

6. $(2 + 2i)^8$ **7.** $\left(\dfrac{\sqrt{3}}{2} + \dfrac{1}{2}i\right)^{10}$

8. $\left(-\dfrac{1}{2} + \dfrac{\sqrt{3}}{2}i\right)^{20}$ **9.** $\left(\dfrac{-1}{\sqrt{2}} + \dfrac{i}{\sqrt{2}}\right)^{14}$

10. $(-1 + \sqrt{3}i)^8$

In Exercises 11 and 12, find the indicated roots of unity and express your answers in the form a + bi.

11. Fourth roots of unity   **12.** Sixth roots of unity

In Exercises 13–22, find the nth roots of the given number in polar form.

13. $64\left(\cos\dfrac{\pi}{5} + i\sin\dfrac{\pi}{5}\right)$; $n = 3$

14. $8\left(\cos\dfrac{\pi}{10} + i\sin\dfrac{\pi}{10}\right)$; $n = 3$

15. $81\left(\cos\dfrac{\pi}{12} + i\sin\dfrac{\pi}{12}\right)$; $n = 4$

16. $16\left(\cos\dfrac{\pi}{7} + i\sin\dfrac{\pi}{7}\right)$; $n = 5$

17. -1; $n = 5$ **18.** 1; $n = 7$

19. i; $n = 5$ **20.** $-i$; $n = 6$

21. $1 + i$; $n = 2$ **22.** $1 - \sqrt{3}i$; $n = 3$

In Exercises 23–30, solve the given equation in the complex number system.

23. $x^6 = -1$ **24.** $x^6 + 64 = 0$ **25.** $x^3 = i$

26. $x^4 = i$ **27.** $x^3 + 27i = 0$ **28.** $x^6 + 729 = 0$

29. $x^4 = -1 + \sqrt{3}i$ **30.** $x^4 = -8 - 8\sqrt{3}i$

In Exercises 31–35, represent the roots of unity graphically. Then use the trace feature to obtain approximations of the form a + bi for each root (round to four places).

31. Seventh roots of unity   **32.** Fifth roots of unity

33. Eighth roots of unity   **34.** Twelfth roots of unity

35. Ninth roots of unity

36. Solve the equation $x^3 + x^2 + x + 1 = 0$. [*Hint:* First find the quotient when $x^4 - 1$ is divided by $x - 1$ and then consider solutions of $x^4 - 1 = 0$.]

37. Solve $x^5 + x^4 + x^3 + x^2 + x + 1 = 0$. [*Hint:* Consider $x^6 - 1$ and $x - 1$ and see Exercise 36.]

38. What do you think are the solutions of $x^{n-1} + x^{n-2} + \cdots + x^3 + x^2 + x + 1 = 0$? (See Exercises 36 and 37.)

Thinkers

39. In the complex plane, identify each point with its complex number label. The unit circle consists of all numbers (points) z such that $|z| = 1$. Suppose v and w are two points (numbers) that move around the unit circle in such a way that $v = w^{12}$ at all times. When w has made one complete trip around the circle, how many trips has v made? [*Hint:* Think polar and DeMoivre.]

40. Suppose u is an nth root of unity. Show that $1/u$ is also an nth root of unity. [*Hint:* Use the definition, *not* polar form.]

41. Let $u_1, u_2, \ldots, u_n$ be the distinct nth roots of unity and suppose v is a nonzero solution of the equation

$$z^n = r(\cos \theta + i \sin \theta).$$

Show that $vu_1, vu_2, \ldots, vu_n$ are n distinct solutions of the equation. [*Remember:* Each u_i is a solution of $x^n = 1$.]

42. Use the formula for nth roots and the identities

$$\cos(x + \pi) = -\cos x \qquad \sin(x + \pi) = -\sin x$$

to show that the nonzero complex number $r(\cos \theta + i \sin \theta)$ has two square roots and that these square roots are negatives of each other.

9.3 Vectors in the Plane

Once a unit of measure has been agreed upon, quantities such as area, length, time, and temperature can be described by a single number. Other quantities, such as an east wind of 10 mph, require two numbers to describe them because they involve both *magnitude* and *direction*. Such quantities are called **vectors** and are represented geometrically by a directed line segment or arrow, as in Figure 9–9.

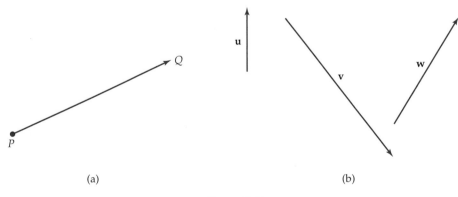

(a)   (b)

Figure 9–9

When a vector extends from a point P to a point Q, as in Figure 9–9(a), P is called the **initial point** of the vector and Q is called the **terminal point,** and the vector is written $\overrightarrow{PQ}$. Its **length** is denoted by $\|\overrightarrow{PQ}\|$. When the endpoints are not specified, as in Figure 9–9(b), vectors are denoted by boldface letters such as **u, v,** and **w.** The length of a vector **u** is denoted by $\|\mathbf{u}\|$ and is called the **magnitude** of **u.**

If **u** and **v** are vectors with the same magnitude and direction, we say that **u** and **v** are **equivalent** and write **u = v.** Some examples are shown in Figure 9–10.

Figure 9–10

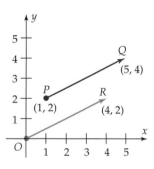

Figure 9–11

EXAMPLE 1

Let $P = (1, 2)$, $Q = (5, 4)$, $O = (0, 0)$, and $R = (4, 2)$, as in Figure 9–11. Show that $\overrightarrow{PQ} = \overrightarrow{OR}$.

SOLUTION The distance formula shows that $\overrightarrow{PQ}$ and $\overrightarrow{OR}$ have the *same length:*

$$\|\overrightarrow{PQ}\| = \sqrt{(5-1)^2 + (4-2)^2} = \sqrt{4^2 + 2^2} = \sqrt{20}.$$
$$\|\overrightarrow{OR}\| = \sqrt{(4-0)^2 + (2-0)^2} = \sqrt{4^2 + 2^2} = \sqrt{20}.$$

Furthermore, the lines through PQ and OR have the same slope:

$$\text{slope } PQ = \frac{4-2}{5-1} = \frac{2}{4} = \frac{1}{2}, \qquad \text{slope } OR = \frac{2-0}{4-0} = \frac{2}{4} = \frac{1}{2}.$$

Since $\overrightarrow{PQ}$ and $\overrightarrow{OR}$ both point to the upper right on lines of the same slope, $\overrightarrow{PQ}$ and $\overrightarrow{OR}$ have the *same direction.* Therefore, $\overrightarrow{PQ} = \overrightarrow{OR}$. ■

According to the definition of equivalence, a vector may be moved from one location to another, provided that its magnitude and direction are not changed. Consequently, we have the following.

Equivalent Vectors

Every vector $\overrightarrow{PQ}$ is equivalent to a vector $\overrightarrow{OR}$ with initial point at the origin: If $P = (x_1, y_1)$ and $Q = (x_2, y_2)$, then

$$\overrightarrow{PQ} = \overrightarrow{OR}, \qquad \text{where } R = (x_2 - x_1, y_2 - y_1).$$

Proof The proof is similar to the one used in Example 1. It follows from the fact that $\overrightarrow{PQ}$ and $\overrightarrow{OR}$ have the same length,

$$\|\overrightarrow{OR}\| = \sqrt{[(x_2 - x_1) - 0]^2 + [(y_2 - y_1) - 0]^2}$$
$$= \sqrt{(x_2 - x_1)^2 + (y_2 - y_1)^2} = \|\overrightarrow{PQ}\|,$$

and that either the line segments PQ and OR are both vertical or they have the same slope,

$$\text{slope } OR = \frac{(y_2 - y_1) - 0}{(x_2 - x_1) - 0} = \frac{y_2 - y_1}{x_2 - x_1} = \text{slope } PQ,$$

as shown in Figure 9–12. ∎

Figure 9–12

The magnitude and direction of a vector with the origin as initial point are completely determined by the coordinates of its terminal point. Consequently, we denote the vector with initial point $(0, 0)$ and terminal point (a, b) by $\langle a, b \rangle$. The numbers a and b are called the **components** of the vector $\langle a, b \rangle$.

Since the length of the vector $\langle a, b \rangle$ is the distance from $(0, 0)$ to (a, b), the distance formula shows that

Magnitude

> The **magnitude** (or **norm**) of the vector $\mathbf{v} = \langle a, b \rangle$ is
> $$\|\mathbf{v}\| = \sqrt{a^2 + b^2}.$$

EXAMPLE 2

Find the components and the magnitude of the vector with initial point $P = (-2, 6)$ and terminal point $Q = (4, -3)$.

SOLUTION According to the fact in the box on page 607 (with $x_1 = -2$, $y_1 = 6$, $x_2 = 4$, $y_2 = -3$):

$$\overrightarrow{PQ} = \overrightarrow{OR}, \qquad \text{where } R = (4 - (-2), -3 - 6) = (6, -9)$$

that is,

$$\overrightarrow{PQ} = \overrightarrow{OR} = \langle 6, -9 \rangle.$$

Therefore,

$$\|\overrightarrow{PQ}\| = \|\overrightarrow{OR}\| = \sqrt{6^2 + (-9)^2} = \sqrt{36 + 81} = \sqrt{117}. \quad \blacksquare$$

▪▪ VECTOR ARITHMETIC

When dealing with vectors, it is customary to refer to ordinary real numbers as **scalars. Scalar multiplication** is an operation in which a scalar k is "multiplied" by a vector $\mathbf{v}$ to produce another *vector* denoted by $k\mathbf{v}$. Here is the formal definition.

Scalar Multiplication

> If k is a real number and $\mathbf{v} = \langle a, b \rangle$ is a vector, then
>
> $$k\mathbf{v} \text{ is the vector } \langle ka, kb \rangle.$$
>
> The vector $k\mathbf{v}$ is called a **scalar multiple** of $\mathbf{v}$.

EXAMPLE 3

If $\mathbf{v} = \langle 3, 1 \rangle$, then

$$3\mathbf{v} = 3\langle 3, 1 \rangle = \langle 3 \cdot 3, 3 \cdot 1 \rangle = \langle 9, 3 \rangle,$$

$$-2\mathbf{v} = -2\langle 3, 1 \rangle = \langle -2 \cdot 3, -2 \cdot 1 \rangle = \langle -6, -2 \rangle,$$

as shown in Figure 9–13:

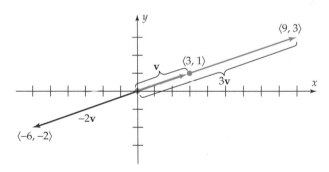

Figure 9–13

Figure 9–13 shows that $3\mathbf{v}$ has the same direction as $\mathbf{v}$, while $-2\mathbf{v}$ has the opposite direction. Also note that

$$\|\mathbf{v}\| = \|\langle 3, 1 \rangle\| = \sqrt{3^2 + 1^2} = \sqrt{10}$$

$$\|-2\mathbf{v}\| = \|\langle -6, -2 \rangle\| = \sqrt{(-6)^2 + (-2)^2} = \sqrt{40} = 2\sqrt{10}.$$

Therefore,

$$\|-2\mathbf{v}\| = 2\sqrt{10} = 2\|\mathbf{v}\| = |-2| \cdot \|\mathbf{v}\|.$$

Similarly, you can verify that $\|3\mathbf{v}\| = |3| \cdot \|\mathbf{v}\| = 3\|\mathbf{v}\|.$ ∎

Example 3 is an illustration of the following facts.

Geometric Interpretation of Scalar Multiplication

The *magnitude* of the vector $k\mathbf{v}$ is $|k|$ times the length of $\mathbf{v}$, that is,

$$\|k\mathbf{v}\| = |k| \cdot \|\mathbf{v}\|.$$

The *direction of $k\mathbf{v}$* is the same as that of $\mathbf{v}$ when k is positive and opposite that of $\mathbf{v}$ when k is negative.

See Exercise 77 for a proof of this statement.

Vector addition is an operation in which two vectors $\mathbf{u}$ and $\mathbf{v}$ are added to produce a new vector denoted $\mathbf{u} + \mathbf{v}$. Formally, we have the following.

Vector Addition

If $\mathbf{u} = \langle a, b \rangle$ and $\mathbf{v} = \langle c, d \rangle$, then

$$\mathbf{u} + \mathbf{v} = \langle a + c, b + d \rangle.$$

EXAMPLE 4

If $\mathbf{u} = \langle -5, 2 \rangle$ and $\mathbf{v} = \langle 3, 1 \rangle$, find $\mathbf{u} + \mathbf{v}$.

SOLUTION

$$\mathbf{u} + \mathbf{v} = \langle -5, 2 \rangle + \langle 3, 1 \rangle = \langle -5 + 3, 2 + 1 \rangle = \langle -2, 3 \rangle$$

as shown in Figure 9–14. ∎

TECHNOLOGY TIP

Vector arithmetic and other vector operations can be done on TI-86/89 and HP-39.

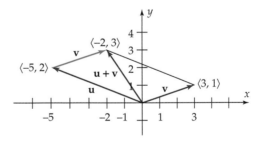

Figure 9–14

Example 4 is an illustration of these facts.

Geometric Interpretations of Vector Addition

1. If $\mathbf{u}$ and $\mathbf{v}$ are vectors with the same initial point P, then $\mathbf{u} + \mathbf{v}$ is the vector $\overrightarrow{PQ}$, where $\overrightarrow{PQ}$ is the diagonal of the parallelogram with adjacent sides $\mathbf{u}$ and $\mathbf{v}$.

2. If the vector $\mathbf{v}$ is moved (without changing its magnitude or direction) so that its initial point lies on the endpoint of the vector $\mathbf{u}$, then $\mathbf{u} + \mathbf{v}$ is the vector with the same initial point P as $\mathbf{u}$ and the same terminal point Q as $\mathbf{v}$.

See Exercise 78 for a proof of these statements.

The **negative** of a vector $\mathbf{v} = \langle c, d \rangle$ is defined to be the vector $(-1)\mathbf{v} = (-1)\langle c, d \rangle = \langle -c, -d \rangle$ and is denoted $-\mathbf{v}$. **Vector subtraction** is then defined as follows.

Vector Subtraction

> If $\mathbf{u} = \langle a, b \rangle$ and $\mathbf{v} = \langle c, d \rangle$, then $\mathbf{u} - \mathbf{v}$ is the vector
>
> $$\mathbf{u} + (-\mathbf{v}) = \langle a, b \rangle + \langle -c, -d \rangle$$
>
> $$= \langle a - c, b - d \rangle.$$

A geometric interpretation of vector subtraction is given in Exercise 79.

EXAMPLE 5

If $\mathbf{u} = \langle 2, 5 \rangle$ and $\mathbf{v} = \langle 6, 1 \rangle$, find $\mathbf{u} - \mathbf{v}$.

SOLUTION

$$\mathbf{u} - \mathbf{v} = \langle 2, 5 \rangle - \langle 6, 1 \rangle = \langle 2 - 6, 5 - 1 \rangle = \langle -4, 4 \rangle,$$

as shown in Figure 9–15. ∎

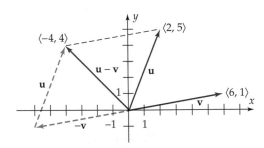

Figure 9–15

The vector $\langle 0, 0 \rangle$ is called the **zero vector** and is denoted $\mathbf{0}$.

EXAMPLE 6

If $\mathbf{u} = \langle -1, 6 \rangle$, $\mathbf{v} = \langle 2/3, -4 \rangle$, and $\mathbf{w} = \langle 2, 5/2 \rangle$, find $2\mathbf{u} + 3\mathbf{v}$ and $4\mathbf{w} - 2\mathbf{u}$.

SOLUTION

$$2\mathbf{u} + 3\mathbf{v} = 2\langle -1, 6 \rangle + 3\left\langle \frac{2}{3}, -4 \right\rangle = \langle -2, 12 \rangle + \langle 2, -12 \rangle$$

$$= \langle 0, 0 \rangle = \mathbf{0},$$

and

$$4\mathbf{w} - 2\mathbf{u} = 4\left\langle 2, \frac{5}{2}\right\rangle - 2\langle -1, 6\rangle$$

$$= \langle 8, 10\rangle - \langle -2, 12\rangle$$

$$= \langle 8 - (-2), 10 - 12\rangle = \langle 10, -2\rangle. \quad \blacksquare$$

Operations on vectors share many of the same properties as arithmetical operations on numbers.

Properties of Vector Addition and Scalar Multiplication

For any vectors **u**, **v**, and **w** and any scalars r and s,

1. $\mathbf{u} + (\mathbf{v} + \mathbf{w}) = (\mathbf{u} + \mathbf{v}) + \mathbf{w}$

2. $\mathbf{u} + \mathbf{v} = \mathbf{v} + \mathbf{u}$

3. $\mathbf{v} + \mathbf{0} = \mathbf{v} = \mathbf{0} + \mathbf{v}$

4. $\mathbf{v} + (-\mathbf{v}) = \mathbf{0}$

5. $r(\mathbf{u} + \mathbf{v}) = r\mathbf{u} + r\mathbf{v}$

6. $(r + s)\mathbf{v} = r\mathbf{v} + s\mathbf{v}$

7. $(rs)\mathbf{v} = r(s\mathbf{v}) = s(r\mathbf{v})$

8. $1\mathbf{v} = \mathbf{v}$

9. $0\mathbf{v} = \mathbf{0}$ and $r\mathbf{0} = \mathbf{0}$

Proof   If $\mathbf{u} = \langle a, b\rangle$ and $\mathbf{v} = \langle c, d\rangle$, then because addition of real numbers is commutative, we have

$$\mathbf{u} + \mathbf{v} = \langle a, b\rangle + \langle c, d\rangle = \langle a + c, b + d\rangle$$

$$= \langle c + a, d + b\rangle = \langle c, d\rangle + \langle a, b\rangle = \mathbf{v} + \mathbf{u}.$$

The other properties are proved similarly; see Exercises 53–58. $\blacksquare$

▉▉ UNIT VECTORS

A vector with length 1 is called a **unit vector.** For instance, $\langle 3/5, 4/5\rangle$ is a unit vector, since

$$\left\|\left\langle \frac{3}{5}, \frac{4}{5}\right\rangle\right\| = \sqrt{\left(\frac{3}{5}\right)^2 + \left(\frac{4}{5}\right)^2} = \sqrt{\frac{9}{25} + \frac{16}{25}} = \sqrt{\frac{25}{25}} = 1.$$

EXAMPLE 7

Find a unit vector that has the same direction as the vector $\mathbf{v} = \langle 5, 12\rangle$.

SOLUTION   The length of **v** is

$$\|\mathbf{v}\| = \|\langle 5, 12\rangle\| = \sqrt{5^2 + 12^2} = \sqrt{169} = 13.$$

The vector

$$\mathbf{u} = \frac{1}{13}\mathbf{v} = \left\langle \frac{5}{13}, \frac{12}{13} \right\rangle$$

has the same direction as **v** (since it is a scalar multiple by a positive number), and **u** is a unit vector because

$$\|\mathbf{u}\| = \left\| \frac{1}{13}\mathbf{v} \right\| = \left| \frac{1}{13} \right| \cdot \|\mathbf{v}\| = \frac{1}{13} \cdot 13 = 1. \quad \blacksquare$$

The procedure used in Example 7 (multiplying a vector by the reciprocal of its length) works in the general case.

Unit Vectors

If **v** is a nonzero vector, then $\dfrac{1}{\|\mathbf{v}\|}$ **v** is a unit vector with the same direction as **v.**

You can easily verify that the vectors $\mathbf{i} = \langle 1, 0 \rangle$ and $\mathbf{j} = \langle 0, 1 \rangle$ are unit vectors. The vectors **i** and **j** play a special role because they lead to a useful alternate notation for vectors. For example, if $\mathbf{u} = \langle 5, -7 \rangle$ then

$$\mathbf{u} = \langle 5, 0 \rangle + \langle 0, -7 \rangle = 5\langle 1, 0 \rangle - 7\langle 0, 1 \rangle = 5\mathbf{i} - 7\mathbf{j}.$$

Similarly, if $\mathbf{v} = \langle a, b \rangle$ is any vector, then

$$\mathbf{v} = \langle a, b \rangle = \langle a, 0 \rangle + \langle 0, b \rangle = a\langle 1, 0 \rangle + b\langle 0, 1 \rangle = a\mathbf{i} + b\mathbf{j}.$$

The vector **v** is said to be a **linear combination** of **i** and **j.** When vectors are written as linear combinations of **i** and **j,** then the properties in the box on page 612 can be used to write the rules for vector addition and scalar multiplication in this form.

$$(a\mathbf{i} + b\mathbf{j}) + (c\mathbf{i} + d\mathbf{j}) = (a + c)\mathbf{i} + (b + d)\mathbf{j}$$

and

$$c(a\mathbf{i} + b\mathbf{j}) = ca\mathbf{i} + cb\mathbf{j}.$$

EXAMPLE 8

If $\mathbf{u} = 2\mathbf{i} - 6\mathbf{j}$ and $\mathbf{v} = -5\mathbf{i} + 2\mathbf{j},$ find $3\mathbf{u} - 2\mathbf{v}.$

SOLUTION

$$3\mathbf{u} - 2\mathbf{v} = 3(2\mathbf{i} - 6\mathbf{j}) - 2(-5\mathbf{i} + 2\mathbf{j}) = 6\mathbf{i} - 18\mathbf{j} + 10\mathbf{i} - 4\mathbf{j}$$

$$= 16\mathbf{i} - 22\mathbf{j}. \quad \blacksquare$$

▉▉ DIRECTION ANGLES

If $\mathbf{v} = \langle a, b \rangle = a\mathbf{i} + b\mathbf{j}$ is a vector, then the direction of **v** is completely determined by the standard position angle θ between $0°$ and $360°$ whose terminal side is **v,** as shown in Figure 9–16. The angle θ is called the **direction angle** of the

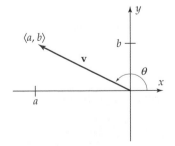

Figure 9–16

vector **v.** According to the point-in-the-plane description of the trigonometric functions,

$$\cos \theta = \frac{a}{\|\mathbf{v}\|} \quad \text{and} \quad \sin \theta = \frac{b}{\|\mathbf{v}\|}.$$

Rewriting each of these equations produces the following result.

Components of the Direction Angle

If $\mathbf{v} = \langle a, b \rangle = a\mathbf{i} + b\mathbf{j}$, then

$$a = \|\mathbf{v}\| \cos \theta \quad \text{and} \quad b = \|\mathbf{v}\| \sin \theta$$

where θ is the direction angle of **v.**

EXAMPLE 9

Find the component form of the vector that represents the velocity of an airplane at the instant its wheels leave the ground if the plane is going 60 mph and the body of the plane makes a 7° angle with the horizontal.

SOLUTION   The velocity vector $\mathbf{v} = a\mathbf{i} + b\mathbf{j}$ has magnitude 60 and direction angle $\theta = 7°$, as shown in Figure 9–17. Hence,

$$\mathbf{v} = (\|\mathbf{v}\| \cos \theta)\mathbf{i} + (\|\mathbf{v}\| \sin \theta)\mathbf{j}$$
$$= (60 \cos 7°)\mathbf{i} + (60 \sin 7°)\mathbf{j}$$
$$\approx (60 \cdot .9925)\mathbf{i} + (60 \cdot .1219)\mathbf{j}$$
$$\approx 59.55\mathbf{i} + 7.31\mathbf{j} = \langle 59.55, 7.31 \rangle. \quad \blacksquare$$

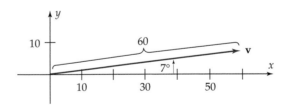

Figure 9–17

If $\mathbf{v} = a\mathbf{i} + b\mathbf{j}$ is a nonzero vector with direction angle θ, then

$$\tan \theta = \frac{\sin \theta}{\cos \theta} = \frac{b/\|\mathbf{v}\|}{a/\|\mathbf{v}\|} = \frac{b}{a}.$$

This fact provides a convenient way to find the direction angle of a vector.

EXAMPLE 10

Find the direction angle of

(a) $\mathbf{u} = 5\mathbf{i} + 13\mathbf{j}$ (b) $\mathbf{v} = -10\mathbf{i} + 7\mathbf{j}.$

SOLUTION

(a) The direction angle θ of $\mathbf{u}$ satisfies $\tan \theta = b/a = 13/5 = 2.6$. Using the TAN^{-1} key on a calculator, we find that $\theta \approx 68.96°$, as shown in Figure 9–18(a).

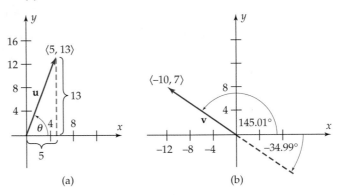

(a)   (b)

Figure 9–18

(b) The direction angle of $\mathbf{v}$ satisfies $\tan \theta = -7/10 = -.7$. Since $\mathbf{v}$ lies in the second quadrant, θ must be between 90° and 180°. A calculator shows that $-34.99°$ is an angle with tangent (approximately) $-.7$. Since tangent has period $\pi (= 180°)$, we know that $\tan t = \tan(t + 180°)$ for every t. Therefore, $\theta = -34.99° + 180° = 145.01°$ is the angle between 90° and 180° such that $\tan \theta \approx -.7$. See Figure 9–18(b). ■

EXAMPLE 11

An object at the origin is acted upon by two forces. A 150-pound force makes an angle of 20° with the positive x-axis, and the other force of 100 pounds makes an angle of 70°, as shown in Figure 9–19. Find the direction and magnitude of the resultant force.

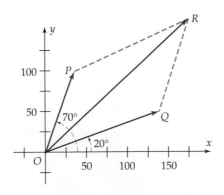

Figure 9–19

SOLUTION   The forces acting on the object are

$$\overrightarrow{OP} = (100 \cos 70°)\mathbf{i} + (100 \sin 70°)\mathbf{j},$$

$$\overrightarrow{OQ} = (150 \cos 20°)\mathbf{i} + (150 \sin 20°)\mathbf{j}.$$

The resultant force $\overrightarrow{OR}$ is the sum of $\overrightarrow{OP}$ and $\overrightarrow{OQ}$. Hence,

$$\overrightarrow{OR} = (100 \cos 70° + 150 \cos 20°)\mathbf{i} + (100 \sin 70° + 150 \sin 20°)\mathbf{j}$$

$$\approx 175.16\mathbf{i} + 145.27\mathbf{j}.$$

Therefore, the magnitude of the resultant force is

$$\|\overrightarrow{OR}\| \approx \sqrt{(175.16)^2 + (145.27)^2} \approx 227.56.$$

The direction angle θ of the resultant force satisfies

$$\tan \theta \approx 145.27/175.16 \approx .8294.$$

A calculator shows that $\theta \approx 39.67°$.   ∎

APPLICATIONS

EXAMPLE 12

A 200-pound box lies on a ramp that makes an angle of 24° with the horizontal. A rope is tied to the box from a post at the top of the ramp to keep it in position. Ignoring friction, how much force is being exerted on the rope by the box?

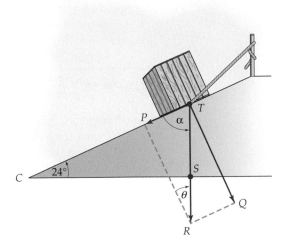

Figure 9–20

SOLUTION   Because of gravity, the box exerts a 200-pound weight straight down (vector $\overrightarrow{TR}$). As Figure 9–20 shows, $\overrightarrow{TR}$ is the sum of $\overrightarrow{TP}$ and $\overrightarrow{TQ}$. The force on the rope is exerted by $\overrightarrow{TP}$, the vector of the force pulling the box down the ramp, so we must find $\|\overrightarrow{TP}\|$. In right triangle TSC, $\alpha + 24° = 90°$, and in right triangle TPR, $\alpha + \theta = 90°$. Hence,

$$\alpha + \theta = \alpha + 24°; \qquad \text{hence,} \qquad \theta = 24°.$$

Therefore,

$$\frac{\|\overrightarrow{TP}\|}{\|\overrightarrow{TR}\|} = \sin \theta$$

$$\frac{\|\overrightarrow{TP}\|}{200} = \sin 24°$$

$$\|\overrightarrow{TP}\| = 200 \sin 24° \approx 81.35.$$

So the force on the rope is 81.35 pounds.   ■

In aerial navigation, directions are given in terms of the angle measured in degrees clockwise from true north. Thus, north is 0°, east is 90°, and so on.

EXAMPLE 13

An airplane is traveling in the direction 50° with an air speed of 300 mph, and there is a 35-mph wind from the direction 120°, as represented by the vectors **p** and **w** in Figure 9–21. Find the *course* and *ground speed* of the plane (that is, its direction and speed relative to the ground).

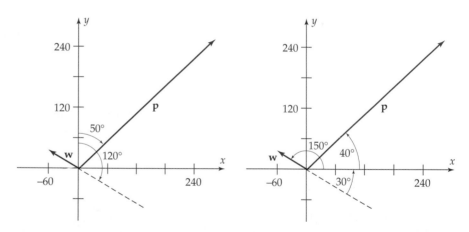

Figure 9–21

SOLUTION   The course of the plane is the direction of the vector **p** + **w,** and its ground speed is the magnitude of **p** + **w.** Figure 9–21(b) shows that the direction angle of **p** (the angle it makes with the positive *x*-axis) is 40° and that the direction angle of **w** is 150°. Therefore,

p + **w** = [(300 cos 40°)**i** + (300 sin 40°)**j**] + [(35 cos 150°)**i** + (35 sin 150°)**j**]

= (300 cos 40° + 35 cos 150°)**i** + (300 sin 40° + 35 sin 150°)**j**

≈ 199.50**i** + 210.34**j.**

The direction angle of **p** + **w** satisfies tan θ = 210.34/199.50 ≈ 1.0543, and a calculator shows that θ ≈ 46.5°. This is the angle **p** + **w** makes with the positive *x*-axis; hence, the course of the plane (the angle between true north and **p** + **w**) is 90° − 46.5° = 43.5°. The ground speed of the plane is

$$\|\mathbf{p} + \mathbf{w}\| \approx \sqrt{(199.5)^2 + (210.34)^2} \approx 289.9 \text{ mph.}$$ ■

EXERCISES 9.3

In Exercises 1–4, find the magnitude of the vector $\overrightarrow{PQ}$.

1. $P = (2, 3), Q = (5, 9)$

2. $P = (-3, 5), Q = (7, -11)$

3. $P = (-7, 0), Q = (-4, -5)$

4. $P = (30, 12), Q = (25, 5)$

In Exercises 5–10, find a vector with the origin as initial point that is equivalent to the vector $\overrightarrow{PQ}$.

5. $P = (1, 5), Q = (7, 11)$ **6.** $P = (2, 7), Q = (-2, 9)$

7. $P = (-4, -8), Q = (-10, 2)$

8. $P = (-5, 6), Q = (-7, -9)$

9. $P = \left(\frac{4}{5}, -2\right), Q = \left(\frac{17}{5}, -\frac{12}{5}\right)$

10. $P = (\sqrt{2}, 4), Q = (\sqrt{3}, -1)$

In Exercises 11–20, find $\mathbf{u} + \mathbf{v}$, $\mathbf{u} - \mathbf{v}$, and $3\mathbf{u} - 2\mathbf{v}$.

11. $\mathbf{u} = \langle -2, 4 \rangle, \mathbf{v} = \langle 6, 1 \rangle$

12. $\mathbf{u} = \langle 4, 0 \rangle, \mathbf{v} = \langle 1, -3 \rangle$

13. $\mathbf{u} = \langle 3, 3\sqrt{2} \rangle, \mathbf{v} = \langle 4\sqrt{2}, 1 \rangle$

14. $\mathbf{u} = \left\langle \frac{2}{3}, 4 \right\rangle, \mathbf{v} = \left\langle -7, \frac{19}{3} \right\rangle$

15. $\mathbf{u} = 2\langle -2, 5 \rangle, \mathbf{v} = \frac{1}{4}\langle -7, 12 \rangle$

16. $\mathbf{u} = \mathbf{i} - \mathbf{j}, \mathbf{v} = 2\mathbf{i} + \mathbf{j}$ **17.** $\mathbf{u} = 8\mathbf{i}, \mathbf{v} = 2(3\mathbf{i} - 2\mathbf{j})$

18. $\mathbf{u} = -4(-\mathbf{i} + \mathbf{j}), \mathbf{v} = -3\mathbf{i}$

19. $\mathbf{u} = -\left(2\mathbf{i} + \frac{3}{2}\mathbf{j}\right), \mathbf{v} = \frac{3}{4}\mathbf{i}$

20. $\mathbf{u} = \sqrt{2}\mathbf{j}, \mathbf{v} = \sqrt{3}\mathbf{i}$

In Exercises 21–26, find the components of the given vector, where $\mathbf{u} = \mathbf{i} - 2\mathbf{j}, \mathbf{v} = 3\mathbf{i} + \mathbf{j}, \mathbf{w} = -4\mathbf{i} + \mathbf{j}$.

21. $\mathbf{u} + 2\mathbf{w}$

22. $\frac{1}{2}(3\mathbf{v} + \mathbf{w})$

23. $\frac{1}{2}\mathbf{w}$

24. $-2\mathbf{u} + 3\mathbf{v}$

25. $\frac{1}{4}(8\mathbf{u} + 4\mathbf{v} - \mathbf{w})$

26. $3(\mathbf{u} - 2\mathbf{v}) - 6\mathbf{w}$

In Exercises 27–34, find the component form of the vector $\mathbf{v}$ whose magnitude and direction angle θ are given.

27. $\|\mathbf{v}\| = 4, \theta = 0°$ **28.** $\|\mathbf{v}\| = 5, \theta = 30°$

29. $\|\mathbf{v}\| = 10, \theta = 225°$ **30.** $\|\mathbf{v}\| = 20, \theta = 120°$

31. $\|\mathbf{v}\| = 6, \theta = 40°$ **32.** $\|\mathbf{v}\| = 8, \theta = 160°$

33. $\|\mathbf{v}\| = 1/2, \theta = 250°$ **34.** $\|\mathbf{v}\| = 3, \theta = 310°$

In Exercises 35–42, find the magnitude and direction angle of the vector $\mathbf{v}$.

35. $\mathbf{v} = \langle 4, 4 \rangle$ **36.** $\mathbf{v} = \langle 5, 5\sqrt{3} \rangle$

37. $\mathbf{v} = \langle -8, 0 \rangle$ **38.** $\mathbf{v} = \langle 4, 5 \rangle$

39. $\mathbf{v} = 6\mathbf{j}$ **40.** $\mathbf{v} = 4\mathbf{i} - 8\mathbf{j}$

41. $\mathbf{v} = -2\mathbf{i} + 8\mathbf{j}$ **42.** $\mathbf{v} = -15\mathbf{i} - 10\mathbf{j}$

In Exercises 43–46, find a unit vector that has the same direction as $\mathbf{v}$.

43. $\langle 4, -5 \rangle$ **44.** $-7\mathbf{i} + 8\mathbf{j}$

45. $5\mathbf{i} + 10\mathbf{j}$ **46.** $-3\mathbf{i} - 9\mathbf{j}$

In Exercises 47–50, an object at the origin is acted upon by two forces, $\mathbf{u}$ and $\mathbf{v}$, with direction angle θ_u and θ_v, respectively. Find the direction and magnitude of the resultant force.

47. $\mathbf{u} = 30$ pounds, $\theta_{\mathbf{u}} = 0°$; $\mathbf{v} = 90$ pounds, $\theta_{\mathbf{v}} = 60°$

48. $\mathbf{u} = 6$ pounds, $\theta_{\mathbf{u}} = 45°$; $\mathbf{v} = 6$ pounds, $\theta_{\mathbf{v}} = 120°$

49. $\mathbf{u} = 12$ newtons, $\theta_{\mathbf{u}} = 130°$; $\mathbf{v} = 20$ newtons $\theta_{\mathbf{v}} = 250°$

50. $\mathbf{u} = 30$ newtons, $\theta_{\mathbf{u}} = 300°$; $\mathbf{v} = 80$ newtons, $\theta_{\mathbf{v}} = 40°$

If forces $\mathbf{u}_1, \mathbf{u}_2, \ldots, \mathbf{u}_k$ act on an object at the origin, the resultant force is the sum $\mathbf{u}_1 + \mathbf{u}_2 + \cdots + \mathbf{u}_k$. The forces are said to be in equilibrium if their resultant force is $\mathbf{0}$. In Exercises 51 and 52, find the resultant force and find an additional force $\mathbf{v}$ that, if added to the system, produces equilibrium.

51. $\mathbf{u}_1 = \langle 2, 5 \rangle, \mathbf{u}_2 = \langle -6, 1 \rangle, \mathbf{u}_3 = \langle -4, -8 \rangle$

52. $\mathbf{u}_1 = \langle 3, 7 \rangle, \mathbf{u}_2 = \langle 8, -2 \rangle, \mathbf{u}_3 = \langle -9, 0 \rangle, \mathbf{u}_4 = \langle -5, 4 \rangle$

In Exercises 53–58, let $\mathbf{u} = \langle a, b \rangle$ and $\mathbf{v} = \langle c, d \rangle$, and let r and s be scalars. Prove that the stated property holds by calculating the vector on each side of the equal sign.

53. $\mathbf{v} + \mathbf{0} = \mathbf{v} = \mathbf{0} + \mathbf{v}$

54. $\mathbf{v} + (-\mathbf{v}) = \mathbf{0}$

55. $r(\mathbf{u} + \mathbf{v}) = r\mathbf{u} + r\mathbf{v}$

56. $(r + s)\mathbf{v} = r\mathbf{v} + s\mathbf{v}$

57. $(rs)\mathbf{v} = r(s\mathbf{v}) = s(r\mathbf{v})$

58. $1\mathbf{v} = \mathbf{v}$ and $0\mathbf{v} = \mathbf{0}$

59. Two ropes are tied to a wagon. A child pulls one with a force of 20 pounds while another child pulls the other with a force of 30 pounds (see figure). If the angle between the two ropes is 28°, how much force must be exerted by a third child, standing behind the wagon, to keep the wagon from moving? [*Hint:* Assume that the wagon is at the origin and one rope runs along the positive *x*-axis. Proceed as in Example 11 to find the resultant force on the wagon from the ropes. The third child must use the same amount in the opposite direction.]

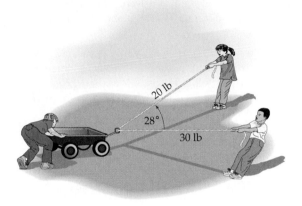

60. Two circus elephants, Bessie and Maybelle, are dragging a large wagon, as shown in the figure. If Bessie pulls with a force of 2200 pounds and Maybelle with a force of 1500 pounds and the wagon moves along the dashed line, what is angle θ?

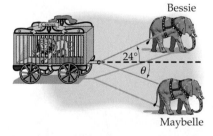

Exercises 61–64 deal with an object on an inclined plane. The situation is similar to that in Figure 9–20 of Example 12, where $\|\overrightarrow{TP}\|$ *is the component of the weight of the object parallel to the plane and* $\|\overrightarrow{TQ}\|$ *is the component of the weight perpendicular to the plane.*

61. An object weighing 50 pounds lies on an inclined plane that makes a 40° angle with the horizontal. Find the components of the weight parallel and perpendicular to the plane. [*Hint:* Solve an appropriate triangle.]

62. Do Exercise 61 when the object weighs 200 pounds and the inclined plane makes a 20° angle with the horizontal.

63. If an object on an inclined plane weighs 150 pounds and the component of the weight perpendicular to the plane is 60 pounds, what angle does the plane make with the horizontal?

64. A force of 500 pounds is needed to pull a cart up a ramp that makes a 15° angle with the ground. Assume that no friction is involved, and find the weight of the cart. [*Hint:* Draw a picture similar to Figure 9–20; the 500-pound force is parallel to the ramp.]

In Exercises 65–70, find the course and ground speed of the plane under the given conditions. (See Example 13.)

65. Air speed 250 mph in the direction 60°; wind speed 40 mph from the direction 330°.

66. Air speed 400 mph in the direction 150°; wind speed 30 mph from the direction 60°.

67. Air speed 300 mph in the direction 300°; wind speed 50 mph in (*not* from) the direction 30°.

68. Air speed 500 mph in the direction 180°; wind speed 70 mph in the direction 40°.

69. The course and ground speed of a plane are 70° and 400 mph, respectively. There is a 60-mph wind blowing south. Find the (approximate) direction and air speed of the plane.

70. A plane is flying in the direction 200° with an air speed of 500 mph. Its course and ground speed are 210° and 450 mph, respectively. What are the direction and speed of the wind?

71. A river flows from east to west. A swimmer on the south bank wants to swim to a point on the opposite shore directly north of her starting point. She can swim at 2.8 mph, and there is a 1-mph current in the river. In what direction should she head so as to travel directly north (that is, what angle should her path make with the south bank of the river)?

72. A river flows from west to east. A swimmer on the north bank swims at 3.1 mph along a straight course that makes a 75° angle with the north bank of the river and reaches the south bank at a point directly south of his starting point. How fast is the current in the river?

73. A 400-pound weight is suspended by two cables (see figure). What is the force (tension) on each cable? [*Hint:* Imagine that the weight is at the origin and that the dashed line is the *x*-axis. Then cable **v** is represented by the vector $(c \cos 65°)\mathbf{i} + (c \sin 65°)\mathbf{j}$, which has magnitude c (why?). Represent cable **u** similarly, denoting its magnitude by d. Use the fact that $\mathbf{u} + \mathbf{v} = 0\mathbf{i} + 400\mathbf{j}$ (why?) to set up a system of two equations in the unknowns c and d.]

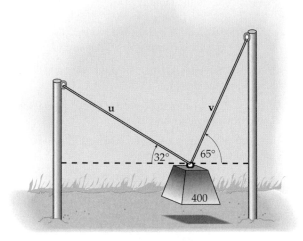

74. An 800-pound weight is suspended from two cables, as shown in the figure. What is the tension (force) on each cable? [See the Hint for Exercise 73.]

75. Do Exercise 74 when the weight is 600 pounds and the angles are 28° and 38°.

76. A 175-pound high-wire artist stands balanced on a tightrope, which sags slightly at the point where he is standing. The rope in front of him makes a 6° angle with the horizontal, and the rope behind him makes a 4° angle with the horizontal. Find the force on each end of the rope. [*Hint:* Use a picture and procedure similar to that in Exercise 73.]

77. Let **v** be the vector with initial point (x_1, y_1) and terminal point (x_2, y_2), and let k be any real number.
  (a) Find the component form of **v** and $k\mathbf{v}$.
  (b) Calculate $\|\mathbf{v}\|$ and $\|k\mathbf{v}\|$.
  (c) Use the fact that $\sqrt{k^2} = |k|$ to verify that $\|k\mathbf{v}\| = |k| \cdot \|\mathbf{v}\|$.
  (d) Show that $\tan \theta = \tan \beta$, where θ is the direction angle of **v** and β is the direction angle of $k\mathbf{v}$. Use the fact that $\tan t = \tan(t + 180°)$ to conclude that **v** and $k\mathbf{v}$ have either the same or opposite directions.
  (e) Use the fact that (c, d) and $(-c, -d)$ lie on the same straight line on opposite sides of the origin (Exercise 85 in Section 1.3) to verify that **v** and $k\mathbf{v}$ have the same direction if $k > 0$ and opposite directions if $k < 0$.

78. Let $\mathbf{u} = \langle a, b \rangle$, $\mathbf{v} = \langle c, d \rangle$. Verify the accuracy of the two geometric interpretations of vector addition given on page 610 as follows.
  (a) Show that the distance from (a, b) to $(a + c, b + d)$ is the same as $\|\mathbf{v}\|$.
  (b) Show that the distance from (c, d) to $(a + c, b + d)$ is the same as $\|\mathbf{u}\|$.
  (c) Show that the line through (a, b) and $(a + c, b + d)$ is parallel to **v** by showing they have the same slope.
  (d) Show that the line through (c, d) and $(a + c, b + d)$ is parallel to **u**.

79. Let $\mathbf{u} = \langle a, b \rangle$ and $\mathbf{v} = \langle c, d \rangle$. Show that $\mathbf{u} - \mathbf{v}$ is equivalent to the vector **w** with initial point (c, d) and terminal point (a, b) as follows. (See the figure.)
  (a) Show that $\|\mathbf{u} - \mathbf{v}\| = \|\mathbf{w}\|$.
  (b) Show that $\mathbf{u} - \mathbf{v}$ and **w** have the same direction.

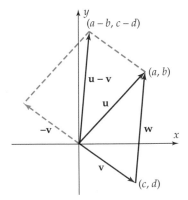

9.4   The Dot Product

When two vectors are added, their sum is another vector, but the situation is different with products. The dot product of two *vectors* is the *real number* defined as follows.

Dot Product

> The **dot product** of vectors $\mathbf{u} = \langle a, b \rangle = a\mathbf{i} + b\mathbf{j}$ and $\mathbf{v} = \langle c, d \rangle = c\mathbf{i} + d\mathbf{j}$ is denoted $\mathbf{u} \cdot \mathbf{v}$ and is defined to be the real number $ac + bd$. Thus,
>
> $$\mathbf{u} \cdot \mathbf{v} = ac + bd.$$

TECHNOLOGY TIP

To compute dot products use DOT or DOTP in this menu/submenu:

TI-86: VECTOR/MATH

TI-89: MATH/MATRIX/VECTOR OPS

HP-39: MATH/MATRIX

EXAMPLE 1

(a) If $\mathbf{u} = \langle 5, 3 \rangle$ and $\mathbf{v} = \langle -2, 6 \rangle$, then

$$\mathbf{u} \cdot \mathbf{v} = 5(-2) + 3 \cdot 6 = 8.$$

(b) If $\mathbf{u} = 4\mathbf{i} - 2\mathbf{j}$ and $\mathbf{v} = 3\mathbf{i} - \mathbf{j}$, then

$$\mathbf{u} \cdot \mathbf{v} = 4 \cdot 3 + (-2)(-1) = 14.$$

(c) $\langle 2, -4 \rangle \cdot \langle 6, 3 \rangle = 2 \cdot 6 + (-4)3 = 0.$ ■

The dot product has a number of useful properties.

Properties of the Dot Product

> If $\mathbf{u}, \mathbf{v}, \mathbf{w}$ are vectors and k is a real number, then
>
> 1. $\mathbf{u} \cdot \mathbf{u} = \|\mathbf{u}\|^2.$
> 2. $\mathbf{u} \cdot \mathbf{v} = \mathbf{v} \cdot \mathbf{u}.$
> 3. $\mathbf{u} \cdot (\mathbf{v} + \mathbf{w}) = \mathbf{u} \cdot \mathbf{v} + \mathbf{u} \cdot \mathbf{w}.$
> 4. $k\mathbf{u} \cdot \mathbf{v} = k(\mathbf{u} \cdot \mathbf{v}) = \mathbf{u} \cdot k\mathbf{v}.$
> 5. $\mathbf{0} \cdot \mathbf{u} = 0.$

Proof

1. If $\mathbf{u} = \langle a, b \rangle$, then

$$\|\mathbf{u}\| = \sqrt{a^2 + b^2}.$$

Hence,

$$\mathbf{u} \cdot \mathbf{u} = \langle a, b \rangle \cdot \langle a, b \rangle = a \cdot a + b \cdot b = a^2 + b^2 = (\sqrt{a^2 + b^2})^2 = \|\mathbf{u}\|^2.$$

2. If $\mathbf{u} = \langle a, b \rangle$ and $\mathbf{v} = \langle c, d \rangle$, then

$$\mathbf{u} \cdot \mathbf{v} = \langle a, b \rangle \cdot \langle c, d \rangle = ac + bd = ca + db = \langle c, d \rangle \cdot \langle a, b \rangle = \mathbf{v} \cdot \mathbf{u}.$$

The last three statements are proved similarly (Exercises 37–39.)   ■

Figure 9–22

▪▪ ANGLES

If $\mathbf{u} = \langle a, b \rangle$ and $\mathbf{v} = \langle c, d \rangle$ are nonzero vectors, then the **angle between u and v** is the smallest angle θ formed by these two line segments, as shown in Figure 9–22. We ignore clockwise or counterclockwise rotation and consider the angle between $\mathbf{v}$ and $\mathbf{u}$ to be the same as the angle between $\mathbf{u}$ and $\mathbf{v}$. Thus, the radian measure of θ ranges from 0 to π.

Nonzero vectors $\mathbf{u}$ and $\mathbf{v}$ are said to be **parallel** if the angle between them is either 0 or π radians (that is, $\mathbf{u}$ and $\mathbf{v}$ lie on the same straight line through the origin and have either the same or opposite directions). The zero vector $\mathbf{0}$ is considered to be parallel to every vector.

Any scalar multiple of $\mathbf{u}$ is parallel to $\mathbf{u}$, since it lies on the same straight line as $\mathbf{u}$ (see Example 3 in Section 9.3). Conversely, if $\mathbf{v}$ is parallel to $\mathbf{u}$, it is easy to show that $\mathbf{v}$ must be a scalar multiple of $\mathbf{u}$ (Exercise 40). Hence, we have the following.

Parallel Vectors

> Vectors $\mathbf{u}$ and $\mathbf{v}$ are parallel exactly when
>
> $$\mathbf{v} = k\mathbf{u} \text{ for some real number } k.$$

EXAMPLE 2

The vectors $\langle 2, 3 \rangle$ and $\langle 8, 12 \rangle$ are parallel because $\langle 8, 12 \rangle = 4\langle 2, 3 \rangle$. ∎

The angle between nonzero vectors $\mathbf{u}$ and $\mathbf{v}$ is closely related to their dot product.

Angle Theorem

> If θ is the angle between the nonzero vectors $\mathbf{u}$ and $\mathbf{v}$, then
>
> $$\mathbf{u} \cdot \mathbf{v} = \|\mathbf{u}\| \, \|\mathbf{v}\| \cos \theta$$
>
> or, equivalently,
>
> $$\cos \theta = \frac{\mathbf{u} \cdot \mathbf{v}}{\,}.$$

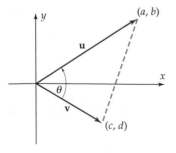

Figure 9–23

Proof   If $\mathbf{u} = \langle a, b \rangle$, $\mathbf{v} = \langle c, d \rangle$, and the angle θ is not 0 or π, then $\mathbf{u}$ and $\mathbf{v}$ form two sides of a triangle, as shown in Figure 9–23.

The lengths of two sides of the triangle are $\|\mathbf{u}\| = \sqrt{a^2 + b^2}$ and $\|\mathbf{v}\| = \sqrt{c^2 + d^2}$. The distance formula shows that the length of the third side (opposite angle θ) is $\sqrt{(a - c)^2 + (b - d)^2}$. Therefore, by the Law of Cosines,

$$[\sqrt{(a - c)^2 + (b - d)^2}]^2 = \|\mathbf{u}\|^2 + \|\mathbf{v}\|^2 - 2\|\mathbf{u}\| \, \|\mathbf{v}\| \cos \theta$$

$$(a - c)^2 + (b - d)^2 = (a^2 + b^2) + (c^2 + d^2) - 2\|\mathbf{u}\| \, \|\mathbf{v}\| \cos \theta$$

$$a^2 - 2ac + c^2 + b^2 - 2bd + d^2 = (a^2 + c^2) + (b^2 + d^2) - 2\|\mathbf{u}\| \, \|\mathbf{v}\| \cos \theta$$

$$-2ac - 2bd = -2\|\mathbf{u}\| \, \|\mathbf{v}\| \cos \theta.$$

Dividing both sides by -2 shows that

$$ac + bd = \|\mathbf{u}\|\,\|\mathbf{v}\|\cos\theta.$$

Since the left side of this equation is precisely $\mathbf{u}\cdot\mathbf{v}$, the proof is complete in this case. The proof when θ is 0 or π is left to the reader (Exercise 41). ∎

EXAMPLE 3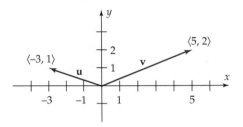

Find the angle θ between the vectors $\langle -3, 1\rangle$ and $\langle 5, 2\rangle$ shown in Figure 9–24.

Figure 9–24

SOLUTION   Apply the formula in the box above with $\mathbf{u} = \langle -3, 1\rangle$ and $\mathbf{v} = \langle 5, 2\rangle$.

$$\cos\theta = \frac{\mathbf{u}\cdot\mathbf{v}}{\|\mathbf{u}\|\,\|\mathbf{v}\|} = \frac{(-3)5 + 1\cdot 2}{\sqrt{(-3)^2 + 1^2}\,\sqrt{5^2 + 2^2}} = \frac{-13}{\sqrt{10}\,\sqrt{29}} = \frac{-13}{\sqrt{290}}.$$

Using the $\cos^{-1}$ key, we see that

$$\theta \approx 2.4393 \text{ radians } (\approx 139.76°). \quad \blacksquare$$

The Angle Theorem has several useful consequences. For instance, by taking absolute values on both sides of $\mathbf{u}\cdot\mathbf{v} = \|\mathbf{u}\|\,\|\mathbf{v}\|\cos\theta$ and using the fact that $\big|\|\mathbf{u}\|\,\|\mathbf{v}\|\big| = \|\mathbf{u}\|\,\|\mathbf{v}\|$ (because $\|\mathbf{u}\|\,\|\mathbf{v}\| \geq 0$), we see that

$$|\mathbf{u}\cdot\mathbf{v}| = \big|\|\mathbf{u}\|\,\|\mathbf{v}\|\cos\theta\big| = \big|\|\mathbf{u}\|\,\|\mathbf{v}\|\big|\,|\cos\theta| = \|\mathbf{u}\|\,\|\mathbf{v}\|\,|\cos\theta|.$$

But for any angle θ, $|\cos\theta| \leq 1$, so

$$|\mathbf{u}\cdot\mathbf{v}| = \|\mathbf{u}\|\,\|\mathbf{v}\|\,|\cos\theta| \leq \|\mathbf{u}\|\,\|\mathbf{v}\|.$$

This proves the Schwarz inequality.

Schwarz Inequality

For any vectors $\mathbf{u}$ and $\mathbf{v}$,

$$|\mathbf{u}\cdot\mathbf{v}| \leq \|\mathbf{u}\|\,\|\mathbf{v}\|.$$

Vectors $\mathbf{u}$ and $\mathbf{v}$ are said to be **orthogonal** (or **perpendicular**) if the angle between them is $\pi/2$ radians (90°), or if at least one of them is **0**. Here is the key fact about orthogonal vectors.

Orthogonal
Vectors

Let **u** and **v** be vectors. Then

u and **v** are orthogonal exactly when **u** · **v** = 0.

Proof If **u** or **v** is **0**, then **u** · **v** = 0, and if **u** and **v** are nonzero orthogonal vectors, then by the Angle Theorem,

$$\mathbf{u} \cdot \mathbf{v} = \|\mathbf{u}\|\,\|\mathbf{v}\| \cos \theta = \|\mathbf{u}\|\,\|\mathbf{v}\| \cos(\pi/2) = \|\mathbf{u}\|\,\|\mathbf{v}\|\,(0) = 0.$$

Conversely, if **u** and **v** are vectors such that **u** · **v** = 0, then Exercise 42 shows that **u** and **v** are orthogonal. ■

EXAMPLE 4

(a) The vectors **u** = ⟨2, −6⟩ and **v** = ⟨9, 3⟩ are orthogonal because

$$\mathbf{u} \cdot \mathbf{v} = \langle 2, -6 \rangle \cdot \langle 9, 3 \rangle = 2 \cdot 9 + (-6)3 = 18 - 18 = 0.$$

(b) The vectors $\frac{1}{2}\mathbf{i} + 5\mathbf{j}$ and $10\mathbf{i} - \mathbf{j}$ are orthogonal, since

$$\left(\frac{1}{2}\mathbf{i} + 5\mathbf{j}\right) \cdot (10\mathbf{i} - \mathbf{j}) = \frac{1}{2}(10) + 5(-1) = 5 - 5 = 0. \quad ■$$

▟ PROJECTIONS AND COMPONENTS

If **u** and **v** are nonzero vectors and θ is the angle between them, construct the perpendicular line segment from the terminal point P of **u** to the straight line on which **v** lies. This perpendicular segment intersects the line at a point Q, as shown in Figure 9–25.

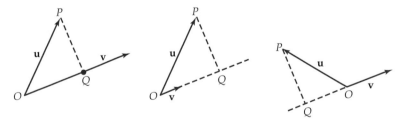

Figure 9–25

The vector $\overrightarrow{OQ}$ is called the **projection of u on v** and is denoted $\mathrm{proj}_{\mathbf{v}}\mathbf{u}.$ Here is a useful description of $\mathrm{proj}_{\mathbf{v}}\mathbf{u}.$

Projection of
u on v

If **u** and **v** are nonzero vectors, then the projection of **u** on **v** is the vector

$$\mathrm{proj}_{\mathbf{v}}\mathbf{u} = \left(\frac{\mathbf{u} \cdot \mathbf{v}}{\|\mathbf{v}\|^2}\right)\mathbf{v}.$$

Proof   Since $\text{proj}_\mathbf{v}\mathbf{u}$ and $\mathbf{v}$ lie on the same straight line, they are parallel, and hence, $\text{proj}_\mathbf{v}\mathbf{u} = k\mathbf{v}$ for some real number k. Let $\mathbf{w}$ be the vector with initial point at the origin and the same length and direction as $\overrightarrow{QP}$, as in the two cases shown in Figure 9–26.

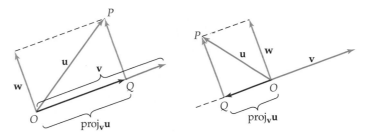

Figure 9–26

Note that $\mathbf{w}$ is parallel to $\overrightarrow{QP}$ and hence is orthogonal to $\mathbf{v}$. As is shown in Figure 9–26, $\mathbf{u} = \text{proj}_\mathbf{v}\mathbf{u} + \mathbf{w} = k\mathbf{v} + \mathbf{w}.$ Consequently, by the properties of the dot product,

$$\mathbf{u} \cdot \mathbf{v} = (k\mathbf{v} + \mathbf{w}) \cdot \mathbf{v} = (k\mathbf{v}) \cdot \mathbf{v} + \mathbf{w} \cdot \mathbf{v}$$

$$= k(\mathbf{v} \cdot \mathbf{v}) + \mathbf{w} \cdot \mathbf{v} = k\|\mathbf{v}\|^2 + \mathbf{w} \cdot \mathbf{v}.$$

But $\mathbf{w} \cdot \mathbf{v} = 0$ because $\mathbf{w}$ and $\mathbf{v}$ are orthogonal. Hence,

$$\mathbf{u} \cdot \mathbf{v} = k\|\mathbf{v}\|^2 \qquad \text{or, equivalently,} \qquad k = \frac{\mathbf{u} \cdot \mathbf{v}}{\|\mathbf{v}\|^2}.$$

Therefore,

$$\text{proj}_\mathbf{v}\mathbf{u} = k\mathbf{v} = \left(\frac{\mathbf{u} \cdot \mathbf{v}}{\|\mathbf{v}\|^2}\right)\mathbf{v},$$

and the proof is complete.   ■

EXAMPLE 5

If $\mathbf{u} = 8\mathbf{i} + 3\mathbf{j}$ and $\mathbf{v} = 4\mathbf{i} - 2\mathbf{j}$, find $\text{proj}_\mathbf{v}\mathbf{u}$ and $\text{proj}_\mathbf{u}\mathbf{v}$.

SOLUTION

$$\mathbf{u} \cdot \mathbf{v} = 8 \cdot 4 + 3(-2) = 26, \qquad \text{and} \qquad \|\mathbf{v}\|^2 = \mathbf{v} \cdot \mathbf{v} = 4^2 + (-2)^2 = 20.$$

Therefore,

$$\text{proj}_\mathbf{v}\mathbf{u} = \left(\frac{\mathbf{u} \cdot \mathbf{v}}{\|\mathbf{v}\|^2}\right)\mathbf{v} = \frac{26}{20}(4\mathbf{i} - 2\mathbf{j}) = \frac{26}{5}\mathbf{i} - \frac{13}{5}\mathbf{j},$$

as is shown in Figure 9–27. We can find the projection of $\mathbf{v}$ on $\mathbf{u}$ by noting that $\|\mathbf{u}\|^2 = \mathbf{u} \cdot \mathbf{u} = 8^2 + 3^2 = 73$, and hence,

$$\text{proj}_{\mathbf{u}}\mathbf{v} = \left(\frac{\mathbf{v} \cdot \mathbf{u}}{\|\mathbf{u}\|^2} \right)\mathbf{u} = \frac{26}{73}(8\mathbf{i} + 3\mathbf{j}) = \frac{208}{73}\mathbf{i} + \frac{78}{73}\mathbf{j}. \qquad \blacksquare$$

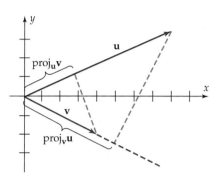

Figure 9–27

Recall that $\dfrac{1}{\|\mathbf{v}\|}\mathbf{v}$ is a unit vector in the direction of $\mathbf{v}$ (see page 613). We can express $\text{proj}_{\mathbf{v}}\mathbf{u}$ as a scalar multiple of this unit vector as follows.

$$\text{proj}_{\mathbf{v}}\mathbf{u} = \left(\frac{\mathbf{u} \cdot \mathbf{v}}{\|\mathbf{v}\|^2} \right)\mathbf{v} = \left(\frac{\mathbf{u} \cdot \mathbf{v}}{\|\mathbf{v}\|} \right)\left(\frac{1}{\|\mathbf{v}\|}\mathbf{v} \right).$$

The scalar $\dfrac{\mathbf{u} \cdot \mathbf{v}}{\|\mathbf{v}\|}$ is called the **component of u along v** and is denoted $\text{comp}_{\mathbf{v}}\mathbf{u}$. Thus,

$$\text{proj}_{\mathbf{v}}\mathbf{u} = \left(\frac{\mathbf{u} \cdot \mathbf{v}}{\|\mathbf{v}\|} \right)\left(\frac{1}{\|\mathbf{v}\|}\mathbf{v} \right) = \text{comp}_{\mathbf{v}}\mathbf{u}\left(\frac{1}{\|\mathbf{v}\|}\mathbf{v} \right).$$

Since $\dfrac{1}{\|\mathbf{v}\|}\mathbf{v}$ is a unit vector, the length of $\text{proj}_{\mathbf{v}}\mathbf{u}$ is

$$\|\text{proj}_{\mathbf{v}}\mathbf{u}\| = \left\| \text{comp}_{\mathbf{v}}\mathbf{u}\left(\frac{1}{\|\mathbf{v}\|}\mathbf{v} \right) \right\| = |\text{comp}_{\mathbf{v}}\mathbf{u}| \left\| \frac{1}{\|\mathbf{v}\|}\mathbf{v} \right\| = |\text{comp}_{\mathbf{v}}\mathbf{u}|.$$

Furthermore, since $\mathbf{u} \cdot \mathbf{v} = \|\mathbf{u}\| \|\mathbf{v}\| \cos \theta$, where θ is the angle between $\mathbf{u}$ and $\mathbf{v}$, we have

$$\text{comp}_{\mathbf{v}}\mathbf{u} = \frac{\mathbf{u} \cdot \mathbf{v}}{\|\mathbf{v}\|} = \frac{\|\mathbf{u}\| \|\mathbf{v}\| \cos \theta}{\|\mathbf{v}\|}.$$

Cancelling $\|\mathbf{v}\|$ on the right side produces this result.

Projections and Components

If $\mathbf{u}$ and $\mathbf{v}$ are nonzero vectors and θ is the angle between them, then

$$\text{comp}_{\mathbf{v}}\mathbf{u} = \frac{\mathbf{u} \cdot \mathbf{v}}{\|\mathbf{v}\|} = \|\mathbf{u}\| \cos \theta$$

and

$$\|\text{proj}_{\mathbf{v}}\mathbf{u}\| = |\text{comp}_{\mathbf{v}}\mathbf{u}|.$$

EXAMPLE 6

If $\mathbf{u} = 2\mathbf{i} + 3\mathbf{j}$ and $\mathbf{v} = -5\mathbf{i} + 2\mathbf{j}$, find $\text{comp}_{\mathbf{v}}\mathbf{u}$ and $\text{comp}_{\mathbf{u}}\mathbf{v}$.

SOLUTION

$$\text{comp}_{\mathbf{v}}\mathbf{u} = \frac{\mathbf{u} \cdot \mathbf{v}}{\|\mathbf{v}\|} = \frac{2(-5) + 3 \cdot 2}{\sqrt{(-5)^2 + 2^2}} = \frac{-4}{\sqrt{29}}.$$

$$\text{comp}_{\mathbf{u}}\mathbf{v} = \frac{\mathbf{v} \cdot \mathbf{u}}{\|\mathbf{u}\|} = \frac{-4}{\sqrt{2^2 + 3^2}} = \frac{-4}{\sqrt{13}}. \qquad \blacksquare$$

■■ APPLICATIONS

Vectors and the dot product can be used to solve a variety of physical problems.

EXAMPLE 7

A 4000-pound automobile is on an inclined ramp that makes a 15° angle with the horizontal. Find the force required to keep it from rolling down the ramp, assuming that the only force that must be overcome is that due to gravity.

SOLUTION   The situation is shown in Figure 9–28, where the coordinate system is chosen so that the car is at the origin, the vector **F** representing the downward force of gravity is on the *y*-axis, and **v** is a unit vector from the origin down the ramp. Since the car weighs 4000 pounds, $\mathbf{F} = -4000\mathbf{j}$. Figure 9–28 shows that the angle between **v** and **F** is 75°. The vector $\text{proj}_{\mathbf{v}}\mathbf{F}$ is the force pulling the car down the ramp, so a force of the same magnitude in the opposite direction is needed to keep the car motionless. As we saw in the preceding box,

$$\|\text{proj}_{\mathbf{v}}\mathbf{F}\| = |\text{comp}_{\mathbf{v}}\mathbf{F}| = \left|\|\mathbf{F}\| \cos 75°\right|$$

$$\approx 4000(.25882) \approx 1035.3.$$

Therefore, a force of 1035.3 pounds is required to hold the car in place.   ■

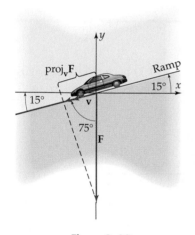

Figure 9–28

If a constant force **F** is applied to an object, pushing or pulling it a distance *d* in the direction of the force as shown in Figure 9–29, the amount of **work** done by the force is defined to be the product

$$W = (\text{magnitude of force})(\text{distance}) = \|\mathbf{F}\| \cdot d.$$

If the magnitude of **F** is measured in pounds and *d* in feet, then the units for *W* are foot-pounds. For example, if you push a car for 35 feet along a level driveway by exerting a constant force of 110 pounds, the amount of work done is $110 \cdot 35 = 3850$ foot-pounds.

Figure 9–29

When a force **F** moves an object in the direction of a vector **d** rather than in the direction of **F**, as shown in Figure 9–30, then the motion of the object can be considered as the result of the vector $\text{proj}_{\mathbf{d}}\mathbf{F}$, which is a force in the same direction as **d**.

Figure 9–30

Therefore, the amount of work done by **F** is the same as the amount of work done by $\text{proj}_\mathbf{d}\mathbf{F}$, namely,

$$W = (\text{magnitude of } \text{proj}_\mathbf{d}\mathbf{F})(\text{length of } \mathbf{d}) = \|\text{proj}_\mathbf{d}\mathbf{F}\| \cdot \|\mathbf{d}\|.$$

The box on page 626 and the Angle Theorem (page 622) show that

$$W = \|\text{proj}_\mathbf{d}\mathbf{F}\| \cdot \|\mathbf{d}\| = |\text{comp}_\mathbf{d}\mathbf{F}| \cdot \|\mathbf{d}\|$$
$$= \|\mathbf{F}\|(\cos\theta)\|\mathbf{d}\|$$
$$= \mathbf{F} \cdot \mathbf{d}.^*$$

Consequently, we have these descriptions of work.

Work

> The work W done by a constant force F as its point of application moves along the vector **d** is
>
> $$W = |\text{comp}_\mathbf{u}\mathbf{F}| \cdot \|\mathbf{d}\| \qquad \text{or, equivalently,} \qquad W = \mathbf{F} \cdot \mathbf{d}.$$

EXAMPLE 8

How much work is done by a child who pulls a sled 100 feet over level ground by exerting a constant 20-pound force on a rope that makes a 45° angle with the ground?

Figure 9–31

SOLUTION   The situation is shown in Figure 9–31, where the force **F** on the rope has magnitude 20 and the sled moves along vector **d** of length 100. The work done is

$$W = \mathbf{F} \cdot \mathbf{d} = \|\mathbf{F}\|\,\|\mathbf{d}\| \cdot \cos\theta = 20 \cdot 100 \cdot \frac{\sqrt{2}}{2}$$
$$= 1000\,\sqrt{2} \approx 1414.2 \text{ foot-pounds.} \quad \blacksquare$$

---

*This formula reduces to the previous one when **F** and **d** have the same direction because in that case, $\cos\theta = \cos 0 = 1$, so $W = \|\mathbf{F}\| \cdot \|\mathbf{d}\| = (\text{magnitude of force})(\text{distance moved})$.

EXERCISES 9.4

In Exercises 1–6, find $\mathbf{u} \cdot \mathbf{v}$, $\mathbf{u} \cdot \mathbf{u}$, and $\mathbf{v} \cdot \mathbf{v}$.

1. $\mathbf{u} = \langle 3, 4 \rangle$, $\mathbf{v} = \langle -5, 2 \rangle$

2. $\mathbf{u} = \langle -1, 6 \rangle$, $\mathbf{v} = \langle -4, 1/3 \rangle$

3. $\mathbf{u} = 2\mathbf{i} + \mathbf{j}$, $\mathbf{v} = 3\mathbf{i}$

4. $\mathbf{u} = \mathbf{i} - \mathbf{j}$, $\mathbf{v} = 5\mathbf{j}$

5. $\mathbf{u} = 3\mathbf{i} + 2\mathbf{j}$, $\mathbf{v} = 2\mathbf{i} + 3\mathbf{j}$

6. $\mathbf{u} = 4\mathbf{i} - \mathbf{j}$, $\mathbf{v} = -\mathbf{i} + 2\mathbf{j}$

In Exercises 7–12, find the dot product when $\mathbf{u} = \langle 2, 5 \rangle$, $\mathbf{v} = \langle -4, 3 \rangle$, and $\mathbf{w} = \langle 2, -1 \rangle$.

7. $\mathbf{u} \cdot (\mathbf{v} + \mathbf{w})$

8. $\mathbf{u} \cdot (\mathbf{v} - \mathbf{w})$

9. $(\mathbf{u} + \mathbf{v}) \cdot (\mathbf{v} + \mathbf{w})$

10. $(\mathbf{u} + \mathbf{v}) \cdot (\mathbf{u} - \mathbf{v})$

11. $(3\mathbf{u} + \mathbf{v}) \cdot (2\mathbf{w})$

12. $(\mathbf{u} + 4\mathbf{v}) \cdot (2\mathbf{u} + \mathbf{w})$

In Exercises 13–18, find the angle between the two vectors.

13. $\langle 4, -3 \rangle$, $\langle 1, 2 \rangle$

14. $\langle 2, 4 \rangle$, $\langle 0, -5 \rangle$

15. $2\mathbf{i} - 3\mathbf{j}$, $-\mathbf{i}$

16. $2\mathbf{j}$, $4\mathbf{i} + \mathbf{j}$

17. $\sqrt{2}\mathbf{i} + \sqrt{2}\mathbf{j}$, $\mathbf{i} - \mathbf{j}$

18. $3\mathbf{i} - 5\mathbf{j}$, $-2\mathbf{i} + 3\mathbf{j}$

In Exercises 19–24, determine whether the given vectors are parallel, orthogonal, or neither.

19. $\langle 2, 6 \rangle$, $\langle 3, -1 \rangle$

20. $\langle -5, 3 \rangle$, $\langle 2, 6 \rangle$

21. $\langle 9, -6 \rangle$, $\langle -6, 4 \rangle$

22. $-\mathbf{i} + 2\mathbf{j}$, $2\mathbf{i} - 4\mathbf{j}$

23. $2\mathbf{i} - 2\mathbf{j}$, $5\mathbf{i} + 8\mathbf{j}$

24. $6\mathbf{i} - 4\mathbf{j}$, $2\mathbf{i} + 3\mathbf{j}$

In Exercises 25–28, find a real number k such that the two vectors are orthogonal.

25. $2\mathbf{i} + 3\mathbf{j}$, $3\mathbf{i} - k\mathbf{j}$

26. $-3\mathbf{i} + \mathbf{j}$, $2k\mathbf{i} - 4\mathbf{j}$

27. $\mathbf{i} - \mathbf{j}$, $k\mathbf{i} + \sqrt{2}\mathbf{j}$

28. $-4\mathbf{i} + 5\mathbf{j}$, $2\mathbf{i} + 2k\mathbf{j}$

In Exercises 29–32, find $\text{proj}_{\mathbf{u}}\mathbf{v}$ and $\text{proj}_{\mathbf{v}}\mathbf{u}$.

29. $\mathbf{u} = 3\mathbf{i} - 5\mathbf{j}$, $\mathbf{v} = 6\mathbf{i} + 2\mathbf{j}$

30. $\mathbf{u} = 2\mathbf{i} - 3\mathbf{j}$, $\mathbf{v} = \mathbf{i} + 2\mathbf{j}$

31. $\mathbf{u} = \mathbf{i} + \mathbf{j}$, $\mathbf{v} = \mathbf{i} - \mathbf{j}$

32. $\mathbf{u} = 5\mathbf{i} + \mathbf{j}$, $\mathbf{v} = -2\mathbf{i} + 3\mathbf{j}$

In Exercises 33–36, find $\text{comp}_{\mathbf{v}}\mathbf{u}$.

33. $\mathbf{u} = 10\mathbf{i} + 4\mathbf{j}$, $\mathbf{v} = 3\mathbf{i} - 2\mathbf{j}$

34. $\mathbf{u} = \mathbf{i} - 2\mathbf{j}$, $\mathbf{v} = 3\mathbf{i} + \mathbf{j}$

35. $\mathbf{u} = 3\mathbf{i} + 2\mathbf{j}$, $\mathbf{v} = -\mathbf{i} + 3\mathbf{j}$

36. $\mathbf{u} = \mathbf{i} + \mathbf{j}$, $\mathbf{v} = -3\mathbf{i} - 2\mathbf{j}$

In Exercises 37–39, let $\mathbf{u} = \langle a, b \rangle$, $\mathbf{v} = \langle c, d \rangle$, and $\mathbf{w} = \langle r, s \rangle$. Verify that the given property of dot products is valid by calculating the quantities on each side of the equal sign.

37. $\mathbf{u} \cdot (\mathbf{v} + \mathbf{w}) = \mathbf{u} \cdot \mathbf{v} + \mathbf{u} \cdot \mathbf{w}$

38. $k\mathbf{u} \cdot \mathbf{v} = k(\mathbf{u} \cdot \mathbf{v}) = \mathbf{u} \cdot k\mathbf{v}$

39. $\mathbf{0} \cdot \mathbf{u} = 0$

40. Suppose $\mathbf{u} = \langle a, b \rangle$ and $\mathbf{v} = \langle c, d \rangle$ are nonzero parallel vectors.
    (a) If $c \neq 0$, show that $\mathbf{u}$ and $\mathbf{v}$ lie on the same nonvertical straight line through the origin.
    (b) If $c \neq 0$, show that $\mathbf{v} = \dfrac{a}{c}\mathbf{u}$ (that is, $\mathbf{v}$ is a scalar multiple of $\mathbf{u}$). [*Hint:* The equation of the line on which $\mathbf{u}$ and $\mathbf{v}$ lie is $y = mx$ for some constant m (why?), which implies that $b = ma$ and $d = mc$.]
    (c) If $c = 0$, show that $\mathbf{v}$ is a scalar multiple of $\mathbf{u}$. [*Hint:* If $c = 0$, then $a = 0$ (why?), and hence, $b \neq 0$ (otherwise, $\mathbf{u} = \mathbf{0}$).]

41. Prove the Angle Theorem in the case when θ is 0 or π.

42. If $\mathbf{u}$ and $\mathbf{v}$ are nonzero vectors such that $\mathbf{u} \cdot \mathbf{v} = 0$, show that $\mathbf{u}$ and $\mathbf{v}$ are orthogonal. [*Hint:* If θ is the angle between $\mathbf{u}$ and $\mathbf{v}$, what is $\cos\theta$ and what does this say about θ?]

43. Show that $(1, 2)$, $(3, 4)$, $(5, 2)$ are the vertices of a right triangle by considering the sides of the triangle as vectors.

44. Find a number x such that the angle between the vectors $\langle 1, 1 \rangle$ and $\langle x, 1 \rangle$ is $\pi/4$ radians.

45. Find nonzero vectors $\mathbf{u}$, $\mathbf{v}$, and $\mathbf{w}$ such that $\mathbf{u} \cdot \mathbf{v} = \mathbf{u} \cdot \mathbf{w}$ and $\mathbf{v} \neq \mathbf{w}$ and neither $\mathbf{v}$ nor $\mathbf{w}$ is orthogonal to $\mathbf{u}$.

46. If $\mathbf{u}$ and $\mathbf{v}$ are nonzero vectors, show that the vectors $\|\mathbf{u}\|\mathbf{v} + \|\mathbf{v}\|\mathbf{u}$ and $\|\mathbf{u}\|\mathbf{v} - \|\mathbf{v}\|\mathbf{u}$ are orthogonal.

47. A 600-pound trailer is on an inclined ramp that makes a 30° angle with the horizontal. Find the force required to keep it from rolling down the ramp, assuming that the only force that must be overcome is that due to gravity.

48. In Example 7, find the vector that represents the force necessary to keep the car motionless.

In Exercises 49–52, find the work done by a constant force F as the point of application of F moves along the vector $\overrightarrow{PQ}$.

49. $F = 2i + 5j$, $P = (0, 0)$, $Q = (4, 1)$

50. $F = i - 2j$, $P = (0, 0)$, $Q = (-5, 2)$

51. $F = 2i + 3j$, $P = (2, 3)$, $Q = (5, 9)$ [*Hint:* Find the component form of $\overrightarrow{PQ}$.]

52. $F = 5i + j$, $P = (-1, 2)$, $Q = (4, -3)$

53. A lawn mower handle makes an angle of 60° with the ground. A woman pushes on the handle with a force of 30 pounds. How much work is done in moving the lawn mower a distance of 75 feet on level ground?

54. A child pulls a wagon along a level sidewalk by exerting a force of 18 pounds on the wagon handle, which makes an angle of 25° with the horizontal. How much work is done in pulling the wagon 200 feet?

55. A 40-pound cart is pushed 100 feet up a ramp that makes a 20° angle with the horizontal (see the figure). How much work is done against gravity? [*Hint:* The amount of work done against gravity is the negative of the amount of work done *by* gravity. Coordinatize the situation so that the cart is at the origin. Then the cart moves along vector $d = (100 \cos 20°)i + (100 \sin 20°)j$, and the downward force of gravity is $F = 0i - 40j$.]

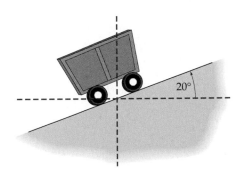

56. Suppose the child in Exercise 54 is pulling the wagon up a hill that makes an angle of 20° with the horizontal and all other facts remain the same. How much work is done in pulling the wagon 150 feet?

Chapter 9 Review

IMPORTANT CONCEPTS _____

Section 9.1

Complex plane   592
Real axis   592
Imaginary axis   592
Absolute value   593
Argument   593
Modulus   593–594
Polar form   594
Multiplication and division in polar
   form   595

Section 9.2

DeMoivre's Theorem   598
Formula for *n*th roots   601

Roots of unity   602–603

Section 9.3

Vector   606
Magnitude   606, 608
Components   608
Scalar multiplication   609
Vector addition and subtraction
   610–611
Unit vector   612
Linear combination of **i** and **j**   613
Direction angle of a vector
   613–614

Section 9.4

Dot product   621
Angle between vectors   622
Parallel vectors   622
Angle Theorem   622
Schwarz inequality   623
Orthogonal vectors   624
Projection of **u** on **v**   624
Component of **u** along **v**   626
Work   628

IMPORTANT FACTS & FORMULAS _____

- $|a + bi| = \sqrt{a^2 + b^2}$

- $a + bi = r(\cos\theta + i\sin\theta)$, where

$$r = \sqrt{a^2 + b^2}, \qquad a = r\cos\theta, \qquad b = r\sin\theta.$$

- $r_1(\cos\theta_1 + i\sin\theta_1) \cdot r_2(\cos\theta_2 + i\sin\theta_2) = r_1 r_2[\cos(\theta_1 + \theta_2) + i\sin(\theta_1 + \theta_2)]$

- $\dfrac{r_1(\cos\theta_1 + i\sin\theta_1)}{r_2(\cos\theta_2 + i\sin\theta_2)} = \dfrac{r_1}{r_2}[\cos(\theta_1 - \theta_2) + i\sin(\theta_1 - \theta_2)]$

- *DeMoivre's Theorem:*

$$[r(\cos\theta + i\sin\theta)]^n = r^n[\cos(n\theta) + i\sin(n\theta)]$$

- The distinct *n*th roots of $r(\cos\theta + i\sin\theta)$ are

$$\sqrt[n]{r}\left[\cos\left(\frac{\theta + 2k\pi}{n}\right) + i\sin\left(\frac{\theta + 2k\pi}{n}\right)\right] \quad (k = 0, 1, 2, \ldots, n-1).$$

- The distinct *n*th roots of unity are

$$\cos\frac{2k\pi}{n} + i\sin\frac{2k\pi}{n} \quad (k = 0, 1, 2, \ldots, n-1).$$

- If $P = (x_1, y_1)$ and $Q = (x_2, y_2)$, then $\overrightarrow{PQ} = \langle x_2 - x_1, y_2 - y_1 \rangle$.

- $\|\langle a, b \rangle\| = \sqrt{a^2 + b^2}$

- If $\mathbf{u} = \langle a, b \rangle$ and k is a scalar, then $k\mathbf{u} = \langle ka, kb \rangle$.

- If $\mathbf{u} = \langle a, b \rangle$ and $\mathbf{v} = \langle c, d \rangle$, then

$$\mathbf{u} + \mathbf{v} = \langle a + c, b + d \rangle \quad \text{and} \quad \mathbf{u} - \mathbf{v} = \langle a - c, b - d \rangle.$$

■ *Properties of Vector Addition and Scalar Multiplication:* For any vectors **u, v,** and **w** and any scalars *r* and *s,*

1. $\mathbf{u} + (\mathbf{v} + \mathbf{w}) = (\mathbf{u} + \mathbf{v}) + \mathbf{w}$
2. $\mathbf{u} + \mathbf{v} = \mathbf{v} + \mathbf{u}$
3. $\mathbf{v} + \mathbf{0} = \mathbf{v} = \mathbf{0} + \mathbf{v}$
4. $\mathbf{v} + (-\mathbf{v}) = \mathbf{0}$
5. $r(\mathbf{u} + \mathbf{v}) = r\mathbf{u} + r\mathbf{v}$
6. $(r + s)\mathbf{v} = r\mathbf{v} + s\mathbf{v}$
7. $(rs)\mathbf{v} = r(s\mathbf{v}) = s(r\mathbf{v})$
8. $1\mathbf{v} = \mathbf{v}$
9. $0\mathbf{v} = \mathbf{0} = r\mathbf{0}$

■ If $\mathbf{v} = \langle a, b \rangle = a\mathbf{i} + b\mathbf{j}$, then

$$a = \|\mathbf{v}\| \cos \theta \qquad \text{and} \qquad b = \|\mathbf{v}\| \sin \theta,$$

where θ is the direction angle of **v**.

■ If $\mathbf{u} = \langle a, b \rangle = a\mathbf{i} + b\mathbf{j}$ and $\mathbf{v} = \langle c, d \rangle = c\mathbf{i} + d\mathbf{j}$, then

$$\mathbf{u} \cdot \mathbf{v} = ac + bd.$$

■ If θ is the angle between nonzero vectors **u** and **v,** then

$$\mathbf{u} \cdot \mathbf{v} = \|\mathbf{u}\| \, \|\mathbf{v}\| \cos \theta.$$

■ *Schwarz Inequality:* $|\mathbf{u} \cdot \mathbf{v}| \le \|\mathbf{u}\| \, \|\mathbf{v}\|.$

■ Vectors **u** and **v** are orthogonal exactly when $\mathbf{u} \cdot \mathbf{v} = 0$.

■ $\text{proj}_{\mathbf{v}}\mathbf{u} = \left(\dfrac{\mathbf{u} \cdot \mathbf{v}}{\|\mathbf{v}\|^2} \right)\mathbf{v}$

■ $\text{comp}_{\mathbf{v}}\mathbf{u} = \dfrac{\mathbf{u} \cdot \mathbf{v}}{\|\mathbf{v}\|} = \|\mathbf{u}\| \cos \theta$, where θ is the angle between **u** and **v.**

REVIEW QUESTIONS

1. Simplify: $|i(4 + 2i)| + |3 - i|$.
2. Simplify: $|3 + 2i| - |1 - 2i|$.
3. Graph the equation $|z| = 2$ in the complex plane.
4. Graph the equation $|z - 3| = 1$ in the complex plane.
5. Express in polar form: $1 + \sqrt{3}i$.
6. Express in polar form: $4 - 5i$.

In Questions 7–11, multiply or divide, and express the answer in the form a + bi.

7. $2\left(\cos \dfrac{\pi}{12} + i \sin \dfrac{\pi}{12} \right) \cdot 4\left(\cos \dfrac{\pi}{6} + i \sin \dfrac{\pi}{6} \right)$

8. $3\left(\cos \dfrac{\pi}{8} + i \sin \dfrac{\pi}{8} \right) \cdot 2\left(\cos \dfrac{3\pi}{8} + i \sin \dfrac{3\pi}{8} \right)$

9. $\dfrac{12\left(\cos \dfrac{7\pi}{12} + i \sin \dfrac{7\pi}{12} \right)}{3\left(\cos \dfrac{5\pi}{12} + i \sin \dfrac{5\pi}{12} \right)}$

10. $\left(\cos \dfrac{\pi}{12} + i \sin \dfrac{\pi}{12} \right)^{18}$

11. $\left[\sqrt[3]{3}\left(\cos \dfrac{5\pi}{36} + i \sin \dfrac{5\pi}{36} \right) \right]^{12}$

In Questions 12–16, solve the given equation in the complex number system, and express your answers in polar form.

12. $x^3 = i$

13. $x^6 = 1$

14. $x^8 = -\sqrt{3} - 3i$

15. $x^4 = i$

16. $x^3 = 1 - i$

In Questions 17–20, let $\mathbf{u} = \langle 3, -2 \rangle$ *and* $\mathbf{v} = \langle 8, 1 \rangle$. *Find*

17. $\mathbf{u} + \mathbf{v}$

18. $\|-3\mathbf{v}\|$

19. $\|2\mathbf{v} - 4\mathbf{u}\|$

20. $3\mathbf{u} - \dfrac{1}{2}\mathbf{v}$

In Questions 21–24, let $\mathbf{u} = -2\mathbf{i} + \mathbf{j}$ *and* $\mathbf{v} = 3\mathbf{i} - 4\mathbf{j}$. *Find*

21. $4\mathbf{u} - \mathbf{v}$

22. $\mathbf{u} + 2\mathbf{v}$

23. $\|\mathbf{u} + \mathbf{v}\|$

24. $\|\mathbf{u}\| + \|\mathbf{v}\|$

25. Find the components of the vector $\mathbf{v}$ such that $\|\mathbf{v}\| = 5$ and the direction angle of $\mathbf{v}$ is $45°$.

26. Find the magnitude and direction angle of $3\mathbf{i} + 4\mathbf{j}$.

27. Find a unit vector whose direction is *opposite* the direction of $3\mathbf{i} - 6\mathbf{j}$.

28. An object at the origin is acted upon by a 10-pound force with direction angle $90°$ and a 20-pound force with direction angle $30°$. Find the magnitude and direction of the resultant force.

29. A plane flies in the direction $120°$ with an air speed of 300 mph. The wind is blowing from north to south at 40 mph. Find the course and ground speed of the plane.

30. An object weighing 40 pounds lies on an inclined plane that makes a $30°$ angle with the horizontal. Find the components of the weight parallel and perpendicular to the plane.

In Questions 31–34, $\mathbf{u} = \langle 3, -4 \rangle$, $\mathbf{v} = \langle -2, 5 \rangle$, *and* $\mathbf{w} = \langle 0, 3 \rangle$. *Find*

31. $\mathbf{u} \cdot \mathbf{v}$

32. $\mathbf{u} \cdot \mathbf{u} - \mathbf{v} \cdot \mathbf{v}$

33. $(\mathbf{u} + \mathbf{v}) \cdot \mathbf{w}$

34. $(\mathbf{u} + \mathbf{w}) \cdot (\mathbf{w} - 3\mathbf{v})$

35. What is the angle between the vectors $5\mathbf{i} - 2\mathbf{j}$ and $3\mathbf{i} + \mathbf{j}$?

36. Is $3\mathbf{i} - 2\mathbf{j}$ orthogonal to $4\mathbf{i} + 6\mathbf{j}$?

In Questions 37 and 38, $\mathbf{u} = 4\mathbf{i} - 3\mathbf{j}$ *and* $\mathbf{v} = 2\mathbf{i} + \mathbf{j}$. *Find*

37. $\text{proj}_{\mathbf{v}}\mathbf{u}$

38. $\text{comp}_{\mathbf{u}}\mathbf{v}$

39. If $\mathbf{u}$ and $\mathbf{v}$ have the same magnitude, show that $\mathbf{u} + \mathbf{v}$ and $\mathbf{u} - \mathbf{v}$ are orthogonal.

40. If $\mathbf{u}$ and $\mathbf{v}$ are nonzero vectors, show that the vector $\mathbf{u} - k\mathbf{v}$ is orthogonal to $\mathbf{v}$, where $k = \dfrac{\mathbf{u} \cdot \mathbf{v}}{\|\mathbf{v}\|^2}$.

41. A 3500-pound automobile is on an inclined ramp that makes a $30°$ angle with the horizontal. Find the force required to keep it from rolling down the ramp, assuming that the only force that must be overcome is that due to gravity.

42. A sled is pulled along level ground by a rope that makes a $50°$ angle with the horizontal. If a force of 40 pounds is used to pull the sled, how much work is done in pulling it 100 feet?

DISCOVERY PROJECT 9　Surveying

You might ask, how far apart are two points *A* and *B*? Under ordinary circumstances, the easiest thing to do would be to measure the distance. However, sometimes it is impractical to make such a direct measurement because of intervening obstacles. Two historical methods for measuring the distance between the two mutually invisible points *A* and *B* are given below.

1. Find the distance from *A* to *B*, using the classic surveyor's method shown in the figure below. A transit is used to measure the angle between two distant objects, and a reasonably short baseline is measured. Angles are reported in degrees. Triangles are drawn, and the law of sines (page 573) is used to calculate the length of unknown edges. To find the distance from *A* to *B*, draw an additional triangle that has *AB* as one of the sides. Caution: The picture is not drawn to scale.

(a)

2. Find the distance from *A* to *B*, using a modern laser range finder. Laser technology removes the necessity of drawing triangles; the tool provides a vector (angle and distance) from point to point, so the distance from *A* to *B* can be found by using vector arithmetic in place of repeated application of the law of sines. Angles are taken from the internal compass and are reported in degrees measured clockwise from North (0°), so you have to convert compass direction angles to angles in standard position. For example, the angle from point *A* (straight South) is reported as 180°, but in standard position, you would list the angle as 270°.

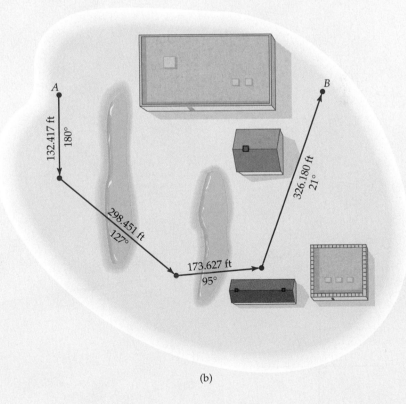

(b)

C H A P T E R 10

Analytic Geometry

Peter Menzel/Stock, Boston Inc./PictureQuest

Calling all ships!

The planets travel in elliptical orbits around the sun, and satellites travel in elliptical orbits around the earth. Parabolic reflectors are used in spotlights, radar antennas, and satellite dishes. The long-range navigation system (loran) uses hyperbolas to enable a ship to determine its exact location. See Exercise 56 on page 662.

Chapter Outline

Interdependence of Sections

Chapters 10–13 are independent of one another and may be read in any order.

Sections 10.1–10.3, 10.5, and 10.6 may be read in any order, but a few exercises in Sections 10.2 and 10.3 assume knowledge of preceding sections. Section 10.4 depends on Sections 10.1–10.3, and Section 10.7 depends on Section 10.6.

10.1   Circles and Ellipses

10.2   Hyperbolas

10.3   Parabolas

10.4   Rotations and Second-Degree Equations

10.4.A   *Special Topics:* Rotation of Axes

10.5   Plane Curves and Parametric Equations

10.6   Polar Coordinates

10.7   Polar Equations of Conics

The discussion of analytic geometry that was begun in Section 1.3 is continued in Sections 10.1–10.4 with an examination of conic sections (which have played a significant role in mathematics since ancient times).

When a right circular cone is cut by a plane, the intersection is a curve called a **conic section,** as shown in Figure 10–1.[*] Conic sections were studied by the ancient Greeks and are still of interest. For instance, planets travel in elliptical orbits, parabolic mirrors are used in telescopes, and certain atomic particles follow hyperbolic paths.

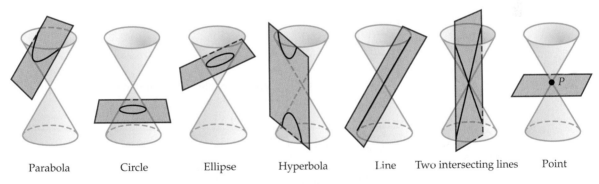

Parabola   Circle   Ellipse   Hyperbola   Line   Two intersecting lines   Point

Figure 10–1

Although the Greeks studied conic sections from a purely geometric point of view, the modern approach is to describe them in terms of the coordinate plane and distance or as the graphs of certain types of equations. This was done for circles in Section 1.3 and will be done here for ellipses, hyperbolas, and parabolas.

Sections 10.5–10.7 present a more thorough treatment of parametric graphing and an alternate method of coordinatizing the plane and graphing.

---

[*]A point, a line, or two intersecting lines are sometimes called **degenerate** conic sections.

10.1 Circles and Ellipses

Let P be a point in the plane, and let r be a positive number. Then the **circle** with center P and radius r consists of all points X in the plane such that

$$\text{Distance from } X \text{ to } P = r,$$

as shown in Figure 10–2.

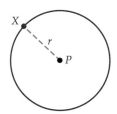

Figure 10–2

In Section 1.3, we proved the following result.

Equation of a Circle

> The circle with center $(0, 0)$ and radius r is the graph of the equation
> $$x^2 + y^2 = r^2.$$
> The circle with center (h, k) and radius r is the graph of
> $$(x - h)^2 + (y - k)^2 = r^2.$$

In addition to the Cartesian equations in the preceding box, circles can also be described by parametric equations.

EXAMPLE 1

Show that the curve given by the parametric equations

$$x = 2 \cos t + 3 \qquad \text{and} \qquad y = 2 \sin t + 1 \quad (0 \le t \le 2\pi)$$

is the circle with center $(3, 1)$ and radius 2.

SOLUTION The circle with center $(3, 1)$ and radius 2 is the graph of the equation

$$(x - 3)^2 + (y - 1)^2 = 2^2.$$

We must show that the coordinates given by the parametric equations satisfy this equation.

$$\begin{aligned}
(x - 3)^2 + (y - 1)^2 &= [(2 \cos t + 3) - 3]^2 + [(2 \sin t + 1) - 1]^2 \\
&= [2 \cos t]^2 + [2 \sin t]^2 \\
&= 4 \cos^2 t + 4 \sin^2 t \\
&= 4(\cos^2 t + \sin^2 t) \\
&= 4(1) = 4. \qquad \text{[Pythagorean Identity]}
\end{aligned}$$

Therefore, the curve given by the parametric equations is the circle with center (3, 1) and radius 2. ∎

If you replace 3, 1, and 2 in Example 1 by *h, k,* and *r,* respectively, then the same computation leads to the following result.

Parametric Equations of a Circle

> The circle with center (h, k) and radius r is given by the parametric equations
>
> $$x = r \cos t + h \qquad \text{and} \qquad y = r \sin t + k \quad (0 \le t \le 2\pi).$$

The preceding discussion of circles provides the model for our discussion of the other conic sections. In each case, the conic is defined in terms of points and distances, and its Cartesian equation is determined. The standard form of the equation of a conic includes the key information necessary for a rough sketch of its graph, just as the standard form of the equation of a circle tells you its center and radius. Finally, appropriate Pythagorean identities are used to develop parametric equations for each conic section.

▪▪ ELLIPSES

Definition. Let *P* and *Q* be points in the plane, and let *r* be a number greater than the distance from *P* to *Q*. The **ellipse** with **foci**[*] *P* and *Q* is the set of all points *X* such that

(Distance from *X* to *P*) + (Distance from *X* to *Q*) = *r*.

To draw this ellipse, take a piece of string of length *r* and pin its ends on *P* and *Q*. Put your pencil point against the string and move it, keeping the string taut. You will trace out the ellipse, as shown in Figure 10–3.

Figure 10–3

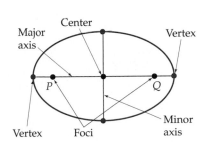

Figure 10–4

The midpoint of the line segment from *P* to *Q* is the **center** of the ellipse. The points where the straight line through the foci intersects the ellipse are its **vertices.** The **major axis** of the ellipse is the line segment joining the vertices; its **minor axis** is the line segment through the center, perpendicular to the major axis, as shown in Figure 10–4.

---

*"Foci" is the plural of "focus."

Equation. Suppose that the foci P and Q are on the x-axis, with coordinates

$$P = (-c, 0) \quad \text{and} \quad Q = (c, 0) \quad \text{for some } c > 0.$$

Let $a = r/2$, so that $2a = r$. Then the point (x, y) is on the ellipse exactly when

[Distance from (x, y) to P] + [Distance from (x, y) to Q] = r

$$\sqrt{(x + c)^2 + (y - 0)^2} + \sqrt{(x - c)^2 + (y - 0)^2} = 2a$$

$$\sqrt{(x + c)^2 + y^2} = 2a - \sqrt{(x - c)^2 + y^2}.$$

Squaring both sides and simplifying (Exercise 60), we obtain

$$a\sqrt{(x - c)^2 + y^2} = a^2 - cx.$$

Again squaring both sides and simplifying, we have

$$(a^2 - c^2)x^2 + a^2y^2 = a^2(a^2 - c^2).$$

To simplify the form of this equation, let $b = \sqrt{a^2 - c^2}$* so that $b^2 = a^2 - c^2$ and the equation becomes

$$b^2x^2 + a^2y^2 = a^2b^2.$$

Dividing both sides by a^2b^2 shows that the coordinates of every point on the ellipse satisfy the equation

$$\frac{x^2}{a^2} + \frac{y^2}{b^2} = 1.$$

Conversely, it can be shown that every point whose coordinates satisfy this equation is on the ellipse. When the equation is in this form the x- and y-intercepts of the graph are easily found. For instance, to find the x-intercepts, we set $y = 0$ and solve

$$\frac{x^2}{a^2} + \frac{0^2}{b^2} = 1$$

$$x^2 = a^2$$

$$x = \pm a.$$

Similarly, the y-intercepts are $\pm b$.

A similar argument applies when the foci are on the y-axis and leads to this conclusion.

---

*The distance between the foci is $2c$. Since $r = 2a$ and $r > 2c$ by definition, we have $2a > 2c$, and hence, $a > c$. Therefore, $a^2 - c^2$ is a positive number and has a real square root.

***Standard Equations
of Ellipses Centered
at the Origin***

Let a and b be real numbers with $a > b > 0$. Then the graph of each of the following equations is an ellipse centered at the origin:

$$\frac{x^2}{a^2} + \frac{y^2}{b^2} = 1 \begin{cases} x\text{-intercepts: } \pm a \qquad y\text{-intercepts: } \pm b \\[4pt] \text{major axis on the } x\text{-axis, with vertices } (a,\, 0) \text{ and } (-a,\, 0) \\[4pt] \text{foci: } (c,\, 0) \text{ and } (-c,\, 0), \text{ where } c = \sqrt{a^2 - b^2} \end{cases}$$

$$\frac{x^2}{b^2} + \frac{y^2}{a^2} = 1 \begin{cases} x\text{-intercepts: } \pm b \qquad y\text{-intercepts: } \pm a \\[4pt] \text{major axis on the } y\text{-axis, with vertices } (0,\, a) \text{ and } (0,\, -a) \\[4pt] \text{foci: } (0,\, c) \text{ and } (0,\, -c), \text{ where } c = \sqrt{a^2 - b^2} \end{cases}$$

In the preceding box $a > b$, but don't let all the letters confuse you: When the equation is in standard form, the denominator of the x term tells you the x-intercepts, the denominator of the y term tells you the y-intercepts, and the major axis is the longer one, as illustrated in the following examples.

EXAMPLE 2

Identify and sketch the graph of the equation $4x^2 + 9y^2 = 36$.

SOLUTION To identify the graph, we put the equation in standard form.

$$4x^2 + 9y^2 = 36$$

Divide both sides by 36:
$$\frac{4x^2}{36} + \frac{9y^2}{36} = \frac{36}{36}$$

Simplify:
$$\frac{x^2}{9} + \frac{y^2}{4} = 1$$

$$\frac{x^2}{3^2} + \frac{y^2}{2^2} = 1.$$

The graph is now in the form of the first equation in the preceding box, with $a = 3$ and $b = 2$. So its graph is an ellipse with x-intercepts ± 3 and y-intercepts ± 2. Its major axis and foci lie on the x-axis, as do its vertices $(3, 0)$ and $(-3, 0)$. A hand-sketched graph is shown in Figure 10–5. To graph this ellipse on a calculator, we first solve its equation for y.

$$4x^2 + 9y^2 = 36$$

Subtract $4x^2$ from both sides: $\qquad 9y^2 = 36 - 4x^2$

Divide both sides by 9: $\qquad y^2 = \dfrac{36 - 4x^2}{9}.$

Taking square roots on both sides, we see that

$$y = \sqrt{\dfrac{36 - 4x^2}{9}} \qquad \text{or} \qquad y = -\sqrt{\dfrac{36 - 4x^2}{9}}.$$

Graphing both of these equations on the same screen, we obtain Figure 10–6. ∎

TECHNOLOGY TIP

On most calculators, you can graph both equations in Example 2 simultaneously by keying in

$$y = \{-1, 1\} \sqrt{\dfrac{36 - 4x^2}{9}}.$$

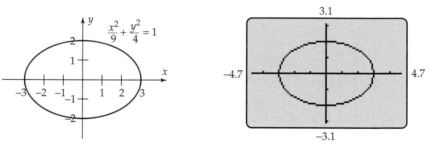

Figure 10–5

Figure 10–6

EXAMPLE 3

Find the equation of the ellipse with vertices $(0, \pm 6)$ and foci $(0, \pm 2\sqrt{6})$, and sketch its graph.

SOLUTION Since the foci are $(0, 2\sqrt{6})$ and $(0, -2\sqrt{6})$, the center of the ellipse is $(0, 0)$, and its major axis lies on the y-axis. Hence, its equation is of the form

$$\dfrac{x^2}{b^2} + \dfrac{y^2}{a^2} = 1.$$

From the box on page 641, we see that $a = 6$ and $c = 2\sqrt{6}$. Since $c = \sqrt{a^2 - b^2}$, we have $c^2 = a^2 - b^2$, so

$$b^2 = a^2 - c^2 = 6^2 - (2\sqrt{6})^2 = 36 - 4 \cdot 6 = 12.$$

Hence, $b = \sqrt{12}$, and the equation of the ellipse is

$$\dfrac{x^2}{(\sqrt{12})^2} + \dfrac{y^2}{6^2} = 1 \qquad \text{or, equivalently,} \qquad \dfrac{x^2}{12} + \dfrac{y^2}{36} = 1.$$

The graph has x-intercepts $\pm\sqrt{12} \approx 3.46$ and y-intercepts ± 6, as sketched in Figure 10–7. ∎

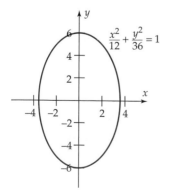

Figure 10–7

▪▪ VERTICAL AND HORIZONTAL SHIFTS

We now examine ellipses that have foci on a line parallel to one of the coordinate axes. Recall that in Section 3.4, we saw that replacing the variable x by $x - 5$ in the rule of the function $y = f(x)$ shifts the graph horizontally 5 units to the right, whereas replacing x by $x + 5$ [that is, $x - (-5)$] shifts the graph horizontally 5 units to the left (see the box on page 168).

Similarly, if the rule of a function is given by $y = f(x)$, then replacing y by $y - 4$ shifts the graph 4 units vertically upward because

$$y - 4 = f(x) \qquad \text{is equivalent to} \qquad y = f(x) + 4$$

(see the box on page 166). For arbitrary equations, we have similar results.

Vertical and Horizontal Shifts

> Let h and k be constants. Replacing x by $x - h$ and y by $y - k$ in an equation shifts the graph of the equation
>
> $|h|$ units horizontally (right for positive h, left for negative h) and
>
> $|k|$ units vertically (upward for positive k, downward for negative k).

EXAMPLE 4

Identify and sketch the graph of

$$\frac{(x - 5)^2}{9} + \frac{(y + 4)^2}{36} = 1.$$

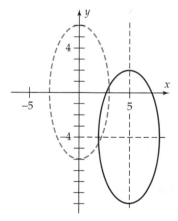

Figure 10–8

SOLUTION This equation can be obtained from the equation $\dfrac{x^2}{9} + \dfrac{y^2}{36} = 1$ (whose graph is known to be an ellipse) as follows.

Replace x by $x - 5$ and replace y by $y - (-4) = y + 4$.

This is the situation described in the previous box with $h = 5$ and $k = -4$. Therefore, the graph is the ellipse $\dfrac{x^2}{9} + \dfrac{y^2}{36} = 1$ shifted horizontally 5 units to the right and vertically 4 units downward, as shown in Figure 10–8. The center of the ellipse is at $(5, -4)$. Its major axis (the longer one) lies on the vertical line $x = 5$, as do its foci. The minor axis is on the horizontal line $y = -4$. ▪

EXAMPLE 5

Identify and sketch the graph of

$$4x^2 + 9y^2 - 32x - 90y + 253 = 0.$$

SOLUTION We first rewrite the equation.

$$(4x^2 - 32x) + (9y^2 - 90y) = -253$$

$$4(x^2 - 8x) + 9(y^2 - 10y) = -253.$$

The factoring in the last equation was done in preparation for completing the square (see the Caution in the margin). To complete the square on $x^2 - 8x$, we add 16 (the square of half the coefficient of x), and to complete the square on $y^2 - 10y$, we add 25 (the square of half the coefficient of y):

$$4(x^2 - 8x + 16) + 9(y^2 - 10y + 25) = -253 + ? + ?.$$

Be careful here: On the left side, we haven't just added 16 and 25. When the left side is multiplied out, we have actually added $4 \cdot 16 = 64$ and $9 \cdot 25 = 225$. To leave the equation unchanged, we must add these numbers on the right:

$$4(x^2 - 8x + 16) + 9(y^2 - 10y + 25) = -253 + 64 + 225$$

Factor and simplify:
$$4(x - 4)^2 + 9(y - 5)^2 = 36$$

Divide both sides by 36:
$$\frac{4(x - 4)^2}{36} + \frac{9(y - 5)^2}{36} = \frac{36}{36}$$

Simplify:
$$\frac{(x - 4)^2}{9} + \frac{(y - 5)^2}{4} = 1.$$

The graph of this equation is the ellipse $\dfrac{x^2}{9} + \dfrac{y^2}{4} = 1$ shifted 4 units to the right and 5 units upward. Its center is at (4, 5). Its major axis lies on the horizontal line $y = 5$, and its minor axis lies on the vertical line $x = 4$, as shown in Figure 10–9. ∎

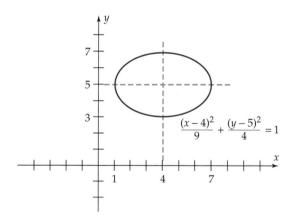

Figure 10–9

By translating the information about ellipses centered at the origin in the box on page 641, we obtain the following.

Standard Equations of Ellipses with Center at (h, k)

Let (h, k) be any point in the plane. If a and b are real numbers with $a > b > 0$, then the graph of each of the following equations is an ellipse with center (h, k).

$$\frac{(x - h)^2}{a^2} + \frac{(y - k)^2}{b^2} = 1 \begin{cases} \text{major axis on the horizontal line } y = k \\ \text{minor axis on the vertical line } x = h \\ \text{foci: } (h - c, k) \text{ and } (h + c, k), \text{ where} \\ \qquad c = \sqrt{a^2 - b^2} \end{cases}$$

$$\frac{(x - h)^2}{b^2} + \frac{(y - k)^2}{a^2} = 1 \begin{cases} \text{major axis on the vertical line } x = h \\ \text{minor axis on the horizontal line } y = k \\ \text{foci: } (h, k - c) \text{ and } (h, k + c), \text{ where} \\ \qquad c = \sqrt{a^2 - b^2} \end{cases}$$

TECHNOLOGY TIP

Casio 9850 and FX 2.0 have conic section graphers (on the main menu) that produce the graphs of equations in standard form when the various coefficients are entered.

EXAMPLE 6

Find an appropriate viewing window and graph the ellipse

$$\frac{(x - 3)^2}{40} + \frac{(y + 2)^2}{120} = 1.$$

SOLUTION   To graph the ellipse, we first solve its equation for y.

Subtract $\dfrac{(x - 3)^2}{40}$ from both sides: $\dfrac{(y + 2)^2}{120} = 1 - \dfrac{(x - 3)^2}{40}$

Multiply both sides by 120: $(y + 2)^2 = 120\left[1 - \dfrac{(x - 3)^2}{40}\right]$

Multiply out right side: $(y + 2)^2 = 120 - 3(x - 3)^2$

Take square roots on both sides: $y + 2 = \pm\sqrt{120 - 3(x - 3)^2}$

$$y = \sqrt{120 - 3(x - 3)^2} - 2 \qquad \text{or} \qquad y = -\sqrt{120 - 3(x - 3)^2} - 2$$

So we should graph both of these last two equations on the same screen.

To determine an appropriate window, look at the original form of the equation:

$$\frac{(x - 3)^2}{40} + \frac{(y + 2)^2}{120} = 1.$$

The center of the ellipse is at $(3, -2)$. Its graph extends a distance of $\sqrt{40}$ to the left and right of the center and a distance of $\sqrt{120}$ above and below the center. Since $\sqrt{40}$ is a bit less than 7 and $\sqrt{120}$ is a bit less than 11 (why?), our window should include x-values from $3 - 7$ to $3 + 7$ (that is, $-4 \le x \le 10$) and y-values from $-2 - 11$ to $-2 + 11$ (that is, $-13 \le y \le 9$). So we first try the window in Figure 10–10 on the next page. In this window, the ellipse looks longer horizontally than vertically. We know from its equation, however, that the major (longer) axis is vertical because the larger constant 120 is the denominator of the y-term. So we change to a square window and obtain the more accurate graph in

Figure 10–11.* Even this graph has gaps that shouldn't be there (an ellipse is a connected figure), but this is unavoidable because of the limited resolution of the calculator screen.   ■

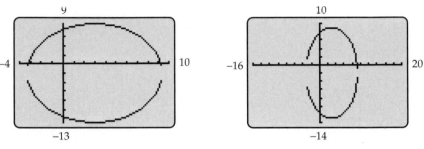

Figure 10–10   **Figure 10–11**

EXAMPLE 7

Graph the equation $x^2 + 8y^2 + 6x + 9y + 4 = 0$ without first putting it in standard form.

SOLUTION   Rewrite it like this:

$$8y^2 + 9y + (x^2 + 6x + 4) = 0.$$

This is a quadratic equation of the form $ay^2 + by + c = 0$, with

$$a = 8, \qquad b = 9, \qquad c = x^2 + 6x + 4$$

and hence can be solved by using the quadratic formula.

$$y = \frac{-b \pm \sqrt{b^2 - 4ac}}{2a}$$

$$y = \frac{-9 \pm \sqrt{9^2 - 4 \cdot 8 \cdot (x^2 + 6x + 4)}}{2 \cdot 8}$$

$$y = \frac{-9 \pm \sqrt{81 - 32(x^2 + 6x + 4)}}{16}$$

TECHNOLOGY TIP

TI users can save keystrokes by entering the first equation in Example 7 as y_1 and then using the RCL key to copy the text of y_1 to y_2. [On TI-89, use COPY and PASTE in place of RCL.] Then only one sign needs to be changed to make y_2 into the second equation of Example 7.

GRAPHING EXPLORATION

Find a complete graph of the original equation by graphing both of the following functions on the same screen. The Technology Tip in the margin may be helpful.

$$y = \frac{-9 + \sqrt{81 - 32(x^2 + 6x + 4)}}{16}$$

$$y = \frac{-9 - \sqrt{81 - 32(x^2 + 6x + 4)}}{16}.$$ ■

---

*Figure 10–11 shows a square window for a TI-83. On wide-screen calculators, a longer x-axis is needed for a square window.

▐▌ PARAMETRIC EQUATIONS

Parametric equations for an ellipse can be obtained in the same way that the ones for the circle were.

EXAMPLE 8

Show that the curve given by the parametric equations

$$x = 2 \cos t + 3 \qquad \text{and} \qquad y = \sqrt{5} \sin t + 6 \quad (0 \le t \le 2\pi)$$

is the ellipse with equation

$$\frac{(x-3)^2}{4} + \frac{(y-6)^2}{5} = 1.$$

SOLUTION   We use the Pythagorean identity to show that the coordinates given by the parametric equations satisfy the equation

$$\frac{(x-3)^2}{4} + \frac{(y-6)^2}{5} = \frac{[(2\cos t + 3) - 3]^2}{4} + \frac{[(\sqrt{5}\sin t + 6) - 6]^2}{5}$$

$$= \frac{[2\cos t]^2}{4} + \frac{[\sqrt{5}\sin t]^2}{5}$$

$$= \frac{4\cos^2 t}{4} + \frac{5\sin^2 t}{5}$$

$$= \cos^2 t + \sin^2 t = 1. \quad ■$$

If you use, *h, k, a,* and *b* in place of 3, 6, 2, and $\sqrt{5}$, respectively, in Example 8, then the same computation leads to this result.

Parametric Equations for Ellipses

> The ellipse with equation
>
> $$\frac{(x-h)^2}{a^2} + \frac{(y-k)^2}{b^2} = 1$$
>
> is given by the parametric equations
>
> $$x = a \cos t + h \qquad \text{and} \qquad y = b \sin t + k \quad (0 \le t \le 2\pi).$$

▐▌ APPLICATIONS

Figure 10–12

Elliptical surfaces have interesting reflective properties. If a sound or light ray passes through one focus and reflects off an ellipse, the ray will pass through the other focus, as shown in Figue 10–12. Exactly this situation occurs under the elliptical dome of the U.S. Capitol. A person who stands at one focus and whispers can be clearly heard by anyone at the other focus. Before this fact was widely known, when Congress used to sit under the dome, several political secrets were inadvertently revealed by congressmen to members of the other party.

The planets and many comets have elliptical orbits, with the sun as one focus. The moon travels in an elliptical orbit with the earth as one focus. Satellites are usually put into elliptical orbits around the earth.

EXAMPLE 9

The earth's orbit around the sun is an ellipse that is almost a circle. The sun is one focus, and the major and minor axes have lengths 186,000,000 miles and 185,974,062 miles, respectively. What are the minimum and maximum distances from the earth to the sun?

SOLUTION   The orbit is shown in Figure 10–13. If we use a coordinate system with the major axis on the x-axis and the sun having coordinates $(c, 0)$, then we obtain Figure 10–14.

Figure 10–13

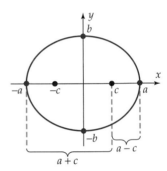

Figure 10–14

The length of the major axis is $2a = 186,000,000$, so that $a = 93,000,000$. Similarly, $2b = 185,974,062$, so $b = 92,987,031$. As was shown earlier, the equation of the orbit is

$$\frac{x^2}{a^2} + \frac{y^2}{b^2} = 1, \qquad \text{where}$$

$$c = \sqrt{a^2 - b^2} = \sqrt{(93,000,000)^2 - (92,987,031)^2} \approx 1,553,083.$$

Figure 10–14 suggests a fact that can also be proven algebraically: The minimum and maximum distances from a point on the ellipse to the focus $(c, 0)$ occur at the endpoints of the major axis:

Minimum distance $= a - c \approx 93,000,000 - 1,553,083 = 91,446,917$ miles

Maximum distance $= a + c \approx 93,000,000 + 1,553,083 = 94,553,083$ miles.

■

EXERCISES 10.1

In Exercises 1–6, determine which of the following equations could possibly have the given graph.

$2x^2 + y^2 = 12$, $(x - 4)^2 + (y - 3)^2 = 4$,

$x^2 + 6y^2 = 18$, $(x + 3)^2 + (y + 4)^2 = 6$,

$(x + 3)^2 + y^2 = 9$, $x^2 + (y - 3)^2 = 4$,

$(x - 3)^2 + (y + 4)^2 = 5$, $(x + 2)^2 + (y - 3)^2 = 2$

1.

2.

3.

4.

5.

6.

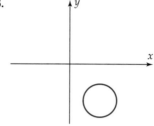

In Exercises 7–14, identify the conic section whose equation is given and find a complete graph of the equation as in Example 2.

7. $\dfrac{x^2}{25} + \dfrac{y^2}{4} = 1$

8. $\dfrac{x^2}{6} + \dfrac{y^2}{16} = 1$

9. $4x^2 + 3y^2 = 12$

10. $9x^2 + 4y^2 = 72$

11. $\dfrac{x^2}{10} - 1 = \dfrac{-y^2}{36}$

12. $\dfrac{y^2}{49} + \dfrac{x^2}{81} = 1$

13. $4x^2 + 4y^2 = 1$

14. $x^2 + 4y^2 = 1$

In Exercises 15–20, find the equation of the ellipse that satisfies the given conditions.

15. Center $(0, 0)$; foci on x-axis; x-intercepts ± 7; y-intercepts ± 2.

16. Center $(0, 0)$; foci on y-axis; x-intercepts ± 1; y-intercepts ± 8.

17. Center $(0, 0)$; foci on x-axis; major axis of length 12; minor axis of length 8.

18. Center $(0, 0)$; foci on y-axis; major axis of length 20; minor axis of length 18.

19. Center $(0, 0)$; endpoints of major and minor axes: $(0, -7)$, $(0, 7)$, $(-3, 0)$, $(3, 0)$.

20. Center $(0, 0)$; vertices $(8, 0)$ and $(-8, 0)$; minor axis of length 8.

Calculus can be used to show that the area of the ellipse with equation $\frac{x^2}{a^2} + \frac{y^2}{b^2} = 1$ is πab. Use this fact to find the area of each ellipse in Exercises 21–26.

21. $\frac{x^2}{16} + \frac{y^2}{4} = 1$ **22.** $\frac{x^2}{9} + \frac{y^2}{5} = 1$

23. $3x^2 + 4y^2 = 12$ **24.** $7x^2 + 5y^2 = 35$

25. $6x^2 + 2y^2 = 14$ **26.** $5x^2 + y^2 = 5$

In Exercises 27–36, identify the conic section whose equation is given, list its center, and find its graph.

27. $\frac{(x-1)^2}{4} + \frac{(y-5)^2}{9} = 1$

28. $\frac{(x-2)^2}{16} + \frac{(y+3)^2}{12} = 1$

29. $\frac{(x+1)^2}{16} + \frac{(y-4)^2}{8} = 1$

30. $\frac{(x+5)^2}{4} + \frac{(y+2)^2}{12} = 1$

31. $9x^2 + 4y^2 + 54x - 8y + 49 = 0$

32. $4x^2 + 5y^2 - 8x + 30y + 29 = 0$

33. $x^2 + y^2 + 6x - 8y + 5 = 0$

34. $x^2 + y^2 - 4x + 2y - 7 = 0$

35. $4x^2 + y^2 + 24x - 4y + 36 = 0$

36. $9x^2 + y^2 - 36x + 10y + 52 = 0$

In Exercises 37–44, find parametric equations for the curve whose equation is given, and use them to find a complete graph of the curve.

37. $\frac{x^2}{10} - 1 = \frac{-y^2}{36}$ **38.** $\frac{y^2}{49} + \frac{x^2}{81} = 1$

39. $4x^2 + 4y^2 = 1$ **40.** $x^2 + 4y^2 = 1$

41. $\frac{(x-1)^2}{4} + \frac{(y-5)^2}{9} = 1$

42. $\frac{(x-2)^2}{16} + \frac{(y+3)^2}{12} = 1$

43. $\frac{(x+1)^2}{16} + \frac{(y-4)^2}{8} = 1$

44. $\frac{(x+5)^2}{4} + \frac{(y+2)^2}{12} = 1$

In Exercises 45–50, find the equation of the ellipse that satisfies the given conditions.

45. Center $(2, 3)$; endpoints of major and minor axes: $(2, -1)$, $(0, 3)$, $(2, 7)$, $(4, 3)$.

46. Center $(-5, 2)$; endpoints of major and minor axes: $(0, 2)$, $(-5, 17)$, $(-10, 2)$, $(-5, -13)$.

47. Center $(7, -4)$; foci on the line $x = 7$; major axis of length 12; minor axis of length 5.

48. Center $(-3, -9)$; foci on the line $y = -9$; major axis of length 15; minor axis of length 7.

49. Center $(3, -2)$; passing through $(3, -6)$ and $(9, -2)$.

50. Center $(2, 5)$; passing through $(2, 4)$ and $(-3, 5)$.

In Exercises 51 and 52, find the equations of two distinct ellipses satisfying the given conditions.

51. Center at $(-5, 3)$; major axis of length 14; minor axis of length 8.

52. Center at $(2, -6)$; major axis of length 15; minor axis of length 6.

In Exercises 53–58, determine which of the following equations could possibly have the given graph.

$\frac{(x+3)^2}{4} + \frac{(y+3)^2}{8} = 1,$ $\frac{(x-3)^2}{9} + \frac{(y+4)^2}{4} = 1,$

$2x^2 + 2y^2 - 8 = 0,$ $4x^2 + 2y^2 - 8 = 0,$

$2x^2 + y^2 - 8x - 6y + 9 = 0,$

$x^2 + 3y^2 + 6x - 12y + 17 = 0.$

53.

54.

55.

56.

57.

58.

59. Consider the ellipse whose equation is $\dfrac{x^2}{a^2} + \dfrac{y^2}{b^2} = 1$. Show that if $a = b$, then the graph is actually a circle.

60. Complete the derivation of the equation of the ellipse on page 640 as follows.
   (a) By squaring both sides, show that the equation
   $$\sqrt{(x + c)^2 + y^2} = 2a - \sqrt{(x - c)^2 + y^2}$$
   may be simplified as
   $$a\sqrt{(x - c)^2 + y^2} = a^2 - cx.$$
   (b) Show that the last equation in part (a) may be further simplified as
   $$(a^2 - c^2)x^2 + a^2y^2 = a^2(a^2 - c^2).$$

61. The orbit of the moon around the earth is an ellipse with the earth as one focus. If the length of the major axis of the orbit is 477,736 miles and the length of the minor axis is 477,078 miles, find the minimum and maximum distances from the earth to the moon.

62. Halley's Comet has an elliptical orbit with the sun as one focus and a major axis that is 1,636,484,848 miles long. The closest the comet comes to the sun is 54,004,000 miles. What is the maximum distance from the comet to the sun?

If $a > b > 0$, then the **eccentricity** of the ellipse
$$\frac{(x - h)^2}{a^2} + \frac{(y - k)^2}{b^2} = 1 \quad \text{or} \quad \frac{(x - h)^2}{b^2} + \frac{(y - k)^2}{a^2} = 1$$
is the number $\dfrac{\sqrt{a^2 - b^2}}{a}$. In Exercises 63–66, find the eccentricity of the ellipse whose equation is given.

63. $\dfrac{x^2}{100} + \dfrac{y^2}{99} = 1$

64. $\dfrac{x^2}{18} + \dfrac{y^2}{25} = 1$

65. $\dfrac{(x - 3)^2}{10} + \dfrac{(y - 9)^2}{40} = 1$

66. $\dfrac{(x + 5)^2}{12} + \dfrac{(y - 4)^2}{8} = 1$

67. On the basis of your answers to Exercises 63–66, how is the eccentricity of an ellipse related to its graph? [*Hint:* What is the shape of the graph when the eccentricity is close to 0? When it is close to 1?]

68. Assuming that the same scale was used for the axes, which of these ellipses has the larger eccentricity?

(a)

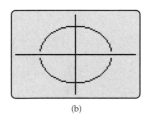
(b)

69. A satellite is to be placed in an elliptical orbit, with the center of the earth as one focus. The satellite's maximum distance from the surface of the earth is to be 22,380 km, and its minimum distance is to be 6540 km. Assume that the radius of the earth is 6400 km, and find the eccentricity of the satellite's orbit.

70. The first step in landing Apollo 11 on the moon was to place the spacecraft in an elliptical orbit such that the minimum distance from the *surface* of the moon to the spacecraft was 110 km and the maximum distance was 314 km. If the radius of the moon is 1740 km, find the eccentricity of the Apollo 11 orbit.

Thinkers

71. The punch bowl and a table holding the punch cups are placed 50 feet apart at a garden party. A portable fence is then set up so that any guest inside the fence can walk straight to the table, then to the punch bowl, and then return to his or her starting point without traveling more than 150 feet. Describe the longest possible such fence that encloses the largest possible area.

72. An arched footbridge over a 100-foot-wide river is shaped like half an ellipse. The maximum height of the bridge over the river is 20 feet. Find the height of the bridge over a point in the river exactly 25 feet from the center of the river.

10.2 Hyperbolas

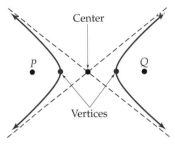

Center

P

Q

Vertices

Figure 10–15

Definition.   Let P and Q be points in the plane, and let r be a positive number. The set of all points X such that

$$|(\text{Distance from } P \text{ to } X) - (\text{Distance from } Q \text{ to } X)| = r$$

is the **hyperbola** with **foci** P and Q; r will be called the **distance difference.** Every hyperbola has the general shape shown by the red curve in Figure 10–15. The dotted straight lines are the **asymptotes** of the hyperbola; it gets closer and closer to the asymptotes, but never touches them. The asymptotes intersect at the midpoint of the line segment from P to Q; this point is called the **center** of the hyperbola. The **vertices** of the hyperbola are the points where it intersects the line segment from P to Q. The line through P and Q is called the **focal axis.**

Equation.   Another complicated exercise in the use of the distance formula, which will be omitted here, leads to the following algebraic description.

Standard Equations of Hyperbolas Centered at the Origin

Let a and b be positive real numbers. Then the graph of each of the following equations is a hyperbola centered at the origin:

$$\frac{x^2}{a^2} - \frac{y^2}{b^2} = 1 \begin{cases} x\text{-intercepts: } \pm a \qquad y\text{-intercepts: none} \\ \text{focal axis on the } x\text{-axis, with vertices } (a, 0) \text{ and } (-a, 0) \\ \text{foci: } (c, 0) \text{ and } (-c, 0), \text{ where } c = \sqrt{a^2 + b^2}. \\ \text{asymptotes: } y = \dfrac{b}{a}x \text{ and } y = -\dfrac{b}{a}x \end{cases}$$

$$\frac{y^2}{a^2} - \frac{x^2}{b^2} = 1 \begin{cases} x\text{-intercepts: none} \qquad y\text{-intercepts: } \pm a \\ \text{focal axis on the } y\text{-axis, with vertices } (0, a) \text{ and } (0, -a) \\ \text{foci: } (0, c) \text{ and } (0, -c), \text{ where } c = \sqrt{a^2 + b^2}. \\ \text{asymptotes: } y = \dfrac{a}{b}x \text{ and } y = -\dfrac{a}{b}x \end{cases}$$

Once again, don't worry about all the letters in the box. When the equation is in standard form with the *x* term positive and *y* term negative, the hyperbola intersects the *x*-axis and opens from side to side. When the *x* term is negative and the *y* term positive, the hyperbola intersects the *y*-axis and opens up and down.

EXAMPLE 1

Identify and sketch the graph of the equation $9x^2 - 4y^2 = 36$.

SOLUTION We first put the equation in standard form:

$$9x^2 - 4y^2 = 36$$

Divide both sides by 36: $$\frac{9x^2}{36} - \frac{4y^2}{36} = \frac{36}{36}$$

Simplify: $$\frac{x^2}{4} - \frac{y^2}{9} = 1$$

$$\frac{x^2}{2^2} - \frac{y^2}{3^2} = 1.$$

Applying the fact in the box with $a = 2$ and $b = 3$ shows that the graph is a hyperbola with vertices $(2, 0)$ and $(-2, 0)$ and asymptotes $y = \frac{3}{2}x$ and $y = -\frac{3}{2}x$. We first plot the vertices and sketch the rectangle determined by the vertical lines $x = \pm 2$ and the horizontal lines $y = \pm 3$. The asymptotes go through the origin and the corners of this rectangle, as shown on the left in Figure 10–16. It is then easy to sketch the hyperbola. ■

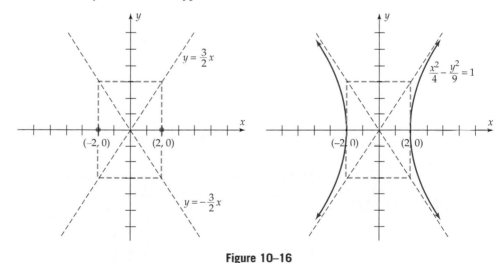

Figure 10–16

EXAMPLE 2

Find the equation of the hyperbola with vertices $(0, 1)$ and $(0, -1)$ that passes through the point $(3, \sqrt{2})$. Then sketch its graph.

SOLUTION The vertices are on the *y*-axis, and the equation is of the form

$$\frac{y^2}{a^2} - \frac{x^2}{b^2} = 1$$

with $a = 1$. Since $(3, \sqrt{2})$ is on the graph, we have

$$\frac{(\sqrt{2})^2}{1^2} - \frac{3^2}{b^2} = 1$$

Simplify: $$2 - \frac{9}{b^2} = 1$$

Subtract 2 from both sides: $$-\frac{9}{b^2} = -1$$

Multiply both sides by $-b^2$: $$9 = b^2.$$

Therefore, $b = 3$, and the equation is

$$\frac{y^2}{1^2} - \frac{x^2}{3^2} = 1 \qquad \text{or, equivalently,} \qquad y^2 - \frac{x^2}{9} = 1.$$

The asymptotes of the hyperbola are the lines $y = \pm\frac{1}{3}x$. To sketch the graph, we first solve the equation for y.

$$y^2 - \frac{x^2}{9} = 1$$

Add $\dfrac{x^2}{9}$ to both sides: $$y^2 = 1 + \frac{x^2}{9}$$

Take square roots on both sides: $$y = \sqrt{1 + \frac{x^2}{9}} \quad \text{or} \quad y = -\sqrt{1 + \frac{x^2}{9}}.$$

Graphing these last two equations on the same screen, we obtain Figure 10–17. ∎

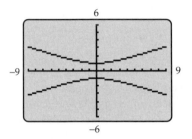

Figure 10–17

▓ VERTICAL AND HORIZONTAL SHIFTS

Let h and k be constants. As we saw on page 643, replacing x by $x - h$ and y by $y - k$ in an equation shifts the graph of the equation h units horizontally and k units vertically. If the graph of the equation is a hyperbola with center at the origin, then the shifted graph has its center at (h, k). Using this fact and translating the information in the box on page 652, we obtain the following.

Standard Equations of Hyperbolas with Center at (h, k)

If a and b are positive real numbers, then the graph of each of the following equations is a hyperbola with center (h, k).

$$\frac{(x - h)^2}{a^2} - \frac{(y - k)^2}{b^2} = 1 \begin{cases} \text{focal axis on the horizontal line } y = k \\ \text{foci: } (h - c, k) \text{ and } (h + c, k), \text{ where} \\ \quad c = \sqrt{a^2 + b^2} \\ \text{vertices: } (h - a, k) \text{ and } (h + a, k) \\ \text{asymptotes: } y = \pm\frac{b}{a}(x - h) + k \end{cases}$$

$$\frac{(y - k)^2}{a^2} - \frac{(x - h)^2}{b^2} = 1 \begin{cases} \text{focal axis on the vertical line } x = h \\ \text{foci: } (h, k - c) \text{ and } (h, k + c), \text{ where} \\ \quad c = \sqrt{a^2 + b^2} \\ \text{vertices: } (h, k - a) \text{ and } (h, k + a) \\ \text{asymptotes: } y = \pm\frac{a}{b}(x - h) + k \end{cases}$$

EXAMPLE 3

Identify and sketch the graph of

$$\frac{(x-3)^2}{4} - \frac{(y+2)^2}{9} = 1.$$

SOLUTION   If we rewrite the equation as

$$\frac{(x-3)^2}{2^2} - \frac{(y-(-2))^2}{3^2} = 1,$$

then it has the first form in the preceding box with $a = 2$, $b = 3$, $h = 3$, and $k = -2$. Its graph is a hyperbola with center $(3, -2)$. There are several ways to obtain the graph.

Method 1. The equation of this hyperbola can be obtained from

$(*)$ $$\frac{x^2}{4} - \frac{y^2}{9} = 1$$

by replacing x by $x - 3$ and y by $y + 2 = y - (-2)$. So its graph is just the graph of $(*)$ (see Figure 10–16) shifted 3 units to the right and 2 units downward, as shown in Figure 10–18.

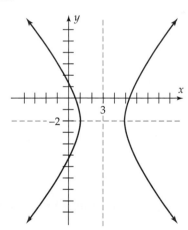

Figure 10–18

Method 2. Solve the original equation for y:

$$\frac{(x-3)^2}{4} - \frac{(y+2)^2}{9} = 1$$

Multiply both sides by 36: $$9(x-3)^2 - 4(y+2)^2 = 36$$

Rearrange terms: $$4(y+2)^2 = 9(x-3)^2 - 36$$

Divide both sides by 4: $$(y+2)^2 = \frac{9(x-3)^2 - 36}{4}$$

Take square roots of both sides: $$y + 2 = \pm\sqrt{\frac{9(x-3)^2 - 36}{4}}$$

$$y = -2 + \sqrt{\frac{9(x-3)^2 - 36}{4}} \quad \text{or} \quad y = -2 - \sqrt{\frac{9(x-3)^2 - 36}{4}}.$$

EXAMPLE 4

Sketch the graph of $6y^2 - 8x^2 - 24y - 48x - 96 = 0$.

SOLUTION   We first solve the equation for y. Begin by rewriting it as

$$6y^2 - 24y + (-8x^2 - 48x - 96) = 0.$$

This is a quadratic equation of the form $ay^2 + by + c = 0$, with

$$a = 6, \qquad b = -24, \qquad c = -8x^2 - 48x - 96.$$

We use the quadratic formula to solve it.

$$y = \frac{-b \pm \sqrt{b^2 - 4ac}}{2a}$$

$$y = \frac{-(-24) \pm \sqrt{(-24)^2 - 4 \cdot 6(-8x^2 - 48x - 96)}}{2 \cdot 6}$$

$$y = \frac{24 \pm \sqrt{576 - 24(-8x^2 - 48x - 96)}}{12}.$$

Now we graph both

$$y = \frac{24 + \sqrt{576 - 24(-8x^2 - 48x - 96)}}{12} \qquad \text{and}$$

$$y = \frac{24 - \sqrt{576 - 24(-8x^2 - 48x - 96)}}{12}$$

on the screen to obtain the hyperbola in Figure 10–19. ■

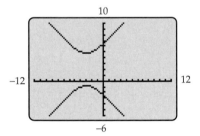

Figure 10–19

EXAMPLE 5

Find the center of the hyperbola in Example 4.

SOLUTION   Begin by rearranging the equation.

$$6y^2 - 8x^2 - 24y - 48x - 96 = 0$$

Add 96 to both sides: $\qquad 6y^2 - 24y - 8x^2 - 48x = 96$

Group x- and y-terms: $\qquad (6y^2 - 24y) - (8x^2 + 48x) = 96$

Factor out coefficients of y^2 and x^2: $\qquad 6(y^2 - 4y) - 8(x^2 + 6x) = 96.$

Complete the square in the expression $y^2 - 4y$ by adding 4 (the square of half the coefficient of y), and complete the square in $x^2 + 6x$ by adding 9 (the square of half the coefficient of x).

$$6(y^2 - 4y + 4) - 8(x^2 + 6x + 9) = 96 + ? + ?$$

CAUTION

Completing the square only works when the coefficient of y^2 is 1. In an expression such as $6y^2 - 24y$, you must first factor out the 6,

$$6(y^2 - 4y),$$

and then complete the square on the expression in parentheses.

On the left side, we have actually added $6 \cdot 4 = 24$ and $-8 \cdot 9 = -72$, so we must add these numbers on the right to keep the equation unchanged.

$$6(y^2 - 4y + 4) - 8(x^2 + 6x + 9) = 96 + 24 - 72$$

Factor and simplify:
$$6(y - 2)^2 - 8(x + 3)^2 = 48$$

Divide both sides by 48:
$$\frac{(y - 2)^2}{8} - \frac{(x + 3)^2}{6} = 1$$

$$\frac{(y - 2)^2}{8} - \frac{(x - (-3))^2}{6} = 1.$$

In this form, we can see that the graph is a hyperbola with center at $(-3, 2)$. ∎

⬛ PARAMETRIC EQUATIONS

Consider the hyperbola with equation

$$\frac{(x - h)^2}{a^2} - \frac{(y - k)^2}{b^2} = 1.$$

Its graph can be obtained from these parametric equations:

$$x = a \sec t + h \qquad \text{and} \qquad y = b \tan t + k \quad (0 \le t \le 2\pi)$$

because, by the Pythagorean identity for secant and tangent,

$$\frac{(x - h)^2}{a^2} - \frac{(y - k)^2}{b^2} = \frac{[(a \sec t + h) - h]^2}{a^2} - \frac{[(b \tan t + k) - k]^2}{b^2}$$

$$= \frac{[a \sec t]^2}{a} - \frac{[b \tan t]^2}{b^2}$$

$$= \frac{a^2 \sec^2 t}{a^2} - \frac{b^2 \tan^2 t}{b^2} = \sec^2 t - \tan^2 t = 1.$$

This proves the first of the following statements; the second is proved similarly.

*Parametric Equations
for Hyperbolas*

Hyperbola	Parametric Equations
$\dfrac{(x - h)^2}{a^2} - \dfrac{(y - k)^2}{b^2} = 1$	$x = a \sec t + h = \dfrac{a}{\cos t} + h \quad$ and $\quad y = b \tan t + k$ $\qquad (0 \le t \le 2\pi)$
$\dfrac{(y - k)^2}{a^2} - \dfrac{(x - h)^2}{b^2} = 1$	$x = b \tan t + h \quad$ and $\quad y = a \sec t + k = \dfrac{a}{\cos t} + k$ $\qquad (0 \le t \le 2\pi)$

EXAMPLE 6

Find parametric equations for the hyperbola with equation

$$\frac{y^2}{9} - \frac{x^2}{16} = 1.$$

SOLUTION   This equation has the form of the second equation in the box above, with $h = 0$, $k = 0$, $a = 3$, and $b = 4$. So the parametric equations are

$$x = 4 \tan t \qquad \text{and} \qquad y = \frac{3}{\cos t} \quad (0 \le t \le 2\pi).$$

EXPLORATION

Use these parametric equations to graph the hyperbola. How is it traced out? Now change the t range so that $-\pi/2 \le t \le 3\pi/2$. Now how is the graph traced out? ∎

▪▪ APPLICATIONS

The reflective properties of hyperbolas are used in the design of camera and telescope lenses. If a light ray passes through one focus of a hyperbola and reflects off the hyperbola at a point P, then the reflected ray moves along the straight line determined by P and the other focus, as shown in Figure 10–20.

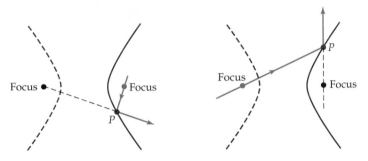

Figure 10–20

Hyperbolas are also the basis of the long-range navigation, system (loran), which enables a ship to determine its exact location by radio, as illustrated in the next example.

EXAMPLE 7

Three loran transmitters Q, P, and R are located 200 miles apart along a straight shoreline and simultaneously transmit signals at regular intervals. These signals travel at a speed of 980 feet per microsecond. A ship S receives a signal from P and 305 microseconds later a signal from R. It also receives a signal from Q 528 microseconds after the one from P. Determine the ship's location.

SOLUTION Take the line through the loran stations as the x-axis, with the origin located midway between Q and P, so that the situation looks like Figure 10–21.

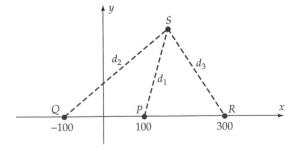

Figure 10–21

If the signal takes t microseconds to go from P to S, then

$$d_1 = 980t \quad \text{and} \quad d_2 = 980(t + 528),$$

so

$$|d_1 - d_2| = |980t - 980(t + 528)| = 980 \cdot 528 = 517{,}440 \text{ feet.}$$

Since 1 mile is 5280 feet, this means that

$$|d_1 - d_2| = 517{,}440/5{,}280 \text{ miles} = 98 \text{ miles.}$$

In other words,

$$|(\text{Distance from } P \text{ to } S) - (\text{Distance from } Q \text{ to } S)| = |d_1 - d_2| = 98.$$

This is precisely the situation described in the definition of "hyperbola" on page 652: S is on the hyperbola with foci $P = (100, 0)$, $Q = (-100, 0)$, and distance difference $r = 98$. This hyperbola has an equation of the form

$$\frac{x^2}{a^2} - \frac{y^2}{b^2} = 1,$$

where $(\pm a, 0)$ are the vertices, $(\pm c, 0) = (\pm 100, 0)$ are the foci, and $c^2 = a^2 + b^2$. Figure 10–22 and the fact that the vertex $(a, 0)$ is on the hyperbola show that

$$|[\text{Distance from } P \text{ to } (a, 0)] - [\text{Distance from } Q \text{ to } (a, 0)]| = r = 98$$

$$|(100 - a) - (100 + a)| = 98$$

$$|-2a| = 98$$

$$|a| = 49.$$

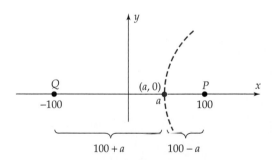

Figure 10–22

Consequently, $a^2 = 49^2 = 2401$, and hence, $b^2 = c^2 - a^2 = 100^2 - 49^2 = 7599$. Thus, the ship lies on the hyperbola

(∗)
$$\frac{x^2}{2401} - \frac{y^2}{7599} = 1.$$

A similar argument using P and R as foci shows that the ship also lies on the hyperbola with foci $P = (100, 0)$ and $R = (300, 0)$ and center $(200, 0)$, whose distance difference r is

$$|d_1 - d_3| = 980 \cdot 305 = 298{,}900 \text{ feet} \approx 56.61 \text{ miles.}$$

As before, you can verify that $a = 56.61/2 = 28.305$, and hence, $a^2 = 28.305^2 = 801.17$. This hyperbola has center $(200, 0)$, and its foci are $(200 - c, k) = (100, 0)$ and $(200 + c, k) = (300, 0)$, which implies that $c = 100$.

Hence, $b^2 = c^2 - a^2 = 100^2 - 801.17 = 9198.83$, and the ship also lies on the hyperbola

(**)
$$\frac{(x - 200)^2}{801.17} - \frac{y^2}{9198.83} = 1.$$

Since the ship lies on both hyperbolas, its coordinates are solutions of both the equations (*) and (**). They can be found algebraically by solving each of the equations for y^2, setting the results equal, and solving for x. They can be found geometrically by graphing both hyperbolas and finding the intersection point. As shown in Figure 10–23, there are actually four points of intersection. However, the two below the x-axis represent points on land in our situation. Furthermore, since the signal from P was received first, the ship is closest to P. So it is located at the point S in Figure 10–23. A graphical intersection finder shows that this point is approximately $(130.48, 215.14)$, where the coordinates are in miles from the origin. ■

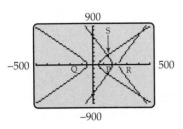

Figure 10–23

EXERCISES 10.2

In Exercises 1–6, determine which of the following equations could possibly have the given graph.

$3x^2 + 3y^2 = 12,$ $6y^2 - x^2 = 6,$
$x^2 + 4y^2 = 1,$ $4x^2 + 4(y + 2)^2 = 12,$
$4(x + 4)^2 + 4y^2 = 12,$ $6x^2 + 2y^2 = 18$
$2x^2 - y^2 = 8,$ $3x^2 - y = 6$

1.

2.

3.

4.

5.

6.

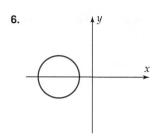

In Exercises 7–14, identify the conic section whose equation is given and find a complete graph of the equation, as in Example 1.

7. $\dfrac{x^2}{6} - \dfrac{y^2}{16} = 1$ **8.** $\dfrac{x^2}{4} - y^2 = 1$

9. $4x^2 - y^2 = 16$ **10.** $3y^2 - 5x^2 = 15$

11. $\dfrac{x^2}{10} - \dfrac{y^2}{36} = 1$ **12.** $\dfrac{y^2}{9} - \dfrac{x^2}{16} = 1$

13. $x^2 - 4y^2 = 1$ **14.** $2x^2 - y^2 = 4$

In Exercises 15–18, find the equation of the hyperbola that satisfies the given conditions.

15. Center $(0, 0)$; x-intercepts ± 3; asymptote $y = 2x$.

16. Center $(0, 0)$; y-intercepts ± 12; asymptote $y = 3x/2$.

17. Center $(0, 0)$; vertex $(2, 0)$; passing through $(4, \sqrt{3})$.

18. Center $(0, 0)$; vertex $(0, \sqrt{12})$; passing through $(2\sqrt{3}, 6)$.

In Exercises 19–32, identify the conic section whose equation is given, list its center, and find its graph.

19. $\dfrac{(y + 3)^2}{25} - \dfrac{(x + 1)^2}{16} = 1$

20. $\dfrac{(y + 1)^2}{9} - \dfrac{(x - 1)^2}{25} = 1$

21. $\dfrac{(x + 3)^2}{1} - \dfrac{(y - 2)^2}{4} = 1$

22. $\dfrac{(y + 5)^2}{9} - \dfrac{(x - 2)^2}{1} = 1$

23. $(y + 4)^2 - 8(x - 1)^2 = 8$

24. $(x - 3)^2 + 12(y - 2)^2 = 24$

25. $4y^2 - x^2 + 6x - 24y + 11 = 0$

26. $x^2 - 16y^2 = 0$

27. $2x^2 + 2y^2 - 12x - 16y + 26 = 0$

28. $3x^2 + 3y^2 + 12x + 6y = 0$

29. $2x^2 + 3y^2 - 12x - 24y + 54 = 0$

30. $x^2 + 2y^2 + 4x - 4y = 8$

31. $x^2 - 3y^2 + 4x + 12y = 20$

32. $2x^2 + 16x = y^2 - 6y - 55$

In Exercises 33–36, find the equation of the hyperbola that satisfies the given conditions.

33. Center $(-2, 3)$; vertex $(-2, 1)$; passing through $(-2 + 3\sqrt{10}, 11)$.

34. Center $(-5, 1)$; vertex $(-3, 1)$; passing through $(-1, 1 - 4\sqrt{3})$.

35. Center $(4, 2)$; vertex $(7, 2)$; asymptote $3y = 4x - 10$.

36. Center $(-3, -5)$; vertex $(-3, 0)$; asymptote $6y = 5x - 15$.

In Exercises 37–42, determine which of the following equations could possibly have the given graph.

$$\dfrac{(y - 2)^2}{4} - \dfrac{(x - 3)^2}{9} = 1, \qquad \dfrac{(x + 3)^2}{3} - \dfrac{(y + 3)^2}{4} = 1,$$

$$4x^2 - 2y^2 = 8, \qquad 9(y - 2)^2 = 36 + 4(x + 3)^2,$$

$$3(y + 3)^2 = 4(x - 3)^2 - 12, \qquad y^2 - 2x^2 = 6.$$

37.

38.

39.

40.

41.

42.

In Exercises 43–50, find parametric equations for the curve whose equation is given, and use them to find a complete graph of the curve.

43. $\dfrac{x^2}{10} - \dfrac{y^2}{36} = 1$
44. $\dfrac{y^2}{9} - \dfrac{x^2}{16} = 1$

45. $x^2 - 4y^2 = 1$
46. $2x^2 - y^2 = 4$

47. $\dfrac{(y+3)^2}{25} - \dfrac{(x+1)^2}{16} = 1$

48. $\dfrac{(y+1)^2}{9} - \dfrac{(x-1)^2}{25} = 1$

49. $\dfrac{(x+3)^2}{1} - \dfrac{(y-2)^2}{4} = 1$

50. $\dfrac{(y+5)^2}{9} - \dfrac{(x-2)^2}{1} = 1$

51. Sketch the graph of $\dfrac{y^2}{4} - \dfrac{x^2}{b^2} = 1$ for $b = 2, b = 4, b = 8,$ $b = 12,$ and $b = 20$. What happens to the hyperbola as b takes larger and larger values? Could the graph ever degenerate into a pair of horizontal lines?

52. Find a number k such that $(-2, 1)$, is on the graph of $3x^2 + ky^2 = 4$. Then graph the equation.

53. Show that the asymptotes of the hyperbola $\dfrac{x^2}{a^2} - \dfrac{y^2}{a^2} = 1$ are perpendicular to each other.

54. Find the approximate coordinates of the points where these hyperbolas intersect:
$$\frac{(x-1)^2}{4} - \frac{(y+1)^2}{8} = 1 \quad \text{and} \quad 4y^2 - x^2 = 1.$$

55 Two listening stations that are 1 mile apart record an explosion. One microphone receives the sound 2 seconds after the other does. Use the line through the microphones as the x-axis, with the origin midway between the microphones, and the fact that sound travels at 1100 feet per second to find the equation of a hyperbola on which the explosion is located. Can you determine the exact location of the explosion?

56. Two transmission stations P and Q are located 200 miles apart on a straight shoreline. A ship 50 miles from shore is moving parallel to the shoreline. A signal from Q reaches the ship 400 microseconds after a signal from P. If the signals travel at 980 feet per microsecond, find the location of the ship (in terms of miles) in the coordinate system with x-axis through P and Q and origin midway between them.

*If $a > 0$ and $b > 0$, then the **eccentricity** of the hyperbola*
$$\frac{(x-h)^2}{a^2} - \frac{(y-k)^2}{b^2} = 1 \quad \text{or} \quad \frac{(y-k)^2}{a^2} - \frac{(x-h)^2}{b^2} = 1$$
is the number $\dfrac{\sqrt{a^2 + b^2}}{a}$. In Exercises 57–61, find the eccentricity of the hyperbola whose equation is given.

57. $\dfrac{(x-6)^2}{10} - \dfrac{y^2}{40} = 1$
58. $\dfrac{y^2}{18} - \dfrac{x^2}{25} = 1$

59. $6(y-2)^2 = 18 + 3(x+2)^2$

60. $16x^2 - 9y^2 - 32x + 36y + 124 = 0$

61. $4x^2 - 5y^2 - 16x - 50y + 71 = 0$

62. (a) Graph these hyperbolas (on the same screen if possible).
$$\frac{y^2}{4} - \frac{x^2}{1} = 1 \qquad \frac{y^2}{4} - \frac{x^2}{12} = 1 \qquad \frac{y^2}{4} - \frac{x^2}{96} = 1$$

(b) Compute the eccentricity of each hyperbola in part (a).

(c) On the basis of parts (a) and (b), how is the shape of a hyperbola related to its eccentricity?

10.3 | Parabolas

Definition. Parabolas appeared in Section 4.1 as the graphs of quadratic functions. Parabolas of this kind are a special case of the following more general definition. Let *L* be a line in the plane, and let *P* be a point not on *L*. If *X* is any point not on *L*, the distance from *X* to *L* is defined to be the length of the perpendicular line segment from *X* to *L*. The **parabola** with focus *P* and **directrix** *L* is the set of all points *X* such that

$$\text{Distance from } X \text{ to } P = \text{Distance from } X \text{ to } L$$

as shown in Figure 10–24.

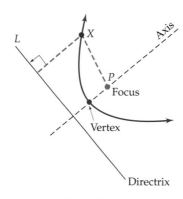

Figure 10–24

The line through *P* perpendicular to *L* is called the **axis.** The intersection of the axis with the parabola (the midpoint of the segment of the axis from *P* to *L*) is the **vertex** of the parabola, as illustrated in Figure 10–24. The parabola is symmetric with respect to its axis.

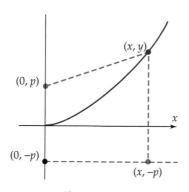

Figure 10–25

Equation. Suppose that the focus is on the *y*-axis at the point $(0, p)$, where *p* is a nonzero constant, and that the directrix is the horizontal line $y = -p$. If (x, y) is any point on the parabola, then the distance from (x, y) to the horizontal line $y = -p$ is the length of the vertical segment from (x, y) to $(x, -p)$ as shown in Figure 10–25.

By the definition of the parabola,

$$\text{Distance from } (x, y) \text{ to } (0, p) = \text{Distance from } (x, y) \text{ to line } y = -p$$

$$\text{Distance from } (x, y) \text{ to } (0, p) = \text{Distance from } (x, y) \text{ to } (x, -p)$$

$$\sqrt{(x - 0)^2 + (y - p)^2} = \sqrt{(x - x)^2 + [y - (-p)]^2}.$$

Squaring both sides and simplifying, we have

$$(x - 0)^2 + (y - p)^2 = (x - x)^2 + (y + p)^2$$

$$x^2 + y^2 - 2py + p^2 = 0^2 + y^2 + 2py + p^2$$

$$x^2 = 4py.$$

Conversely, it can be shown that every point whose coordinates satisfy this equation is on the parabola.

A similar argument works for the parabola with focus $(p, 0)$ on the x-axis and directrix the vertical line $x = -p$, and leads to this conclusion.

Standard Equations of Parabolas with Vertex at the Origin

Let p be a nonzero real number. Then the graph of each of the following equations is a parabola with vertex at the origin.

$x^2 = 4py$ $\begin{cases} \text{focus: } (0, p) \\ \text{directrix: } y = -p \\ \text{axis: } y\text{-axis} \\ \text{opens upward if } p > 0, \\ \text{downward if } p < 0 \end{cases}$

$y^2 = 4px$ $\begin{cases} \text{focus: } (p, 0) \\ \text{directrix: } x = -p \\ \text{axis: } x\text{-axis} \\ \text{opens to right if } p > 0, \text{ to left if } p < 0 \end{cases}$

EXAMPLE 1

Show that the graph of the equation is a parabola, and find its focus and directrix; then sketch the graph:

(a) $y = -x^2/8$ (b) $x = 3y^2$

SOLUTION

(a) We rewrite the equation so that it matches one of the forms in the preceding box:

$$y = -\frac{x^2}{8}$$

Multiply both sides by -8: $-8y = x^2$.

The equation $x^2 = -8y$ is of the form $x^2 = 4py$, with $4p = -8$, so $p = -2$. Hence, the graph is a downward-opening parabola with focus $(0, p) = (0, -2)$ and directrix $y = -p = -(-2) = 2$, as shown in Figure 10–26.

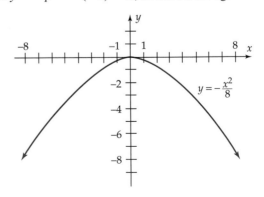

Figure 10–26

(b) Divide both sides of $x = 3y^2$ by 3 so that the equation becomes

$$y^2 = \frac{x}{3}$$

$$y^2 = \frac{1}{3}x.$$

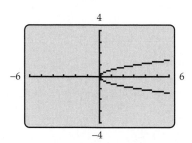

This equation is of the form $y^2 = 4px$, with $4p = 1/3$, so $p = 1/12$. Therefore, the graph is a parabola with focus $(1/12, 0)$ and directrix $x = -1/12$ that opens to the right. To sketch its graph, we solve the equation $y^2 = x/3$ for y.

$$y = \sqrt{\frac{x}{3}} \qquad \text{or} \qquad y = -\sqrt{\frac{x}{3}}.$$

Figure 10–27

Graphing both of these equations on the same screen produces Figure 10–27.

■

EXAMPLE 2

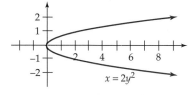

Figure 10–28

Find the focus, directrix, and equation of the parabola that passes through the point $(8, 2)$, has vertex $(0, 0)$, and focus on the x-axis.

SOLUTION The equation is of the form $y^2 = 4px$. Since $(8, 2)$ is on the graph, we have $2^2 = 4p \cdot 8$, so $p = \frac{1}{8}$. Therefore, the focus is $\left(\frac{1}{8}, 0\right)$, and the directrix is the vertical line $x = -\frac{1}{8}$. The equation is $y^2 = 4\left(\frac{1}{8}\right)x = \frac{1}{2}x$ or, equivalently, $x = 2y^2$. Its graph is sketched in Figure 10–28. ■

▪▪ VERTICAL AND HORIZONTAL SHIFTS

Let h and k be constants. If we replace x by $x - h$ and y by $y - k$ in the equation of a parabola with vertex at the origin, then the graph of the new equation can be obtained by shifting the parabola vertically and horizontally so that its vertex is at (h, k), as explained on page 643. Using this fact and translating the information in the box on page 644, we obtain the following.

Standard Equations of Parabolas with Vertex at (h, k)

If p is a nonzero real number, then the graph of each of the following equations is a parabola with vertex (h, k).

$$(x - h)^2 = 4p(y - k) \begin{cases} \text{focus: } (h, k + p) \\ \text{directrix: the horizontal line } y = k - p \\ \text{axis: the vertical line } x = h \\ \text{opens upward if } p > 0, \text{ downward if } p < 0 \end{cases}$$

$$(y - k)^2 = 4p(x - h) \begin{cases} \text{focus: } (h + p, k) \\ \text{directrix: the vertical line } x = h - p \\ \text{axis: the horizontal line } y = k \\ \text{opens to right if } p > 0, \text{ to left if } p < 0 \end{cases}$$

EXAMPLE 3

Identify and sketch the graph of $y = 2(x - 3)^2 + 1$.

SOLUTION   We first rewrite the equation.

$$y = 2(x - 3)^2 + 1$$

Subtract 1 from both sides $$y - 1 = 2(x - 3)^2$$

Divide both sides by 2: $$\frac{1}{2}(y - 1) = (x - 3)^2$$

$$(x - 3)^2 = \frac{1}{2}(y - 1).$$

This is the first form in the preceding box, with $h = 3$, $k = 1$, and $4p = 1/2$. Hence, $p = 1/8$, and the graph is an upward-opening parabola with vertex $(3, 1)$, focus $(3, 1 + 1/8) = (3, 9/8)$, and directrix the horizontal line $y = 1 - 1/8 = 7/8$. The graph of

$$(x - 3)^2 = \frac{1}{2}(y - 1) \qquad \text{or, equivalently,} \qquad y = 2(x - 3)^2 + 1$$

is the graph of $y = 2x^2$ shifted 3 units to the right and 1 unit upward, as shown in Figure 10–29. ∎

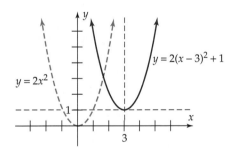

Figure 10–29

EXAMPLE 4

Graph the equation $x = 5y^2 + 30y + 41$ without putting it in standard form.

SOLUTION   There are two methods of graphing this equation on a calculator.

Method 1. Rewrite the equation as

$$5y^2 + 30y + 41 - x = 0.$$

This is a quadratic equation of the form $ay^2 + by + c = 0$, with $a = 5$, $b = 30$, and $c = 41 - x$. It can be solved by using the quadratic formula.

$$y = \frac{-30 \pm \sqrt{30^2 - 4 \cdot 5(41 - x)}}{2 \cdot 5} = \frac{-30 \pm \sqrt{900 - 20(41 - x)}}{10}.$$

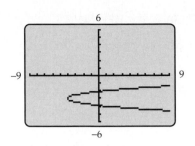

Figure 10–30

Now graph both

$$y = \frac{-30 + \sqrt{900 - 20(41 - x)}}{10} \quad \text{and} \quad y = \frac{-30 - \sqrt{900 - 20(41 - x)}}{10}$$

on the same screen to obtain the parabola in Figure 10–30.

Method 2. Since the equation $x = 5y^2 + 30y + 41$ defines x as a function of y, we can use parametric graphing, as explained in Example 12 of Section 3.3. The parametric equations are

$$x = 5t^2 + 30t + 41,$$

$$y = t.$$

Graphing these equations in parametric mode also produces Figure 10–30. ∎

The parametric graphing technique used in Example 4 can be applied to the parabola with equation $(y - k)^2 = 4p(x - h)$ by solving the equation for x and letting $y = t$.

Parametric Equations for Parabolas

> The parabola with vertex (h, k) and equation
>
> $$(y - k)^2 = 4p(x - h)$$
>
> is given by the parametric equations
>
> $$x = \frac{(t - k)^2}{4p} + h \quad \text{and} \quad y = t \quad (t \text{ any real number}).$$

EXAMPLE 5

Find the vertex, focus, and directrix of the parabola

$$x = 5y^2 + 30y + 41$$

that was graphed in Example 4.

SOLUTION   We first rewrite the equation.

$$5y^2 + 30y + 41 = x$$

Subtract 41 from both sides: $5y^2 + 30y = x - 41$

Factor out 5 on left side: $5(y^2 + 6y) = x - 41.$

Complete the square on the expression $y^2 + 6y$ by adding 9 (the square of half the coefficient of y):

$$5(y^2 + 6y + 9) = x - 41 + \;?.$$

On the left side, we have actually added $5 \cdot 9 = 45$, so we must add the same amount to the right side.

$$5(y^2 + 6y + 9) = x - 41 + 45$$

Factor left side: $\qquad\qquad 5(y + 3)^2 = x + 4$

Divide both sides by 5: $\qquad\qquad (y + 3)^2 = \dfrac{1}{5}(x + 4)$

$$[y - (-3)]^2 = \dfrac{1}{5}[x - (-4)].$$

Thus, the graph is a parabola with vertex $(-4, -3)$. In this case, $4p = 1/5$, so $p = 1/20 = .05$. Hence, the focus is $(-4 + .05, -3) = (-3.95, -3)$, and the directrix is $x = -4 - .05 = -4.05$.

 EXPLORATION

Use the parametric equations in the preceding box, with $h = -4$, $k = -3$, and $p = 1/20$, to graph this parabola in the window with

$$-9 \leq x \leq 9 \qquad \text{and} \qquad -6 \leq y \leq 6 \quad (-6 \leq t \leq 6).$$

Your graph should be identical to Figure 10–30. ∎

■■ APPLICATIONS

Certain laws of physics show that sound waves or light rays from a source at the focus of a parabola will reflect off the parabola in rays parallel to the axis of the parabola, as shown in Figure 10–31. This is the reason that parabolic reflectors are used in automobile headlights and searchlights.

Conversely, a light ray coming toward a parabola will be reflected into the focus, as shown in Figure 10–32. This fact is used in the design of radar antennas, satellite dishes, and field microphones used at outdoor sporting events to pick up conversation on the field.

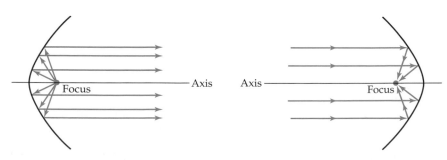

Figure 10–31   **Figure 10–32**

Projectiles follow a parabolic curve, a fact that is used in the design of water slides in which the rider slides down a sharp incline, then up and over a hill, before plunging downward into a pool. At the peak of the hill, the rider shoots up along a parabolic arc several inches above the slide, experiencing a sensation of weightlessness.

EXAMPLE 6

The radio telescope in Figure 10–33 has the shape of a parabolic dish (a cross section through the center of the dish is a parabola). It is 30 feet deep at the center and has a diameter of 200 feet. How far from the vertex of the parabolic dish should the receiver be placed to catch all the rays that hit the dish?

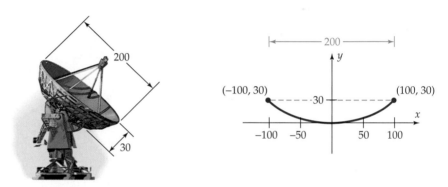

Figure 10–33   **Figure 10–34**

SOLUTION   Rays hitting the dish are reflected into the focus, as explained above. So the radio receiver must be located at the focus. To find the focus, draw a cross section of the dish, with vertex at the origin, as in Figure 10–34. The equation of this parabola is of the form $x^2 = 4py$. Since the point $(100, 30)$ is on the parabola, we have

$$x^2 = 4py$$

Substitute: $$100^2 = 4p(30)$$

Simplify: $$120p = 100^2$$

Divide both sides by 120: $$p = \frac{100^2}{120}$$

Simplify: $$p = \frac{250}{3}.$$

As we saw in the box on page 664, the focus is the point $(0, p)$, which is p units from the vertex $(0, 0)$. Therefore, the receiver should be placed $250/3 \approx 83.33$ feet from the vertex.   ■

✓ EXERCISES 10.3

In Exercises 1–6, determine which of the following equations could possibly have the given graph.

$y = x^2/4,$ $x^? = -8y,$ $6x - y^2,$ $y^2 = 4x,$

$2x^2 + y^2 = 12,$ $x^2 + 6y^2 = 18,$

$6y^2 - x^2 = 6,$ $2x^2 - y^2 = 8$

1. 1

2.

3.

4.

5.

6.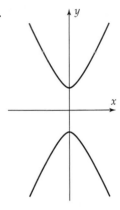

In Exercises 7–10, find the focus and directrix of the parabola.

7. $y = 3x^2$ **8.** $x = .5y^2$

9. $y = .25x^2$ **10.** $x = -6y^2$

In Exercises 11–22, determine the vertex, focus, and directrix of the parabola without graphing and state whether it opens upward, downward, left, or right.

11. $x - y^2 = 2$ **12.** $y - 3 = x^2$

13. $x + (y + 1)^2 = 2$ **14.** $y + (x + 2)^2 = 3$

15. $3x - 2 = (y + 3)^2$ **16.** $2x - 1 = -6(y + 1)^2$

17. $x = y^2 - 9y$ **18.** $x = y^2 + y + 1$

19. $y = 3x^2 + x - 4$ **20.** $y = -3x^2 + 4x - 1$

21. $y = -3x^2 + 4x + 5$ **22.** $y = 2x^2 - x - 1$

In Exercises 23–30, sketch the graph of the equation and label the vertex.

23. $y = 4(x - 1)^2 + 2$ **24.** $y = 3(x - 2)^2 - 3$

25. $x = 2(y - 2)^2$ **26.** $x = -3(y - 1)^2 - 2$

27. $y = x^2 - 4x - 1$ **28.** $y = x^2 + 8x + 6$

29. $y = x^2 + 2x$ **30.** $x = y^2 - 3y$

In Exercises 31–42, find the equation of the parabola satisfying the given conditions.

31. Vertex (0, 0); axis $x = 0$; (2, 12) on graph.

32. Vertex (0, 1); axis $x = 0$; (2, −7) on graph.

33. Vertex (1, 0); axis $x = 1$; (2, 13) on graph.

34. Vertex (−3, 0); axis $y = 0$; (−1, 1) on graph.

35. Vertex (2, 1); axis $y = 1$; (5, 0) on graph.

36. Vertex (1, −3); axis $y = -3$; (−1, −4) on graph.

37. Vertex $(-3, -2)$; focus $(-47/16, -2)$.

38. Vertex $(-5, -5)$; focus $(-5, -99/20)$.

39. Vertex $(1, 1)$; focus $(1, 9/8)$.

40. Vertex $(-4, -3)$; $(-6, -2)$ and $(-6, -4)$ on graph.

41. Vertex $(-1, 3)$; $(8, 0)$ and $(0, 4)$ on graph.

42. Vertex $(1, -3)$; $(0, -1)$ and $(-1, 5)$ on graph.

In Exercises 43–48, find parametric equations for the curve whose equation is given, and use them to find a complete graph of the curve.

43. $8x = 2y^2$

44. $4y = x^2$

45. $y = 4(x - 1)^2 + 2$

46. $y = 3(x - 2)^2 - 3$

47. $x = 2(y - 2)^2$

48. $x = -3(y - 1)^2 - 2$

In Exercises 49–56, identify the conic section whose equation is given, list its vertex or vertices, if any, and find its graph.

49. $x^2 = 6x - y - 5$

50. $y^2 = x - 2y - 2$

51. $3y^2 = x - 1 + 2y$

52. $2y^2 = x - 4y - 5$

53. $3x^2 + 3y^2 - 6x - 12y - 6 = 0$

54. $2x^2 + 3y^2 + 12x - 6y + 9 = 0$

55. $2x^2 - y^2 + 16x + 4y + 24 = 0$

56. $4x^2 - 40x - 2y + 105 = 0$

In Exercises 57–62, determine which of the following equations could possibly have the given graph.

$$y = (x + 5)^2 - 3, \qquad x = (y + 3)^2 + 2,$$

$$y = (x - 4)^2 + 2, \qquad x = -(y - 3)^2 - 2.$$

$$y^2 = 4y + x - 1, \qquad y = -x^2 - 8x - 18$$

57.

58.

59.

60.

61.

62.

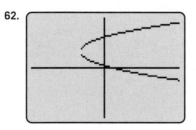

63. Find the number b such that the vertex of the parabola $y = x^2 + bx + c$ lies on the y-axis.

64. Find the number d such that the parabola $(y + 1)^2 = dx + 4$ passes through $(-6, 3)$.

65. Find the points of intersection of the parabola $4y^2 + 4y = 5x - 12$ and the line $x = 9$.

66. Find the points of intersection of the parabola $4x^2 - 8x = 2y + 5$ and the line $y = 15$.

67. A parabolic satellite dish is 4 feet in diameter and 1.5 feet deep. How far from the vertex should the receiver be placed to catch all the signals that hit the dish?

68. A radio telescope has a parabolic dish with a diameter of 300 feet. Its receiver (focus) is located 130 feet from the vertex. How deep is the dish at its center? [*Hint:* Position the dish as in Figure 10–34, and find the equation of the parabola.]

69. A flashlight has a parabolic reflector that is 3 inches in diameter and 1.5 inches deep. For the light from the bulb to reflect in beams that are parallel to the center axis of the flashlight, how far from the vertex of the reflector should the bulb be located? [*Hint:* See Figure 10–31 and the preceding discussion.]

70. A large spotlight has a parabolic reflector that is 3 feet deep at its center. The light source is located $1\frac{1}{3}$ feet from the vertex. What is the diameter of the reflector?

71. The cables of a suspension bridge are shaped like parabolas. The cables are attached to the towers 100 feet from the bridge surface, and the towers are 420 feet apart. The cables touch the bridge surface at the center (midway between the towers). At a point on the bridge 100 feet from one of the towers, how far is the cable from the bridge surface?

72. At a point 120 feet from the center of a suspension bridge, the cables are 24 feet above the bridge surface. Assume that the cables are shaped like parabolas and touch the bridge surface at the center (which is midway between the towers). If the towers are 600 feet apart, how far above the surface of the bridge are the cables attached to the towers?

10.4   Rotations and Second-Degree Equations

A **second-degree equation** in x and y is one that can be written in the form

$$Ax^2 + Bxy + Cy^2 + Dx + Ey + F = 0$$

for some constants A, B, C, D, E, F, with at least one of A, B, C nonzero.

EXAMPLE 1

Show that each of the following conic sections is the graph of a second-degree equation.

(a) Ellipse: $\dfrac{x^2}{6} + \dfrac{y^2}{5} = 1$

(b) Hyperbola: $\dfrac{(x + 1)^2}{4} - \dfrac{(y - 3)^2}{6} = 1$

SOLUTION   We need only show that each of these equations is in fact a second-degree equation. In each case, eliminate denominators, multiply out all terms, and gather them on one side of the equal sign.

(a)

$$\frac{x^2}{6} + \frac{y^2}{5} = 1$$

Multiply both sides by 30: $5x^2 + 6y^2 = 30$

Rearrange terms: $5x^2 + 6y^2 - 30 = 0.$

This equation is a second-degree equation because it has the form

$$Ax^2 + Bxy + Cy^2 + Dx + Ey + F = 0$$

with $A = 5$, $B = 0$, $C = 6$, $D = 0$, $E = 0$, and $F = -30$.

(b)

$$\frac{(x+1)^2}{4} - \frac{(y-3)^2}{6} = 1$$

Multiply both sides by 12: $3(x+1)^2 - 2(y-3)^2 = 12$

Multiply out left side: $3(x^2 + 2x + 1) - 2(y^2 - 6y + 9) = 12$

$$3x^2 + 6x + 3 - 2y^2 + 12y - 18 = 12$$

Rearrange terms: $3x^2 - 2y^2 + 6x + 12y - 27 = 0.$

This is a second-degree equation with $A = 3$, $B = 0$, $C = -2$, $D = 6$, $E = -12$, and $F = -27$.   ■

Calculations like those in Example 1 can be used on the equation of any conic section to show that it is the graph of a second-degree equation. Conversely, it can be shown that

The graph of every second-degree equation is a conic section

(possibly degenerate). When the second-degree equation has no xy-term (that is, $B = 0$), as was the case in Example 1, the graph is a conic section in standard position (axis or axes parallel to the coordinate axes). When $B \neq 0$, however, the conic is rotated from standard position, so its axis or axes are not parallel to the coordinate axes.

EXAMPLE 2

Graph the equation

$$3x^2 + 6xy + y^2 + x - 2y + 7 = 0.$$

SOLUTION   We first rewrite it as

$$y^2 + 6xy - 2y + 3x^2 + x + 7 = 0$$

$$y^2 + (6x - 2)y + (3x^2 + x + 7) = 0.$$

This equation has the form $ay^2 + by + c = 0$, with $a = 1$, $b = 6x - 2$, and $c = 3x^2 + x + 7$. It can be solved by using the quadratic formula.

$$y = \frac{-b \pm \sqrt{b^2 - 4ac}}{2a} = \frac{-(6x - 2) \pm \sqrt{(6x - 2)^2 - 4 \cdot 1 \cdot (3x^2 + x + 7)}}{2 \cdot 1}.$$

The top half of the graph is obtained by graphing

$$y = \frac{-6x + 2 + \sqrt{(6x - 2)^2 - 4(3x^2 + x + 7)}}{2},$$

and the bottom half is obtained by graphing

$$y = \frac{-6x + 2 - \sqrt{(6x - 2)^2 - 4(3x^2 + x + 7)}}{2}.$$

The graph is a hyperbola whose focal axis tilts upward to the left, as shown in Figure 10–35. ■

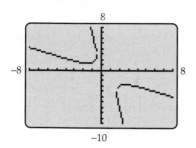

Figure 10–35

▪▪ THE DISCRIMINANT

The following fact, whose proof is outlined in Exercise 15 of Special Topics 10.4.A, makes it easy to identify the graphs of second-degree equations without graphing them.

Graphs of Second-Degree Equations

If the equation

$$Ax^2 + Bxy + Cy^2 + Dx + Ey + F = 0 \quad (A,\ B,\ C \text{ not all } 0)$$

has a graph, then that graph is

A circle or an ellipse (or a point), if $B^2 - 4AC < 0$;

A parabola (or a line or two parallel lines), if $B^2 - 4AC = 0$;

A hyperbola (or two intersecting lines), if $B^2 - 4AC > 0$.

The expression $B^2 - 4AC$ is called the **discriminant** of the equation.

EXAMPLE 3

Identify the graph of

$$2x^2 - 4xy + 3y^2 + 5x + 6y - 8 = 0$$

and sketch the graph.

SOLUTION   We compute the discriminant with $A = 2$, $B = -4$, and $C = 3$.

$$B^2 - 4AC = (-4)^2 - 4 \cdot 2 \cdot 3 = 16 - 24 = -8.$$

Hence, the graph is an ellipse (possibly a circle or a single point). To find the graph, we rewrite the equation as

$$3y^2 - 4xy + 6y + 2x^2 + 5x - 8 = 0$$

$$3y^2 + (-4x + 6)y + (2x^2 + 5x - 8) = 0.$$

The equation has the form $ay^2 + by + c = 0$ and can be solved by the quadratic formula.

$$y = \frac{-b \pm \sqrt{b^2 - 4ac}}{2a}$$

$$= \frac{-(-4x + 6) \pm \sqrt{(-4x + 6)^2 - 4 \cdot 3 \cdot (2x^2 + 5x - 8)}}{2 \cdot 3}.$$

The graph can now be found by graphing the last two equations on the same screen.

GRAPHING EXPLORATION

Find a viewing window that shows a complete graph of the equation. In what direction does the major axis run?   ■

EXAMPLE 4

Does Figure 10–36 show a complete graph of

$$3x^2 + 5xy + 2y^2 - 8y - 1 = 0?$$

SOLUTION   Although the graph in the figure looks like a parabola, appearances are deceiving. The discriminant of the equation is

$$B^2 - 4AC = 5^2 - 4 \cdot 3 \cdot 2 = 1,$$

which means that the graph is a hyperbola. So Figure 10–36 cannot possibly be a complete graph of the equation.*

GRAPHING EXPLORATION

Solve the equation for y, as in Example 3. Then find a viewing window that clearly shows both branches of the hyperbola.   ■

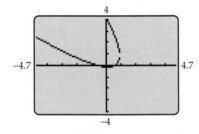

Figure 10–36

EXAMPLE 5

Sketch the graph of

$$3x^2 + 6xy + 3y^2 + 13x + 9y + 53 = 0.$$

SOLUTION   The discriminant is $B^2 - 4AC = 6^2 - 4 \cdot 3 \cdot 3 = 0$. Hence, the graph is a parabola (or a line or parallel lines in the degenerate case).

GRAPHING EXPLORATION

Find a viewing window that shows a complete graph of the equation.   ■

---

*When I asked students to graph this equation on an examination, many of them did not bother to compute the discriminant and produced something similar to Figure 10–36. They didn't get much credit for this answer. *Moral:* Use the discriminant to identify the conic so that you know the shape of the graph you are looking for.

EXERCISES 10.4

In Exercises 1–6, assume that the graph of the equation is a nondegenerate conic section. Without graphing, determine whether the graph an ellipse, hyperbola, or parabola.

1. $x^2 - 2xy + 3y^2 - 1 = 0$

2. $xy - 1 = 0$

3. $x^2 + 2xy + y^2 + 2\sqrt{2}x - 2\sqrt{2}y = 0$

4. $2x^2 - 4xy + 5y^2 - 6 = 0$

5. $17x^2 - 48xy + 31y^2 + 50 = 0$

6. $2x^2 - 4xy - 2y^2 + 3x + 5y - 10 = 0$

In Exercises 7–24, use the discriminant to identify the conic section whose equation is given, and find a viewing window that shows a complete graph.

7. $9x^2 + 4y^2 + 54x - 8y + 49 = 0$

8. $4x^2 + 5y^2 - 8x + 30y + 29 = 0$

9. $4y^2 - x^2 + 6x - 24y + 11 = 0$

10. $x^2 - 16y^2 = 0$

11. $3y^2 - x - 2y + 1 = 0$

12. $x^2 - 6x + y + 5 = 0$

13. $41x^2 - 24xy + 34y^2 - 25 = 0$

14. $x^2 + 2\sqrt{3}xy + 3y^2 + 8\sqrt{3}x - 8y + 32 = 0$

15. $17x^2 - 48xy + 31y^2 + 49 = 0$

16. $52x^2 - 72xy + 73y^2 = 200$

17. $9x^2 + 24xy + 16y^2 + 90x - 130y = 0$

18. $x^2 + 10xy + y^2 + 1 = 0$

19. $23x^2 + 26\sqrt{3}xy - 3y^2 - 16x + 16\sqrt{3}y + 128 = 0$

20. $x^2 + 2xy + y^2 + 12\sqrt{2}x - 12\sqrt{2}y = 0$

21. $17x^2 - 12xy + 8y^2 - 80 = 0$

22. $11x^2 - 24xy + 4y^2 + 30x + 40y - 45 = 0$

23. $3x^2 + 2\sqrt{3}xy + y^2 + 4x - 4\sqrt{3}y - 16 = 0$

24. $3x^2 + 2\sqrt{2}xy + 2y^2 - 12 = 0$

10.4.A *SPECIAL TOPICS* Rotation of Axes

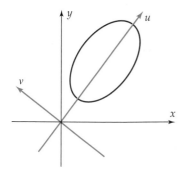

Figure 10–37

We have seen that the graph of the second-degree equation

$$Ax^2 + Bxy + Cy^2 + Dx + Ey + F = 0 \quad (B \neq 0)$$

is a conic section whose axes are not parallel to the coordinate axes, as in Figure 10–37. Although we can graph the equation on a calculator (as in Section 10.4), we cannot read off useful information about the center, vertices, etc. from the equation, as we can from an equation in standard form.

The key is to replace the xy coordinate system by a new coordinate system, as indicated by the blue u- and v-axes in Figure 10–37. The conic is not rotated in the new coordinate system, so it has an equation (in u and v) in standard form that will provide the desired information.

To use this approach, we must first determine the relationship between the xy coordinates of a point and its coordinates in the uv system. Suppose the uv coordinate system is obtained by rotating the xy axes about the origin, counterclockwise through an angle θ.[*] If a point P has coordinates (x, y) in the xy system, we

---

[*] All rotations in this section are counterclockwise about the origin, with $0° < \theta < 90°$.

can find its coordinates (u, v) in the rotated coordinate system by using Figure 10–38.

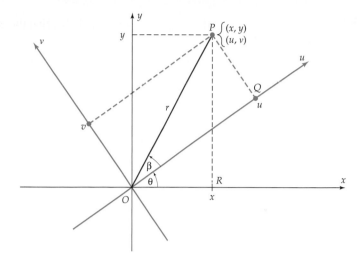

Figure 10–38

Triangle OPQ shows that

$$\cos \beta = \frac{OQ}{OP} = \frac{u}{r} \qquad \text{and} \qquad \sin \beta = \frac{PQ}{OP} = \frac{v}{r}.$$

Therefore,

$$u = r \cos \beta \qquad \text{and} \qquad v = r \sin \beta.$$

Similarly, triangle OPR shows that

$$\cos(\theta + \beta) = \frac{OR}{OP} = \frac{x}{r} \qquad \text{and} \qquad \sin(\theta + \beta) = \frac{PR}{OP} = \frac{y}{r},$$

so

$$x = r \cos(\theta + \beta) \qquad \text{and} \qquad y = r \sin(\theta + \beta).$$

Applying the addition identity for cosine (page 499) shows that

$$x = r \cos(\theta + \beta)$$
$$= r(\cos \theta \cos \beta - \sin \theta \sin \beta)$$
$$= (r \cos \beta) \cos \theta - (r \sin \beta) \sin \theta$$
$$= u \cos \theta - v \sin \theta.$$

A similar argument with $y = r \sin(\theta + \beta)$ and the addition identity for sine leads to this result.

The Rotation Equations

If the xy coordinate axes are rotated through an angle θ to produce the uv coordinate axes, then the coordinates (x, y) and (u, v) of a point are related by these equations:

$$x = u \cos \theta - v \sin \theta,$$

$$y = u \sin \theta + v \cos \theta.$$

EXAMPLE 1

If the *xy* axes are rotated 30°, find the equation relative to the *uv* axes of the graph of

$$3x^2 + 2\sqrt{3}xy + y^2 + x - \sqrt{3}y = 0,$$

and graph the equation.

SOLUTION   Since $\sin 30° = 1/2$ and $\cos 30° = \sqrt{3}/2$, the rotation equations are

$$x = u \cos 30° - v \sin 30° = \frac{\sqrt{3}}{2}u - \frac{1}{2}v,$$

$$y = u \sin 30° + v \cos 30° = \frac{1}{2}u + \frac{\sqrt{3}}{2}v.$$

Substitute these expressions in the original equation.

$$3x^2 + 2\sqrt{3}xy + y^2 + x - \sqrt{3}y = 0$$

$$3\left(\frac{\sqrt{3}}{2}u - \frac{1}{2}v\right)^2 + 2\sqrt{3}\left(\frac{\sqrt{3}}{2}u - \frac{1}{2}v\right)\left(\frac{1}{2}u + \frac{\sqrt{3}}{2}v\right)$$

$$+ \left(\frac{1}{2}u + \frac{\sqrt{3}}{2}v\right)^2 + \left(\frac{\sqrt{3}}{2}u - \frac{1}{2}v\right) - \sqrt{3}\left(\frac{1}{2}u + \frac{\sqrt{3}}{2}v\right) = 0$$

$$3\left(\frac{3}{4}u^2 - \frac{\sqrt{3}}{2}uv + \frac{1}{4}v^2\right) + 2\sqrt{3}\left(\frac{\sqrt{3}}{4}u^2 + \frac{1}{2}uv - \frac{\sqrt{3}}{4}v^2\right)$$

$$+ \left(\frac{1}{4}u^2 + \frac{\sqrt{3}}{2}uv + \frac{3}{4}v^2\right) + \left(\frac{\sqrt{3}}{2}u - \frac{1}{2}v\right) - \sqrt{3}\left(\frac{1}{2}u + \frac{\sqrt{3}}{2}v\right) = 0.$$

Verify that the last equation simplifies as

$$4u^2 - 2v = 0 \qquad \text{or, equivalently,} \qquad v = 2u^2.$$

In the *uv* system, $v = 2u^2$ is the equation of an upward-opening parabola with vertex at (0, 0) as shown in Figure 10–39.   ■

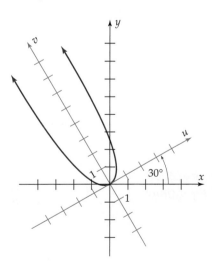

Figure 10–39

Rotating the axes in the preceding example changed the original equation, which included an *xy* term, to an equation that had no *uv* term. That enabled us to identify the graph of the new equation as a conic section. This can be done for any second-degree equation by choosing an angle of rotation that will eliminate the *xy* term. The choice of the angle is determined by this fact, which is proved in Exercise 13.

Rotation Angle

> The equation $Ax^2 + Bxy + Cy^2 + Dx + Ey + F = 0$ $(B \neq 0)$ can be rewritten as $A'u^2 + C'v^2 + D'u + E'v + F' = 0$ by rotating the *xy* axes through an angle θ such that
>
> $$\cot 2\theta = \frac{A - C}{B} \quad (0° < \theta < 90°)$$
>
> and using the rotation equations.

EXAMPLE 2

What angle of rotation will eliminate the *xy* term in the equation

$$153x^2 + 192xy + 97y^2 - 1710x - 1470y + 5625 = 0,$$

and what are the rotation equations in this case?

SOLUTION   According to the fact in the box with $A = 153$, $B = 192$, and $C = 97$, we should rotate through an angle of θ, where

$$\cot 2\theta = \frac{153 - 97}{192} = \frac{56}{192} = \frac{7}{24}.$$

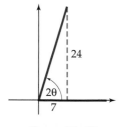

Figure 10–40

Since $0° < 2\theta < 180°$ and $\cot 2\theta$ is positive, the terminal side of the angle 2θ lies in the first quadrant, as shown in Figure 10–40. The hypotenuse of this triangle has length $\sqrt{7^2 + 24^2} = \sqrt{625} = 25$. Hence, $\cos 2\theta = 7/25$. The half-angle identities (page 513) show that

$$\sin \theta = \sqrt{\frac{1 - \cos 2\theta}{2}} = \sqrt{\frac{1 - 7/25}{2}} = \sqrt{\frac{9}{25}} = \frac{3}{5},$$

$$\cos \theta = \sqrt{\frac{1 + \cos 2\theta}{2}} = \sqrt{\frac{1 + 7/25}{2}} = \sqrt{\frac{16}{25}} = \frac{4}{5}.$$

Using $\sin \theta = 3/5$ and the SIN^{-1} key on a calculator, we find that the angle θ of rotation is approximately $36.87°$. The rotation equations are

$$x = u \cos \theta - v \sin \theta = \frac{4}{5}u - \frac{3}{5}v,$$

$$y = u \sin \theta + v \cos \theta = \frac{3}{5}u + \frac{4}{5}v. \quad \blacksquare$$

EXAMPLE 3

Graph the equation without using a calculator.

$$153x^2 + 192xy + 97y^2 - 1710x - 1470y + 5625 = 0.$$

SOLUTION   The angle θ and the rotation equations for eliminating the xy term were found in the preceding example. Substitute the rotation equations in the given equation.

$$153x^2 + 192xy + 97y^2 - 1710x - 1470y + 5625 = 0$$

$$153\left(\frac{4}{5}u - \frac{3}{5}v\right)^2 + 192\left(\frac{4}{5}u - \frac{3}{5}v\right)\left(\frac{3}{5}u + \frac{4}{5}v\right)$$

$$+ 97\left(\frac{3}{5}u + \frac{4}{5}v\right)^2 - 1710\left(\frac{4}{5}u - \frac{3}{5}v\right) - 1470\left(\frac{3}{5}u + \frac{4}{5}v\right) + 5625 = 0$$

$$153\left(\frac{16}{25}u^2 - \frac{24}{25}uv + \frac{9}{25}v^2\right) + 192\left(\frac{12}{25}u^2 + \frac{7}{25}uv - \frac{12}{25}v^2\right)$$

$$+ 97\left(\frac{9}{25}u^2 + \frac{24}{25}uv + \frac{16}{25}v^2\right) - 2250u - 150v + 5625 = 0$$

$$225u^2 + 25v^2 - 2250u - 150v + 5625 = 0$$

$$9u^2 + v^2 - 90u - 6v + 225 = 0$$

$$9(u^2 - 10u) + (v^2 - 6v) = -225.$$

Finally, complete the square in u and v (adding the appropriate amounts to the right side so as not to change the equation).

$$9(u^2 - 10u + 25) + (v^2 - 6v + 9) = -225 + 9 \cdot 25 + 9$$

$$9(u - 5)^2 + (v - 3)^2 = 9$$

$$\frac{(u - 5)^2}{1} + \frac{(v - 3)^2}{9} = 1.$$

Therefore the graph is an ellipse centered at $(5, 3)$ in the uv coordinate system, as shown in Figure 10–41.   ■

Figure 10–41

EXERCISES 10.4.A

In Exercises 1–4, find the new coordinates of the point when the coordinate axes are rotated through the given angle.

1. $(3, 2)$; $\theta = 45°$ **2.** $(-2, 4)$; $\theta = 60°$

3. $(1, 0)$; $\theta = 30°$ **4.** $(3, 3)$; $\sin \theta = 5/13$

In Exercises 5–8, rotate the axes through the given angle to form the uv coordinate system. Express the given equation in terms of the uv coordinate system.

5. $\theta = 45°$; $xy = 1$

6. $\theta = 45°$; $13x^2 + 10xy + 13y^2 = 72$

7. $\theta = 30°$; $7x^2 - 6\sqrt{3}xy + 13y^2 - 16 = 0$

8. $\sin \theta = 1/\sqrt{5}$; $x^2 - 4xy + 4y^2 + 5\sqrt{5}y + 1 = 0$

In Exercises 9–12, find the angle of rotation that will eliminate the xy term of the equation and list the rotation equations in this case.

9. $41x^2 - 24xy + 34y^2 - 25 = 0$

10. $x^2 + 2\sqrt{3}xy + 3y^2 + 8\sqrt{3}x - 8y + 32 = 0$

11. $17x^2 - 48xy + 31y^2 + 49 = 0$

12. $52x^2 - 72xy + 73y^2 = 200$

Thinkers

13. (a) Given an equation
$Ax^2 + Bxy + Cy^2 + Dx + Ey + F = 0$, with $B \neq 0$. and an angle θ, use the rotation equations in the box on page 677 to rewrite the equation in the form

$$A'u^2 + B'uv + C'v^2 + D'u + E'v + F' = 0,$$

where $A', \ldots, F'$ are expressions involving $\sin \theta$, $\cos \theta$, and the constants $A, \ldots, F$.

(b) Verify that
$$B' = 2(C - A) \sin \theta \cos \theta + B(\cos^2\theta - \sin^2\theta).$$

(c) Use the double-angle identities to show that
$$B' = (C - A) \sin 2\theta + B \cos 2\theta.$$

(d) If θ is chosen so that $\cot 2\theta = (A - C)/B$, show that $B' = 0$. This proves the statement in the box on page 679.

14. Assume that the graph of
$A'u^2 + C'v^2 + D'u + E'v + F' = 0$ (with at least one of A', C' nonzero) in the uv coordinate system is a nondegenerate conic. Show that its graph is an ellipse if $A'C' > 0$ (A' and C' have the same sign), a hyperbola if $A'C' < 0$ (A' and C' have opposite signs), or a parabola if $A'C' = 0$.

15. Assume the graph of
$Ax^2 + Bxy + Cy^2 + Dx + Ey + F = 0$ is a nondegenerate conic section. Prove the statement in the box on page 674 as follows.

(a) In Exercise 13(a), show that
$(B')^2 - 4A'C' = B^2 - 4AC.$

(b) Assume that θ has been chosen so that $B' = 0$. Use Exercise 14 to show that the graph of the original equation is an ellipse if $B^2 - 4AC < 0$, a parabola if $B^2 - 4AC = 0$, and a hyperbola if $B^2 - 4AC > 0$.

10.5 Plane Curves and Parametric Equations

There are many curves in the plane that cannot be represented as the graph of a function $y = f(x)$. Parametric graphing enables us to represent such curves in terms of functions and also provides a formal definition of a curve in the plane.

Consider, for example, an object moving in the plane during a particular time interval. To describe both the path of the object and its location at any particular time, three variables are needed: the time t and the coordinates (x, y) of the object at time t. For instance, the coordinates x and y might be given by

$$x = 4 \cos t + 5 \cos(3t) \qquad \text{and} \qquad y = \sin(3t) + t.$$

From $t = 0$ to $t = 12.5$, the object traces out the curve shown in Figure 10–42. The points marked on the graph show the location of the object at various times. Note that the object may be at the same location at different times (the points where the graph crosses itself).

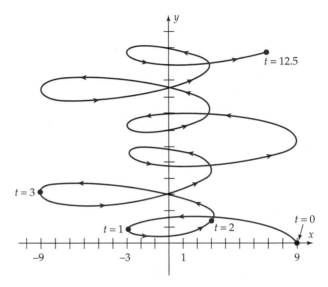

Figure 10–42

In the preceding example, both x and y were determined by continuous functions of t, with t taking values in the interval $[0, 12.5]$.[*] The example suggests the following definition.

Definition of Plane Curve

Let f and g be continuous functions of t on an interval I. The set of all points (x, y) where

$$x = f(t) \qquad \text{and} \qquad y = g(t)$$

is called a **plane curve**. The variable t is called a **parameter,** and the equations defining x and y are **parametric equations.**

In this general definition of "curve," the variable t need not represent time. As the following examples illustrate, different pairs of parametric equations may produce the same curve. Each such pair of parametric equations is called a **parameterization** of the curve.

A curve given by parametric equations can be graphed by hand [choose values of t, plot the corresponding points (x, y), and make an educated guess about the shape of the curve]. A calculator or computer in parametric mode is more likely to produce an accurate graph, however.

---

[*]Intuitively, "continuous" means that the graph of the function that determines x, namely, $f(t) = 4 \cos t + 5 \cos 3t$, is a connected curve with no gaps or holes and similarly for the function that determines y. Continuous functions are defined more precisely in Chapter 13.

The basics of parametric graphing were introduced in Section 3.3. We saw that when you set the viewing window in parametric graphing, you must specify both

> the range of values for *t*   *and*   the *t*-step (or *t*-pitch).

Failure to make appropriate choices for these may result in an inaccurate graph, as the next example illustrates.

EXAMPLE 1

In the window with $-10 \leq x \leq 10$ and $0 \leq y \leq 15$, graph the curve given by

$$x = 4 \cos t + 5 \cos(3t) \qquad \text{and} \qquad y = \sin(3t) + t$$

(a) with $0 \leq t \leq 12.5$ and each of these *t*-steps: 2, 1, and .5;

(b) with *t*-step = .1 and each of these ranges: $2.4 \leq t \leq 3.9$, $4.6 \leq t \leq 8$, and $0 \leq t \leq 12.5$.

How do these graphs compare with Figure 10–42?

SOLUTION

(a) The graphs are shown in Figure 10–43. Only the last one looks even remotely like Figure 10–42.

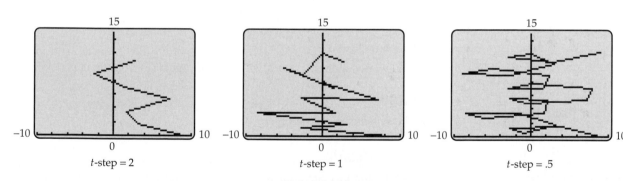

t-step = 2   *t*-step = 1   *t*-step = .5

Figure 10–43

(b) The graphs are shown in Figure 10–44. The first two show only a small portion of Figure 10–42, but the last one is a complete graph that closely resembles Figure 10–42. ∎

$2.4 \leq t \leq 3.9$ $4.6 \leq t \leq 8$ $0 \leq t \leq 12.5$

Figure 10–44

As Example 1 suggests, a *t*-step between .05 and .15 usually produces a reasonably smooth graph. Larger *t*-steps may produce jagged or totally inaccurate graphs. Smaller *t*-steps may involve very long graphing times and usually don't increase smoothness very much.

EXAMPLE 2

16

−24   24

−16

Figure 10–45

Graph the curve given by

$$x = t + 5 \cos t \qquad \text{and} \qquad y = 1 - 3 \sin t$$

in the viewing window with

$$-24 \le x \le 24, \qquad -16 \le y \le 16, \qquad -20 \le t \le 20.$$

SOLUTION   See Figure 10–45. Reproduce this graph yourself to see how the curve spirals along the *x*-axis.   ∎

EXAMPLE 3

14

−10   10

−6

Figure 10–46

Consider the curve given by

$$x = -2t \qquad \text{and} \qquad y = 4t^2 - 4 \quad (-1 \le t \le 2).$$

(a) Graph the curve.

(b) Find an equation in *x* and *y* whose graph includes the graph in part (a).

SOLUTION

(a) The graph is shown in Figure 10–46.

GRAPHING EXPLORATION

Graph this curve on your calculator, using the same viewing window and range of *t* values as in Figure 10–46. In what direction is the curve traced out (that is, what is the first point graphed (corresponding to $t = -1$) and what is the last point graphed (corresponding to $t = 2$))?

(b) We shall solve one of the parametric equations for *t* and substitute the result in the other one. Solving $x = -2t$ for *t*, we have $t = -x/2$. Substituting this in the equation for *y*, we have

$$y = 4t^2 - 4 = 4\left(-\frac{x}{2}\right)^2 - 4 = 4\left(\frac{x^2}{4}\right) - 4 = x^2 - 4.$$

Therefore, every point on the curve is also on the graph of $y = x^2 - 4$. From Section 3.4 or 4.1 we know that the graph of $y = x^2 - 4$ is the parabola in Figure 10–47. However, the curve given by the parametric equations is *not* the entire parabola, but ony the part shown in red, which joins the points $(2, 0)$ and $(-4, 12)$ that correspond to the minimum and maximum values of *t*, namely, $t = -1$ and $t = 2$.   ∎

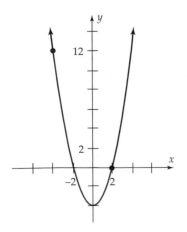

Figure 10–47

Expressing a parametric curve as (part of) the graph of an equation in x and y is called **eliminating the parameter.** As illustrated in Example 3,

> *To eliminate the parameter, solve one of the parametric equations for t and substitute this result in the other parametric equation.*

██ FINDING PARAMETRIC EQUATIONS FOR CURVES

Having seen how to graph a curve given by parametric equations, we now consider the reverse problem: finding a parametric respresentation for the graph of an equation in x and y. This is easy when the equation defines y as a function of x, such as

$$y = x^3 + 5x^2 - 3x + 4.$$

A parametric description of this function can be obtained by changing the variable:

$$x = t \qquad \text{and} \qquad y = t^3 + 5t^2 - 3t + 4.$$

The same thing is true for equations that define x as a function of y, such as $x = y^2 + 3$, which can be parameterized by letting

$$x = t^2 + 3 \qquad \text{and} \qquad y = t.$$

In other cases, different techniques may be needed.

EXAMPLE 4

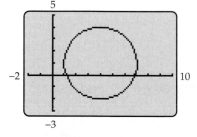

Figure 10–48

In Section 10.1 we saw that a parameterization of the circle

$$(x - 4)^2 + (y - 1)^2 = 9$$

is given by

(∗) $x = 3 \cos t + 4 \qquad$ and $\qquad y = 3 \sin t + 1 \quad (0 \le t \le 2\pi).$

With this parameterization, the circle is traced out in a counterclockwise direction from the point (7, 1), as shown in Figure 10–48. Another parameterization is given by

$$x = 3 \cos 2t + 4 \qquad \text{and} \qquad y = -3 \sin 2t + 1 \quad (0 \le t \le \pi).$$

GRAPHING EXPLORATION

Verify that this parameterization traces out the circle in a clockwise direction, twice as fast as the parameterization given by (∗), since t runs from 0 to π rather than 2π. ■

EXAMPLE 5

Find three parameterizations of the straight line through (1, −3) with slope −2.

SOLUTION   The point-slope form of the equation of this line is

$$y - (-3) = -2(x - 1) \qquad \text{or, equivalently,} \qquad y = -2x - 1.$$

Figure 10–49

Its graph is shown in Figure 10–49. Since this equation defines y as a function of x, one parameterization is

$$x = t \qquad \text{and} \qquad y = -2t - 1 \quad (t \text{ any real number}).$$

A second parameterization is given by letting $x = t + 1$; then

$$y = -2x - 1 = -2(t + 1) - 1 = -2t - 3 \quad (t \text{ any real number}).$$

A third parameterization can be obtained by letting

$$x = \tan t \qquad \text{and} \qquad y = -2x - 1 = -2\tan t - 1 \quad (-\pi/2 < t < \pi/2).$$

When t runs from $-\pi/2$ to $\pi/2$, then $x = \tan t$ takes all possible real number values, and hence so does y.   ■

CAUTION

Some substitutions in an equation $y = f(x)$ do *not* lead to a parameterization of the entire graph. For instance, in Example 5, letting $x = t^2$ and substituting in the equation $y = -2x - 1$ lead to

$$x = t^2 \qquad \text{and} \qquad y = -2t^2 - 1 \quad (\text{any real number } t).$$

Thus, x is always nonnegative, and y is always negative. So the parameterization produces only the half of the line $y = -2x - 1$ to the right of the y-axis in Figure 10–49.

▆▆ APPLICATIONS

In the following applications, we ignore air resistance and assume some facts about gravity that are proved in physics.

EXAMPLE 6

A golfer hits the ball with an initial velocity of 140 feet per second so that its path as it leaves the ground makes an angle of 31° with the horizontal.

(a) When does the ball hit the ground?

(b) How far from its starting point does it land?

(c) What is the maximum height of the ball during its flight?

SOLUTION   Imagine that the golf ball starts at the origin and travels in the direction of the positive x-axis. If there were no gravity, the distance traveled by the ball in t seconds would be $140t$ feet. As shown in Figure 10–50, the coordinates (x, y) of the ball would satisfy

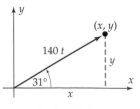

Figure 10–50

$$\frac{x}{140t} = \cos 31° \qquad\qquad \frac{y}{140t} = \sin 31°,$$

$$x = (140 \cos 31°)t \qquad\qquad y = (140 \sin 31°)t.$$

However, there *is* gravity, and at time t, it exerts a force of $16t^2$ downward (that is, in the negative direction on the y-coordinate). Consequently, the coordinates of the golf ball at time t are

$$x = (140 \cos 31°)t \qquad \text{and} \qquad y = (140 \sin 31°)t - 16t^2.$$

The path given by these parametric equations is graphed in Figure 10–51.[*]

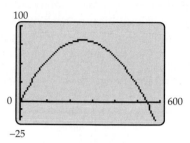

Figure 10–51

(a) The ball is on the ground when $y = 0$, that is, at the x-intercepts of the graph. They can be found geometrically by using trace and zoom-in (the graphical root finder does not operate in parametric mode), but this is very time-consuming. To find the intercepts algebraically we need only set $y = 0$ and solve for t.

$$(140 \sin 31°)t - 16t^2 = 0$$

$$t(140 \sin 31° - 16t) = 0$$

$$t = 0 \qquad \text{or} \qquad 140 \sin 31° - 16t = 0$$

$$t = \frac{140 \sin 31°}{16} \approx 4.5066.$$

Thus, the ball hits the ground after approximately 4.5066 seconds.

(b) The horizontal distance traveled by the ball is given by the x-coordinate of the intercept. The x-coordinate when $t \approx 4.5066$ is

$$x = (140 \cos 31°)(4.5066) \approx 540.81 \text{ feet.}$$

(c) The graph in Figure 10–51 looks like a parabola, and it is, as you can verify by eliminating the parameter t (Exercise 40). The y-coordinate of the vertex is the maximum height of the ball. It can be found geometrically by using trace and zoom-in (the maximum finder doesn't work in parametric mode) or algebraically as follows. The vertex occurs halfway between its two x-intercepts ($x = 0$ and $x \approx 540.81$), that is, when $x \approx 270.405$. Hence,

$$(140 \cos 31°)t = x = 270.405,$$

so

$$t = \frac{270.45}{140 \cos 31°} \approx 2.2533.$$

Therefore, the y-coordinate of the vertex (the maximum height of the ball) is

$$y = (140 \sin 31°)(2.2533) - 16(2.2533)^2 \approx 81.237 \text{ feet.} \quad ■$$

The argument used in Example 6 also applies when the initial position of the golf ball is k feet above the ground (for instance, if the golfer were on a platform at a driving range). In that case, the ball begins at $(0, k)$ instead of $(0, 0)$, and its position at time t is k feet higher than before, so its coordinates are

$$x = (140 \cos 31°)t \qquad \text{and} \qquad y = (140 \sin 31°)t - 16t^2 + k.$$

---

[*]Only the part of the graph on or above the x-axis represents the ball's path, since the ball does not go underground after it lands.

Then replacing 140 by v and 31° by θ in Example 6 leads to this conclusion.

Projectile Motion

> When a projectile is fired from the position $(0, k)$ on the positive y-axis at an angle θ with the horizontal, in the direction of the positive x-axis, with initial velocity v feet per second, with negligible air resistance, then its position at time t seconds is given by the parametric equations
>
> $$x = (v \cos \theta)t \qquad \text{and} \qquad y = (v \sin \theta)t - 16t^2 + k.$$

EXAMPLE 7

A batter hits a ball that is 3 feet above the ground and leaves the bat with an initial velocity of 138 feet per second, making an angle of 26° with the horizontal and heading toward a 25-foot-high fence that is 400 feet away. Will the ball go over the fence?

SOLUTION   According to the preceding box (with $v = 138$, $\theta = 26°$, and $k = 3$), the path of the ball is given by the parametric equations

$$x = (138 \cos 26°)t \qquad \text{and} \qquad y = (138 \sin 26°)t - 16t^2 + 3.$$

The graph of the ball's path in Figure 10–52 was made with the grid-on feature and vertical tic marks 25 units apart. It shows that the y-coordinate of the ball is greater than 25 when its x-coordinate is 400. So the ball goes over the fence.   ∎

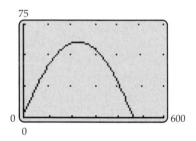

Figure 10–52

TECHNOLOGY TIP

In parametric graphing, zoom-in can be very time-consuming. It's often more effective to limit the t range to the values near the points you are interested in and set the t step very small. The picture may be hard to read, but trace can be used to determine coordinates.

GRAPHING EXPLORATION

Will the ball in Example 7 go over the fence if its initial velocity is 135 feet per second? Use degree mode and the viewing window of Figure 10–52 (with $0 \le t \le 4$ and t step = .1) to graph the ball's path. You might need to use trace if the graph is hard to read. If the answer still isn't clear, try changing the t step to .02.

Our final example is a curve that has several interesting applications.

EXAMPLE 8

Choose a point P on a circle of radius 3, and find a parametric description of the curve that is traced out by P as the circle rolls along the x-axis, as shown in Figure 10–53.

Figure 10–53

SOLUTION This curve is called a **cycloid.** Begin with P at the origin and the center C of the circle at $(0, 3)$. As the circle rolls along the x-axis, the line segment CP moves from vertical through an angle of t radians, as shown in Figure 10–54.

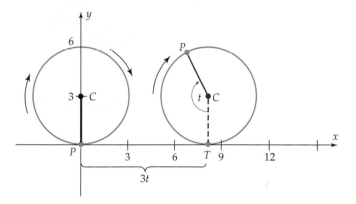

Figure 10–54

The distance from point T to the origin is the length of arc of the circle from T to P. As shown on page 419, this arc has length $3t$. Therefore the center C has coordinates $(3t, 3)$. When $0 < t < \pi/2$, the situation looks like Figure 10–55.

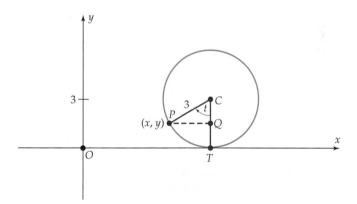

Figure 10–55

Right triangle PQC shows that

$$\sin t = \frac{PQ}{3} \qquad \text{or, equivalently,} \qquad PQ = 3 \sin t$$

and

$$\cos t = \frac{CQ}{3} \qquad \text{or, equivalently,} \qquad CQ = 3 \cos t.$$

In Figure 10–55, P has coordinates (x, y) and we have

$$x = OT - PQ = 3t - 3 \sin t = 3(t - \sin t),$$
$$y = CT - CQ = 3 - 3 \cos t = 3(1 - \cos t).$$

A similar analysis for other values of t (Exercises 49–51) shows that these equations are valid for every t. Therefore, the parametric equations of this cycloid are

$$x = 3(t - \sin t) \quad \text{and} \quad y = 3(1 - \cos t) \quad (t \text{ any real number}). \quad ■$$

If a cycloid is traced out by a circle of radius r, then the argument given in Example 8, with r in place of 3, shows that the parametric equations of the cycloid are

$$x = r(t - \sin t) \quad \text{and} \quad y = r(1 - \cos t) \quad (t \text{ any real number}).$$

Cycloids have a number of interesting applications. For example, among all the possible paths joining points P and Q in Figure 10–56, an arch of an inverted cycloid (shown in red) is the curve along which a particle (subject only to gravity) will slide from P to Q in the shortest possible time. This fact was first proved by J. Bernoulli in 1696.

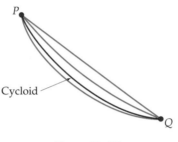

Figure 10–56   Figure 10–57

The Dutch physicist Christiaan Huygens (who invented the pendulum clock) proved that a particle takes the same time to slide to the bottom point Q of an inverted cycloid arch (as in Figure 10–57) from *any* point P on the curve.

✓ EXERCISES 10.5

In Exercises 1–14, find a viewing window that shows a complete graph of the curve.

1. $x = t^2 - 4, \quad y = t/2, \quad -2 \le t \le 3$

2. $x = 3t^2, \quad y = 2 + 5t, \quad 0 \le t \le 2$

3. $x = 2t, \quad y = t^2 - 1, \quad -1 \le t \le 2$

4. $x = t - 1, \quad y = \dfrac{t + 1}{t - 1}, \quad t \ge 1$

5. $x = 4 \sin 2t + 9, \quad y = 6 \cos t - 8, \quad 0 \le t \le 2\pi$

6. $x = t^3 - 3t - 8, \quad y = 3t^2 - 15, \quad -4 \le t \le 4$

7. $x = 6 \cos t + 12 \cos^2 t, \quad y = 8 \sin t + 8 \sin t \cos t, \\ \quad 0 \le t \le 2\pi$

8. $x = 12 \cos t, \quad y = 12 \sin 2t, \quad 0 \le t \le 2\pi$

9. $x = 6 \cos t + 5 \cos 3t, \quad y = 6 \sin t - 5 \sin 3t, \\ \quad 0 \le t \le 2\pi$

10. $x = 3t^2 + 10, \quad y = 4t^3, \quad \text{any real number } t$

11. $x = 12 \cos 3t \cos t + 6, \quad y = 12 \cos 3t \sin t - 7, \\ \quad 0 \le t \le 2\pi$

12. $x = 2 \cos 3t - 6, \quad y = 2 \cos 3t \sin t + 7, \\ \quad 0 \le t \le 2\pi$

13. $x = t \sin t, \quad y = t \cos t, \quad 0 \le t \le 8\pi$

14. $x = 9 \sin t, \quad y = 9t \cos t, \quad 0 \le t \le 20$

In Exercises 15–22, the given curve is part of the graph of an equation in x and y. Find the equation by eliminating the parameter.

15. $x = t - 3, \quad y = 2t + 1, \quad t \ge 0$

16. $x = t + 5, \quad y = \sqrt{t}, \quad t \ge 0$

17. $x = -2 + t^2, \quad y = 1 + 2t^2, \quad \text{any real number } t$

18. $x = t^2 + 1$, $\quad y = t^2 - 1$, $\quad$ any real number t

19. $x = e^t$, $\quad y = t$, $\quad$ any real number t

20. $x = 2e^t$, $\quad y = 1 - e^t$, $\quad t \geq 0$

21. $x = 3\cos t$, $\quad y = 3\sin t$, $\quad 0 \leq t \leq 2\pi$

22. $x = 4\sin 2t$, $\quad y = 2\cos 2t$, $\quad 0 \leq t \leq 2\pi$

In Exercises 23 and 24, sketch the graphs of the given curves and compare them. Do they differ and if so, how?

23. (a) $x = -4 + 6t$, $\quad y = 7 - 12t$, $\quad 0 \leq t \leq 1$
    (b) $x = 2 - 6t$, $\quad y = -5 + 12t$, $\quad 0 \leq t \leq 1$

24. (a) $x = t$, $\quad y = t^2$
    (b) $x = \sqrt{t}$, $\quad y = t$
    (c) $x = e^t$, $\quad y = e^{2t}$

In Exercises 25–28, find a parameterization of the given curve. Confirm your answer by graphing.

25. circle with center $(7, -4)$ and radius 6

26. circle with center $(9, 12)$ and radius 5

27. $x^2 + y^2 - 14x + 8y + 29 = 0$ $\quad$ [*Hint:* Example 11 in Section 1.3.]

28. $x^2 + y^2 - 4x - 6y + 9 = 0$

29. (a) What is the slope of the line through (a, b) and (c, d)?
    (b) Use the slope from part (a) and the point (a, b) to write the equation of the line. Do not simplify.
    (c) Show that the curve with parametric equations

$$x = a + (c - a)t \quad \text{and} \quad y = b + (d - b)t$$
$$(t \text{ any real number})$$

is the line through (a, b) and (c, d). [*Hint:* Solve both equations for t, and set the results equal to each other; compare with the equation in part (b).]

30. Find parametric equations whose graph is the line segment joining the points (a, b) and (c, d). [*Hint:* Adjust the range of t values in Exercise 29(c).]

In Exercises 31–33, use Exercise 30 to find a parameterization of the line segment joining the two points. Confirm your answer by graphing.

31. $(-6, 12)$ and $(12, -10)$

32. $(14, -5)$ and $(5, -14)$

33. $(18, 4)$ and $(-16, 14)$

34. (a) Find a parameterization of the line segment joining $(-5, -3)$ and $(7, 4)$, as in Exercises 30–33.
    (b) Explain why another parameterization of this line segment is given by

$$x = -5 + 12\sin t \quad \text{and}$$
$$y = -3 + 7\sin t \quad (0 \leq t \leq \pi/2).$$

(c) Use the trace feature to verify that the segment is traced out twice when the t-range in part (b) is changed to $0 \leq t \leq \pi$ (use t-step $= \pi/20$). Explain why.
(d) What happens when $0 \leq t \leq 2\pi$?

35. (a) Graph the curve given by

$$x = \sin kt \quad \text{and} \quad y = \cos t \quad (0 \leq t \leq 2\pi)$$

when $k - 1, 2, 3$, and 4. Use the window with

$$-1.5 \leq x \leq 1.5 \quad \text{and} \quad -1.5 \leq y \leq 1.5$$

and t-step $= \pi/30$.
    (b) Without graphing, predict the shape of the graph when $k = 5$ and $k = 6$. Then verify your predictions graphically.

36. (a) Graph the curve given by

$$x = 3\sin 2t \quad \text{and} \quad y = 2\cos kt \quad (0 \leq t \leq 2\pi)$$

when $k = 1, 2, 3, 4$. Use the window with $-3.5 \leq x \leq 3.5$ and $-2.5 \leq y \leq 2.5$ and t-step $= \pi/30$.
    (b) Predict the shape of the graph when $k = 5, 6, 7, 8$. Verify your predictions graphically.

In Exercises 37–39, locate all local maxima and minima (other than endpoints) of the curve.

37. $x = 4t - 6$, $\quad y = 3t^2 + 2$, $\quad -10 \leq t \leq 10$

38. $x = t^3 + \sin t + 4$, $\quad y = \cos t$, $\quad -1.5 \leq t \leq 2$

39. $x = 4t^3 - t + 4$, $\quad y = -3t^2 + 5$, $\quad -2 \leq t \leq 2$

40. Show that the ball's path in Example 6 is a parabola by eliminating the parameter in the parametric equations

$$x = (140\cos 31°)t \quad \text{and} \quad y = (140\sin 31°)t - 16t^2.$$

[*Hint:* Solve the first equation for t, and substitute the result in the second equation.]

In Exercises 41–48, use a calculator in degree mode and assume that air resistance is negligible.

41. A skeet is fired from the ground with an initial velocity of 110 feet per second at an angle of 28°.
    (a) Graph the skeet's path.
    (b) How long is the skeet in the air?
    (c) How high does it go?

42. A ball is thrown from a height of 5 feet above the ground with an initial velocity of 60 feet per second at an angle of 50° with the horizontal.
    (a) Graph the ball's path.
    (b) When and where does the ball hit ground?

43. A medieval bowman shoots an arrow which leaves the bow 4 feet above the ground with an initial velocity of 88 feet per second at an angle of 48° with the horizontal.
  (a) Graph the arrow's path.
  (b) Will the arrow go over the 40-foot-high castle wall that is 200 feet from the archer?

44. A golfer at a driving range stands on a platform 2 feet above the ground and hits the ball with an initial velocity of 120 feet per second at an angle of 39° with the horizontal. There is a 32-foot-high fence 400 feet away. Will the ball fall short, hit the fence, or go over it?

45. A golf ball is hit off the tee at an angle of 30° and lands 300 feet away. What was its initial velocity? [*Hint:* The ball lands when $x = 300$ and $y = 0$. Use this fact and the parametric equations for the ball's path to find two equations in the variables t and v. Solve for v.]

46. A football kicked from the ground has an initial velocity of 75 feet per second.
  (a) Set up the parametric equations that describe the ball's path. Experiment graphically with different angles to find the smallest angle (within one degree) needed so that the ball travels at least 150 feet.
  (b) Use algebra and trigonometry to find the angle needed for the ball to travel exactly 150 feet. [*Hint:* The ball lands when $x = 150$ and $y = 0$. Use this fact and the parametric equations for the ball's path to find two equations in the variables t and θ. Solve the "x equation" for t, and substitute this result in the other one; then solve for θ. The double-angle identity may be helpful for putting this equation in a form that is easy to solve.]

47. A golf ball is hit off the ground at an angle of θ degrees with an initial velocity of 100 feet per second.
  (a) Graph the path of the ball when $\theta = 20°$, $\theta = 40°$, $\theta = 60°$, and $\theta = 80°$.
  (b) For what angle in part (a) does the ball land farthest from where it started?
  (c) Experiment with different angles, as in parts (a) and (b), and make a conjecture as to which angle results in the ball landing farthest from its starting point.

48. A golf ball is hit off the ground at an angle of θ degrees with an initial velocity of 100 feet per second.
  (a) Graph the path of the ball when $\theta = 30°$ and when $\theta = 60°$. In which case does the ball land farthest away?
  (b) Do part (a) when $\theta = 25°$ and $\theta = 65°$.
  (c) Experiment further, and make a conjecture as to the results when the sum of the two angles is 90°.
  (d) Prove your conjecture algebraically. [*Hint:* Find the value of t at which a ball hit at angle θ hits the ground (which occurs when $y = 0$); this value of t will be an expression involving θ. Find the corresponding value of x (which is the distance of the ball from the starting point). Then do the same for an angle of $90° - \theta$ and use the cofunction identities (in degrees) to show that you get the same value of x.]

In Exercises 49–51, complete the derivation of the parametric equations of the cycloid in Example 8.

49. (a) If $\pi/2 < t < \pi$, verify that angle θ in the figure has measure $t - \pi/2$ and that

$$x = OT \quad CQ = 3t - 3\cos\left(t - \frac{\pi}{2}\right),$$

$$y = CT + PQ = 3 + 3\sin\left(t - \frac{\pi}{2}\right).$$

  (b) Use the addition and subtraction identities for sine and cosine to show that in this case,

$$x = 3(t - \sin t) \quad \text{and} \quad y = 3(1 - \cos t).$$

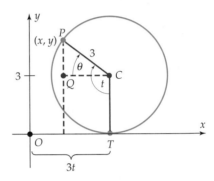

50. (a) If $\pi < t < 3\pi/2$, verify that angle θ in the figure measures $3\pi/2 - t$ and that

$$x = OT + CQ = 3t + 3\cos\left(\frac{3\pi}{2} - t\right),$$

$$y = CT + PQ = 3 + 3\sin\left(\frac{3\pi}{2} - t\right).$$

  (b) Use the addition and subtraction identities for sine and cosine to show that in this case

$$x = 3(t - \sin t) \quad \text{and} \quad y = 3(1 - \cos t).$$

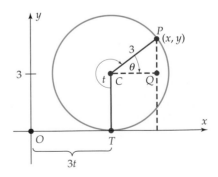

51. (a) If $3\pi/2 < t < 2\pi$, verify that angle θ in the figure has measure $t - 3\pi/2$ and that

$$x = OT + CQ = 3t + 3\cos\left(t - \frac{3\pi}{2}\right),$$

$$y = CT - PQ = 3 - 3\sin\left(y - \frac{3\pi}{2}\right).$$

(b) Use the addition and subtraction identities for sine and cosine to show that in this case

$$x = 3(t - \sin t) \qquad \text{and} \qquad y = 3(1 - \cos t).$$

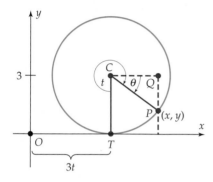

Thinkers

52. A particle moves on the horizontal line $y = 3$. Its x-coordinate at time t seconds is given by

$$x = 2t^3 - 13t^2 + 23t - 8.$$

This exercise explores the motion of the particle.

(a) Graph the path of the particle in the viewing window with $-10 \le x \le 10$, $-2 \le y \le 4$, $0 \le t \le 4.3$, and t step $= .05$. Note that the calculator seems to pause before completing the graph.

(b) Use trace (starting with $t = 0$) and watch the path of the particle as you press the right arrow key at regular intervals. How many times does it change direction? When does it appear to be moving the fastest?

(c) At what times t does the particle change direction? What are its x-coordinates at these times?

53. Set your calculator for radian mode and for simultaneous graphing mode [check your instruction manual for how to do this]. Particles A, B, and C are moving in the plane, with their positions at time t seconds given by:

$$A: \quad x = 8\cos t \quad \text{and} \quad y = 5\sin t,$$

$$B: \quad x = 3t \qquad \text{and} \quad y = 5t,$$

$$C: \quad x = 3t \qquad \text{and} \quad y = 4t.$$

(a) Graph the paths of A and B in the window with $0 \le x \le 12$, $0 \le y \le 6$, and $0 \le t \le 2$. The paths intersect, but do the particles actually collide? That is, are they at the same point at the same time? [For slow motion, choose a very small t step, such as .01.]

(b) Set t step $= .05$ and use trace to estimate the time at which A and B are closest to each other.

(c) Graph the paths of A and C and determine geometrically [as in part (b)] whether they collide. Approximately when are they closest?

(d) Confirm your answers in part (c) as follows. Explain why the distance between particles A and C at time t is given by

$$d = \sqrt{(8\cos t - 3t)^2 + (5\sin t - 4t)^2}.$$

A and C will collide if $d = 0$ at some time. Using function graphing mode, graph this distance function when $0 \le t \le 2$, and using zoom-in if necessary, show that d is always positive. Find the value of t for which d is smallest.

54. Let P be a point at distance k from the center of a circle of radius r. As the circle rolls along the x-axis, P traces out a curve called a **trochoid**. [When $k \le d$, it might help to think of the circle as a bicycle wheel and P as a point on one of the spokes.]

(a) Assume that P is on the y-axis as close as possible to the x-axis when $t = 0$, and show that the parametric equations of the trochoid are

$$x = rt - k\sin t \qquad \text{and} \qquad y = r - k\cos t.$$

Note that when $k = r$, these are the equations of a cycloid.

(b) Sketch the graph of the trochoid with $r = 3$ and $k = 2$.

(c) Sketch the graph of the trochoid with $r = 3$ and $k = 4$.

55. A circle of radius b rolls along the inside of a larger circle of radius a. The curve traced out by a fixed point P on the smaller circle is called a **hypocycloid.**

(a) Assume that the larger circle has center at the origin and that the smaller circle starts with P located at $(a, 0)$. Use the figure to show that the parametric equations of the hypocycloid are

$$x = (a - b) \cos t + b \cos\left(\frac{a - b}{b}t\right),$$

$$y = (a - b) \sin t - b \sin\left(\frac{a - b}{b}t\right).$$

(b) Sketch the graph of the hypocycloid with $a = 5$, $b = 1$, and $0 \le t \le 2\pi$.

(c) Sketch the graph of the hypocycloid with $a = 5$, $b = 2$, and $0 \le t \le 4\pi$.

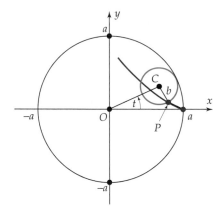

56. Jill is on a ferris wheel of radius 20 feet whose bottom just grazes the ground. The wheel is rotating counterclockwise at the rate of .7 radians per second. Jack is standing 100 feet from the bottom of the ferris wheel. When Jill is at point P, he throws a ball toward the wheel (see the figure). The ball leaves Jack's hand 5 feet above the ground with an initial velocity of 62 feet/second at an angle of 1.2 radians with the horizontal. Will Jill be able to catch the ball? Follow the steps below to answer the question.

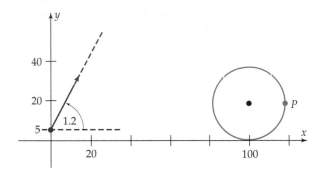

(a) Imagine that Jack is at the origin and the bottom of the ferris wheel is at $(100, 0)$. Then the ball leaves his hand from the point $(0, 5)$, and the wheel is a circle with center $(100, 20)$ and radius 20 (see the figure). Therefore, the wheel is the graph of the equation

$$(x - 100)^2 + (y - 20)^2 = 20^2.$$

Show that Jill's movement around the wheel is given by the parametric equations

$$x = 20 \cos(.7t) + 100,$$

$$y = 20 \sin(.7t) + 20$$

by verifying that these equations give a parameterization of the circle [as in Example 4].

(b) Find the parametric equations that describe the position of the ball at time t.

(c) Set your calculator for parametric mode, radian mode, and simultaneous graphing mode [check your instruction manual for how to do this]. Using a square viewing window with $0 \le x \le 130$, $0 \le t \le 9$, and t step $= .1$ to graph both sets of parametric equations (Jill's motion and the ball's) simultaneously. [For slow motion, make the t step smaller.] Assuming that Jill can reach 2 feet in any direction, can she catch the ball? If not, use the trace to estimate the time at which Jill is closest to the ball.

(d) Experiment by changing the angle or initial velocity (or both) of the ball to find values that will allow Jill to catch the ball.

10.6 Polar Coordinates

In the past, we used a rectangular coordinate system in the plane, based on two perpendicular coordinate axes. Now we introduce another coordinate system for the plane, based on angles.

Choose a point O in the plane (called the **origin** or **pole**) and a half-line extending from O (called the **polar axis**). As shown in Figure 10–58, a point P is given **polar coordinates** (r, θ), where

r = Distance from P to O

θ = Angle with polar axis as initial side and OP as terminal side.

Figure 10–58

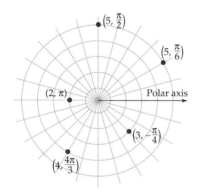

Figure 10–59

We shall usually measure the angle θ in radians; it may be either positive or negative, depending on whether it is generated by a clockwise or counterclockwise rotation. Some typical points are shown in Figure 10–59, which also illustrates the "circular grid" that a polar coordinate system imposes on the plane.

The polar coordinates of a point P are *not* unique. The angle θ may be replaced by any angle that has the same terminal side as θ, such as $\theta \pm 2\pi$. For instance, the coordinates $(2, \pi/3)$, $(2, 7\pi/3)$, and $(2, -5\pi/3)$ all represent the same point, as shown in Figure 10–60.

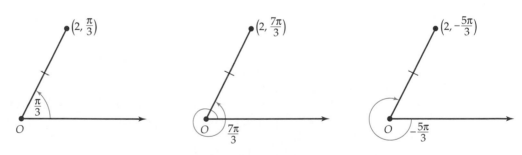

Figure 10–60

We shall consider the coordinates of the origin to be $(0, \theta)$, where θ is *any* angle.

Negative values for the first coordinate will be allowed according to this convention: For each positive r, the point $(-r, \theta)$ lies on the straight line containing the terminal side of θ, at distance r from the origin, on the *opposite* side of the origin from the point (r, θ), as shown in Figure 10–61.

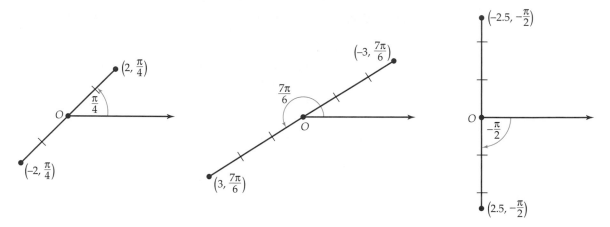

Figure 10–61

It is sometimes convenient to use both a rectangular and a polar coordinate system in the plane, with the polar axis coinciding with the positive x-axis. Then the y-axis is the polar line $\theta = \pi/2$. Suppose P has rectangular coordinates (x, y) and polar coordinates (r, θ), with $r > 0$, as in Figure 10–62.

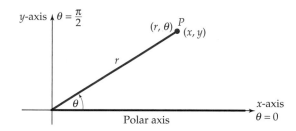

Figure 10–62

Since r is the distance from (x, y) to $(0, 0)$, the distance formula shows that $r = \sqrt{x^2 + y^2}$, and hence, $r^2 = x^2 + y^2$. The point-in-the-plane description of the trigonometric functions shows that

$$\cos \theta = \frac{x}{r}, \qquad \sin \theta = \frac{y}{r}, \qquad \tan \theta = \frac{y}{x}.$$

Solving the first two equations for x and y, we obtain the relationship between polar and rectangular coordinates.[*]

---

[*]The conclusions in the next box are also true when $r < 0$ (Exercise 50).

Coordinate Conversion Formulas

If a point has rectangular coordinates (x, y) and polar coordinates (r, θ), then

$$x = r \cos \theta \qquad \text{and} \qquad y = r \sin \theta,$$

$$r^2 = x^2 + y^2 \qquad \text{and} \qquad \tan \theta = \frac{y}{x}.$$

EXAMPLE 1

Convert each of the following points in polar coordinates to rectangular coordinates.

(a) $(2, \pi/6)$ (b) $(3, 4)$.

SOLUTION

(a) Apply the first set of equations in the box with $r = 2$ and $\theta = \pi/6$.

$$x = 2 \cos \frac{\pi}{6} = 2 \cdot \frac{\sqrt{3}}{2} = \sqrt{3} \qquad \text{and} \qquad y = 2 \sin \frac{\pi}{6} = 2 \cdot \frac{1}{2} = 1.$$

So the rectangular coordinates are $(\sqrt{3}, 1)$.

(b) The point with polar coordinates $(3, 4)$ has $r = 3$ and $\theta = 4$ radians. Therefore, its rectangular coordinates are

$$(r \cos \theta, r \sin \theta) = (3 \cos 4, 3 \sin 4) \approx (-1.9609, -2.2704). \qquad \blacksquare$$

EXAMPLE 2

Convert each of the following points in rectangular coordinates to polar coordinates.

(a) $(2, -2)$ (b) $(3, 5)$ (c) $(-2, 4)$.

SOLUTION

(a) The second set of equations in the box, with $x = 2$, $y = -2$, shows that

$$r = \sqrt{2^2 + (-2)^2} = \sqrt{8} = 2\sqrt{2} \qquad \text{and} \qquad \tan \theta = -2/2 = -1.$$

We must find an angle θ whose terminal side passes through $(2, -2)$ and whose tangent is -1. Figure 10–63 shows that two of the many possibilities are

$$\theta = -\frac{\pi}{4} \qquad \text{and} \qquad \theta = \frac{7\pi}{4}.$$

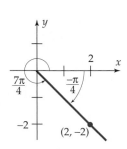

Figure 10–63

So one pair of polar coordinates is $\left(2\sqrt{2}, -\frac{\pi}{4}\right)$, and another is $\left(2\sqrt{2}, \frac{7\pi}{4}\right)$.

(b) Applying the second set of equations in the box, with $x = 3$, $y = 5$, we have

$$r = \sqrt{3^2 + 5^2} = \sqrt{34} \qquad \text{and} \qquad \tan \theta = 5/3.$$

The TAN^{-1} key on a calculator shows that $\theta \approx 1.0304$ radians is an angle between 0 and $\pi/2$ with tangent $5/3$. Since $(3, 5)$ is in the first quadrant, one pair of (approximate) polar coordinates is $(\sqrt{34}, 1.0304)$.

(c) In this case,

$$r = \sqrt{(-2)^2 + 4^2} = \sqrt{20} = 2\sqrt{5} \qquad \text{and} \qquad \tan\theta = \frac{4}{-2} = -2.$$

Using the TAN^{-1} key, we find that $\theta \approx -1.1071$ is an angle between $-\pi/2$ and 0 with tangent -2. However, we want an angle between $\pi/2$ and π because $(-2, 4)$ is in the second quadrant. Since tangent has period π,

$$\tan(-1.1071 + \pi) = \tan(-1.1071) = -2.$$

Thus, $-1.1071 + \pi \approx 2.0344$ is an angle between $\pi/2$ and π whose tangent is -2. Therefore, one pair of polar coordinates is $(2\sqrt{5}, 2.0344)$. ∎

▗▖ POLAR GRAPHS

The graphs of a few polar coordinate equations can be easily determined from the appropriate definitions.

EXAMPLE 3

Graph the equations

(a) $r = 3^*$ (b) $\theta = \pi/6.^*$

SOLUTION

(a) The graph consists of all points (r, θ) with first coordinate 3, that is, all points whose distance from the origin is 3. So the graph is a circle with center O and radius 3, as shown in Figure 10–64.

(b) The graph consists of all points $(r, \pi/6)$. If $r \geq 0$, then $(r, \pi/6)$ lies on the terminal side of an angle of $\pi/6$ radians, whose initial side is the polar axis. If $r < 0$, then $(r, \pi/6)$ lies on the extension of this terminal side across the origin. So the graph is the straight line in Figure 10–65. ∎

Figure 10–64

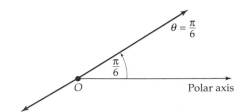

Figure 10–65

Some polar graphs can be sketched by hand by using basic facts about trigonometric functions.

---

*Every equation is understood to involve two variables, but one may have coefficient 0, as is the case here: $r = 3 + 0 \cdot \theta$ and $\theta = 0 \cdot r + \pi/6$. This is analogous to equations such as $y = 5$ and $x = 2$ in rectangular coordinates.

EXAMPLE 4

Graph $r = 1 + \sin\theta$.

SOLUTION Remember the behavior of $\sin\theta$ between 0 and 2π:

As θ takes values larger than 2π, $\sin\theta$ repeats the same pattern, and hence, so does $r = 1 + \sin\theta$. The same is true for negative values of θ. The full graph (called a **cardioid**) is at the lower right in Figure 10–66. ∎

As θ increases from 0 to $\pi/2$, $\sin\theta$ increases from 0 to 1. So $r = 1 + \sin\theta$ increases from 1 to 2.

As θ increases from $\pi/2$ to π, $\sin\theta$ decreases from 1 to 0. So $r = 1 + \sin\theta$ decreases from 2 to 1.

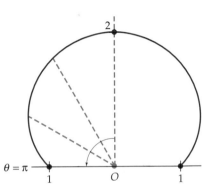

As θ increases from π to $3\pi/2$, $\sin\theta$ decreases from 0 to -1. So $r = 1 + \sin\theta$ decreases from 1 to 0.

As θ increases from $3\pi/2$ to 2π, $\sin\theta$ increases from -1 to 0. So $r = 1 + \sin\theta$ increases from 0 to 1.

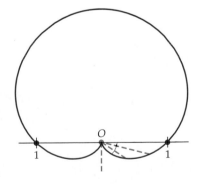

Figure 10–66

The easiest way to graph polar equation $r = f(\theta)$ is to use a calculator in polar graphing mode. A second way is to use parametric graphing mode, with the coordinate converison formulas as a parameterization.

$$x = r\cos\theta = f(\theta)\cos\theta,$$

$$y = r\sin\theta = f(\theta)\sin\theta.$$

Figure 10–67

Figure 10–68

Graph $r = 2 + 4 \cos \theta$.

SOLUTION *Polar Method:* Put your calculator in polar graphing mode and enter $r = 2 + 4 \cos \theta$ in the function memory. Set the viewing window by entering minimum and maximum values for x, y, and θ. Since cosine has period 2π, a complete graph can be obtained by taking $0 \le \theta \le 2\pi$. You must also set the θ step (or θ pitch), which determines how many values of θ the calculator uses to plot the graph. With an appropriate θ step, the graph should look like Figure 10–67.

Parametric Method: Put your calculator in parametric graphing mode. The parametric equations for $r = 2 + 4 \cos \theta$ are as follows (using t as the variable instead of θ with $0 \le t \le 2\pi$):

$$x = r \cos t = (2 + 4 \cos t) \cos t = 2 \cos t + 4 \cos^2 t$$

$$y = r \sin t = (2 + 4 \cos t) \sin t = 2 \sin t + 4 \sin t \cos t.$$

They also produce the graph in Figure 10–67. ∎

EXAMPLE 6

The graph of $r = \sin 2\theta$ in Figure 10–68 can be obtained either by graphing directly in polar mode or by using parametric mode and the equations

$$x = r \cos t = \sin 2t \cos t \quad \text{and} \quad y = r \sin t = \sin 2t \sin t \quad (0 \le t \le 2\pi).$$ ∎

Here is a summary of commonly encountered polar graphs (in each case, *a* and *b* are constants).

Equation	Name of Graph	Shape of Graph*	
$r = a\theta \; (\theta \geq 0)$ $r = a\theta \; (\theta \leq 0)$	Archimedean spiral	 $r = a\theta \; (\theta \geq 0)$	 $r = a\theta \; (\theta \leq 0)$
$r = a(1 \pm \sin \theta)$ $r = a(1 \pm \cos \theta)$	cardioid	 $r = a(1 + \cos \theta)$	 $r = a(1 - \sin \theta)$
$r = a \sin n\theta$ $r = a \cos n\theta$ $(n \geq 2)$	rose (There are *n* petals when *n* is odd and 2*n* petals when *n* is even.)	 $r = a \cos n\theta$	 $r = a \sin n\theta$

---

*Depending on the plus or minus sign and whether sine or cosine is involved, the basic shape of a specific graph may differ from those shown by a rotation, reversal, or horizontal or vertical shift.

Equation	Name of Graph	Shape of Graph
$r = a \sin \theta$ $r = a \cos \theta$	circle	 $r = a \cos \theta$ $r = a \sin \theta$
$r^2 = \pm a^2 \sin 2\theta$ $r^2 = \pm a^2 \cos 2\theta$	lemniscate	 $r^2 = a^2 \sin 2\theta$ $r^2 = a^2 \cos 2\theta$
$r = a \pm b \sin \theta$ $r = a \pm b \cos \theta$ $(a, b > 0;\ a \neq b)$	limaçon	 $a < b$ $b < a < 2b$ $a \geq 2b$ $r = a + b \cos \theta$ $r = a + b \sin \theta$ $r = a - b \sin \theta$

✓ EXERCISES 10.6

1. What are the polar coordinates of the points P, Q, R, S, T, U, V in the figure?

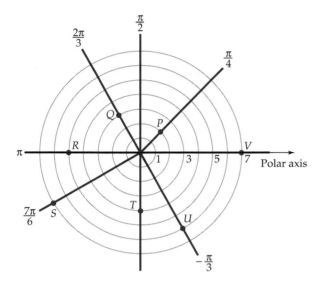

In Exercises 2–6, list four other pairs of polar coordinates for the given point, each with a different combination of signs (that is, $r > 0$, $\theta > 0$; $r > 0$, $\theta < 0$; $r < 0$, $\theta > 0$; $r < 0$, $\theta < 0$).

2. $(3, \pi/3)$ 3. $(-5, \pi)$ 4. $(2, -2\pi/3)$

5. $(-1, -\pi/6)$ 6. $(\sqrt{3}, 3\pi/4)$

In Exercises 7–10, convert the polar coordinates to rectangular coordinates.

7. $(3, \pi/3)$ 8. $(-2, \pi/4)$ 9. $(-1, 5\pi/6)$

10. $(2, 0)$

In Exercises 11–16, convert the rectangular coordinates to polar coordinates.

11. $(3\sqrt{3}, -3)$ 12. $(2\sqrt{3}, -2)$ 13. $(2, 4)$

14. $(3, -2)$ 15. $(-5, 2.5)$ 16. $(-6.2, -3)$

In Exercises 17–22, sketch the graph of the equation without using a calculator.

17. $r = 4$ 18. $r = -1$ 19. $\theta = -\pi/3$

20. $\theta = 5\pi/6$ 21. $\theta = 1$ 22. $\theta = -4$

In Exercises 23–46, sketch the graph of the equation.

23. $r = \theta$ $(\theta \le 0)$ 24. $r = 3\theta$ $(\theta \ge 0)$

25. $r = 1 - \sin\theta$ 26. $r = 3 - 3\cos\theta$

27. $r = -2\cos\theta$ 28. $r = -6\sin\theta$

29. $r = \cos 2\theta$ 30. $r = \cos 3\theta$

31. $r = \sin 3\theta$ 32. $r = \sin 4\theta$

33. $r^2 = 4\cos 2\theta$ 34. $r^2 = \sin 2\theta$

35. $r = 2 + 4\cos\theta$ 36. $r = 1 + 2\cos\theta$

37. $r = \sin\theta + \cos\theta$ 38. $r = 4\cos\theta + 4\sin\theta$

39. $r = \sin(\theta/2)$ 40. $r = 4\tan\theta$

41. $r = \sin\theta\tan\theta$ (cissoid)

42. $r = 4 + 2\sec\theta$ (conchoid)

43. $r = e^{\theta}$ (logarithmic spiral)

44. $r^2 = 1/\theta$ 45. $r = 1/\theta$ $(\theta > 0)$

46. $r^2 = \theta$

47. (a) Find a complete graph of $r = 1 - 2\sin 3\theta$.
    (b) Predict what the graph of $r = 1 - 2\sin 4\theta$ will look like. Then check your prediction with a calculator.
    (c) Predict what the graph of $r = 1 - 2\sin 5\theta$ will look like. Then check your prediction with a calculator.

48. (a) Find a complete graph of $r = 1 - 3\sin 2\theta$.
    (b) Predict what the graph of $r = 1 - 3\sin 3\theta$ will look like. Then check your prediction with a calculator.
    (c) Predict what the graph of $r = 1 - 3\sin 4\theta$ will look like. Then check your prediction with a calculator.

49. If a, b are constants such that $ab \ne 0$, show that the graph of $r = a\sin\theta + b\cos\theta$ is a circle. [*Hint:* Multiply both sides by r and convert to rectangular coordinates.]

50. Prove that the coordinate conversion formulas are valid when $r < 0$. [*Hint:* If P has coordinates (x, y) and (r, θ), with $r < 0$, verify that the point Q with rectangular coordinates $(-x, -y)$ has polar coordinates $(-r, \theta)$. Since $r < 0$, $-r$ is positive and the conversion formulas proved in the text apply to Q. For instance, $-x = -r\cos\theta$, which implies that $x = r\cos\theta$.]

51. **Distance Formula for Polar Coordinates:** Prove that the distance from (r, θ) to (s, β) is

$$\sqrt{r^2 + s^2 - 2rs\cos(\theta - \beta)}$$

[*Hint:* If $r > 0$, $s > 0$, and $\theta > \beta$, then the triangle with vertices (r, θ), (s, β), $(0, 0)$ has an angle of $\theta - \beta$, whose sides have lengths r and s. Use the Law of Cosines.]

52. Explain why the following symmetry tests for the graphs of polar equations are valid.

(a) If replacing θ by $-\theta$ produces an equivalent equation, then the graph is symmetric with respect to the line $\theta = 0$ (the x-axis).

(b) If replacing θ by $\pi - \theta$ produces an equivalent equation, then the graph is symmetric with respect to the line $\theta = \pi/2$ (the y-axis).

(c) If replacing r by $-r$ produces an equivalent equation, then the graph is symmetric with respect to the origin.

10.7 | Polar Equations of Conics

In a rectangular coordinate system, each type of conic section has a different definition. By using polar coordinates, it is possible to give a unified treatment of conics and their equations. Before doing this, we must first introduce a concept that will play a key role in the development.

Recall that both ellipses and hyperbolas are defined in terms of two foci and both have two vertices that lie on the line through the foci (see pages 639 and 652). The **eccentricity** of an ellipse or a hyperbola is denoted e and is defined to be the ratio

$$e = \frac{\text{distance between the foci}}{\text{distance between the vertices}}.$$

For conics centered at the origin, with foci on the x-axis, the situation is as follows.

Ellipse	**Hyperbola**
$$\frac{x^2}{a^2} + \frac{y^2}{b^2} = 1 \quad (a > b)$$	$$\frac{x^2}{a^2} - \frac{y^2}{b^2} = 1$$
foci: $(\pm c, 0)$ vertices: $(\pm a, 0)$	foci: $(\pm c, 0)$ vertices: $(\pm a, 0)$
$$c = \sqrt{a^2 - b^2}$$	$$c = \sqrt{a^2 + b^2}$$

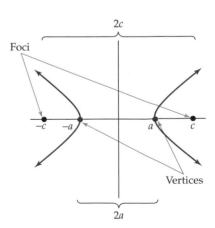

$$e = \frac{2c}{2a} = \frac{c}{a} = \frac{\sqrt{a^2 - b^2}}{a}$$

$$e = \frac{2c}{2a} = \frac{c}{a} = \frac{\sqrt{a^2 + b^2}}{a}$$

A similar analysis shows that the formulas for e are also valid for conics whose equations are of the form

$$\frac{x^2}{b^2} + \frac{y^2}{a^2} = 1 \quad (a > b) \qquad \text{or} \qquad \frac{y^2}{a^2} - \frac{x^2}{b^2} = 1.$$

These formulas can be used to compute the eccentricity of any ellipse or hyperbola whose equation can be put in standard form.

EXAMPLE 1

Find the eccentricity of the conic with equation

(a) $\dfrac{y^2}{4} - \dfrac{x^2}{21} = 1$;

(b) $4x^2 + 9y^2 - 32x - 90y + 253 = 0.$

SOLUTION

(a) In this case, $a^2 = 4$ (so $a = 2$) and $b^2 = 21$. Hence, the eccentricity is

$$e = \frac{\sqrt{a^2 + b^2}}{a} = \frac{\sqrt{4 + 21}}{2} = \frac{\sqrt{25}}{2} = \frac{5}{2} = 2.5.$$

(b) In Example 5 of Section 10.1, we saw that the equation can be put into this standard form:

$$\frac{(x-4)^2}{9} + \frac{(y-5)^2}{4} = 1.$$

Hence, its graph is just the ellipse

$$\frac{x^2}{9} + \frac{y^2}{4} = 1$$

shifted vertically and horizontally. Since the shifting does not change the distances between foci or vertices, both ellipses have the same eccentricity, which can be computed by using $a^2 = 9$ and $b^2 = 4$.

$$e = \frac{\sqrt{a^2 - b^2}}{a} = \frac{\sqrt{9 - 4}}{3} = \frac{\sqrt{5}}{3} \approx .745. \quad \blacksquare$$

Example 1 and the preceding pictures illustrate the following fact. For ellipses, the distance between the foci (numerator of e) is less than the distance between the vertices (denominator), so $e < 1$. For hyperbolas, however, $e > 1$ because the distance between the foci is greater than that between the vertices.

The eccentricity of an ellipse measures its "roundness." An ellipse whose eccentricity is close to 0 is almost circular (Exercise 19). The eccentricity of a hyperbola measures how "flat" its branches are. The branches of a hyperbola with large eccentricity look almost like parallel lines (Exercise 20).

■■ CONICS AND POLAR EQUATIONS

The polar analogues of the standard equations of ellipses, parabolas, and hyperbolas are given in the following chart. The proof of these statements is given at the end of the section. In the chart, e and d are constants, with $e > 0$. Remember that in a rectangular coordinate system whose positive x-axis coincides with the polar axis, a point with polar coordinates (r, θ) is on the x-axis when $\theta = 0$ or π and on the y-axis when $\theta = \pi/2$ or $3\pi/2$.

Polar Equations for Conic Sections

Equation	Graph	
$r = \dfrac{ed}{1 + e\cos\theta}$ or $r = \dfrac{ed}{1 - e\cos\theta}$	$0 < e < 1$	*Ellipse* with eccentricity e One of the foci: $(0, 0)$ Vertices at $\theta = 0$ and $\theta = \pi$
	$e = 1$	*Parabola* with focus $(0, 0)$ Vertex at $\theta = 0$ or $\theta = \pi$; (r is not defined for the other value of θ)
	$e > 1$	*Hyperbola* with eccentricity e One of the foci: $(0, 0)$ Vertices at $\theta = 0$ and $\theta = \pi$
$r = \dfrac{ed}{1 + e\sin\theta}$ or $r = \dfrac{ed}{1 - e\sin\theta}$	$0 < e < 1$	*Ellipse* with eccentricity e One of the foci: $(0, 0)$ Vertices at $\theta = \pi/2$ and $\theta = 3\pi/2$
	$e = 1$	*Parabola* with focus $(0, 0)$ Vertex at $\theta = \pi/2$ or $\theta = 3\pi/2$; (r is not defined for the other value of θ)
	$e > 1$	*Hyperbola* with eccentricity e One of the foci: $(0, 0)$ Vertices at $\theta = \pi/2$ and $\theta = 3\pi/2$

EXAMPLE 2

Find a complete graph of

$$r = \frac{3e}{1 + e\cos\theta}$$

when

(a) $e = .7$ (b) $e = 1$ (c) $e = 2$.

SOLUTION   From the first equation in the preceding chart (with $d = 3$), we know that the graphs are an ellipse, a parabola, and a hyperbola, respectively, as shown in Figure 10–69.   ■

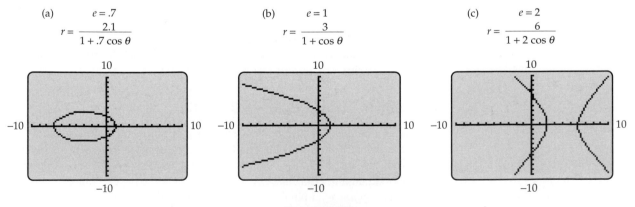

(a) $e = .7$
$r = \dfrac{2.1}{1 + .7\cos\theta}$

(b) $e = 1$
$r = \dfrac{3}{1 + \cos\theta}$

(c) $e = 2$
$r = \dfrac{6}{1 + 2\cos\theta}$

Figure 10–69

EXAMPLE 3

Identify the conic section that is the graph of

$$r = \frac{20}{4 - 10 \sin \theta},$$

and find its vertices and eccentricity.

SOLUTION   First, rewrite the equation in one of the forms listed in the preceding box.

$$r = \frac{20}{4 - 10 \sin \theta} = \frac{20}{4\left(1 - \dfrac{10}{4} \sin \theta\right)} = \frac{5}{1 - 2.5 \sin \theta}.$$

This is one such form, with $e = 2.5$ and $ed = 5$ (so $d = 2$). Consequently, the graph is a hyperbola with eccentricity $e = 2.5$ whose vertices are at

$$\theta = \frac{\pi}{2}, \qquad r = \frac{20}{4 - 10 \sin \dfrac{\pi}{2}} = \frac{20}{4 - 10 \cdot 1} = -\frac{20}{6} = -\frac{10}{3}$$

and

$$\theta = \frac{3\pi}{2}, \qquad r = \frac{20}{4 - 10 \sin \dfrac{3\pi}{2}} = \frac{20}{4 - 10(-1)} = \frac{20}{14} = \frac{10}{7}.$$

GRAPHING EXPLORATION

Find a viewing window that shows a complete graph of this hyperbola. ■

EXAMPLE 4

Find a polar equation of the ellipse with $(0, 0)$ as a focus and vertices $(3, 0)$ and $(6, \pi)$.

SOLUTION   Because of the location of the vertices, the polar equation is of the form $r = ed/(1 \pm e \cos \theta)$. We first consider the equation

$$r = \frac{ed}{1 + e \cos \theta}.$$

Since the coordinates of the vertices satisfy the equation, we must have

$$3 = \frac{ed}{1 + e \cos 0} = \frac{ed}{1 + e} \qquad \text{and} \qquad 6 = \frac{ed}{1 + e \cos \pi} = \frac{ed}{1 - e},$$

which imply that

$$3(1 + e) = ed \qquad \text{and} \qquad 6(1 - e) = ed.$$

Therefore,

$$3(1 + e) = 6(1 - e)$$

$$3 + 3e = 6 - 6e$$

$$9e = 3$$

$$e = 1/3.$$

Substituting $e = 1/3$ in either of the original equations shows that $d = 12$. So an equation of the ellipse is

$$r = \frac{ed}{1 + e \cos \theta} = \frac{\frac{1}{3} \cdot 12}{1 + \frac{1}{3} \cos \theta} = \frac{12}{3 + \cos \theta}.$$

If we had started instead with the equation $r = ed/(1 - e \cos \theta)$ and solved for e as above, we would have obtained $e = -1/3$, which is impossible, since $e > 0$.

ALTERNATE SOLUTION   Verify that the vertex (3, 0) also has polar coordinates $(-3, \pi)$. Similarly, $(6, \pi)$ also has polar coordinates $(-6, 0)$. If you begin with the equation $r = ed/(1 - e \cos \theta)$ and the vertices $(-3, \pi)$ and $(-6, 0)$ and proceed as before to find e and $d,$ you obtain the equation

$$r = \frac{-12}{3 - \cos \theta}. \quad\blacksquare$$

■■ ALTERNATE DEFINITION OF CONICS

The theorem stated in the following box is sometimes used as a definition of the conic sections because it provides a unified approach instead of the variety of descriptions given in Sections 10.1–10.3. Its proof also provides a proof of the statements in the box on page 706.

The basic idea is to describe every conic in terms of a straight line L (the **directrix**) and a point P not on L (the **focus**), in much the same way that parabolas were defined in Section 10.3. The number e in the theorem turns out to be the eccentricity of the conic.

Conic Section Theorem

> Let L be a straight line, P a point not on L, and e a positive constant. The set of all points X in the plane such that
>
> $$\frac{\text{distance from } X \text{ to } P}{\text{distance from } X \text{ to } L} = e$$
>
> is a conic section with P as one of the foci.[*] The conic is an ellipse if $0 < e < 1$, a parabola if $e = 1$,[†] and a hyperbola if $e > 1$.

---

[*]The distance from X to L is measured along the line through X that is perpendicular to L.
[†]When $e = 1$, the given condition is equivalent to

$$\text{distance from } X \text{ to } P = \text{distance from } X \text{ to } L$$

which is the definition of a parabola given in Section 10.3.

Proof   Coordinatize the plane so that the pole is the point *P*, the polar axis is horizontal, and the directrix *L* is a vertical line to the left of the pole, as in Figure 10–70. Let *d* be the distance from *P* to *L*, and let (*r*, *θ*) be the polar coordinates of *X*.

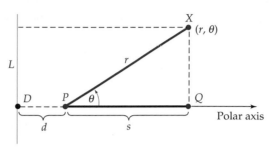

Figure 10–70

If *X* satisfies the condition

$$\frac{\text{distance from } X \text{ to } P}{\text{distance from } X \text{ to } L} = e,$$

then

(∗)   distance from *X* to *P* = *e*(distance from *X* to *L*).

Figure 10–70 shows that *r* is the distance from *P* to *X* and that the distance from *X* to *L* is the same as that from *D* to *Q*, namely, *d* + *s*. Furthermore, cos *θ* = *s*/*r*, so *s* = *r* cos *θ*. Consequently, equation (∗) can be written in polar coordinates as follows.

$$\text{distance from } X \text{ to } P = e(\text{distance from } X \text{ to } L)$$

$$r = e(d + s)$$

$$r = e(d + r\cos\theta)$$

$$r - er\cos\theta = ed$$

$$r(1 - e\cos\theta) = ed$$

$$r = \frac{ed}{1 - e\cos\theta}.$$

To show that this is actually the equation of a conic, we translate it into rectangular coordinates using the conversion formulas from Section 10.6.

$$r^2 = x^2 + y^2 \qquad \text{and} \qquad \cos\theta = \frac{x}{r} = \frac{x}{\pm\sqrt{x^2 + y^2}}.$$

Then the polar coordinate equation becomes

$$\pm\sqrt{x^2 + y^2} = \frac{ed}{1 - e\left(\dfrac{x}{\pm\sqrt{x^2 + y^2}}\right)}$$

$$\pm\sqrt{x^2 + y^2}\left(1 - \frac{ex}{\pm\sqrt{x^2 + y^2}}\right) = ed$$

$$\pm\sqrt{x^2 + y^2} - ex = ed$$

$$\pm\sqrt{x^2 + y^2} = ed + ex.$$

Squaring both sides and rearranging terms, we have

$$x^2 + y^2 = e^2d^2 + 2de^2x + e^2x^2$$

(∗∗) $$(1 - e^2)x^2 - 2de^2x + y^2 = e^2d^2.$$

Now we consider the two possibilities $e = 1$ and $e \neq 1$.

Case 1. If $e = 1$, then equation (∗∗) becomes

$$-2dx + y^2 = d^2$$

$$y^2 = 2dx + d^2$$

$$(y - 0)^2 = 2d\left(x + \frac{d}{2}\right)$$

$$(y - 0)^2 = 4\left(\frac{d}{2}\right)\left(x - \left(-\frac{d}{2}\right)\right).$$

The box on page 665 (with $k = 0$, $p = d/2$, $h = -d/2$) shows that this is the standard equation of a parabola with

$$\text{vertex}\left(-\frac{d}{2}, 0\right), \qquad \text{focus } (0, 0), \qquad \text{directrix } x = -d.$$

Case 2. If $e \neq 1$, then we can divide both sides of equation (∗∗) by the nonzero number $1 - e^2$.

$$\left(x^2 - \frac{2de^2}{1 - e^2}x\right) + \frac{y^2}{1 - e^2} = \frac{e^2d^2}{1 - e^2}.$$

Next, we complete the square on the expression in parentheses by adding the square of half of the coefficient of x to both sides of the equation and simplify the result.

$$\left[x^2 - \frac{2de^2}{1 - e^2}x + \left(\frac{de^2}{1 - e^2}\right)^2\right] + \frac{y^2}{1 - e^2} = \frac{e^2d^2}{1 - e^2} + \left(\frac{de^2}{1 - e^2}\right)^2$$

$$\left(x - \frac{de^2}{1 - e^2}\right)^2 + \frac{y^2}{1 - e^2} = \frac{(1 - e^2)e^2d^2 + (de^2)^2}{(1 - e^2)^2}$$

$$\left(x - \frac{de^2}{1 - e^2}\right)^2 + \frac{y^2}{1 - e^2} = \frac{e^2d^2}{(1 - e^2)^2}.$$

Dividing both sides of the last equation by $e^2d^2/(1 - e^2)^2$ produces the equation

(∗∗∗) $$\frac{\left(x - \dfrac{de^2}{1 - e^2}\right)^2}{\dfrac{e^2d^2}{(1 - e^2)^2}} + \frac{y^2}{\dfrac{e^2d^2}{1 - e^2}} = 1.$$

Now we consider the two possibilities $e < 1$ and $e > 1$.

Case 2A.   If $e < 1$, then $1 - e^2 > 0$ and the constants in the denominators on the left side of equation (∗∗∗) are positive. Therefore, equation (∗∗∗) is of the form

$$\frac{(x - h)^2}{a^2} + \frac{(y - k)^2}{b^2} = 1$$

with $h = de^2/(1 - e^2)$, $k = 0$, and a and b positive numbers such that

$$a^2 = \frac{e^2 d^2}{(1 - e^2)^2} \qquad \text{and} \qquad b^2 = \frac{e^2 d^2}{1 - e^2}.$$

In this case, $a > b$ by Exercise 47. According to the box on page 645, this is the standard equation of an ellipse with center $(h, 0)$ and foci $(h - c, 0)$ and $(h + c, 0)$, where $c^2 = a^2 - b^2$. Its eccentricity is the number

$$\frac{c}{a} = \sqrt{\frac{c^2}{a^2}} = \sqrt{\frac{a^2 - b^2}{a^2}} = \sqrt{1 - b^2 \cdot \frac{1}{a^2}}$$

$$= \sqrt{1 - \frac{e^2 d^2}{1 - e^2} \cdot \frac{(1 - e^2)^2}{e^2 d^2}} = \sqrt{e^2} = e.$$

Case 2B.   If $e > 1$, then $1 - e^2 < 0$. Therefore,

$$e^2 - 1 = -(1 - e^2) > 0,$$

so equation (∗∗∗) may be written as

$$\frac{\left(x - \dfrac{de^2}{1 - e^2}\right)^2}{\dfrac{e^2 d^2}{(1 - e^2)^2}} - \frac{y^2}{\dfrac{e^2 d^2}{e^2 - 1}} = 1.$$

This is an equation of the form

$$\frac{(x - h)^2}{a^2} - \frac{(y - k)^2}{b^2} = 1$$

with a and b positive. The box on page 654 shows that it is the standard equation of a hyperbola with foci $(h - c, 0)$ and $(h + c, 0)$, where $c^2 = a^2 + b^2$. Exercise 48 shows that its eccentricity is e.

   The preceding argument depends on coordinatizing the plane in a certain way and taking d to be the distance from the pole to L. Similar arguments, in which d is the distance from the pole to L and the plane is coordinatized so that L is to the right of the pole or parallel to the polar axis, lead to the other polar equations shown in Figure 10–71 on the next page.

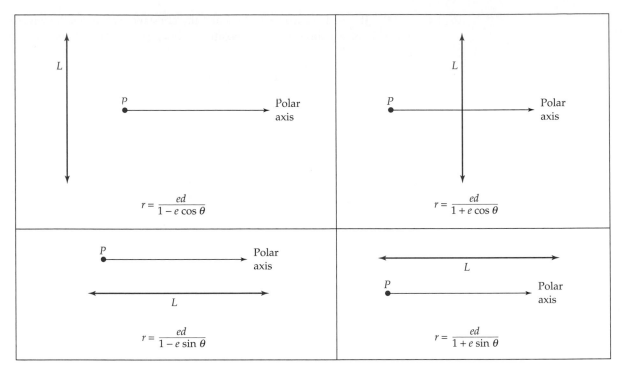

Figure 10–71

Analogous arguments when $d < 0$ (using $-d$ as the distance from the pole to L) then complete the proof. ∎

EXAMPLE 5

Find the polar equation of the hyperbola with focus at the pole, directrix

$$r = -4 \csc \theta,$$

and eccentricity 3.

SOLUTION   The equation of the directrix can be written as

$$r = \frac{-4}{\sin \theta} \qquad \text{or, equivalently,} \qquad r \sin \theta = -4.$$

With the conversion formulas for a rectangular coordinate system whose positive x-axis coincides with the polar axis, this equation becomes $y = -4$. So the directrix is a line parallel to the polar axis and 4 units below it. Using Figure 10–71, we see that $d = 4$, and the equation is

$$r = \frac{ed}{1 - e \sin \theta} = \frac{3 \cdot 4}{1 - 3 \sin \theta} = \frac{12}{1 - 3 \sin \theta}. ∎$$

EXERCISES 10.7

In Exercises 1–6, which of the graphs (a)–(f) at the bottom of the page could possibly be the graph of the equation?

1. $r = \dfrac{3}{1 - \cos \theta}$

2. $r = \dfrac{6}{2 + \cos \theta}$

3. $r = \dfrac{6}{2 - 4 \sin \theta}$

4. $r = \dfrac{15}{1 + 4 \cos \theta}$

5. $r = \dfrac{6}{3 - 2 \sin \theta}$

6. $r = \dfrac{6}{\dfrac{3}{2} + \dfrac{3}{2} \sin \theta}$

In Exercises 7–12, identify the conic section whose equation is given; if it is an ellipse or hyperbola, state its eccentricity.

7. $r = \dfrac{12}{3 + 4 \sin \theta}$

8. $r = \dfrac{-10}{2 + 3 \cos \theta}$

9. $r = \dfrac{8}{3 + 3 \sin \theta}$

10. $r = \dfrac{20}{5 - 10 \sin \theta}$

11. $r = \dfrac{2}{6 - 4 \cos \theta}$

12. $r = \dfrac{-6}{5 + 2 \cos \theta}$

In Exercises 13–18, find the eccentricity of the conic whose equation is given.

13. $\dfrac{x^2}{100} + \dfrac{y^2}{99} = 1$

14. $\dfrac{(x - 4)^2}{18} + \dfrac{(y + 5)^2}{25} = 1$

15. $\dfrac{(x - 6)^2}{10} - \dfrac{y^2}{40} = 1$

16. $4x^2 + 9y^2 - 24x + 36y + 36 = 0$

17. $16x^2 - 9y^2 - 32x + 36y + 124 = 0$

18. $4x^2 - 5y^2 - 16x - 50y + 71 = 0$

19. (a) Using a square viewing window (so that circles look like circles), graph these ellipses (on the same screen if possible):

$$\frac{x^2}{16} + \frac{y^2}{1} = 1, \qquad \frac{x^2}{16} + \frac{y^2}{6} = 1, \qquad \frac{x^2}{16} + \frac{y^2}{14} = 1$$

(b) Compute the eccentricity of each ellipse in part (a).

(c) On the basis of parts (a) and (b), how is the shape of an ellipse related to its eccentricity?

20. (a) Graph these hyperbolas (on the same screen if possible):

$$\frac{y^2}{4} - \frac{x^2}{1} = 1, \qquad \frac{y^2}{4} - \frac{x^2}{12} = 1, \qquad \frac{y^2}{4} - \frac{x^2}{96} = 1$$

(b) Compute the eccentricity of each hyperbola in part (a).

(c) On the basis of parts (a) and (b), how is the shape of a hyperbola related to its eccentricity?

In Exercises 21–32, sketch the graph of the equation and label the vertices.

21. $r = \dfrac{8}{1 - \cos \theta}$

22. $r = \dfrac{5}{3 + 2 \sin \theta}$

23. $r = \dfrac{4}{2 - 4 \cos \theta}$

24. $r = \dfrac{5}{1 + \cos \theta}$

25. $r = \dfrac{10}{4 - 3 \sin \theta}$

26. $r = \dfrac{12}{3 + 4 \sin \theta}$

(a)

(b)

(c)

(d)   (e)

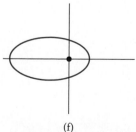

(f)

27. $r = \dfrac{15}{3 - 2 \cos \theta}$

28. $r = \dfrac{32}{3 + 5 \sin \theta}$

29. $r = \dfrac{3}{1 + \sin \theta}$

30. $r = \dfrac{10}{3 + 2 \cos \theta}$

31. $r = \dfrac{10}{2 + 3 \sin \theta}$

32. $r = \dfrac{15}{4 - 4 \cos \theta}$

In Exercises 33–46, find the polar equation of the conic section that has focus $(0, 0)$ and satisfies the given conditions.

33. Parabola; vertex $(3, \pi)$

34. Parabola; vertex $(2, \pi/2)$

35. Ellipse; vertices $(2, \pi/2)$ and $(8, 3\pi/2)$

36. Ellipse; vertices $(2, 0)$ and $(4, \pi)$

37. Hyperbola; vertices $(1, 0)$ and $(-3, \pi)$

38. Hyperbola; vertices $(-2, \pi/2)$ and $(4, 3\pi/2)$

39. Eccentricity 4; directrix; $r = -2 \sec \theta$

40. Eccentricity 2; directrix: $r = 4 \csc \theta$

41. Eccentricity 1; directrix: $r = -3 \csc \theta$

42. Eccentricity 1; directrix: $r = 5 \sec \theta$

43. Eccentricity $1/2$; directrix: $r = 2 \sec \theta$

44. Eccentricity $4/5$; directrix: $r = 3 \csc \theta$

45. Hyperbola; vertical directrix to the left of the pole; eccentricity 2; $(1, 2\pi/3)$ is on the graph.

46. Hyperbola; horizontal directrix above the pole; eccentricity 2; $(1, 2\pi/3)$ is on the graph.

47. In Case 2A of the proof of the Conic Section Theorem, show that $a > b$.

48. In Case 2B of the proof of the Conic Section Theorem, show that the hyperbola has eccentricity e.

49. A comet travels in a parabolic orbit with the sun as focus. When the comet is 60 million miles from the sun, the line segment from the sun to the comet makes an angle of $\pi/3$ radians with the axis of the parabolic orbit. Using the sun as the pole and assuming the axis of the orbit lies along the polar axis, find a polar equation for the orbit.

50. Halley's Comet has an elliptical orbit, with eccentricity .97 and the sun as a focus. The length of the major axis of the orbit is 3364.74 million miles. Using the sun as the pole and assuming the major axis of the orbit is perpendic–ular to the polar axis, find a polar equation for the orbit.

Chapter 10 Review

IMPORTANT CONCEPTS

Section 10.1
Circle   638
Equations of circles   638–639
Ellipse   639
Foci   639
Center   639
Major and minor axes   639
Vertical and horizontal shifts   643
Equations of ellipses   641, 645, 647

Section 10.2
Hyperbola   652
Foci   652
Center   652
Vertices   652
Asymptotes   652
Equations of hyperbolas   652, 654, 657

Section 10.3
Parabola   663
Focus   663
Directrix   663
Vertex   663
Equations of parabolas   664, 665, 667

Section 10.4
Graphs of rotated conics   673–675
Discriminant   674

Special Topics 10.4.A
Rotation equations   677
Eliminating the xy term   679

Section 10.5
Plane curve   682
Parameter   682

Parametric equations   682
Eliminating the parameter   685
Finding parameterizations   685
Cycloid   689

Section 10.6
Polar coordinates   695
Pole   695
Polar axis   695
Polar/rectangular coordinate conversion   697
Polar graphs   699–702

Section 10.7
Eccentricity   704
Polar equations for conics   706
Focus   708
Directrix   708

IMPORTANT FACTS & FORMULAS

■ Equations of conic sections

Conic Section	Cartesian Equation	Parametric Equations
Circle center (h, k) radius r	$(x - h)^2 + (y - k)^2 = r^2$	$x = r \cos t + h$ $y = r \sin t + k$ $(0 \le t \le 2\pi)$
Ellipse center (h, k) axes on the lines $x = h, y = k$	$\dfrac{(x - h)^2}{a^2} + \dfrac{(y - k)^2}{b^2} = 1$	$x = a \cos t + h$ $y = b \sin t + k$ $(0 \le t \le 2\pi)$
Hyperbola center (h, k) vertices on the line $y = k$	$\dfrac{(x - h)^2}{a^2} - \dfrac{(y - k)^2}{b^2} = 1$	$x = \dfrac{a}{\cos t} + h$ $y = b \tan t + k$ $(0 \le t \le 2\pi)$

Hyperbola
  center (h, k)
  vertices on the line
      $r = h$

$$\frac{(y-k)^2}{a^2} - \frac{(x-h)^2}{b^2} = 1$$

$$x = b \tan t + h$$
$$y = \frac{a}{\cos t} + k$$
$$(0 \leq t \leq 2\pi)$$

Parabola
  vertex (h, k)
  axis $x = h$

$$(x-h)^2 = 4p(y-k)$$

$$x = t$$
$$y = \frac{(t-h)^2}{4p} + k$$
$$(t \text{ any real})$$

Parabola
  vertex (h, k)
  axis $y = k$

$$(y-k)^2 = 4p(x-h)$$

$$x = \frac{(t-k)^2}{4p} + h$$
$$y = t$$
$$(t \text{ any real})$$

■ The discriminant of the equation $Ax^2 + Bxy + Cy^2 + Dx + Ey + F = 0$ (where A, B, C are not all zero) is $B^2 - 4AC$.

    If $B^2 - 4AC < 0$, the graph is a circle or ellipse (or a point).

    If $B^2 - 4AC = 0$, the graph is a parabola (or a line or two parallel lines).

    If $B^2 - 4AC > 0$, the graph is a hyperbola (or two intersecting lines).

■ Rotation Equations:

$$x = u \cos \theta - v \sin \theta,$$

$$y = u \sin \theta + v \cos \theta.$$

■ To eliminate the xy term in $Ax^2 + Bxy + Cy^2 + Dx + Ey + F = 0$, rotate the axes through an angle θ such that $\cot 2\theta = \dfrac{A - C}{B}$.

■ The rectangular and polar coordinates of a point are related by

$$x = r \cos \theta \quad \text{and} \quad y = r \sin \theta;$$

$$r^2 = x^2 + y^2 \quad \text{and} \quad \tan \theta = \frac{y}{x}.$$

■ If e and d are constants with $e > 0$, then the graph of a polar equation of the form

$$r = \frac{ed}{1 \pm e \cos \theta} \quad \text{or} \quad r = \frac{ed}{1 \pm e \sin \theta}$$

is an ellipse if $0 < e < 1$, a parabola if $e = 1$, and a hyperbola if $e > 1$.

REVIEW QUESTIONS

In Questions 1–4, find the foci and vertices of the conic, and state whether it is an ellipse or a hyperbola.

1. $\dfrac{x^2}{16} + \dfrac{y^2}{20} = 1$

2. $\dfrac{x^2}{9} - \dfrac{y^2}{16} = 1$

3. $\dfrac{(x-1)^2}{7} + \dfrac{(y-3)^2}{16} = 1$

4. $3x^2 = 1 + 2y^2$

5. Find the focus and directrix of the parabola $10y = 7x^2$.

6. Find the focus and directrix of the parabola

$$3y^2 - x - 4y + 4 = 0.$$

In Questions 7–20, sketch the graph of the equation. If there are asymptotes, give their equations.

7. $\dfrac{x^2}{4} + \dfrac{y^2}{25} = 1$

8. $25x^2 + 4y^2 = 100$

9. $\dfrac{(x-3)^2}{9} + \dfrac{(y+5)^2}{4} = 1$

10. $\dfrac{x^2}{9} - \dfrac{y^2}{16} = 1$

11. $\dfrac{(y+4)^2}{25} - \dfrac{(x-1)^2}{4} = 1$

12. $4x^2 - 9y^2 = 144$

13. $x^2 + 4y^2 - 10x + 9 = 0$

14. $9x^2 - 4y^2 - 36x + 24y - 36 = 0$

15. $2y = 4(x-3)^2 + 6$

16. $3y = 6(x+1)^2 - 9$

17. $x = y^2 + 2y + 2$

18. $y = x^2 - 2x + 3$

19. $x^2 + y^2 - 6x + 5 = 0$

20. $x^2 + y^2 - 4x + 6y + 4 = 0$

21. Find the center and radius of the circle whose equation is

$$x^2 + y^2 + 8x + 10y + 33 = 0.$$

22. Find the equation of the circle with center $(-2, 3)$ that passes through the point $(1, 7)$.

23. What is the center of the ellipse

$$4x^2 + 3y^2 - 32x + 36y + 124 = 0?$$

24. Find the equation of the ellipse with center at the origin, one vertex at $(0, 4)$, passing through $(\sqrt{3}, 2\sqrt{3})$.

25. Find the equation of the ellipse with center at $(3, 1)$, one vertex at $(1, 1)$, passing through $(2, 1 + \sqrt{3}/2)$.

26. Find the equation of the hyperbola with center at the origin, one vertex at $(0, 5)$, pass through $(1, 3\sqrt{5})$.

27. Find the equation of the hyperbola with center at $(3, 0)$, one vertex at $(3, 2)$, passing through $(1, \sqrt{5})$.

28. Find the equation of the parabola with vertex $(2, 5)$, axis $x = 2$, and passing through $(3, 12)$.

29. Find the equation of the parabola with vertex $(3/2, -1/2)$, axis $y = -1/2$, and passing through $(-3, 1)$.

30. Find the equation of the parabola with vertex $(5, 2)$ that passes through the points $(7, 3)$ and $(9, 6)$.

In Questions 31–36, use parametric graphing to sketch the graph of the equation.

31. $\dfrac{x^2}{4} + \dfrac{y^2}{25} = 1$

32. $25x^2 + 4y^2 = 100$

33. $\dfrac{(x-3)^2}{9} + \dfrac{(y+5)^2}{4} = 1$

34. $\dfrac{x^2}{9} - \dfrac{y^2}{16} = 1$

35. $\dfrac{(y+4)^2}{25} - \dfrac{(x-1)^2}{4} = 1$

36. $4x^2 - 9y^2 = 144$

37. The arch shown in the figure has the shape of half of an ellipse. How wide is the arch at a point 8 feet above the ground? [*Hint:* Think of the arch as sitting on the *x*-axis, with its center at the origin, and find its equation.]

38. A satellite dish with a parabolic cross section is 6 feet in diameter. The receiver is located on the center axis, 1 foot from the base of the dish. How deep is the dish at its center?

In Questions 39–42, assume that the graph of the equation is a nondegenerate conic. Use the discriminant to identify the graph.

39. $3x^2 + 2\sqrt{2}xy + 2y^2 - 12 = 0$

40. $x^2 + y^2 - xy - 4y = 0$

41. $4xy - 3x^2 - 20 = 0$

42. $4x^2 - 4xy + y^2 - \sqrt{5}x - 2\sqrt{5}y = 0$

In Questions 43–48, find a viewing window that shows a complete graph of the equation.

43. $x^2 - xy + y^2 - 6 = 0$

44. $x^2 + xy + y^2 - 3y - 6 = 0$

REVIEW QUESTIONS

45. $x^2 + xy - 2 = 0$

46. $x^2 - 4xy + y^2 + 5 = 0$

47. $x^2 + 3xy + y^2 - 2\sqrt{2}x + 2\sqrt{2}y = 0$

48. $x^2 + 2xy + y^2 - 4\sqrt{2}y = 0$

In Questions 49–50, find the rotation equations when the x-y axes are rotated through the given angle.

49. $60°$ **50.** $45°$

In Questions 51–52, find the angle through which the x-y axes should be rotated to eliminate the xy term in the equation.

51. $x^2 + xy + y^2 - 3y - 6 = 0$

52. $x^2 - 4xy + y^2 + 5 = 0$

In Questions 53–56, find a viewing window that shows a complete graph of the curve whose parametric equations are given.

53. $x = 8 \cos t + \cos 8t$
$y = 8 \sin t - \sin 8t$ $(0 \le t \le 2\pi)$

54. $x = [64 \cos(\pi/6)]t$
$y = -16t^2 + [64 \sin(\pi/6)]t$ $(0 \le t \le \pi)$

55. $x = t^3 + t + 1$ and $y = t^2 + 2t$ $(-3 \le t \le 3)$

56. $x = t^2 - t + 3$ and $y = t^3 - 5t$ $(-3 \le t \le 3)$

In Questions 57–60, sketch the graph of the curve whose parametric equations are given and find an equation in x and y whose graph contains the given curve.

57. $x = 2t - 1$, $y = 2 - t$, $-3 \le t \le 3$

58. $x = 3 \cos t$, $y = 5 \sin t$, $0 \le t \le 2\pi$

59. $x = \cos t$, $y = 2 \sin^2 t$, $0 \le t \le 2\pi$

60. $x = e^t$, $y = \sqrt{t + 1}$, $t \ge 1$

61. Which of the following are *not* parameterizations of the curve $x = y^2 + 1$?
(a) $x = t^2 + 1$, $y = t$,   any real number t
(b) $x = \sin^2 t + 1$, $y = \sin t$,   any real number t
(c) $x = t^4 + 1$, $y = t^2$,   any real number t
(d) $x = t^6 + 1$, $y = t^3$,   any real number t

62. Which of the curves in Questions 53–56 appear to be the graphs of functions of the form $y = f(x)$?

63. Plot the points $(2, 3\pi/4)$ and $(-3, -2\pi/3)$ on a polar coordinate graph.

64. List four other pairs of polar coordinates for the point $(-2, \pi/4)$.

In Questions 65–74, sketch the graph of the equation in a polar coordinate system.

65. $r = 5$ **66.** $r = -2$

67. $\theta = 2\pi/3$ **68.** $\theta = -5\pi/6$

69. $r = 2\theta$ $(\theta \le 0)$ **70.** $r = 4 \cos \theta$

71. $r = 2 - 2 \sin \theta$ **72.** $r = \cos 3\theta$

73. $r^2 = \cos 2\theta$ **74.** $r = 1 + 2 \sin \theta$

75. Convert $(3, -2\pi/3)$ from polar to rectangular coordinates.

76. Convert $(3, \sqrt{3})$ from rectangular to polar coordinates.

77. What is the eccentricity of the ellipse $3x^2 + y^2 = 84$?

78. What is the eccentricity of the ellipse $24x^2 + 30y^2 = 120$?

In Questions 79–82, sketch the graph of the equation, labeling the vertices and identifying the conic.

79. $r = \dfrac{12}{2 - \sin \theta}$ **80.** $r = \dfrac{14}{7 + 7 \cos \theta}$

81. $r = \dfrac{-24}{3 - 9 \cos \theta}$ **82.** $r = \dfrac{10}{3 + 4 \sin \theta}$

In Questions 83–86, find a polar equation of the conic that has focus (0, 0) and satisfies the given conditions.

83. Ellipse; vertices $(4, 0)$ and $(6, \pi)$

84. Hyperbola; vertices $(5, \pi/2)$ and $(-3, 3\pi/2)$

85. Eccentricity 1; directrix $r = 2 \sec \theta$

86. Eccentricity .75; directrix $r = -3 \csc \theta$

DISCOVERY PROJECT 10   Designing Machines to Make Designs

The Studio Dog/Getty Images

Parametric equations are extremely useful for modeling the behavior of moving parts, particularly when the action can be decomposed into two or more discrete movements. You saw this in Section 10.5 in the form of the cycloid curves—the rolling of a circle was decomposed into the spinning of the circle and the linear motion of the center of the circle. Similarly, the motion of machine parts can be examined from the reverse point of view. That is, you can look at how two or more discrete parts act in concert to create a desirable pattern. In the case of the cycloid, the parametric equations $x = r(t - \sin t)$ and $y = r(1 - \cos t)$ model the motion of a point on the rim of a wheel of radius r which is rolling across the x-axis. The first component of each formula, the t and the 1, show the linear movement of the axle or center of the circle, while the $\sin t$ and $\cos t$ take care of the rotation of the rim around the axle. What happens if the motion of the center and rim are no longer synchronized?

1. Graph the parametric equations $x = t + \cos(kt)$ and $y = 5 + \sin(kt)$ for various values of k greater than 1. What does it mean physically when k is larger than 1?

2. Graph the parametric equations $x = t + \cos(kt)$ and $y = 5 + 5\sin(kt)$ for various values of k less than 1. What does it mean physically when k is less than 1?

3. The formulas from the previous questions can also be used to design a machine in which objects move at a constant speed below a device that spins in a circle above the objects. This is a popular method for decorating mass-produced pastries. Find a value of k so that the graph of the parametric system looks like the picture to the right. Interpret what the value of k means in this instance.

4. On the basis of your answer to the previous question, modify the parametric equations so that the pastries will have a strip of decorative icing approximately 1 cm wide, measured perpendicularly to the linear motion. Be sure to give time and length units in your answer.

DISCOVERY PROJECT 10

5. Sewing machines use similar devices, called feed dogs, to move the cloth beneath the presser foot. The motion of the feed dogs can be simulated using functions of the type $\dfrac{|\sin kt|}{\sin kt}$ in place of sines and cosines. Use this to design a set of parametric equations that would produce stitches like the ones shown to the right.

6. How would you change your answer to question 5 if the width of the stitches, measured perpendicularly to the direction of motion, needs to be 5 times as wide but the length when measured from peak to peak in the direction of motion remains the same?

C H A P T E R 11

Systems of Equations

S. Stammer/Photo Researchers, Inc.

Is this a diamond in the rough?

The structure of certain crystals can be described by a large system of linear equations (more than a hundred equations and variables). A variety of resource allocation problems involving many variables can be handled by solving an appropriate system of equations. The fastest solution methods involve matrices and are easily implemented on a computer or calculator. See Exercise 51 on page 755.

Chapter Outline

Interdependence of Sections

Chapters 11–13 are independent of each other and may be read in any order.

In this chapter, the interdependence of sections is

11.1 ⟶ 11.2 ⟶ 11.3

11.1   Systems of Linear Equations in Two Variables
11.1.A *Special Topics:* Systems of Nonlinear Equations
11.2   Large Systems of Linear Equations
11.3   Matrix Methods for Square Systems

This chapter deals with *systems of equations*, such as

$$2x - 5y + 3z = 1$$
$$x + 2y - z = 2$$
$$3x + y + 2z = 11$$

Three equations in
three variables

$$2x + 5y + z + w = 0$$
$$2y - 4z + 41w = 5$$
$$3x + 7y + 5z - 8w = -6$$

Three equations in
four variables

$$x^2 + y^2 = 25$$
$$x^2 - y = 7$$

Two equations in
two variables

Sections 11.1–11.3 deal with systems of linear equations (such as the first two shown above). Systems involving nonlinear equations are considered in Special Topics 11.1.A.

A **solution of a system** is a solution that satisfies *all* the equations in the system. For instance, in the first system of equations above, $x = 1$, $y = 2$, $z = 3$ is a solution of all three equations (check it) and hence is a solution of the system. On the other hand, $x = 0$, $y = 7$, $z = 12$ is a solution of the first two equations but not of the third (check it). So $x = 0$, $y = 7$, $z = 12$ is not a solution of the system.

11.1   Systems of Linear Equations in Two Variables

Systems of linear equations in two variables may be solved geometrically or algebraically. The geometric method is similar to what we have done previously.

EXAMPLE 1

Solve this system geometrically.

$$2x - y = 1$$
$$3x + 2y = 4.$$

SOLUTION   First, we solve each equation for y.

$$2x - y = 1 \qquad 3x + 2y = 4$$
$$-y = -2x + 1 \qquad 2y = -3x + 4$$
$$y = 2x - 1 \qquad y = \frac{-3x + 4}{2}.$$

Next, we graph both equations on the same screen (Figure 11–1). As we saw in Section 1.4, each graph is a straight line, and every point on the graph represents a solution of the equation. Therefore, the solution of the system is given by the coordinates of the point that lies on both lines. An intersection finder (Figure 11–2) shows that the approximate coordinates of this point are

$$x \approx .85714286 \qquad \text{and} \qquad y \approx .71428571. \qquad \blacksquare$$

Figure 11–1

Figure 11–2

As is shown in Example 1, the solutions of a system of linear equations are determined by the points where their graphs intersect. There are exactly three geometric possibilities for two lines in the plane: They are parallel, they intersect at a single point, or they coincide, as illustrated in Figure 11–3. Each of these possibilities leads to a different number of solutions for the system.

Number of Solutions of a System

A system of two linear equations in two variables must have

No solutions *or*

Exactly one solution *or*

An infinite number of solutions.

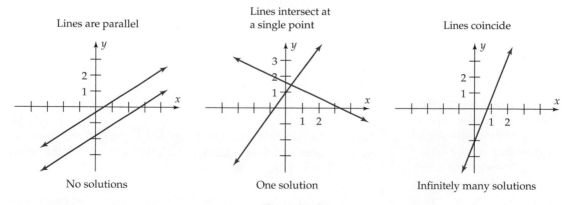

Figure 11–3

:: THE SUBSTITUTION METHOD

When you use a calculator to solve systems geometrically, you may have to settle for an approximate solution, as we did in Example 1. Algebraic methods, however, produce exact solutions. Furthermore, algebraic methods are often as easy to implement as geometric ones, so we shall use them, whenever practical, to obtain exact solutions. One algebraic method is **substitution,** which is explained in the next example.

EXAMPLE 2

Use substitution to find the exact solution of the system from Example 1.

$$2x - y = 1$$
$$3x + 2y = 4.$$

SOLUTION   Any solution of this system must satisfy the first equation, $2x - y = 1$. Solving this equation for y, as in Example 1, shows that

$$y = 2x - 1.$$

Substituting this expression for y in the second equation, we have

$$3x + 2y = 4$$
$$3x + 2(2x - 1) = 4$$
$$3x + 4x - 2 = 4$$
$$7x = 6$$
$$x = 6/7.$$

Therefore, every solution of the original system must have $x = 6/7$. But when $x = 6/7$, we see from the first equation that

$$2x - y = 1$$
$$2\left(\frac{6}{7}\right) - y = 1$$
$$\frac{12}{7} - y = 1$$
$$-y = -\frac{12}{7} + 1$$
$$y = \frac{12}{7} - 1 = \frac{5}{7}.$$

(We would also have found that $y = 5/7$ if we had substituted $x = 6/7$ in the second equation.) Consequently, the exact solution of the original system is $x = 6/7, y = 5/7$.   ■

When using the substitution method, you may solve either of the given equations for either one of the variables and then substitute that result in the other equation. In Example 2, we solved for y in the first equation because that avoided the fractional expression that would have occurred if we had solved for x or had solved the second equation for x or y.

▪▪ THE ELIMINATION METHOD

The **elimination method** of solving systems of linear equations is often more convenient than substitution. It depends on this fact.

> **Multiplying both sides of an equation by a nonzero constant does not change the solutions of the equation.**

For example, the equation $x + 3 = 5$ has the same solution as $2x + 6 = 10$ (the first equation multiplied by 2). The elimination method also uses this fact from basic algebra:

> If $A = B$ and $C = D$, then $A + C = B + D$ and $A - C = B - D$.

EXAMPLE 3

To solve this system

$$x - 3y = 4$$
$$2x + y = 1,$$

we replace the first equation by an equivalent one (that is, one with the same solutions).

$$-2x + 6y = -8 \quad \text{[First equation multiplied by } -2\text{]}$$
$$2x + y = 1.$$

The multiplier -2 was chosen so that the coefficients of x in the two equations would be negatives of each other. Any solution of this last system must also be a solution of the sum of the two equations.

$$-2x + 6y = -8$$
$$\underline{2x + y = 1}$$
$$7y = -7. \quad \text{[The first variable has been eliminated]}$$

Solving this last equation, we see that $y = -1$. Substituting this value in the first of the original equations shows that

$$x - 3(-1) = 4$$
$$x = 1.$$

Therefore, $x = 1$, $y = -1$ is the solution of the original system. ▪

EXAMPLE 4

Any solution of the system

$$5x - 3y = 3$$
$$3x - 2y = 1$$

must also be a solution of this system:

$$10x - 6y = 6 \qquad \text{[First equation multiplied by 2]}$$

$$-9x + 6y = -3. \qquad \text{[Second equation multiplied by } -3]$$

The multipliers 2 and -3 were chosen so that the coefficients of y in the new equations would be negatives of each other. Any solution of this last system must also be a solution of the equation obtained by adding these two equations.

$$\begin{array}{r} 10x - 6y = 6 \\ \underline{-9x + 6y = -3} \\ x = 3. \end{array} \qquad \text{[The second variable has been eliminated]}$$

Substituting $x = 3$ in the first of the original equations shows that

$$5(3) - 3y = 3$$

$$-3y = -12$$

$$y = 4.$$

Therefore, the solution of the original system is $x = 3$, $y = 4$. ∎

EXAMPLE 5

To solve the system

$$2x - 3y = 5$$

$$4x - 6y = 1,$$

we multiply the first equation by -2 and add.

$$\begin{array}{r} -4x + 6y = -10 \\ \underline{4x - 6y = 1} \\ 0 = -9. \end{array}$$

Since $0 = -9$ is always false, the original system cannot possibly have any solutions. A system with no solution is said to be **inconsistent.** ∎

GRAPHING EXPLORATION

Confirm the result of Example 5 geometrically by graphing the two equations in the system. Do these lines intersect, or are they parallel?

EXAMPLE 6

To solve the system

$$3x - y = 2$$

$$6x - 2y = 4$$

we multiply the first equation by 2 to obtain the following system.

$$6x - 2y = 4$$
$$6x - 2y = 4.$$

The two equations are identical. So the solutions of this system are the same as the solutions of the single equation $6x - 2y = 4$, which can be rewritten as

$$2y = 6x - 4$$
$$y = 3x - 2.$$

This equation, and hence the original system, has infinitely many solutions. They can be described as follows: Choose any real number for x, say, $x = b$. Then $y = 3x - 2 = 3b - 2$. So the solutions of the system are all pairs of numbers of the form

$$x = b, \qquad y = 3b - 2, \quad \text{where } b \text{ is any real number.}$$

A system such as this is said to be **dependent.**   ∎

   Some nonlinear systems can be solved by replacing them with equivalent linear systems.

EXAMPLE 7

To solve the system

$$\frac{1}{x} + \frac{3}{y} = -1$$
$$\frac{2}{x} - \frac{1}{y} = 5,$$

we let $u = 1/x$ and $v = 1/y$ so that the system becomes

$$u + 3v = -1$$
$$2u - v = 5.$$

We can solve this system by multiplying the first equation by -2 and adding it to the second equation.

$$-2u - 6v = 2$$
$$\underline{2u - v = 5}$$
$$-7v = 7$$
$$v = -1.$$

Substituting $v = -1$ in the equation $u + 3v = -1$, we see that

$$u = -3(-1) - 1 = 2.$$

Consequently, the possible solution of the original system is

$$x = \frac{1}{u} = \frac{1}{2} \qquad \text{and} \qquad y = \frac{1}{v} = \frac{1}{(-1)} = -1.$$

You should substitute this possible solution in both equations of the original system to check that it is actually a solution of the system.   ∎

▪▪ APPLICATIONS

Producers are happy to supply items at a high price, but if they do, the consumers' demand for them may be low. At lower prices, more items will be demanded, but if the price is too low, producers may not want to supply the items. The **equilibrium price** is the price at which the number of items demanded by consumers is the same as the number supplied by producers. The number of items demanded and supplied at the equilibrium price is the **equilibrium quantity.**

EXAMPLE 8

The consumer demand for a certain type of floor tile is related to its price by the equation $p = 60 - .75x$, where p is the price (in dollars) at which x thousand boxes of tile will be demanded. The supply of these tiles is related to their price by the equation $p = .8x + 5.75$, where p is the price at which x thousand boxes will be supplied by the producer. Find the equilibrium quantity and the equilibrium price.

SOLUTION   We must find the values of x and p that satisfy both the supply and demand equations. In other words, we must solve this system:

$$p = 60 - .75x$$

$$p = .8x + 5.75.$$

Algebraic Method.   Since both equations are already solved for p, we use substitution. Substituting the value of p given by the first equation into the second, we obtain

$$60 - .75x = .8x + 5.75$$

Subtract $.8x$ from both sides: $60 - 1.55x = 5.75$

Subtract 60 from both sides: $-1.55x = -54.25$

Divide both sides by -1.55: $x = \dfrac{-54.25}{-1.55} = 35.$

We can determine p by substituting $x = 35$ in either equation, say, the first one.

$$p = 60 - .75x$$

$$p = 60 - .75(35) = 33.75.$$

Therefore, the equilibrium quantity is 35,000 boxes (x is measured in thousands), and the equilibrium price is $33.75 per box.

Graphical Method.   We graph the two equations in the form $y = 60 - .75x$ and $y = .8x + 5.75$ on the same screen and find their intersection point (Figure 11–4). The intersection point (35, 33.75) is called the **equilibrium point.** Its first coordinate is the equilibrium quantity, and its second coordinate is the equilibrium price.   ▪

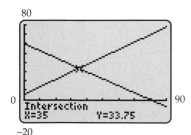

Figure 11–4

EXAMPLE 9

575 people attend a ball game, and total ticket sales are $2575. If adult tickets cost $5 and children's tickets cost $3, how many adults attended the game? How many children?

SOLUTION Let x be the number of adults, and let y be the number of children. Then,

Number of adults + Number of children = Total attendance

$$x + y = 575.$$

We can obtain a second equation by using the information about ticket sales.

Adult ticket sales + Children ticket sales = Total ticket sales

$$\left(\begin{matrix} \text{Price} \\ \text{per} \\ \text{ticket} \end{matrix}\right) \times \left(\begin{matrix} \text{Number} \\ \text{of} \\ \text{adults} \end{matrix}\right) + \left(\begin{matrix} \text{Price} \\ \text{per} \\ \text{ticket} \end{matrix}\right) \times \left(\begin{matrix} \text{Number} \\ \text{of} \\ \text{children} \end{matrix}\right) = 2575$$

$$5x \quad + \quad 3y \quad\quad\quad = 2575.$$

To find x and y, we need only solve this system of equations:

$$x + \ y = \ \ 575$$
$$5x + 3y = 2575.$$

Multiplying the first equation by -3 and adding, we have

$$-3x - 3y = -1725$$
$$\underline{5x + 3y = \ \ \ 2575}$$
$$2x \quad\quad = \quad\ \ 850$$
$$x = \quad\ \ 425.$$

So 425 adults attended the game. The number of children was

$$y = 575 - x = 575 - 425 = 150. \quad\blacksquare$$

EXAMPLE 10

A plane flies 3000 miles from San Francisco to Boston in 5 hours, with a tailwind all the way. The return trip on the same route, now with a headwind, takes 6 hours. Assuming that both remain constant, find the speed of the plane and the speed of the wind.

SOLUTION Let x be the plane's speed, and let y be the wind speed (both in miles per hour). Then on the trip to Boston with a tailwind,

$x + y =$ actual speed of the plane (wind and plane go in same direction).

On the return trip against a headwind,

$x - y$ = actual speed of plane (wind and plane go in opposite directions).

Using the basic rate/distance equation, we have

Trip to Boston

time × rate = distance

$$5(x + y) = 3000$$
$$5x + 5y = 3000$$
$$x + y = 600$$

Return Trip

time × rate = distance

$$6(x - y) = 3000$$
$$6x - 6y = 3000$$
$$x - y = 500$$

Thus, we need only solve this system of equations:

$$x + y = 600$$
$$x - y = 500.$$

Adding the two equations shows that

$$2x = 1100$$
$$x = 550.$$

Substituting this result in the first equation, we have

$$550 + y = 600$$
$$y = 50.$$

Thus, the plane's speed is 550 mph, and the wind speed is 50 mph.   ■

EXAMPLE 11

How many pounds of tin and how many pounds of copper should be added to 1000 pounds of an alloy that is 10% tin and 30% copper to produce a new alloy that is 27.5% tin and 35% copper?

SOLUTION   Let x be the number of pounds of tin and y the number of pounds of copper to be added to the 1000 pounds of the old alloy. Then there will be $1000 + x + y$ pounds of the new alloy. We first find the *amounts* of tin and copper in the new alloy.

	Pounds in old alloy	+ Pounds added =	Pounds in new alloy
Tin	10% of 1000 +	x	= 100 + x
Copper	30% of 1000 +	y	= 300 + y

Now consider the *percentages* of tin and copper in the new alloy.

	Percentage in new alloy	×	Total weight of new alloy	=	Pounds in new alloy
Tin	27.5%	of	$1000 + x + y$	=	$.275(1000 + x + y)$
Copper	35%	of	$1000 + x + y$	=	$.35(1000 + x + y)$

The two ways of computing the weight of each metal in the alloy must produce the same result, that is,

$$100 + x = .275(1000 + x + y) \quad \text{[pounds of tin]}$$

$$300 + y = .35(1000 + x + y). \quad \text{[pounds of copper]}$$

Multiplying out the right sides and rearranging terms produces this system of equations:

$$.725x - .275y = 175$$

$$-.35x + .65y = 50.$$

Multiplying the first equation by .65 and the second by .275 and adding the results, we have

$$.47125x - .17875y = 113.75$$

$$\underline{-.09625x + .17875y = 13.75}$$

$$.37500x \qquad\qquad = 127.50$$

$$x = 340.$$

Substituting this in the first equation and solving for y shows that $y = 260$. Therefore, 340 pounds of tin and 260 pounds of copper should be added. ∎

✓ EXERCISES 11.1

In Exercises 1–6, determine whether the given values of x, y, and z are a solution of the system of equations.

1. $x = -1, y = 3$

$\quad 2x + y = 1$

$\quad -3x + 2y = 9$

2. $x = 3, y = 4$

$\quad 2x + 6y = 30$

$\quad x + 2y = 11$

3. $x = 2, y = -1$

$\quad \dfrac{1}{3}x + \dfrac{1}{2}y = \dfrac{1}{6}$

$\quad \dfrac{1}{2}x + \dfrac{1}{3}y = \dfrac{2}{3}$

4. $x = .4, \quad y = .7$

$\quad 3.1x - 2y = -.16$

$\quad 5x - 3.5y = -.48$

5. $x = \dfrac{1}{2}, y = 3, z = -1$

$\quad 2x - y + 4z = -6$

$\quad 3y + 3z = 6$

$\quad 2z = 2$

6. $x = 2, y = \dfrac{3}{2}, z = -\dfrac{1}{2}$

$\quad 3x + 4y - 2z = 13$

$\quad \dfrac{1}{2}x \qquad + 8z = -3$

$\quad x - 3y + 5z = -5$

In Exercises 7–14, use substitution to solve the system.

7. $\quad x - 2y = 5$

$\quad 2x + y = 3$

8. $3x - y = 1$

$\quad -x + 2y = 4$

9. $3x - 2y = 4$

$\quad 2x + y = -1$

10. $5x - 3y = -2$

$\quad -x + 2y = 3$

11. $r + s = 0$

  $r - s = 5$

12. $t = 3u + 5$

  $t = u + 5$

13. $x + y = c + d$ (where c, d are constants)

  $x - y = 2c - d$

14. $x + 3y = c - d$ (where c, d are constants)

  $2x - y = c + d$

In Exercises 15–30, use the elimination method to solve the system.

15. $2x - 2y = 12$

  $-2x + 3y = 10$

16. $3x + 2y = -4$

  $4x - 2y = -10$

17. $x + 3y = -1$

  $2x - y = 5$

18. $4x - 3y = -1$

  $x + 2y = 19$

19. $2x + 3y = 15$

  $8x + 12y = 40$

20. $2x + 5y = 8$

  $6x + 15y = 18$

21. $3x - 2y = 4$

  $6x - 4y = 8$

22. $2x - 8y = 2$

  $3x - 12y = 3$

23. $\dfrac{x}{3} - \dfrac{y}{2} = -3$

  $\dfrac{2x}{5} + \dfrac{y}{5} = -2$

24. $\dfrac{x}{3} + \dfrac{3y}{5} = 4$

  $\dfrac{x}{6} - \dfrac{y}{2} = -3$

25. $\dfrac{x + y}{4} - \dfrac{x - y}{3} = 1$

  $\dfrac{x + y}{4} + \dfrac{x - y}{2} = 9$

26. $\dfrac{x - y}{4} + \dfrac{x + y}{3} = 1$

  $\dfrac{x + 2y}{3} + \dfrac{3x - y}{2} = -2$

27. $\dfrac{1}{x} - \dfrac{3}{y} = 2$

  $\dfrac{2}{x} + \dfrac{1}{y} = 3$

28. $\dfrac{5}{x} + \dfrac{2}{y} = 0$

  $\dfrac{6}{x} + \dfrac{4}{y} = 3$

29. $\dfrac{2}{x} + \dfrac{3}{y} = 8$

  $\dfrac{3}{x} - \dfrac{1}{y} = 1$

30. $\dfrac{3}{x + 1} - \dfrac{4}{y - 2} = 2$

  $\dfrac{1}{x + 1} + \dfrac{4}{y - 2} = 5$

  $\left[\text{\textit{Hint:} Let } u = \dfrac{1}{x + 1} \text{ and } v = \dfrac{1}{y - 2}. \right]$

31. The population y in year x of Philadelphia and Houston is approximated by these equations:

  Philadelphia: $12.04x + y = 2099.64$

  Houston: $-26.47x + y = 660.64$,

  where $x = 0$ corresponds to 1950 and y is in thousands.*

(a) How can you tell from the equation whether a city's population was increasing or decreasing since 1950?

(b) In what year do the two cities have the same population?

32. The population y in year x of Long Beach and New Orleans is approximated by these equations:

  Long Beach: $-6.1x + 2y = 660$

  New Orleans: $3.82x + y = 628.4$,

  where $x = 0$ corresponds to 1960 and y is in thousands.* In what year do the two cities have the same population?

33. On the basis of data from 1995–2000, the median income y in year x for men and women is approximated by these equations:

  Men: $-2106x + 3y = 74,148$

  Women: $-1190x + 2y = 26,366$,

  where $x = 0$ corresponds to 1995 and y is in constant 1999 dollars.* If the equations remain valid in the future, when will the median income of men and women be the same?

34. The death rate per 100,000 population y in year x for heart disease and cancer is approximated by these equations:

  Heart Disease: $6.9x + 2y = 728.4$

  Cancer: $-1.3x + y = 167.5$,

  where $x = 0$ corresponds to 1970.† If the equations remain accurate, when will the death rates for heart disease and cancer be the same?

In Exercises 35 and 36, solve the system of equations.

35. $ax + by = r$ (where a, b, c, d, r, s are

  $cx + dy = s$ constants and $ad - bc \neq 0$)

36. $ax + by = ab$ (where a, b are nonzero

  $bx - ay = ab$ constants)

37. Let c be any real number. Show that this system has exactly one solution.

  $$x + 2y = c$$

  $$6x - 3y = 4.$$

38. (a) Find the values of c for which this system has an infinite number of solutions.

  $$2x - 4y = 6$$

  $$-3x + 6y = c.$$

(b) Find the values of c for which the system in part (a) has no solutions.

---

*U.S. Census Bureau

†U.S. Department of Health and Human Services.

In Exercises 39 and 40, find the values of c and d for which both given points lie on the given straight line.

39. $cx + dy = 2$; (0, 4) and (2, 16)
40. $cx + dy = -6$; (1, 3) and (-2, 12)

In Exercises 41–44, find the equilibrium quantity and the equilibrium price. In the supply and demand equations, p is price (in dollars) and x is quantity (in thousands).

41. Supply: $p = .75x$
Demand: $p = 16 - 1.25x$

42. Supply: $p = 1.4x - .6$
Demand: $p = -2x + 3.2$

43. Supply: $p = 300 - 30x$
Demand: $p = 80 + 25x$

44. Supply: $p = 181 - .01x$
Demand: $p = 52 + .02x$

45. A 200-seat theater charges $3 for adults and $1.50 for children. If all seats were filled and the total ticket income was $510, how many adults and how many children were in the audience?

46. A theater charges $4 for main floor seats and $2.50 for balcony seats. If all seats are sold, the ticket income is $2100. At one show, 25% of the main floor seats and 40% of the balcony seats were sold, and ticket income was $600. How many seats are on the main floor and how many are in the balcony?

47. A boat made a 4-mile trip upstream against a constant current in 15 minutes. The return trip at the same constant speed with the same current took 12 minutes. What is the speed of the boat and what is the speed of the current?

48. A plane flying into a headwind travels 2000 miles in 4 hours and 24 minutes. The return flight along the same route with a tailwind takes 4 hours. Find the wind speed and the plane's speed (assuming that both are constant).

49. At a certain store, cashews cost $4.40/pound and peanuts cost $1.20/pound. If you want to buy exactly 3 pounds of nuts for $6.00, how many pounds of each kind of nuts should you buy? [*Hint:* If you buy x pounds of cashews and y pounds of peanuts, then $x + y = 3$. Find a second equation by considering cost, and solve the resulting system.]

50. How many cubic centimeters (cm^3) of a solution that is 20% acid and of another solution that is 45% acid should be mixed to produce 100 cm^3 of a solution that is 30% acid?

51. How many grams of a 50%-silver alloy should be mixed with a 75%-silver alloy to obtain 40 grams of a 60%-silver alloy?

52. A winemaker has two large casks of wine. One wine is 8% alcohol, and the other is 18% alcohol. How many liters of each wine should be mixed to produce 30 liters of wine that is 12% alcohol?

53. Bill and Ann plan to install a heating system for their swimming pool. Since gas is not available, they have a choice of electric or solar heat. They have gathered the following cost information.

System	Installation Costs	Monthly Operational Cost
Electric	$2,000	$80
Solar	$14,000	$9.50

(a) Ignoring changes in fuel prices, write a linear equation for each heating system that expresses its total cost y in terms of the number of *years x* of operation.
(b) What is the five-year total cost of electric heat? Of solar heat?
(c) In what year will the total cost of the two heating systems be the same? Which is the cheaper system before that time? After that time?

54. A toy company makes Boomie Babies, as well as collector cases for each Boomie Baby. To make x cases cost the company $5000 in fixed overhead, plus $7.50 per case. An outside supplier has offered to produce any desired volume of cases for $8.20 per case.
(a) Write an equation that expresses the company's cost to make x cases itself.
(b) Write an equation that expresses the cost of buying x cases from the outside supplier.
(c) Graph both equations on the same axes and determine when the two costs are the same.
(d) When should the company make the cases themselves and when should they buy them from the outside supplier?

55. When Neil Simon planned to open his play *London Suite,* his producer, Emanuel Azenberg, made the following cost and revenue estimates for opening on Broadway or off Broadway.[*]

	On Broadway	Off Broadway
Initial cost to open[†]	$1,295,000	$440,000
Weekly costs[†]	$206,500	$82,000
Weekly revenue	$250,500	$109,000

For a production on Broadway that runs x weeks, find a linear equation that gives the
 (a) total revenue R;
 (b) total cost C;
 (c) total profit P.
 (d) Do parts (a)–(c) for an off-Broadway production.
 (e) After how many weeks will the on-Broadway profit equal the off-Broadway profit? What is the profit then?
 (f) When would it be better to open off Broadway rather than on Broadway?

56. A store sells deluxe tape recorders for $150. The regular model costs $120. The total tape recorder inventory would sell for $43,800. But during a recent month, the store actually sold half of its deluxe models and two-thirds of the regular models and took in a total of $26,700. How many of each kind of recorder did the store have at the beginning of the month?

*Exercises 57–60 deal with the **break-even point,** which is the point at which revenue equals cost.*

57. The cost C of making x hedge trimmers is given by $C = 45x + 6000$. Each hedge trimmer can be sold for $60.
 (a) Find an equation that expresses the revenue R from selling x hedge trimmers.
 (b) How many hedge trimmers must be sold for the company to break even?

58. The cost C of making x cases of pet food is given by $C = 100x + 2600$. Each case of 100 boxes sells for $120.
 (a) Find an equation that expresses the revenue from selling x cases.
 (b) How many cases must be sold for the company to break even?

59. A firm is developing a new product. The marketing department estimates that no more than 450 units can be sold. Costs for producing x units are expected to be given by $C = 80x + 7500$, and revenues are given by $R = 95x$. You are the manager who must determine whether or not to produce the product. What is your decision? Why?

60. Do Exercise 59 when costs are given by

$$C = 1750x + 96{,}175,$$

and revenue is given by $R = 1975x$; no more than 800 units can be produced.

61. An investor has part of her money in an account that pays 9% annual interest and the rest in an account that pays 11% annual interest. If she has $8000 less in the higher-paying account than in the lower-paying one and her total annual interest income is $2010, how much does she have invested in each account?

62. Joyce has money in two investment funds. Last year, the first fund paid a dividend of 8%, the second paid a dividend of 2%, and Joyce received a total of $780. This year, the first fund paid a 10% dividend, the second paid only 1%, and Joyce received $810. How much money does she have invested in each fund?

63. The table shows the percentages of men and women in the labor force who are college graduates in selected years.[‡]

Year	1992	1995	1997	1998	1999	2000
Men	27.5	29.7	29.2	29.6	30.5	30.9
Women	25.0	26.6	28.0	28.6	29.5	29.8

 (a) Use linear regression (with $x = 0$ corresponding to 1990) to find an equation that gives the percentage of men in the labor force in year x who are college graduates.
 (b) Do part (a) for women.
 (c) If these models remain accurate, when will the percentage of women equal that of men?

---

[*]Albert Goetz, "Basic Economics: Calculating against Theatrical Disaster," *Mathematics Teacher* 89, no. 1 (January 1996) © 1996. Reprinted with permission of the National Councils of Teachers of Mathematics. All rights reserved.
[†]Initial cost includes sets, costumes, rehearsals, etc. Weekly expenses include theater rent, salaries, advertising, etc.

[‡]U.S. Bureau of Labor Statistics.

64. The table shows the number of passenger cars (in thousands) imported into the United States from Japan and Canada in selected years.[*]

Year	1995	1996	1997	1998	1999	2000
Japan	1114.4	1190.9	1387.8	1456.1	1707.3	1839.1
Canada	1552.7	1690.7	1731.2	1837.6	2170.4	2138.8

(a) Use linear regression to find an equation that approximates the number y of cars imported from Japan in year x, with $x = 5$ corresponding to 1995.

(b) Do part (a) for Canada.

(c) If these models remain accurate, in what year will the imports from Japan and Canada be the same? Approximately how many cars will be imported that year?

65. A machine in a pottery factory takes 3 minutes to form a bowl and 2 minutes to form a plate. The material for a bowl costs \$.25, and the material for a plate costs \$.20. If the machine runs for 8 hours straight and exactly \$44 is spent for material, how many bowls and plates can be produced?

66. Because Chevrolet and Saturn produce cars in the same price range, Chevrolet's sales are a function not only of Chevy prices (x), but of Saturn prices (y) as well. Saturn prices are related similarly to both Saturn and Chevy prices. Suppose General Motors forecasts the demand z_1 for Chevrolets and the demand z_2 for Saturns to be given by

$$z_1 = 68,000 - 6x + 4y \quad \text{and} \quad z_2 = 42,000 + 3x - 3y.$$

Solve this system of equations and express
(a) the price x of Chevrolets as a function of z_1 and z_2;
(b) the price y of Saturns as a function of z_1 and z_2.

---

[*]U.S. Census Bureau, Foreign Trade Division.

11.1.A   SPECIAL TOPICS Systems of Nonlinear Equations

Some systems that include nonlinear equations can be solved algebraically.

EXAMPLE 1

Solve the system

$$-2x + y = -1$$
$$xy = 3.$$

SOLUTION   Solve the first equation for y:

$$y = 2x - 1$$

and substitute this into the second equation:

$$xy = 3$$
$$x(2x - 1) = 3$$
$$2x^2 - x = 3$$
$$2x^2 - x - 3 = 0$$
$$(2x - 3)(x + 1) = 0$$
$$2x - 3 = 0 \quad \text{or} \quad x + 1 = 0$$
$$x = 3/2 \qquad\qquad x = -1.$$

Using the equation $y = 2x - 1$ to find the corresponding values of y, we see that

$$\text{If } x = \frac{3}{2}, \quad \text{then } y = 2\left(\frac{3}{2}\right) - 1 = 2.$$

$$\text{If } x = -1, \quad \text{then } y = 2(-1) - 1 = -3.$$

Therefore, the solutions of the system are $x = 3/2, y = 2$, and $x = -1, y = -3$. ∎

EXAMPLE 2

Solve the system

$$x^2 + y^2 = 8$$
$$x^2 - y = 6.$$

SOLUTION   Solve the second equation for y, obtaining $y = x^2 - 6$, and substitute this into the first equation.

$$x^2 + y^2 = 8$$
$$x^2 + (x^2 - 6)^2 = 8$$
$$x^2 + x^4 - 12x^2 + 36 = 8$$
$$x^4 - 11x^2 + 28 = 0$$
$$(x^2 - 4)(x^2 - 7) = 0$$
$$x^2 - 4 = 0 \quad \text{or} \quad x^2 - 7 = 0$$
$$x^2 = 4 \qquad\qquad x^2 = 7$$
$$x = \pm 2 \qquad\qquad x = \pm\sqrt{7}.$$

Using the equation $y = x^2 - 6$ to find the corresponding values of y, we find that the solutions of the system are

$$x = 2, y = -2; \quad x = -2, y = -2; \quad x = \sqrt{7}, y = 1;$$
$$x = -\sqrt{7}, y = 1. \quad ∎$$

Algebraic techniques were successful in Examples 1 and 2 because substitution led to equations whose solutions could be found exactly. When this is not the case, graphical methods are needed. The solutions of a system of equations are the points that are on the graphs of all the equations in the system. They can be approximated with a graphical intersection finder.

EXAMPLE 3

Solve the system

$$y = x^4 - 4x^3 + 9x - 1$$
$$y = 3x^2 - 3x - 7.$$

SOLUTION   If you try substitution on this system, say, by substituting the expression for y from the first equation into the second, you obtain

$$x^4 - 4x^3 + 9x - 1 = 3x^2 - 3x - 7$$
$$x^4 - 4x^3 - 3x^2 + 12x + 6 = 0.$$

This fourth-degree equation cannot be readily solved algebraically, so a graphical approach is appropriate.

    Graph both equations of the original system on the same screen. In the viewing window of Figure 11–5, the graphs intersect at three points. However, the graphs seem to be getting closer together as they run off the screen at the top right, which suggests that there may be another intersection point.

Figure 11–5

Figure 11–6

Figure 11–7

The larger window in Figure 11–6 shows four intersection points. There cannot be any more because, as we saw in the previous paragraph, the intersection points (solutions of the system) correspond to the solutions of a fourth-degree polynomial equation, which has a maximum of four solutions. An intersection finder (Figure 11–7) shows that one of the approximate solutions of the system is

$$x = -1.509484, \qquad y = 4.3640718.$$

EXPLORATION

GRAPHING

Graph the two equations in the viewing window of Figure 11–6 and use your intersection finder to approximate the other three solutions of this system. ■

▟▙ SYSTEMS WITH SECOND-DEGREE EQUATIONS

Solving systems of equations graphically depends on our ability to graph each equation in the system. With some equations of higher degree, this may require special techniques.

EXAMPLE 4

Solve this system graphically.

$$x^2 - 4x - y + 1 = 0$$
$$10x^2 + 25y^2 = 100.$$

SOLUTION   It's easy to graph the first equation, since it can be rewritten as $y = x^2 - 4x + 1$. To graph the second equation, we must first solve for y.

$$10x^2 + 25y^2 = 100$$
$$25y^2 = 100 - 10x^2$$
$$y^2 = \frac{100 - 10x^2}{25}$$
$$y = \sqrt{\frac{100 - 10x^2}{25}} \quad \text{or} \quad y = -\sqrt{\frac{100 - 10x^2}{25}}.$$

TECHNOLOGY TIP

On most calculators, you can graph both of these functions at once by keying in

$$y = \{1, -1\}\sqrt{\frac{100 - 10x^2}{25}}.$$

By graphing both these functions on the same screen (see the Technology Tip), we obtain the complete graph of the equation $10x^2 + 25y^2 = 100$, as shown in Figure 11–8. The graphs of both equations in the system are shown in Figure 11–9.

Figure 11–8

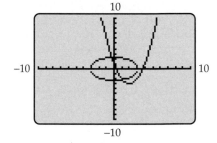

Figure 11–9

The two intersection points (solutions of the system) in Figure 11–9 can now be determined by an intersection finder.

$$x = -.2348, \, y = 1.9945 \quad \text{and} \quad x = .9544, \, y = -1.9067. \quad ∎$$

EXAMPLE 5

To solve the system

$$-4x^2 + 24xy + 3y^2 - 48 = 0$$
$$16x^2 + 24xy + 9y^2 + 100x + 50y + 100 = 0,$$

we must express each equation in terms of functions in order to graph it. The first equation may be rewritten as

$$3y^2 + (24x)y + (-4x^2 - 48) = 0.$$

This is a quadratic equation of the form $ay^2 + by + c = 0$, with

$$a = 3, \qquad b = 24x, \qquad c = -4x^2 - 48,$$

and hence can be solved by using the quadratic formula.

$$y = \frac{-b \pm \sqrt{b^2 - 4ac}}{2a}$$

$$y = \frac{-24x \pm \sqrt{(24x)^2 - 4 \cdot 3 \cdot (-4x^2 - 48)}}{2 \cdot 3}$$

$$= \frac{-24x \pm \sqrt{(24x)^2 - 12(-4x^2 - 48)}}{6}.$$

Figure 11–10

Consequently, the graph of the first equation can be obtained by graphing both of these functions on the same screen (Figure 11–10).

$$(*) \qquad \begin{aligned} y &= \frac{-24x + \sqrt{(24x)^2 - 12(-4x^2 - 48)}}{6}. \\ y &= \frac{-24x - \sqrt{(24x)^2 - 12(-4x^2 - 48)}}{6}. \end{aligned}$$

TECHNOLOGY TIP

TI-83/86 users can save keystrokes by entering the first equation of (∗) as y_1 and then using the RCL key to copy the text of y_1 to y_2. On TI-89, use COPY and PASTE in place of RCL. Then only one sign needs to be changed to make y_2 into the second equation to be graphed.

The second equation can also be solved for y by rewriting it as follows.

$$16x^2 + 24xy + 9y^2 + 100x + 50y + 100 = 0$$

$$9y^2 + (24xy + 50y) + (16x^2 + 100x + 100) = 0$$

$$9y^2 + (24x + 50)y + (16x^2 + 100x + 100) = 0.$$

Now apply the quadratic formula with $a = 9$, $b = 24x + 50$, and $c = 16x^2 + 100x + 100$.

$$y = \frac{-b \pm \sqrt{b^2 - 4ac}}{2a}$$

$$= \frac{-(24x + 50) \pm \sqrt{(24x + 50)^2 - 4 \cdot 9(16x^2 + 100x + 100)}}{2 \cdot 9}.$$

Figure 11–11

Thus, the graph of the second equation (Figure 11–11) consists of the graphs of these two functions.

$$(**) \qquad \begin{aligned} y &= \frac{-(24x + 50) + \sqrt{(24x + 50)^2 - 36(16x^2 + 100x + 100)}}{18} \\ y &= \frac{-(24x + 50) - \sqrt{(24x + 50)^2 - 36(16x^2 + 100x + 100)}}{18}. \end{aligned}$$

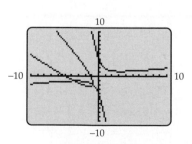

Figure 11–12

By graphing both equations (that is, all four functions shown in (∗) and (∗∗)), we obtain Figure 11–12. Then an intersection finder shows that the solutions of the system (points of intersection) are

$$x = -3.623, y = -1.113 \qquad \text{and} \qquad x = -.943, y = -1.833. \quad \blacksquare$$

EXAMPLE 6

A 52-foot-long piece of wire is to be cut into three pieces, two of which are the same length. The two equal pieces are to be bent into squares and the third piece into a circle. What should the length of each piece be if the total area enclosed by the two squares and the circle is 100 square feet?

SOLUTION   Let x be the length of each piece of wire that is to be bent into a square, and let y be the length of the piece that is to be bent into a circle. Since the original wire is 52 feet long,

$$x + x + y = 52 \qquad \text{or, equivalently,} \qquad y = 52 - 2x.$$

If a piece of wire of length x is bent into a square, the side of the square will have length $x/4$ and hence the area of the square will be $(x/4)^2 = x^2/16$. The remaining piece of wire will be made into a circle of circumference (length) y. Since the circumference is 2π times the radius (that is, $y = 2\pi r$), the circle has radius $r = y/2\pi$. Therefore, the area of the circle is

$$\pi r^2 = \pi \left(\frac{y}{2\pi} \right)^2 = \frac{\pi y^2}{4\pi^2} = \frac{y^2}{4\pi}.$$

The sum of the areas of the two squares and the circle is 100, that is,

$$\frac{x^2}{16} + \frac{x^2}{16} + \frac{y^2}{4\pi} = 100$$

$$\frac{y^2}{4\pi} = 100 - \frac{2x^2}{16}$$

$$y^2 = 4\pi \left(100 - \frac{x^2}{8} \right).$$

Therefore, the lengths x and y are solutions of this system of equations.

$$y = 52 - 2x$$

$$y^2 = 4\pi \left(100 - \frac{x^2}{8} \right).$$

The system may be solved either algebraically or graphically, using Figure 11–13. Since x and y are lengths, both must be positive. Consequently, we need only consider the intersection point in the first quadrant. An intersection finder shows that its coordinates are $x \approx 9.25$, $y \approx 33.50$. Therefore, the wire should be cut into two 9.25-foot pieces and one 33.5-foot piece. ∎

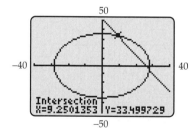

Figure 11–13

✓ EXERCISES 11.1.A

In Exercises 1–12, solve the system algebraically.

1. $x^2 - y = 0$
$-2x + y = 3$

2. $x^2 - y = 0$
$-3x + y = -2$

3. $x^2 - y = 0$
$x + 3y = 6$

4. $x^2 - y = 0$
$x + 4y = 4$

5. $x + y = 10$
$xy = 21$

6. $2x + y = 4$
$xy = 2$

7. $xy + 2y^2 = 8$
$x - 2y = 4$

8. $xy + 4x^2 = 3$
$3x + y = 2$

9. $x^2 + y^2 - 4x - 4y = -4$
$$x - y = 2$$

10. $x^2 + y^2 - 4x - 2y = -1$
$$x + 2y = 2$$

11. $x^2 + y^2 = 25$
$$x^2 + y = 19$$

12. $x^2 + y^2 = 1$
$$x^2 - y = 5$$

In Exercises 13–28, solve the system by any means.

13. $y = x^3 - 3x^2 + 4$
$$y = -.5x^2 + 3x - 2$$

14. $y = -x^3 + 3x^2 + x - 3$
$$y = -2x^2 + 5$$

15. $y = x^3 - 3x + 2$
$$y = \frac{3}{x^2 + 3}$$

16. $y = .25x^4 - 2x^2 + 4$
$$y = x^3 - x^2 - 2x + 1$$

17. $y = x^3 + x + 1$
$$y = \sin x$$

18. $y = x^2 - 4$
$$y = \cos x$$

19. $25x^2 - 16y^2 = 400$
$$-9x^2 + 4y^2 = -36$$

20. $9x^2 + 16y^2 = 140$
$$-x^2 + 4y^2 = -4$$

21. $5x^2 + 3y^2 - 20x + 6y = -8$
$$x - y = 2$$

22. $4x^2 + 9y^2 = 36$
$$2x - y = -1$$

23. $x^2 + 4xy + 4y^2 - 30x - 90y + 450 = 0$
$$x^2 + x - y + 1 = 0$$

24. $3x^2 + 4xy + 3y^2 - 12x + 2y + 7 = 0$
$$x^2 - 10x - y + 21 = 0$$

25. $4x^2 - 6xy + 2y^2 - 3x + 10y = 6$
$$4x^2 + y^2 = 64$$

26. $5x^2 + xy + 6y^2 - 79x - 73y + 196 = 0$
$$x^2 - 2xy + y^2 - 8x - 8y + 48 = 0$$

27. $x^2 + 3xy + y^2 = 2$
$$3x^2 - 5xy + 3y^2 = 7$$

28. $2x^2 - 8xy + 8y^2 + 2x - 5 = 0$
$$16x^2 - 24xy + 9y^2 + 100x - 200y + 100 = 0$$

29. What would be the answer in Example 6 if the original piece of wire were 70 feet long? Explain.

30. A 52-foot-long piece of wire is to be cut into three pieces, two of which are the same length. The two equal pieces are to be bent into circles and the third piece into a square. What should the length of each piece be if the total area enclosed by the two circles and the square is 100 square feet?

31. A rectangular box (including top) with square ends and a volume of 16 cubic meters is to be constructed from 40 square meters of cardboard. What should its dimensions be?

32. A rectangular sheet of metal is to be rolled into a circular tube. If the tube is to have a surface area (excluding ends) of 210 square inches and a volume of 252 cubic inches, what size sheet of metal should be used?

33. Find two real numbers whose sum is -16 and whose product is 48.

34. Find two real numbers whose sum is 34.5 and whose product is 297.

35. Find two positive real numbers whose difference is 1 and whose product is 4.16.

36. Find two real numbers whose difference is 25.75 and whose product is 127.5.

37. Find two real numbers whose sum is 3 such that the sum of their squares is 369.

38. Find two real numbers whose sum is 2 such that the difference of their squares is 60.

39. Find the dimensions of a rectangular room whose perimeter is 58 feet and whose area is 204 square feet.

40. Find the dimensions of a rectangular room whose perimeter is 53 feet and whose area is 165 square feet.

41. A rectangle has area 120 square inches and a diagonal of length 17 inches. What are its dimensions?

42. A right triangle has area 225 square centimeters and a hypotenuse of length 35 centimeters. To the nearest tenth of a centimeter, how long are the legs of the triangle?

43. Find the equation of the straight line that intersects the parabola $y = x^2$ *only* at the point $(3, 9)$. [*Hint:* What condition on the discriminant guarantees that a quadratic equation has exactly one real solution?]

11.2   Large Systems of Linear Equations

Systems of linear equations in three variables can be interpreted geometrically as the intersection of planes. However, algebraic methods are the only practical means to solve such systems or ones with more variables. Large systems can be solved by **Gaussian elimination,**[*] which is an extension of the elimination method used in Section 11.1.

Two systems of equations are said to be **equivalent** if they have the same solutions. The basic idea of Gaussian elimination is to transform a given system into an equivalent system that can easily be solved. There are several operations on a system of equations that leave the solutions to the system unchanged and, hence, produce an equivalent system. The first one is to

Interchange any two equations in the system,

which obviously won't affect the solutions of the system. The second is to

Multiply an equation in the system by a nonzero constant.

This does not change the solutions of the equation and therefore does not change the solutions of the system. To understand how the next operation works, we shall examine an earlier example from a different viewpoint.

EXAMPLE 1

In Example 3 of Section 11.1, we solved the system

$$x - 3y = 4$$
$$2x + y = 1$$

by multiplying the first equation by -2 and adding it to the second to eliminate the variable x.

$$-2x + 6y = -8 \qquad \text{[}-2 \text{ times first equation]}$$
$$\underline{2x + y = 1} \qquad \text{[Second equation]}$$
$$7y = -7. \qquad \text{[Sum of second equation and } -2 \text{ times first equation]}$$

We then solved this last equation for y and substituted the answer, $y = -1$, in the original first equation to find that $x = 1$. What we did, in effect, was

Replace the original system by the following system, in which x has been eliminated from the second equation; then solve this new system.

$$(*) \qquad x - 3y = 4 \qquad \text{[First equation]}$$
$$7y = -7. \qquad \text{[Sum of second equation and } -2 \text{ times first equation]}$$

The solution of system $(*)$ is easily seen to be $y = -1$, $x = 1$, and you can readily verify that this is the solution of the original system. So the two systems are equivalent. [*Note:* We are not claiming that the second equations in the two systems have the same solutions—they don't—but only that the two *systems* have the same solutions.]   ∎

---

*Named after the German mathematician K. F. Gauss (1777–1855).

Example 1 is an illustration of the third of the following operations.

Elementary Operations

> Performing any of the following operations on a system of equations produces an equivalent system.
>
> 1. Interchange any two equations in the system.
>
> 2. Replace an equation in the system by a nonzero constant multiple of itself.
>
> 3. Replace an equation in the system by the sum of itself and a constant multiple of another equation in the system.

The next example shows how elementary operations can be used to transform a system into an equivalent system that can be solved.

EXAMPLE 2

Solve the system

$$x + 4y - 3z = 1 \qquad \text{[Equation } A\text{]}$$
$$-3x - 6y + z = 3 \qquad \text{[Equation } B\text{]}$$
$$2x + 11y - 5z = 0. \qquad \text{[Equation } C\text{]}$$

SOLUTION   We first use elementary operations to produce an equivalent system in which the variable x has been eliminated from the second and third equations.

To eliminate x from equation B, replace equation B by the sum of itself and 3 times equation A.

$$\begin{aligned} \text{[3 times } A\text{]} \qquad 3x + 12y - 9z &= 3 \\ \text{[}B\text{]} \qquad -3x - 6y + z &= 3 \\ \hline 6y - 8z &= 6 \end{aligned}$$

$$\begin{aligned} x + 4y - 3z &= 1 \qquad \text{[}A\text{]} \\ 6y - 8z &= 6 \qquad \text{[Sum of } B \text{ and 3 times } A\text{]} \\ 2x + 11y - 5z &= 0. \qquad \text{[}C\text{]} \end{aligned}$$

To eliminate x from equation C, we replace equation C by the sum of itself and -2 times equation A.

$$\begin{aligned} \text{[}-2 \text{ times } A\text{]} \qquad -2x - 8y + 6z &= -2 \\ \text{[}C\text{]} \qquad 2x + 11y - 5z &= 0 \\ \hline 3y + z &= -2 \end{aligned}$$

$$\begin{aligned} x + 4y - 3z &= 1 \\ 6y - 8z &= 6 \\ 3y + z &= -2. \qquad \text{[Sum of } C \text{ and } -2 \text{ times } A\text{]} \end{aligned}$$

The next step is to eliminate the y term in one of the last two equations. This can be done by replacing the second equation by the sum of itself and -2 times the third equation.

$$
\begin{aligned}
x + 4y - 3z &= 1 \\
-10z &= 10 \qquad \text{[Sum of second equation and} \\
&\qquad\qquad\quad -2 \text{ times third equation]} \\
3y + z &= -2.
\end{aligned}
$$

Finally, interchange the last two equations.

$$
(\ast) \qquad
\begin{aligned}
x + 4y - 3z &= 1 \\
3y + z &= -2 \\
-10z &= 10.
\end{aligned}
$$

This last system, which is equivalent to the original one, is easily solved. The last equation shows that

$$
-10z = 10 \qquad \text{or, equivalently,} \qquad z = -1.
$$

Substituting $z = -1$ in the second equation shows that

$$
\begin{aligned}
3y + z &= -2 \\
3y + (-1) &= -2 \\
3y &= -1 \\
y &= -\frac{1}{3}.
\end{aligned}
$$

Substituting $y = -1/3$ and $z = -1$ in the first equation yields

$$
\begin{aligned}
x + 4y - 3z &= 1 \\
x + 4\left(-\frac{1}{3}\right) - 3(-1) &= 1 \\
x = 1 + \frac{4}{3} - 3 &= -\frac{2}{3}.
\end{aligned}
$$

Therefore, the original system has just one solution: $x = -2/3$, $y = -1/3$, $z = -1$. ∎

The process used to solve the final system $(\ast)$ in Example 2 is called **back substitution** because you begin with the last equation and work back to the first. It works because system $(\ast)$ is in **triangular form:** The first variable in the first equation, x, does not appear in any subsequent equation; the first variable in the second equation, y, does not appear in any subsequent equation, and so on. It can be shown that the procedure in Example 2 works in every case.

Gaussian Elimination

Any system of linear equations can be transformed into an equivalent system in triangular form by using a finite number of elementary operations. If the system has solutions, they can then be found by back substitution in the triangular form system.

In terms of convenience, most people prefer to use a calculator or computer to solve large systems of equations. However, the system solvers on calculators (see the Technology Tip in the margin) work only when the system has the same number of equations as variables *and* has a unique solution. When the solver gives an error message, you can't tell whether the system has many solutions or none at all. So we now develop a version of Gaussian elimination that also handles these situations and can be implemented on a calculator.

⠿ MATRIX METHODS

When solving systems by hand, a lot of time is wasted copying the x's, y's, z's, and so on. This fact suggests a shorthand system for representing a system of equations. For example, the system

$$(*) \qquad \begin{aligned} x + 2y + 3z &= -2 \\ 2x + 6y + z &= 2 \\ 3x + 3y + 10z &= -2 \end{aligned}$$

can be represented by the following rectangular array of numbers, consisting of the coefficients of the variables and the constants on the right of the equal sign, arranged in the same order in which they appear in the system.

$$\left(\begin{array}{ccc|c} 1 & 2 & 3 & -2 \\ 2 & 6 & 1 & 2 \\ 3 & 3 & 10 & -2 \end{array} \right)$$

This array is called the **augmented matrix** of the system. It has three horizontal **rows** and four vertical **columns.**

EXAMPLE 3

Use the matrix form of the preceding system (*) to solve the system.

SOLUTION   To solve the system in its original equation form, we would use elementary operations to eliminate the x terms from the last two equations, and then eliminate the y term from the last equation. With matrices, we do essentially the same thing, with the elementary operations on equations being replaced by the corresponding **row operations** on the augmented matrix in order to make certain entries in the first and second columns 0. Here is a side-by-side development of the two solution methods.

Equation Method	**Matrix Method**

Replace the second equation by the sum of itself and -2 times the first equation:

Replace the second row by the sum of itself and -2 times the first row:

$$x + 2y + 3z = -2$$
$$2y - 5z = 6$$
$$3x + 3y + 10z = -2$$

$$\begin{pmatrix} 1 & 2 & 3 & 2 \\ 0 & 2 & -5 & 6 \\ 3 & 3 & 10 & -2 \end{pmatrix}$$

Replace the third equation by the sum of itself and -3 times the first equation:

Replace the third row by the sum of itself and -3 times the first row:

$$x + 2y + 3z = -2$$
$$2y - 5z = 6$$
$$-3y + z = 4$$

$$\begin{pmatrix} 1 & 2 & 3 & -2 \\ 0 & 2 & -5 & 6 \\ 0 & -3 & 1 & 4 \end{pmatrix}$$

Multiply the second equation by $1/2$ (so that y has coefficient 1):

Multiply the second row by $1/2$:

$$x + 2y + 3z = -2$$
$$y - \frac{5}{2}z = 3$$
$$-3y + z = 4$$

$$\begin{pmatrix} 1 & 2 & 3 & -2 \\ 0 & 1 & -\frac{5}{2} & 3 \\ 0 & -3 & 1 & 4 \end{pmatrix}$$

Replace the third equation by the sum of itself and 3 times the second equation:

Replace the third row by the sum of itself and 3 times the second row:

$$x + 2y + 3z = -2$$
$$y - \frac{5}{2}z = 3$$
$$-\frac{13}{2}z = 13$$

$$\begin{pmatrix} 1 & 2 & 3 & -2 \\ 0 & 1 & -\frac{5}{2} & 3 \\ 0 & 0 & -\frac{13}{2} & 13 \end{pmatrix}$$

Finally, multiply the last equation by $-2/13$:[†]

Finally, multiply the last row by $-2/13$:

$$x + 2y + 3z = -2$$
$$(**) \quad y - \frac{5}{2}z = 3$$
$$z = -2$$

$$\begin{pmatrix} 1 & 2 & 3 & -2 \\ 0 & 1 & -\frac{5}{2} & 3 \\ 0 & 0 & 1 & -2 \end{pmatrix}$$

[†]This step isn't necessary, but it is often convenient to have 1 as the coefficient of the first variable in each equation.

System (∗∗) is easily solved. The third equation shows that $z = -2$, and substituting this in the second equation shows that

$$y - \frac{5}{2}(-2) = 3$$

$$y = 3 - 5 = -2.$$

Substituting $y = -2$ and $z = -2$ in the first equation yields

$$x + 2(-2) + 3(-2) = -2$$

$$x = -2 + 4 + 6 = 8.$$

Therefore, the only solution of the original system is $x = 8$, $y = -2$, $z = -2$. ∎

When using matrix notation, row operations replace elementary operations on equations, as shown in Example 3. The solution process ends when you reach a matrix, (and corresponding system), such as the final matrix in Example 1:

(∗∗)
$$\begin{pmatrix} 1 & 2 & 3 & -2 \\ 0 & 1 & -\dfrac{5}{2} & 3 \\ 0 & 0 & 1 & -2 \end{pmatrix} \qquad \begin{aligned} x + 2y + 3z &= -2 \\ y - \frac{5}{2}z &= 3 \\ z &= -2. \end{aligned}$$

The final matrix should satisfy these conditions:

All rows consisting entirely of zeros (if any) are at the bottom.

The first nonzero entry in each nonzero row is a 1 (called a **leading 1**).

Each leading 1 appears to the right of leading 1's in any preceding rows.

Such a matrix is said to be in **row echelon form.**

Most calculators have a key that uses row operations to put a given matrix into row echelon form (see the Technology Tip in the margin). For example, using the TI-83 REF key on the first matrix in Example 3 produced this row echelon matrix and corresponding system of equations.

$$\begin{pmatrix} 1 & 1 & \dfrac{10}{3} & -\dfrac{2}{3} \\ 0 & 1 & -\dfrac{17}{12} & \dfrac{5}{6} \\ 0 & 0 & 1 & -2 \end{pmatrix} \qquad \begin{aligned} x + y + \frac{10}{3}z &= -\frac{2}{3} \\ y - \frac{17}{12}z &= \frac{5}{6} \\ z &= -2. \end{aligned}$$

Because the calculator used a different sequence of row operations than was used in Example 3, it produced a row echelon matrix (and corresponding system) that differs from matrix (∗∗) above. You can easily verify, however, that the preceding system has the same solutions as system (∗∗), namely,

$$x = 8, \quad y = -2, \quad z = -2.$$

TECHNOLOGY TIP

To put a matrix in row echelon form, use REF in this menu/submenu:

TI-83: MATRIX/MATH

TI-86: MATRIX/OPS

TI-89: MATH/MATRIX

Casio and HP-39 users should consult the Technology Tip after Example 5.

EXAMPLE 4

Solve the system

$$x + y + 2z = 1$$
$$2x + 4y + 5z = 2$$
$$3x + 5y + 7z = 4.$$

SOLUTION If you try to use a systems equation solver on a calculator, you will get an error message. So we form the augmented matrix and reduce it to row echelon form either by hand or by using the REF key. A TI-86 produced the row echelon matrix in Figure 11–14.

TECHNOLOGY TIP

On TI calculators, using the FRAC key in conjunction with the REF key, as in Figure 11–14, usually eliminates long decimals and makes the matrix easier to read.

Figure 11–14

Look at the last row of the matrix in Figure 11–14; it represents the equation

$$0x + 0y + 0z = 1.$$

Since this equation has no solutions (the left side is always 0 and the right side is always 1), neither does the original system. Such a system is said to be **inconsistent.** ■

■■ THE GAUSS-JORDAN METHOD

Gaussian elimination on a calculator is an efficient method of solving systems of equations but may involve some messy calculations when you solve the final triangular form system by hand. Most hand computations can be eliminated by using a slight variation, known as the **Gauss-Jordan method,**[*] which is illustrated in the next example.

EXAMPLE 5

Use the Gauss-Jordan method to solve this system.

$$x - y + 5z = -6$$
$$3x + 3y - z = 10$$
$$x - 5y + 8z = -17$$
$$x + 3y + 2z = 5.$$

---

[*]This method was developed by the German engineer Wilhelm Jordan (1842–1899).

SOLUTION   The augmented matrix of the system is shown in Figure 11–15, and an equivalent row echelon matrix (obtained by using the REF and FRAC keys) is shown in Figure 11–16.

$$\begin{pmatrix} 1 & -1 & 5 & -6 \\ 3 & 3 & -1 & 10 \\ 1 & -5 & 8 & -17 \\ 1 & 3 & 2 & 5 \end{pmatrix}$$

Figure 11–15   **Figure 11–16**

At this point in Gaussian elimination, we would use back substitution to solve the triangular form system represented by the last matrix in Figure 11–16. In the Gauss-Jordan method, however, additional elimination of variables replaces back substitution. Look at the leading 1 in the third row (shown in color).

$$\begin{pmatrix} 1 & 1 & -\dfrac{1}{3} & \dfrac{10}{3} \\ 0 & 1 & -\dfrac{25}{18} & \dfrac{61}{18} \\ 0 & 0 & 1 & -1 \\ 0 & 0 & 0 & 0 \end{pmatrix}.$$

In Gauss-Jordan elimination, we make the entries above this leading 1 into 0's.

Replace the second row by the sum of itself and 25/18 times the third row:

$$\begin{pmatrix} 1 & 1 & -\dfrac{1}{3} & \dfrac{10}{3} \\ 0 & 1 & 0 & 2 \\ 0 & 0 & 1 & -1 \\ 0 & 0 & 0 & 0 \end{pmatrix}$$

Replace the first row by the sum of itself and 1/3 times the third row:

$$\begin{pmatrix} 1 & 1 & 0 & 3 \\ 0 & 1 & 0 & 2 \\ 0 & 0 & 1 & -1 \\ 0 & 0 & 0 & 0 \end{pmatrix}$$

Now consider the leading 1 in the second row (shown in color), and make the entry above it 0.

Replace the first row by the sum of itself and -1 times the second row:

$$\begin{pmatrix} 1 & 0 & 0 & 1 \\ 0 & 1 & 0 & 2 \\ 0 & 0 & 1 & -1 \\ 0 & 0 & 0 & 0 \end{pmatrix}$$

This last matrix represents the following system, whose solution is obvious:

$$x = 1$$
$$y = 2$$
$$z = -1. \quad \blacksquare$$

A row echelon form matrix, such as the last one in Example 5, in which any column containing a leading 1 has 0's in all other positions, is said to be in **reduced row echelon form.** The goal in the Gauss-Jordan method is to use row operations to put a given augmented matrix into reduced row echelon form (from which the solutions can be read immediately, as in Example 5).

As a general rule, Gaussian elimination (matrix version) is the method of choice when working by hand, (the additional row operations needed to put a matrix in reduced row echelon form are usually more time-consuming—and error-prone—than back substitution). With a calculator or computer, however, it's better to find a reduced row echelon matrix for the system. You can do this in one step on a calculator by using the RREF key (see the Technology Tip in the margin).

EXAMPLE 6

Solve this system.

$$2x + 5y + z + 3w = 0$$
$$2y - 4z + 6w = 0$$
$$2x + 17y - 23z + 40w = 0.$$

SOLUTION   A system such as this, in which all the constants on the right side are zero, is called a **homogeneous system.** Every homogeneous system has at least one solution, namely, $x = 0$, $y = 0$, $z = 0$, $w = 0$, which is called the **trivial solution.** The issue with homogeneous systems is whether or not they have any nonzero solutions. The augmented matrix of the system is shown in Figure 11–17, and an equivalent reduced row echelon form matrix is shown in Figure 11–18.*

Figure 11–17

Figure 11–18

The system corresponding to the reduced echelon form matrix in Figure 11–18 is

$$x + \frac{11}{2}z \quad = 0$$
$$y - 2z \quad = 0$$
$$w = 0.$$

---

*When dealing with homogeneous systems, it's not really necessary to include the last column of zeros, as is done here, because row operations do not change this column.

The second equation shows that

$$y = 2z.$$

This equation has an infinite number of solutions, for instance,

$$z = 1, y = 2 \quad \text{or} \quad z = 3, y = 6 \quad \text{or} \quad z = -2.5, y = -5.$$

In fact, for each real number t, there is a solution: $z = t$, $y = 2t$. Substituting $z = t$ into the first equation shows that

$$x + \frac{11}{2}t = 0$$

$$x = -\frac{11}{2}t.$$

Therefore, this system, and hence the original one, has an infinite number of solutions, one for each real number t.

$$x = -\frac{11}{2}t, \qquad y = 2t, \qquad z = t, \qquad w = 0.$$

A system with infinitely many solutions, such as this one, is said to be **dependent.** ∎

NOTE

Every system that has more variables than equations (as in Example 6) is dependent, but other systems may be dependent as well.

▟ APPLICATIONS

In calculus, it is sometimes necessary to write a complicated rational expression as the sum of simpler ones. One technique for doing this involves systems of equations.

EXAMPLE 7

Find constants A, B, and C such that

$$\frac{2x^2 + 15x + 10}{(x-1)(x+2)^2} = \frac{A}{x-1} + \frac{B}{x+2} + \frac{C}{(x+2)^2}.$$

SOLUTION   Multiply both sides of the equation by the common denominator $(x-1)(x+2)^2$ and collect like terms on the right side.

$$2x^2 + 15x + 10 = A(x+2)^2 + B(x-1)(x+2) + C(x-1)$$

$$= A(x^2 + 4x + 4) + B(x^2 + x - 2) + C(x-1)$$

$$= Ax^2 + 4Ax + 4A + Bx^2 + Bx - 2B + Cx - C$$

$$= (A + B)x^2 + (4A + B + C)x + (4A - 2B - C).$$

Since the polynomials on the left and right sides of the last equation are equal, their coefficients must be equal term by term, that is,

$$A + B = 2 \qquad \text{[Coefficients of } x^2 \text{]}$$

$$4A + B + C = 15 \qquad \text{[Coefficients of } x \text{]}$$

$$4A - 2B - C = 10. \qquad \text{[Constant terms]}$$

We can consider this as a system of equations with unknowns A, B, C. The augmented matrix of the system is shown in Figure 11–19, and an equivalent reduced row echelon form matrix is shown in Figure 11–20.

Figure 11–19   Figure 11–20

The solutions of the system can be read from the reduced row echelon form matrix in Figure 11–20.

$$A = 3, \qquad B = -1, \qquad C = 4.$$

Therefore,

$$\frac{2x^2 + 15x + 10}{(x-1)(x+2)^2} = \frac{3}{x-1} + \frac{-1}{x+2} + \frac{4}{(x+2)^2}.$$

The right side of this equation is called the **partial fraction decomposition** of the fraction on the left side. ∎

EXAMPLE 8

Charlie is starting a small business and borrows $10,000 on three different credit cards, with annual interest rates of 18%, 15%, and 9%, respectively. He borrows three times as much as on the 15% card as on the 18% card, and his total annual interest on all three cards is $1244.25. How much did he borrow on each credit card?

SOLUTION   Let x be the amount on the 18% card, let y be the amount on the 15% card, and let z be the amount on the 9% card. Then $x + y + z = 10,000$. Furthermore,

$$\underset{\begin{array}{c}\text{Interest on}\\\text{18\% card}\end{array}}{} + \underset{\begin{array}{c}\text{Interest on}\\\text{15\% card}\end{array}}{} + \underset{\begin{array}{c}\text{Interest on}\\\text{9\% card}\end{array}}{} = \underset{\begin{array}{c}\text{Total}\\\text{interest}\end{array}}{}$$

$$.18x \quad + \quad .15y \quad + \quad .09z \quad = 1244.25.$$

Finally, we have

$$\underset{\begin{array}{c}\text{Amount on}\\\text{15\% card}\end{array}}{} = \underset{\begin{array}{c}\text{3 times amount}\\\text{on 18\% card}\end{array}}{}$$

$$y = 3x,$$

which is equivalent to $3x - y = 0$. Therefore, we must solve this system of equations.

$$x + y + z = 10{,}000$$
$$.18x + .15y + .09z = 1{,}244.25$$
$$3x - y = 0,$$

whose augmented matrix is

$$\begin{pmatrix} 1 & 1 & 1 & 10{,}000 \\ .18 & .15 & .09 & 1{,}244.25 \\ 3 & -1 & 0 & 0 \end{pmatrix}.$$

CALCULATOR EXPLORATION

Enter this matrix in your calculator. Use row operations or the RREF key to put it in reduced row echelon form. Read the solutions of the system from this last matrix.

The Calculator Exploration shows that Charlie borrowed $1275 on the 18% card, $3825 on the 15% card, and $4900 on the 9% card. ■

The preceding examples illustrate the following fact, whose proof is omitted.

Number of Solutions of a System

Any system of linear equations must have

No solutions (an inconsistent system) *or*

Exactly one solution *or*

An infinite number of solutions (a dependent system).

✓ **EXERCISES 11.2**

In Exercises 1–4, write the augmented matrix of the system.

1. $2x - 3y + 4z = 1$
$x + 2y - 6z = 0$
$3x - 7y + 4z = -3$

2. $x + 2y - 3w + 7z = -5$
$2x - y + 2z = 4$
$3x + 7w - 6z = 0$

3. $x - \dfrac{1}{2}y + \dfrac{7}{4}z = 0$
$2x - \dfrac{3}{2}y + 5z = 0$
$- 2y + \dfrac{1}{3}z = 0$

4. $2x - \dfrac{1}{2}y + \dfrac{7}{2}w - 6z = 1$
$\dfrac{1}{4}x - 6y + 2w - z = 2$
$4y - \dfrac{1}{2}w + z = 3$
$2x + 3y + \dfrac{1}{2}z = 4$

In Exercises 5–8, the augmented matrix of a system of equations is given. Express the system in equation notation.

5. $\begin{pmatrix} 2 & -3 & 1 \\ 4 & 7 & 2 \end{pmatrix}$

6. $\begin{pmatrix} 2 & 3 & 5 & 2 \\ 1 & 6 & 9 & 0 \end{pmatrix}$

7. $\begin{pmatrix} 1 & 0 & 1 & 0 & 1 \\ 1 & -1 & 4 & -2 & 3 \\ 4 & 2 & 5 & 0 & 2 \end{pmatrix}$

8. $\begin{pmatrix} 1 & 7 & 0 & 4 \\ 2 & 3 & 1 & 6 \\ -1 & 0 & 2 & 3 \end{pmatrix}$

In Exercises 9–12, the reduced row echelon form of the augmented matrix of a system of equations is given. Find the solutions of the system.

9. $\begin{pmatrix} 1 & 0 & 0 & 0 & 3/2 \\ 0 & 1 & 0 & 0 & 5 \\ 0 & 0 & 1 & 0 & -2 \\ 0 & 0 & 0 & 1 & 0 \end{pmatrix}$

10. $\begin{pmatrix} 1 & 0 & 0 & 0 & 0 & 5 \\ 0 & 1 & 0 & 0 & 0 & 4 \\ 0 & 0 & 1 & 0 & 0 & 3 \\ 0 & 0 & 0 & 0 & 1 & 2 \\ 0 & 0 & 0 & 0 & 0 & 1 \end{pmatrix}$

11. $\begin{pmatrix} 1 & 0 & 0 & 1 & 2 \\ 0 & 1 & 0 & 2 & -3 \\ 0 & 0 & 1 & 0 & 4 \\ 0 & 0 & 0 & 0 & 0 \end{pmatrix}$

12. $\begin{pmatrix} 1 & 0 & 0 & 0 & 7 \\ 0 & 1 & 0 & 0 & 1 \\ 0 & 0 & 1 & 0 & -5 \\ 0 & 0 & 0 & 1 & 4 \\ 0 & 0 & 0 & 0 & 0 \\ 0 & 0 & 0 & 0 & 0 \end{pmatrix}$

In Exercises 13–16, use Gaussian elimination to solve the system.

13. $\begin{aligned} -x + 3y + 2z &= 0 \\ 2x - y - z &= 3 \\ x + 2y + 3z &= 0 \end{aligned}$

14. $\begin{aligned} 3x + 7y + 9z &= 0 \\ x + 2y + 3z &= 2 \\ x + 4y + z &= 2 \end{aligned}$

15. $\begin{aligned} x + y + z &= 1 \\ x - 2y + 2z &= 4 \\ 2x - y + 3z &= 5 \end{aligned}$

16. $\begin{aligned} 2x - y + z &= 1 \\ 3x + y + z &= 0 \\ 7x - y + 3z &= 2 \end{aligned}$

In Exercises 17–20, use the Gauss-Jordan method to solve the system.

17. $\begin{aligned} x - 2y + 4z &= 6 \\ x + y + 13z &= 6 \\ -2x + 6y - z &= -10 \end{aligned}$

18. $\begin{aligned} x - y + 5z &= -6 \\ 3x + 3y - z &= 10 \\ x + 3y + 2z &= 5 \end{aligned}$

19. $\begin{aligned} x + y + z &= 200 \\ x - 2y &= 0 \\ 2x + 3y + 5z &= 600 \\ 2x - y + z &= 200 \end{aligned}$

20. $\begin{aligned} 3x - y + z &= 6 \\ x + 2y - z &= 0 \end{aligned}$

In Exercises 21–36, solve the system.

21. $\begin{aligned} 11x + 10y + 9z &= 5 \\ x + 2y + 3z &= 1 \\ 3x + 2y + z &= 1 \end{aligned}$

22. $\begin{aligned} -x + 2y - 3z + 4w &= 8 \\ 2x - 4y + z + 2w &= -3 \\ 5x - 4y + z + 2w &= -3 \end{aligned}$

23. $\begin{aligned} x + y &= 3 \\ 5x - y &= 3 \\ 9x - 4y &= 1 \end{aligned}$

24. $\begin{aligned} 2x - y + 2z &= 3 \\ -x + 2y - z &= 0 \\ x + y - z &= 1 \end{aligned}$

25. $\begin{aligned} x - 4y - 13z &= 4 \\ x - 2y - 3z &= 2 \\ -3x + 5y + 4z &= 2 \end{aligned}$

26. $\begin{aligned} 2x - 4y + z &= 3 \\ x + 3y - 7z &= 1 \\ -2x + 4y - z &= 10 \end{aligned}$

27. $\begin{aligned} 4x + y + 3z &= 7 \\ x - y + 2z &= 3 \\ 3x + 2y + z &= 4 \end{aligned}$

28. $\begin{aligned} x + 4y + z &= 3 \\ -x + 2y + 2z &= 0 \\ 2x + 2y - z &= 3 \end{aligned}$

29. $\begin{aligned} x + y + z &= 0 \\ 3x - y + z &= 0 \\ -5x - y + z &= 0 \end{aligned}$

30. $\begin{aligned} x + y + z &= 0 \\ x - y - z &= 0 \\ x - y + z &= 0 \end{aligned}$

31. $\begin{aligned} 2x + y + 3z - 2w &= -6 \\ 4x + 3y + z - w &= -2 \\ x + y + z + w &= -5 \\ -2x - 2y + 2z + 2w &= -10 \end{aligned}$

32. $\begin{aligned} x + y + z + w &= -1 \\ -x + 4y + z - w &= 0 \\ x - 2y + z - 2w &= 11 \\ -x - 2y + z + 2w &= -3 \end{aligned}$

33. $\begin{aligned} x - 2y - z - 3w &= -3 \\ -x + y + z &= 0 \\ 4y + 3z - 2w &= -1 \\ 2x - 2y + w &= 1 \end{aligned}$

34. $\begin{aligned} 3x - y + 2z &= 0 \\ -x + 3y + 2z + 5w &= 0 \\ x + 2y + 5z - 4w &= 0 \\ 2x - y + 3w &= 0 \end{aligned}$

35. $\begin{aligned} \frac{3}{x} - \frac{1}{y} + \frac{4}{z} &= -13 \\ \frac{1}{x} + \frac{2}{y} - \frac{1}{z} &= 12 \\ \frac{4}{x} - \frac{1}{y} + \frac{3}{z} &= -7 \end{aligned}$

[*Hint:* Let $u = 1/x$, $v = 1/y$, $w = 1/z$.]

36.
$$\frac{1}{x+1} - \frac{2}{y-3} + \frac{3}{z-2} = 4$$
$$\frac{5}{y-3} - \frac{10}{z-2} = -5$$
$$\frac{-3}{x+1} + \frac{4}{y-3} - \frac{1}{z-2} = -2$$

[*Hint:* Let $u = 1/(x+1)$, $v = 1/(y-3)$, $w = 1/(z-2)$.]

Exercises 37–40, solve the system. [Note: The REF and RREF keys on some calculators produce an error message when there are more rows than columns in a matrix, in which case you will have to solve the system by some other means.]

37. $2x - y = 1$
$3x + y = 2$
$4x - 2y = 2$
$5x + 5y = 4$

38. $x + y = 3$
$-x + 2y = 3$
$5x - y = 3$
$-7x + 5y = 3$

39. $x + 2y = 3$
$2x + 3y = 4$
$3x + 4y = 5$
$4x + 5y = 6$

40. $x - y = 2$
$x + y = 4$
$2x + 3y = 9$
$3x - 2y = 6$

In Exercises 41–46, find the constants A, B, and C.

41. $\dfrac{x}{(x+1)(x+2)} = \dfrac{A}{x+1} + \dfrac{B}{x+2}$

42. $\dfrac{1}{(x+1)(x-1)} = \dfrac{A}{x+1} + \dfrac{B}{x-1}$

43. $\dfrac{2x+1}{(x+2)(x-3)^2} = \dfrac{A}{x+2} + \dfrac{B}{x-3} + \dfrac{C}{(x-3)^2}$

44. $\dfrac{x^2 - x - 21}{(2x-1)(x^2+4)} = \dfrac{A}{2x-1} + \dfrac{Bx+C}{x^2+4}$

45. $\dfrac{5x^2 + 1}{(x+1)(x^2-x+1)} = \dfrac{A}{x+1} + \dfrac{Bx+C}{x^2-x+1}$

46. $\dfrac{x-2}{(x+4)(x^2+2x+2)} = \dfrac{A}{x+4} + \dfrac{Bx+C}{x^2+2x+2}$

47. Lillian borrows $10,000. She borrows some from her friend at 8% annual interest, twice as much as that from her bank at 9%, and the remainder from her insurance company at 5%. She pays a total of $830 in interest for the first year. How much did she borrow from each source?

48. An investor puts a total of $25,000 into three very speculative stocks. She invests some of it in Crystalcomp and $2000 more than one-half that amount in Flyboys. The remainder is invested in Zumcorp. Crystalcomp rises 16% in value, Flyboys rises 20%, and Zumcorp rises 18%. Her investment in the three stocks is now worth $29,440. How much was originally invested in each stock?

49. An investor has $70,000 invested in a mutual fund, bonds, and a fast food franchise. She has twice as much invested in bonds as in the mutual fund. Last year the mutual fund paid a 2% dividend, the bonds paid 10%, and the fast food franchise paid 6%; her dividend income was $4800. How much is invested in each of the three investments?

50. Tickets to a band concert cost $2 for children, $3 for teenagers, and $5 for adults. 570 people attended the concert and total ticket receipts were $1950. Three-fourths as many teenagers as children attended. How many children, adults, and teenagers attended?

51. The table shows the calories, sodium, and protein in one cup of various kinds of soup.

	Progresso Roasted Chicken Rotini	Healthy Choice Hearty Chicken	Campbell's Chunky Chicken Noodle
Calories	100	130	130
Sodium (mg)	970	480	880
Protein (g)	6	8	8

How many cups of each kind of soup should be mixed together to produce ten servings of soup, each of which provides 203 calories, 1190 milligrams of sodium, and 12.4 grams of protein? What is the serving size (in cups)? (*Hint:* In ten servings, there must be 2030 calories, 11,900 milligrams of sodium, and 124 grams of protein.)

52. The table shows the calories, sodium, and fat in one ounce of various snack foods (all produced by Planters).

	Cashews	Dry Roasted Honey Peanuts	Cajun Crunch Trail Mix
Calories	170	150	130
Sodium (mg)	120	95	10
Fat (g)	6	7	3

How many ounces of each kind of snack should be combined to produce ten servings, each of which provides 288 calories, 115 milligrams of sodium, and 9.6 grams of fat? What is the serving size?

53. Comfort Systems, Inc., sells three models of humidifiers. The bedroom model weighs 10 pounds and comes in an 8-cubic-foot box; the living room model weighs 20 pounds and comes in an 8-cubic-foot box; the whole-house model weighs 60 pounds and comes in a 28-cubic-foot box. Each of the company's delivery vans has 248 cubic feet of space and can hold a maximum of 440 pounds. For a van to be as fully loaded as possible, how many of each model should it carry?

54. Peanuts cost $3 per pound, almonds cost $4 per pound, and cashews costs $8 per pound. How many pounds of each should be used to produce 140 pounds of a mixture costing $6 per pound, in which there are twice as many peanuts as almonds?

Exercises 55 and 56 deal with computer-aided tomography (CAT) scanners that take X-rays of body parts from different directions to create a picture of a cross section of the body.[*] The amount by which the X-ray energy decreases (measured in linear-attenuation units) indicates whether the X-ray has passed through healthy tissue, tumorous tissue, or bone, according to the following table.

Tissue Type	Linear-Attenuation Units
Healthy	.1625–.2977
Tumorous	.2679–.3930
Bone	.3857–.5108

The body part being scanned is divided into cells. The total linear-attenuation value is the sum of the values for each cell the X-ray passes through. In the figure for Exercise 55, for example, let a, b, and c be the values for cells A, B, and C, respectively; then the attenuation value for X-ray 3 is b + c.

55. (a) In the figure, find the linear-attenuation values for X-rays 1 and 2.
    (b) If the total linear-attenuation values for X-rays 1, 2, and 3 are .75, .60, and .54, respectively, set up and solve a system of three equations in a, b, c.
    (c) What kind of tissue are cells A, B, and C?

56. Four X-ray beams are aimed at four cells, as shown in the figure.

(a) If the total linear-attenuation values for X-rays 1, 2, 3, and 4 are .60, .75, .65, and .70, respectively, is there enough information to determine the values of a, b, c, and d? Explain.
(b) If an additional X-ray beam is added, with a linear-attenuation value of .85, as shown in the figure below, can the values of a, b, c, and d be determined? If so, what are they? What can be said about cells A, B, C, and D?

57. A furniture manufacturer has 1950 machine hours available each week in the cutting department, 1490 hours in the assembly department, and 2160 in the finishing department. Manufacturing a chair requires .2 hours of cutting, .3 hours of assembly, and .1 hours of finishing. A chest requires .5 hours of cutting, .4 hours of assembly, and .6 hours of finishing. A table requires .3 hours of cutting, .1 hours of assembly, and .4 hours of finishing. How many chairs, chests, and tables should be produced to use all the available production capacity?

58. A stereo equipment manufacturer produces three models of speakers, R, S, and T, and has three kinds of delivery vehicles: trucks, vans, and station wagons. A truck holds two boxes of model R, one of model S, and three of model T. A van holds one box of model R, three of model S, and two of model T. A station wagon holds one box of model R, three of model S, and one of model T. If 15 boxes of model R, 20 boxes of model S, and 22 boxes of model T are to be delivered, how many vehicles of each type should be used so that all operate at full capacity?

---

[*]Exercises 55 and 56 are based on D. Jabon, G. Nord, B. W. Wilson, and P. Coffman, "Medical Applications of Linear Equations," *Mathematics Teacher 89*, no. 5 (May 1996).

59. The diagram shows the traffic flow at four intersections during a typical one-hour period. The streets are all one-way, as indicated by the arrows. To adjust the traffic lights to avoid congestion, engineers must determine the possible values of x, y, z, and t.

(a) Write a system of linear equations that describes congestion-free traffic flow. [*Hint:* 600 cars per hour come down Euclid to intersection A, and 400 come down 4th Avenue to intersection A. Also, x cars leave intersection A on Euclid, and t cars leave on 4th Avenue. To avoid congestion, the number of cars leaving the intersection must be the same as the number entering, that is, $x + t = 600 + 400$. Use intersections B, C, and D to find three more equations.]

(b) Solve the system in part (a), which is dependent. Express your answers in terms of the variable t.

(c) Find the largest and smallest number of cars that can leave the given intersection on the given street: A on 4th Avenue, A on Euclid, C on 5th Avenue, and C on Chester.

60. The diagram shows the traffic flow at four intersections during rush hour, as in Exercise 59.

(a) What are the possible values of x, y, z, and t in order to avoid any congestion? [Express your answers in terms of t.]

(b) What are the possible values of t?

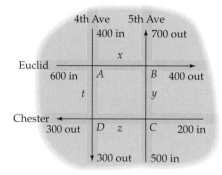

11.3 Matrix Methods for Square Systems

Matrices were used in Section 11.2 as a convenient shorthand for solving systems of linear equations. We now consider matrices in a more general setting and show how the algebra of matrices provides an alternative method for solving systems of equations that are not dependent and have the same number of equations as variables.

Let m and n be positive integers. An $m \times n$ **matrix** (read "m by n matrix") is a rectangular array of numbers, with m horizontal rows and n vertical columns. For example,

$$\begin{pmatrix} 3 & 2 & -5 \\ 6 & 1 & 7 \\ -2 & 5 & 0 \end{pmatrix} \qquad \begin{pmatrix} -3 & 4 \\ 2 & 0 \\ 0 & 1 \\ 7 & 3 \\ 1 & -6 \end{pmatrix} \qquad \begin{pmatrix} 3 & 0 & 1 & 0 \\ \sqrt{2} & -\frac{1}{2} & 4 & \frac{8}{3} \\ 10 & 2 & -\frac{3}{4} & 12 \end{pmatrix} \qquad \begin{pmatrix} \sqrt{3} \\ 2 \\ 0 \\ 11 \end{pmatrix}$$

3×3 matrix	5×2 matrix	3×4 matrix	4×1 matrix
3 rows	5 rows	3 rows	4 rows
3 columns	2 columns	4 columns	1 column

In a matrix, the rows are horizontal and are numbered from top to bottom. The columns are vertical and are numbered from left to right. For example,

$$\begin{array}{c} \text{Row } 1 \rightarrow \\ \text{Row } 2 \rightarrow \\ \text{Row } 3 \rightarrow \end{array} \begin{pmatrix} 11 & 3 & 14 \\ -2 & 0 & -5 \\ \frac{1}{3} & 6 & 7 \end{pmatrix}$$

$$\begin{array}{ccc} \uparrow & \uparrow & \uparrow \\ \text{Column 1} & \text{Column 2} & \text{Column 3} \end{array}$$

Each entry in a matrix can be located by stating the row and column in which it appears. For instance, in the preceding 3×3 matrix, 14 is the entry in row 1, column 3, and 0 is the entry in row 2, column 2. When you enter a matrix on a calculator, the words "row" and "column" won't be displayed, but the row numbers will always be listed before the column number. Thus, a display such as "$A[3, 2]$," or simply "3, 2," indicates the entry in row 3, column 2.

Two matrices are said to be **equal** if they have the same size (same number of rows and columns) and the corresponding entries are equal. For example,

$$\begin{pmatrix} 3 & (-1)^2 \\ 6 & 12 \end{pmatrix} = \begin{pmatrix} 3 & 1 \\ \sqrt{36} & 12 \end{pmatrix}, \qquad \text{but} \qquad \begin{pmatrix} 6 & 4 \\ 5 & 1 \end{pmatrix} \neq \begin{pmatrix} 6 & 5 \\ 4 & 1 \end{pmatrix}.$$

▋▋ MATRIX MULTIPLICATION

Although there is an extensive arithmetic of matrices, we shall need only matrix multiplication. The simplest case is the product of a matrix with a single row and a matrix with a single column, where the row and column have the same number of entries. This is done by multiplying corresponding entries (first by first, second by second, and so on) and then adding the results. An example is shown in Figure 11–21.

$$(3 \quad 1 \quad 2)\begin{pmatrix} 2 \\ 0 \\ 1 \end{pmatrix} = 3 \cdot 2 + 1 \cdot 0 + 2 \cdot 1 = 8$$

$$\uparrow \qquad \uparrow \qquad \uparrow$$
$$\text{First} \quad \text{Second} \quad \text{Third}$$
$$\text{Terms} \quad \text{Terms} \quad \text{Terms}$$

Figure 11–21

Note that the product of a row and a column is a single number.

Now let A be an $m \times n$ matrix, and let B be an $n \times p$ matrix, so that the number of columns of A is the same as the number of rows of B (namely, n). The product matrix AB is defined to be an $m \times p$ matrix (same number of rows as A and same number of columns as B). The product AB is defined as follows.

Matrix Multiplication

> If A is an $m \times n$ and B is an $n \times p$ matrix, then AB is the $m \times p$ matrix whose entry in the ith row and jth column is
>
> the product of the ith row of A and the jth column of B.

EXAMPLE 1

If it is defined, find the product AB, where

$$A = \begin{pmatrix} 3 & 1 & 2 \\ -1 & 0 & 4 \end{pmatrix} \quad \text{and} \quad B = \begin{pmatrix} 2 & -3 & 0 & 1 \\ 0 & 5 & 2 & 7 \\ 1 & 8 & -4 & 1 \end{pmatrix}.$$

SOLUTION A has 3 columns, and B has 3 rows. So the product matrix AB is defined. AB has 2 rows (same as A) and 4 columns (same as B). Its entries are calculated as follows. The entry in row 1, column 1 of AB is the product of row 1 of A and column 1 of B, which is the number 8, as shown in Figure 11–21 and indicated at the right on the next page.

row 1, column 1 $\begin{pmatrix} \boxed{3 \ 1 \ 2} \\ -1 \ 0 \ 4 \end{pmatrix} \begin{pmatrix} 2 & -3 & 0 & 1 \\ 0 & 5 & 2 & 7 \\ 1 & 8 & -4 & 1 \end{pmatrix} = \begin{pmatrix} \boxed{8} \end{pmatrix}$ $3 \cdot 2 + 1 \cdot 0 + 2 \cdot 1 = 8$

The other entries in AB are obtained similarly.

row 1, column 2 $\begin{pmatrix} \boxed{3 \ 1 \ 2} \\ -1 \ 0 \ 4 \end{pmatrix} \begin{pmatrix} 2 & -3 & 0 & 1 \\ 0 & 5 & 2 & 7 \\ 1 & 8 & -4 & 1 \end{pmatrix} = \begin{pmatrix} 8 & 12 \end{pmatrix}$ $3(-3) + 1 \cdot 5 + 2 \cdot 8 = 12$

row 1, column 3 $\begin{pmatrix} \boxed{3 \ 1 \ 2} \\ -1 \ 0 \ 4 \end{pmatrix} \begin{pmatrix} 2 & -3 & 0 & 1 \\ 0 & 5 & 2 & 7 \\ 1 & 8 & -4 & 1 \end{pmatrix} = \begin{pmatrix} 8 & 12 & -6 \end{pmatrix}$ $3 \cdot 0 + 1 \cdot 2 + 2(-4) = -6$

row 1, column 4 $\begin{pmatrix} \boxed{3 \ 1 \ 2} \\ -1 \ 0 \ 4 \end{pmatrix} \begin{pmatrix} 2 & -3 & 0 & 1 \\ 0 & 5 & 2 & 7 \\ 1 & 8 & -4 & 1 \end{pmatrix} = \begin{pmatrix} 8 & 12 & -6 & 12 \end{pmatrix}$ $3 \cdot 1 + 1 \cdot 7 + 2 \cdot 1 = 12$

row 2, column 1 $\begin{pmatrix} 3 \ 1 \ 2 \\ \boxed{-1 \ 0 \ 4} \end{pmatrix} \begin{pmatrix} 2 & -3 & 0 & 1 \\ 0 & 5 & 2 & 7 \\ 1 & 8 & -4 & 1 \end{pmatrix} = \begin{pmatrix} 8 & 12 & -6 & 12 \\ 2 \end{pmatrix}$ $(-1)2 + 0 \cdot 0 + 4 \cdot 1 = 2$

row 2, column 2 $\begin{pmatrix} 3 \ 1 \ 2 \\ \boxed{-1 \ 0 \ 4} \end{pmatrix} \begin{pmatrix} 2 & -3 & 0 & 1 \\ 0 & 5 & 2 & 7 \\ 1 & 8 & -4 & 1 \end{pmatrix} = \begin{pmatrix} 8 & 12 & -6 & 12 \\ 2 & 35 \end{pmatrix}$ $(-1)(-3) + 0 \cdot 5 + 4 \cdot 8 = 35$

row 2, column 3 $\begin{pmatrix} 3 \ 1 \ 2 \\ \boxed{-1 \ 0 \ 4} \end{pmatrix} \begin{pmatrix} 2 & -3 & 0 & 1 \\ 0 & 5 & 2 & 7 \\ 1 & 8 & -4 & 1 \end{pmatrix} = \begin{pmatrix} 8 & 12 & -6 & 12 \\ 2 & 35 & -16 \end{pmatrix}$ $(-1)0 + 0 \cdot 2 + 4(-4) = -16$

row 2, column 4 $\begin{pmatrix} 3 \ 1 \ 2 \\ \boxed{-1 \ 0 \ 4} \end{pmatrix} \begin{pmatrix} 2 & -3 & 0 & 1 \\ 0 & 5 & 2 & 7 \\ 1 & 8 & -4 & 1 \end{pmatrix} = \begin{pmatrix} 8 & 12 & -6 & 12 \\ 2 & 35 & -16 & 3 \end{pmatrix}$ $(-1)1 + 0 \cdot 7 + 4 \cdot 1 = 3$

The last matrix on the right is the product AB. ■

EXAMPLE 2

Let A, B, C, and D be the following matrices.

$$A = \begin{pmatrix} 3 & 2 & 1 \\ -2 & 0 & 4 \\ 1 & -2 & 5 \end{pmatrix}, \quad B = \begin{pmatrix} 5 & -2 \\ 1 & -1 \\ 4 & 2 \end{pmatrix}, \quad C = \begin{pmatrix} 4 & -2 & 7 \\ 6 & 3 & 1 \\ 2 & 1 & 4 \end{pmatrix},$$

$$D = \begin{pmatrix} 1 & 0 & -5 \\ 2 & 3 & -4 \\ 3 & 7 & 2 \end{pmatrix}$$

Find each of the following matrices, if possible.

(a) *AB*   (b) *BC*   (c) *CD* and *DC*

SOLUTION

(a) Following the same procedure as in Example 1, we have

$$AB = \begin{pmatrix} 3 & 2 & 1 \\ -2 & 0 & 4 \\ 1 & -2 & 5 \end{pmatrix} \begin{pmatrix} 5 & -2 \\ 1 & -1 \\ 4 & 2 \end{pmatrix}$$

$$= \begin{pmatrix} 3 \cdot 5 + 2 \cdot 1 + 1 \cdot 4 & 3(-2) + 2(-1) + 1 \cdot 2 \\ (-2) \cdot 5 + 0 \cdot 1 + 4 \cdot 4 & (-2)(-2) + 0(-1) + 4 \cdot 2 \\ 1 \cdot 5 + (-2) \cdot 1 + 5 \cdot 4 & 1(-2) + (-2)(-1) + 5 \cdot 2 \end{pmatrix}$$

$$= \begin{pmatrix} 21 & -6 \\ 6 & 12 \\ 23 & 10 \end{pmatrix}.$$

(b) Matrix *B* is 3 × 2, and *C* is 3 × 3. Since the number of columns in *B* is different from the number of rows in *C*, the product is not defined.

(c) We use the matrix editor of a calculator to enter the matrices (Figure 11–22).

Figure 11–22

We then use the calculator to compute both products (Figure 11–23).

Figure 11–23

Note that *DC* is not equal to *CD*.   ■

It can be shown that matrix multiplication is associative, meaning that *A*(*BC*) = (*AB*)*C* for all matrices *A, B, C* for which the products are defined. As we saw in Example 2, however, matrix multiplication is *not* commutative, that is, *AB* might not be equal to *BA*, even when both products are defined.

▪▪ IDENTITY MATRICES AND INVERSES

The $n \times n$ **identity matrix** I_n is the matrix with 1's on the diagonal from upper left to lower right and 0's everywhere else; for example,

$$I_2 = \begin{pmatrix} 1 & 0 \\ 0 & 1 \end{pmatrix}, \qquad I_3 = \begin{pmatrix} 1 & 0 & 0 \\ 0 & 1 & 0 \\ 0 & 0 & 1 \end{pmatrix}, \qquad I_4 = \begin{pmatrix} 1 & 0 & 0 & 0 \\ 0 & 1 & 0 & 0 \\ 0 & 0 & 1 & 0 \\ 0 & 0 & 0 & 1 \end{pmatrix}.$$

The number 1 is the multiplicative identity of the number system because $a \cdot 1 = a = 1 \cdot a$ for every number a. The identity matrix I_n is the multiplicative identity for $n \times n$ matrices.

Identity Matrix

For any $n \times n$ matrix A,

$$AI_n = A = I_nA.$$

For example, in the 2×2 case,

$$\begin{pmatrix} a & b \\ c & d \end{pmatrix}\begin{pmatrix} 1 & 0 \\ 0 & 1 \end{pmatrix} = \begin{pmatrix} a \cdot 1 + b \cdot 0 & a \cdot 0 + b \cdot 1 \\ c \cdot 1 + d \cdot 0 & c \cdot 0 + d \cdot 1 \end{pmatrix} = \begin{pmatrix} a & b \\ c & d \end{pmatrix}.$$

Verify that the same answer results if you reverse the order of multiplication.

Every nonzero number c has a multiplicative inverse $c^{-1} = 1/c$ with the property that $cc^{-1} = 1$. The analogous statement for matrix multiplication does not always hold, and special terminology is used when it does. An $n \times n$ matrix A is said to be **invertible** (or **nonsingular**) if there is an $n \times n$ matrix B such that $AB = I_n$. In this case, it can be proved that $BA = I_n$ also. The matrix B is called the **inverse** of A and is sometimes denoted A^{-1}.

EXAMPLE 3

You can readily verify that

$$\begin{pmatrix} 2 & 1 \\ 3 & 1 \end{pmatrix}\begin{pmatrix} -1 & 1 \\ 3 & -2 \end{pmatrix} = \begin{pmatrix} 1 & 0 \\ 0 & 1 \end{pmatrix} = \begin{pmatrix} -1 & 1 \\ 3 & -2 \end{pmatrix}\begin{pmatrix} 2 & 1 \\ 3 & 1 \end{pmatrix}.$$

Therefore, $A = \begin{pmatrix} 2 & 1 \\ 3 & 1 \end{pmatrix}$ is an invertible matrix with inverse

$$A^{-1} = \begin{pmatrix} -1 & 1 \\ 3 & -2 \end{pmatrix}. \qquad ■$$

EXAMPLE 4

Find the inverse of the matrix $\begin{pmatrix} 2 & 6 \\ 1 & 4 \end{pmatrix}$.

SOLUTION   We must find numbers x, y, u, v such that

$$\begin{pmatrix} 2 & 6 \\ 1 & 4 \end{pmatrix}\begin{pmatrix} x & u \\ y & v \end{pmatrix} = \begin{pmatrix} 1 & 0 \\ 0 & 1 \end{pmatrix},$$

which is the same as

$$\begin{pmatrix} 2x + 6y & 2u + 6v \\ x + 4y & u + 4v \end{pmatrix} = \begin{pmatrix} 1 & 0 \\ 0 & 1 \end{pmatrix}.$$

Since corresponding entries in these last two matrices are equal, finding x, y, u, v amounts to solving these systems of equations:

$$\begin{array}{ccc} \begin{aligned} 2x + 6y &= 1 \\ x + 4y &= 0 \end{aligned} & \text{and} & \begin{aligned} 2u + 6v &= 0 \\ u + 4v &= 1. \end{aligned} \end{array}$$

We shall solve the systems by the Gauss-Jordan method of Section 11.2. The augmented matrices for the two systems are

$$A = \begin{pmatrix} 2 & 6 & \vdots & 1 \\ 1 & 4 & \vdots & 0 \end{pmatrix} \quad \text{and} \quad B = \begin{pmatrix} 2 & 6 & \vdots & 0 \\ 1 & 4 & \vdots & 1 \end{pmatrix}.$$

Note that the row operations that are needed for both A and B will be the same (because the first two columns are the same in both A and B). Consequently, we can save space and time by combining both of these matrices into this single matrix.

$$\begin{pmatrix} 2 & 6 & \vdots & 1 & 0 \\ 1 & 4 & \vdots & 0 & 1 \end{pmatrix}.$$

The first three columns of the last matrix are matrix A, and the first two and last columns are matrix B. Performing row operations on this matrix amounts to doing the operations simultaneously on both A and B.

Multiply row 1 by 1/2:

$$\begin{pmatrix} 1 & 3 & \vdots & \frac{1}{2} & 0 \\ 1 & 4 & \vdots & 0 & 1 \end{pmatrix}$$

Replace row 2 by the sum of itself and -1 times row 1:

$$\begin{pmatrix} 1 & 3 & \vdots & \frac{1}{2} & 0 \\ 0 & 1 & \vdots & -\frac{1}{2} & 1 \end{pmatrix}$$

Replace row 1 by the sum of itself and -3 times row 2:

$$\begin{pmatrix} 1 & 0 & \vdots & 2 & -3 \\ 0 & 1 & \vdots & -\frac{1}{2} & 1 \end{pmatrix}$$

The first three columns of the last matrix show that $x = 2$ and $y = -1/2$. Similarly, the first two and last columns show that $u = -3$ and $v = 1$. Therefore,

$$A^{-1} = \begin{pmatrix} 2 & -3 \\ -\frac{1}{2} & 1 \end{pmatrix}.$$

Observe that A^{-1} is just the right half of the final form of the preceding augmented matrix and that the left half is the identity matrix I_2.   ∎

Although the technique in Example 4 can be used to find the inverse of any matrix that has one, it's quicker to use a calculator. Any calculator with matrix capabilities can find the inverse of an invertible matrix (see the Technology Tip in the margin).

TECHNOLOGY TIP

On calculators other than TI-89, you can find the inverse A^{-1} of matrix A by keying in A (or Mat A) and using the x^{-1} key. Using the $\wedge$ key and -1 produces A^{-1} on TI-89 and HP-39 but leads to an error message on other calculators.

CAUTION

A calculator should produce an error message when asked for the inverse of a matrix A that does not have one. However, because of round-off errors, it may sometimes display a matrix that it says is A^{-1}. As an accuracy check when finding inverses, multiply A by A^{-1} to see whether the product is the identity matrix. If it isn't, A does not have an inverse.

▰▰ INVERSE MATRICES AND SYSTEMS OF EQUATIONS

Any system of linear equations can be expressed in matrix form, as shown in the next example.

EXAMPLE 5

Use matrix multiplication to express this system of equations in matrix form.

$$
\begin{aligned}
x + \ y + z &= \ \ 2 \\
2x + 3y \ \ \ \ \ &= \ \ 5 \\
x + 2y + z &= -1.
\end{aligned}
$$

SOLUTION   Let A be the 3×3 matrix of coefficients on the left side of the equations, let B be the column matrix of constants on the right side, and let X be the column matrix of unknowns.

$$
A = \begin{pmatrix} 1 & 1 & 1 \\ 2 & 3 & 0 \\ 1 & 2 & 1 \end{pmatrix}, \qquad X = \begin{pmatrix} x \\ y \\ z \end{pmatrix}, \qquad B = \begin{pmatrix} 2 \\ 5 \\ -1 \end{pmatrix}.
$$

Then AX is a matrix with three rows and one column, as is B.

$$
AX = \begin{pmatrix} 1 & 1 & 1 \\ 2 & 3 & 0 \\ 1 & 2 & 1 \end{pmatrix} \begin{pmatrix} x \\ y \\ z \end{pmatrix} = \begin{pmatrix} x + \ \ y + \ \ z \\ 2x + \ 3y + \ 0z \\ x + \ 2y + \ \ z \end{pmatrix} \quad \text{and} \quad B = \begin{pmatrix} 2 \\ 5 \\ -1 \end{pmatrix}.
$$

The entries in AX are just the left sides of the equations of the system, and the entries in B are the constants on the right sides. Therefore, the system can be expressed as the matrix equation $AX = B$.   ■

Suppose a system of equations is written in matrix form $AX = B$ and that the matrix A has an inverse. Then we can solve $AX = B$ by multiplying both sides of A^{-1}.

$$
A^{-1}(AX) = A^{-1}B
$$

$$
(A^{-1}A)X = A^{-1}B \qquad \text{[Matrix multiplication is associative]}
$$

$$
I_n X = A^{-1}B \qquad \text{[$A^{-1}A$ is the identity matrix]}
$$

$$
X = A^{-1}B \qquad \text{[Product of identity matrix and X is X]}
$$

The next example shows how this works in practice.

EXAMPLE 6

Solve the system

$$x + y + z = 2$$
$$2x + 3y = 5$$
$$x + 2y + z = -1.$$

SOLUTION   As we saw in Example 5, this system is equivalent to the matrix equation

$$AX = B$$

$$\begin{pmatrix} 1 & 1 & 1 \\ 2 & 3 & 0 \\ 1 & 2 & 1 \end{pmatrix} \begin{pmatrix} x \\ y \\ z \end{pmatrix} = \begin{pmatrix} 2 \\ 5 \\ -1 \end{pmatrix}.$$

Use a calculator to find the inverse of the coefficient matrix A and multiply both sides of the equation by A^{-1}.

$$A^{-1}AX = A^{-1}B$$

$$\begin{pmatrix} 1.5 & .5 & -1.5 \\ -1 & 0 & 1 \\ .5 & -.5 & .5 \end{pmatrix} \begin{pmatrix} 1 & 1 & 1 \\ 2 & 3 & 0 \\ 1 & 2 & 1 \end{pmatrix} \begin{pmatrix} x \\ y \\ z \end{pmatrix} = \begin{pmatrix} 1.5 & .5 & -1.5 \\ -1 & 0 & 1 \\ .5 & -.5 & .5 \end{pmatrix} \begin{pmatrix} 2 \\ 5 \\ -1 \end{pmatrix}$$

$$\begin{pmatrix} 1 & 0 & 0 \\ 0 & 1 & 0 \\ 0 & 0 & 1 \end{pmatrix} \begin{pmatrix} x \\ y \\ z \end{pmatrix} = \begin{pmatrix} 1.5 & .5 & -1.5 \\ -1 & 0 & 1 \\ .5 & -.5 & .5 \end{pmatrix} \begin{pmatrix} 2 \\ 5 \\ -1 \end{pmatrix} \quad \text{[Since } A^{-1}A = I_3]$$

$$\begin{pmatrix} x \\ y \\ z \end{pmatrix} = \begin{pmatrix} 7 \\ -3 \\ -2 \end{pmatrix} \quad \text{[Since } I_3X = X]$$

Therefore, the solution of the original system is $x = 7$, $y = -3$, $z = -2$.   ∎

Only a matrix with the same number of rows as columns can possibly have an inverse. Consequently, the method of Example 6 can be tried only when the system has the same number of equations as unknowns. In this case, you should use your calculator to verify that the coefficient matrix actually has an inverse (see the Caution on page 764). If it does not, other methods must be used. Here is a summary of the possibilities.

Matrix Solution of a System of Equations

Suppose a system with the same number of equations as unknowns is written in matrix form as $AX = B$.

If the matrix A has an inverse, then the unique solution of the system is

$$X = A^{-1}B.$$

If A does not have an inverse, then the system either has no solutions or has infinitely many solutions. Its solutions (if any) may be found by using Gauss-Jordan elimination (Section 11.2).

EXAMPLE 7

Solve the system

$$2x + y - z = 2$$
$$x + 3y + 2z = 1$$
$$x + y + z = 2.$$

SOLUTION   Since there are the same number of equations as unknowns, we can try the method of Example 6. In this case, we have

$$A = \begin{pmatrix} 2 & 1 & -1 \\ 1 & 3 & 2 \\ 1 & 1 & 1 \end{pmatrix} \quad \text{and} \quad B = \begin{pmatrix} 2 \\ 1 \\ 2 \end{pmatrix}.$$

CALCULATOR EXPLORATION

Verify that the matrix A does have an inverse. Show that the solutions of the system are $x = 2$, $y = -1$, $z = 1$ by computing $A^{-1}B$.   ∎

▐▌ APPLICATIONS

Just as two points determine a unique line, three points (that aren't on the same line) determine a unique parabola, as the next example demonstrates.

Figure 11–24

EXAMPLE 8

A batter hits a baseball, and special measuring devices locate its position at various times during its flight. If the path of the ball is drawn on a coordinate plane, with the batter at the origin, it looks like Figure 11–24. According to the measuring devices, the ball passes through the points $(7, 9)$, $(47, 38)$, and $(136, 70)$.

(a) What is the equation of the path of the ball?

(b) How far from the batter does the ball hit the ground?

SOLUTION

(a) The path of the ball appears to be part of a parabola (a fact that will not be proved here) and hence has an equation of the form

$$y = ax^2 + bx + c$$

for some constants a, b, and c. Since $(7, 9)$ is on the graph, we know that

$$\text{when } x = 7, \quad \text{then } y = 9,$$

so

$$9 = a(7^2) + b(7) + c \qquad \text{or, equivalently,} \qquad 49a + 7b + c = 9.$$

Similarly, since (47, 38) and (136, 70) are on the graph, we have

$$38 = a(47^2) + b(47) + c \qquad \text{or, equivalently,} \qquad 2209a + 47b + c = 38$$

and

$$70 = a(136^2) + b(136) + c \qquad \text{or, equivalently,} \qquad 18{,}496a + 136b + c = 70.$$

We can determine a, b, and c by solving this system of equations.

$$49a + \quad 7b + c = 9$$
$$2209a + \quad 47b + c = 38$$
$$18{,}496a + 136b + c = 70$$

or, in matrix form,

$$\begin{pmatrix} 49 & 7 & 1 \\ 2209 & 47 & 1 \\ 18{,}496 & 136 & 1 \end{pmatrix} \begin{pmatrix} a \\ b \\ c \end{pmatrix} = \begin{pmatrix} 9 \\ 38 \\ 70 \end{pmatrix}.$$

A calculator shows that the solution is

$$\begin{pmatrix} a \\ b \\ c \end{pmatrix} = \begin{pmatrix} 49 & 7 & 1 \\ 2209 & 47 & 1 \\ 18{,}496 & 136 & 1 \end{pmatrix}^{-1} \begin{pmatrix} 9 \\ 38 \\ 70 \end{pmatrix} \approx \begin{pmatrix} -.00283 \\ .87798 \\ 2.99296 \end{pmatrix}.$$

Therefore, the approximate equation for the ball's path is

$$y = -.00283x^2 + .87798x + 2.99296.$$

(b) The ball hits the ground at a point where its height y is 0, that is, when

$$-.00283x^2 + .87798x + 2.99296 = 0.$$

Use the quadratic formula or an equation solver to verify that the solutions of this equation are $x \approx -3.37$ (which is not applicable here) and $x \approx 313.61$. Therefore, the ball hits the ground approximately 314 feet from the batter. ■

Given any three points not on a straight line, the method in Example 8 can be used to find the unique parabola that passes through these points. This parabola can also be found by quadratic regression.[*]

CALCULATOR EXPLORATION

Use quadratic regression on the three points in Example 8 and verify that the equation obtained is the same as the one in Example 8.

To see why regression produces the parabola that actually passes through the three points, recall that the error in a quadratic model is measured by taking the difference between the y-coordinate of each data point and the y-coordinate of the corresponding point on the model and summing the squares of these errors. If the data points actually lie on a parabola (which is always the case with three points that are not on a line), then the error for that parabola will be 0 (the smallest possible error). Hence, it will be the parabola produced by the least squares quadratic regression procedure.

---

[*]If you have not read Section 2.5 and Special Topics 4.4.A, you may skip this discussion.

EXERCISES 11.3

In Exercises 1–6, determine whether the product AB or BA is defined. If a product is defined, state its size (number of rows and columns). Do not calculate any products.

1. $A = \begin{pmatrix} 3 & 6 & 7 \\ 8 & 0 & 1 \end{pmatrix}, \qquad B = \begin{pmatrix} 2 & 5 & 9 & 1 \\ 7 & 0 & 0 & 6 \\ -1 & 3 & 8 & 7 \end{pmatrix}$

2. $A = \begin{pmatrix} -1 & -2 & -5 \\ 9 & 2 & -1 \\ 10 & 34 & 5 \end{pmatrix}, \quad B = \begin{pmatrix} 17 & -9 \\ -6 & 12 \\ 3 & 5 \end{pmatrix}$

3. $A = \begin{pmatrix} 1 & 0 \\ 1 & 1 \\ 0 & 1 \end{pmatrix}, \qquad B = \begin{pmatrix} 5 & 6 & 11 \\ 7 & 8 & 15 \end{pmatrix}$

4. $A = \begin{pmatrix} 1 & -5 & 7 \\ 2 & 4 & 8 \\ 1 & -1 & 2 \end{pmatrix}, \qquad B = \begin{pmatrix} -2 & 4 & 9 \\ 13 & -2 & 1 \\ 5 & 25 & 0 \end{pmatrix}$

5. $A = \begin{pmatrix} -4 & 15 \\ 3 & -7 \\ 2 & 10 \end{pmatrix}, \qquad B = \begin{pmatrix} 1 & 2 \\ 3 & 4 \end{pmatrix}$

6. $A = \begin{pmatrix} 10 & 12 \\ -6 & 0 \\ 1 & 23 \\ -4 & 0 \end{pmatrix}, \qquad B = \begin{pmatrix} 1 & 2 & 3 \\ 3 & 2 & 1 \end{pmatrix}$

In Exercises 7–13, find AB.

7. $A = \begin{pmatrix} 3 & 2 \\ 2 & 4 \end{pmatrix}, \qquad B = \begin{pmatrix} 1 & -2 & 3 \\ 0 & 3 & 1 \end{pmatrix}$

8. $A = \begin{pmatrix} -1 & 2 & 3 \\ 0 & -1 & 2 \\ 1 & 2 & 0 \end{pmatrix}, \qquad B = \begin{pmatrix} 3 & -2 & -1 \\ 1 & 0 & 5 \\ 1 & -1 & -1 \end{pmatrix}$

9. $A = \begin{pmatrix} 1 & 0 & -4 \\ 0 & 2 & -1 \\ 2 & 3 & 4 \end{pmatrix}, \qquad B = \begin{pmatrix} 1 & 1 \\ 1 & 0 \\ 0 & 1 \end{pmatrix}$

10. $A = \begin{pmatrix} 1 & -2 \\ 3 & 0 \\ 0 & -1 \\ 2 & 1 \end{pmatrix}, \qquad B = \begin{pmatrix} -1 & 3 & -2 & 0 \\ 6 & 1 & 0 & -2 \end{pmatrix}$

11. $A = \begin{pmatrix} 2 & 0 & -1 \\ 1 & 1 & 2 \\ 0 & 2 & -3 \\ 2 & 3 & 0 \end{pmatrix}, \qquad B = \begin{pmatrix} 1 & 0 & 1 & 1 \\ 1 & 1 & 0 & 1 \\ 1 & 1 & 1 & 0 \end{pmatrix}$

12. $A = \begin{pmatrix} 10 & 0 & 1 & 0 \\ -1 & 1 & 0 & 1 \end{pmatrix}, \qquad B = \begin{pmatrix} 2 & -1 & 0 & 1 \\ -2 & 3 & 1 & -4 \\ 3 & 5 & 2 & -5 \end{pmatrix}$

In Exercises 13–16, show that AB is not equal to BA by computing both products.

13. $A = \begin{pmatrix} 3 & 2 \\ 5 & 1 \end{pmatrix}, \qquad B = \begin{pmatrix} 7 & -5 \\ -2 & 6 \end{pmatrix}$

14. $A = \begin{pmatrix} 3/2 & 2 \\ 4 & 7/2 \end{pmatrix}, \qquad B = \begin{pmatrix} 1 & -3/2 \\ 5/2 & 1 \end{pmatrix}$

15. $A = \begin{pmatrix} 4 & 2 & -1 \\ 0 & 1 & 2 \\ -3 & 0 & 1 \end{pmatrix}, \qquad B = \begin{pmatrix} 1 & 7 & -5 \\ 2 & -2 & 6 \\ 0 & 0 & 0 \end{pmatrix}$

16. $A = \begin{pmatrix} 1 & 1 & -1 & 1 \\ 2 & 0 & 3 & 2 \\ -3 & 0 & 0 & 1 \\ 1 & -1 & 1 & 2 \end{pmatrix}, \qquad B = \begin{pmatrix} 0 & 1 & 7 & 7 \\ 2 & 3 & -2 & 1 \\ 5 & 0 & 1 & 0 \\ -1 & 0 & 1 & 0 \end{pmatrix}$

In Exercises 17–24, find the inverse of the matrix, if it exists.

17. $\begin{pmatrix} 1 & 2 \\ 3 & 4 \end{pmatrix}$

18. $\begin{pmatrix} 3 & 5 \\ 1 & 4 \end{pmatrix}$

19. $\begin{pmatrix} 3 & -1 \\ -6 & 2 \end{pmatrix}$

20. $\begin{pmatrix} 1 & -1 & 0 \\ 1 & 0 & -1 \\ 6 & -2 & -3 \end{pmatrix}$

21. $\begin{pmatrix} 1 & 2 & 0 \\ 3 & -1 & 2 \\ -2 & 3 & -2 \end{pmatrix}$

22. $\begin{pmatrix} 1 & -3 & 4 \\ 2 & -5 & 7 \\ 0 & -1 & 1 \end{pmatrix}$

23. $\begin{pmatrix} 5 & 0 & 2 \\ 2 & 2 & 1 \\ -3 & 1 & -1 \end{pmatrix}$

24. $\begin{pmatrix} -1 & 3 & 1 \\ 2 & 5 & 0 \\ 3 & 1 & -2 \end{pmatrix}$

In Exercises 25–28, solve the system of equations by using the method of Example 6.

25.
$$\begin{aligned} -x + y &= 1 \\ -x \phantom{{}+y} + z &= -2 \\ 6x - 2y - 3z &= 3 \end{aligned}$$

26.
$$\begin{aligned} x + 2y + 3z &= 1 \\ 2x + 5y + 3z &= 0 \\ x + 8z &= -1 \end{aligned}$$

27.
$$\begin{aligned} 2x + y &= 0 \\ -4x - y - 3z &= 1 \\ 3x + y + 2z &= 2 \end{aligned}$$

28.
$$\begin{aligned} -3x - 3y - 4z &= 2 \\ y + z &= 1 \\ 4x + 3y + 4z &= 3 \end{aligned}$$

In Exercises 29–39, solve the system by any method.

29.
$$x + y + 2w = 3$$
$$2x - y + z - w = 5$$
$$3x + 3y + 2z - 2w = 0$$
$$x + 2y + z = 2$$

30.
$$x - 2y + 3z = 1$$
$$y - z + w = -2$$
$$-2x + 2y - 2z + 4w = 5$$
$$2y - 3z + w = 8$$

31.
$$x + y + 6z + 2v = 1.5$$
$$x + 5z + 2v - 3w = 2$$
$$3x + 2y + 17z + 6v - 4w = 2.5$$
$$4x + 3y + 21z + 7v - 2w = 3$$
$$-6x - 5y - 36z - 12v + 3w = 3.5$$

32.
$$x - 1.5y + 1.5v - 4w = 3$$
$$-1.5y + .5w = 0$$
$$-x + 2y + .5z - 2v + 4.5w = 2$$
$$-.5y - 2.5z + .5v - .75w = 8$$
$$y - .5z + .5w = 4$$

33.
$$x + 2y + 2z - 2w = -23$$
$$4x + 4y - z + 5w = 7$$
$$-2x + 5y + 6z + 4w = 0$$
$$5x + 13y + 7z + 12w = -7$$

34.
$$x + 4y + 5z + 2w = 0$$
$$2x + y + 4z - 2w = 0$$
$$-x + 7y + 10z + 5w = 0$$
$$-4x + 2y + z + 5w = 0$$

35.
$$x + 2y + 5z - 2v + 4w = 0$$
$$2x - 4y + 6z + v + 4w = 0$$
$$5x + 2y - 3z + 2v + 3w = 0$$
$$6x - 5y - 2z + 5v + 3w = 0$$
$$x + 2y - z - 2v + 4w = 0$$

36.
$$4x + 2y + 3z + 3v + 2w = 1$$
$$2x + 2z + 2u + v - w = 2$$
$$10x + 2y + 10z + 10u + 3v + 4w = 5$$
$$16x + 4y + 16z + 18u + 7v - 2w = -3$$
$$x + 2y + 4z - 6u + 2v + w = -2$$
$$6x + 2y + 6z + 6u + 3v = 1$$

37.
$$x + 2y + 3z = 1$$
$$3x + 2y + 4z = -1$$
$$2x + 6y + 8z + w = 3$$
$$2x + 2z - 2w = 3$$

38.
$$x + 2y + 4z = 6$$
$$y + z = 1$$
$$x + 3y + 5z = 10$$

39.
$$x + 3w = -2$$
$$x - 4y - z + 3w = -7$$
$$4y + z = 5$$
$$-x + 12y + 3z - 3w = 17$$

In Exercises 40–42, find constants a, b, c such that the three given points lie on the parabola $y = ax^2 + bx + c$. See Example 8.

40. $(-3, 2), (1, 1), (2, -1)$

41. $(-3, 15), (1, -7), (5, 111)$

42. $(1, -2), (3, 1), (4, -1)$

43. Concentrations of the greenhouse gas carbon dioxide (CO_2) have increased quadratically over the past half-century.[*] So the concentration y of CO_2 (in parts per million) in year x is given by an equation of the form

$$y = ax^2 + bx + c.$$

(a) Let $x = 0$ correspond to 1958 and use the following data to find a, b, and c.

Year	1958	1979	2001
CO_2 Concentration	315	337	371

(b) Use this equation to estimate the CO_2 concentration in the years 1983, 1993, and 2003. [For comparison purposes, the actual concentrations in 1983 and 1993 were 343 ppm and 357 ppm respectively.]

44. The table shows the per capita consumption of chicken (in pounds) in selected years.[†]

Year	1980	1990	1999
Chicken Consumption	32.7	42.4	54.2

(a) Let $x = 0$ correspond to 1980. Find a quadratic equation that models this data.

(b) Use the equation to estimate the consumption of chicken in 1985, 1995, and 2005. [The actual consumption in 1985 and 1995 was 36.4 and 49.0 pounds, respectively.]

---

[*]C. D. Keeling and T. P. Whorf, Scripps Institution of Oceanography.
[†]U.S. Department of Agriculture.

45. The table shows the per capita sales (in dollars) of electronics and appliances in selected years.[*]

Year	1992	1996	2000
Sales	$169	$260	$320

(a) Find a quadratic equation that models this data, with $x = 2$ corresponding to 1992.

(b) Use the equation to estimate sales of electronics and appliances in 1993, 1998, and 2002.

46. The table shows the percentage y of total consumer spending on sound recordings that is spent on pop music in year x.[†]

Year	1990	1994	1996	1998	2000
% Pop	13.7	10.3	9.3	10	11

(a) Assume that the data can be modeled by an equation of the form

$$y = ax^4 + bx^3 + cx^2 + dx + k.$$

Use the same method used in Exercises 43–45 (with five equations instead of three) to find the constants a, b, c, d, and k.

(b) Use the equation in part (a) to estimate the percentage of consumer spending spent on pop music in 1993 and 1999.

47. Find constants a, b, c such that the points $(0, -2)$, $(\ln 2, 1)$, and $(\ln 4, 4)$ lie on the graph of $f(x) = ae^x + be^{-x} + c$. [*Hint:* Proceed as in Example 8.]

48. Find constants a, b, c such that the points $(0, -1)$, $(\ln 2, 4)$, and $(\ln 3, 7)$ lie on the graph of $f(x) = ae^x + be^{-x} + c$.

49. New Army Stores, a national chain of casual clothing stores, recently sent shipments of jeans, jackets, sweaters, and shirts to its stores in various cities. The number of items shipped to each city and their total wholesale cost are shown in the table below. Find the wholesale price of each of the following: one pair of jeans, one jacket, one sweater, and one shirt.

City	Jeans	Jackets	Sweaters	Shirts	Total Cost
Cleveland	3000	3000	2200	4200	$507,650
St. Louis	2700	2500	2100	4300	459,075
Seattle	5000	2000	1400	7500	541,225
Phoenix	7000	1800	600	8000	571,500

50. A 100-bed nursing home provides two levels of long-term care: regular and maximum. Patients at each level have a choice of a private room or a less expensive semiprivate room. The table below shows the number of patients in each category at various times last year and the total daily cost for these patients. Find the daily cost of each of the following: a private room (regular care), a private room (maximum care), a semiprivate room (regular care), and a semiprivate room (maximum care).

Month	Regular Care Patients		Maximum Care Patients		Daily Cost
	Semi-private	Private	Semi-private	Private	
January	22	8	60	10	$18,824
April	26	8	54	12	18,738
July	24	14	56	6	18,606
October	20	10	62	8	18,824

51. A candy company produces three types of gift boxes: A, B, and C. A box of variety A contains .6 pound of chocolates and .4 pound of mints. A box of variety B contains .3 pound of chocolates, .4 pound of mints, and .3 pound of caramels. A box of variety C contains .5 pound of chocolates, .3 pound of mints, and .2 pound of caramels. The company has 41,400 pounds of chocolates, 29,400 pounds of mints, and 16,200 pounds of caramels in stock. How many boxes of each variety should be made to use up all the stock?

52. Certain circus animals are fed the same three food mixes: R, S, and T. Lions receive 1.1 units of R, 2.4 units of S, and 3.7 units of T each day. Horses receive 8.1 units of R, 2.9 units of S, and 5.1 units of T each day. Bears receive 1.3 units of R, 1.3 units of S, and 2.3 units of T each day. If 16,000 units of R, 28,000 units of S, and 44,000 units of T are available each day, how many of each type of animal can be supported?

---

[*]U.S. Census Bureau.
[†]Recording Industry Association of America, Inc.

Chapter 11 Review

IMPORTANT CONCEPTS

Section 11.1

System of equations 722
Solution of a system 722
Number of solutions of a
    system 723
Substitution method 724
Elimination method 725
Inconsistent system 726
Dependent system 727

Special Topics 11.1.A

Algebraic solution of systems of
    nonlinear equations 735–736
Graphical solution of systems of
    nonlinear equations 737–740

Section 11.2

Equivalent systems 742
Elementary operations 743
Back substitution 744
Triangular form system 744
Gaussian elimination 744
Augmented matrix 745
Row operations 745
Row echelon form matrix 747
Inconsistent system 748
Gauss-Jordan method 748
Reduced row echelon form matrix
    750
Homogeneous system 750
Dependent system 751
Partial fraction decomposition
    751–752

Number of solutions of a system
    753

Section 11.3

$m \times n$ matrix 758
Rows and columns 758
Equality of matrices 758
Multiplication of matrices 759
Identity matrix 762
Invertible matrix 762
Inverse of a matrix 762
Matrix form of a system of
    equations 764
Inverses and solutions of a
    system 765

REVIEW QUESTIONS

In Questions 1–4, solve the system of linear equations by any means you want.

1. $-5x + 3y = 4$
    $2x - y = -3$

2. $3x - y = 6$
    $2x + 3y = 7$

3. $3x - 5y = 10$
    $4x - 3y = 6$

4. $\dfrac{1}{4}x - \dfrac{1}{3}y = -\dfrac{1}{4}$
    $\dfrac{1}{10}x + \dfrac{2}{5}y = \dfrac{2}{5}$

5. The number of days y in year x in which acceptable air quality standards were not met in two metropolitan areas are approximated by these equations (in which $x = 4$ corresponds to 1994).[*]

Riverside–San Bernadino, CA: $10.89x + y = 184.92$

Atlanta, GA: $16.06x - 2y = 32.04$

In what year is the number of unacceptable days the same in both areas? How many such days are there in that year?

6. The sum of one number and three times a second number is -20. The sum of the second number and two times the first number is 55. Find the two numbers.

7. An alloy containing 40% gold and an alloy containing 70% gold arc to be mixed to produce 50 pounds of an alloy containing 60% gold. How much of each alloy is needed?

8. The table shows the population (in thousands) of the cities of Miami and Cleveland in selected years.[†]

Year	1950	1960	1970	1980	1990	2000
Miami	249.3	291.7	335.0	347.0	358.5	362.5
Cleveland	914.8	876.1	751.0	574.0	505.6	478.4

(a) Use linear regression to find an equation that gives the population y of Miami in year x, with $x = 0$ corresponding to 1950.

(b) Do part (a) for Cleveland.

(c) If these models remain accurate, when do the two cities have the same population and what is that population?

---

[*]U.S. Environmental Protection Agency.

[†]U.S. Census Bureau.

In Questions 9–12, solve the nonlinear system.

9. $x^2 - y = 0$

$y - 2x = 3$

10. $x^2 + y^2 = 25$

$x^2 + y = 19$

11. $x^2 + y^2 = 16$

$x + y = 2$

12. $6x^2 + 4xy + 3y^2 = 36$

$x^2 - xy + y^2 = 9$

13. Write the augmented matrix of the system

$$x - 2y + 3z = 4$$

$$2x + y - 4z = 3$$

$$-3x + 4y - z = -2$$

14. Use matrix methods to solve the system in Question 13.

15. Write the coefficient matrix of the system

$$2x - y - 2z + 2u = 0$$

$$x + 3y - 2z + u = 0$$

$$-x + 4y + 2z - 3u = 0$$

16. Use matrix methods to solve the system in Question 15.

In Questions 17–22, solve the system.

17. $3x + y - z = 13$

$x + 2z = 9$

$-3x - y + 2z = 9$

18. $x + 2y + 3z = 1$

$4x + 4y + 4z = 2$

$10x + 8y + 6z = 4$

19. $4x + 3y - 3z = 2$

$5x - 3y + 2z = 10$

$2x - 2y + 3z = 14$

20. $x + y - 4z = 0$

$2x + y - 3z = 2$

$-3x - y + 2z = -4$

21. $x - 2y - 3z = 1$

$5y + 10z = 0$

$8x - 6y - 4z = 8$

22. $4x - y - 2z = 4$

$x - y - \frac{1}{2}z = 1$

$2x - y - z = 8$

23. Let L be the line with equation $4x - 2y = 6$, and let M be the line with equation $-10x + 5y = -15$. Which of the following statements is true?

(a) L amd M do not intersect.

(b) L and M intersect at a single point.

(c) L and M are the same line.

(d) All of the above are true.

(e) None of the above are true.

24. Which of the following statements about this system of equations are *false*?

$$x + z = 2$$

$$6x + 4y + 14z = 24$$

$$2x + y + 4z = 7$$

(a) $x = 2, y = 3, z = 0$ is a solution.

(b) $x = 1, y = 1, z = 1$ is a solution.

(c) $x = 1, y = -3, z = 3$ is a solution.

(d) The system has an infinite number of solutions.

(e) $x = 2, y = 5, z = -1$ is not a solution.

In Questions 25 and 26, find the constants A, B, C that make the statement true.

25. $\dfrac{4x - 7}{x^2 - x - 6} = \dfrac{A}{x - 3} + \dfrac{B}{x + 2}$

26. $\dfrac{6x^2 + 6x - 6}{(x^2 - 1)(x + 2)} = \dfrac{A}{x + 1} + \dfrac{B}{x - 1} + \dfrac{C}{x + 2}$

In Questions 27–30, perform the indicated matrix multiplication or state that the product is not defined. Use these matrices.

$$A = \begin{pmatrix} -1 & 0 \\ 0 & -1 \end{pmatrix}, \quad B = \begin{pmatrix} 2 & -3 \\ 4 & 1 \end{pmatrix}, \quad C = \begin{pmatrix} 3 & 2 \\ 2 & 4 \end{pmatrix},$$

$$D = \begin{pmatrix} -3 & 1 & 2 \\ 1 & 0 & 4 \end{pmatrix}, \quad E = \begin{pmatrix} 1 & 2 \\ -3 & 4 \\ 0 & 5 \end{pmatrix}, \quad F = \begin{pmatrix} 2 & 3 \\ 6 & 3 \\ 6 & 1 \end{pmatrix}$$

27. AB **28.** CD **29.** AE **30.** DF

In Questions 31–34, find the inverse of the matrix, if it exists.

31. $\begin{pmatrix} 3 & -7 \\ 4 & -9 \end{pmatrix}$

32. $\begin{pmatrix} 2 & 6 \\ 1 & 3 \end{pmatrix}$

33. $\begin{pmatrix} 3 & 2 & 6 \\ 1 & 1 & 2 \\ 2 & 2 & 5 \end{pmatrix}$

34. $\begin{pmatrix} 1 & -1 & 1 \\ 2 & -3 & 2 \\ -4 & 6 & 1 \end{pmatrix}$

In Questions 35 and 36, use matrix inverses to solve the system.

35. $x + 2z + 6w = 2$

$3x + 4y - 2z - w = 0$

$5x + 2z - 5w = -4$

$4x - 4y + 2z + 3w = 1$

REVIEW QUESTIONS

36. $2x + y + 2z \quad\quad + u = 2$

$x + 3y - 4z - 2u + 2v = -2$

$2x + 3y + 5z - 4u + v = 1$

$x \quad\quad - 2z \quad\quad + 4v = 4$

$2x \quad\quad + 6z \quad\quad - 5v = 0$

In Questions 37 and 38, find the equation of the parabola passing through the given points.

37. $(-3, 52)$, $(2, 17)$, $(8, 305)$

38. $(-2, -18)$, $(2, 6)$, $(4, -12)$

39. The table shows the number of hours spent per person per year on home video games.[*]

Year	1996	2000	2004
Hours	25	76	161

(a) Find a quadratic equation that models this data, with $x = 6$ corresponding to 1996.

(b) Use the equation to estimate the number of hours spent in 1998, 2002, and 2006.

40. The table shows the per capita consumption of eggs in selected years.[†]

Year	1980	1990	1999
Egg Consumption	271	234	255

(a) Find a quadratic equation that models this data, with $x = 0$ corresponding to 1980.

(b) Use the equation to estimate egg consumption in 1985, 1995, and 2005.

41. Tickets to a lecture cost $1 for students, $1.50 for faculty, and $2 for others. Total attendance at the lecture was 460, and the total income from tickets was $570. Three times as many students as faculty attended. How many faculty members attended the lecture?

42. If Tom, Dick, and Harry work together, they can paint a large room in 4 hours. When only Dick and Harry work together, it takes 8 hours to paint the room. Tom and Dick, working together, take 6 hours to paint the room. How long would it take each of them to paint the room alone? [*Hint:* If x is the amount of the room painted in 1 hour by Tom, y is the amount painted by Dick, and z is the amount painted by Harry, then $x + y + z = 1/4$.]

43. An animal feed is to be made from corn, soybeans, and meat by-products. One bag is to supply 1800 units of fiber, 2800 units of fat, and 2200 units of protein. Each pound of corn has 10 units of fiber, 30 units of fat, and 20 units of protein. Each pound of soybeans has 20 units of fiber, 20 units of fat, and 40 units of protein. Each pound of by-products has 30 units of fiber, 40 units of fat, and 25 units of protein. How many pounds of corn, soybeans, and by-products should each bag contain?

44. A company produces three camera models: A, B, and C. Each model A requires 3 hours of lens polishing, 2 hours of assembly time, and 2 hours of finishing time. Each model B requires 2 hours of lens polishing, 2 hours of assembly time, and 1 hour of finishing time. Each model C requires 1, 3, and 1 hours of lens polishing, assembly, and finishing time, respectively. There are 100 hours available for lens polishing, 100 hours for assembly, and 65 hours for finishig each week. How many of each model should be produced if all available time is to be used?

---

[*]Data and projections from *Statistical Abstract of the United States: 2001.*

[†]U.S. Department of Agriculture.

DISCOVERY PROJECT 11 Input-Output Analysis

Wassily Leontief won the Nobel Prize in economics in 1973 for his method of input-output analysis of the economies of industrialized nations. This method has become a permanent part of production planning and forecasting by both national governments and private corporations. During the Arab oil boycott in 1973, for example, General Electric used input-output analysis on 184 sectors of the economy (such as energy, agriculture, and transportation) to predict the effect of the energy crisis on public demand for its products. The key to Leontief's method is knowing how much each sector of the economy needs from other sectors in order to do its job.

A simple economic model will illustrate the basic ideas behind input-output analysis. Suppose the country Hypothetica has just two sectors in its economy: agriculture and manufacturing. The production of a ton of agricultural products requires the use of .1 ton of agricultural products and .1 ton of manufactured products. Similarly, the production of a ton of manufactured products consumes .1 ton of agricultural products and .3 ton of manufactured products. The key economic question is: How many tons of each sector must Hypothetica produce in order to have enough surplus to export 10,000 tons of agricultural products and 10,000 tons of manufactured goods?

Let A be the total amount of agricultural goods, and let M be the total amount of manufactured goods produced. The total agricultural production A is the sum of the amount needed for producing the agricultural products plus the amount needed for producing manufactured products plus the amount targeted for export. In other words,

agricultural needs	+	manufacturing needs	+	export needs	=	total agricultural production
$.1A$	+	$.1M$	+	$100{,}000$	=	$A.$

1. Write a similar equation for manufactured goods.

2. Solve the system of equations given by the agricultural equation above and the manufacturing equation of Problem 1. What level of production of agricultural and manufactured goods should Hypothetica work toward to reach its export goals?

3. The same kind of analysis can be used to determine the surplus available for exports in an economy. Suppose that Hypothetica can produce a total of 120,000 tons of agricultural products and 28,000 tons of manufactured goods. How much of each is available for export?

4. Input-output analysis can also be used for allocating resources in smaller-scale situations. Suppose a horse outfitter is hired by the State of Washington Department of Fish and Wildlife to haul 20 salt blocks into a game area for winter feeding. Each horse can carry a 200-pound load. A single person can handle three horses. Each person requires one horse to ride and one-half horse to carry his or her personal gear. A single horse can carry the feed and equipment for five horses. How many people and horses must go on the trip? [*Note:* Each horse load of salt must consist of an even number of blocks. Each salt block weighs 50 pounds. Also, fractional numbers of horses and people are not allowed. Make sure to adjust your answer to integer values, and be sure that the answer still provides sufficient carrying capacity for the salt.]

12

Discrete Algebra

Bob Daemmrich/Stock, Boston Inc./PictureQuest

$$\sum_{m=1}^{k} a_m = \frac{k}{2}\left(a_1 + a_k\right)$$

$$\sum_{m=1}^{k} a_m = a_1\left(\frac{1-r^k}{1-r}\right)$$

What's next?

CD and CD-ROM players, fax machines, cameras, and other devices incorporate digital technology, which uses sequences of 0's and 1's to send signals. Determining the monthly payment on a car loan involves the sum of a geometric sequence. Problems involving the action of bouncing balls, vacuum pumps, and other devices can sometimes be solved by using sequences. See Exercises 53, 57, and 58 on page 800.

Chapter Outline

Interdependence of Sections

Sections 12.1, 12.4, and 12.5 are independent of one another and may be read in any order. The interdependence of sections is

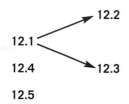

12.5

12.1	**Sequences and Sums**
12.2	**Arithmetic Sequences**
12.3	**Geometric Sequences**
12.3.A	*Special Topics:* **Infinite Series**
12.4	**The Binomial Theorem**
12.5	**Mathematical Induction**

This chapter deals with a variety of subjects involving counting processes and the nonnegative integers 0, 1, 2, 3, ...

12.1 Sequences and Sums

A **sequence** is an ordered list of numbers, such as

$$2, 4, 6, 8, 10, 12, \ldots$$

$$1, -3, 5, -7, 9, -11, 13, \ldots$$

$$1, 0, 1, 0, 1, 0, 1, 0, \ldots$$

$$2, 1, \frac{2}{3}, \frac{3}{4}, \frac{4}{5}, \frac{5}{6}, \frac{6}{7}, \ldots,$$

where the dots indicate that the same pattern continues forever.[*] Each number on the list is called a **term** of the sequence.

$$
\begin{array}{cccccc}
2, & 1, & 0, & 1, & 2, & 3, \quad 2, 1, 0, 1, 2, 3, 2, \ldots . \\
\uparrow & \uparrow & \uparrow & \uparrow & \uparrow & \uparrow \\
\text{1st} & \text{2nd} & \text{3rd} & \text{4th} & \text{5th} & \text{6th} \\
\text{term} & \text{term} & \text{term} & \text{term} & \text{term} & \text{term}
\end{array}
$$

When the pattern isn't obvious, as in the preceding examples, sequences are usually described in terms of a formula.

---

[*]Such a list defines a function f whose domain is the set of positive integers. The rule is $f(1) =$ first number on the list, $f(2) =$ second number on the list, and so on. Conversely, any function g whose domain is the set of positive integers leads to an ordered list of numbers, namely, $g(1)$, $g(2)$, $g(3)$, ... So a sequence is formally defined to be a function whose domain is the set of positive integers.

EXAMPLE 1

Consider the sequence $a_1, a_2, a_3, \ldots, a_n, \ldots$ where a_n is given by the formula

$$a_n = \frac{n^2 - 3n + 1}{2n + 5}.$$

To find a_1, we substitute $n = 1$ in the formula for a_n; to find a_2, we substitute $n = 2$ in the formula; and so on.

$$a_1 = \frac{1^2 - 3 \cdot 1 + 1}{2 \cdot 1 + 5} = -\frac{1}{7},$$

$$a_2 = \frac{2^2 - 3 \cdot 2 + 1}{2 \cdot 2 + 5} = -\frac{1}{9},$$

$$a_3 = \frac{3^2 - 3 \cdot 3 + 1}{2 \cdot 3 + 5} = \frac{1}{11}.$$

Thus, the sequence begins $-1/7, -1/9, 1/11, \ldots$. The 39th term is

$$a_{39} = \frac{39^2 - 3 \cdot 39 + 1}{2 \cdot 39 + 5} = \frac{1405}{83}. \qquad \blacksquare$$

The subscript notation for sequences is sometimes abbreviated by writing $\{a_n\}$ in place of $a_1, a_2, a_3, \ldots$.

EXAMPLE 2

Find the first three terms, the 41st term, and the 206th term of the sequence

$$\left\{ \frac{(-1)^n}{n + 2} \right\}.$$

SOLUTION   The formula is

$$a_n = \frac{(-1)^n}{n + 2}.$$

Substituting $n = 1$, $n = 2$, and $n = 3$ shows that

$$a_1 = \frac{(-1)^1}{1 + 2} = -\frac{1}{3}, \qquad a_2 = \frac{(-1)^2}{2 + 2} = \frac{1}{4}, \qquad a_3 = \frac{(-1)^3}{3 + 2} = -\frac{1}{5}.$$

Similarly,

$$a_{41} = \frac{(-1)^{41}}{41 + 2} = -\frac{1}{43} \qquad \text{and} \qquad a_{206} = \frac{(-1)^{206}}{206 + 2} = \frac{1}{208}. \qquad \blacksquare$$

EXAMPLE 3

Here are some other sequences whose nth term can be described by a formula.

Sequence	nth Term	First 5 Terms
$a_1, a_2, a_3, \ldots$	$a_n = n^2 + 1$	$2, 5, 10, 17, 26$
$b_1, b_2, b_3, \ldots$	$b_n = \dfrac{1}{n}$	$1, \dfrac{1}{2}, \dfrac{1}{3}, \dfrac{1}{4}, \dfrac{1}{5}$
$c_1, c_2, c_3, \ldots$	$c_n = \dfrac{(-1)^{n+1} 2n}{(n+1)(n+2)}$	$\dfrac{1}{3}, -\dfrac{1}{3}, \dfrac{3}{10}, -\dfrac{4}{15}, \dfrac{5}{21}$
$x_1, x_2, x_3, \ldots$	$x_n = 3 + \dfrac{1}{10^n}$	$3.1, 3.01, 3.001, 3.0001, 3.00001.$
$a_1, a_2, a_3, \ldots$	$a_n = 7$	$7, 7, 7, 7, 7.$ ■

A sequence in which every term is the same, such as the last one in Example 3, is called a **constant sequence.** A calculator is often useful for computing and displaying the terms of more complicated sequences.

EXAMPLE 4

(a) Display the first five terms of the sequence $\{a_n = n^2 - n - 3\}$ on your calculator screen.

(b) Display the first, fifth, ninth, and thirteenth terms of this sequence.

SOLUTION

Method 1: Enter the sequence in the function memory as $y_1 = x^2 - x - 3$. In the table set-up screen, begin the table at $x = 1$ and set the increment at 1 and display a table of values (Figure 12–1). To display the first, fifth, ninth, and thirteenth terms, set the table increment at 4 (Figure 12–2).

Figure 12–1

Figure 12–2

Method 2: Using the Technology Tip in the margin, enter the following, which produces Figure 12–3 on the next page.

$$\text{SEQ}(x^2 - x - 3, x, 1, 5, 1).$$

To display the first, fifth, ninth and thirteenth terms, enter

$$\text{SEQ}(x^2 - x - 3, x, 1, 13, 4),$$

which tells the calculator to look at every fourth term from 1 to 13 and produces Figure 12–4.

Figure 12–3   Figure 12–4

TECHNOLOGY TIP

To change to sequence graphing mode, choose SEQUENCE in this menu:

TI-83/89: MODE

HP-39: APLET

On Casio, choose RECUR in the main menu.

Method 3: If possible, put your calculator in sequence graphing mode (see the Technology Tip in the margin). Enter the formula in the equation memory (Figure 12–5). Now you can either construct a table (as in *Method 1*) or graph the sequence and use the trace feature to determine its terms, as in Figure 12–6.[*] ∎

Figure 12–5   Figure 12–6

EXAMPLE 5

Display the first ten terms of the sequence $\left\{\dfrac{n}{n+1}\right\}$ on your calculator screen in fractional form, if possible.

SOLUTION   Creating a table always produces decimal approximations, as you can easily verify. The same is usually true of the SEQ key, unless you take special steps. On HP-39, change the number format mode to "fraction" (MODE menu). On TI calculators (other than TI-89), either use the FRAC key after obtaining the sequence (Figure 12–7) or use parentheses and the FRAC key as part of the function: Entering

$$\text{SEQ}(x/(x+1) \blacktriangleright \text{FRAC}, x, 1, 10, 1)$$

```
seq(X/(X+1),X,1,
10,1)
{.5 .6666666667...
Ans▶Frac
{1/2 2/3 3/4 4/...
```

Figure 12–7

---

*Set TI calculators for DOT instead of CONNECTED graphing in the MODE menu. This is not necessary on Casio and not available on HP-39 when it is in sequence mode.

```
seq((X/(X+1))▶Fr
ac,X,1,10,1)
{1/2 2/3 3/4 4/…
```

Figure 12–8

produces Figure 12–8. In each figure, you must use the arrow key to scroll to the right to see all the terms. ■

A sequence is said to be defined **recursively** (or **inductively**) if the first term is given (or the first several terms) and there is a method of determining the nth term by using the terms that precede it.

EXAMPLE 6

Consider the sequence whose first two terms are

$$a_1 = 1 \qquad \text{and} \qquad a_2 = 1$$

and whose nth term (for $n \geq 3$) is the sum of the two preceding terms.

$$a_3 = a_2 + a_1 = 1 + 1 = 2,$$
$$a_4 = a_3 + a_2 = 2 + 1 = 3,$$
$$a_5 = a_4 + a_3 = 3 + 2 = 5.$$

For each integer n, the two preceding integers are $n - 1$ and $n - 2$. So

$$a_n = a_{n-1} + a_{n-2} \quad (n \geq 3).$$

This sequence 1, 1, 2, 3, 5, 8, 13, . . . is called the **Fibonacci sequence,** and the numbers that appear in it are called **Fibonacci numbers.** Fibonacci numbers have many surprising and interesting properties. See Exercises 60–66 for details. ■

EXAMPLE 7

The sequence given by

$$a_1 = -7 \qquad \text{and} \qquad a_n = a_{n-1} + 3 \quad \text{for } n \geq 2$$

is defined recursively. Its first three terms are

$$a_1 = -7, \qquad a_2 = a_1 + 3 = -7 + 3 = -4,$$
$$a_3 = a_2 + 3 = -4 + 3 = -1. \quad ■$$

Entering a recursively defined sequence in a calculator (in sequence mode) may require the use of special keys; check your instruction manual. Figure 12–9 shows the Fibonacci sequence in the function memory of a TI-83; the entry "u(nMin) = {1, 1}" indicates that the first two terms of the sequence are 1, 1. On other calculators, these terms are entered directly as a_1 and a_2, either in the function memory (HP-39) or in the RANG menu (Casio).

```
Plot1  Plot2  Plot3
 nMin=1
·.u(n)◻u(n-1)+u(n
-2)
  u(nMin)◻{1,1}
·.v(n)=
  v(nMin)=
·.w(n)=
```

Figure 12–9

Sometimes, it is convenient or more natural to begin numbering the terms of a sequence with a number other than 1. So we may consider sequences such as

$$b_4, b_5, b_6, \ldots \qquad \text{or} \qquad c_0, c_1, c_2, \ldots .$$

EXAMPLE 8

The sequence $4, 5, 6, 7, \ldots$ can be conveniently described by saying $b_n = n$, with $n \geq 4$. In the brackets notation, we write $\{n\}_{n \geq 4}$. Similarly, the sequence

$$2^0, 2^1, 2^2, 2^3, \ldots$$

may be described as $\{2^n\}_{n \geq 0}$ or by saying $c_n = 2^n$, with $n \geq 0$.   ■

EXAMPLE 9

To buy a car, Leslie borrows \$14,000 at 7% annual interest. Her monthly payment is \$277.22 for 60 months.

(a) Find a formula for a recursively defined sequence $\{u_n\}$ such that u_n is the balance due on the loan after the nth payment.

(b) Find the balance after 30 months.

SOLUTION   With car loans and home mortgages, interest is computed monthly on the unpaid balance. The monthly interest rate is understood to be one-twelfth of the annual rate, that is, $.07/12$.

(a) Let $u_0 = 14,000$, the balance after 0 payments. When the first payment is made, the loan balance is

$$\$14,000 + \text{one month's interest} = \$14,000 + \frac{.07}{12}(14,000).$$

Subtracting the first loan payment, gives the balance after one month (rounded to the nearest penny).

$$u_1 = \$14,000 + \frac{.07}{12}(14,000) - 277.22 = \$13,804.45.$$

Similarly,

$$u_n = u_{n-1} + \frac{.07}{12}u_{n-1} - 277.22,$$

Balance after | Balance after | Interest | nth payment
n payments | $n-1$ payments | on u_{n-1}

which can be written as

$$u_n = \left(1 + \frac{.07}{12}\right)u_{n-1} - 277.22.$$

(b) Using the table feature of a calculator in sequence mode, we find that $u_{30} = \$7609.07$, as shown in Figure 12–10.   ■

n	$u(n)$
24	8978
25	8753.1
26	8527
27	8299.5
28	8070.7
29	7840.6
30	7609.1

$u(n)=7609.070348$

Figure 12–10

⠿ SUMMATION NOTATION

It is sometimes necessary to find the sum of various terms in a sequence. For instance, we might want to find the sum of the first nine terms of the sequence $\{a_n\}$. Mathematicians often use the Greek letter sigma (Σ) to abbreviate such a sum:[*]

$$\sum_{k=1}^{9} a_k = a_1 + a_2 + a_3 + a_4 + a_5 + a_6 + a_7 + a_8 + a_9.$$

Similarly, for any positive integer m and numbers $c_1, c_2, \ldots, c_m$, we have the following.

Summation Notation

$$\sum_{k=1}^{m} c_k \qquad \text{means} \qquad c_1 + c_2 + c_3 + \cdots + c_m.$$

EXAMPLE 10

Compute each of these sums.

(a) $\displaystyle\sum_{k=1}^{5} k^2$ (b) $\displaystyle\sum_{k=1}^{4} k^2(k-2)$ (c) $\displaystyle\sum_{k=1}^{6} (-1)^k k.$

SOLUTION

(a) We successively substitute 1, 2, 3, 4, 5 for k in the expression k^2 and add up the results.

$$\sum_{k=1}^{5} k^2 = 1^2 + 2^2 + 3^2 + 4^2 + 5^2 = 55.$$

(b) Successively substituting 1, 2, 3, 4 for k in $k^2(k-2)$ and adding the results, we have

$$\sum_{k=1}^{4} k^2(k-2) = 1^2(1-2) + 2^2(2-2) + 3^2(3-2) + 4^2(4-2)$$
$$= 1(-1) + 4(0) + 9(1) + 16(2) = 40.$$

(c) $\displaystyle\sum_{k=1}^{6} (-1)^k k =$
$$(-1)^1 \cdot 1 + (-1)^2 \cdot 2 + (-1)^3 \cdot 3 + (-1)^4 \cdot 4 + (-1)^5 \cdot 5 + (-1)^6 \cdot 6$$
$$= -1 + 2 - 3 + 4 - 5 + 6 = 3. \qquad ■$$

In sums such as $\displaystyle\sum_{k=1}^{5} k^2$ and $\displaystyle\sum_{k=1}^{6} (-1)^k k$, The letter k is called the **summation index.** Any letter may be used for the summation index, just as the rule of a function f may be denoted by $f(x)$ or $f(t)$ or $f(k)$. For example, $\displaystyle\sum_{n=1}^{5} n^2$ means: Take the sum of the terms n^2 as n takes values from 1 to 5. In other words,

$$\sum_{n=1}^{5} n^2 = \sum_{k=1}^{5} k^2.$$

---

[*]Σ is the letter S in the Greek alphabet, the first letter in *Sum*.

Similarly,

$$\sum_{k=1}^{4} k^2(k-2) = \sum_{j=1}^{4} j^2(j-2) = \sum_{n=1}^{4} n^2(n-2).$$

The Σ notation for sums can also be used for sums that don't begin with $k = 1$. For instance,

$$\sum_{k=4}^{10} k^2 = 4^2 + 5^2 + 6^2 + 7^2 + 8^2 + 9^2 + 10^2 = 371$$

$$\sum_{j=0}^{3} j^2(2j+5) = 0^2(2 \cdot 0 + 5) + 1^2(2 \cdot 1 + 5) + 2^2(2 \cdot 2 + 5) + 3^2(2 \cdot 3 + 5).$$

EXAMPLE 11

Use a calculator to compute these sums.

(a) $\displaystyle\sum_{k=1}^{50} k^2$ (b) $\displaystyle\sum_{k=38}^{75} k^2.$

SOLUTION   In each case, use SUM together with SEQ (or ΣLIST and MAKELIST on HP-39), with the same syntax for SEQ as in Example 4:

$$\sum_{k=1}^{50} k^2 = \text{SUM SEQ}(x^2, x, 1, 50, 1) = 42{,}925$$

and

$$\sum_{k=38}^{75} k^2 = \text{SUM SEQ}(x^2, x, 38, 75, 1) = 125{,}875,$$

as shown in Figure 12–11. ■

TECHNOLOGY TIP

SUM (or ΣLIST on HP-39) is in this menu/submenu:

TI-83: LIST/MATH

TI-86: LIST/OPS

TI-89: MATH/LIST

Casio: OPTN/LIST

HP-39: MATH/LIST

(a)

(b)

Figure 12–11

▪▪ PARTIAL SUMS

Suppose $\{a_n\}$ is a sequence and k is a positive integer. The sum of the first k terms of the sequence is called the **kth partial sum** of the sequence. Thus, we have the following.

Partial Sums

> The **kth partial sum** of $\{a_n\}$ is $\displaystyle\sum_{n=1}^{k} a_n = a_1 + a_2 + a_3 + \cdots + a_k.$

EXAMPLE 12

Here are some partial sums of the sequence $\{n^3\}$:

First partial sum: $\displaystyle\sum_{n=1}^{1} n^3 = 1^3 = 1,$

Second partial sum: $\displaystyle\sum_{n=1}^{2} n^3 = 1^3 + 2^3 = 9,$

Sixth partial sum: $\displaystyle\sum_{n=1}^{6} n^3 = 1^3 + 2^3 + 3^3 + 4^3 + 5^3 + 6^3 = 441.$ ■

EXAMPLE 13

The sequence $\{2^n\}_{n \geq 0}$ begins with the 0th term, so the fourth partial sum (the sum of the first four terms) is

$$2^0 + 2^1 + 2^2 + 2^3 = \sum_{n=0}^{3} 2^n.$$

Similarly, the fifth partial sum of the sequence $\left\{\dfrac{1}{n(n-2)}\right\}_{n \geq 3}$ is the sum of the first five terms.

$$\frac{1}{3(3-2)} + \frac{1}{4(4-2)} + \frac{1}{5(5-2)} + \frac{1}{6(6-2)} + \frac{1}{7(7-2)} = \sum_{n=3}^{7} \frac{1}{n(n-2)}.$$ ■

Certain calculations can be written very compactly in summation notation. For example, the distributive law shows that

$$ca_1 + ca_2 + ca_3 + \cdots + ca_r = c(a_1 + a_2 + a_3 + \cdots + a_r).$$

In summation notation, this becomes

$$\sum_{n=1}^{r} ca_n = c\left(\sum_{n=1}^{r} a_n\right).$$

This proves the first of the following statements.

Properties of Sums

1. $\displaystyle\sum_{n=1}^{r} ca_n = c\left(\sum_{n=1}^{r} a_n\right)$ for any number c.

2. $\displaystyle\sum_{n=1}^{r} (a_n + b_n) = \sum_{n=1}^{r} a_n + \sum_{n=1}^{r} b_n$

3. $\displaystyle\sum_{n=1}^{r} (a_n - b_n) = \sum_{n=1}^{r} a_n - \sum_{n=1}^{r} b_n$

To prove statement 2, use the commutative and associative laws repeatedly to show that

$$(a_1 + b_1) + (a_2 + b_2) + (a_3 + b_3) + \cdots + (a_r + b_r)$$
$$= (a_1 + a_2 + a_3 + \cdots + a_r) + (b_1 + b_2 + b_3 + \cdots + b_r),$$

which can be written in summation notation as

$$\sum_{n=1}^{r} (a_n + b_n) = \sum_{n=1}^{r} a_n + \sum_{n=1}^{r} b_n.$$

The last statement is proved similarly.

✓ EXERCISES 12.1

In Exercises 1–10, find the first five terms of the sequence $\{a_n\}$.

1. $a_n = 2n + 6$

2. $a_n = 2^n - 7$

3. $a_n = \dfrac{1}{n^3}$

4. $a_n = \dfrac{1}{(n+3)(n+1)}$

5. $a_n = (-1)^n \sqrt{n+2}$

6. $a_n = (-1)^{n+1} n(n-1)$

7. $a_n = 4 + (-.1)^n$

8. $a_n = 5 - (.1)^n$

9. $a_n = (-1)^n + 3n$

10. $a_n = (-1)^{n+2} - (n+1)$

In Exercises 11–14, express the sum in Σ notation.

11. $1 + 2 + 3 + 4 + 5 + 6 + 7 + 8 + 9 + 10 + 11$

12. $1^1 + 2^2 + 3^3 + 4^4 + 5^5$

13. $\dfrac{1}{2^7} + \dfrac{1}{2^8} + \dfrac{1}{2^9} + \dfrac{1}{2^{10}} + \dfrac{1}{2^{11}} + \dfrac{1}{2^{12}} + \dfrac{1}{2^{13}}$

14. $(-6)^{11} + (-6)^{12} + (-6)^{13} + (-6)^{14} + (-6)^{15}$

In Exercises 15–20, find the sum.

15. $\displaystyle\sum_{i=1}^{5} 3i$

16. $\displaystyle\sum_{i=1}^{4} \dfrac{1}{2^i}$

17. $\displaystyle\sum_{n=1}^{16} (2n - 3)$

18. $\displaystyle\sum_{n=1}^{75} (-1)^n(3n + 1)$

19. $\displaystyle\sum_{n=15}^{36} (n^2 - 8)$

20. $\displaystyle\sum_{k=0}^{25} (2k^2 - 5k + 1)$

In Exercises 21–26, find a formula for the nth term of the sequence whose first few terms are given.

21. $-1, 1, -1, 1, -1, 1, \ldots$

22. $2, -2, 2, -2, 2, -2, \ldots$

23. $\dfrac{1}{2}, \dfrac{2}{3}, \dfrac{3}{4}, \dfrac{4}{5}, \dfrac{5}{6}, \ldots$

24. $\dfrac{1}{2 \cdot 3}, \dfrac{1}{3 \cdot 4}, \dfrac{1}{4 \cdot 5}, \dfrac{1}{5 \cdot 6}, \dfrac{1}{6 \cdot 7}, \ldots$

25. $2, 7, 12, 17, 22, 27, \ldots$

26. $8, -5, 2, -11, -4, -17, -10, \ldots$

In Exercises 27–34, find the first five terms of the given sequence.

27. $a_1 = 4$ and $a_n = 2a_{n-1} + 3$ for $n \geq 2$

28. $a_1 = -3$ and $a_n = (-1)^n 4a_{n-1} - 5$ for $n \geq 2$

29. $a_1 = 1, a_2 = -2, a_3 = 3,$ and
$a_n = a_{n-1} + a_{n-2} + a_{n-3}$ for $n \geq 4$

30. $a_1 = 1, a_2 = 3,$ and $a_n = 2a_{n-1} + 3a_{n-2}$ for $n \geq 3$

31. $a_0 = 2, a_1 = 3,$ and $a_n = (a_{n-1})\left(\dfrac{1}{2} a_{n-2}\right)$ for $n \geq 2$

32. $a_0 = 1, a_1 = 1,$ and $a_n = na_{n-1}$ for $n \geq 2$

33. a_n is the nth digit in the decimal expansion of π.

34. a_n is the nth digit in the decimal expansion of $1/13$.

In Exercises 35–38, find the third and the sixth partial sums of the sequence.

35. $\{n^2 - 5n + 2\}$

36. $\{(2n - 3n^2)^2\}$

37. $\{(-1)^{n+1} 5\}$

38. $\{2^n(2 - n^2)\}_{n \geq 0}$

In Exercises 39–42, express the given sum in Σ notation.

39. $\dfrac{1}{3} + \dfrac{1}{5} + \dfrac{1}{7} + \dfrac{1}{9} + \dfrac{1}{11} + \dfrac{1}{13}$

40. $2 + 1 + \dfrac{4}{5} + \dfrac{5}{7} + \dfrac{2}{3} + \dfrac{7}{11} + \dfrac{8}{13}$

41. $\dfrac{1}{8} - \dfrac{2}{9} + \dfrac{3}{10} - \dfrac{4}{11} + \dfrac{5}{12}$

42. $\dfrac{2}{3 \cdot 5} + \dfrac{4}{5 \cdot 7} + \dfrac{8}{7 \cdot 9} + \dfrac{16}{9 \cdot 11}$
$+ \dfrac{32}{11 \cdot 13} + \dfrac{64}{13 \cdot 15} + \dfrac{128}{15 \cdot 17}$

In Exercises 43–48, use a calculator to approximate the required term or sum.

43. a_{12} where $a_n = \left(1 + \dfrac{1}{n}\right)^n$

44. a_{50} where $a_n = \dfrac{\ln n}{n^2}$

45. a_{102} where $a_n = \dfrac{n^3 - n^2 + 5n}{3n^2 + 2n - 1}$

46. a_{125} where $a_n = \sqrt[n]{n}$

47. $\displaystyle\sum_{k=1}^{14} \frac{1}{k^2}$ **48.** $\displaystyle\sum_{n=8}^{22} \frac{1}{n}$

In Exercises 49–50, use a recursively defined sequence, as in Example 9.

49. Suppose you take out a car loan for $9500 at 5% annual interest for 48 months. Your monthly payment is $218.78. How much do you owe after one year (12 payments)? After three years?

50. Phil is buying a condo. He takes out an $80,000 mortgage for 30 years at 6% annual interest. His monthly payment is $479.64.
   (a) How much does he owe after one year (12 payments)?
   (b) How much interest does Phil pay during the first five years? [*Hint:* The interest paid is the difference between his total payments and the amount of the loan paid off after five years

$$(\$80,000 - \text{remaining balance}).]$$

   (c) After 15 years, Phil sells the condo and pays off the remaining mortgage balance. How much does he pay?

51. The number of hours that the average person spends listening to the radio in year n is approximated by the sequence $\{a_n\}$, where $n = 1$ corresponds to 1998 and $a_n = 937.84 + 37.55 \ln n$.[*]
   (a) For how many hours did the average person listen to the radio in 2000? In 2003?
   (b) How much total time did the average person spend listening from 1998 to 2003 (inclusive)?

52. The budget of the National Institutes of Health (in billions of dollars) in year n is approximated by the sequence $\{b_n\}$, where $n = 0$ corresponds to 1998 and $b_n = 2.72n + 12.96$.[†]
   (a) What was the NIH budget in 2002?
   (b) Assuming that NIH kept to its budget and spent the entire amount each year, how much did NIH spend from 1998 to 2002?

53. Consumer spending per person (in dollars) on home video games each year is approximated by the sequence $\{c_n\}$, where $n = 0$ is 1995 and $c_n = .42n^2 + 1.528n + 10.396$.[*]
   (a) What is the per person spending in 2002? In 2004?
   (b) What was the total amount spent per person on home video games from 1995 to 2003?

54. Book sales (in billions of dollars) in the United States are approximated by the sequence $\{d_n\}$, where $n = 0$ corresponds to 1990 and $d_n = 1.26n + 18.83$.[‡]
   (a) What were the sales in 2000?
   (b) What was the total spent on books from 1990 to 2000?

---

[*]Based on data and projections from the *Statistical Abstract of the United States: 2001.*
[†]Federal budget.
[‡]Book Industry Study Group, Inc.

Thinkers

Exercises 55–59 deal with prime numbers. A positive integer greater than 1 is prime if its only positive integer factors are itself and 1. For example, 7 is prime because its only factors are 7 and 1, but 15 is not prime because it has factors other than 15 and 1 (namely, 3 and 5).

55. (a) Let $\{a_n\}$ be the sequence of prime integers in their usual ordering. Verify that the first ten terms are 2, 3, 5, 7, 11, 13, 17, 19, 23, 29.
   (b) Find $a_{17}, a_{18}, a_{19}, a_{20}$.

In Exercises 56–59, find the first five terms of the sequence.

56. a_n is the nth prime integer larger than 10. [*Hint:* $a_1 = 11$.]

57. a_n is the square of the nth prime integer.

58. a_n is the number of prime integers less than n.

59. a_n is the largest prime integer less than $5n$.

Exercises 60–66 deal with the Fibonacci sequence $\{a_n\}$ that was discussed in Example 6.

60. Leonardo Fibonacci discovered the sequence in the thirteenth century in connection with this problem: A rabbit colony begins with one pair of adult rabbits (one male, one female). Each adult pair produces one pair of babies (one male, one female) every month. Each pair of baby rabbits becomes adult and produces the first offspring at age two months. Assuming that no rabbits die, how many adult pairs of rabbits are in the colony at the end of n months $(n = 1, 2, 3, \ldots)$? [*Hint:* It may be helpful to make up a chart listing for each month the number of adult pairs, the number of one-month-old pairs, and the number of baby pairs.]

61. (a) List the first ten terms of the Fibonacci sequence.
   (b) List the first ten partial sums of the sequence.
   (c) Do the partial sums follow an identifiable pattern?

62. Verify that every positive integer less than or equal to 15 can be written as a sum of Fibonacci numbers, with none used more than once.

63. Verify that $5(a_n)^2 + 4(-1)^n$ is always a perfect square for $n = 1, 2, \ldots, 10$.

64. Verify that $(a_n)^2 = a_{n+1}a_{n-1} + (-1)^{n-1}$ for $n = 2, \ldots, 10$.

65. Show that $\displaystyle\sum_{n=1}^{k} a_n = a_{k+2} - 1$. [*Hint:* $a_1 = a_3 - a_2$; $a_2 = a_4 - a_3$; etc.]

66. Show that $\displaystyle\sum_{n=1}^{k} a_{2n-1} = a_{2k}$, that is, the sum of the first k odd-numbered terms is the kth even-numbered term. [*Hint:* $a_3 = a_4 - a_2$; $a_5 = a_6 - a_4$; etc.]

12.2   Arithmetic Sequences

An **arithmetic sequence** (sometimes called an **arithmetic progression**) is a sequence in which the difference between each term and the preceding one is always the same constant.

EXAMPLE 1

In the sequence 3, 8, 13, 18, 23, 28, . . . the difference between each term and the preceding one is always 5. So this is an arithmetic sequence.   ■

EXAMPLE 2

Which of the following sequences are arithmetic?

(a) 14, 10, 6, 2, -2, -6, -10, -14, . . .
(b) $7, 7^2, 7^3, 7^4, 7^5, \ldots$
(c) $\log 5, \log 5^2, \log 5^3, \log 5^4, \log 5^5, \ldots$

SOLUTION

(a) The difference between each term and the preceding one is -4.

$$10 - 14 = -4 \qquad\qquad -2 - 2 = -4,$$
$$6 - 10 = -4 \qquad -6 - (-2) = -4,$$
$$2 - 6 = -4 \qquad -10 - (-6) = -4,$$

and so on. Hence, the sequence is arithmetic.

(b) Note that

$$7^2 - 7 = 49 - 7 = 42, \qquad \text{but} \qquad 7^3 - 7^2 = 343 - 49 = 294.$$

Since the difference between consecutive terms is not always the same, this sequence is not arithmetic.

(c) Using the Power Law for logarithms (Section 5.4), we see that

$$\log 5^2 - \log 5 = 2 \log 5 - \log 5 = \log 5,$$
$$\log 5^3 - \log 5^2 = 3 \log 5 - 2 \log 5 = \log 5,$$
$$\log 5^4 - \log 5^3 = 4 \log 5 - 3 \log 5 = \log 5,$$

and so on. The difference between each term and the preceding one is always $\log 5 \approx .69897$, so the sequence is arithmetic.   ■

If $\{a_n\}$ is an arithmetic sequence, then for each $n \geq 2$, the term preceding a_n is a_{n-1}, and the difference $a_n - a_{n-1}$ is some constant—call it d. Therefore, $a_n - a_{n-1} = d$ or, equivalently,

***Arithmetic
Sequences***

In an arithmetic sequence $\{a_n\}$,

$$a_n = a_{n-1} + d$$

for some constant d and all $n \geq 2$.

The number d is called the **common difference** of the arithmetic sequence.

EXAMPLE 3

If $\{a_n\}$ is an arithmetic sequence with $a_1 = 3$ and $a_2 = 4.5$, find the common difference and list the first eight terms.

SOLUTION The common difference is

$$d = a_2 - a_1 = 4.5 - 3 = 1.5.$$

Therefore, the sequence begins 3, 4.5, 6, 7.5, 9, 10.5, 12, 13.5, . . . ∎

EXAMPLE 4

Show that the sequence $\{-7 + 4n\}$ is arithmetic and find the common difference.

SOLUTION The sequence is arithmetic because for each $n \geq 2$,

$$a_n - a_{n-1} = (-7 + 4n) - [-7 + 4(n-1)]$$
$$= (-7 + 4n) - (-7 + 4n - 4) = 4.$$

Therefore, the common difference is $d = 4$. ∎

If $\{a_n\}$ is an arithmetic sequence with common difference d, then for each $n \geq 2$, we know that $a_n = a_{n-1} + d$. Applying this fact repeatedly shows that

$$a_2 = a_1 + d$$
$$a_3 = a_2 + d = (a_1 + d) + d = a_1 + 2d$$
$$a_4 = a_3 + d = (a_1 + 2d) + d = a_1 + 3d$$
$$a_5 = a_4 + d = (a_1 + 3d) + d = a_1 + 4d,$$

and in general, we have the following.

***nth Term of an
Arithmetic Sequence***

In an arithmetic sequence $\{a_n\}$ with common difference d,

$$a_n = a_1 + (n-1)d \quad \text{for every } n \geq 1.$$

EXAMPLE 5

Find the nth term of the arithmetic sequence with first term -5 and common difference 3.

SOLUTION Since $a_1 = -5$ and $d = 3$, the formula in the preceding box shows that

$$a_n = a_1 + (n - 1)d = -5 + (n - 1)3 = 3n - 8. \quad \blacksquare$$

EXAMPLE 6

What is the 45th term of the arithmetic sequence whose first three terms are 5, 9, 13?

SOLUTION The first three terms show that $a_1 = 5$ and that the common difference d is 4. Applying the formula in the box with $n = 45$, we have

$$a_{45} = a_1 + (45 - 1)d = 5 + (44)4 = 181. \quad \blacksquare$$

EXAMPLE 7

If $\{a_n\}$ is an arithmetic sequence with $a_6 = 57$ and $a_{10} = 93$, find a_1 and a formula for a_n.

SOLUTION Apply the formula $a_n = a_1 + (n - 1)d$ with $n = 6$ and $n = 10$.

$$a_6 = a_1 + (6 - 1)d \qquad \text{and} \qquad a_{10} = a_1 + (10 - 1)d$$

$$57 = a_1 + 5d \qquad\qquad\qquad 93 = a_1 + 9d$$

$$a_1 = 57 - 5d \qquad\qquad\qquad a_1 = 93 - 9d.$$

These two expressions for a_1 must be equal.

$$57 - 5d = 93 - 9d$$

$$4d = 36$$

$$d = 9.$$

Substituting $d = 9$ in either of the equations above shows that $a_1 = 12$.

$$a_1 = 57 - 5(9) = 12 \qquad \text{or} \qquad a_1 = 93 - 9(9) = 12.$$

So the formula for a_n is

$$a_n = a_1 + (n - 1)d = 12 + (n - 1)9 = 9n + 3. \quad \blacksquare$$

▪▪ PARTIAL SUMS

It's easy to compute partial sums of arithmetic sequences by using the following formulas.

Partial Sums of an Arithmetic Sequence

If $\{a_n\}$ is an arithmetic sequence with common difference d, then for each positive integer k, the kth partial sum can be found by usuing *either* of these formulas.

$$1. \sum_{n=1}^{k} a_n = \frac{k}{2}(a_1 + a_k) \qquad \text{or}$$

$$2. \sum_{n=1}^{k} a_n = ka_1 + \frac{k(k-1)}{2}d.$$

Proof   Let S denote the kth partial sum $a_1 + a_2 + \cdots + a_k$. For reasons that will become apparent later, we shall calculate the number $2S$.

$$2S = S + S = (a_1 + a_2 + \cdots + a_k) + (a_1 + a_2 + \cdots + a_k).$$

Now we rearrange the terms on the right by grouping the first and last terms together, then the first and last of the remaining terms, and so on.

$$2S = (a_1 + a_k) + (a_2 + a_{k-1}) + (a_3 + 2_{k-2}) + \cdots + (a_k + a_1).$$

Since adjacent terms of the sequence differ by d, we have

$$a_2 + a_{k-1} = (a_1 + d) + (a_k - d) = a_1 + a_k.$$

Using this fact, we have

$$a_3 + a_{k-2} = (a_2 + d) + (a_{k-1} - d) = a_2 + a_{k-1} = a_1 + a_k.$$

Continuing in this manner, we see that every pair in the sum for $2S$ is equal to $a_1 + a_k$. Therefore,

$$2S = (a_1 + a_k) + (a_2 + a_{k-1}) + (a_3 + a_{k-2}) + \cdots + (a_k + a_1)$$

$$= (a_1 + a_k) + (a_1 + a_k) + (a_1 + a_k) + \cdots + (a_1 + a_k) \quad (k \text{ terms})$$

$$= k(a_1 + a_k).$$

Dividing both sides of this last equation by 2 shows that $S = \frac{k}{2}(a_1 + a_k)$. This proves the first formula. To obtain the second one, note that

$$a_1 + a_k = a_1 + [a_1 + (k-1)d] = 2a_1 + (k-1)d.$$

Substituting the right side of this equation in the first formula for S shows that

$$S = \frac{k}{2}(a_1 + a_k) = \frac{k}{2}[2a_1 + (k-1)d] = ka_1 + \frac{k(k-1)}{2}d.$$

This proves the second formula.   ■

EXAMPLE 8

Find the 12th partial sum of the arithmetic sequence that begins $-8, -3, 2, 7, \ldots$.

SOLUTION   We first note that the common difference d is 5. Since $a_1 = -8$ and $d = 5$, the second formula in the box with $k = 12$ shows that

$$\sum_{n=1}^{12} a_n = 12(-8) + \frac{12(11)}{2}5 = -96 + 330 = 234. \quad \blacksquare$$

EXAMPLE 9

Find the sum of all multiples of 3 from 3 to 333.

SOLUTION   The answer can be found in two ways.

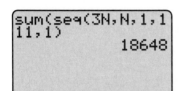

Figure 12–12

Algebraic Method: This sum is a partial sum of the arithmetic sequence 3, 6, 9, 12, ... Since this sequence can be written in the form

$$3 \cdot 1, 3 \cdot 2, 3 \cdot 3, 3 \cdot 4, 3 \cdot 5, 3 \cdot 6, \ldots,$$

we see that $333 = 3 \cdot 111$ is the 111th term. The 111th partial sum of this sequence can be found by using the first formula in the box with $k = 111$, and $a_1 = 3$, and $a_{111} = 333$.

$$\sum_{n=1}^{111} a_n = \frac{111}{2}(3 + 333) = \frac{111}{2}(336) = 18{,}648.$$

Calculator Method: Note that the nth term of the sequence 3, 6, 9, . . . is $3n$. Consequently, we can compute the 111th partial sum as in Figure 12–12. $\blacksquare$

EXAMPLE 10

If the starting salary for a job is \$20,000 and you get a \$2000 raise at the beginning of each subsequent year, what will your salary be during the tenth year? How much will you earn during the first ten years?

SOLUTION   Your yearly salary rates form a sequence: 20,000, 22,000, 24,000, 26,000, and so on. It is an arithmetic sequence with $a_1 = 20{,}000$ and $d = 2000$. Your tenth-year salary is

$$a_{10} = a_1 + (10 - 1)d = 20{,}000 + 9 \cdot 2000 = \$38{,}000.$$

Your ten-year total earnings are the tenth partial sum of the sequence.

$$\frac{10}{2}(a_1 + a_{10}) + \frac{10}{2}(20{,}000 + 38{,}000) = 5(58{,}000) = \$290{,}000. \quad \blacksquare$$

✔ **EXERCISES 12.2**

In Exercises 1–6, determine whether the sequence is arithmetic or not. If it is, find the common difference.

1. $1, 3, 5, 7, 9, \ldots$

2. $\dfrac{1}{3}, \dfrac{2}{3}, \dfrac{3}{3}, \dfrac{4}{3}, \dfrac{5}{3}, \ldots$

3. $1, 2, 4, 8, 16, \ldots$

4. $\log 1, \log 2, \log 4, \log 8, \log 16, \ldots$

5. $\log 3, \log 6, \log 9, \log 12, \log, 15, \ldots$

6. $\dfrac{1}{3}, -\dfrac{4}{6}, -\dfrac{15}{9}, -\dfrac{32}{12}, -\dfrac{55}{15}, -\dfrac{84}{18}, \ldots$

In Exercises 7–12, write the first five terms of the sequence whose nth term is given. Use them to decide whether the sequence is arithmetic. If it is, list the common difference.

7. $a_n = 5 + 4n$

8. $b_n = n - \dfrac{5}{4}$

9. $c_n = (-1)^n$

10. $d_n = 2n + \dfrac{1}{n}$

11. $a_n = e^{\ln n}$

12. $b_n = 3\sqrt{7n^2}$

In Exercises 13–18, show that the sequence is arithmetic and find its common difference.

13. $\{3 - 2n\}$

14. $\left\{4 + \dfrac{n}{3}\right\}$

15. $\left\{\dfrac{5 + 3n}{2}\right\}$

16. $\left\{\dfrac{\pi - n}{2}\right\}$

17. $\{c + 2n\}$ (*c* constant)

18. $\{2b + 3nc\}$ (*b, c* constants)

In Exercises 19–24, the first term a_1 and the common difference d of an arithmetic sequence are given. Find the fifth term and the formula for the nth term.

19. $a_1 = 5, d = 2$

20. $a_1 = -4, d = 5$

21. $a_1 = 4, d = \dfrac{1}{4}$

22. $a_1 = -6, d = \dfrac{2}{3}$

23. $a_1 = 10, d = -\dfrac{1}{2}$

24. $a_1 = \pi, d = \dfrac{1}{5}$

In Exercises 25–30, use the given information about the arithmetic sequence with common difference d to find a_1 and a formula for a_n.

25. $a_4 = 12, d = 2$

26. $a_7 = -8, d = 3$

27. $a_2 = 4, a_6 = 32$

28. $a_7 = 6, a_{12} = -4$

29. $a_5 = 0, a_9 = 6$

30. $a_5 = -3, a_9 = -18$

In Exercises 31–36, find the kth partial sum of the arithmetic sequence $\{a_n\}$ with common difference d.

31. $k = 6, a_1 = 2, d = 5$

32. $k = 8, a_1 = \dfrac{2}{3}, d = -\dfrac{4}{3}$

33. $k = 7, a_1 = \dfrac{3}{4}, d = -\dfrac{1}{2}$

34. $k = 9, a_1 = 6, a_9 = -24$

35. $k = 6, a_1 = -4, a_6 = 14$

36. $k = 10, a_1 = 0, a_{10} = 30$

In Exercises 37–40, find the sum.

37. $\displaystyle\sum_{n=1}^{20} (3n + 4)$

38. $\displaystyle\sum_{n=1}^{25} \left(\dfrac{n}{4} + 5\right)$

39. $\displaystyle\sum_{n=1}^{40} \dfrac{n+3}{6}$

40. $\displaystyle\sum_{n=1}^{30} \dfrac{4 - 6n}{3}$

41. Find the sum of all the even integers from 2 to 100.

42. Find the sum of all the integer multiples of 7 from 7 to 700.

43. Find the sum of the first 200 positive integers.

44. Find the sum of the positive integers from 101 to 200 (inclusive). [*Hint:* What's the sum from 1 to 100? Use it and Exercise 43.]

45. A business makes a $10,000 profit during its first year. If the yearly profit increases by $7500 in each subsequent year, what will the profit be in the tenth year and what will the total profit for the first ten years be?

46. If a man's starting salary is $15,000 and he receives a $1000 increase every six months, what will his salary be during the last six months of the sixth year? How much will he earn during the first six years?

47. The sequence in which

a_n = per capita amount spent on health care in year *n*,

with $n = 1$ corresponding to 1981, is approximately arithmetic.[*]
(a) If the per capita amount was $1223.50 in 1981 and $2786.80 in 1990, find a formula for a_n.
(b) Use the sequence to estimate the per capita amount in 2000 and 2003.

48. The sequence in which

b_n = remaining life expectancy of a man at age *n*

is approximately arithmetic.[*]
(a) Use the fact that a man's remaining life expectancy is 59.3 years at age 15 and 55 years at age 20 to find a formula for b_n.
(b) Determine the remaining life expectancy of a man at these ages: 18, 22, 25, and 30.

49. Let $n = 1$ correspond to 1981. The arithmetic sequence in which

$c_n = $ average yearly earnings of a production worker in manufacturing

has $c_1 = \$15,828.80$ and $c_{19} = \$28,932.80$.[†]
(a) Find a formula for c_n.
(b) Find the total earnings of an average production worker from 1981 to 2002.

---

[*]Based on data from the U.S. Department of Health and Human Services.
[†]Based on data from the U.S. Bureau of Labor Statistics.

50. Let $n = 1$ correspond to 1997. The arithmetic sequence in which

$$d_n = \text{hours per person spent on the Internet in year } n$$

has $d_1 = 39.9$ and $d_6 = 181.9.$*

(a) Find a formula for d_n.

(b) How many hours per person were spent on the Internet in 2002?

(c) Find the total number of hours per person on the Internet from 1997 to 2002.

51. A lecture hall has six seats in the first row, eight in the second, ten in the third, and so on, through row 12. Rows 12 through 20 (the last row) all have the same number of seats. Find the number of seats in the lecture hall.

52. A monument is constructed by first laying a row of 60 bricks at ground level. A second row, with two fewer bricks, is centered on that; a third row, with two fewer bricks, is centered on the second; and so on. The top row contains ten bricks. How many bricks are there in the monument?

53. A ladder with nine rungs is to be built, with the bottom rung 24 inches wide and the top rung 18 inches wide. If the lengths of the rungs decrease uniformly from bottom to top, how long should each of the seven intermediate rungs be?

54. Find the first eight numbers in an arithmetic sequence in which the sum of the first and seventh term is 40 and the product of the first and fourth terms is 160.

---

*Based on data and projections in *Statistical Abstract of the United States: 2001.*

12.3   Geometric Sequences

A **geometric sequence** (sometimes called a **geometric progression**) is a sequence in which the quotient of each term and the preceding one is the same constant r. This constant r is called the **common ratio** of the geometric sequence.

EXAMPLE 1

The sequence $3, 9, 27, \ldots 3^n, \ldots$ is geometric with common ratio 3. For instance, $a_2/a_1 = 9/3 = 3$ and $a_3/a_2 = 27/9 = 3$. If 3^n is any term ($n \geq 2$), then the preceding term is 3^{n-1}, and

$$\frac{3^n}{3^{n-1}} = \frac{3 \cdot 3^{n-1}}{3^{n-1}} = 3. \qquad \blacksquare$$

EXAMPLE 2

Which of the following sequences are geometric?

(a) $\dfrac{1}{2}, \dfrac{2}{3}, \dfrac{3}{4}, \dfrac{4}{5}, \dfrac{5}{6}, \ldots$

(b) $e, -e^3, e^5, -e^7, e^9, \ldots$

(c) $\left\{ \dfrac{5}{2^n} \right\}$

SOLUTION

(a) Note that

$$\frac{a_2}{a_1} = \frac{2/3}{1/2} = \frac{2}{3} \cdot \frac{2}{1} = \frac{4}{3}, \qquad \text{but} \qquad \frac{a_3}{a_2} = \frac{3/4}{2/3} = \frac{3}{4} \cdot \frac{3}{2} = \frac{9}{8}.$$

Since the quotient of consecutive terms is not always the same, this sequence is not geometric.

(b) Compute the quotient of each term and the preceding one.

$$\frac{a_2}{a_1} = \frac{-e^3}{e} = -e^2, \qquad \frac{a_4}{a_3} = \frac{-e^7}{e^5} = -e^2,$$

$$\frac{a_3}{a_2} = \frac{e^5}{-e^3} = -e^2, \qquad \frac{a_5}{a_4} = \frac{e^9}{-e^7} = -e^2,$$

and so on. The consecutive quotients are always the number $-e^2$, so the sequence is geometric with common ratio $-e^2$.

(c) For each $n \geq 2$, the term preceding a_n is a_{n-1}. Here,

$$a_n = \frac{5}{2^n} \qquad \text{and} \qquad a_{n-1} = \frac{5}{2^{n-1}},$$

so

$$\frac{a_n}{a_{n-1}} = \frac{5/2^n}{5/2^{n-1}} = \frac{5}{2^n} \cdot \frac{2^{n-1}}{5} = \frac{2^{n-1}}{2^n} = \frac{1}{2}.$$

Since the quotient of each term and the preceding one is always $1/2$, the sequence is geometric with common ratio $1/2$. ∎

Suppose $\{a_n\}$ is a geometric sequence with common ratio r. For each $n \geq 2$, the quotient of the term a_n and the preceding one is $\dfrac{a_n}{a_{n-1}} = r$. Multiplying both sides of the equation by a_{n-1} produces this fact.

Geometric Sequences

In a geometric sequence $\{a_n\}$ with common ratio r,

$$a_n = ra_{n-1} \qquad \text{for each } n \geq 2.$$

Applying this last formula for $n = 2, 3, 4, \ldots$, we have

$$a_2 = ra_1,$$

$$a_3 = ra_2 = r(ra_1) = r^2 a_1,$$

$$a_4 = ra_3 = r(r^2 a_1) = r^3 a_1,$$

$$a_5 = ra_4 = r(r^3 a_1) = r^4 a_1,$$

and in general we have the following.

nth Term of a Geometric Sequence

If $\{a_n\}$ is a geometric sequence with common ratio r, then for all $n \geq 1$,

$$a_n = r^{n-1} a_1.$$

EXAMPLE 3

Find a formula for the nth term of the geometric sequence $\{a_n\}$ that satisfies the given conditions.

(a) $a_1 = 7$ and $r - 2$ (b) $a_3 = 8$ and $r = 1/2$

SOLUTION

(a) The equation in the preceding box shows that

$$a_n = r^{n-1}a_1 = 2^{n-1} \cdot 7.$$

(b) We must first find a_1.

$$a_n = r^{n-1}a_1$$

Let $n = 3$: $a_3 = r^{3-1}a_1$

Substitute $a_3 = 8$ and $r = 1/2$: $8 = \left(\dfrac{1}{2}\right)^2 a_1$

$$8 = \dfrac{1}{4}a_1$$

Multiply both sides by 4: $32 = a_1.$

Therefore,

$$a_n = r^{n-1}a_1 = \left(\dfrac{1}{2}\right)^{n-1} \cdot 32. \quad \blacksquare$$

EXAMPLE 4

Find a formula for the nth term of the geometric sequence whose first two terms are 2 and $-2/5$.

SOLUTION   The common ratio is

$$r = \frac{a_2}{a_1} = \frac{-2/5}{2} = \frac{-2}{5} \cdot \frac{1}{2} = -\frac{1}{5}.$$

Using the equation in the box, we now see that the formula for the nth term is

$$a_n = r^{n-1}a_1 = \left(-\frac{1}{5}\right)^{n-1}(2) = \frac{(1)^{n-1}}{(-5)^{n-1}}(2) = \frac{2}{(-5)^{n-1}}.$$

So, the sequence begins $2, -2/5, 2/5^2, -2/5^3, 2/5^4, \ldots . \quad \blacksquare$

EXAMPLE 5

Find a formula for the nth term of the geometric sequence $\{a_n\}$ in which $a_2 = 20/9$ and $a_5 = 160/243$.

SOLUTION   By the equation in the box above, we have

$$\frac{160/243}{20/9} = \frac{a_5}{a_2} = \frac{r^4 a_1}{r a_1} = r^3.$$

Consequently,

$$r = \sqrt[3]{\frac{160/243}{20/9}} = \sqrt[3]{\frac{160}{243} \cdot \frac{9}{20}} = \sqrt[3]{\frac{8 \cdot 9}{243}} = \sqrt[3]{\frac{8}{27}} = \frac{2}{3}.$$

Since $a_2 = r a_1$, we see that

$$a_1 = \frac{a_2}{r} = \frac{20/9}{2/3} = \frac{20}{9} \cdot \frac{3}{2} = \frac{10}{3}.$$

Therefore,

$$a_n = r^{n-1} a_1 = \left(\frac{2}{3}\right)^{n-1} \cdot \frac{10}{3} = \frac{2^{n-1} \cdot 2 \cdot 5}{3^{n-1} \cdot 3} = \frac{2^n \cdot 5}{3^n} = 5\left(\frac{2}{3}\right)^n. \quad \blacksquare$$

⠿ PARTIAL SUMS

If the common ratio r of a geometric sequence is the number 1, then we have

$$a_n = 1^{n-1} a_1 \quad \text{for every } n \geq 1.$$

Therefore, the sequence is just the constant sequence $a_1, a_1, a_1, \ldots$. For any positive integer k, the kth partial sum of this constant sequence is

$$\underbrace{a_1 + a_1 + \cdots + a_1}_{k \text{ terms}} = k a_1$$

In other words, the kth partial sum of a constant sequence is just k times the constant. If a geometric sequence is not constant (that is, $r \neq 1$), then its partial sums are given by the following formula.

*Partial Sums of a
Geometric Sequence*

> The kth partial sum of the geometric sequence $\{a_n\}$ with common ratio $r \neq 1$ is
> $$\sum_{n=1}^{k} a_n = a_1 \left(\frac{1 - r^k}{1 - r}\right).$$

Proof   If S denotes the kth partial sum, then the formula for the nth term of a geometric sequence shows that

$$S = a_1 + a_2 + \cdots + a_k = a_1 + a_1 r + a_1 r^2 + a_1 r^3 + \cdots + a_1 r^{k-1}.$$

Use this equation to compute $S - rS$.

$$S = a_1 + a_1 r + a_1 r^2 + a_1 r^3 + \cdots + a_1 r^{k-1}$$
$$rS = \quad\quad a_1 r + a_1 r^2 + a_1 r^3 + \cdots + a_1 r^{k-1} + a_1 r^k$$
$$\overline{\rule{0pt}{1em}\quad\quad\quad\quad\quad\quad\quad\quad\quad\quad\quad\quad\quad\quad\quad\quad\quad\quad\quad}$$
$$S - rS = a_1 \quad\quad\quad\quad\quad\quad\quad\quad\quad\quad\quad\quad - a_1 r^k$$
$$(1 - r)S = a_1(1 - r^k).$$

Since $r \neq 1$, we can divide both sides of this last equation by $1 - r$ to complete the proof.

$$S = \frac{a_1(1 - r^k)}{1 - r} = a_1\left(\frac{1 - r^k}{1 - r}\right). \quad \blacksquare$$

EXAMPLE 6

Find the sum

$$-\frac{3}{2} + \frac{3}{4} - \frac{3}{8} + \frac{3}{16} - \frac{3}{32} + \frac{3}{64} - \frac{3}{128} + \frac{3}{256} - \frac{3}{512}.$$

SOLUTION   Note that this is the ninth partial sum of the geometric sequence $\left\{3\left(\frac{-1}{2}\right)^n\right\}$. The common ratio is $r = -1/2$. The formula in the box shows that

$$\sum_{n=1}^{9} 3\left(\frac{-1}{2}\right)^n = a_1\left(\frac{1 - r^9}{1 - r}\right) = \left(\frac{-3}{2}\right)\left[\frac{1 - (-1/2)^9}{1 - (-1/2)}\right]$$

$$= \left(\frac{-3}{2}\right)\left(\frac{1 + 1/2^9}{3/2}\right) = \left(\frac{-3}{2}\right)\left(\frac{2}{3}\right)\left(1 + \frac{1}{2^9}\right)$$

$$= -1 - \frac{1}{2^9} = -1 - \frac{1}{512} = -\frac{513}{512}.$$

Figure 12–13

The sum can also be found with a calculator (Figure 12–13) and expressed in fractional form.[*] $\blacksquare$

EXAMPLE 7

A superball is dropped from a height of 9 feet. It hits the ground and bounces to a height of 6 feet. It continues to bounce up and down. On each bounce, it rises to 2/3 of the height of the previous bounce. How far has the ball traveled (both up and down) when it hits the ground for the seventh time?

SOLUTION   We first consider how far the ball travels on each bounce. On the first bounce, it rises 6 feet and falls 6 feet for a total of 12 feet. On the second bounce, it rises and falls 2/3 of the previous height and hence travels 2/3 of 12 feet. If a_n denotes the distance traveled on the nth bounce, then

$$a_1 = 12, \qquad a_2 = \left(\frac{2}{3}\right)a_1, \qquad a_3 = \left(\frac{2}{3}\right)a_2 = \left(\frac{2}{3}\right)^2 a_1,$$

and in general

$$a_n = \left(\frac{2}{3}\right)a_{n-1} = \left(\frac{2}{3}\right)^{n-1} a_1.$$

---

[*]For fractional form on TI, use the FRAC key, as in Figure 12–13. On HP-39, change the number format in the MODE menu to "fraction." On Casio 9850, use the FRAC program in the Program Appendix; on Casio FX 2.0, use the a b/c key.

So $\{a_n\}$ is a geometric sequence with common ratio $r = 2/3$. When the ball hits the ground for the seventh time, it has completed six bounces. Therefore, the total distance it has traveled is the distance it was originally dropped (9 feet) plus the distance traveled in six bounces, namely,

$$9 + a_1 + a_2 + a_3 + a_4 + a_5 + a_6 = 9 + \sum_{n=1}^{6} a_n = 9 + a_1\left(\frac{1 - r^6}{1 - r}\right)$$

$$= 9 + 12\left[\frac{1 - (2/3)^6}{1 - (2/3)}\right] \approx 41.84 \text{ feet.} \quad \blacksquare$$

✓ EXERCISES 12.3

In Exercises 1–12, determine whether the sequence is arithmetic, geometric, or neither.

1. $2, 7, 12, 17, 22, \ldots$ **2.** $2, 6, 18, 54, 162, \ldots$

3. $13, 13/2, 13/4, 13/8, \ldots$

4. $-1, -\frac{1}{2}, 0, \frac{1}{2}, \ldots$ **5.** $50, 48, 46, 44, \ldots$

6. $2, -3, 9/2, -27/4, -81/8, \ldots$

7. $3, -3/2, 3/4, -3/8, 3/16, \ldots$

8. $-6, -3.7, -1.4, .9, 3.2, \ldots$

9. $3, 3\sqrt{2}, 6, 6\sqrt{2}, 12, 12\sqrt{2}, \ldots$

10. $\ln e, \ln e^2, \ln e^3, \ln e^4, \ln e^5, \ldots$

11. $6, 6, 6, 6, 6, \ldots$

12. $1, \sqrt[3]{3}, \sqrt[3]{9}, 3, 3\sqrt[3]{3}, 3\sqrt[3]{9}, 9, \ldots$

In Exercises 13–22, one term and the common ratio r of a geometric sequence are given. Find the sixth term and a formula for the nth term.

13. $a_1 = 5, r = 2$ **14.** $a_1 = 1, r = -2$

15. $a_1 = 4, r = \frac{1}{4}$ **16.** $a_1 = -6, r = \frac{2}{3}$

17. $a_1 = 10, r = -\frac{1}{2}$ **18.** $a_1 = \pi, r = \frac{1}{5}$

19. $a_2 = 12, r = 1/3$ **20.** $a_3 = 1/2, r = 3$

21. $a_4 = -4/5, r = 2/5$ **22.** $a_5 = 2/3, r = -1/3$

In Exercises 23–28, show that the given sequence is geometric and find the common ratio.

23. $\left\{\left(-\frac{1}{2}\right)^n\right\}$ **24.** $\{2^{3n}\}$

25. $\{5^{n+2}\}$ **26.** $\{3^{n/2}\}$

27. $\{(\sqrt{5})^n\}$ **28.** $\{4^{n-4}\}$

In Exercises 29–36, use the given information about the geometric sequence $\{a_n\}$ to find a_5 and a formula for a_n.

29. $a_1 = 256, a_2 = -64$ **30.** $a_1 = 1/6, a_2 = -1/18$

31. $a_1 = 1/2, a_2 = 5$ **32.** $a_1 = \sqrt{7}, a_2 = \sqrt{42}$

33. $a_3 = 4, a_6 = 1/16$ **34.** $a_3 - 4, a_6 = -32$

35. $a_1 = 5, a_7 = 20$ (assume that $r > 0$)

36. $a_2 = 6, a_7 = 192$

In Exercises 37–40, find the kth partial sum of the geometric sequence $\{a_n\}$ with common ratio r.

37. $k = 6, a_1 = 5, r = \frac{1}{2}$ **38.** $k = 8, a_1 = 9, r = \frac{1}{3}$

39. $k = 7, a_2 = 6, r = 2$ **40.** $k = 9, a_2 = 6, r = \frac{1}{4}$

In Exercises 41–46, find the sum.

41. $\displaystyle\sum_{n=1}^{7} 2^n$ **42.** $\displaystyle\sum_{k=1}^{6} 3\left(\frac{1}{2}\right)^k$

43. $\displaystyle\sum_{n=1}^{9}\left(-\frac{1}{3}\right)^n$ **44.** $\displaystyle\sum_{n=1}^{5} 5 \cdot 3^{n-1}$

45. $\displaystyle\sum_{j=1}^{6} 4\left(\frac{3}{2}\right)^{j-1}$ **46.** $\displaystyle\sum_{t=1}^{8} 6(.9)^{t-1}$

In Exercises 47–52, you are asked to find a geometric sequence. In each case, round the common ratio r to four decimal places.

47. The number of students per computer in U.S. schools in year n (with $n = 1$ corresponding to 1985) can be approximated by a geometric sequence whose first two terms are $a_1 = 44.4313$ and $a_2 = 38.0154$.[*]
(a) Find a formula for a_n.
(b) What is the number of students per computer in 2003?
(c) In what year will there be a computer for each student?

---

*Based on data from Quality Education Data, Inc.

48. According to data from the U.S. Census Bureau, the population of the United States (in millions) in year n can be approximated by a geometric sequence $\{b_n\}$, where $n = 1$ corresponds to 1971.
   (a) If $b_1 = 207.5489$ and $b_3 = 211.7207$, find a formula for b_n.
   (b) Estimate the U.S. population in 2005 and 2010.

49. The value of all group life insurance (in billions of dollars) in year n can be approximated by a geometric sequence $\{c_n\}$, where $n = 1$ corresponds to 1991.[*]
   (a) If there was \$3.9631 billion in effect in 1991 and \$4.1672 billion in 1992, find a formula for c_n.
   (b) How much group life insurance is in effect in 1998? In 2002? In 2004?

50. Data from the U.S. Centers for Disease Control and Prevention indicates that the number of new cases of AIDS each year can be approximated by a geometric sequence $\{a_n\}$, where $n = 1$ corresponds in 1994.
   (a) If there were 80,691 new cases in 1994 and 71,605 in 1995, find a formula for a_n.
   (b) About how many new cases were reported in 2000?
   (c) Find the total number of new AIDS cases reported from 1994 to 2000.

51. The amount spent per person per year on cable and satellite TV can be approximated by a geometric sequence $\{b_n\}$, where $n = 1$ corresponds to 1995.[†]
   (a) If \$130.76 was spent in 1995 and \$151.39 was spent in 1997, find a formula for b_n.
   (b) Find the total spent per person from 1995 to 2003.

52. According to data from the U.S. National Center for Education Statistics, the number of bachelor's degrees earned by women can be approximated by a geometric sequence $\{c_n\}$, where $n = 1$ corresponds to 1970.
   (a) If 365,916 degrees were earned in 1970 and 373,454 in 1971, find a formula for c_n.
   (b) How many degrees were earned in 1997? In 2000? In 2003?
   (c) Find the total number of degrees earned from 1995 to 2003.

53. A ball is dropped from a height of 8 feet. On each bounce, it rises to half its previous height. When the ball hits the ground for the seventh time, how far has it traveled?

54. A ball is dropped from a height of 10 feet. On each bounce, it rises to 45% of its previous height. When it hits the ground for the tenth time, how far has it traveled?

55. If you are paid a salary of 1¢ on the first day of March and 2¢ on the second day, and your salary continues to double each day, how much will you earn in the month of March?

56. Starting with your parents, how many ancestors do you have for the preceding ten generations?

57. A car that sold for \$8000 depreciates in value 25% each year. What is it worth after five years?

58. A vacuum pump removes 60% of the air in a container at each stroke. What percentage of the original amount of air remains after six strokes?

Thinkers

59. Suppose $\{a_n\}$ is a geometric sequence with common ratio $r > 0$ and each $a_n > 0$. Show that the sequence $\{\log a_n\}$ is an arithmetic sequence with common difference $\log r$.

60. Suppose $\{a_n\}$ is an arithmetic sequence with common difference d. Let C be any positive number. Show that the sequence $\{C^{a_n}\}$ is a geometric sequence with common ratio C^d.

61. In the geometric sequence 1, 2, 4, 8, 16, . . . , show that each term is 1 plus the sum of all preceding terms.

62. In the geometric sequence 2, 6, 18, 54, . . . , show that each term is twice the sum of 1 and all preceding terms.

63. The minimum monthly payment for a certain bank credit card is the larger of 1/25 of the outstanding balance or \$5. If the balance is less than \$5, the entire balance is due. If you make only the minimum payment each month, how long will it take to pay off a balance of \$200 (excluding any interest that might be due)?

---

[*]Based on data from the American Council of Life Insurance.
[†]Based on data in the *Statistical Abstract of the United States: 2001*.

12.3.A SPECIAL TOPICS Infinite Series

An **infinite series** (or simply **series**) is an expression of the form

$$a_1 + a_2 + a_3 + a_4 + a_5 + \cdots,$$

where each a_n is a real number. This notation suggests the "sum" of the infinite sequence $a_1, a_2, a_3, a_4, a_5, \ldots$ However, ordinary addition is defined only for a finite list of numbers. So we must *define* what is meant by the sum of an infinite series. Consider, for example, the infinite series

$$\frac{3}{10} + \frac{3}{10^2} + \frac{3}{10^3} + \frac{3}{10^4} + \cdots.$$

We begin with the *sequence* $3/10, 3/10^2, 3/10^3, 3/10^4, \ldots$ and compute its partial sums.

$$S_1 = \frac{3}{10},$$

$$S_2 = \frac{3}{10} + \frac{3}{10^2} = \frac{33}{100},$$

$$S_3 = \frac{3}{10} + \frac{3}{10^2} + \frac{3}{10^3} = \frac{333}{1000},$$

$$S_4 = \frac{3}{10} + \frac{3}{10^2} + \frac{3}{10^3} + \frac{3}{10^4} = \frac{3333}{10,000}.$$

These partial sums $S_1, S_2, S_3, S_4, \ldots$ themselves form a sequence.

$$\frac{3}{10}, \frac{33}{100}, \frac{333}{1000}, \frac{3333}{10,000}, \ldots$$

The terms in the sequence of partial sums appear to be getting closer and closer to $1/3$. In other words, as k gets larger and larger, the corresponding partial sum S_k gets closer and closer to $1/3$. So we say that $1/3$ is the *sum* of the infinite series

$$\frac{3}{10} + \frac{3}{10^2} + \frac{3}{10^3} + \frac{3}{10^4} + \cdots,$$

or that this series *converges* to $1/3$, and we write

$$\frac{3}{10} + \frac{3}{10^2} + \frac{3}{10^3} + \frac{3}{10^4} + \cdots = \frac{1}{3}.$$

The general case is handled similarly. If $a_1 + a_2 + a_3 + a_4 + \cdots$ is an infinite series, then its **partial sums** are the numbers

$$S_1 = a_1,$$
$$S_2 = a_1 + a_2,$$
$$S_3 = a_1 + a_2 + a_3,$$

and in general, for any $k \geq 1$,

$$S_k = a_1 + a_2 + a_3 + a_4 + \cdots + a_k.$$

If it happens that the terms $S_1, S_2, S_3, S_4, \ldots$ of the *sequence* of partial sums get closer and closer to a particular real number S in such a way that the partial sum S_k is arbitrarily close to S when k is large enough, then we say that the series **converges** and that S is the **sum of the convergent series.**

Not every series has a sum. For instance, the partial sums of the series

$$1 + 2 + 3 + 4 + \cdots$$

get larger and larger and never get closer and closer to a single real number. So this series is not convergent and does not have a sum.

EXAMPLE 1

Although no proof will be given here, it is intuitively clear that every infinite decimal may be thought of as the sum of a convergent series. For instance,

$$\pi = 3.1415926 \cdots = 3 + .1 + .04 + .001 + .0005 + .00009 + \cdots.$$

Note that the third partial sum is $3 + .1 + .04 = 3.14$, which is π to two decimal places. Similarly, the kth partial sum of this series is just π to $k - 1$ decimal places. ■

▪▪ INFINITE GEOMETRIC SERIES

If $\{a_n\}$ is a geometric sequence with common ratio r, then the corresponding infinite series

$$a_1 + a_2 + a_3 + a_4 + a_5 + \cdots$$

is called an **infinite geometric series.** By using the formula for the nth term of a geometric sequence, we can also express the corresponding geometric series in the form

$$a_1 + ra_1 + r^2a_1 + r^3a_1 + r^4a_1 + \cdots.$$

Under certain circumstances, an infinite geometric series is convergent and has a sum.

Sum of an Infinite Geometric Series

If $|r| < 1$, then the infinite geometric series

$$a_1 + ra_1 + r^2a_1 + r^3a_1 + r^4a_1 + \cdots$$

converges, and its sum is

$$\frac{a_1}{1 - r}.$$

Although we cannot prove this fact rigorously here, we can make it highly plausible both geometrically and algebraically.

EXAMPLE 2

$\dfrac{8}{5} + \dfrac{8}{5^2} + \dfrac{8}{5^3} + \cdots$ is an infinite geometric series with $a_1 = 8/5$ and $r = 1/5$. The kth partial sum of this series is the same as the kth partial sum of the sequence $\{8/5^n\}$, and hence, from the box on page 797, we know that

$$S_k = a_1\left(\frac{1 - r^k}{1 - r}\right) = \frac{8}{5}\left[\frac{1 - \left(\frac{1}{5}\right)^k}{1 - \frac{1}{5}}\right]$$

$$= \frac{8}{5}\left(\frac{1 - \frac{1}{5^k}}{\frac{4}{5}}\right) = \frac{8}{5} \cdot \frac{5}{4}\left(1 - \frac{1}{5^k}\right)$$

$$= 2\left(1 - \frac{1}{5^k}\right) = 2 - \frac{2}{5^k}.$$

The function $f(x) = 2 - 2/5^x$ is defined for all real numbers. When $x = k$ is a positive integer, then $f(k)$ is S_k, the kth partial sum of the series. Using a calculator, we obtain the graph of $f(x)$ in Figure 12–14.

Figure 12–14

EXPLORATION

Graph $f(x)$ in the same viewing window as in Figure 12–14. Use the trace feature to move the cursor along the graph. As x gets larger, what is the apparent value of $f(x)$ (that is, the value of the partial sum)?

Your calculator will probably tell you that every partial sum is 2, once you move beyond approximately $x = 15$. Actually, the partial sums are slightly smaller than 2 but are rounded to 2 by the calculator. In any case, the horizontal line through 2 is a horizontal asymptote of the graph (meaning that the graph gets very close to the line as x gets larger), so it is very plausible that the sequence converges to the number 2. But 2 is exactly what the preceding box says the sum should be.

$$\frac{a_1}{1 - r} = \frac{\frac{8}{5}}{1 - \frac{1}{5}} = \frac{\frac{8}{5}}{\frac{4}{5}} = \frac{8}{4} = 2. \quad \blacksquare$$

Example 2 is typical of the general case, as can be seen algebraically. Consider the geometric series $a_1 + a_2 + a_3 + \cdots$ with common ratio r such that $|r| < 1$. The kth partial sum S_k is the same as the kth partial sum of the geometric sequence $\{a_n\}$, and hence,

$$S_k = a_1\left(\frac{1 - r^k}{1 - r}\right).$$

As k gets larger and larger, the number r^k gets very close to 0 because $|r| < 1$ (for instance, $(-.6)^{20} \approx .0000366$ and $.2^9 \approx .000000512$). Consequently, when k is very large, $1 - r^k$ is very close to $1 - 0$ so that

$$S_k = a_1\left(\frac{1 - r^k}{1 - r}\right) \qquad \text{is very close to} \qquad a_1\left(\frac{1 - 0}{1 - r}\right) = \frac{a_1}{1 - r}.$$

EXAMPLE 3

$-\dfrac{1}{2} + \dfrac{1}{4} - \dfrac{1}{8} + \dfrac{1}{16} + \cdots$ is an infinite geometric series with $a_1 = -1/2$ and $r = -1/2$. Since $|r| < 1$, this series converges, and its sum is

$$\frac{a_1}{1 - r} = \frac{-\dfrac{1}{2}}{1 - \left(-\dfrac{1}{2}\right)} = \frac{-\dfrac{1}{2}}{\dfrac{3}{2}} = -\frac{1}{3}. \qquad \blacksquare$$

Infinite geometric series provide another way of writing an infinite repeating decimal as a rational number.

EXAMPLE 4

To express $6.8573573573\cdots$ as a rational number, we first write it as $6.8 + .0573573573\cdots$. Consider $.0573573573\cdots$ as an infinite series.

$$.0573 + .0000573 + .0000000573 + .0000000000573 + \cdots,$$

which is the same as

$$.0573 + (.001)(.0573) + (.001)^2(.0573) + (.001)^3(.0573) + \cdots.$$

This is a convergent geometric series with $a_1 = .0573$ and $r = .001$. Its sum is

$$\frac{a_1}{1 - r} = \frac{.0573}{1 - .001} = \frac{.0573}{.999} = \frac{573}{9990}.$$

Therefore,

$$6.8573573573\cdots = 6.8 + [.0573 + .0000573 + \cdots]$$

$$= 6.8 + \frac{573}{9990}$$

$$= \frac{68}{10} + \frac{573}{9990}$$

$$= \frac{68{,}505}{9990} = \frac{4567}{666}. \qquad \blacksquare$$

We close this section by mentioning an alternative notation for infinite series. The series

$$a_1 + a_2 + a_3 + a_4 + \cdots \qquad \text{is denoted by} \qquad \sum_{n=1}^{\infty} a_n.$$

For example, $\displaystyle\sum_{n=1}^{\infty} 2(.6^n)$ denotes the series

$$2(.6) + 2(.6^2) + 2(.6^3) + 2(.6^4) + \cdots,$$

and $\displaystyle\sum_{n=1}^{\infty} \left(\frac{-1}{2}\right)^n$ denotes

$$-\frac{1}{2} + \left(\frac{-1}{2}\right)^2 + \left(\frac{-1}{2}\right)^3 + \left(\frac{-1}{2}\right)^4 + \cdots - \frac{1}{2} + \frac{1}{4} - \frac{1}{8} + \frac{1}{16} - \frac{1}{32} + \cdots.$$

✓ EXERCISES 12.3.A

In Exercises 1–8, find the sum of the infinite series, if it has one.

1. $\displaystyle\sum_{n=1}^{\infty} \frac{1}{2^n}$ **2.** $\displaystyle\sum_{n=1}^{\infty} \left(-\frac{3}{4}\right)^n$ **3.** $\displaystyle\sum_{n=1}^{\infty} (.06)^n$

4. $1 - .5 + .25 - .125 + .0625 - \cdots$

5. $500 + 200 + 80 + 32 + \cdots$

6. $9 - 3\sqrt{3} + 3 - \sqrt{3} + 1 - \dfrac{1}{\sqrt{3}} + \cdots$

7. $2 + \sqrt{2} + 1 + \dfrac{1}{\sqrt{2}} + \dfrac{1}{2} + \cdots$

8. $\displaystyle\sum_{n=1}^{\infty} \left(\frac{1}{2^n} - \frac{1}{3^n}\right)$

In Exercises 9–15, express the repeating decimal as a rational number.

9. $.22222\cdots$ **10.** $.37373737\cdots$

11. $5.4272727\cdots$ **12.** $85.131313\cdots$

13. $2.1425425425\cdots$ **14.** $3.7165165165\cdots$

15. $1.74241241241\cdots$

16. If $\{a_n\}$ is an arithmetic sequence with common difference $d > 0$ and each $a_i > 0$, explain why the infinite series $a_1 + a_2 + a_3 + a_4 + \cdots$ is not convergent.

17. (a) Verify that $\displaystyle\sum_{n=1}^{\infty} 2(1.5)^n$ is a geometric series with $a_1 = 3$ and $r = 1.5$.

(b) Find the kth partial sum of the series and use this expression to define a function f, as in Example 2.

(c) Graph the function f in a viewing window with $0 \le x \le 30$. As x gets very large, what happens to the corresponding value of $f(x)$? Does the graph get closer and closer to some horizontal line, as in Example 2? What does this say about the convergence of the series?

18. Use the graphical approach illustrated in Example 2 to find the sum of the series in Example 3. Does the graph get very close to the horizontal line through $-1/3$? What's going on?

In Exercises 19 and 20, use your calculator to compute partial sums and estimate the sum of the convergent infinite series.

19. $\dfrac{\pi}{2} - \dfrac{(\pi/2)^3}{3!} + \dfrac{(\pi/2)^5}{5!} - \dfrac{(\pi/2)^7}{7!} + \cdots$, where $n!$ is the product $1 \cdot 2 \cdot 3 \cdot 4 \cdot \cdots \cdot n$.*

20. $1 + 1 + \dfrac{1}{2!} + \dfrac{1}{3!} + \dfrac{1}{4!} + \cdots$ [*Hint:* What is the decimal expansion of e?]

---

*To compute $n!$ quickly, look for ! in the PROB submenu of the MATH or OPTN menu.

12.4   The Binomial Theorem

The Binomial Theorem provides a formula for calculating the product $(x + y)^n$ for any positive integer n. Before we state the theorem, some preliminaries are needed.

Let n be a positive integer. The symbol $n!$ (read **n factorial**) denotes the product of all the integers from 1 to n. For example,

$$2! = 1 \cdot 2 = 2, \quad 3! = 1 \cdot 2 \cdot 3 = 6, \quad 4! = 1 \cdot 2 \cdot 3 \cdot 4 = 24,$$

$$5! = 1 \cdot 2 \cdot 3 \cdot 4 \cdot 5 = 120,$$

$$10! = 1 \cdot 2 \cdot 3 \cdot 4 \cdot 5 \cdot 6 \cdot 7 \cdot 8 \cdot 9 \cdot 10 = 3,628,800.$$

In general, we have this result.

n Factorial

Let n be a positive integer. Then
$$n! = 1 \cdot 2 \cdot 3 \cdot 4 \cdots (n - 2)(n - 1)n.$$

$0!$ is defined to be the number 1.

Learn to use your calculator to compute factorials. You will find ! in the PROB (or PRB) submenu of the MATH or OPTN menu.

CALCULATOR EXPLORATION

15! is such a large number your calculator will switch to scientific notation to express it. What is this approximation? Many calculators cannot compute factorials larger than 69! If yours does compute larger ones, how large a one can you compute without getting an error message [or on HP-39, getting the number 9.9999 · · · E499]?

If r and n are integers with $0 \le r \le n$, then we have the following.

Binomial Coefficients

Either of the symbols $\binom{n}{r}$ or $_nC_r$ denotes the number $\dfrac{n!}{r!(n - r)!}$.

$\binom{n}{r}$ is called a **binomial coefficient.**

For example,

$$_5C_3 = \binom{5}{3} = \frac{5!}{3!(5 - 3)!} = \frac{5!}{3!2!} = \frac{1 \cdot 2 \cdot 3 \cdot 4 \cdot 5}{(1 \cdot 2 \cdot 3)(1 \cdot 2)} = \frac{4 \cdot 5}{2} = 10,$$

$$_4C_2 = \binom{4}{2} = \frac{4!}{2!(4 - 2)!} = \frac{4!}{2!2!} = \frac{1 \cdot 2 \cdot 3 \cdot 4}{(1 \cdot 2)(1 \cdot 2)} = \frac{3 \cdot 4}{2} = 6.$$

Binomial coefficients can be computed on a calculator by using $_nC_r$ or Comb in the PROB (or PRB) submenu of the MATH or OPTN menu.

CALCULATOR EXPLORATION

Compute $_{56}C_{47} = \begin{pmatrix} 56 \\ 47 \end{pmatrix}$. Although calculators cannot compute 475!, they can compute many binomial coefficients, such as $\begin{pmatrix} 475 \\ 400 \end{pmatrix}$, because most of the factors cancel out (as in the previous example). Check yours. Will it also compute $\begin{pmatrix} 475 \\ 50 \end{pmatrix}$?

The preceding examples illustrate a fact whose proof will be omitted: *Every binomial coefficient is an integer.* Furthermore, for every nonnegative integer n,

$$\begin{pmatrix} n \\ 0 \end{pmatrix} = 1 \qquad \text{and} \qquad \begin{pmatrix} n \\ n \end{pmatrix} = 1$$

because

$$\begin{pmatrix} n \\ 0 \end{pmatrix} = \frac{n!}{0!(n-0)!} + \frac{n!}{0!n!} = \frac{n!}{n!} = 1 \qquad \text{and}$$

$$\begin{pmatrix} n \\ n \end{pmatrix} = \frac{n!}{n!(n-n)!} = \frac{n!}{n!0!} = \frac{n!}{n!} = 1.$$

If we list the binomial coefficients for each value of n in this manner:

$n = 0$ $\qquad\qquad\qquad\qquad\qquad \begin{pmatrix} 0 \\ 0 \end{pmatrix}$

$n = 1$ $\qquad\qquad\qquad\qquad \begin{pmatrix} 1 \\ 0 \end{pmatrix} \qquad \begin{pmatrix} 1 \\ 1 \end{pmatrix}$

$n = 2$ $\qquad\qquad\quad \begin{pmatrix} 2 \\ 0 \end{pmatrix} \qquad \begin{pmatrix} 2 \\ 1 \end{pmatrix} \qquad \begin{pmatrix} 2 \\ 2 \end{pmatrix}$

$n = 3$ $\qquad \begin{pmatrix} 3 \\ 0 \end{pmatrix} \qquad \begin{pmatrix} 3 \\ 1 \end{pmatrix} \qquad \begin{pmatrix} 3 \\ 2 \end{pmatrix} \qquad \begin{pmatrix} 3 \\ 3 \end{pmatrix}$

$n = 4$ $\begin{pmatrix} 4 \\ 0 \end{pmatrix} \qquad \begin{pmatrix} 4 \\ 1 \end{pmatrix} \qquad \begin{pmatrix} 4 \\ 2 \end{pmatrix} \qquad \begin{pmatrix} 4 \\ 3 \end{pmatrix} \qquad \begin{pmatrix} 4 \\ 4 \end{pmatrix}$

$\vdots \qquad \ddots$ $\qquad\qquad\qquad\qquad\qquad\qquad\qquad\qquad\qquad\qquad \ddots$

and then calculate each of them, we obtain the following array of numbers.

row 0					1				
row 1				1		1			
row 2			1		2		1		
row 3		1		3		3		1	
row 4	1		4		6		4		1

$\vdots \qquad \ddots$ $\qquad\qquad\qquad\qquad\qquad\qquad\qquad\qquad \ddots$

This array is called **Pascal's triangle.** Its pattern is easy to remember: Each entry (except the 1's at the beginning or end of a row) is the sum of the two closest entries in the row above it. In the fourth row, for instance, 6 is the sum of the two 3's above it, and each 4 is the sum of the 1 and 3 above it. See Exercise 47 for a proof.

To develop a formula for calculating $(x + y)^n$, we first calculate these products for small values of n to see whether we can find some kind of pattern.

$$n = 0 \qquad (x + y)^0 = \qquad\qquad 1$$
$$n = 1 \qquad (x + y)^1 = \qquad\qquad 1x + 1y$$
$$(*) \qquad n = 2 \qquad (x + y)^2 = \qquad\qquad 1x^2 + 2xy + 1y^2$$
$$n = 3 \qquad (x + y)^3 = \qquad 1x^3 + 3x^2y + 3xy^2 + 1y^3$$
$$n = 4 \qquad (x + y)^4 = 1x^4 + 4x^3y + 6x^2y^2 + 4xy^3 + 1y^4$$

One pattern is immediately obvious: The coefficients here (shown in color) are the top part of Pascal's triangle! In the case $n = 4$, for example, this means that the coefficients are the numbers

$$1 \qquad 4 \qquad 6 \qquad 4 \qquad 1$$
$$\binom{4}{0}, \; \binom{4}{1}, \; \binom{4}{2}, \; \binom{4}{3}, \; \binom{4}{4}.$$

If this pattern holds for larger n, then the coefficients in the expansion of $(x + y)^n$ are

$$\binom{n}{0}, \binom{n}{1}, \binom{n}{2}, \binom{n}{3}, \ldots, \binom{n}{n-1}, \binom{n}{n}.$$

As for the xy-terms associated with each of these coefficients, look at the pattern in $(*)$ above: The exponent of x goes down by 1 and the exponent of y goes up by 1 as you go from term to term, which suggests that the terms of the expansion of $(x + y)^n$ (without the coefficients) are

$$x^n, \qquad x^{n-1}y, \qquad x^{n-2}y^2, \qquad x^{n-3}y^3, \ldots, xy^{n-1}, \quad y^n.$$

Combining the patterns of coefficients and xy-terms and using the fact that

$$\binom{n}{0} = 1 \qquad \text{and} \qquad \binom{n}{n} = 1$$

suggests that the following result is true about the expansion of $(x + y)^n$.

The Binomial Theorem

For each positive integer n,

$$(x + y)^n = x^n + \binom{n}{1}x^{n-1}y + \binom{n}{2}x^{n-2}y^2 +$$
$$\binom{n}{3}x^{n-3}y^3 + \cdots + \binom{n}{n-1}xy^{n-1} + y^n.$$

Using summation notation and the fact that

$$\binom{n}{0} = 1 = \binom{n}{n},$$

we can write the Binomial Theorem compactly as

$$(x + y)^n = \sum_{j=0}^{n} \binom{n}{j} x^{n-j} y^j.$$

The Binomial Theorem will be proved in Section 12.5 by means of mathematical induction. We shall assume its truth for now and illustrate some of its uses.

EXAMPLE 1

Expand $(x + y)^8$.

SOLUTION   We apply the Binomial Theorem in the case $n = 8$.

$$(x + y)^8 = x^8 + \binom{8}{1} x^7 y + \binom{8}{2} x^6 y^2 + \binom{8}{3} x^5 y^3$$

$$+ \binom{8}{4} x^4 y^4 + \binom{8}{5} x^3 y^5 + \binom{8}{6} x^2 y^6 + \binom{8}{7} x y^7 + y^8.$$

The coefficients can be computed individually by hand or by using $_nC_r$ (or COMB) on a calculator; for instance,

$$_8C_2 = \binom{8}{2} = \frac{8!}{2!6!} = 28 \qquad \text{or} \qquad _8C_3 = \binom{8}{3} = \frac{8!}{3!5!} = 56.$$

Alternatively, you can display all the coefficients at once by making a table of values for the function $f(x) = {}_8C_x$, as shown in Figure 12–15.[*]

Figure 12–15

Substituting these values in the preceding expansion, we have

$$(x + y)^8 = x^8 + 8x^7 y + 28x^6 y^2$$

$$+ 56x^5 y^3 + 70x^4 y^4 + 56x^3 y^5 + 28x^2 y^6 + 8xy^7 + y^8. \qquad ■$$

EXAMPLE 2

Expand $(1 - z)^6$.

---

*Thanks to Nick Goodbody for suggesting this.

SOLUTION   Note that $1 - z = 1 + (-z)$ and apply the Binomial Theorem with $x = 1$, $y = -z$, and $n = 6$.

$$(1 - z)^6 = 1^6 + \binom{6}{1}1^5(-z) + \binom{6}{2}1^4(-z)^2 + \binom{6}{3}1^3(-z)^3$$

$$+ \binom{6}{4}1^2(-z)^4 + \binom{6}{5}1(-z)^5 + (-z)^6$$

$$= 1 - \binom{6}{1}z + \binom{6}{2}z^2 - \binom{6}{3}z^3 + \binom{6}{4}z^4 - \binom{6}{5}z^5 + z^6$$

$$= 1 - 6z + 15z^2 - 20z^3 + 15z^4 - 6z^5 + z^6. \quad \blacksquare$$

EXAMPLE 3

Expand $(x^2 + x^{-1})^4$.

SOLUTION   Use the Binomial Theorem with x^2 in place of x and x^{-1} in place of y.

$$(x^2 + x^{-1})^4 = (x^2)^4 + \binom{4}{1}(x^2)^3(x^{-1}) + \binom{4}{2}(x^2)^2(x^{-1})^2 + \binom{4}{3}(x^2)(x^{-1})^3 + (x^{-1})^4$$

$$= x^8 + 4x^6x^{-1} + 6x^4x^{-2} + 4x^2x^{-3} + x^{-4}$$

$$= x^8 + 4x^5 + 6x^2 + 4x^{-1} + x^{-4}. \quad \blacksquare$$

EXAMPLE 4

Show that $(1.001)^{1000} > 2$ without using a calculator.

SOLUTION   We write 1.001 as $1 + .001$ and apply the Binomial Theorem with $x = 1$, $y = .001$, and $n = 1000$.

$$(1.001)^{1000} = (1 + .001)^{1000}$$

$$= 1^{1000} + \binom{1000}{1}1^{999}(.001) + \text{other positive terms}$$

$$= 1 + \binom{1000}{1}(.001) + \text{other positive terms}.$$

But

$$\binom{1000}{1} = \frac{1000!}{1!999!} = \frac{1000 \cdot 999!}{999!} = 1000.$$

Therefore,

$$\binom{1000}{1}(.001) = 1,000(.001) = 1$$

and

$$(1.001)^{1000} = 1 + 1 + \text{other positive terms} = 2 + \text{other positive terms}.$$

Hence, $(1.001)^{1000} > 2$. $\blacksquare$

Sometimes we need to know only one term in the expansion of $(x + y)^n$. If you examine the expansion given by the Binomial Theorem, you will see that in the second term y has exponent 1, in the third term y has exponent 2, and so on. Thus, we have the following.

Properties of the Binomial Expansion

> In the binomial expansion of $(x + y)^n$,
>
> The exponent of y is always one less than the number of the term.
>
> Furthermore, in each of the middle terms of the expansion,
>
> The coefficient of the term containing y^r is $\binom{n}{r}$.
>
> The sum of the x exponent and the y exponent is n.

For instance, in the *ninth* term of the expansion of $(x + y)^{13}$, y has exponent 8, the coefficient is $\binom{13}{8}$, and x must have exponent 5 (since $8 + 5 = 13$). Thus, the ninth term is $\binom{13}{8} x^5 y^8$.

EXAMPLE 5

Find the ninth term of the expansion of $\left(2x^2 + \dfrac{\sqrt[4]{y}}{\sqrt{6}} \right)^{13}$.

SOLUTION   We use the Binomial Theorem with $n = 13$ and with $2x^2$ in place of x and $\sqrt[4]{y}/\sqrt{6}$ in place of y. The remarks above show that the ninth term is

$$\binom{13}{8}(2x^2)^5 \left(\frac{\sqrt[4]{y}}{\sqrt{6}}\right)^8.$$

Since $\sqrt[4]{y} = y^{1/4}$ and $\sqrt{6} = \sqrt{3}\sqrt{2} = 3^{1/2}2^{1/2}$, we can simplify as follows.

$$\binom{13}{8}(2x^2)^5 \left(\frac{\sqrt[4]{y}}{\sqrt{6}}\right)^8 = \binom{13}{8}2^5(x^2)^5 \frac{(y^{1/4})^8}{(3^{1/2})^8(2^{1/2})^8} = \binom{13}{8}2^5 x^{10} \frac{y^2}{3^4 \cdot 2^4}$$

$$= \binom{13}{8}\frac{2}{3^4} x^{10} y^2 = \frac{13 \cdot 12 \cdot 11 \cdot 10 \cdot 9}{5 \cdot 4 \cdot 3 \cdot 2} \cdot \frac{2}{3^4} x^{10} y^2$$

$$= \frac{286}{9} x^{10} y^2. \quad \blacksquare$$

✓ **EXERCISES 12.4**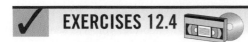

In Exercises 1–10, evaluate the expression.

1. $6!$ **2.** $\dfrac{11!}{8!}$ **3.** $\dfrac{12!}{9!3!}$ **4.** $\dfrac{9! - 8!}{7!}$

5. $\dbinom{5}{3} + \dbinom{5}{2} - \dbinom{6}{3}$ **6.** $\dbinom{12}{11} - \dbinom{11}{10} + \dbinom{7}{0}$

7. $\dbinom{6}{0} + \dbinom{6}{1} + \dbinom{6}{2} + \dbinom{6}{3} + \dbinom{6}{4} + \dbinom{6}{5} + \dbinom{6}{6}$

8. $\dbinom{6}{0} - \dbinom{6}{1} + \dbinom{6}{2} - \dbinom{6}{3} + \dbinom{6}{4} - \dbinom{6}{5} + \dbinom{6}{6}$

9. $\dbinom{100}{96}$ **10.** $\dbinom{75}{72}$

In Exercises 11–16, expand the expression.

11. $(x + y)^5$ **12.** $(a + b)^7$ **13.** $(a - b)^5$

14. $(c - d)^8$ **15.** $(2x + y^2)^5$ **16.** $(3u - v^3)^6$

In Exercises 17–26, use the Binomial Theorem to expand and (where possible) simplify the expression.

17. $(\sqrt{x} + 1)^6$ **18.** $(2 - \sqrt{y})^5$

19. $(1 - c)^{10}$ **20.** $\left(\sqrt{c} + \dfrac{1}{\sqrt{c}}\right)^7$

21. $(x^{-3} + x)^4$ **22.** $(3x^{-2} - x^2)^6$

23. $(1 + \sqrt{3})^4 + (1 - \sqrt{3})^4$

24. $(\sqrt{3} + 1)^6 - (\sqrt{3} - 1)^6$

25. $(1 + i)^6$, where $i^2 = -1$

26. $(\sqrt{2} - i)^4$, where $i^2 = -1$

In Exercises 27–32, find the indicated term of the expansion of the given expression.

27. third, $(x + y)^5$ **28.** fourth, $(a + b)^6$

29. fifth, $(c - d)^7$ **30.** third, $(a + 2)^8$

31. fourth, $\left(u^{-2} + \dfrac{u}{2}\right)^7$ **32.** fifth, $(\sqrt{x} - \sqrt{2})^7$

33. Find the coefficient of $x^5 y^8$ in the expansion of $(2x - y^2)^9$.

34. Find the coefficient of $x^{12}y^6$ in the expansion of $(x^3 - 3y)^{10}$.

35. Find the coefficient of $1/x^3$ in the expansion of $\left(2x + \dfrac{1}{x^2}\right)^6$.

36. Find the constant term in the expansion of $\left(y - \dfrac{1}{2y}\right)^{10}$.

37. (a) Verify that $\dbinom{9}{1} = 9$ and $\dbinom{9}{8} = 9$.

(b) Prove that for each positive integer n, $\dbinom{n}{1} = n$ and $\dbinom{n}{n-1} = n$. [*Note:* Part (a) is just the case when $n = 9$ and $n - 1 = 8$.]

38. (a) Verify that $\dbinom{7}{2} = \dbinom{7}{5}$.

(b) Let r and n be integers with $0 \le r \le n$. Prove that $\dbinom{n}{r} = \dbinom{n}{n-r}$. [*Note:* Part (a) is just the case when $n = 7$ and $r = 2$.]

39. Prove that for any positive integer n,
$$2^n = \dbinom{n}{0} + \dbinom{n}{1} + \dbinom{n}{2} + \cdots + \dbinom{n}{n}.$$
[*Hint:* $2 = 1 + 1$.]

40. Prove that for any positive integer n,
$$\dbinom{n}{0} - \dbinom{n}{1} + \dbinom{n}{2} - \dbinom{n}{3} + \dbinom{n}{4} - \cdots$$
$$+ (-1)^k \dbinom{n}{k} + \cdots + (-1)^n \dbinom{n}{n} = 0.$$

41. Use the Binomial Theorem with $y = i \sin \theta$ and $x = \cos \theta$ to find $(\cos \theta + i \sin \theta)^4$, where $i^2 = -1$.

42. (a) Use DeMoivre's Theorem to find
$$(\cos \theta + i \sin \theta)^4.$$

(b) Use the fact that the two expressions obtained in part (a) and in Exercise 41 must be equal to express $\cos 4\theta$ and $\sin 4\theta$ in terms of $\sin \theta$ and $\cos \theta$.

43. (a) Let f be the function given by $f(x) = x^5$. Let h be a nonzero number and compute $f(x + h) - f(x)$ (but leave all binomial coefficients in the form $\dbinom{5}{r}$ here and below).

(b) Use part (a) to show that h is a factor of $f(x + h) - f(x)$ and find $\dfrac{f(x + h) - f(x)}{h}$.

(c) If h is *very* close to 0, find a simple approximation of the quantity $\dfrac{f(x + h) - f(x)}{h}$. [See part (b).]

44. Do Exercise 43 with $f(x) = x^8$ in place of $f(x) = x^5$.

45. Do Exercise 43 with $f(x) = x^{12}$ in place of $f(x) = x^5$.

46. Let n be a fixed positive integer. Do Exercise 43 with $f(x) = x^n$ in place of $f(x) = x^5$.

Thinkers

47. Let r and n be integers such that $0 \le r \le n$.
   (a) Verify that $(n - r)! = (n - r)[n - (r + 1)]!$
   (b) Verify that $(n - r)! = [(n + 1) - (r + 1)]!$
   (c) Prove that $\begin{pmatrix} n \\ r + 1 \end{pmatrix} + \begin{pmatrix} n \\ r \end{pmatrix} = \begin{pmatrix} n + 1 \\ r + 1 \end{pmatrix}$ for any $r \le n - 1$.
   [*Hint:* Write out the terms on the left side and use parts (a) and (b) to express each of them as a fraction with denominator $(r + 1)!(n - r)!$. Then add these two fractions, simplify the numerator, and compare the result with $\begin{pmatrix} n + 1 \\ r + 1 \end{pmatrix}$.]
   (d) Use part (c) to explain why each entry in Pascal's triangle (except the 1's at the beginning or end of a row) is the sum of the two closest entries in the row above it.

48. (a) Find these numbers and write them one *below* the next: 11^0, 11^1, 11^2, 11^3, 11^4.
   (b) Compare the list in part (a) with rows 0 to 4 of Pascal's triangle. What's the explanation?
   (c) What can be said about 11^5 and row 5 of Pascal's triangle?
   (d) Calculate all integer powers of 101 from 101^0 to 101^8, list the results one under the other, and compare the list with rows 0 to 8 of Pascal's triangle. What's the explanation? What happens with 101^9?

12.5 | Mathematical Induction

Mathematical induction is a method of proof that can be used to prove a wide variety of mathematical facts, including the Binomial Theorem, DeMoivre's Theorem, and statements such as the following.

The sum of the first n positive integers is the number $\dfrac{n(n + 1)}{2}$.

$2^n > n$ for every positive integer n.

For each positive integer n, 4 is a factor of $7^n - 3^n$.

All of the preceding statements have a common property. For example, a statement such as

The sum of the first n positive integers is the number $\dfrac{n(n + 1)}{2}$

or, in symbols,

$$1 + 2 + 3 + \cdots + n = \frac{n(n + 1)}{2}$$

is really an infinite sequence of statements, one for each possible value of n.

$$n = 1: \qquad 1 = \frac{1(2)}{2},$$

$$n = 2: \qquad 1 + 2 = \frac{2(3)}{2},$$

$$n = 3: \qquad 1 + 2 + 3 = \frac{3(4)}{2},$$

and so on. Obviously, there isn't time enough to verify every one of the statements on this list, one at a time. But we can find a workable method of proof by examining how each statement on the list is *related* to the *next* statement on the list.

For example, for $n = 50$, the statement is

$$1 + 2 + 3 + \cdots + 50 = \frac{50(51)}{2}.$$

At the moment, we don't know whether or not this statement is true. But just *suppose* that it were true. What could then be said about the next statement, the one for $n = 51$?

$$1 + 2 + 3 + \cdots + 50 + 51 = \frac{51(52)}{2}.$$

Well, *if* it is true that

$$1 + 2 + 3 + \cdots + 50 = \frac{50(51)}{2},$$

then adding 51 to both sides and simplifying the right side would yield these equalities.

$$1 + 2 + 3 + \cdots + 50 + 51 = \frac{50(51)}{2} + 51,$$

$$1 + 2 + 3 + \cdots + 50 + 51 = \frac{50(51)}{2} + \frac{2(51)}{2} = \frac{50(51) + 2(51)}{2},$$

$$1 + 2 + 3 + \cdots + 50 + 51 = \frac{(50 + 2)51}{2},$$

$$1 + 2 + 3 + \cdots + 50 + 51 = \frac{51(52)}{2}.$$

Since this last equality is just the original statement for $n = 51$, we conclude that

If the statement is true for $n = 50$, *then* it is also true for $n = 51$.

We have *not* proved that the statement actually *is* true for $n = 50$, but only that *if* it is, then it is also true for $n = 51$.

We claim that this same conditional relationship holds for any two consecutive values of n. In other words, we claim that for any positive integer k,

① ***If* the statement is true for $n = k$, *then* it is also true for $n = k + 1$.**

The proof of this claim is the same argument used earlier (with k and $k + 1$ in place of 50 and 51): *If* it is true that

$$1 + 2 + 3 + \cdots + k = \frac{k(k + 1)}{2}, \qquad \text{[Original statement for } n = k\text{]}$$

then adding $k + 1$ to both sides and simplifying the right side produces these equalities.

$$1 + 2 + 3 + \cdots + k + (k + 1) = \frac{k(k + 1)}{2} + (k + 1),$$

$$1 + 2 + 3 + \cdots + k + (k + 1)$$
$$= \frac{k(k + 1)}{2} + \frac{2(k + 1)}{2} = \frac{k(k + 1) + 2(k + 1)}{2},$$

$$1 + 2 + 3 + \cdots + k + (k + 1) = \frac{(k + 2)(k + 1)}{2},$$

$$1 + 2 + 3 + \cdots + k + (k + 1) = \frac{(k + 1)[(k + 1) + 1]}{2}.$$

[Original statement for $n = k + 1$]

We have proved that claim ① is valid for each positive integer k. We have *not* proved that the original statement is true for any value of n, but only that *if* it is true for $n = k$, then it is also true for $n = k + 1$. Applying this fact when $k = 1, 2, 3, \ldots$, we see that

② *If* the statement is true for $n = 1$,   *then* it is also true for $n = 1 + 1 = 2$;

If the statement is true for $n = 2$,   *then* it is also true for $n = 2 + 1 = 3$;

If the statement is true for $n = 3$,   *then* it is also true for $n = 3 + 1 = 4$;
$\vdots$

If the statement is true for $n = 50$,   *then* it is also true for $n = 50 + 1 = 51$;

If the statement is true for $n = 51$,   *then* it is also true for $n = 51 + 1 = 52$;
$\vdots$

and so on.

We are finally in a position to *prove* the original statement:

$$1 + 2 + 3 + \cdots + n = n(n + 1)/2.$$

Obviously, it *is true* for $n = 1$, since $1 = 1(2)/2$. Now apply in turn each of the propositions on list ②. Since the statement *is* true for $n = 1$, it must also be true for $n = 2$, and hence for $n = 3$, and hence for $n = 4$, and so on, for every value of n. Therefore, the original statement is true for *every* positive integer n.

The preceding proof is an illustration of the following principle.

Principle of Mathematical Induction

Suppose there is given a statement involving the positive integer n and that

(i) The statement is true for $n = 1$.

(ii) If the statement is true for $n = k$ (where k is any positive integer), then the statement is also true for $n = k + 1$.

Then the statement is true for every positive integer n.

Property (i) is simply a statement of fact. To verify that it holds, you must prove the given statement is true for $n = 1$. This is usually easy, as in the preceding example.

Property (ii) is a *conditional* property. It does not assert that the given statement *is* true for $n = k$, but only that *if* it is true for $n = k$, then it is also true for $n = k + 1$. So to verify that property (ii) holds, you need only prove this conditional proposition.

If the statement is true for $n = k$, *then* it is also true for $n = k + 1$.

To prove this or any conditional proposition, you must proceed as in the previous example: Assume the "if" part and use this assumption to prove the "then" part. As we saw earlier, the same argument will usually work for any possible k. Once this conditional proposition has been proved, you can use it *together* with property (i) to conclude that the given statement is necessarily true for every n, just as in the preceding example.

Thus proof by mathematical induction reduces to two steps.

Step 1   Prove that the given statement is true for $n = 1$.

Step 2   Let k be a positive integer. Assume that the given statement is true for $n = k$. Use this assumption to prove that the statement is true for $n = k + 1$.

Step 2 may be performed before step 1 if you wish. Step 2 is sometimes referred to as the **inductive step.** The assumption that the given statement is true for $n = k$ in this inductive step is called the **induction hypothesis.**

EXAMPLE 1

Prove that $2^n > n$ for every positive integer n.

SOLUTION   Here, the statement involving n is $2^n > n$.

Step 1   When $n = 1$, we have the statement $2^1 > 1$. This is obviously true.

Step 2   Let k be any positive integer. We assume that the statement is true for $n = k$, that is, we assume that $2^k > k$. We shall use this assumption to prove that the statement is true for $n = k + 1$, that is, that $2^{k+1} > k + 1$.

We begin with the induction hypothesis:* $2^k > k$. Multiplying both sides of this inequality by 2 yields

③

$$2 \cdot 2^k > 2k$$

$$2^{k+1} > 2k.$$

---

*This is the point at which you usually must do some work. Remember that what follows is the "finished proof." It does not include all the thought, scratch work, false starts, and so on that were done before this proof was actually found.

Since k is a positive integer, we know that $k \geq 1$. Adding k to each side of the inequality $k \geq 1$, we have

$$k + k \geq k + 1$$

$$2k \geq k + 1.$$

Combining this result with inequality ③, we see that

$$2^{k+1} > 2k \geq k + 1.$$

The first and last terms of this inequality show that $2^{k+1} > k + 1$. Therefore, the statement is true for $n = k + 1$. This argument works for any positive integer k. Thus, we have completed the inductive step. By the Principle of Mathematical Induction, we conclude that $2^n > n$ for every positive integer n. ■

EXAMPLE 2

Simple arithmetic shows that

$$7^2 - 3^2 = 49 - 9 = 40 = 4 \cdot 10$$

and

$$7^3 - 3^3 = 343 - 27 = 316 = 4 \cdot 79.$$

In each case, 4 is a factor. These examples suggest that

For each positive integer n, 4 is a factor of $7^n - 3^n$.

This conjecture can be proved by induction as follows.

Step 1   When $n = 1$, the statement is "4 is a factor of $7^1 - 3^1$." Since $7^1 - 3^1 = 4 = 4 \cdot 1$, the statement is true for $n = 1$.

Step 2   Let k be a positive integer and assume that the statement is true for $n = k$, that is, that 4 is a factor of $7^k - 3^k$. Let us denote the other factor by D, so that the induction hypothesis is: $7^k - 3^k = 4D$. We must use this assumption to prove that the statement is true for $n = k + 1$, that is, that 4 is a factor of $7^{k+1} - 3^{k+1}$. Here is the proof.

$$7^{k+1} - 3^{k+1} = 7^{k+1} - 7 \cdot 3^k + 7 \cdot 3^k - 3^{k+1} \qquad \text{[Since } -7 \cdot 3^k + 7 \cdot 3^k = 0]$$

$$= 7(7^k - 3^k) + (7 - 3)3^k \qquad \text{[Factor]}$$

$$= 7(4D) + (7 - 3)3^k \qquad \text{[Induction hypothesis]}$$

$$= 7(4D) + 4 \cdot 3^k \qquad \text{[7} - 3 = 4]$$

$$= 4(7D + 3^k). \qquad \text{[Factor out 4]}$$

From this last line, we see that 4 is a factor of $7^{k+1} - 3^{k+1}$. Thus, the statement is true for $n = k + 1$, and the inductive step is complete. Therefore, by the Principle of Mathematical Induction, the conjecture is actually true for every positive integer n. ■

Another example of mathematical induction, the proof of the Binomial The-orem, is given at the end of this section.

Sometimes a statement involving the integer n may be false for $n = 1$ and (possibly) other small values of n but true for all values of n beyond a particular number. For instance, the statement $2^n > n^2$ is false for $n = 1, 2, 3,$ and 4. But it is true for $n = 5$ and all larger values of n. A variation on the Principle of Mathematical Induction can be used to prove this fact and similar statements. See Exercise 28 for details.

■■ A COMMON MISTAKE WITH INDUCTION

It is sometimes tempting to omit step 2 of an inductive proof when the given statement can easily be verified for small values of n, especially if a clear pattern seems to be developing. As the next example shows, however, *omitting step 2 may lead to error.*

EXAMPLE 3

An integer (> 1) is said to be *prime* if its only positive integer factors are itself and 1. For instance, 11 is prime, since its only positive integer factors are 11 and 1. But 15 is not prime because it has factors other than 15 and 1 (namely, 3 and 5). For each positive integer n, consider the number

$$f(n) = n^2 - n + 11.$$

You can readily verify that

$$f(1) = 11, \quad f(2) = 13, \quad f(3) = 17, \quad f(4) = 23, \quad f(5) = 31$$

and that *each of these numbers is prime.* Furthermore, there is a clear pattern: The first two numbers (11 and 13) differ by 2; the next two (13 and 17) differ by 4; the next two (17 and 23) differ by 6; and so on. On the basis of this evidence, we might conjecture that

For each positive integer n, the number $f(n) = n^2 - n + 11$ is prime.

We have seen that this conjecture is true for $n = 1, 2, 3, 4, 5$. Unfortunately, however, it is *false* for some values of n. For instance, when $n = 11$,

$$f(11) = 11^2 - 11 + 11 + 11^2 = 121.$$

But 121 is obviously *not* prime, since it has a factor other than 121 and 1, namely, 11. You can verify that the statement is also false for $n = 12$ but true for $n = 13$. ■

In the preceding example, the proposition

If the statement is true for $n = k$, then it is true for $n = k + 1$

is false when $k = 10$ and $k + 1 = 11$. If you were not aware of this and tried to complete step 2 of an inductive proof, you would not have been able to find a valid proof for it. Of course, the fact that you can't find a proof of a proposition doesn't always mean that no proof exists. But when you are unable to complete step 2, you are warned that there is a possibility that the given statement may be false for some values of n. This warning should prevent you from drawing any wrong conclusions.

▪▪ PROOF OF THE BINOMIAL THEOREM

We shall use induction to prove that for every positive integer n,

$$(x + y)^n = x^n + \binom{n}{1}x^{n-1}y$$

$$+ \binom{n}{2}x^{n-2}y^2 + \binom{n}{3}x^{n-3}y^3 + \cdots + \binom{n}{n-1}xy^{n-1} + y^n.$$

This theorem was discussed and its notation explained in Section 12.4.

Step 1 When $n = 1$, there are only two terms on the right side of the preceding equation, and the statement reads $(x + y)^1 = x^1 + y^1$. This is certainly true.

Step 2 Let k be any positive integer and assume that the theorem is true for $n = k$, that is, that

$$(x + y)^k = x^k + \binom{k}{1}x^{k-1}y + \binom{k}{2}x^{k-2}y^2 + \cdots$$

$$+ \binom{k}{r}x^{k-r}y^r + \cdots + \binom{k}{k-1}xy^{k-1} + y^k.$$

[On the right side of this equation, we have included a typical middle term $\binom{k}{r}x^{k-r}y^r$. The sum of the exponents is k, and the bottom part of the binomial coefficient is the same as the y exponent.] We shall use this assumption to prove that the theorem is true for $n = k + 1$, that is, that

$$(x + y)^{k+1} = x^{k+1} + \binom{k+1}{1}x^k y + \binom{k+1}{2}x^{k-1}y^2 + \cdots$$

$$+ \binom{k+1}{r+1}x^{k-r}y^{r+1} + \cdots + \binom{k+1}{k}xy^k + y^{k+1}.$$

We have simplified some of the terms on the right side; for instance, $(k + 1) - 1 = k$ and $(k + 1) - (r + 1) = k - r$. But this is the correct statement for $n = k + 1$: The coefficients of the middle terms are $\binom{k+1}{1}$, $\binom{k+1}{2}$, $\binom{k+1}{3}$, and so on; the sum of the exponents of each middle term is $k + 1$, and the bottom part of each binomial coefficient is the same as the y exponent.

To prove the theorem for $n = k + 1$, we shall need this fact about binomial coefficients: For any integers r and k with $0 \leq r < k$,

④ $$\binom{k}{r+1} + \binom{k}{r} = \binom{k+1}{r+1}.$$

A proof of this fact is outlined in Exercise 47 on page 813.

To prove the theorem for $n = k + 1$, we first note that

$$(x + y)^{k+1} = (x + y)(x + y)^k.$$

Applying the induction hypothesis to $(x + y)^k$, we see that

$$(x + y)^{k+1} = (x + y)\left[x^k + \binom{k}{1}x^{k-1}y + \binom{k}{2}x^{k-2}y^2 + \cdots + \binom{k}{r}x^{k-r}y^r \right.$$
$$\left. + \binom{k}{r+1}x^{k-(r+1)}y^{r+1} + \cdots + \binom{k}{k-1}xy^{k-1} + y^k \right]$$
$$= x\left[x^k + \binom{k}{1}x^{k-1}y + \cdots + y^k \right] + y\left[x^k + \binom{k}{1}x^{k-1}y + \cdots + y^k \right].$$

Next we multiply out the right side. Remember that multiplying by x increases the x exponent by 1 and multiplying by y increases the y exponent by 1.

$$(x + y)^{k+1} = \left[x^{k+1} + \binom{k}{1}x^k y + \binom{k}{2}x^{k-1}y^2 + \cdots + \binom{k}{r}x^{k-r+1}y^r \right.$$
$$\left. + \binom{k}{r+1}x^{k-r}y^{r+1} + \cdots + \binom{k}{k-1}x^2 y^{k-1} + xy^k \right]$$
$$+ \left[x^k y + \binom{k}{1}x^{k-1}y^2 + \binom{k}{2}x^{k-2}y^3 + \cdots + \binom{k}{r}x^{k-r}y^{r+1} \right.$$
$$\left. + \binom{k}{r+1}x^{k-(r+1)}y^{r+2} + \cdots + \binom{k}{k-1}xy^k + y^{k+1} \right]$$
$$= x^{k+1} + \left[\binom{k}{1} + 1 \right]x^k y + \left[\binom{k}{2} + \binom{k}{1} \right]x^{k-1}y^2 + \cdots$$
$$+ \left[\binom{k}{r+1} + \binom{k}{r} \right]x^{k-r}y^{r+1} + \cdots + \left[1 + \binom{k}{k-1} \right]xy^k + y^{k+1}.$$

Now apply statement ④ to each of the coefficients of the middle terms. For instance, with $r = 1$, statement ④ shows that

$$\binom{k}{2} + \binom{k}{1} = \binom{k+1}{2}.$$

Similarly, with $r = 0$,

$$\binom{k}{1} + 1 = \binom{k}{1} + \binom{k}{0} = \binom{k+1}{1},$$

and so on. Then the expression above for $(x + y)^{k+1}$ becomes

$$(x + y)^{k+1} = x^{k+1} + \binom{k+1}{1}x^k y + \binom{k+1}{2}x^{k-1}y^2 + \cdots$$
$$+ \binom{k+1}{r+1}x^{k-r}y^{r+1} + \cdots + \binom{k+1}{k}xy^k + y^{k+1}.$$

Since this last statement says that the theorem is true for $n = k + 1$, the inductive step is complete. By the Principle of Mathematical Induction, the theorem is true for every positive integer n.

EXERCISES 12.5

In Exercises 1–18, use mathematical induction to prove that each of the given statements is true for every positive integer n.

1. $1 + 2 + 2^2 + 2^3 + 2^4 + \cdots + 2^{n-1} = 2^n - 1$

2. $1 + 3 + 3^2 + 3^3 + 3^4 + \cdots + 3^{n-1} = \dfrac{3^n - 1}{2}$

3. $1 + 3 + 5 + 7 + \cdots + (2n - 1) = n^2$

4. $2 + 4 + 6 + 8 + \cdots + 2n = n^2 + n$

5. $1^2 + 2^2 + 3^2 + \cdots + n^2 = \dfrac{n(n + 1)(2n + 1)}{6}$

6. $\dfrac{1}{2} + \dfrac{1}{4} + \dfrac{1}{8} + \cdots + \dfrac{1}{2^n} = 1 - \dfrac{1}{2^n}$

7. $\dfrac{1}{1 \cdot 2} + \dfrac{1}{2 \cdot 3} + \dfrac{1}{3 \cdot 4} + \cdots + \dfrac{1}{n(n + 1)} = \dfrac{n}{n + 1}$

8. $\left(1 + \dfrac{1}{1}\right)\left(1 + \dfrac{1}{2}\right)\left(1 + \dfrac{1}{3}\right) \cdots \left(1 + \dfrac{1}{n}\right) = n + 1$

9. $n + 2 > n$ 10. $2n + 2 > n$

11. $3^n \geq 3n$ 12. $3^n \geq 1 + 2n$

13. $3n > n + 1$ 14. $\left(\dfrac{3}{2}\right)^n > n$

15. 3 is a factor of $2^{2n+1} + 1$

16. 5 is a factor of $2^{4n-2} + 1$

17. 64 is a factor of $3^{2n+2} - 8n - 9$

18. 64 is a factor of $9^n - 8n - 1$

19. Let c and d be fixed real numbers. Prove that
$$c + (c + d) + (c + 2d) + (c + 3d) + \cdots$$
$$+ [c + (n - 1)d] = \frac{n[2c + (n - 1)d]}{2}$$

20. Let r be a fixed real number with $r \neq 1$. Prove that
$$1 + r + r^2 + r^3 + \cdots + r^{n-1} = \frac{r^n - 1}{r - 1}.$$
    [*Remember:* $1 = r^0$; so when $n = 1$, the left side reduces to $r^0 = 1$.]

21. (a) Write *each* of $x^2 - y^2$, $x^3 - y^3$, and $x^4 - y^4$ as a product of $x - y$ and another factor.
    (b) Make a conjecture as to how $x^n - y^n$ can be written as a product of $x - y$ and another factor. Use induction to prove your conjecture.

22. Let $x_1 = \sqrt{2}$; $x_2 = \sqrt{2 + \sqrt{2}}$;
    $x_3 = \sqrt{2 + \sqrt{2 + \sqrt{2}}}$; and so on. Prove that $x_n < 2$ for every positive integer n.

In Exercises 23–27, if the given statement is true, prove it. If it is false, give a counterexample.

23. Every odd positive integer is prime.

24. The number $n^2 + n + 17$ is prime for every positive integer n.

25. $(n + 1)^2 > n^2 + 1$ for every positive integer n.

26. 3 is a factor of the number $n^3 - n + 3$ for every positive integer n.

27. 4 is a factor of the number $n^4 - n + 4$ for every positive integer n.

28. Let q be a *fixed* integer. Suppose a statement involving the integer n has these two properties:
    (i) The statement is true for $n = q$.
    (ii) *If* the statement is true for $n = k$ (where k is any integer with $k \geq q$), then the statement is also true for $n = k + 1$.
    Then we claim that the statement is true for every integer n greater than or equal to q.
    (a) Give an informal explanation that shows why this claim should be valid. Note that when $q = 1$, this claim is precisely the Principle of Mathematical Induction.
    (b) The claim made before part (a) will be called the *Extended Principle of Mathematical Induction.* State the two steps necessary to use this principle to prove that a given statement is true for all $n \geq q$. (See the discussion on page 816.)

In Exercises 29–34, use the Extended Principle of Mathematical Induction (Exercise 28) to prove the given statement.

29. $2n - 4 > n$ for every $n \geq 5$. (Use 5 for q here.)

30. Let r be a fixed real number with $r > 1$. Then $(1 + r)^n > 1 + nr$ for every integer $n \geq 2$. (Use 2 for q here.)

31. $n^2 > n$ for all $n \geq 2$

32. $2^n > n^2$ for all $n \geq 5$

33. $3^n > 2^n + 10n$ for all $n \geq 4$

34. $2n < n!$ for all $n \geq 4$

Thinkers

35. Let n be a positive integer. Suppose that there are three pegs and on one of them, n rings are stacked, each ring being smaller in diameter than the one below it (see the figure). We want to transfer the stack of rings to another peg according to these rules: (i) Only one ring may be moved at a time; (ii) a ring can be moved to any peg, provided that it is never placed on top of a smaller ring; (iii) the final order of the rings on the new peg must be the same as the original order on the first peg.

(a) What is the smallest possible number of moves when $n = 2$? $n = 3$? $n = 4$?

(b) Make a conjecture as to the smallest possible number of moves required for any n. Prove your conjecture by induction.

36. The basic formula for compound interest $T(x) = P(1 + r)^x$ was discussed on page 354. Prove by induction that the formula is valid whenever x is a positive integer. [*Note: P* and r are assumed to be constant.]

37. Use induction to prove DeMoivre's Theorem: For any complex number $z = r(\cos \theta + i \sin \theta)$ and any positive integer n,

$$z^n = r^n[\cos(n\theta) + i \sin(n\theta)].$$

Chapter 12 Review

IMPORTANT CONCEPTS

Section 12.1

Sequence   777
Term   777
Constant sequence   779
Recursively defined sequence   781
Fibonacci sequence   781
Summation notation   783
Summation index   783
Partial sum   784

Section 12.2

Arithmetic sequence   788
Common difference   789
nth term formula   789
Partial sum formulas   791

Section 12.3

Geometric sequence   794
Common ratio   794
nth term formula   795
Partial sum formula   797

Special Topics: 12.3.A

Infinite series   801
Infinite geometric series   802

Section 12.4

n factorial ($n!$)   806
Binomial coefficient   806
Pascal's triangle   807–808
Binomial Theorem   808

Section 12.5

Principle of Mathematical Induction
    815
Inductive step   816
Induction hypothesis   816

IMPORTANT FACTS & FORMULAS

- In an arithmetic sequence $\{a_n\}$ with common difference d,

$$a_n = a_1 + (n-1)d, \qquad \sum_{n=1}^{k} a_n = \frac{k}{2}(a_1 + a_k),$$

$$\sum_{n=1}^{k} a_n = ka_1 + \frac{k(k-1)}{2}d.$$

- In a geometric sequence $\{a_n\}$ with common ratio $r \neq 1$,

$$a_n = r^{n-1}a_1, \qquad \sum_{n=1}^{k} a_n = a_1\left(\frac{1-r^k}{1-r}\right).$$

- $n! = 1 \cdot 2 \cdot 3 \cdots (n-2)(n-1)n$

- $\dbinom{n}{r} = \dfrac{n!}{r!(n-r)!} = {}_nC_r$

- *The Binomial Theorem:*

$$(x+y)^n = x^n + \binom{n}{1}x^{n-1}y + \binom{n}{2}x^{n-2}y^2 + \binom{n}{3}x^{n-3}y^3 + \cdots + \binom{n}{n-1}xy^{n-1} + y^n$$

$$= \sum_{j=0}^{n} \binom{n}{j}x^{n-j}y^j$$

REVIEW QUESTIONS

In Questions 1–4, find the first four terms of the sequence $\{a_n\}$.

1. $a_n = 2n - 5$

2. $a_n = 3^n - 27$

3. $u_n = \left(\dfrac{-1}{n}\right)^2$

4. $a_n = (-1)^{n+1}(n - 1)$

5. The number of people employed at Lucent Technologies over a three-year period can be approximated by the sequence $\{a_n\}$, where n is measured in months (with $n = 1$ corresponding to September 1999) and

$$a_n = -.153n^4 + 18.09n^3 - 668.19n^2 + 4980.5n + 148{,}270.^*$$

   (a) About how many were employed in September 1999? In July 2001? In August 2002?

   (b) Approximately when did employment drop below 60,000?

6. Find the fifth partial sum of the sequence $\{a_n\}$, where $a_1 = -4$ and $a_n = 3a_{n-1} + 2$.

7. Find the fourth partial sum of the sequence $\{a_n\}$, where $a_1 = 1/9$ and $a_n = 3a_{n-1}$.

8. $\displaystyle\sum_{n=0}^{4} 2^n(n + 1) = ?$

9. $\displaystyle\sum_{n=2}^{4} (3n^2 - n + 1) = ?$

10. Average tuition and fees each year in private colleges can be approximated by the sequence given by

$$b_n = 9401.63 \cdot e^{.05467n},$$

where $n = 0$ corresponds to the 1990–1991 school year.[†]

   (a) What were the average tuition and fees in 2000–2001? What are they this year?

   (b) If you started at a private college with average tuition and fees in Fall 1990 and graduated after four years, what was the total that you (or your parents) spent on tuition and fees?

   (c) Do part (b), assuming that you start in Fall 2002.

In Questions 11–14, find a formula for a_n; assume that the sequence is arithmetic.

11. $a_1 = 3$ and the common difference is -6.

12. $a_2 = 4$ and the common difference is 3.

13. $a_1 = -5$ and $a_3 = 7$.

14. $a_3 = 2$ and $a_7 = -1$.

15. Find the 11th partial sum of the arithmetic sequence with $a_1 = 5$ and common difference -2.

16. Find the 12th partial sum of the arithmetic sequence with $a_1 = -3$ and $a_{12} = 16$.

In Questions 17–20, find a formula for a_n; assume that the sequence is geometric.

17. $a_1 = 2$ and the common ratio is 3.

18. $a_1 = 5$ and the common ratio is $-1/2$.

19. $a_2 = 192$ and $a_7 = 6$.

20. $a_3 = 9/2$ and $a_6 = -243/16$.

21. Find the fifth partial sum of the geometric sequence with $a_1 = 1/4$ and common ratio 3.

22. Find the sixth partial sum of the geometric sequence with $a_1 = 5$ and common ratio $1/2$.

23. Find numbers b, c, d such that 4, b, c, d, 23 are the first five terms of an arithmetic sequence.

24. Find numbers c and d such that 8, c, d, 27 are the first four terms of a geometric sequence.

25. The amount spent on movie theater admissions (in billions of dollars) in year n is approximated by a geometric sequence $\{a_n\}$ in which $n = 1$ corresponds to 1994.[‡]

   (a) If \$5.123 billion was spent in 1994 and \$5.5067 billion was spent in 1995, find a formula for a_n.

   (b) How much was spent in 2002?

   (c) What was the total amount spent from 1994 to 2002?

26. Attendance at college football games in year n is approximated by an arithmetic sequence $\{b_n\}$ in which $n = 1$ corresponds to 1995.[§]

   (a) If attendance was 35,410,000 in 1995 and 36,320,000 in 1996, find the attendance in 2003.

   (b) What was the total attendance from 1995 to 2000?

In Questions 27 and 28, find the sums, if they exist.

27. $\displaystyle\sum_{n=1}^{\infty} \dfrac{1}{2^{n-1}}$

28. $\displaystyle\sum_{n=1}^{\infty} \left(\dfrac{-1}{4^n}\right)$

29. Use the Binomial Theorem to show that $(1.02)^{51} > 2.5$.

30. What is the coefficient of u^3v^2 in the expansion of $(u + 5v)^5$?

31. $\dbinom{15}{12} = ?$

32. $\dbinom{18}{3} = ?$

33. Let n be a positive integer. Simplify $\dbinom{n + 1}{n}$.

---

*Based on data from Lucent Technologies.

[†]Based on data from the College Board.

[‡]Based on data from the U.S. Bureau of Economic Analysis.

[§]Based on data from the NCAA.

34. Use the Binomial Theorem to expand $(\sqrt{x} + 1)^5$. Simplify your answer.

35. $\dfrac{20!5!}{6!17!} = ?$

36. $\dfrac{7! - 5!}{4!} = ?$

37. Find the coefficient of x^2y^4 in the expansion of $(2y + x^2)^5$.

38. Prove that for every positive integer n,

$$1^3 + 2^3 + 3^3 + \cdots + n^3 = \frac{n^2(n + 1)^2}{4}.$$

39. Prove that for every positive integer n,

$$1 + 5 + 5^2 + 5^3 + \cdots + 5^{n-1} = \frac{5^n - 1}{4}.$$

40. Prove that $2^n \geq 2n$ for every positive integer n.

41. If x is a real number with $|x| < 1$, then prove that $|x^n| < 1$ for all $n \geq 1$.

42. Prove that for any positive integer n,

$$1 + 5 + 9 + \cdots + (4n - 3) = n(2n - 1).$$

43. Prove that for any positive integer n,

$$1 + 4 + 4^2 + 4^3 + \cdots + 4^{n-1} = \frac{1}{3}(4^n - 1).$$

44. Prove that $3n < n!$ for every $n \geq 4$.

45. Prove that for every positive integer n, 8 is a factor of $9^n - 8n - 1$.

DISCOVERY PROJECT 12   Taking Your Chances

Henryk T. Kaiser/Rex Instock/PictureQuest

Games of chance have been played in many societies throughout history, so it should not be surprising that as mathematics becomes more sophisticated, its principles are applied to the study of games of chance. Many common games can be analyzed by using the techniques of discrete mathematics, particularly the Binomial Theorem. The Binomial Theorem comes into play because of the nature of gaming; each play is either a win or a loss, one of two choices. Consider the simple game of flipping a coin. What does a sequence of two games look like?

The tree diagram indicates four paths through the sequence of two games: head-head, head-tail, tail-head, and tail-tail.

If you think of heads and tails as the variables H and T, then you have three final results: H^2, T^2, and two HT's if you consider $HT = TH$. Indeed, you can think of the last column of the tree as the polynomial of two variables

$$H^2 + 2HT + T^2.$$

The coefficients of the terms are the binomial coefficients you learned about in the Binomial Theorem. You can use the theorem to help you describe many such sequential games.

1. Write a polynomial that describes the results of a game in which you flip a coin eight times in a row.

The variables H and T can also be used to represent the probability that you will get heads or tails, respectively, when you flip a coin. If the coin you are using is a fair one, then the probability of heads or tails is equally likely; that is, you would expect to see heads half the time and tails half the time. These values of $\frac{1}{2}$ are the probabilities that heads or tails will appear on a given flip.

2. What is the probability that you will get exactly four heads and four tails when you flip a coin eight times? Answer this by evaluating the H^4T^4 term of the probability polynomial using the values $H = \frac{1}{2}$ and $T = \frac{1}{2}$.

3. Show that the probability polynomial for flipping eight coins adds to exactly 1.

There is no magic to the values of $\frac{1}{2}$ for winning or losing a game. The only constraint is that the probability values for winning and losing must add to 1.

4. Suppose that you play a game in which the probability that you win is 0.47 and the probability that you lose is 0.53. What is the probability that you win exactly four out of seven games? Evaluate the $W^4 L^3$ term of the probability polynomial $(W + L)^7$.

C H A P T E R 13

Limits and Continuity

David Ryan/Photo 20-20/PictureQuest

Where did that number come from?

Doctors use a formula to determine when a dialysis patient has been adequately dialyzed. Engineers used a function to determine the shape of the Gateway Arch in St. Louis. Economists and scientists estimate the capacity of the earth to support its population. Geologists and physicists compute the amount of energy released by an earthquake. What do these disparate formulas, functions, estimates, and computations have in common? They all involve e, a number that is given by a limit. See Exercise 46 on page 839 and Exercise 51 on page 875.

Chapter Outline

Interdependence of Sections

13.1 ⟶ 13.2 ⟶ 13.3
                ⟶ 13.4

13.1 **Limits of Functions**
13.2 **Properties of Limits**
13.2.A *Special Topics:* **The Formal Definition of Limit**
13.3 **Continuity**
13.4 **Limits Involving Infinity**

In earlier chapters, we dealt with *static* situations, such as the following:
    What is the size of the angle?
    How far did the rock fall?
    How old is the skeleton?
Calculus, by contrast, deals with *dynamic* questions:
    At what rate is the angle changing size?
    How fast is the rock falling at time $t = 2$ seconds?
    What production rate will increase profits the most?
The key to dealing with such dynamic problems is the concept of limit, which is introduced in this chapter.

13.1 Limits of Functions

You have often dealt with questions such as this:

> What is the value of the function $f(x)$ when $x = 4$?

Underlying the idea of "limit," however, is the question:

> How does the function $f(x)$ behave *near* $x = 4$?

EXAMPLE 1

The function

$$f(x) = \frac{.1x^4 - .8x^3 + 1.6x^2 + 2x - 8}{x - 4}$$

is not defined at $x = 4$ (why?). What happens to the values of $f(x)$ when x is *very close* to 4?

SOLUTION   We evaluate f at several numbers that are very close to 4, including numbers that are smaller than 4 and numbers that are larger than 4, that is, numbers that lie on the left of 4 and on the right of 4 on the number line, as shown on the next page.

x gets closer and closer to 4 from the left   *x* gets closer and closer to 4 from the right

x	3.9	3.99	3.999	3.9999	4	4.0001	4.001	4.01	4.1
f(x)	1.8479	1.9841	1.9984	1.9998	*	2.0002	2.0016	2.0161	2.1681

The table suggests the following.

> As *x* gets very close to 4 (from either side), the corresponding values of *f(x)* get very close to 2.

Furthermore, by taking *x* close enough to 4, the corresponding values of *f(x)* can be made *as close as you want* to 2. For instance,

$$f(3.999999) = 1.999998 \quad \text{and} \quad f(4.000001) = 2.000002.$$

This situation is usually expressed by saying that "the *limit* of *f(x)* as *x* approaches 4 is the number 2," which is written symbolically as

$$\lim_{x \to 4} f(x) = 2 \quad \text{or} \quad \lim_{x \to 4} \frac{.1x^4 - .8x^3 + 1.6x^2 + 2x - 8}{x - 4} = 2. \quad ■$$

Example 1 is an illustration of the following definition of "limit." Now *f* is any function, and *c* and *L* are fixed real numbers (in Example 1, *c* = 4 and *L* = 2). The phrase "arbitrarily close" means "as close as you want."

Informal Definition of Limit

Let *f* be a function, and let *c* be a real number such that *f(x)* is defined for all values of *x* near *x* = *c*, except possibly at *x* = *c* itself. Suppose that

> as *x* takes values very close (but not equal) to *c* (on both sides of *c*), the corresponding values of *f(x)* are very close (and possibly equal) to the real number *L*

and that

> the values of *f(x)* can be made arbitrarily close to *L* by taking values of *x* that are sufficiently close (but not equal) to *c*.

Then we say that

> the **limit** of the function *f(x)* as *x* approaches *c* is the number *L*,

which is written

$$\lim_{x \to c} f(x) = L.$$

EXAMPLE 2

Find $\lim_{x \to 2} 3x^2$.

SOLUTION   In this case, $f(x) = 3x^2$. When *x* is a number very close to 2, then x^2 is very close to 2 · 2 = 4. For instance,

$$x = 1.9995 \text{ is very close to 2} \quad \text{and} \quad x^2 = 1.9995^2 = 3.99800025.$$

X	Y2
1.99	11.88
1.999	11.988
1.9999	11.999
2	12
2.0001	12.001
2.001	12.012
2.01	12.12

Y2=3X²

Figure 13–1

Similarly, when x^2 is very close to 4, then $3x^2$ is very close to $3 \cdot 4 = 12$. This reasoning suggests that

$$\lim_{x \to 2} 3x^2 = 12.$$

The table in Figure 13–1 supports this conclusion (values of x close to 2 produce values of $f(x) = 3x^2$ close to 12). Furthermore, the *limit* of f as x approaches 2 is the *value* of f at $x = 2$ because $f(2) = 3(2^2) = 12$. ■

EXAMPLE 3

Find $\lim_{x \to 2} f(x)$, where f is the function given by this two-part rule:

$$f(x) = \begin{cases} 0 & \text{if } x \text{ is an integer,} \\ 1 & \text{if } x \text{ is not an integer.} \end{cases}$$

SOLUTION A calculator is not much help here, but the function is easily graphed by hand, as shown in Figure 13–2.

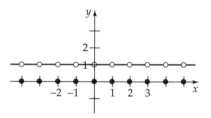

Figure 13–2

When x is a number very close (but not equal) to 2 (on either side of 2), the corresponding value of $f(x)$ is the number 1, and this is true no matter how close x is to 2. Thus,

$$\lim_{x \to 2} f(x) = 1.$$

Since $f(2) = 0$ by definition, we see that the limit of f as x approaches 2 is *not* the same as the value of the function f at $x = 2$. ■

Examples 1–3 illustrate the following facts.

Limits and Functional Values

If the function $f(x)$ has a limit as x approaches c, then there are three possibilities:

1. $f(c)$ is not defined, but $\lim_{x \to c} f(x)$ is defined.   [Example 1]

2. $f(c)$ is defined and $\lim_{x \to c} f(x) = f(c)$.   [Example 2]

3. $f(c)$ is defined, but $\lim_{x \to c} f(x) \neq f(c)$.   [Example 3]

EXAMPLE 4

The graph of $f(x) = \dfrac{\sin x}{x}$ is shown in Figure 13–3. Find $\lim\limits_{x \to 0} f(x)$.

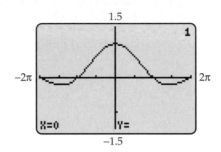

Figure 13–3

SOLUTION   The function is not defined at $x = 0$ (which is indicated by the trace feature in Figure 13–3). To see what's going on very close to $x = 0$, we graph f in the narrow viewing window of Figure 13–4, in which the graph looks like a straight line.

Figure 13–4

GRAPHING EXPLORATION

Graph f in the viewing window of Figure 13–4. Use the trace feature to move closer and closer to the y-axis from both the left and the right, and note the values of $f(x)$ as x gets closer and closer to 0.

The Graphing Exploration strongly suggests that

$$\lim_{x \to 0} f(x) = 1 \qquad \text{or, equivalently,} \qquad \lim_{x \to 0} \frac{\sin x}{x} = 1.$$

This is indeed the case, as will be proved in calculus.   ∎

EXAMPLE 5

Use numerical methods to find

$$\lim_{x \to -3} \frac{x + 3}{e^{x+3} - 1}.$$

Confirm your answer graphically.

SOLUTION   Create a table of values for $f(x) = \dfrac{x + 3}{e^{x+3} - 1}$ when x is very close to -3, as in Figure 13–5 (which shows a TI-86 screen). The table indicates that f is not defined when $x = -3$. Nevertheless, the table suggests that

$$\lim_{x \to -3} \frac{x + 3}{e^{x+3} - 1} = 1.$$

Figure 13–5

We can confirm this conclusion by graphing f in a very narrow window around $x = -3$ and using the trace feature to examine the values of $f(x)$ when x is very close to -3. Figure 13–6 also suggests that the limit is 1. ■

Figure 13–6

◼◼ NONEXISTENCE OF LIMITS

Not every function has a limit at every number. Limits can fail to exist for a variety of reasons, some of which are illustrated in the following examples.

EXAMPLE 6

Figure 13–7 shows the graph of

$$f(x) = \frac{1}{(x + 1)^2}.$$

Find $\lim\limits_{x \to -1} f(x)$, if it exists.

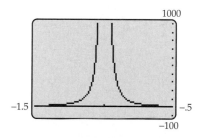

Figure 13–7

SOLUTION

GRAPHING EXPLORATION

Graph $f(x)$ in the viewing window of Figure 13–7. Use the trace feature to show that as x approaches -1 from the left or right, the corresponding values of $f(x)$ become larger and larger, without bound, rather than getting closer and closer to one particular number.

Therefore, the limit of $f(x)$ as x approaches -1 does not exist. ■

EXAMPLE 7

As was explained in Example 5 on page 472, the graph of $f(x) = \sin(\pi/x)$ oscillates infinitely often between -1 and 1 as x approaches 0, as shown in Figure 13–8.

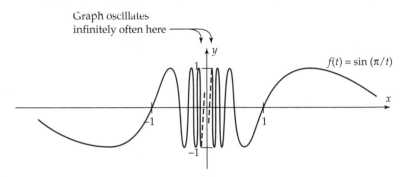

Figure 13–8

As x approaches 0, the function takes every value between -1 and 1 infinitely many times. In particular, $f(x)$ does not get closer and closer to one particular number. Therefore, the limit of $f(x) = \sin(\pi/x)$ as x approaches 0 does not exist.

■

EXAMPLE 8

Find $\lim\limits_{x\to 0} \dfrac{|x|}{x}$ if it exists.

SOLUTION   The function $f(x) = |x|/x$ is not defined when $x = 0$. According to the definition of absolute value, $|x| = x$ when $x > 0$ and $|x| = -x$ when $x < 0$. Therefore,

When $x > 0$, then $f(x) = \dfrac{|x|}{x} = \dfrac{x}{x} = 1$;

When $x < 0$, then $f(x) = \dfrac{|x|}{x} = \dfrac{-x}{x} = -1$.

Consequently, the graph of f looks like Figure 13–9.

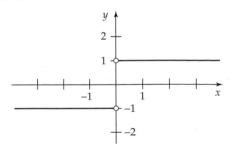

Figure 13–9

If x takes positive values closer and closer to 0, then the corresponding value of $f(x)$ is always 1. If x takes negative values closer and closer to 0, however, the corresponding value of $f(x)$ is always -1. Thus, as x takes values close to 0 on *both* sides of 0, as required by the definition of limit, the corresponding values of $f(x)$ do not get closer and closer to the same real number. Therefore, the limit does not exist.   ■

■■ ONE-SIDED LIMITS

Although the limit does not exist in Example 8, we can adapt the limit concept to describe the behavior of the function $f(x) = \dfrac{|x|}{x}$ near $x = 0$. If you consider only *positive* values of x, then the graph in Figure 13–9 shows that

> as x takes values very close to 0 (with $x > 0$), the corresponding values of $f(x)$ are very close (in fact, equal) to 1.

We express this fact by writing

$$\lim_{x \to 0^+} f(x) = 1$$

and saying

> the limit of $f(x)$ as x approaches 0 from the right is 1.

Similarly, if you consider only *negative* values of x, then

> as x takes values very close to 0 (with $x < 0$), the corresponding values of $f(x)$ are very close (in fact, equal) to -1.

We express this fact by writing

$$\lim_{x \to 0^-} f(x) = -1$$

and saying

> the limit of $f(x)$ as x approaches 0 from the left is -1.

The same idea carries over to the general case, in which f is any function and c and L are real numbers.

One-Sided Limits

Suppose that

> as x takes values very close to c (with $x > c$), the corresponding values of $f(x)$ are very close (and possibly equal) to the real number L

and that

> the values of $f(x)$ can be made arbitrarily close to L by taking values of x (with $x > c$) that are sufficiently close to c.

Then we say that

> the limit of $f(x)$ as x approaches c from the right is L,

which is written

$$\lim_{x \to c^+} f(x) = L.$$

The symbol

$$\lim_{x \to c^-} f(x) = L,$$

which is read

> the limit of $f(x)$ as x approaches c from the left is L,

is defined analogously (just replace "$x > c$" by "$x < c$" in the preceding definition).

EXAMPLE 9

The function $f(x) = \sqrt{x - 3} + 1$ is defined only when $x \geq 3$, that is, for values of x to the right of 3. Find the limit of $f(x)$ as x approaches 3 from the right.

SOLUTION   The graph of f in Figure 13–10 suggests that the limit is 1. You can confirm this by using a very narrow window.

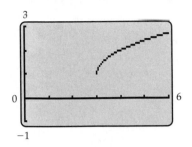

Figure 13–10

GRAPHING EXPLORATION

Graph f in the viewing window with $2.9999 \leq x \leq 3.0001$ and $.98 \leq y \leq 1.02$. Use the trace feature to move along the graph. Verify that as x approaches 3 from the right, the corresponding values of $f(x)$ get closer and closer to 1.

Therefore, we conclude that $\lim\limits_{x \to 3^+} (\sqrt{x - 3} + 1) = 1$.   ■

We now have three kinds of limits: the left-hand limits and right-hand limits defined on the preceding page and the two-sided limits defined on page 830. Example 9 exhibits a function that has a right-hand limit at $x = 3$ but no left-hand or two-sided limit. The function in Example 8 has both a left- and a right-hand limit but no two-sided limit.

When a function has a two-sided limit L at $x = c$, then it automatically has L as both the left- and right-hand limit at $x = c$. Example 5 illustrates this clearly. Figures 13–5 and 13–6 show that as x gets closer and closer to -3 (on *both* sides of -3), the corresponding values of

$$f(x) = \frac{x + 3}{e^{x+3} - 1}$$

get closer and closer to 1. So it is certainly true that when x takes values on the right of -3 (that is, $x > -3$), the corresponding values of $f(x)$ get closer and closer to 1—and similarly, when x takes values on the left of -3. A similar argument applies in the general case.

Two-Sided Limits

Let f be a function and let c and L be real numbers. Then

$$\lim_{x \to c} f(x) = L \quad \text{exactly when} \quad \lim_{x \to c^-} f(x) = L \quad \text{and} \quad \lim_{x \to c^+} f(x) = L.$$

EXAMPLE 10

The graph of the function f in Figure 13–11 shows that $f(2) = 1$ and that

$$\lim_{x \to 2} f(x) = 3, \qquad \lim_{x \to 2^-} f(x) = 3, \qquad \lim_{x \to 2^+} f(x) = 3.$$

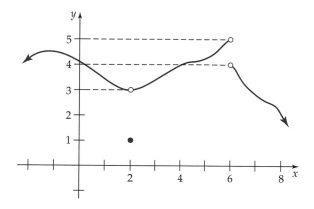

Figure 13–11

CAUTION

Whenever a calculator was used in the preceding examples, we were careful to say that the information it provided *suggested* that the limit was a particular number. Although calculators can provide strong numerical and graphical evidence for the existence of a limit, they do *not* provide a proof of this fact. In fact, calculators can occasionally be very misleading (see Exercise 59).

The graph also shows that

$$\lim_{x \to 6^-} f(x) = 5 \qquad \text{and} \qquad \lim_{x \to 6^+} f(x) = 4,$$

so the two-sided limit, $\lim_{x \to 6} f(x)$, is not defined. ■

EXERCISES 13.1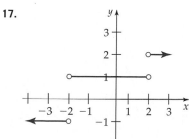

Note: In the following exercises, "find the limit" means "find the exact limit, if possible, and otherwise the best possible approximation."

In Exercises 1–16, use the table feature of your calculator to find the limit.

1. $\displaystyle\lim_{x \to -1} \frac{x^6 - 1}{x^4 - 1}$

2. $\displaystyle\lim_{x \to 2} \frac{x^5 - 32}{x^3 - 8}$

3. $\displaystyle\lim_{x \to 0} \frac{\tan x - x}{x^3}$

4. $\displaystyle\lim_{x \to 0} \frac{\tan x}{x + \sin x}$

5. $\displaystyle\lim_{x \to 0} \frac{x - \tan x}{x - \sin x}$

6. $\displaystyle\lim_{x \to 0} \frac{x + \sin 2x}{x - \sin 2x}$

7. $\displaystyle\lim_{x \to 0} \frac{x}{\ln |x|}$

8. $\displaystyle\lim_{x \to 0} \frac{e^{2x} - 1}{x}$

9. $\displaystyle\lim_{x \to 0} \frac{e^x - 1}{\sin x}$

10. $\displaystyle\lim_{x \to 0} \left(x \sin \frac{1}{x}\right)$

11. $\displaystyle\lim_{x \to \pi^-} \frac{\sin x}{1 - \cos x}$

12. $\displaystyle\lim_{x \to \frac{\pi}{2}^-} (\sec x - \tan x)$

13. $\displaystyle\lim_{x \to 0^+} \sqrt{x}\,(\ln x)$

14. $\displaystyle\lim_{x \to 0^+} \left(\frac{1}{x}\right)^x$

15. $\displaystyle\lim_{x \to 0^-} \frac{\sin(6x)}{x}$

16. $\displaystyle\lim_{x \to 0^+} \frac{\sin(3x)}{1 + \sin(4x)}$

In Exercises 17–22, use the graph of the function f to determine the following limits.

$$\lim_{x \to -3} f(x) \qquad \lim_{x \to 0} f(x), \qquad \lim_{x \to 2} f(x).$$

17.

18.

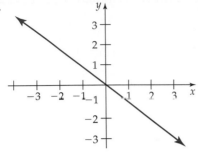

In Exercises 23–26, use the graph of the function f to determine

(a) $\lim\limits_{x \to -2^-} f(x)$ (b) $\lim\limits_{x \to 0^+} f(x)$

(c) $\lim\limits_{x \to 3^-} f(x)$ (d) $\lim\limits_{x \to 3^+} f(x)$

23.

19.

24.

20.

25.

21.

26.

22.

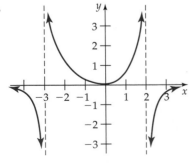

In Exercises 27–45, use numerical or graphical means to find the limit, if it exists. If the limit of f as x approaches c does exist, answer this question: Is it equal to $f(c)$?

27. $\lim\limits_{x \to 3} \dfrac{5x - 4}{2x - 1}$ **28.** $\lim\limits_{x \to -2} \dfrac{2x + 3}{2x - 5}$

29. $\lim\limits_{x \to 2} \dfrac{x^2 - 5x + 6}{x^2 - 6x + 8}$ **30.** $\lim\limits_{x \to -2} \dfrac{x^2 + 3x + 2}{x^2 - x - 6}$

31. $\lim\limits_{x \to 4} \dfrac{x - 6}{x - 4}$

32. $\lim\limits_{x \to -3} \dfrac{(x - 3)(x + 3)(x + 4)}{(x - 4)(x + 3)(x + 8)}$

33. $\lim\limits_{x \to 2^+} \sqrt{x^2 - 4}$

34. $\lim\limits_{x \to 3^-} \sqrt{9 - x^2}$

35. $\lim\limits_{x \to 4} \dfrac{-6}{(x - 4)^2}$

36. $\lim\limits_{x \to -2} \dfrac{2x}{(x + 2)^3}$

37. $\lim\limits_{x \to 0} \dfrac{x}{e^x - 1}$

38. $\lim\limits_{x \to 0} \dfrac{e^{2x} + e^x - 2}{e^x - 1}$

39. $\lim\limits_{x \to 1} \dfrac{\ln x}{x - 1}$

40. $\lim\limits_{x \to 5} \dfrac{\ln x - \ln 5}{x - 5}$

41. $\lim\limits_{x \to 0} (x \ln |x|)$

42. $\lim\limits_{x \to \pi/2} x \sin x$

43. $\lim\limits_{x \to \pi/2} x \cos x$

44. $\lim\limits_{x \to 1} \tan \pi x$

45. $\lim\limits_{x \to 1} e^x \sin \dfrac{\pi x}{2}$

46. (a) Approximate $\lim\limits_{x \to 0} (1 + x)^{1/x}$ to seven decimal places. (Evaluate the function at numbers closer and closer to 0 until successive approximations agree in the first seven decimal places.)

(b) Find the decimal expansion of e to at least nine decimal places.

(c) On the basis of the results of parts (a) and (b), what do you think is the exact value of $\lim\limits_{x \to 0} (1 + x)^{1/x}$?

47. (a) Graph the function f whose rule is

$$f(x) = \begin{cases} 3 - x & \text{if } x < -2, \\ x + 2 & \text{if } -2 \le x < 2, \\ 1 & \text{if } x = 2, \\ 4 - x & \text{if } x > 2. \end{cases}$$

Use the graph in part (a) to evaluate these limits:

(b) $\lim\limits_{x \to -2} f(x)$ (c) $\lim\limits_{x \to 1} f(x)$ (d) $\lim\limits_{x \to 2} f(x)$

48. In Exercise 47, find

(a) $\lim\limits_{x \to -2^-} f(x)$ (b) $\lim\limits_{x \to -2^+} f(x)$

(c) $\lim\limits_{x \to 2^-} f(x)$ (d) $\lim\limits_{x \to 2^+} f(x)$

49. (a) Graph the function g whose rule is

$$g(x) = \begin{cases} x^2 & \text{if } x < -1, \\ x + 2 & \text{if } -1 \le x < 1, \\ 3 - x & \text{if } x \ge 1. \end{cases}$$

Use the graph in part (a) to evaluate these limits:

(b) $\lim\limits_{x \to -1} g(x)$ (c) $\lim\limits_{x \to 0} g(x)$ (d) $\lim\limits_{x \to 1} g(x)$

50. In Exercise 49, find

(a) $\lim\limits_{x \to -1^-} g(x)$ (b) $\lim\limits_{x \to -1^+} g(x)$

(c) $\lim\limits_{x \to 1^-} g(x)$ (d) $\lim\limits_{x \to 1^+} g(x)$

In Exercises 51 and 52, use a unit circle diagram to explain why the given statement is true.

51. $\lim\limits_{t \to \pi/2} \sin t = 1$

52. $\lim\limits_{t \to \pi/2} \cos t = 0$

Exercises 53–58, involve the greatest integer function $f(x) = [x]$, which was defined in Example 5 on pages 127–128. Use your calculator as an aid in analyzing these problems.

53. Let $h(x) = [x] + [-x]$; find $\lim\limits_{x \to 2} h(x)$, if this limit exists.

54. Let $g(x) = x - [-x]$; find $\lim\limits_{x \to 2} g(x)$, if this limit exists.

55. Find $\lim\limits_{x \to 2^+} [x]$ and $\lim\limits_{x \to 2^-} [x]$.

56. Find $\lim\limits_{x \to 3^-} (x - [x])$ and $\lim\limits_{x \to 3^+} (x - [x])$.

57. Let $r(x) = \dfrac{[x] + [-x]}{x}$; find $\lim\limits_{x \to 3} r(x)$, if this limit exists.

58. Let $k(x) = \dfrac{x}{[x] + [-x]}$; find $\lim\limits_{x \to 1} k(x)$, if this limit exists.

59. If $f(x) = \dfrac{1 - \cos(x^6)}{x^{12}}$, then $\lim\limits_{x \to 0} f(x) = 1/2$, as is shown in calculus. A calculator or computer, however, may indicate otherwise. Graph $f(x)$ in a viewing window with

$$-.1 \le x \le .1,$$

and use the trace feature to determine the values of $f(x)$ when x is very close to 0. What does this suggest that the limit is?

60. Consider the function t whose rule is

$$t(x) = \begin{cases} 0 & \text{if } x \text{ is rational}, \\ 1 & \text{if } x \text{ is irrational}. \end{cases}$$

Explain why $\lim\limits_{x \to 4} t(x)$ does not exist.

13.2   Properties of Limits

We now consider a number of facts that greatly simplify the computation of limits.

Properties of Limits

If f and g are functions and c, L, M are numbers such that

$$\lim_{x \to c} f(x) = L \qquad \text{and} \qquad \lim_{x \to c} g(x) = M,$$

then

1. $\lim\limits_{x \to c} (f(x) + g(x)) = L + M$

2. $\lim\limits_{x \to c} (f(x) - g(x)) = L - M$

3. $\lim\limits_{x \to c} (f(x) \cdot g(x)) = L \cdot M$

4. $\lim\limits_{x \to c} \left(\dfrac{f(x)}{g(x)} \right) = \dfrac{L}{M}$ (provided that $M \neq 0$)

5. $\lim\limits_{x \to c} \sqrt{f(x)} = \sqrt{L}$ (provided that $f(x) \geq 0$ for all x near c)

All of these properties remain valid for one-sided limits (that is, when "$x \to c$" is replaced throughout either by "$x \to c^{+}$" or by "$x \to c^{-}$").

Although formal proofs of these properties won't be given here, the central idea is easily understood.[*] Consider, for example, Properties 1 and 3. We are given that

$$\lim_{x \to c} f(x) = L \qquad \text{and} \qquad \lim_{x \to c} g(x) = M.$$

Consequently, as x gets very close to c, the corresponding values of $f(x)$ are very close to L, and the corresponding values of $g(x)$ are very close to M. So it seems plausible that $f(x) + g(x)$ is very close to $L + M$ and $f(x)\,g(x)$ is very close to LM.

These properties are often stated somewhat differently. Since $\lim\limits_{x \to c} f(x) = L$ and $\lim\limits_{x \to c} g(x) = M$, Properties 1–5 can be written as follows.

1. $\lim\limits_{x \to c} [f(x) + g(x)] = \lim\limits_{x \to c} f(x) + \lim\limits_{x \to c} g(x)$

2. $\lim\limits_{x \to c} [f(x) - g(x)] = \lim\limits_{x \to c} f(x) - \lim\limits_{x \to c} g(x)$

3. $\lim\limits_{x \to c} [f(x) \cdot g(x)] = \lim\limits_{x \to c} f(x) \cdot \lim\limits_{x \to c} g(x)$

4. $\lim\limits_{x \to c} \left(\dfrac{f(x)}{g(x)} \right) = \dfrac{\lim\limits_{x \to c} f(x)}{\lim\limits_{x \to c} g(x)}$ (provided that $\lim\limits_{x \to c} g(x) \neq 0$)

5. $\lim\limits_{x \to c} \sqrt{f(x)} = \sqrt{\lim\limits_{x \to c} f(x)}$ (provided that $f(x) \geq 0$ for all x near c)

---

[*]A formal proof requires a rigorous definition of "limit." This rigorous definition and proofs of several of these properties are discussed in *Special Topics* 13.2.A.

▚ LIMITS OF POLYNOMIAL FUNCTIONS

We begin with the most simple polynomial functions, the constant functions, such as

$$f(x) = 5.$$

To compute $\lim_{x \to c} f(x)$, for example, you must ask, "When x is very close to c, what is the value of $f(x)$?" The answer is easy, because

no matter what x is, the value of $f(x)$ is *always* the number 5.

So as x gets closer and closer to c, the value of $f(x)$ is always 5. Hence,

$$\lim_{x \to c} f(x) = 5, \qquad \text{which is usually written} \qquad \lim_{x \to c} 5 = 5.$$

The same thing is true for any constant function.

Constant Limits

> If d is a constant, then $\lim_{x \to c} d = d$.

Now consider the *identity function,* whose rule is $f(x) = x$, and the limit $\lim_{x \to c} f(x)$. When x is very close to c, the corresponding value of $f(x)$ (namely, x itself) obviously is very close to c. So we have this conclusion.

Identity Function Limit

> For every real number c, $\lim_{x \to c} x = c$.

The preceding facts, together with Limit Properties 1–3, now make it easy to find the limit of any polynomial function.

EXAMPLE 1

If $f(x) = x^2 - 2x + 3$, find $\lim_{x \to -4} f(x)$.

SOLUTION By Properties 1 and 2,

$$\lim_{x \to -4} f(x) = \lim_{x \to -4} (x^2 - 2x + 3)$$
$$= \lim_{x \to -4} x^2 - \lim_{x \to -4} 2x + \lim_{x \to -4} 3.$$

Consequently, by Property 3,

$$\lim_{x \to -4} f(x) = (\lim_{x \to -4} x)(\lim_{x \to -4} x) - (\lim_{x \to -4} 2)(\lim_{x \to -4} x) + \lim_{x \to -4} 3.$$

Using the facts about the limits of constant functions and the identity function in the preceding boxes, we see that

$$\lim_{x \to -4} f(x) = (\lim_{x \to -4} x)(\lim_{x \to -4} x) - (\lim_{x \to -4} 2)(\lim_{x \to -4} x) + \lim_{x \to -4} 3$$
$$= (\lim_{x \to -4} x)(\lim_{x \to -4} x) - 2(\lim_{x \to -4} x) + 3$$
$$= (-4)(-4) - 2(-4) + 3 = 27.$$

Note that the limit of $f(x) = x^2 - 2x + 3$ at $x = -4$ is the same as the value of the function at $x = -4$, namely, $f(-4) = 27$. ■

Since any polynomial function consists of sums and products of constants and x, the argument used in Example 1 works for any polynomial function and leads to the following conclusion.

Polynomial Limits

> If $f(x)$ is a polynomial function and c is any real number, then
> $$\lim_{x \to c} f(x) = f(c).$$
> In other words, the limit is the value of the polynomial function at $x = c$.

This result also applies to one-sided limits because the left- and right-hand limits at $x = c$ must be the same as the two-sided limit (see the box on page 836).

EXAMPLE 2

The function $g(x) = \sqrt{9 - x^2}$ is defined only when $-3 \le x \le 3$ (why?). Find $\lim\limits_{x \to 3^-} \sqrt{9 - x^2}$.

SOLUTION   By Property 5,
$$\lim_{x \to 3^-} \sqrt{9 - x^2} = \sqrt{\lim_{x \to 3^-} (9 - x^2)}.$$
The limit under the right-side radical is the limit of the polynomial function $f(x) = 9 - x^2$. According to the preceding box (and the remark after it), the limit can be found by evaluating $f(x)$ at 3. Therefore,
$$\lim_{x \to 3^-} \sqrt{9 - x^2} = \sqrt{\lim_{x \to 3^-} (9 - x^2)} = \sqrt{9 - 3^2} = 0. \quad \blacksquare$$

▪▪ LIMITS OF RATIONAL FUNCTIONS

Property 4 and the fact that polynomial limits can be found by evaluation make it easy to compute the limits of rational functions.

EXAMPLE 3

If
$$f(x) = \frac{x^3 - 3x^2 + 10}{x^2 - 6x + 1},$$

find $\lim\limits_{x \to 2} f(x)$.

SOLUTION   The graph of $f(x)$ near $x = 2$ (Figure 13–12) suggests that the limit is approximately $-.86$. We can determine the limit exactly by noting that $f(x)$ is the quotient of two polynomial functions
$$g(x) = x^3 - 3x^2 + 10 \quad \text{and} \quad h(x) = x^2 - 6x + 1,$$

Figure 13–12

each of whose limits, as x approaches 2, can be found by evaluation of the functions at $x = 2$. Therefore,

$$
\begin{aligned}
\lim_{x \to 2} f(x) &= \lim_{x \to 2} \frac{x^3 - 3x^2 + 10}{x^2 - 6x + 1} \\
&= \frac{\lim_{x \to 2} (x^3 - 3x^2 + 10)}{\lim_{x \to 2} (x^2 - 6x + 1)} \qquad \text{[Property 4]} \\
&= \frac{2^3 - 3 \cdot 2^2 + 10}{2^2 - 6 \cdot 2 + 1} = \frac{6}{-7} = -\frac{6}{7} \qquad \text{[Limits of Polynomial Functions]}
\end{aligned}
$$

Note that the limit of $f(x)$, as x approaches 2, is the number $f(2)$. ■

The procedure in Example 3 works for other rational functions as well.

Rational Limits

> Let $f(x)$ be a rational function, and let c be a real number such that $f(c)$ is defined. Then
>
> $$\lim_{x \to c} f(x) = f(c).$$
>
> In other words, the limit is the value of the function at $x = c$.

This result is also valid for one-sided limits (see the box on page 836). When a rational function is not defined at a number, different techniques must be used to find the limit (if there is one).

EXAMPLE 4

If

$$
f(x) = \frac{x^2 - 2x - 3}{x - 3},
$$

find $\lim_{x \to 3} f(x)$.

SOLUTION Since $f(x)$ is not defined when $x = 3$, we cannot find the limit by evaluation. Nor can we use Property 4 because the limit of the denominator as x approaches 3 is 0. To find the limit, we first factor the numerator.

$$
\frac{x^2 - 2x - 3}{x - 3} = \frac{(x + 1)(x - 3)}{x - 3}.
$$

The factor $x - 3$ on the right side may be canceled *provided that $x - 3 \neq 0$*, that is, provided that $x \neq 3$. In other words,

$$
\frac{x^2 - 2x - 3}{x - 3} = \frac{(x + 1)(x - 3)}{x - 3} = x + 1 \qquad \text{for all } x \neq 3.
$$

The definition of limit, as x approaches 3, involves only the behavior of a function *near* $x = 3$ and not *at* $x = 3$. The preceding equation shows that both $f(x)$

and the function $g(x) = x + 1$ have exactly the same values at all numbers near $x = 3$. Therefore, they must have the same limit as x approaches 3. Therefore,

$$\lim_{x \to 3} \left(\frac{x^2 - 2x - 3}{x - 3} \right) = \lim_{x \to 3} (x + 1) = 3 + 1 = 4. \quad \blacksquare$$

The technique in Example 4 applies in many cases. When two functions have identical behavior, except possibly as $x = c$, they will have the same limit as x approaches c. More precisely,

Limit Theorem

> If f and g are functions that have limits as x approaches c and
>
> $$f(x) = g(x) \qquad \text{for all } x \neq c,$$
>
> then
>
> $$\lim_{x \to c} f(x) = \lim_{x \to c} g(x).$$

EXAMPLE 5

Find $\displaystyle\lim_{x \to 4} \frac{\sqrt{x} - 2}{x - 4}$.

SOLUTION   Property 4 does not apply here since the limit of the denominator is 0 when x approaches 4. However, the denominator can be factored.

$$x - 4 = (\sqrt{x} + 2)(\sqrt{x} - 2).$$

Consequently,

$$\frac{\sqrt{x} - 2}{x - 4} = \frac{\sqrt{x} - 2}{(\sqrt{x} + 2)(\sqrt{x} - 2)} = \frac{1}{\sqrt{x} + 2} \cdot \frac{\sqrt{x} - 2}{\sqrt{x} - 2}.$$

When $x \neq 4$, then $\sqrt{x} - 2 \neq 0$ and $\dfrac{\sqrt{x} - 2}{\sqrt{x} - 2} = 1$, so

$$\frac{\sqrt{x} - 2}{x - 4} = \frac{1}{\sqrt{x} + 2} \cdot \frac{\sqrt{x} - 2}{\sqrt{x} - 2} = \frac{1}{\sqrt{x} + 2} \qquad \text{for all } x \neq 4.$$

By the Limit Theorem,

$$\lim_{x \to 4} \frac{\sqrt{x} - 2}{x - 4} = \lim_{x \to 4} \frac{1}{\sqrt{x} + 2}$$

$$= \frac{\displaystyle\lim_{x \to 4} 1}{\displaystyle\lim_{x \to 4} (\sqrt{x} + 2)} \qquad \text{[Property 4]}$$

$$= \frac{\displaystyle\lim_{x \to 4} 1}{\displaystyle\lim_{x \to 4} \sqrt{x} + \lim_{x \to 4} 2} \qquad \text{[Property 1]}$$

$$= \frac{\displaystyle\lim_{x \to 4} 1}{\sqrt{\displaystyle\lim_{x \to 4} x} + \lim_{x \to 4} 2} \qquad \text{[Property 5]}$$

$$= \frac{1}{\sqrt{4} + 2} = \frac{1}{4}. \qquad \text{[Constant Limits; Polynomial Limit]} \quad \blacksquare$$

⬛ LIMITS OF DIFFERENCE QUOTIENTS

Limits involving the difference quotient of a function play an important role in calculus. Recall from Section 3.2 that the difference quotient of a function f is the quantity

$$\frac{f(x + h) - f(x)}{h}.$$

In limit computations with difference quotients, the variable is the quantity h and x is treated as a constant.

EXAMPLE 6

Suppose $f(x) = x^2$.

(a) Find and simplify the difference quotient of f, when $x = 5$.
(b) Find the limit of this difference quotient as h approaches 0.

SOLUTION

(a) $\dfrac{f(5 + h) - f(5)}{h} = \dfrac{(5 + h)^2 - 5^2}{h} = \dfrac{(25 + 10h + h^2) - 25}{h}$

$$= \frac{10h + h^2}{h}.$$

(b) $\displaystyle\lim_{h\to 0} \frac{f(5 + h) - f(5)}{h} = \lim_{h\to 0} \frac{10h + h^2}{h}$

$$= \lim_{h\to 0} \frac{h(10 + h)}{h} \qquad \text{[Factor numerator]}$$

$$= \lim_{h\to 0} (10 + h) \qquad \text{[Limit Theorem]}$$

$$= 10 + 0 = 10. \qquad \text{[Limit of a polynomial function]} \qquad ■$$

EXAMPLE 7

Find the limit of the difference quotient of $f(x) = \sqrt{x}$ as h approaches 0.

SOLUTION   The difference quotient is

$$\frac{f(x + h) - f(x)}{h} = \frac{\sqrt{x + h} - \sqrt{x}}{h}.$$

Rationalizing the numerator of the last fraction (as was done in Example 8 of Section 5.1) shows that the fraction on the right above is equal to

$$\frac{1}{\sqrt{x + h} + \sqrt{x}}.$$

Therefore,

$$\lim_{h \to 0} \frac{f(x+h) - f(x)}{h} = \lim_{h \to 0} \frac{\sqrt{x+h} - \sqrt{x}}{h}$$

$$= \lim_{h \to 0} \frac{1}{\sqrt{x+h} + \sqrt{x}}$$

$$= \frac{\lim\limits_{h \to 0} 1}{\lim\limits_{h \to 0} \sqrt{x+h} + \lim\limits_{h \to 0} \sqrt{x}} \qquad \text{[Property 4]}$$

$$= \frac{1}{\lim\limits_{h \to 0} \sqrt{x+h} + \sqrt{x}} \qquad \text{[Constant Limits}^*\text{]}$$

$$= \frac{1}{\sqrt{\lim\limits_{h \to 0} (x+h)} + \sqrt{x}} \qquad \text{[Property 5]}$$

$$= \frac{1}{\sqrt{x} + \sqrt{x}} = \frac{1}{2\sqrt{x}}. \qquad \text{[Polynomial Limit}^*\text{]} \qquad \blacksquare$$

Limits such as the one in Example 7 play a central role in calculus and have a special name. The **derivative** of the function f is defined to be the function that is denoted f' and whose rule is

$$f'(x) = \lim_{h \to 0} \frac{f(x+h) - f(x)}{h}.$$

Example 7 shows that the derivative of $f(x) = \sqrt{x}$ is the function f' given by

$$f'(x) = \frac{1}{2\sqrt{x}}.$$

---

*Remember that x is treated as a constant here. Hence, $x + h$ is a polynomial in h with constant term x.

EXERCISES 13.2

In Exercises 1–8, use the following facts about the functions f, g, and h to find the required limit.

$$\lim_{x \to 4} f(x) = 5 \qquad \lim_{x \to 4} g(x) = 0 \qquad \lim_{x \to 4} h(x) = -2$$

1. $\lim\limits_{x \to 4} (f(x) + g(x))$

2. $\lim\limits_{x \to 4} (g(x) - h(x))$

3. $\lim\limits_{x \to 4} \dfrac{f(x)}{g(x)}$

4. $\lim\limits_{x \to 4} \dfrac{g(x)}{h(x)}$

5. $\lim\limits_{x \to 4} f(x)g(x)$

6. $\lim\limits_{x \to 4} h(x)^2$

7. $\lim\limits_{x \to 4} \dfrac{3h(x)}{2f(x) + g(x)}$

8. $\lim\limits_{x \to 4} \dfrac{f(x) - 2g(x)}{4h(x)}$

In Exercises 9–40, find the limit if it exists. If the limit does not exist, explain why.

9. $\lim\limits_{x \to 2} (6x^3 - 2x^2 + 5x - 3)$

10. $\lim\limits_{x \to -1} (x^7 + 2x^5 - x^4 + 3x + 4)$

11. $\lim\limits_{x \to -2} \dfrac{3x - 1}{2x + 3}$

12. $\lim\limits_{x \to 3} \dfrac{x^2 + x + 1}{x^2 - 2x}$

13. $\lim\limits_{x \to 3} \dfrac{x^2 - x - 6}{x^2 - 2x - 3}$

14. $\lim\limits_{x \to 1} \dfrac{x^2 - 1}{x^2 + x - 2}$

15. $\lim\limits_{x \to 1} \dfrac{x^3 - 1}{x^2 - 1}$

16. $\lim\limits_{x \to -2} \dfrac{x^2 + 5x + 6}{x^2 - x - 6}$

17. $\lim\limits_{x \to 4^-} \dfrac{x - 4}{x^2 - 16}$

18. $\lim\limits_{x \to 2^+} \dfrac{|x - 2|}{x - 2}$

19. $\lim\limits_{x \to 3^+} \dfrac{3}{x^2 - 9}$

20. $\lim\limits_{x \to 2^-} \dfrac{x + 1}{x^2 - x - 2}$

21. $\lim\limits_{x \to 1} \sqrt{x^3 + 6x^2 + 2x + 5}$

22. $\lim\limits_{x \to 2} \sqrt{x^2 + x + 3}$

23. $\lim\limits_{x \to 1^+} (\sqrt{x - 1} + 3)$

24. $\lim\limits_{x \to 3^-} \sqrt{-3 - x}$

25. $\lim\limits_{x \to -2.5^+} (\sqrt{5 + 2x} + x)$

26. $\lim\limits_{x \to 3^+} (\sqrt{x - 3} + \sqrt{3x})$

27. $\lim\limits_{x \to 3} \dfrac{\sqrt{x} - \sqrt{3}}{x - 3}$

28. $\lim\limits_{x \to 25} \dfrac{\sqrt{x} - 5}{x - 25}$

29. $\lim\limits_{x \to 0} \left(\dfrac{1/(x + 5) - 1/5}{x} \right)$ [*Hint:* Write the expression in parentheses as a single fraction.]

30. $\lim\limits_{x \to 0} \left(\dfrac{2/(x + 6) - 1/3}{x} \right)$

31. $\lim\limits_{x \to 1} \left[\dfrac{1}{x - 1} - \dfrac{2}{x^2 - 1} \right]$

32. $\lim\limits_{x \to -2} \left[\dfrac{x^2}{x + 2} + \dfrac{2x}{x + 2} \right]$

33. $\lim\limits_{x \to 0} \dfrac{x^2}{|x|}$

34. $\lim\limits_{x \to -2} |x + 2|$

35. $\lim\limits_{x \to -3^+} \left(\dfrac{|x + 3|}{x + 3} + \sqrt{x + 3} + 1 \right)$

36. $\lim\limits_{x \to -4^-} (\sqrt{|x| - 4} + x^2)$

37. $\lim\limits_{x \to 0} \dfrac{\sqrt{2 - x} - \sqrt{2}}{x}$ [*Hint:* Rationalize the numerator; see Section 5.1.]

38. $\lim\limits_{x \to 0} \left[\dfrac{1}{x \sqrt{x + 1}} - \dfrac{1}{x} \right]$

39. $\lim\limits_{x \to 0} \left[\dfrac{|x|}{x} - \dfrac{x}{|x|} \right]$

40. $\lim\limits_{x \to -3} \dfrac{|x + 3|}{x + 3}$

In Exercises 41–44, find
$$\lim_{h \to 0} \frac{f(2 + h) - f(2)}{h}.$$

41. $f(x) = x^2$

42. $f(x) = x^3$

43. $f(x) = x^2 + x + 1$

44. $f(x) = \sqrt{x}$

In Exercises 45 and 46, find $\lim\limits_{h \to 0} \dfrac{f(0 + h) - f(0)}{h}$, *if it exists.*

45. $f(x) = |x|$

46. $f(x) = x|x|$

In Exercises 47–50, find the rule of the derivative of the function f. [See Example 7 and the remarks following it.]

47. $f(x) = 2x + 3$

48. $f(x) = 3x - 5$

49. $f(x) = x^2 + x$

50. $f(x) = x^2 - x + 1$

In Exercises 51–54, find the rule of the derivative of the function f. [The difference quotients of these functions were found and simplified in Exercises 49–52 of Section 5.1.]

51. $f(x) = \sqrt{x + 1}$

52. $f(x) = 2\sqrt{x + 3}$

53. $f(x) = \sqrt{x^2 + 1}$

54. $f(x) = \sqrt{x^2 - x}$

55. Give an example of functions f and g and a number c such that neither $\lim\limits_{x \to c} f(x)$ nor $\lim\limits_{x \to c} g(x)$ exists, but $\lim\limits_{x \to c} (f(x) + g(x))$ does exist.

56. Give an example of functions f and g and a number c such that neither $\lim\limits_{x \to c} f(x)$ nor $\lim\limits_{x \to c} g(x)$ exists, but $\lim\limits_{x \to c} f(x)g(x)$ does exist.

13.2.A   *SPECIAL TOPICS* The Formal Definition of Limit

The informal definition of limit in Section 13.1 is adequate for understanding the properties of limits in Section 13.2 and using them to calculate the limits of many familiar functions. Indeed, this definition (or one very much like it) was used for more than a century and played a crucial role in the development of calculus. To avoid error, however, mathematical concepts such as limit must be based on precise definitions and theorems. In this section, we take the first step in building this rigorous foundation by developing the formal definition of limit.

To keep the discussion as concrete as possible, suppose we have a function f such that $\lim\limits_{x \to 5} f(x) = 12$. You don't need to know the rule of f or anything else about it, except the informal definition of limit, to understand the following discussion.

The informal definition of the statement "$\lim\limits_{x \to 5} f(x) = 12$" that was given in Section 13.1 has two components:

A. As x takes values very close (but not equal) to 5, the corresponding values of $f(x)$ are very close (and possibly equal) to 12.

B. The value of $f(x)$ can be made arbitrarily close (as close as you want) to 12 by taking x sufficiently close (but not equal) to 5.

Strictly speaking, only component B is necessary, since it implies component A: If the values of $f(x)$ can be made *arbitrarily close* to 12 by taking x close enough to 5, then it is certainly true that when x is *very* close to 5, $f(x)$ must be very close to 12.* Consequently, we begin with component B of the informal definition.

① | $\lim\limits_{x \to 5} f(x) = 12$ means that the values of $f(x)$ can be made as close as you want to 12 by taking x sufficiently close to 5.

We shall modify this definition step by step until we reach the desired formal definition.

Definition ① says, in effect, that a two-step process is involved: Whenever you tell us how close you want $f(x)$ to 12, we can tell you how close x must be to 5 to guarantee this. So it can be restated like this.

② | $\lim\limits_{x \to 5} f(x) = 12$ means that whenever you specify how close $f(x)$ should be to 12, we can tell you how close x must be to 5 to guarantee this.

For example, if you want $f(x)$ to be within .01 of 12 (that is, between 11.99 and 12.01), we must tell you how close x must be to 5 to guarantee that the corresponding values of $f(x)$ are between 11.99 and 12.01. But "arbitrarily close" implies much more. We must be able to answer your demands, regardless of how close you want $f(x)$ to be to 12. If you want $f(x)$ to be within .002 of 12, or within .0001 of 12, or within any distance of 12, we must be able to tell you how close x must be to 5 in each case to accomplish this. So Definition ② can be restated as follows.

③ | $\lim\limits_{x \to 5} f(x) = 12$ means that no matter what positive number you specify (measuring how close you want $f(x)$ to 12), we can tell you how close x must be to 5 to guarantee that $f(x)$ is that close to 12.

---

*Component A was included in the informal definition because it describes the procedures that are actually used to estimate a limit. In this case, we would compute values of $f(x)$ when $x = 4.9, 4.99, 4.999, 5.1, 5.01, 5.001$, etc. to see whether they are very close to 12.

Hereafter, whatever small positive number you specify (measuring how close $f(x)$ should be to 12) will be denoted by the Greek letter ϵ (epsilon).[*] When we tell you how close x should be to 5 to accomplish this, we must give some small positive number that measures "how close" (for instance, x must be within .0003 of 5, or x must be within .00002 of 5). We denote the number that measures the closeness of x to 5 by the Greek letter δ (delta).[†] Presumably the number δ, which measures how close x must be to 5, will depend on the number ϵ, which measures how close you want $f(x)$ to be to 12. In this language, Definition ③ becomes the following.

④

> $\lim_{x \to 5} f(x) = 12$ means that for every positive number ϵ, there is a positive number δ (depending on ϵ) with this property:
>
> If x is within δ of 5 (but not equal to 5), then $f(x)$ is within ϵ of 12 (and possibly equal to 12).

Although Definition ④ is essentially the formal definition, somewhat briefer notation is usually used, as we now explain.

If we think of $f(x)$ and 12 as numbers on the number line, then the statement

$$f(x) \text{ is within } \epsilon \text{ of } 12$$

means that

$$\text{the distance from } f(x) \text{ to } 12 \text{ is less than } \epsilon.$$

Since distance on the number line is measured by absolute value (see Section 1.1), this last statement means

$$|f(x) - 12| < \epsilon.$$

Similarly, saying that x is within δ of 5 and not equal to 5 means that the distance from x to 5 is less than δ but greater than 0, that is,

$$0 < |x - 5| < \delta.$$

In this notation, Definition ④ becomes the desired formal definition.

⑤

> $\lim_{x \to 5} f(x) = 12$ means that for each positive number ϵ, there is a positive number δ (depending on ϵ) with this property:
>
> If $\quad 0 < |x - 5| < \delta, \quad$ then $\quad |f(x) - 12| < \epsilon.$

---

[*]Mathematicians have used Greek letters in this context since the formal definition was first developed in the nineteenth century. Epsilon is the Greek letter e, the first letter in the word "error." You can think of ϵ as measuring the allowable degree of error—the amount by which you will permit $f(x)$ to differ from 12.

[†]Delta is the Greek letter d, the first letter of "difference." We must state the difference between x and 5 that will guarantee that $f(x)$ is within ϵ of 12.

Definition ⑤ is sufficiently rigorous because the imprecise terms such as "arbitrarily close" and "sufficiently close" in the informal definition have been replaced by a precise statement about inequalities that can be verified in specific cases, as will be shown in the examples below. There is nothing special about 5 and 12 in the preceding discussion; the entire analysis applies equally well in the general case and leads to this formal definition of limit (which is just Definition ⑤ with c in place of 5, L in place of 12, and f any function).

***Definition
of Limit***

Let f be a function and let c be a real number such that $f(x)$ is defined for all x (except possibly $x = c$) in some open interval containing c.* We say "the limit of $f(x)$, as x approaches c, is L" and write

$$\lim_{x \to c} f(x) = L,$$

provided that for each positive number ϵ, there is a positive number δ (depending on ϵ) with this property:

$$\text{If} \quad 0 < |x - c| < \delta, \quad \text{then} \quad |f(x) - L| < \epsilon.$$

EXAMPLE 1

Let $f(x) = 4x - 8$, and prove that $\lim_{x \to 5} f(x) = 12$.

SOLUTION   We must apply the definition of limit with $c = 5$, $L = 12$, and $f(x) = 4x - 8$. Suppose that ϵ is any positive number. We must find a positive number δ with this property.

$$\text{If} \quad 0 < |x - 5| < \delta, \quad \text{then} \quad |f(x) - 12| < \epsilon.$$

Let δ be the number $\epsilon/4$; we claim that this δ will work. For now, don't worry about how we figured out that $\delta = \epsilon/4$ will work; just verify that the following argument is valid: If $|x - 5| < \delta$, then

$$|x - 5| < \epsilon/4 \qquad \text{[Since } \delta = \epsilon/4 \text{]}$$

$$4|x - 5| < \epsilon \qquad \text{[Both sides multiplied by 4]}$$

$$|4||x - 5| < \epsilon \qquad \text{[Since } 4 = |4|\text{]}$$

$$|4(x - 5)| < \epsilon. \qquad \text{[Property of absolute value (page 9)]}$$

$$|4x - 20| < \epsilon$$

$$|(4x - 8) - 12| < \epsilon \qquad \text{[Rewrite } -20 \text{ as } -8 - 12\text{]}$$

$$|f(x) - 12| < \epsilon. \qquad \text{[Because } f(x) = 4x - 8\text{]}$$

We have verified that for each positive ϵ, there is a positive δ (namely, $\epsilon/4$) with this property:

$$\text{If} \quad 0 < |x - 5| < \delta, \quad \text{then} \quad |f(x) - 12| < \epsilon.$$

This completes the proof that $\lim_{x \to 5} f(x) = 12$. ■

---

*This means that there are numbers a and b with $a < c < b$ such that $f(x)$ is defined for all x with $a < x < b$, except possibly $x = c$.

Proofs like the one in Example 1 often seem mysterious to beginners. Although they can follow the argument after the appropriate δ has been announced, they don't see how anyone found that particular δ in the first place. So here's an example that gives a somewhat fuller picture of the mental processes that are used in proving statements about limits.

EXAMPLE 2

Prove that $\lim\limits_{x \to 1} (2x + 7) = 9$.

SOLUTION   In this case, $f(x) = 2x + 7$, $c = 1$, and $L = 9$. Let ϵ be any positive number. We must find a δ with this property.

$$\text{If} \quad 0 < |x - 1| < \delta, \quad \text{then} \quad |(2x + 7) - 9| < \epsilon.$$

To get some idea which δ might have this property, we *work backwards*. The conclusion we want to reach, namely,

$$|(2x + 7) - 9| < \epsilon,$$

is equivalent to

$$|2x - 2| < \epsilon,$$

which in turn is equivalent to each of these statements.

$$|2(x - 1)| < \epsilon$$
$$|2||x - 1| < \epsilon$$
$$2|x - 1| < \epsilon$$
$$|x - 1| < \epsilon/2.$$

When the conclusion is written this way, it suggests that number $\epsilon/2$ would be a good choice for δ.

Everything up to here has been "scratch work" (the similar scratch work was not included in Example 1). Now we must give the actual proof.

Given a positive number ϵ, let δ be the number $\epsilon/2$. If $0 < |x - 1| < \delta$, then

$$|x - 1| < \epsilon/2 \qquad \text{[Since } \delta = \epsilon/2\text{]}$$
$$2|x - 1| < \epsilon \qquad \text{[Both sides multiplied by 2]}$$
$$|2||x - 1| < \epsilon \qquad \text{[Since } 2 = |2|\text{]}$$
$$|2(x - 1)| < \epsilon \qquad \text{[Property of absolute value]}$$
$$|2x - 2| < \epsilon$$
$$|(2x + 7) - 9| < \epsilon \qquad \text{[Rewrite } -2 \text{ as } 7 - 9\text{]}$$
$$|f(x) - 9| < \epsilon. \qquad \text{[Since } f(x) = 2x + 7\text{]}$$

Therefore, $\delta = \epsilon/2$ has the required property, and the proof is complete. ∎

Once we did some algebraic scratch work in Examples 1 and 2, the limit proofs were relatively easy. In most cases, however, a more involved argument is required. In fact, it can be quite difficult to prove directly from the definition, for example, that

$$\lim_{x \to 3} \frac{x^2 - 4x + 1}{x^3 - 2x^2 - x} = -\frac{1}{3}.$$

Fortunately, such complicated calculations can often be avoided by using the various limit properties given in Section 13.2. Of course, these properties must first be proved by using the definition. Surprisingly, these proofs are comparatively easy.

EXAMPLE 3

Let f and g be functions such that

$$\lim_{x \to c} f(x) = L \quad \text{and} \quad \lim_{x \to c} g(x) = M.$$

Prove that

$$\lim_{x \to c} (f(x) + g(x)) = L + M.$$

SOLUTION   *Scratch Work*   If ϵ is any positive number, we must find a positive δ with this property:

If $0 < |x - c| < \delta$,   then $|(f(x) + g(x)) - (L + M)| < \epsilon$.

We first note that by the triangle inequality (Section 1.1),

$$|(f(x) + g(x)) - (L + M)| = |(f(x) - L) + (g(x) - M)|$$
$$\leq |f(x) - L| + |g(x) - M|.$$

If we can find a δ with this property:

If $0 < |x - c| < \delta$,   then $|f(x) - L| + |g(x) - M| < \epsilon$,

then the smaller quantity $|(f(x) + g(x)) - (L + M)|$ will also be less than ϵ when $0 < |x - c| < \delta$. We can find such a δ as follows.

Proof   Let ϵ be any positive number. Apply the definition of $\lim_{x \to c} f(x) = L$ with $\epsilon/2$ in place of ϵ. There is a positive number δ_1 with the following property.

If $0 < |x - c| < \delta_1$,   then $|f(x) - L| < \epsilon/2$.

Similarly, by the definition of $\lim_{x \to c} g(x) = M$ with $\epsilon/2$ in place of ϵ, there is a positive number δ_2 with the following property.

If $0 < |x - c| < \delta_2$,   then $|g(x) - M| < \epsilon/2$.

Now let δ be the smaller of the two numbers δ_1 and δ_2, so that $\delta \leq \delta_1$ and $\delta \leq \delta_2$. Then if $0 < |x - c| < \delta$, we must have

$$0 < |x - c| < \delta_1 \quad \text{and} \quad 0 < |x - c| < \delta_2,$$

and therefore,

$$|f(x) - L| < \epsilon/2 \quad \text{and} \quad |g(x) - M| < \epsilon/2.$$

Consequently, if $0 < |x - c| < \delta$, then

$$|(f(x) + g(x)) - (L + M)| = |(f(x) - L) + (g(x)) - M)|$$
$$\leq |f(x) - L| + |g(x) - M|$$
$$< \frac{\epsilon}{2} + \frac{\epsilon}{2} = \epsilon.$$

Thus, we have shown that for any $\epsilon > 0$, there is a $\delta > 0$ with the following property.

If $0 < |x - c| < \delta$,   then $|(f(x) + g(x)) - (L + M)| < \epsilon.$

Therefore, $\lim\limits_{x \to c} (f(x) + g(x)) = L + M.$ ■

The proofs of the other limit properties and theorems of Section 13.2 are dealt with in Exercise 15 and in calculus.

Finally, the formal definition of limit may easily be carried over to one-sided limits by using this fact (see Special Topics 4.6.A).

$$|x - c| < \delta \quad \text{exactly when} \quad -\delta < x - c < \delta,$$

which is equivalent to

$$|x - c| < \delta \quad \text{exactly when} \quad c - \delta < x < c + \delta.$$

Thus, the numbers between c and $c + \delta$ lie to the right of c, within distance δ of c, and the numbers between $c - \delta$ and c lie to the left of c, within distance δ of c.

Consequently, the formal definition of right-hand limits can be obtained from the definition above by replacing the phrase "if $0 < |x - c| < \delta$" by "$c < x < c + \delta$." For a formal definition of left-hand limits, replace the phrase "if $0 < |x - c| < \delta$" by "$c - \delta < x < c$."

✓ EXERCISES 13.2.A

In Exercises 1–12, use the formal definition of limit to prove the statement, as in Examples 1 and 2.

1. $\lim\limits_{x \to 3} (3x - 2) = 7$

2. $\lim\limits_{x \to 1} (4x + 2) = 6$

3. $\lim\limits_{x \to 5} x = 5$

4. $\lim\limits_{x \to 0} (x + 2) = 2$

5. $\lim\limits_{x \to 2} (6x + 3) = 15$

6. $\lim\limits_{x \to 7} (-2x + 19) = 5$

7. $\lim\limits_{x \to 1} 4 = 4$

8. $\lim\limits_{x \to 2} \pi = \pi$

9. $\lim\limits_{x \to 4} (x - 6) = -2$

10. $\lim\limits_{x \to 1} (2x - 7) = -5$

11. $\lim\limits_{x \to -2} (2x + 5) = 1$

12. $\lim\limits_{x \to -3} (2x - 4) = -10$

In Exercises 13 and 14, use the formal definition of limit to prove the statement.

13. $\lim\limits_{x \to 0} x^2 = 0$

14. $\lim\limits_{x \to 0} x^3 = 0$

In Exercises 15 and 16, let f and g be functions such that

$$\lim\limits_{x \to c} f(x) = L \quad \text{and} \quad \lim\limits_{x \to c} g(x) = M.$$

15. Prove that $\lim\limits_{x \to c} (f(x) - g(x)) = L - M.$

16. If k is a constant, prove that $\lim\limits_{x \to c} kf(x) = kL.$

13.3   Continuity

Let c be a real number in the domain of a function f. Intuitively speaking, the function f is **continuous** at $x = c$ if you can draw the graph of f *at and near* the point $(c, f(c))$ without lifting your pencil from the paper. For example, each of the four graphs in Figure 13–13 is the graph of a function that is continuous at $x = c$.

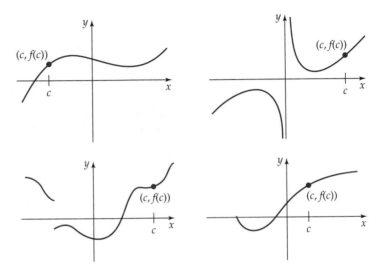

Figure 13–13

On the other hand, none of the functions whose graphs are shown in Figure 13–14 is continuous at $x = c$. If you don't believe it, just try to draw one of these graphs near $x = c$ without lifting your pencil from the paper.

Figure 13–14

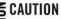
When a calculator graphs in "connected" mode, it plots a number of points and then connects them with line segments to produce a curve. Thus, a calculator *assumes* that the function is continuous at every point it plots, which sometimes results in faulty or misleading graphs. For example, a calculator may not show the hole in graph (d) of Figure 13–14, or it may insert a vertical line segment in graph (b) at the point where the graph jumps.

Our goal is to find an analytical description of continuity at a point that does not depend on having the graph given in advance. So let us consider the two possibilities when a function is continuous at $x = c$, as shown in Figure 13–15.

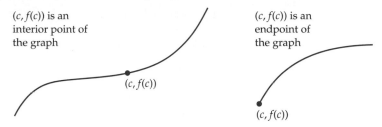

$(c, f(c))$ is an interior point of the graph

$(c, f(c))$ is an endpoint of the graph

$(c, f(c))$

$(c, f(c))$

Figure 13–15

We begin with the case of an interior point.[*]

If the graph can be drawn around the point $(c, f(c))$ without lifting pencil from paper, then at the very least, $f(x)$ must be defined for $x = c$ and for $x = t$ when t is any number near c. We can describe this more precisely by saying the following.

①

> $f(x)$ is defined for all x in some open interval containing c. In other words, there are numbers a and b with $a < c < b$ such that $f(x)$ is defined for all x with $a < x < b$. In particular, $f(c)$ is defined.

Although condition ① is necessary for f to be continuous at $x = c$, this condition by itself does not *guarantee* continuity. For instance, the functions whose graphs are shown in Figure 13–14(a) and 13–14(b) on the facing page, are defined for all values of x near c but are *not* continuous at $x = c$.

Consider Figure 13–16, which shows a typical function f that is continuous at $x = c$. The point $(x, f(x))$ represents the pencil point as it moves along the graph near the point $(c, f(c))$.

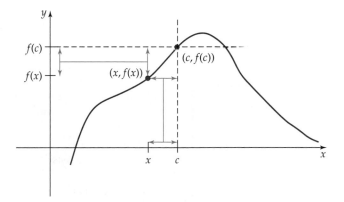

Figure 13–16

---

[*]The endpoint case is considered on page 858.

Note that

②
> As the pencil point $(x, f(x))$ gets closer to the vertical line through $x = c$ (from the left or the right), it also gets closer and closer to the horizontal line through $y = f(c)$.

Furthermore,

③
> The pencil point $(x, f(x))$ can be brought as close as you want to the horizontal line by moving it close enough to the vertical line.

As shown in Figure 13–16, the distance from the point $(x, f(x))$ to the dashed vertical line is the same as the distance from x to c on the x-axis. Similarly, the distance from the point $(x, f(x))$ to the dashed horizontal line is the same as the distance from $f(x)$ to $f(c)$ on the y-axis. Consequently, statements ② and ③ mean that

> as x takes values closer and closer to c (on both sides of c),
> the corresponding values of $f(x)$ get closer and closer to $f(c)$

and that

> the values of $f(x)$ can be made as close as you want
> to $f(c)$ by taking x close enough to c.

This last statement, together with statement ① above, looks suspiciously like the definition of limit (with $f(c)$ in place of L) and suggests that the idea of "not lifting the pencil" can be expressed mathematically by saying

④
$$\lim_{x \to c} f(x) = f(c).$$

The preceding analysis leads to the following.

Definition of Continuity

> Let f be a function that is defined for all x in some open interval containing c. Then f is said to be **continuous** at $x = c$ provided that
>
> (i) $f(c)$ is defined;
>
> (ii) $\lim_{x \to c} f(x)$ exists; and
>
> (iii) $\lim_{x \to c} f(x) = f(c)$.

Strictly speaking, it isn't necessary to list all three of the conditions in the definition, since condition (iii) implies that the other two conditions hold.

EXAMPLE 1

Without graphing, show that the function

$$f(x) = \frac{\sqrt{x^2 - x + 1}}{x - 5}$$

is continuous at $x = -3$.

SOLUTION

$$f(-3) = \frac{\sqrt{(-3)^2 - (-3) + 1}}{(-3) - 5} = -\frac{\sqrt{13}}{8},$$

and by the properties of limits (page 840),

$$\lim_{x \to -3} f(x) = \lim_{x \to -3} \frac{\sqrt{x^2 - x + 1}}{x - 5} = \frac{\lim\limits_{x \to -3} \sqrt{x^2 - x + 1}}{\lim\limits_{x \to -3} (x - 5)}$$

$$= \frac{\sqrt{\lim\limits_{x \to -3} (x^2 - x + 1)}}{\lim\limits_{x \to -3} (x - 5)} = \frac{\sqrt{(-3)^2 - (-3) + 1}}{(-3) - 5}$$

$$= -\frac{\sqrt{13}}{8}.$$

Therefore, $\lim\limits_{x \to -3} f(x) = f(-3)$, and hence, f is continuous at $x = -3$. ■

The properties of limits (Section 13.2) and the definition of continuity provide justification for assumptions made about various graphs in earlier chapters. For example, in Section 13.2, we saw that for any polynomial or rational function f and any number c such that $f(c)$ is defined,

$$\lim_{x \to c} f(x) = f(c).$$

Thus, every polynomial and every rational function is continuous wherever it is defined. That proves the first two of the following statements.

Continuous Functions

Each of the following functions is continuous at every number in its domain.

1. Polynomial functions

2. Rational functions

3. The exponential functions

$$f(x) = 10^x, \qquad g(x) = e^x, \qquad h(x) = b^x$$

4. The logarithmic functions

$$f(x) = \log x, \qquad g(x) = \ln x, \qquad h(x) = \log_b x$$

5. The trigonometric functions

$$f(x) = \sin x, \qquad g(x) = \cos x, \qquad h(x) = \tan x,$$

$$k(x) = \csc x, \qquad r(x) = \sec x, \qquad s(x) = \cot x$$

6. The inverse trigonometric functions

$$f(x) = \sin^{-1}x, \qquad g(x) = \cos^{-1}x, \qquad h(x) = \tan^{-1}x$$

The facts in the preceding box justify the graphing techniques used in this book. Plotting points (a few when graphing by hand, many when graphing with technology) and connecting them with an unbroken curve is a reasonable procedure when graphing continuous functions.

▪▪ CONTINUITY ON AN INTERVAL

Now let us consider continuity at an endpoint of the graph of a function f, such as $(a, f(a))$ or $(b, f(b))$ in Figure 13–17.

$(a, f(a))$ $(b, f(b))$

Figure 13–17

The intuitive idea of continuity at $(a, f(a))$ is that we can draw the graph of f at, and to the right of, the point $(a, f(a))$ without lifting the pencil from the paper. Essentially the same analysis that was given above can be made here, provided that we consider only values of x to the right of $x = a$. It shows that the intuitive idea of continuity at $x = a$ is equivalent to a statement about right-hand limits, namely, $\lim_{x \to a^+} f(x) = f(a)$. An analogous discussion (with left-hand limits) applies to the endpoint $(b, f(b))$ and leads to this definition.

Continuity from the Left and Right

> A function f is **continuous from the right** at $x = a$ provided that
> $$\lim_{x \to a^+} f(x) = f(a).$$
> A function f is **continuous from the left** at $x = b$ provided that
> $$\lim_{x \to b^-} f(x) = f(b).$$

EXAMPLE 2

The function $f(x) = \sqrt{x}$, which is not defined when $x < 0$, is continuous from the right at $x = 0$ because $f(0) = \sqrt{0} = 0$ and
$$\lim_{x \to 0^+} f(x) = \lim_{x \to 0^+} \sqrt{x} = 0 = f(0). \qquad ■$$

The most useful functions are those that are continuous at every point in an interval. Intuitively, this means that their graphs can be drawn over the entire interval without lifting the pencil from the paper. As we saw above, most of the functions in this book are of this type. Here is a formal definition.

*Continuity on
an Interval*

A function *f* is said to be **continuous on an open interval** (*a*, *b*) provided that *f* is continuous at every number in the interval.

A function *f* is said to be **continuous on a closed interval** [*a*, *b*] provided that *f* is continuous from the right at *x* = *a*, continuous from the left at *x* = *b*, and continuous at every number in the open interval (*a*, *b*).

Analogous definitions may be given for continuity on intervals of the form [*a*, *b*), (*a*, *b*], (*a*, ∞), [*a*, ∞), (−∞, *b*), and (−∞, *b*].

EXAMPLE 3

The function *f* whose graph is shown in Figure 13–18 is discontinuous at $x = -3$ and $x = 2$ but is continuous on each of the intervals $(-5, -3)$, $[-3, 2]$, and $(2, \infty)$. ■

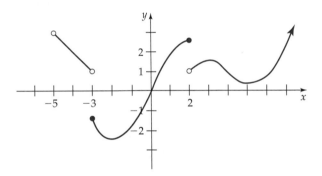

Figure 13–18

▪▪ PROPERTIES OF CONTINUOUS FUNCTIONS

Using the definition is not always the most convenient way to show that a particular function is continuous. It is often easier to establish continuity by using the following facts.

*Properties of
Continuous Functions*

If the functions *f* and *g* are continuous at $x = c$, then each of the following functions is also continuous at $x = c$.

1. The sum function $f + g$

2. The difference function $f - g$

3. The product function fg

4. The quotient function f/g (provided that $g(c) \neq 0$)

Proof   **1.** By the definition of the sum function, $(f + g)(x) = f(x) + g(x)$. Since f and g are continuous at $x = c$, we know that

$$\lim_{x \to c} f(x) = f(c) \qquad \text{and} \qquad \lim_{x \to c} g(x) = g(c).$$

Therefore, by the first property of limits (see page 840),

$$\lim_{x \to c} (f + g)(x) = \lim_{x \to c} (f(x) + g(x)) = \lim_{x \to c} f(x) + \lim_{x \to c} g(x)$$

$$= f(c) + g(c)$$

$$= (f + g)(c).$$

This says that $f + g$ is continuous at $x = c$.

The remaining statements are proved similarly, using limit properties 2, 3, and 4. ∎

EXAMPLE 4

We noted above that the function $f(x) = \sin x$ is continuous at $x = 0$. We also know that the polynomial function $g(x) = x^3 + 5x - 2$ is continuous at $x = 0$. Therefore, each of the following functions is continuous at $x = 0$.

$$(f + g)(x) = \sin x + x^3 + 5x - 2$$

$$(f - g)(x) = \sin x - (x^3 + 5x - 2) = \sin x - x^3 - 5x + 2$$

$$(fg)(x) = (\sin x)(x^3 + 5x - 2)$$

$$\left(\frac{f}{g}\right)(x) = \frac{\sin x}{x^3 + 5x - 2}. \qquad ∎$$

Composition of functions is often used to construct new functions from given ones.

Continuity of Composite Functions

> If the function f is continuous at $x = c$ and the function g is continuous at $x = f(c)$, then the composite function $g \circ f$ is continuous at $x = c$.

EXAMPLE 5

The polynomial function $f(x) = x^3 - 3x^2 + x + 7$ is continuous at $x = 2$, and $f(2) = 2^3 - 3 \cdot 2^2 + 2 + 7 = 5$. The function $g(x) = \sqrt{x}$ is continuous at $x = 5$ because by Limit Property 5 (page 840).

$$\lim_{x \to 5} \sqrt{x} = \sqrt{\lim_{x \to 5} x} = \sqrt{5} = g(5).$$

By the box above (with $c = 2$ and $f(c) = f(2) = 5$), the composite function $g \circ f$, which is given by

$$(g \circ f)(x) = g(f(x)) = g(x^3 - 3x^2 + x + 7) = \sqrt{x^3 - 3x^2 + x + 7},$$

is also continuous at $x = 2$. ∎

◨◨ THE INTERMEDIATE VALUE THEOREM

The following property of continuous functions has several important applications.

Intermediate Value Theorem

> If the function f is continuous on the closed interval $[a, b]$ and k is any number between $f(a)$ and $f(b)$, then there exists at least one number c between a and b such that $f(c) = k$.

Although we shall not prove the Intermediate Value Theorem, we can demonstrate its plausibility by a simple geometric argument. Figure 13–19 shows the graph of a function f that is continuous on the interval $[a, b]$ and a number k that is between $f(a)$ and $f(b)$.

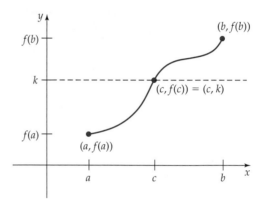

Figure 13–19

Because f is continuous on the interval, its graph can be drawn from $(a, f(a))$ to $(b, f(b))$ without lifting the pencil from the paper. As Figure 13–19 suggests, there is no way to do this unless the graph crosses the horizontal line $y = k$ at least once between $x = a$ and $x = b$. The coordinates of a point where the graph crosses the line can be described in two ways:

$(c, f(c))$ because the point is on the graph of f;

(c, k) because the point is on the line $y = k$.

Thus, this point satisfies the conclusion of the Intermediate Value Theorem:

$$a < c < b \qquad \text{and} \qquad f(c) = k.$$

The Intermediate Value Theorem further explains why the graph of a continuous function is connected and unbroken. If f is continuous on the interval $[a, b]$, then its graph cannot go from the point $(a, f(a))$ to the point $(b, f(b))$ without moving through all the y values between $f(a)$ and $f(b)$.

Here is an important special case of the Intermediate Value Theorem.

Equation Theorem

> If the function f is continuous on the interval $[a, b]$ and $f(a)$ and $f(b)$ have opposite signs, then the equation $f(x) = 0$ has a solution between a and b.

Proof   Since one of $f(a)$ and $f(b)$ is positive and the other is negative, the number 0 is between them. By the Intermediate Value Theorem, there is a number c between a and b such that $f(c) = 0$.   ∎

Many root finders on calculators are based on the Equation Theorem. To see why, consider the equation

$$x^3 - x - 1 = 0.$$

Figure 13–20

The graph of $f(x) = x^3 - x - 1$ in Figure 13–20 shows that there is a solution between 1 and 2. So we evaluate $f(x)$, beginning at $x = 1$ and moving by tenths, as in Figure 13–21. Since $f(1.3)$ and $f(1.4)$ have opposite signs, the solution is between 1.3 and 1.4 by the Equation Theorem. Hence, the decimal expansion of the solution begins with 1.3.

Figure 13–21	Figure 13–22	Figure 13–23

Now evaluate $f(x)$ beginning at 1.3 and moving by hundredths, as in Figure 13–22. Since $f(1.32)$ and $f(1.33)$ have opposite signs, the solution is between 1.32 and 1.33. So its decimal expansion begins with 1.32. Next evaluate $f(x)$ beginning at 1.32 and moving by thousandths, as in Figure 13–23, which shows that the solution is between 1.324 and 1.325. Thus, its decimal approximation begins 1.324. Each time the evaluation procedure is carried out, one more decimal place in the solution is found.

Root finders use this process (or a variation of it) to approximate solutions to as many decimal places as the calculator can handle. The Equation Theorem is also the reason that a root finder that cannot find a solution may display an error message that reads "no sign change."

EXERCISES 13.3

1. Use the graph of the function *f* in the figure to find all the numbers at which *f* is not continuous.

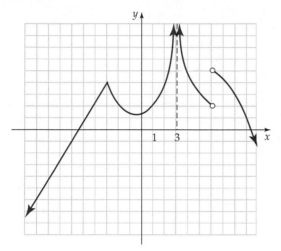

2. Use the graph of the function *g* in the figure to find all the numbers at which *g*(*x*) is defined but not continuous.

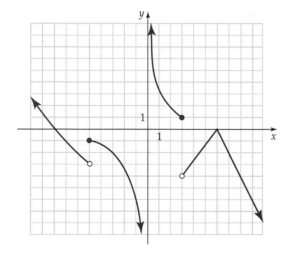

In Exercises 3–6, determine whether the function whose graph is given is continuous at x = −2, at x = 0, and at x = 3.

3.

4.

5.

6.

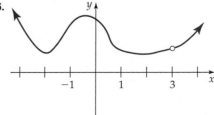

In Exercises 7–12, use the definition of continuity and the properties of limits to show that the function is continuous at the given number.

7. $f(x) = x^2 + 5(x-2)^7$ at $x = 3$

8. $g(x) = (x^2 - 3x - 10)(x^3 + 2x^2 - 5x + 4)$ at $x = -1$

9. $f(x) = \dfrac{x^2 - 9}{(x^2 - x - 6)(x^2 + 6x + 9)}$ at $x = 2$

10. $h(x) = \dfrac{x + 3}{(x^2 - x - 1)(x^2 + 1)}$ at $x = -2$

11. $f(x) = \dfrac{x\sqrt{x}}{(x-6)^2}$ at $x = 36$

12. $k(x) = \dfrac{\sqrt{8 - x^2}}{2x^2 - 5}$ at $x = -2$

In Exercises 13–18, explain why the function is not continuous at the given number.

13. $f(x) = 1/(x-3)^3$ at $x = 3$

14. $h(x) = \dfrac{x^2 + 4}{x^2 - x - 2}$ at $x = 2$

15. $f(x) = \dfrac{x^2 + 4x + 3}{x^2 - x - 2}$ at $x = -1$

16. $g(x) = \begin{cases} \sin(\pi/x) & \text{if } x \neq 0 \\ 1 & \text{if } x = 0 \end{cases}$ at $x = 0$

17. $f(x) = \begin{cases} x^2 & \text{if } x \neq 0 \\ 1 & \text{if } x = 0 \end{cases}$ at $x = 0$

18. $f(x) = \dfrac{\sqrt{2 + x} - \sqrt{2}}{x}$ at $x = 0$

In Exercises 19–24, determine whether or not the function is continuous at the given number.

19. $f(x) = \begin{cases} -2x + 4 & \text{if } x \leq 2 \\ 2x - 4 & \text{if } x > 2 \end{cases}$ at $x = 2$

20. $g(x) = \begin{cases} 2x + 5 & \text{if } x < -1 \\ -2x + 1 & \text{if } x \geq -1 \end{cases}$ at $x = -1$

21. $f(x) = \begin{cases} x^2 - x & \text{if } x \leq 0 \\ 2x^2 & \text{if } x > 0 \end{cases}$ at $x = 0$

22. $g(x) = \begin{cases} x^3 - x + 1 & \text{if } x < 2 \\ 3x^2 - 2x - 1 & \text{if } x \geq 2 \end{cases}$ at $x = 2$

23. $f(x) = |x - 3|$ at $x = 3$

24. $k(x) = -|x + 2| + 3$ at $x = -2$

In Exercises 25–28, determine all numbers at which the function is continuous.

25. $f(x) = \begin{cases} \dfrac{x^2 + x - 2}{x^2 - 4x + 3} & \text{if } x \neq 1 \\ -3/2 & \text{if } x = 1 \end{cases}$

26. $g(x) = \begin{cases} \dfrac{x^2 - x - 6}{x^2 - 4} & \text{if } x \neq -2 \\ 4/5 & \text{if } x = -2 \end{cases}$

27. $f(x) = \begin{cases} x^2 + 1 & \text{if } x < 0 \\ x & \text{if } 0 < x \leq 2 \\ -2x + 3 & \text{if } x > 2 \end{cases}$

28. $h(x) = \begin{cases} 1/x & \text{if } x < 1 \text{ and } x \neq 0 \\ x^2 & \text{if } x \geq 1 \end{cases}$

In Exercises 29–32, justify your answers.

29. Taxis in New York City cost $2 plus 30 cents for each $1/5$ of a mile (or portion thereof). Let $f(x)$ be the cost of traveling x miles. Is f continuous on the interval $[0, 3]$?

30. On a four and a half hour flight from Chicago to Seattle, let $h(x)$ be the height of the plane above the ground at time x hours. Is h continuous on the interval $[0, 4.5]$?

31. The U.S. Weather Bureau at Hopkins Airport in Cleveland continuously records the temperature. If $g(x)$ is the temperature at time x hours, where $x = 0$ corresponds to midnight, at what points on the interval $[0, 24]$ is g continuous?

32. Postage on a letter from the United States to Germany is 80 cents for each ounce (or fraction thereof) for letters weighing up to 8 ounces. Let $f(x)$ be the postage for a letter weighing x ounces. At what points on the interval $(0, 8]$ is f continuous.

33. If you don't have a calculator and you don't remember how to find square roots by hand, explain how you could use the Equation Theorem (and a lot of pencil and paper multiplication and addition) to find the decimal expansion of $\sqrt{7}$. [*Hint:* What are the solutions of $x^2 - 7 = 0$?]

34. (a) The function whose graph is shown in Figure 13–2 on page 831 is discontinuous at an infinite number of places. Where is it discontinuous?
    (b) Explain why a rational function cannot be discontinuous at an infinite number of places.

Thinkers

35. Show that the function
$$f(x) = \frac{x^4 - 5x^2 + 4}{x - 1}$$
is *not* continuous on $[-3, 3]$ but *does* satisfy the conclusion of the Intermediate Value Theorem (that is, if k is a number between $f(-3)$ and $f(3)$, there is a number c between -3 and 3 such that $f(c) = k$). [*Hint:* What can be said about f on the intervals $[-3, -2]$ and $[2, 3]$?]

36. Show that the function

$$f(x) = \frac{x^4 - 2x^3}{x - 2}$$

is not continuous on $[-3, 3]$ and does not satisfy the conclusion of the Intermediate Value Theorem (that is, there is a number k between $f(-3)$ and $f(3)$ for which there is no number c between -3 and 3 such that $f(c) = k$).

37. For what values of b is the function

$$f(x) - \begin{cases} bx + 4 & \text{if } x \le 3 \\ bx^2 - 2 & \text{if } x > 3 \end{cases}$$

continuous at $x = 3$?

38. Show that $f(x) = \sqrt{|x|}$ is continuous at $x = 0$.

A function f that is not defined as x = c is said to have a removable discontinuity at x = c, if there is a function g such that g(c) is defined, g is continuous at x = c, and g(x) = f(x) for x ≠ c. In Exercises 39–43, show that the function f has a removable discontinuity by finding an appropriate function g.

39. $f(x) = \dfrac{x - 1}{x^2 - 1}$ **40.** $f(x) = \dfrac{x^2}{|x|}$

41. $f(x) = \dfrac{2 - \sqrt{x}}{4 - x}$

42. $f(x) = \dfrac{\sin x}{x}$ [*Hint:* See Example 4 on page 832.]

43. Show that the function $f(x) = \dfrac{|x|}{x}$ has a discontinuity at $x = 0$ that is *not* removable.

44. A ranger leaves his truck at a parking lot at the trail head at 8:00 A.M. and hikes 11 miles to a fire tower, arriving there at noon. He stays overnight and starts back along the same trail at 8:00 A.M., arriving at the parking lot at noon. Show that there is a point on the trail that he passes at exactly the same time on both days. [*Hint:* Let $f(t)$ be his distance from the parking lot at time t on the first day, and let $g(t)$ be his distance from the parking lot at time t on the second day. Use an appropriate theorem to solve the equation $f(t) = g(t)$.]

13.4 Limits Involving Infinity

In the following discussion, it is important to remember that

> **There is no real number called "infinity," and the symbol ∞, which is usually read "infinity," does not represent any real number.**

Nevertheless, the word "infinity" and the symbol ∞ are often used as a convenient shorthand to describe the way some functions behave under certain circumstances. General speaking, "infinity" indicates a situation in which some numerical quantity gets larger and larger without bound, meaning that it can be made larger than any given number. Similarly, "negative infinity" (in symbols $-\infty$) indicates a situation in which a numerical quantity gets smaller and smaller without bound, meaning that it can be made smaller than any given negative number.

Consider the function f whose graph is shown in Figure 13–24.

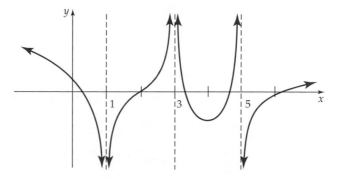

Figure 13–24

The graph shows that as x approaches 3 from the left or right, the corresponding values of $f(x)$ do not get closer and closer to a particular number. Instead, they become larger and larger without bound. Although there is no limit in the sense defined in Section 13.1, it is convenient to describe this situation symbolically by writing

$$\lim_{x \to 3} f(x) = \infty$$

which is read "the limit of $f(x)$ as x approaches 3 is infinity." Similarly, f does not have a limit as x approaches 1 (from the left or right) because the corresponding values of $f(x)$ get smaller and smaller without bound. We say that "the limit of $f(x)$ as x approaches 1 is negative infinity" and write

$$\lim_{x \to 1} f(x) = -\infty.$$

Near $x = 5$, the values of $f(x)$ get very large on the left side of 5 and very small on the right side of 5, so we write

$$\lim_{x \to 5^-} f(x) = \infty \qquad \text{and} \qquad \lim_{x \to 5^+} f(x) = -\infty$$

and say "the limit as x approaches 5 from the left is infinity" and "the limit as x approaches 5 from the right is negative infinity."

There are many cases like the ones illustrated above in which the language of limits and the word "infinity" is useful for describing the behavior of a function that actually does not have a limit in the sense of Section 13.1.

EXAMPLE 1

How does $f(x) = -5/x^4$ behave near $x = 0$?

SOLUTION

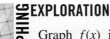**EXPLORATION**

Graph $f(x)$ in the viewing window with $-.5 \le x \le .5$ and $-500{,}000 \le y \le 0$. Use the trace feature to move along the graph on both sides of $x = 0$ and confirm that

$$\lim_{x \to 0} \frac{-5}{x^4} = -\infty. \qquad \blacksquare$$

EXAMPLE 2

Describe the behavior of the function

$$g(x) = \frac{8}{x^2 - 2x - 8}$$

near $x = -2$.

SOLUTION

GRAPHING EXPLORATION

Graph $g(x)$ in a suitable viewing window around $x = -2$, and verify that

$$\lim_{x \to -2^-} \frac{8}{x^2 - 2x - 8} = \infty \quad \text{and} \quad \lim_{x \to -2^+} \frac{8}{x^2 - 2x - 8} = -\infty.$$

The "infinite limits" considered in Figure 13–24 and Examples 1 and 2 can be interpreted geometrically. Each such limit corresponds to a vertical asymptote of the graph. Consequently, we have this formal definition.

Vertical Asymptotes

> The vertical line $x = c$ is a **vertical asymptote** of the graph of the function f if at least one of the following is true:
>
> $$\lim_{x \to c^-} f(x) = \infty, \qquad \lim_{x \to c^+} f(x) = \infty, \qquad \lim_{x \to c} f(x) = \infty,$$
>
> $$\lim_{x \to c^-} f(x) = -\infty, \qquad \lim_{x \to c^+} f(x) = -\infty, \qquad \lim_{x \to c} f(x) = -\infty.$$

LIMITS AT INFINITY

The word "limit" has been used thus far to describe the behavior of a function f when x is near a particular number c. Now we consider the behavior of a function when x is very large or very small.

EXAMPLE 3

Figure 13–25 shows the graph of

$$f(x) = \frac{5}{1 + 24e^{-x/4}} + 1.$$

Describe the behavior of f when x is very large and when x is very small.

Figure 13–25   **Figure 13–26**

Figure 13–27

SOLUTION   As you move to the right on the graph (that is, as x gets very large), the graph gets very close to the horizontal line $y = 6$, as shown in Figure 13–26 on the preceding page. In other words, as x gets larger and larger, the corresponding values of $f(x)$ get very close to 6. We express this fact symbolically by writing

$$\lim_{x \to \infty} f(x) = 6,$$

which is read "the limit of $f(x)$ as x approaches infinity is the number 6."

Toward the left, the graph gets very close to the horizontal line $y = 1$. As x gets smaller and smaller, the corresponding values of $f(x)$ get very close to 1, as shown in Figure 13–27.* We say that "the limit of $f(x)$ as x approaches negative infinity is 1" and write

$$\lim_{x \to -\infty} f(x) = 1. \quad \blacksquare$$

The limits illustrated in Example 3 are similar to those in Section 13.1 in that the values of the function do approach a fixed number. The definition in the general case is similar.

Limits
at Infinity

Let f be a function that is defined for all $x > a$, for some number a. If

as x takes larger and larger positive values, increasing without bound, the corresponding values of $f(x)$ get very close (and possibly equal) to the real number L

and

the values of $f(x)$ can be made arbitrarily close (as close as you want) to L by taking large enough values of x,

then we say that

the limit of $f(x)$ as x approaches infinity is L,

which is written

$$\lim_{x \to \infty} f(x) = L.$$

Similarly, if f is defined for all $x < b$, for some number b, then

$$\lim_{x \to -\infty} f(x) = L,$$

which is read

the limit of $f(x)$ as x approaches negative infinity is L,

means that

as x takes smaller and smaller negative values, decreasing without bound, the corresponding values of $f(x)$ get very close (and possibly equal) to the real number L,

and

the values of $f(x)$ can be made arbitrarily close (as close as you want) to L by taking small enough values of x.

---

*Because of rounding off, the trace feature on most calculators will say that $f(x) = 1$ when x is smaller than approximately -60. However, the value of the function is always greater than 1 (why?).

This is an informal definition, as was the one in Section 13.1. Rigorous definitions (similar to those in *Special Topics* 13.2.A for ordinary limits) are discussed in Exercises 49 and 50.

As Example 3 shows, limits as x approaches infinity or negative infinity correspond to horizontal asymptotes of the graph of the function. Consequently, we have this formal definition.

Horizontal Asymptotes

> The line $y = L$ is a **horizontal asymptote** of the graph of the function f if either
>
> $$\lim_{x \to \infty} f(x) = L \qquad \text{or} \qquad \lim_{x \to -\infty} f(x) = L.$$

EXAMPLE 4

When x is a very large positive number, then $1/x$ is a positive number that is very close to 0. Similarly, when x is a negative number that is large in absolute value (such as $-5,000,000$), $1/x$ is a negative number that is very close to 0. These facts suggest that

$$\lim_{x \to \infty} \frac{1}{x} = 0 \qquad \text{and} \qquad \lim_{x \to -\infty} \frac{1}{x} = 0.$$

GRAPHING EXPLORATION

Confirm that fact geometrically by graphing the function $f(x) = 1/x$ and verifying that the x-axis (the horizontal line $y = 0$) is a horizontal asymptote of the graph in both directions. ∎

No polynomial function has a limit as x approaches infinity or negative infinity, because no polynomial graph has a horizontal asymptote.

EXAMPLE 5

Describe the behavior of the polynomial function

$$f(x) = x^3 - 10x + 5$$

as x approaches infinity and as x approaches negative infinity.

SOLUTION   Look what happens when x takes larger and larger positive values.

GRAPHING EXPLORATION

Graph $f(x)$ in the viewing window with

$$0 \le x \le 5000 \text{ and } 0 \le y \le 100,000,000,000.$$

Use the trace feature to verify that as you move to the right along the graph, the values of $f(x)$ get larger and larger, increasing without bound.

Thus, $\lim\limits_{x \to \infty} f(x)$, in the sense defined on the preceding page, does not exist. Nevertheless, we can describe this situation by writing

$$\lim_{x \to \infty} f(x) = \infty.$$

A similar situation occurs when x takes smaller and smaller negative values.

GRAPHING EXPLORATION

Graph $f(x)$ in the viewing window with $-5000 \le x \le 0$ and $-100{,}000{,}000{,}000 \le y \le 0$. Use the trace feature to verify that as you move to the left along the graph, the values of $f(x)$ get smaller and smaller, decreasing without bound.

We describe this situation by writing

$$\lim_{x \to -\infty} f(x) = -\infty. \qquad \blacksquare$$

■■ PROPERTIES OF LIMITS

The limits of constant functions are easily found. Consider, for example, the function $f(x) = 5$. As x gets larger and larger (or smaller and smaller), the corresponding value of $f(x)$ is *always* the number 5, so that $\lim\limits_{x \to \infty} f(x) = 5$ and $\lim\limits_{x \to -\infty} f(x) = 5$. A similar argument works for any constant function.

Limit of a Constant

> If c is a constant, then
>
> $$\lim_{x \to \infty} c = c \qquad \text{and} \qquad \lim_{x \to -\infty} c = c.$$

Infinite limits have the same useful properties that ordinary limits have. For instance, suppose that as x approaches infinity, the values of a function f get close to a number L and the values of a function g get close to a number M. Then it is plausible that the values of $f(x) + g(x)$ get close to $L + M$, the values of $f(x)g(x)$ get close to $L \cdot M$, and so forth. Similar remarks apply when x approaches negative infinity. More formally, we have the following.

Properties of Limits at Infinity

> If f and g are functions and L and M are numbers such that
>
> $$\lim_{x \to \infty} f(x) = L \qquad \text{and} \qquad \lim_{x \to \infty} g(x) = M,$$
>
> then
>
> 1. $\lim\limits_{x \to \infty} (f(x) + g(x)) = L + M$;
> 2. $\lim\limits_{x \to \infty} (f(x) - g(x)) = L - M$;
> 3. $\lim\limits_{x \to \infty} (f(x) \cdot g(x)) = L \cdot M$;
> 4. $\lim\limits_{x \to \infty} \left(\dfrac{f(x)}{g(x)} \right) = \dfrac{L}{M}$; (provided that $M \neq 0$)
> 5. $\lim\limits_{x \to \infty} \sqrt{f(x)} = \sqrt{L}$. (provided that $f(x) \geq 0$ for all large x)
>
> Properties 1–5 also hold with $-\infty$ in place of ∞ (provided that for property 5, $f(x) \geq 0$ for all small x).

These properties are often stated somewhat differently. For example, since $\lim\limits_{x\to\infty} f(x) = L$ and $\lim\limits_{x\to\infty} g(x) = M$, Property 4 can be written as

$$\lim_{x\to\infty} \frac{f(x)}{g(x)} = \frac{\lim\limits_{x\to\infty} f(x)}{\lim\limits_{x\to\infty} g(x)},$$

and similarly for the others.

If c is a constant, then Property 3 and Example 4 show that

$$\lim_{x\to\infty} \frac{c}{x} = \lim_{x\to\infty} \left(c \cdot \frac{1}{x} \right) = \left(\lim_{x\to\infty} c \right) \left(\lim_{x\to\infty} \frac{1}{x} \right) = c \cdot 0 = 0.$$

Repeatedly using Property 3 with this fact and Example 4, we see that for any integer $n \geq 2$,

$$\lim_{x\to\infty} \frac{c}{x^n} = \lim_{x\to\infty} \left(\frac{c}{x} \cdot \frac{1}{x} \cdot \frac{1}{x} \cdots \frac{1}{x} \right)$$

$$= \left(\lim_{x\to\infty} \frac{c}{x} \right)\left(\lim_{x\to\infty} \frac{1}{x} \right)\left(\lim_{x\to\infty} \frac{1}{x} \right) \cdots \left(\lim_{x\to\infty} \frac{1}{x} \right)$$

$$= 0 \cdot 0 \cdot 0 \cdots 0 = 0.$$

A similar argument works with $-\infty$ in place of ∞ and produces this useful result (which is essentially a formal statement of half of the Big-Little Principle on page 272):

**Infinite
Limit Theorem**

If c is a constant, then for each positive integer n,

$$\lim_{x\to\infty} \frac{c}{x^n} = 0 \qquad \text{and} \qquad \lim_{x\to-\infty} \frac{c}{x^n} = 0.$$

This theorem and the limit properties above now make it possible to determine the limit (if it exists) of any rational function as x approaches infinity or negative infinity.

EXAMPLE 6

If you graph

$$f(x) = \frac{3x^2 - 2x + 1}{2x^2 + 4x - 5}$$

to the right of the y-axis, you will see that there appears to be a horizontal asymptote close to $y = 1.5$. We can confirm this fact algebraically by computing

$$\lim_{x\to\infty} \frac{3x^2 - 2x + 1}{2x^2 + 4x - 5}.$$

Property 4 cannot be used directly because neither the numerator nor the denominator has a finite limit as x approaches infinity (see Example 5). So we first divide both numerator and denominator by the highest power of x that appears, namely, x^2. When $x > 0$, this does not change the value of the fraction, and we have

$$\lim_{x\to\infty} \frac{3x^2 - 2x + 1}{2x^2 + 4x - 5} = \lim_{x\to\infty} \frac{\dfrac{3x^2 - 2x + 1}{x^2}}{\dfrac{2x^2 + 4x - 5}{x^2}}$$

$$= \lim_{x\to\infty} \frac{\dfrac{3x^2}{x^2} - \dfrac{2x}{x^2} + \dfrac{1}{x^2}}{\dfrac{2x^2}{x^2} + \dfrac{4x}{x^2} - \dfrac{5}{x^2}}$$

$$= \lim_{x\to\infty} \frac{3 - \dfrac{2}{x} + \dfrac{1}{x^2}}{2 + \dfrac{4}{x} - \dfrac{5}{x^2}}$$

$$= \frac{\lim_{x\to\infty} \left(3 - \dfrac{2}{x} + \dfrac{1}{x^2}\right)}{\lim_{x\to\infty} \left(2 + \dfrac{4}{x} - \dfrac{5}{x^2}\right)}$$ [Property 4]

$$= \frac{\lim_{x\to\infty} 3 - \lim_{x\to\infty} \dfrac{2}{x} + \lim_{x\to\infty} \dfrac{1}{x^2}}{\lim_{x\to\infty} 2 + \lim_{x\to\infty} \dfrac{4}{x} - \lim_{x\to\infty} \dfrac{5}{x^2}}$$ [Properties 1, 2]

$$= \frac{3 - \lim_{x\to\infty} \dfrac{2}{x} + \lim_{x\to\infty} \dfrac{1}{x^2}}{2 + \lim_{x\to\infty} \dfrac{4}{x} - \lim_{x\to\infty} \dfrac{5}{x^2}}$$ [Limit of constant]

$$= \frac{3 - 0 + 0}{2 + 0 - 0} = \frac{3}{2}.$$ [Limit Theorem]   ∎

The technique in Example 6 carries over to any rational function $f(x) = g(x)/h(x)$, where g and h have the same degree, and provides the mathematical justification for the treatment of horizontal asymptotes in Section 4.5 (see Exercise 48). A slight variation of the procedure can be used to compute certain limits involving square roots.

EXAMPLE 7

Find

(a) $\displaystyle\lim_{x\to\infty} \frac{\sqrt{3x^2 + 1}}{2x + 3}$

(b) $\displaystyle\lim_{x\to-\infty} \frac{\sqrt{3x^2 + 1}}{2x + 3}$

SOLUTION

(a) We need only consider positive values of x. When x is positive, $\sqrt{x^2} = x$. Therefore,

$$\lim_{x \to \infty} \frac{\sqrt{3x^2 + 1}}{2x + 3} = \lim_{x \to \infty} \frac{\dfrac{\sqrt{3x^2 + 1}}{x}}{\dfrac{2x + 3}{x}} = \lim_{x \to \infty} \frac{\dfrac{\sqrt{3x^2 + 1}}{\sqrt{x^2}}}{\dfrac{2x + 3}{x}}$$

$$= \lim_{x \to \infty} \frac{\sqrt{\dfrac{3x^2 + 1}{x^2}}}{\dfrac{2x + 3}{x}} = \lim_{x \to \infty} \frac{\sqrt{3 + \dfrac{1}{x^2}}}{2 + \dfrac{3}{x}}$$

$$= \frac{\displaystyle\lim_{x \to \infty} \sqrt{3 + \dfrac{1}{x^2}}}{\displaystyle\lim_{x \to \infty} \left(2 + \dfrac{3}{x}\right)} \qquad \text{[Property 4]}$$

$$= \frac{\sqrt{\displaystyle\lim_{x \to \infty} \left(3 + \dfrac{1}{x^2}\right)}}{\displaystyle\lim_{x \to \infty} \left(2 + \dfrac{3}{x}\right)} \qquad \text{[Property 5]}$$

$$= \frac{\sqrt{\displaystyle\lim_{x \to \infty} 3 + \lim_{x \to \infty} \dfrac{1}{x^2}}}{\displaystyle\lim_{x \to \infty} 2 + \lim_{x \to \infty} \dfrac{3}{x}} \qquad \text{[Property 1]}$$

$$= \frac{\sqrt{3 + 0}}{2 + 0} = \frac{\sqrt{3}}{2}. \qquad \text{[Constant limit and Limit Theorem]}$$

(b) To compute the limit as x approaches negative infinity, we need only consider negative values of x and use the fact that when x is negative, $x = -\sqrt{x^2}$ (for instance, $-2 = -\sqrt{(-2)^2}$). Then an argument similar to the one in part (a) shows that

$$\lim_{x \to -\infty} \frac{\sqrt{3x^2 + 1}}{2x + 3} = -\frac{\sqrt{3}}{2}. \qquad \blacksquare$$

EXERCISES 13.4

In Exercises 1–8, use a calculator to estimate the limit.

1. $\lim\limits_{x \to \infty} \left[\sqrt{x^2 + 1} - (x + 1) \right]$

2. $\lim\limits_{x \to \infty} \left[\sqrt{x^2 + x + 1} + x \right]$

3. $\lim\limits_{x \to -\infty} \dfrac{x^{2/3} - x^{4/3}}{x^3}$

4. $\lim\limits_{x \to \infty} \dfrac{x^{5/4} + x}{2x - x^{5/4}}$

5. $\lim\limits_{x \to -\infty} \sin \dfrac{1}{x}$

6. $\lim\limits_{x \to \infty} \dfrac{\sin x}{x}$

7. $\lim\limits_{x \to \infty} \dfrac{\ln x}{x}$

8. $\lim\limits_{x \to \infty} \dfrac{5}{1 + (1.1)^{-x/20}}$

In Exercises 9–14, list the vertical asymptotes of the graph. Then use the graph of the function f to find

$$\lim_{x \to \infty} f(x) \quad \text{and} \quad \lim_{x \to -\infty} f(x).$$

9.

10.

11.

12.

13.

14.

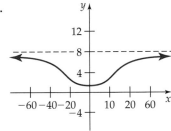

In Exercises 15–20, use the Infinite Limit Theorem and the properties of limits as in Example 6 to find the horizontal asymptotes (if any) of the graph of the given function.

15. $f(x) = \dfrac{3x^2 + 5}{4x^2 - 6x + 2}$

16. $g(x) = \dfrac{x^2}{x^2 - 2x + 1}$

17. $h(x) = \dfrac{2x^2 - 6x + 1}{2 + x - x^2}$

18. $k(x) = \dfrac{3x + x^2 - 4}{2x - x^3 + x^2}$

19. $f(x) = \dfrac{3x^4 - 2x^3 + 5x^2 - x + 1}{7x^3 - 4x^2 + 6x - 12}$

20. $g(x) = \dfrac{2x^5 - x^3 + 2x - 9}{5 - x^5}$

In Exercises 21–39, use the Infinite Limit Theorem and the properties of limits to find the limit.

21. $\lim\limits_{x \to -\infty} \dfrac{(x-3)(x+2)}{2x^2 + x + 1}$

22. $\lim\limits_{x \to \infty} \dfrac{(2x+1)(3x-2)}{3x^2 + 2x - 5}$

23. $\lim\limits_{x \to \infty} \left(3x - \dfrac{1}{x^2}\right)$

24. $\lim\limits_{x \to -\infty} (3x^2 + 1)^{-2}$

25. $\lim\limits_{x \to -\infty} \left(\dfrac{3x}{x+2} + \dfrac{2x}{x-1}\right)$

26. $\lim\limits_{x \to \infty} \left(\dfrac{x}{x^2 + 1} + \dfrac{2x^2}{x^3 + x}\right)$

27. $\lim\limits_{x \to \infty} \dfrac{2x}{\sqrt{x^2 - 2x}}$

28. $\lim\limits_{x \to -\infty} \dfrac{x}{\sqrt{x^2 - 1}}$

29. $\lim\limits_{x \to -\infty} \dfrac{3x - 2}{\sqrt{2x^2 + 1}}$

30. $\lim\limits_{x \to \infty} \dfrac{3x - 2}{\sqrt{2x^2 + 1}}$

31. $\lim\limits_{x \to \infty} \dfrac{\sqrt{2x^2 + 1}}{3x - 5}$

32. $\lim\limits_{x \to -\infty} \dfrac{\sqrt{2x^2 + 1}}{3x - 5}$

33. $\lim\limits_{x \to -\infty} \dfrac{\sqrt{3x^2 + 3}}{x + 3}$

34. $\lim\limits_{x \to \infty} \dfrac{\sqrt{3x^2 + 2x}}{2x + 1}$

35. $\lim\limits_{x \to \infty} \dfrac{x^2 + 2x + 1}{\sqrt{x^4 + 2x}}$

36. $\lim\limits_{x \to \infty} \dfrac{\sqrt{x^6 - x^2}}{2x^3}$

37. $\lim\limits_{x \to \infty} \dfrac{1 - \sqrt{x}}{1 + \sqrt{x}}$ [*Hint:* Rationalize the denominator.]

38. $\lim\limits_{x \to \infty} \dfrac{\sqrt{x} + 2}{\sqrt{x} - 3}$

39. $\lim\limits_{x \to \infty} \left(\sqrt{x^2 + 1} - x\right)$ $\left[\textit{Hint: Multiply by } \dfrac{\sqrt{x^2 + 1} + x}{\sqrt{x^2 + 1} + x}.\right]$

In Exercises 40–42, find the limit by adapting the hint for Exercise 39.

40. $\lim\limits_{x \to -\infty} \left(x + \sqrt{x^2 + 4}\right)$

41. $\lim\limits_{x \to \infty} \left(\sqrt{x^2 - 1} - \sqrt{x^2 + 1}\right)$

42. $\lim\limits_{x \to -\infty} \left(\sqrt{x^2 + 5x + 5} + x + 1\right)$

Thinkers

In Exercises 43–44, find the limit.

43. $\lim\limits_{x \to \infty} \dfrac{x}{|x|}$

44. $\lim\limits_{x \to -\infty} \dfrac{|x|}{|x| + 1}$

45. Let $[x]$ denote the greatest integer function (see Example 5 on pages 127–128) and find:

(a) $\lim\limits_{x \to \infty} \dfrac{[x]}{x}$

(b) $\lim\limits_{x \to -\infty} \dfrac{[x]}{x}$

46. Use the change of base formula for logarithms (*Special Topics 5.4.A*) to show that $\lim\limits_{x \to \infty} \dfrac{\ln x}{\log x} = \ln 10$.

47. Find $\lim\limits_{x \to \infty} \dfrac{\sqrt{x + \sqrt{x + \sqrt{x}}}}{\sqrt{x + 1}}$.

48. Let $f(x)$ be a nonzero polynomial with leading coefficient a, and let $g(x)$ be a nonzero polynomial with leading coefficient c. Prove that

(a) If $\deg f(x) < \deg g(x)$, then $\lim\limits_{x \to \infty} \dfrac{f(x)}{g(x)} = 0$.

(b) If $\deg f(x) = \deg g(x)$, then $\lim\limits_{x \to \infty} \dfrac{f(x)}{g(x)} = \dfrac{a}{c}$.

(c) If $\deg f(x) > \deg g(x)$, then $\lim\limits_{x \to \infty} \dfrac{f(x)}{g(x)}$ does not exist.

Formal definitions of limits at infinity and negative infinity are given in Exercises 49 and 50. Adapt the discussion in Special Topics 13.2.A to explain how these definitions are derived from the informal definitions given in this section.

49. Let f be a function, and let L be a real number. Then the statement $\lim\limits_{x \to \infty} f(x) = L$ means that for each positive number ϵ, there is a positive real number k (depending on ϵ) with this property:

If $x > k$, then $|f(x) - L| < \epsilon$.

[*Hint:* Concentrate on the second part of the informal definition. The number k measures "large enough," that is, how large the values of x must be to guarantee that $f(x)$ is as close as you want to L.]

50. Let f be a function, and let L be a real number. Then the statement $\lim\limits_{x \to -\infty} f(x) = L$ means that for each positive number ϵ, there is a negative real number n (depending on ϵ) with this property:

If $x < n$, then $|f(x) - L| < \epsilon$.

51. (a) Approximate $\lim\limits_{x \to \infty} \left(1 + \dfrac{1}{x}\right)^x$ to seven decimal places. (Evaluate the function at larger and larger values of x until successive approximations agree in the first seven decimal places.)

(b) Find the decimal expansion of e to at least nine decimal places.

(c) On the basis of the results in parts (a) and (b), what do you think is the exact value of $\lim\limits_{x \to \infty} \left(1 + \dfrac{1}{x}\right)^x$?

(d) Compare this limit to the one in Exercise 46 of Section 13.1. How are the two related?

Chapter 13 Review

IMPORTANT CONCEPTS

Section 13.1

Limit of a function   830
Nonexistence of limits   833–834
Limit of a function from the right
   835
Limit of a function from the left
   835

Section 13.2

Properties of limits   840
Limits of polynomial and rational
   functions   841–843
Limit Theorem   844
Limits of difference quotients   845
Derivative   846

Special Topics 13.2.A

The formal definition of limit   850

Section 13.3

Continuity of a function at
   $x = c$ 856
Continuity from the left and right
   858
Continuity on an interval   859
Properties of continuous functions
   859
Continuity of composite functions
   860
Intermediate Value Theorem   861
Equation Theorem   861

Section 13.4

Infinite limits   865–866
Vertical asymptotes   867
Limit of a function as x approaches
   infinity or negative infinity
   868
Horizontal asymptotes   869
Properties of limits at infinity   870

REVIEW QUESTIONS

In Questions 1–4, use a calculator to estimate the limit.

1. $\lim\limits_{x \to 0} \dfrac{3x - \sin x}{x}$

2. $\lim\limits_{x \to \pi/2} \dfrac{1 - \sin x}{1 + \cos 2x}$

3. $\lim\limits_{x \to -1} \dfrac{10^x - .1}{x + 1}$

4. $\lim\limits_{x \to 0} \dfrac{\sqrt{x - 3} - \sqrt{3}}{x}$

In Questions 5 and 6, use the graph of the function to determine the limit.

5. $\lim\limits_{x \to 2} f(x)$

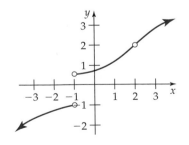

6. $\lim\limits_{x \to -1} f(x)$

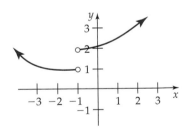

In Questions 7–10, assume that $\lim\limits_{x \to 3} f(x) = 5$ and $\lim\limits_{x \to 3} g(x) = -2$. Find the limit.

7. $\lim\limits_{x \to 3} [3f(x) - 15]$

8. $\lim\limits_{x \to 3} (f(x)[g(x) - 2])$

9. $\lim\limits_{x \to 3} \dfrac{f(x)g(x) - 2f(x)}{[g(x)]^2}$

10. $\lim\limits_{x \to 3} \dfrac{\sqrt{f(x) - 2g(x)}}{f(x) + g(x)}$

In Questions 11–20, find the limit if it exists. If the limit does not exist, explain why.

11. $\lim\limits_{x \to -2} (x^3 - 3x + 1)$

12. $\lim\limits_{x \to 4} \dfrac{2x + 1}{x - 2}$

REVIEW QUESTIONS

13. $\lim\limits_{x \to 1} \dfrac{x^2 - 1}{x^2 - 3x + 2}$

14. $\lim\limits_{x \to -2} \dfrac{x^2 - x - 6}{x^2 + x - 2}$

15. $\lim\limits_{x \to 0} \dfrac{\sqrt{1 + x} - 1}{x}$

16. $\lim\limits_{x \to 2} \dfrac{x^2 - 2x - 3}{x^2 - 6x + 9}$

17. $\lim\limits_{x \to -1^+} \sqrt{9 + 8x - x^2}$

18. $\lim\limits_{x \to 8^+} \dfrac{x^2 - 64}{x - 8}$

19. $\lim\limits_{x \to -5^+} \dfrac{|x + 5|}{x + 5}$

20. $\lim\limits_{x \to 7^-} \left(\sqrt{7 - x^2 + 6x} + 2 \right)$

21. If $f(x) = x^2 + 1$, find

$$\lim_{h \to 0} \frac{f(2 + h) - f(2)}{h}.$$

22. If $f(x) = 3x - 2$ and c is a constant, find

$$\lim_{h \to 0} \frac{f(c + h) - f(c)}{h}.$$

In Questions 23 and 24, find the rule of the derivative of the function f.

23. $f(x) = 4x - 2$

24. $f(x) = x^2$

In Questions 25 and 26, use the formal definition of limit to prove the statement.

25. $\lim\limits_{x \to 3} (2x + 1) = 7$

26. $\lim\limits_{x \to 2} \left(\dfrac{1}{2}x + 3 \right) = 4$

In Questions 27 and 28, determine whether the function whose graph is given is continuous at $x = -3$ and $x = 2$.

27.

28.

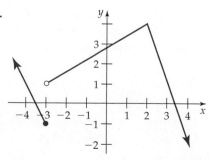

In Questions 29–32, determine if the function is continuous at the given points.

29. $f(x) = \sqrt{x^2 - 8}$; $x = 4, x = \sqrt{8}$

30. $g(x) = |x - 3|$; $x = 3, x = -3$

31. $h(x) = \dfrac{x - 7}{x + 4}$; $x = -4, x = 0$

32. $f(x) = \dfrac{x^2 - 1}{x + 1}$; $x = -1, x = 0, x = 1$

33. Show that $f(x) = \dfrac{x^2 - x - 6}{x^2 - 9}$ is

    (a) continuous at $x = 2$

    (b) discontinuous at $x = 3$.

34. Is the function given by

$$f(x) = \begin{cases} 3x - 2 & \text{if } x \le 3 \\ 10 - x & \text{if } x > 3 \end{cases}$$

    continuous at $x = 3$? Justify your answer.

In Questions 35 and 36, find the vertical asymptotes of the graph of the given function and state whether the graph moves upward or downward on each side of each asymptote.

35. $f(x) = \dfrac{x^2 + 1}{x^2 - x - 2}$

36. $g(x) = \dfrac{x^2 - 1}{x^2 - 3x + 2}$

In Questions 37–44, find the limit.

37. $\lim\limits_{x \to 3} \dfrac{1}{(x - 3)^2}$

38. $\lim\limits_{x \to -2^-} \dfrac{x^2 + x - 2}{x - 2}$

39. $\lim\limits_{x \to \infty} \dfrac{\sin x}{x^2}$

40. $\lim\limits_{x \to \infty} \dfrac{x}{(\ln x)^2}$

41. $\lim\limits_{x \to \infty} \dfrac{2x^3 - 3x^2 + 5x - 1}{4x^3 + 2x^2 - x + 10}$

42. $\lim\limits_{x \to -\infty} \dfrac{4 - 3x - 2x^2}{x^3 + 2x + 5}$

43. $\lim\limits_{x \to -\infty} \left(\dfrac{2x + 1}{x - 3} + \dfrac{4x - 1}{3x} \right)$

44. $\lim\limits_{x \to \infty} \dfrac{\sqrt{3x^2 + 2}}{4x + 1}$

In Questions 45 and 46, find the horizontal asymptotes of the graph of the given function algebraically and verify your results graphically.

45. $f(x) = \dfrac{x^2 - x + 7}{2x^2 + 5x + 7}$

46. $f(x) = \dfrac{x - 9}{\sqrt{4x^2 + 3x + 2}}$

DISCOVERY PROJECT 13   Black Holes

Chad Baker/Getty Images

You can't get here from there. Or, as the Eagles said in their 1976 song *Hotel California,* "You can never leave." A black hole is the ultimate "gotcha." Nothing can escape the gravitational attraction of a black hole—not even light. To understand why this is so and to understand why the entire universe isn't sucked into some black hole, we need to use a little algebra and two ideas from physics: the speed of light and escape velocity.

The **speed of light** (c in the famous equation $E = mc^2$) is approximately 3.00×10^8 meters per second. The speed of light is also the absolute maximum velocity in this universe. Nothing moves faster. The **escape velocity** is the velocity required for an object to escape the gravity of a planet or some other object.

Suppose that an object of mass m kilograms is moving with velocity v meters per second. Its **kinetic energy** (energy of motion) is given by

$$KE = \frac{1}{2} mv^2.$$

Near the surface of a planet with mass M kilograms, the gravitational **potential energy** (stored energy) of the object is given by

$$PE = \frac{GmM}{r^2} \cdot r = \frac{GmM}{r},$$

where $G \approx 6.67 \times 10^{-11}$ is the gravitational constant and r is the distance (in meters) from the center of the planet. If the velocity is large enough, the kinetic energy equals (or exceeds) the potential energy, and the object will escape. To find the escape velocity, we solve the equation $KE = PE$.

$$\frac{1}{2} mv^2 = \frac{GmM}{r}$$

Divide both sides by m and multiply both sides by 2:

$$v^2 = \frac{2GM}{r}$$

(1)   Take the square root of both sides: $v = \sqrt{\dfrac{2GM}{r}}$ meters per second.

Notice that the escape velocity v depends only on the mass M of the planet and the distance r of the object from the planet's center. The mass of the object itself is irrelevant. For an atom or an ant or a rocket or the entire city of Atlanta, the escape velocity is the same!

Under the preceding assumptions, find the escape velocity from the following locations. Express all your answers in terms of kilograms (kg), meters (m), and seconds. Express your final answers (but not any intermediate results) in scientific notation (with the number between 1 and 10 rounded to two decimal places).

1. Surface of the earth. The mass of the earth is 5.97×10^{24} kg, and its radius is 6.38×10^6 m.

2. Surface of the moon. The mass of the moon is 7.36×10^{22} kg, and its radius is 1.74×10^6 m.

3. A satellite at an altitude of 35,786 kilometers orbits the earth at the same rate as the earth rotates. Satellites in such *geosynchronous orbits* stay "parked" above the same spot. [*Hint:* A kilometer is 10^3 meters. Add the altitude to the earth's radius to find r.]

Gravity works in all stars to pull matter toward the center. The energy released by fusion counteracts gravity. But when a star dies, gravity eventually prevails. Depending on its mass, the star becomes a white dwarf, a neutron star, or a black hole. In the case of a black hole, the mass is so large that gravity overcomes even neutron repulsion, and the mass contracts to a singularity (a point).

As a dying star contracts, it reaches its **Schwarzschild radius,** R_S, also known as the **event horizon.** At the event horizon the escape velocity is c, the speed of light. To find the Schwarzschild radius, we first solve the escape velocity equation (1) for r:

(2) $$r = \frac{2GM}{v^2}$$

Then set $v = c$, so that $R_S = 2GM/c^2$. For points closer to the center/singularity, the escape velocity is greater than c. But since nothing moves faster than light, nothing escapes. Nothing. At greater distances, outside the imaginary sphere defined by the Schwarzschild radius, everything is pretty much normal. The gravitational attraction and escape velocity are no different than they would be for any regular star of the same mass. You just can't see what is attracting you.

Use $c = 3.00 \times 10^8$ meters per second for the speed of light, and denote the mass of the sun by M_{sun}. Assume that $M_{sun} = 1.99 \times 10^{30}$ kg. Find the Schwarzschild radius for black holes with the following masses.

4. Mass M equal to five times the mass of the sun $(5 \times M_{sun})$

5. $M = 8.4 \times M_{sun}$

6. $M = 20 \times M_{sun}$

DISCOVERY PROJECT 13

The figure below is a depiction of the space-time fabric in the vicinity of a black hole.

A cross-section by a perpendicular plane through the singularity will resemble the graph of $f(x) = 1/x^2$.

Although you may have thought that we would never make the connection, there is a relationship between the physics of black holes and the algebra of limits. As we saw in Section 13.4, the behavior of $f(x) = 1/x^2$ near $x = 0$ is described by writing

$$\lim_{x \to 0} \frac{1}{x^2} = \infty.$$

Infinite limits such as this one can be rigorously defined (as ordinary limits were in Special Topics 13.2.A) as follows.

Let f be a function and let a be a real number such that $f(x)$ is defined for all x in some open interval containing a (except possibly at $x = a$). We say that **the limit of $f(x)$, as x approaches a, is infinity,** and write

$$\lim_{x \to a} f(x) = \infty,$$

provided that for each positive number N, there is a positive number δ (depending on N) with this property:

$$\text{If } 0 < |x - a| < \delta, \qquad \text{then} \qquad f(x) > N.$$

To relate this definition to the preceding discussion, note that equation **(1)** defines the velocity v as a function of the distance r; the rule of the function is

$$f(r) = \sqrt{\frac{2GM}{r}}.$$

Finding the Schwarzschild radius amounts to finding δ for this function f when $N = c$ and a is the center/singularity of the black hole (so that $r = |x - a|$), as in Exercises 4–6.

More about this topic can be found at the following Web sites:

http://www.brookscole.com/astronomy_d/resources/universe/index.html
http://antwrp.gsfc.nasa.gov/htmltest/rjn_bht.html
http://archive.ncsa.uiuc.edu/Cyberia/NumRel/BlackHoles.html

Appendix 1 ALGEBRA REVIEW

This appendix reviews the fundamental algebraic facts that are used frequently in this book. You must be able to handle these algebraic manipulations to succeed in this course and in calculus.

1.A Integral Exponents

Exponents provide a convenient shorthand for certain products. If c is a real number, then c^2 denotes cc, and c^3 denotes ccc. More generally, for any positive integer n,

$$c^n \text{ denotes the product } ccc \cdots c \quad (n \text{ factors}).$$

In this notation, c^1 is just c, so we usually omit the exponent 1.

EXAMPLE 1

$3^4 = 3 \cdot 3 \cdot 3 \cdot 3 = 81$, and

$$(-2)^5 = (-2)(-2)(-2)(-2)(-2) = -32.$$

For every positive integer n, $0^n = 0 \cdots 0 = 0$. ∎

EXAMPLE 2

To find $(2.4)^9$, use the $\wedge$ (or a^b or x^y) key on your calculator:

$$2.4 \wedge 9 \text{ ENTER}^*$$

which produces the (approximate) answer 2641.80754. ∎

CAUTION

Be careful with negative bases. For instance, if you want to compute $(-12)^4$, which is a positive number, but you key in $(-)\ 12 \wedge 4$ ENTER, the calculator will interpret this as $-(12^4)$ and produce a negative answer. To get the correct answer, you must key in the parentheses:

$$((-)\ 12) \wedge 4 \text{ ENTER.}$$

---

*The ENTER key is labeled EXE on Casio calculators.

Because exponents are just shorthand for multiplication, it is easy to determine the rules they obey. For instance,

$$c^3 c^5 = (ccc)(ccccc) = c^8, \qquad \text{that is,} \qquad c^3 c^5 = c^{3+5}.$$

$$\frac{c^7}{c^4} = \frac{ccccccc}{cccc} = \frac{\cancel{ccccccc}}{\cancel{cccc}} = ccc = c^3, \qquad \text{that is,} \qquad \frac{c^7}{c^4} = c^{7-4}.$$

Similar arguments work in the general case.

Multiplication and Division with Exponents

To multiply c^m by c^n, add the exponents.

$$c^m \cdot c^n = c^{m+n}.$$

To divide c^m by c^n, subtract the exponents.

$$\frac{c^m}{c^n} = c^{m-n}.$$

EXAMPLE 3

$$4^2 \cdot 4^7 = 4^{2+7} = 4^9 \qquad \text{and} \qquad 2^8/2^3 = 2^{8-3} = 2^5. \qquad \blacksquare$$

The notation c^n can be extended to the cases in which n is zero or negative as follows.

Zero and Negative Exponents

If $c \neq 0$ and n is a positive integer, then

$$c^0 \text{ is defined to be the number } 1;$$

$$c^{-n} \text{ is defined to the number } \frac{1}{c^n}.$$

Note that 0^0 and negative powers of 0 are *not* defined (negative powers of 0 would involve division by 0). The reason for choosing these definitions is that the multiplication and division rules for exponents remain valid. For instance,

$$c^5 \cdot c^0 = c^5 \cdot 1 = c^5, \qquad \text{so} \qquad c^5 c^0 = c^{5+0}.$$

$$c^7 c^{-7} = c^7 \frac{1}{c^7} = 1 = c^0, \qquad \text{so} \qquad c^7 c^{-7} = c^{7-7}.$$

EXAMPLE 4

$6^{-3} = 1/6^3 = 1/216$ and $(-2)^{-5} = 1/(-2)^5 = -1/32$. A calculator shows that $(.287)^{-12} \approx 3{,}201{,}969.857.$* $\qquad \blacksquare$

---

*$\approx$ means "approximately equal to."

If c and d are nonzero real numbers and m and n are integers (positive, negative, or zero), then we have these laws.

Exponent Laws

1. $c^m c^n = c^{m+n}$

2. $\dfrac{c^m}{c^n} = c^{m-n}$

3. $(c^m)^n = c^{mn}$

4. $(cd)^n = c^n d^n$

5. $\left(\dfrac{c}{d}\right)^n = \dfrac{c^n}{d^n}$

6. $\dfrac{1}{c^{-n}} = c^n$

EXAMPLE 5

Here are examples of the six exponent laws.

1. $\pi^{-5}\pi^2 = \pi^{-5+2} = \pi^{-3} = 1/\pi^3$.

2. $x^9/x^4 = x^{9-4} = x^5$.

3. $(5^{-3})^2 = 5^{(-3)2} = 5^{-6}$.

4. $(2x)^5 = 2^5 x^5 = 32x^5$. ∎

5. $\left(\dfrac{7}{3}\right)^{10} = \dfrac{7^{10}}{3^{10}}$.

6. $1/x^{-5} = \dfrac{1}{\left(\dfrac{1}{x^5}\right)} = x^5$.

 CAUTION

$(2x)^5$ is *not* the same as $2x^5$. Part 4 of Example 5 shows that $(2x)^5 = 32x^5$ and *not* $2x^5$.

The exponent laws can often be used to simplify complicated expressions.

EXAMPLE 6

(a) $(2x^2 y^3 z)^4 = 2^4 (x^2)^4 (y^3)^4 z^4 = 16 x^8 y^{12} z^4$.

     ↑ Law (4)   ↑ Law (3)

(b) $(r^{-3} s^2)^{-2} = (r^{-3})^{-2}(s^2)^{-2} = r^6 s^{-4} = r^6/s^4$.

     ↑ Law (4)   ↑ Law (3)

(c) $\dfrac{x^5(y^2)^3}{(x^2 y)^2} = \dfrac{x^5 y^6}{(x^2 y)^2} = \dfrac{x^5 y^6}{(x^2)^2 y^2} = \dfrac{x^5 y^6}{x^4 y^2} = x^{5-4} y^{6-2} = xy^4$. ∎

     ↑ Law (3) ↑ Law (4)   ↑ Law (3) ↑ Law (2)

It is usually more efficient to use the exponent laws with the negative exponents rather than first converting to positive exponents. If positive exponents are required, the conversion can be made in the last step.

EXAMPLE 7

Simplify and express without negative exponents

$$\frac{a^{-2}(b^2c^3)^{-2}}{(a^{-3}b^{-5})^2c}.$$

SOLUTION

$$\frac{a^{-2}(b^2c^3)^{-2}}{(a^{-3}b^{-5})^2c} = \frac{a^{-2}(b^2)^{-2}(c^3)^{-2}}{(a^{-3})^2(b^{-5})^2c} = \frac{a^{-2}b^{-4}c^{-6}}{a^{-6}b^{-10}c}$$

$$\underset{\text{Law (4)}}{\uparrow} \qquad\qquad \underset{\text{Law (3)}}{\uparrow}$$

$$= a^{-2-(-6)}b^{-4-(-10)}c^{-6-1} = a^4b^6c^{-7} = \frac{a^4b^6}{c^7}. \qquad \blacksquare$$

$$\underset{\text{Law (2)}}{\uparrow}$$

Since $(-1)(-1) = +1$, any even power of -1, such as $(-1)^4$ or $(-1)^{12}$, will be equal to 1. Every odd power of -1 is equal to -1; for instance, $(-1)^5 = (-1)^4(-1) = 1(-1) = -1$. Consequently, for every positive number c,

$$(-c)^n = [(-1)c]^n = (-1)^nc^n = \begin{cases} c^n & \text{if } n \text{ is even} \\ -c^n & \text{if } n \text{ is odd} \end{cases}.$$

EXAMPLE 8

$(-3)^4 = 3^4 = 81$ and $(-5)^3 = -5^3 = -125.$ ■

✔ EXERCISES 1.A

In Exercises 1–18, evaluate the expression.

1. $(-6)^2$

2. -6^2

3. $5 + 4(3^2 + 2^3)$

4. $(-3)2^2 + 4^2 - 1$

5. $\dfrac{(-3)^2 + (-2)^4}{-2^2 - 1}$

6. $\dfrac{(-4)^2 + 2}{(-4)^2 - 7} + 1$

7. $\left(\dfrac{-5}{4}\right)^3$

8. $-\left(\dfrac{7}{4} + \dfrac{3}{4}\right)^2$

9. $\left(\dfrac{1}{3}\right)^3 + \left(\dfrac{2}{3}\right)^3$

10. $\left(\dfrac{5}{7}\right)^2 + \left(\dfrac{2}{7}\right)^2$

11. $2^4 - 2^7$

12. $3^3 - 3^{-7}$

13. $(2^{-2} + 2)^2$

14. $(3^{-1} - 3^3)^2$

15. $2^2 \cdot 3^{-3} - 3^2 \cdot 2^{-3}$

16. $4^3 \cdot 5^{-2} + 4^2 \cdot 5^{-1}$

17. $\dfrac{1}{2^3} + \dfrac{1}{2^{-4}}$

18. $3^2\left(\dfrac{1}{3} + \dfrac{1}{3^{-2}}\right)$

In Exercises 19–38, simplify the expression. Each letter represents a nonzero real number and should appear at most once in your answer.

19. $x^2 \cdot x^3 \cdot x^5$

20. $y \cdot y^4 \cdot y^6$

21. $(.03)y^2 \cdot y^7$

22. $(1.3)z^3 \cdot z^5$

23. $(2x^2)^3 3x$

24. $(3y^3)^4 5y^2$

25. $(3x^2y)^2$

26. $(2xy^3)^3$

27. $(a^2)(7a)(-3a^3)$

28. $(b^3)(-b^2)(3b)$

29. $(2w)^3(3w)(4w)^2$

30. $(3d)^4(2d)^2(5d)$

31. $a^{-2}b^3a^3$

32. $c^4d^5c^{-3}$

33. $(2x)^{-2}(2y)^3(4x)$

34. $(3x)^{-3}(2y)^{-2}(2x)$

35. $(-3a^4)^2(9x^3)^{-1}$

36. $(2y^3)^3(3y^2)^{-2}$

37. $(2x^2y)^0(3xy)$

38. $(3x^2y^4)^0$

In Exercises 39–42, express the given number as a power of 2.

39. $(64)^2$

40. $(1/8)^3$

41. $(2^4 \cdot 16^{-2})^3$

42. $(1/2)^{-8}(1/4)^4(1/16)^{-3}$

In Exercises 43–60, simplify and write the given expression without negative exponents. All letters represent nonzero real numbers.

43. $\dfrac{x^4(x^2)^3}{x^3}$

44. $\left(\dfrac{z^2}{t^3}\right)^4 \cdot \left(\dfrac{z^3}{t}\right)^5$

45. $\left(\dfrac{e^6}{c^4}\right)^2 \cdot \left(\dfrac{c^3}{e}\right)^3$

46. $\left(\dfrac{x^7}{y^6}\right)^2 \cdot \left(\dfrac{y^2}{x}\right)^4$

47. $\left(\dfrac{ab^2c^3d^4}{abc^2d}\right)^2$

48. $\dfrac{(3x)^2(y^2)^3x^2}{(2xy^2)^3}$

49. $\left(\dfrac{a^6}{b^{-4}}\right)^2$

50. $\left(\dfrac{x^{-2}}{y^{-2}}\right)^2$

51. $\left(\dfrac{c^5}{d^{-3}}\right)^{-2}$

52. $\left(\dfrac{x^{-1}}{2y^{-1}}\right)\left(\dfrac{2y}{x}\right)^{-2}$

53. $\left(\dfrac{3x}{y^2}\right)^{-3}\left(\dfrac{-x}{2y^3}\right)^2$

54. $\left(\dfrac{5u^2v}{2uv^2}\right)^2\left(\dfrac{-3uv}{2u^2v}\right)^{-3}$

55. $\dfrac{(a^{-3}b^2c)^{-2}}{(ab^{-2}c^3)^{-1}}$

56. $\dfrac{(-2cd^2e^{-1})^3}{(5c^{-3}de)^{-2}}$

57. $(c^{-1}d^{-2})^{-3}$

58. $[(x^2y^{-1})^2]^{-3}$

59. $a^2(a^{-1} + a^{-3})$

60. $\dfrac{a^{-2}}{b^{-2}} + \dfrac{b^2}{a^2}$

In Exercises 61–66, determine the sign of the given number without calculating the product.

61. $(-2.6)^3(-4.3)^{-2}$

62. $(4.1)^{-2}(2.5)^{-3}$

63. $(-1)^9(6.7)^5$

64. $(-4)^{12}6^9$

65. $(-3.1)^{-3}(4.6)^{-6}(7.2)^7$

66. $(45.8)^{-7}(-7.9)^{-9}(-8.5)^{-4}$

In Exercises 67–72, r, s, and t are positive integers and a, b, and c are nonzero real numbers. Simplify and write the given expression without negative exponents.

67. $\dfrac{3^{-r}}{3^{-s-r}}$

68. $\dfrac{4^{-(t+1)}}{4^{2-t}}$

69. $\left(\dfrac{a^6}{b^{-4}}\right)^t$

70. $\dfrac{c^{-t}}{(6b)^{-s}}$

71. $\dfrac{(c^{-r}b^s)^t}{(c^t b^{-s})^r}$

72. $\dfrac{(a^r b^{-s})^{-t}}{(b^t c^r)^{-s}}$

Errors to Avoid

In Exercises 73–80, give an example to show that the statement may be false for some numbers.

73. $a^r + b^r = (a + b)^r$

74. $a^r a^s = a^{rs}$

75. $a^r b^s = (ab)^{r+s}$

76. $c^{-r} = -c^r$

77. $\dfrac{c^r}{c^s} = c^{r/s}$

78. $(a + 1)(b + 1) = ab + 1$

79. $(-a)^2 = -a^2$

80. $(-a)(-b) = -ab$

1.B   Arithmetic of Algebraic Expressions

Expressions such as

$$b + 3c^2, \qquad 3x^2 - 5x + 4, \qquad \sqrt{x^3 + z}, \qquad \dfrac{x^3 + 4xy - \pi}{x^2 + xy}$$

are called **algebraic expressions.** The letters in algebraic expressions represent numbers. A letter that represents a number whose value remains unchanged throughout the discussion is called a **constant.** Constants are usually denoted by letters near the beginning of the alphabet. A letter that may represent *any* number is called a **variable.** Letters near the end of the alphabet usually denote variables. Thus, in the expression $3x + 5 + c$, it is understood that x is a variable and c is a constant. The value of this expression depends on the value of the variable x. If $x = 4$, then

$$3x + 5 + c = 3(4) + 5 + c = 17 + c,$$

and if $x = \frac{1}{3}$, then

$$3x + 5 + c = 3\left(\tfrac{1}{3}\right) + 5 + c = 6 + c,$$

and so on.*

---

*We assume any conditions on the constants and variables necessary to guarantee that an algebraic expression does represent a real number. For instance, in $\sqrt{z}$, we assume $z \geq 0$, and in $1/c$, we assume $c \neq 0$.

The usual rules of arithmetic are valid for algebraic expressions:

Commutative Laws:

$$a + b = b + a \quad \text{and} \quad ab = ba.$$

Associative Laws:

$$(a + b) + c = a + (b + c) \quad \text{and} \quad (ab)c = a(bc).$$

Distributive Laws:

$$a(b + c) = ab + ac \quad \text{and} \quad (b + c)a = ba + ca.$$

EXAMPLE 1

Use the distributive law to *combine like terms;* for instance,

$$3x + 5x + 4x = (3 + 5 + 4)x = 12x.$$

In practice, you do the middle part in your head and simply write

$$3x + 5x + 4x = 12x. \quad \blacksquare$$

EXAMPLE 2

In more complicated expressions, eliminate parentheses, use the commutative law to group like terms together, and then combine them.

$$(a^2b - 3\sqrt{c}) + (5ab + 7\sqrt{c}) + 7a^2b = a^2b - 3\sqrt{c} + 5ab + 7\sqrt{c} + 7a^2b$$

$$\text{Regroup} \qquad = \underbrace{a^2b + 7a^2b}_{} - \underbrace{3\sqrt{c} + 7\sqrt{c}}_{} + 5ab$$

$$\text{Combine like terms:} \quad = \quad 8a^2b \quad + \quad 4\sqrt{c} \quad + 5ab. \quad \blacksquare$$

CAUTION

Be careful when parentheses are preceded by a minus sign: $-(b + 3) = -b - 3$ and *not* $-b + 3$. Here's the reason: $-(b + 3)$ means $(-1)(b + 3)$; hence by the distributive law,

$$-(b + 3) = (-1)(b + 3) = (-1)b + (-1)3 = -b - 3.$$

Similarly, $-(7 - y) = -7 - (-y) = -7 + y$.

The examples in the Caution Box illustrate the following.

Rules for Eliminating Parentheses

> Parentheses preceded by a plus sign (or no sign) may be deleted.
>
> Parentheses preceded by a minus sign may be deleted *if* the sign of every term within the parentheses is changed.

The usual method of multiplying algebraic expressions is to use the distributive laws repeatedly, as shown in the following examples. The net result is to *multiply every term in the first sum by every term in the second sum.*

EXAMPLE 3

Compute $(y - 2)(3y^2 - 7y + 4)$.

SOLUTION   We first apply the distributive law, treating $(3y^2 - 7y + 4)$ as a single number.

$$(y - 2)(3y^2 - 7y + 4) = y(3y^2 - 7y + 4) - 2(3y^2 - 7y + 4)$$

Distributive law: $\qquad = 3y^3 - 7y^2 + 4y - 6y^2 + 14y - 8$

Regroup: $\qquad = 3y^3 - \underline{7y^2 - 6y^2} + \underline{4y + 14y} - 8$

Combine like terms: $\qquad = 3y^3 - \quad 13y^2 \quad + \quad 18y \quad - 8.$ ■

EXAMPLE 4

Compute $(2x - 5y)(3x + 4y)$.

SOLUTION   Treat $3x + 4y$ as a single number, and use the distributive law.

$$(2x - 5y)(3x + 4y) = 2x(3x + 4y) - 5y(3x + 4y)$$

Distributive law: $\qquad = 2x \cdot 3x + 2x \cdot 4y + (-5y) \cdot 3x + (-5y) \cdot 4y$

Regroup $\qquad = 6x^2 + \underline{8xy - 15xy} - 20y^2$

Combine like terms: $\qquad = 6x^2 - \quad 7xy \quad - 20y^2.$ ■

Observe the pattern in the second line of Example 4 and its relationship to the terms being multiplied.

$$(2x - 5y)(3x + 4y) = 2x \cdot 3x + 2x \cdot 4y + (-5y) \cdot 3x + (-5y) \cdot 4y$$

First terms

$$(2x - 5y)(3x + 4y)$$

Outside terms

$$(2x - 5y)(3x + 4y)$$

Inside terms

$$(2x - 5y)(3x + 4y)$$

Last terms

CAUTION

The FOIL method can be used only when multiplying two expressions that each have two terms.

This pattern is easy to remember by using the acronym FOIL (**F**irst, **O**utside, **I**nside, **L**ast). The FOIL method makes it easy to find products such as this one mentallly, without the necessity of writing out the intermediate steps.

EXAMPLE 5

$$(3x + 2)(x + 5) = 3x^2 + 15x + 2x + 10 = 3x^2 + 17x + 10.$$ ■

$\qquad\qquad\qquad$ First $\quad$ Outside $\quad$ Inside $\quad$ Last

✓ EXERCISES 1.B

In Exercises 1–54, perform the indicated operations and simplify your answer.

1. $x + 7x$ **2.** $5w + 7w - 3w$

3. $6a^2b + (-8b)a^2$ **4.** $-6x^3\sqrt{t} + 7x^3\sqrt{t} + 15x^3\sqrt{t}$

5. $(x^2 + 2x + 1) - (x^3 - 3x^2 + 4)$

6. $\left[u^4 - (-3)u^3 + \dfrac{u}{2} + 1\right] + \left(u^4 - 2u^3 + 5 - \dfrac{u}{2}\right)$

7. $\left[u^4 - (-3)u^3 + \dfrac{u}{2} + 1\right] - \left(u^4 - 2u^3 + 5 - \dfrac{u}{2}\right)$

8. $(6a^2b + 3a\sqrt{c} - 5ab\sqrt{c}) + (-6ab^2 - 3ab + 6ab\sqrt{c})$

9. $[4z - 6z^2w - (-2)z^3w^2] + (8 - 6z^2w - zw^3 + 4z^3w^2)$

10. $(x^5y - 2x + 3xy^3) - (-2x - x^5y + 2xy^3)$

11. $(9x - x^3 + 1) - [2x^3 + (-6)x + (-7)]$

12. $(x - \sqrt{y} - z) - (x + \sqrt{y} - z) - (\sqrt{y} + z - x)$

13. $(x^2 - 3xy) - (x + xy) - (x^2 + xy)$

14. $2x(x^2 + 2)$ **15.** $(-5y)(-3y^2 + 1)$

16. $x^2y(xy - 6xy^2)$

17. $3ax(4ax - 2a^2y + 2ay)$ **18.** $2x(x^2 - 3xy + 2y^2)$

19. $6z^3(2z + 5)$ **20.** $-3x^2(12x^6 - 7x^5)$

21. $3ab(4a - 6b + 2a^2b)$ **22.** $(-3ay)(4ay - 5y)$

23. $(x + 1)(x - 2)$ **24.** $(x + 2)(2x - 5)$

25. $(-2x + 4)(-x - 3)$ **26.** $(y - 6)(2y + 2)$

27. $(y + 3)(y + 4)$ **28.** $(w - 2)(3w + 1)$

29. $(3x + 7)(-2x + 5)$ **30.** $(ab + 1)(a - 2)$

31. $(y - 3)(3y^2 + 4)$ **32.** $(y + 8)(y - 8)$

33. $(x + 4)(x - 4)$ **34.** $(3x - y)(3x + y)$

35. $(4a + 5b)(4a - 5b)$ **36.** $(x + 6)^2$

37. $(y - 11)^2$ **38.** $(2x + 3y)^2$

39. $(5x - b)^2$ **40.** $(2s^2 - 9y)(2s^2 + 9y)$

41. $(4x^3 - y^4)^2$ **42.** $(4x^3 - 5y^2)(4x^3 + 5y^2)$

43. $(-3x^2 + 2y^4)^2$ **44.** $(c - 2)(2c^2 - 3c + 1)$

45. $(2y + 3)(y^2 + 3y - 1)$ **46.** $(x + 2y)(2x^2 - xy + y^2)$

47. $(5w + 6)(-3w^2 + 4w - 3)$

48. $(5x - 2y)(x^2 - 2xy + 3y^2)$

49. $2x(3x + 1)(4x - 2)$ **50.** $3y(-y + 2)(3y + 1)$

51. $(x - 1)(x - 2)(x - 3)$

52. $(y - 2)(3y + 2)(y + 2)$

53. $(x + 4y)(2y - x)(3x - y)$

54. $(2x - y)(3x + 2y)(y - x)$

In Exercises 55–64, find the coefficient of x^2 in the given product. Avoid doing any more multiplying than necessary.

55. $(x^2 + 3x + 1)(2x - 3)$ **56.** $(x^2 - 1)(x + 1)$

57. $(x^3 + 2x - 6)(x^2 + 1)$ **58.** $(\sqrt{3} + x)(\sqrt{3} - x)$

59. $(x + 2)^3$ **60.** $(x^2 + x + 1)(x - 1)$

61. $(x^2 + x + 1)(x^2 - x + 1)$

62. $(2x^2 + 1)(2x^2 - 1)$

63. $(2x - 1)(x^2 + 3x + 2)$

64. $(1 - 2x)(4x^2 + x - 1)$

In Exercises 65–70, perform the indicated multiplication, and simplify your answer if possible.

65. $(\sqrt{x} + 5)(\sqrt{x} - 5)$

66. $(2\sqrt{x} + \sqrt{2y})(2\sqrt{x} - \sqrt{2y})$

67. $(3 + \sqrt{y})^2$ **68.** $(7w - \sqrt{2x})^2$

69. $(1 + \sqrt{3}x)(x + \sqrt{3})$ **70.** $(2y + \sqrt{3})(\sqrt{5}y - 1)$

In Exercises 71–76, compute the product, and arrange the terms of your answer according to decreasing powers of x, with each power of x appearing at most once.

Example: $(ax + b)(4x - c) = 4ax^2 + (4b - ac)x - bc$.

71. $(ax + b)(3x + 2)$ **72.** $(4x - c)(dx + c)$

73. $(ax + b)(bx + a)$ **74.** $rx(3rx + 1)(4x - r)$

75. $(x - a)(x - b)(x - c)$ **76.** $(2dx - c)(3cx + d)$

In Exercises 77–79, assume that all exponents are nonnegative integers, and find the product.

Example: $2x^k(3x + x^{n+1}) = (2x^k)(3x) + (2x^k)(x^{n+1})$

$\qquad\qquad\qquad = 6x^{k+1} + 2x^{k+n+1}$.

77. $(2x^n)(8x^k)$

78. $(x^m + 2)(x^n - 3)$ **79.** $(y^r + 1)(y^s - 4)$

In Exercises 80–82, express the area of the shaded region as an algebraic expression in x.

80.

81.

82.

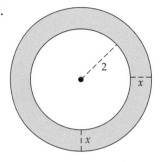

84. $x - (3y + 4) = x - 3y + 4$

85. $(x + y)^2 = x + y^2$ **86.** $(2x)^3 = 2x^3$

87. $(7x)(7y) = 7xy$ **88.** $(x + y)^2 = x^2 + y^2$

89. $y + y + y = y^3$ **90.** $(a - b)^2 = a^2 - b^2$

91. $(x - 3)(x - 2) = x^2 - 5x - 6$

92. $(a + b)(a^2 + b^2) = a^3 + b^3$

Thinkers

In Exercises 93 and 94, explain algebraically why each of these parlor tricks always works.

93. Write down a nonzero number. Add 1 to it, and square the result. Subtract 1 from the original number, and square the result. Subtract this second square from the first one. Divide by the number with which you started. The answer is 4.

94. Write down a positive number. Add 4 to it. Multiply the result by the original number. Add 4 to this result, and then take the square root. Subtract the number with which you started. The answer is 2.

95. Invent a similar parlor trick in which the answer is always the number with which you started.

Errors to Avoid

In Exercises 83–92, find a numerical example to show that the given statement is false. *Then find the mistake in the statement and correct it.*

Example: The statement $-(b + 2) = -b + 2$ *is false when* $b = 5$, *since* $-(5 + 2) = -7$ *but* $-5 + 2 = -3$. *The mistake is the sign on the 2. The correct statement is* $-(b + 2) = -b - 2$.

83. $3(y + 2) = 3y + 2$

1.C   Factoring

Factoring is the reverse of multiplication: We begin with a product and find the factors that multiply together to produce this product. Factoring skills are necessary to simplify expressions, to do arithmetic with fractional expressions, and to solve equations and inequalities.

Here is the first general rule for factoring.

Common Factors

> If there is a common factor in every term of the expression, factor out the common factor of highest degree.

EXAMPLE 1

In $4x^6 - 8x$, for example, each term contains a factor of $4x$, so

$$4x^6 - 8x = 4x(x^5 - 2).$$

Similarly, the common factor of highest degree in $x^3y^2 + 2xy^3 - 3x^2y^4$ is xy^2, and

$$x^3y^2 + 2xy^3 - 3x^2y^4 = xy^2(x^2 + 2y - 3xy^2). \quad \blacksquare$$

You can greatly increase your factoring proficiency by learning to recognize multiplication patterns that appear frequently. Here are the most common ones.

Quadratic Factoring Patterns

Difference of Squares: $u^2 - v^2 = (u + v)(u - v);$

Perfect Squares: $u^2 + 2uv + v^2 = (u + v)^2;$

$u^2 - 2uv + v^2 = (u - v)^2.$

EXAMPLE 2

(a) $x^2 - 9y^2$ can be written $x^2 - (3y)^2$, a difference of squares. Therefore, $x^2 - 9y^2 = (x + 3y)(x - 3y).$

(b) $y^2 - 7 = y^2 - (\sqrt{7})^2 = (y + \sqrt{7})(y - \sqrt{7}).$*

(c) $36r^2 - 64s^2 = (6r)^2 - (8s)^2 = (6r + 8s)(6r - 8s)$
$= 2(3r + 4s)2(3r - 4s) = 4(3r + 4s)(3r - 4s).$ ∎

EXAMPLE 3

Since the first and last terms of $4x^2 - 36x + 81$ are perfect squares, we try to use the perfect square pattern with $u = 2x$ and $v = 9$.

$$4x^2 - 36x + 81 = (2x)^2 - 36x + 9^2$$
$$= (2x)^2 - 2 \cdot 2x \cdot 9 + 9^2 = (2x - 9)^2.$$ ∎

Cubic Factoring Patterns

Difference of Cubes: $u^3 - v^3 = (u - v)(u^2 + uv + v^2);$

Sum of Cubes: $u^3 + v^3 = (u + v)(u^2 - uv + v^2);$

Perfect Cubes: $u^3 + 3u^2v + 3uv^2 + v^3 = (u + v)^3;$

$u^3 - 3u^2v + 3uv^2 - v^3 = (u - v)^3.$

EXAMPLE 4

(a) $x^3 - 125 = x^3 - 5^3 = (x - 5)(x^2 + 5x + 5^2)$
$= (x - 5)(x^2 + 5x + 25).$

(b) $x^3 + 8y^3 = x^3 + (2y)^3 = (x + 2y)[x^2 - x \cdot 2y + (2y)^2]$
$= (x + 2y)(x^2 - 2xy + 4y^2).$

(c) $x^3 - 12x^2 + 48x - 64 = x^3 - 12x^2 + 48x - 4^3$
$= x^3 - 3x^2 \cdot 4 + 3x \cdot 4^2 - 4^3$
$= (x - 4)^3.$ ∎

*When a polynomial has integer coefficients, we normally look only for factors with integer coefficients. But when it is easy to find other factors, as here, we shall do so.

When none of the multiplication patterns applies, use trial and error to factor quadratic polynomials. If a quadratic has two first-degree factors, then the factors must be of the form $ax + b$ and $cx + d$ for some constants a, b, c, and d. The product of such factors is

$$(ax + b)(cx + d) = acx^2 + adx + bcx + bd$$
$$= acx^2 + (ad + bc)x + bd.$$

Note that ac *is the coefficient of* x^2 and bd *is the constant term* of the product polynomial. This pattern can be used to factor quadratics by reversing the FOIL process.

EXAMPLE 5

Factor $x^2 + 9x + 18$.

SOLUTION   If $x^2 + 9x + 18$ factors as $(ax + b)(cx + d)$, then we must have $ac = 1$ (coefficient of x^2) and $bd = 18$ (constant term). Thus, $a = \pm 1$ and $c = \pm 1$ (the only integer factors of 1). The only possibilities for b and d are

$$\pm 1, \pm 18 \quad \text{or} \quad \pm 2, \pm 9 \quad \text{or} \quad \pm 3, \pm 6.$$

We mentally try the various possibilities, using FOIL as our guide. For example, we try $b = 2$, $d = 9$ and check this factorization: $(x + 2)(x + 9)$. The sum of the outside and inside terms is $9x + 2x = 11x$, so this product can't be $x^2 + 9x + 18$. By trying other possibilities we find that $b = 3$, $d = 6$ leads to the correct factorization.

$$x^2 + 9x + 18 = (x + 3)(x + 6). \quad \blacksquare$$

EXAMPLE 6

Factor $6x^2 + 11x + 4$ as $(ax + b)(cx + d)$.

SOLUTION   We must find numbers a and c whose product is 6, the coefficient of x^2, and numbers b and d whose product is the constant term 4. Here are some possibilities.

$ac = 6$	a	± 1	± 2	± 3	± 6
	c	± 6	± 3	± 2	± 1

$bd = 4$	b	± 1	± 2	± 4
	d	± 4	± 2	± 1

Using trial and error shows that $(2x + 1)(3x + 4) = 6x^2 + 11x + 4$. $\blacksquare$

Occasionally, the patterns above can be used to factor expressions involving larger exponents than 2.

EXAMPLE 7

(a) $x^6 - y^6 = (x^3)^2 - (y^3)^2 = (x^3 + y^3)(x^3 - y^3)$
$$= (x + y)(x^2 - xy + y^2)(x - y)(x^2 + xy + y^2).$$
(b) $x^8 - 1 = (x^4)^2 - 1 = (x^4 + 1)(x^4 - 1)$
$$= (x^4 + 1)(x^2 + 1)(x^2 - 1)$$
$$= (x^4 + 1)(x^2 + 1)(x + 1)(x - 1). \quad \blacksquare$$

EXAMPLE 8

To factor $x^4 - 2x^2 - 3$, let $u = x^2$. Then,
$$x^4 - 2x^2 - 3 = (x^2)^2 - 2x^2 - 3$$
$$= u^2 - 2u - 3 = (u + 1)(u - 3)$$
$$= (x^2 + 1)(x^2 - 3)$$
$$= (x^2 + 1)(x + \sqrt{3})(x - \sqrt{3}). \quad \blacksquare$$

EXAMPLE 9

$3x^3 + 3x^2 + 2x + 2$ can be factored by regrouping and using the distributive law to factor out a common factor.
$$(3x^3 + 3x^2) + (2x + 2) = 3x^2(x + 1) + 2(x + 1)$$
$$= (3x^2 + 2)(x + 1). \quad \blacksquare$$

 ## EXERCISES 1.C

In Exercises 1–58, factor the expression.

1. $x^2 - 4$
2. $x^2 + 6x + 9$
3. $9y^2 - 25$
4. $y^2 - 4y + 4$
5. $81x^2 + 36x + 4$
6. $4x^2 - 12x + 9$
7. $5 - x^2$
8. $1 - 36u^2$
9. $49 + 28z + 4z^2$
10. $25u^2 - 20uv + 4v^2$
11. $x^4 - y^4$
12. $x^2 - 1/9$
13. $x^2 + x - 6$
14. $y^2 + 11y + 30$
15. $z^2 + 4z + 3$
16. $x^2 - 8x + 15$
17. $y^2 + 5y - 36$
18. $z^2 - 9z + 14$
19. $x^2 - 6x + 9$
20. $4y^2 - 81$
21. $x^2 + 7x + 10$
22. $w^2 - 6w - 16$
23. $x^2 + 11x + 18$
24. $x^2 + 3xy - 28y^2$
25. $3x^2 + 4x + 1$
26. $4y^2 + 4y + 1$
27. $2z^2 + 11z + 12$
28. $10x^2 - 17x + 3$
29. $9x^2 - 72x$
30. $4x^2 - 4x - 3$
31. $10x^2 - 8x - 2$
32. $7z^2 + 23z + 6$
33. $8u^2 + 6u - 9$
34. $2y^2 - 4y + 2$
35. $4x^2 + 20xy + 25y^2$
36. $63u^2 - 46uv + 8v^2$
37. $x^3 - 125$
38. $y^3 + 64$
39. $x^3 + 6x^2 + 12x + 8$
40. $y^3 - 3y^2 + 3y - 1$
41. $8 + x^3$
42. $z^3 - 9z^2 + 27z - 27$
43. $-x^3 + 15x^2 - 75x + 125$
44. $27 - t^3$
45. $x^3 + 1$
46. $x^3 - 1$
47. $8x^3 - y^3$
48. $(x - 1)^3 + 1$
49. $x^6 - 64$

50. $x^5 - 8x^2$

51. $y^4 + 7y^2 + 10$

52. $z^4 - 5z^2 + 6$

53. $81 - y^4$

54. $x^6 + 16x^3 + 64$

55. $z^6 - 1$

56. $y^6 + 26y^3 - 27$

57. $x^4 + 2x^2y - 3y^2$

58. $x^8 - 17x^4 + 16$

61. $a^3 - 2b^2 + 2a^2b - ab$ **62.** $u^2v - 2w^2 - 2uvw + uw$

63. $x^3 + 4x^2 - 8x - 32$ **64.** $z^8 - 5z^7 + 2z - 10$

Thinker

65. Show that there do *not* exist real numbers c and d such that $x^2 + 1 = (x + c)(x + d)$.

In Exercises 59–64, factor by regrouping and using the distributive law (as in Example 9).

59. $x^2 - yz + xz - xy$ **60.** $x^6 - 2x^4 - 8x^2 + 16$

1.D   Fractional Expressions

Quotients of algebraic expressions, such as

$$\frac{x^2 - 2}{x - 1}, \quad \frac{1}{x^3 + 4x^2 - x}, \quad \text{or} \quad \frac{3\sqrt{x} - 7}{x^3(x + 1)}$$

are called **fractional expressions.** Throughout this section, we assume that all denominators are nonzero.

The basic rules for dealing with fractional expressions are essentially the same as those for ordinary numerical fractions. In particular, the usual cancellation property holds.

Cancellation

> If $k \neq 0$, then $\dfrac{ka}{kb} = \dfrac{a}{b}$.

EXAMPLE 1

Simplify $\dfrac{x^4 - 1}{x^2 + 1}$.

SOLUTION

Factor the numerator:
$$\frac{x^4 - 1}{x^2 + 1} = \frac{(x^2 + 1)(x^2 - 1)}{x^2 + 1}$$

Cancel the common factor $x^2 + 1$:
$$= \frac{x^2 - 1}{1} = x^2 - 1. \quad \blacksquare$$

From arithmetic, we know that

$$\frac{2}{4} \neq \frac{3}{6}.$$

Note that the cross-products are equal: $2 \cdot 6 = 4 \cdot 3$. The same thing is true in the general case.

Equality Rule

$$\frac{a}{b} = \frac{c}{d} \qquad \text{exactly when} \qquad ad = bc.$$

EXAMPLE 2

We know that

$$\frac{x^2 + 2x}{x^2 + x - 2} = \frac{x}{x - 1}$$

because the cross-products are equal.

$$(x^2 + 2x)(x - 1) = x^3 + x^2 - 2x \qquad \text{and} \qquad (x^2 + x - 2)x = x^3 + x^2 - 2x. \quad \blacksquare$$

A fraction is in **lowest terms** if its numerator and denominator have no common factors except ± 1.

EXAMPLE 3

Express $\dfrac{x^2 + x - 6}{x^2 - 3x + 2}$ in lowest terms.

SOLUTION   Proceed as we did in Example 1: Factor numerator and denominator, and cancel common factors.

$$\frac{x^2 + x - 6}{x^2 - 3x + 2} = \frac{(x - 2)(x + 3)}{(x - 2)(x - 1)} = \frac{x + 3}{x - 1}. \quad \blacksquare$$

To add or subtract two fractions with the same denominator, simply add or subtract the numerators.

$$\frac{a}{b} + \frac{c}{b} = \frac{a + c}{b} \qquad \text{and} \qquad \frac{a}{b} - \frac{c}{b} = \frac{a - c}{b}.$$

EXAMPLE 4

$$\begin{aligned}
\frac{7x^2 + 2}{x^2 + 3} - \frac{4x^2 + 2x - 5}{x^2 + 3} &= \frac{(7x^2 + 2) - (4x^2 + 2x - 5)}{x^2 + 3} \\
&= \frac{7x^2 + 2 - 4x^2 - 2x + 5}{x^2 + 3} \\
&= \frac{3x^2 - 2x + 7}{x^2 + 3}. \quad \blacksquare
\end{aligned}$$

To add or subtract fractions with different denominators, you must use a **common denominator,** that is, replace the given fractions by equivalent ones that have the same denominator. For instance, to add 1/3 and 1/4, we use the common denominator 12 (the product of 3 and 4).

$$\frac{1}{3} = \frac{4}{12} \quad \text{and} \quad \frac{1}{4} = \frac{3}{12},$$

so

$$\frac{1}{3} + \frac{1}{4} = \frac{4}{12} + \frac{3}{12} = \frac{4+3}{12} = \frac{7}{12}.$$

EXAMPLE 5

Compute $\dfrac{2x+1}{3x} - \dfrac{x^2-2}{x-1}$.

SOLUTION   We use the product of the denominators $3x(x-1)$ as the common denominator. To rewrite the first fraction in terms of this denominator, we must find the right-side numerator here.

$$\frac{2x+1}{3x} = \frac{?}{3x(x-1)}.$$

Note that to get from the left-side denominator to the right-side one, you multiply by $x-1$. To keep the fractions equal, you must multiply the left-hand numerator by $x-1$ also.

$$\frac{2x+1}{3x} = \frac{(2x+1)(x-1)}{3x(x-1)}.$$

A similar procedure works with the other fraction.

$$\frac{x^2-2}{x-1} = \frac{?}{3x(x-1)} \qquad \left(\frac{x^2-2}{x-1} = \frac{3x(x^2-2)}{3x(x-1)}. \right.$$

Multiply by $3x$

Now that the fractions have the same denominator, it's easy to subtract them.

$$\frac{2x+1}{3x} - \frac{x^2-2}{x-1} = \frac{(2x+1)(x-1)}{3x(x-1)} - \frac{3x(x^2-2)}{3x(x-1)}$$

$$= \frac{(2x+1)(x-1) - 3x(x^2-2)}{3x(x-1)}$$

$$= \frac{2x^2 - x - 1 - 3x^3 + 6x}{3x^2 - 3x}$$

$$= \frac{-3x^3 + 2x^2 + 5x - 1}{3x^2 - 3x}. \qquad \blacksquare$$

Although the product of the denominators can always be used as a common denominator, it's usually more efficient to use the **least common denominator,** that is, the smallest quantity that can be the denominator of both fractions. The next example illustrates an easy way to find the least common denominator when dealing with numbers.

EXAMPLE 6

Find the least common denominator for $\dfrac{1}{100}$ and $\dfrac{1}{120}$.

SOLUTION The two denominators can be factored as

$$100 = 2^2 \cdot 5^2 \qquad \text{and} \qquad 120 = 2^3 \cdot 3 \cdot 5.$$

The distinct factors that appear in both numbers are 2, 3, and 5. For each factor, take the highest power to which it appears in either number (namely, 2^3, 3^1, and 5^2). The product of these highest powers is the least common denominator.

$$2^3 \cdot 3 \cdot 5^2 = 600.$$

Note that $600 = 100 \cdot 6$ and $600 = 120 \cdot 5$, so

$$\frac{1}{100} = \frac{1 \cdot 6}{100 \cdot 6} = \frac{6}{600} \qquad \text{and} \qquad \frac{1}{120} = \frac{1 \cdot 5}{120 \cdot 5} = \frac{5}{600}. \qquad \blacksquare$$

The technique of Example 6 can also be used to find the least common denominator of rational expressions.

EXAMPLE 7

Find the least common denominator for

$$\frac{1}{x^2 + 2x + 1}, \qquad \frac{5x}{x^2 - x}, \qquad \frac{3x - 7}{x^4 + x^3},$$

and express each fraction in terms of this common denominator.

SOLUTION Begin by factoring each denominator completely.

$$x^2 + 2x + 1 = (x + 1)^2 \qquad x^2 - x - x(x - 1) \qquad x^4 + x^3 = x^3(x + 1).$$

The distinct factors are x, $x + 1$, and $x - 1$. The least common denominator is

$$x^3(x + 1)^2(x - 1).$$

Therefore,

$$\frac{1}{(x + 1)^2} = \frac{1}{(x + 1)^2} \cdot \frac{x^3(x - 1)}{x^3(x - 1)} = \frac{x^3(x - 1)}{x^3(x + 1)^2(x - 1)}$$

$$\frac{5x}{x(x - 1)} = \frac{5x}{x(x - 1)} \cdot \frac{x^2(x + 1)^2}{x^2(x + 1)^2} = \frac{5x^3(x + 1)^2}{x^3(x + 1)^2(x - 1)}$$

$$\frac{3x - 7}{x^3(x + 1)} = \frac{3x - 7}{x^3(x + 1)} \cdot \frac{(x + 1)(x - 1)}{(x + 1)(x - 1)} = \frac{(3x - 7)(x + 1)(x - 1)}{x^3(x + 1)^2(x - 1)}. \qquad \blacksquare$$

EXAMPLE 8

To find $\dfrac{1}{z} + \dfrac{3z}{z+1} - \dfrac{z^2}{(z+1)^2}$, we use the least common denominator $z(z+1)^2$.

$$\frac{1}{z} + \frac{3z}{z+1} - \frac{z^2}{(z+1)^2} = \frac{(z+1)^2}{z(z+1)^2} + \frac{3z^2(z+1)}{z(z+1)^2} - \frac{z^3}{z(z+1)^2}$$

$$= \frac{(z+1)^2 + 3z^2(z+1) - z^3}{z(z+1)^2}$$

$$= \frac{z^2 + 2z + 1 + 3z^3 + 3z^2 - z^3}{z(z+1)^2}$$

$$= \frac{2z^3 + 4z^2 + 2z + 1}{z(z+1)^2}. \quad\blacksquare$$

Multiplication of fractions is easy: Multiply corresponding numerators and denominators, then simplify your answer.

EXAMPLE 9

$$\frac{x^2 - 1}{x^2 + 2} \cdot \frac{3x - 4}{x+1} = \frac{(x^2-1)(3x-4)}{(x^2+2)(x+1)}$$

$$= \frac{(x-1)(x+1)(3x-4)}{(x^2+2)(x+1)} = \frac{(x-1)(3x-4)}{x^2+2}. \quad\blacksquare$$

Division of fractions is given by the following rule.

Invert the divisor and multiply: $\quad \dfrac{a}{b} \div \dfrac{c}{d} = \dfrac{a}{b} \cdot \dfrac{d}{c} = \dfrac{ad}{bc}.$

EXAMPLE 10

$$\frac{x^2 + x - 2}{x^2 - 6x + 9} \div \frac{x^2 - 1}{x - 3} = \frac{x^2 + x - 2}{x^2 - 6x + 9} \cdot \frac{x-3}{x^2 - 1}$$

$$= \frac{(x+2)(x-1)}{(x-3)^2} \cdot \frac{x-3}{(x-1)(x+1)}$$

$$= \frac{x+2}{(x-3)(x+1)}. \quad\blacksquare$$

Division problems can also be written in fractional form. For instance, 8/2 means $8 \div 2$. Similarly, a compound fraction such as

$$\frac{\dfrac{x+1}{x^2-4}}{\dfrac{3x+2}{x+2}}$$

means

$$\frac{x+1}{x^2-4} \div \frac{3x+2}{x+2}, \qquad \text{which is equivalent to} \qquad \frac{x+1}{x^2-4} \cdot \frac{x+2}{3x+2}.$$

Consequently, the basic rule for simplifying a compound fraction is as follows.

Invert the denominator, and multiply it by the numerator.

EXAMPLE 11

Compute

(a) $\quad \dfrac{\dfrac{16y^2z}{8yz^2}}{\dfrac{yz}{6y^3z^3}}$

(b) $\quad \dfrac{\dfrac{y^2}{y+2}}{y^3+y}$

SOLUTION

(a) $\dfrac{16y^2z/8yz^2}{yz/6y^3z^3} = \dfrac{16y^2z}{8yz^2} \cdot \dfrac{6y^3z^3}{yz} = \dfrac{16 \cdot 6 \cdot y^5z^4}{8y^2z^3}$

$$= 2 \cdot 6 \cdot y^{5-2}z^{4-3} = 12y^3z.$$

(b) $\dfrac{\dfrac{y^2}{y+2}}{y^3+y} = \dfrac{y^2}{y+2} \cdot \dfrac{1}{y^3+y} = \dfrac{y^2}{(y+2)(y^3+y)}$

$$= \dfrac{y^2}{(y+2)y(y^2+1)}$$

$$= \dfrac{y}{(y+2)(y^2+1)}. \qquad \blacksquare$$

EXAMPLE 12

Simplify the compound fraction

$$\frac{\dfrac{1}{x+h} - \dfrac{1}{x}}{h}.$$

SOLUTION   First write the numerator as a single fraction, then invert and multiply.

$$\frac{\dfrac{1}{x+h} - \dfrac{1}{x}}{h} = \frac{\dfrac{x}{x(x+h)} - \dfrac{x+h}{x(x+h)}}{h}$$

$$= \frac{\dfrac{x-(x+h)}{x(x+h)}}{h} = \frac{\dfrac{-h}{x(x+h)}}{h}$$

$$= \frac{-h}{x(x+h)} \cdot \frac{1}{h} = \frac{-1}{x(x+h)}. \qquad \blacksquare$$

✓ EXERCISES 1.D

In Exercises 1–10, express the fraction in lowest terms.

1. $\dfrac{63}{49}$

2. $\dfrac{121}{33}$

3. $\dfrac{13 \cdot 27 \cdot 22 \cdot 10}{6 \cdot 4 \cdot 11 \cdot 12}$

4. $\dfrac{x^2 - 4}{x + 2}$

5. $\dfrac{x^2 - x - 2}{x^2 + 2x + 1}$

6. $\dfrac{z + 1}{z^3 + 1}$

7. $\dfrac{a^2 - b^2}{a^3 - b^3}$

8. $\dfrac{x^4 - 3x^2}{x^3}$

9. $\dfrac{(x + c)(x^2 - cx + c^2)}{x^4 + c^3 x}$

10. $\dfrac{x^4 - y^4}{(x^2 + y^2)(x^2 - xy)}$

In Exercises 11–28, perform the indicated operations.

11. $\dfrac{3}{7} + \dfrac{2}{5}$

12. $\dfrac{7}{8} - \dfrac{5}{6}$

13. $\left(\dfrac{19}{7} + \dfrac{1}{2}\right) - \dfrac{1}{3}$

14. $\dfrac{1}{a} - \dfrac{2a}{b}$

15. $\dfrac{c}{d} + \dfrac{3c}{e}$

16. $\dfrac{r}{s} + \dfrac{s}{t} + \dfrac{t}{r}$

17. $\dfrac{b}{c} - \dfrac{c}{b}$

18. $\dfrac{a}{b} + \dfrac{2a}{b^2} + \dfrac{3a}{b^3}$

19. $\dfrac{1}{x + 1} - \dfrac{1}{x}$

20. $\dfrac{1}{2x + 1} + \dfrac{1}{2x - 1}$

21. $\dfrac{1}{x + 4} + \dfrac{2}{(x + 4)^2} - \dfrac{3}{x^2 + 8x + 16}$

22. $\dfrac{1}{x} + \dfrac{1}{xy} + \dfrac{1}{xy^2}$

23. $\dfrac{1}{x} - \dfrac{1}{3x - 4}$

24. $\dfrac{3}{x - 1} + \dfrac{4}{x + 1}$

25. $\dfrac{1}{x + y} + \dfrac{x + y}{x^3 + y^3}$

26. $\dfrac{6}{5(x - 1)(x - 2)^2} + \dfrac{x}{3(x - 1)^2(x - 2)}$

27. $\dfrac{1}{4x(x + 1)(x + 2)^3} - \dfrac{6x + 2}{4(x + 1)^3}$

28. $\dfrac{x + y}{(x^2 - xy)(x - y)^2} - \dfrac{2}{(x^2 - y^2)^2}$

In Exercises 29–42, express in lowest terms.

29. $\dfrac{3}{4} \cdot \dfrac{12}{5} \cdot \dfrac{10}{9}$

30. $\dfrac{10}{45} \cdot \dfrac{6}{14} \cdot \dfrac{1}{2}$

31. $\dfrac{3a^2 c}{4ac} \cdot \dfrac{8ac^3}{9a^2 c^4}$

32. $\dfrac{6x^2 y}{2x} \cdot \dfrac{y}{21xy}$

33. $\dfrac{7x}{11y} \cdot \dfrac{66y^2}{14x^3}$

34. $\dfrac{ab}{c^2} \cdot \dfrac{cd}{a^2 b} \cdot \dfrac{ad}{bc^2}$

35. $\dfrac{3x + 9}{2x} \cdot \dfrac{8x^2}{x^2 - 9}$

36. $\dfrac{4x + 16}{3x + 15} \cdot \dfrac{2x + 10}{x + 4}$

37. $\dfrac{5y - 25}{3} \cdot \dfrac{y^2}{y^2 - 25}$

38. $\dfrac{6x - 12}{6x} \cdot \dfrac{8x^2}{x - 2}$

39. $\dfrac{u}{u - 1} \cdot \dfrac{u^2 - 1}{u^2}$

40. $\dfrac{t^2 - t - 6}{t^2 - 6t + 9} \cdot \dfrac{t^2 + 4t - 5}{t^2 - 25}$

41. $\dfrac{2u^2 + uv - v^2}{4u^2 - 4uv + v^2} \cdot \dfrac{8u^2 + 6uv - 9v^2}{4u^2 - 9v^2}$

42. $\dfrac{2x^2 - 3xy - 2y^2}{6x^2 - 5xy - 4y^2} \cdot \dfrac{6x^2 + 6xy}{x^2 - xy - 2y^2}$

In Exercises 43–60, compute the quotient and express in lowest terms.

43. $\dfrac{5}{12} \div \dfrac{4}{14}$

44. $\dfrac{\dfrac{100}{52}}{\dfrac{27}{26}}$

45. $\dfrac{uv}{v^2 w} \div \dfrac{uv}{u^2 v}$

46. $\dfrac{3x^2 y}{(xy)^2} \div \dfrac{3xyz}{x^2 y}$

47. $\dfrac{\dfrac{x + 3}{x + 4}}{\dfrac{2x}{x + 4}}$

48. $\dfrac{\dfrac{(x + 2)^2}{(x - 2)^2}}{\dfrac{x^2 + 2x}{x^2 - 4}}$

49. $\dfrac{x + y}{x + 2y} \div \left(\dfrac{x + y}{xy}\right)^2$

50. $\dfrac{\dfrac{u^3 + v^3}{u^2 - v^2}}{\dfrac{u^2 - uv + v^2}{u + v}}$

51. $\dfrac{(c + d)^2}{\dfrac{c^2 - d^2}{cd}}$

52. $\dfrac{\dfrac{1}{x} - \dfrac{3}{2}}{\dfrac{2}{x - 2} + \dfrac{5}{x}}$

53. $\dfrac{\dfrac{1}{x^2} - \dfrac{1}{y^2}}{\dfrac{1}{x} + \dfrac{1}{y}}$

54. $\dfrac{\dfrac{x}{x + 1} + \dfrac{1}{x}}{\dfrac{1}{x} + \dfrac{1}{x + 1}}$

55. $\dfrac{\dfrac{6}{y} - 3}{1 - \dfrac{1}{y - 1}}$

56. $\dfrac{\dfrac{1}{3x} - \dfrac{1}{4y}}{\dfrac{5}{6x^2} + \dfrac{1}{y}}$

57. $\dfrac{\dfrac{1}{2x + 2h} - \dfrac{1}{2x}}{h}$

58. $\dfrac{\dfrac{1}{(x + h)^2} - \dfrac{1}{x^2}}{h}$

59. $(x^{-1} + y^{-1})^{-1}$

60. $\dfrac{(x + y)^{-1}}{x^{-1} + y^{-1}}$

In Exercises 61–64, a dartboard is shown. Assuming that every dart actually hits the board, find the probability that the dart will land in the shaded area. This probability is the quotient.

$$\frac{\text{Area of shaded region}}{\text{Area of dartboard}}.$$

61.

62.

63.

64.

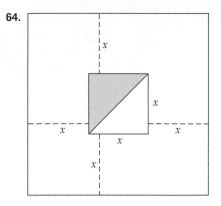

Errors to Avoid

In Exercises 65–71, find a numerical example to show that the given statement is false. *Then find the mistake in the statement and correct it.*

65. $\dfrac{1}{a} + \dfrac{1}{b} = \dfrac{1}{a+b}$

66. $\dfrac{x^2}{x^2 + x^6} = 1 + x^3$

67. $\left(\dfrac{1}{\sqrt{a} + \sqrt{b}}\right)^2 = \dfrac{1}{a+b}$

68. $\dfrac{r+s}{r+t} = 1 + \dfrac{s}{t}$

69. $\dfrac{u}{v} + \dfrac{v}{u} = 1$

70. $\dfrac{\frac{1}{x}}{\frac{1}{y}} = \dfrac{1}{xy}$

71. $(\sqrt{x} + \sqrt{y})\,\dfrac{1}{\sqrt{x} + \sqrt{y}} = x + y$

Appendix 2 GEOMETRY REVIEW

An **angle** consists of two half-lines that begin at the same point *P,* as in Figure 1. The point *P* is called the **vertex** of the angle, and the half-lines are called the **sides** of the angle.

Figure 1

An angle may be labeled by a Greek letter, such as angle θ in Figure 1(a), or by listing three points (a point on one side, the vertex, a point on the other side), such as angle *QPM* in Figure 1(b).

To measure the size of an angle, we must assign a number to each angle. Here is the classical method for doing this.

1. Construct a circle whose center is the vertex of the angle.

2. Divide the circumference of the circle into 360 equal parts (called **degrees**) by marking 360 points on the circumference, beginning with the point where one side of the angle intersects the circle. Label these points $0°$, $1°$, $2°$, $3°$, and so on.

3. The label of the point where the second side of the angle intersects the circle is the degree measure of the angle.

For example, Figure 2 shows an angle θ of measure 25 degrees (in symbols, $25°$) and an angle β of measure $135°$.

Figure 2

An **acute angle** is an angle whose measure is strictly between $0°$ and $90°$, such as angle θ in Figure 2. A **right angle** is an angle that measures $90°$. An **obtuse angle** is an angle whose measure is strictly between $90°$ and $180°$, such as angle β in Figure 2.

▪▪ TRIANGLES

A **triangle** has three sides (straight line segments) and three angles, formed at the points where the various sides meet. When angles are measured in degrees,

the sum of the measures of all three angles of a triangle is *always* 180°.

For instance, see Figure 3. This fact is proved in Exercise 37.

Figure 3

A **right triangle** is a triangle, one of whose angles is a right angle, such as the first two triangles shown in Figure 3. The side of a right triangle that lies opposite the right angle is called the **hypotenuse.** In each of the right triangles in Figure 3, side AC is the hypotenuse.

Pythagorean Theorem

If the sides of a right triangle have lengths a and b and the hypotenuse has length c, then

$$c^2 = a^2 + b^2.$$

See Exercise 38 for a proof of the Pythagorean Theorem.

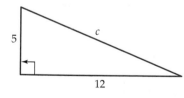

Figure 4

EXAMPLE 1

Consider the right triangle with sides of lengths 5 and 12, as shown in Figure 4.

According to the Pythagorean Theorem, the length c of the hypotenuse satisfies the equation $c^2 = 5^2 + 12^2 = 25 + 144 = 169$. Since $169 = 13^2$, we see that c must be 13. ∎

Isosceles Triangle Theorem

If two angles of a triangle are equal, then the two sides opposite these angles have the same length.

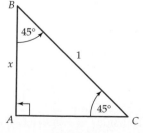

Figure 5

EXAMPLE 2

Suppose the hypotenuse of the right triangle shown in Figure 5 has length 1 and that angles B and C measure 45° each.

Then by the Isosceles Triangle Theorem, sides AB and AC have the same length. If x is the length of side AB, then by the Pythagorean Theorem,

$$x^2 + x^2 = 1^2$$

$$2x^2 = 1$$

$$x^2 = \frac{1}{2}$$

$$x = \sqrt{\frac{1}{2}} = \frac{1}{\sqrt{2}} = \frac{\sqrt{2}}{2}.$$

(We ignore the other solution of this equation, namely, $x = -\sqrt{1/2}$, since x represents a length here and therefore must be nonnegative.) Therefore, the sides of a $90°-45°-45°$ triangle with hypotenuse 1 are each of length $\sqrt{2}/2$. ∎

30-60-90
Triangle Theorem

In a right triangle that has an angle of 30°, the length of the side opposite the 30° angle is one-half the length of the hypotenuse.

Figure 6

EXAMPLE 3

Suppose that in the right triangle shown in Figure 6 angle B is 30° and the length of hypotenuse BC is 2.

By the 30-60-90 Triangle Theorem, the side opposite the 30° angle, namely, side AC, has length 1. If x denotes the length of side AB, then by the Pythagorean Theorem,

$$1^2 + x^2 = 2^2$$

$$x^2 = 3$$

$$x = \sqrt{3} \qquad ∎$$

Figure 7

EXAMPLE 4

The right triangle shown in Figure 7 has a 30° angle at C, and side AC has length $\sqrt{3}/2$.

Let x denote the length of the hypotenuse BC. By the 30-60-90 Triangle Theorem, side AB has length $\frac{1}{2}x$. By the Pythagorean Theorem,

$$\left(\frac{1}{2}x\right)^2 + \left(\frac{\sqrt{3}}{2}\right)^2 = x^2$$

$$\frac{x^2}{4} + \frac{3}{4} = x^2$$

$$\frac{3}{4} = \frac{3}{4}x^2$$

$$x^2 = 1$$

$$x = 1.$$

Therefore, the triangle has hypotenuse of length 1 and sides of lengths $1/2$ and $\sqrt{3}/2$. ∎

Figures 6 and 7 illustrate properties shared by all triangles.

The longest side of a triangle is always opposite its largest angle.

The shortest side of a triangle is always opposite its smallest angle.

You may find these facts helpful when checking solutions to triangle problems.

▐▌ CONGRUENT TRIANGLES

Two angles are said to be **congruent** if they have the same measure, and two line segments are **congruent** if they have the same length. We say that two triangles are **congruent** if the three sides and three angles of one are congruent respectively to the corresponding sides and angles of the other, as illustrated in Figure 8.

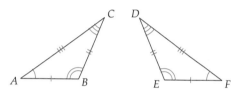

Congruent Triangles

Corresponding Congruent Sides*	Corresponding Congruent Angles
$AB \cong FE$	$\angle A \cong \angle F$
$BC \cong ED$	$\angle B \cong \angle E$
$CA \cong DF$	$\angle C \cong \angle D$

Figure 8

When two triangles are congruent, you can place one on top of the other so that they coincide. (You might have to rotate or flip one of the triangles over to do this, as is the case in Figure 8.) Thus,

Congruent triangles have the same size and same shape.

The key facts about congruent triangles are proved in high school geometry.

Congruent Triangles Theorem

If two triangles are given, then any one of the following conditions guarantees that they are congruent.

1. SAS: Two sides and the angle between them in one triangle are congruent to the corresponding sides and angle in the other triangle.

2. ASA: Two angles and the side between them in one triangle are congruent to the corresponding angles and side in the other triangle.

3. SSS: The three sides of one triangle are congruent to the corresponding sides of the other triangle.

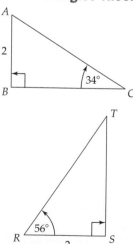

Figure 9

EXAMPLE 5

Show that the right triangles in Figure 9 are congruent.

*The symbol $\cong$ means "is congruent to."

SOLUTION   Triangle ABC has angles of 90° and 34°. Since the sum of all three angles is 180°, angle A must measure 56°. Hence, we have these congruences.

$$\angle A \cong \angle R \quad \text{[Both measure 56°]}$$

$$\angle B \cong \angle S \quad \text{[Both measure 90°]}$$

$$AB \cong RS \quad \text{[Both have length 2]}$$

Since two angles and the side between them in triangle ABC are congruent to the corresponding angles and side in triangle DEF, the triangles are congruent by ASA.   ■

▤ SIMILAR TRIANGLES

Two triangles are said to be **similar** if the three angles of one are congruent respectively to the three angles of the other, as illustrated in Figure 10.

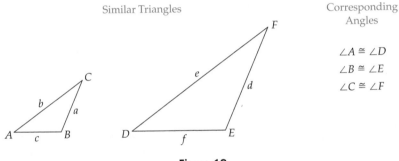

Similar Triangles

Corresponding Angles

$$\angle A \cong \angle D$$
$$\angle B \cong \angle E$$
$$\angle C \cong \angle F$$

Figure 10

Thus,

Similar triangles have the same shape, but not necessarily the same size.

In Figure 10, for example, the sides of triangle DEF are twice as long as the corresponding sides of triangle ABC:

$$d = 2a, \qquad e = 2b, \qquad f = 2c,$$

which is equivalent to:

$$\frac{d}{a} = 2, \qquad \frac{e}{b} = 2, \qquad \frac{f}{c} = 2.$$

Hence,

$$\frac{d}{a} = \frac{e}{b} = \frac{f}{c}.$$

A similar fact holds in the general case.

Ratios Theorem

Suppose triangle ABC is similar to triangle DEF (with $\angle A \cong \angle D$; $\angle B \cong \angle E$; $\angle C \cong \angle F$).

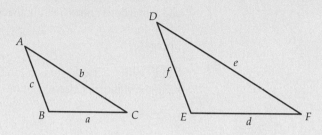

Then

$$\frac{d}{a} = \frac{e}{b} = \frac{f}{c}.$$

Figure 11

EXAMPLE 6

Suppose the triangles in Figure 11 are similar and that the sides have the lengths indicated. Find r and s.

SOLUTION By the Ratios Theorem,

$$\frac{\text{length } DF}{\text{length } AC} = \frac{\text{length } EF}{\text{length } BC}.$$

In other words,

$$\frac{s}{\sqrt{8}} = \frac{10}{3},$$

so

$$3s = 10\sqrt{8}$$

$$s = \left(\frac{10}{3}\right)\sqrt{8}.$$

Similarly,

$$\frac{\text{length } DE}{\text{length } AB} = \frac{\text{length } EF}{\text{length } BC},$$

so

$$\frac{r}{1} = \frac{10}{3}$$

$$r = \frac{10}{3}.$$

Therefore, the sides of triangle DEF are of lengths 10, $\frac{10}{3}$, and $\frac{10}{3}\sqrt{8}$. ∎

EXAMPLE 7

Tom, who is 6 feet tall, stands in the sunlight near the main library on campus, and Anne determines that his shadow is 7 feet long. At the same time, their friend Mary finds that the library casts a 100-foot-long shadow. How high is the main library?

SOLUTION   Both Tom and the library are perpendicular to the ground. So we have the right triangles in Figure 12 (which is not to scale).

 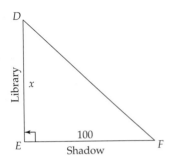

Figure 12

Angles B and E each measure 90°; hence,

$$\angle B \cong \angle E.$$

Because the sun hits both Tom and the library at the same angle,

$$\angle A \cong \angle D.$$

If each of these angles measures $k°$, then angle C measures $180° - 90° - k°$ (because the sum of the angles in triangle ABC must be 180°). Similarly, angle F must also measure $180° - 90° - k°$, so

$$\angle C \cong \angle F.$$

Therefore, triangle ABC is similar to triangle DEF. By the Ratios Theorem,

$$\frac{\text{length } DE}{\text{length } AB} = \frac{\text{length } EF}{\text{length } BC}$$

$$\frac{x}{6} = \frac{100}{7}$$

$$x = 6\left(\frac{100}{7}\right) \approx 85.7.$$

So the library is about 85.7 feet high.   ■

In addition to the definition, there are several other ways to show that two triangles are similar.

Similar Triangles Theorem

Given triangles *ABC* and *DEF*,

then any one of the following conditions guarantees that they are similar.

1. AA: Two angles of one are congruent to two angles of the other.

2. S/S/S: The ratios of corresponding sides are all equal:

$$\frac{d}{a} = \frac{e}{b} = \frac{f}{c}.$$

3. S/A/S: An angle in one is congruent to an angle in the other, and the ratios of the corresponding sides of these angles are the same:

$$\angle A \cong \angle D \qquad \text{or} \qquad \angle B \cong \angle E \qquad \text{or} \qquad \angle C \cong \angle F$$

$$\frac{e}{b} = \frac{f}{c}, \qquad\qquad \frac{d}{a} = \frac{f}{c} \qquad\qquad \frac{d}{a} = \frac{e}{b}.$$

Example 7 shows why the first condition guarantees similarity: If two pairs of angles are congruent, then the third angles are also congruent because the sum of the angles in each triangle is 180°. The proofs of the other conditions will be omitted.

EXAMPLE 8

Show that the triangles in Figure 13 are similar, and find angles *C*, *D*, and *E* and side *f*.

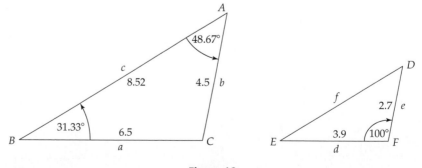

Figure 13

SOLUTION   Since the sum of the measures of the angles in triangle *ABC* is 180°, angle *C* must measure

$$180° - 31.33° - 48.67° = 100°.$$

Hence, $\angle C \cong \angle F$. The ratios of the corresponding sides of these angles are

$$\frac{d}{a} = \frac{3.9}{6.5} = .6 \quad \text{and} \quad \frac{e}{b} = \frac{2.7}{4.5} = .6$$

Thus,

$$\angle C \cong \angle F \quad \text{and} \quad \frac{d}{a} = \frac{e}{b}.$$

Therefore, triangle *ABC* is similar to triangle *DEF* by S/A/S of the Similar Triangles Theorem. Consequently, corresponding angles are congruent. In particular, angle *D* measures 48.67° and angle *E* measures 31.33° because $\angle D \cong \angle A$ and $\angle E \cong \angle B$. Since corresponding side ratios are equal,

$$\frac{f}{c} = \frac{d}{a}$$

$$\frac{f}{8.52} = \frac{3.9}{6.5} = .6$$

$$f = .6(8.52) = 5.112. \quad \blacksquare$$

✓ EXERCISES 2

In Exercises 1–8, find the missing side(s) of the triangle.

1.

2.

3.

4.

5.

6.

7.

8.

In Exercises 9 and 10, two congruent triangles are given. Find the missing sides and angles.

9.

10.

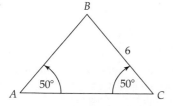

In Exercises 11 and 12, prove that the two triangles are congruent.

11.

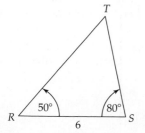

12.

13. If $\angle u \cong \angle v$ and $OP \cong OQ$, show that $PD \cong DQ$ and that PQ is perpendicular to OD. [*Hint:* Prove that triangles POD and DOQ are congruent.]

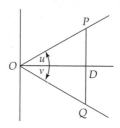

14. If $AB \cong AD$, $BC \cong DC$, $\angle x \cong \angle y$, and $\angle u \cong \angle v$, prove that triangles ADC and ABC are congruent.

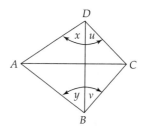

15. If $AB \cong DC$ and $\angle u \cong \angle v$, prove that triangles ABC and ACD are congruent.

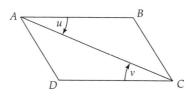

16. If $AB \cong CB$ and $\angle u \cong \angle v$, prove that triangles ABD and BCD are congruent.

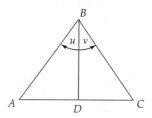

17. If $\angle x \cong \angle y$ and $\angle u \cong \angle v$, prove that triangles ABC and ADC are congruent.

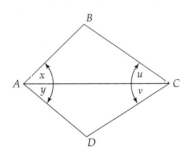

18. If $\angle x \cong \angle y$ and $\angle u \cong \angle v$, prove that triangles ABD and DBC are congruent.

In Exercises 19–22, triangle ABC is similar to triangle RST. The following notation is used: The side opposite angle A is denoted a, the side opposite angle B is denoted b, the side opposite angle R is denoted r, and so on.

19. If $\angle B = 37°$, $\angle C = 90°$, $a = 6$, $b = 4.5$ and $\angle R = 53°$, $\angle T = 90°$, $s = 21$, find c, r, t.

20. If $\angle A = 90°$, $c = 220$ and $\angle T = 90°$, $\angle S \cong \angle B$, $r = 100$, $s = 50$, find a, b, t.

21. If $a = 7$, $b = 5$, $c = 6$, and $\angle A \cong \angle T$, $\angle C \cong \angle R$, $s = 2.5$, find r and t.

22. If $a = 12$, $b = 9$, $c = 4$, and $\angle B \cong \angle T$, $\angle C \cong \angle R$, $t = 6.3$, find r and s.

In Exercises 23–26, prove that the triangles are similar, and find the missing sides and angles.

23.

24.

25.

26.

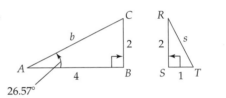

In Exercises 27–30, prove that triangles ABC and RST are similar, and find the missing sides and angles. The notation is the same as in Exercises 19–22.

27. $\angle C = 90°$, $a = 5$, $b = 5$ and $\angle T = 90°$, $r = 17.5$, $s = 17.5$.

28. $\angle B = (160/3)°$, $\angle C = 90°$, $a = 3.5$, $b = 4.7$ and $\angle R = (110/3)°$, $\angle T = 90°$, $r = 7$, $s = 9.4$

29. $\angle A = 105°$, $\angle B = 33.5°$, $a = 7$, $b = 4$, $c = 4.8$ and $\angle R = 105°$, $\angle T = 33.5°$, $r = 3.5$, $t = 2$.

30. $\angle B = 25°$, $\angle C = 142.8°$, $a = 5$, $b = 10$, $c = 14.3$, and $\angle S = 25°$, $\angle T = 12.2°$, $t = 10$

31. A flagpole casts a 99-foot shadow at the same time as a 5-foot-long stick (perpendicular to the ground) casts a 9-foot shadow. How high is the flagpole?

32. If a 6-foot-tall person casts a 10-foot shadow, how long is the shadow cast by a 28.3-foot-high tree that is next to the person?

33. A scale model of a pyramid is 1/20 the size of the original. If the model pyramid is 4 feet tall, how tall is the original pyramid?

34. A scale model of a ship is 8 inches long, and its mast is 4.5 inches high. If the real ship is 65 feet long, how high is the real mast?

35. On a map of Ohio, Columbus is 9.2 centimeters from Cincinnati and 8.5 centimeters from Marietta. On the map, Cincinnati is 15.75 centimeters from Marietta. If the actual distance from Columbus to Cincinnati is 100 miles, how far is Marietta from Columbus and from Cincinnati?

36. A small mirror is placed on level ground 4 feet from the base of a flagpole. A person whose eyes are 66 inches above the ground sees the top of the pole in the mirror when he is 2 feet from the mirror. The base of the pole, the mirror, and the person are aligned on a straight line. How high is the flagpole?

Thinkers

37. If *ABC* is a triangle, prove that the sum of its angles is 180°. [*Hint:* In the figure below, use congruent triangles to show that $\angle u \cong \angle x$ and $\angle v \cong \angle y$. You may assume that opposite sides of a rectangle have the same length.]

38. Prove the Pythagorean Theorem. [*Hint:* Use the figure below, in which triangle *ABC* has a right angle at *C*. Find three similar triangles in the figure, and use the properties of similar triangles to show that $a^2 + b^2 = c^2$.]

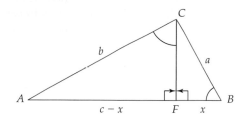

Appendix 3 PROGRAMS

Each program is preceded by a *Description,* which describes, in general terms, how the program operates and what it does. Some programs require that certain things be done before the program is run (such as entering a function in the function memory); these requirements are listed as *Preliminaries.* Occasionally, italic remarks appear in brackets after a program step; they are *not* part of the program but are intended to provide assistance when you are entering the program into your calculator. A remark such as "[*MATH NUM menu*]" means that the symbols or commands needed for that step of the program are in the NUM submenu of the MATH menu.

Fraction Conversion for Casio (Built-in on other calculators)

Description:　Enter a positive repeating decimal; the program converts it into a fraction.

Note:　When two submenus are given, the first is for 9850 and the second is for FX 2.0.

Fix 7　[*SETUP DISPLAY menu*]

"N ="?→N　[*PROG REL or LOGIC menu*]

0→D

Lbl 1　[*PROG JUMP menu*]

D + 1→D

N × D

Rnd　[*OPTN NUM menu*]

If (Frac Ans) ≠ 0　[*PROG COM or IF; OPTN NUM; PROG REL or LOGIC menus*]

Then Goto 1　[*PROG COM or IF; PROG JUMP menus*]

If End　[*PROG COM or IF menu*]

(Ans + .5)→N

Norm　[*SETUP DISP menu*]

Locate 3, 4, Int N　[*PROG IO; OPTN NUM menus*]

Locate 3, 5, "—"

Locate 3, 6, D◢

TI-82/83 Quadratic Formula (Built-in on other calculators)

Description:　Enter the coefficients of the quadratic equation $ax^2 + bx + c = 0$; the program finds all real solutions.

:ClrHome [*Optional*]

:Disp "AX^2 + BX + C = 0" [*Optional*]

:Prompt A

:Prompt B

:Prompt C

:(B^2 − 4AC) → S

:If S < 0

:Goto 1

:Disp $(-B + \sqrt{S})/(2A)$

:Disp $(-B - \sqrt{S})/(2A)$

:Stop

:Lbl 1

:Disp "NO REAL ROOTS"

TI-85 Table Maker (Built-in on other calculators)

Preliminaries: Enter the function to be evaluated in the function memory as Y_1.

Description: Select a starting point and an increment (the amount by which adjacent x entries differ); the program displays a table of function values. To scroll through the table a page at a time, press "down" or "up." Press "quit" to end the program. [*Note:* An error message will occur if the calculator attempts to evaluate the function at a number for which it is not defined. In this case, change the starting point or increment to avoid the undefined point.]

:Lbl SETUP

:ClLCD [*Optional*]

Disp " "TABLE SETUP"

:Input "TblMin =", tblmin

:Input "ΔTbl =", dtbl   [*CHAR GREEK menu*]

:tblmin → x   [*Use the x-var key for x.*]

:Lbl CONTD

:ClLCD

:Output(1,1,"x")   [*I/O menu*]

:Output(1,10,"y₁")

:For (cnt,2,7,1)

:Output(cnt,1,x)

Output(cnt,8," ")

:Output(cnt,9,y₁)

:x + dtbl → x

:End

:Menu(1, "Down",CONTD, 2,"UP",CONTU, 4, "Setup," SETUP, 5, "quit", TQUIT)

:Lbl CONTU

:x − 12∗dtbl → x

:Goto CONTD

:Lbl TQUIT

:ClLCD

:Stop

Synthetic Division (Built-in on TI-89/92)

Preliminaries: Enter the coefficients of the dividend $F(x)$ (in order of decreasing powers of x, putting in zeros for missing coefficients) as list L_1 (or List 1). If the coefficients are 1, 2, 3, for example, key in {1, 2, 3,} and store it in L_1. The symbols { } are on the keyboard or in the LIST menu. The list name L_1 is on the keyboard of TI-82/83. On TI-85/86 and HP-39, type in L1. On Casio, use "List" in the LIST submenu of the OPTN menu to type in List 1.

Description: Write the divisor in the form $x − a$ and enter a. The program displays the degree of the quotient $Q(x)$, the coefficients of $Q(x)$ (in order of decreasing powers of x), and the remainder. If the program pauses before it has displayed all these items, you can use the arrow keys to scroll through the display; then press ENTER (or OK) to continue.

TI

:ClrHome [ClLCD on TI-85/86]

:Disp "DIVISOR IS X $-$ A"

:Prompt A

:$L_1 \rightarrow L_2$ [*See Preliminaries for how to enter list names*]

:dim $L_1 \rightarrow$ N [dimL L_1 on TI-85/86]   [*LIST OPS menu*]

:For (K, 2, N)

:$(L_1(K) + A \times L_2(K - 1)) \rightarrow L_2(K)$

:End

:round$(L_2(N),9) \rightarrow$ R [*MATH NUM menu*]

:$(N - 1) \rightarrow$ dim L_2

:Disp "DEGREE OF QUOTIENT"

:Disp dim $L_2 - 1$

:Disp "COEFFICIENTS"

:Pause L_2

:Disp "REMAINDER"

:Disp R

Casio

"DIVISOR IS X $-$ A"

"A $=$ "? $\rightarrow$ A

List 1 $\rightarrow$ List 2   [*See Preliminaries for how to enter list names*]

dim List 1 $\rightarrow$ N   [*OPTN LIST menu*]

2 $\rightarrow$ K

Lbl 1

(List 1[K] + A $\times$ List 2 [K $-$ 1]) $\rightarrow$ List 2 [K]

K + 1 $\rightarrow$ K

If K $\leq$ N

Then Goto 1

If End

Fix 9   [*SETUP DISP menu*]

List 2 [N]

Rnd   [*OPTN MATH NUM menu*]

Ans $\rightarrow$ R

Norm   [*SETUP DISP menu*]

Seq(List 2 [X], X, 1, N $-$ 1, 1) $\rightarrow$ List 2

      [*OPTN LIST menu*]

"DEGREE OF QUOTIENT"

dim List 2 $-$ 1◀

"COEFFICIENTS"

List 2 ◀

"REMAINDER"

R

HP-38/39

Input A; "SYNDIV", "X $-$ A"; "ENTER A"; 0:

$L_1 \rightarrow L_2$:   [*See Preliminaries for how to enter list names*]

Size$(L_2) \rightarrow$ N: [*MATH LIST menu*]

For C = 1 to (N $-$ 1)   Step 1; $L_2(C) \times A + L_2(C + 1) \rightarrow L_2(C + 1)$ End:

Makelist $(L_2(J), J, 1, N - 1, 1) \rightarrow L_3$:   [*MATH LIST menu*]

Msgbox "DEGREE OF QUOTIENT" Size$(L_3) - 1$:

Msgbox "COEFFICIENTS" L_3:

Msgbox "REMAINDER" $L_2(N)$:

RREF Program for TI-82 and Casio (Built-in on other calculators)

Preliminaries: Enter a matrix in the matrix memory as matrix A.

Description: When the program is run, it uses row operations to put matrix A in reduced row echelon form. The reduced matrix is displayed; you may scroll through it, using the arrow keys. When the program is finished, the original matrix is in the matrix memory as matrix A, and the reduced matrix is in the matrix memory as matrix B.

TI

:ClrHome
:[A] → [B]
:dim([B]) → L$_1$
:L$_1$(1) → M
:L$_1$(2) → N
:1 → R
:1 → J
:Lbl 6
:abs(seq([B](I, J), I, R, M, 1)) → L$_2$ [*LIST menu*]
:max(L$_2$) → T
:If T = 0
:Goto 3
:R → I
:Lbl 1
:If abs([B](I, J)) = T
:Goto 2
:I + 1 → I
:Goto 1
:Lbl 2
:*row([B](I, J)$^{-1}$, [B], I) → [B] [*MATRIX MATH menu*]
:For(K, 1, I − 1, 1)
:*row+(−[B](K, J), [B], I, K) → [B]
:End
:For(K, I + 1, M, 1)
:*row+(−[B](K, J), [B], I, K) → [B]
:End
:rowswap([B], I, R) → [B]
:round([B], 9) → [B]
:If R = min(M, N)
:Then
:Lbl 7
:Pause round([B], 9)▶Frac

:Stop
:End
:R + 1 → R
:J + 1 → J
:If J = N + 1
:Goto 7
:Goto 6
:Lbl 3
:J + 1 → J
:If J = N + 1
:Goto 7
:Goto 6

Casio

Mat A → Mat B
"NUMBER OF ROWS"? → M
"NUMBER OF COLUMNS"? → N
{M, N} → List 3
1 → R
1 → J
Lbl 6
Mat → List(Mat B, J) → List 1
seq(List 1[X], X, R, M, 1) → List 1 [*OPTN LIST menu*]
abs(List 1) → List 1
Max(List 1) → T [*OPTN LIST menu*]
If T = 0
Then Goto 3
If End
R → I
Lbl 1
abs(Mat B[I, J]) → S
If S = T
Then Goto 2
If End

I + 1 → I
Goto 1
Lbl 2
*Row Mat B[I, J]$^{-1}$, B, I [(*MENU*) *MAT menu*]
For 1 → K To I − 1
*Row+ −Mat B[K, J], B, I, K
Next
For I + 1 → K To M
*Row+ −Mat B[K, J], B, I, K
Next
Swap B, I, R
If R = Min(List 3)
Then Goto 7
Else Goto 8
If End
Lbl 7

Mat B ◢
Stop
Lbl 8
R + 1 → R
J + 1 → J
If J = N + 1
Then Goto 7
If End
Goto 6
Lbl 3
J + 1 → J
If J = N + 1
Then Goto 7
If End
Goto 6

HP-39 Rectangular/Polar Conversion Program

Description: Enter the rectangular coordinates of a point in the plane; the program displays the polar coordinates of the point.

Input X; "RECTANGULAR TO POLAR"; "X ="; "ENTER X"; 0:

Input Y: "RECTANGULAR TO POLAR"; "Y ="; "ENTER Y"; 0:

If X > 0

Then (ATAN(Y/X)) → C

Else (ATAN(X, Y) + π) → C:

End:

MSGBOX "R =" $\sqrt{X^2 + Y^2}$ "θ =" C:

HP-39 Polar/Rectangular Conversion Program

Description: Enter the polar coordinates of a point in the plane; the program displays the rectangular coordinates of the point.

Input R; "POLAR TO RECTANGULAR"; "R = "; "ENTER R"; 0:

Input θ; "POLAR TO RECTANGULAR"; "θ = "; "ENTER θ"; 0: [*CHARS menu*]

MSGBOX "X = " R(cos θ) "Y = " R(sin θ):

CHAPTER 1

Section 1.1, page 13

1.

3. $6.785 \cdot 10^9$ **5.** $5.91 \cdot 10^{12}$ **7.** $2 \cdot 10^{-9}$ m

9. 150,000,000,000 m

11. .0000000000000000000016726 kg

13. (a) 5.626×10^{12}; 2.813×10^8
(b) $20,000

15. $-17 < 14$ **17.** $x \geq 0$

19. $t > 0$ **21.** $c \leq 3$ **23.** $<$

25. $=$ **27.** $>$ **29.** $b + c = a$

31. a lies to the right of b.

33. $a < b$

35.

37.

39.

41. $[5, 8]$ **43.** $(-3, 14)$ **45.** $[-8, \infty)$

47. 11   **49.** 0   **51.** 169

53. $\pi - \sqrt{2}$ **55.** π **57.** $<$

59. $>$ **61.** $<$ **63.** 7

65. 14.5   **67.** $\pi - 3$ **69.** $\sqrt{3} - \sqrt{2}$

71. $\dfrac{4 \cdot \pi \cdot \sqrt{805}}{161} \approx 2.21$

73. $|x - 143| \leq 21$; $|x - 163| \leq 26$

75. t^2 **77.** $b - 3$

79. $-(c - d) = d - c$ **81.** 0

83. $|(c - d)^2| = (c - d)^2 = c^2 - 2cd + d^2$

85. $|x - 5| < 4$ **87.** $|x + 4| \leq 17$

89. $|c| < |b|$

91. The distance from x to 3 is less than 2 units.

93. The distance from x to -7 is at most 3 units.

95. (a) iii   (b) i   (c) ii   (d) v   (e) iv

97. $x = 1$ or -1 **99.** $x = 1$ or 3

101. $x = -\pi + 4$ or $-\pi - 4$

103. $-7 < x < 7$ **105.** $3 < x < 7$

107. $x \leq -5$ or $x \geq 1$

109. Since $|a| \geq 0$, $|b| \geq 0$, and $|c| \geq 0$, the sum $|a| + |b| + |c|$ is positive only when one or more of $|a|$, $|b|$, $|c|$ is positive. But $|a|$ is positive only when $a \neq 0$; similarly for b, c.

Special Topics 1.1.A, page 17

1. $.7777 \cdots$ **3.** $1.6428571428571 \cdots$

5. $.052631578947368421052 \cdots$

7. $\dfrac{37}{99}$ **9.** $\dfrac{758,679}{9900} = \dfrac{252,893}{3300}$

11. $\dfrac{5}{37}$ **13.** $\dfrac{517,896}{9900} = \dfrac{14,386}{275}$

15. no; $\dfrac{2}{3} = .6666 \cdots$ **17.** yes; $\dfrac{1}{64} = .015625$

19. no   **21.** yes; $\dfrac{1}{0.625} = 1.6$

23. If $d = .74999 \cdots$, then $1000d - 100d = (749.999 \cdots) - (74.999 \cdots) = 675$. Hence $900d = 675$ so that $d = \dfrac{675}{900} = \dfrac{3}{4}$. Also $.75000 \cdots = .75 = \dfrac{75}{100} = \dfrac{3}{4}$.

25. $.05882352941176470588823$

27. $\dfrac{1}{29} = .03448275862068965517241379310344 \cdots$

29. $\dfrac{283}{47} =$
$6.0212765957446808510638297872340425531914893617021\overline{2} \cdots$

31. (a) One of many possible ways is to use the nonrepeating decimal expansion of π. For instance, with .75, associate $.7531415926 \cdots$; with 6.593 associate $6.59331415926 \cdots$, etc. Thus different terminating decimals correspond to different nonrepeating ones.
(b) As suggested in the *Hint*, associate with $.134134134 \cdots$ the irrational number $.134013400134000134 \cdots$. With $6.17398419841 \cdots$ associate $6.173984109841009841000 9841 \cdots$, etc. Thus, different repeating decimals lead to different nonrepeating ones.

Section 1.2, page 26

1. $x = 8$

3. $x = \dfrac{-5}{6}$

5. $y = -32$

7. $y = \dfrac{x + 5}{3}$

9. $b = \dfrac{2A - hc}{h}$ $(h \neq 0)$

11. $h = \dfrac{4V}{\pi d^2}$ $(d \neq 0)$

13. $x = 3$ or 5

15. $x = -2$ or 7

17. $y = \dfrac{1}{2}$ or -3

19. $t = -2$ or $\dfrac{-1}{4}$

21. $u = 1$ or $\dfrac{-4}{3}$

23. $x = 1 \pm \sqrt{13}$

25. $x = \dfrac{1 + \sqrt{5}}{2}$ or $\dfrac{1 - \sqrt{5}}{2}$

27. 2

29. 2

31. 1

33. $x = 2 \pm \sqrt{3}$

35. $x = -3 \pm \sqrt{2}$

37. No real solutions

39. $x = \dfrac{1}{2} \pm \sqrt{2}$

41. $x = \dfrac{2 \pm \sqrt{3}}{2}$

43. $x = -3$ or -6

45. $x = \dfrac{-1 \pm \sqrt{2}}{2}$

47. $x = 5$ or $\dfrac{-3}{2}$

49. No real solutions

51. No real solutions

53. $x \approx 1.824$ or $.470$

55. $x = 13.79$

57. $y = \pm 1$ or $\pm \sqrt{6}$

59. $x = \pm \sqrt{7}$

61. $y = \pm 2$ or $\pm 1/\sqrt{2}$

63. $x = \pm 1/\sqrt{5}$

65. $x = -5$ or 8

67. $x = \dfrac{-5 \pm \sqrt{57}}{8}$

69. $Y = 1500$

71. About 6.3 sec

73. (a) After about 4.38 sec
(b) After 50 sec

75. (a) 2116.1 lb/ft^2; 670.5 lb/ft^2
(b) About 14,410 ft

77. 1960

79. $k = 10$ or -10

81. $k = 16$

83. Yes

85. No

87. $k = 4$

Special Topics 1.2.A, page 30

1. $x = -6$ or 3

3. $x = 3/2$

5. $x = 2$

7. $x = 3/2$

9. $x = -5$ or 1 or -3 or -1

11. $x = 1$ or 4 or $\dfrac{5 + \sqrt{33}}{2}$ or $\dfrac{5 - \sqrt{33}}{2}$

13. $CL = .02 + \sqrt{.000882}$ or $.02 - \sqrt{.000882}$

Special Topics 1.2.B, page 34

1. A varies directly as the square of r; the constant of variation is π.

3. A varies jointly as l and w; the constant of variation is 1.

5. V varies jointly as the square of r and h; the constant of variation is $\dfrac{\pi}{3}$.

7. $a = \dfrac{k}{b}$

9. $z = kxyw$

11. $d = k\sqrt{h}$

13. $v = ku$; $k = 4$

15. $v = k/u$; $k = 16$

17. $t = krs$; $k = 4$

19. $w = kxy^2$; $k = 2$

21. $T = kpv^3/u^2$; $k = 16$

23. $r = 4$

25. $b = 9/4$

27. $u = 50$

29. $r = 3$

31. \$1638

33. 12 lb

35. 80 kg/cm^2

37. 400 ft

39. 30 in.

41. 3750 kg

43. (a) 2400 lb
(b) No more than 4 ft

Section 1.3, page 44

1. $A(-3, \ 3)$; $B(-1.5, \ 3)$; $C(-2.5, \ 0)$; $D(-1.5, \ -3)$; $E(0, 2)$; $F(0, 0)$; $G(2, 0)$; $H(3, 1)$; $I(3, -1)$

3. $P(-6, 3)$

5. $P(4, 2)$

7.

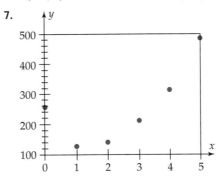

9. (a) Quadrant IV   (b) Quadrants III or IV

11. (a)

(b)

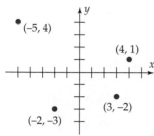

(c) They are mirror images of each other, with the *x*-axis being the mirror. In other words, they lie on the same vertical line, on opposite sides of the *x*-axis, the same distance from the axis.

13. $13; \left(\dfrac{-1}{2}, -1\right)$ **15.** $\sqrt{17}; \left(\dfrac{3}{2}, -3\right)$

17. $\sqrt{6 - 2\sqrt{6}} \approx 1.05; \dfrac{\sqrt{2} + \sqrt{3}}{2}, \dfrac{3}{2}$

19. $\sqrt{2} \cdot |a - b|; \left(\dfrac{a + b}{2}, \dfrac{a + b}{2}\right)$

21. $13 + \sqrt{13} + \sqrt{8} \approx 19.434$

23. Hypotenuse from $(1, 1)$ to $(2, -2)$ has length $\sqrt{10}$; other sides have lengths $\sqrt{2}$ and $\sqrt{8}$. Since $(\sqrt{2})^2 + (\sqrt{8})^2 = (\sqrt{10})^2$, this is a right triangle.

25. (a)

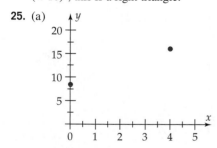

(b) Midpoint: $(2, 11.9)$
(c) If sales increase at a constant rate, then the midpoint suggests what sales will be in 2002.

27. (a) $\sqrt{\left(28\dfrac{1}{3}\right)^2 + 35^2} \approx 45.03$ yd

(b) $\left(34\dfrac{1}{6}, 27\dfrac{1}{2}\right)$

29. Yes   **31.** Yes   **33.** No

35. $x = 1, x = 5, y = -5$

37. $x = -1, x = 5, y = \pm\sqrt{5}$

39. $x = 0, x = -10, y = 0, y = 8$

41. $(x + 3)^2 + (y - 4)^2 = 4$

43. $x^2 + y^2 = 2$

45.

47.

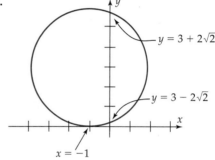

49. Center $(-4, 3)$; radius $2\sqrt{10}$

51. Center $(-3, 2)$; radius $2\sqrt{7}$

53. Center $(-12.5, -5)$; radius $\sqrt{169.25}$

55. (a) About 7.25 million
(b) 2007; about 7.75 million
(a) 2000–2002; 2011–2015

57. *B*   **59.** *C*

61. $(x - 2)^2 + (y - 2)^2 = 8$

63. $(x - 1)^2 + (y - 2)^2 = 8$

65. $(x + 5)^2 + (y - 4)^2 = 16$

67. $x^2 + y^2 - 4x - 2y = 0$

69. $(-3, -4)$ and $(2, 1)$

71. Assume $k > d$. The other two vertices of one possible square are $(c + k - d, d), (c + k - d, k)$; those of another square are $(c - (k - d), d), (c - (k - d), k)$; those of a third square are $\left(c + \dfrac{k - d}{2}, \dfrac{k + d}{2}\right), \left(c - \dfrac{k - d}{2}, \dfrac{k + d}{2}\right)$.

73. $(0, 0), (6, 0)$

75. $(3, -5 + \sqrt{11}), (3, -5 - \sqrt{11})$ **77.** $x = 6$

79. M has coordinates $(s/2, r/2)$ by the midpoint formula. Hence the distance from M to $(0, 0)$ is

$$\sqrt{\left(\frac{s}{2}-0\right)^2+\left(\frac{r}{2}-0\right)^2}=\sqrt{\frac{s^2}{4}+\frac{r^2}{4}},$$

and the distance from M to $(0, r)$ is the same:

$$\sqrt{\left(\frac{s}{2}-0\right)^2+\left(\frac{r}{2}-r\right)^2}=\sqrt{\left(\frac{s}{2}\right)^2+\left(-\frac{r}{2}\right)^2}$$
$$=\sqrt{\frac{s^2}{4}+\frac{r^2}{4}}$$

as is the distance from M to $(s, 0)$:

$$\sqrt{\left(\frac{s}{2}-s\right)^2+\left(\frac{r}{2}-0\right)^2}=\sqrt{\left(-\frac{s}{2}\right)^2+\left(\frac{r}{s}\right)^2}$$
$$=\sqrt{\frac{s^2}{4}+\frac{r^2}{4}}.$$

81. Place one vertex of the rectangle at the origin, with one side on the positive x-axis and another on the positive y-axis. Let $(a, 0)$ be the coordinates of the vertex on the x-axis and $(0, b)$ the coordinates of the vertex on the y-axis. Then the fourth vertex has coordinates (a, b) (draw a picture!). One diagonal has endpoints $(0, b)$ and $(a, 0)$, so that its length is $\sqrt{(0-a)^2+(b-0)^2}=\sqrt{a^2+b^2}$. The other diagonal has endpoints $(0, 0)$ and (a, b) and hence has the same length:
$$\sqrt{(0-a)^2+(0-b)^2}=\sqrt{a^2+b^2}.$$

83. The circle $(x-k)^2+y^2=k^2$ has center $(k, 0)$ and radius $|k|$ (the distance from $(k, 0)$ to $(0, 0)$). So the family consists of every circle that is tangent to the y-axis *and* has center on the x-axis.

85. The points are on opposite sides of the origin because one first coordinate is positive and one is negative. They are equidistant from the origin because the midpoint on the line segment joining them is

$$\left(\frac{c+(-c)}{2}, \frac{d+(-d)}{2}\right)=(0, 0).$$

Section 1.4, page 58

1. (a) C (b) B (c) B (d) D

3. $\dfrac{5}{2}$ **5.** 4   **7.** $t=22$ **9.** $t=\dfrac{12}{5}$

11.

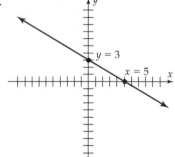

Slope of $L=\dfrac{C-0}{1-0}=C$

13. Slope 2; y-intercept 5

15. Slope, $\dfrac{-3}{7}$; y-intercept $\dfrac{-11}{7}$

17. $y=x+2$ **19.** $y=-x+8$

21. $y=-x-5$ **23.** $y=\dfrac{-7}{3}x+\dfrac{34}{9}$

25.

27.

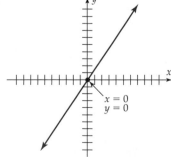

29.

31. Perpendicular **33.** Parallel **35.** Parallel

37. Perpendicular **39.** Yes

41. $y = 3x + 7$ **43.** $y = \dfrac{3}{2}x$

45. $y = x - 5$ **47.** $y = -x + 2$

49. $k = \dfrac{-11}{3}$ **51.** $y = \dfrac{-3}{4}x + \dfrac{25}{4}$

53. $y = \dfrac{-x}{2} + 6$ **55.** Both have slope $\dfrac{-A}{B}$.

57. (a) $y = \dfrac{9}{8}x + 54$
  (b) 57,375,000
  (c) 2005

59. (a) $y = 4.09x + 136.2$
  (b) 193.46
  (c) 4.09 per year

61. (a) $y = 5x + 115$
  (b) 125 lb; 185 lb
  (c) 6 ft 7 in.

63. (a) About 6,230,000   (b) In 2009

65. (a) $\dfrac{145x + 120,000}{x}$ (b) 4,000

67. (a) $c(x) = 110,000 + 50x$
  (b) $r(x) = 72x$
  (c) $p(x) = 22x - 110,000$
  (d) 5000 books

69. (a) 12.5 gal/min; 25/3 gal/min; 25 gal/min
  (b) $y = 75 - 10x$

71. (a) $y = 60 - 2x$ (b) $y = 80 - 2x$
  (c) $y = 160 - 4x$

73. (a) No more than \$20   (b) No more than \$40

75. Let $y = mx + b$ and $y = mx + c$ be equations of lines with same slope m, and $b \neq c$. Suppose (x_1, y_1) is an arbitrary point lying on both lines. Then, $y_1 = mx_1 + b$ and $y_1 = mx_1 + c$. So, $mx_1 + b = mx_1 + c$ and, $b = c$, a contradiction. Thus, the lines share no point in common so must be parallel.

77.

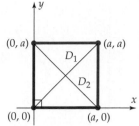

Equation of D_1; $y = x$, with slope 1
Equation of D_2; $y = -x + a$, with slope -1
(slope D_1)(slope D_2) = $(1)(-1) = -1$

Chapter 1 Review, page 64

1. (a) $>$ (b) $<$ (c) $<$ (d) $>$ (e) $=$

3. (a) 1.232×10^{16} (b) 7.89×10^{-11}

5. (a) $-10 < y < 0$ (b) $0 \leq x \leq 10$

7. (a) $|x + 7| < 3$ (b) $|y| > |x - 3|$

9. $x = 2$ or 8 **11.** $x = -1/2$ or $-11/2$

13. $-4 \leq x \leq 0$

15. (a) $7 - \pi$ (b) $\sqrt{23} - \sqrt{3}$

17. (a) $(-8, \infty)$ (b) $(-\infty, 5]$

19. $28/99$ **21.** $x = 44/7$

23. No real solutions **25.** $z = \dfrac{-3 \pm 2\sqrt{11}}{5}$

27. 2.25 times as large

29. 2 **31.** 6 **33.** 1

35. $x = 3$ or -3 or $\sqrt{2}$ or $-\sqrt{2}$

37. $x = -1$ or $5/3$ **39.** $\sqrt{58}$

41. $\sqrt{c^2 + d^2}$ **43.** $\left(d, \dfrac{c + 2d}{2}\right)$

45. (a) $\sqrt{17}$ (b) $(x - 2)^2 + (y + 3)^2 = 17$

47.

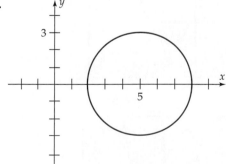

49. b and d **51.** c

53.

55. (a) 1   (b) 4/5   **57.** $y = 3x - 7$

59. $y = -2x + 1$ **61.** $x - 5y = -29$ **63.** 25,000 ft

65. False   **67.** False   **69.** False   **71.** False

73. (d)   **75.** (e)   **77.** 5/3

79. (a) $y = 0.2x + 70.8$
     (b) 73.6 years
     (c) In 2016

81. (c)   **83.** (d)

CHAPTER 2

Section 2.1, page 80

1–5. Answers vary

7.

9.

11.

13. (a)
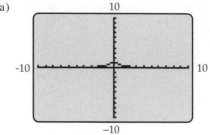

    (c) $-10 \leq x \leq 10$ and $0 \leq y \leq 1$

15.
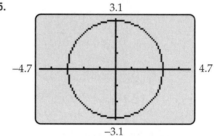

17. The two halves of the graph should be connected, but the ends of the two pieces are almost vertical so the calculator could not plot enough points near them to make the graph appear connected.

19.

21.

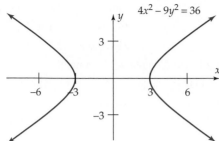

$4x^2 - 9y^2 = 36$

23.

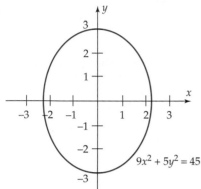

$9x^2 + 5y^2 = 45$

25. Highest at about (.7612, 3.9358); lowest at about (4.0812, −3.3356)

27. 1961   **29.** c   **31.** d   **33.** e

35. Not the same

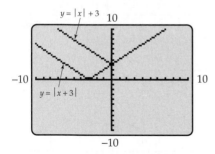

$y = |x| + 3$

$y = |x + 3|$

37. Same

39. Not the same

$y = \sqrt{x^2 + 9}$

$y = x + 3$

41. (a) Graphs appear identical.

(b) Graphs are different.

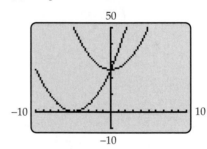

43. Possibly true   **45.** (c)

47. An appropriate viewing window should be large enough to contain the points (0, 4000), (20, 104000), and (25.080869, 0), and should not show negative values.

49. x should range from 0 to 48, and y should range from 0 to at least 31.5.

51. (a) 1917; 1929   (b) 1925; 104,000

53. (a) After about 6.3 hours (b) After about 18.7 hours

55. $-5 \le x \le 5$ and $-100 \le y \le 100$

57. $-10 \le x \le 10$ and $-2 \le y \le 20$ [Where is the right half of the graph?]

59. $-6 \le x \le 12$ and $-100 \le y \le 250$

61. All four graphs have the same shape. Graph (b) is graph (a) shifted 5 units vertically upward, graph (c) is graph (a) shifted 5 units vertically downward, and graph (d) is graph (a) shifted 2 units vertically downward.

63. Graph (b) is graph (a) stretched by a factor of 2 vertically away from the x-axis; similarly for graph (c), except the stretching factor is 3. Graph (d) is graph (a) shrunk vertically toward the x-axis by a factor of $\frac{1}{2}$.

65. The graphs are "mirror images" of each other, with the straight line $y = x$ being the mirror.

67. Same answer as Exercise 65

Section 2.2, page 89

1. 3   **3.** 3   **5.** 2

7. $x = -2.42645$ **9.** $x = 1.1640$ **11.** $x = 1.23725$

13. $x = 1.1921$ **15.** $x = -1.3794$ **17.** $x = -1.6005$

19. $x = 1.8608059$ **21.** $x = 2.1017$

23. $x = -1.7521$ **25.** $x = .9505$

27. $x = 0$ or 2.2074 **29.** $x = 2.3901454$

31. $x = -.6513878188$ or 1.151387819

33. $x = 7.033393042$ **35.** $x = 2/3$

37. $x = 1/12$ **39.** $x = \sqrt{3}$

41. $x = 1.4528$ or 112.00 **43.** $x = .0499$ or 1.9097

45. 2004   **47.** (a) 1999   (b) 2004

Section 2.3, page 98

1. The two numbers: x and y; their sum is 15: $x + y = 15$; the difference of their squares is 5: $x^2 - y^2 = 5$.

3.

English Language	Mathematical Language
Length of rectangle	x
Width of rectangle	y
Perimeter is 45	$x + y + x + y = 45$ or $2x + 2y = 45$
Area is 112.5	$xy = 112.5$

5. Let x be the old salary. Then the raise is 8% of x. Hence,

$$\text{old salary} + \text{raise} = \$1600$$

$$x + (8\% \text{ of } x) = 1600$$

$$x + .08x = 1600.$$

7. The circle has radius $r = 16/2 = 8$, so its area is $\pi r^2 = \pi \cdot 8^2 = 64\pi$. Let x be the amount by which the radius is to be reduced. Then $r = 8 - x$ and the new area is $\pi(8 - x)^2$, which must be 48π less than the original area, that is,
$\pi(8 - x)^2 = 64\pi - 48\pi$, or equivalently,
$\pi(8 - x)^2 = 16\pi$.

9. \$366.67 at 12% and \$733.33 at 6%

11. $2\frac{2}{3}$ qt   **13.** 60 mph   **15.** 44, 38.25

17. 65 mph   **19.** About 1.753 ft

21. 2 meters   **23.** About 132.7 ft

25. Red Riding Hood, 54 mph; wolf, 48 mph

27. 23 cm by 24 cm by 25 cm

29. About 4.658 in.   **31.** Less than 6 in.

33. 2.2 by 4.4 by 4 ft high

Section 2.4, page 103

1. $(-1, 8)$ **3.** $(1.1428562, -2.0625)$

5. $(-.3409, .0003222)$

7. (a) $(-1, 4)$ (b) $(-1, 4)$ and $(2, 4)$ (c) $(3, 20)$

9. $34.2 \times 34.2 \times 17.1$

11. (a) 4.4267 by 4.4267 inches.
(b) $10/3$ by $10/3$ inches

13. (a) Approximately 206
(b) Approximately 269; approximately \$577

15. (a) 600   (b) 958

17. $x = 9.306$; area ≈ 220.18 ft^2

19. Approximately $(1.871, 1.5)$ **21.** 12 times

Section 2.5, page 113

1.

MODEL: $y = x$			
Data Point	Model Point	Residual	Squared Residual
$(1, 2)$	$(1, 1)$	1	1
$(2, 2)$	$(2, 2)$	0	0
$(3, 3)$	$(3, 3)$	0	0
$(4, 3)$	$(4, 4)$	-1	1
$(5, 5)$	$(5, 5)$	0	0
		Sum: 0	Sum: 2

MODEL: $y = .5x + 1.5$			
Data Point	Model Point	Residual	Squared Residual
$(1, 2)$	$(1, 2)$	0	0
$(2, 2)$	$(2, 2.5)$	-0.5	0.25
$(3, 3)$	$(3, 3)$	0	0
$(4, 3)$	$(4, 3.5)$	-0.5	0.25
$(5, 5)$	$(5, 4)$	1	1
		Sum: 0	Sum: 1.5

The model $y = .5x + 1.5$ is better.

3.

MODEL: $y = 11x + 121$			
Data Point	Model Point	Residual	Squared Residual
(0, 129)	(0, 121)	8	64
(5, 161)	(5, 176)	−15	225
(9, 215)	(9, 220)	−5	25
(10, 244)	(10, 231)	13	169
		Sum: −1	Sum: 483

MODEL: $y = 11.5x + 129$			
Data Point	Model Point	Residual	Squared Residual
(0, 129)	(0, 129)	0	0
(5, 161)	(5, 186.5)	−25.5	650.25
(9, 215)	(9, 232.5)	−17.5	306.25
(10, 244)	(10, 244)	0	0
		Sum: −43	Sum: 956.5

The model $y = 11x + 121$ is better.

5. Positive   **7.** Very little

9.

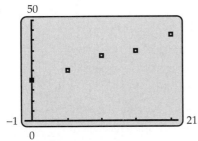

(a) The data appear to be linear.
(b) Positive

11.

(a) Nonlinear
(b) Positive

13. (a) Linear   (b) Negative

15. (a)

(b) $y = 398.24x − 417.32$
(c) $1176 million; $2769 million
(d) 5158

17. (a) $67.7x + 660.6$
(b) The y intercept is the expenditures for 1990. The slope is the amount the expenditures increase per year.
(c) $1811.5 billion

19. (a) $y = 12.3x + 238.5$; $y = 15.2x + 354.2$
(b) 435.3; 597.4

21. (a) $y = 18.5x + 63.7$; 193.2 million
(b) If one doesn't expect more than 80% of people to use the Internet, then the model loses its accuracy in 2006.

23. (a) $y = 227.2x + 1142.7$
(b) 2007

Chapter 2 Review, page 117

1. (a) a, d
(b) b, c do not show peaks and valleys near the origin; e crowds the graph onto the y-axis.
(c) a or $−4 \le x \le 6$ and $−10 \le y \le 10$

3. (a) None of them
(b) a, b, d do not show any peaks or valleys; c does not show the valleys; e shows only one point on the graph (which can't be distinguished because it's on the y-axis).
(c) $−7 \le x \le 11$ and $−1000 \le y \le 500$

5. (a) b, c
(b) a shows no peaks or valleys; d is too crowded horizontally; e shows only one point.
(c) $−10 \le x \le 10$ and $−150 \le y \le 150$

7. As x moves left or right from the origin, x^2 grows larger and larger, and hence so does $x^2 - 10$. Therefore, the graph always rises as it moves away from the y-axis and is complete.

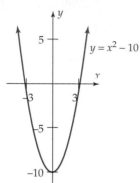

9. Note that y is defined only when $x \geq 5$ (why?). Also, $x - 5$ grows larger as x gets larger (that is, as you move to the right) and hence the same is true of $\sqrt{x - 5}$. Therefore, the graph always rises as you move to the right and is complete.

11.

13.

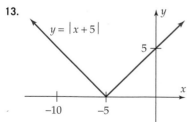

15. $x = 2.7644$ **17.** $x = 3.2678$

19. $x = -3.2843$ **21.** $x = 1.6511$

23. Gold, $\frac{3}{11}$ oz; silver, $\frac{8}{11}$ oz

25. $2\frac{2}{9}$ hrs   **27.** 9.6 ft

29. 4 ft   **31.** 25

33. 20 yd by 30 yd (interior fence is 20 yd)

35. $x = \sqrt{3} \approx 1.732$

37. (a)   140

(b) Nonlinear

39. (a) $y = 0.37x + 10.7$
(b) 1993: model: \$11.81; data: \$11.74; residual: -0.07
2000: model: \$14.40; data: \$14.38; residual: -0.02
(c) \$15.88

41. (a) $y = 12.1817x + 74.5656$
(b) \$220.8 billion; \$257.3 billion
(c) late 2001

CHAPTER 3

Section 3.1, page 132

1. Yes. Each input produces only one output.

3. No. The value -5 produces two outputs.

5. 6   **7.** -2 **9.** -17

11. y is a function of x.   **13.** x is a function of y.

15. y is a function of x and x is a function of y.

17. Neither x nor y is a function of the other.

19.

X	Y1
-2	-2
-1.5	-3.25
-1	-4
-.5	-4.25
0	-4
.5	-3.25
1	-2

X=-2

X	Y1
1.5	-.25
2	2
2.5	4.75
3	8
3.5	11.75
4	16
4.5	20.75

X=4.5

21.

23. $500 \to 0$
$1509 \to 0$
$3754 \to 0$
$6783 \to 45.66$
$12,500 \to 210$
$55,342 \to 2030.52$

25. The definition requires that a given input have only one output, not that a given output be produced by at most one input.

27. Postage is a function of weight since each weight determines one and only one postage amount. But weight is *not* a function of postage since a given postage amount may apply to several different weights. For instance, *all* letters under 1 oz use just one first-class stamp.

29. (a)

AVERAGE MAN	
Drinks in 1 hour	**Blood Alcohol Level**
2	.03
3	.05
4	.07
5	.10

AVERAGE WOMAN	
Drinks in 1 hour	**Blood Alcohol Level**
2	.05
3	.08

(b) Yes. For average man chart, the domain consists of all x such that $2 \le x \le 5$ and the range of all y such that $.03 \le y \le .10$. For the average woman the domain consists of all x such that $2 \le x \le 3$ and the range of all y such that $.05 \le y \le .08$.

31. (a) $A = \pi r^2$ (b) $A = \frac{1}{4}\pi d^2$

33. $V = 4x^3$

35. $D = 400 - 16t^2$; range $= [0, 400]$

37. (a)

Year	Percent
1960	12.82
1970	13.24
1980	17.53
1990	20.85
2000	23.21

(b) The model seems reasonable.
(c) 21.96; 25.38
(d) 2010

39. Domain: $[-3, 3]$; range: $[-4, 3]$

41. Domain: $[-3, 4]$; range: $[-1.2, 3]$

43. Domain: $[-2.5, 2]$; range: $[-3, 3]$

45. In all cases, there is an input that produces more than one output.

47. (a) All real numbers except numbers of the form $\frac{\pi}{2} + k \cdot \pi$, where k is an integer (assuming the calculator is in radian mode).
(b) $(0, \infty)$

Section 3.2, page 143

1. (a) $-4/5$ (b) $-3/4$ (c) $-2/5$
(d) $-1/8$ (e) 0

3. $\sqrt{3} + 1$

5. $\sqrt{\sqrt{2} + 3} - \sqrt{2} + 1$ **7.** 4 **9.** $34/3$

11. $59/12$ **13.** $(a + k)^2 + \dfrac{1}{a + k} + 2$

15. $(2 - x)^2 + \dfrac{1}{2 - x} + 2 = 6 - 4x + x^2 + \dfrac{1}{2 - x}$

17. 8

19. $(s + 1)^2 - 1 = s^2 + 2s$ **21.** $t^2 - 1$

23. (a) $3r + 7$ (b) $3(r - x)$ (c) 3

25. (a) $r - r^2$ (b) $r - r^2 - x + x^2$ (c) $1 - r - x$

27. 1 **29.** 3 **31.** $-2x - h + 1$

33. $\dfrac{1}{\sqrt{x + h} + \sqrt{x}}$ **35.** $2x + h$

37. (a) False (b) True **39.** $c = -2$

41. (a) $[-3, 4]$ (b) $[-2, 3]$ (c) -2
(d) $.5$ (e) 1 (f) -1

43. (a) $(-\infty, 20]$ (b) 3 (c) -1
(d) 1 (e) 2

45. All real numbers **47.** All real numbers

49. All nonnegative real numbers

51. All nonzero real numbers **53.** All real numbers

55. All real numbers except -2 and 3

57. $[6, 12]$

59. Two examples are $f(x) = x^2$ and $g(x) = |x|$.

61. $4 \pm \sqrt{26}$ **63.** $3, \dfrac{-7}{2}$

65. $\sqrt{2x - 5}$ **67.** $\dfrac{x^3 + 6}{5}$

69. (a) $f(x) = 5x + 200$; $g(x) = 10x$
(b) 300 and 200; 375 and 350; 450 and 500
(c) 45

71.
(a) $T(x) = \begin{cases} .10x & 0 \le x \le 6000 \\ 600 + .15(x - 6000) & 6000 < x \le 27{,}950 \\ 3892.50 + .27(x - 27{,}950) & 27{,}950 < x \le 67{,}700 \\ 14{,}625 + .3(x - 67{,}700) & 67{,}700 < x \le 141{,}250 \\ 36{,}690 + .35(x - 141{,}250) & 141{,}250 < x \le 307{,}050 \\ 94{,}720 + .386(x - 307{,}050) & 307{,}050 < x \end{cases}$
(b) 3300; 5796; 24,315

73.
$f(x) = \begin{cases} 3x & 0 \le x \le 0.75 \\ 2.25 + 5(x - .75) & .75 < x \le 2 \\ 8.5 & 2 < x \le 2.5 \\ 8.5 + 3(x - 2.5) & 2.5 < x \le 4 \end{cases}$

75. $C(x) = 5.75x + \dfrac{45{,}000}{x}$

77. (a) $f(x) = 806.1714x + 8177.0476$
(b) 13,014; 14,626; 16,239
(c) $18,657

Section 3.3, page 156

1. Yes; 0 **3.** No

5.

$y = 2[x]$

7.

$y = [-x]$

9.

11.

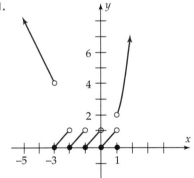

13. (a) Several possibilities, including

$$p(x) = \begin{cases} [x] & \text{if } x \text{ is an integer} \\ [x] + 1 & \text{if } x \text{ is not an integer} \end{cases}$$

or $p(x) = -[-x]$, with $x > 0$ in all cases, where x is the weight in ounces
(b) Graph for $0 < x \le 4$:

Weight in ounces

(c)

15. (a) $f(x) = \begin{cases} x + 2 & x \ge 0 \\ -x + 2 & x < 0 \end{cases}$

(b)

17. (a) $h(x) = \begin{cases} \frac{x}{2} - 2 & x \ge 0 \\ -\frac{x}{2} - 2 & x < 0 \end{cases}$

(b)
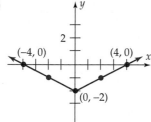

19. (a) $f(x) = \begin{cases} x - 5 & x \ge 5 \\ 5 - x & x < 5 \end{cases}$

(b)
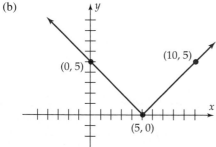

21. Minimum at $x = .57735$; maximum at $x = -.57735$

23. Minimum at $x = -1$; maximum at $x = 1$

25. Minimum at $x = 0.7633$; maximum at $x = 0.4367$

27. Increasing on $(-2.5, 0)$ and $(1.7, 4)$; decreasing on $(-6, -2.5)$ and $(0, 1.7)$

29. Constant on $(-\infty, -1]$ and $[1, \infty)$; decreasing on $(-1, 1)$.

31. Increasing on $[-5.7936, .46028]$; decreasing on $(-\infty, -5.7936]$ and $[.46028, \infty)$

33. Increasing on $(0, .867)$ and $(2.883, \infty)$; decreasing on $(-\infty, 0)$ and $(.867, 2.883)$

35. (a) $2x + 2z = 100$
(b) $A(x) = 50x - x^2$
(c) $x = 25, z = 25$

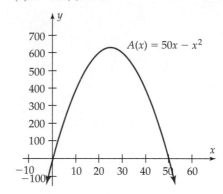

37. (a) $S = 2x^2 + 4xh$
(b) $867 = hx^2$
(c) $S(x) = 2x^2 + \dfrac{3468}{x}$
(d) $x = h = 9.5354$

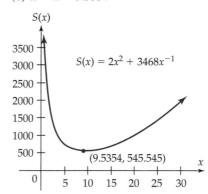

39. (a) iv   (b) i   (c) v   (d) iii   (e) ii

41. Many correct answers, including:

Domain $\approx [0, 150]$
Range $\approx [0, 300]$

43. Many correct answers, including

45.

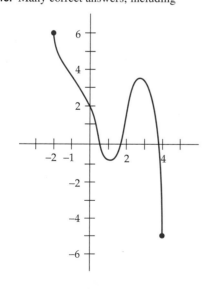

47. (a) 9
  (b) 15
  (c) 8
  (d) 1993; 1990
  (e) 1978–1981; this can be determined from the slope of the graph.

49. Many correct answers, including

51. $-15 \le x \le 30$ and $-10 \le y \le 10$ ($-10 \le t \le 10$)

53. $-10 \le x \le 6$ and $-7 \le y \le 7$ ($-7 \le t \le 7$)

55. $-15 \le x \le 0$ and $0 \le y \le 4$ ($0 \le t \le 4$)

57. $-5 \le x \le 45$ and $-65 \le y \le 65$

59. Entire graph: $-2 \le x \le 32$ and $-10 \le y \le 75$; near the origin: $-2 \le x \le 5$ and $-10 \le y \le 10$

61. Entire graph: $-16 \le x \le 2$ and $-62 \le y \le 60$; near the origin: $-2 \le x \le 2$ and $-4 \le y \le 2$

63. (a)

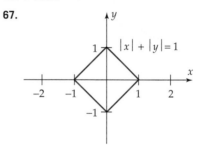

  (b) $f(x) = 12.361x + 36.7764$
  (c) \$110.955 billion
  (d) 2005

65. 7 times

67.

Special Topics 3.3.A, page 163

  1. Domain: all real numbers
      Range: all real numbers

  3. Domain: real numbers that are ≤ -2 or ≥ 2
      Range: nonnegative real numbers

  5. $[-3, 5]$ **7.** 4   **9.** 3.5

  11. 4.5   **13.** 4   **15.** 1, 5

  17. $[-8, 9]$ **19.** Approx. $-7, -3, 3, 6.8$

  21. 1, 3, 5, and others   **23.** 1 and -9

  25. The gross domestic product must be growing faster than the gross federal debt.

  27. 39 and 75   **29.** noon–8 P.M.

  31. Domain f: $[-6, 7]$
      Domain g: $[-8, 9]$

  33. Approx. -1.6 and 0   **35.** $x = 3$

  37. $[-2, -1]$ and $[3, 7]$

  39. Approx. $-\$13,000$ (a loss of \$13,000)

  41. Approx. 12,300   **43.** $h(x) = \sqrt{4 - x^2}$

  45. $h(x) = x - x^3$

47. At 15 min she stopped for 5 min, then continued running steadily for 10 min. At 30 min she turned back and jogged home without stopping, arriving home 55 min after her start.

Section 3.4, page 174

1. H **3.** F **5.** K **7.** C

9.

11.

13. Window: $-10 \le x \le 10$
$\qquad -36 \le y \le 42$

15. Window: $-13 \le x \le 12$
$\qquad -2 \le y \le 14$

17.

19. Shift the graph of f horizontally 3 units to the right; then shift it 2 units vertically upward.

21. Reflect the graph of f in the x-axis; shrink it toward the x-axis by a factor of $1/2$; shift it vertically 6 units downward.

23. Shift the graph of f horizontally 2 units to the right; stretch it away from the x-axis by a factor of 2; then reflect it in the x-axis.

25. $g(x) = f(x + 5) + 4 =$
$\qquad\qquad (x + 5)^2 + 2 + 4 = (x + 5)^2 + 6$

27. $g(x) = 2f(x - 6) - 3 = 2\sqrt{x - 6} - 3$

29. (a) $g(x) = x^2 + 3x + 2$
(b) $2x + h + 3; 2x + h + 3$
They are equal.

31.

33.

35.

37.

39.

41.

43.

45.

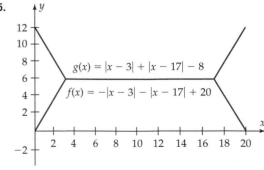

$$g(x) = |x - 3| + |x - 17| - 8$$

$$f(x) = -|x - 3| - |x - 17| + 20$$

47.

49.

51.

53.

55.

57.

59.

61.

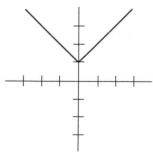

Special Topics 3.4.A, page 180

1. Symmetric with respect to the *y*-axis

3. The graph has none of the listed symmetries.

5. Odd   **7.** Even   **9.** Even   **11.** Even

13. Neither   **15.** Yes   **17.** No   **19.** Origin

21. Origin   **23.** *y*-axis

25.

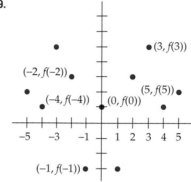

27.

29.

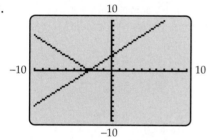

31. Many correct graphs, including the one shown here:

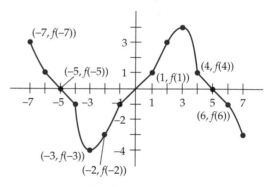

33. Suppose the graph is symmetric to the *x*-axis and the *y*-axis. If (*x*, *y*) is on the graph, then (*x*, −*y*) is on the graph by *x*-axis symmetry. Hence, (−*x*, −*y*) is on the graph of *y*-axis symmetry. Therefore, (*x*, *y*) on the graph implies that (−*x*, −*y*) is on the graph, so the graph is symmetric with respect to the origin. Next suppose that the graph is symmetric to the *y*-axis and the origin. If (*x*, *y*) is on the graph, then (−*x*, *y*) is on the graph by *y*-axis symmetry. Hence, (−(−*x*), −*y*) = (*x*, −*y*) is on the graph by origin symmetry. Therefore, (*x*, *y*) on the graph implies that (*x*, −*y*) is on the graph, so the graph is symmetric with respect to the *x*-axis. The proof of the third case is similar to that of the second case.

Section 3.5, page 189

1. $(f+g)(x) = x^3 - 3x + 2;$
$(f-g)(x) = -x^3 - 3x + 2;$
$(g-f)(x) = x^3 + 3x - 2$

3. $(f+g)(x) = \dfrac{1}{x} + x^2 + 2x - 5; \; (f-g) =$
$\dfrac{1}{x} - x^2 - 2x + 5; \; (g-f)(x) = x^2 + 2x - 5 - \dfrac{1}{x}$

5. $(fg)(x) = -3x^4 + 2x^3;$
$\left(\dfrac{f}{g}\right)(x) = \dfrac{-3x+2}{x^3}; \left(\dfrac{g}{f}\right)(x) = \dfrac{x^3}{-3x+2}$

7. $(fg)(x) = x^2 \sqrt{x-3} - 3\sqrt{x-3}$
$(f/g)(x) = \dfrac{x^2 - 3}{\sqrt{x-3}}$
$(g/f)(x) = \dfrac{\sqrt{x-3}}{x^2 - 3}$

9. Domain of fg: all real numbers except 0; domain of f/g: all real numbers except 0

11. Domain of fg: $[-4/3, 2]$; domain of f/g: $(-4/3, 2]$

13. 0   **15.** 30   **17.** 49; 1; -8 **19.** $-3; -3; 0$

21. $(f \circ g)(x) = (x+3)^2; (g \circ f)(x) = x^2 + 3;$ domain of $f \circ g$ and $g \circ f$ is all real numbers.

23. $(f \circ g)(x) = 1/\sqrt{x}; (g \circ f)(x) = \sqrt{\dfrac{1}{x}}$ domain of $f \circ g$ and $g \circ f$ is $(0, \infty)$.

25. $(ff)(x) = x^6; (f \circ f)(x) = x^9$

27. $(ff)(x) = \dfrac{1}{x^2}; (f \circ f)(x) = x$

29. $(f \circ g)(x) = f\left(\dfrac{x-2}{9}\right) = 9\left(\dfrac{x-2}{9}\right) + 2 = x$ and
$(g \circ f)(x) = g(9x + 2) = \dfrac{(9x + 2) - 2}{9} = x$

31. $(f \circ g)(x) = f(x-2)^3 = \sqrt[3]{(x-2)^3} + 2 = x$ and
$(g \circ f)(x) = g(\sqrt[3]{x} + 2) = (\sqrt[3]{x} + 2 - 2)^3 = x$

33.

x	-4	-3	-2	-1	0
$f(x)$	-3	-1	0	$1/2$	1
$g(x) = f(f(x))$	-1	$1/2$	1	1.2	1.5

x	1	2	3	4
$f(x)$	1.5	1	-2	-2
$g(x) = f(f(x))$	2	1.5	0	0

35.

x	1	2	3	4	5
$(g \circ f)(x)$	4	2	5	4	4

37.

x	1	2	3	4	5
$(f \circ f)(x)$	1	3	3	5	1

In Answers 39–45, the given function is $B \circ A$, where A and B are the functions listed here. In some cases other correct answers are possible.

39. $A(x) = x^2 + 2, B(x) = \sqrt[3]{x}$

41. $A(x) = 7x^3 - 10x + 17, B(x) = x^7$

43. $A(x) = 3x^2 + 5x - 7, B(x) = \dfrac{1}{x}$

45. Several possible answers, including
$h(x) = x^2 + 2x; k(t) = t + 1$

47. $(f \circ g)(x) = (\sqrt{x})^3$, domain $[0, \infty); (g \circ f)(x) = \sqrt{x^3}$, domain $[0, \infty)$

49. $(f \circ g)(x) = \sqrt{5x + 10}$, domain $[-2, \infty);$
$(g \circ f)(x) = 5\sqrt{x + 10}$, domain $[-10, \infty)$

51. (a) $f(x^2) = 2x^6 + 5x^2 - 1$
(b) $(f(x))^2 = (2x^3 + 5x - 1)^2$
$= 4x^6 + 20x^4 - 4x^3 + 25x^2 - 10x + 1$
(c) No; $f(x^2) \neq (f(x))^2$ in general

53. Not the same

55. $(g \circ f)(x) = x^2 + 6x + 10$
Difference quotient $= 2x + h + 6$

57. $(g \circ f)(x) = \dfrac{1}{x + 1}$
Difference quotient $= \dfrac{-1}{(x + 1 + h)(x + 1)}$

59. (a) One day, .00012246 sq in.; one week, .0000025 sq in.; one 31-day month, .00000013 sq in.
(b) No; no. The model is probably reasonable until the puddle is about the size of a period, with a radius of approximately .01. This occurs after approximately 15 hours.

61. (a) $A = \pi\left(\dfrac{d}{2}\right)^2 = \pi \cdot \dfrac{d^2}{4} = \dfrac{\pi}{4}\left(6 - \dfrac{50}{t^2 + 10}\right)^2$
(b) $\pi/4 \approx .7854$ sq in.; 22.2648 sq in.
(c) In approximately 11.39 weeks

63. $V = 256\pi \cdot t^3/3; 17{,}157.28$ cm^3

65. $s = 10t/3$

67. One such function is $f(x) = \dfrac{x-1}{x}$.

Section 3.6, page 200

1. (a) 14 ft/sec   (b) 54 ft/sec
(c) 112 ft/sec   (d) $93\frac{1}{3}$ ft/sec

3. (a) About 2086 people per year
(b) About -2756 people per year

5. (a) A decrease of 245,800 students/year
(b) An increase of 552,400 students/year
(c) An increase of 296,200 students/year
(d) A decrease of 89,600 students/year
(e) 1985–1995; 1995–2005
(f) 1980–1985; 2005–2010

7. (a) 10.475 billion shares per year
(b) 29.175 billion shares per year
(c) 51.8 billion shares per year
(d) 26.22 billion shares per year
(e) 1999–2001

9. (a) $5,000 per page
(b) $1875 per page
(c) $625 per page
(d) About $1750 per page
(e) No; no; no

11. -2
13. -1
15. 1.5858

17. 1
19. $2x + h$
21. $2t + h - 8000$

23. $2\pi r + \pi h$

25. (a) The water is leaving the tank at a rate of 7979.9 gal/min.
(b) -7979.99
(c) -7980

27. (a) $6.5\,\pi$
(b) $6.2\,\pi$
(c) $6.1\,\pi$
(d) $6\,\pi$
(e) It's the same.

29. (a) C, 62.5 ft/sec; D, 75 ft/sec
(b) Approximately $t = 4$ to $t = 9.8$ sec
(c) The average speed of car D from $t = 4$ to $t = 10$ sec is the slope of the secant line joining the (approximate) points (4, 100) and (10, 600), namely,
$\dfrac{600 - 100}{10 - 4} \approx 83.33$ ft/sec. The average speed of car C is the slope of the secant line joining the (approximate) points (4, 475) and (10, 800), namely,
$\dfrac{800 - 475}{10 - 4} \approx 54.17$ ft/sec.

31. (a) From day 0 until any day up to day 94, the average growth rate is positive.
(b) From day 0 to day 95
(c) -28, meaning that the population is decreasing at a rate of 28 chipmunks per day
(d) 20, -20, and 0 chipmunks per day

33. (a) $f(x) = 12.76x + 190.4$
(b) $12.76 per year
(c) 1980–1990: $12.80; 1990–2000: $13.50. The first is close; the second is not.
(d) 2008

Section 3.7, page 211

1. No
3. Yes
5. Yes
7. No

9. $g(x) = -x$
11. $g(x) = \dfrac{x + 4}{5}$

13. $g(x) = \sqrt[3]{\dfrac{5 - x}{2}}$
15. $g(x) = \dfrac{x^2 + 7}{4}, \quad (x \geq 0)$

17. $g(x) = \dfrac{1}{x}$
19. $g(x) = \dfrac{1}{2x} - \dfrac{1}{2}$

21. $g(x) = \sqrt[3]{\dfrac{5x + 1}{1 - x}}$

23. $(f \circ g)(x) = f(g(x)) = f(x - 1) = (x - 1) + 1 = x$ and
$(g \circ f)(x) = g(f(x)) = g(x + 1) = (x + 1) - 1 = x$

25. $(f \circ g)(x) = f\left(\dfrac{1 - x}{x}\right) = \dfrac{1}{\left(\dfrac{1 - x}{x}\right) + 1} =$

$\dfrac{1}{\dfrac{(1 - x) + x}{x}} = x$ and $(g \circ f)(x) = g\left(\dfrac{1}{x + 1}\right) =$

$\dfrac{1 - \dfrac{1}{x + 1}}{\dfrac{1}{x + 1}} = \dfrac{\dfrac{(x + 1) - 1}{x + 1}}{\dfrac{1}{x + 1}} = x$

27. $(f \circ g)(x) = f(\sqrt[5]{x}) = (\sqrt[5]{x})^5 = x$ and
$(g \circ f)(x) = g(x^5) = \sqrt[5]{x^5} = x$

29. $(f \circ f)(x) = f(f(x)) = \dfrac{2f(x) + 1}{3f(x) - 2} =$

$\dfrac{2\left[\dfrac{2x + 1}{3x - 2}\right] + 1}{3\left[\dfrac{2x + 1}{3x - 2}\right] - 2} = \dfrac{\dfrac{2(2x + 1) + (3x - 2)}{3x - 2}}{\dfrac{3(2x + 1) - 2(3x - 2)}{3x - 2}} =$

$\dfrac{\dfrac{7x}{3x - 2}}{\dfrac{7}{3x - 2}} = x$

31.

33.

35.

37.

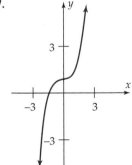

In Exercises 39–45, there are several correct answers for each, including these:

39. One restricted function is $h(x) = |x|$ with $x \geq 0$ (so that $h(x) = x$); inverse function $g(y) = y$ with $y \geq 0$.

41. One restricted function is $h(x) = -x^2$ with $x \leq 0$; inverse function $g(y) = -\sqrt{-y}$ with $y \leq 0$. Another restricted function is $h(x) = -x^2$ with $x \geq 0$; inverse function $g(y) = \sqrt{-y}$ with $y \leq 0$.

43. One restricted function is $h(x) = \dfrac{x^2 + 6}{2}$ with $x \geq 0$; inverse function $g(y) = \sqrt{2y - 6}$ with $y \geq 3$.

45. One restricted function is $f(x) = \dfrac{1}{x^2 + 1}$ with $x \leq 0$; inverse function $g(y) = -\sqrt{\dfrac{1}{y} - 1} = -\sqrt{\dfrac{1 - y}{y}}$ with $0 < y \leq 1$.

47. (a) $f^{-1}(x) = \dfrac{x - 2}{3}$ (b) $f^{-1}(1) = -1/3$ and $1/f(1) = 1/5$

49. Let $y = f(x) = mx + b$. Since $m \neq 0$, we can solve for x and obtain $x = \dfrac{y - b}{m}$. Hence, the rule of the inverse function g is $g(y) = \dfrac{y - b}{m}$, and we have:

$(f \circ g)(y) = f(g(y)) = f\left(\dfrac{y - b}{m}\right) = m\left(\dfrac{y - b}{m}\right) + b = y$ and

$(g \circ f)(x) = g(f(x)) = g(mx + b) = \dfrac{(mx + b) - b}{m} = x.$

51. (a) Slope $= \dfrac{a - b}{b - a} = \dfrac{-(b - a)}{b - a} = -1.$

(b) The line $y = x$ has slope 1 (why?) and by (a), line PQ has slope -1. Since the product of their slopes is -1, the lines are perpendicular.

(c) Length $PR = \sqrt{(a - c)^2 + (b - c)^2}$
$= \sqrt{a^2 - 2ac + c^2 + b^2 - 2bc + c^2}$
$= \sqrt{a^2 + b^2 + 2c^2 - 2ac - 2bc};$

Length $RQ = \sqrt{(c - b)^2 + (c - a)^2}$
$= \sqrt{c^2 - 2bc + b^2 + c^2 - 2ac + a^2}$
$= \sqrt{a^2 + b^2 + 2c^2 - 2ac - 2bc}.$

Since the two lengths are the same, $y = x$ is the perpendicular bisector of segment PQ.

53. (a) Suppose $a \neq b$. If $f(a) = f(b)$, then $g(f(a)) = g(f(b))$. But $a = g(f(a))$ by (1) and $b = g(f(b))$. Hence, $a = b$, contrary to our hypothesis. Therefore, it cannot happen that $f(a) = f(b)$, that is, $f(a) \neq f(b)$. Hence, f is one-to-one.

(b) If $g(y) = x$, then $f(g(y)) = f(x)$. But $f(g(y)) = y$ by (2). Hence, $y = f(g(y)) = f(x)$.

(c) If $f(x) = y$, then $g(f(x)) = g(y)$. But $g(f(x)) = x$ by (1). Hence, $x = g(f(x)) = g(y)$.

Chapter 3 Review, page 215

1. (a) -3 (b) 1755   (c) 2   (d) -14

3.

x	0	1	2	-4	t	k
$f(x)$	7	5	3	15	$7 - 2t$	$7 - 2k$

x	$b - 1$	$1 - b$	$6 - 2u$
$f(x)$	$9 - 2b$	$5 + 2b$	$-5 + 4u$

5. Many possible answers, including:
(a) $f(x) = x^2$, $a = 2$, $b = 3$; $f(a + b) = f(2 + 3) = 5^2 = 25$, but $f(a) + f(b) = f(2) + f(3) = 2^2 + 3^2 = 13$, so the statement is false.

(b) $f(x) = x + 1$, $a = 0$, $b = 1$; $f(ab) = f(0) = 1$, but $f(a)f(b) = f(0)f(1) = 1 \cdot 2 = 2$, so the statement is false.

7. $[4, \infty)$ **9.** $(t + 2)^2 - 3(t + 2) = t^2 + t - 2$

11. $2\left(\dfrac{x}{2}\right)^3 + \left(\dfrac{x}{2}\right) + 1 = \dfrac{x^3}{4} + \dfrac{x}{2} + 1$

13. (a) $f(t) = 50\sqrt{t}$
(b) $g(t) = 2500\pi t$
(c) radius: 150 meters; area: 70,685.83 square meters
(d) 12.73 hr

15.

17. (b)

19. No local maxima; minimum at $x = -.5$; increasing on $[-.5, \infty)$; decreasing on $(-\infty, -.5]$

21. Maximum at $x \approx -5.0704$; minimum at $x \approx -.2629$. Increasing on $(-\infty, -5.0704]$ and $[-.2629, \infty)$; decreasing on $[-5.0704, -.2629]$

23.

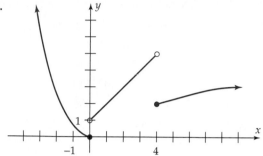

$x = t^2 - 4$
$y = 2t + 1$ $(-3 \le t \le 3)$

25. Many correct answers, including

27. x-axis, y-axis, origin   **29.** Even

31. Odd   **33.** y-axis

35.

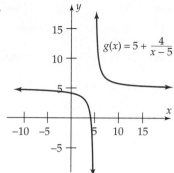

$g(x) = 5 + \dfrac{4}{x - 5}$

37. Approximately $[-3, 3.8]$

39. Many correct answers, including $x = -2$; all x in the interval $(2.5, 3.8)$; all x in the interval $[5, 6]$

41. 1   **43.** -3 **45.** True

47. $x = 4$ **49.** $x \le 3$ **51.** $x < 3$

53. (a) King Richard's car
(b) King Richard's
(c) Fireball Bob's

55. Shrink the graph of g toward the x-axis by a factor of .25, then shift the graph vertically 2 units upward.

57. Shift the graph of g horizontally 7 units to the right; then stretch it away from the x-axis by a factor of 3; then reflect it in the x-axis; finally, shift the graph vertically 2 units upward.

59. (e)

61. (a) $1/3$ (b) $(x - 1)\sqrt{x^2 + 5}$ $(x \ne 1)$
(c) $\dfrac{\sqrt{c^2 + 2c + 6}}{c}$

63.

x	-4	-3	-2	-1	0
$g(x)$	1	4	3	1	-1
$h(x)$	-3	-3	-4	-3	1

x	1	2	3	4
$g(x)$	-3	-2	-4	-3
$h(x)$	4	3	1	4

65. $\dfrac{82}{27}$ **67.** $\dfrac{1}{x^3} + 3$ **69.** $\dfrac{1}{4}$

71. $(f \circ g)(x) = f(x^2 - 1) = \dfrac{1}{x^2 - 1}$;
$(g \circ f)(x) = g\left(\dfrac{1}{x}\right) = \dfrac{1}{x^2} - 1$

73. All nonnegative numbers except 1

75. (a) $-1/3$ (b) $5/8$

77. 6   **79.** 3   **81.** $2x + h$

83. (a) \$290/ton (b) \$230/ton (c) \$212/ton

85. (a) Approximately day 45 to day 50 (population decreasing)
(b) Approximately any 10-day interval between day 20 and day 35
(c) Approximately day 30 to day 40

87. (a) \$5338; \$9900; \$6194 dollars per year
(b) \$234,100

89. $g(x) = 5 - (x - 7)^2 = -x^2 + 14x - 44; \quad x \geq 7$

91.

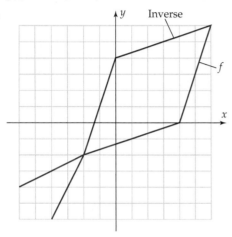

93. The graph of f passes the horizontal line test and hence has an inverse function. It is easy to verify either geometrically [by reflecting the graph of f in the line $y = x$] or algebraically [by calculating $f(f(x))$] that f is its own inverse function.

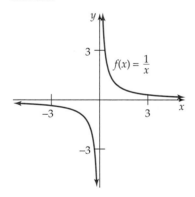

95. There is no inverse function because the graph of f fails the horizontal line test (use the viewing window with $-10 \leq x \leq 20$ and $-200 \leq y \leq 100$).

CHAPTER 4

Section 4.1, page 232

1. I **3.** K **5.** J **7.** F

9. (5, 2), upward **11.** (1, 2), downward

13. (3, −6), upward **15.** (−3/2, 15/4), upward

17. (−3, −21), upward **19.** (4, 14), downward

21. (a) $-2x - h + 1$ (b) (1/2, 1/4) (c) $-h$

23. (a) $-4x - 2h + 2$ (b) $\left(\dfrac{1}{2}, -\dfrac{1}{2}\right)$ (c) $-2h$

25. $g(x) = 2x^2 - 5$; vertex $(0, -5)$

27. $h(x) = 2(x - 3)^2 + 4$; vertex (3, 4)

29. $F(x) - 3x^2$ **31.** $f(x) = 3(x - 2)^2 + 5$

33. $b = 0$ **35.** $b = -4, c = 8$

37. (a) 20,000; 100,000 (b) 200; 550

39. Height = 841.92 ft. Distance = 13, 986.52 ft.

41. (a) At 30 km/h, 30 meters;
At 100 km/h, 170 meters
(b) 50 km/h

43. $t = 2.5$ sec; $h = 196$ ft

45. 22 ft **47.** $h = 15, b = 15$

49. Two 50-ft sides and one 100-ft side

51. Area = 704.17 ft^2

53. (a) $f(t) = \dfrac{1}{1152}x^2 - 2$
(b) Approx. 33.94 in. from center

55. \$3.50

57. \$3.67 (But if tickets must be priced in multiples of \$.20, \$3.60 is best.)

Section 4.2, page 241

1. Polynomial of degree 3; leading coefficient 1; constant term 1

3. Polynomial of degree 3; leading coefficient 1; constant term −1

5. Polynomial of degree 2; leading coefficient 1; constant term −3

7. Not a polynomial

9. Quotient $3x^3 - 3x^2 + 5x - 11$; remainder 12

11. Quotient $x^2 + 2x - 6$; remainder $-7x + 7$

13. Quotient $5x^2 + 5x + 5$; remainder 0

15. No **17.** Yes **19.** 0, 2 **21.** $2\sqrt{2}, -1$

23. 2 **25.** 6 **27.** −30 **29.** 170,802

31. 5,935,832 **33.** No **35.** No **37.** Yes

39. $(x + 4)(2x - 7)(3x - 5)$

41. $(x - 3)(x + 3)(2x + 1)^2$

43. $f(x) = (x + 2)(x + 1)(x - 1)(x - 2)(x - 3)$
$= x^5 - 3x^4 - 5x^3 + 15x^2 + 4x - 12$

45. $f(x) = x(x + 1)(x - 1)(x - 2)(x - 3)$
$= x^5 - 5x^4 + 5x^3 + 5x^2 - 6x$

47. Many correct answers, including
$(x - 1)(x - 7)(x + 4)$

49. Many correct answers including
$(x - 1)(x - 2)^2(x - \pi)^3$

51. $f(x) = \dfrac{17}{100}(x - 5)(x - 8)x$

53. $k = 1$ **55.** $k = 1$

57. If $x - c$ were a factor of $x^4 + x^2 + 1$, then c would be a solution of $x^4 + x^2 + 1 = 0$, that is, c would satisfy $c^4 + c^2 = -1$. But $c^4 \geq 0$ and $c^2 \geq 0$, so that is impossible. Hence, $x - c$ is not a factor.

59. (a) Many possible answers, including: if $n = 3$ and $c = 1$, then $x + 1 = x - (-1)$ is not a factor of $x^3 - 1$ since -1 is not a solution of $x^3 - 1 = 0$.
(b) Since n is odd $(-c)^n = -c^n$ and hence $-c$ is a solution of $x^n + c^n = 0$. Thus, $x - (-c) = x + c$ is a factor of $x^n + c^n$ by the Factor Theorem.

61. $k = 5$

Special Topics 4.2.A, page 246

1. $\underline{2|}\ \ 3 \ \ -8 \ \ \ \ 0 \ \ \ \ 9 \ \ \ 5$
$\phantom{\underline{2|}\ \ 3\ \ }6 \ \ -4 \ \ -8 \ \ \ 2$
$\overline{\phantom{\underline{2|}\ } 3 \ \ -2 \ \ -4 \ \ \ \ 1 \ \ \underline{|7}}$

quotient $3x^3 - 2x^2 - 4x + 1$;
remainder 7

3. $\underline{-3|}\ \ 2 \ \ \ \ 5 \ \ \ \ 0 \ \ -2 \ \ -8$
$\phantom{\underline{-3|}\ \ 2\ }-6 \ \ \ \ 3 \ \ -9 \ \ \ \ 33$
$\overline{\phantom{\underline{-3|}\ } 2 \ \ -1 \ \ \ \ 3 \ \ -11 \ \ \underline{|25}}$

quotient $2x^3 - x^2 + 3x - 11$; remainder 25

5. $\underline{7|}\ \ 5 \ \ \ \ 0 \ \ \ \ -3 \ \ \ \ -4 \ \ \ \ \ \ 6$
$\phantom{\underline{7|}\ \ 5\ }35 \ \ \ 245 \ \ 1{,}694 \ \ 11{,}830$
$\overline{\phantom{\underline{7|}\ } 5 \ \ \ 35 \ \ \ 242 \ \ 1{,}690 \ \ \underline{|11{,}836}}$

quotient $5x^3 + 35x^2 + 242x + 1690$; remainder 11,836

7. $\underline{2|}\ \ 1 \ \ -6 \ \ \ \ 4 \ \ \ \ 2 \ \ -7$
$\phantom{\underline{2|}\ \ 1\ }2 \ \ -8 \ \ -8 \ \ -12$
$\overline{\phantom{\underline{2|}\ } 1 \ \ -4 \ \ -4 \ \ -6 \ \ \underline{|{-19}}}$

quotient $x^3 - 4x^2 - 4x - 6$; remainder -19

9. Quotient $3x^3 + \dfrac{3}{4}x^2 - \dfrac{29}{16}x - \dfrac{29}{64}$; remainder $\dfrac{483}{256}$

11. Quotient $2x^3 - 6x^2 + 2x + 2$; remainder 1

13. $g(x) = (x + 4)(3x^2 - 3x + 1)$

15. $g(x) = \left(x - \dfrac{1}{2}\right)(2x^4 - 6x^3 + 12x^2 - 10)$

17. Quotient $x^2 - 2.15x + 4$; remainder 2.25

19. $c = -4$

Section 4.3, page 252

1. $x = \pm 1$ or -3 **3.** $x = \pm 1$ or -5

5. $x = -4, 0, 1$, or $1/2$ **7.** $x = -3$ or 2

9. $x = 2$ **11.** $x = -5, 2$, or 3

13. $(x - 2)(2x^2 + 1)$ **15.** $x^3(x^2 + 3)(x + 2)$

17. $(x - 2)(x - 1)^2(x^2 + 3)$ **19.** Lower -5; upper 2

21. Lower -7; upper 3   **23.** $x = 1, 2$, or $-1/2$

25. $x = 1, 1/2$, or $1/3$ **27.** $x = 2$ or $\dfrac{-5 \pm \sqrt{37}}{2}$

29. $x = 1/2$ or $\pm\sqrt{2}$ or $\pm\sqrt{3}$

31. $x = -1, 5$, or $\pm\sqrt{3}$ **33.** $x = 1/3$ or -1.8393

35. $x = -2.2470$ or $-.5550$ or $.8019$ or 50

37. (a) The only possible rational roots of $f(x) = x^2 - 2$ are ± 1 or ± 2 (why?). But $\sqrt{2}$ is a root of $f(x)$ and $\sqrt{2} \neq \pm 1$ or ± 2. Hence, $\sqrt{2}$ is irrational.
(b) $\sqrt{3}$ is a root of $x^2 - 3$ whose only possible rational roots are ± 1 or ± 3 (why?). But $\sqrt{3} \neq \pm 1$ or ± 3.

39. (a) 8.2395; 5.519 people per 100,000
(b) 1996
(c) 1991

41. 2 by 2 in.

43. (a) $6°$ per day at the beginning; $6.6435°$ per day at the end
(b) Day 2.0330 and day 10.7069
(c) Day 5.0763 and day 9.6126
(d) Day 7.6813

Section 4.4, page 261

1. Yes   **3.** Yes   **5.** No

7. Degree 3, yes; degree 4, no; degree 5, yes

9. No

11. Degree 3, no; degree 4, no; degree 5, yes

13. The graphs have the same *shape* in the window with $-40 \leq x \leq 40$ and $-1000 \leq y \leq 5000$ but don't look identical.

15. -2 is a root of odd multiplicity, as are 1 and 3

17. -2 and -1 are roots of odd multiplicity; 2 is a root of even multiplicity.

19. (e)   **21.** (f)   **23.** (c)

25. The graph in the standard viewing window does not rise at the far right as does the graph of the highest degree term x^3, so it is not complete.

27. The graph in the standard viewing window does not rise at the far left and far right as does the graph of the highest degree term $.005x^4$, so it is not complete.

29. $-9 \le x \le 3$ and $-20 \le y \le 40$

31. $-6 \le x \le 6$ and $-60 \le y \le 320$

33. $-3 \le x \le 4$ and $-35 \le y \le 20$

35. Left half: $-33 \le x \le -2$ and $-50{,}000 \le y \le 260{,}000$; right half: $-2 \le x \le 3$ and $-20 \le y \le 30$

37. $-90 \le x \le 120$ and $-15{,}000 \le y \le 5000$

39. Overall: $-3 \le x \le 3$ and $-20 \le y \le 20$; near y-axis: $-.1 \le x \le .2$ and $4.997 \le y \le 5.001$

41. (a) The graph of a cubic polynomial (degree 3) has at most $3 - 1 = 2$ local extrema. When $|x|$ is large the graph resembles the graph of ax^3, that is, one end shoots upward and the other end downward. If the graph had only one local extremum, both ends of the graph would go in the same direction (both up or down). Thus, the graph of a cubic polynomial has either two local extrema or none.

   (b) These are the only possible shapes for a graph that has 0 or 2 local extrema, 1 point of inflection, and resembles the graph of ax^3 when $|x|$ is large.

43. (a) Odd   (b) Positive
    (c) $-2, 0, 4,$ and 6 (d) 5

45. (d)

47.

49.

51.

53.

55. 180,000

   (b) 4950; 3750
   (c) The name "point of diminishing returns" refers to the slowing rate of sales increases.

57. (a) $R(x) = 347x - c(x) = -.001x^5 + .29x^3 - 8x^2 + 347x - 1600$
   (b) About 5101
   (c) About 50,517
   (d) About 37,167

59. (a) 82,794; $1546.39   (b) 147,200

61. 2.7085 by 2.7085 in.

63. (a) The solutions are roots of
$g(x) - 4 = .01x^3 - .06x^2 + .12x - .08$. This polynomial has degree 3 and hence has at most 3 roots.

(b) $1 \le x \le 3$ and $3.99 \le y \le 4.01$

(c) Suppose $f(x)$ has degree n. If the graph of $f(x)$ had a horizontal segment lying on the line $y = k$ for some constant k, then the equation $f(x) = k$ would have infinitely many solutions (why?). But the polynomial $f(x) - k$ has degree n (why?) and thus has at most n roots. Hence the equation $f(x) = k$ has at most n solutions which means the graph cannot have a horizontal segment.

65. (a) No, except between $x = 0$ and $x = 2$

(b) Yes

(c) Approximately .0007399

67.

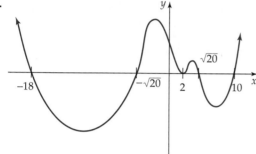

(b)–(d) Depends on the calculator

Special Topics 4.4.A, page 270

1. Cubic   **3.** Quartic

5. (a) $.76893x^3 - 40.565x^2 + 545.09x + 2896.9$

(b) 4988.6; 4817.0 per 100,000

(c) 4541.2 per 100,000

(d) Assuming the post-1990 downward trend continues, the model may remain accurate until about 2006.

7.

(b) $y = -.5179820180x^2 + 14.65684316x - 20.88711289$

(c) Noon: 80°; 9 A.M.: 69°; 2 P.M.: 83°

9. (a)

(b) Quartic; $y = -.0016x^4 + .05804x^3 - .63003x^2 + 2.13144x + 36.46698$

(c) \$42,546

(d) No

11. (a)

50

0 ⌞_____14
0

(b) $y = -.32545x^2 + 4.2625x + 22.338$
$y = -.04109x^3 + .41414x^2 + .65498x + 26.233$
$y = .01237x^4 - .33796x^3 + 2.7503x^2 - 6.0035x + 31.179$

(c) Quartic seems to fit best, but quadratic may be most reasonable for future years.

13. (a) $y = .0139x^3 - .6438x^2 + 9.3748x - 35.1114$;
$y = .00127x^4 - .0622x^3 + 1.0098x^2 - 5.9605x + 15.9725$

(b) No

Section 4.5, page 284

1. All real numbers except $-5/2$

3. All real numbers except $3 + \sqrt{5}$ and $3 - \sqrt{5}$

5. All real numbers except $-\sqrt{2}$, 1, and $\sqrt{2}$

7. F   **9.** A

11. Vertical asymptotes $x = -1$ and $x = 6$

13. Hole at $x = 0$; vertical asymptote $x = -1$

15. Vertical asymptotes: $x = -2$ and $x = 2$

17. $y = 3$; any window with $-115 \le x \le 110$

19. $y = -1$; any window with $-31 \le x \le 35$

21. $y = 5/2$; any window with $-40 \le x \le 42$

23.

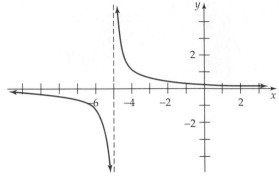

Vertical asymptote $x = -5$
Horizontal asymptote $y = 0$

25.

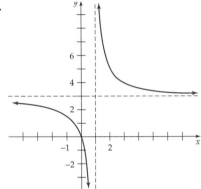

Vertical asymptote $x = 1$
Horizontal asymptote $y = 3$

27.

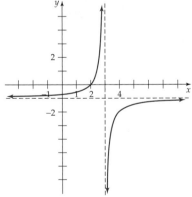

Vertical asymptote $x = 3$
Horizontal asymptote $y = -1$

29.

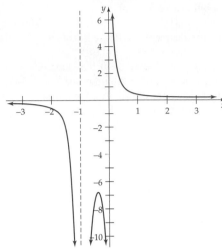

Vertical asymptotes $x = -1$, $x = 0$
Horizontal asymptote $y = 0$

31.

Vertical asymptotes $x = -2$, $x = 1$
Horizontal asymptote $y = 0$

33.

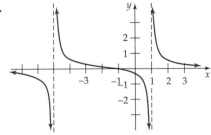

Vertical asymptotes $x = -5$, $x = 1$
Horizontal asymptote $y = 0$

35.

Vertical asymptote $x = 0$
Horizontal asymptote $y = -4$

37.

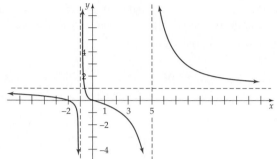

Vertical asymptotes $x = -1$, $x = 5$
Horizontal asymptote $y = 1$

39.

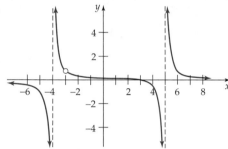

Vertical asymptotes $x = -4$, $x = 5$
Hole at $x = -3$
Horizontal asymptote $y = 0$

41. Overall: $-5 \le x \le 4.4$ and $-8 \le y \le 4$; hidden area near origin: $-2 \le x \le 2$ and $-.5 \le y \le .5$; hidden area near $x = -5$; $-15 \le x \le -3$ and $-.07 \le y \le .02$

43. $-9.4 \le x \le 9.4$ and $-4 \le y \le 4$; there is a hole at $x = 2$.

45. For vertical asymptotes and x-intercepts:
$-4.7 \le x \le 4.7$ and $-8 \le y \le 8$; to see graph get close to the horizontal asymptote: $-40 \le x \le 35$ and $-2 \le y \le 3$

47. Overall: $-4.7 \le x \le 4.7$ and $-3 \le y \le 3$; hidden area near $x = 4$: $3 \le x \le 15$ and $-.02 \le y \le .01$

49. $\dfrac{-1}{x(x + h)}$ **51.** $\dfrac{-2x - h}{x^2(x + h)^2}$

53. (a) $-1/4.2 \approx -.2381$; $-1/4.02 \approx -.2488$;
$-1/4.002 \approx -.2499$; instantaneous rate of change $-1/4 = -.25$
(b) $-1/9.3 \approx -.1075$; $-1/9.03 \approx -.1107$;
$-1/9.003 \approx -.1111$; instantaneous rate of change $-1/9 = -.1111 \cdots$
(c) They are the same.

55. (a) $y = \dfrac{x - 1}{x - 2}$ (b) $y = \dfrac{x - 1}{-x - 2}$
(c) Graph (a) has a vertical asymptote at $x = 2$ and graph (b) has a vertical asymptote at $x = -2$.

57. 8.4343 in. $\times$ 8.4343 in. $\times$ 14.057 in.

59. (a) $c(x) = \dfrac{2800 + 3.5x^2}{x}$ (b) $13.91 \le x \le 57.52$
(c) 28.28 mph

61. (a) $p(x) = \dfrac{500 + x^2}{x}$ (b) $10 \le x \le 50$
(c) $x = 22.36$; 22.36 m by 11.18 m

63. (a) $h_1 = h - 2$
(b) $h_1 = \dfrac{150}{\pi r^2} - 2$ (because $\pi r^2 h = 150$)
(c) $V = \pi(r - 1)^2 \left(\dfrac{150}{\pi r^2} - 2 \right)$
(d) The walls are 1 ft thick.
(e) $r \approx 2.88$ ft; $h \approx 5.76$ ft

65. (a) $g(0) = 9.801$ m/sec^2
(b)

(c) There are no r-intercepts because the numerator is never zero. So you can never completely escape the pull of gravity.

Special Topics 4.5.A, page 291

1. Asymptote: $y = x$; window: $-14 \le x \le 14$ and $-15 \le y \le 15$

3. Asymptote: $y = x^2 - x$; window: $-15 \le x \le 6$ and $-40 \le y \le 240$

5.

Vertical asymptote $x = 2$
Oblique asymptote $y = x + 1$

7.

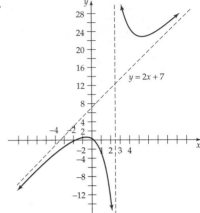

Vertical asymptote $x = 5/2$
Oblique asymptote $y = 2x + 7$

9.

Vertical asymptote $x = 1$
Parabolic asymptote $y = x^2 + x + 1$

11.

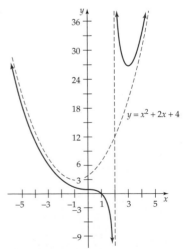

Vertical asymptote $x = 2$
Parabolic asymptote $y = x^2 + 2x + 4$

13. $-15.5 \leq x \leq 8.5$ and $-16 \leq y \leq 8$

15. $-4.7 \leq x \leq 4.7$ and $-12 \leq y \leq 8$

17. Overall: $-13 \leq x \leq 7$ and $-20 \leq y \leq 20$; hidden area near the origin: $-2.5 \leq x \leq 1$ and $-.02 \leq y \leq .02$

19. (b) Approx. $.06 \leq x \leq 2.78$

Section 4.6, page 300

1. $(-\infty, 3/2]$　　**3.** $(-2, \infty)$　　**5.** $(-\infty, -8/5]$

7. $(1, \infty)$　　**9.** $(2, 4)$　　**11.** $[-3, 5/2)$

13. $(-\infty, 1/7)$　　**15.** $[-7/17, \infty)$　　**17.** $[-1, 1/8)$

19. $[5, \infty)$　　**21.** $x < \dfrac{b + c}{a}$　　**23.** $c < x < a + c$

25. $1 \leq x \leq 3$

27. $x \leq \dfrac{-9 - \sqrt{21}}{2}$ or $x \geq \dfrac{-9 + \sqrt{21}}{2}$

29. $x \leq \dfrac{1 - \sqrt{33}}{2}$ or $x \geq \dfrac{1 + \sqrt{33}}{2}$

31. $-1 \leq x \leq 0$ or $x \geq 1$

33. $x < -1$ or $0 < x < 3$

35. $-2 < x < -1$ or $1 < x < 2$

37. $-2.26 \leq x \leq 0.76$ or $x \geq 3.51$

39. $.5 < x < .84$　　**41.** $x < -1/3$ or $x > 2$

43. $-2 < x < -1$ or $1 < x < 3$　　**45.** $x > 1$

47. $x \leq -9/2$ or $x > -3$　　**49.** $-3 < x < 1$ or $x \geq 5$

51. $-\sqrt{7} < x < \sqrt{7}$ or $x > 5.34$

53. $x < -3$ or $1/2 < x < 5$

55. $x > -1.43$

57. $x \leq -3.79$ or $x \geq .79$

59. Approximately 8.608 cents per kwh

61. More than \$12,500　　**63.** Between \$4000 and \$5400

65. $1 < x < 19$ and $y = 20 - x$　　**67.** $10 < x < 35$

69. $1 \leq t \leq 4$　　**71.** $2 < t < 2.25$

73. (a) $x^2 < x$ when $0 < x < 1$ and $x^2 > x$ when $x < 0$ or $x > 1$.
(b) If c is nonzero and $|c| < 1$, then either $0 < c < 1$ or $-1 < c < 0$ (which is equivalent to $1 > -c > 0$). If $0 < c < 1$, then $|c| = c$ and c is a solution of $x^2 < x$ by part (a), so that $c^2 < c = |c|$. If $1 > -c > 0$, then $|c| = -c$, which is a solution of $x^2 < x$ by part (a), so that $c^2 = (-c)^2 < (-c) = |c|$.
(c) If $|c| > 1$, then either $c < -1$ or $c > 1$. In either case, c is a solution of $x^2 > x$ by part (a).

Special Topics 4.6.A, page 305

1. $-4/3 \leq x \leq 0$　　**3.** $7/6 < x < 11/6$

5. $x < -2$ or $x > -1$

7. $x \leq -11/20$ or $x \geq -1/4$

9. $x < -53/40$ or $x > -43/40$

11. $x \le -7/2$ or $x \ge -5/4$

13. $x < -5$ or $-5 < x < -4/3$ or $x > 6$

15. $-1/7 < x < 3$

17. $-\sqrt{3} < x < -1$ or $1 < x < \sqrt{3}$

19. $x < -\sqrt{6}$ or $x > \sqrt{6}$

21. $x \le 2$ or $-1 \le x \le 0$ or $x \ge 1$

23. $0 < x < 2/3$ or $2 < x < 8/3$

25. $-1.43 < x < 1.24$ **27.** $x < -.89$ or $x > 1.56$

29. $x \le 2$ or $x \ge 14/3$

31. $-1.13 < x < 1.35$ or $1.35 < x < 1.67$

33. If $|x - 3| < E/5$, then multiplying both sides by 5 shows that $5|x - 3| < E$. But
$5|x - 3| = |5| \cdot |x - 3| = |5(x - 3)| =$
$|5x - 15| = |(5x - 4) - 11|$. Thus,
$|(5x - 4) - 11| < E$.

Section 4.7, page 310

1. $8 + 2i$ **3.** $-2 - 10i$ **5.** $-\dfrac{1}{2} - 2i$

7. $\left(\dfrac{\sqrt{2} - \sqrt{3}}{2}\right) + 2i$ **9.** $1 + 13i$

11. $-10 + 11i$ **13.** $-21 - 20i$ **15.** 4

17. $-i$ **19.** i **21.** i

23. $\dfrac{5}{29} + \dfrac{2}{29}i$ **25.** $-\dfrac{1}{3}i$ **27.** $\dfrac{12}{41} - \dfrac{15}{41}i$

29. $\dfrac{-5}{41} - \dfrac{4}{41}i$ **31.** $\dfrac{10}{17} - \dfrac{11}{17}i$ **33.** $\dfrac{7}{10} + \dfrac{11}{10}i$

35. $-\dfrac{113}{170} + \dfrac{41}{170}i$ **37.** $6i$ **39.** $\sqrt{14}i$

41. $-4i$ **43.** $11i$ **45.** $(\sqrt{15} - 3\sqrt{2})i$

47. $\dfrac{2}{3}$ **49.** $-41 - i$

51. $(2 + 5\sqrt{2}) + (\sqrt{5} - 2\sqrt{10})i$

53. $\dfrac{1}{3} - \dfrac{\sqrt{2}}{3}i$ **55.** $x = 2, y = -2$

57. $x = -3/4, y = 3/2$ **59.** $x = \dfrac{1}{3} \pm \dfrac{\sqrt{14}}{3}i$

61. $x = -\dfrac{1}{2} \pm \dfrac{\sqrt{7}}{2}i$ **63.** $x = \dfrac{1}{4} \pm \dfrac{\sqrt{31}}{4}i$

65. $x = \dfrac{3 \pm \sqrt{3}}{2}$

67. $x = 2, -1 + \sqrt{3}i, -1 - \sqrt{3}i$

69. $x = 1, -1, i, -i$ **71.** -1

73. $z + w = (a + bi) + (c + di)$
$\qquad = (a + c) + (b + d)i$ and hence
$\overline{z + w} = (a + c) - (b + d)i$
$\qquad = a + c - bi - di$
$\qquad = (a - bi) + (c - di) = \bar{z} + \bar{w}$

75. We first express z/w in standard form:
$$\frac{z}{w} = \frac{a + bi}{c + di} = \frac{a + bi}{c + di} \cdot \frac{c - di}{c - di} =$$
$$\frac{(ac + bd) + (bc - ad)i}{c^2 + d^2}. \text{ Hence,}$$
$$\overline{\left(\frac{z}{w}\right)} = \frac{(ac + bd) - (bc - ad)i}{c^2 + d^2} =$$
$$\frac{ac + bd - bci + adi}{c^2 + d^2}. \text{ On the other hand,}$$
$$\frac{\bar{z}}{\bar{w}} = \frac{a - bi}{c - di} = \frac{a - bi}{c - di} \cdot \frac{c + di}{c + di} = \frac{ac + bd - bci + adi}{c^2 + d^2}.$$

77. If $z = a + bi$, with a, b real numbers, then
$z - \bar{z} = (a + bi) - (a - bi) = 2bi$. If $z = a + bi$ is real,
then $b = 0$ and hence, $z - \bar{z} = 2bi = 0$. Therefore,
$z = \bar{z}$. Conversely, if $z = \bar{z}$, then $0 = z - \bar{z} = 2bi$, which
implies that $b = 0$. Hence, $z = a$ is real.

79. $\dfrac{1}{z} = \left(\dfrac{a}{a^2 + b^2}\right) + \left(\dfrac{-b}{a^2 + b^2}\right)i$

Section 4.8, page 316

1. 2 **3.** 6 **5.** -30

7. $x = 0$ (multiplicity 54); $x = -4/5$ (multiplicity 1)

9. $x = 0$ (multiplicity 15); $x = \pi$ (multiplicity 14);
$x = \pi + 1$ (multiplicity 13)

11. $x = 1 + 2i$ or $1 - 2i$;
$f(x) = (x - 1 - 2i)(x - 1 + 2i)$

13. $x = -\dfrac{1}{3} + \dfrac{2\sqrt{5}}{3}i$ or $-\dfrac{1}{3} - \dfrac{2\sqrt{5}}{3}i$;
$f(x) = \left(x + \dfrac{1}{3} - \dfrac{2\sqrt{5}}{3}i\right)\left(x + \dfrac{1}{3} + \dfrac{2\sqrt{5}}{3}i\right)$

15. $x = 3$ or $-\dfrac{3}{2} + \dfrac{3\sqrt{3}}{2}i$ or $-\dfrac{3}{2} - \dfrac{3\sqrt{3}}{2}i$;
$f(x) = (x - 3)\left(x + \dfrac{3}{2} - \dfrac{3\sqrt{3}}{2}i\right)\left(x + \dfrac{3}{2} + \dfrac{3\sqrt{3}}{2}i\right)$

17. $x = -2$ or $1 + \sqrt{3}i$ or $1 - \sqrt{3}i$;
$f(x) = (x + 2)(x - 1 - \sqrt{3}i)(x - 1 + \sqrt{3}i)$

19. $x = 1$ or i or -1 or $-i$;
$f(x) = (x - 1)(x - i)(x + 1)(x + i)$

21. $x = \sqrt{5}$ or $-\sqrt{5}$ or $\sqrt{2}i$ or $-\sqrt{2}i$;
$f(x) = (x - \sqrt{5})(x + \sqrt{5})(x - \sqrt{2}i)(x + \sqrt{2}i)$

23. Many correct answers, including
$(x - 1)(x - 7)(x + 4)$

25. Many correct answers, including
$(x - 1)(x - 2)^2(x - \pi)^3$

27. $f(x) = 2x(x - 4)(x + 3)$

In Exercises 29–40, there are many correct answers, including the following.

29. $x^2 - 4x + 5$ **31.** $(x - 2)(x^2 - 4x + 5)$

33. $(x + 3)(x^2 - 2x + 2)(x^2 - 2x + 5)$

35. $x^2 - 2x + 5$ **37.** $(x - 4)^2(x^2 - 6x + 10)$

39. $(x^4 - 3x^3)(x^2 - 2x + 2)$

41. $3x^2 - 6x + 6$ **43.** $-2x^3 + 2x^2 - 2x + 2$

45. Many correct answers, including

$$x^2 - (1 - i)x + (2 + i)$$

47. Many correct answers, including

$$x^3 - 5x^2 + (7 + 2i)x - (3 + 6i)$$

49. $3, -\dfrac{1}{2} + \dfrac{\sqrt{3}}{2}i, -\dfrac{1}{2} - \dfrac{\sqrt{3}}{2}i$ **51.** $i, -i, -1, -2$

53. $1, 2i, -2i$ **55.** $i, -i, 2 + i, 2 - i$

57. (a) Since $z + w = (a + c) + (b + d)i$,
$\overline{z + w} = (a + c) - (b + d)i$. Since $\bar{z} = a - bi$ and
$\bar{w} = c - di, \bar{z} + \bar{w} = (a - bi) + (c - di) =$
$(a + c) - (b + d)i$. Hence $\overline{z + w} = \bar{z} + \bar{w}$.
(b) Since $zw = (ac - bd) + (ad + bc)i$,
$\overline{zw} = (ac - bd) - (ad + bc)i$. Since $\bar{z} = a - bi$ and
$\bar{w} = c - di, \bar{z}\,\bar{w} = (a - bi)(c - di) =$
$(ac - bd) - (ad + bc)i$. Hence $\overline{zw} = \bar{z} \cdot \bar{w}$.

59. (a) $\overline{f(z)} = \overline{az^3 + bz^2 + cz + d}$ (definition of $f(z)$)
$= \overline{az^3} + \overline{bz^2} + \overline{cz} + \bar{d}$ (Exercise 57(a))
$= \bar{a}\overline{z^3} + \bar{b}\overline{z^2} + \bar{c}\,\bar{z} + \bar{d}$ (Exercise 57(b))
$= a\overline{z^3} + b\overline{z^2} + c\bar{z} + d$ ($r = \bar{r}$ for r real)
$= a\bar{z}^3 + b\bar{z}^2 + c\bar{z} + d$ (Exercise 57(b))
$= f(\bar{z})$ (definition of f)
(b) Since $f(z) = 0$, we have $0 = \bar{0} = \overline{f(z)} = f(\bar{z})$. Hence
$\bar{z}$ is a root of $f(x)$.

61. If $f(z)$ is a polynomial with real coefficients, then $f(z)$ can
be factored as $g_1(z)g_2(z)g_3(z) \cdots g_k(z)$, where each $g_i(z)$ is
a polynomial with real coefficients and degree 1 or 2. The
rules of polynomial multiplication show that the degree of
$f(z)$ is the sum: degree $g_1(z)$ + degree $g_2(z)$ + degree $g_3(z)$
$+ \cdots +$ degree $g_k(z)$. If all of the $g_i(z)$ have degree 2, then
this last sum is an even number. But $f(z)$ has odd degree,
so this can't occur. Therefore, at least one of the $g_i(z)$ is a
first-degree polynomial and hence must have a real root.
This root is also a root of $f(z)$.

Chapter 4 Review, page 319

1. $(2, 3)$ **3.** $(4, -4)$ **5.** $(1.5, -5.75)$

7. (a) $y = 260 - x$
(b) $A = -x^2 + 260x - 3500$
(c) $x = 130$ ft; $y = 130$ ft

9. 30 ft, 60 ft, 30 ft

11. (a), (c), (e), (f)   **13.** 0

15.

2|	1	−5	8	1	−17	16	−4
		2	−6	4	10	−14	4
	1	−3	2	5	−7	2	0

other factor: $x^5 - 3x^4 + 2x^3 + 5x^2 - 7x + 2$

17. Many correct answers, including
$f(x) = 5(x - 1)^2(x + 1) = 5x^3 - 5x^2 - 5x + 5$

19. -1 and $5/3$ **21.** $\sqrt[3]{2}$ **23.** $0, \pm\sqrt{\dfrac{2 + \sqrt{21}}{2}}$

25. (a) $1, -1, 3, -3, \dfrac{1}{2}, -\dfrac{1}{2}, \dfrac{3}{2}, -\dfrac{3}{2}$ (b) 3

(c) $3, \dfrac{1 + \sqrt{3}}{2}, \dfrac{1 - \sqrt{3}}{2}$

27. 1   **29.** $3, -3, \sqrt{2}, -\sqrt{2}$

31. $(x^4 - 4x^3 + 16x - 16) \div (x - 5)$ is
$x^3 + x^2 + 5x + 41$ with remainder 189. Since all coefficients and the remainder are positive, 5 is an upper
bound for the roots.

33. $1, -1, 1.867, -.867$ **35.** $2x + h + 1$

37. Many correct answers

39. (c)

41. $-3 \le x \le 9$ and $-35 \le y \le 15$

43. $-2 \le x \le 18$ and $-500 \le y \le 1100$

45. (a)

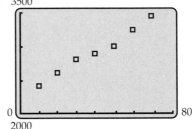

(b) $c(x) \approx 0.008515x^3 - 1.1094x^2 + 56.2583x + 2017.2576$

(c) \$26.80

(d) Average cost of 35: \$85.49; average cost of 75:
\$47.84

47.

49.

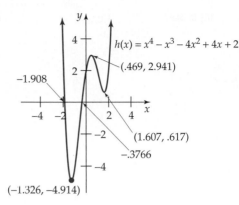

$h(x) = x^4 - x^3 - 4x^2 + 4x + 2$

(.469, 2.941)

−1.908

(1.607, .617)

−.3766

(−1.326, −4.914)

51.

53.

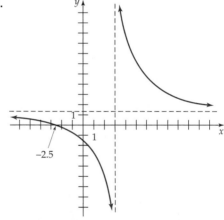

55. Vertical asymptotes $x = -2$ and $x = 3$; horizontal asymptote $y = 0$; hole at $x = 1$

57. Overall: $-4.7 \le x \le 4.7$ and $-5 \le y \le 5$ (Adjust x range to avoid erroneous vertical lines on widescreen calculators.) Near $x = 3$: $1 \le x \le 25$ and $-.5 \le y \le .1$

59. $-19 \le x \le 19$ and $-8 \le y \le 8$

61. At least 400 bags (priced at \$1.75 each) and at most 2500 bags (priced at \$.70 each)

63. (a) $T = \dfrac{40}{x} + \dfrac{110}{x + 25}$, where x is the speed of the car and $0 < x \le 55$ (speed limit)
(b) At least 44.08 mph

65. $\dfrac{1}{(x + h + 1)(x + 1)}$ **67.** (d)

69. $y \le -17$ or $y \ge 13$

71. $(-\infty, -2)$ and $\left(-\dfrac{1}{3}, \infty\right)$

73. $-\sqrt{3} \le x < -1$ or $-1 < x < 1$ or $1 < x \le \sqrt{3}$

75. $x \le -1$ or $0 \le x \le 1$

77. (e)   **79.** $x \le -4/3$ or $x \ge 0$

81. $x \le -7$ or $x > -4$

83. $x < -2\sqrt{3}$ or $-3 < x < 2\sqrt{3}$

85. $x < \dfrac{1 - \sqrt{13}}{6}$ or $x > \dfrac{1 + \sqrt{13}}{6}$

87. $x = \dfrac{-3 \pm \sqrt{31}i}{2}$ **89.** $x = \dfrac{3 \pm \sqrt{31}i}{10}$

91. $x = \sqrt{2/3}$ or $-\sqrt{2/3}$ or i or $-i$

93. $x = -2$ or $1 + \sqrt{3}i$ or $1 - \sqrt{3}i$

95. $i, -i, 2, -1$

97. Many correct answers, including $x^4 - 2x^3 + 2x^2$

CHAPTER 5

Section 5.1, page 332

1. $1000k^{13/4}$ **3.** $c^{42/5} \cdot d^{10/3}$ **5.** $\dfrac{1}{3y^{2/3}}$

7. $\dfrac{a^{1/2}}{49b^{5/2}}$ **9.** a^x **11.** $x^{7/6} - x^{11/6}$ **13.** $x - y$

15. $x + y - (x - y)^{3/2}$ **17.** $(x^{1/3} + 3)(x^{1/3} - 2)$

19. $(x^{1/2} + 3)(x^{1/2} + 1)$

21. $(x^{2/5} + 9)(x^{1/5} + 3)(x^{1/5} - 3)$

23. $(a^2 + b^2)^{1/3}$ **25.** $a^{3/16}$

27. $4t^{27/10}$ **29.** $4\sqrt{5}$ **31.** $6\sqrt{2}$

33. $\dfrac{-2 + \sqrt{11}}{5}$ **35.** $-\sqrt{2}$ **37.** $15\sqrt{5}$

39. $\dfrac{4a^4}{|b|}$ **41.** $\dfrac{|d^5|}{2\sqrt{c}}$

43. $\dfrac{3\sqrt{2}}{4}$ **45.** $\dfrac{3\sqrt{3} - 3}{4}$ **47.** $\dfrac{2\sqrt{x} - 4}{x - 4}$

49. $\dfrac{1}{\sqrt{x + h + 1} + \sqrt{x + 1}}$

51. $\dfrac{2x + h}{\sqrt{(x + h)^2 + 1} + \sqrt{x^2 + 1}}$

53. 36 million miles   **55.** 886.8 million miles

57. 5.313 million   **59.** 20.004 million

61. (a) Any even power of a real number is never negative.
(b) $\sqrt[3]{-8} = -2$, but $\sqrt[6]{(-8)^2} = 2$

63. (a) Since its graph passes the horizontal test, f is one-to-one and hence has an inverse.

(b) $(g \circ f)(x) = g(f(x)) = (f(x))^{1/5} = (x^5)^{1/5} = x$ and
$(f \circ g)(x) = f(g(x)) = (g(x))^5 = (x^{1/5})^5 = x$

65. The graph of g is the graph of $f(x) = \sqrt{x}$ shifted horizontally 3 units to the left.

67. The graph of k is the graph of $f(x) = \sqrt{x}$ shifted horizontally 4 units to the left and vertically 4 units downward.

69. (a) 7.65, 15.31, 22.96, 30.61, 45.92
(b) Doubling capital and labor doubles output. Tripling capital and labor triples output.

71. (a) 13.61, 32.37, 53.73, 76.98, 127.79
(b) Doubling capital and labor increases output by a factor of $2^{5/4}$. Tripling capital and labor increases output by a factor of $3^{5/4}$.

Special Topics 5.1.A, page 339

1. $x = 7$ **3.** $x = 4$ **5.** $x = -2$

7. $x = \pm 3$ **9.** $x = -1$ or 2 **11.** $x = 9$

13. $x = 1/2$ **15.** $x = 1/2$ or -4 **17.** $x \approx \pm.73$

19. $x \approx -1.17$ or $x \approx 2.59$ or $x = -1$

21. $x \approx 1.658$ **23.** $x = 6$ **25.** $x = 3$ or 7

27. $r = 4.658$ **29.** $r \approx 8.019$

31. $b = \sqrt{\dfrac{a^2}{A^2 - 1}}$ **33.** $u = \sqrt{\dfrac{x^2}{1 - K^2}}$

35. $x = 4$ **37.** $x = 4$ **39.** $x = -1$ or -8

41. $x = 16$ **43.** $x \approx 105.236$ **45.** $x \approx -.283$

47. (a) 11.47 ft or 29.91 ft   (b) 21 ft

49. Approx. 6.205 mi

Section 5.2, page 350

1.

3.

5.

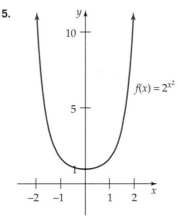

$f(x) = 2^{x^2}$

7. Shift the graph of h vertically 5 units downward.

9. Stretch the graph of h away from the x-axis by a factor of 3.

11. Shift the graph of h horizontally 2 units to the left, then vertically 5 units downward.

13. $f(x)$: C; $g(x)$: A; $h(x)$: B

15. Neither   **17.** Even

19. Even   **21.** 11   **23.** .8

25. $\dfrac{10^{x+h} - 10^x}{h}$ **27.** $\dfrac{(2^{x+h} + 2^{-x-h}) - 2^x - 2^{-x}}{h}$

29. $-3 \le x \le 3$ and $0 \le y \le 12$

31. $-4 \le x \le 4$ and $0 \le y \le 10$

33. $-10 \le x \le 10$ and $0 \le y \le 20$

35. $-10 \le x \le 10$ and $0 \le y \le 6$

37. The x-axis is a horizontal asymptote for the left side of the graph; local minimum at $(-1.44, -.53)$.

39. No asymptotes; local minimum at $(0, 1)$

41. The x-axis is a horizontal asymptote; local maximum at $(0, 1)$.

43. (a) About 520 in 15 days; about 1559 in 25 days
(b) In 29.3 days

45. (a) $k = 15$ (b) 12.128 lb/in^2 (c) $.01671$ lb/in^2

47. (a) 1980: 74.06; 2000: 76.34   (b) 1930

49. (a) The current population is 10, and in 5 years it will be about 149.
(b) After about 9.55 years

51. (a) $t = .75, c = 8; t = 1, c = 16$
(b) $c(t) = 2^{4t}$

53. (a) $f(x) = 6(3^x)$ or $f(x) = 18(3^{x-1})$
(b) 3
(c) No; yes

55. (a)

Folds	0	1	2	3	4
Thickness	.002	.004	.008	.016	.032

(b) $f(x) = .002(2^x)$ (c) 2097.15 in. = 174.76 ft
(d) 43

57. (a) $f(x) = 1200(1.04)^x$
(b) 1349.84; 1503.57
(c) After approximately 11 years and 1 month

59. (a) $100.4(1.014)^x$ (b) 115.3 million   (c) 2016

61. (a) $32.44e^{.02216x}$
(b) 40.49 million; 56.45 million   (c) 2024

63. About 256; about 654

65. (a) $f(x) = .75^x$ (b) About 8 ft

67. (a) $M(x) = 5(0.5^{x/5730})$ (b) 3.08 g; 1.90 g
(c) After about 13,305 years

69. $f(x) = a^x$ for any nonnegative constant a.

71. (a) The graph of f is the mirror image of the graph of g.
(b) $k(x) = f(x)$; see part (a).

73. (a) Not entirely   (b) $f_8(x)$
(c) Not at the right side; $f_{12}(x)$

Special Topics 5.2.A, page 359

1. Annually: $1469.33
Quarterly: $1485.95
Monthly: $1489.85
Weekly: $1491.37

3. $585.83   **5.** $610.40   **7.** $639.76

9. $563.75   **11.** $582.02   **13.** $3325.29

15. $3359.59   **17.** Fund C   **19.** $385.18

21. $1,162,003.14   **23.** About 5%   **25.** About 5.92%

27. (a) 9 years   (b) 9 years   (c) 9 years
(d) Investment amount and doubling time are independent.

29. 9.9 years

Section 5.3, page 367

1. 4   **3.** -2.5 **5.** $10^3 = 1000$

7. $10^{2.88} = 750$ **9.** $e^{1.0986} = 3$ **11.** $e^{-4.6052} = .01$

13. $x^2 + 2y = e^{z+w}$ **15.** $\log .01 = -2$

17. $\log 3 = .4771$ **19.** $\ln 25.79 = 3.25$

21. $\ln 5.5527 = 12/7$ **23.** $\ln w = 2/r$ **25.** $\sqrt{43}$

27. 15   **29.** $1/2$ **31.** 931

33. $x + y$ **35.** x^2

37. $4e^{(\ln 25)x}$ **39.** $-16e^{(\ln 30.5)x}$

41. $x > -1$ **43.** $x < 0$

45. (a) for all $x > 0$ (b) $e^{\ln x} = x$ for $x > 0$

47. They are the same for $x > 0$. If $x < 0$, $g(x)$ does not exist, while $f(x)$ is symmetric with respect to the y-axis.

49. Stretch away from x-axis by a factor of 2.

51. Horizontal shift to the right by 4

53. Horizontal shift to the left by 3
Vertical shift downward by 4

55.

57.

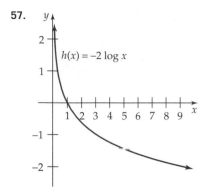

$h(x) = -2 \log x$

59. $0 \le x \le 9.4$ and $-6 \le y \le 6$ (vertical asymptote at $x = 1$)

61. $-10 \le x \le 10$ and $-3 \le y \le 3$

63. $0 \le x \le 20$ and $-6 \le y \le 3$

65. .5493   **67.** $-.2386$

69. (a) $\dfrac{\ln(3 + h) - \ln 3}{h}$ (b) $h \approx 2.2$

71. $(f \circ g)(x) = \dfrac{1}{1 + e^{-\ln[x/(1-x)]}} = \dfrac{1}{1 + \dfrac{1}{e^{\ln[x/(1-x)]}}}$

$= \dfrac{1}{1 + \dfrac{1}{\dfrac{x}{1-x}}} = \dfrac{1}{1 + \dfrac{1-x}{x}} = \dfrac{1}{\dfrac{1}{x}} = x;$

$(g \circ f)(x) = \ln\left(\dfrac{\dfrac{1}{1 + e^{-x}}}{1 - \dfrac{1}{1 + e^{-x}}}\right)$

$= \ln\left(\dfrac{\dfrac{1}{1 + e^{-x}}}{1 - \dfrac{1}{1 + e^{-x}}} \cdot \dfrac{1 + e^{-x}}{1 + e^{-x}}\right)$

$= \ln\left(\dfrac{1}{1 + e^{-x} - 1}\right) = \ln(e^x) = x$

73. $A = -9$; $B = 10$ **75.** About 4392 meters

77. (a) 26.55 billion pounds, 27.72 billion pounds
   (b) 2008

79. (a) 9.9 days   (b) About 6986

81. (a) No ads: about 120 bikes; $1000: about 299 bikes;
     $10,000: about 513 bikes
   (b) $1000, yes; $10,000, no
   (c) Both yes, but spending $1000 is more worthwhile.

83. $n = 30$ gives an approximation with a maximum error of
   .00001 when $-.7 \le x \le .7$.

Section 5.4, page 374

1. $\ln(x^2 y^3)$ **3.** $\log(x - 3)$ **5.** $\ln(x^{-7})$

7. $\ln(e - 1)^3$ **9.** $\log(20xy)$ **11.** $2u + 5v$

13. $\dfrac{1}{2}u + 2v$ **15.** $\dfrac{2}{3}u + \dfrac{1}{6}v$

17. False; the right side is not defined when $x < 0$, but the left
   side is.

19. True by the Power Law

21. False; the graph of the left side differs from the graph of
   the right side.

23. Many possible answers. For example $a = 2, b = 3$.

25. e **27.** 2

29. Approximately 2.54

31. 20 decibels

33. Approximately 66 decibels

35. Twice as loud

37. (a) 1.255; 1.255
   (b) 2.659; 2.659
   (c) 3.952; 3.952
   (d) $\log x = \dfrac{\ln x}{\ln 10}$ for all $x > 0$

39. (a) .9422
   (b) 1.9422
   (c) 2.9422
   (d) 3.9422
   (e) 4.9422
   (f) If we have x_n, then $x_{n+1} = 10x_n$; if we have $\log x_n$,
     then $\log x_{n+1} = \log 10x_n = 1 + \log x_n$; each increase
     of the logarithmic scale by 1 corresponds to multiply-
     ing by 10. Also, an increase by y in the logarithmic
     scale corresponds to multiplying by 10^y.

Special Topics 5.4.A, page 381

1.

x	0	1	2	4
$f(x) = \log_4 x$	Not defined	0	.5	1

3.

x	1/36	1/6	1	216
$h(x) = \log_6 x$	-2	-1	0	3

5.

x	0	1/7	$\sqrt{7}$	49
$f(x) = 2\log_7 x$	Not defined	-2	1	4

7.

x	-2.75	-1	1	29
$h(x) = 3\log_2(x + 3)$	-6	3	6	15

9. $\log .01 = -2$ **11.** $\log \sqrt[3]{10} = 1/3$

13. $\log r = 7k$ **15.** $\log_7 5{,}764{,}801 = 8$

17. $\log_3(1/9) = -2$ **19.** $10^4 = 10{,}000$

21. $10^{2.88} \approx 750$ **23.** $5^3 = 125$ **25.** $2^{-2} = \dfrac{1}{4}$

27. $10^{z+w} = x^2 + 2y$ **29.** $\sqrt{97}$ **31.** $x^2 + y^2$

33. 1/2   **35.** 6   **37.** $b = 3$ **39.** $b = 20$

41. 5   **43.** 3   **45.** 4

47. $\log \dfrac{x^2 y^3}{z^6}$

49. $\log (x^2 - 3x)$

51. $\log_2 (5|c|)$

53. $\log_4\left(\dfrac{1}{49c^2}\right)$

55. $\ln\left[\dfrac{(x+1)^2}{x+2}\right]$

57. $\log_2(x)$

59. $\ln(e^2 - 2e + 1) = \ln(e-1)^2 = 2\ln(e-1)$

61. 3.3219 **63.** $.8271$ **65.** 1.1115 **67.** 1.6199

69. True   **71.** True   **73.** False   **75.** 397^{398}

77. $\log_b u = \dfrac{\log_a u}{\log_a b}$

79. $\log_{10} u = 2\log_{100} u$

81. $\log_b x = \dfrac{1}{2}\log_b v + 3 = \log_b \sqrt{v} + \log_b b^3 = \log_b(b^3 \sqrt{v})$; hence $x = b^3 \sqrt{v}$.

83. $f(x) = g(x)$ only when $x \approx .123$, so the statement is false.

85.

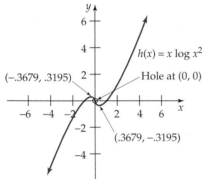

$h(x) = x \log x^2$

Hole at $(0, 0)$

$(-.3679, .3195)$

$(.3679, -.3195)$

Section 5.5, page 390

1. $x = 4$ **3.** $x = 1/9$ **5.** $x = \dfrac{1}{2}$ or -3

7. $x = -2$ or $-1/2$ **9.** $x = \ln 5/\ln 3 \approx 1.465$

11. $x = \ln 3/\ln 1.5 \approx 2.7095$

13. $x = \dfrac{\ln 3 - 5\ln 5}{\ln 5 + 2\ln 3} \approx -1.825$

15. $x = \dfrac{\ln 2 - \ln 3}{3\ln 2 + \ln 3} \approx -.1276$

17. $x = (\ln 5)/2 \approx .805$

19. $x = (-\ln 3.5)/1.4 \approx -.895$

21. $x = 2\ln(5/2.1)/\ln 3 \approx 1.579$ **23.** $x = 0$ or 1

25. $x = \ln 2 \approx .693$ or $x = \ln 3 \approx 1.099$

27. $x = \ln 3 \approx 1.099$

29. $x = \ln 2/\ln 4 = 1/2$ or $x = \ln 3/\ln 4 \approx .792$

31. $x = \ln(t + \sqrt{t^2 + 1})$

33. If $\ln u = \ln v$, then $e^{\ln u} = e^{\ln v}$, so $u = v$

35. $x = 9$ **37.** $x = 5$ **39.** $x = 6$ **41.** $x = 3$

43. $x = \dfrac{-5 + \sqrt{37}}{2}$ **45.** $x = 9/(e-1)$ **47.** $x = 5$

49. $x = \pm\sqrt{10001}$ **51.** $x = \sqrt{\dfrac{e+1}{e-1}}$

53. .444 billion years   **55.** 14.95 days

57. 3.8134 days   **59.** About 3689 years old

61. About 2534 years ago   **63.** About 6.99%

65. (a) About 22.5 years   (b) About 22.1 years

67. $3197.05   **69.** 79.36 years

71. (a) 1.36%   (b) 2027

73. (a) $k \approx 21.459$ (b) $t \approx .182$

75. (a) 20 at the beginning and 2500 3 hours later

(b) $\dfrac{\ln 2}{\ln 5} \approx .43$

77. $t = 0$: 200 people
$t = 3$: about 2795 people
(b) About 6.02 weeks

79. (a) $k \approx .229$, $c \approx 83.3$ (b) 12.43 weeks

Section 5.6, page 398

1. Cubic, logistic

3. Exponential, quadratic, cubic, power

5. Exponential, logarithmic, quadratic, cubic

7. Quadratic, cubic

9. Quadratic, cubic

11. Ratios range from approximately 5.059 to 5.076; exponential is appropriate

13. (a) For large values of x the term $54e^{-.0228x}$ is close to zero so the quantity $(1 + 54e^{-.0228x})$ is slightly larger than 1, which means $\dfrac{384.57}{1 + 54e^{-.0228x}}$ is always less than (but very close to) 384.57.

(b)

15. (a)

(b)

(c)

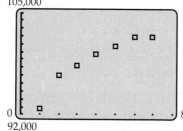

17. Power model

19. Power or logarithmic model

21. (a) 105,000

(b) 105,000

(c)

23. (a) 80

(b) 5

(c) An exponential model would be best:
$f(x) = 152.22 \cdot e^{-0.03x}$.

25. (a) 40

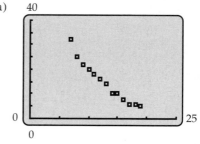

(b) $y = 91.1630 \cdot .85919^x$ or, equivalently,
$y = 91.1630e^{-.15176x}$

(c) About 2.8

(d) In late 2009 ($x \approx 29.74$)

(e) Any projection of future trends becomes tenuous with passing time. The availability of new technology could have unforeseen effects.

27. (a)

(b) Ratio
‾‾‾‾
1.01
1.01   An exponential model
1.01   would be appropriate.
1.01
1.01
1.01

(c) $f(t) = 98.74e^{0.012x}$

(d) Predicted number of Households
‾‾‾‾‾‾‾‾‾‾‾‾‾‾‾‾‾‾‾‾‾‾‾‾‾‾‾
98.7
99.9
101.1
102.3
103.6
104.8

(e) 108.6 million

29. (a) $f(x) = 17.6 + 13.42 \ln x$
(b) 77.4 years
(c) 2013

Chapter 5 Review, page 405

1. c^2 **3.** $a^{10/3}b^{42/5}$ **5.** $u^{1/2} - v^{1/2}$ **7.** $c^2d^4/2$

9. $\dfrac{2}{\sqrt{2x + 2h + 1} + \sqrt{2x + 1}}$

11. $x = \dfrac{5 - \sqrt{5}}{2}$ **13.** No solutions

15. $x = -1.733$ or 5.521

17. $-3 \le x \le 3$ and $0 \le y \le 2$

19.

21.

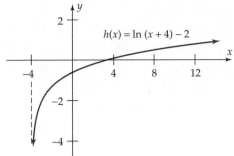

23. (a) About 2.03
(b) About 31.97
(c) Approximately 6 to 10 months
(d) Never; however, at the end of 18 months about 99.6% of the program will be mastered.

25. $\ln 756 = 6.628$ **27.** $\ln (u + v) = r^2 - 1$

29. $\log 756 = 2.8785$ **31.** $e^{7.118} = 1234$

33. $e^t = rs$ **35.** $5^u = cd - k$ **37.** 3   **39.** 3/4

41. $2 \ln x$ **43.** $\ln (9y/x^2)$ **45.** $\ln (1/x^{10})$

47. 2   **49.** (c)

51. (c)   **53.** $x = \pm\sqrt{2}$ **55.** $x = \dfrac{3 \pm \sqrt{57}}{4}$

57. $x = -1/2$ **59.** $x = e^{(u-c)/d}$ **61.** $x = 2$

63. $x = 101$ **65.** About 1.64 mg

67. Approximately 12 years   **69.** $452.89   **71.** 7.6

73. (a) 11°F

(b)

(c) The ratios of successive entries are approxiately constant.
(d) $f(t) = 22.42e^{-.034x}$.
(e) About 10°F

CHAPTER 6

Section 6.1, page 421

1. $2\pi/9$ **3.** $\pi/9$ **5.** $\pi/18$

7. $9\pi/4, 17\pi/4, -7\pi/4, -15\pi/4$

9. $11\pi/6, 23\pi/6, -13\pi/6, -25\pi/6$

11. $5\pi/3$ **13.** $3\pi/4$ **15.** $3\pi/5$ **17.** $7 - 2\pi$

19. $\pi/30$ **21.** $-\pi/15$ **23.** $5\pi/12$ **25.** $3\pi/4$

27. $-5\pi/4$ **29.** $31\pi/6$ **31.** $36°$ **33.** $-18°$

35. $135°$ **37.** $4°$ **39.** $-75°$ **41.** $972°$

43. $4\pi/3$ **45.** $7\pi/6$ **47.** $41\pi/6$ **49.** 8π cm

51. $17/4$ **53.** $50/9$ **55.** 2000 **57.** 5

59. 8.75 **61.** 3490.66 mi   **63.** 942.48 mi

65. 7π **67.** 4π **69.** 42.5π **71.** $2\pi k$

73. 3 radians ($\approx 171.9°$)

75. (a) 400π radians per min

(b) 800π in. per min or $\dfrac{200\pi}{3}$ ft per min

77. (a) 5π radians per sec
(b) 6.69 mph

79. 15.92 ft

Section 6.2, page 429

1. -1 **3.** 0 **5.** 0 **7.** 0 **9.** 0

11. $\sin t = 1/\sqrt{5}$, $\cos t = -2/\sqrt{5}$, $\tan t = -1/2$

13. $\sin t = -4/5$, $\cos t = -3/5$, $\tan t = 4/3$

15. $\sqrt{3}/2$, $1/2$, $\sqrt{3}$ **17.** $\sqrt{2}/2$, $\sqrt{2}/2$, 1

19. $\sqrt{2}/2$, $-\sqrt{2}/2$, -1 **21.** $1/2$, $-\sqrt{3}/2$, $\dfrac{-\sqrt{3}}{3}$

23. $\sin\left(-\dfrac{23\pi}{6}\right) = \dfrac{1}{2}$, $\cos\left(-\dfrac{23\pi}{6}\right) = \dfrac{\sqrt{3}}{2}$, $\tan\left(-\dfrac{23\pi}{6}\right) = \dfrac{\sqrt{3}}{3}$

25. $\sin\left(-\dfrac{19\pi}{3}\right) = -\dfrac{\sqrt{3}}{2}$, $\cos\left(-\dfrac{19\pi}{3}\right) = \dfrac{1}{2}$,
$\tan\left(-\dfrac{19\pi}{3}\right) = -\sqrt{3}$

27. $\sin\left(-\dfrac{15\pi}{4}\right) = \dfrac{\sqrt{2}}{2}$, $\cos\left(-\dfrac{15\pi}{4}\right) = \dfrac{\sqrt{2}}{2}$, $\tan\left(-\dfrac{15\pi}{4}\right) = 1$

29. $\sin\left(-\dfrac{17\pi}{2}\right) = -1$, $\cos\left(-\dfrac{17\pi}{2}\right) = 0$,
$\tan\left(-\dfrac{17\pi}{2}\right)$ is not defined

31. $-\sqrt{3}/2$ **33.** $-\sqrt{2}/2$ **35.** $\dfrac{\sqrt{2}}{4}(1 - \sqrt{3})$

37. $\sin t = 7/\sqrt{53}$, $\cos t = 2/\sqrt{53}$, $\tan t = 7/2$

39. $\sin t = -6/\sqrt{71}$, $\cos t = -5/\sqrt{71}$, $\tan t = 6/5$

41. $\sin t = -10/\sqrt{103}$, $\cos t = \sqrt{3}/\sqrt{103}$, $\tan t = -10/\sqrt{3}$

43. $\sin t = -3/\sqrt{10}$, $\cos t = 1/\sqrt{10}$, $\tan t = -3$

45. $\sin t = -5/\sqrt{34}$, $\cos t = 3/\sqrt{34}$, $\tan t = -5/3$

47. $\sin t = 1/\sqrt{5}$, $\cos t = -2/\sqrt{5}$, $\tan t = -1/2$

49. Quadrant I: $\sin t(+)$, $\cos t(+)$, $\tan t(+)$;
Quadrant II: $\sin t(+)$, $\cos t(-)$, $\tan t(-)$;
Quadrant III: $\sin t(-)$, $\cos t(-)$; $\tan t(+)$;
Quadrant IV: $\sin t(-)$, $\cos t(+)$, $\tan t(-)$

51. Positive since $0 < 1 < \pi/2$

53. Negative since $\pi/2 < 3 < \pi$

55. Positive since $0 < 1.5 < \pi/2$

57. $t = \dfrac{\pi}{2} + 2\pi n$, n any integer

59. $t = \pi n$, n any integer

61. $t = \dfrac{\pi}{2} + \pi n$, n any integer

63. $\sin(\cos 0) = \sin 1$, while $\cos(\sin 0) = \cos 0 = 1$. Since $\sin 1 < 1$ (draw a picture!), $\cos(\sin 0)$ is larger than $\sin(\cos 0)$.

65. (a) Each horse moves through an angle of 2π radians in 1 min. The angle between horses A and B is $\pi/4$ radians ($= \frac{1}{8}$ of 2π radians). It takes $\frac{1}{8}$ min for each horse to move through an angle of $\pi/4$ radians. Thus the position occupied by B at time t will be occupied by A $\frac{1}{8}$ min later, that is, at time $t + \frac{1}{8}$. Therefore, $B(t) = A\left(t + \frac{1}{8}\right)$.

(b) $C(t) = A\left(t + \frac{1}{3}\right)$.

(c) $E(t) = D\left(t + \frac{1}{8}\right)$; $F(t) = D\left(t + \frac{1}{3}\right)$.

(d) The triangles in Figure B are similar, so that $\dfrac{5}{1} = \dfrac{D(t)}{A(t)}$. Therefore, $D(t) = 5A(t)$.

(e) $E(t) = 5B(t) = 5A\left(t + \frac{1}{8}\right)$; $F(t) = 5C(t) = 5A\left(t + \frac{1}{3}\right)$.

(f) Since horse A travels through an angle of 2π radians each minute and its starting angle is 0 radians, then at the end of t min horse A will be on the terminal side of an angle of $2\pi t$ radians, at the point where it intersects the unit circle. $A(t)$ is the second coordinate of this point; hence, $A(t) = \sin(2\pi t)$.

(g) $B(t) = A\left(t + \frac{1}{8}\right) = \sin[2\pi\left(t + \frac{1}{8}\right)] = \sin(2\pi t + \pi/4)$; $C(t) = \sin(2\pi t + 2\pi/3)$

(h) $D(t) = 5\sin(2\pi t)$; $E(t) = 5\sin(2\pi t + \pi/4)$;
$f(t) = 5\sin(2\pi t + 2\pi/3)$

Section 6.3, page 443

1. $(fg)(t) = 3\sin^2 t + 6\sin t\cos t$

3. $3\sin^3 t + 3\sin^2 t\tan t$ **5.** $(\cos t - 2)(\cos t + 2)$

7. $(\sin t - \cos t)(\sin t + \cos t)$ **9.** $(\tan t + 3)^2$

11. $(3\sin t + 1)(2\sin t - 1)$

13. $(\cos^2 t + 5)(\cos t + 1)(\cos t - 1)$

15. $(f \circ g)(t) = \cos(2t + 4)$, $(g \circ f)(t) = 2\cos t + 4$

17. $(f \circ g)(t) = \tan(t^2 + 2)$, $(g \circ f)(t) = \tan^2(t + 3) - 1$

19. Yes   **21.** No   **23.** No   **25.** $\sin t = -\sqrt{3}/2$

27. $\sin t = \sqrt{3}/2$ **29.** $-3/5$ **31.** $-3/5$

33. $3/4$ **35.** $-3/4$

37. $-\sqrt{21}/5$ **39.** $-2/5$ **41.** $-\sqrt{21}/5$

43. $\dfrac{\sqrt{2 + \sqrt{2}}}{2}$ **45.** $\dfrac{\sqrt{2 - \sqrt{2}}}{2}$

47. $\sin^2 t - \cos^2 t$

49. $\sin t$

51. $|\sin t \cos t| \sqrt{\sin t}$

53. $1/4$

55. $\cos t + 2$

57. $\cos t$

59. $f(t + \pi) = \sin 2(t + \pi) = \sin(2t + 2\pi) = \sin 2t = f(t)$

61. $f(t + \pi/2) = \sin 4(t + \pi/2) = \sin(4t + 2\pi)$
$\qquad = \sin 4t = f(t)$

63. $f(t + \pi/2) = \tan 2(t + \pi/2) = \tan(2t + \pi)$
$\qquad = \tan 2t = f(t)$

65. (a) There is no such number k.
  (b) If we substitute $t = 0$ in $\cos(t + k) = \cos t$, we get $\cos k = \cos 0 = 1$.
  (c) If there were such a number k, then by part (b), $\cos k = 1$, which is impossible by part (a). Therefore, there is no such number k, and the period is 2π.

Section 6.4, page 453

1. $t = \ldots, -2\pi, -\pi, 0, \pi, 2\pi, \ldots$; or $t = \pi k$, where k is any integer

3. $t = \ldots, -7\pi/2, -3\pi/2, \pi/2, 5\pi/2, 9\pi/2, \ldots$; or $t = \pi/2 + 2\pi k$, where k is any integer

5. $t = \ldots, -3\pi, -\pi, \pi, 3\pi, \ldots$; or $t = \pi + 2k\pi$, where k is any integer

7. 11

9. 1.4

11. Shift the graph of f vertically 3 units upward.

13. Reflect the graph of f in the horizontal axis.

15. Shift the graph of f vertically 5 units upward.

17. Stretch the graph of f away from the horizontal axis by a factor of 3.

19. Stretch the graph of f away from the horizontal axis by a factor of 3, then shift the resulting graph vertically 2 units upward.

21. Shift the graph of f horizontally 2 units to the right.

23. 2 solutions

25. 2 solutions

27. 2 solutions

29. 2 solutions

31. Possibly an identity

33. Possibly an identity

35. Possibly an identity

37. Not an identity

39. Possibly an identity

41. Not an identity

43. (a) Yes if proper value of k is used; no
  (b) $0, 2\pi, 4\pi, 6\pi$, etc. So why do the graphs look identical?

45. (a) 80
  (b) 14 or 15 on 96-pixel-wide screens; up to 40–50 on wider screens; quite different from part (a). Explain what's going on. [*Hint:* How many points have to be plotted in order to get even a rough approximation of one full wave? How many points is the calculator plotting for the entire graph?]

47. (a) $-\pi \le t \le \pi$
  (b) $n = 15$; $f_{15}(2)$ and $g(2)$ are identical in the first nine decimal places and differ in the tenth, a very good approximation.

49. $r(t)/s(t)$, where $r(t) = f_{15}(t)$ in Exercise 47 and $s(t) = f_{16}(t)$ in Exercise 48.

51. The y-coordinate of the new point is the same as the x-coordinate of the point on the unit circle. To explain what's going on, look at the definition of the cosine function.

Section 6.5, page 464

1. Amplitude: 3; period: π, phase shift: $+\dfrac{\pi}{2}$

3. Amplitude: 7; period: $\dfrac{2\pi}{7}$, phase shift: $-\dfrac{1}{49}$

5. Amplitude: 1; period: 1; phase shift: 0

7. Amplitude: 6; period: $\dfrac{2}{3}$; phase shift: $-\dfrac{1}{3\pi}$

9. $f(t) = 3 \sin\left(8t - \dfrac{8\pi}{5}\right)$, (other answers possible)

11. $f(t) = \dfrac{2}{3}\sin(2\pi t)$

13. $f(t) = 7 \sin\left(\dfrac{6\pi}{5}t + \dfrac{3\pi^2}{5}\right)$

15. $f(t) = 2 \sin 4t$

17. $f(t) = 1.5 \cos\dfrac{t}{2}$

19. (a) $\dfrac{2\pi}{300}$
  (b) The graph makes 300 complete waves between 0 and 2π.
  (c) $0 \le x \le \dfrac{2\pi}{75}$; $-2 \le y \le 2$

21. (a) $\dfrac{2\pi}{900}$
  (b) The graph makes 900 complete waves between 0 and 2π.
  (c) $0 \le x \le \dfrac{2\pi}{225}$; $-2 \le y \le 2$

23. (a) $f(t) = -12 \sin\left(10t + \dfrac{\pi}{2}\right)$
  (b) $g(t) = -12 \cos 10t$

25. (a) $f(t) = -\sin 2t$
  (b) $g(t) = -\cos\left(2t - \dfrac{\pi}{2}\right)$

27.

29.

31.

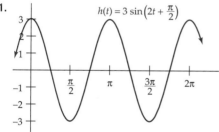

33. Local maximum at $t = 5\pi/6 \approx 2.6180$; local minimum at $t = 11\pi/6 \approx 5.7596$

35. Local maxima at $t = \pi/6 \approx .5236$, $t = 5\pi/6 \approx 2.6180$, $t = 3\pi/2 \approx 4.7124$; local minima at $t = \pi/2 \approx 1.5708$, $t = 7\pi/6 \approx 3.6652$, $t = 11\pi/6 \approx 5.7596$

37. $A \approx 5.3852$, $b = 1$, $c \approx 1.1903$

39. $A \approx 3.8332$, $b = 4$, $c \approx 1.4572$

41. All waves in the graph of g are of equal height, which is not the case with the graph of f.

43. $1/980,000$; $980,000$ **45.** $f(t) = 125 \sin(\pi t/5)$

47. $f(t) = \cos 20\pi t + \sqrt{16 - \sin^2(20\pi t)}$

49. $h(t) = 6 \sin(\pi t/2)$ **51.** $h(t) = 6 \cos(\pi t/2)$

53. $d(t) = 10 \sin(\pi t/2)$

55. (a) At least four (starting point, high point, low point, ending point)
  (b) At least 301 (4 points for the first wave and 3 for each of the remaining 99 waves because the starting point of one is the ending point of the preceding one)
  (c) Answers vary from 95 to 239.

57. (a)

  (b) A periodic model would be
      $f(x) = 1.28 \cdot \sin(.61x - .24) + 7.59$.
  (c) The model gives only a rough approximation of the data, and it is unlikely to be of much use for making predictions.

59. (a) $f(t) = 2.77 \cdot \sin(.39x + 1.91) + 2.52$
  (b) $f(x)$ has period ≈ 16.1. One would expect a period of 12.
  (c)

  (d) $g(x) = 2.17 \sin(.51x + 1.04) + 1.71$
  (e) $g(x)$ has period ≈ 12.3, and it seems to fit better than $f(x)$.

61. (a) $k = 9.8/\pi^2$
  (b) When k is replaced by $(k + .01\%$ of $k)$, the value of ω changes and the period of the pendulum becomes approximately 2.000099998 sec, meaning that the clock loses .000099998 sec every 2 sec, for a total of approximately 397.43 sec (6.62 min) during the three months.

Special Topics 6.5.A, page 473

  1. $A \approx 2.2361$, $b = 1$, $c \approx 1.1071$

  3. $A \approx 5.3852$, $b = 4$, $c \approx -1.1903$

  5. $A \approx 5.1164$, $b = 3$, $c \approx -.7442$

  7. $0 \le t \le 2\pi$ and $-5 \le y \le 5$ (one period)

  9. $-10 \le t \le 10$ and $-10 \le y \le 10$

11. $0 \le t \le \pi/50$ and $-2 \le y \le 2$ (one period)

13. $0 \le t \le .04$ and $-7 \le y \le 7$ (one period)

15. $0 \le t \le 10$ and $-6 \le y \le 10$ (one period)

17. To the left of the y-axis, the graph lies above the t-axis, which is a horizontal asymptote of the graph. To the right of the y-axis, the graph makes waves of amplitude 1, of shorter and shorter period.
Window: $-3 \le t \le 3.2$ and $-2 \le y \le 2$

19. The graph is symmetric with respect to the y-axis and consists of waves along the t-axis, whose amplitude slowly increases as you move farther from the origin in either direction.
Window: $-30 \le t \le 30$ and $-6 \le y \le 6$

21. The graph which is not defined at $t = 0$, is symmetric with respect to the y-axis and consists of waves along the t-axis whose amplitude rapidly decreases as you move farther from the origin in either direction.
Window: $-30 \le t \le 30$ and $-.3 \le y \le 1$

23. The function is periodic with period π. (Why?) The graph lies on or below the t-axis because the logarithmic function is negative for numbers between 0 and 1 and $|\cos t|$ is always between 0 and 1. The graph has vertical asymptotes when $t = \pm\pi/2, \pm3\pi/2, \pm5\pi/2, \pm7\pi/2, \ldots$ ($\cos t = 0$ at these points and $\ln 0$ is not defined).
Window: $-2\pi \le t \le 2\pi$ and $-3 \le y \le 1$ (two periods).

Section 6.6, page 479

1. Fourth quadrant

3. Second quadrant

5. Fourth quadrant

7. $\sin t = 4/5$, $\cos t = 3/5$, $\tan t = 4/3$, $\cot t = 3/4$, $\sec t = 5/3$, $\csc t = 5/4$

9. $\sin t = 12/13$, $\cos t = -5/13$, $\tan t = -12/5$, $\cot t = -5/12$, $\sec t = -13/5$, $\csc t = 13/12$

11. $\sin t = 5/\sqrt{26}$, $\cos t = -1/\sqrt{26}$, $\tan t = -5$, $\cot t = -1/5$, $\sec t = -\sqrt{26}$, $\csc t = \sqrt{26}/5$

13. $\sin t = \sqrt{3}/\sqrt{5}$, $\cos t = \sqrt{2}/\sqrt{5}$, $\tan t = \sqrt{3}/\sqrt{2}$, $\cot t = \sqrt{2}/\sqrt{3}$, $\sec t = \sqrt{5}/\sqrt{2}$, $\csc t = \sqrt{5}/\sqrt{3}$

15. $\sin t = \dfrac{3}{\sqrt{12 + 2\sqrt{2}}}$, $\cos t = \dfrac{1 + \sqrt{2}}{\sqrt{12 + 2\sqrt{2}}}$,
$\tan t = \dfrac{3}{1 + \sqrt{2}}$, $\cot t = \dfrac{1 + \sqrt{2}}{3}$,
$\sec t = \dfrac{\sqrt{12 + 2\sqrt{2}}}{1 + \sqrt{2}}$, $\csc t = \dfrac{\sqrt{12 + 2\sqrt{2}}}{3}$

17. $\sin\left(\dfrac{4\pi}{3}\right) = -\dfrac{\sqrt{3}}{2}$, $\cos\left(\dfrac{4\pi}{3}\right) = -\dfrac{1}{2}$, $\tan\left(\dfrac{4\pi}{3}\right) = \sqrt{3}$,
$\cot\left(\dfrac{4\pi}{3}\right) = \dfrac{1}{\sqrt{3}}$, $\sec\left(\dfrac{4\pi}{3}\right) = -2$, $\csc\left(\dfrac{4\pi}{3}\right) = \dfrac{-2}{\sqrt{3}}$

19. $\sin\left(\dfrac{7\pi}{4}\right) = -\dfrac{\sqrt{2}}{2}$, $\cos\left(\dfrac{7\pi}{4}\right) = \dfrac{\sqrt{2}}{2}$, $\tan\left(\dfrac{7\pi}{4}\right) = -1$,
$\cot\left(\dfrac{7\pi}{4}\right) = -1$, $\sec\left(\dfrac{7\pi}{4}\right) = \sqrt{2}$, $\csc\left(\dfrac{7\pi}{4}\right) = -\sqrt{2}$

21. -3.8287

23. (a) 5.6511; 5.7618; 5.7731; 5.7743
(b) $(\sec 2)^2$

25. $\cos t + \sin t$

27. $1 - 2\sec t + \sec^2 t$

29. $\cot^3 t - \tan^3 t$

31. $\csc t(\sec t - \csc t)$

33. $(-1)(\tan^2 t + \sec^2 t)$

35. $(\cos^2 t + 1 + \sec^2 t)(\cos t - \sec t)$ **37.** $\cot t$

39. $\dfrac{2\tan t + 1}{3\sin t + 1}$

41. $4 - \tan t$

43. $1 + \cot^2 t = 1 + \dfrac{\cos^2 t}{\sin^2 t} = \dfrac{\sin^2 t + \cos^2 t}{\sin^2 t} = \dfrac{1}{\sin^2 t} = \csc^2 t$

45. $\sec(-t) = \dfrac{1}{\cos(-t)} = \dfrac{1}{\cos t} = \sec t$

47. $\sin t = \dfrac{\sqrt{3}}{2}$, $\cos t = -\dfrac{1}{2}$, $\tan t = -\sqrt{3}$,
$\csc t = \dfrac{2\sqrt{3}}{3}$, $\sec t = -2$, $\cot t = -\dfrac{\sqrt{3}}{3}$

49. $\sin t = 1$, $\cos t = 0$, $\tan t$ is undefined, $\cot t = 0$, $\sec t$ is undefined, $\csc t = 1$

51. $\sin t = \dfrac{12}{13}$, $\cos t = -\dfrac{5}{13}$, $\tan t = -\dfrac{12}{5}$,
$\cot t = -\dfrac{5}{12}$, $\sec t = -\dfrac{13}{5}$, $\csc t = \dfrac{13}{12}$

53. Possibly an identity   **55.** Not an identity

57. Look at the graph of $y = \sec t$ in Figure 6–76 on page 478. If you draw in the line $y = t$, it will pass through $(-\pi/2, -\pi/2)$ and $(\pi/2, \pi/2)$, and obviously will not intersect the graph of $y = \sec t$ when $-\pi/2 \le t \le \pi/2$. But it will intersect each part of the graph that lies above the horizontal axis, to the right of $t = \pi/2$; it will also intersect those parts that lie below the horizontal axis, to the left of $-\pi/2$. The first coordinate of each of these infinitely many intersection points will be a solution of $\sec t = t$.

59. (a) $\dfrac{\cos\theta\sin\theta}{2}$ (b) $\dfrac{\tan\theta}{2}$ (c) $\dfrac{\theta}{2}$

Chapter 6 Review, page 483

1. $\dfrac{\pi}{3}$ **3.** $324°$ **5.** $\dfrac{11\pi}{9}$

7. $-495°$ **9.** $\cos v = -\dfrac{1}{3}$ **11.** 0

13. $.809$ **15.** $-\dfrac{\sqrt{3}}{2}$ **17.** $\dfrac{\sqrt{3}}{2}$

19.

t	0	$\pi/6$	$\pi/4$	$\pi/3$	$\pi/2$
$\sin t$	0	$1/2$	$\sqrt{2}/2$	$\sqrt{3}/2$	1
$\cos t$	1	$\sqrt{3}/2$	$\sqrt{2}/2$	$1/2$	0

21. $9/4$ **23.** 0   **25.** $-\sqrt{\dfrac{2}{3}}$ **27.** $-\dfrac{3}{5}$

29. $-\dfrac{12}{13}$ **31.** $-\dfrac{1}{2}$ **33.** $\dfrac{\sqrt{3}}{2}$ **35.** (c)

37. $\dfrac{7}{\sqrt{58}}$ **39.** $-\dfrac{7}{3}$ **41.** $y = -\sqrt{3}x$ **43.** $-\dfrac{3}{2}$

45. $\dfrac{2}{\sqrt{13}}$ **47.** See Figure 6–77 on page 479.

49. (d)   **51.** $-\dfrac{3}{2}$ **53.** 0

55. (d)   **57.** $-\dfrac{1}{\sqrt{3}}$

59. (a) $\dfrac{3}{2}$ (b) $t = \dfrac{\pi}{5}$

61.

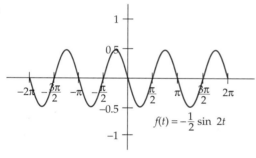

$$f(t) = -\tfrac{1}{2}\sin 2t$$

63. Not an identity   **65.** Possibly an identity

67. $1/2$ **69.** $2\cos(5t/2)$

71. $f(t) = 8\sin\left(\dfrac{2\pi t - 28\pi}{5}\right)$ is one possibility.

73. $A \approx 10.5588,\ b = 4,\ c = .4581$

75. $0 \le t \le \pi/50$ and $-5 \le y \le 5$ (one period)

CHAPTER 7

Section 7.1, page 497

1. Possibly an identity

3. Possibly an identity

5. B   **7.** E

9. $\tan x \cos x = \left(\dfrac{\sin x}{\cos x}\right)\cos x = \sin x$

11. $\cos x \sec x = \cos x\left(\dfrac{1}{\cos x}\right) = 1$

13. $\tan x \csc x = \left(\dfrac{\sin x}{\cos x}\right)\left(\dfrac{1}{\sin x}\right) = \dfrac{1}{\cos x} = \sec x$

15. $\dfrac{\tan x}{\sec x} = \dfrac{\sin x/\cos x}{1/\cos x} = \sin x$

17. $(1 + \cos x)(1 - \cos x) = 1 - \cos^2 x = \sin^2 x$

19. Not an identity

21. $\dfrac{\sin(-x)}{\cos(-x)} = \dfrac{-\sin x}{\cos x} = -\tan x$

23. $\cot(-x) = \dfrac{\cos(-x)}{\sin(-x)} = \dfrac{\cos x}{-\sin x} = -\cot x$

25. Not an identity

27. $\sec^2 x - \csc^2 x = (1 + \tan^2 x) - (1 + \cot^2 x) = \tan^2 x - \cot^2 x$

29. $\sin^2 x(\cot x + 1)^2 = [\sin x(\cot x + 1)]^2 = (\cos x + \sin x)^2 = [\cos x(1 + \tan x)]^2 = \cos^2 x(\tan x + 1)^2$

31. $\sin^2 x - \tan^2 x = \sin^2 x - \dfrac{\sin^2 x}{\cos^2 x} = \sin^2 x\left(1 - \dfrac{1}{\cos^2 x}\right) = \sin^2 x(1 - \sec^2 x) = \sin^2 x(-\tan^2 x) = -\sin^2 x \tan^2 x$

33. $(\cos^2 x - 1)(\tan^2 x + 1) = (-\sin^2 x)(\sec^2 x) = (-\sin^2 x)\left(\dfrac{1}{\cos^2 x}\right) = -\tan^2 x$

35. $\dfrac{\sec x}{\csc x} = \dfrac{1/\cos x}{1/\sin x} = \dfrac{\sin x}{\cos x} = \tan x$

37. $\cos^4 x - \sin^4 x = (\cos^2 x - \sin^2 x)(\cos^2 x + \sin^2 x) = \cos^2 x - \sin^2 x$

39. Not an identity

41. $\dfrac{\sec x}{\csc x} + \dfrac{\sin x}{\cos x} = \dfrac{1/\cos x}{1/\sin x} + \tan x = \dfrac{\sin x}{\cos x} + \tan x = \tan x + \tan x = 2\tan x$

43. $\dfrac{\sec x + \csc x}{1 + \tan x} = \dfrac{\dfrac{1}{\cos x} + \dfrac{1}{\sin x}}{1 + \dfrac{\sin x}{\cos x}} \cdot \dfrac{\sin x \cos x}{\sin x \cos x} = \dfrac{\sin x + \cos x}{\sin x \cos x + \sin^2 x} = \dfrac{\sin x + \cos x}{\sin x(\cos x + \sin x)} = \dfrac{1}{\sin x} = \csc x$

45. $\dfrac{1}{\csc x - \sin x} = \dfrac{1}{\dfrac{1}{\sin x} - \sin x} \cdot \dfrac{\sin x}{\sin x} = \dfrac{\sin x}{1 - \sin^2 x} = \dfrac{\sin x}{\cos^2 x} = \left(\dfrac{1}{\cos x}\right)\left(\dfrac{\sin x}{\cos x}\right) = \sec x \tan x$

47. Not an identity

49. Conjecture: $\cos x$. Proof: $1 - \dfrac{\sin^2 x}{1 + \cos x} = \dfrac{1 + \cos x - \sin^2 x}{1 + \cos x} = \dfrac{\cos x + (1 - \sin^2 x)}{1 + \cos x} = \dfrac{\cos x + \cos^2 x}{1 + \cos x} = \dfrac{\cos x(1 + \cos x)}{1 + \cos x} = \cos x$

51. Conjecture: $\tan x$. Proof:
$(\sin x + \cos x)(\sec x + \csc x) - \cot x - 2 =$
$\sin x \sec x + \sin x \csc x + \cos x \sec x +$

$\cos x \csc x - \cot x - 2 = \sin x \cdot \dfrac{1}{\cos x} +$

$\sin x \cdot \dfrac{1}{\sin x} + \cos x \cdot \dfrac{1}{\cos x} + \cos x \cdot \dfrac{1}{\sin x} -$

$\cot x - 2 = \dfrac{\sin x}{\cos x} + 1 + 1 + \dfrac{\cos x}{\sin x} - \cot x - 2 =$

$\tan x + \cot x - \cot x = \tan x$

53. $\dfrac{1 - \sin x}{\sec x} = \dfrac{1 - \sin x}{\sec x} \cdot \dfrac{(1 + \sin x)}{(1 + \sin x)} =$

$\dfrac{1 - \sin^2 x}{\sec x(1 + \sin x)} = \dfrac{\cos^2 x}{\dfrac{1}{\cos x}(1 + \sin x)} = \dfrac{\cos^3 x}{1 + \sin x}$

55. $\dfrac{\cos x}{1 - \sin x} \cdot \dfrac{1 + \sin x}{1 + \sin x} = \dfrac{\cos x(1 + \sin x)}{1 - \sin^2 x}$

$= \dfrac{\cos x(1 + \sin x)}{\cos^2 x} = \dfrac{1 + \sin x}{\cos x} = \dfrac{1}{\cos x} + \dfrac{\sin x}{\cos x} =$

$\sec x + \tan x$

57. $\dfrac{\cos x \cot x}{\cot x - \cos x} \cdot \dfrac{\sin x}{\sin x} = \dfrac{\cos x \cos x}{\cos x - \cos x \sin x} =$

$\dfrac{\cos^2 x}{\cos x(1 - \sin x)} = \dfrac{\cos x}{1 - \sin x} \cdot \dfrac{1 + \sin x}{1 + \sin x} =$

$\dfrac{\cos x(1 + \sin x)}{1 - \sin^2 x} = \dfrac{\cos x + \sin x \cos x}{\cos^2 x} \cdot \dfrac{\dfrac{1}{\sin x}}{\dfrac{1}{\sin x}} =$

$= \dfrac{\dfrac{\cos x}{\sin x} + \cos x}{\cos x\left(\dfrac{\cos x}{\sin x}\right)} = \dfrac{\cot x + \cos x}{\cos x \cot x}$

59. $\cot x = \dfrac{1}{\tan x}$, so $\log_{10}(\cot x) = \log_{10}\left(\dfrac{1}{\tan x}\right)$
$= -\log_{10}(\tan x)$

61. $\csc x + \cot x = (\csc x + \cot x) \cdot \dfrac{(\csc x - \cot x)}{(\csc x - \cot x)}$

$= \dfrac{\csc^2 x - \cot^2 x}{\csc x - \cot x} = \dfrac{1}{\csc x - \cot x}$;

so, $\log_{10}(\csc x + \cot x) = \log_{10}\left(\dfrac{1}{\csc x - \cot x}\right)$
$= -\log_{10}(\csc x - \cot x)$

63. $-\tan x \tan y(\cot x - \cot y)$
$\qquad = -\tan y(\tan x \cot x) + \tan x(\tan y \cot y)$
$\qquad = -\tan y + \tan x = \tan x - \tan y$

65. $\dfrac{\cos x - \sin y}{\cos y - \sin x} \cdot \dfrac{\cos x + \sin y}{\cos x + \sin y}$

$= \dfrac{\cos^2 x - \sin^2 y}{(\cos y - \sin x)(\cos x + \sin y)}$

$= \dfrac{(1 - \sin^2 x) - (1 - \cos^2 y)}{(\cos y - \sin x)(\cos x + \sin y)}$

$= \dfrac{\cos^2 y - \sin^2 x}{(\cos y - \sin x)(\cos x + \sin y)}$

$= \dfrac{(\cos y - \sin x)(\cos y + \sin x)}{(\cos y - \sin x)(\cos x + \sin y)} = \dfrac{\cos y + \sin x}{\cos x + \sin y}$

Section 7.2, page 505

1. $\dfrac{\sqrt{6} - \sqrt{2}}{4}$ **3.** $2 - \sqrt{3}$ **5.** $2 - \sqrt{3}$

7. $-2 - \sqrt{3}$ **9.** $-2 - \sqrt{3}$ **11.** $\dfrac{\sqrt{6} + \sqrt{2}}{4}$

13. $\cos x$ **15.** $-\sin x$ **17.** $-1/\cos x$

19. $-\sin 2$ **21.** $\cos x$ **23.** $-2 \sin x \sin y$

25. $\dfrac{4 + \sqrt{2}}{6}$ **27.** $\dfrac{2\sqrt{6} - \sqrt{3}}{10}$

29. $\dfrac{4 + 3\sqrt{3}}{10}$ **31.** $\dfrac{4 - 3\sqrt{3}}{10}$

33. Since $\sin t = \dfrac{4}{5}$ and $\cos t = \dfrac{3}{5}$,

$5 \sin(x + t) = 5(\sin x \cos t + \cos x \sin t)$
$\qquad = 5\left(\sin x \cdot \dfrac{3}{5} + \cos x \cdot \dfrac{4}{5}\right)$
$\qquad = 3 \sin x + 4 \cos x.$

35. $\dfrac{f(x + h) - f(x)}{h} = \dfrac{\cos(x + h) - \cos x}{h}$

$= \dfrac{\cos x \cos h - \sin x \sin h - \cos x}{h}$

$= \dfrac{\cos x \cos h - \cos x}{h} - \dfrac{\sin x \sin h}{h}$

$= \cos x\left(\dfrac{\cos h - 1}{h}\right) - \sin x\left(\dfrac{\sin h}{h}\right)$

37. $\sin(x + y) = -44/125$; $\tan(x + y) = 44/117$; $x + y$ is in the third quadrant.

39. $\cos(x + y) = -56/65$; $\tan(x + y) = 33/56$; $x + y$ is in the third quadrant.

41. $\sin(u + v + w) = \sin u \cos v \cos w$
$\qquad\qquad + \cos u \sin v \cos w$
$\qquad\qquad + \cos u \cos v \sin w - \sin u \sin v \sin w$

43. Since $y = \pi/2 - x$, $\sin y = \sin(\pi/2 - x) = \cos x$.
Hence, $\sin^2 x + \sin^2 y = \sin^2 x + \cos^2 x = 1$

45. $\sin(x - \pi) = \sin x \cos \pi - \cos x \sin \pi$
$\qquad = (\sin x)(-1) - (\cos x)(0) = -\sin x$

47. $\cos(\pi - x) = \cos \pi \cos x + \sin \pi \sin x$
$$= (-1)\cos x + (0)\sin x = -\cos x$$

49. $\sin(x + \pi) = \sin x \cos \pi + \cos x \sin \pi$
$$= (\sin x)(-1) + (\cos x)(0) = -\sin x$$

51. By Exercises 49 and 50, $\tan(x + \pi) = \dfrac{\sin(x + \pi)}{\cos(x + \pi)}$
$$= \dfrac{-\sin x}{-\cos x} = \tan x$$

53. $\dfrac{1}{2}[\cos(x - y) - \cos(x + y)] = \dfrac{1}{2}[\cos x \cos y +$
$\sin x \sin y - (\cos x \cos y - \sin x \sin y)]$
$$= \dfrac{1}{2}(2 \sin x \sin y) = \sin x \sin y$$

55. $\cos(x + y)(x - y)$
$$= (\cos x \cos y - \sin x \sin y)(\cos x \cos y + \sin x \sin y)$$
$$= (\cos x \cos y)^2 - (\sin x \sin y)^2$$
$$= \cos^2 x \cos^2 y - \sin^2 x \sin^2 y$$

57. $\dfrac{\cos(x - y)}{\sin x \cos y} = \dfrac{\cos x \cos y + \sin x \sin y}{\sin x \cos y}$
$$= \dfrac{\cos x}{\sin x} + \dfrac{\sin y}{\cos y} = \cot x + \tan y$$

59. Not an identity

61. $\dfrac{\sin(x + y)}{\sin(x - y)}$
$$= \dfrac{\sin x \cos y + \cos x \sin y}{\sin x \cos y - \cos x \sin y} \cdot \dfrac{1/\cos x \cos y}{1/\cos x \cos y}$$
$$= \dfrac{\dfrac{\sin x}{\cos x} + \dfrac{\sin y}{\cos y}}{\dfrac{\sin x}{\cos x} - \dfrac{\sin y}{\cos y}} = \dfrac{\tan x + \tan y}{\tan x - \tan y}$$

63. Not an identity

65. Not an identity

Special Topics 7.2.A, page 509

1. .64 **3.** 2.47 **5.** $\pi/2$

7. 1.37 or 1.77 **9.** $\pi/4$ or $3\pi/4$

11. 1.39 or 1.75 **13.** 0.142

Section 7.3, page 516

1. $\dfrac{\sqrt{2 + \sqrt{2}}}{2}$ **3.** $\dfrac{\sqrt{2 + \sqrt{2}}}{2}$

5. $2 - \sqrt{3}$ **7.** $\dfrac{\sqrt{2 + \sqrt{3}}}{2}$

9. $\dfrac{\sqrt{2 - \sqrt{2}}}{2}$ **11.** $-\sqrt{2} + 1$

13. $\dfrac{1}{2} \sin 10x - \dfrac{1}{2} \sin 2x$ **15.** $\dfrac{1}{2} \cos 6x + \dfrac{1}{2} \cos 2x$

17. $\dfrac{1}{2} \cos 20x - \dfrac{1}{2} \cos 14x$ **19.** $2 \sin 4x \cos x$

21. $2 \sin 2x \cos 7x$

23. $\sin 2x = \dfrac{120}{169}$, $\cos 2x = \dfrac{119}{169}$, $\tan 2x = \dfrac{120}{119}$

25. $\sin 2x = \dfrac{24}{25}$, $\cos 2x = -\dfrac{7}{25}$, $\tan 2x = -\dfrac{24}{7}$

27. $\sin 2x = \dfrac{24}{25}$, $\cos 2x = \dfrac{7}{25}$, $\tan 2x = \dfrac{24}{7}$

29. $\sin 2x = \dfrac{\sqrt{15}}{8}$, $\cos 2x = \dfrac{7}{8}$, $\tan 2x = \dfrac{\sqrt{15}}{7}$

31. $\sin \dfrac{x}{2} = .5477$, $\cos \dfrac{x}{2} = .8367$, $\tan \dfrac{x}{2} = .6547$

33. $\sin \dfrac{x}{2} = \dfrac{1}{\sqrt{10}}$, $\cos \dfrac{x}{2} = \dfrac{-3}{\sqrt{10}}$, $\tan \dfrac{x}{2} = \dfrac{-1}{3}$

35. $\sin \dfrac{x}{2} = \sqrt{\dfrac{\sqrt{5} + 2}{2\sqrt{5}}}$, $\cos \dfrac{x}{2} = -\sqrt{\dfrac{\sqrt{5} - 2}{2\sqrt{5}}}$,
$\tan \dfrac{x}{2} = -\sqrt{9 + 4\sqrt{5}} = -\sqrt{5} - 2$

37. $\sin 2x = .96$ **39.** $\cos 2x = .28$

41. $\sin \dfrac{x}{2} = .3162$

43. $\cos 3x = 4 \cos^3 x - 3 \cos x$

45. $\cos x$ **47.** $\sin 4y$ **49.** 1

51. $\sin 16x = \sin[2(8x)] = 2 \sin 8x \cos 8x$

53. $\cos^4 x - \sin^4 x = (\cos^2 x - \sin^2 x)(\cos^2 x + \sin^2 x)$
$$= \cos^2 x - \sin^2 x = \cos 2x$$

55. Not an identity

57. $\dfrac{1 + \cos 2x}{\sin 2x} = \dfrac{1}{\dfrac{\sin 2x}{1 + \cos 2x}} = \dfrac{1}{\tan\left(\dfrac{2x}{2}\right)} = \cot x$

59. $\sin 3x = \sin(2x + x)$
$$= \sin 2x \cos x + \cos 2x \sin x$$
$$= (2 \sin x \cos x)\cos x + (1 - 2 \sin^2 x)\sin x$$
$$= 2 \sin x \cos^2 x + \sin x - 2 \sin^3 x$$
$$= 2 \sin x(1 - \sin^2 x) + \sin x - 2 \sin^3 x$$
$$= \sin x(2 - 2 \sin^2 x + 1 - 2 \sin^2 x)$$
$$= \sin x(3 - 4 \sin^2 x)$$

61. Not an identity

63. $\csc^2\left(\dfrac{x}{2}\right) = \dfrac{1}{\sin^2\left(\dfrac{x}{2}\right)} = \dfrac{1}{\dfrac{1 - \cos x}{2}} = \dfrac{2}{1 - \cos x}$

65. $\dfrac{\sin x - \sin 3x}{\cos x + \cos 3x} = \dfrac{-2 \cos 2x \sin x}{2 \cos 2x \cos x} = \dfrac{-\sin x}{\cos x} = -\tan x$

67. $\dfrac{\sin 4x + \sin 6x}{\cos 4x - \cos 6x} = \dfrac{2 \sin 5x \cos x}{2 \sin 5x \sin x} = \dfrac{\cos x}{\sin x} = \cot x$

69. $\dfrac{\sin x + \sin y}{\cos x - \cos y} = \dfrac{2 \sin\left(\dfrac{x+y}{2}\right)\cos\left(\dfrac{x-y}{2}\right)}{-2 \sin\left(\dfrac{x+y}{2}\right)\sin\left(\dfrac{x-y}{2}\right)}$

$= \dfrac{-\cos\left(\dfrac{x-y}{2}\right)}{\sin\left(\dfrac{x-y}{2}\right)} = -\cot\left(\dfrac{x-y}{2}\right)$

71. (a) It suffices to prove the equivalent identity
$(1 - \cos x)(1 + \cos x) = (\sin x)(\sin x)$. We have
$(1 - \cos x)(1 + \cos x) = 1 - \cos^2 x =$
$\sin^2 x = (\sin x)(\sin x)$.

(b) By the half-angle identity proved in the text and
part (a), $\tan\dfrac{x}{2} = \dfrac{1 - \cos x}{\sin x} = \dfrac{\sin x}{1 + \cos x}$

Section 7.4, page 525

1. $\dfrac{\pi}{2}$ **3.** $-\dfrac{\pi}{4}$ **5.** 0 **7.** $\dfrac{\pi}{6}$ **9.** $-\dfrac{\pi}{4}$

11. $-\dfrac{\pi}{3}$ **13.** $\dfrac{2\pi}{3}$ **15.** $.3576$ **17.** -1.2728

19. $.7168$ **21.** $-.8584$ **23.** 2.2168

25. $\cos u = 1/2;\ \tan u = -\sqrt{3}$

27. $\dfrac{\pi}{2}$ **29.** $\dfrac{5\pi}{6}$ **31.** $-\dfrac{\pi}{3}$ **33.** $\dfrac{\pi}{3}$

35. $\dfrac{\pi}{6}$ **37.** $\dfrac{4}{5}$ **39.** $\dfrac{4}{5}$ **41.** $\dfrac{5}{12}$

43. $\cos(\sin^{-1}v) = \sqrt{1 - v^2}$ $(-1 \le v \le 1)$

45. $\tan(\sin^{-1}v) = \dfrac{v}{\sqrt{1 - v^2}}$ $(-1 < v < 1)$

47.

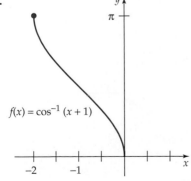

$f(x) = \cos^{-1}(x + 1)$

49.

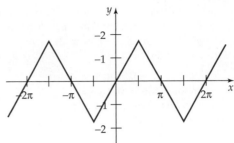

51. (a) $t = \dfrac{1}{2\pi f}\sin^{-1}\left(\dfrac{V}{V_{\max}}\right) + \dfrac{k}{f}$ or

$t = \dfrac{1}{2\pi f}\left[\pi - \sin^{-1}\left(\dfrac{V}{V_{\max}}\right)\right] + \dfrac{k}{f}$

$(k = 0, \pm1, \pm2, \dots)$

(b) $t = .0005822$

53. (a) $\theta = \tan^{-1}\left(\dfrac{25}{x}\right) - \tan^{-1}\left(\dfrac{10}{x}\right)$ or

$\theta = \cos^{-1}\left(\dfrac{\sqrt{x^2 + 10^2} + \sqrt{x^2 + 25^2} - 15^2}{2\sqrt{x^2 + 10^2}\,\sqrt{x^2 + 25^2}}\right)$

(b) 15.8 ft

55. No horizontal line intersects the graph of $g(x) = \sec x$ more than once when $0 \le x \le \pi$ (see Figure 6–76). Hence, the restricted secant function has an inverse function, as explained in Section 3.7.

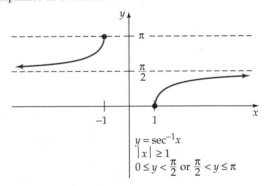

$y = \sec^{-1}x$
$|x| \ge 1$
$0 \le y < \dfrac{\pi}{2}$ or $\dfrac{\pi}{2} < y \le \pi$

57. No horizontal line intersects the graph of $f(x) = \cot x$ more than once when $0 < x < \pi$ (see Figure 6–77). Hence, the restricted cotangent function has an inverse function, as explained in Section 3.7.

$y = \cot^{-1}x$

x any real number
$0 < y < \pi$

59. Let $u = \sin^{-1}(-x)$. Then $-\pi/2 \le u \le \pi/2$ and $\sin u = -x$; hence, $x = -\sin u = \sin(-u)$. Therefore, $\sin^{-1} x = -u$ (since $-\pi/2 \le -u \le \pi/2$) so that $\sin^{-1}(-x) = u = -\sin^{-1}x$.

61. Let $u = \cos^{-1}(-x)$. Then $0 \le u \le \pi$ and $\cos u = -x$. Since $0 \le u \le \pi$, we have $0 \le \pi - u \le \pi$. Since $\cos(\pi - u) = -\cos u = x$, $\cos^{-1}x = \pi - u = \pi - \cos^{-1}(-x)$. Therefore, $\cos^{-1}(-x) = \pi - \cos^{-1}x$.

63. If $0 < x < \pi$, then $-\dfrac{\pi}{2} < \dfrac{\pi}{2} - x < \dfrac{\pi}{2}$. Since $\tan\left(\dfrac{\pi}{2} - x\right) = \cot x$, $\tan^{-1}(\cot x) = \dfrac{\pi}{2} - x$.

65. Let $u = \sin^{-1}x$. Then $-\pi/2 \le u \le \pi/2$ and $\sin u = x$. Then $\cos u = \pm\sqrt{1 - \sin^2 u} = \pm\sqrt{1 - x^2}$. Since $-\pi/2 \le u \le \pi/2$, $\cos u$ is positive, so $\cos u = \sqrt{1 - x^2}$. Therefore,

$\tan u = \dfrac{\sin u}{\cos u} = \dfrac{x}{\sqrt{1 - x^2}}$. Hence,

$u = \tan^{-1}\left(\dfrac{x}{\sqrt{1 - x^2}}\right)$, so that

$\sin^{-1}x = u = \tan^{-1}\left(\dfrac{x}{\sqrt{1 - x^2}}\right)$.

67. No, the graphs differ.

Section 7.5, page 534

In this section, $k = 0, \pm 1, \pm 2, \pm 3, \ldots$.

1. $x = -.4836$ or $3.6252 + 2k\pi$

3. $x = \pm 2.1700 + 2k\pi$ **5.** $x = -.2327 + k\pi$

7. $x = .4101 + k\pi$ **9.** $x = \pm 1.9577 + 2k\pi$

11. $x = .1193$ or 3.0223 **13.** $x = 1.3734$ or 4.5150

15. $x = \pi/3$ or $\dfrac{2\pi}{3} + 2k\pi$ **17.** $x = \dfrac{-\pi}{3} + k\pi$

19. $x = \pm\dfrac{5\pi}{6} + 2k\pi$ **21.** $x = \dfrac{-\pi}{6}$ or $\dfrac{7\pi}{6} + 2k\pi$

23. $x = \dfrac{-\pi}{6}$ or $\dfrac{2\pi}{3} + k\pi$ **25.** $x = \pm\dfrac{\pi}{2} + 4k\pi$

27. $x - \dfrac{-\pi}{9} + \dfrac{k\pi}{3}$ **29.** $x = \pm.7381 + \dfrac{2k\pi}{3}$

31. $x = 2.2143 + 2k\pi$ **33.** $65.38°$

35. $30°$ **37.** $27.57°$ **39.** $14.18°$

41. $x = 3.4814, 5.9433$

43. $x = \dfrac{3\pi}{4}, \dfrac{7\pi}{4}, 2.1588, 5.3004$

45. $x = \dfrac{\pi}{4}, \dfrac{\pi}{2}, \dfrac{5\pi}{4}, \dfrac{3\pi}{2}$ **47.** $x = \dfrac{\pi}{6}, \dfrac{\pi}{2}, \dfrac{5\pi}{6}, \dfrac{3\pi}{2}$

49. $x = .8481, 1.7682, 2.2935, 4.9098$

51. $x = .8213, 2.3203$

53. $x = .3649, 1.2059, 3.5065, 4.3475$

55. $x = 1.0591, 2.8679, 4.2007, 6.0095$

57. No solution   **59.** $x = \dfrac{\pi}{2}, \dfrac{7\pi}{6}, \dfrac{3\pi}{2}, \dfrac{11\pi}{6}$

61. $x = \dfrac{\pi}{6}, \dfrac{5\pi}{6}$ **63.** $x = \dfrac{\pi}{2}, \dfrac{7\pi}{6}, \dfrac{11\pi}{6}$

65. $x = 0, \dfrac{\pi}{3}, \dfrac{5\pi}{3}$ **67.** $\theta = 82.83°, 262.83°$

69. $\theta = 114.83°, 245.17°$ **71.** $\theta = 210°, 270°, 330°$

73. $\theta = 60°, 120°, 240°, 300°$

75. $\theta = 120°, 240°$

77. $x = .5725$ or $1.6868 + k\pi$

79. $x = .4959$ or 1.2538 or 1.5708 or 1.8877 or 2.6957 or $4.7124 + 2k\pi$

81. $x = .1671$ or 1.8256 or 2.8876 or $4.5453 + 2k\pi$

83. $x = 1.2161$ or $5.0671 + 2k\pi$

85. $x = 2.4620$ or $3.8212 + 2k\pi$

87. $x = .5166$ or $5.6766 + 2k\pi$

89. (a) March 1 (day 60); October 9 (day 282)
(b) June 20 (day 171)

91. $.5475$ or 1.0233 **93.** $13.25°$ or $76.75°$

95. $\sin x = k$ and $\cos x = k$ have no solutions when $k > 1$ or $k < -1$.

97. The solutions $x = 0, \pi$ are missed due to dividing by $\sin x$.

Chapter 7 Review, page 538

1. $\dfrac{1}{3} + \cot t$ **3.** $\sin^4 x$

5. $\sin^4 t - \cos^4 t = (\sin^2 t - \cos^2 t)(\sin^2 t + \cos^2 t) = [\sin^2 t - (1 - \sin^2 t)](1) = 2\sin^2 t - 1$

7. $\dfrac{\sin t}{1 - \cos t} = \dfrac{\sin t}{1 - \cos t} \cdot \dfrac{(1 + \cos t)}{(1 + \cos t)}$

$= \dfrac{\sin t(1 + \cos t)}{1 - \cos^2 t} = \dfrac{\sin t(1 + \cos t)}{\sin^2 t} = \dfrac{1 + \cos t}{\sin t}$

9. Not an identity

11. $(\sin x + \cos x)^2 - \sin 2x$
$= \sin^2 x + 2\sin x \cos x + \cos^2 x - 2\sin x \cos x$
$= \sin^2 x + \cos^2 x = 1$

13. $\dfrac{\tan x - \sin x}{2 \tan x} \cdot \dfrac{\cos x}{\cos x} = \dfrac{\sin x - \sin x \cos x}{2 \sin x}$

$= \dfrac{\sin x(1 - \cos x)}{2 \sin x} = \dfrac{1 - \cos x}{2} = \sin^2\left(\dfrac{x}{2}\right)$

15. $\cos(x + y)\cos(x - y)$
$= [\cos x \cos y - \sin x \sin y][\cos x \cos y + \sin x \sin y]$
$= \cos^2 x \cos^2 y - \sin^2 x \sin^2 y$
$= \cos^2 x(1 - \sin^2 y) - (1 - \cos^2 x)\sin^2 y$
$= \cos^2 x - \cos^2 x \sin^2 y - \sin^2 y + \cos^2 x \sin^2 y$
$= \cos^2 x - \sin^2 y$

17. $\dfrac{\sec x + 1}{\tan x} = \dfrac{(\sec x + 1)(\sec x - 1)}{\tan x(\sec x - 1)}$

$= \dfrac{\sec^2 x - 1}{\tan x(\sec x - 1)} = \dfrac{\tan^2 x}{\tan x(\sec x - 1)} = \dfrac{\tan x}{\sec x - 1}$

19. $\dfrac{1 + \tan^2 x}{\tan^2 x} = \dfrac{1}{\tan^2 x} + 1 = \cot^2 x + 1 = \csc^2 x$

21. $\tan^2 x - \sec^2 x = -(\sec^2 x - \tan^2 x) = -1 =$
$-(\csc^2 x - \cot^2 x) = \cot^2 x - \csc^2 x$

23. $\dfrac{120}{169}$ **25.** $-\dfrac{56}{65}$ **27.** $\dfrac{3\sqrt{5} + 1}{8}$

29. Yes. $\sin 2x = 2 \sin x \cos x = 2(0) \cos x = 0$

31. $\cos\left(\dfrac{\pi}{12}\right) = \cos\left[\dfrac{1}{2}\left(\dfrac{\pi}{6}\right)\right]$

$= \sqrt{\dfrac{1 + \cos \dfrac{\pi}{6}}{2}} = \sqrt{\dfrac{1 + \sqrt{3}/2}{2}}$

$= \sqrt{\dfrac{2 + \sqrt{3}}{4}} = \dfrac{\sqrt{2 + \sqrt{3}}}{2}; \cos\left(\dfrac{\pi}{12}\right)$

$= \cos\left(\dfrac{\pi}{4} - \dfrac{\pi}{6}\right) = \cos\dfrac{\pi}{4}\cos\dfrac{\pi}{6} + \sin\dfrac{\pi}{4}\sin\dfrac{\pi}{6}$

$= \dfrac{\sqrt{2}}{2}\left(\dfrac{\sqrt{3}}{2}\right) + \dfrac{\sqrt{2}}{2}\left(\dfrac{1}{2}\right) = \dfrac{\sqrt{6} + \sqrt{2}}{4}.$ So,

$\dfrac{\sqrt{2 + \sqrt{3}}}{2} = \dfrac{\sqrt{6} + \sqrt{2}}{4}$ or $\sqrt{2 + \sqrt{3}} = \dfrac{\sqrt{2} + \sqrt{6}}{2}$

33. $\dfrac{\sqrt{6} + \sqrt{2}}{4}$ **35.** (a)   **37.** .96

39. $\pi/4$ **41.** $\pi/4$ **43.** $\pi/3$

45. $5\pi/6$ **47.** .75   **49.** $\pi/3$

51.

53. $\dfrac{\sqrt{15}}{4}$ **55.** $x = \dfrac{\pi}{6}$ or $\dfrac{5\pi}{6} + 2k\pi$

57. $x = -\dfrac{\pi}{4} + k\pi$

59. $x = .7754$ or $2.3662 + 2k\pi$

61. $x = 1.4940 + k\pi$ **63.** $x = \dfrac{\pi}{6}$ or $\dfrac{5\pi}{6} + 2k\pi$

65. $x = -\dfrac{\pi}{6}$ or $\dfrac{7\pi}{6} + 2k\pi$ **67.** $x = \pm\dfrac{\pi}{3} + k\pi$

69. $x = .8959$ or $2.2457 + 2k\pi$

71. $x = k\pi$

73. $x = .8419$ or 2.2997 or 4.1784 or $5.2463 + 2k\pi$

75. $\theta = 225.50°, 314.50°$ **77.** $9.06°$ or $80.94°$

CHAPTER 8

Section 8.1, page 551

	sin	cos	tan	csc	sec	cot
1.	$\dfrac{3\sqrt{13}}{13}$	$\dfrac{2\sqrt{13}}{13}$	$\dfrac{3}{2}$	$\dfrac{\sqrt{13}}{3}$	$\dfrac{\sqrt{13}}{2}$	$\dfrac{2}{3}$
3.	$\dfrac{6\sqrt{61}}{61}$	$\dfrac{-5\sqrt{61}}{61}$	$\dfrac{-6}{5}$	$\dfrac{\sqrt{61}}{6}$	$\dfrac{-\sqrt{61}}{5}$	$\dfrac{-5}{6}$
5.	$\dfrac{-\sqrt{22}}{11}$	$\dfrac{-3\sqrt{11}}{11}$	$\dfrac{\sqrt{2}}{3}$	$\dfrac{-\sqrt{22}}{2}$	$\dfrac{-\sqrt{11}}{3}$	$\dfrac{3\sqrt{2}}{2}$

7. $\sin \theta = \sqrt{\dfrac{2}{11}}, \cos \theta = \dfrac{3}{\sqrt{11}}, \tan \theta = \dfrac{\sqrt{2}}{3}$

9. $\sin \theta = \sqrt{\dfrac{3}{7}}, \cos \theta = \dfrac{2}{\sqrt{7}}, \tan \theta = \dfrac{\sqrt{3}}{2}$

11. $\sin \theta = \dfrac{h}{m}, \cos \theta = \dfrac{d}{m}, \tan \theta = \dfrac{h}{d}$

13. $c = 36$ **15.** $c = 36$ **17.** $c = 8.4$

19. $h = 25\sqrt{2}/2$ **21.** $h = 300$ **23.** $h = 50\sqrt{3}$

25. $c = 4\sqrt{3}/3$ **27.** $a = 10\sqrt{3}/3$

29. $\angle A = 40°, a = 10 \cos 50° = 6.4, c = 10 \sin 50° = 7.7$

31. $\angle C = 76°, b = 6/\sin 14° = 24.8, c - 6/\tan 14° = 24.1$

33. $\angle C = 25°, a = 5 \tan 65° = 10.7, b = 5/\cos 65° = 11.8$

35. $\angle C = 18°, a = 3.5 \sin 72° = 3.3, c = 3.5 \cos 72° = 1.1$

37. About $48.59°$ **39.** About $48.19°$

41. $\angle A = 33.7°, \angle C = 56.3°$

43. $\angle A = 44.4°, \angle C = 45.6°$

45. $\angle A = 48.2°, \angle C = 41.8°$

47. $\angle A = 60.8°, \angle C = 29.2°$

49. (a) $\sin \theta = \dfrac{a}{c}; \cos \alpha = \dfrac{a}{c}$

(b) The third angle is a right angle.

(c) $\cos(90° - \theta) = \cos(\alpha) = \sin \theta$

Section 8.2, page 562

1. 100.7 ft   **3.** About 68°   **5.** 71 ft

7. About 15.32 ft   **9.** 30°   **11.** 460.2 ft

13. 8598.3 ft   **15.** No   **17.** 27.5 ft; no

19. 19.25 ft; 53.97°   **21.** 351.1 m   **23.** 10.1 ft

25. 1.6 mi

27. (a) 56.7 ft   (b) 9.7 ft

29. 205.7 ft   **31.** 173.2 mi   **33.** 52.5 mph

35. (a) $A(t) = 200 \cos t \sin t$
       (b) $t \approx .7854$; approx. 14.1421 by 7.0711

37. 449.1 ft

Section 8.3, page 570

1. $a = 4.2$, $\angle B = 125.0°$, $\angle C = 35.0°$

3. $c = 13.9$, $\angle A = 22.5°$, $\angle B = 39.5°$

5. $a = 24.4$, $\angle B = 18.4°$, $\angle C = 21.6°$

7. $c = 21.5$, $\angle A = 33.5°$, $\angle B = 67.9°$

9. $\angle A = 120°$, $\angle B = 21.8°$, $\angle C = 38.2°$

11. $\angle A = 24.1°$, $\angle B = 30.8°$, $\angle C = 125.1°$

13. $\angle A = 38.8°$, $\angle B = 34.5°$, $\angle C = 106.7°$

15. $\angle A = 34.1°$, $\angle B = 50.5°$, $\angle C = 95.4°$

17. 54.2° at vertex (0, 0); 48.4° at vertex (5, −2); 77.5° at vertex (1, −4)

19. 57.72 miles   **21.** 425.67 ft, 469.38 ft

23. 63.7 ft   **25.** 84.9°

27. 8.4 km   **29.** 154.5 ft

31. 4.7 cm and 9.0 cm   **33.** 33.44°   **35.** 978.7 mi

37. $AB = 24.27$, $AC = 21.23$, $BC = 19.5$, $\angle A = 50.2°$, $\angle B = 56.8°$, $\angle C = 73.0°$

39. 16.99 m

Section 8.4, page 580

1. $\angle C = 110°$, $b = 2.5$, $c = 6.3$

3. $\angle B = 14°$, $b = 2.2$, $c = 6.8$

5. $\angle A = 88°$, $a = 17.3$, $c = 12.8$

7. $\angle C = 41.5°$, $b = 9.7$, $c = 10.9$

9. No solution

11. $\angle A_1 = 55.2°$, $\angle C_1 = 104.8°$, $c_1 = 14.1$
       $\angle A_2 = 124.8°$, $\angle C_2 = 35.2°$, $c_2 = 8.4$

13. No solution

15. $\angle B_1 = 65.8$, $\angle A_1 = 58.2°$, $a_1 = 10.3$
       $\angle B_2 = 114.2°$, $\angle A_2 = 9.8°$, $a_2 = 2.1$

17. $\angle C = 72°$, $b = 14.7$, $c = 15.2$

19. $a = 9.8$, $\angle B = 23.3°$, $\angle C = 81.7$

21. $\angle A = 18.6°$, $\angle B = 39.6°$, $\angle C = 121.9°$

23. $c = 13.9$, $\angle A = 60.1°$, $\angle B = 72.9°$

25. $\angle C = 39.8°$, $\angle A = 77.7°$, $a = 18.9$

27. No solution

29. $a = 18.5$, $\angle C = 78.3°$, $\angle B = 36.7°$

31. $\angle C = 126°$, $b = 193.8$, $c = 273.3$

33. 135.5 m   **35.** 5.4°   **37.** 5 ft

39. 5.3°   **41.** 30.1 km   **43.** About 9642 ft

45. (a) Use the law of cosines in triangle *ABD* to find angle *ABD*, then angle $EBA = 180°$ − angle *ABD*. Use the law of cosines in triangle *ABC* to find angle *CAB*, then angle $EAB = 180°$ − angle *CAB*. Then two of the angles in triangle *EAB* have been found, and the third can easily be found. Use these angles, side *AB*, and the law of sines to find *AE*.
       (b) 94.24 ft

47. 13.36 m

Special Topics 8.4.A, page 584

1. 7.3   **3.** 32.5   **5.** 82.3

7. 31.4   **9.** 6.5   **11.** About 7691 square units

13. About 235.9 square feet

15. 5.8 gal   **17.** 11.18 units²

19. It is an impossible triangle, $A + B < C$.

Chapter 8 Review, page 587

1. (d)   **3.** (e)

5. $b = \sqrt{313} \approx 17.7$, $A \approx 42.7°$, $C \approx 47.3°$

7. $b \approx 14.6$, $c \approx 8.40$, $A = 55°$

9. 225.9 ft   **11.** 1.5°

13. $A = 52.9°$, $B = 41.6°$, $C = 85.5°$

15. $A = 20.6°$, $b = 21.8$, $C = 29.4°$

17. Approximately 301 mi

19. $A = 25°$, $a = 2.9$, $b = 5.6$

21. $A = 52.03°$, $B = 65.97°$, $b = 86.9$

23. $B = 81.8°$, $C = 38.2°$, $c = 2.5$ and $B = 98.2°$, $C = 21.8°$, $c = 1.5$

25. 147.4   **27.** 13.4 km

29. $a = 41.6$; $C = 75°$, $c = 54.1$

31. $A = 35.5°$, $b = 8.3$, $C = 68.5°$

33. Joe is 217.9 m from the pole and Alice is 240 m from the pole.

35. (a) 3617.65 ft   (b) 4018.71 ft   (c) 3642.19 ft

37. 71.89°   **39.** 10   **41.** 37.95

CHAPTER 9

Section 9.1, page 596

1–7.

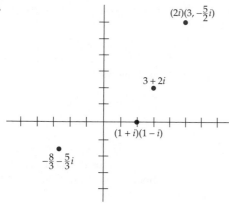

9. 13   **11.** $\sqrt{3}$ **13.** 12

15. Many correct answers, including $z = 1$, $w = i$

17.

19.

21.

23.

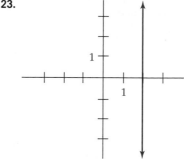

25. $5(\cos .9273 + i \sin .9273)$

27. $13(\cos 5.1072 + i \sin 5.1072)$

29. $\sqrt{5}(\cos 1.1071 + i \sin 1.1071)$

31. $\sqrt{18.5}(\cos 2.1910 + i \sin 2.1910)$

33. $2\left(\cos\dfrac{2\pi}{3} + i \sin\dfrac{2\pi}{3}\right) = -1 + \sqrt{3}\,i$

35. $42\left(\cos\dfrac{7\pi}{6} + i \sin\dfrac{7\pi}{6}\right) = -21\sqrt{3} - 21i$

37. $\dfrac{3}{2}\left(\cos\dfrac{\pi}{4} + i \sin\dfrac{\pi}{4}\right) = \left(\dfrac{3\sqrt{2}}{4}\right) + \left(\dfrac{3\sqrt{2}}{4}\right)i$

39. $2\sqrt{2}\left(\cos\dfrac{7\pi}{12} + i \sin\dfrac{7\pi}{12}\right)$

41. $\cos\dfrac{\pi}{2} + i \sin\dfrac{\pi}{2}$ **43.** $12\left(\cos\dfrac{2\pi}{3} + i \sin\dfrac{2\pi}{3}\right)$

45. $2\sqrt{2}\left(\cos\dfrac{19\pi}{12} + i \sin\dfrac{19\pi}{12}\right)$

47. The polar form of i is $1(\cos 90° + i \sin 90°)$. Hence, by the Polar Multiplication Rule

$$zi = r \cdot 1(\cos(\theta + 90°) + i \sin(\theta + 90°)).$$

You can think of z as lying on a circle with center at the origin and radius r. Then zi lies on the same circle (since it too is r units from the origin), but 90° farther around the circle (in a counterclockwise direction).

49. (a) $\dfrac{b}{a}$ (b) $\dfrac{d}{c}$ (c) $y - b = \left(\dfrac{d}{c}\right)(x - a)$

(d) $y - d = \dfrac{b}{a}(x - c)$

(f) $(a + c, b + d)$ lies on L since $(b + d) - b =$

$\dfrac{d}{c}[(a + c) - a]$ and $(a + c, b + d)$

lies on M since $(b + d) - d = \dfrac{b}{a}[(a + c) - c]$.

51. (a) $r_2(\cos \theta_2 + i \sin \theta_2)(\cos \theta_2 - i \sin \theta_2)$
$= r_2(\cos^2\theta_2 + \sin^2\theta_2) = r_2$

(b) $r_1(\cos \theta_1 + i \sin \theta_1)(\cos \theta_2 - i \sin \theta_2)$
$= r_1[(\cos \theta_1 \cos \theta_2 + \sin \theta_1 \sin \theta_2) +$
$i(\sin \theta_1 \cos \theta_2 - \cos \theta_1 \sin \theta_2)]$
$= r_1[\cos(\theta_1 - \theta_2) + i \sin(\theta_1 - \theta_2)]$

Section 9.2, page 605

1. i **3.** $\dfrac{-243\sqrt{3}}{2} - \dfrac{243}{2}i$ **5.** -64

7. $\dfrac{1}{2} - \dfrac{\sqrt{3}}{2}i$ **9.** i **11.** $1, -1, i, -i$

13. $4\left(\cos\dfrac{\pi}{15} + i\sin\dfrac{\pi}{15}\right), 4\left(\cos\dfrac{11\pi}{15} + i\sin\dfrac{11\pi}{15}\right),$
$4\left(\cos\dfrac{7\pi}{5} + i\sin\dfrac{7\pi}{5}\right)$

15. $3\left(\cos\dfrac{\pi}{48} + i\sin\dfrac{\pi}{48}\right), 3\left(\cos\dfrac{25\pi}{48} + i\sin\dfrac{25\pi}{48}\right),$
$3\left(\cos\dfrac{49\pi}{48} + i\sin\dfrac{49\pi}{48}\right), 3\left(\cos\dfrac{73\pi}{48} + i\sin\dfrac{73\pi}{48}\right)$

17. $\left(\cos\dfrac{\pi}{5} + i\sin\dfrac{\pi}{5}\right), \left(\cos\dfrac{3\pi}{5} + i\sin\dfrac{3\pi}{5}\right), (\cos \pi + i \sin \pi),$
$\left(\cos\dfrac{7\pi}{5} + i\sin\dfrac{7\pi}{5}\right), \left(\cos\dfrac{9\pi}{5} + i\sin\dfrac{9\pi}{5}\right)$

19. $\left(\cos\dfrac{\pi}{10} + i\sin\dfrac{\pi}{10}\right), \left(\cos\dfrac{\pi}{2} + i\sin\dfrac{\pi}{2}\right),$
$\left(\cos\dfrac{9\pi}{10} + i\sin\dfrac{9\pi}{10}\right), \left(\cos\dfrac{13\pi}{10} + i\sin\dfrac{13\pi}{10}\right),$
$\left(\cos\dfrac{17\pi}{10} + i\sin\dfrac{17\pi}{10}\right)$

21. $\sqrt[4]{2}\left(\cos\dfrac{\pi}{8} + i\sin\dfrac{\pi}{8}\right), \sqrt[4]{2}\left(\cos\dfrac{9\pi}{8} + i\sin\dfrac{9\pi}{8}\right)$

23. $x = \dfrac{\sqrt{3}}{2} + \dfrac{1}{2}i$ or $\dfrac{\sqrt{3}}{2} - \dfrac{1}{2}i$ or $-\dfrac{\sqrt{3}}{2} + \dfrac{1}{2}i$ or
$-\dfrac{\sqrt{3}}{2} - \dfrac{1}{2}i$ or i or $-i$

25. $\dfrac{\sqrt{3}}{2} + \dfrac{1}{2}i$ or $-\dfrac{\sqrt{3}}{2} + \dfrac{1}{2}i$ or $-i$

27. $x = 3i$ or $\dfrac{3\sqrt{3}}{2} - \dfrac{3}{2}i$ or $-\dfrac{3\sqrt{3}}{2} - \dfrac{3}{2}i$

29. $\sqrt[4]{2}\left(\dfrac{\sqrt{3}}{2} + \dfrac{1}{2}i\right)$ or $\sqrt[4]{2}\left(-\dfrac{1}{2} + \dfrac{\sqrt{3}}{2}i\right)$ or
$\sqrt[4]{2}\left(-\dfrac{\sqrt{3}}{2} - \dfrac{1}{2}i\right)$ or $\sqrt[4]{2}\left(\dfrac{1}{2} - \dfrac{\sqrt{3}}{2}i\right)$

31. $1, .6235 \pm .7818i, -.2225 \pm .9749i, -.9010 \pm .4339i$

33. $\pm 1, \pm i, .7071 \pm .7071i, -.7071 \pm .7071i$

35. $1, .7660 \pm .6428i, .1736 \pm .9848i, -.5 \pm .8660i,$
$-.9397 \pm .3420i$

37. $x^6 - 1 = (x - 1)(x^5 + x^4 + x^3 + x^2 + x + 1)$, so the
solutions of $x^5 + x^4 + x^3 + x^2 + x + 1 = 0$ are the sixth
roots of unity other than 1; namely,

$-1, \dfrac{1}{2} + \dfrac{\sqrt{3}}{2}i, \dfrac{1}{2} - \dfrac{\sqrt{3}}{2}i, -\dfrac{1}{2} + \dfrac{\sqrt{3}}{2}i, -\dfrac{1}{2} - \dfrac{\sqrt{3}}{2}i.$

39. 12

41. For each i, u_i is an nth root of unity, so$(u_i)^n = 1$. Hence
$(vu_i)^n = (v^n)(u_i)^n = v^n \cdot 1 = r(\cos \theta + i \sin \theta)$ and vu_i is
a solution of the equation. If $vu_i = vu_j$, then multiplying
both sides by $1/v$ shows that $u_i = u_j$. In other words, if
u_i is not equal to u_j, then $vu_i \neq vu_j$. Thus, the solutions
$vu_1, \ldots, vu_n$ are all distinct.

Section 9.3, page 618

1. $3\sqrt{5}$ **3.** $\sqrt{34}$ **5.** $\langle 6, 6\rangle$

7. $\langle -6, 10\rangle$ **9.** $\langle 13/5, -2/5\rangle$

11. $\mathbf{u} + \mathbf{v} = \langle 4, 5\rangle$; $\mathbf{u} - \mathbf{v} = \langle -8, 3\rangle$; $3\mathbf{u} - 2\mathbf{v} = \langle -18, 10\rangle$

13. $\mathbf{u} + \mathbf{v} = \langle 3 + 4\sqrt{2}, 1 + 3\sqrt{2}\rangle$; $\mathbf{u} - \mathbf{v} =$
$\langle 3 - 4\sqrt{2}, -1 + 3\sqrt{2}\rangle$; $3\mathbf{u} - 2\mathbf{v} =$
$\langle 9 - 8\sqrt{2}, -2 + 9\sqrt{2}\rangle$

15. $\mathbf{u} + \mathbf{v} = \langle -23/4, 13\rangle$; $\mathbf{u} - \mathbf{v} = \langle -9/4, 7\rangle$;
$3\mathbf{u} - 2\mathbf{v} = \langle -17/2, 24\rangle$

17. $\mathbf{u} + \mathbf{v} = 14\mathbf{i} - 4\mathbf{j}$; $\mathbf{u} - \mathbf{v} = 2\mathbf{i} + 4\mathbf{j}$; $3\mathbf{u} - 2\mathbf{v} = 12\mathbf{i} + 8\mathbf{j}$

19. $\mathbf{u} + \mathbf{v} = -\dfrac{5}{4}\mathbf{i} - \dfrac{3}{2}\mathbf{j}$; $\mathbf{u} - \mathbf{v} = -\dfrac{11}{4}\mathbf{i} - \dfrac{3}{2}\mathbf{j}$;

$3\mathbf{u} - 2\mathbf{v} = -\dfrac{15}{2}\mathbf{i} - \dfrac{9}{2}\mathbf{j}$

21. $-7\mathbf{i}$ **23.** $-2\mathbf{i} + \dfrac{1}{2}\mathbf{j}$ **25.** $6\mathbf{i} - \dfrac{13}{4}\mathbf{j}$

27. $\mathbf{v} = \langle 4, 0\rangle$ **29.** $\mathbf{v} = \langle -5\sqrt{2}, -5\sqrt{2}\rangle$

31. $\mathbf{v} = \langle 4.5963, 3.8567\rangle$

33. $\mathbf{v} = \langle -.1710, -.4698\rangle$

35. $\|\mathbf{v}\| = 4\sqrt{2}, \theta = 45°$ **37.** $\|\mathbf{v}\| = 8, \theta = 180°$

39. $\|\mathbf{v}\| = 6, \theta = 90°$

41. $\|\mathbf{v}\| = 2\sqrt{17}, \theta = 104.04°$

43. $\left\langle \dfrac{4}{\sqrt{41}}, \dfrac{-5}{\sqrt{41}}\right\rangle$ **45.** $\dfrac{1}{\sqrt{5}}\mathbf{i} + \dfrac{2}{\sqrt{5}}\mathbf{j}$

47. Direction: 46.1°; magnitude: 108.2 lb

49. Direction: 213.4°; magnitude: 17.4 kg

51. Resultant force = $\langle -8, -2 \rangle$; $\mathbf{v} = \langle 8, 2 \rangle$

53. $\mathbf{v} + \mathbf{0} = \langle c, d \rangle + \langle 0, 0 \rangle = \langle c, d \rangle = \mathbf{v}$, and
$\mathbf{0} + \mathbf{v} = \langle 0, 0 \rangle + \langle c, d \rangle = \langle c, d \rangle = \mathbf{v}$

55. $r(\mathbf{u} + \mathbf{v}) = r(\langle a, b \rangle + \langle c, d \rangle) = r\langle a + c, b + d \rangle =$
$\langle r(a + c), r(b + d) \rangle = \langle ra + rc, rb + rd \rangle$, and
$r\mathbf{u} + r\mathbf{v} = r\langle a, b \rangle + r\langle c, d \rangle = \langle ra, rb \rangle + \langle rc, rd \rangle =$
$\langle ra + rc, rb + rd \rangle$

57. $(rs)\mathbf{v} = (rs)\langle c, d \rangle = \langle rsc, rsd \rangle$; $r(s\mathbf{v}) = r(s\langle c, d \rangle) =$
$r\langle sc, sd \rangle = \langle rsc, rsd \rangle$; and $s(r\mathbf{v}) = s\langle rc, rd \rangle =$
$\langle src, srd \rangle = \langle rsc, rsd \rangle$

59. 48.58 lb

61. Parallel to plane: 32.1 lb; perpendicular to plane: 38.3 lb

63. 66.4°

65. Ground speed: 253.2 mph; course: 69.1°

67. Ground speed: 304.1 mph; course: 309.5°

69. Air speed: 424.3 mph; direction: 62.4°

71. 69.08°

73. 170.32 lb on $\mathbf{u}$; 341.77 lb on $\mathbf{v}$

75. 517.55 for the 28° rope and 579.90 for the 38° rope

77. (a) $\mathbf{v} = \langle x_2 - x_1, y_2 - y_1 \rangle$; $k\mathbf{v} = \langle kx_2 - kx_1, ky_2 - ky_1 \rangle$
(b) $\|\mathbf{v}\| = \sqrt{(x_2 - x_1)^2 + (y_2 - y_1)^2}$;
$\|k\mathbf{v}\| = \sqrt{(kx_2 - kx_1)^2 + (ky_2 - ky_1)^2}$
(c) $\|k\mathbf{v}\| = \sqrt{k^2(x_2 - x_1)^2 + k^2(y_2 - y_1)^2}$
$= \sqrt{k^2}\sqrt{(x_2 - x_1)^2 + (y_2 - y_1)^2} = |k|\,\|\mathbf{v}\|$
(d) $\tan\theta = \dfrac{y_2 - y_1}{x_2 - x_1}$; $\tan\beta = \dfrac{ky_2 - ky_1}{kx_2 - kx_1} =$
$\dfrac{k(y_2 - y_1)}{k(x_2 - x_1)} = \dfrac{y_2 - y_1}{x_2 - x_1}$. Since the values of the tangent
function are repeated only at intervals of π, it follows
that θ and β differ by a multiple of π, and so represent
either the same direction or opposite directions.
(e) If $k > 0$, the signs of the components of $k\mathbf{v}$ are the
same as those of $\mathbf{v}$, so $\mathbf{v}$ and $k\mathbf{v}$ lie in the same quadrant. Therefore $\mathbf{v}$ and $k\mathbf{v}$ must have the same direction
rather than opposite directions. If $k < 0$, the signs of
$k\mathbf{v}$ are opposite those of $\mathbf{v}$, so $\mathbf{v}$ and $k\mathbf{v}$ do not lie in the
same quadrant. They cannot have the same direction,
so must have opposite directions.

79. (a) Since $\mathbf{u} - \mathbf{v} = \langle a - c, b - s \rangle$, the box on page 608
shows that its magnitude is $\sqrt{(a - c)^2 + (b - d)^2}$.
On the other hand, the distance formula shows that the
length (magnitude) of $\mathbf{w}$ is $\sqrt{(a - c)^2 + (b - d)^2}$.
Hence, $\|\mathbf{u} - \mathbf{v}\| = \|\mathbf{w}\|$.

(b) $\mathbf{u} - \mathbf{v}$ lies on the straight line through $(0, 0)$ and
$(a - c, b - d)$ which has slope
$\dfrac{(a - c) - 0}{(b - d) - 0} = \dfrac{a - c}{b - d}$. Similarly, $\mathbf{w}$ lies on the line
through (a, b) and (c, d), which also has slope
$\dfrac{a - c}{b - d}$. So, $\mathbf{u} - \mathbf{v}$ and $\mathbf{w}$ are parallel. Verify that
they actually have the same direction by considering
the relative positions of (a, b), (c, d), and
$(a - c, b - d)$. For instance, if $\mathbf{u} - \mathbf{v}$ points upward
to the right, then (a, b) lies to the right and above
(c, d). Hence $c < a$ and $d < b$, so that $a - c > 0$
and $b - d > 0$, which means that the endpoint of $\mathbf{w}$
lies in the first quadrant, that is, $\mathbf{w}$ points upward to
the right.

Section 9.4, page 629

1. $\mathbf{u} \cdot \mathbf{v} = -7$, $\mathbf{u} \cdot \mathbf{u} = 25$, $\mathbf{v} \cdot \mathbf{v} = 29$

3. $\mathbf{u} \cdot \mathbf{v} = 6$, $\mathbf{u} \cdot \mathbf{u} = 5$, $\mathbf{v} \cdot \mathbf{v} = 9$

5. $\mathbf{u} \cdot \mathbf{v} = 12$, $\mathbf{u} \cdot \mathbf{u} = 13$, $\mathbf{v} \cdot \mathbf{v} = 13$

7. 6   **9.** 20   **11.** -28 **13.** 1.75065 radians

15. 2.1588 radians   **17.** $\pi/2$ radians

19. Orthogonal   **21.** Parallel   **23.** Neither

25. $k = 2$ **27.** $k = \sqrt{2}$

29. $\text{proj}_{\mathbf{u}}\mathbf{v} = \langle 12/17, -20/17 \rangle$; $\text{proj}_{\mathbf{v}}\mathbf{u} = \langle 6/5, 2/5 \rangle$

31. $\text{proj}_{\mathbf{u}}\mathbf{v} = \langle 0, 0 \rangle$; $\text{proj}_{\mathbf{v}}\mathbf{u} = \langle 0, 0 \rangle$

33. $\text{comp}_{\mathbf{v}}\mathbf{u} = 22/\sqrt{13}$ **35.** $\text{comp}_{\mathbf{v}}\mathbf{u} = 3/\sqrt{10}$

37. $\mathbf{u} \cdot (\mathbf{v} + \mathbf{w}) = \langle a, b \rangle \cdot (\langle c, d \rangle + \langle r, s \rangle)$
$= \langle a, b \rangle \cdot \langle c + r, d + s \rangle = a(c + r) + b(d + s)$
$= ac + ar + bd + bs$
$\mathbf{u} \cdot \mathbf{v} + \mathbf{u} \cdot \mathbf{w} = \langle a, b \rangle \cdot \langle c, d \rangle + \langle a, b \rangle \cdot \langle r, s \rangle$
$= (ac + bd) + (ar + bs) = ac + ar + bd + bs$

39. $\mathbf{0} \cdot \mathbf{u} = \langle 0, 0 \rangle \cdot \langle a, b \rangle = 0a + 0b = 0$

41. If $\theta = 0$ or π, then $\mathbf{u}$ and $\mathbf{v}$ are parallel, so $\mathbf{v} = k\mathbf{u}$ for
some real number k. We know that $\|\mathbf{v}\| = |k|\,\|\mathbf{u}\|$
(Exercise 77, Section 9.3). If $\theta = 0$, then $\cos\theta = 1$ and
$k > 0$. Since $k > 0$, $|k| = k$ and so $\|\mathbf{v}\| = k\|\mathbf{u}\|$. Therefore,
$\mathbf{u} \cdot \mathbf{v} = \mathbf{u} \cdot k\mathbf{u} = k\mathbf{u} \cdot \mathbf{u} = k\|\mathbf{u}\|^2 = \|\mathbf{u}\|(k\|\mathbf{u}\|) = \|\mathbf{u}\|\|\mathbf{v}\| =$
$\|\mathbf{u}\|\|\mathbf{v}\|\cos\theta$. On the other hand, if $\theta = \pi$, then
$\cos\theta = -1$ and $k < 0$. Since $k < 0$, $|k| = -k$ and so
$\|\mathbf{v}\| = -k\|\mathbf{u}\|$. Then $\mathbf{u} \cdot \mathbf{v} = \mathbf{u} \cdot k\mathbf{u}, = k\mathbf{u} \cdot \mathbf{u} = k\|\mathbf{u}\|^2 =$
$\|\mathbf{u}\|(-k\|\mathbf{u}\|) = -\|\mathbf{u}\|\|\mathbf{v}\| = \|\mathbf{u}\|\|\mathbf{v}\|\cos\theta$. In both cases we
have shown $\mathbf{u} \cdot \mathbf{v} = \|\mathbf{u}\|\|\mathbf{v}\|\cos\theta$.

43. If $A = (1, 2)$, $B = (3, 4)$, and $C = (5, 2)$, then the vector
$\overrightarrow{AB} = \langle 2, 2 \rangle$, $\overrightarrow{AC} = \langle 4, 0 \rangle$, and $\overrightarrow{BC} = \langle 2, -2 \rangle$. Since
$\overrightarrow{AB} \cdot \overrightarrow{BC} = 0$, $\overrightarrow{AB}$ and $\overrightarrow{BC}$ are perpendicular, so the
angle at vertex B is a right angle.

45. Many possible answers: One is $\mathbf{u} = \langle 1, 0 \rangle$, $\mathbf{v} = \langle 1, 1 \rangle$, and
$\mathbf{w} = \langle 1, -1 \rangle$.

47. 300 lb ($= 600 \sin 30°$)

49. 13   **51.** 24

53. The force in the direction of the lawnmower's motion is $30 \cos 60° = 15$ lb. Thus, the work done is

$$15(75) = 1125 \text{ ft-lb.}$$

55. 1368 ft-lb

Chapter 9 Review, page 632

1. $\sqrt{10} + \sqrt{20}$

3. The graph is a circle of radius 2 centered at the origin.

5. $2\left(\cos\dfrac{\pi}{3} + i \sin\dfrac{\pi}{3}\right)$ **7.** $4\sqrt{2} + 4\sqrt{2}i$

9. $2\sqrt{3} + 2i$ **11.** $\dfrac{81}{2} - \dfrac{81\sqrt{3}}{2}i$

13. $1, \cos\dfrac{\pi}{3} + i \sin\dfrac{\pi}{3}, \cos\dfrac{2\pi}{3} + i \sin\dfrac{2\pi}{3}, -1,$

$\cos\dfrac{4\pi}{3} + i \sin\dfrac{4\pi}{3}, \cos\dfrac{5\pi}{3} + i \sin\dfrac{5\pi}{3}$

15. $\cos\dfrac{\pi}{8} + i \sin\dfrac{\pi}{8}, \cos\dfrac{5\pi}{8} + i \sin\dfrac{5\pi}{8},$

$\cos\dfrac{9\pi}{8} + i \sin\dfrac{9\pi}{8}, \cos\dfrac{13\pi}{8} + i \sin\dfrac{13\pi}{8}$

17. $\langle 11, -1 \rangle$ **19.** $2\sqrt{29}$ **21.** $-11\mathbf{i} + 8\mathbf{j}$

23. $\sqrt{10}$ **25.** $\left\langle \dfrac{5\sqrt{2}}{2}, \dfrac{5\sqrt{2}}{2} \right\rangle$

27. $-\dfrac{1}{\sqrt{5}}\mathbf{i} + \dfrac{2}{\sqrt{5}}\mathbf{j}$

29. Ground speed: 321.87 mph; course: 126.18°

31. -26 **33.** 3   **35.** .70 radians

37. $\text{proj}_{\mathbf{v}}\mathbf{u} = \mathbf{v} = 2\mathbf{i} + \mathbf{j}$

39. $(\mathbf{u} + \mathbf{v}) \cdot (\mathbf{u} - \mathbf{v}) = \mathbf{u} \cdot \mathbf{u} - \mathbf{u} \cdot \mathbf{v} + \mathbf{v} \cdot \mathbf{u} - \mathbf{v} \cdot \mathbf{v} = \mathbf{u} \cdot \mathbf{u} - \mathbf{v} \cdot \mathbf{v} = \|\mathbf{u}\|^2 - \|\mathbf{v}\|^2 = 0$ since $\mathbf{u}$ and $\mathbf{v}$ have the same magnitude.

41. 1750 lb

CHAPTER 10

Section 10.1, page 649

1. $x^2 + (y - 3)^2 = 4$

3. $x^2 + 6y^2 = 18$

5. $(x - 4)^2 + (y - 3)^2 = 4$

7. Ellipse

9. Ellipse

11. Ellipse

13. Circle

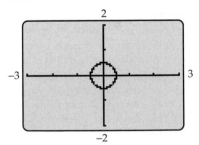

15. $\dfrac{x^2}{49} + \dfrac{y^2}{4} = 1$ **17.** $\dfrac{x^2}{36} + \dfrac{y^2}{16} = 1$

19. $\dfrac{x^2}{9} + \dfrac{y^2}{49} = 1$

21. 8π **23.** $2\sqrt{3}\pi$ **25.** $\dfrac{7\pi}{\sqrt{3}}$

27. Ellipse; center (1, 5)

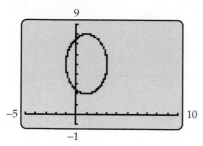

29. Ellipse; center (−1, 4)

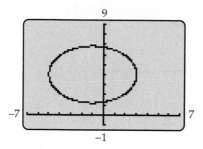

31. Ellipse; center (−3, 1)

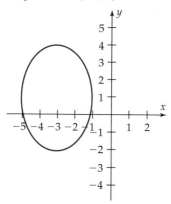

33. Circle; center (−3, 4)

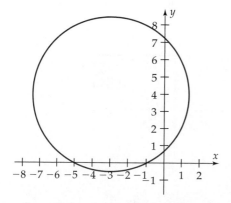

35. Ellipse; center (−3, 2)

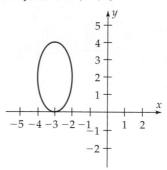

37. $x = \sqrt{10} \cos t, y = 6 \sin t \quad (0 \leq t \leq 2\pi)$

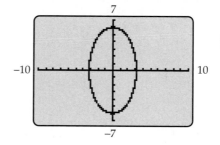

39. $x = \frac{1}{2} \cos t, y = \frac{1}{2} \sin t \quad (0 \leq t \leq 2\pi)$

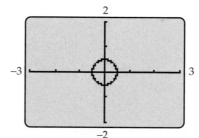

41. $x = 2 \cos t + 1, y = 3 \sin t + 5 \quad (0 \leq t \leq 2\pi)$

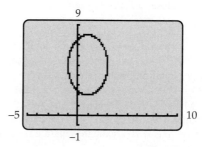

43. $x = 4 \cos t - 1,\ y = \sqrt{8} \sin t + 4 \quad (0 \le t \le 2\pi)$

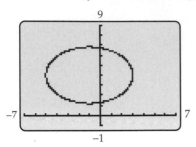

45. $\dfrac{(x-2)^2}{4} + \dfrac{(y-3)^2}{16} = 1$

47. $\dfrac{4(x-7)^2}{25} + \dfrac{(y+4)^2}{36} = 1$

49. $\dfrac{(x-3)^2}{36} + \dfrac{(y+2)^2}{16} = 1$

51. $\dfrac{(x+5)^2}{49} + \dfrac{(y-3)^2}{16} = 1 \quad$ or

$\dfrac{(x+5)^2}{16} + \dfrac{(y-3)^2}{49} = 1$

53. $2x^2 + 2y^2 - 8 = 0$

55. $2x^2 + y^2 - 8x - 6y + 9 = 0$

57. $\dfrac{(x+3)^2}{4} + \dfrac{(y+3)^2}{8} = 1$

59. If $a = b$, then $\dfrac{x^2}{a^2} + \dfrac{y^2}{a^2} = 1$. Multiplying both sides by a^2 gives $x^2 + y^2 = a^2$, the equation of a circle of radius a with center at the origin.

61. Approximately 226,335 miles and 251,401 miles

63. Eccentricity $= .1$ **65.** Eccentricity $= \dfrac{\sqrt{3}}{2} \approx .87$

67. If eccentricity is close to 0, the ellipse is almost circular. If the eccentricity is close to 1, the ellipse is elongated.

69. Eccentricity $\approx .38$

71. Let P denote the punch bowl and Q the table. In the longest possible trip starting at point X, the sum of the distance from X to Q and the distance from X to P must be 100 (since the distance from Q to P is 50). Thus, the fence should be an ellipse with foci P and Q and $r = 100$, as described on pages 639–640 (with $c = 25$). Verify that the length of its major axis is 100 ft and the length of its minor axis is approximately 86.6 ft.

Section 10.2, page 660

1. $x^2 + 4y^2 = 1$ **3.** $2x^2 - y^2 = 8$

5. $6x^2 + 2y^2 = 18$

7. Hyperbola

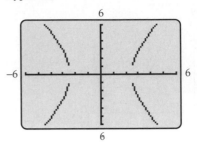

Because of limited resolution, this calculator-generated graph does not show that the top and bottom halves of the graph are connected.

9. Hyperbola

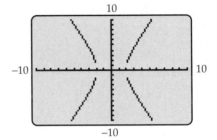

Because of limited resolution, this calculator-generated graph does not show that the top and bottom halves of the graph are connected.

11. Hyperbola

13. Hyperbola

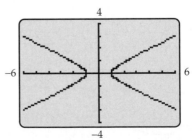

15. $\dfrac{x^2}{9} - \dfrac{y^2}{36} = 1$ **17.** $\dfrac{x^2}{4} - y^2 = 1$

19. Hyperbola; center $(-1, -3)$

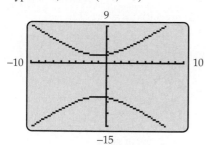

21. Hyperbola; center $(-3, 2)$

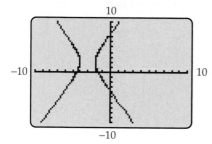

23. Hyperbola; center $(1, -4)$

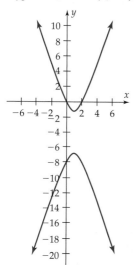

25. Hyperbola; center $(3, 3)$

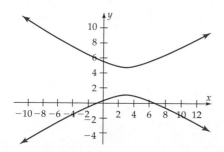

27. Circle; center $(3, 4)$

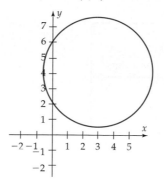

29. Ellipse; center $(3, 4)$

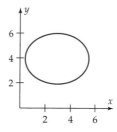

31. Hyperbola; center $(-2, 2)$

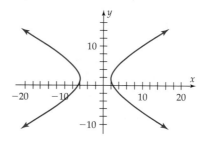

33. $\dfrac{(y-3)^2}{4} - \dfrac{(x+2)^2}{6} = 1$

35. $\dfrac{(x-4)^2}{9} - \dfrac{(y-2)^2}{16} = 1$

37. $y^2 - 2x^2 = 6$

39. $\dfrac{(y-2)^2}{4} - \dfrac{(x-3)^2}{9} = 1$

41. $\dfrac{(x+3)^2}{3} - \dfrac{(y+3)^2}{4} = 1$

43. $x = \sqrt{10}/\cos t,\ y = 6 \tan t \quad (0 \le t \le 2\pi)$

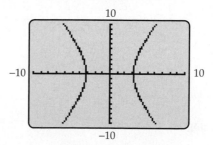

45. $x = 1/\cos t$, $y = \frac{1}{2}\tan t$ $(0 \le t \le 2\pi)$

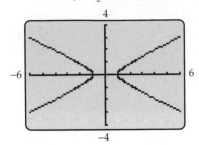

47. $x = 4\tan t - 1$, $y = 5/\cos t - 3$ $(0 \le t \le 2\pi)$

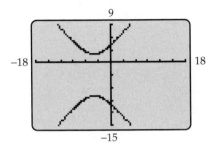

49. $x = 1/\cos t - 3$, $y = 2\tan t + 2$ $(0 \le t \le 2\pi)$

51. The two branches of the hyperbola are very "flat" when b is large. With very large b and a small viewing window, the hyperbola may look like two horizontal lines, but its asymptotes $y = \pm\frac{2}{b}x$ are not horizontal (their slopes, $\pm 2/b$, are close to, but not equal to, 0 when b is large).

53. The asymptotes of $\dfrac{x^2}{a^2} - \dfrac{y^2}{a^2} = 1$ are $y = \pm\dfrac{a}{a}x$ or $y = \pm x$, with slopes 1 and -1. Since $(1)(-1) = -1$, these lines are perpendicular.

55. $\dfrac{x^2}{1,210,000} - \dfrac{y^2}{5,759,600} = 1$ (measurement in feet). The exact location cannot be determined from the given information.

57. $\sqrt{5} \approx 2.24$ **59.** $\sqrt{3} \approx 1.73$ **61.** 1.5

Section 10.3, page 670

1. $6x = y^2$ **3.** $2x^2 - y^2 = 8$

5. $x^2 + 6y^2 = 18$

7. Focus: $(0, 1/12)$, directrix: $y = -1/12$.

9. Focus: $(0, 1)$, directrix: $y = -1$.

11. Vertex: $(2, 0)$, focus: $(9/4, 0)$, directrix: $x = 7/4$, opens right

13. Vertex: $(2, -1)$, focus: $(7/4, -1)$, directrix: $x = 9/4$, opens left

15. Vertex: $(2/3, -3)$, focus: $(17/12, -3)$, directrix: $x = -1/12$, opens right

17. Vertex: $(-81/4, 9/2)$, focus: $(-20, 9/2)$, directrix: $x = -41/2$, opens right

19. Vertex: $(-1/6, -49/12)$, focus: $(-1/6, -4)$, directrix: $y = -25/6$, opens up

21. Vertex: $(2/3, 19/3)$, focus: $(2/3, 25/4)$, directrix: $y = 77/12$, opens down

23.

25.

27.

29.

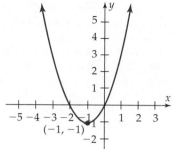

31. $y = 3x^2$

33. $y = 13(x - 1)^2$

35. $x - 2 = 3(y - 1)^2$

37. $x + 3 = 4(y + 2)^2$

39. $2(x - 1)^2 = y - 1$

41. $(y - 3)^2 = x + 1$

43. $x = t^2/4$, $y = t$ (*t* any real number)

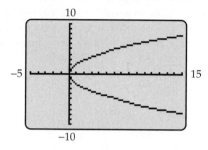

45. $x = t$, $y = 4(t - 1)^2 + 2$ (any real number *t*)

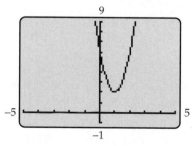

47. $x = 2(t - 2)^2$, $y = t$ (any real number *t*)

49. Parabola

51. Parabola

53. Circle

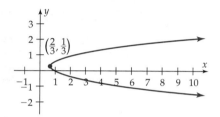

55. Hyperbola; vertices $(-4 - \sqrt{2}, 2)$, $(-4 + \sqrt{2}, 2)$

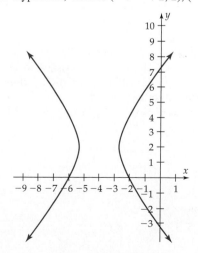

57. $y = (x - 4)^2 + 2$ **59.** $y = -x^2 - 8x - 18$

61. $y = (x + 5)^2 - 3$ **63.** $b = 0$

65. $\left(9, -\dfrac{1}{2} + \dfrac{1}{2}\sqrt{34}\right), \left(9, -\dfrac{1}{2} - \dfrac{1}{2}\sqrt{34}\right)$

67. 2/3 of a foot from the vertex

69. .375 inches from the vertex

71. 27.44 feet above the bridge surface

Section 10.4, page 676

1. Ellipse   **3.** Paraboa   **5.** Hyperbola

7. Ellipse; $-6 \le x \le 3$ and $-2 \le y \le 4$

9. Hyperbola; $-7 \le x \le 13$ and $-3 \le y \le 9$

11. Parabola; $-1 \le x \le 8$ and $-3 \le y \le 3$

13. Ellipse; $-1.5 \le x \le 1.5$ and $-1 \le y \le 1$

15. Hyperbola; $-15 \le x \le 15$ and $-10 \le y \le 10$

17. Parabola; $-19 \le x \le 2$ and $-1 \le y \le 13$

19. Hyperbola; $-15 \le x \le 15$ and $-15 \le y \le 15$

21. Ellipse; $-6 \le x \le 6$ and $-4 \le y \le 4$

23. Parabola; $-9 \le x \le 4$ and $-2 \le y \le 14$

Special Topics 10.4.A, page 681

1. $\left(\dfrac{5\sqrt{2}}{2}, -\dfrac{\sqrt{2}}{2}\right)$ **3.** $\left(\dfrac{\sqrt{3}}{2}, -\dfrac{1}{2}\right)$

5. $\dfrac{u^2}{2} - \dfrac{v^2}{2} = 1$ **7.** $\dfrac{u^2}{4} + v^2 = 1$

9. $\theta \approx 53.13°$; $x = \dfrac{3}{5}u - \dfrac{4}{5}v$; $y = \dfrac{4}{5}u + \dfrac{3}{5}v$

11. $\theta \approx 36.87°$; $x = \dfrac{4}{5}u - \dfrac{3}{5}v$; $y = \dfrac{3}{5}u + \dfrac{4}{5}v$

13. (a) $(A\cos^2\theta + B\cos\theta\sin\theta + C\sin^2\theta)u^2 + (B\cos^2\theta - 2A\cos\theta\sin\theta + 2C\cos\theta\sin\theta - B\sin^2\theta)uv + (C\cos^2\theta - B\cos\theta\sin\theta + A\sin^2\theta)v^2 + (D\cos\theta + E\sin\theta)u + (E\cos\theta - D\sin\theta)v + F = 0$
(b) $B' = B\cos^2\theta - 2A\cos\theta\sin\theta + 2C\cos\theta\sin\theta - B\sin^2\theta = 2(C - A)\sin\theta\cos\theta + B(\cos^2\theta - \sin^2\theta)$ since B' is the coefficient of uv
(c) Since $\sin 2\theta = 2\sin\theta\cos\theta$ and $\cos 2\theta = \cos^2\theta - \sin^2\theta$, $B' = (C - A)\sin 2\theta + B\cos 2\theta$.

15. (a) From Exercise 13(a) we have $(B')^2 - 4A'C' = (B\cos^2\theta - 2A\cos\theta\sin\theta + 2C\cos\theta\sin\theta - B\sin^2\theta)^2 - 4(A\cos^2\theta + B\cos\theta\sin\theta + C\sin^2\theta)(C\sin^2\theta - B\cos\theta\sin\theta + A\sin^2\theta) = [B(\cos^2\theta - \sin^2\theta) + 2(C - A)\cos\theta\sin\theta]^2 - 4(A\cos^2\theta + B\cos\theta\sin\theta + C\sin^2\theta)(C\cos^2\theta - B\cos\theta\sin\theta + A\sin^2\theta) = B^2(\cos^2\theta - \sin^2\theta)^2 + 4(C - A)^2\cos^2\theta\sin^2\theta + 4B(C - A)(\cos^2\theta - \sin^2\theta)\cos\theta\sin\theta - [4AC(\cos^4\theta + \sin^4\theta) + 4(A^2 + C^2 - B^2)\cos^2\theta\sin^2\theta - 4AB(\cos^3\theta\sin\theta - \cos\theta\sin^3\theta) + 4BC(\cos^3\theta\sin\theta - \cos\theta\sin^3\theta)] = B^2(\cos^4\theta - 2\cos^2\theta\sin^2\theta + \sin^4\theta + 4\cos^2\theta\sin^2\theta) - 4AC(2\cos^2\theta\sin^2\theta + \cos^4\theta + \sin^4\theta)$ (everything else cancels) $= (B^2 - 4AC)(\cos^4\theta + 2\cos^2\theta\sin^2\theta + \sin^4\theta) = (B^2 - 4AC)(\cos^2\theta + \sin^2\theta)^2 = B^2 - 4AC$
(b) If $B^2 - 4AC < 0$, then also $(B')^2 - 4A'C' < 0$. Since $B' = 0$, $-4A'C' < 0$ and so $A'C' > 0$. By Exercise 14, the graph is an ellipse. The other two cases are proved in the same way.

Section 10.5, page 690

1. $-5 \le x \le 6$ and $-2 \le y \le 2$

3. $-3 \le x \le 4$ and $-2 \le y \le 3$

5. $0 \le x \le 14$ and $-15 \le y \le 0$

7. $-2 \le x \le 20$ and $-11 \le y \le 11$

9. $-12 \le x \le 12$ and $-12 \le y \le 12$

11. $-2 \le x \le 20$ and $-20 \le y \le 4$

13. $-25 \le x \le 22$ and $-25 \le y \le 26$

15. $y = 2x + 7$ **17.** $y = 2x + 5$ **19.** $y = \ln x$

21. $x^2 + y^2 = 9$

23. Both give a straight line segment between $P = (-4, 7)$ and $Q = (2, -5)$. The parametric equations in (a) move from P to Q, and the parametric equations in (b) move from Q to P.

25. $x = 6\cos t + 7$, $y = 6\sin t - 4$ $0 \le t < 2\pi$

27. $x = 6\cos t + 7$, $y = 6\sin t - 4$ $0 \le t < 2\pi$

29. (a) $\dfrac{d - b}{c - a}$
(b) $y = \left(\dfrac{d - b}{c - a}\right)(x - a) + b$

31. $x = -6 + 18t$, $y = 12 - 22t$ $0 \le t \le 1$

33. $x = 18 - 34t$, $y = 4 + 10t$ $0 \le t \le 1$

35. (a)

(b)

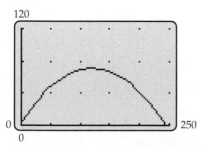

37. Local minimum at $(-6, 2)$

39. Local maximum at $(4, 5)$

41. (a)

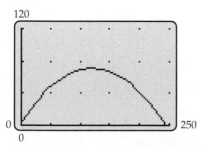

$x = (110 \cos 28°)t$
$y = (110 \sin 28°)t - 16t^2$
$0 \le t \le 3.5$

(b) About 3.2 sec
(c) 41.67 ft

43. (a)

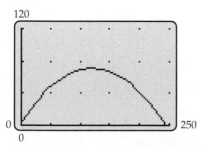

$x = (88 \cos 48°)t$
$y = (88 \sin 48°)t - 16t^2 + 4$
$0 \le t \le 4.5$

(b) Yes

45. $v = 80\sqrt[4]{3} \approx 105.29$ ft/sec

47. (a)

(b) 40°
(c)

An angle of 45° seems to result in the longest distance.

49. (b) $\cos(t - \pi/2) = \cos t \cos \pi/2 + \sin t \sin \pi/2 = (\cos t)(0) + (\sin t)(1) = \sin t$. Therefore, $3t - 3 \cos(t - \pi/2) = 3t - 3 \sin t = 3(t - \sin t)$.
$\sin(t - \pi/2) = \sin t \cos \pi/2 - \cos t \sin \pi/2 = (\sin t)(0) - (\cos t)(1) = -\cos t$. Therefore, $3 + 3 \sin(t - \pi/2) = 3 - 3 \cos t = 3(1 - \cos t)$.

51. (b) $\cos(t - 3\pi/2) = \cos t \cos 3\pi/2 + \sin t \sin 3\pi/2 = (\cos t)(0) + (\sin t)(-1) = -\sin t$. Therefore, $3t + 3 \cos(t - 3\pi/2) = 3t - 3 \sin t = 3(t - \sin t)$.
$\sin(t - 3\pi/2) = \sin t \cos 3\pi/2 - \cos t \sin 3\pi/2 = (\sin t)(0) - (\cos t)(-1) = \cos t$. Therefore, $3 - 3 \sin(t - 3\pi/2) = 3 - 3 \cos t = 3(1 - \cos t)$.

53. (a)

The particles do not collide

(b) $t \approx 1.1$

(c)

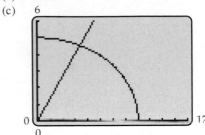

The particles do not collide; they are closest when $t \approx 1.13$.

(d)

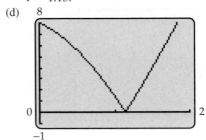

d is smallest when $t \approx 1.1322$.

55. (a) As shown at the top of the next column, the center Q of the small circle is always at distance $a - b$ from the origin O. Suppose t is the angle that OQ makes with the x-axis. Then the coordinates of Q are $x = (a - b) \cos t$, $y = (a - b) \sin t$. Examining the smaller circle in detail, we see that the change in x-coordinate from Q to P is $b \cos u$, where u is the angle that PQ makes with the positive x-axis. Likewise, the change in y-coordinate from Q to P is $-b \sin u$. Therefore, the coordinates of P are

$$x = (a - b) \cos t + b \cos u$$

$$y = (a - b) \sin t - b \sin u.$$

The angles t and u are related by the fact that the inner circle must roll without "slipping." This means the arc length that P has moved around the inner circle must equal the arc length that the inner circle has moved along the circumference of the larger circle. In other words, the arc length from P to W must equal the arc length from S to V. Since the length of a circular arc is the radius times the angle, this means

$$bu = (a - b)t, \text{ or } u = (a - b)t/b.$$

Substituting this for u in the above equations will give the desired parametric equations.

(b)

(c)

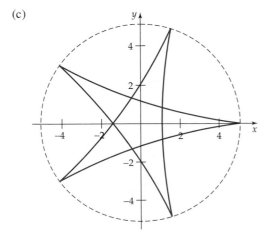

Section 10.6, page 703

1. $P = (2, \pi/4)$, $Q = (3, 2\pi/3)$, $R = (5, \pi)$, $S = (7, 7\pi/6)$, $T = (4, 3\pi/2)$, $U = (6, -\pi/3)$ or $(6, 5\pi/3)$, $V = (7, 0)$

3. $(5, 2\pi)$, $(5, -2\pi)$, $(-5, 3\pi)$, $(-5, -\pi)$, and others

5. $(1, 5\pi/6)$, $(1, -7\pi/6)$, $(-1, 11\pi/6)$, $(-1, -13\pi/6)$, and others

7. $(3/2, 3\sqrt{3}/2)$ **9.** $(\sqrt{3}/2, -1/2)$

11. $(6, -\pi/6)$ **13.** $(2\sqrt{5}, 1.1071)$

15. $(\sqrt{31.25}, 2.6779)$

17.

19.

21.

23.

25.

27.

29.

31.

33.

35.

37.

39.

41.

43.

45.

47. (a)   (b)

49. $r = a \sin \theta + b \cos \theta \Rightarrow r^2 = ar \sin \theta + br \cos \theta \Rightarrow$
$x^2 + y^2 = ay + bx \Rightarrow x^2 - bx + y^2 - ay = 0 \Rightarrow$
$(x^2 - bx + b^2/4) + (y^2 - ay + a^2/4) = (a^2 + b^2)/4$
$\Rightarrow (x - b/2)^2 + (y - a/2)^2 = (a^2 + b^2)/4$, a circle
with center $(b/2, a/2)$ and radius $\sqrt{a^2 + b^2}/2$.

51. Using the Law of Cosines in the following diagram,
$d^2 = r^2 + s^2 - 2rs \cos(\theta - \beta)$,
so $d = \sqrt{r^2 + s^2 - 2rs \cos(\theta - \beta)}$.

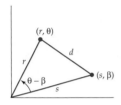

Section 10.7, page 713

1. (d)   **3.** (c)   **5.** (a)

7. Hyperbola, $e = 4/3$ **9.** Parabola, $e = 1$

11. Ellipse, $e = 2/3$

13. .1   **15.** $\sqrt{5}$ **17.** $5/4$

19. (b) $\sqrt{15}/4, \sqrt{10}/4, \sqrt{2}/4$
(c) The smaller the eccentricity, the closer the shape is to circular.

In the graphs for Exercises 21–31, the x- and y-axes with scales are given for convenience, but coordinates of points are in polar *coordinates.*

21.

23.

25.

27.

29.

31.

33. $r = \dfrac{6}{1 - \cos \theta}$ **35.** $r = \dfrac{16}{5 + 3 \sin \theta}$

37. $r = \dfrac{3}{1 + 2 \cos \theta}$ **39.** $r = \dfrac{8}{1 - 4 \cos \theta}$

41. $r = \dfrac{3}{1 - \sin \theta}$ **43.** $r = \dfrac{2}{2 + \cos \theta}$

45. $r = \dfrac{2}{1 - 2 \cos \theta}$

47. Since $0 < e < 1, 0 < 1 - e^2 < 1$ as well. The formulas for a^2 and b^2 show that $a^2 = b^2/(1 - e^2)$, so $a^2 > b^2$. Since a and b are both positive, $a > b$.

49. $r = \dfrac{3 \cdot 10^7}{1 - \cos \theta}$

Chapter 10 Review, page 717

1. Ellipse, foci: $(0, 2), (0, -2)$, vertices: $(0, 2\sqrt{5})$, $(0, -2\sqrt{5})$

3. Ellipse, foci: $(1, 6), (1, 0)$, vertices: $(1, 7), (1, -1)$

5. Focus: $(0, 5/14)$, directrix: $y = -5/14$

7.

9.

11.

13.

15.

17.

19.

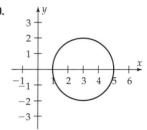

21. Center: $(-4, -5)$; radius: $\sqrt{8}$

23. Center: $(4, -6)$

25. $\dfrac{(x-3)^2}{4} + \dfrac{(y-1)^2}{2} = 1$

27. $\dfrac{y^2}{4} - \dfrac{(x-3)^2}{16} = 1$

29. $\left(y + \dfrac{1}{2}\right)^2 = -\dfrac{1}{2}\left(x - \dfrac{3}{2}\right)$

31.   **33.**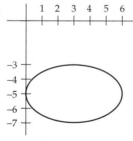

35. Asymptotes

$$y + 4 = \pm\dfrac{5}{2}(x - 1)$$

37. 6 ft   **39.** Ellipse   **41.** Hyperbola

43. $-6 \le x \le 6$ and $-4 \le y \le 4$

45. $-9 \le x \le 9$ and $-6 \le y \le 6$

47. $-15 \le x \le 10$ and $-10 \le y \le 20$

49. $x = \dfrac{1}{2}u - \dfrac{\sqrt{3}}{2}v,\ y = \dfrac{\sqrt{3}}{2}u + \dfrac{1}{2}v$

51. $45°$

53. $-15 \le x \le 15$ and $-10 \le y \le 10$

55. $-35 \le x \le 35$ and $-2 \le y \le 16$

57. $x = 3 - 2y$ or $y = -\dfrac{1}{2}x + \dfrac{3}{2}$

59. $y = -2x^2 + 2$

Point moves from $(1, 0)$ to $(-1, 0)$ as t goes from 0 to π. Then point retraces its path, moving from $(-1, 0)$ to $(1, 0)$ as t goes from π to 2π.

61. (b) and (c)

63.

65.

67.

69.

71.

73.

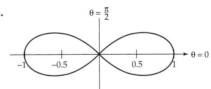

75. $\left(-\dfrac{3}{2}, -\dfrac{3\sqrt{3}}{2}\right)$ **77.** Eccentricity $= \sqrt{\dfrac{2}{3}} \approx 0.8165$

79. Ellipse

81. Hyperbola

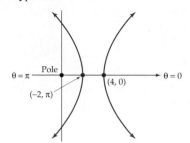

83. $r = \dfrac{24}{5 + \cos\theta}$ **85.** $r = \dfrac{2}{1 + \cos\theta}$

CHAPTER 11

Section 11.1, page 731

1. Yes **3.** Yes **5.** No

7. $x = \dfrac{11}{5},\ y = -\dfrac{7}{5}$ **9.** $x = \dfrac{2}{7},\ y = -\dfrac{11}{7}$

11. $r = \dfrac{5}{2},\ s = -\dfrac{5}{2}$ **13.** $x = \dfrac{3c}{2},\ y = \dfrac{-c + 2d}{2}$

15. $x = 28,\ y = 22$ **17.** $x = 2,\ y = -1$

19. Inconsistent

21. $x = b,\ y = \dfrac{3b - 4}{2}$, where b is any real number

23. $x = -6,\ y = 2$ **25.** $x = \dfrac{66}{5},\ y = \dfrac{18}{5}$

27. $x = \dfrac{7}{11},\ y = -7$ **29.** $x = 1,\ y = 1/2$

31. (a) When the equation is written in the form $y = mx + b$, the sign of m determines whether the population is increasing or decreasing.
  (b) In 1987

33. Never, according to this model

35. $x = \dfrac{rd - sb}{ad - bc},\ y = \dfrac{as - cr}{ad - bc}$

37. $x = \dfrac{3c + 8}{15},\ y = \dfrac{6c - 4}{15}$ is the only solution.

39. $c = -3,\ d = 1/2$ **41.** $p = 6,\ x = 8$

43. $p = 180,\ x = 4$ **45.** 140 adults, 60 children

47. Boat: 18 mph; current: 2 mph

49. $\frac{3}{4}$ lb cashews and $2\frac{1}{4}$ lb peanuts

51. 24 g of 50% alloy; 16 g of 75% alloy

53. (a) Electric: $y = 960x + 2000$; solar: $y = 114x + 14{,}000$
  (b) Electric: $6800; solar: $14,570
  (c) Costs same in 14th year; electric; solar

55. (a) $R = 250{,}500x$

(b) $C = 1{,}295{,}000 + 206{,}500x$
(c) $P = 44{,}000x - 1{,}295{,}000$
(d) $R = 109{,}000x$
  $C = 440{,}000 + 82{,}000x$
  $P = 27{,}000x,\ -440{,}000$
(e) After 50 weeks; approx. $900,000
(f) If the show runs less than 50 weeks, it would be better to open off Broadway.

57. (a) $R(x) = 60x$ (b) 400

59. No

61. $14,450 at 9% and $6450 at 11%

63. (a) $f(x) = 0.373x + 27.02$
  (b) $f(x) = 0.624x + 23.65$
  (c) 2003

65. 80 bowls and 120 plates

Special Topics 11.1.A, page 740

1. $x = 3,\ y = 9$ or $x = -1,\ y = 1$

3. $x = \dfrac{-1 + \sqrt{73}}{6},\ y = \dfrac{37 - \sqrt{73}}{18}$ or
  $x = \dfrac{-1 - \sqrt{73}}{6},\ y = \dfrac{37 + \sqrt{73}}{18}$

5. $x = 7,\ y = 3$ or $x = 3,\ y = 7$

7. $x = 0,\ y = -2$ or $x = 6,\ y = 1$

9. $x = 2,\ y = 0$ or $x = 4,\ y = 2$

11. $x = 4,\ y = 3$ or $x = -4,\ y = 3$ or $x = \sqrt{21},\ y = -2$ or $x = -\sqrt{21},\ y = -2$

13. $x = -1.6237,\ y = -8.1891$ or $x = 1.3163,\ y = 1.0826$ or $x = 2.8073,\ y = 2.4814$

15. $x = -1.9493,\ y = .4412$ or $x = .3634,\ y = .9578$ or $x = 1.4184,\ y = .5986$

17. $x = -.9519,\ y = -.8145$ **19.** No solutions

21. $x = \dfrac{13 - \sqrt{105}}{8},\ y = \dfrac{-3 - \sqrt{105}}{8}$ or
  $x = \dfrac{13 + \sqrt{105}}{8},\ y = \dfrac{-3 + \sqrt{105}}{8}$

23. $x = -4.8093,\ y = 19.3201$ or $x = -3.1434,\ y = 7.7374$ or $x = 2.1407,\ y = 7.7230$ or $x = 2.8120,\ y = 11.7195$

25. $x = -3.8371,\ y = -2.2596$ or $x = -.9324,\ y = -7.7796$

27. $x = -1.4873,\ y = .0480$ or $x = -.0480,\ y = 1.4873$ or $x = .0480,\ y = -1.4873$ or $x = 1.4873,\ y = -.0480$

29. There is no solution when the wire is 70 ft long because the graphs of $y = 70 - 2x$ and $y^2 = 4\pi(100 - x^2/8)$ do not intersect.

31. Two possible boxes: one is 2 by 2 by 4 m and the other is approximately 3.123 by 3.123 by 1.640 m.

33. -4 and -12 **35.** 1.6 and 2.6

37. 15 and -12 **39.** 12 ft by 17 ft

41. 8×15 inches   **43.** $y = 6x - 9$

Section 11.2, page 753

1. $\begin{pmatrix} 2 & -3 & 4 & 1 \\ 1 & 2 & -6 & 0 \\ 3 & -7 & 4 & -3 \end{pmatrix}$ **3.** $\begin{pmatrix} 1 & -\frac{1}{2} & \frac{7}{4} & 0 \\ 2 & -\frac{3}{2} & 5 & 0 \\ 0 & -2 & \frac{1}{3} & 0 \end{pmatrix}$

5. $2x - 3y = 1$
$4x + 7y = 2$

7. $x \quad\ \ + z \qquad\ = 1$
$x - \ y + 4z - 2w = 3$
$4x + 2y + 5z \qquad = 2$

9. $x = 3/2, y = 5, z = -2, w = 0$

11. $x = 2 - t, y = -3 - 2t, z = 4, w = t$, where t is any real number

13. $x = \frac{3}{2}, y = \frac{3}{2}, z = -\frac{3}{2}$

15. $z = t, y = -1 + \frac{1}{3}t, x = 2 - \frac{4}{3}t$, where t is any real number

17. $x = -14, y = -6, z = 2$

19. $x = 100, y = 50, z = 50$

21. $z = t, y = \frac{1}{2} - 2t, x = t$, where t is any real number

23. $x = 1, y = 2$ **25.** No solutions

27. $z = t, y = t - 1, x = -t + 2$, for any real number t

29. $x = 0, y = 0, z = 0$

31. $x = -1, y = 1, z = -3, w = -2$

33. $x = \frac{7}{31}, y = \frac{6}{31}, z = \frac{1}{31}, w = \frac{29}{31}$

35. $x = 1/2, y = 1/3, z = -1/4$

37. $x = 3/5, y = 1/5$

39. $x = -1, y = 2$

41. $A = -1, B = 2$

43. $A = -3/25, B = 3/25, C = 7/5$

45. $A = 2, B = 3, C = -1$

47. $3000 from her friend, $6000 from the bank, $1000 from the insurance company

49. $15,000 from the mutual fund; $30,000 in bonds; $25,000 in food franchise

51. 6 cups Progresso
9 cups Healthy Choice
2 cups Campbell's
Serving size: 1.7 cups

53. 3 possible solutions:
18 bedroom, 13 living room
16 bedroom, 8 living room, 2 whole house
14 bedroom, 3 living room, 4 whole house

55. a) Ray 1: $a + b$
Ray 2: $a + c$
(b) $a = 0.405$
$b = 0.345$
$c = 0.195$
(c) A is bone; B is tumorous; C is healthy.

57. 2000 chairs, 1600 chests, 2500 tables

59. (a) $x + t = 1000$
$x + y = 1100$
$y + z = 700$
$t + z = 600$
(b) $x = 1000 - t$
$y = 100 + t$
$z = 600 - t$
(c)

Intersection	Maximum number of cars	Minimum number of cars
A on 4th	600	0
A on Euclid	1000	400
C on 5th	700	100
C on Chester	600	0

Section 11.3, page 768

1. AB defined, 2×4; BA not defined

3. AB defined, 3×3; BA defined, 2×2

5. AB defined, 3×2; BA not defined

7. $\begin{pmatrix} 3 & 0 & 11 \\ 2 & 8 & 10 \end{pmatrix}$ **9.** $\begin{pmatrix} 1 & -3 \\ 2 & -1 \\ 5 & 6 \end{pmatrix}$

11. $\begin{pmatrix} 1 & -1 & 1 & 2 \\ 4 & 3 & 3 & 2 \\ -1 & -1 & -3 & 2 \\ 5 & 3 & 2 & 5 \end{pmatrix}$

13. $AB = \begin{pmatrix} 17 & -3 \\ 33 & -19 \end{pmatrix}$; $BA = \begin{pmatrix} -4 & 9 \\ 24 & 2 \end{pmatrix}$

15. $AB = \begin{pmatrix} 8 & 24 & -8 \\ 2 & -2 & 6 \\ -3 & -21 & 15 \end{pmatrix}$; $BA = \begin{pmatrix} 19 & 9 & 8 \\ -10 & 2 & 0 \\ 0 & 0 & 0 \end{pmatrix}$

17. $\begin{pmatrix} -2 & 1 \\ 3/2 & -1/2 \end{pmatrix}$

19. No inverse

21. No inverse   **23.** $\begin{pmatrix} -3 & 2 & -4 \\ -1 & 1 & -1 \\ 8 & -5 & 10 \end{pmatrix}$

25. $x = -1, y = 0, z = -3$

27. $x = -8, y = 16, z = 5$

29. $x = -.5, y = -2.1, z = 6.7, w = 2.8$

31. $x = 10.5, y = 5, z = -13, v = 32, w = 2.5$

33. $x = -1149/161, y = 426/161, z = -1124/161,$
$w = 579/161$

35. $x = 0, y = 0, z = 0, v = 0, w = 0$

37. Inconsistent system; no solutions

39. $x = -2 - 3w, y = 5/4 - (1/4)z$, where z and w are any real numbers.

41. $a = \dfrac{35}{8}, b = \dfrac{13}{4}, c = \dfrac{-117}{8}$

43. (a) $a \approx .01158$
$b \approx .80449$
$c = 315$
(b) 342.35, 357.34, 374.65

45. (a) $-0.969x^2 + 30.5x + 111.875$
(b) 194.7; 293.9; 338.4

47. $a = 1, b = -4, c = 1$

49. Jeans are \$34.50; jackets are \$72; sweaters are \$44; shirts are \$21.75.

51. 15,000 of A, 18,000 of B, 54,000 of C

Chapter 11 Review, page 771

1. $x = -5, y = -7$ **3.** $x = 0, y = -2$

5. 2000; approximately 70

7. About 16.67 lb of the 40% alloy and 33.33 lb of the 70% alloy

9. $x = 3, y = 9$ or $x = -1, y = 1$

11. $x = 1 - \sqrt{7}, y = 1 + \sqrt{7}$ or $x = 1 + \sqrt{7}, y = 1 - \sqrt{7}$

13. $\begin{pmatrix} 1 & -2 & 3 & 4 \\ 2 & 1 & -4 & 3 \\ 3 & 4 & -1 & -2 \end{pmatrix}$ **15.** $\begin{pmatrix} 2 & -1 & -2 & 2 \\ 1 & 3 & -2 & 1 \\ -1 & 4 & 2 & -3 \end{pmatrix}$

17. $x = -35, y = 140, z = 22$

19. $x = 2, y = 4, z = 6$

21. $x = -t + 1, y = -2t, z = t$ for any real number t

23. (c)   **25.** $\dfrac{3}{x+2} + \dfrac{1}{x-3}$

27. $\begin{pmatrix} -2 & 3 \\ -4 & -1 \end{pmatrix}$ **29.** Not defined

31. $\begin{pmatrix} -9 & 7 \\ -4 & 3 \end{pmatrix}$ **33.** $\begin{pmatrix} 1 & 2 & -2 \\ -1 & 3 & 0 \\ 0 & -2 & 1 \end{pmatrix}$

35. $x = -1/85, y = -14/85, z = -21/34, w = 46/85$

37. $y = 5x^2 - 2x + 1$

39. (a) $f(x) = 1.0625x^2 - 4.25x + 12.25$
(b) 46.25; 114.25; 216.25

41. 100

43. 30 lb corn, 15 lb soybeans, 40 lb by-products

CHAPTER 12

Section 12.1, page 786

1. 8, 10, 12, 14, 16   **3.** $1, \dfrac{1}{8}, \dfrac{1}{27}, \dfrac{1}{64}, \dfrac{1}{125}$

5. $-\sqrt{3}, 2, -\sqrt{5}, \sqrt{6}, -\sqrt{7}$

7. 3.9, 4.01, 3.999, 4.0001, 3.99999

9. 2, 7, 8, 13, 14   **11.** $\displaystyle\sum_{i=1}^{11} i$

13. $\displaystyle\sum_{i=7}^{13} \dfrac{1}{2^i}$ (other answers possible)

15. 45   **17.** 224   **19.** 15,015

21. $a_n = (-1)^n$ **23.** $a_n = \dfrac{n}{n+1}$ **25.** $a_n = 5n - 3$

27. 4, 11, 25, 53, 109   **29.** $1, -2, 3, 2, 3$

31. $2, 3, 3, \dfrac{9}{2}, \dfrac{27}{4}$ **33.** 3, 1, 4, 1, 5

35. Third -10; sixth -2 **37.** Third 5; sixth 0

39. $\displaystyle\sum_{n=1}^{6} \dfrac{1}{2n+1}$ (other answers possible)

41. $\displaystyle\sum_{n=8}^{12} (-1)^n \left(\dfrac{n-7}{n} \right)$ (other answers possible)

43. 2.613035   **45.** $\dfrac{1051314}{31415} \approx 33.465$

47. 1.5759958   **49.** \$7300; \$2556

51. (a) 979.1, 1005.1   (b) 5874.1

53. (a) 41.67; 58.17   (b) 234.25

55. (b) 59, 61, 67, 71

57. 4, 9, 25, 49, 121   **59.** 3, 7, 13, 19, 23

61. (a) 1, 1, 2, 3, 5, 8, 13, 21, 34, 55
(b) 1, 2, 4, 7, 12, 20, 33, 54, 88, 143
(c) nth partial sum: $a_{n+2} - 1$

63. $n = 1: 5(1)^2 + 4(-1)^1 = 1 = 1^2$;
$\quad n = 2: 5(1)^2 + 4(-1)^2 = 9 = 3^2$;
$\quad n = 3: 5(2)^2 + 4(-1)^3 = 16 = 4^2$;
$\quad n = 4: 5(3)^2 + 4(-1)^4 = 49 = 7^2$; etc.

65. We have $a_1 = a_3 - a_2$; $a_2 = a_4 - a_3$; $a_3 = a_5 - a_4$; ...
$a_{k-1} = a_{k+1} - a_k$; $a_k = a_{k+2} - a_{k+1}$. If these equations
are listed vertically, the sum of the left-side terms is
$\sum_{n=1}^{k} a_k$. On the right side, one term in each line is the same
as a term in the next line, except for sign. So the sum of
the right-side terms is $a_{k+2} - a_2$. Since $a_2 = 1$, we conclude that $\sum_{n=1}^{k} a_n = a_{k+2} - 1$.

Section 12.2, page 792

1. 2   **3.** Not arithmetic

5. Not arithmetic

7. 9, 13, 17, 21, 25   Common difference = 4

9. $-1, 1, -1, 1, -1$

11. 1, 2, 3, 4, 5   Common difference = 1

13. $a_n - a_{n-1} = -2 = d$ **15.** $a_n - a_{n-1} = \frac{3}{2} = d$

17. $a_n - a_{n-1} = 2 = d$ **19.** 13; $a_n = 2n + 3$

21. 5; $a_n = n/4 + 15/4$ **23.** 8; $a_n = -n/2 + 21/2$

25. $a_1 = 6$; $a_n = 2n + 4$ **27.** $a_1 = -3$; $a_n = 7n - 10$

29. $a_1 = -6$; $a_n = \frac{-15}{2} + 3n/2$

31. 87   **33.** $-21/4$ **35.** 30   **37.** 710

39. $156\frac{2}{3}$ **41.** 2550   **43.** 20,100

45. $77,500 in tenth year; $437,500 over ten years

47. (a) $a_n = 1049.8 + 173.7\,n$ (b) $4523.80, 5044.90

49. (a) $c_n = 15100.8 + 728n$ (b) 516,401.6

51. 428

53. 23.25, 22.5, 21.75, 21, 20.25, 19.5, 18.75

Section 12.3, page 799

1. Arithmetic   **3.** Geometric   **5.** Arithmetic

7. Geometric   **9.** Geometric   **11.** Both

13. $a_6 = 160$; $a_n = 2^{n-1} \cdot 5$

15. $a_6 = \frac{1}{256}$; $a_n = \frac{1}{4^{n-2}}$

17. $a_6 = \frac{-5}{16}$; $a_n = \frac{(-1)^{n-1} \cdot 5}{2^{n-2}}$

19. $a_6 = \frac{4}{27}$; $a_n = 36 \cdot \left(\frac{1}{3}\right)^{n-1}$

21. $a_6 = \frac{-16}{125}$; $a_n = -5 \cdot \left(\frac{2}{5}\right)^{n-2}$

23. $\frac{a_n}{a_{n-1}} = \frac{-1}{2} = r$ **25.** $\frac{a_n}{a_{n-1}} = 5 = r$

27. $\frac{a_n}{a_{n-1}} = \sqrt{5} = r$ **29.** $a_5 = 1$; $a_n = \left(\frac{-1}{4}\right)^{n-5}$

31. $a_5 = 5000$; $a_n = \frac{1}{2} \cdot 10^{n-1}$

33. $a_5 = \frac{1}{4}$; $a_n = \left(\frac{1}{4}\right)^{n-4}$

35. $a_5 = 10 \cdot 2^{1/3}$; $a_n = 5 \cdot 2^{\frac{n-1}{3}}$ **37.** $\frac{315}{32}$

39. 381   **41.** 254

43. $\frac{-4921}{19{,}683}$ **45.** $\frac{665}{8}$

47. (a) $a_n = 44.4313 \cdot 0.8556^{n-1}$
$\quad$ (b) 2.6826   (c) 2009

49. (a) $c_n = 3.9631 \cdot 1.0515^{n-1}$
$\quad$ (b) $5.6325 billion, $6.8855 billion, $7.6130 billion

51. (a) $b_n = 130.76 \cdot 1.076^{n-1}$ (b) $1605.85

53. 23.75 ft   **55.** $21,474, 836.47   **57.** $1898.44

59. $\log a_n - \log a_{n-1} = \log \frac{a_n}{a_{n-1}} = \log_r$

61. The sequence is $\{2^{n-1}\}$ and $r = 2$. So for any k, the kth
term is 2^{k-1}, and the sum of the preceding terms is the
$(k - 1)$th partial sum of the sequence,
$\sum_{n=1}^{k-1} 2^{n-1} = \frac{1 - 2^{k-1}}{1 - 2} = 2^{k-1} - 1$.

63. 37 payments

Special Topics 12.3.A, page 805

1. 1   **3.** $.06/.94 = 3/47$ **5.** $\frac{500}{.6} = 833\frac{1}{3}$

7. $4 + 2\sqrt{2}$ **9.** $2/9$ **11.** $597/110$

13. $10{,}702/4995$ **15.** $174{,}067/99{,}900$

17. (a) $a_1 = 2(1.5)^1 = 3$. For each $n \geq 2$,
$\quad a_{n-1} = 2(1.5)^{n-1}$ and $a_n = 2(1.5)^n$, so that the ratio
$\quad \frac{a_n}{a_{n-1}}$ is $\frac{2(1.5)^n}{2(1.5)^{n-1}} = 1.5$. Therefore, this is a
$\quad$ geometric series.
$\quad$ (b) $S_k = -6 + 6(1.5)^k$ and $f(x) = -6 + 6(1.5)^x$
$\quad$ (c) The graph shows that as x gets large, $f(x)$ gets huge,
$\quad$ so there is no horizontal asymptote. Hence, the series
$\quad$ does not converge.

19. 1

Section 12.4, page 812

1. 720   **3.** 220   **5.** 0   **7.** 64   **9.** 3,921,225

11. $x^5 + 5x^4y + 10x^3y^2 + 10x^2y^3 + 5xy^4 + y^5$

13. $a^5 - 5a^4b + 10a^3b^2 - 10a^2b^3 + 5ab^4 - b^5$

15. $32x^5 + 80x^4y^2 + 80x^3y^4 + 40x^2y^6 + 10xy^8 + y^{10}$

17. $x^3 + 6x^2\sqrt{x} + 15x^2 + 20x\sqrt{x} + 15x + 6\sqrt{x} + 1$

19. $1 - 10c + 45c^2 - 120c^3 + 210c^4 - 252c^5 + 210c^6 - 120c^7 + 45c^8 - 10c^9 + c^{10}$

21. $x^{-12} + 4x^{-8} + 6x^{-4} + 4 + x^4$ **23.** 56

25. $-8i$ **27.** $10x^3y^2$ **29.** $35c^3d^4$

31. $\dfrac{35}{8} u^{-5}$ **33.** 4032   **35.** 160

37. (a) $\binom{9}{1} = \dfrac{9!}{1!8!} = 9; \binom{9}{8} = \dfrac{9!}{8!1!} = 9$

(b) $\binom{n}{1} = \binom{n}{n-1} = \dfrac{n!}{1!(n-1)!} = \dfrac{n(n-1)!}{(n-1)!} = n$

39. $2^n = (1 + 1)^n = 1^n + \binom{n}{1}1^{n-1} \cdot 1$
$+ \binom{n}{2}1^{n-2} \cdot 1^2 + \binom{n}{3}1^{n-3} \cdot 1^3 + \cdots +$
$\binom{n}{n-1}1^1 \cdot 1^{n-1} + 1^n = \binom{n}{0} + \binom{n}{1} +$
$\binom{n}{2} + \binom{n}{3} + \cdots + \binom{n}{n-1} + \binom{n}{n}$

41. $\cos^4 \theta + 4i \cos^3 \theta \sin \theta - 6 \cos^2 \theta \sin^2 \theta - 4i \cos \theta \sin^3 \theta + \sin^4 \theta$

43. (a) $f(x + h) - f(x) = (x + h)^5 - x^5 =$
$\left[x^5 + \binom{5}{1}x^4h + \binom{5}{2}x^3h^2 + \binom{5}{3}x^2h^3 + \binom{5}{4}xh^4 + h^5 \right] - x^5 = \binom{5}{1}x^4h + \binom{5}{2}x^3h^2 + \binom{5}{3}x^2h^3 + \binom{5}{4}xh^4 + h^5$

(b) $\dfrac{f(x + h) - f(x)}{h} = \binom{5}{1}x^4 + \binom{5}{2}x^3h + \binom{5}{3}x^2h^2 + \binom{5}{4}xh^3 + h^4$

(c) When h is *very* close to 0, so are the last four terms in part (b), so $\dfrac{f(x + h) - f(x)}{h} \approx \binom{5}{1}x^4 = 5x^4$.

45. $\dfrac{f(x + h) - f(x)}{h} = \dfrac{(x + h)^{12} - x^{12}}{h} = \binom{12}{1}x^{11} + \binom{12}{2}x^{10}h + \binom{12}{3}x^9h^2 + \binom{12}{4}x^8h^3 + \cdots + \binom{12}{10}x^2h^9 + \binom{12}{11}xh^{10} + h^{11} \approx \binom{12}{1}x^{11}$
$= 12x^{11}$, when h is very close to 0.

47. (a) $(n - r)! =$
$(n - r)(n - r - 1)(n - r - 2)(n - r - 3) \cdots 2 \cdot 1$
$= (n - r)[n - (r + 1)](n - (r + 1) - 1) \cdots 2 \cdot 1$
$= (n - r)[n - (r + 1)]!$

(b) Since $(n + 1) - (r + 1) = n - r$,
$[(n + 1) - (r + 1)]! = (n - r)!$

(c) $\binom{n}{r + 1} + \binom{n}{r} = \dfrac{n!}{(r + 1)![n - (r + 1)]!} + \dfrac{n!}{r!(n - r)!} = \dfrac{n!(n - r) + n!(r + 1)}{(r + 1)!(n - r)!} = \dfrac{n!(n + 1)}{(r + 1)!(n - r)!} = \dfrac{(n + 1)!}{(r + 1)![(n + 1) - (r + 1)]!} = \binom{n + 1}{r + 1}$

(d) For example, rows 2 and 3 of Pascal's triangle are

1   2   1

1   ③   3   1

that is,

$\binom{2}{0}$ $\binom{2}{1}$ $\binom{2}{2}$

$\binom{3}{0}$ $\binom{3}{1}$ $\binom{3}{2}$ $\binom{3}{3}$

The circled 3 is the sum of the two closest entries in the row above: $1 + 2$. But this just says that $\binom{3}{1} = \binom{2}{0} + \binom{2}{1}$, which is part (c) with $n = 2$ and $r = 0$. Similarly, in the general case, verify that the two closest entries in the row above $\binom{n + 1}{r + 1}$ are $\binom{n}{r}$ and $\binom{n}{r + 1}$ and use part (c).

Section 12.5, page 821

1. *Step 1:* For $n = 1$ the statement is $1 = 2^1 - 1$, which is true. *Step 2:* Assume that the statement is true for $n = k$: that is,
$1 + 2 + 2^2 + 2^3 + \cdots + 2^{k-1} = 2^k - 1$. Add 2^k to both sides, and rearrange terms:
$1 + 2 + 2^2 + 2^3 + \cdots + 2^{k-1} + 2^k = 2^k - 1 + 2^k$
$1 + 2 + 2^2 + 2^3 + \cdots + 2^{k-1} + 2^{(k+1)-1} = 2(2^k) - 1$
$1 + 2 + 2^2 + 2^3 + \cdots + 2^{k-1} + 2^{(k+1)-1} = 2^{k+1} - 1$
But this last line says that the statement is true for $n = k + 1$. Therefore, by the Principle of Mathematical Induction the statement is true for every positive integer n.

Note: **Hereafter, in these answers, step 1 will be omitted if it is trivial (as in Exercise 1), and only the essential parts of step 2 will be given.**

3. Assume that the statement is true for $n = k$:
$1 + 3 + 5 + \cdots + (2k - 1) = k^2$.
Add $2(k + 1) - 1$ to both sides:
$1 + 3 + 5 + \cdots + (2k - 1) + [2(k + 1) - 1] = k^2 + 2(k + 1) - 1 = k^2 + 2k + 1 = (k + 1)^2$.
The first and last parts of this equation say that the statement is true for $n = k + 1$.

5. Assume that the statement is true for $n = k$:

$$1^2 + 2^2 + 3^2 + \cdots + k^2 = \frac{k(k + 1)(2k + 1)}{6}$$

Add $(k + 1)^2$ to both sides:
$1^2 + 2^2 + 3^2 + \cdots + k^2 + (k + 1)^2$

$$= \frac{k(k + 1)(2k + 1)}{6} + (k + 1)^2$$

$$= \frac{k(k + 1)(2k + 1) + 6(k + 1)^2}{6}$$

$$= \frac{(k + 1)[k(2k + 1) + 6(k + 1)]}{6}$$

$$= \frac{(k + 1)(2k^2 + 7k + 6)}{6}$$

$$= \frac{(k + 1)(k + 2)(2k + 3)}{6}$$

$$= \frac{(k + 1)[(k + 1) + 1][2(k + 1) + 1]}{6}$$

The first and last parts of this equation say that the statement is true for $n = k + 1$.

7. Assume that the statement is true for $n = k$:

$$\frac{1}{1 \cdot 2} + \frac{1}{2 \cdot 3} + \cdots + \frac{1}{k(k + 1)} = \frac{k}{k + 1}.$$

Adding $\dfrac{1}{(k + 1)[(k + 1) + 1]} = \dfrac{1}{(k + 1)(k + 2)}$ to both sides yields:

$$\frac{1}{1 \cdot 2} + \frac{1}{2 \cdot 3} + \cdots + \frac{1}{k(k + 1)} + \frac{1}{(k + 1)(k + 2)}$$

$$= \frac{k}{k + 1} + \frac{1}{(k + 1)(k + 2)}$$

$$= \frac{k(k + 2) + 1}{(k + 1)(k + 2)} = \frac{k^2 + 2k + 1}{(k + 1)(k + 2)}$$

$$= \frac{(k + 1)^2}{(k + 1)(k + 2)} = \frac{k + 1}{k + 2} = \frac{k + 1}{(k + 1) + 1}$$

The first and last parts of this equation show that the statement is true for $n = k + 1$.

9. Assume the statement is true for $n = k$: $k + 2 > k$. Adding 1 to both sides, we have: $k + 2 + 1 > k + 1$, or equivalently, $(k + 1) + 2 > (k + 1)$. Therefore, the statement is true for $n = k + 1$.

11. Assume the statement is true for $n = k$: $3^k \geq 3k$. Multiplying both sides by 3 yields: $3 \cdot 3^k \geq 3 \cdot 3k$, or equiva-

lently, $3^{k+1} \geq 3 \cdot 3k$. Now since $k \geq 1$, we know that $3k \geq 3$ and hence that $2 \cdot 3k \geq 3$.
Therefore, $2 \cdot 3k + 3k \geq 3 + 3k$, or equivalently, $3 \cdot 3k \geq 3k + 3$. Combining this last inequality with the fact that $3^{k+1} \geq 3 \cdot 3k$, we see that $3^{k+1} \geq 3k + 3$, or equivalently, $3^{k+1} \geq 3(k + 1)$. Therefore, the statement is true for $n = k + 1$.

13. Assume the statement is true for $n = k$: $3k > k + 1$. Adding 3 to both sides yields: $3k + 3 > 1 + 3$, or equivalently, $3(k + 1) > (k + 1) + 3$. Since $(k + 1) + 3$ is certainly greater than $(k + 1) + 1$, we conclude that $3(k + 1) > (k + 1) + 1$. Therefore, the statement is true for $n = k + 1$.

15. Assume the statement is true for $n = k$; then 3 is a factor of $2^{2k+1} + 1$; that is, $2^{2k+1} + 1 = 3M$ for some integer M. Thus, $2^{2k+1} = 3M - 1$. Now $2^{2(k+1)+1} = 2^{2k+2+1} = 2^{2+2k+1} = 2^2 \cdot 2^{2k+1} = 4(3M - 1) = 12M - 4 = 3(4M) - 3 - 1 = 3(4M - 1) - 1$. From the first and last terms of this equation we see that $2^{2(k+1)+1} + 1 = 3(4M - 1)$. Hence, 3 is a factor of $2^{2(k+1)+1} + 1$. Therefore, the statement is true for $n = k + 1$.

17. Assume the statement is true for $n = k$: 64 is a factor of $3^{2k+2} - 8k - 9$. Then $3^{2k+2} - 8k - 9 = 64N$ for some integer N so that $3^{2k+2} = 8k + 9 + 64N$.
Now $3^{2(k+1)+2} = 3^{2k+2+2} = 3^{2+(2k+2)} = 3^2 \cdot 3^{2k+2} = 9(8k + 9 + 64N)$. Consequently,
$3^{2(k+1)+2} - 8(k + 1) - 9 = 3^{2(k+1)+2} - 8k - 8 - 9$
$= 3^{2(k+1)+2} - 8k - 17$
$= [9(8k + 9 + 64N)] - 8k - 17$
$= 72k + 81 + 9 \cdot 64N - 8k - 17$
$= 64k + 64 + 9 \cdot 64N = 64(k + 1 + 9N)$.
From the first and last parts of this equation we see that 64 is a factor of $3^{2(k+1)+2} - 8(k + 1) - 9$. Therefore, the statement is true for $n = k + 1$.

19. Assuming that the statement is true for $n = k$:
$c + (c + d) + (c + 2d) + \cdots + [c + (k - 1)d] = \frac{k[2c + (k - 1)d]}{2}$. Adding $c + kd$ to both sides, we have
$c + (c + d) + (c + 2d) + \cdots +$
$\qquad\qquad\qquad [c + (k - 1)d] + (c + kd)$
$= \frac{k[2c + (k - 1)d]}{2} + c + kd$
$= \frac{k[2c + (k - 1)d] + 2(c + kd)}{2}$
$= \frac{2ck + k(k - 1)d + 2c + 2kd}{2}$
$= \frac{2ck + 2c + kd(k - 1) + 2kd}{2}$
$= \frac{(k + 1)2c + kd(k - 1 + 2)}{2}$

$$= \frac{(k+1)2c + kd(k+1)}{2} = \frac{(k+1)(2c+kd)}{2}$$

$$= \frac{(k+1)(2c + [(k+1)-1]d)}{2}$$

Therefore, the statement is true for $n = k + 1$.

21. (a) $x^2 - y^2 = (x-y)(x+y)$;
$x^3 - y^3 = (x-y)(x^2 + xy + y^2)$;
$x^4 - y^4 = (x-y)(x^3 + x^2y + xy^2 + y^3)$

(b) *Conjecture:* $x^n - y^n = (x-y)(x^{n-1} + x^{n-2}y + x^{n-3}y^2 + \cdots + x^2y^{n-3} + xy^{n-2} + y^{n-1})$.
Proof: The statement is true for $n = 2, 3, 4$, by part (a). Assume that the statement is true for $n = k$:
$x^k - y^k =$
$\qquad (x-y)(x^{k-1} + x^{k-2}y + \cdots + xy^{k-2} + y^{k-1})$.
Now use the fact that $-yx^k + yx^k = 0$ to write $x^{k+1} - y^{k+1}$ as follows:
$x^{k+1} - y^{k+1} = x^{k+1} - yx^k + yx^k - y^{k+1}$
$= (x^{k+1} - yx^k) + (yx^k - y^{k+1})$
$= (x-y)x^k + y(x^k - y^k)$
$= (x-y)x^k + y(x-y)(x^{k-1} + x^{k-2}y + x^{k-3}y^2 + \cdots + xy^{k-2} + y^{k-1})$
$= (x-y)x^k + (x-y)(x^{k-1}y + x^{k-2}y^2 + x^{k-3}y^3 + \cdots + xy^{k-1} + y^k)$
$= (x-y)[x^k + x^{k-1}y + x^{k-2}y^2 + x^{k-3}y^3 + \cdots + xy^{k-1} + y^k]$

The first and last parts of this equation show that the conjecture is true for $n = k + 1$. Therefore, by mathematical induction, the conjecture is true for every integer $n \geq 2$.

23. False, counterexample: $n = 9$

25. True: *Proof:* Since $(1+1)^2 > 1^2 + 1$, the statement is true for $n = 1$. Assume the statement is true for $n = k$: $(k+1)^2 > k^2 + 1$: Then $[(k+1)+1]^2 = (k+1)^2 + 2(k+1) + 1 > k^2 + 1 + 2(k+1) + 1 = k^2 + 2k + 2 + 2 > k^2 + 2k + 2 = k^2 + 2k + 1 + 1 = (k+1)^2 + 1$. The first and last terms of this inequality say that the statement is true for $n = k + 1$. Therefore, by induction the statement is true for every positive integer n.

27. False, counterexample: $n = 2$

29. Since $2 \cdot 5 - 4 > 5$, the statement is true for $n = 5$. Assume the statement is true for $n = k$ (with $k \geq 5$): $2k - 4 > k$. Adding 2 to both sides shows that $2k - 4 + 2 > k + 2$, or equivalently, $2(k+1) - 4 > k + 2$. Since $k + 2 > k + 1$, we see that $2(k+1) - 4 > k + 1$. So the statement is true for $n = k + 1$. Therefore, by the Extended Principle of Mathematical Induction, the statement is true for all $n \geq 5$.

31. Since $2^2 > 2$, the statement is true for $n = 2$. Assume that $k \geq 2$ and that the statement is true for $n = k$: $k^2 > k$. Then $(k+1)^2 = k^2 + 2k + 1 > k^2 + 1 > k + 1$. The first and

last terms of this inequality show that the statement is true for $n = k + 1$. Therefore, by induction, the statement is true for all $n \geq 2$.

33. Sinc $3^4 = 81$ and $2^4 + 10 \cdot 4 = 16 + 40 = 56$, we see that $3^4 > 2^4 + 10 \cdot 4$. So the statement is true for $n = 4$. Assume that $k \geq 4$ and that the statement is true for $n = k$: $3^k > 2^k + 10k$. Multiplying both sides by 3 yields: $3 \cdot 3^k > 3(2^k + 10k)$, or equivalently, $3^{k+1} > 3 \cdot 2^k + 30k$. But

$$3 \cdot 2^k + 30k > 2 \cdot 2^k + 30k = 2^{k+1} + 30k.$$

Therefore, $3^{k+1} > 2^{k+1} + 30k$. Now we shall show that $30k > 10(k+1)$. Since $k \geq 4$, we have $20k \geq 20 \cdot 4$, so that $20k > 80 > 10$. Adding $10k$ to both sides of $20k > 10$ yields: $30k > 10k + 10$, or equivalently, $30k > 10(k+1)$. Consequently,

$$3^{k+1} > 2^{k+1} + 30k > 2^{k+1} + 10(k+1).$$

The first and last terms of this inequality show that the statement is true for $n = k + 1$. Therefore, the statement is true for all $n \geq 4$ by induction.

35. (a) 3 (that is, $2^2 - 1$) for $n = 2$; 7 (that is, $2^3 - 1$) for $n = 3$; 15 (that is, $2^4 - 1$) for $n = 4$.

(b) *Conjecture:* The smallest possible number of moves for n rings is $2^n - 1$. *Proof:* This conjecture is easily seen to be true for $n = 1$ or $n = 2$. Assume it is true for $n = k$ and that we have $k + 1$ rings to move. In order to move the *bottom* ring from the first peg to another peg (say, the second one), it is first necessary to move the top k rings off the first peg *and* leave the second peg vacant at the end (the second peg will have to be used *during* this moving process). If this is to be done according to the rules, we will end up with the top k rings on the third peg in the *same* order they were on the first peg. According to the induction assumption, the least possible number of moves needed to do this is $2^k - 1$. It now takes one move to transfer the bottom ring [the $(k+1)$st] from the first to the second peg. Finally, the top k rings now on the third peg must be moved to the second peg. Once again by the induction hypothesis, the least number of moves for doing this is $2^k - 1$. Therefore, the smallest total number of moves needed to transfer all $k + 1$ rings from the first to the second peg is $(2^k - 1) + 1 + (2^k - 1) = (2^k + 2^k) - 1 = 2 \cdot 2^k - 1 = 2^{k+1} - 1$. Hence, the conjecture is true for $n = k + 1$. Therefore, by induction it is true for all positive integers n.

37. *De Moivre's Theorem:* For any complex number $z = r(\cos\theta + i\sin\theta)$ and any positive integer n, $z^n = r^n[\cos(n\theta) + i\sin(n\theta)]$. *Proof:* The theorem is obviously true when $n = 1$. Assume that the theorem is true for $n = k$, that is, $z^k = r^k[\cos(k\theta) + i \sin(k\theta)]$. Then

$z^{k+1} = z \cdot z^k =$
$$[r(\cos\theta + i\sin\theta)](r^k[\cos(k\theta) + i\sin(k\theta)]).$$

According to the multiplication rule for complex numbers in polar form (multiply the moduli and add the arguments) we have:

$$z^{k+1} = r \cdot r^k[\cos(\theta + k\theta) + i\sin(\theta + k\theta)]$$
$$= r^{k+1}\{\cos[(k+1)\theta] + i\sin[(k+1)\theta]\}.$$

This statement says the theorem is true for $n = k + 1$. Therefore, by induction, the theorem is true for every positive integer n.

Chapter 12 Review, page 824

1. $-3, -1, 1, 3$ 3. $1, \dfrac{1}{4}, \dfrac{1}{9}, \dfrac{1}{16}$

5. (a) 152,600, 86,634, 48,620   (b) Feb 2002

7. $40/9$ 9. 81 11. $a_n = 9 - 6n$

13. $a_n = 6n - 11$ 15. -55

17. $a_n = 2 \cdot 3^{n-1}$ 19. $a_n = \dfrac{3}{2^{n-8}}$

21. $\dfrac{121}{4}$ 23. $8.75, 13.5, 18.25$

25. (a) $a_n = 5.123 \cdot 1.0749^{x-1}$ (b) About \$9.13 billion
    (c) About \$62.63 billion

27. 2

29. $(1.02)^{51} = (1 + 0.02)^{51} = 1^{51} + \dbinom{51}{1}1^{50} \cdot (.02) +$
$\dbinom{51}{2}1^{49} \cdot (.02)^2 +$ other positive terms $=$
$2.53 +$ other positive terms > 2.5

31. 455 33. $n + 1$ 35. 1140
37. 80

39. True for $n = 1$. If the statement is true for $n = k$,
then $1 + 5 + \cdots + 5^{k-1} = \dfrac{5^k - 1}{4}$ so that
$1 + 5 + \cdots + 5^{k-1} + 5^k = \dfrac{5^k - 1}{4} + 5^k =$
$\dfrac{5^k - 1 + 4 \cdot 5^k}{4} = \dfrac{5 \cdot 5^k - 1}{4} = \dfrac{5^{k+1} - 1}{4}$. Hence,
the statement is true for $n = k + 1$ and therefore true for all n by induction.

41. Since the statement is obviously true for $x = 0$, assume $x \ne 0$. Then the statement is true for $n = 1$. If the statement is true for $n = k$, then $|x^k| < 1$. Then $|x^k| \cdot |x| < |x|$. Thus, $|x^{k+1}| = |x^k| \cdot |x| < |x| < 1$. Hence, the statement is true for $n = k + 1$ and therefore true for all n by induction.

43. True for $n = 1$. If the statement is true for $n = k$, then
$1 + 4 + \cdots + 4^{k-1} = \dfrac{1}{3}(4^k - 1)$. Hence,
$1 + 4 + \cdots + 4^{k-1} + 4^k = \dfrac{1}{3}(4^k - 1) + 4^k =$
$\dfrac{1}{3}(4^k - 1) + \dfrac{3 \cdot 4^k}{3} = \dfrac{4^k - 1 + 3 \cdot 4^k}{3} =$
$\dfrac{4 \cdot 4^k - 1}{3} = \dfrac{1}{3}(4^{k+1} - 1)$. Hence the statement is
true for $n = k + 1$ and therefore for all n by induction.

45. If $n = 1$, then $9^n - 8n - 1 = 0$. Since $0 = 0 \cdot 8$, the statement is true for $n = 1$. If the statement is true for $n = k$, then $9^k - 8k - 1 = 8D$, so that
$9^k - 1 = 8k + 8D = 8(k + D)$. Consequently,
$9^{k+1} - 8(k + 1) - 1 = 9^{k+1} - 8k - 8 - 1 =$
$9^{k+1} - 9 - 8k = 9(9^k - 1) - 8k =$
$9[8(k + D)] - 8k = 8[9(k + D) - k]$. Thus, 8 is a factor of $9^{k+1} - 8(k + 1) - 1$ and the statement is true for $n = k + 1$. Therefore it is true for all n by induction.

CHAPTER 13

Section 13.1, page 837

1. $3/2$ 3. $1/3$ 5. -2 7. 0

9. 1 11. 0 13. 0 15. 6

17. $-1, 1,$ does not exist

19. $2,$ does not exist, 0 21. $2, 1, 1$

23. (a) 0 (b) 1 (c) -1 (d) -1

25. (a) 1 (b) 0 (c) Limit does not exist.
    (d) Limit does not exist.

27. $\displaystyle\lim_{x \to 3} f(x) = \dfrac{11}{5} = f(3)$

29. $\displaystyle\lim_{x \to 2} f(x) = \dfrac{1}{2}$, but $f(x)$ is not defined at $x = 2$.

31. The limit does not exist.

33. $\displaystyle\lim_{x \to 2^+} f(x) = 0 = f(2)$

35. The limit does not exist.

37. $\displaystyle\lim_{x \to 0} f(x) = 1$, but $f(x)$ is not defined at $x = 0$.

39. $\displaystyle\lim_{x \to 1} f(x) = 1$, but $f(x)$ is not defined at $x = 1$.

41. $\displaystyle\lim_{x \to 0} f(x) = 0$, but $f(x)$ is not defined at $x = 0$.

43. $\displaystyle\lim_{x \to \pi/2} f(x) = 0 = f(\pi/2)$

45. $\displaystyle\lim_{x \to 1} f(x) = e = f(1)$

47. (a)

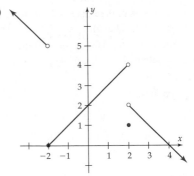

(b) $\lim_{x \to -2} f(x)$ does not exist.

(c) $\lim_{x \to 1} f(x) = 3$ (d) $\lim_{x \to 2} f(x)$ does not exist.

49. (a)

(b) 1   (c) 2
(d) Does not exist

51. The point P, where the terminal side of an angle of t radians in standard position meets the unit circle, has coordinates $(\cos t, \sin t)$. The picture below shows that as t gets closer and closer to $\pi/2$, the terminal side gets very close to the y-axis and the point P moves along the unit circle toward the point $(0, 1)$. Hence the second coordinate of P, $\sin t$, gets closer to 1.

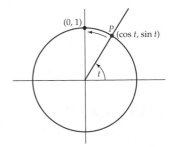

53. -1 **55.** 2; 1   **57.** $-1/3$

59. A viewing window may suggest that the limit is 0.

Section 13.2, page 846

1. 5   **3.** The limit does not exist.

5. 0   **7.** $-3/5$ **9.** 47   **11.** 7

13. $5/4$ **15.** $3/2$ **17.** $1/8$

19. The limit does not exist.

21. $\sqrt{14}$ **23.** 3   **25.** -2.5 **27.** $\dfrac{1}{2\sqrt{3}}$

29. $-1/25$ **31.** $1/2$ **33.** 0   **35.** 2

37. $\dfrac{-1}{2\sqrt{2}}$ **39.** 0   **41.** 4   **43.** 5

45. The right hand and left hand limits are not the same, hence the limit does not exist.

47. 2   **49.** $2x + 1$

51. $1/(2\sqrt{x + 1})$ **53.** $x/\sqrt{x^2 + 1}$

55. $f(x) = \dfrac{|x|}{x}$, $g(x) = \dfrac{-x}{|x|}$, $c = 0$

Special Topics 13.2.A, page 853

The symbol $\Rightarrow$ means "implies."

1. Given $\epsilon > 0$, let $\delta = \epsilon/3$. Then $0 < |x - 3| < \delta \Rightarrow$ $|x - 3| < \epsilon/3 \Rightarrow 3|x - 3| < \epsilon \Rightarrow |3(x - 3)| < \epsilon \Rightarrow$ $|3x - 9| < \epsilon \Rightarrow |(3x - 2) - 7| < \epsilon \Rightarrow |f(x) - 7| < \epsilon$

3. Given $\epsilon > 0$, let $\delta = \epsilon$. Then $0 < |x - 5| < \delta \Rightarrow$ $|x - 5| < \epsilon \Rightarrow |f(x) - 5| < \epsilon$

5. Given $\epsilon > 0$, let $\delta = \epsilon/6$. Then $0 < |x - 2| < \delta \Rightarrow$ $|x - 2| < \epsilon/6 \Rightarrow 6|x - 2| < \epsilon \Rightarrow |6(x - 2)| < \epsilon \Rightarrow$ $|6x - 12| < \epsilon \Rightarrow |(6x + 3) - 15| < \epsilon \Rightarrow |f(x) - 15| < \epsilon$

7. Given $\epsilon > 0$, let δ be any positive number. Then for every number x (including those satisfying $0 < |x - 1| < \delta$), $|f(x) - 4| = |4 - 4| = 0 < \epsilon$

9. Given $\epsilon > 0$, let $\delta = \epsilon$. Then $0 < |x - 4| < \delta \Rightarrow$ $|x - 4| < \epsilon \Rightarrow |(x - 6) - (-2)| < \epsilon \Rightarrow$ $|f(x) - (-2)| < \epsilon$

11. Given $\epsilon > 0$, let $\delta = \epsilon/2$. Then $0 < |x - (-2)| < \delta \Rightarrow$ $|x - (-2)| < \epsilon/2 \Rightarrow 2|x - (-2)| < \epsilon \Rightarrow$ $|2(x - (-2))| < \epsilon \Rightarrow |2x + 4| < \epsilon \Rightarrow$ $|(2x + 5) - 1| < \epsilon \Rightarrow |f(x) - 1| < \epsilon$

13. Given $\epsilon > 0$, let $\delta = \sqrt{\epsilon}$. Then $0 < |x - 0| < \delta \Rightarrow$ $0 < |x| < \sqrt{\epsilon} \Rightarrow |x|^2 < \epsilon \Rightarrow |x^2| < \epsilon \Rightarrow$ $|x^2 - 0| < \epsilon \Rightarrow |f(x) - 0| < \epsilon$

15. Apply Example 3 with $-g(x)$ in place of $g(x)$.

Section 13.3, page 863

1. $x = 3, x = 6$

3. Continuous at $x = -2$ and $x = 3$; discontinuous at $x = 0$

5. Continuous at $x = 0$ and $x = 3$; discontinuous at $x = -2$

7. $\lim_{x \to 3} f(x) = \lim_{x \to 3} (x^2 + 5(x - 2)^7) = \lim_{x \to 3} x^2 +$ $\lim_{x \to 3} (5(x - 2)^7) = \lim_{x \to 3} x^2 + (\lim_{x \to 3} 5)(\lim_{x \to 3} (x - 2))^7 =$ $9 + 5(3 - 2)^7 = 14 = f(3)$

9. $\lim\limits_{x \to 2} f(x) = \lim\limits_{x \to 2} \dfrac{x^2 - 9}{(x^2 - x - 6)(x^2 + 6x + 9)} =$

$$\dfrac{\lim\limits_{x \to 2} (x^2 - 9)}{\lim\limits_{x \to 2} (x^2 - x - 6) \cdot \lim\limits_{x \to 2} (x^2 + 6x + 9)} =$$

$$\dfrac{2^2 - 9}{(2^2 - 2 - 6)(2^2 + 6 \cdot 2 + 9)} = \dfrac{-5}{(-4)(25)} = \dfrac{1}{20} = f(2)$$

11. $\lim\limits_{x \to 36} f(x) = \dfrac{\lim\limits_{x \to 36} (x\sqrt{x})}{\lim\limits_{x \to 36} (x - 6)^2} =$

$$\dfrac{\lim\limits_{x \to 36} x \cdot \lim\limits_{x \to 36} \sqrt{x}}{\lim\limits_{x \to 36} (x - 6) \cdot \lim\limits_{x \to 36} (x - 6)} = \dfrac{36\sqrt{36}}{(36 - 6)(36 - 6)} =$$

$$\dfrac{36 \cdot 6}{30 \cdot 30} = \dfrac{216}{900} = \dfrac{6}{25} = f(36)$$

13. f is not defined at $x = 3$.

15. f is not defined at $x = -1$.

17. $\lim\limits_{x \to 0} f(x) = 0$, but $f(0) = 1$; hence $\lim\limits_{x \to 0} f(x) \neq f(0)$.

19. Continuous   **21.** Continuous

23. Continuous

25. Continuous at every real number except $x = 3$

27. Continuous at every real number except $x = 0$ and $x = 2$

29. f is discontinuous at $1/5, 2/5, 3/5 \cdots$ etc.

31. Generally, temperature is continuous, hence $g(x)$ is continuous everywhere on $[0, 24]$.

33. Since $\sqrt{7}$ is a solution of $x^2 - 7 = 0$, the root-finding algorithm explained after the Equation Theorem can be used to find the decimal expansion of $\sqrt{7}$.

35. Because f is not defined at $x = 1$, it is not continuous on $[-3, 3]$. Note that $f(-3) = -10$ and $f(3) = 20$. We must show that f satisfies the conclusion of the Intermediate Value Theorem: For any number k between -10 and 20, there is a number c between -3 and 3 such that $f(c) = k$.

Let k be a number between -10 and 20. First, suppose that $-10 \leq k \leq 0$ and note that $f(-2) = 0$. Since f is continuous on $[-3, -2]$ (why?), the Intermediate Value Theorem applies on that interval. Hence, there is a number c between -3 and -2 such that $f(c) = k$. Now suppose that $0 < k \leq 20$ and note that $f(2) = 0$. Since f is continuous on $[2, 3]$ (why?), the Intermediate Value Theorem applies on that interval. Hence, there is a number c between 2 and 3 such that $f(c) = k$. Thus, in all cases, there is a number c between -3 and 3 such that $f(c) = k$.

37. $b = 1$

39. $g(x) = \dfrac{1}{x + 1}$

41. $g(x) = \dfrac{1}{2 + \sqrt{x}}$

43. If $g(x) = f(x)$ for all $x \neq 0$, then $g(x) = 1$ for all $x > 0$ and $g(x) = -1$ for all $x < 0$. Thus $\lim\limits_{x \to 0} g(x)$ does not exist (see Example 8 of Section 13.1). Hence the definition of continuity cannot be satisfied, no matter what $g(0)$ is.

Section 13.4, page 874

1. -1 **3.** 0 **5.** 0 **7.** 0

9. No vertical asymptotes; $\lim\limits_{x \to \infty} f(x) = \infty$; $\lim\limits_{x \to -\infty} f(x) = 0$

11. No vertical asymptotes; $\lim\limits_{x \to \infty} f(x) = 2$; $\lim\limits_{x \to -\infty} f(x) = -1$

13. Vertical asymptote $x = -10$; $\lim\limits_{x \to \infty} f(x) = \infty$; $\lim\limits_{x \to -\infty} f(x) = -\infty$

15. $y = 3/4$ **17.** $y = -2$

19. No horizontal asymptote   **21.** $1/2$

23. ∞ **25.** 5 **27.** 2 **29.** $-\sqrt{3}/2$

31. $\sqrt{2}/3$ **33.** $-\sqrt{3}$ **35.** 1 **37.** -1

39. 0 **41.** 0 **43.** 1

45. (a) 1 **47.** 1
(b) 1

49. The first part of the informal definition is included in the second part, which says "the values of $f(x)$ can be made arbitrarily close to L by taking large enough values of x." This means that whenever you specify how close $f(x)$ should be to L, we can tell you how large x must be to guarantee this. In other words, you specify how close you want $f(x)$ to be to L by giving a positive number ϵ and we tell you how large x must be to guarantee that $f(x)$ is within ϵ of L, that is, to guarantee that $|f(x) - L| < \epsilon$. We do this by giving a positive number k such that $|f(x) - L| < \epsilon$ whenever $x > k$. This can be reworded as follows: For every positive number ϵ, there is a positive number k (depending on ϵ) with this property:

$$\text{If } x > k, \text{ then } |f(x) - L| < \epsilon.$$

51. (a) 2.7182818 (b) 2.718281828 (c) e

(d) They are the same; substitute $z = \dfrac{1}{x}$ for x.

Chapter 13 Review, page 876

1. 2 **3.** About $.23026$ **5.** 2

7. 0 **9.** -5 **11.** -1

13. -2 **15.** $1/2$ **17.** 0

19. 1 **21.** 4 **23.** 4

25. Given $\epsilon > 0$, let $\delta = \epsilon/2$. Then $0 < |x - 3| < \delta \Rightarrow |x - 3| < \epsilon/2 \Rightarrow 2|x - 3| < \epsilon \Rightarrow |2(x - 3)| < \epsilon \Rightarrow |2x - 6| < \epsilon \Rightarrow |(2x + 1) - 7| < \epsilon \Rightarrow |f(x) - 7| < \epsilon$

27. Continuous at $x = -3$; Discontinuous at $x = 2$

29. Continuous at $x = 4$; Continuous at $x = \sqrt{8}$

31. Discontinuous at $x = -4$; Continuous at $x = 0$

33. (a) $\lim\limits_{x \to 2} f(x) = \lim\limits_{x \to 2} \dfrac{x^2 - x - 6}{x^2 - 9} =$
$\dfrac{\lim\limits_{x \to 2}(x^2 - x - 6)}{\lim\limits_{x \to 3}(x^2 - 9)} = \dfrac{2^2 - 2 - 6}{2^2 - 9} = \dfrac{-4}{-5} = \dfrac{4}{5} = f(2)$

(b) f is not defined at $x = 3$ and hence is discontinuous there.

35. Vertical asymptotes at $x = -1$ and $x = 2$. The graph moves upward as x approaches -1 from the left and downward as x approaches -1 from the right. The graph moves downward as x approaches 2 from the left and upward as x approaches 2 from the right.

37. The limit does not exist.

39. 0   **41.** 1/2   **43.** 10/3   **45.** $y = 1/2$

APPENDIX 1

Section 1.A, page 885

1. 36   **3.** 73   **5.** -5
7. $-125/64$ **9.** 1/3   **11.** -112
13. $81/16$ **15.** $-211/216$ **17.** $129/8$
19. x^{10} **21.** $.03y^9$ **23.** $24x^7$
25. $9x^4y^2$ **27.** $-21a^6$ **29.** $384w^6$
31. ab^3 **33.** $8x^{-1}y^3$ **35.** a^8x^{-3}
37. $3xy$ **39.** 2^{12} **41.** 2^{-12}
43. x^7 **45.** ce^9 **47.** $b^2c^2d^6$
49. $a^{12}b^8$ **51.** $1/(c^{10}d^6)$ **53.** $1/(108x)$
55. a^7c/b^6 **57.** c^3d^6 **59.** $a + \dfrac{1}{a}$
61. Negative   **63.** Negative   **65.** Negative
67. 3^s **69.** $a^{6t}b^{4t}$ **71.** b^{rs+st}/c^{2rt}

73. Many possible examples, including $3^2 + 4^2 = 9 + 16 = 25$, but $(3 + 4)^2 = 7^2 = 49$

75. Many possible examples, including $3^2 \cdot 2^3 = 9 \cdot 8 = 72$; but $(3 \cdot 2)^{2+3} = 6^5 = 7776$

77. Many possible examples, including $2^6/2^3 = 64/8 = 8$, but $2^{6/3} = 2^2 = 4$

79. False for all nonzero a; for instance, $(-3)^2 = (-3)(-3) = 9$, but $-3^2 = -9$

Section 1.B, page 889

1. $8x$ **3.** $-2a^2b$
5. $-x^3 + 4x^2 + 2x - 3$ **7.** $5u^3 + u - 4$
9. $4z - 12z^2w + 6z^3w^2 - zw^3 + 8$
11. $-3x^3 + 15x + 8$ **13.** $-5xy - x$

15. $15y^3 - 5y$
17. $12a^2x^2 - 6a^3xy + 6a^2xy$
19. $12z^4 + 30z^3$ **21.** $12a^2b - 18ab^2 + 6a^3b^2$
23. $x^2 - x - 2$
25. $2x^2 + 2x - 12$ **27.** $y^2 + 7y + 12$
29. $-6x^2 + x + 35$ **31.** $3y^3 - 9y^2 + 4y - 12$
33. $x^2 - 16$ **35.** $16a^2 - 25b^2$
37. $y^2 - 22y + 121$ **39.** $25x^2 - 10bx + b^2$
41. $16x^6 - 8x^3y^4 + y^8$ **43.** $9x^4 - 12x^2y^4 + 4y^8$
45. $2y^3 + 9y^2 + 7y - 3$
47. $-15w^3 + 2w^2 + 9w - 18$
49. $24x^3 - 4x^2 - 4x$ **51.** $x^3 - 6x^2 + 11x - 6$
53. $-3x^3 - 5x^2y + 26xy^2 - 8y^3$
55. 3   **57.** -6 **59.** 6   **61.** 1   **63.** 5
65. $x - 25$ **67.** $9 + 6\sqrt{y} + y$
69. $\sqrt{3}x^2 + 4x + \sqrt{3}$ **71.** $3ax^2 + (3b + 2a)x + 2b$
73. $abx^2 + (a^2 + b^2)x + ab$
75. $x^3 - (a + b + c)x^2 + (ab + ac + bc)x - abc$
77. $16x^{n+k}$
79. $y^{r+s} - 4y^r + y^s - 4$ **81.** $4x^2 + 32x$

83. Example: if $y = 4$, then $3(4 + 2) \neq (3 \cdot 4) + 2$; correct statement: $3(y + 2) = 3y + 6$

85. Example: if $x = 2$, $y = 3$, then $(2 + 3)^2 \neq 2 + 3^2$; correct statement: $(x + y)^2 = x^2 + 2xy + y^2$

87. Example: if $x = 2$, $y = 3$, then $(7 \cdot 2)(7 \cdot 3) \neq 7 \cdot 2 \cdot 3$; correct statement: $(7x)(7y) = 49xy$

89. Example: if $y = 2$, then $2 + 2 + 2 \neq 2^3$; correct statement: $y + y + y = 3y$

91. Example: if $x = 4$, then $(4 - 3)(4 - 2) \neq 4^2 - 5 \cdot 4 - 6$; correct statement: $(x - 3)(x - 2) = x^2 - 5x + 6$

93. If x is the chosen number, then adding 1 and squaring the result gives $(x + 1)^2$. Subtracting 1 from the original number x and squaring the result gives $(x - 1)^2$. Subtracting the second of these squares from the first yields: $(x + 1)^2 - (x - 1)^2 = (x^2 + 2x + 1) - (x^2 - 2x + 1) = 4x$. Dividing by the original number x now gives $\dfrac{4x}{x} = 4$. So the answer is always 4, no matter what number x is chosen.

95. Many correct answers.

Section 1.C, page 893

1. $(x + 2)(x - 2)$ **3.** $(3y + 5)(3y - 5)$
5. $(9x + 2)^2$ **7.** $(\sqrt{5} + x)(\sqrt{5} - x)$
9. $(7 + 2z)^2$ **11.** $(x^2 + y^2)(x + y)(x - y)$

13. $(x + 3)(x - 2)$ **15.** $(z + 3)(z + 1)$

17. $(y + 9)(y - 4)$ **19.** $(x - 3)^2$

21. $(x + 5)(x + 2)$ **23.** $(x + 9)(x + 2)$

25. $(3x + 1)(x + 1)$ **27.** $(2z + 3)(z + 4)$

29. $9x(x - 8)$ **31.** $2(x - 1)(5x + 1)$

33. $(4u - 3)(2u + 3)$ **35.** $(2x + 5y)^2$

37. $(x - 5)(x^2 + 5x + 25)$

39. $(x + 2)^3$ **41.** $(2 + x)(4 - 2x + x^2)$

43. $(-x + 5)^3$ **45.** $(x + 1)(x^2 - x + 1)$

47. $(2x - y)(4x^2 + 2xy + y^2)$

49. $(x^3 + 2^3)(x^3 - 2^3) =$
$(x + 2)(x^2 - 2x + 4)(x - 2)(x^2 + 2x + 4)$

51. $(y^2 + 5)(y^2 + 2)$ **53.** $(9 + y^2)(3 + y)(3 - y)$

55. $(z + 1)(z^2 - z + 1)(z - 1)(z^2 + z + 1)$

57. $(x^2 + 3y)(x^2 - y)$ **59.** $(x + z)(x - y)$

61. $(a + 2b)(a^2 - b)$

63. $(x^2 - 8)(x + 4) = (x + \sqrt{8})(x - \sqrt{8})(x + 4)$

65. If $x^2 + 1 = (x + c)(x + d) = x^2 + (c + d)x + cd$, then $c + d = 0$ and $cd = 1$. But $c + d = 0$ implies that $c = -d$ and hence that $1 = cd = (-d)d = -d^2$, or equivalently, that $d^2 = -1$. Since there is no real number with this property, $x^2 + 1$ cannot possibly factor in this way.

Section 1.D, page 900

1. $\dfrac{9}{7}$ **3.** $\dfrac{195}{8}$ **5.** $\dfrac{x - 2}{x + 1}$

7. $\dfrac{a + b}{a^2 + ab + b^2}$ **9.** $\dfrac{1}{x}$ **11.** $\dfrac{29}{35}$

13. $\dfrac{121}{42}$ **15.** $\dfrac{ce + 3cd}{de}$ **17.** $\dfrac{b^2 - c^2}{bc}$

19. $\dfrac{-1}{x(x + 1)}$ **21.** $\dfrac{x + 3}{(x + 4)^2}$ **23.** $\dfrac{2x - 4}{x(3x - 4)}$

25. $\dfrac{x^2 - xy + y^2 + x + y}{x^3 + y^3}$

27. $\dfrac{-6x^5 - 38x^4 - 84x^3 - 71x^2 - 14x + 1}{4x(x + 1)^3(x + 2)^3}$

29. 2 **31.** $2/(3c)$ **33.** $3y/x^2$ **35.** $\dfrac{12x}{x - 3}$

37. $\dfrac{5y^2}{3(y + 5)}$ **39.** $\dfrac{u + 1}{u}$ **41.** $\dfrac{(u + v)(4u - 3v)}{(2u - v)(2u - 3v)}$

43. $\dfrac{35}{24}$ **45.** $\dfrac{u^2}{vw}$ **47.** $\dfrac{x + 3}{2x}$

49. $\dfrac{x^2 y^2}{(x + y)(x + 2y)}$ **51.** $\dfrac{cd(c + d)}{c - d}$

53. $\dfrac{y - x}{xy}$ **55.** $\dfrac{-3y + 3}{y}$

57. $\dfrac{-1}{x(2x + 2h)}$ **59.** $\dfrac{xy}{x + y}$

61. $\dfrac{\pi}{4} \approx 0.785$, or a 78.5% chance

63. $\dfrac{1}{25} = 0.04$, or 4% chance

65. Example: if $a = 1$, $b = 2$, then $\dfrac{1}{1} + \dfrac{1}{2} \neq \dfrac{1}{1 + 2}$;
correct statement: $\dfrac{1}{a} + \dfrac{1}{b} = \dfrac{b + a}{ab}$

67. Example: if $a = 4$, $b = 9$, then
$\left(\dfrac{1}{\sqrt{4} + \sqrt{9}}\right)^2 \neq \dfrac{1}{4 + 9}$; correct statement:
$\left(\dfrac{1}{\sqrt{a} + \sqrt{b}}\right)^2 = \dfrac{1}{a + 2\sqrt{ab} + b}$

69. Example: if $u = 1$, $v = 2$, then $\dfrac{1}{2} + \dfrac{2}{1} \neq 1$; correct
statement: $\dfrac{u}{v} + \dfrac{v}{u} = \dfrac{u^2 + v^2}{vu}$

71. Example: if $x = 4$, $y = 9$, then
$(\sqrt{4} + \sqrt{9}) \cdot \dfrac{1}{\sqrt{4} + \sqrt{9}} \neq 4 + 9$; correct statement:
$(\sqrt{x} + \sqrt{y}) \cdot \dfrac{1}{\sqrt{x} + \sqrt{y}} = 1$

APPENDIX 2, PAGE 910

1. $b = 2\sqrt{5}$ **3.** $c = 3\sqrt{7}$

5. $c = 5$, $b = 5\sqrt{2}$ **7.** $c = 5$, $b = 5\sqrt{3}$

9. $\angle A = 53.13°$, $\angle R = 36.87°$, $r = 3$, $s = 4$, $t = 5$

11. Since $\angle A = 65° = \angle T$, $\angle R = 25° = \angle C$, and $AC = RT = 5$, the triangle ABC is congruent to triangle TSR by ASA.

13. It is given that $\angle POD = \angle QOD$, evidently $OD = OD$, and since $\angle PDO = \angle QDO = 90°$, triangle POD is congruent to triangle QOD by ASA. Hence, $PD \cong DQ$

15. Since it is given that $AB = DC$ and $\angle BAC = \angle ACD$, and since evidently $AC = AC$, triangle BAC is congruent to triangle ACD by SAS.

17. Since $AC = AC$, the triangles ABC and ADC are congruent by ASA.

19. $c = 7.5$, $r = 28$, $t = 35$

21. $r = 3.5$, $t = 3$

23. Since $AB = AC$, $\angle B = \angle C = 60°$. Similarly $\angle S = \angle T = 60°$. Hence the triangles are similar by AAA. Since triangle ABC is equiangular, it must also be equilateral, and so $a = 15$.

25. Since $BC/ST = AC/SR$ and $\angle C = \angle S$, triangle ABC is similar to triangle RTS. Then $\angle A = \angle R = 42.54°$, and $\angle T = 62.46°$.

27. Since $a/b = r/s$ and $\angle C = \angle T$, triangle ABC is similar to triangle RTS. Then $\angle A = \angle B = \angle R = \angle S = 45°$, $c = 5\sqrt{2}$, and $t = 17.5\sqrt{2}$.

29. Since two angles in triangles ABC and RTS are equal, the third must also be equal, and the triangles must be similar. In particular, $\angle C = \angle S = 41.5°$, and $s = 2.4$.

31. 55 ft

33. 80 ft

35. Marietta is 92.4 miles from Columbus and 171.2 miles from Cincinnati.

37. Since $CD = SB$, $CS = DB$, and $\angle S = \angle D$, triangle CSB is congruent to triangle BDC. Therefore $\angle y = \angle v$. Similarly, $\angle u = \angle x$. Now since RCS is a straight line, $\angle x + \angle w + \angle y = 180°$. Hence $\angle u + \angle w + \angle v = 180°$.

INDEX
of Applications

Biology and Life Science

Animal feed, 773
Apple orchard yield, 265
Bacteria population, 158, 190, 253, 346, 353, 386, 392
Beaver population, 352, 392
Beef and poultry consumption, 369, 769
Blood alcohol content, 133, 391
Blood flow, 201
Carbon-14, half-life, 353, 385
CAT scans, 756
Cell multiplication, 352
Chipmunk population, 202
Death rate, 732
Disease rate, 116, 409–410
Flu virus, spread of, 369, 392
Fruit fly population, 219, 351
Fungus growth, 190
Heart disease, 60, 733
HIV/AIDS, 27, 46, 203, 268–269, 800
Infant mortality rates, 400
Life expectancy, 67, 351, 402, 793
Medicine, concentration in bloodstream, 82, 286
Population growth, 27, 67, 81–82, 85–86, 114, 194, 200, 267, 333, 345, 348, 349, 353, 369, 387, 391, 392, 393–395, 397, 400, 407, 732, 771, 800
Precipitation, 467
Rabbit colony population, 787
Seedling survival rate, 322
Single-parent families, 134
Smoking, 160, 271, 408
Tree frog population, 352
Water boiling, 60, 68
Weather bureau, 864
Weight, 15, 35, 60, 423

Business and Manufacturing

Advertising expenses, 200, 201, 265, 271, 369
Apartment rentals, 230–231
Book sales, 787
Calculator sales, 118, 320
Can design, 103, 265
Candy gift boxes, 770
Car importing, 735

Business and Manufacturing
(*continued*)

Car production, 59
Car sales, 735
Cost function, 52–53, 60, 99
Departmental expenses, 15
Depreciation, 52, 60, 61
Digital devices, 353
Drug approval, FDA, 114–115
Economics, 27, 774–775
Electricity, 33, 35
Express Mail rates, 215
Gross domestic product, 114
Jewelry production, 118
Manufacturing costs, 104, 118, 134, 141–142, 144, 165, 176, 265, 286, 287, 301, 322, 733, 734
Manufacturing revenue, 104
Marketing costs of prescription drugs, 119
Personal digital video sales, 44
Pottery, 235, 735
Production time, 756, 773
Profits, 77, 81, 202, 233, 265, 793
Research and development costs, 114
Retail costs, 734, 770
Revenues, 130
Sales calls, 234
Sales of electronics and appliances, 770
Sales related to advertising, 265, 369
Sewing machines, 720
Train design, 487–488
Truck expenses, 286–287
Vans, loading, 755
Winemaking, 82

Chemistry and Physics

AC current, 525
Air volume, 800
Alloy mixture, 730–731, 733, 771
Antifreeze, 94
Atmospheric pressure, 27, 60, 351, 368–369, 392
Black holes, 878–881
Cable length, 571
Carbon dioxide, 351, 391, 769
Comets, 714
Concentration of alcohol, 287

Chemistry and Physics
(*continued*)

Earthquakes, 375, 407
Evaporation of a puddle, 187, 190
Fahrenheit vs. Celsius, 68
Flashlight, 672
Focal length of camera, 287
Folding paper, 352
Force, 35
Gas concentration, 59
Gas pressure, 35
Gas volume, 35
Gravity, 287
Half-lives, 347–348, 353–354, 407
Loran transmitters, 658–660
Object on inclined plane, 616–617, 619, 627–628, 630, 633
Orbit of planets, 333, 396, 540–541, 648, 651
Pendulum, 14, 35
Purifying water, 346–347
Radioactive decay, 391
Radio telescope, 669, 672
Sea level, height above, 368–369, 558–559
Solutions, mixing, 94, 99, 301, 733, 755
Stack of rings puzzle, 822
Vapor pressure of water, 114
Wagon pulling, 630
Water balloon volume, 199–200
Water pressure, 34
Weather balloon volume, 191, 193
Weight hanging from cable, 620
Weight hanging from spring, 35, 462–463, 466, 536, 539
Winemaking, 733

Construction

Antenna tower, 569
Arch, 717
Bridge design, 590
Building a monument, 794
Building height, 909
Bus shelter, 100
Cable lengths, 233
Cement walk, 95–96, 99
Drawbridge, 562–563, 564
Fencing, 104, 118, 145, 230, 234, 287, 320, 581, 651

Construction

(*continued*)

Floor tile, 728
Footbridge, 651
Gate, swinging, 571
Gutters, 234
Hinged crane, 582
Ladder angle, 99, 559, 562
Ladder rung length, 794
Painting, 585, 773
Pedestrian overpass, 564
Pole, 581, 588
Ramps, 58, 562, 627–628, 633
Road construction, 581, 587, 588
Sinkhole, 572
Subdivision, 301
Surveying, 580, 588, 634–635
Suspension bridge, 672
Towers, 571, 580, 587
Traffic flow, 757
Tunnel, 565
Window size, 118
Wire, cutting into three sections, 741
Wire, stretched, 560, 562, 563
Wiring costs, 341

Consumer Affairs

Cable TV subscribers, 352, 800
Car leases, 301
Car loans, 787
College graduates as percentage of labor
    force, 734, 800
Commission income, 144, 301
Computer monitor size, 65
Computer sales, 77, 299
Consumer confidence level, 161–162
Consumer debt, 391
Consumer price index (CPI), 164
Consumer spending, 787
Credit cards, 353, 752–753
Depreciation, 800
Disposable income, 114
Eggs, consumption of, 773
Food sold by the pound, 733, 756
Freezer costs, 301
Health care, 115, 787, 793
Health care costs, 54–55
Home sale prices, 65, 219
Home video games, 773
Income levels, men and women, 732
Internet use, 115, 353, 794
IRA, 44
Life insurance, 114, 118–119, 800
Loans, 99, 755
Nursing home costs, 770
Postage stamps, 156, 864
Poverty levels, 60, 108–109

Consumer Affairs

(*continued*)

Prescription drug cost, 116
Prices, 60
Salaries, 99, 115, 301, 352, 392, 405, 793,
    800
Sales, 200
Satellite TV, 672, 717, 800
Taxes, 34, 125, 133, 136, 141, 144
Taxi fares, 864
Ticket costs, 61, 235, 733, 734, 755, 824
Unemployment rates, 111, 467
Wages, 119, 200, 203
Wireless telephone, 160

Financial

Car leases, 301
Compound interest, 353, 355–359, 391
Inflation, 353
Input-output analysis, 774–775
Interest income, 301, 359, 822
Interest rates, 99, 133, 344, 355–359, 391,
    407, 755
Investments, 118, 201, 355–358, 368, 734,
    755
IRA, 44
Mortgages, 159
National debt, 163, 196, 271, 401
Stock dividends, 93–94
Stock prices, 81
Stock returns, 99

Geometry

Baseball, 23, 38–39, 45, 234, 517, 688,
    766–767
Baseball diamond, 571
Boxes, 92, 96–98, 100, 102, 104, 118, 133,
    145, 157, 234, 253–254, 265, 283–284,
    286, 741
Building height, 909
Cylinders, 104, 133, 158, 286
Flagpole, 560–561, 566, 587, 913, 914
Football, 45, 692, 824
Gateway Arch, 350, 352
Glass dimensions, 565
Gravity, 23
Hands of a clock, 419, 421
Hinged crane, 582
Horizon, 589–590
Kite flying, 563
Latitudes, 422
Lot size, 585
Ohio Turnpike, 563
Pyramid, 340, 913
Rectangles, 157, 301
Rope stretching, 340

Geometry

(*continued*)

Sears Tower and Smith Tower, 8
Statue, 564, 581
Swimming pool, 563, 733
Wheelchair ramp, 58
Wheels, 422–423, 461, 465

Miscellaneous

Accidental deaths, 110, 391
AM radio transmitter, 465
Ancestors, 800
Audio feedback, 220–223
Ball game ticket sales, 729
Balloon launches, 66
Basketball tickets, 61
Camera angle, 526
Cannon, 539
Car accident, 46, 391, 402
Car racing, 66, 202, 217
Charitable donations, 119–120
Circus animals, 619, 770
Civil War, 233
Computers in schools, 401, 799
Crime rate, 270
Crutches, length of, 144
Daylight hours, 536
Ferris wheel, 465, 694
Forest fire, 580
Games of chance, 826–827
Golfing, 686–687, 692
Guns, 233
Household number, 402
Joggling board, 234–235
Kerosene, filtering, 353
Lawn mower, pushing, 630
Lighting, 32–33, 341, 526, 535, 561–562,
    563, 564, 672
Lightning and thunder, 35
Military budget, 159
Military troop levels, 151
Murder rates, 253
Probability theory, 826–827
Radio listening hours, 787
Radio station formats, 145
Rumor spreading, 407
Skeet shooting, 691
Sledding, 628, 633
Snowboarding, 66, 216
Superfund Trust balance, 82
Tank, emptying, 61, 202
Temperatures, 126, 164, 194–195, 203, 212,
    270, 408, 463, 467
Typing speed, 392
Unmarried couples, 60
Wind-chill factor, 408

Student Issues

College enrollment, 115, 402
Exam average, 369
Experimental scores, 406
Financial aid, 37, 116
Home schooling, 401
Lecture tickets, 773
SAT scores, 120
Scholarships, 134–135
School costs, 322
School enrollment, 200, 270, 271
Seats in lecture hall, 794
Tuition rates, 44, 145

Time and Distance

Airplane travel, 30, 94–95, 99, 158, 535,
    563, 564, 565, 566, 571, 572, 577, 579,
    582, 614, 617, 619, 633, 729–730, 733

Time and Distance
(*continued*)

Arrow, 692
Balls, 301, 691, 694, 798–799, 800
Batman, 563
Bicycles, 99, 420–421, 423
Boat travel, 572
Bullets, 234
Buoy, 564
Car brakes, 233
Car travel, 65, 99, 118, 129, 145, 165, 188,
    200, 322
Falling objects, 35, 100, 126, 134, 136, 191,
    192, 197, 199, 219, 301, 800
Helicopter, 340
Hiking time, 865
Jogging, 165
Maps, 914
Merry-go-round, 420, 423, 430–431, 466
Model plane, 525

Rivers, 337–339, 558, 562, 582, 619
Rockets, 27, 81, 234, 301, 320, 564
Rope and pulley, 573
Satellite orbit, 572, 651
Ships, 571, 587, 588, 733, 914
Sound waves, traveling, 662
Spotlight, 672
Train travel, 99, 322, 568, 587
Trampoline, 158
Tree tops, 581, 913
Walking, 145, 191

SUBJECT INDEX

A

Absolute value, 8–10
  of complex number, 593
  equations, 28–30
  function, 149–150, 155, 214
  inequalities, 302–305
Acute angle, 553–554, 902
Addition
  identities, 498–506
  of vectors, 610–611, 612
Adjacent side, 545–546
Algebra
  fundamental theorem of, 312
  identities and, 437–444
  review, 882–901
  of trigonometric equations, 531–533
  of trigonometric functions, 476–478
Algebraic
  equations, solving, 18–26, 89
  exponential and logarithmic equations,
    solving, 382–392
  expression, 886–890
  inverse functions, finding, 206
  linear and quadratic inequalities, methods
    of solving, 302–305
  linear equations in two variables, meth-
    ods of solving, 724–727
  nonlinear equations, methods of solving,
    735–736
  test for symmetry, 179
Algorithm
  division, 237
  solution for trigonometric equations,
    529–531
Ambiguous case, 575
Amplitude, 457–458, 460
Analytic geometry, 636–720
Angle(s), 412–423
  acute, 553–554, 902
  central, 418
  coterminal, 413, 417
  degree measure, 413, 544–545
  direction, 613–616
  double—identities, 510–511
  of elevation, 559–562
  half—identities, 512–514
  of inclination, 506–507
  initial side, 412, 555
  between lines, 507–509
  negative, 413
  negative—identities, 442–443, 545

Angle(s) (*continued*)
  obtuse, 902
  positive, 413
  radian measure, 414–417
  right, 553, 902
  rotation, 413, 679
  in standard position, 413, 555
  supplementary—identity, 576
  terminal side, 412, 555
  theorem, 622
  between vectors, 622–624
  vertex, 412, 902
Angular speed, 419–421
Applications. *See* Applications Index
Applied problems, setting up, 90–93
Arc
  length, 418–419
  subtended by angle, 418
Arccosine function, 521–522
Arches, 324
Archimedean spiral, 701
Arcsine function, 518–520
Arctangent function, 523–524
Argument, 593
Arithmetic
  algebraic expressions, 886–890
  described, 6
  polynomial, 236–238
  progression, 788–794
  vector, 609–613
Associative law, 887
Asymptote
  horizontal, 273, 280, 869
  of hyperbola, 652
  vertical, 273, 278, 867
Attracting fixed point, 221
Augmented matrix, 745
Axis
  coordinate, 36
  of ellipse, 639
  of hyperbola, 652
  imaginary, 592
  of parabola, 226, 663
  polar, 695
  real, 592
  rotation of, 676–681

B

Back substitution, 744
Base
  change of—formula, 381

Base (*continued*)
  logarithmic functions to, 375–383
Basic
  principles for solving equations, 18–26
  trigonometric equations, 527–531
  trigonometric identities, 490–498,
    510–517
Bifurcation, 222
Big-Little Principle, 272
Binomial
  coefficient, 806–807
  theorem, 806–813, 819–820, 826–827
Bound(s), 249–252
  lower, 251
  test, 251
  upper, 251

C

Cancellation property, 864
Cardioid, 699, 701
Cartesian coordinate system, 36
Catenary, 349–350
Celsius, 68
Center
  of circle, 42
  of ellipse, 639
  of hyperbola, 652
Central angle, 418
Change
  average rate of, 192–196
  of base formula, 381
  difference quotient and rate of, 198
  instantaneous rate of, 198–200
  in x or y, 48
Circle, 42–43
  as conic section, 637
  equations of, 638–639
  parametric equations for, 639
  polar equations for, 702
  unit, 433–435
Closed interval, 4, 859
Coefficient
  binomial, 806–807
  of correlation, 109
  leading, 236
  of polynomial, 235, 314–316
Cofunction identities, 501–502
Column of matrix, 745
Common
  denominator, 895–896
  difference, 789

Common (*continued*)
  factor, 890
  logarithm, 360–362, 365
  ratio, 794
Common logarithm, 377
Commutative law, 887
Complete graph, 77–80
Completing the square, 20–21
Complex number(s), 305–311
  absolute value of, 593
  conjugate of, 307
  factorization over, 312–313
  imaginary part of, 311
  modulus of, 593–594
  parallelogram rule, 597
  polar form, 593–598
  real part of, 311
  standard form, 306
  trigonometric form, 593–598
Complex plane, 592
Component(s)
  of **u** along **v**, 626–627
  of vector, 608
Composite function, 184, 860
Composition of functions, 183–188
Compound interest formula, 354–360
Compressed trigonometric graphs, 470–473
Comp$_\mathbf{v}\mathbf{u}$, 626
Congruent triangles, 905–907
Conic section(s), 637, 673
  degenerate, 637
  parametric equations of, 719–720
  polar equations for, 704–714
  standard equations for, 705–706,
    708–709
Conjugate
  complex numbers, 307
  roots theorem, 314
  solutions, 310
Constant
  algebraic expressions, 886
  function, 151–152, 214
  limits, 841, 870
  polynomial, 235
  of proportionality, 30
  sequence, 779
  term, 235
  of variation, 30
Continuous
  compounding interest, 357–358
  of functions, 854–865
  graph, 255–256, 278
  of rational graphs, 278
Contraction of graph, 169–170
Convention, domain, 127
Convergent series, 802
Conversion
  degree/radian, 417–418
  polar/rectangular coordinates, 697
Coordinate(s)
  conversion, 697

Coordinate(s) (*continued*)
  plane, 36–48
  of point, 36
  polar, 695–704
  rectangular, 36
  test for symmetry, 179
Correlation
  coefficient, 109
  negative, 110
  positive, 110
Cosecant function, 474–475
  graph, 478–479
  period, 476
Cosine function, 423, 424, 431–433, 434
  amplitude, 457–458
  graph, 447–449
  inverse, 521–522
  period, 441, 455–457
  phase shift, 458–461
  point-in-plane description, 475–476,
    544
  range, 449
  right triangle description, 554
Cosines, law of, 566–573
Cos *t*. *See* Cosine function
Cotangent function, 474–475
  period, 476
Coterminal angles, 413, 417
Cot *t*. *See* Cotangent function
Csc *t*. *See* Cosecant function
Cube root of 1/unity, 310
Cubic factoring patterns, 891
Cubic function, graphing, 147, 155, 214
Cubic regression, 267, 269
Curve, plane, 681–694
Cycloid, 689

D

Damped trigonometric graphs, 470–473
Data points, 112
Decay
  exponential, 346–348
  radioactive, 347
Decimal
  repeating, 16
  representation of real numbers, 16–18
  window, 74, 282
Decreasing
  on an interval, 151–152
  function, 205
Degenerate conic section, 637
Degree, 413, 902
  conversion to radians, 417–418
  measuring angles, 544–545
  of polynomial, 236
Demand curve, 121–123
DeMoivre's Theorem, 598–599
Denominator
  common, 895–897
  rationalizing the, 331–332
Dependent system, 727, 751

Depression, angle of, 559–562
Derivative, 846
Difference
  common, 789
  of cubes, 891
  of functions, 182
  quotient, 139, 196–198, 845–846
  of squares, 891
Direction angle, 613–616
Directrix, 708
Direct variation, 30
Discriminant, 23–24, 674–675
Distance
  difference, 652
  formula, 38–40
  on number line, 10–13
  -rate-time formula, 94–95
Distributive law, 6, 887, 888
Dividend, 238
Division
  algorithm, 237
  with exponents, 883
  polar, 595–596
  of polynomials, 238–241, 243–246
  synthetic, 239, 243–246
Divisor, 238
Domain, 127
  of composite function, 185
  convention, 140–141
  of cotangent, secant, and cosecant func-
    tions, 474
  of rational function, 272
  of tangent function, 425
Dot
  graphing mode, 80
  product, 621–630
Double angle identities, 510–511

E

Eccentricity, 651, 704
Echelon form matrix, 747, 750
E (irrational number)
  continuous compounding and, 357–358
  natural exponent function, 348
Elementary operations, 743
Elevation, angle of, 559–562
Eliminating
  the parameter, 685
  parentheses, 887–888
Elimination method, 725–727
Ellipse, 637
  eccentricity of, 651, 704
  equation of, 639–647
  polar equation of, 704, 706
  vertical and horizontal shifts,
    643–646
Endpoint of interval, 4
Equality
  of fractions, 895
  of matrices, 758
  rule for polar form, 598

Equation memory, 72–73
Equation(s)
   absolute value, 28–30
   applications of, 90–100
   of circle, 42, 638–639
   of ellipse, 639–647
   equivalent systems of, 742
   exponential, 383–387
   first-degree, 19
   functions defined by, 128–129
   graphical solution, 70
   graph of second-degree, 673–675
   of hyperbola, 652–657
   intersection method solution, 84–86
   large systems of linear equations,
     746
   linear, 19
   logarithmic, 387–390
   memory, 72–73
   nonlinear systems of, 735–741
   of parabola, 663–672
   parametric, 154, 639
   point-slope, 53–55
   polar, 704–714
   polynomial, 24–26
   principles for solving, 18–26
   quadratic, 20–24
   radical, 334–341
   rotation, 676–681
   slope-intercept, 50–53
   solver, 86–89
   theory of, 312–317
   trigonometric, 527–536
Equation theorem, 861
Equilibrium point, 122, 728
Equilibrium price, 122, 728
Equilibrium quantity, 728
Equivalent
   equation, 18
   system of equations, 742
   vectors, 607–608
Escape velocity, 878–881
EVAL key, 138
Even function, 179–180
Event horizon, 879
Expansion
   binomial, 811
   of graph, 169–170
Exponent
   integral, 882–886
   irrational, 332
   laws, 329, 438, 884
   negative, 883
   rational, 328–330
Exponential
   decay, 346–348
   equations, 383–387
   function, 341–354, 365, 404
   growth, 344–346
   model, 392–396, 409–410
   natural, function, 348

Expression
   algebraic, 886–890
   fractional, 894–901
Extended principle of mathematical induction, 821
Extremum, local, 258

F
Factor
   common, 890
   theorem, 240, 248–249
Factorial, n, 806
Factoring, 890–894
   identities, 515
   patterns, 891
   quadratic equations, 20
Factorization
   over complex numbers, 312–313
   over real numbers, 315
Factors and remainders, 238
Fahrenheit, 68
Fibonacci sequence, 781, 787
Finder
   graphical intersection, 85
   graphical root, 83, 89
   maximum/minimum, 76–77, 79
First-degree equation, 19
Fixed point, 221
Flywheels, 487–488
Focus (foci)
   of conic section, 708
   of ellipse, 639
   of hyperbola, 652
   of parabola, 663
FOIL, 888
Formula
   compound interest, 354–360
   distance, 38–40
   Herron's, 583–584
   midpoint, 40
   quadratic, 22
   simple interest, 93–94
FRAC key, 17
Fraction
   conversion program, 915
   denominator, 331–332, 895–897
   identity, proving, 496–497
   numerator, 331–332
   partial, 752
   rewriting, 493
Fractional expression, 894–901
Functional notation, 136–145
Function(s), 124–135
   absolute value, 149–150, 155, 214
   amplitude, 458
   arccosine, 521–522
   arcsine, 518–520
   arctangent, 523–524
   average rate of change, 192–193
   composite, 184, 860
   composition of, 183–188

Function(s) (*continued*)
   constant on an interval, 151–152, 214
   continuous, 854–865
   cosecant, 474–475, 476, 478–479
   cosine, 423, 424, 447–449, 521–522
   cotangent, 474–475
   cube, 147, 155, 214
   decreasing, 205
   decreasing on an interval, 151–152
   defined by equation, 128–129
   defined by graph, 129–132
   difference of, 182
   domain of, 185, 272, 474
   evaluation of, 131–132, 138
   even, 179–180
   exponential, 341–354, 365, 376, 404
   graph of, 146–161, 214
   greatest integer, 147, 155, 214
   identity, 155, 214
   increasing, 205
   inverse, 203–212
   inverse trigonometric, 518–526
   limits of, 829–839
   limits of polynomial, 841–842
   limits of rational, 842–844
   linear rational, 274–277
   logarithmic, 360–369, 375–383, 404
   logistic, 222
   odd, 180
   one-to-one, 203–204
   operations on, 182–191
   periodic, 441, 455–457, 458
   piecewise-defined, 140–141, 148–150
   polynomial, 235–243
   power, 319
   product of, 183
   quadratic, 225–235
   quotient of, 183
   range of, 127
   rational, 272–291
   reciprocal, 319
   rule of, 127
   secant, 474–475, 476, 478–479
   sine, 423–425, 431–433, 434
   square root, 147, 155, 214
   step, 148
   sum of, 182
   table of values of, 137
   tangent, 424–425, 523–524
   trigonometric, 426–429, 435–436, 438,
     474–476, 543–557
   Vertical Line Test, 152–153
Fundamental Theorem of Algebra, 312–313

G
Gateway Arch, 350
Gauss, K.F., 742
Gaussian elimination, 742–757
Gauss-Jordan method, 748–751
Geometric
   interpretation of vector addition, 610–611

Geometric (*continued*)
  progression, 794–800
  series, 802–805
Geometry
  analytic, 636–720
  review, 902–914
Geosynchronous orbits, 879
Graphical
  intersection finder, 85
  maximum/minimum finder, 76–77, 79
  root finder, 83, 89
Graphing
  conventions, 70, 79
  dot mode for, 80
  supply and demand curves, 121–123
  technology, 71–74
Graph(s)
  amplitude, 457–458
  of average rate of change, 195–196
  basic, 445–455
  complete, 77–80
  of conic sections, 695–704, 705–706,
    708–709
  continuous, 255–256
  contraction, 169–170
  of cosine function, 447–449
  damped, 470–473
  equation "by hand," 70
  of equations, 41
  expansion, 169–170
  exponential function, 404
  function defined by, 129–132
  of functions, 146–161
  horizontal (phase) shifts, 458–461
  and identities, 452–453
  of inverse function, 209–210
  line, 37
  of linear function, 146
  of linear rational functions, 274–277
  logarithmic function, 404
  of nonlinear equations, 736–737
  parametric, 153–154
  periodic, 455–457
  polar, 698–702
  of polynomial functions, 254–266
  of quadratic function, 225
  of rational functions, 277–284
  reading, 161–165
  reflections, 171–172
  secant function, 478–479
  of second-degree equations, 673–675
  sinusoidal, 446–447
  symmetric with respect to axes and ori-
    gin, 177–182
  of tangent function, 450–452
  transformations of, 166–176
  trigonometric, 468–473, 533–534
  vertical line test, 152–153
  vertical shifts, 166–167, 643–646,
    654–657
Greater than, 4

Greatest integer function, 147, 155, 214
Growth, exponential, 344–346

H
Half-angle identities, 512–514
Half-life, 347
Harmonic motion, 463
Herron's formula, 583–584
Hidden behavior, 152
Hole, 279
Homogenous system, 750
Horizontal
  asymptote, 273, 280, 869
  line, 51
  line test, 204, 518
  shift, 168–169, 458–461, 643–646,
    654–657, 665–668
Hyperbola, 637
  equations of, 652–657
  polar equations for, 704, 706
Hypotenuse, 903
Hypothesis, induction, 816

I
I. *See* Vector
I (complex number), 306
Identities, 437–444, 490
  addition, 498–506
  angles measured in degrees, 545
  cofunction, 501–502
  double angle, 510–511
  factoring, 515
  function, 155, 214
  function limit, 841
  graphs and, 452–453
  half-angle, 512–514
  matrix, 762–763
  negative angle, 442–443, 492, 545
  periodicity, 441–442, 492, 545
  power-reducing, 512
  product, 515
  proof of, 491–497
  Pythagorean, 439–440, 477, 492
  reciprocal, 476, 492
  subtraction, 498–506
  supplementary angle, 576
Imaginary
  axis, 592
  number, 306
  part of complex number, 311
Inclination, angle of, 506–507
Inconsistent system, 726
Increasing
  on an interval, 151–152
  function, 205
Index, summation, 783–784
Induction
  common mistake, 818–819
  hypothesis, 816
  mathematical, 813–822
Inductively defined sequence, 781

Inductive step, 816
Inequalities
  absolute value, 302–305
  linear, 292–293
  polynomial, 293–295
  principles for solving, 292
  quadratic, 295–297
  rational, 297–299
  Schwarz, 623
  triangle, 10
Infinite
  geometric series, 802–805
  limits, 865–875
Infinite limit theorem, 871
Inflection, point of, 258
Initial
  point of vector, 606
  side of angle, 412, 555
Input-output analysis, 774–775
Instantaneous rate of change, 198–200
Integer, 2
  prime, 818
Integer part function, 135
Integral exponent, 882–886
Intercepts, 41, 84
  of polynomial graphs, 257
  of rational graphs, 278
Interest, 93–94
  compound, 354–360
  continuous compounding, 357–358
  simple, 93–94
Intermediate value theorem, 861–862
Intersection
  finder, 89
  method of solving equations, 84–86
Interval
  continuity on, 859
  increasing or decreasing on, 151–152
  notation, 4–5
Inverse
  cosine function, 521–522
  function, 203–212
  of matrix, 762–763
  sine function, 518–520
  tangent function, 523–524
  variation, 31
Invertible matrix, 762
Irrational exponent, 332
Irrational number (*e*), 17
  continuous compounding and, 357–358
  natural exponent function, 348
Isosceles triangle theorem, 903–904
Iteration, 221

J
J. *See* Vector
Jordan, W.J., 748

K
Keynes, J.M., 27
*k*th partial sum, 784–786

L

Latitude, 422
Law(s)
  of cosines, 566–573
  exponent, 329, 438, 884
  logarithm, 379–381
  of sines, 573–582
Leading
  coefficient, 236
  one, 747
Least
  common denominator, 896–897
  squares regression line,
      108, 112
Left, continuity from, 858
Lemniscate, 702
Length
  arc, 418–419
  of vector, 606
Leontieff, W., 774
Less than, 4
Limaçon, 702
Limits
  constant, 841, 870
  difference quotients, 845–846
  formal definition, 847–853
  of functions, 829–839
  infinity, 865–875
  informal definition, 830
  nonexistence, 833–835
  one-sided, 835–837
  properties of, 840–847, 870–873
  rational functions, 842–844
Limit theorem, 844
Linear
  combination, 613
  equation, 19
  function, 146, 155, 214, 236
  inequality, 292–293
  model, 105–116
  rate of change, slope of a line, 53,
      195–196
  rational function, 274–277
  regression, 108
  system of—equations, 721–775
Linear speed, 419
Line(s), 48–62, 637
  angle between, 507–509
  angle of inclination, 506–507
  equations of, 50–58
  graph, 37
  horizontal, 51
  horizontal—test, 204
  least squares regression, 108, 112
  number, 3, 10
  parallel, 56–57
  perpendicular, 57–58
  real world models, 68
  secant, 195–196
  slope of, 49
  vertical, 55–56

Line(s) (*continued*)
  vertical—test, 152–153
Local
  extremum, 258
  maximum, 150–151
  minimum, 150–151
Logarithmic
  equations, 387–390
  functions, 366–367, 375–383, 404
  models, 392, 397–398
Logarithm(s)
  to base *b*, 375–383
  change of base, 381
  common, 360–362, 365, 377
  laws, 379–381
  natural, 362–363, 365
  properties of, 363–365, 369–375, 378
Logistic function, 222
Logistic model, 392, 395, 409–410
Lower bound, 251
Lowest terms, 895

M

Magnitude of vector, 606, 608
Major axis of ellipse, 639
Manual zoom-in, 74
Mathematical induction, 813–822
Mathematical model, 105
Matrix (matrices)
  augmented, 745
  identity, 762–763
  inverse, 764–766
  invertible, 762
  multiplication, 758–761
  nonsingular, 762
  reduced row echelon form, 750
  row echelon form, 747
  solution of system, 745–748
  for square systems, 758–770
Maximum
  finder, 76–77, 79
  local, 150–151
Memory, equation, 72–73
Method
  Gaussian, 742–757
  solution, choosing, 88–89
Midpoint formula, 40
Minimum
  finder, 76–77, 79
  local, 150–151
Minor axis of ellipse, 639
Mode
  dot graphing, 80
  RADIAN, 424
Model
  exponential, 392–396, 409–410
  linear, 105–116
  logarithmic, 392, 397–398
  logistic, 392, 395, 409–410
  mathematical, 105
  polynomial, 267–269, 395

Model (*continued*)
  power, 392, 396
Modulus, 593–594
Motion
  harmonic, 463
  projectile, 688
Multiplication
  with exponents, 883
  matrix, 758–761
  in polar form, 595–596
  scalar, 609–610, 612
Multiplicity of root, 257

N

Natural
  logarithm, 362–363, 365
  number, 2
$_NC_r$ (binomial coefficient), 806–807
Negative
  angle, 413
  angle identities, 442–443, 492, 545
  correlation, 110
  exponent, 883
  number, 7
  square root of—number, 308
  of vector, 611
N (*n* factorial), 806
Nonlinear system of equations, 735–741
Nonnegative number, 7
Nonrepeating decimal, 17
Nonsingular matrix, 762
Norman arch, 324
Norm of vector, 608
*n*th root, 327–328, 599–602
*n*th term
  of arithmetic sequence, 789
  of geometric sequence, 795
Number
  complex, 305–311
  Fibonacci, 781
  imaginary, 306
  irrational, 17, 348
  line, 3, 10–13
  natural, 2
  negative, 7, 308
  positive, 7
  rational, 2
  real, 2, 16–18, 306, 412
  of roots, 241, 313
  of solutions of system, 723, 753
Numerator, rationalizing, 331–332
Numerical equation finder, 89

O

Oblique triangles, 573–582
Obtuse angle, 902
Odd function, 180
One-sided limits, 835–837
One-step solver, 89
One-to-one function, 203–204
Open interval, 4–5, 859

Operations
  elementary, 743
  on functions, 182–191
  order, 6
  row, 745–747
Opposite side, 545–546
Orbit, 221
Order, 4–5
Ordered pair, 37
Origin, 36, 695
  circle at, 43
  symmetry with respect to, 178–179
Orthogonal vectors, 623–624
Output, 774–775

P
Parabola, 225, 637
  axis of, 226
  equations of, 663–672
  polar equations for, 706
  vertex of, 226
Parabolic arch, 324
Parallel
  lines, 56–57
  vectors, 622
Parallelogram rule, 597
Parameter, 682–683
  eliminating the, 685
Parametric equations, 154
  of circle, 639
  of conic sections, 719–720
  of curve, 682–683, 685–686
  of ellipse, 647
  graphing, 153–154
  of hyperbola, 657
  of parabola, 667–668
Parentheses
  functional notation, 136
  rules, 6, 437, 887–888
Partial fraction decomposition, 752
Partial sum, 790–792
  of arithmetic sequence, 791
  of geometric sequence, 797–799
  of infinite series, 801
Pascal's triangle, 808
Periodicity identities, 441–442, 492,
    545
Period
  of cosecant, 476
  of cosine, 441, 455–457
  of cotangent, 476
  of function, 441, 455–457, 458, 460
  of secant, 450, 476
  of sine, 441, 455–457
  of tangent, 451
Periodic function, 441
Perpendicular
  lines, 57–58, 509
  vectors, 623–624
Phase shift, 458–461
Piecewise-defined function, 140–141
  graph of, 148–150

Pistons, 487–488
Plane
  complex, 592
  coordinate, 36–48
  curve, 681–694
Plot, scatter, 37
Point, 637
  coordinates of, 36
  of inflection, 258
  -in-the-plane description, 427–429,
    475–476
  -slope form, 53–55
Point of diminishing returns, 265
Polar
  axis, 695
  conversion, 697
  coordinates, 695–704
  division rules, 595–596
  equations of conic sections,
    704–714
  form of complex numbers, 593–598
  graphs, 698–702
  multiplication rules, 595–596
Pole, 695
Polynomial, 235
  approximation, 369
  arithmetic, 236–238
  bounds, 249–252
  constant, 235
  degree of, 236
  division, 238–241, 243–246
  equation solver, 24–26
  function, 235–243
  graph of—function, 254–266
  inequality, 293–295
  limits of—function, 841–842
  model, 267–269, 395
  number of roots of, 241, 313
  with real coefficients, 314–316
  real roots of, 240, 247–254
  regression, 267
  solvers, 86, 89
  zero, 235
Positive
  angle, 413
  correlation, 110
  integer, 2
  number, 7
Power
  functions, 319
  law for logarithms, 372–373
  model, 392, 396
  principle, 334
  reducing identities, 512
Present value, 391
Prime number, 818
Principle(s)
  Big-Little, 272
  of mathematical induction, 813–822
  power, 334
  for solving equations, 18–26
  for solving inequalities, 292

PROB, 806
Probability, 826–827
Product
  dot, 621–630
  of functions, 182
  identities, 515
  law for logarithms, 369–370,
    379–380
  zero, 20
Programs
  fraction conversion, 915
  quadratic formula, 915
  rectangular/polar conversion, 920
  RREF, 918–920
  synthetic division, 916–917
  table maker, 916
Progression
  arithmetic, 788–794
  geometric, 794–800
Projectile motion, 688
Projection of **u** on **v**, 624–627
Proj$_v$**u**, 624–627
Proportionality, constant of, 30
Proving identities, 624–627
Pythagorean
  identities, 439–440, 477, 492, 545
  Theorem, 903

Q
Quadrant, 36
Quadratic
  equation, 20–24
  factoring patterns, 891
  formula, 22
  formula program, 915
  function, 225–235, 236
  inequality, 295–297
  regression, 267, 269
Quartic regression, 267, 269
Quotient
  of algebraic expressions, 894
  difference, 139, 196–198, 845–846
  law for logarithms, 371, 379–380
  in polynomial division, 238

R
Radian
  -degree conversion, 417–418
  measure, 414–417
RADIAN mode, 424
Radical, 330–331
  equations, 334–341
Radioactive decay, 347
Radius of circle, 42
Range, 127
  of sine and cosine, 449
Rate of change
  average, 192–196
  instantaneous, 198–200
Ratio, common, 794
Rational
  exponent, 328–330

Rational (*continued*)
function, 272–291
function limit, 842–844
inequality, 297–299
linear—function, 274–277
number, 2, 16
root, 247–248
Rationalizing denominators and numerators, 331–332
Ratios theorem, 908
Reading graphs, 161–165
Real
axis, 592
coefficients, polynomials with, 314–316
number, 2, 16–18, 306, 412, 432
part of complex number, 311
root, 240
Reciprocal functions, 319
graphing, 273
Reciprocal identities, 476, 492
Rectangular
conversion, 697
coordinate system, 36
Recursively defined sequence, 781
Reduced row echelon form, 750
Reflection of graph, 171–172
Regression
commands, 267
least squares line, 108, 112
linear, 108
polynomial, 267
Remainder(s), 238–241
and factors, 238
theorem, 239
Repeating decimal, 16
Repelling fixed point, 221
Residual, 107
Richter scale, 374
Right, continuity from, 858
Right angle, 553, 902
Right triangle, 903
description of trigonometric functions, 545–548, 553–557
30-60-90 triangle theorem, 904–905
trigonometry, 426–427, 436, 548–551, 558–566
Root(s)
bounds test for, 249–252
conjugate, 314
factor theorem, 248–249
finder, 83, 89
multiplicity of, 257
nth, 327–328, 599–602
number of, 241, 313
of polynomial, 240, 247–254
rational, 247–248
real, 240, 247–254
square, 7–8, 9, 87, 147
of unity, 310, 602–605
Rose, 701
Rotation
of axes, 676–681

Rotation (*continued*)
equations, 672–676
Round-Trip Theorem, 207–209
Row
echelon form, 747
of matrix, 745, 750
operation, 745–747
reduced—echelon form, 750
RREF program, 918–920
Rule
equality for polar form, 598
of function(s), 127
parallelogram, 597
parentheses, 6, 437, 887–888
polar multiplication and division, 595–596

S

SAS, 567
Scalar multiplication, 609–610, 612, 623–624
Scatter plot, 37
Schwarz inequality, 623
Schwarzschild radius, 879
Scientific notation, 3–4
Sears Tower, 1, 8
Secant function, 474–475
graph of, 478–479
period of, 476
Secant line, 195–196
Second-degree equation(s), 20–24, 672
graph of, 673–675
systems of, 738–740
Sec t. *See* Secant function
Section, conic, 637
Semicircular arch, 324
SEQ, 780
Sequence, 777
arithmetic, 788–794
constant, 779
Fibonacci, 781, 787
geometric, 794–800
inductively defined, 781
partial sum of, 784–786, 791, 797–799
recursively defined, 781
summation notation, 783–784
Series
convergent, 802
geometric, 802–805
infinite, 801–805
Setting up applied problems, 90–93
Shift, horizontal, 168–169, 458–461
Shift, vertical, 166–167
Shift, vertical and horizontal
of ellipses, 643–646
of hyperbolas, 654–657
of parabolas, 665–668
Shortage, 122
Side
adjacent, 545–546
of angle, 412, 902
opposite, 545–546

Signs of trigonometric functions, 431
Similar triangles, 907–911
Simple
harmonic motion, 463
interest, 93–94
Sine function, 423–425, 431–433, 434
amplitude, 457–458
graph, 446–447
inverse, 518–520
period, 441, 455–457
phase shift, 458–461
point-in-plane description, 475–476, 544
range, 449
restricted, 518, 520
right triangle description, 554
Sines, law of, 573–582
Sin t. *See* Sine function
Sinusoidal graphs, 468–470
Slope(s)
-intercept form, 50–53
of line, 49
of perpendicular lines, 509
point—form, 53–55
properties of, 50
and tangent function, 450
Solution(s)
of equation, 18
and intercepts, 84
method, choosing, 88–89
number of, 23–24, 723, 753
of quadratic equation, 24
real, 18
of system of equations, 722, 723, 735–741, 746
trivial, 750
Solver
equation, 86–87
polynomial, 24–26, 86
Solving
equations, 18–26, 83–90
by factoring, 20
triangles, 548–551
trigonometric equations, 527–536
Special values of trigonometric functions, 435–436
Speed, angular, 419–421
Speed of light, 878
Spiral, Archimedean, 701
Square root, 7–8
function, graphing, 147, 155, 214
of negative number, 308
of squares, 9
of zero, 87
Square(s)
completing the, 20–21
difference of, 891
function, graphing, 147, 155, 214
matrix for, 758–770
perfect, 891
viewing window, 75, 80
SSA, 575, 576
SSS, 567

Standard
  equations of conic sections, 705–706, 708–709
  form for complex numbers, 306
  notation for triangles, 566
  position, 413, 555
  viewing window, 74
Statistics editor, 112
Step function, 148
Substitution
  back, 744
  method, 724–725
Subtraction
  of functions, 182
  identities, 498–506
  of vectors, 611
Sum
  of convergent series, 802
  of cubes, 891
  of first n positive integers, 813
  of functions, 182
  of infinite geometric series, 802
  partial, 784–786
Summation index or notation, 783–784
Supplementary angle identity, 576
Supply curve, 121–123
Surveying, 588, 634–635
Symmetry
  origin, 178–179
  tests, 179
  x-axis, 178
  y-axis, 177
Synthetic division, 239, 243–246
  program, 916–917
System of equations, 721–775
  dependent, 727, 751
  equivalent, 742
  homogenous, 750
  inconsistent, 726
  large systems, 742–757
  matrix solutions of, 745–748, 764–766
  nonlinear, 735–741
  number of solutions, 723, 753
  triangular form, 744
  in two variables, 722–735

T
Table, 129–130
  program, 916
Tangent function, 424–425, 431–433
  addition and subtraction identities, 500–501
  graph, 450–452
  half-angle identities, 514
  inverse, 523–524
  period, 451
  point-in-plane description, 475–476
  restricted, 524
  right triangle description, 554
  and slope, 450

Term
  constant, 235
  of sequence, 777
Terminal
  point of vector, 606
  side of angle, 412
Tick marks, 79
TRACE, 73–74
Transformations
  combining, 172–173
  expansions and contractions, 169–170
  graphs, 166–176
  horizontal shifts, 168–169
  reflections, 171–172
  vertical shifts, 166–167, 643–646, 654–657
Triangle(s), 902–905. *See also* Right triangle
  area of, 582–585
  inequality, 10
  Pascal's, 808
  similar, 907–911
  solving, 548–551
  standard notation for, 566
  trigonometry, 542–590
Triangular form system, 744
Trigonometric
  equations, 527–536
  forms of complex numbers, 593–598
  functions, 426–429, 435–436, 438, 474–476, 543–557
  graphs, 468–473
  identities, 490–498, 510–517
Trigonometry, triangle, 542–590
Trivial solution, 750
Trochoid, 693
Two-sided limits, 836–837

U
Unit
  circle, 433–435
  vector, 612–613
Unity, roots of, 310, 602–605
Upper bound, 251

V
Value
  absolute, 28–30, 149–150
  absolute—of complex number, 593
  of function, 137, 831
  intermediate—theorem, 861–862
  present, 391
Variable, 887
Variation
  direct, 30
  inverse, 31
Vector(s), 606–620
  addition, 610–611, 612
  angle between, 622–624
  component of **u** alongside **v**, 626–627
  components of, 608
  direction angle of, 613–616

Vector(s) (*continued*)
  dot product of, 621–630
  equivalent, 607–608
  initial point of, 606
  length of, 606
  magnitude of, 606
  negative of, 611
  norm of, 608
  orthogonal, 623–624
  parallel, 622
  perpendicular, 623–624
  projection of **u** on **v**, 624–627
  scalar multiplication of, 609–610
  subtraction, 611
  unit, 612–613
  zero, 611
Vertex
  of angle, 412, 902
  of ellipse, 639
  of hyperbola, 652
  of parabola, 226, 663
Vertical
  asymptote, 273, 278, 867
  line, 55–56
  line test, 152–153
  shifts, 166–167, 643–646, 654–657, 665–668
Vertices. *See* Vertex
Viewing window, 74–76

W
Window
  decimal, 74
  square, 75, 80
  viewing, 74–76, 80
Work, 628

X
X as a function of y, 128
X-axis, 36
  symmetry, 178, 179
X-coordinate, 36
X-intercepts, 41, 84
  polynomial function, 257
  rational function, 278

Y
Y as a function of x, 128
Y-axis, 36
  symmetry, 177, 179
Y-coordinate, 36
Y-intercept, 41

Z
Zero
  distance to, 11
  exponent, 883
  fraction, 88
  of polynomial, 235
  products, 20
  square root, 87
  vector, 611
ZOOM, 74

Trigonometry

If t is a real number and P is the point where the terminal side of an angle of t radians in standard position meets the unit circle, then

$$\cos t = x\text{-coordinate of } P \qquad \sin t = y\text{-coordinate of } P$$

$$\tan t = \frac{\sin t}{\cos t} \qquad \csc t = \frac{1}{\sin t} \qquad \sec t = \frac{1}{\cos t} \qquad \cot t = \frac{\cos t}{\sin t}$$

Point-in-the-Plane Description

For any real number t and point (x, y) on the terminal side of an angle of t radians in standard position:

$$\sin t = \frac{y}{r} \qquad \cos t = \frac{x}{r} \qquad \tan t = \frac{y}{x} \quad (x \neq 0)$$

$$\csc t = \frac{r}{y} \quad (y \neq 0) \qquad \sec t = \frac{r}{x} \quad (x \neq 0) \qquad \cot t = \frac{x}{y} \quad (y \neq 0)$$

Periodic Graphs

If $A \neq 0$ and $b > 0$, then each of $f(t) = A \sin(bt + c)$ and $g(t) = A \cos(bt + c)$ has

$$\text{amplitude } |A|, \qquad \text{period } 2\pi/b, \qquad \text{phase shift } -c/b.$$

Right Triangle Trigonometry

$$\sin \theta = \frac{\text{opposite}}{\text{hypotenuse}}$$

$$\cos \theta = \frac{\text{adjacent}}{\text{hypotenuse}}$$

$$\tan \theta = \frac{\text{opposite}}{\text{adjacent}}$$

Special Values

Degrees	Radians	$\sin \theta$	$\cos \theta$	$\tan \theta$
0°	0	0	1	0
30°	$\dfrac{\pi}{6}$	$\dfrac{1}{2}$	$\dfrac{\sqrt{3}}{2}$	$\dfrac{\sqrt{3}}{3}$
45°	$\dfrac{\pi}{4}$	$\dfrac{\sqrt{2}}{2}$	$\dfrac{\sqrt{2}}{2}$	1
60°	$\dfrac{\pi}{3}$	$\dfrac{\sqrt{3}}{2}$	$\dfrac{1}{2}$	$\sqrt{3}$
90°	$\dfrac{\pi}{2}$	1	0	undefined

The top header of the table reads **θ** spanning Degrees and Radians.

Special Right Triangles

Law of Cosines

$$a^2 = b^2 + c^2 - 2bc \cos A$$

$$b^2 = a^2 + c^2 - 2ac \cos B$$

$$c^2 = a^2 + b^2 - 2ab \cos C$$

Law of Sines

$$\frac{a}{\sin A} = \frac{b}{\sin B} = \frac{c}{\sin C}$$

Area Formulas

Herron's Formula: Area $= \sqrt{s(s-a)(s-b)(s-c)}$, where $s = \dfrac{1}{2}(a + b + c)$ $\qquad$ Area $= \dfrac{1}{2}ab \sin C$